Gross Anatomy
in the Practice of Medicine

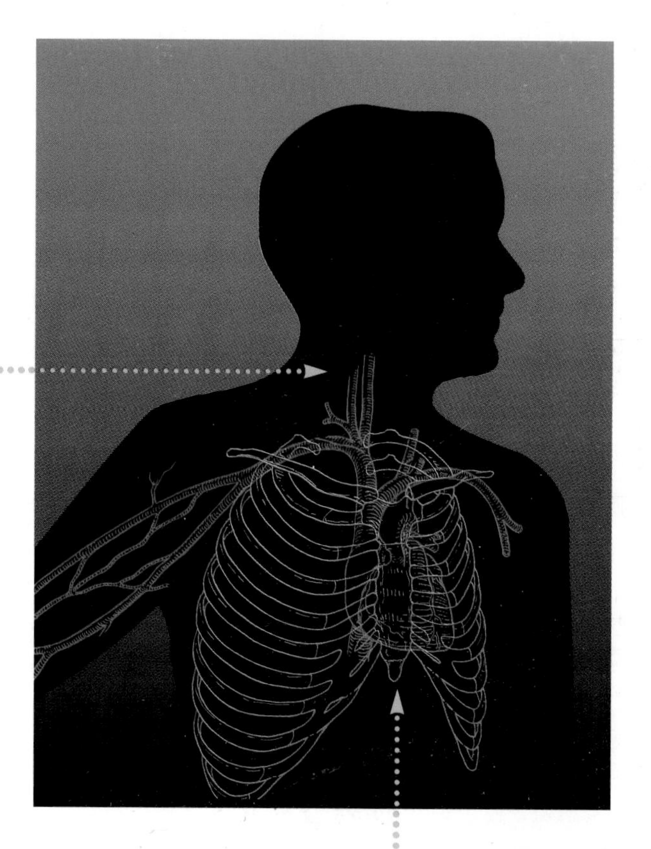

Gross Anatomy
in the Practice of Medicine

Frank J. Slaby, Ph.D.

Associate Professor of Anatomy
Department of Anatomy
George Washington University School of Medicine
Washington, D.C.

Susan K. McCune, M.D.

Assistant Professor, Department of Pediatrics
George Washington University School of Medicine
Washington, D.C.

Robert W. Summers, M.D.

Professor, Department of Internal Medicine
Division of Gastroenterology/Hepatology
University of Iowa College of Medicine
University of Iowa Hospitals and Clinics and
Iowa City Veterans Affairs Medical Center
Iowa City, Iowa

Lea & Febiger

PHILADELPHIA • BALTIMORE • HONG KONG
LONDON • MUNICH • SYDNEY • TOKYO

A WAVERLY COMPANY
1994

Lea & Febiger
Box 3024
200 Chester Field Parkway
Malvern, PA 19355-9725
U.S.A.
(215) 251-2230

Executive Editor—J. Matthew Harris
Development Editor—Lisa Stead
Project Editor—Raymond Lukens
Production Coordinator—MaryClare Beaulieu

Slaby, Frank.
Gross anatomy in the practice of medicine / Frank J. Slaby, Susan K. McCune, Robert W. Summers.
p. cm.
ISBN 0-8121-1664-X
1. Human anatomy. 2. Physical diagnosis. 3. Medical history taking. 4. Medical logic. I. McCune, Susan K. II. Summers, Robert W. (Robert Wendell), 1938- . III. Title.
[DNLM: 1. Diagnosis—case studies. 2. Clinical Medicine—methods. 3. Anatomy.
WB 141 S161g 1994]
QM23.2.S525 1994
611'. 002461—dc20
DNLM/DLC
for Library of Congress 93-5249
CIP

Reprints of chapters may be purchased from Lea & Febiger in quantities of 100 or more. Contact Sally Grande in the Sales Department.

NOTE: Although the author(s) and the publisher have taken reasonable steps to ensure the accuracy of the drug information included in this text before publication, drug information may change without notice and readers are advised to consult the manufacturer's packaging inserts before prescribing medications.

PRINTED IN THE UNITED STATES OF AMERICA

Print No. 5 4 3 2 1

Dedication

To my mother and late father.

Frank J. Slaby

To my parents, the late Doctors Violet and Wallace McCune, whose support and encouragement I miss.

Susan K. McCune

To my wife, Edith, whose unfailing support is a continuing inspiration and encouragement for all my efforts.

Robert W. Summers

Preface

We have written this book to provide an understanding of how anatomic knowledge is actually used in the practice of medicine. To achieve this objective, we have integrated a traditional approach to gross anatomy with a new case-based approach. Section I presents a traditional introduction to gross anatomy; it provides overviews of the cardiovascular, lymphatic, musculoskeletal, and nervous systems, a description of the vertebral column and the muscles of the back, and a discussion of the physical basis of the radiologic methods used to image the body. The parts of the chapters of Sections II to VI that present the gross anatomy of the human body on a regional basis give the traditional approach to gross anatomy. This approach has proved highly successful in teaching and learning basic gross anatomy because it emphasizes the major anatomic relationships in each body region and correlates both surface anatomy and radiologic anatomy with gross anatomy.

The patient cases in Sections II to VI give a new practical approach to gross anatomy that provides insight into the process by which physicians incorporate anatomic knowledge into their clinical reasoning skills. Gross anatomy is most commonly applied by physicians to the analysis of patient data gathered from the history, physical examination, and radiologic procedures. Accordingly, cases with major anatomic components have been prepared to demonstrate by example how information gathered from the history, physical examination, and radiologic procedures is evaluated to determine the anatomic basis of a patient's chief medical problem or problems.

Most of the patient cases are presented in nine parts:

(1) **Initial Presentation and Appearance of the Patient** (a description of the patient's physical appearance and chief medical complaint or complaints)

(2) **Questions Asked of the Patient** [a transcription of the dialogue (how questions were asked and answered) during the history taking]

(3) **Physical Examination of the Patient** (a summary of the physical exam that emphasizes the pertinent normal and abnormal findings)

(4) **Initial Assessment of the Patient's Condition** (a first, general description of the patient's apparent chief medical problem or problems)

(5) **Anatomic Basis of the Patient's History and Physical Examination** (a discussion of what is known or can be deduced from the patient's history and physical exam)

(6) **Intermediate Evaluation of the Patient's Condition** (a secondary evaluation of the patient's chief medical problem or problems, based largely on anatomic and physiologic information acquired from the history and physical examination)

(7) **Clinical Reasoning Process** (a discussion of the clinical reasoning process used to evaluate further the patient's condition)

(8) **Radiologic Evaluation and Resolution of the Patient's Condition** (a discussion of the radiologic findings that lead to a resolution of the patient's condition)

(9) **Chronology of the Patient's Condition** (a chronology of the patient's condition told in terms of the anatomic basis of the patient's signs and symptoms).

A brief, general discussion of the appropriate medical and/or surgical treatment for the patients and their prognoses is presented near the end of each case presentation.

The case presentations in this book are unique among medical texts, and gross anatomy texts in particular. The following points highlight the clinical problem-solving features that have been incorporated into these presentations:

Each case outlines an efficient yet thorough evaluation of the patient's condition. The individual examiner in each case has extensive knowledge and experience in the evaluation of the patient's chief medical problem or problems.

The history is not a complete record of the patient's present and past medical history, but focuses instead on clarifying the anatomic (and occasionally physiologic) basis of the patient's present medical condition. Each history has italicized comments to explain how the examiner evaluates the information received from one or more questions and then uses this information to determine the next set of questions.

The physical examination is not an exhaustive examination of systems, but instead focuses primarily on the anatomic regions involved with the patient's chief medical problems and only secondarily on other anatomic regions. In each case, the examiner documents findings from seven basic exams. The **HEENT Examination** covers the **H**ead, **E**ars, **E**yes, **N**ose, and **T**hroat. Examination of the **Lungs** involves auscultation and percussion of the lungs. The **Cardiovascular Examination** is essentially the auscultation of the heart sounds and a survey of the body for signs and symptoms of cardiovascular disease. Examination of the **Abdomen** involves visual inspection, auscultation, percussion, and palpation of the abdomen. The **Genitourinary Examination** comprises basically inspection and palpation of the pelvis and external genitalia. The **Musculoskeletal Examination** involves visual inspection, tests of movements, isometric exercises, and palpation of the upper and lower limbs. The **Neurologic Examination** is a survey of the body for signs and symptoms of central and peripheral nervous system disorders.

Each case documents how the examiner uses the history, physical exam, and radiologic procedures to gather anatomic information about the patient. Many medical students tend to associate gross anatomy with the end of life, because the gift of the bodies of unknown individuals for dissection in the laboratory first reveals to students the organization of the human body. It is our hope, however, that what you read in this book will show you how the facts of gross anatomy apply to life itself. This is because gross anatomy is used in medical practice to understand what you see, hear, and touch on an individual who has come to you for help. The initial facts you learn about each patient are the features of his or her gross anatomy. When you first talk with a patient and then touch the person during the physical exam, it is your first and best opportunity to verbally inform and physically assure the individual that you are there to evaluate and help if possible. So, as you read each case, imagine that you are beside the examiner and that it is the examiner who is teaching you how gross anatomy comes alive in the examination room.

Each case demonstrates how gross anatomic knowledge is used in the clinical reasoning process. Case presentations in other medical texts begin with a synopsis of the history, physical examination, and lab findings, then state a definitive diagnosis, and finally end with a description of the manner in which an injury, disease, or disorder accounts for the patient's signs and symptoms. This format adequately describes how the defined altered condition of a patient's anatomy and physiology accounts for the patient's signs and symptoms. But this line of reasoning is opposite to that which occurs in the real world.

In the real world of medical practice, the patient's condition is unknown and defined in the beginning by only signs and symptoms. The clinical reasoning process is a line of reasoning which enables the examiner to transform the signs and symptoms into reasonable deductions about the altered condition of the patient's anatomy and physiology. Understanding the anatomic basis of the history and physical exam is key to understanding the role of anatomic knowledge in the clinical reasoning process.

The vast majority of signs and symptoms are not pathognomonic. In other words, most signs and symptoms do not point to a specific diagnosis. In general, a sign or symptom can be interpreted in a number of ways. Even when all of a patient's signs and symptoms are considered together, they generally point not to one but to a small number of likely injuries, diseases, or disorders. These considerations explain why the words "suggest" and "indicate" are used to describe the implications of signs and symptoms in a case. If a sign or symptom is said to **suggest** a particular condition, this means that there is roughly a 50% likelihood that this is the correct implication of the sign or symptom. If a sign or symptom is said to **indicate** a particular condition, this means that there is roughly a 70 to 90% likelihood that this is the correct implication of the sign or symptom.

Each case illustrates the importance of considering the mechanism or process responsible for a patient's condition. The history and physical exam are used not only to gather anatomic and physiologic information about the patient, but also to initiate definition of the mechanism or process responsible for the patient's ill health.

Many physicians use the acronym **VINDICATE** to consider the pathologic basis of a patient's condition. The letters of the word **VINDICATE** represent nine major causes of pain and suffering: **V**ascular disorders, **I**nfection, **N**eoplasm, **D**egeneration/deficiency, **I**atrogenic/inherited disorders, disorders of **C**ollagen metabolism, **A**llergy, **T**rauma, and **E**ndocrine disorders. An iatrogenic disorder is an unfavorable response to a medical treatment or surgical procedure.

In each case, the findings from the history, physical examination, and radiologic procedures provide, for the most part, a fairly accurate resolution of the patient's condition. Anatomic information is the key to the resolution of almost every case in this book; however, anatomic information is generally no more important than other types of medical information in the consideration of most cases encountered in medical practice. Therefore, anatomic information has to be generally processed with other medical information in the clinical reasoning process.

Although this book has been written to be used primarily for the study of gross anatomy during the first year of medical school, the broad scope of its content and its emphasis on practical matters make it a book useful to medical, allied health, and nursing students in all the years of training. Second-year medical students will find the cases helpful in understanding the progression of diseases and illnesses. Third- and fourth-year medical students will find the cases useful as practical aids for the refinement of history-taking and physical diagnostic skills. Allied health and nursing students may find the depth of the gross anatomy sections beyond your instruction; however, you will find that the cases address the practical aspects of your training and will help you understand at a deeper level the anatomic basis of the history and physical examination.

Washington, D.C. — Frank J. Slaby
Washington, D.C. — Susan K. McCune
Iowa City, Iowa — Robert W. Summers

Acknowledgments

The authors want to thank, first and foremost, J. Matthew Harris, Senior Textbook Acquisitions Editor for Lea & Febiger, for his support and contributions to this book. His efforts in refining the format of the case presentations and in helping us achieve our curricular objectives are greatly appreciated. We also want to thank Richard Perry, President, Waverly, U.S., for his interest and support in the project; Lisa Stead, Development Editor, for her considerable help in coordinating the assembly of the text and its illustrations; Raymond Lukens, Project Editor, for his skill in correcting and clarifying the text; and Samuel Rondinelli, Director of Production, in managing the final production of the book. Their combined assistance has made this project a pleasurable and exciting endeavor.

The authors want to thank the following faculty reviewers: Robert T. Binhammer, University of Nebraska Medical Center; Bruce I. Bogart, New York University Medical Center; Catherine G. Cusick, Tulane University School of Medicine; M. Lynn Palmer, Simmons College and Harvard Medical School, and Robert E. Waterman, University of New Mexico School of Medicine.

The author and publisher acknowledge with thanks the considerable use of illustrations from three Urban & Schwarzenberg publications, specifically, Clemente's Anatomy, 3rd edition (1987), Hall-Cragg's Anatomy, 2nd edition (1990), and Pernkopf's Anatomy, 3rd edition (1989).

F. J. S.
S. K. M.
R. W. S.

Contents

Notes to the Student

DEFINITIONS OF BASIC GROSS ANATOMY TERMS

Anatomists and physicians always refer to a standard body position called the **anatomic position** (Fig. A) when describing spatial relationships among regions or parts of the body. The patient assumes the anatomic position when standing erect with face and eyes forward, the feet close together, the upper limbs hanging by the side of the body, and the hands turned so that the palms face anteriorly (forward).

The vertical plane which divides the body into its left and right halves is called the **median plane** (Fig. B). Planes parallel to the median plane are called **sagittal planes.** Vertical planes perpendicular to the median or sagittal planes are called **coronal planes** (Fig. B). Coronal planes divide the body into **anterior** (front) and **posterior** (back) portions. **Horizontal planes** are alternatively called **transverse planes** (Fig. B). Horizontal planes divide the body into **superior** (upper) and **inferior** (lower) portions.

A cross sectional view of a body region is named according to the imaginary plane that cuts through the body region to provide the cross sectional view. For example, the cross sectional view of the head and neck provided by the cut of the median plane is called the median sectional view.

The terms **medial** and **lateral** are used to compare the proximity of structures to the median plane. The following statements exemplify how these terms may be used: (1) in the anatomic position, the thumb is the most lateral digit of the hand, and the little finger is the most medial digit (Fig. A); (2) the big toe is the most medial digit of the foot, and the little toe is the most lateral digit.

The terms **proximal** and **distal** are used to compare the closeness of structures to the origin of a reference structure or to the regions of attachment of the limbs to the trunk of the body. The following statements exemplify how these terms may be used: (1) the aorta is the largest artery of the body and arises from the heart and ends in the lower region of the abdomen. Therefore, the most proximal segment of the aorta is the segment which arises from the heart, and the most distal segment is the segment which ends in the abdomen. (2) In the upper limb, the arm is proximal to

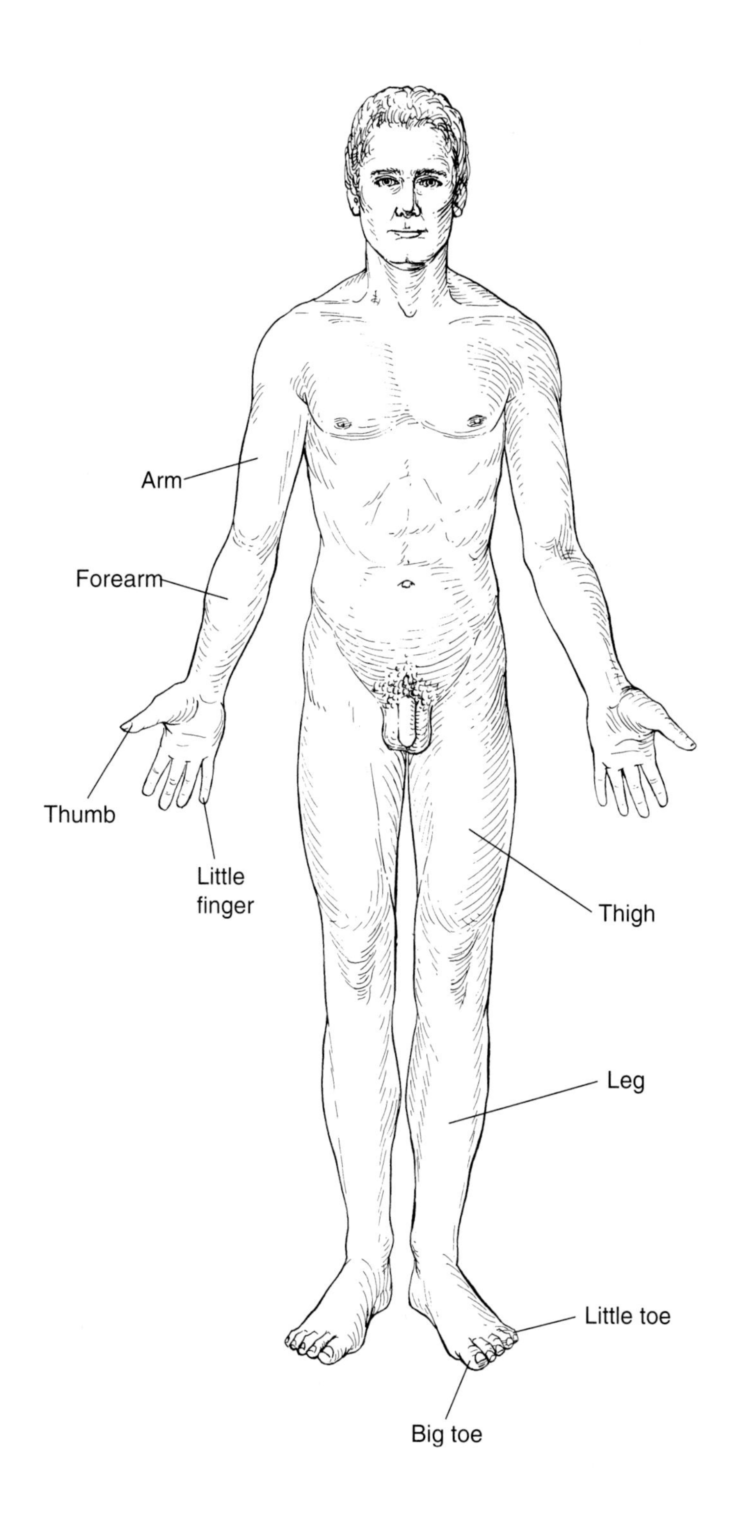

Fig. A: Anterior view of the anatomic position of the human body.

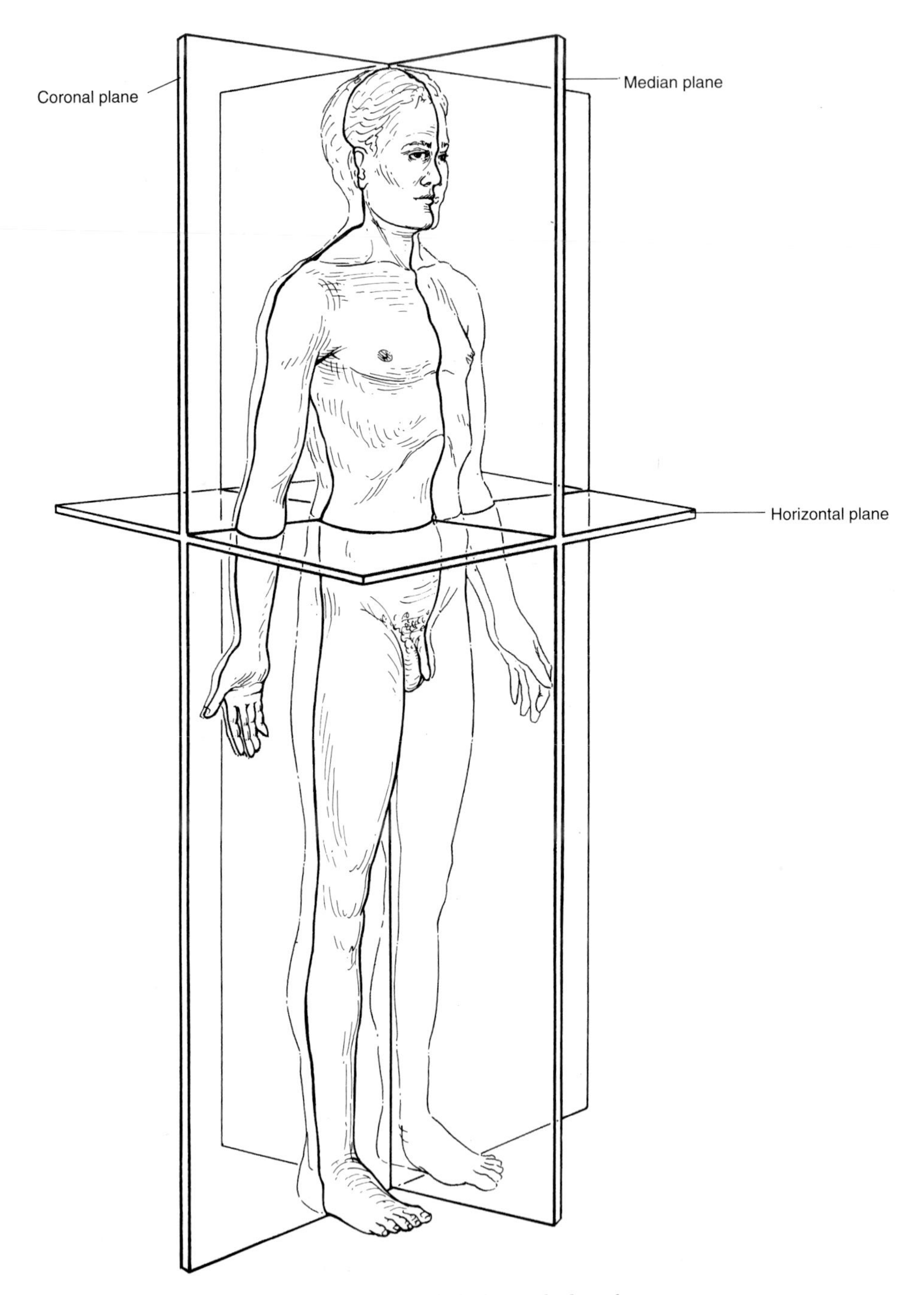

Fig. B: Planes of division through the human body in the anatomic position.

the forearm (Fig. A). (3) In the lower limb, the leg is distal to the thigh.

The terms **cranial** and **caudal** are used to compare the proximity of structures to the head versus the tail region of the body. Reference to the vertebrae of the vertebral column exemplifies how these terms may be used. The most superior vertebrae in the vertebral column are the cervical vertebrae in the neck, and the most inferior vertebrae are the coccygeal vertebrae in the pelvis. Therefore, the cervical vertebrae are the most cranial vertebrae, and the coccygeal vertebrae are the most caudal vertebrae.

The terms **superficial** and **deep** are used to compare the proximity of structures to the surface of the body. In particular, the terms are used to identify the two layers of fascia that underlie the skin in most regions of the body (fascia is a term which generally refers to any collection of connective tissue large enough to be described by the unaided eye). For example, in the arm, the two layers of fascia occur as two coaxial, tubular sheaths that extend along the length of the arm (see Fig. 9–11). The outer layer of fascia that directly underlies the skin is called the superficial fascia of the arm, and the inner layer of fascia that directly ensheaths the muscles of the arm is called the deep fascia of the arm.

The term **ipsilateral** is used to refer to two structures or regions which are both on either the left or right side of the body. The term **contralateral** is used to refer to two structures or regions which are on opposite sides of the body.

THE TISSUES OF THE BODY

There are four basic classes of tissues that comprise the human body: (1) epithelial, (2) connective, (3) muscular, and (4) nervous tissue.

Epithelial tissues line the outer and inner surfaces of the body. The cells of epithelial tissues have very little extracellular material between them. Examples of epithelial tissues include the skin of the body, the endothelial cells that line blood and lymphatic vessels, the mucosa that lines the lumen of the digestive tract, the cells that line the airways of the lungs, and the secretory cells of glands. Epithelial tissues also line the body's internal cavities, such as the peritoneal cavity around the organs of the abdomen and pelvis, the pleural spaces about the lungs, the pericardial cavity about the heart, and the synovial cavities within synovial joints.

Connective tissues are characterized by the relative abundance of extracellular material among their constituent cells. The various types of connective tissues are distinguished by the nature of their extracellular material. There are many types of connective tissues in which protein fibers comprise most of the extracellular matrix; examples of such connective tissues include the tendons of muscles and the ligaments that join bones. The cartilaginous and bony tissues of the body represent connective tissues in which the extracellular matrix consists of a mixture of protein fibers and ground substance. The ground substance of cartilaginous tissues consists of organic molecules and that of bones consists of both organic molecules and inorganic crystals. Other connective tissues include adipose (fat) tissue, bone marrow, the reticular framework of lymph nodes, and blood. Blood is a connective tissue in which fluid is the chief extracellular material.

Muscle tissues are characterized by the contractility of their constituent cells. There are three types of muscle tissues in the body: the **cardiac muscle tissue** of the heart, the **skeletal muscle tissue** of the muscles that move the bones of the body, and the **smooth muscle tissue** that resides in the walls of hollow organs and blood vessels.

Nervous tissues consist of cells called **neurons** that conduct electrical activity and supportive cells called **neuroglia.** Neurons have a **cell body** (in which is located the nucleus, Golgi complex, and most of the mitochondria) from which project fine, cell membrane-lined extensions of the cytoplasm called dendrites and axons. Neurons may have several dendrites but only one axon. **Dendrites** are short cellular projections which generally conduct electrical signals to the cell body of the neuron. **Axons** are long cellular extensions which conduct electrical activity away from the cell body of the neuron.

SECTION I

Introduction

CHAPTER **1**

Overview of the Cardiovascular System

The cardiovascular system is so named because it is a vascular system (a system composed of vessels) in which the heart is the central organ (the prefix cardio- is a term derived from *kardia,* the word for heart in ancient Greek). The vessels form closed passageways through which the blood continuously circulates between the heart and the body's tissues. Vessels that conduct blood from the heart to the tissues are called **arteries.** Vessels that conduct blood from the tissues back to the heart are called **veins.** The smallest arteries conducting blood into a tissue are called **arterioles,** and the smallest veins which conduct blood from a tissue are called **venules.** The minute vessels that transmit blood through a tissue are called **blood capillaries,** and they arise directly from arterioles and end directly with venules. The network of blood capillaries within a tissue is commonly called the tissue's **blood capillary bed.**

OUTLINE OF HEART ANATOMY AND BLOOD FLOW THROUGH ITS CHAMBERS

When viewed in vertical section, the heart is seen to be a hollowed-out organ with a right side and a left side (Fig. 1–1). Each side has two chambers, a smaller one called an atrium and a larger called a ventricle. The two chambers on the right side are called the **right atrium** and **right ventricle,** and the two chambers on the left side are called the **left atrium** and **left ventricle.**

The anatomic division of the heart into two sides underscores the fact that the heart functions as two pumps working side-by-side. On each side of the heart, the atrium and ventricle work together as an atrioventricular pump. Each atrioventricular pump works by first receiving blood from veins which open directly into the atrium (Fig. 1–1). The atrium then directs the blood to flow through an opening into the ventricle, and the ventricle finally ejects the blood from the heart into a large artery arising directly from the ventricle (Fig. 1–1). The electrical activity conducted through the heart during each heartbeat synchronizes the pumping activities of the two atrioventricular pumps.

On each side of the heart, the atrial opening into the ventricle is guarded by a valve that permits sustained blood flow only from the atrium into the ventricle. The valve guarding the opening of the right atrium into the right ventricle is formed from three leaf-like cusps, and thus is called the **tricuspid valve** (a less commonly used name is the **right atrioventricular valve**). The valve guarding the opening of the left atrium into the left ventricle is formed from a pair of cusps in the shape of a miter, and thus is called the **mitral valve** (less commonly used names are the **biscuspid valve** or **left atrioventricular valve**) (a miter is a Catholic or Anglican bishop's hat with pointed, beveled ends) (Fig. 1–1).

On each side of the heart, there is a valve that guards the origin of the large arterial trunk arising from the ventricle; this valve permits sustained blood flow only from the ventricle into the artery. The artery that arises from the right ventricle is called the **pulmonary trunk,** and the valve at its origin is called the **pulmonary valve.** The artery that arises from the left ventricle is called the **aorta,** and the valve at its origin is called the **aortic valve.** The aortic and pulmonary valves are each formed from three cup-shaped cusps.

The synchronous pumping activities of the heart's two atrioventricular pumps can be described by a single cycle of events called the **cardiac cycle.** The cardiac cycle is divided into two periods: the cycle begins with a period of ventricular relaxation called diastole, and ends with a period of ventricular contraction called **systole.** When the heart beats at its resting rate of once every 0.8 seconds (72 times per minute), diastole occupies the first 0.6 seconds of the cardiac cycle and systole occupies the last 0.2 seconds.

The diastolic period of the cardiac cycle begins with

the entire heart's musculature relaxed (Fig. 1–2A). Both ventricles have just ejected most of their blood, and both atria are nearly filled with blood that has been flowing into them during the terminal moments of the preceding cardiac cycle. All valves (the tricuspid, mitral, pulmonary, and aortic) are closed.

On the right side of the heart, the atrium is filled with oxygen-poor blood received mainly through two large veins: the **superior vena cava** (which conducts oxygen-poor blood drained from the head, neck, thorax, and upper limbs) and the **inferior vena cava** (which conducts oxygen-poor blood drained from the abdomen, pelvis, perineum, and lower limbs) (Fig. 1–1). On the left side of the heart, the atrium is filled with oxygen-rich blood received via four large **pulmonary veins,** a pair of which comes from each lung (Fig. 1–1).

In the early moments of diastole, the increasing blood pressure in each filled atrium soon exceeds that in the adjoining ventricle. The pressure difference between the two chambers forces open the cusps of the atrioventricular valve, and there ensues a flood of

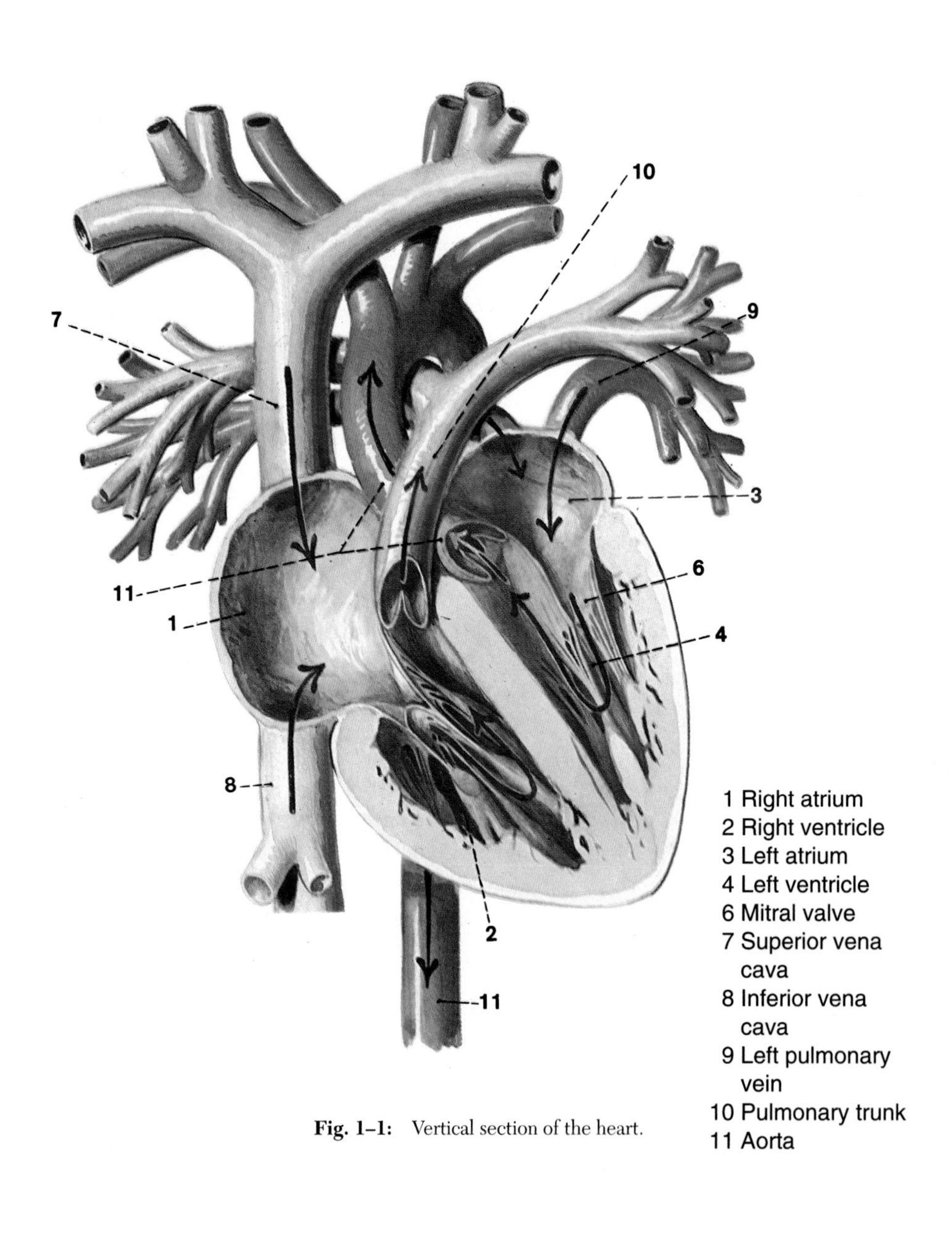

Fig. 1–1: Vertical section of the heart.

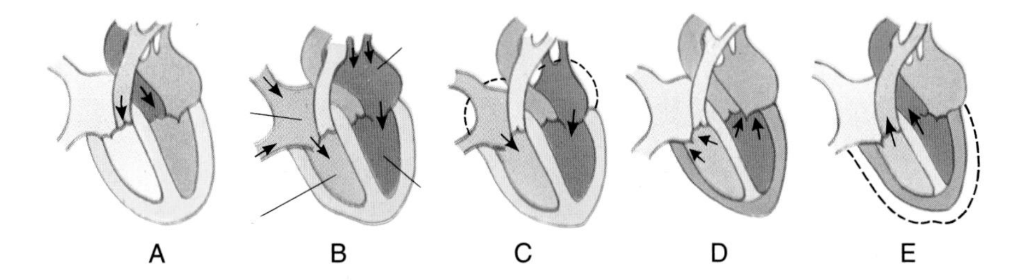

Fig. 1–2: The cardiac cycle. (A) Beginning of diastole upon closure of aortic and pulmonary valves. (B) Opening of the atrioventricular valves during the early moments of diastole. (C) Atrial contraction during the final moments of diastole. (D) Closure of the atrioventricular valves very soon after the beginning of systole. (E) Opening of the aortic and pulmonary valves during systole.

blood flow from the atrium into the ventricle (Fig. 1–2B). The blood flow soon ebbs, and throughout much of the remainder of diastole, each atrium serves essentially as a conduit for transporting blood into the ventricle.

Blood flow into both sides of the heart abruptly ends during the final moments of diastole with the beginning of contraction of the cardiac muscle tissue in the walls of both atria (Fig. 1–2C). During atrial contraction, blood flow into each atrium is momentarily blocked as all of the blood in the atrium is squeezed into the adjoining ventricle. The final moments of diastole are thus characterized by the near-complete emptying of both atria and the final filling of both ventricles.

The beginning of systole is marked by ventricular contraction concurring with atrial relaxation. These concurrent events on both sides of the heart promptly generate a blood pressure in each contracting ventricle that exceeds blood pressure in the adjoining, relaxing atrium. This pressure difference across each atrioventricular valve closes the valve shut (Fig. 1–2D). The tricuspid and mitral valves close shut almost simultaneously, producing immediately after their closure vibrations in the ventricular walls and in the blood confined to the ventricular chambers. When these vibrations reach the chest wall, they collectively form **S1,** the **first heart sound** (the "lub" sound) of each heartbeat.

Continuing ventricular contraction now markedly elevates the blood pressure in each ventricle until it distends the root of the arterial trunk emanating from the ventricle. There occurs a sudden ejection of the high-pressured, ventricular blood into the artery at the moment that the apposed cusps of the valve at the artery's origin become separated by the artery's distention (Fig. 1–2E). Consequently, soon after the onset of systole, oxygen-poor blood is ejected from the right ventricle into the pulmonary trunk and oxygen-rich blood is ejected from the left ventricle into the aorta. The pulmonary trunk and its two terminal branches, the **left and right pulmonary arteries,** conduct the oxygen-poor blood to the lungs for replenishment with oxygen (Fig. 1–1). The aorta and its major branches conduct the oxygen-rich blood to all the regions of the body for their nourishment.

The relaxation of the atrial musculature and the closure of the atrioventricular valves during the early moments of systole initiate the filling of both atria again with blood from their venous sources (Figs. 1–2D and 1–2E). The atria serve as reservoirs of this blood until the atrioventricular valves reopen during the early moments of the diastolic period of the next cardiac cycle.

Toward the end of systole, after the ventricles have ejected most of their blood, there begins a backward flow of blood from the pulmonary trunk and aorta into the ventricles (as a consequence of the blood pressure in each arterial trunk being greater than that in the ventricle). This backward blood flow snaps the pulmonary and aortic valves shut (Fig. 1–2A). The pulmonary and aortic valves snap shut almost simultaneously, generating immediately after their closure vibrations in the walls of both arterial trunks and in the blood they bear. When these vibrations reach the chest wall, they collectively produce **S2,** the **second heart sound** (the "dup" sound) of the heartbeat. At this time, the cardiac cycle begins anew.

OUTLINE OF THE PULMONARY AND SYSTEMIC CIRCULATIONS

During each heartbeat, the heart pumps oxygen-poor blood to the lungs and receives oxygen-rich blood from the lungs. The heart thus sustains a circulation of blood flow between itself and the lungs which serves to replenish the blood's oxygen supply. This circulation of blood flow is called the **pulmonary circulation** (Fig. 1–3).

During each heartbeat, the heart also pumps oxygen-rich blood to all the regions of the body and receives oxygen-poor blood from these regions. The

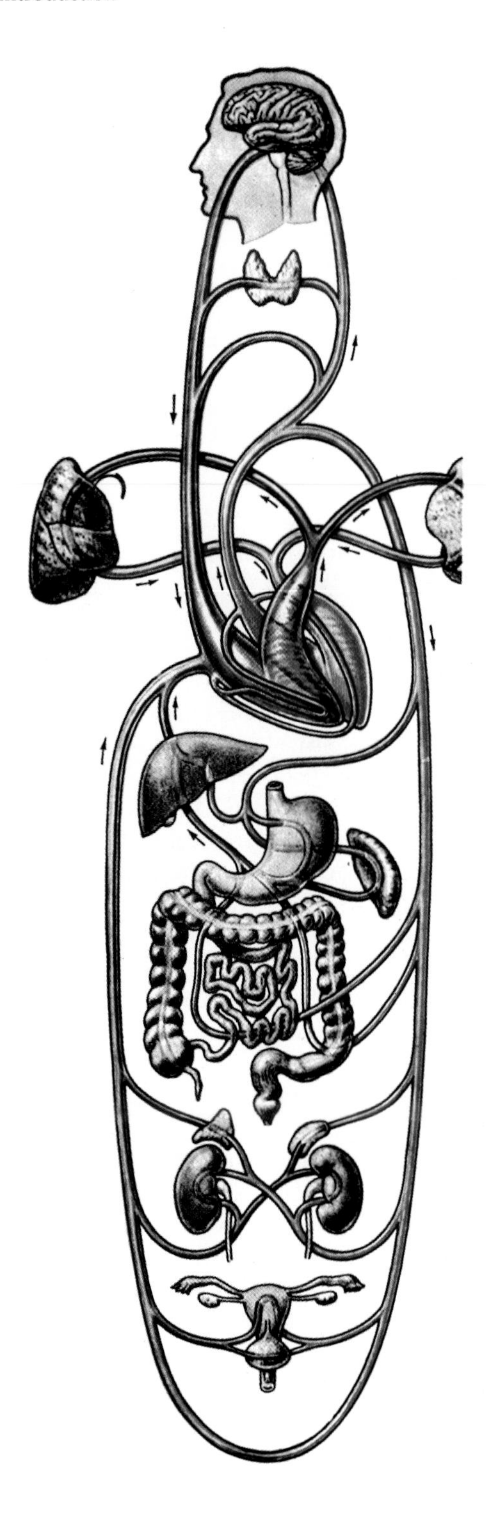

heart, therefore, also sustains a circulation of blood flow between itself and all the regions of the body which serves to deliver nutrients to and remove metabolic waste products from all the tissues and organs of the body. This circulation of blood flow is called the **systemic circulation** (Fig. 1–3).

The blood flow through the cardiovascular system is thus divisible into two circulatory pathways, the larger systemic circulation and the smaller pulmonary circulation. The general arrangement of blood vessels in the pulmonary circulation is basically the same as that in the systemic circulation. The oxygen contents of arterial and venous blood in the pulmonary circulation, however, are opposite to those in the systemic circulation. In the pulmonary circulation, arteries conduct oxygen-poor blood from the heart to the lungs, and veins conduct oxygen-rich blood back to the heart from the lungs. By contrast, in the systemic circulation, arteries conduct oxygen-rich blood from the heart to the body's tissues, and veins conduct oxygen-poor blood from the tissues back to the heart.

Fig. 1–3: General organization of the pulmonary and systemic circulations.

CHAPTER **2**

Overview of the Lymphatic System

The lymphatic system is the second most extensive vascular system of the body. Lymphatic vessels drain excess interstitial fluid from the body's tissues, conduct the fluid to the base of the neck, and there, near the origins of the brachiocephalic veins, return the fluid to the circulating blood. The lymphatic system also consists of tissues and organs which provide immunologic responses: these tissues and organs include the thymus, the spleen, aggregates of lymphoid tissue in the respiratory and gastrointestinal tracts, and lymph nodes.

As blood flows through the capillary bed of a tissue, there is a net flow of fluid out from the capillaries at their arteriolar origins and a net flow of fluid into the capillaries at their venular ends. The net outward flow at the arteriolar origins generates interstitial fluid from **blood plasma.** The net inward flow at the venular ends absorbs interstitial fluid into the blood plasma.

The venular ends of a tissue's blood capillary bed reabsorb all but about 10% of the interstitial fluid formed at the bed's arteriolar origins. There are also some high molecular weight plasma proteins which exit the bloodstream at the bed's arteriolar origins but are not returned to the bloodstream at the bed's venular ends. The excess interstitial fluid and unrecovered high molecular weight plasma proteins are removed from the interstitial spaces by drainage into minute lymphatic vessels called **lymphatic capillaries.** The lymphatic capillaries begin blindly in the tissue's intercellular spaces, and end by anastomosing with each other to form slightly larger lymphatic vessels that exit the tissue. The interstitial fluid drained into lymphatic capillaries is called **lymph,** and the lymphatic vessels that collect lymph from lymphatic capillaries and conduct it away from the tissue are called simply **lymphatics.**

There are some tissues that do not have lymphatic capillaries for the drainage of excess interstitial fluid and unrecovered high molecular weight plasma proteins. Such exceptions include the central nervous system, bone marrow, articular cartilage, epidermis, and the cornea. The interstitial fluid within these tissues flows through minute interstitial channels before ultimately being collected by lymphatics located outside the tissues.

The lymph collected from a tissue generally passes through at least one cluster of **lymph nodes** before finally being emptied into the bloodstream near the origins of the brachiocephalic veins. Lymphatics conducting lymph into a cluster of lymph nodes are called **afferent lymphatics;** the lymphatics draining lymph from the nodes are called **efferent lymphatics.** As lymph passes through lymph nodes, immunologic reactions occur against any foreign agents or neoplastic cells being borne by the lymph.

OUTLINE OF LYMPH FLOW FROM THE ABDOMEN, PELVIS, PERINEUM, AND LOWER LIMBS

Almost all lymph drained from the abdomen, pelvis, perineum, and lower limbs passes through a terminal chain of nodes interspersed along the abdominal aorta and the common iliac, external iliac, and internal iliac arteries (Fig. 2–1). These nodes are grouped and labeled according to the artery along which they lie. On each side of the body, the **external iliac nodes** receive almost all the lymph drained from the corresponding lower limb and the skin and superficial fascia of the ipsilateral side of the trunk below the umbilicus. The external and internal iliac nodes together receive most of the lymph drained from the pelvis and the deep tissues of the perineum. The efferent lymphatics from the external and internal iliac nodes on each side

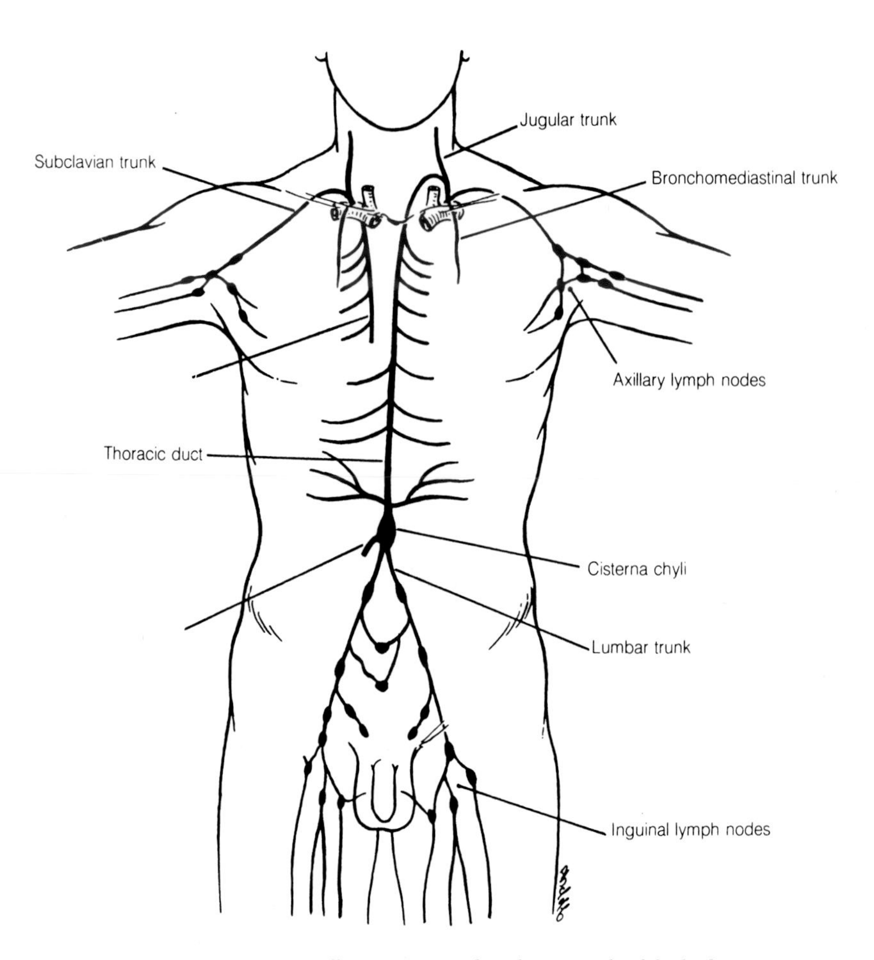

Fig. 2–1: Diagram illustrating major lymphatic vessels of the body.

of the body converge and drain into the **common iliac nodes.** The efferent lymphatics from both sets of common iliac nodes drain into the **aortic (lumbar) nodes.** In addition to receiving efferent lymphatics from the common iliac nodes, the aortic nodes also receive efferent lymphatics from all abdominal viscera and some pelvic viscera.

The efferent lymphatics from the aortic nodes unite with each other in the abdomen at the level of the first or second lumbar vertebra, immediately to the right of the aorta. The union of these vessels forms a dilated sac called the **cisterna chyli** (Fig. 2–1). The cisterna chyli is named for its role of serving as a cistern (reservoir) of **chyle,** the triglyceride-rich lymph drained from the small intestine. The superior margin of the cisterna chyli gives rise to a large lymphatic vessel called the **thoracic duct.** The thoracic duct extends superiorly for only a short distance in the abdomen before it enters the thorax by passing through the aortic opening of the diaphragm. It then ascends within the thorax to the left side of the base of the neck. The thoracic duct ends by uniting with the **left brachiocephalic vein** (near its origin at the juncture of the left internal jugular and left subclavian veins).

OUTLINE OF LYMPH FLOW FROM THE HEAD, NECK, THORAX, AND UPPER LIMBS

On each side of the body, the deep cervical nodes (a group of nodes that lie in the neck along the internal jugular vein) receive nearly all the lymph drained from the corresponding side of the head and neck. The efferent lymphatics from the deep cervical nodes on each side unite with each other to form a large lymphatic vessel called a **jugular trunk** (Fig. 2–1).

On each side of the body, the **apical group of axillary nodes** (a group of nodes that lie in the axilla immediately lateral to the first rib) receive almost all the lymph drained from the corresponding upper limb and the skin and superficial fascia of the ipsilateral side of the trunk above the umbilicus. The efferent lymphatics from the apical group of axillary nodes on each side unite with each other to form a large lymphatic vessel called a **subclavian trunk** (Fig. 2–1).

On each side of the body, nodes located in the mediastinum receive almost all the lymph drained from the corresponding side of the thorax. The efferent lymphatics from the mediastinal nodes on each side unite with each other to form a large lymphatic vessel called a **bronchomediastinal trunk** (Fig. 2–1).

The left jugular, subclavian, and bronchomediastinal trunks end in union with either the left brachiocephalic vein or the thoracic duct (Fig. 2–1). The left brachiocephalic vein thus receives lymph drained from the lower limbs, perineum, pelvis, abdomen, left side of the thorax, left upper limb, and left side of the head and neck.

The right jugular, subclavian, and bronchomediastinal trunks end most commonly by uniting individually with the **right brachiocephalic vein** (near its origin at the juncture of the right internal jugular and right subclavian veins). In about 20% of all individuals, the three trunks join together to form a large but short lymphatic vessel called the **right lymphatic duct,** which ends by directly joining the right brachiocephalic vein near its origin. The right brachiocephalic vein thus receives lymph drained from the right side of the thorax, right upper limb, and right side of the head and neck.

Some of the lymph borne by lymphatic vessels returns to the circulating blood at sites other than those near the origins of the left and right brachiocephalic veins. These sites include passageways between many lymphatic vessels and adjoining veins. The existence of these alternate routes of lymphatic return to the systemic venous system explains the absence of any signs or symptoms when the thoracic duct becomes blocked by disease.

CHAPTER 3

Overview of the Musculoskeletal System

The musculoskeletal system consists of the bones, joints, and skeletal muscles of the body. The skeletal muscles, which are named for their attachment to the bones of the skeleton, move the body.

BONES

Structure of Bones

All bones show in cross section an outer layer of dense mineralization called compact bone and a central region of spongelike mineralization called **cancellous (spongy) bone** (Fig. 3–1). The anastomotic spicules that form the mineralized meshwork of spongy bone are called **trabeculae.** Bone marrow fills the maze of interconnected spaces surrounding the trabeculae of cancellous bone. The bone marrow of adults bears either hematopoietic (blood-forming) or fatty tissue.

Classification of Bones

The skull, sternum, ribs, and vertebrae are called the bones of the **axial skeleton.** The bones of the upper and lower limbs (upper and lower extremities) are called the bones of the **appendicular skeleton.**

Axial and appendicular bones are commonly categorized by shape. The bones of the calvaria (roof of the skull) are termed **flat bones.** The bones of the extremities which have relatively long shafts and enlarged ends are called **long bones.** The roughly cuboidal bones of the extremities (such as the carpals of the wrist and the tarsals of the hindfoot) are called **short bones.** Bones which develop in tendons are shaped similar to sesame seeds, and hence are called **sesamoid bones;** the patellae are the largest sesamoid bones of the body. The vertebrae and the coxal bone are termed **irregular bones.**

Ossification of Bones

All bones develop from membranous condensations of mesenchyme (embryonic connective tissue). The mesenchymal model of the bone is ossified either by a direct mode called **intramembranous ossification** or an indirect mode called **intracartilaginous (endochondral) ossification.** Intramembranous ossification is so named because the deposition of bony tissue occurs within the membranous, mesenchymal model of the bone. The flat bones of the calvaria, the mandible, and most of the clavicles all develop by intramembranous ossification. In intracartilaginous ossification, a cartilaginous model of the bone first replaces the initial mesenchymal model, and the deposition of bony tissue then subsequently occurs within the cartilaginous model. The long bones of the limbs all develop via intracartilaginous ossification.

Intracartilaginous ossification of the long bones of the limbs begins during early fetal development at a single primary site in the middle of the **diaphysis** (shaft) of each long bone (Fig. 3–2). At times characteristic for each long bone, one or more secondary centers of ossification appear at one or both of the cartilaginous ends of the bone (Fig. 3–2). Almost all of these secondary centers of ossification appear after birth. The end part of a long bone which develops from a secondary center of ossification is called an **epiphysis.** At the ends of a long bone which bear one or more growing epiphyses, a zone of hyaline cartilage called an **epiphyseal plate** intervenes between the diaphysis and the one or more epiphyses (Fig. 3–2). The end part of a diaphysis that is in contact with an epiphyseal plate is called a **metaphysis** (Fig. 3–2). The axial growth of long bones during childhood and adolescence occurs primarily by ossification of cartilage continuously formed at the metaphyseal side of epiphyseal plates. When the adult dimensions of a long bone are attained, bony tissue replaces the epiphyseal plates, and the diaphysis fuses with the epiphyses to form a mature bone. Bony tissue replaces almost all the epiphyseal plates in the long bones of the limbs by 25 years of age.

JOINTS

Joints are the sites of union between bones or bodies of cartilage, and consist of the tissues which unite the bones or cartilages at these sites. The **articular surfaces** of a joint are, in general, the surface areas which appose each other at the joint.

Fibrous Joints

In fibrous joints, bones are united by fibrous tissue. The distal tibiofibular joint of the leg and the joints between the flat bones of the calvaria are fibrous joints.

Cartilaginous Joints

In cartilaginous joints, bones are united by cartilage. Cartilaginous joints may be divided into two classes: primary and secondary.

Primary cartilaginous joints (synchondroses) are joints in which bones are united by hyaline cartilage. The union of the diaphysis of a growing long bone with an epiphysis and the union of the first rib with the manubrium of the sternum are two examples of primary cartilaginous joints.

Secondary cartilaginous joints (symphyses) are joints in which the articular surfaces of the bones are covered with plates of hyaline cartilage, and the hyaline cartilage plates are united by a plate of fibrous cartilage. The manubriosternal joint, the pubic symphysis, and the intervertebral joints in which the intervertebral discs unite the bodies of adjacent vertebrae are all examples of secondary cartilaginous joints.

Synovial Joints

A synovial joint is an articulation in which the articular surfaces of the bones are separated from each other within the confines of a membrane-lined, fluid-filled cavity (Fig. 3–3). Synovial joints are, in general, the joints that provide the greatest range and freedom of movement.

The articular surfaces of the bones in a synovial joint are covered with cartilage, which is usually hyaline in type. The fluid-filled cavity is directly lined by a **synovial membrane,** except where the cavity is bordered by articular cartilage (Fig. 3–3). The synovial membrane secretes the synovial fluid that fills the **synovial cavity** of the joint. The entire joint is enclosed within a capsule of fibrous connective tissue.

Some synovial joints have rings, wedges, or discs of fibrous cartilage that are interposed between and articulate with the articular surfaces of the bones. These bodies of articular cartilage enhance the stability of the joint by increasing the congruency of the articular surfaces within the joint. The glenoid labrum of the shoul-

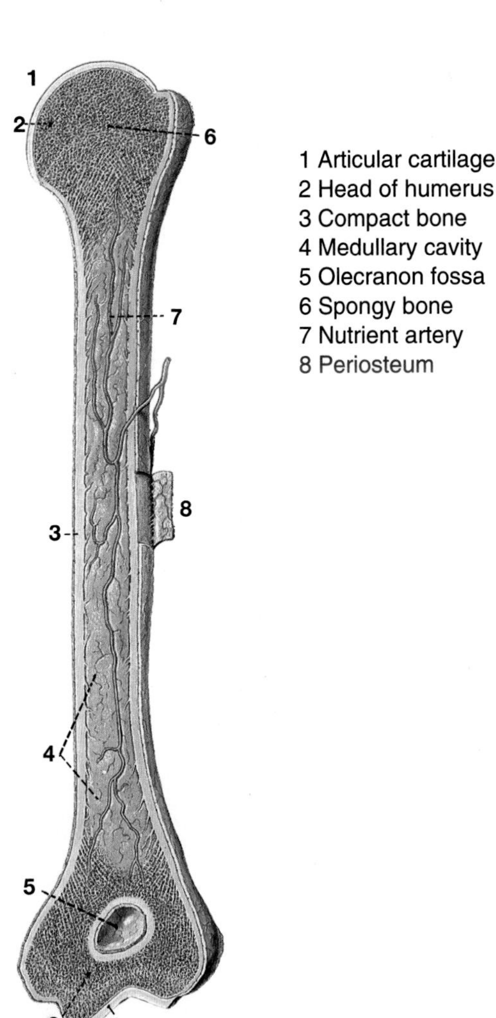

1 Articular cartilage
2 Head of humerus
3 Compact bone
4 Medullary cavity
5 Olecranon fossa
6 Spongy bone
7 Nutrient artery
8 Periosteum

Fig. 3–1: Vertical section of the humerus.

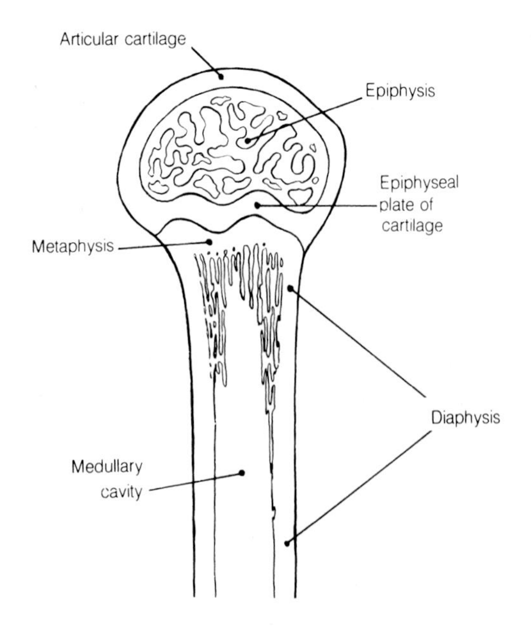

Fig. 3–2: Diagram illustrating the parts of a developing long bone.

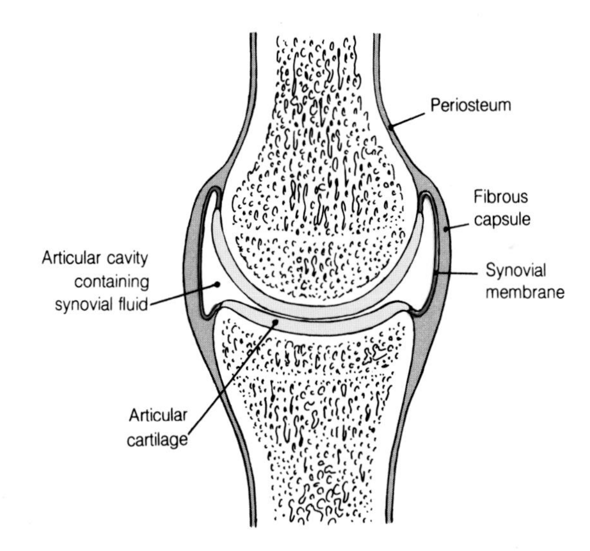

Fig. 3–3: Diagram illustrating major features of a typical synovial joint.

der joint, the acetabular labrum of the hip joint, and the medial and lateral menisci of the knee joint are all examples of bodies of articular fibrous cartilage in a synovial joint.

Stability of Joints

The tone of the muscles acting across a joint is the most important factor in stabilizing the joint. The ligaments that bind together the bones of a joint also contribute to the joint's stability. In most instances, ligaments resist disruptive forces only at the limits of the movements at the joint. Muscles, however, can serve as extensile ligaments that resist disruptive forces throughout the entire range of movements at the joint. It is this capacity of muscles to exert cohesive forces across a joint throughout its entire range of movements that maintains the articular surfaces in their close and proper apposition.

Innervation of Joints

Hilton's Law (which is named in honor of the nineteenth-century English surgeon, John Hilton) best describes the sensory innervation of joints: The nerve supplying a joint also supplies muscles that act across the joint and the skin covering the articular insertions of these muscles.

SKELETAL MUSCLES

The muscles responsible for the gross movements of the upper and lower limbs all consist of skeletal muscle tissue. The connective tissue framework of each skeletal muscle is the mechanical transducer for the forces

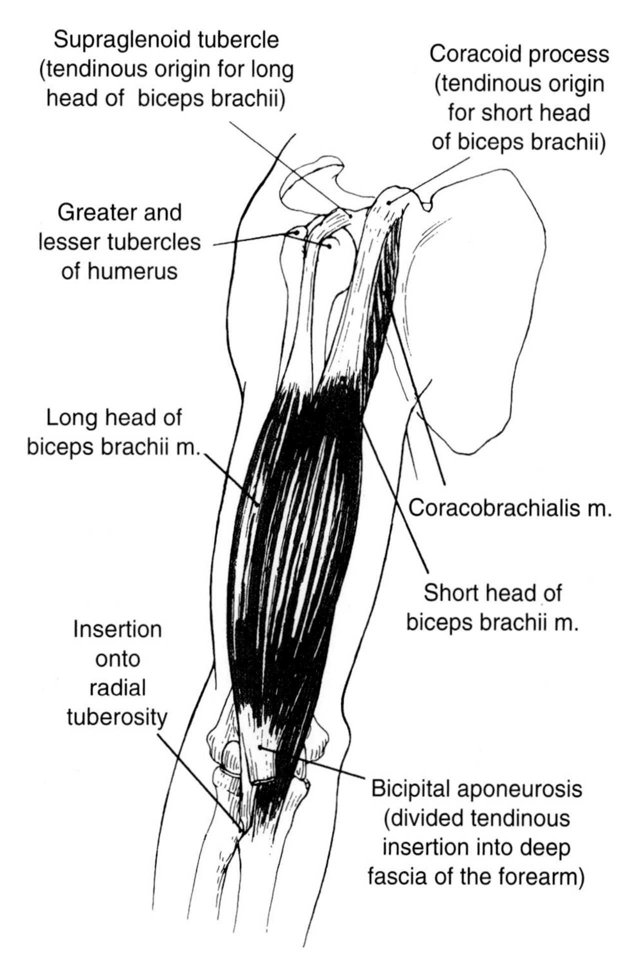

Fig. 3–4: The biceps brachii muscle of the arm. Biceps brachii has two heads of origin and two insertion sites.

generated by the contracting muscle fibers. In other words, the connective tissue elements of a muscle represent the medium which transmits the muscle's tractive forces to the muscle's attachment sites.

Skeletal muscles are attached to structures via bands of connective tissue elements. In some instances, the connective tissue elements of a muscle fuse together to form a connective tissue entity called a **tendon** that extends between the muscle fiber-bearing belly of the muscle and the structure(s) to which it is attached (Fig. 3–4). Since tendons consist mainly of closely packed bundles of collagen fibers, they are very strong, unvarying in length, and relatively flexible. Therefore, tendons can be smoothly curved around bony surfaces to change the final direction of pull of a muscle. Tendons appear whitish because of their low density of vascular networks. Tendons commonly appear as straps or cords. A tendon which is in the shape of a thin, flat sheet is called an **aponeurosis.**

Each skeletal muscle is considered to have two basic attachment sites, called the **origin** and the **insertion** (Fig. 3–4). In many instances, a muscle is attached to

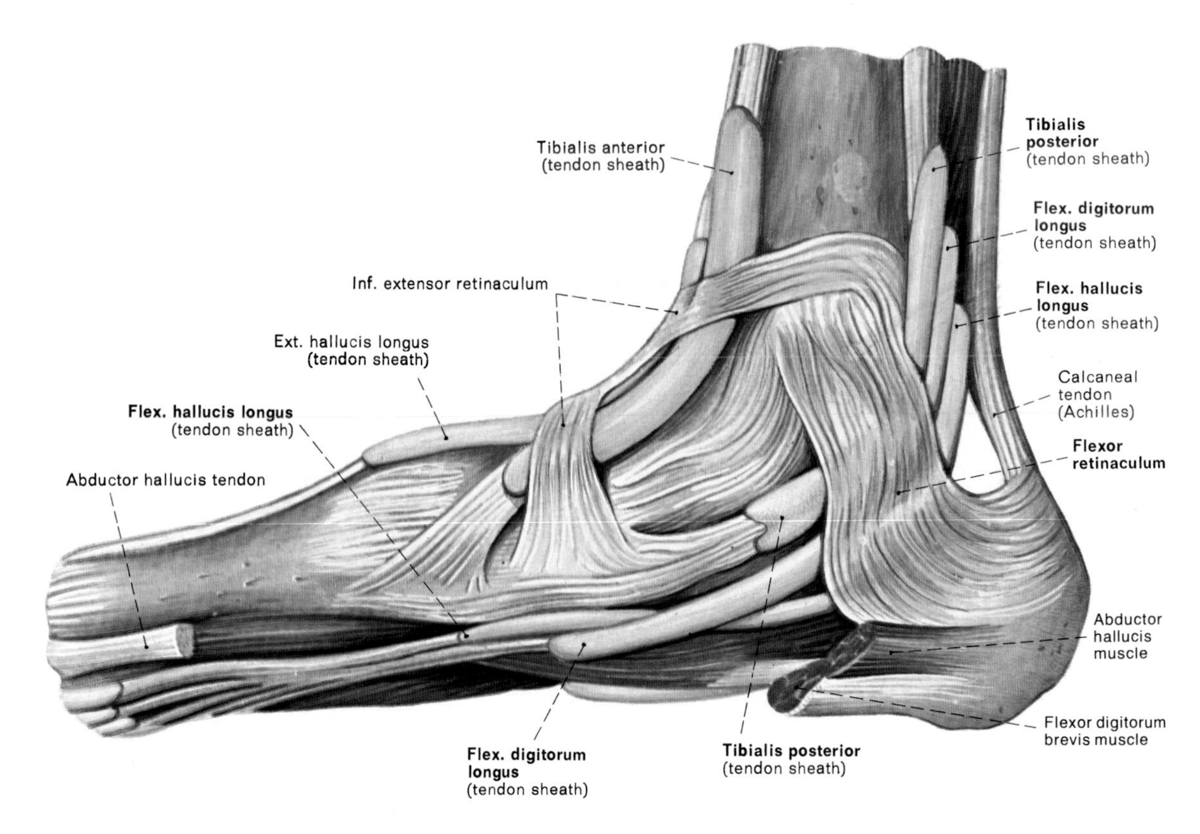

Fig. 3–5: Medial view of the tendons, synovial sheaths, and retinacula of the right ankle region.

more than one structure at its origin and/or insertion. In the limbs, each muscle's origin is the more proximal attachment site, and its insertion is the more distal attachment site. In general, the origin of a muscle is the attachment site of lesser movement. If there are distinct intervals along the boundary of a muscle's origin, the muscle is said to have two or more heads of origin (Fig. 3–4).

Muscle Action

When a skeletal muscle is stimulated and its fibers contract, the muscle's tension and length can be independently increased, decreased, or maintained constant. In other words, muscle contraction does not always produce muscle shortening and/or an increase in muscle tension. If muscle contraction generates a change in the tension of the muscle but not its length, the muscle's contraction is labelled **isometric.** If, on the other hand, muscle contraction generates a change in the length of the muscle but not its tension, the muscle's contraction is labelled **isotonic.** Muscle contraction which produces muscle shortening is called **concentric** contraction. Muscle contraction which occurs in association with muscle lengthening is called **eccentric** contraction.

When a muscle concentrically contracts, and, in so doing, approximates its insertion toward its origin, it exerts its action(s). In other words, the expression of a muscle's action(s) requires that the muscle's origin remain fixed as the muscle pulls upon its insertion. The one or more actions of a muscle are the effects that are produced on the part or parts of the body that are moved. The actions of a muscle necessarily depend on the muscle's relationship with the joint(s) that it crosses and also on the types of movements permissible at the joint(s).

Functional Classification of Skeletal Muscles

Skeletal muscles generally collaborate in the production of movements. Because muscles act as members of functional associations, there is a functional classification of the role that individual muscles or muscle groups play in particular movements.

A **prime mover** is the muscle or a member of the group of muscles chiefly responsible for a particular

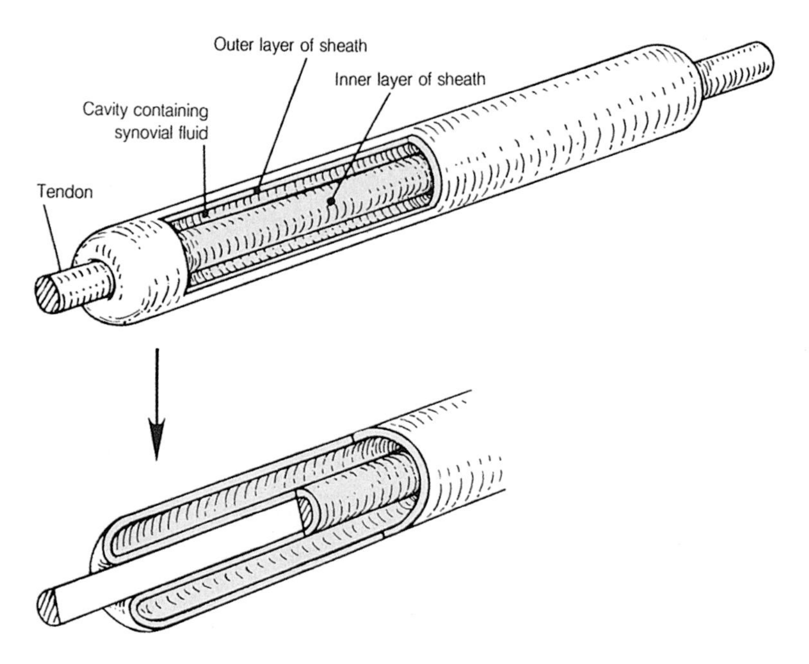

Fig. 3–6: Diagram illustrating the arrangement of a synovial sheath about a tendon.

movement. An **antagonist** is any muscle that opposes the action(s) of a prime mover. A **fixator** is any muscle whose isometric contraction stabilizes the origin of a prime mover. A **synergist** is any muscle whose contraction opposes an unwanted movement produced by the unrestricted contraction of a prime mover (the overview of the actions of the anterior and posterior forearm muscles in Chapter 10 discusses how the carpi muscles serve as synergists of the forearm muscles which can flex or extend the digits).

ACCESSORY STRUCTURES RELATED TO MUSCLE ACTION

Fascia

Fascia is a term that generally refers to any collection of connective tissue large enough to be described by the unaided eye. **Superficial fascia** is the layer of loose areolar connective tissue which lies immediately beneath the skin. **Deep fascia** consists of the layer or layers of much denser fibrous connective tissue which underlie the superficial fascia.

In the wrist and ankle regions, the deep fascia is thickened into retention bands called **retinacula** (Fig. 3–5). The retinacula are attached along their sides to underlying bones. Together the retinacula and bones form tunnel-like corridors through which pass the tendons of muscles in route to their insertion sites. The retinacula retain the tendons in constant, close association with the underlying bones during movements of the hands and feet.

Bursae and Synovial Sheaths

A bursa is a flattened, fluid-filled sac. The wall of a bursa consists of an internal synovial membrane supported externally by dense connective tissue. The opposed walls of a bursa are separated only by a capillary film of synovial fluid. Bursae establish discontinuities between tissues which provide for much freedom of movement over a limited range .

A synovial sheath is an elongated bursa wrapped around a tendon (Fig. 3–6). Synovial sheaths envelop the tendons which traverse beneath the retinacula at the wrist and ankle (Fig. 3–5).

CHAPTER 4

The Back

This chapter focus on the principal structures of the back: (a) the vertebral column and the intrinsic back muscles that support it, (b) the spinal cord and the membranes, or meninges, that envelop it, and (c) the spinal nerves that arise from the spinal cord.

THE VERTEBRAL COLUMN

Names of the Vertebrae

The vertebral column extends inferiorly from the base of the skull through the back of the neck and the back of the trunk of the body (Fig. 4–1). It ends at the level of the floor of the pelvis. The vertebral column transmits the spinal cord and the roots of the spinal nerves.

The 33 vertebrae are subdivided into

7 **cervical vertebrae** in the neck,
12 **thoracic vertebrae** in the thorax,
5 **lumbar vertebrae** in the abdomen, and
5 **sacral vertebrae** and
4 **coccygeal vertebrae** in the pelvis.

The vertebrae in each subdivision are numbered, with the first vertebra in each subdivision being the most superior vertebra. The most superior thoracic vertebra is thus called the first thoracic vertebra, and the most inferior lumbar vertebra is called the fifth lumbar vertebra. By adulthood, the five sacral vertebrae fuse with each other to form a wedge-shaped bone called the **sacrum,** and the four coccygeal vertebrae fuse with each other to form an arrowhead-shaped bone called the **coccyx** (Fig. 4–1).

Each vertebra from the third cervical to the fifth lumbar is separated from the vertebra immediately above and the vertebra immediately below by a disc of fibrous cartilage called an **intervertebral disc** (Fig. 4–2). There is no intervertebral disc between the first and second cervical vertebrae. Remnants of intervertebral discs may be found within the sacral and coccygeal regions of the vertebral column.

Vertebral Column Curvatures

The vertebral column in a midgestational fetus is concave anteriorly (Fig. 4–3A). At some time between late intrauterine life and 3 to 4 months after birth, a secondary curvature which is concave posteriorly appears

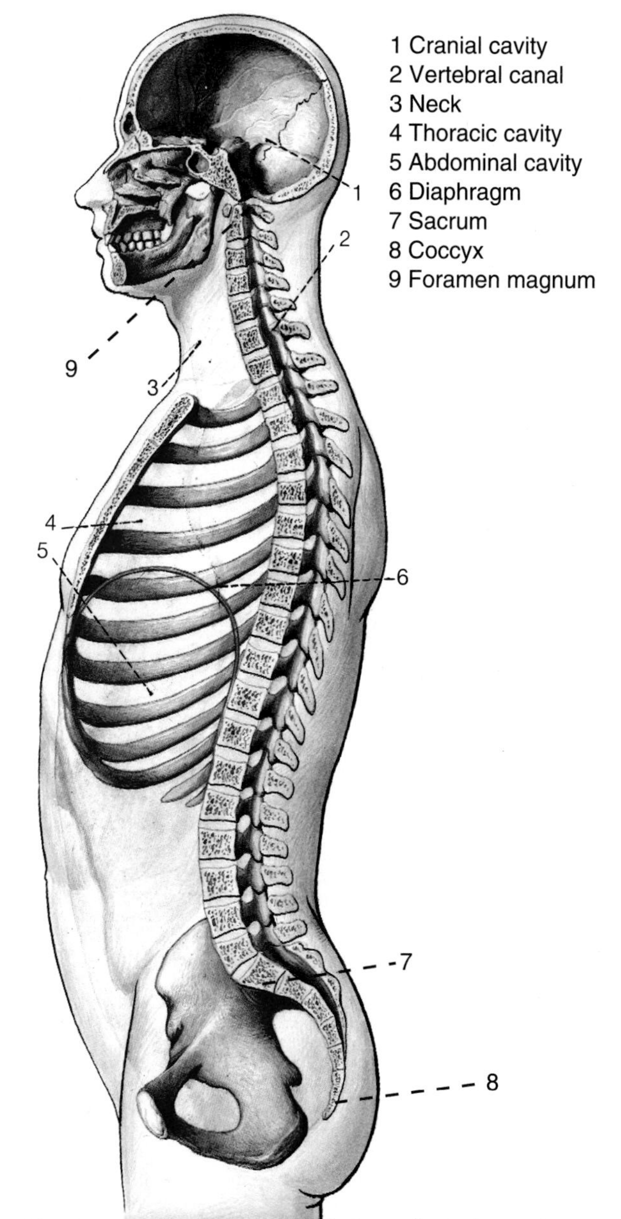

Fig. 4–1: Median section of the skull, vertebral column, rib cage, and bony pelvis.

in the neck region. This cervical curvature is an adaptation by the cervical region of the vertebral column to support the head. At some time between 12 and 18 months after birth, another secondary curvature which is concave posteriorly appears in the lumbar region. This lumbar curvature is an adaptation by the lumbar region of the vertebral column to support the upper body in the upright posture. There are thus four normal curvatures in the vertebral column of each individual after one and a half years of age (Fig. 4–3B):

(secondary) cervical curvature,
(primary) thoracic curvature,
(secondary) lumbar curvature, and
(primary) sacral curvature.

The primary curvatures are concave anteriorly, and the secondary curvatures are concave posteriorly. The formations of the secondary curvatures arise primarily from changes in the shapes of intervertebral discs.

An exaggerated thoracic curvature is called a **kyphosis** (Fig. 4–4). An exaggerated lumbar curvature is called a **lordosis.** An abnormal lateral curvature is called a **scoliosis.**

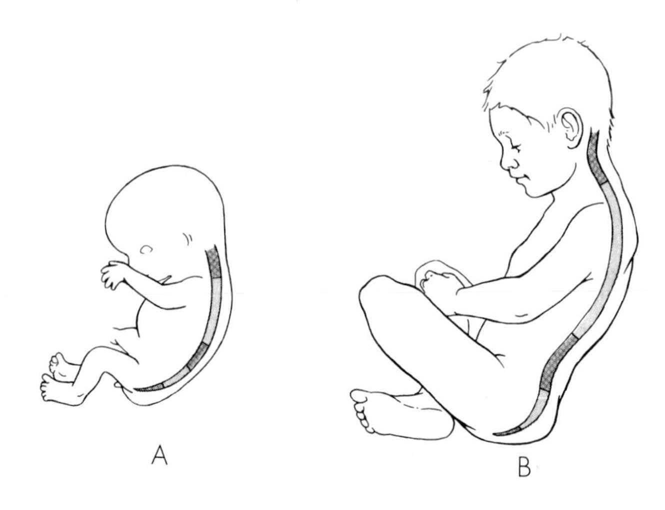

Fig. 4–3: Diagrams showing the normal curvatures of the vertebral column in (A) a midgestational fetus and (B) a 4-year old child.

Structure of Typical Vertebrae

A typical vertebra is composed of a **body** anteriorly and a **vertebral arch** posteriorly (Fig. 4–5). The space encircled by the body and vertebral arch is called the **vertebral foramen;** the spinal cord descends through this foramen. The vertebral arch is composed of paired **pedicles** and paired **laminae.** The pedicles, which are the most anterior parts of the vertebral arch, project posterolaterally from the vertebral body; the laminae, which are the most posterior parts of the vertebral arch, fuse with each other in the midline.

Seven processes project outward from the vertebral arch (Fig. 4–5). A **spinous process (spine)** projects posteriorly from the region of union of the laminae. A transverse process projects laterally on each side from the region of union between the pedicle and lamina. **Superior and inferior articular processes** on each side respectively project superiorly and inferiorly from the pedicle.

The spinous and transverse processes of the cervical, thoracic, and lumbar vertebrae bear the attachment sites for most of the intrinsic muscles of the back. The spinous and transverse processes serve as levers which enhance the mechanical advantage of these deep back muscles in their control of vertebral movements.

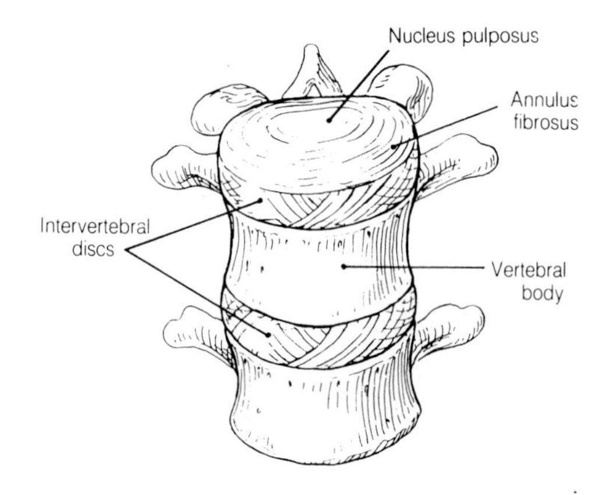

Fig. 4–2: Diagram illustrating the separation of vertebral bodies in the vertebral column by intervertebral discs.

Articulations of Typical Vertebrae

Each typical vertebra forms two synovial joints and one secondary cartilaginous joint with (a) the vertebra above and (b) the vertebra below. Between each adjacent pair of typical vertebrae, the inferior articular processes of the vertebra above form a pair of synovial joints with the superior articular processes of the vertebra below, and the body of the vertebra above forms a secondary cartilaginous joint with the body of the vertebra below. The intervertebral discs are the discs of fibrous cartilage which unite the bodies of adjacent vertebrae in their secondary cartilaginous joints.

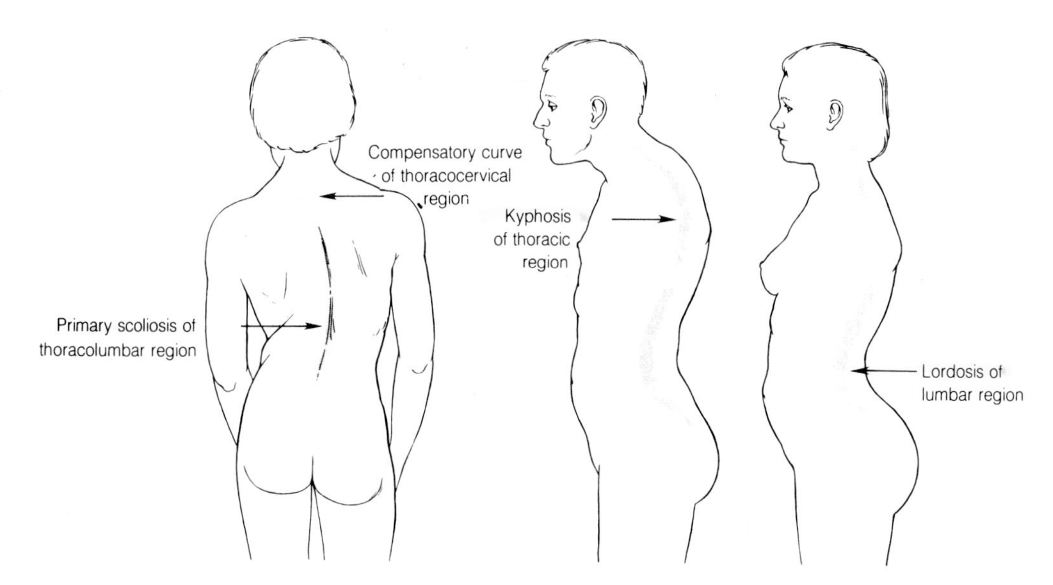

Fig. 4–4: Diagrams illustrating abnormal curvatures of the vertebral column.

The intervertebral discs account for about one-quarter of the length of the vertebral column. Each intervertebral disc consists of a central region called the **nucleus pulposus** surrounded by a circumferential region called the **anulus fibrosis** (Fig. 4–2). The nucleus pulposus is a gelatinous mass with a very high water content (70 to 80%); it functions as a noncompressible but deformable pad of tissue between the bodies of two vertebrae. Some water is lost from the nuclei pulposi of the vertebral column during the waking hours of each day; although the loss can shorten a person's height by as much as 1 cm, almost all the water is recovered during the night's sleep. The anulus fibrosis is a fibrocartilaginous band of tissue which retains in place the more fluid nucleus pulposus.

Ligaments of the Vertebral Column

The ligaments of the vertebral column stabilize the alignment between adjacent vertebrae. The **supraspinous ligaments** connect the tips of the spinous processes (Fig. 12–9). The **interspinous ligaments** stretch between adjacent spinous processes. The **flaval ligaments** extend between adjacent laminae. The **anterior and posterior longitudinal ligaments** connect the vertebral bodies and intervertebral discs anteriorly and posteriorly, respectively (Fig. 12–5).

INTRINSIC MUSCLES OF THE BACK

The intrinsic muscles of the back are the muscles chiefly responsible for the posture of the vertebral column. They lie for the most part in the posterolateral grooves of the vertebral column that on each side are bordered anteriorly by the transverse processes of the vertebrae and medially by their spinous processes. In the lumbar and lower thoracic regions, the intrinsic muscles of the back are all covered superficially by a collection of deep fascial layers collectively called the **thoracolumbar fascia.** The intrinsic muscles of the back are all innervated by posterior rami of spinal nerves.

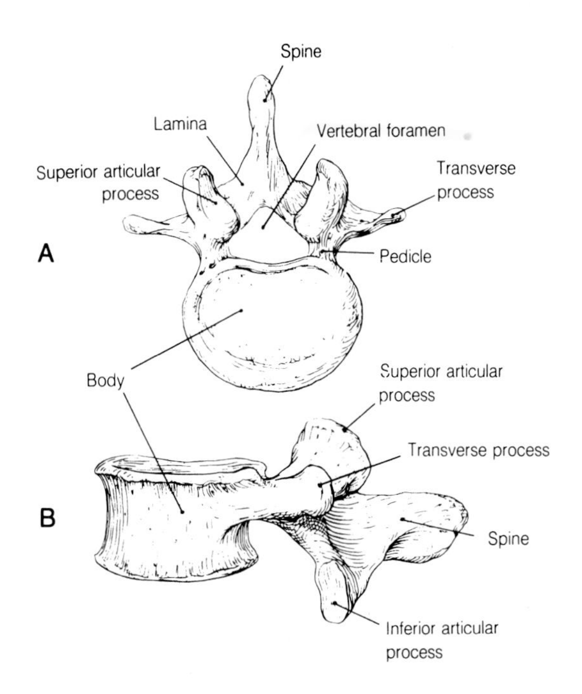

Fig. 4–5: Diagrams illustrating (A) superior and (B) left lateral views of a typical lumbar vertebra.

Erector Spinae

Erector spinae is the principal intrinsic muscle of the back (Fig. 4–6). Its chief actions are extension and lateral flexion of the vertebral column. Its muscle fibers have multiple attachments as they collectively extend superiorly from the sacrum to the back of the head. The muscle as a whole originates from the sacrum, iliac crest, and the spinous processes of all five lumbar vertebrae and the two lowest thoracic vertebrae. It ascends from the upper lumbar region as three vertical columns of muscle fibers.

The most lateral column of muscle fibers is called **iliocostalis** (Fig. 4–6). Iliocostalis fibers are divisible into three groups: (i) fibers which extend from the origin of erector spinae to the lower ribs, (ii) fibers which extend for the most part from the lower ribs to the upper ribs, and (iii) fibers which extend from the upper ribs to the transverse processes of the lower cervical vertebrae.

The intermediate column of muscle fibers is called **longissimus** (Fig. 4–6). Longissimus fibers insert for the most part onto the transverse processes of lumbar, thoracic, and cervical vertebrae. Some fibers extend from the upper thoracic and lower cervical vertebrae to the mastoid process of the temporal bone of the skull.

The most medial column of muscle fibers is called **spinalis** (Fig. 4–6). Spinalis fibers insert mainly onto the spinous processes of upper thoracic vertebrae.

Transversospinal Muscles

The transversospinal muscles lie deep to erector spinae in the posterolateral grooves of the vertebral column (Fig. 4–7). Their chief actions are extension and rotation of the vertebral column.

The transversospinal muscles are so named because, for the most part, they extend from the transverse process of one vertebra to the spinous process of a vertebra above. There are three groups of transversospinal muscles: the semispinales, multifidus, and rotatores groups. Each of these groups consists of numerous individual muscles. The semispinales muscles ascend the greatest number of vertebral levels (generally 5 or 6) between their origin and insertion; the rotatores muscles ascend the smallest number of vertebral levels (either 1 or 2) between their origin and insertion.

SPINAL CORD

The major components of the human nervous system are the brain, the spinal cord, and the nerves which arise from the brain and spinal cord. The 12 pairs of nerves which arise from the brain are called the **cranial nerves,** and the 31 pairs of nerves which arise from the spinal cord are called the **spinal nerves.**

The brain and spinal cord together are commonly referred to as the **central nervous system;** the cranial and spinal nerves collectively are referred to as the **peripheral nervous system.** It is important to understand, however, that the expressions "central nervous system" and "peripheral nervous system" have only artificial descriptive value, since there is neither an anatomic nor a functional basis for the division of the human nervous system into two such subsidiary nervous systems.

Gross Morphology of the Spinal Cord and Its Inferior Limits in Neonates and Adults

The spinal cord begins just below the **foramen magnum** as a continuation of the medulla oblongata [the foramen magnum is the largest opening in the base of the skull (Fig. 4–1), and the medulla oblongata is the most inferior part of the brainstem]. The spinal cord descends for its entire length within the **vertebral canal,** which is the passageway that serially extends through the vertebral foramina of the vertebral column (Fig. 4–1).

The spinal cord is roughly in the shape of a cylinder that is slightly flattened both anteriorly and posteriorly (Fig. 4–8). It has anterior and posterior midline clefts respectively called the **anterior median fissure** and the **posterior median septum.** It exhibits slight enlargements of its diameter near its upper and lower ends respectively called the **cervical and lumbosacral enlargements** (Fig. 4–9). The inferior end of the spinal cord is tapered into the shape of a cone, and is called the **conus medullaris.**

During early embryogenesis, the spinal cord initially extends throughout the entire length of the vertebral canal. However, during subsequent embryonic and fetal development, the vertebral column grows longitudinally faster than the spinal cord. The consequence of this difference in growth is that at birth the conus medullaris generally lies at the level of the second or third lumbar vertebra. During childhood and adolescence, the vertebral column continues to grow longitudinally faster than the spinal cord. The consequence of this final growth differential is that in an adult the conus medullaris generally lies at the level of the lower border of the first lumbar vertebra (Fig. 4–10).

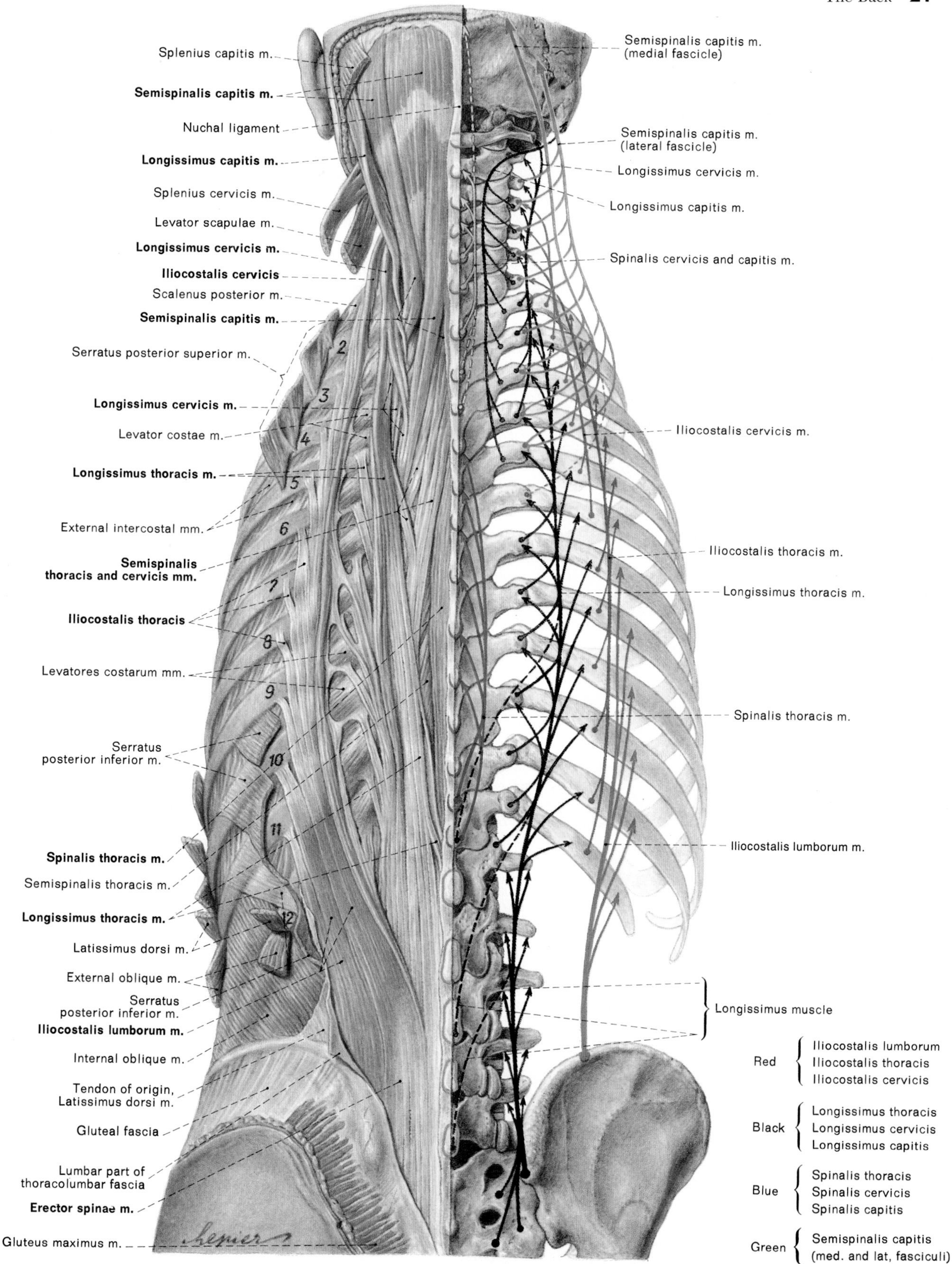

Fig. 4–6: Deep muscles of the back: Erector spinae. NOTE: 1) *on the left,* the erector spinae (sacrospinalis) muscle is separated into its iliocostalis, longissimus and spinalis portions. In the neck observe the semispinalis capitis which has both medial and lateral fascicles. The semispinalis cervicis and thoracis extend inferiorly from above, and lie deep to the sacrospinalis layer of musculature. 2) *on the right,* all of the muscles have been removed and their attachments have been diagrammed by means of colored lines.

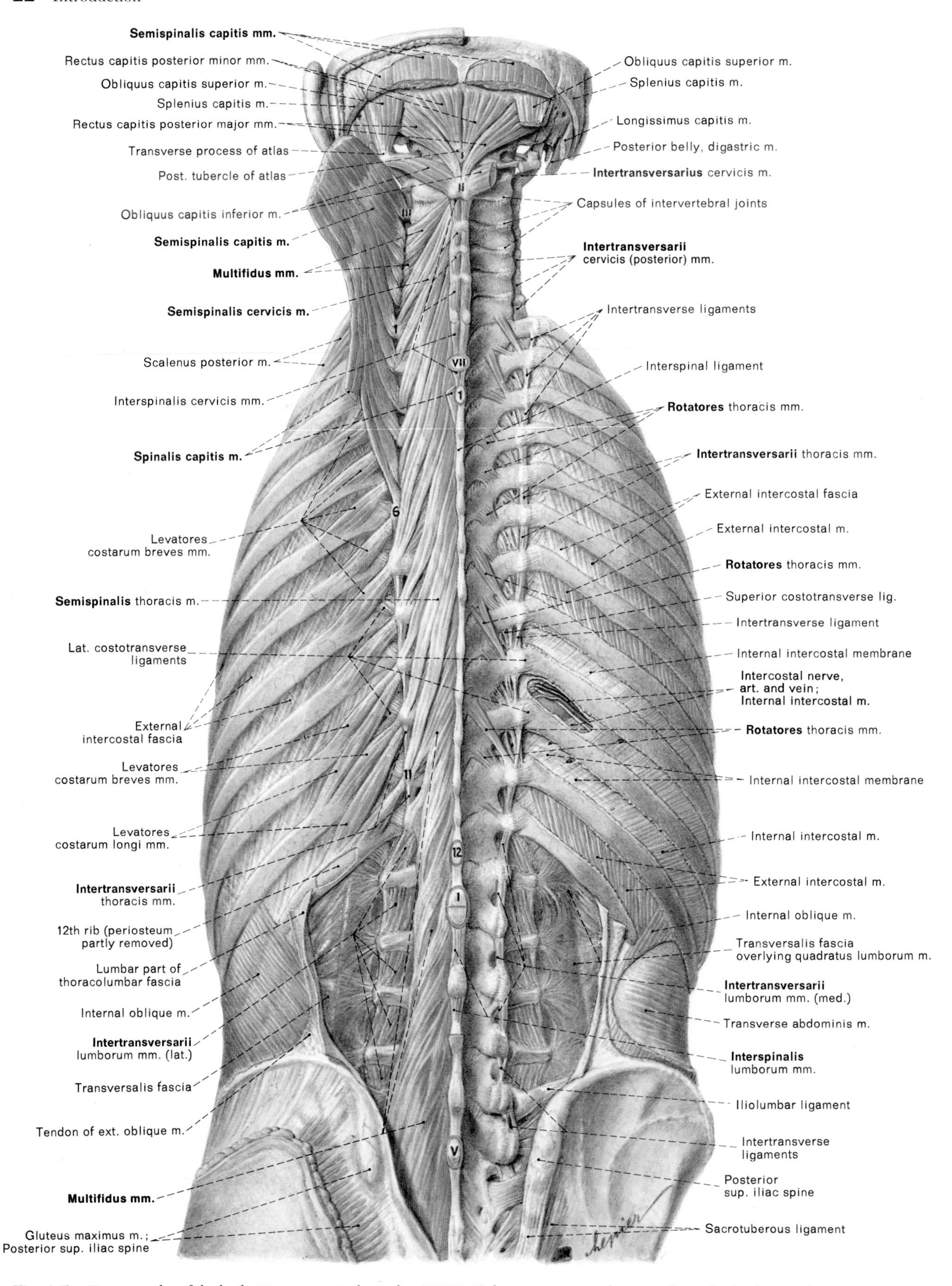

Fig. 4–7: Deep muscles of the back: Transversospinal muscles. NOTE: 1) the transversospinal groups of muscles lie deep to the erector spinae muscles and generally extend between the transverse processes of the vertebrae to the spinous processes of more superior vertebrae. Their actions extend the vertebral column, or upon acting individually and on one side, they bend and rotate the vertebrae. 2) within this group of muscles are the semispinalis (thoracis, cervivis, and capitis), the multifidus, the rotatores (lumborum, thoracis, cervicis), the interspinales (lumborum, thoracis, cervicis) and the intertransversarii.

Names of the Spinal Nerves

31 pairs of spinal nerves emerge from the spinal cord as it descends through the vertebral canal. They consist of

8 pairs of **cervical spinal neves,**
12 pairs of **thoracic spinal nerves,**
5 pairs of **lumbar spinal nerves,**
5 pairs of **sacral spinal nerves,** and
1 pair of **coccygeal spinal nerves.**

The spinal nerves in each group are numbered, with the first pair in each group being the most superior pair. The most superior pair of thoracic spinal nerves is thus called the first thoracic pair, and the most inferior pair of lumbar spinal nerves is called the fifth lumbar pair.

It is common practice to refer to all the spinal nerves except for the first coccygeal pair via an abbreviated notation which designates first the first letter of the name of the spinal nerve and then secondly the number of the spinal nerve (Fig. 4–10). For example, the first cervical spinal nerves may be referred to as C1, the second thoracic as T2, the third lumbar as L3, and the fourth sacral as S4. The first coccygeal spinal nerves may be designated by Coc1.

Roots and Rami of the Spinal Nerves

Each spinal nerve begins as a series of **anterior and posterior rootlets** which respectively emerge from the anterolateral and posterolateral aspects of the spinal cord (Fig. 4–8). As the rootlets extend laterally from the spinal cord, the anterior rootlets merge with each other to form an **anterior root,** and the posterior rootlets merge with each other to form a **posterior root.** As the roots exit the vertebral canal via an **intervertebral foramen,** they unite with each other to form a **spinal nerve trunk** (the intervertebral foramina of the vertebral column are the spaces on each side between the pedicles of adjacent vertebrae). The spinal nerve trunk extends for only a few millimeters before branching into two nerve bundles called the **anterior and posterior rami.**

The thoracic, lumbar, sacral, and coccyeal spinal nerves each exit the vertebral column via the interver-

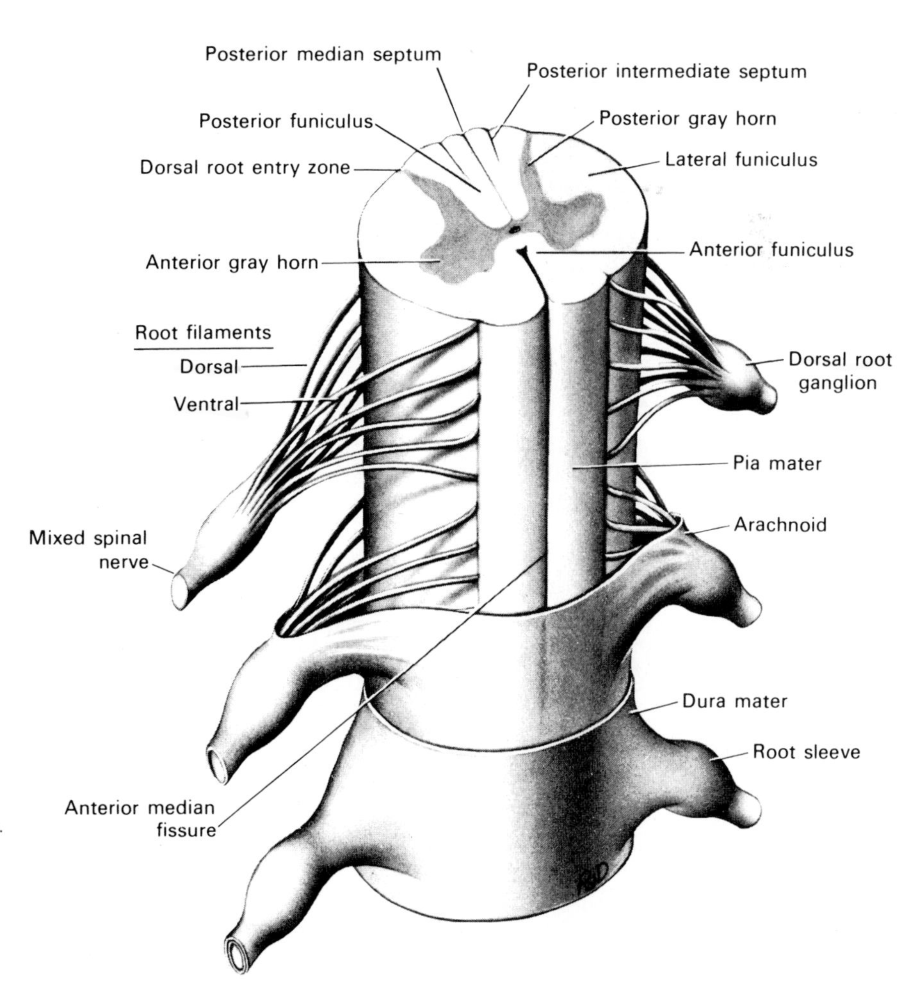

Fig. 4–8: Diagram of the spinal cord, roots of the spinal nerves, and meninges.

tebral foramen immediately below the vertebra of the same name (Fig. 4–10). The cervical spinal nerves, except for the eighth cervical pair, each exit via the intervertebral foramen immediately above the vertebra of the same name. The eighth cervical spinal nerves exit via the intervertebral foramina between the seventh cervical and first thoracic vertebrae.

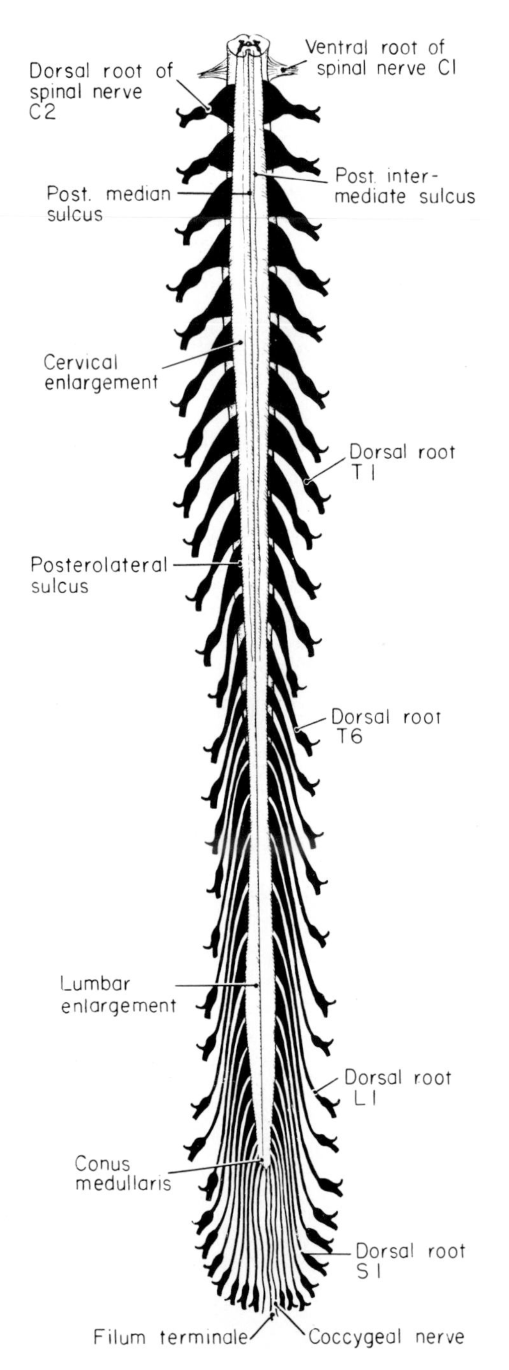

Fig. 4–9: Posterior view of the spinal cord showing the dorsal roots and dorsal root ganglia of the spinal nerves. Letters and numbers indicate corresponding spinal nerves.

The transverse segment of the spinal cord which gives rise to the anterior and posterior rootlets for each pair of spinal nerves is called the **spinal cord segment** for the spinal nerve pair. The spinal cord is thus divisible, for descriptive purposes, into 31 segments. The spinal cord segments which give rise to the spinal nerves that form the **brachial plexus** collectively constitute the cervical enlargement of the spinal cord (the brachial plexus is the nerve plexus which gives rise to almost all the sensory and motor fibers for the upper limb; the brachial plexus is formed from the anterior rami of C5-T1). The spinal cord segments which give rise to the spinal nerves that form the sacral plexus collectively constitute the lumbosacral enlargement of the spinal cord (the sacral plexus is the nerve plexus which gives rise to many of the sensory and motor fibers for the lower limb; the sacral plexus is formed from the anterior rami of L4–S4).

The cervical and thoracic spinal cord segments lie at or near the vertebral levels at which their spinal nerves exit the vertebral column (Fig. 4–10). The lumbar, sacral, and coccygeal spinal cord segments, however, lie at vertebral levels markedly higher than those levels at which their spinal nerves exit the vertebral column. In an adult, the lumbar, sacral, and coccygeal segments reside between the levels of the eleventh thoracic and first lumbar vertebrae (Fig. 4–10). The relatively high placement of the lumbar, sacral, and coccygeal segments in the vertebral column is a result of the difference in longitudinal growth between the spinal cord and the vertebral column during intrauterine life, childhood, and adolescence.

The roots of the cervical spinal nerves extend almost horizontally from the spinal cord to their respective intervertebral foramina (Fig. 4–10). The roots of the thoracic spinal nerves descend 1 to 2 vertebral levels before entering their respective intervertebral foramina. In an adult, the lumbar, sacral, and coccygeal spinal nerve roots descend to or below the level of the conus medullaris before entering their respective intervertebral foramina. The collection of spinal nerve roots that descends below the conus medullaris resembles the collection of hairs at the end of a horse's tail, and thus the collection is called the **cauda equina** (which is the Latin expression for horse's tail).

Organization of Neuronal Elements within the Spinal Cord and the Roots and Rami of the Spinal Nerves

Each spinal cord segment transmits motor nerve fibers to and receives sensory nerve fibers from its pair of spinal nerves. The posterior rami of the 31 spinal nerves collectively innervate the musculoskeletal tissues (bones, joints, and skeletal muscles) and skin of the midline region of the back of the neck and body trunk. The anterior rami of the 31 spinal nerves collectively innervate almost all the musculoskeletal tissues and skin of the remainder of the neck and body trunk and almost all of the musculoskeletal tissues and skin of the upper and lower limbs.

The spinal cord may be divided, for descriptive purposes, not only transversely into its segments, but also longitudinally into central and peripheral columnar regions. Whereas the central columnar regions bear neuronal cell bodies and unmyelinated nerve fibers at relatively high densities, the peripheral columnar regions bear myelinated nerve fibers at relatively high density. The markedly high density of whitish myelin in the peripheral columnar regions renders them lighter than the central columnar regions. The central columnar regions, which collectively present an H-shaped outline when viewed in horizontal cross section, are thus called the **gray matter** of the spinal cord, and the peripheral columnar regions its **white matter** (Fig. 4–8). The vast majority of myelinated fibers within the peripheral columnar regions are either ascending or descending within the spinal cord.

The crossbar of the H-shaped gray matter is called the **gray commissure.** On each side, the anterior and posterior parts of the post of the H-shaped gray matter are respectively called the **anterior and posterior gray columns.** The parts of the anterior and posterior gray columns which lie at the level of a spinal cord segment are called the anterior and posterior gray horns of that segment (Fig. 4–8). The T1 to L2 segments collectively bear a gray matter region on each side called a **lateral gray** column which projects laterally from between the anterior and posterior gray columns. The white matter on each side is grossly divided into three cordlike columns called the **anterior, lateral, and posterior funiculi** (Fig. 4–8).

Each spinal cord segment bears the cell bodies of the motor fibers which the segment's spinal nerve pair transmits to the periphery. The motor neurons are called **lower motor neurons,** and their cell bodies reside within

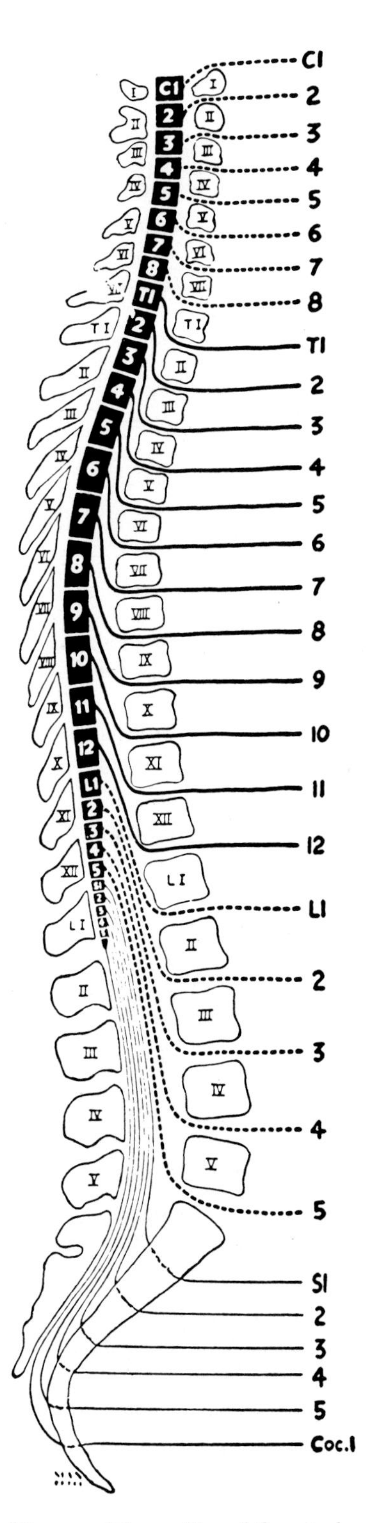

Fig. 4–10: Diagram of the position of the spinal cord segments with reference to the bodies and spinous processes of the vertebrae. Locations of emergence and entrance of spinal roots are indicated.

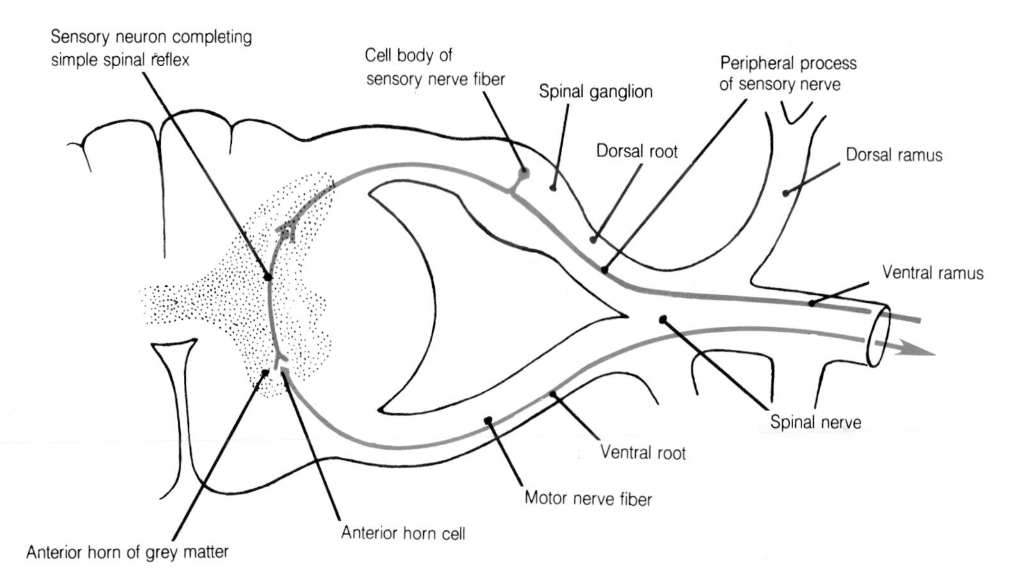

Fig. 4–11: Diagram illustrating the course of motor and sensory nerve fibers between the spinal cord and a spinal nerve.

the anterior gray horns of the segment (Fig. 4–11).

The cell bodies of the sensory fibers which each segment's spinal nerve pair transmits from the periphery reside, on each side, within the enlargement, called the **spinal ganglion,** that occurs along the length of the posterior root. A spinal ganglion may also be called a **dorsal (posterior) root ganglion.**

All the motor fibers that exit each spinal cord segment exit via the anterior roots only of that segment's spinal nerve pair (Fig. 4–11). All the sensory fibers that enter each spinal cord segment enter via the posterior roots only of that segment's spinal nerve pair. The roots and rami of spinal nerves thus differ in that whereas rami transmit both motor and sensory fibers, anterior roots transmit motor fibers only and posterior roots transmit sensory fibers only.

Dermatomes and Myotomes

The segmental derivation of the spinal nerves is manifested on the surface of the body by the segmental innervation of the skin. In other words, the cutaneous sensory neurons of the spinal nerves innervate the body's skin in a segmental, strip-like fashion. The strip-like area of skin innervated by the cutaneous sensory neurons of a spinal nerve is called the dermatome of that spinal nerve. Figure 4–12 shows the dermatomes of the body. The dermatomes have been determined by mapping the cutaneous areas that lose sensation upon complete transection of individual or multiple cutaneous nerves.

The dermatomes exhibit three major features: (1) Every spinal nerve pair except for the C1 pair has a dermatome. The dermatomes collectively cover all the body surface except for most of the face (Fig. 4–12). (2) Adjacent dermatomes are generally but not always derived from consecutive spinal nerves. (3) There is variable overlap between adjacent dermatomes. In general, the extent of overlap is greater on the body trunk than on the limbs. This difference explains why an individual commonly appreciates a greater loss of sensation when a cutaneous nerve of a limb is completely transected than when a cutaneous nerve of the body trunk is completely transected. Furthermore, in general, the extent of overlap of pain and temperature fibers between adjacent dermatomes is greater than that of tactile fibers.

The skeletal muscles innervated by the lower motor neurons of a spinal nerve constitute the myotome of that spinal nerve. Individuals exhibit moderate variability in their myotomes.

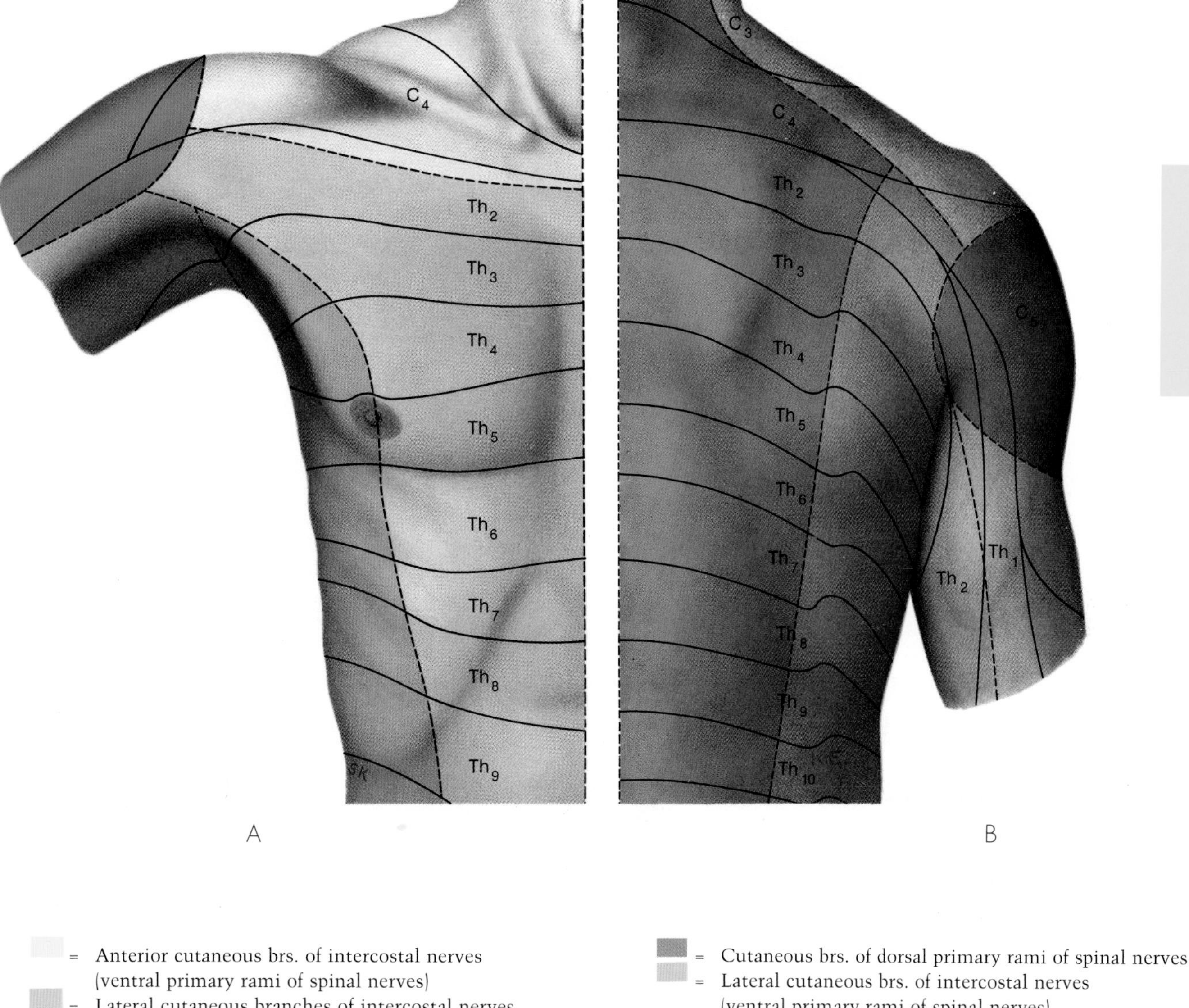

= Anterior cutaneous brs. of intercostal nerves (ventral primary rami of spinal nerves)
= Lateral cutaneous branches of intercostal nerves (ventral primary rami of spinal nerves)
= Superior lateral brachial cutaneous n. (axillary n.)
= Medial brachial cutaneous n. and intercostobrachial n.

= Cutaneous brs. of dorsal primary rami of spinal nerves
= Lateral cutaneous brs. of intercostal nerves (ventral primary rami of spinal nerves)
= Superior lateral brachial cutaneous n. (axillary n.)
= Posterior brachial cutaneous n. (radial n.)
= Medial brachial cutaneous n. and intercostobrachial n.

C_4 Th_2–Th_9 = Dermatomes
Solid lines = Boundaries of dermatomes
Broken lines = Fields of innervation

C_3–C_5, Th_2–Th_{10} = Dermatomes
Solid lines = Boundaries of dermatomes
Broken lines = Fields of innervation

Fig. 4–12: (**A**) Anterior view and (**B**) posterior view of the cutaneous fields of innervation and dermatomes of a male thorax.

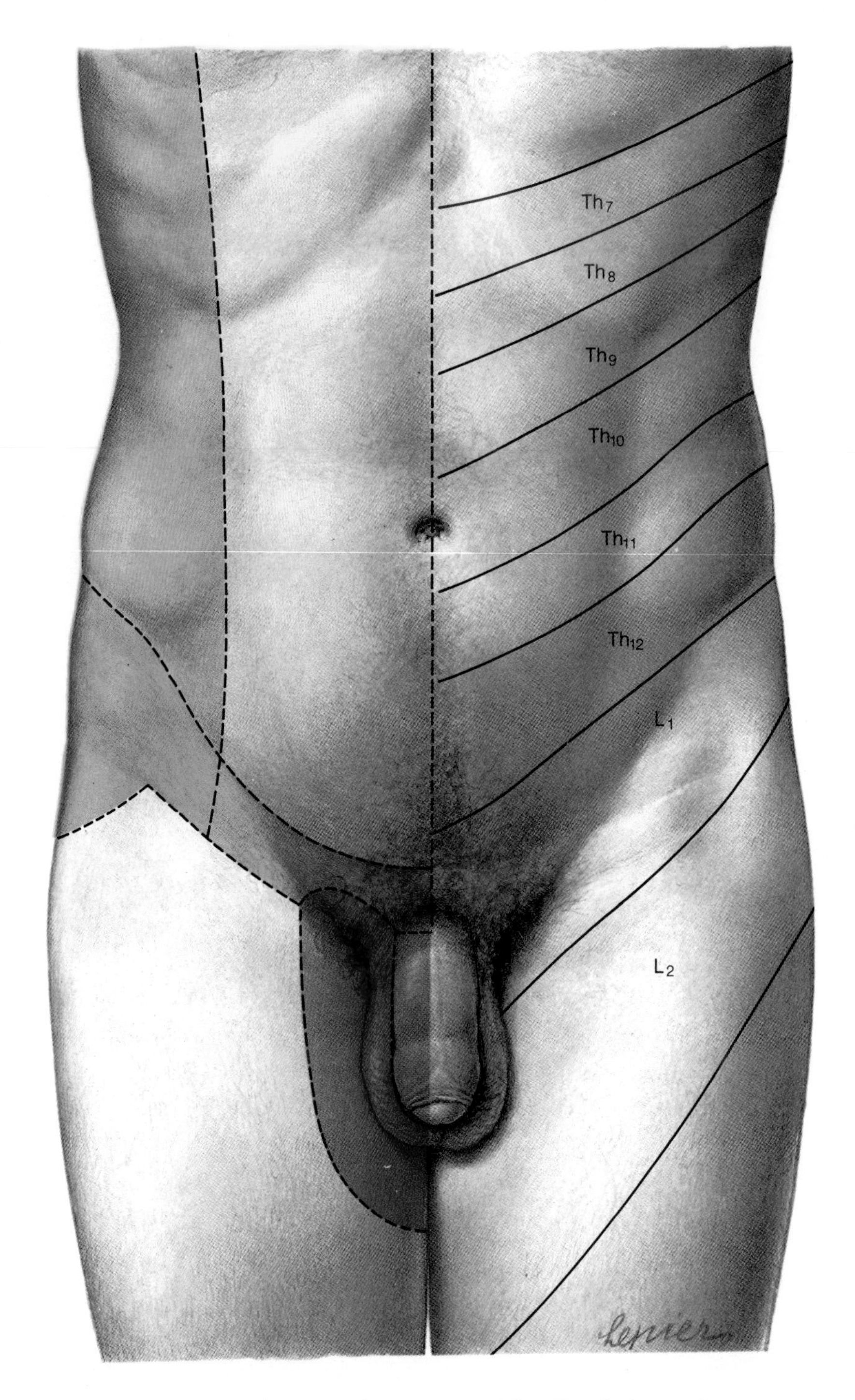

= Anterior cutaneous brs. of intercostal nerves
= Lateral cutaneous brs. of intercostal nerves
= Anterior cutaneous br. of iliohypogastric n.
= Lateral cutaneous br. of iliohypogastric n.
= Anterior scrotal brs. of ilioinguinal n. and genital br. of genitofemoral n.
= Dorsal n. of penis (pudendal n.)

Th_7–Th_{12}, L_1–L_2 = Dermatomes

Solid lines = Dermatome boundaries
Dashed lines = Fields of innervation

Fig. 4–12: (**C**) Anterior view of the cutaneous fields of innervation and dermatomes of a male abdomen.

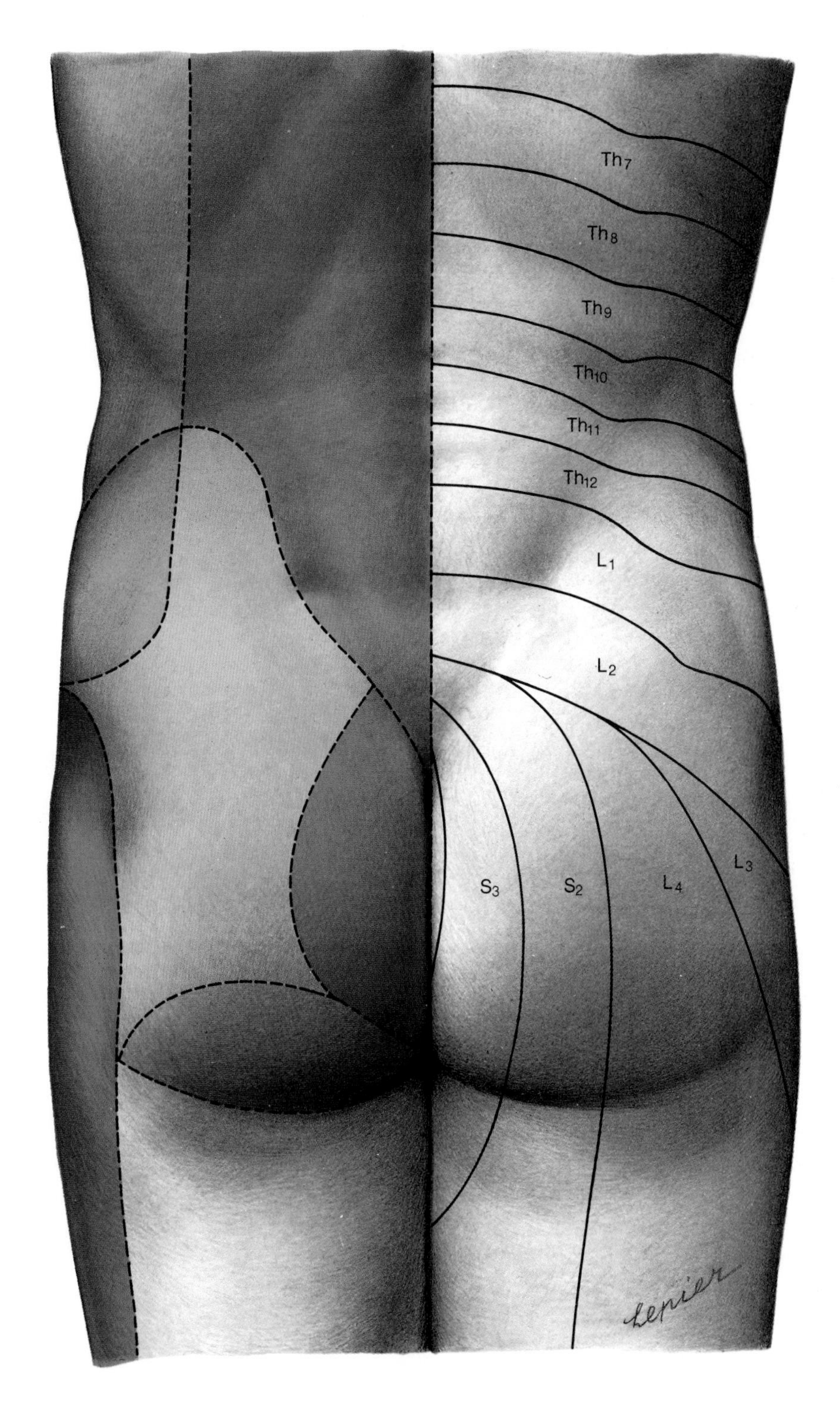

= Lateral cutaneous brs. of intercostal nerves (ventral brs. of spinal nerves)

= Cutaneous brs. of dorsal spinal nerves

= Superior clunial nerves (dorsal brs. of spinal nerves L_1–L_3)

= Middle clunial nerves (dorsal brs. of spinal nerves S_1–S_3)

= Lateral cutaneous br. of iliohypogastric n.

= Lateral femoral cutaneous n.

= Inferior clunial nerves (sacral plexus)

Th_7–Th_{12}, L_1–L_4, S_2–S_3 = Dermatomes

Solid lines = Dermatome boundaries
Dashed lines = Fields of innervation

Fig. 4–12: (**D**) Posterior view of the cutaneous fields of innervation and dermatomes of the lower back and buttock regions of a male.

Spinal Cord Reflexes

A spinal cord reflex is, in general, any reflex response which involves only the sensory and motor fibers of spinal nerves and their connections in the spinal cord. Although pathways between the spinal nerves and higher centers in the brain can and do participate in many reflexes, there is no such participation in a true spinal cord reflex. All spinal cord reflexes begin with the stimulation of sensory neurons in one or more spinal nerves. The sensory neurons then stimulate lower motor neurons in the anterior gray horns of the corresponding and/or other spinal cord segments. The sensory neurons stimulate the lower motor neurons either directly or via interconnecting neurons. The activities of the lower motor neurons elicit an unwilled reaction (the reflex) to the initial stimulus.

Spinal cord reflexes can be monosynapatic or polysynaptic. Monosynaptic reflexes are the simplest spinal cord reflexes; they are produced by a single sensory neuron in synaptic contact with a single lower motor neuron, and they elicit contraction of a single muscle. Polysynaptic reflexes elicit contraction of certain muscle groups in coordination with relaxation of other muscle groups.

Three general types of spinal cord reflex activity are always checked during a neurologic examination: (1) A **phasic stretch reflex** is seen when the abrupt stretching of a muscle's fibers elicits a phasic (abrupt and transient) contraction of the muscle fibers. The knee jerk

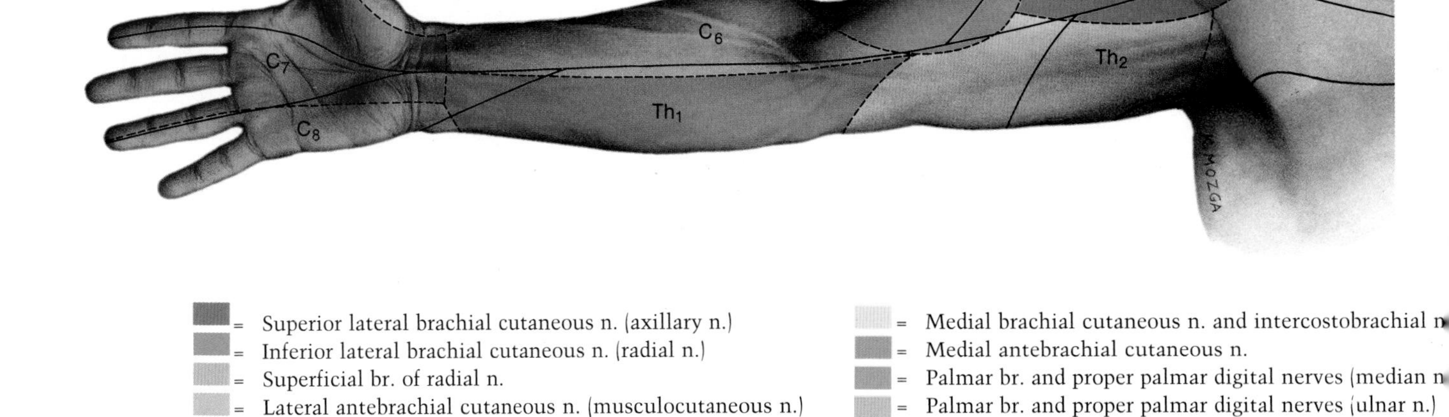

C_5–C_8, Th_1–Th_2 = Dermatomes
Solid lines = Boundaries between dermatomes
Dashed lines = Fields of innervation

E

Fig. 4–12: (**E**) Anterior view of the cutaneous fields of innervation and dermatomes of the upper limb.

reflex is an example of a phasic stretch reflex. (2) A **tonic stretch reflex** occurs if the slow, continuous stretching of a muscle's fibers elicits a sustained, low level of contractile activity in the muscle fibers. Tonic stretch reflexes are responsible for **muscle tone,** which is the resistive force muscles exert when passively and slowly stretched. (3) A **withdrawal reflex** is polysynaptic, one in which a noxious stimulus elicits contraction of muscles whose actions move body parts away from the source of irritation.

Voluntary control of skeletal muscle activity is provided by motor neurons called **upper motor neurons.** The cell bodies of upper motor neurons are located in the cerebral cortex of the brain, and their fibers descend within the corticospinal tracts of the spinal cord's anterior and lateral funiculi to lower motor neurons located in the anterior gray columns of the spinal cord. Conduction of impulses from upper motor neurons to lower motor neurons, and then from lower motor neurons to a skeletal muscle's fibers provide for the willful regulation of the muscle's activity. If a lesion strikes and destroys all the upper motor neurons for a skeletal muscle, the muscle initially does not exhibit any activity. Following this transient period of

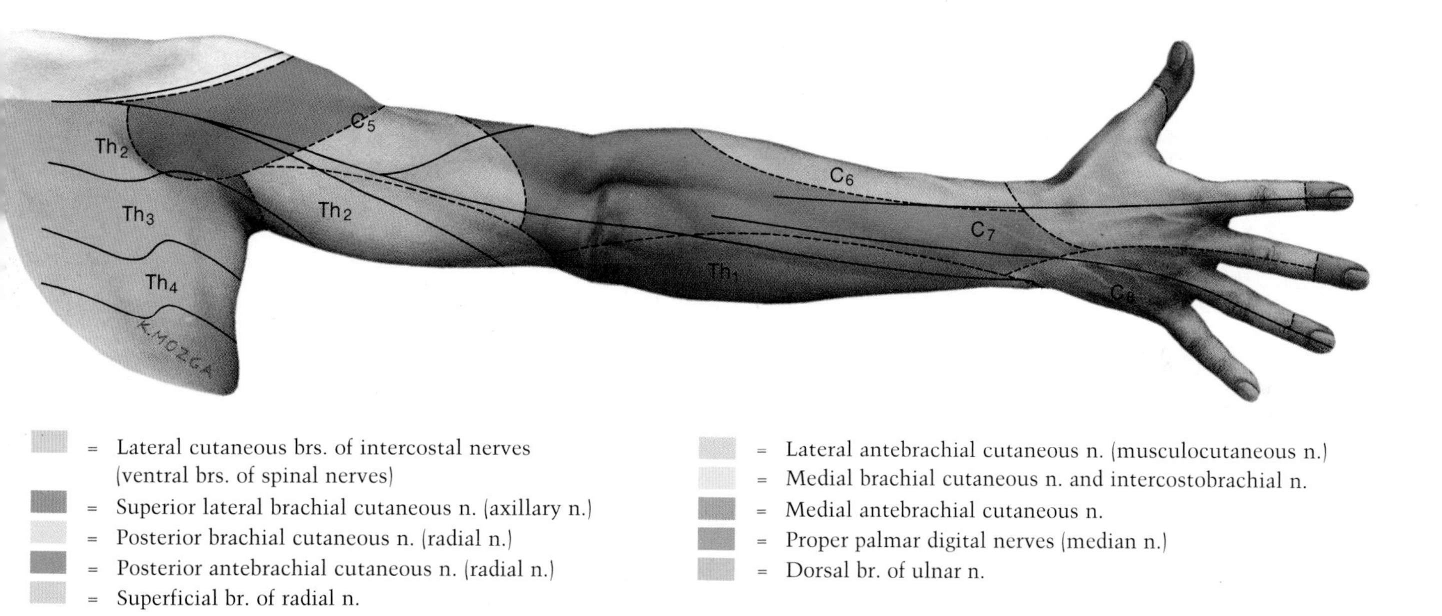

= Lateral cutaneous brs. of intercostal nerves (ventral brs. of spinal nerves)
= Superior lateral brachial cutaneous n. (axillary n.)
= Posterior brachial cutaneous n. (radial n.)
= Posterior antebrachial cutaneous n. (radial n.)
= Superficial br. of radial n.
= Lateral antebrachial cutaneous n. (musculocutaneous n.)
= Medial brachial cutaneous n. and intercostobrachial n.
= Medial antebrachial cutaneous n.
= Proper palmar digital nerves (median n.)
= Dorsal br. of ulnar n.

C_4–C_8, Th_1–Th_4 = Dermatomes
Solid lines = Boundaries between dermatomes
Dashed lines = Fields of innervation

F

Fig. 4–12: (F) Posterior view of the cutaneous fields of innervation and dermatomes of the upper limb.

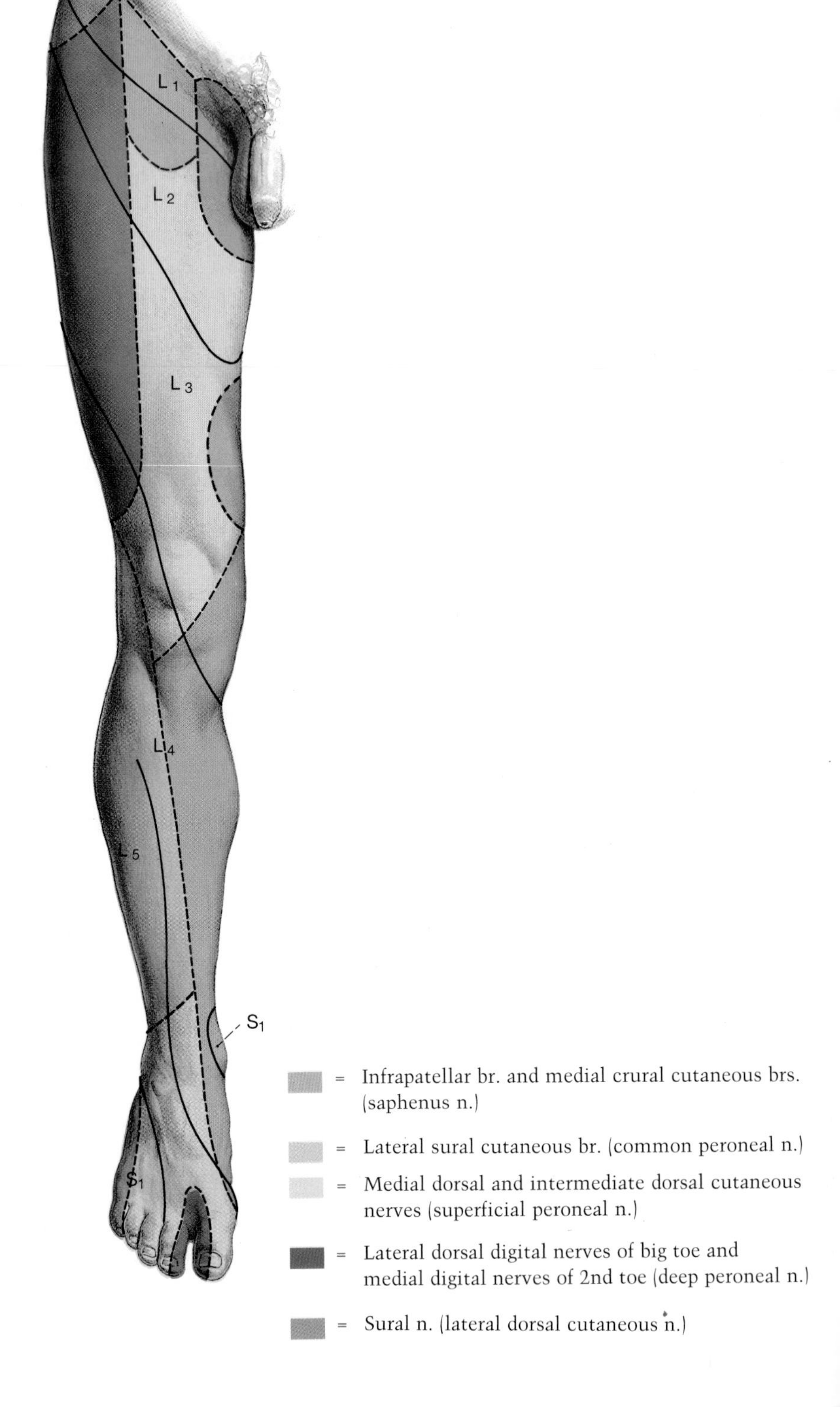

= Lateral cutaneous br. of iliohypogastric n.
= Lateral femoral cutaneous n.
= Femoral br. of genitofemoral n.
= Anterior scrotal brs. of ilioinguinal n. and genital br. of genitofemoral n.
= Cutaneous br. of obturator n.
= Anterior cutaneous brs. of femoral n.

G

L_1–L_5, S1 = Dermatomes
Solid lines = Dermatome boundaries
Dashed lines = Fields of innervation

Fig. 4–12: (**G**) Anterior view of the cutaneous fields of innervation and dermatomes of the lower limb.

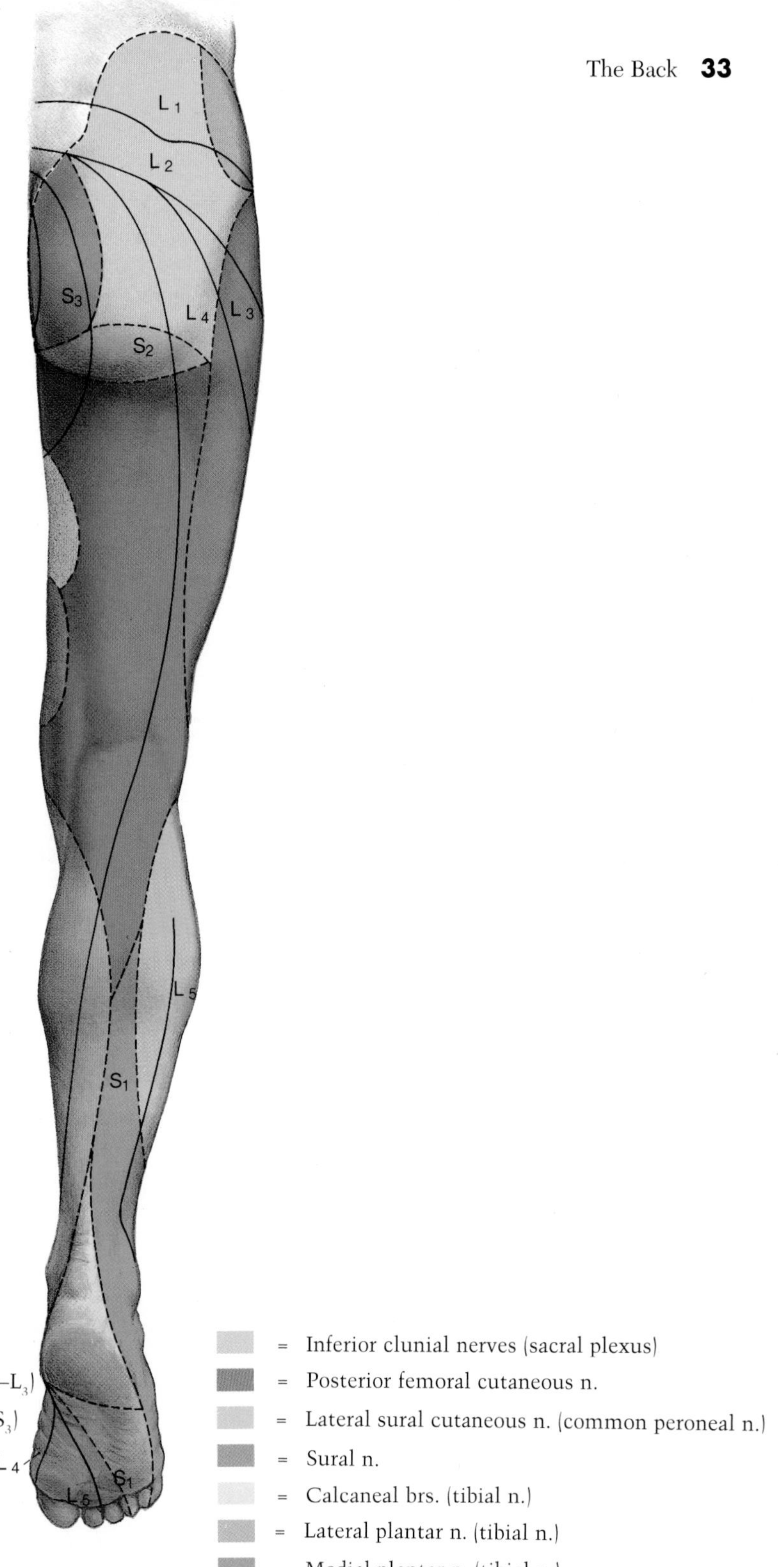

- = Lateral cutaneous br. of iliohypogastric n.
- = Superior clunial nerves (dorsal brs. of spinal nerves L_1–L_3)
- = Middle clunial nerves (dorsal brs. of spinal nerves S_1–S_3)
- = Lateral femoral cutaneous n.
- = Anterior cutaneous brs. of femoral n.
- = Cutaneous br. of obturator n.
- = Infrapatellar and medial crural cutaneous brs. (saphenous n.)
- = Inferior clunial nerves (sacral plexus)
- = Posterior femoral cutaneous n.
- = Lateral sural cutaneous n. (common peroneal n.)
- = Sural n.
- = Calcaneal brs. (tibial n.)
- = Lateral plantar n. (tibial n.)
- = Medial plantar n. (tibial n.)

L_1–L_5, S_1–S_3 = Dermatomes
Solid lines = Dermatome boundaries
Dashed lines = Field of innervation

H

Fig. 4–12: **(H)** posterior view of the cutaneous fields of innervation and dermatomes of the lower limb.

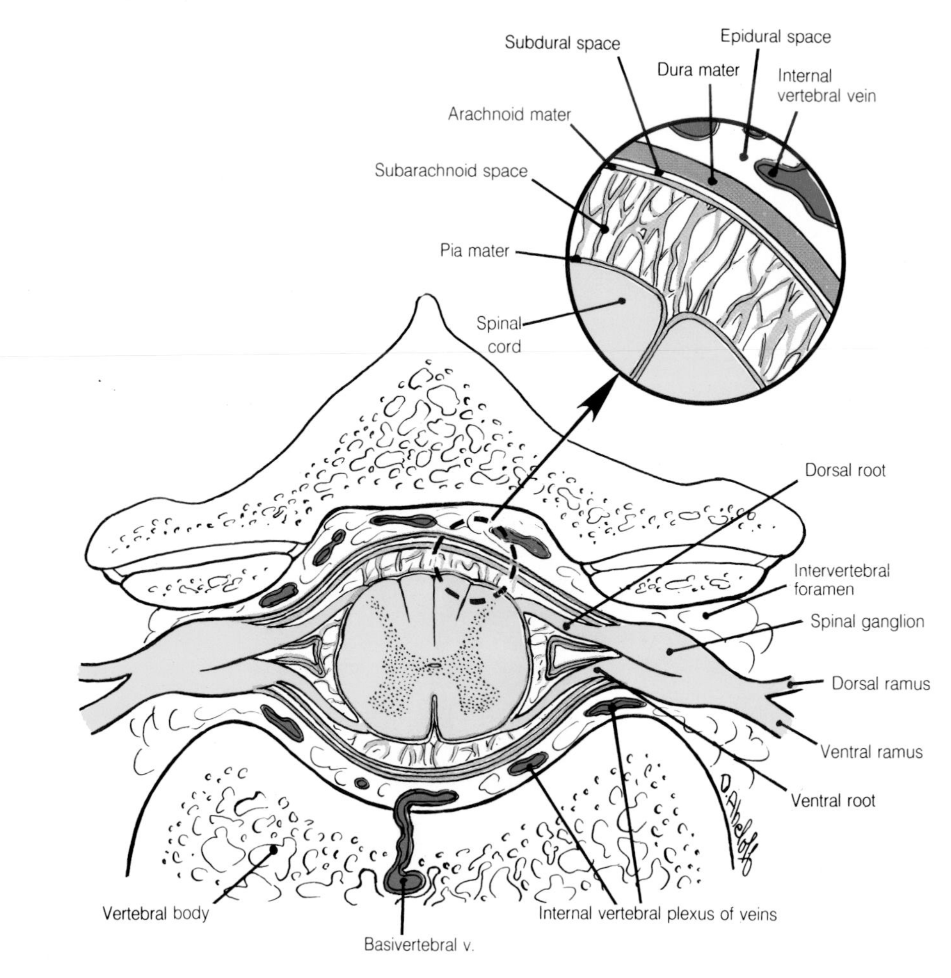

Fig. 4–13: Horizontal section through the spinal cord and vertebral column.

inactivity, the muscle exhibits spastic activity when spinal cord reflexes designed to test its activity are conducted. If, by contrast, a lesion strikes and destroys all the lower motor neurons that innervate a skeletal muscle, the muscle permanently loses its tone, becomes flaccid and ultimately atrophies; the muscle exhibits greatly diminished or no activity when spinal cord reflexes designed to test its activity are conducted.

Meninges of the Spinal Cord

The brain and the spinal cord are enveloped by three membranes called meninges. The spinal meninges (the meninges which envelop the spinal cord) are direct continuations of the cranial meninges (the meninges which envelop the brain). For both the brain and spinal cord, the innermost meningeal layer is called the **pia mater,** the middle layer the **arachnoid mater,** and the outermost layer the **dura mater.**

Each meningeal layer is named for its structure and its protection of the brain and spinal cord. The pia mater (which is the Latin expression for "delicate mother") is an epithelial lining with a fine, richly vascularized layer of loose connective tissue. The spinal pia mater closely encloaks the spinal cord (Fig. 4–13). The spinal pia mater which encloaks the conus medullaris continues inferiorly from the tip of the conus medullaris as a very fine connective tissue strand (Fig. 4–14). The strand is centrally located within the cauda equina, and is called the **filum terminale** since it is the terminal filament to extend from the spinal cord (Fig. 4–14). The filum terminale extends inferiorly through the vertebral canal from the tip of the conus medullaris to the posterior surface of the body of the first coccygeal vertebra.

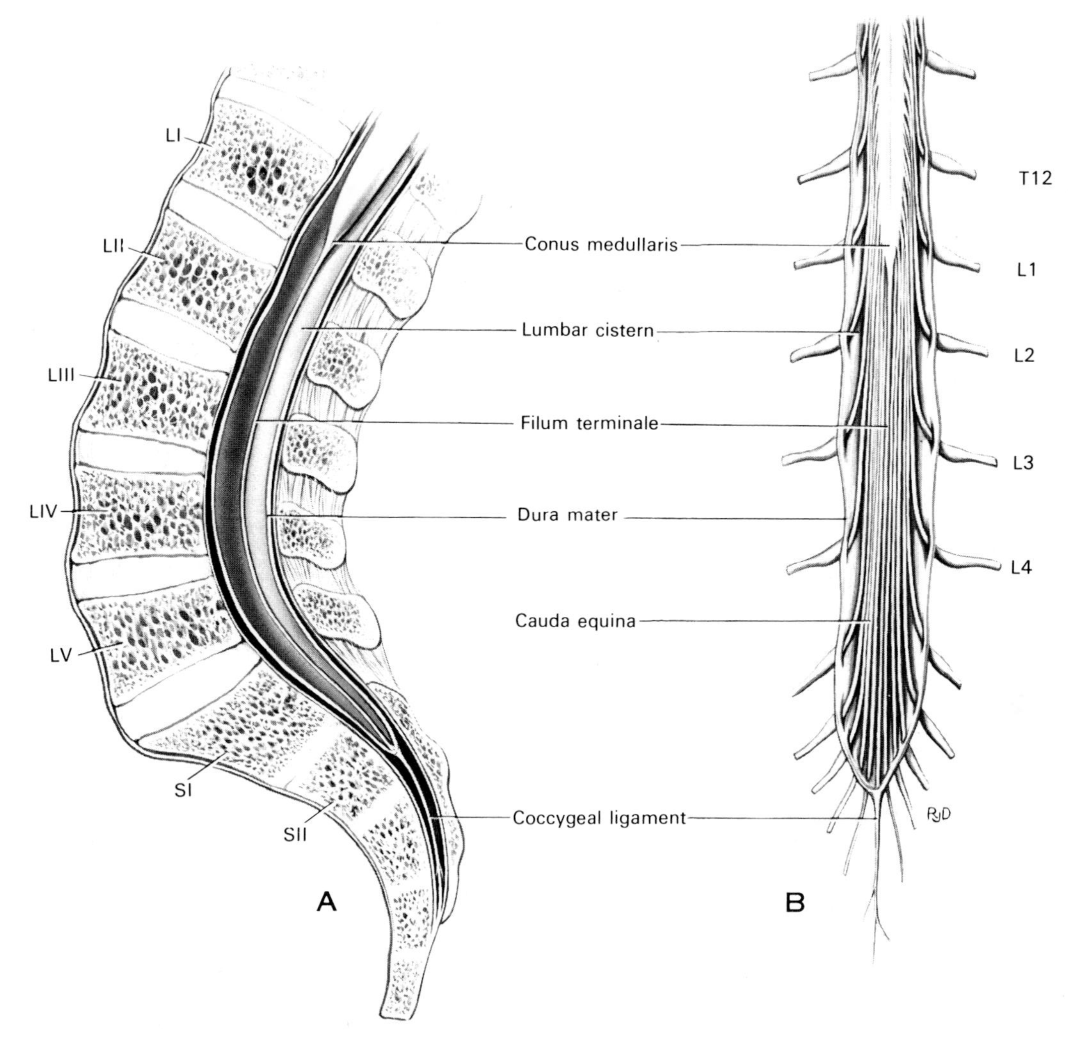

Fig. 4–14: Diagrams illustrating the lowest parts of the spinal cord and vertebral column. (A) Median section of the lumbar and sacral vertebrae, subarachnoid space, conus medullaris, and filum terminale. (B) Posterior view of the conus medullaris, cauda equina, and associated spinal nerve roots.

The arachnoid mater (which is the Latin expression for "spidery mother") is a layer of loose connective tissue. The spinal arachnoid mater forms a loose, tubular sheath around the spinal cord (Fig. 4–13) and cauda equina down to the lower border of the body of the second sacral vertebra (Fig. 4–14). The spinal arachnoid mater ends abruptly at this level as it encircles the filum terminale.

A thin fibrous sheet called a **denticulate ligament** extends on each side of the spinal cord from the spinal pia mater to the spinal arachnoid mater (Fig. 4–15). The denticulate ligament is so named because its attachment to the spinal arachnoid mater occurs via 21 minute, tooth-like processes. The left and right denticulate ligaments help suspend the spinal cord along the central axis of the vertebral column.

The dura mater (Latin for "tough mother") is a layer of dense, fibrous connective tissue. The spinal dura mater lines the outer surface of the spinal arachnoid mater down to the latter's blind ending at the lower border of the body of the second sacral vertebra (Figs. 4–13 and 4–14). The spinal dura mater tapers at this level to form a tight connective tissue sheath around the filum terminale. The spinal dura mater thus invests the terminal segment of the filum terminale, and tethers it inferiorly to the posterior surface of the body of the first coccygeal vertebra.

The spinal meninges bound three spaces within the vertebral canal. The space surrounding the spinal dura mater within the vertebral canal is called the **epidural space,** and it is filled with loose connective tissue and an extensive venous plexus called the **internal**

venous plexus (Fig. 4–13). The space between the spinal dura mater and the spinal arachnoid mater is a microscopically-thin space called the subdural space, and it bears lymph (Fig. 4–13). The space between the spinal arachnoid mater and spinal pia mater is called the **subarachnoid space** of the spinal cord, and it is filled with **cerebrospinal fluid (CSF)** (Fig. 4–13).

Origin and Drainage of CSF and Its Support of the Brain and Spinal Cord

CSF is a clear, colorless, watery fluid derived from blood plasma; it normally contains only a few lymphocytes. Although its electrolyte composition closely resembles that of blood plasma, its protein concentration is markedly lower. CSF is secreted into a series of four chambers within the brain (the **ventricles** of the brain) by the **choroid plexuses** (the highly vascularized plexuses which line the ventricles) (Fig. 4–16). The CSF flows from the ventricles into the subarachnoid space which surrounds the brain in the cranial cavity of the skull and into the corresponding subarachnoid space which surrounds the spinal cord in the vertebral canal. CSF drains into the systemic venous circulation mainly via the **dural venous sinuses** (venous channels which course within the dura mater of the brain). Some CSF is also drained by the internal venous plexus of the epidural space in the vertebral column.

The brain in the cranial cavity and the spinal cord in the vertebral column thus both lie immersed in the CSF. It is in this fashion that the CSF serves to support the brain and spinal cord in a fluid environment of uniform pressure.

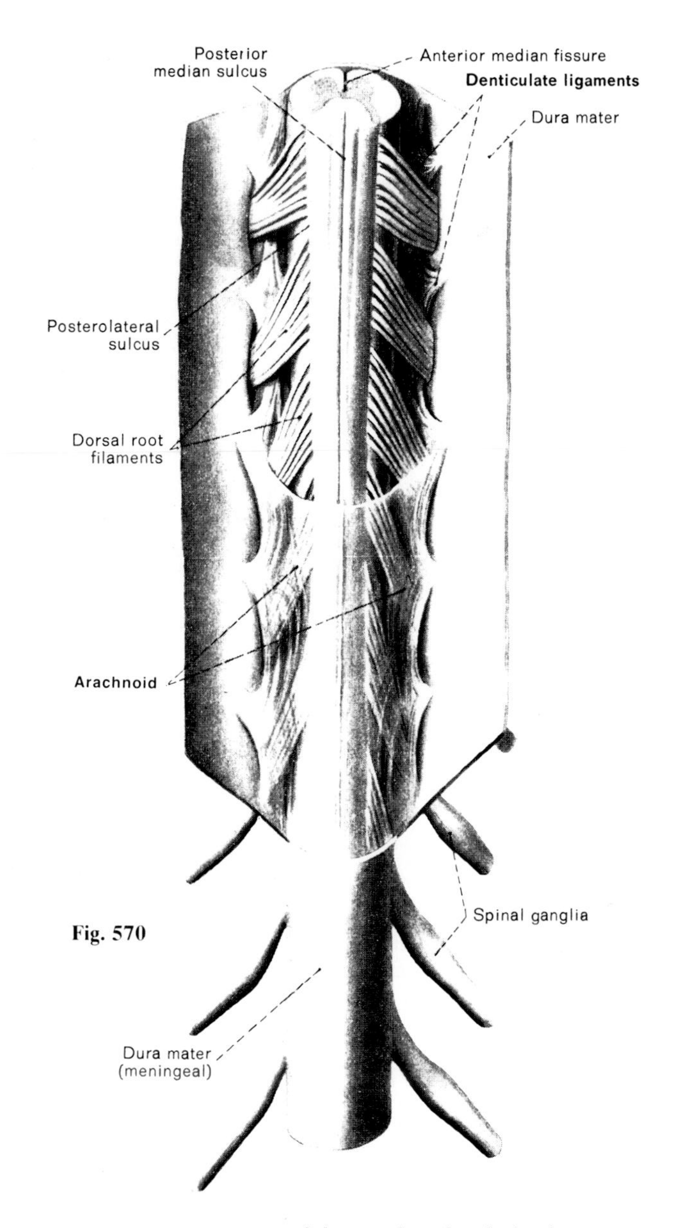

Fig. 4–15: Posterior view of the spinal cord with the dura mater dissected and reflected laterally.

CLINICAL PANEL I-1

MENINGITIS

Definition Meningitis is inflammation of the meninges. In most cases, inflammation is the result of infection by bacterial, viral, fungal, or parasitic agents. The infectious process almost invariably extends throughout the subarachnoid space to involve both the cranial and spinal meninges. Viral meningitides (cases of viral meningitis) are generally benign and self-limiting in duration. Bacterial meningitides have serious sequelae (serious conditions resulting as a consequence of the disease) or a fatal outcome if not promptly treated with appropriate antibiotics.

Common Symptoms The symptoms of meningitis include fever, headache, alterations in mental status (such as lethargy or confusion), photophobia (enhanced sensitivity to light), stiff neck, and nausea and vomiting. Some patients may present with only one or two symptoms. It is believed that most of these symptoms are a consequence of increased intracranial pressure and inflammation of sensory nerves. The pathophysiologic basis of the photophobia is not known.

Anatomic Basis of the Examination *Brudzinski's and Kernig's tests* are the physical tests commonly used to elicit signs of meningitis. Each test is conducted as the patient lies supine on an examination table with the hands folded behind the head.

In Brudzinski's test, the examiner requests the patient to raise the head up, an act which stretches the spinal meninges from the cervical part of the vertebral column. If the meninges are inflamed, such stretching produces pain in the head, neck or back and elicits involuntary flexion of the lower limbs at the hip and knee joints. The involuntary movements at the hip and knee joints constitute a positive sign for Brudzinski's test, since they represent an attempt by the patient to minimize tension in the spinal meninges through flexion of the thighs and legs.

The mechanism by which flexion of the thighs and legs minimizes tension in the spinal meninges is related to the course of the sciatic nerve in the lower limb. The sciatic nerve is one of the major nerves of the lower limb (Fig. 4–17). It extends from the buttock into the thigh by passing behind the hip joint, and its two large terminal branches, the tibial and common peroneal nerves, extend from the thigh into the leg by passing behind the knee joint. Whereas flexion of the thigh at the hip stretches the sciatic nerve along its course behind the hip joint, flexion of the leg at the knee slackens the tibial and common peroneal nerves along their course behind the knee joint. Combined flexion at both joints produces a net slack in the sciatic nerve and its tibial and common peroneal branches; this slack extends back to the spinal meninges that envelop the roots of the spinal nerves (L4, L5, S1, S2, and S3) which contribute nerve fibers to the sciatic nerve. It is in this fashion that flexion of the thighs and legs slackens the spinal meninges from the lumbosacral part of the vertebral column.

In Kernig's test, the patient slowly raises one of the lower limbs, taking care to keep the leg fully extended at the knee. This action stretches the spinal meninges from the lumbosacral part of the vertebral column. If such stretching produces pain in the head, neck, or back, and the pain can be relieved by flexion of the leg at the knee, then the test is judged positive.

In cases of suspected meningitis, a sample of CSF is collected for cell count, bacterial culture, and chemical analysis. The collection procedure is called lumbar puncture, since the CSF is drawn through a needle that punctures the spinal dural sheath (and its inner lining of arachnoid mater) in the lower lumbar region of the vertebral column (Fig. 4–14). The lower lumbar region is selected because of the following anatomic relationships: The spinal subarachnoid space extends inferiorly within the vertebral canal farther than the spinal cord. Whereas the conus medullaris ends in an adult at the lower border of the body of the first lumbar vertebra, the subarachnoid space extends inferiorly to the lower border of the body of the second sacral vertebra (Fig. 4–14). This difference between the inferior extents of the spinal cord and spinal subarachnoid space makes it possible to collect samples of CSF from the lower lumbar region of the spinal subarachnoid space without risking direct damage to the spinal cord.

To perform a lumbar puncture, the patient is either seated or lain in a lateral decubitus position (a sidelying position), with the lumbar portion of the vertebral column maximally flexed (the flexion increases the heights of the spaces between the spinous processes of the lumbar vertebrae) (Fig. 4–18). The highest point of the iliac crest is used to mark the level of the spinous process of the fourth lumbar

CLINICAL PANEL I-1 (continued from p. 37)

vertebra. In an adult, the needle may be inserted into either the interspinous space (the space between the spinous processes) of the third and fourth lumbar vertebrae or the interspinous space between the fourth and fifth lumbar vertebrae. In a child, it is advisable to insert the needle into the lower interspinous space because the spinal cord may extend below the level of the body of the third lumbar vertebra. After application of a local anesthetic, the spinal needle gains entrance into the subarachnoid space by passing through the following series of tissues in the median plane of the body: skin, superficial fascia, supraspinous ligament, interspinous ligament, the fused, posterior margins of the paired flaval ligaments, the connective tissue of the epidural space, the dura mater, and, finally, the arachnoid mater.

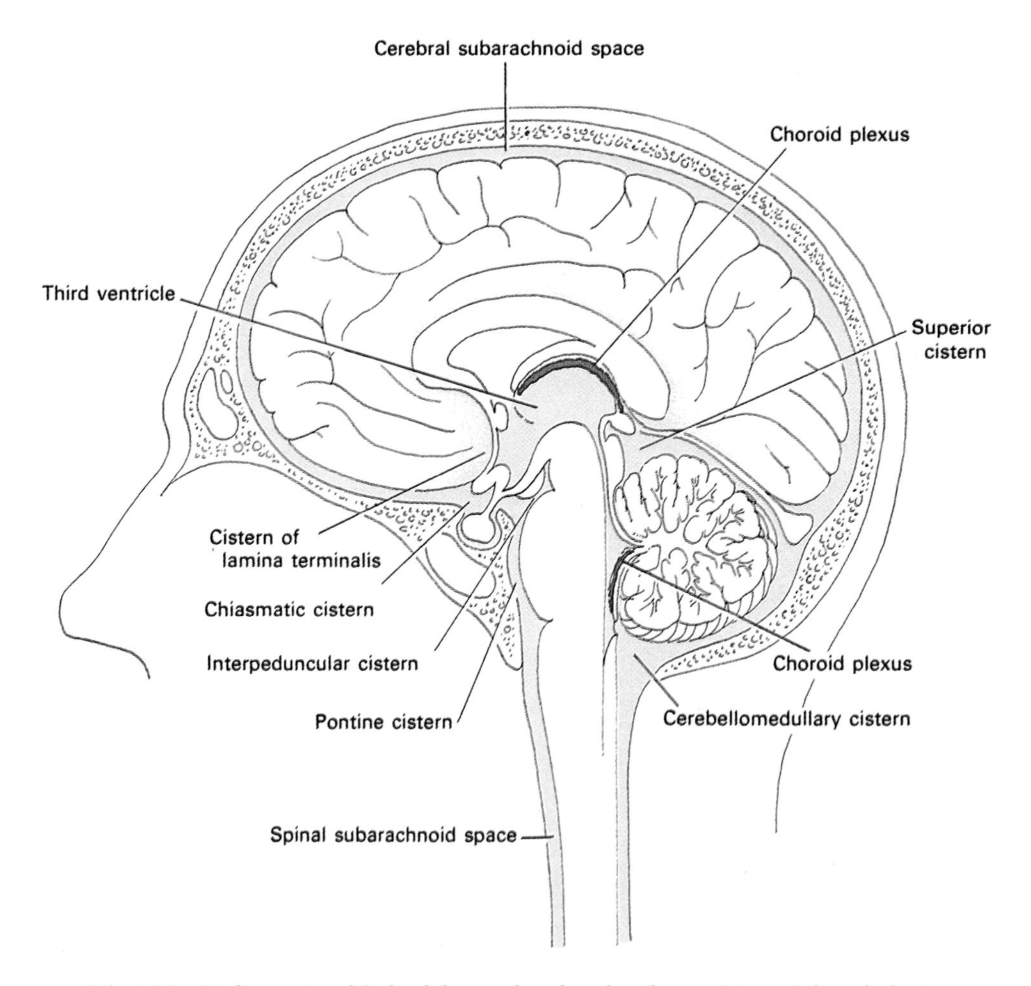

Fig. 4–16: Median section of the head showing the subarachnoid spaces (cisterns) about the brain.

Arterial Supply and Venous Drainage of the Spinal Cord

Three arteries extend vertically along the length of the spinal cord: a single **anterior spinal artery** and a **pair of posterior spinal arteries** (Fig. 4–19). The anterior spinal artery supplies the longitudinal columnar regions in the anterior two-thirds of the spinal cord (anterior gray columns, gray commissure, anterior funiculi, and lateral funiculi). The posterior spinal arteries supply the remainder of the longitudinal columnar regions (posterior gray columns and posterior funiculi).

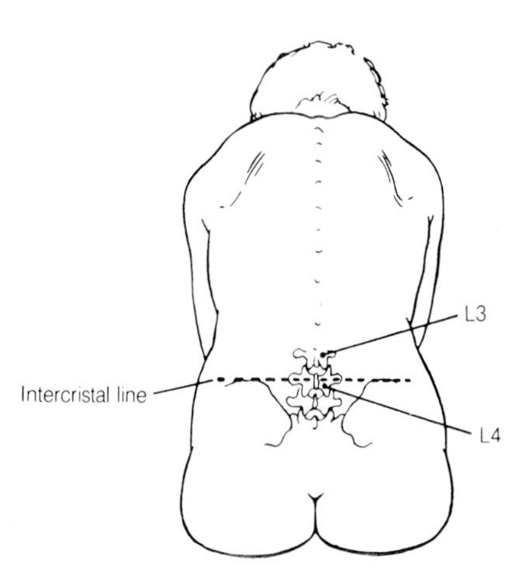

Fig. 4–18: Diagram illustrating the surface features used in performing a lumbar puncture.

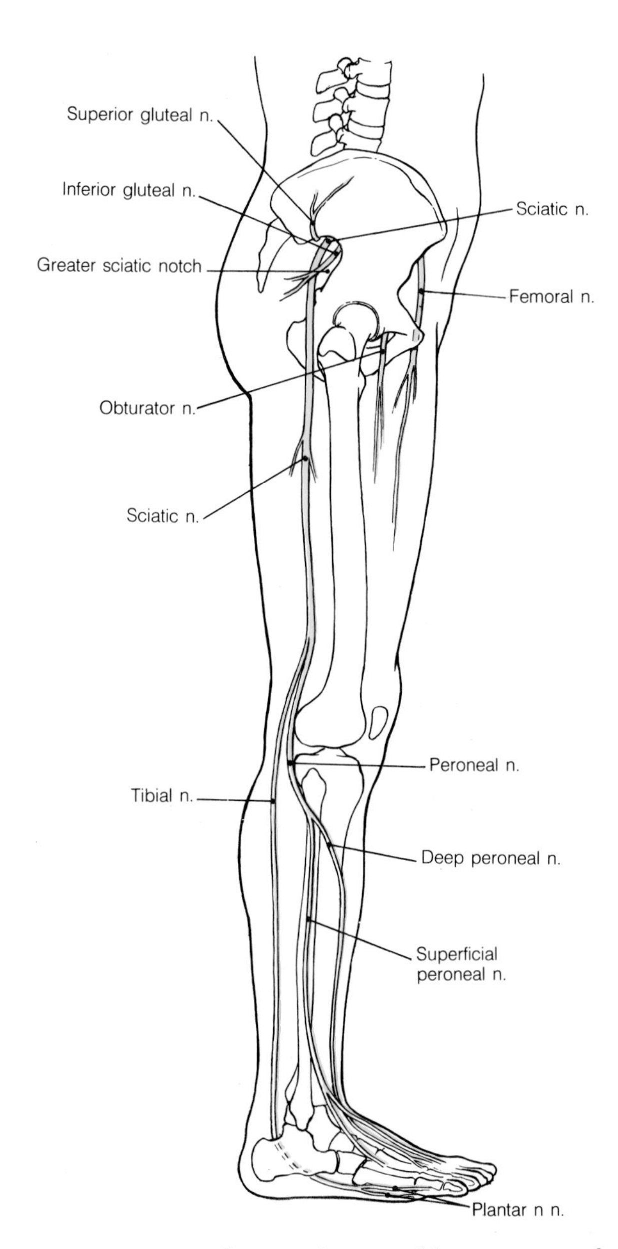

Fig. 4–17: Diagram illustrating the course of the sciatic nerve and its tibial and common peroneal branches in the lower limb.

Branches of the vertebral arteries give rise to the superior origins of the anterior and posterior spinal arteries (Fig. 4–19). Most of the blood supply to the upper cervical segments of the spinal cord is provided by these vertebral artery branches. **Anterior and posterior radicular arteries** respectively give rise to the lengths of the anterior and posterior spinal arteries which extend inferiorly from the lower cervical segments to the conus medullaris (Fig. 4–19). Most of the anterior and posterior radicular arteries arise from spinal branches of intercostal and lumbar arteries. One of the anterior radicular arteries that arises in the lower thoracic or upper lumbar region is frequently so large that it is called the **arteria radicularis magna** (an arteria radicularis magna is not shown in Fig. 4–19). An arteria radicularis magna generally supplies most of the blood to the lower two-thirds of the anterior spinal artery.

The anterior and posterior spinal arteries do not always occur as continuous arterial vessels. Discontinuities in the anterior spinal artery are most common at the levels of the mid-thoracic and upper lumbar segments. Discontinuities in the posterior spinal arteries are most common at the levels of the upper thoracic segments.

Anterior and posterior spinal veins extend vertically along the spinal cord (Fig. 4–20). These spinal veins drain mainly into the internal vertebral plexus in the epidural space and into the **external vertebral plexus** (the extensive venous plexus that extends longitudinally along the length of the external surface of the vertebral column). The vertebral venous plexuses drain primarily into the **azygos system of veins.**

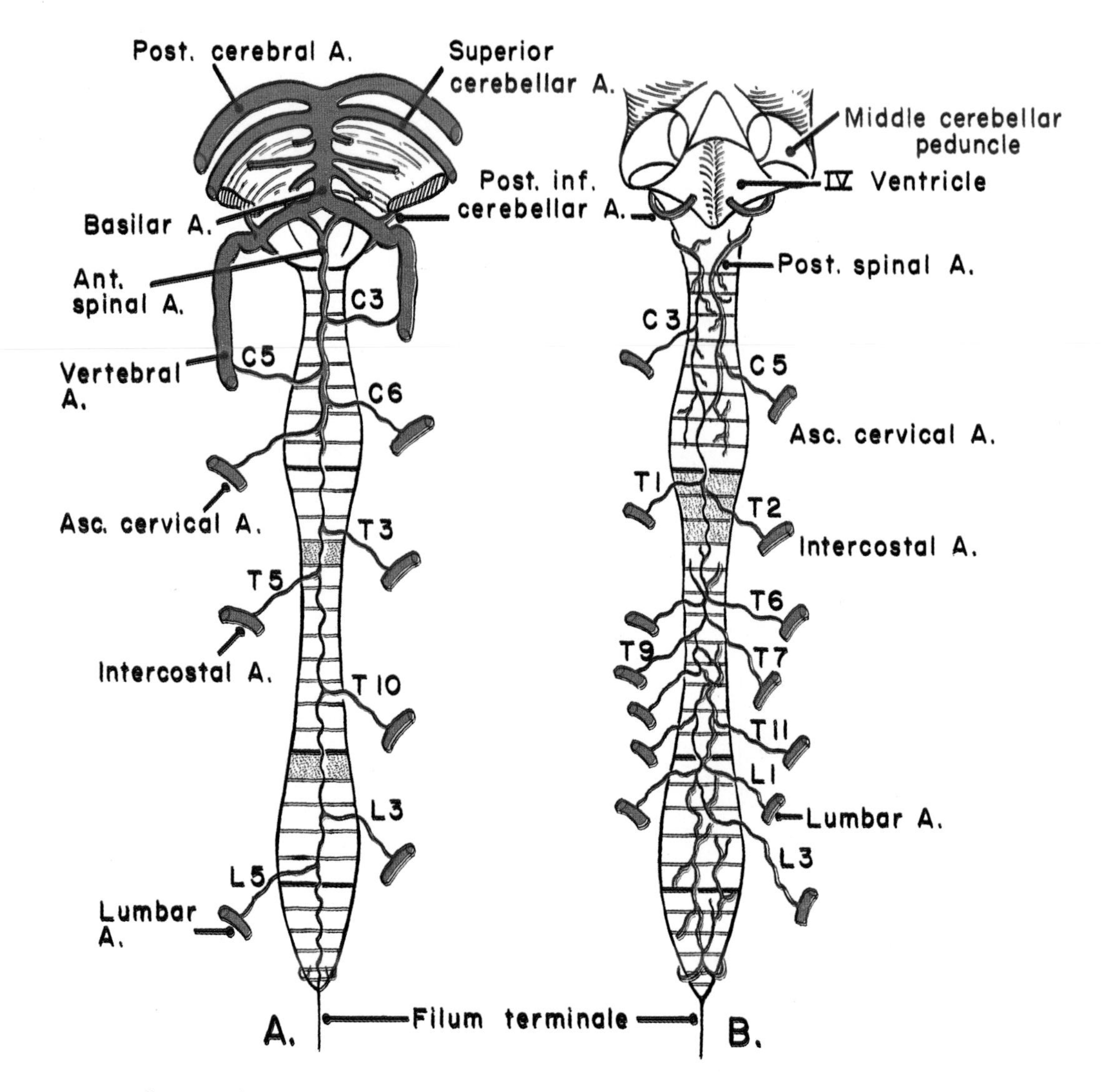

Fig. 4–19: Diagrams illustrating the arterial supply of the spinal cord. (A) Anterior surface. (B) Posterior surface. Vulnerable spinal cord segments are stippled.

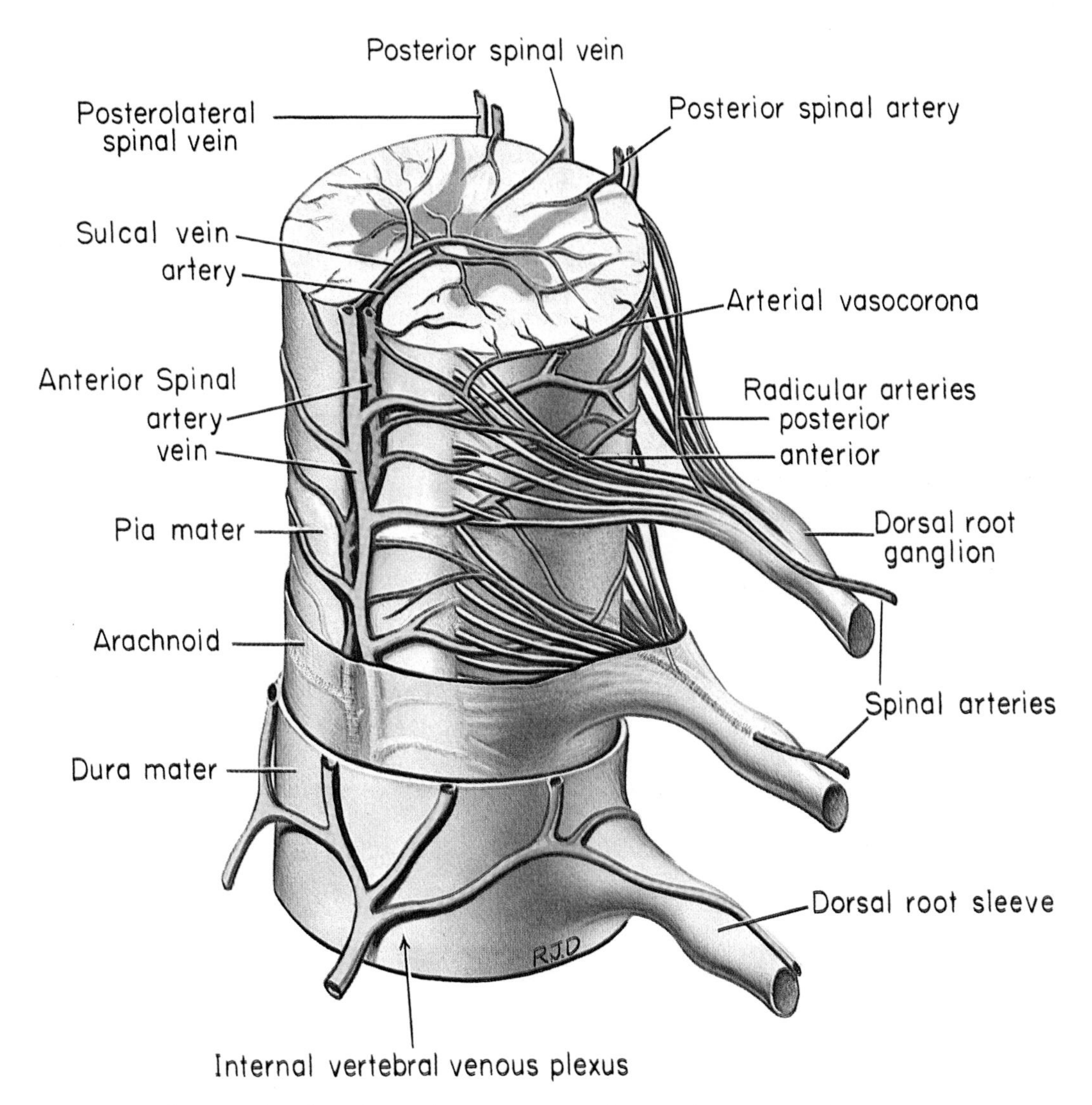

Fig. 4–20: Diagram illustrating the relationships of the spinal arteries and veins with the spinal cord, meninges, and spinal nerve roots.

CHAPTER 5

The Autonomic Nervous System

The autonomic nervous system is the visceral motor component of the human nervous system. In other words, it consists of the neuronal populations that convey efferent impulses from the central nervous system to the body's internal organs and glands and its smooth muscle tissue. The autonomic nervous system is basically responsible for homeostasis (the regulation and maintenance of the body states). The autonomic nervous system's activities are primarily involuntary, occurring reflexively in response to sensory input. The autonomic nervous system is named for the substantial independence of its activities from one's volition.

TWO MAJOR DIVISIONS OF THE AUTONOMIC NERVOUS SYSTEM

The autonomic nervous system has two major divisions: the **sympathetic division** and the **parasympathetic division.** The divisions were named for the early belief that the sympathetic division conveys the "feelings" or "sympathies" of the internal organs, and that the parasympathetic division acts alongside the sympathetic division in this respect.

Organs that are innervated by the autonomic nervous system are generally innervated by both of its divisions, and the divisions generally exert opposite effects on a tissue. However, there are some exceptions to these generalizations. There are some tissues upon which the sympathetic and parasympathetic divisions exert similar effects, and there are other tissues that are innervated by one division only. For example, the upper and lower limbs, the walls of the thoracic and abdominopelvic cavities, and the cutaneous tissues of the body as a whole receive sympathetic but no parasympathetic innervation.

In both divisions, two neurons transmit every efferent impulse from the central nervous system to a target tissue. The first neuron is called the **preganglionic neuron,** as it transmits impulses from either the brain or the spinal cord to a ganglion. The second neuron is called the **postganglionic neuron,** as it transmits impulses from the ganglion directly to one or more cells in a target tissue.

In both divisions, there are more postganglionic than preganglionic neurons. Almost all the preganglionic neurons in each division synapse with two or more (commonly severalfold more) postganglionic neurons. In the sympathetic division, a preganglionic neuron may synapse with as many as 10 to 20 postganglionic neurons. The disproportion between postganglionic and preganglionic neurons is greater in the sympathetic division than in the parasympathetic division.

GENERAL ORGANIZATION OF THE SYMPATHETIC DIVISION

Locations of the Cell Bodies of the Preganglionic and Postganglionic Neurons

In the sympathetic division, the cell bodies of the preganglionic neurons are all located in the lateral gray columns of the spinal cord from T1 to L2 (or L3) (Fig. 5–1A). The cell bodies of the postganglionic neurons are primarily congregated in major ganglia lying relatively close to the vertebral column (Fig. 5–1A). All but six of these major sympathetic ganglia are called the **paravertebral ganglia** since they form chains on the sides of the vertebral column; these chains are called the **left and right sympathetic chains (trunks)** (Fig. 5–2). The sympathetic chains extend along the entire length of the vertebral column, from the uppermost cervical vertebrae to the lowest sacral vertebrae.

Initially during embryogenesis, there is, on each side, a separate paravertebral ganglion which develops in association with each spinal nerve. However, subsequent fusion of some of the ganglia with each other abolishes this direct correspondence between paravertebral ganglia and spinal nerves, and reduces the ganglia in each sympathetic chain to generally 3 **cervical ganglia,** 11 to 12 **thoracic ganglia,** 4 **lumbar ganglia,** and 4 **sacral ganglia,** for a total of 22 to 23 ganglia (Fig. 5–2). The ganglia which derive from single

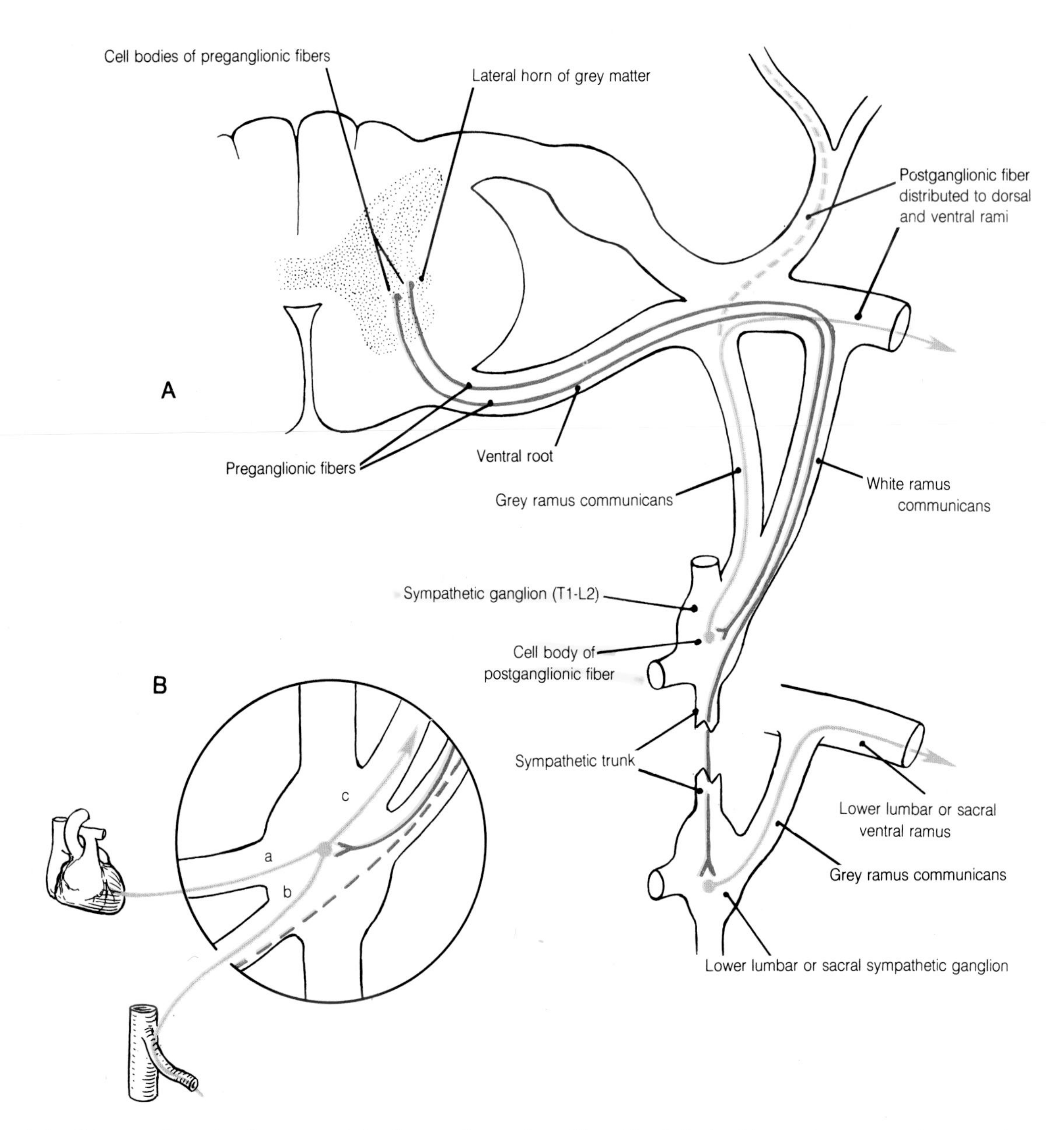

Fig. 5–1: Diagrams illustrating the origin and distribution of sympathetic fibers. The dashed red line in B represents the preganglionic sympathetic fibers which do not end in the paravertebral ganglia of the sympathetic chains, but instead pass through the paravertebral ganglia and exit the sympathetic chains via splanchnic nerves to end in the prevertebral ganglia and their associated nerve plexuses. The yellow line labeled "a" in B represents a postganglionic sympathetic fiber exiting the sympathetic chain via a plexus branch. The yellow line labeled "b" in B represents a postganglionic sympathetic fiber exiting the sympathetic chain via a vascular branch. The yellow line labeled "c" in B represents a postganglionic sympathetic fiber exiting the sympathetic chain via a gray ramus communicans.

embryonic ganglia each remain attached to a single spinal nerve; those ganglia which arise from the fusion of two or more embryonic ganglia become attached to the corresponding number of spinal nerves.

The six remaining major sympathetic ganglia are found near the origins of the abdominal aorta's five largest visceral branches. The **left and right celiac ganglia** lie near the origin of the celiac artery, the **superior mesenteric ganglion** lies near the origin of the superior mesenteric artery, the **left and right aorticorenal ganglia** lie near the origins of the renal arteries, and the **inferior mesenteric ganglion** lies near the origin of the inferior mesenteric artery. Since these six major sympathetic ganglia lie anterior to both the vertebral column and abdominal aorta, they are commonly referred to as the **prevertebral ganglia** or **preaortic ganglia.**

Courses of Preganglionic Fibers in the Sympathetic Division

The thoracic and uppermost lumbar ganglia of the sympathetic chains receive all of the fibers that emanate from the preganglionic sympathetic neurons in the lateral gray columns of T1-L2 (Fig. 5–1A). At each of these spinal cord levels, the preganglionic sympathetic fibers exit the spinal cord segment via the anterior (ventral) root, nerve trunk, and anterior (ventral) ramus of its spinal nerve. A branch of the anterior ramus called a **white ramus communicans** transmits the preganglionic sympathetic fibers to the paravertebral ganglion attached to the spinal nerve. The white rami communicantes of the sympathetic chains are so named because they are the branches, or rami, which communicate the "whitish" preganglionic sympathetic fibers of the spinal nerves to the paravertebral ganglia of the sympathetic chains. Almost all preganglionic sympathetic fibers are myelinated, and hence whitish in color. It is important to recognize that T1-L2 are the only spinal nerves that are each attached to their paravertebral ganglia via a white ramus communicans; this is because the T1-L2 spinal cord segments are the only segments to bear cell bodies of preganglionic sympathetic neurons, and, therefore, to give rise to preganglionic sympathetic fibers.

Upon entering the thoracic and uppermost lumbar ganglia of the sympathetic chains, the preganglionic sympathetic fibers take one of two courses: they either end in the paravertebral ganglia of the sympathetic chains (Fig. 5–1A) or exit the chains and end in the prevertebral ganglia and their associated nerve plexuses (Fig. 5–1B). The preganglionic sympathetic fibers which end in the paravertebral ganglia synapse with postganglionic sympathetic neurons in these ganglia (Fig. 5–1A). Some, but not all, of these fibers end in the paravertebral ganglion that they initially enter. The remainder of the fibers pass through their initial paravertebral ganglion, then extend within the sympathetic chain to either higher or lower ganglia in the chain, and there finally end by synapsing with postganglionic sympathetic neurons. The preganglionic sympathetic fibers which ascend and descend within the sympathetic chain form a large portion of the segments of the chain that link the ganglia together on each side.

The preganglionic sympathetic fibers which end in the prevertebral ganglia and their associated plexuses synapse with postganglionic sympathetic neurons in these ganglia and plexuses. Nerves called **splanchnic nerves** transmit these preganglionic sympathetic fibers from the paravertebral ganglia to the prevertebral ganglia and their plexuses. The thoracic and lumbar ganglia are the only paravertebral ganglia to give rise to splanchnic nerves; collectively, these splanchnic nerves provide almost all the preganglionic sympathetic fibers for the abdominal and pelvic viscera.

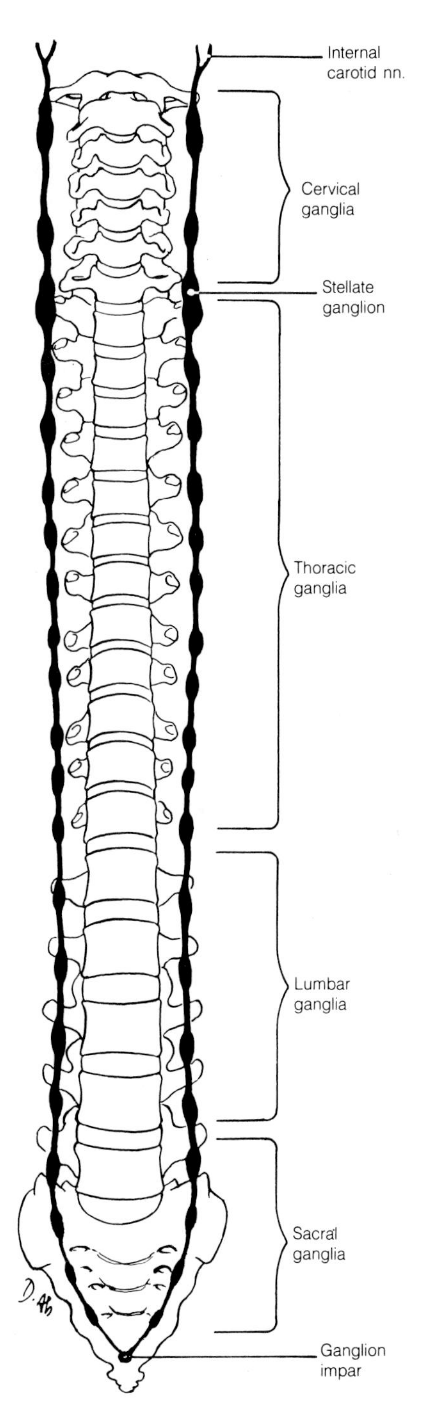

Fig. 5–2: Anterior view of the relationship of the sympathetic chains to the vertebral column.

The **greater splanchnic nerve** is the most prominent splanchnic nerve to arise from each sympathetic chain. It transmits preganglionic sympathetic fibers from the T5–T9 paravertebral ganglia (that is, from the paravertebral ganglia attached to T5–T9) to mainly the celiac ganglion. The **lesser splanchnic nerve** is the next most prominent splanchnic nerve; it transmits preganglionic sympathetic fibers from the T10 and T11 paravertebral ganglia to mainly the aorticorenal ganglion. The **least splanchnic nerve** transmits preganglionic sympathetic fibers from the T12 paravertebral ganglion to mainly the renal plexus (the nerve plexus around the renal artery). The **lumbar splanchnic nerves** arise from the lumbar ganglia; they transmit preganglionic sympathetic fibers to mainly the intermesenteric plexus (the nerve plexus around the abdominal aorta that lies between the origins of the superior and inferior mesenteric arteries) and the inferior mesenteric plexus. It is important to recognize that although the greater, lesser, least, and lumbar splanchnic nerves all arise from the sympathetic chains, the preganglionic sympathetic fibers in these splanchnic nerves collectively emanate from the preganglionic sympathetic neurons in the lateral gray columns of T5-L2.

Courses and Target Tissues of Postganglionic Fibers in the Sympathetic Division

The vast majority of the fibers that arise from postganglionic sympathetic neurons in the paravertebral ganglia exit the sympathetic chains via one of three principal types of branches: spinal nerve branches, vascular branches, and plexus branches.

The spinal nerve branches are called the **gray rami communicantes** of the sympathetic chains (Fig. 5–1A). They are so named because they are the branches which communicate "grayish" postganglionic sympathetic fibers from the paravertebral ganglia to the spinal nerves. Almost all postganglionic sympathetic fibers are not myelinated, and hence grayish in color. Each of the 31 spinal nerves receives a gray ramus communicans from one of the paravertebral ganglia. Since the number of ganglia in each sympathetic chain is less than the number of spinal nerves, it follows that some of the paravertebral ganglia extend a gray ramus communicans to each of two or more spinal nerves.

The postganglionic sympathetic fibers which enter each spinal nerve trunk are distributed into both the anterior and posterior rami of the spinal nerve, and conveyed to their target tissues by branches of the rami (Fig. 5–1A). The postganglionic sympathetic fibers borne by the 31 spinal nerves collectively innervate the smooth muscle tissue of blood vessels, the arrector pili muscles of hair follicles, and the fluid-secreting cells of sweat glands in (a) the cutaneous tissues of the neck, (b) the entirety of the upper and lower limbs, and (c) the walls of the thorax, abdomen, and pelvis. In these body regions, the fibers provide stimulation for vasoconstriction of blood vessels, erection of body hair, and sweat secretion.

The vascular branches of the paravertebral ganglia transmit their postganglionic sympathetic fibers onto arteries located in the vicinity of the sympathetic chains (Fig. 5–1B). These arteries and their branches convey the fibers for either part or all of their course to target tissues.

The postganglionic sympathetic fibers for the head exit the sympathetic chains almost exclusively via vascular branches of the cervical ganglia. The fibers arise from postganglionic sympathetic neurons in the cervical ganglia. Vascular branches of the cervical ganglia transmit the fibers onto the major arterial routes for the head (the carotid system of arteries and the vertebral arteries).

The plexus branches of the paravertebral ganglia transmit their postganglionic sympathetic fibers to autonomic plexuses in the neck and trunk of the body (Fig. 5–1B). The most notable plexus branches are those of the cervical and upper five thoracic ganglia which transmit postganglionic sympathetic fibers to the autonomic plexuses of the heart and lung (the cardiac and pulmonary plexuses). The vast majority of sympathetic fibers entering the cardiac and pulmonary plexuses are postganglionic sympathetic fibers transmitted via plexus branches of the cervical and upper five thoracic ganglia.

Almost all the fibers that arise from postganglionic sympathetic neurons in the prevertebral ganglia and their associated nerve plexuses exit their ganglia or plexuses via vascular branches. The fibers reach their target tissues by coursing alongside the arteries that supply the tissues. The postganglionic sympathetic fibers that arise from the prevertebral ganglia and their associated nerve plexuses collectively innervate all abdominal and pelvic viscera.

GENERAL ORGANIZATION OF THE PARASYMPATHETIC DIVISION

Locations of the Cell Bodies of the Preganglionic and Postganglionic Neurons

The parasympathetic division of the autonomic nervous system is divisible into two parts: a cranial part and a sacral part (Fig. 5–3). The cell bodies of the preganglionic neurons in the cranial part are all located in the brain. The cell bodies of the preganglionic neurons in the sacral part are all located in the lateral gray matter of the S2-S4 spinal cord segments.

In the parasympathetic division, the cell bodies of the postganglionic neurons are all located in minute ganglia distributed throughout the head, neck, thorax, abdomen, pelvis, and perineum (Fig. 5–3). Each ganglion lies either near to or embedded within its target tissue or organ. Only four pairs of these minute ganglia

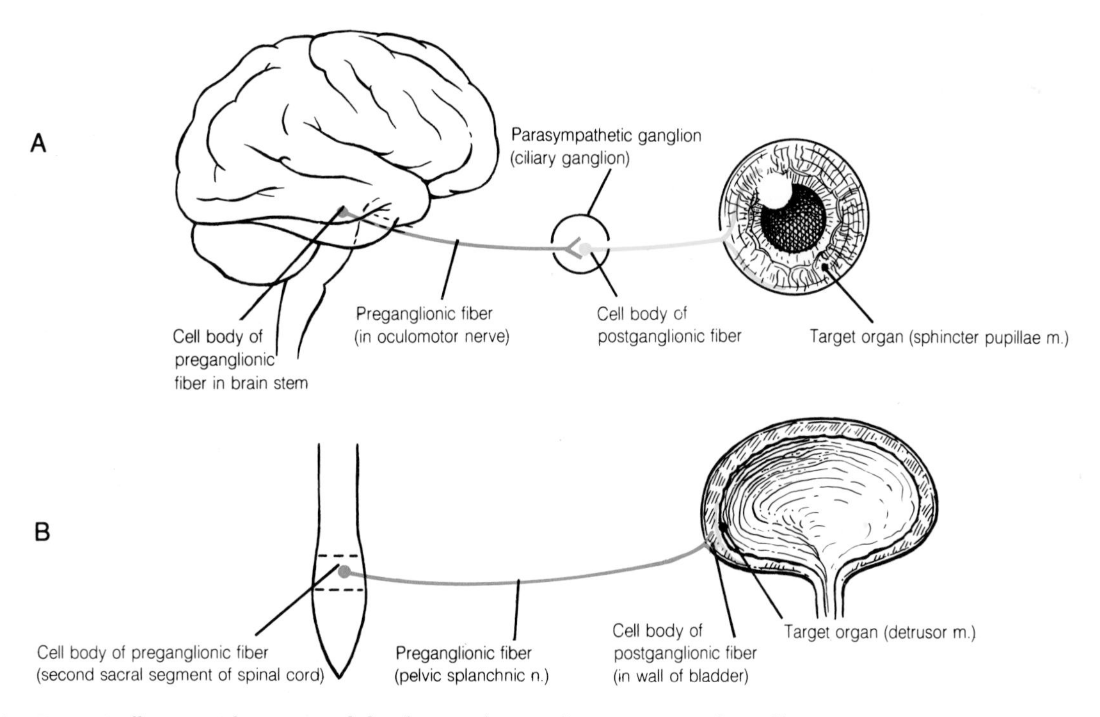

Fig. 5–3: Diagrams illustrating the origin and distribution of preganglionic parasympathetic fibers in (A) a cranial nerve and (B) a pelvic splanchnic nerve.

have specific, gross anatomic names, and they are all located in the head and neck. The **ciliary ganglion** lies in the orbital cavity, the **pterygopalatine and otic ganglia** lie deep in the head, and the **submandibular ganglion** lies below the floor of the mouth.

Courses of Preganglionic Fibers in the Parasympathetic Division

The fibers that emanate from the preganglionic parasympathetic neurons in the brain exit the brain via four of the cranial nerve pairs, specifically, the **oculomotor nerves (cranial nerves III),** the **facial nerves (cranial nerves VII),** the **glossopharyngeal nerves (cranial nerves IX),** and the **vagus nerves (cranial nerves X).** The preganglionic parasympathetic fibers of the oculomotor nerve end in the ciliary ganglion, those of the facial nerve end in either the pterygopalatine or submandibular ganglion, and those of the glossopharyngeal nerve end in the otic ganglion. The preganglionic parasympathetic fibers of the vagus nerves end in minute ganglia distributed throughout the neck, thorax, and abdomen. The paired vagus nerves transmit about 75% of the body's preganglionic parasympathetic fibers.

The fibers that emanate from the preganglionic parasympathetic neurons in the S2-S4 spinal cord segments exit the spinal cord via the anterior roots, nerve trunks, and anterior rami of S2-S4. The fibers exit the anterior rami of S2-S4 via small branches collectively called the **pelvic splanchnic nerves** (Fig. 5–3). The pelvic splanchnic nerves transmit their preganglionic parasympathetic fibers to minute ganglia in abdominal and pelvic viscera.

Target Tissues of Postganglionic Fibers in the Parasympathetic Division

The postganglionic parasympathetic neurons in the ciliary ganglion innervate two muscles in the eye: sphincter pupillae and ciliaris. These muscles are respectively responsible for constriction of the pupil and accommodation. The postganglionic parasympathetic neurons in the pterygopalatine ganglion are secretomotor for the lacrimal gland and for mucosal glands in the nasal cavities, palate, and pharynx. The postganglionic parasympathetic neurons in the submandibular ganglion are secretomotor for the submandibular and sublingual salivary glands and also for the smaller salivary glands of the oral cavity. The postganglionic parasympathetic neurons in the otic ganglion are secretomotor for the parotid salivary gland.

The postganglionic parasympathetic neurons in the minute ganglia which receive their preganglionic fibers from the vagus nerves distribute fibers within the following viscera: the glands of the larynx, the trachea, lungs, heart, liver, pancreas, spleen, the tubular digestive tract from the esophagus to the juncture between the proximal two-thirds and distal one-third of the transverse colon, and the upper portions of the ureters.

The postganglionic parasympathetic fibers in the minute ganglia which receive their preganglionic fibers from the pelvic splanchnic nerves distribute fibers within the following viscera: the lower portions of the ureters, bladder, the erectile tissues of the penis or clitoris, testes or ovaries, the uterus and uterine tubes in a female, and the tubular digestive tract from the distal third of the transverse colon to the upper half of the anal canal.

Neurotransmitters of the Autonomic Nervous System

Almost all the preganglionic and postganglionic neurons of the autonomic nervous system release either **acetylcholine** or **norepinephrine** at their axonal endings. Neurons which release acetylcholine are generically labelled **cholinergic neurons,** and those which release norepinephrine are generically labelled **adrenergic neurons.**

In the sympathetic division, almost all the preganglionic neurons are cholinergic and most of the postganglionic neurons are adrenergic. The postganglionic neurons which are cholinergic innervate almost exclusively sweat glands (small numbers also innervate blood vessels). In the parasympathetic division, almost all the preganglionic and postganglionic neurons are cholinergic.

There are some autonomic neurons (particularly those of the enteric nervous system in the gastrointestinal tract) which release neurotransmitters other than acetylcholine or norepinephrine. Neurotransmitters known to be active include serotonin, somatostatin, enkephalins and endorphins, vasoactive intestinal peptide (VIP), cholecystokinin (CCK), and the purines ATP and adenosine. Some of these substances may act both as neurotransmitters and hormones.

There is also a neurotransmitter of the autonomic nervous system whose chemical identity is unknown but whose effects have been demonstrated via electrical stimulation. This neurotransmitter is called the **nonadrenergic, noncholinergic (NANC) transmitter.** It is likely that nitric oxide (NO) or a closely related substance is the NANC transmitter. In any instance, NO should also be included as an autonomic neurotransmitter because it has been demonstrated to affect arteriolar blood flow and gastrointestinal motility.

BASIC CHARACTERISTICS OF SYMPATHETIC AND PARASYMPATHETIC ACTIVITY

The sympathetic and parasympathetic divisions are each responsive to sensory input from both somatic and visceral afferent neurons. Somatic afferent neurons convey sensory input from the skin, skeletal muscles, bones, and joints. Visceral afferent neurons convey sensory input from internal organs and glands, blood vessels, smooth muscle fibers, and the heart's cardiac muscle fibers.

In the sympathetic division, there are times when only certain portions of the division are selectively stimulated. For example, when a small skin area becomes overheated, sensory input from that area elicits stimulation of just those limited portions of the sympathetic division responsible for the sympathetic innervation of the skin area. The limited stimulation increases sweat secretion and also produces local vasodilation. However, stimulated activity in the sympathetic division is much more commonly a mass response (a response in which many portions of the division are simultaneously stimulated). This mass response, which is elicited generally by physical and/or emotional stress, produces reactions that prepare the entire body for increased mental and physical activity. The evoked reactions include dilation of the pupils, acceleration of the heartbeat, an increase in arterial blood pressure, and elevated blood plasma glucose and free fatty acid concentrations. This mass response by the sympathetic division is commonly called the "flight or fight response," as it is a response which occurs when an individual anticipates that he/she will have to either flee from or struggle with a very threatening situation.

By contrast, stimulated activity in the parasympathetic division is commonly discrete and specific (localized to specific viscera) and elicited by quiet and relaxed conditions. The different portions of the parasympathetic division evoke reactions such as lens accommodation, constriction of the pupil, slowing of the heartbeat, secretion of gastric juice by the stomach, peristaltic contractions of the rectum during defecation, and contraction of the urinary bladder during micturition. The physiological reactions evoked by the preganglionic parasympathetic fibers of the vagus nerves are commonly regarded as the basic effects of stimulated activity in the parasympathetic division.

CHAPTER **6**

Radiologic Anatomy

Physicians use radiologic techniques to image the internal anatomy of their patients. Each radiologic technique furnishes its own characteristic images of internal anatomy. Internal injury or pathology can be evaluated at the gross anatomic level of resolution by one or more of five major radiologic methods: conventional radiography, computerized tomography, nuclear medicine imaging, magnetic resonance imaging, and ultrasonography.

CONVENTIONAL RADIOGRAPHY

Conventional radiography is the oldest and most commonly employed radiologic method. In conventional radiography, an X-ray beam is directed through a body region for projection onto a flat X-ray film (Fig. 6–1). The X-ray film is differentially exposed because the various tissues of the body region differentially attenuate (absorb and scatter) the X-rays impinging upon them. Variation in film exposure produces an image of the body region's internal anatomy. The X-ray film and the image cast upon it together are called a **radiograph.** The parts of the body region that collectively attenuate the X-rays the least yield the darkest areas in the radiograph; the parts that collectively attenuate the X-rays the most provide the lightest areas.

The greater the radiodensity of a tissue, the more it attenuates X-rays (the **radiodensity** of a tissue is a function of its physical density and effective atomic number; the **effective atomic number** of a tissue is a measure of the average atomic number of its constituent atoms). Tissues which minimally attenuate X-rays are said to be **radiolucent.** The lungs are among the most radiolucent tissues of the body; their radiolucency is due to their high air content (which gives them a low physical density). Tissues which greatly attenuate X-rays are said to be **radiopaque.** Bones are among the most radiopaque tissues of the body; their high calcium content gives them an effective atomic number greater than that of all other body tissues except teeth.

When radiologic examination is warranted, radiographs are almost always the first radiologic images requested. This is because radiographs are comparatively inexpensive, can be quickly and easily obtained, and convey a relatively large amount of information. All radiographs exhibit the following general features:

(1) Every radiograph is a two-dimensional representation of a three-dimensional set of structures. Every radiograph thus consists of the superimposition of the images cast by all the structures which lie along the path of the X-ray beam. The more radiopaque structures tend to obscure the outlines of the more radiolucent structures. Therefore, in order to view more clearly the more radiolucent structures and to locate more accurately the position of any structural abnormality, it is generally necessary to examine radiographs of a body region from at least two or three different perspectives.

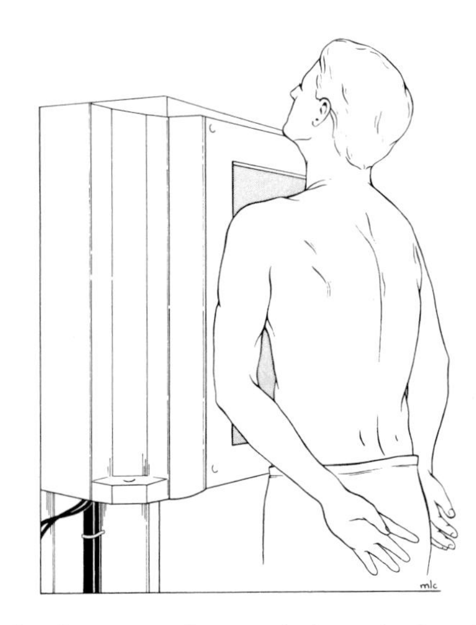

Fig. 6–1: Orientation of a patient's chest relative to the film cassette for a posteroanterior (PA) chest film.

(2) Since the X-ray beam is divergent, the images of all structures are, to some extent, magnified in radiographs. The parts of a body region that lie closest to the X-ray film cassette are magnified the least and cast the sharpest images.

(3) Almost all the tissues of the body exhibit a radiodensity similar to that of air, fat, water, or bone. The body's solid or fluid-filled soft tissues (such as its

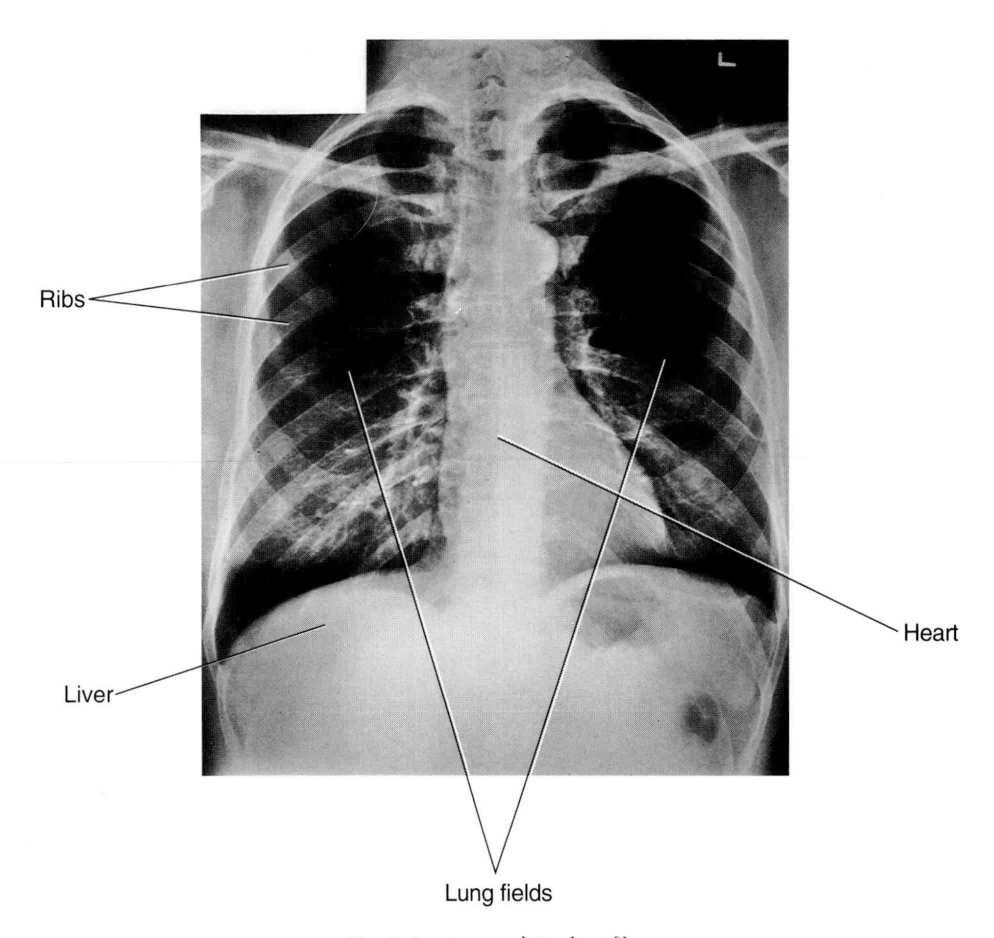

Fig. 6–2: A normal PA chest film.

skeletal muscles, the heart, the liver, and the kidneys) all have a radiodensity similar to that of water. A border of a solid or fluid-filled soft tissue is outlined in a radiograph only if that border faces an air-filled soft tissue (such as the lungs), a layer of fat, or a bone.

The **posteroanterior chest radiograph (PA chest film)** illustrates three of the four basic radiographic densities of body tissues (Fig. 6–2). A PA chest film is taken by pressing the anterior surface of the chest against an X-ray film cassette and exposing the film by passing an X-ray beam sagitally through the thorax from the back to the front of the chest (Fig. 6–1). The heart casts most of the central radiopaque shadow in a PA chest film, and the liver casts the radiopaque shadow below and to the right of the heart shadow; both organs exhibit a radiographic density similar to that of water. The lungs exhibit a radiodensity similar to that of air, and thus produce the paired radiolucent areas in the radiograph called the **lung fields.** The ribs, with their radiodensity of bone, are easily identified in the radiolucent lung fields.

COMPUTERIZED TOMOGRAPHY

Computerized tomography, or **CT,** gives physicians the capability to examine the X-ray attenuation pattern of a transverse (horizontal) section of a patient's body. This is accomplished as follows: A thin, fan-shaped X-ray beam is transversely directed toward the patient's body, and a curved, linear array of X-ray detectors on the opposite side of the body measures the amount of radiation that passes through the horizontal section of the patient (Fig. 6–3). Measurements are taken as the X-ray beam source is rotated (scanned) 360° around the patient's longitudinal axis. A computer divides the horizontal section into tissue volume elements called **voxels,** employs a reconstruction algorithm to determine the X-ray attenuation values of the voxels, and then transforms the X-ray attenuation values into CT numbers. **CT numbers** are expressed in terms of **Hounsfield units (HU)** (in honor of one of the inventors of computerized tomography). Hounsfield units span a scale whose limits extend from −1000 to +1000; the CT numbers −1000 HU, 0 HU, and +1000 HU are arbitrarily assigned to voxels consisting exclusively of air, water, and compact bone, respectively. Voxels of fat exhibit CT numbers which average approximately −110 HU.

A range of CT numbers is finally matched to the spectrum of gray shades that can be displayed by the picture elements **(pixels)** of a video monitor (a video monitor's pixels can display shades of gray from white to black). This final step permits the X-ray attenuation pattern of the array of voxels to be imaged on the video monitor by a matrix of gray-shaded pixels. In the CT scan, each voxel is represented by a single pixel, and, in general, the greater the X-ray attenuation value (and CT number) of the voxel, the lighter the shade of gray of the pixel.

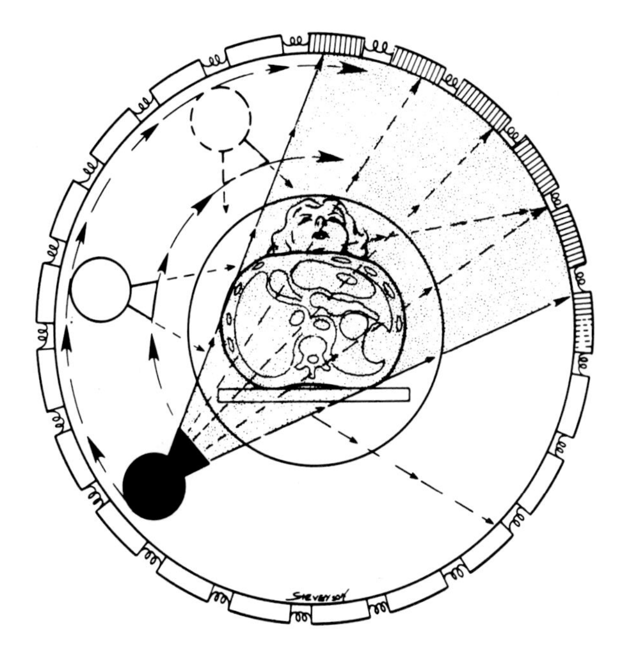

Fig. 6–3: Diagram of a CT scanner. The X-ray source rotates around the patient, directing multiple beams of X-rays through the patient to a ring of detectors.

CT scans are always oriented to show how a horizontal section of a patient's body would appear to a physician who stands at the foot of a hospital bed and examines the section as it is "removed" from the supine patient in the bed. In other words, physicians always examine CT scans as though they are looking toward the head of a supine patient from the patient's feet. The superior edge of the image in a CT scan therefore represents the anterior surface of the body region, and the right lateral edge of the image represents the left lateral surface of the body region (Fig. 6–4).

Computerized tomography is used principally as a diagnostic adjunct to conventional radiography. CT scans are superior to radiographs mainly in the extent to which they reveal radiodensity differences among and within soft tissues. This high image contrast sensitivity in CT scans is a consequence of primarily three factors:

(1) CT scanners use a relatively thin, collimated X-ray beam. The thinness of the beam reduces X-ray scatter, which, in turn, permits greater accuracy in the determination of the X-ray attenuation values of the voxels.

(2) Every voxel consists of a thin cross-sectional slice of generally just a single major structure (such as a bone, muscle, organ, gland, or large blood vessel). The images of most structures in CT scans are thus free of the superimposition of the images of other structures. Slice thickness of CT scans can be increased from 1 to 10 mm.

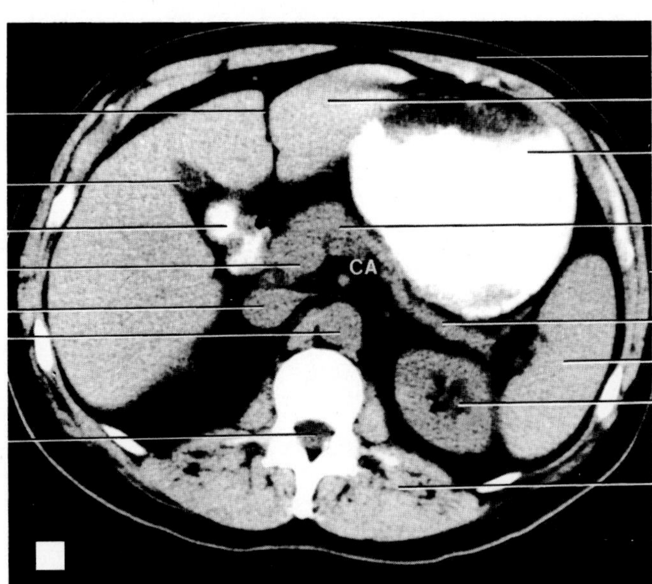

Fig. 6–4: A normal CT scan of the upper abdomen.

(3) The image contrast among tissues of different radiodensity can be enhanced by adjusting (windowing) the range of CT numbers that corresponds to the gray shade spectrum of the pixels on the video monitor. Every CT scan has an **upper window setting** and a **lower window setting.** All pixels with CT numbers greater than the upper widow setting appear white, all pixels with CT numbers lesser than the lower window setting appear black, and all pixels with CT numbers between the upper and lower window settings appear as a shade of gray. Adjustment of the upper and lower window settings to a relatively narrow range of the CT number scale enhances the contrast among contiguous pixels whose CT numbers lie in the narrow range.

NUCLEAR MEDICINE IMAGING

Nuclear medicine imaging techniques provide information about the distribution of trace amounts of radioactive substances introduced into the body. In most instances, the radioactive substance is either inhaled by and intravenously injected into a patient. Substances which have proven useful in nuclear medicine imaging are substances whose chemical and/or physical properties result in either preferential accumulation in or selective exclusion from certain body regions.

Substances used in nuclear medicine imaging are radiolabelled with isotopes whose nuclear disintegration either directly or indirectly generates gamma rays. Gamma rays have sufficient energy to escape from any region of the body for detection by external devices. Radioisotopes which indirectly generate gamma rays emit positrons upon their nuclear disintegration; the emitted positrons travel a few millimeters before each produces a pair of gamma rays via self-annihilation with an electron.

In most cases, radiologists use devices which provide projection images of emitted gamma rays (in a manner analogous to the way that an X-ray film provides a projection image of X-rays that pass through a body region). In other cases, radiologists use devices that provide tomographic images of emitted gamma rays (in a manner analogous to the way that a CT scanner provides tomographic images of X-rays that pass through a body region). Devices which provide tomographic images of the emission pattern from gamma

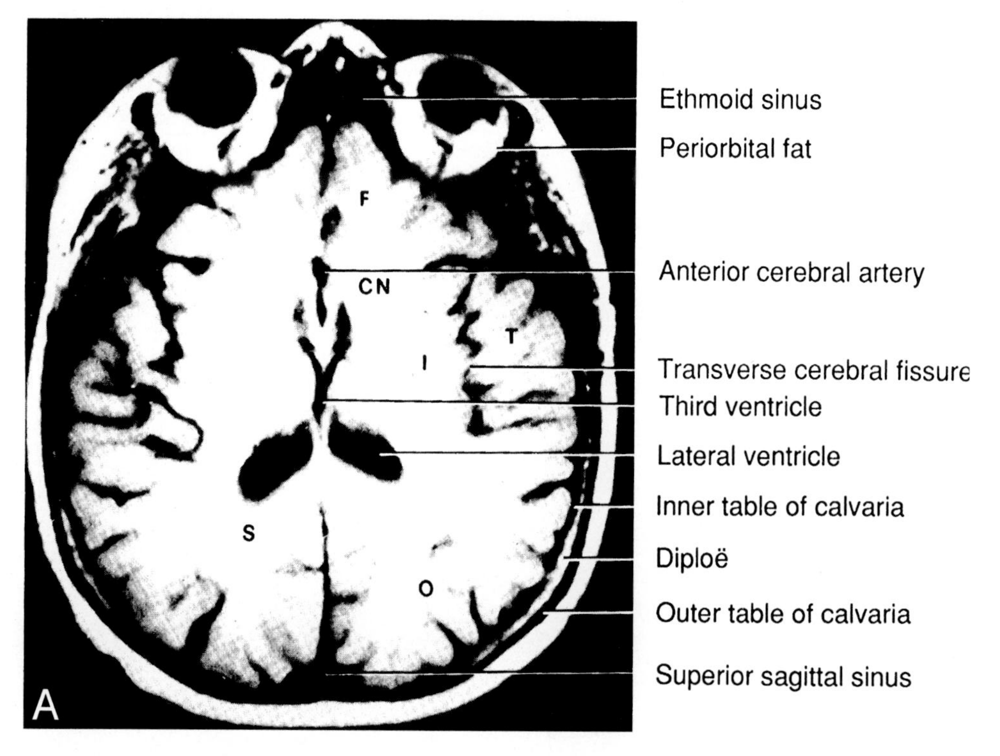

Fig. 6–5: (A) Transverse, (B) sagittal, and (C) coronal magnetic resonance images of the head.

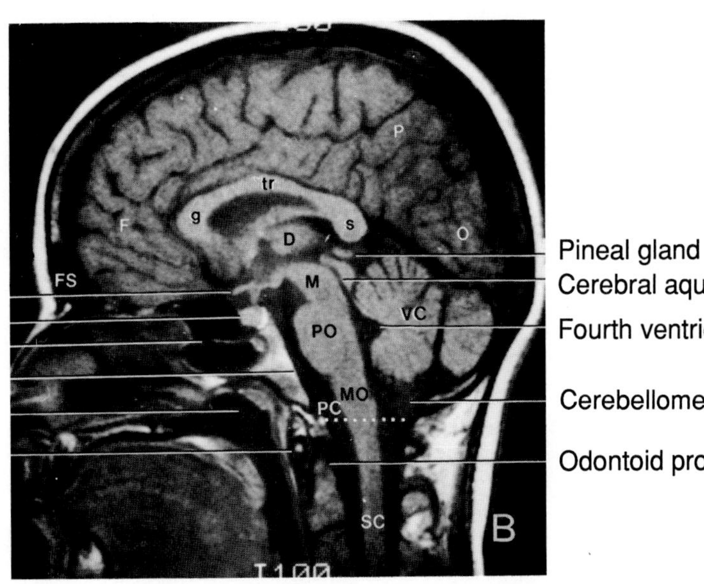

Fig. 6–5: Continued

ray-emitting radioisotopes are called **single photon emission tomography scanners (SPET scanners).** Devices which produce tomographic images of the emission pattern from positron-emitting radioisotopes are called **positron emitting tomography scanners (PET scanners).**

MAGNETIC RESONANCE IMAGING

Magnetic resonance imaging (MRI) is based upon the interaction of very strong magnetic fields and radiowaves with the magnetic atomic nuclei in a person's tissues. Magnetic atomic nuclei have an odd number of protons (such as hydrogen nuclei with their single proton). Magnetic atomic nuclei behave as if they are minute bar magnets. Magnetic atomic nuclei can also be regarded as spinning around their north-south polar axis.

MRI systems place patients first in a very strong, static magnetic field. The polar axes of a very small fraction of the magnetic atomic nuclei in the person's tissues tend to become angularly aligned parallel to the axis of the magnetic field. The polar axes of the angularly aligned nuclei precess, or rotate, around the axis of the magnetic field at frequencies within the radiowave frequency range. The patient at this time is exposed to a short pulse of radiowaves. The angularly aligned nuclei whose precessional frequency matches that of the radiowaves exactly can absorb energy from the microwaves. The nuclei that become excited to a higher energy state are angularly aligned in a new direction and spin in phase with each other. Upon cessation of radiowave exposure, the excited nuclei rapidly realign with the static magnetic field and their spins progressively shift out of phase. This angular realignment and spin dephasing generate radiowaves that are detected by the MRI system and processed by a computer to generate magnetic resonance images.

MRI systems can construct images of transverse, sagittal, or coronal sections of body regions (Fig. 6–5). The interactions of the hydrogen nuclei in a patient's body with the static magnetic field and radiowave pulses are generally selected to construct the images because hydrogen nuclei are the most abundant type of magnetic atomic nuclei in the body.

ULTRASONOGRAPHY

Ultrasonic vibrations are physical vibrations whose frequency range is immediately above that audible to humans. Ultrasonic vibrations can penetrate and pass through certain tissues of the body. When ultrasonic vibrations pass through a body region, they are partially reflected at interfaces between materials of different acoustic impedance (the acoustic impedance of a material is the product of its physical density and the velocity at which it transmits physical vibrations). Ultrasonography uses this reflective property of ultrasonic vibrations to image structural interfaces within the

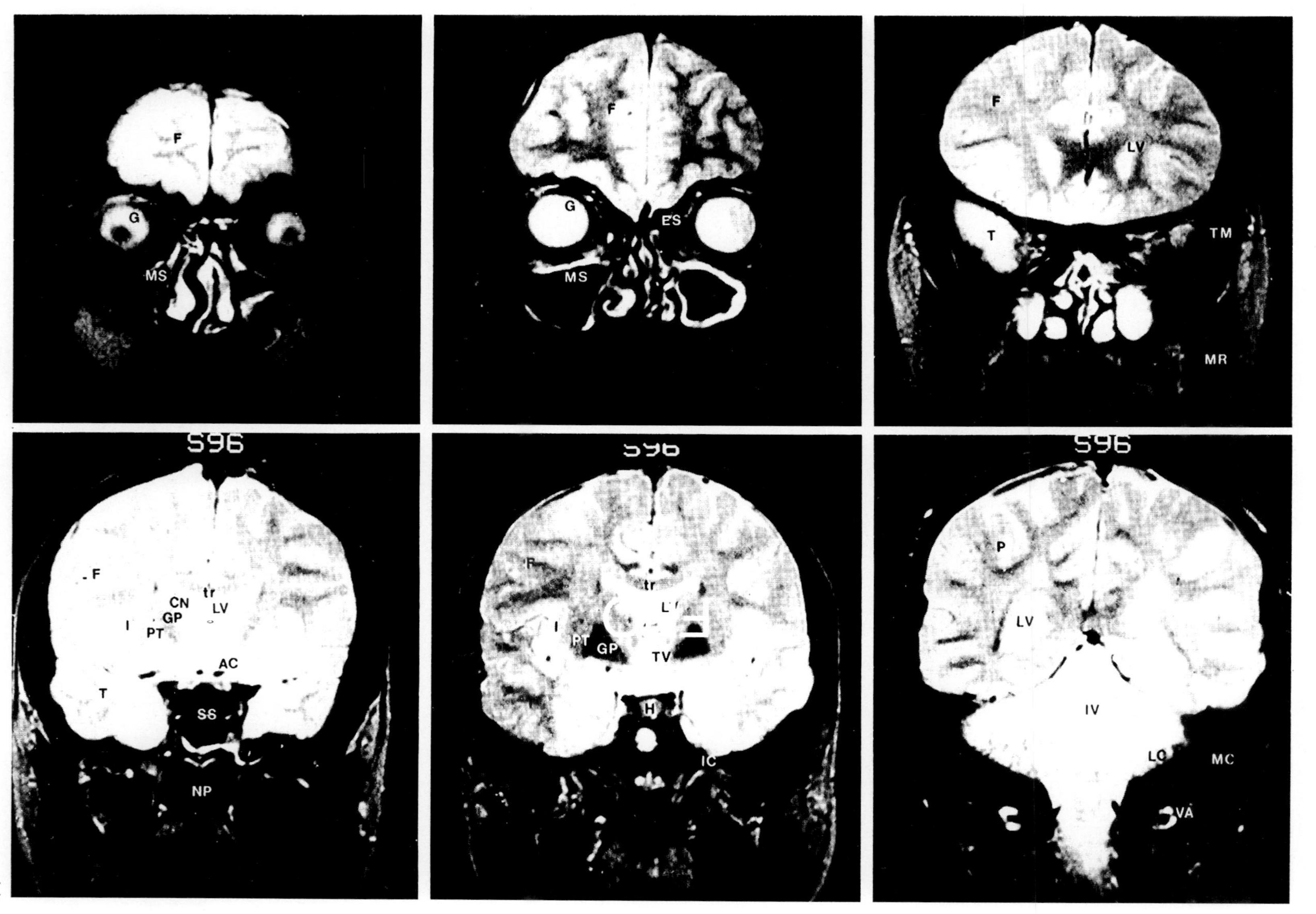

Fig. 6–5: Continued

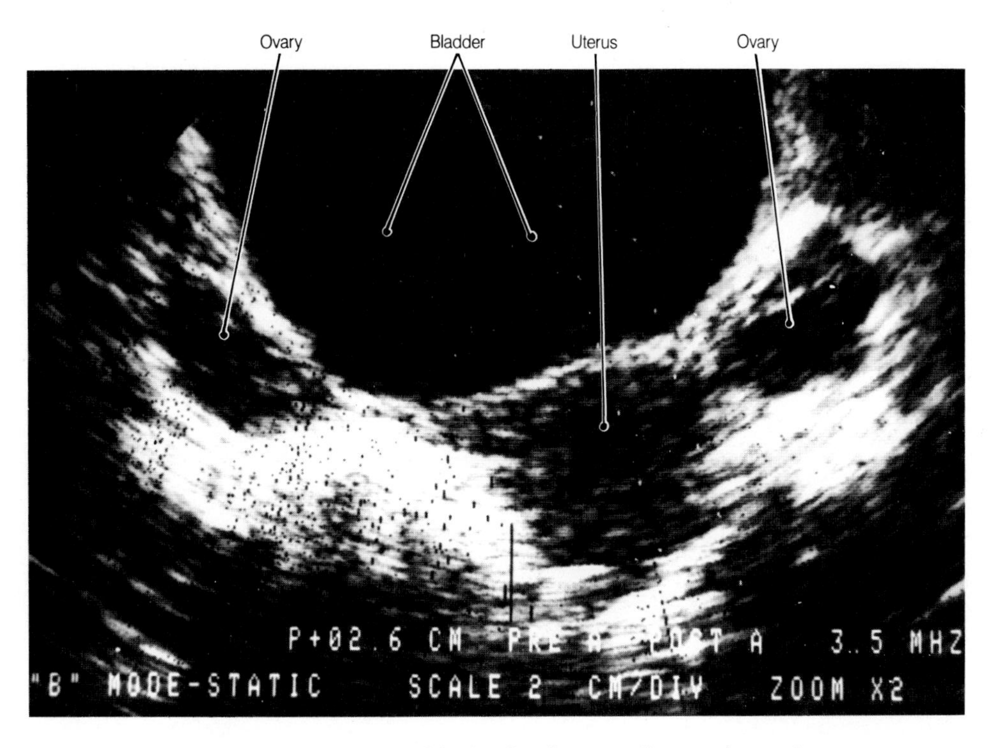

Fig. 6-6: Sonogram of the female pelvic cavity (horizontal section).

body. Ultrasonography is safe and inexpensive, and thus can be repetitively used to monitor certain medical conditions.

The various tissues of the body differentially attenuate ultrasonic vibrations. Bone and air-filled, soft tissues (such as the lungs and air-filled segments of the intestines) have the greatest attenuation rates; such tissues shield deeper tissues from ultrasonographic imaging. Water is the material within the body that has the smallest attenuation rate; this feature accounts for the use of a urine-filled bladder as a sonolucent window during ultrasonography of pelvic tissues and organs (Fig. 6–6).

SECTION II

Upper Limb

CHAPTER **7**

The Shoulder

This chapter focuses on the innervation and actions of the muscles that move the upper limb at its most proximal joints. The discussion begins with an examination of the bones of the shoulder and the joints associated with shoulder and arm movements.

THE SKELETAL FRAMEWORK OF THE SHOULDER

The clavicle, scapula and proximal end of the humerus form the skeletal framework of the shoulder (Figs. 7–1 and 7–2). The crank-shaped **clavicle** is palpable along its entire length in the upper anterior part of the shoulder (Fig 7–1). The clavicle is the first bone to undergo ossification, which occurs in the intramembranous mode.

The **scapula** is a triangular bone in the upper posterior part of the shoulder. The principal palpable parts of the scapula are the **acromion, coracoid process, spine** and **inferior angle** (Figs. 7–1 and 7–2). The **point of the shoulder** acquires its shape from the subcutaneous acromion (Fig. 7–1). The coracoid process can be palpated in the upper anterior region of the shoulder beneath the lateral third of the clavicle (Fig. 7–1). The spine and inferior angle are both palpable in the upper part of the back.

The **humerus** is the bone of the arm. The most important parts of the proximal half of the humerus include the **head, greater and lesser tuberosities, intertubercular groove, anatomic neck** and **surgical neck** (Fig. 7–3). The anatomic neck is the region at the upper end of the shaft of the humerus immediately adjoining the margin of the head. The surgical neck is the region of the shaft that is immediately proximal to the insertion of the pectoralis major, teres major, and latissimus dorsi muscles. Longitudinal growth of the humeral shaft during childhood and adolescence occurs more actively at the proximal than at the distal end of the bone. Growth ends at the distal end of the shaft at about 14 to 16 years of age and at the proximal end at about 18 to 20 years of age.

THE JOINTS OF THE SHOULDER

The sternoclavicular, acromioclavicular, and shoulder joints are the most proximal joints of the upper limb. They are associated with all shoulder and arm movements. All of the upper limb joints in an adult (except for the middle radioulnar joint) are of the synovial type.

The **sternoclavicular (SC) joint** joins the medial end of the clavicle to the **manubrium of the sternum** (Fig. 7–4). It is the only joint between a bone of the upper limb and a bone of the axial skeleton. An articular disc of fibrous cartilage lies between the clavicular

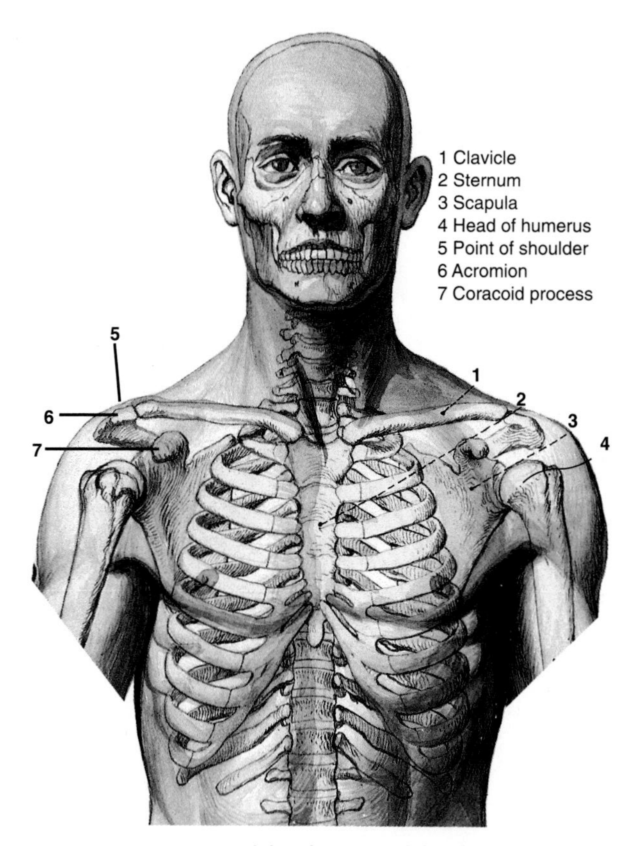

Fig. 7–1: Anterior view of the placement of the clavicle, scapula, and proximal end of the humerus in the shoulder.

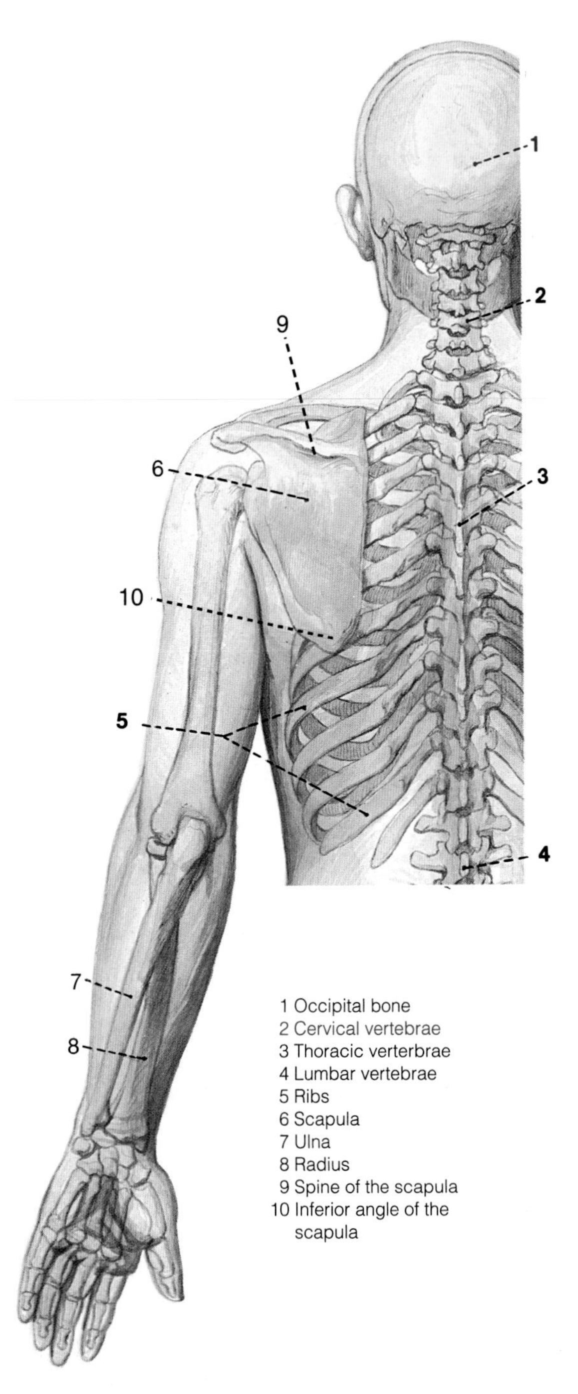

Fig. 7–2: Posterior view of the placement of the scapula and proximal end of the humerus in the shoulder.

and sternal articular surfaces in the joint (Fig. 7–4). The **medial supraclavicular nerve** (a nerve derived from the cervical plexus) and the **nerve to subclavius** (a nerve derived from the brachial plexus) innervate the SC joint. The **costoclavicular ligament,** which attaches the clavicle to the **first costal cartilage** (the first costal cartilage is the bar of hyaline cartilage which joins the first rib to the sternum), stabilizes the SC joint by serving as the strongest non-muscular structure binding the clavicle to the rib cage (Fig. 7–4).

The **acromioclavicular (AC) joint** joins the lateral end of the clavicle to the medial margin of the acromion of the scapula (Fig. 7–4). An articular disc of fibrous cartilage lies between the clavicular and scapular articular surfaces in the joint. The **suprascapular and lateral pectoral nerves** innervate the AC joint. The **coracoclavicular ligament,** which attaches the clavicle to the coracoid process of the scapula, stabilizes the AC joint by serving as the strongest non-muscular structure suspending the scapula from the clavicle (Fig. 7–4).

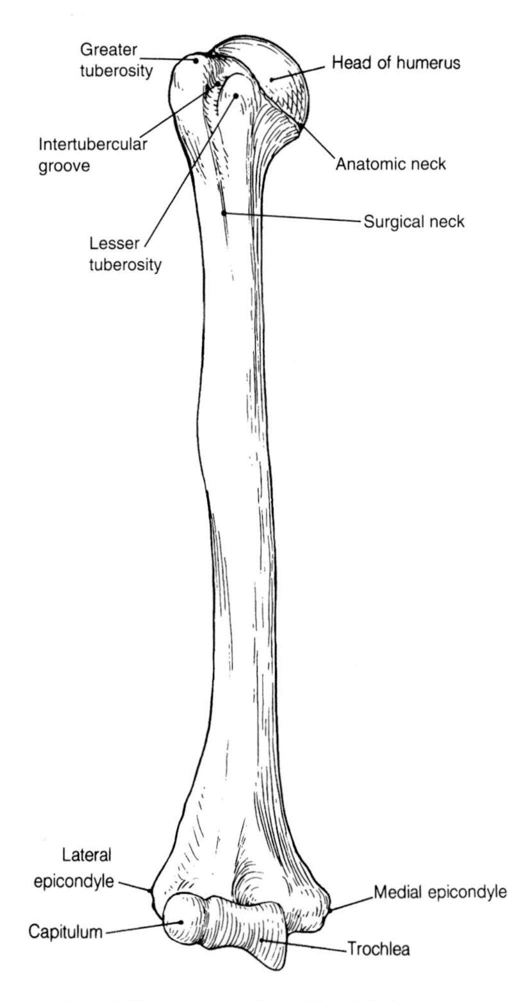

Fig. 7–3: Anterior view of the right humerus.

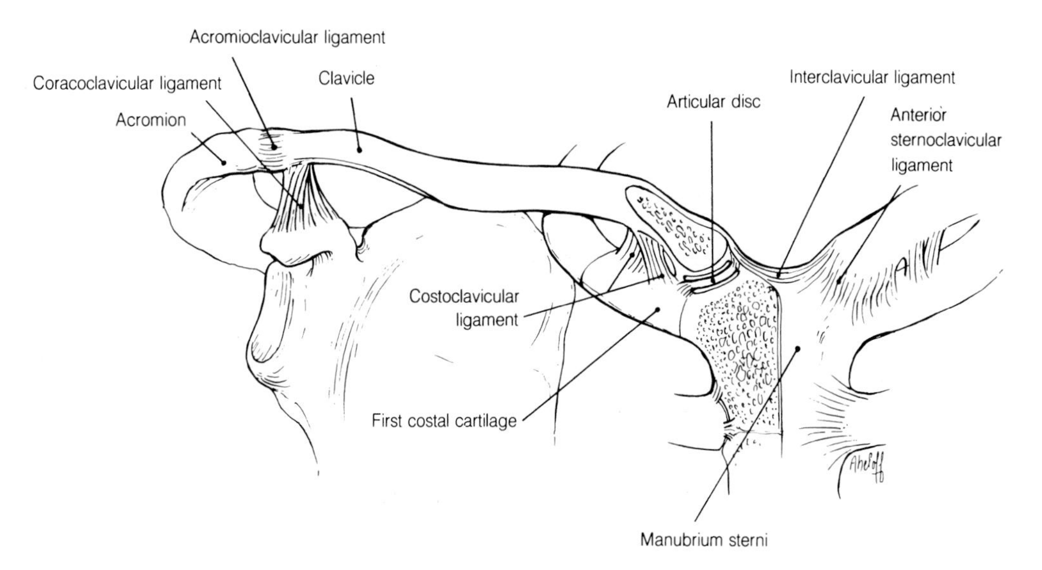

Fig. 7–4: Diagram illustrating coronal section of the right sternoclavicular joint and anterior view of the right acromioclavicular joint.

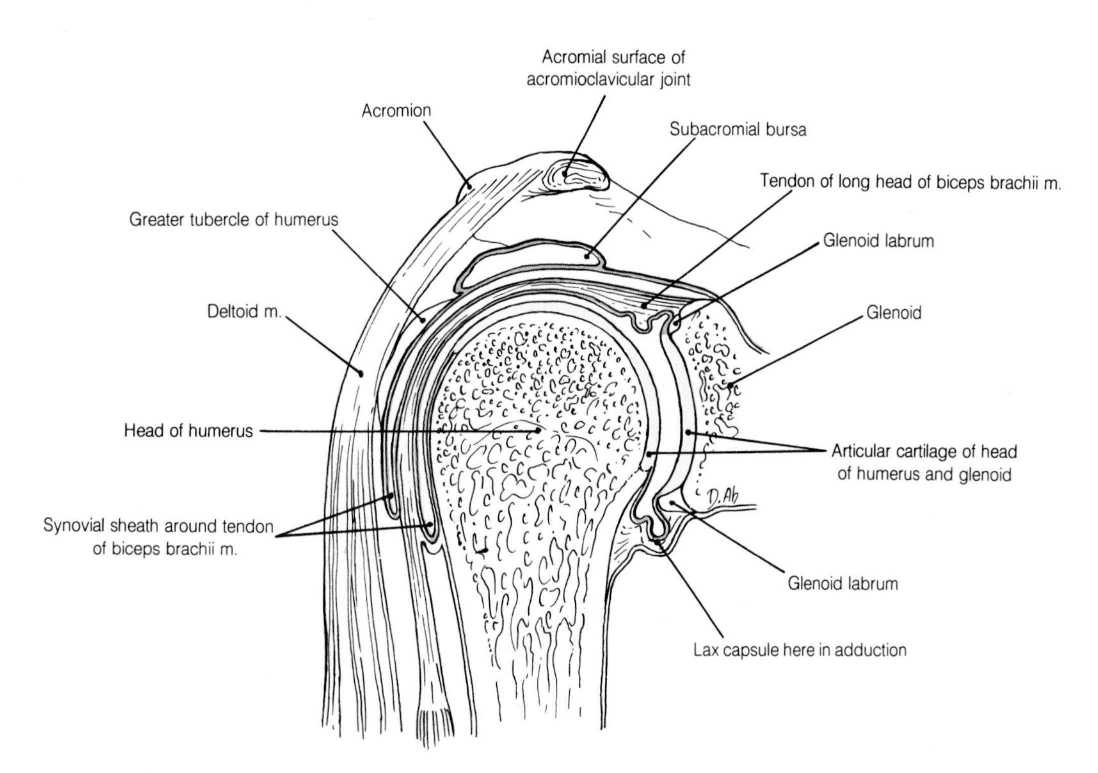

Fig. 7–5: Coronal section of the right shoulder joint.

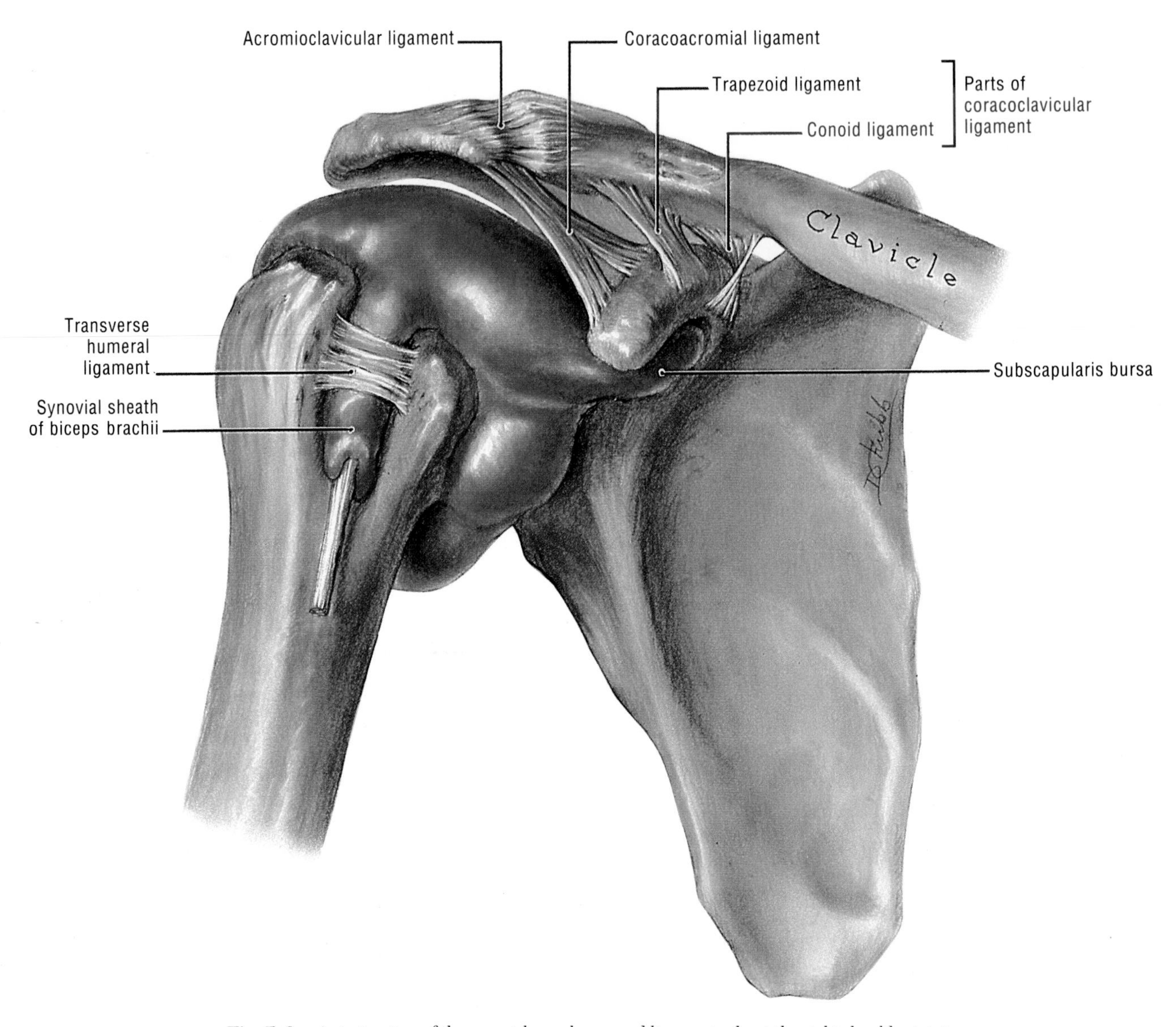

Fig. 7–6: Anterior view of the synovial membrane and ligaments about the right shoulder joint.

The **shoulder joint** joins the head of the humerus with the **glenoid cavity of the scapula** (Fig. 7–5). The humeral head also articulates in the joint with the **glenoid labrum,** a ring of fibrous cartilage that rims the glenoid cavity. The **suprascapular, axillary, and lateral pectoral nerves** innervate the joint. The **coracoacromial ligament,** which extends between the coracoid process and acromion of the scapula, arches over the shoulder joint (Fig. 7–6). The coracoacromial ligament resists superior dislocation of the head of the humerus from its articulation in the shoulder joint.

The fibrous capsule of the shoulder joint is attached proximally to the outer margin of the glenoid cavity and distally to the upper end of the shaft of the humerus (mostly the anatomic neck). There is a gap in the humeral attachment which spans over the intertubercular groove and transmits the **tendon of the long head of biceps brachii** (Fig. 7–6). The thickened part of the fibrous capsule that spans over the intertubercular groove is called the **transverse humeral ligament.** The tendon of the long head of biceps brachii originates within the shoulder joint from the **supraglenoid tubercle of the scapula** and arches over the head of the humerus in route to its extension within the intertubercular groove (Fig. 7–5). The synovial membrane of the shoulder joint envelops the tendon along its course through the joint.

The subscapular and subacromion-subdeltoid bursae are associated with the shoulder joint. The **subscapular bursa,** which is an extracapsular extension of the shoulder joint's synovial lining (Fig. 7–6), intervenes between the scapular fossa and the tendinous insertion of subscapularis onto the shoulder joint's capsule. The **subacromion-subdeltoid bursa** separates the tendinous insertion of supraspinatus onto the joint capsule from the overlying acromion and deltoid muscle. Figure 7–5 depicts the bursa and the overlying acromion and deltoid muscle but not the underlying, tendinous insertion of supraspinatus. The synovial lining of the subacromion-subdeltoid bursa is not continuous with that of the shoulder joint's synovial cavity.

THE MECHANICAL ROLES OF THE BONES AND JOINTS OF THE SHOULDER

The clavicle serves as a strut for the shoulder because of its position, shape, and ligamentous attachments to the rib cage and scapula. If the clavicle were not present, the shoulder would (by virtue of its mass and lack of support) hang more inferiorly, anteriorly and medially. With the clavicle in place, however, there is a bone (securely attached to the rib cage by the costoclavicular ligament) positioned and shaped to brace the shoulder superiorly, posteriorly and laterally. The forthcoming section on the muscles of the shoulder explains how the clavicle acts as a strut that provides for trapezius's capacity to both elevate and posterolaterally position the shoulder.

The strut role served by the clavicles is emphasized by the effect of their absence in individuals afflicted with an hereditary condition called **cleidocranial dysostosis** (Fig. 7–7). This condition is characterized by defective ossification of cranial bones and the clavicles; the defect can cause the complete absence of the clavicles. Individuals who do not have their clavicles

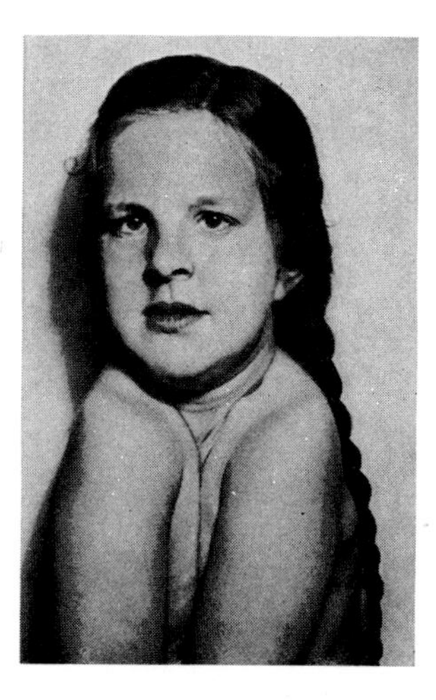

Fig. 7–7: A young girl who can bring her shoulders almost into contact with each other in front of the chest because of the absence of the clavicles. The clavicles are absent because of cleidocranial dysostosis.

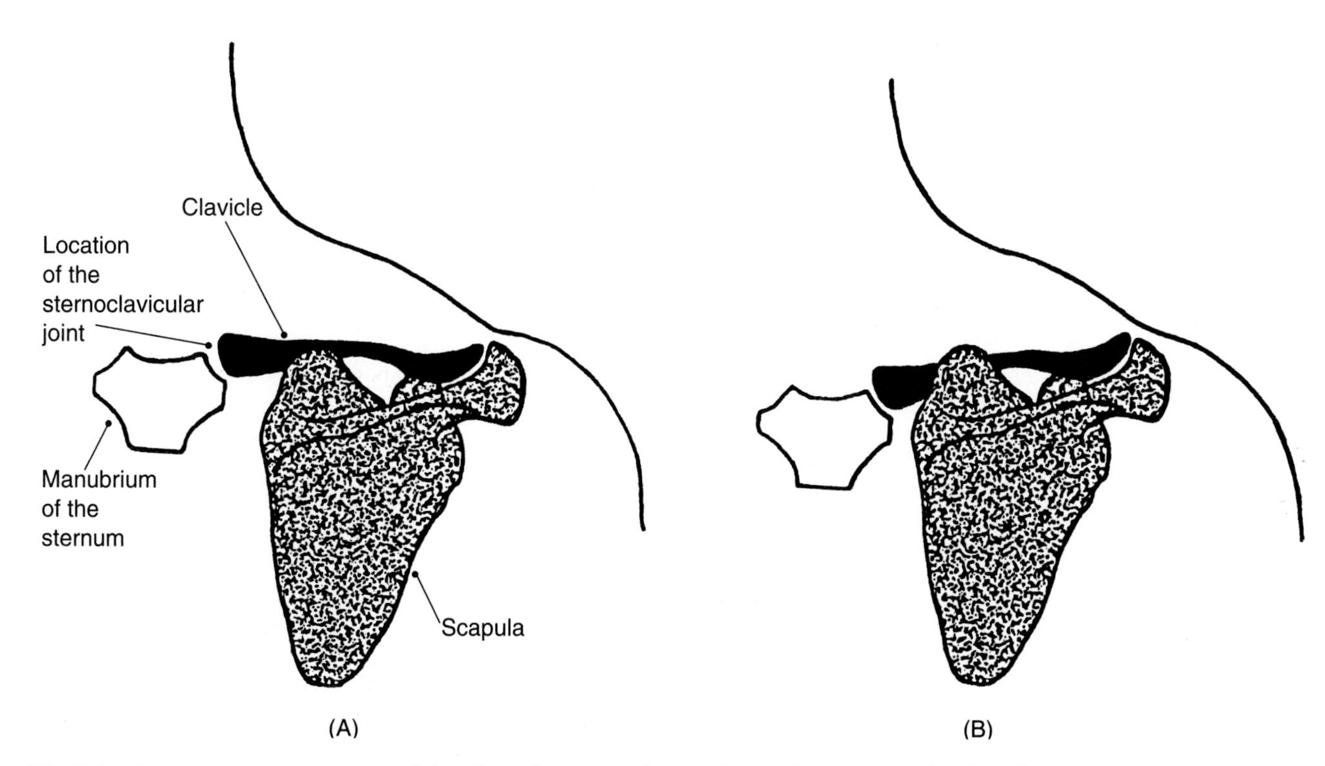

Fig. 7–8: Diagrammatic posterior views of the relationships among the manubrium of the sternum, clavicle, and scapula in (A) the anatomic position and (B) a position in which the pectoral girdle has been raised from its anatomic position. Notice that elevation of the pectoral girdle involves upward rotation of the clavicle at the sternoclavicular joint.

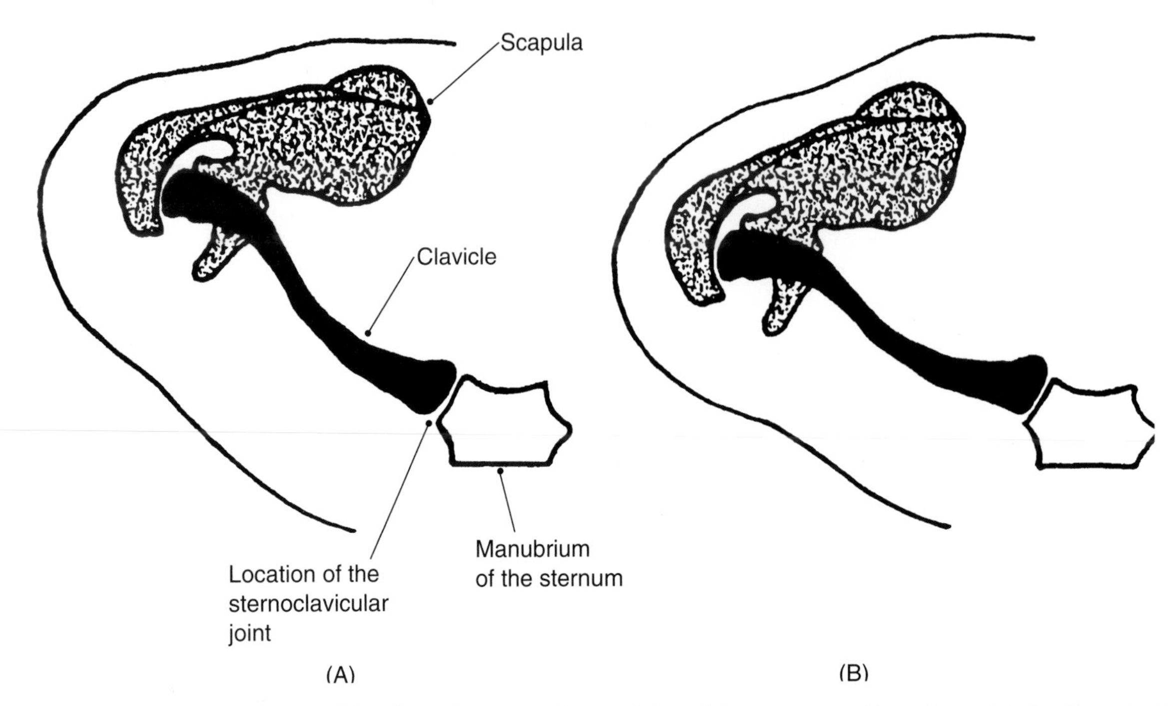

Fig. 7–9: Diagrammatic superior views of the relationships among the manubrium of the sternum, clavicle, and scapula in (A) the anatomic position and (B) a position in which the pectoral girdle has been protracted from its anatomic position. Notice that protraction of the pectoral girdle involves forward (anterior) rotation of the clavicle at the sternoclavicular joint. Retraction of the pectoral girdle is the movement opposite to that of protraction.

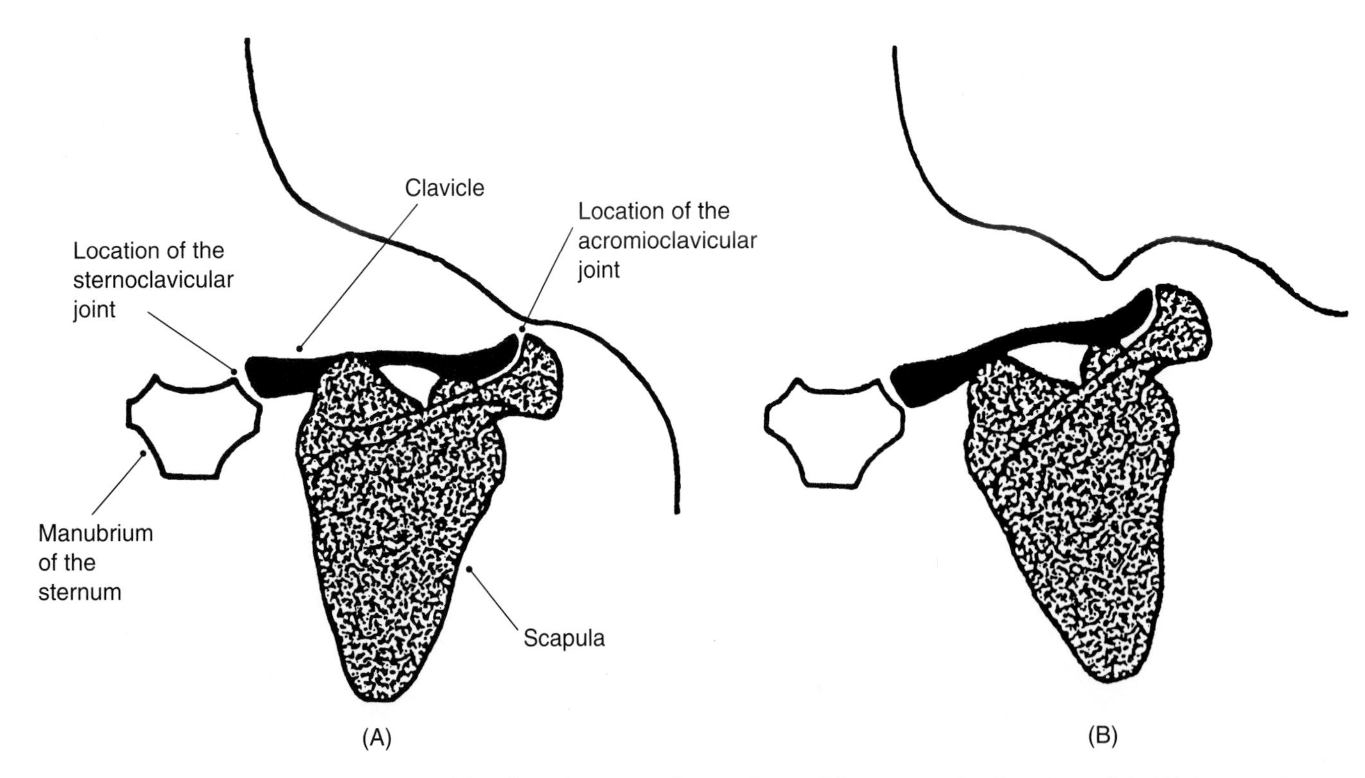

Fig. 7–10: Diagrammatic posterior views of the relationships among the manubrium of the sternum, clavicle, and scapula in (A) the anatomic position and (B) a position in which the pectoral girdle has been laterally rotated from its anatomic position. Notice that lateral rotation of the pectoral girdle involves upward rotation of the clavicle at the sternoclavicular joint and rotation of the scapula at the acromioclavicular joint (the scapular rotation at the AC joint increases the angle between the shaft of the clavicle and the spine of the scapula). Medial rotation of the pectoral girdle is the movement opposite to that of lateral rotation.

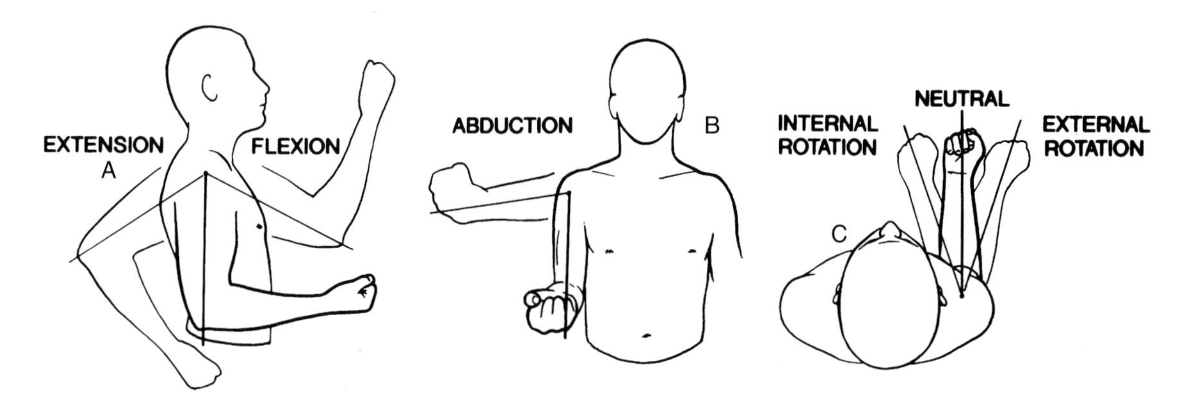

Fig. 7–11: Diagrams illustrating (A) flexion and extension of the arm, (B) abduction and adduction of the arm, and (C) external rotation and internal rotation of the arm. The forearm is flexed 90° at the elbow in all the diagrams.

can bring their shoulders almost into contact with each other in front of the chest.

The clavicle and scapula together form a mobile boom for the upper limb. This boom is called the **pectoral girdle** because it girds the upper part of the chest. The pectoral girdle is moved about by the muscles that suspend it from the head, neck, back, and chest wall. The muscles that **raise and lower the pectoral girdle** (Fig. 7–8), in effect, raise and lower the shoulder. Other muscles **protract and retract the pectoral girdle** (Fig. 7–9) and, in doing so, pull the shoulder horizontally forward and backward around the rib cage. Finally, there are muscles which **laterally rotate and medially rotate the scapula** (Fig. 7–10). The lateral rotators of the scapula rotate the clavicle at the SC joint and rotate the scapula about the AC joint in such a fashion that the inferior angle of the scapula moves upward and laterally; the medial rotators of the scapula exert opposite actions. The lateral rotators of the scapula provide much of the force required to raise the arm above the shoulder.

The pectoral girdle serves as a boom for the upper limb because its movements at the SC and AC joints greatly extend the range of movements of the arm. There are limits to the extent to which the humeral head can be moved in certain directions on the surface of the glenoid cavity, and these limits define the range of movements of the arm at the shoulder joint. The muscles which move the pectoral girdle, however, can change the position and orientation of the glenoid cavity, and thus extend the range of movements for the arm. For example, abduction of the humerus at the shoulder joint alone cannot abduct the arm much above the level of the shoulder. However, abduction of the humerus at the shoulder joint can also be performed in tandem with lateral rotation of the scapula at the SC and AC joints. The coupling of the abduction range available at the shoulder joint with that available at the SC and AC joints extends the range of arm abduction to 180°.

The shoulder joint provides more range and freedom of movement than any other synovial joint in the body. The shoulder joint provides for **flexion, extension, abduction, adduction, external (lateral) rotation** and **internal (medial) rotation** of the arm (Fig. 7–11). The combination of the flexion, extension, abduction and adduction movements provides for **circumduction** of the arm. Three anatomic features account for this great range and freedom of movement:

(1) The shoulder joint has a ball-and-socket configuration in which the head of the humerus is free to rotate in any direction on the surface of the glenoid cavity (starting from the anatomic position).
(2) The comparatively small surface area of the glenoid cavity enhances the extent to which the humeral head can rotate in any direction on the glenoid cavity (Fig. 7–5).
(3) The joint has a comparatively lax fibrous capsule (Fig. 7–5).

Most arm movements involve movements of the pectoral girdle and humerus at the SC, AC, and shoulder joints. These movements occur in a concomitant and smoothly coordinated pattern called **scapulohumeral rhythm.** Scapulohumeral rhythm is particularly evident during abduction of the arm through the full 180° range of motion. Lateral rotation of the scapula at the SC and AC joints accompanies upward rotation of the humerus at the shoulder joint through most of the 180° range (Fig. 7–12). On average, lateral rotation of the scapula contributes 1° abduction for every 2° provided by upward rotation of the humerus at the shoulder joint.

CLINICAL PANEL II-1

FRACTURE OF THE CLAVICLE PROXIMAL TO THE CORACOCLAVICULAR LIGAMENT

Common Mechanism of the Injury: When the upper limb is used to brace the body, forces are transmitted along the skeletal framework of the shoulder to the trunk of the body. In particular, forces are transmitted from the humerus to the scapula across the shoulder joint, from the scapula to the clavicle via the coracoclavicular ligament, then medially along the clavicle and finally from the clavicle to the body trunk, generally by the costoclavicular ligament. Because bracing the body with the upper limb commonly occurs before accidental collisions and falls, such bracing can impose sufficient stresses on the clavicle to fracture it between the clavicular attachments of the coracoclavicular and costoclavicular ligaments. These considerations explain why the clavicle is the most commonly fractured bone of the body.

Common Symptoms: Because a fracture of the clavicle proximal to the coracoclavicular ligament abolishes the clavicle's strut role for the shoulder, a person with such a fracture presents with a shoulder that (by virtue of its mass and lack of support) is displaced inferiorly, anteriorly and medially. The fracture site is generally evident because of angulation of the fragments.

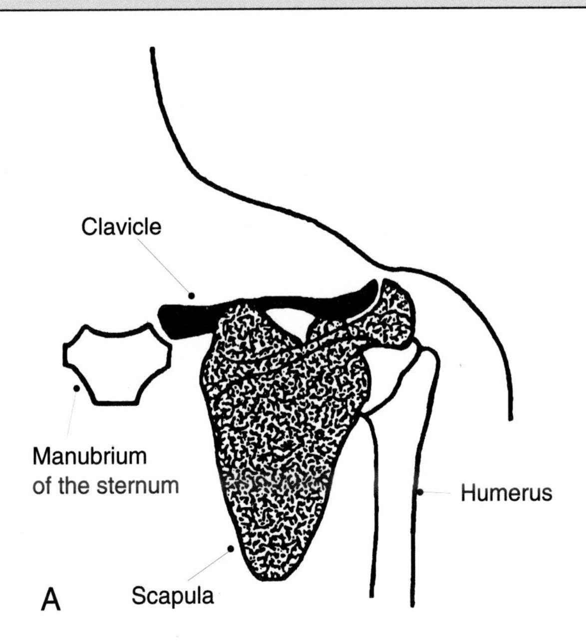

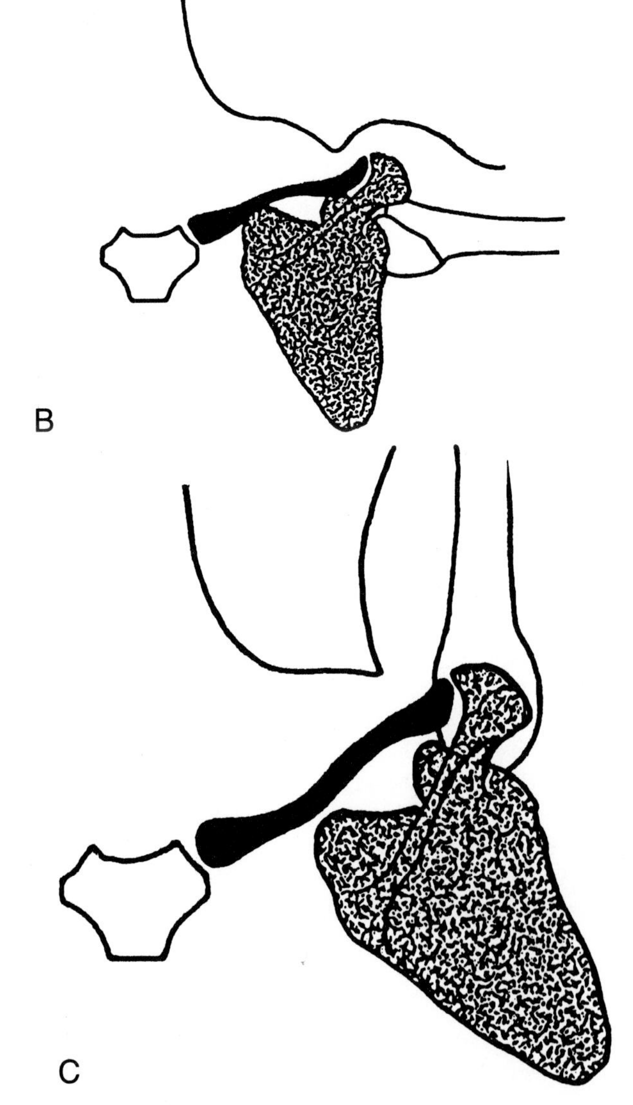

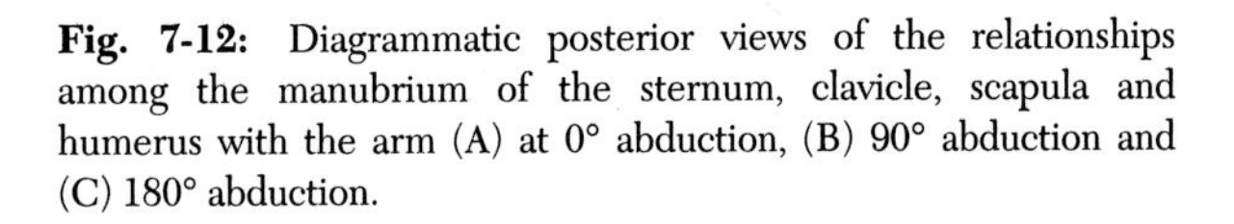

Fig. 7-12: Diagrammatic posterior views of the relationships among the manubrium of the sternum, clavicle, scapula and humerus with the arm (A) at 0° abduction, (B) 90° abduction and (C) 180° abduction.

THE MUSCLES OF THE SHOULDER

The shoulder muscles may be classified functionally into four groups: muscles that suspend the pectoral girdle from the vertebral column, muscles that pull on the pectoral girdle from the anterior chest wall, the muscles of the rotator cuff, and the prime movers of abduction and adduction of the arm at the shoulder joint. Tables 7–1 to 7–4 give a reference for the origin, insertion, nerve supply, and actions of the shoulder muscles.

Trapezius is the major muscle that suspends the upper limb from the vertebral column (Fig. 7–13). Its upper muscle fibers, which descend from the back of the head and neck to insert onto the lateral third of the clavicle and the acromion, can raise the pectoral girdle. Its middle muscle fibers, which extend horizontally from the upper thoracic vertebrae to insert onto the spine of the scapula, can retract the pectoral girdle. Its lower muscle fibers, which ascend from the lower thoracic vertebrae to insert onto the medial end of the spine of the scapula, lower the pectoral girdle. The upper and lower fibers of trapezius together laterally rotate the scapula.

The action of trapezius results from its upper, middle, and lower muscle fibers. The tractive forces generated by the upper, middle, and lower sets of muscle fibers can be varied independently of each other. This independence of action is possible because of the manner by which skeletal muscles (of which trapezius is an example) are innervated. An explanation of this process follows:

Table 7–1
Shoulder Muscles that Suspend the Pectoral Girdle from the Vertebral Column

Muscle	Origin	Insertion	Nerve Supply	Actions
Trapezius	Medial third of the superior nuchal line of the occipital bone, external occipital protuberance, ligamentum nuchae, and spinous processes of vertebrae C7 to T12	Spine of the scapula, the acromion, and lateral third of the clavicle	Spinal root of the accessory nerve provides motor supply, fibers from spinal nerves C3 and C4 provide for proprioception	Upper fibers raise the pectoral girdle, middle fibers retract the pectoral girdle, and lower fibers lower the pectoral girdle; upper and lower fibers act together to laterally rotate the scapula
Levator scapulae	Transverse processes of vertebrae C1 to C4	Medial border of the supraspinous fossa of the scapula	Dorsal scapular nerve (C5) and fibers from spinal nerves C3 and C4	Raises the pectoral girdle and medially rotates the scapula
Rhomboid minor	Ligamentum nuchae and spinous processes of vertebrae C7 and T1	Medial border of the scapula near the medial end of the spine of the scapula	Dorsal scapular nerve (C5)	Raises and retracts the pectoral girdle and medially rotates the scapula
Rhomboid major	Spinous processes of vertebrae T2 to T5	Medial border of the infraspinous fossa of the scapula	Dorsal scapular nerve (C5)	Raises and retracts the pectoral girdle and medially rotates the scapula

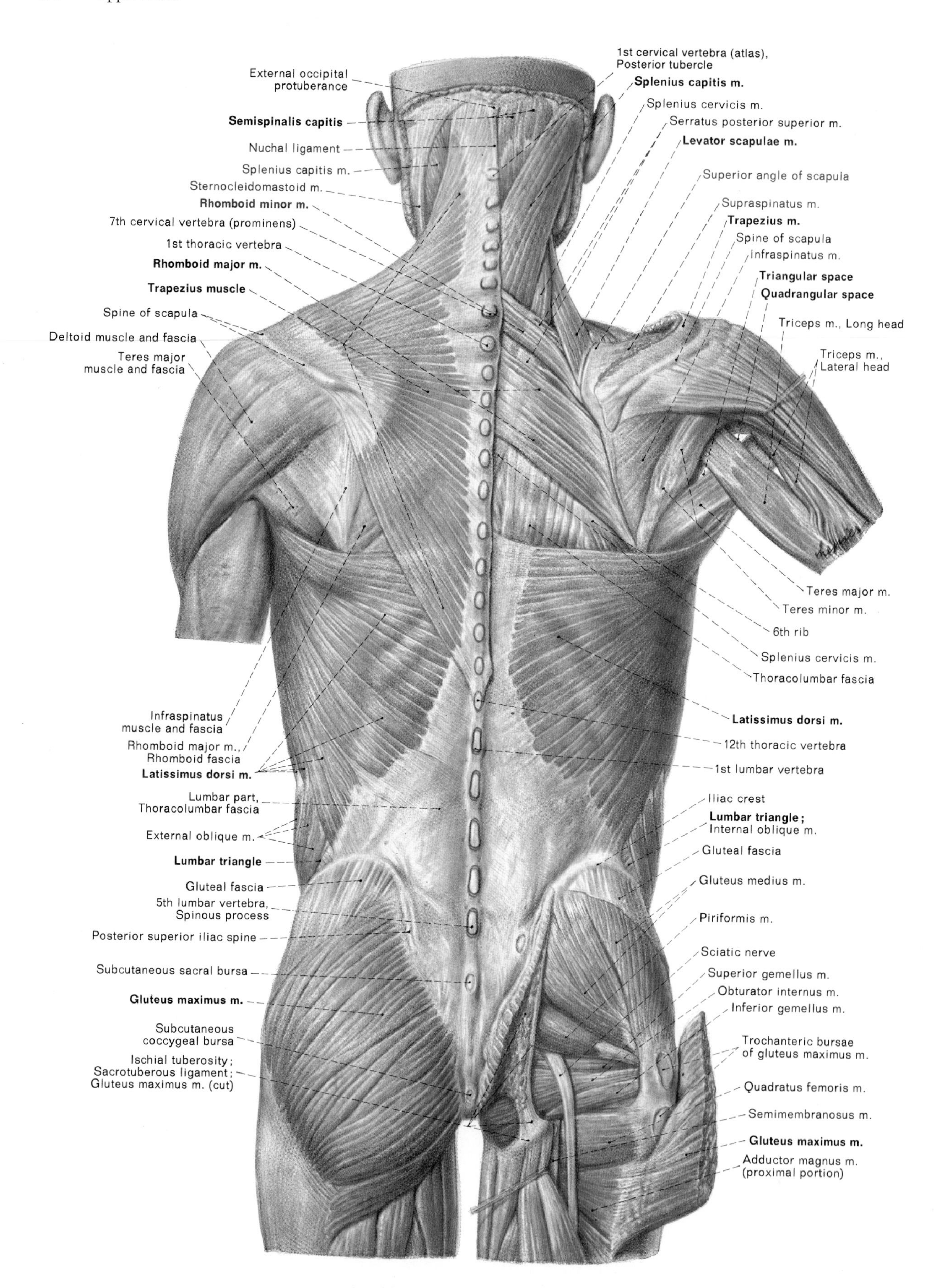

Fig. 7–13: Muscles of the posterior neck, shoulder, back, and gluteal region.

Table 7–2
Shoulder Muscles that Pull on the Pectoral Girdle from the Anterior Chest Wall

Muscle	Origin	Insertion	Nerve Supply	Actions
Serratus anterior	Outer surfaces of the anterolateral portions of ribs 1-8	Anterior aspect of the medial border of the scapula	Long thoracic nerve (C5, **C6,** and **C7**)	Protracts the pectoral girdle and laterally rotates the scapula
Pectoralis minor	Outer surfaces of ribs 3-5	Coracoid process of the scapula	Medial and lateral pectoral nerves (C6, **C7,** and C8)	Lowers and protracts the pectoral girdle and medially rotates the scapula
Subclavius	First costal cartilage	Inferior surface of the clavicle	Nerve to subclavius (**C5** and C6)	Lowers the pectoral girdle

Every skeletal muscle is innervated by a set of motor nerve fibers. These motor nerve fibers are axons that extend from the cell bodies of lower motor neurons to the muscle fibers of the skeletal muscle. Within the muscle, each motor nerve fiber forms cell-to-cell junctions (motor end plates) with a group of muscle fibers called a **motor unit.** When the motor nerve fiber transmits an impulse sufficient to generate the action potential of the muscle fibers in its motor unit, the impulse elicits simultaneous contraction of all the muscle fibers within the motor unit. Because all or almost all the muscle fibers within a motor unit are innervated by just a single motor nerve fiber and that motor nerve fiber can be stimulated independently of the motor nerve fibers that innervate other motor units within the muscle, each motor unit of a skeletal muscle can be stimulated independently of the other motor units. Thus, tractive forces of the upper, middle, and lower fibers of trapezius can be varied independently of each other.

The average size of motor units varies among the skeletal muscles of the body. In general, the smaller a skeletal muscle and the finer its actions, the smaller the average size of its motor units. The largest skeletal muscles of the body have motor units ranging in size from several hundred to over one thousand muscle fibers; the smallest skeletal muscles have motor units with as few as three muscle fibers.

The trapezius muscle is chiefly responsible for the ability to raise or shrug the shoulder. It is a prime mover for lateral rotation of the scapula, and thus provides much of the force required to raise the arm above the shoulder. Consequently, loss of trapezius's actions results in a lowering of the shoulder and weakness in shrugging the shoulder and raising the arm above the shoulder. Isolated trapezius palsy also results in flaring of the vertebral border and inferior angle of the scapula when the upper limb is at rest in the anatomic position. Abduction of the arm against resis-

Table 7–3
Muscles of the Rotator Cuff

Muscle	Origin	Insertion	Nerve Supply	Actions
Supraspinatus	Supraspinous fossa of the scapula	Capsule of the shoulder joint and greater tuberosity of the humerus	Suprascapular nerve (**C5** and C6)	Abducts the arm
Infraspinatus	Infraspinous fossa of the scapula	Capsule of the shoulder joint and greater tuberosity of the humerus	Suprascapular nerve (**C5** and C6)	Externally rotates the arm
Teres minor	Upper lateral border of the scapula	Capsule of the shoulder joint and greater tuberosity of the humerus	Axillary nerve (**C5** and C6)	Externally rotates the arm
Subscapularis	Subscapular fossa of the scapula	Capsule of the shoulder joint and lesser tuberosity of the humerus	Upper and lower subscapular nerves (C5, **C6,** and C7)	Internally rotates the arm

tance accentuates the flaring, but flexion of the arm minimizes it.

Levator scapulae, rhomboid major and **rhomboid minor** lie immediately deep to trapezius (Fig. 7–13). They descend from the cervical and uppermost thoracic vertebrae to insert onto the medial border of the scapula. Levator scapulae assists trapezius in raising the pectoral girdle and is a medial rotator of the scapula. The rhomboids assist trapezius in raising and retracting the pectoral girdle and are medial rotators of the scapula.

Serratus anterior originates from the anterolateral surface of the rib cage, curves around the lateral surface of the rib cage, and inserts onto the anteromedial border of the scapula (Fig. 7–14). It is the most powerful protractor of the pectoral girdle and a prime mover for lateral rotation of the scapula. Isolated serratus anterior palsy results in pronounced flaring of the scapula (in particular, the inferior angle) when the outstretched arm is thrust forward (the pronounced flaring is frequently referred to as "winging" of the scapula). The flaring is less evident if the upper limb is at rest or abducted against resistance. Loss of serratus anterior's actions weakens the ability to raise the arm above the shoulder.

Pectoralis minor ascends from upper ribs to insert onto the coracoid process of the scapula (Fig. 7–15). It can lower and protract the pectoral girdle and medially rotate the scapula.

Subclavius is a comparatively minor muscle that extends from the first costal cartilage to the undersurface of the clavicle (Figs. 7–16 and 7–17). It can lower the pectoral girdle.

Supraspinatus, infraspinatus, teres minor and **subscapularis** are called the **muscles of the rotator cuff** because their tendons of insertion form a musculotendinous cuff about the shoulder joint. Supraspinatus originates from the upper posterior surface of the scapula and its tendon of insertion passes deep to the coracoacromial ligament and acromion and over the shoulder joint before inserting onto the capsule of the shoulder joint and the greater tuberosity of the humerus (Fig. 7–16). The tendons of insertion of infraspinatus and teres minor pass posterior to the shoulder joint before inserting onto the capsule of the shoulder joint and the greater tuberosity (Fig. 7–16). The tendon of insertion of subscapularis passes anterior to the shoulder joint before inserting onto the capsule of the shoulder joint and the lesser tuberosity (Fig. 7–17). The muscles of the rotator cuff are not powerful movers of the arm. When the powerful arm movers exert their actions, however, the muscles of the rotator cuff maintain the humeral head in close and proper apposition to the glenoid cavity. In effect, the muscles of the rotator cuff serve to stabilize the

Table 7–4
Prime Movers of Abduction and Adduction of the Arm at the Shoulder Joint

Muscle	Origin	Insertion	Nerve Supply	Actions
Deltoid	Lateral third of the clavicle, acromion, and spine of the scapula	Deltoid tuberosity of the humerus	Axillary nerve (**C5** and C6)	Anterior fibers flex and internally rotate the arm, middle fibers abduct the arm, and posterior fibers extend and externally rotate the arm
Pectoralis major	Medial half of the clavicle, sternum, and costal cartilages of ribs 2-6	Lateral lip of the intertubercular groove of the humerus	Medial and lateral pectoral nerves; clavicular head (C5 and **C6**); sternocostal head (C7, **C8** and T1)	Adducts and internally rotates the arm; clavicular fibers flex the arm
Teres major	Lower lateral border of the scapula	Medial lip of the intertubercular groove of the humerus	Lower subscapular nerve (**C6** and C7)	Adducts and internally rotates the arm
Latissimus dorsi	Posterior part of the iliac crest, thoracolumbar fascia, spinous processes of the lower 6 thoracic vertebrae, and ribs 9 or 10–12	Floor of the intertubercular groove of the humerus	Thoracodorsal nerve (**C6, C7,** and C8)	Extends, adducts, and internally rotates the arm

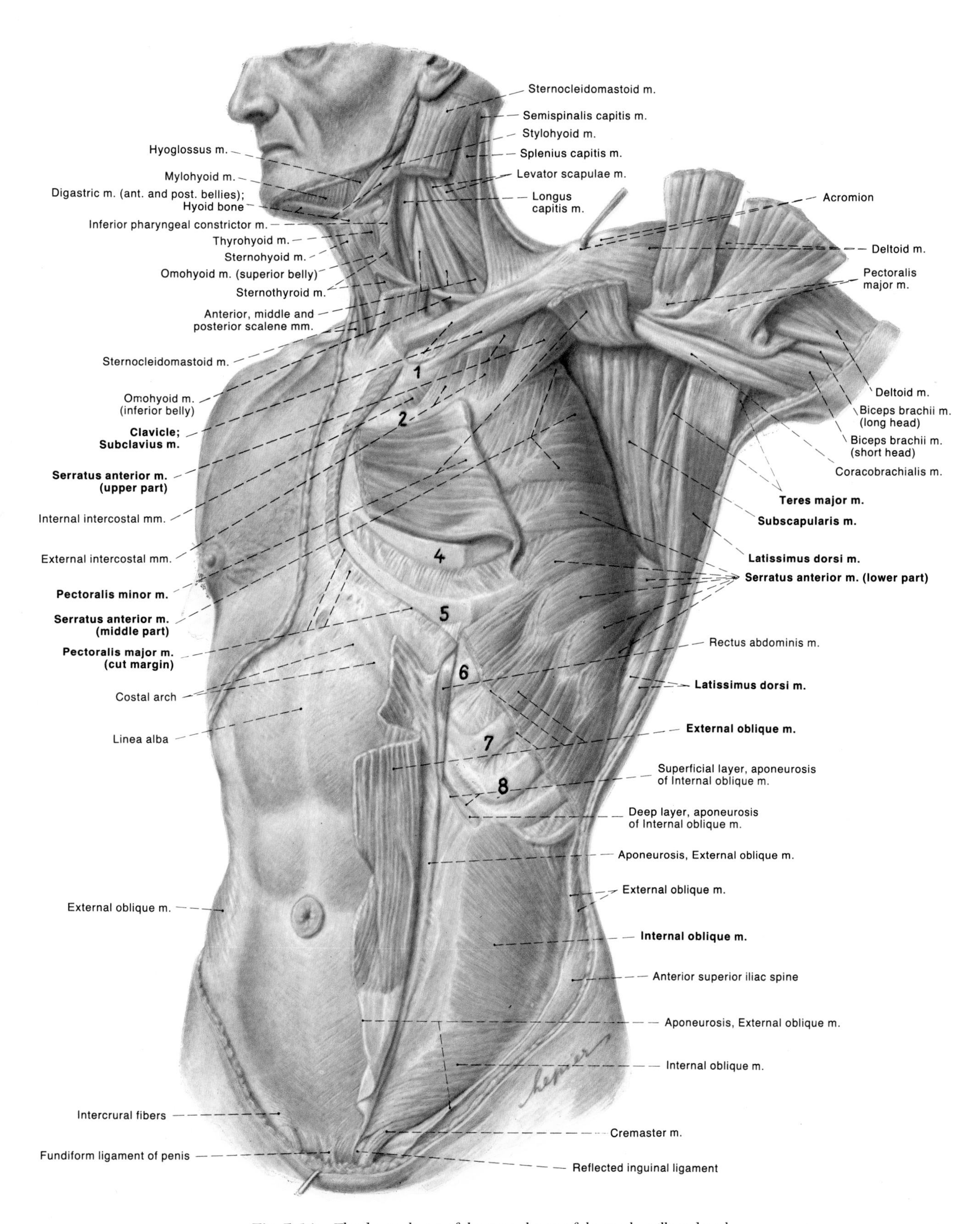

Fig. 7–14: The deeper layers of the musculature of the trunk, axilla and neck.

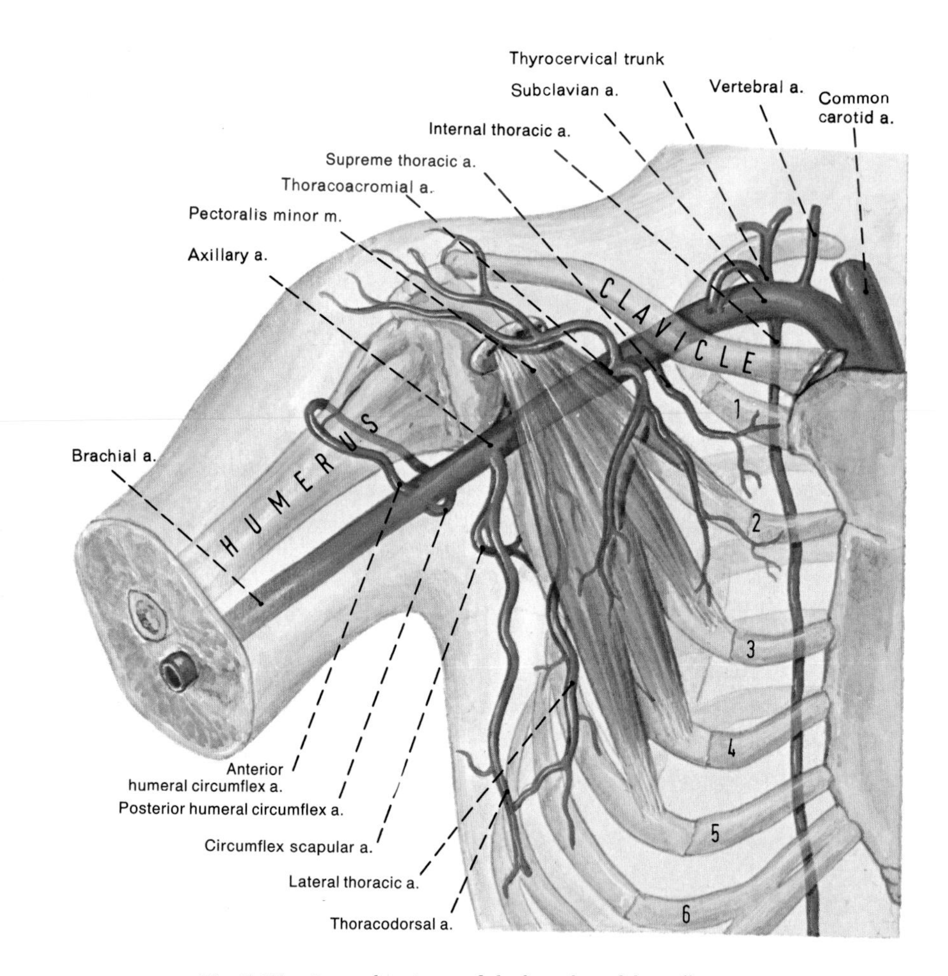

Fig. 7–15: Pectoralis minor and the branches of the axillary artery.

dynamic integrity of the shoulder joint when the prime movers of the arm exert their forces across the joint.

Deltoid forms the prominent muscle mass in the uppermost lateral aspect of the arm. The action of deltoid is the result of the actions of its anterior, middle and posterior fibers, all of which converge to insert onto the lateral side of the shaft of the humerus (Fig. 7–18). The anterior fibers, which originate from the lateral third of the clavicle and pass in front of the shoulder joint, flex and internally rotate the arm (Fig. 7–18). The middle fibers, which originate from the acromion and pass over the shoulder joint, abduct the arm at the shoulder joint (Fig. 7–16). The posterior fibers, which originate from the spine of the scapula and pass behind the shoulder joint, extend and externally rotate the arm (Fig. 7–16).

Deltoid is the prime mover for abduction of the arm at the shoulder joint. If the upper limb is abducted from the anatomic position, supraspinatus assists deltoid in initiating abduction.

Pectoralis major represents the muscle mass in the **anterior axillary fold** (Fig. 7–19). Its muscle fibers originate from the medial half of the clavicle and the sternocostal surface (the sternum and upper costal cartilages) of the rib cage (Fig. 7–20). The fibers converge to insert onto the anterior aspect of the upper humerus (Fig. 7–14). It adducts and internally rotates the arm. The clavicular fibers can flex the arm.

Pectoralis major activity can be assessed by palpation of the anterior axillary fold as the patient presses his or her hand upon the hip. The attempt to apply pressure should elicit isometric contraction of pectoralis major and thus a distinct increase in the tone of the muscle mass in the anterior axillary fold.

Teres major is part of the muscle mass in the upper part of the **posterior axillary fold** (Fig. 7–19).

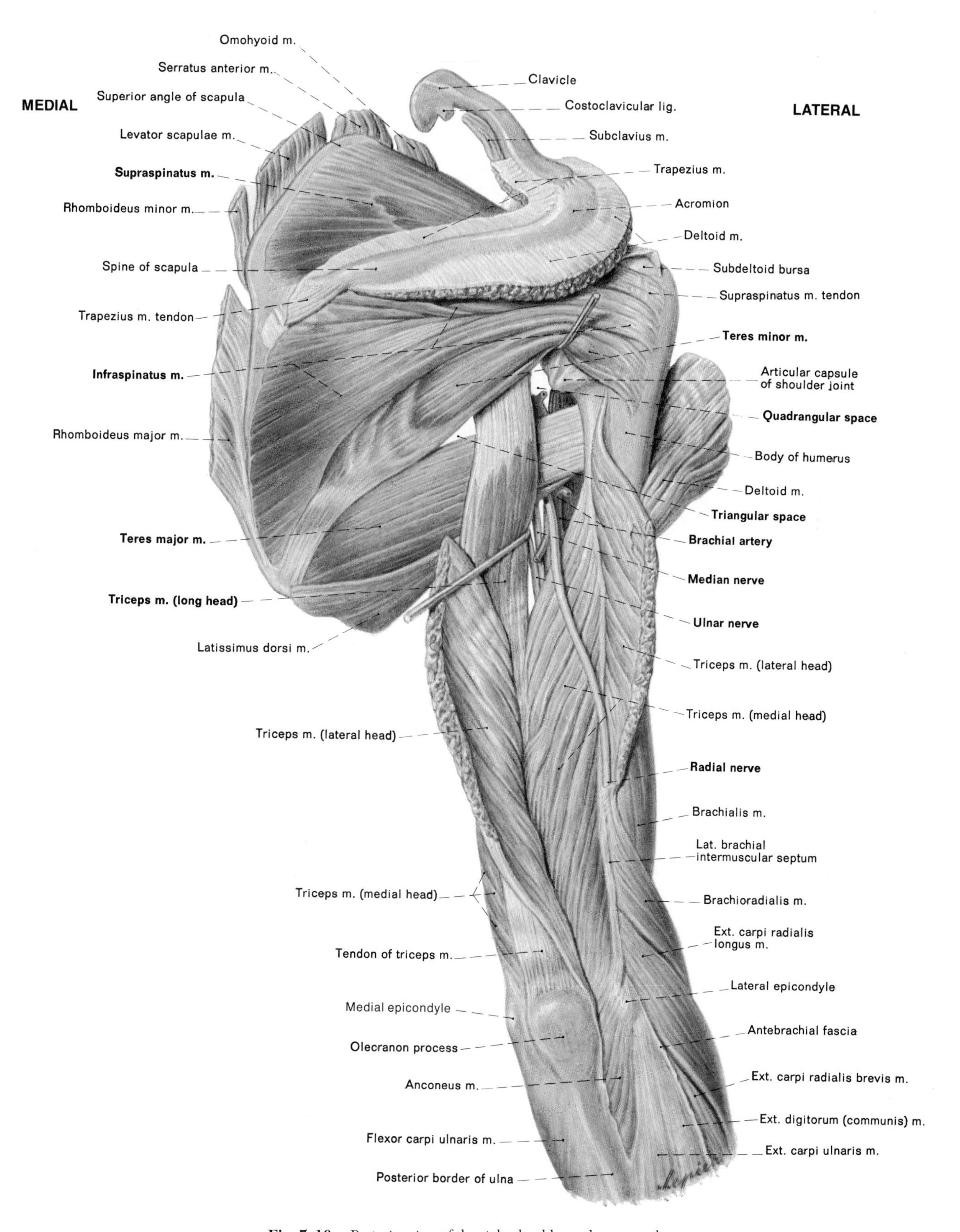

Fig. 7–16: Posterior view of the right shoulder and arm muscles.

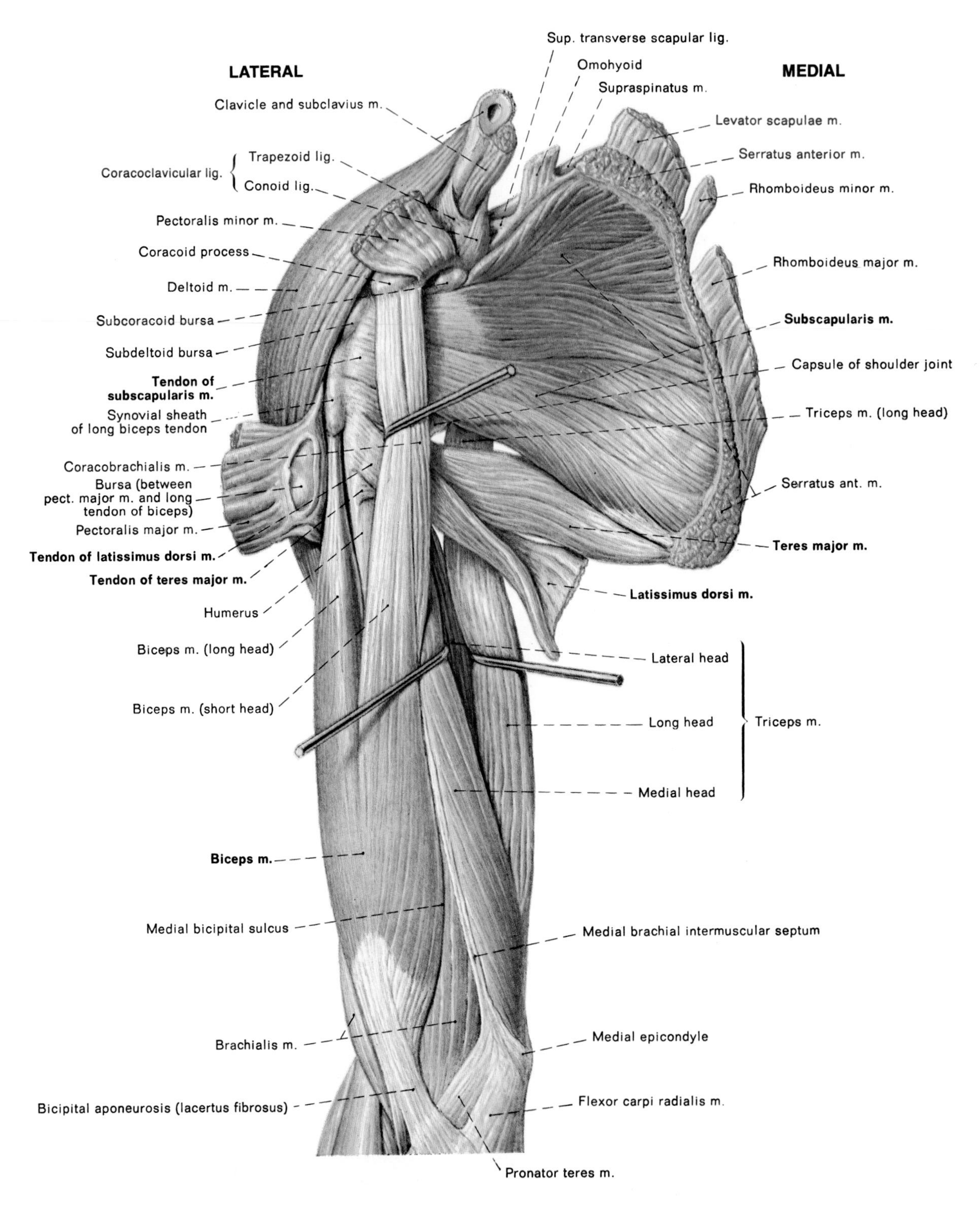

Fig. 7–17: Anterior view of the right shoulder and arm muscles.

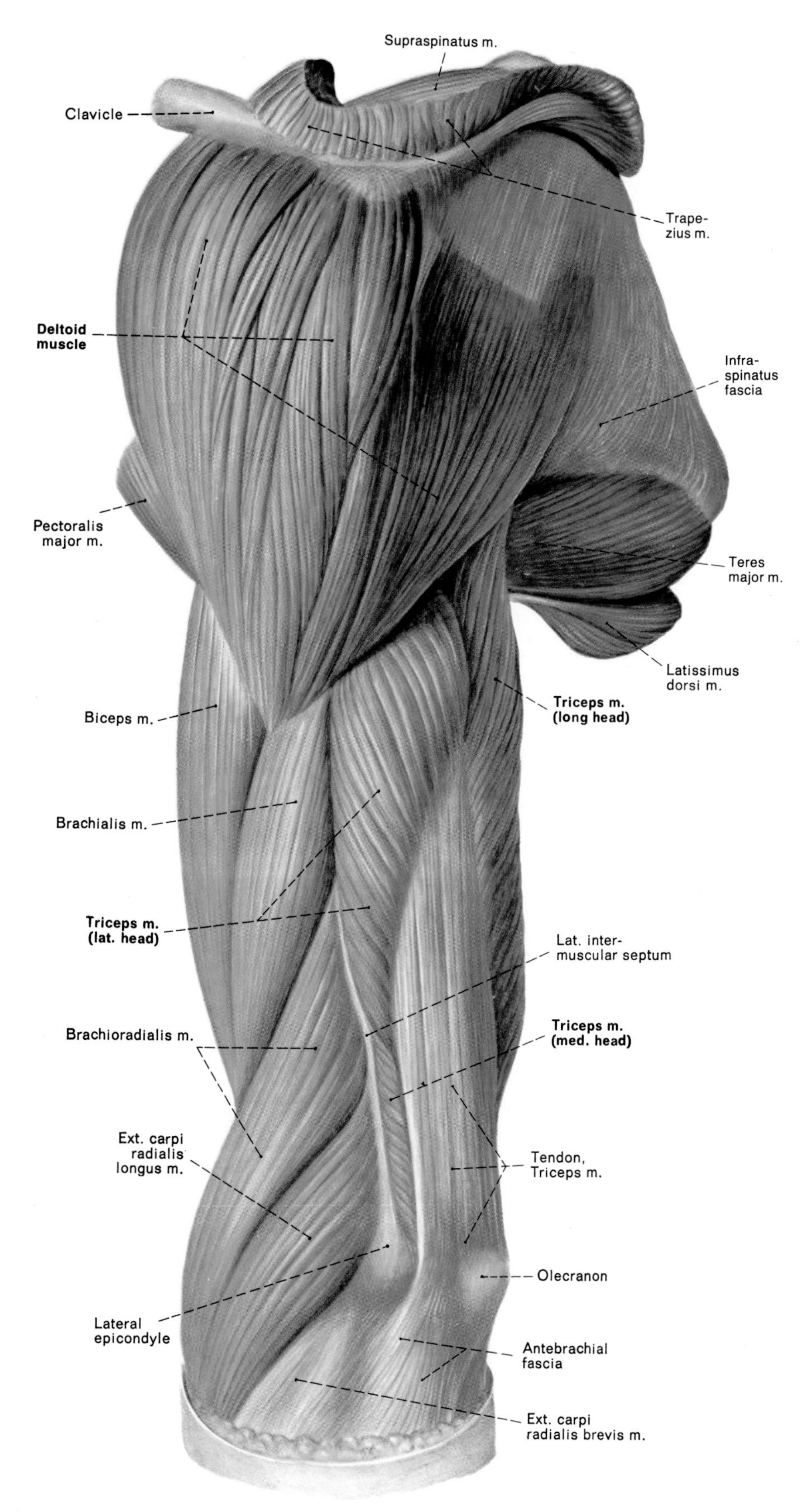

Fig. 7–18: Lateral view of left arm muscles.

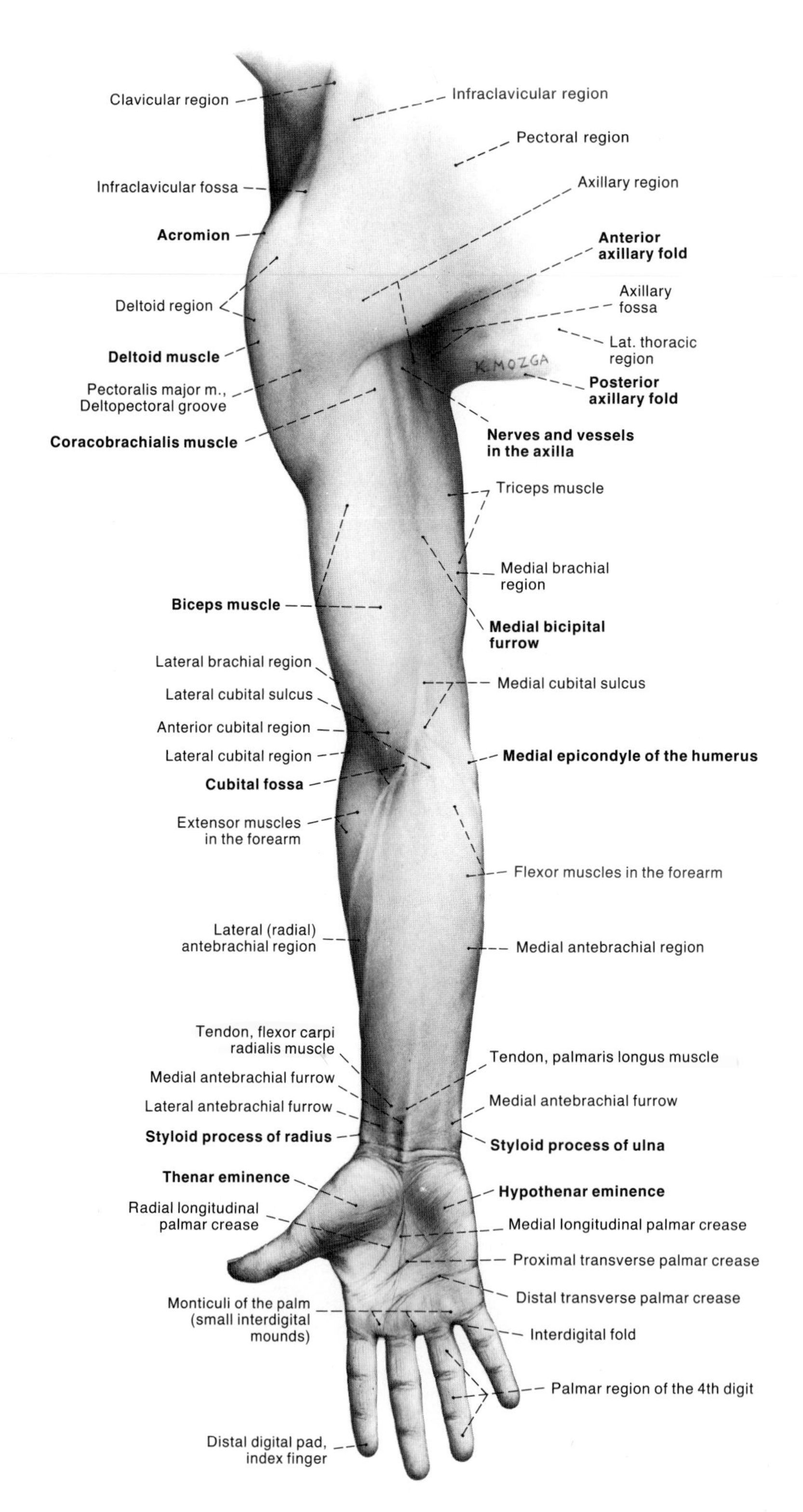

Fig. 7–19: Anterior view of the surface anatomy of the right upper limb.

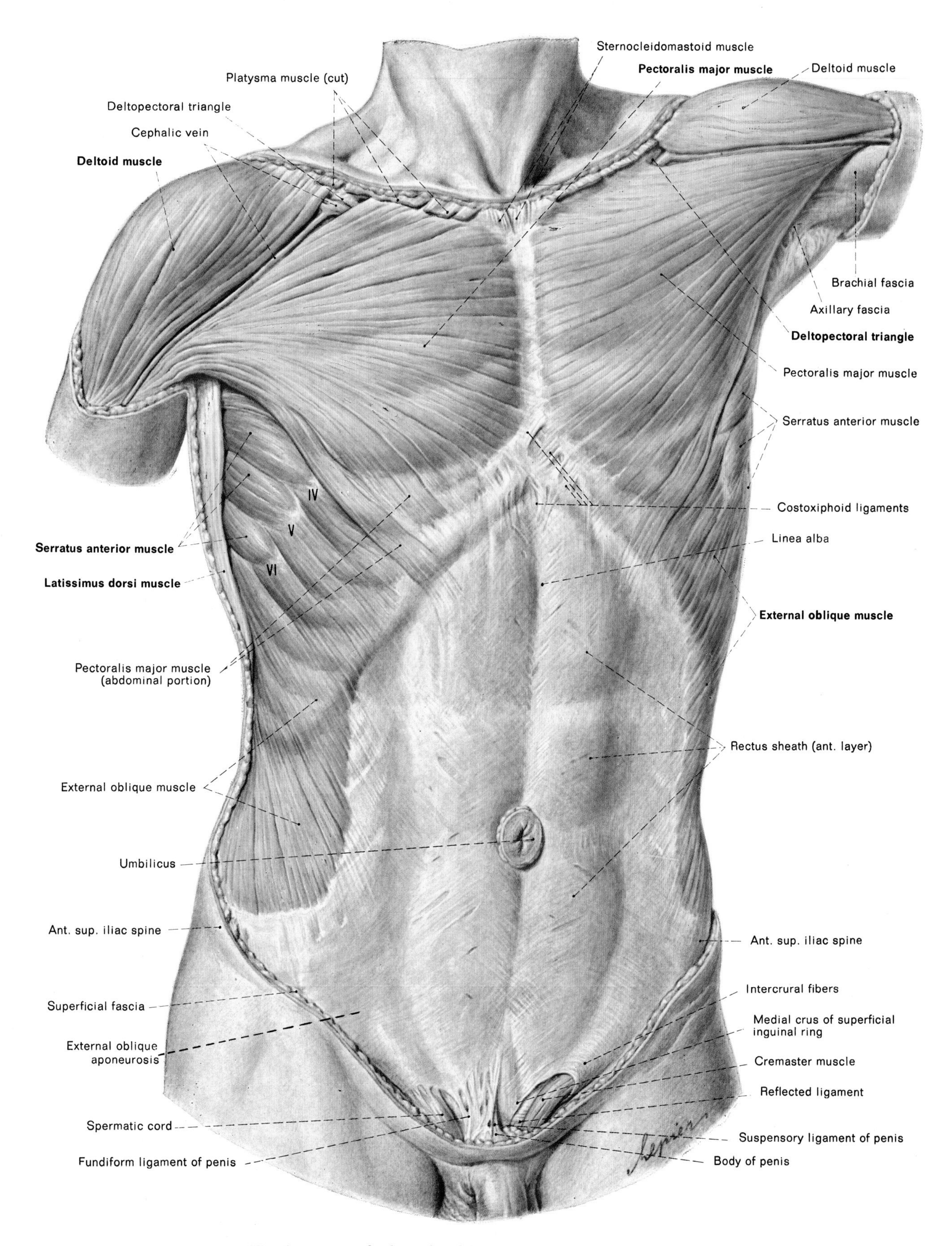

Fig. 7–20: Superficial muscles of the anterior chest and abdominal walls.

CLINICAL PANEL II-2

ANTEROINFERIOR DISLOCATION OF THE HEAD OF THE HUMERUS

Common Mechanism of the Injury: The anatomic features that account for the great range and freedom of movement at the shoulder joint are also the basis for the joint's relative instability to potentially disruptive forces. The inferior aspect of the joint is particularly deficient in ligamentous and muscular supports. The comparative laxity of the inferior aspect of the fibrous capsule (Fig. 7–5) is the basis for this deficiency in ligamentous support. The rotator cuff muscles, which serve to stabilize the dynamic integrity of the joint, provide extensile support across the superior, anterior, and posterior aspects of the joint. The lack of a rotator cuff-like muscle across the inferior aspect of the joint is the basis for the deficiency in muscular support. These anatomic factors and that the upper limb is commonly used to brace the body against accidental collisions and falls show why the shoulder joint is the most commonly dislocated large joint in adults.

Consequently, if a person braces the body with the upper limb in such a fashion that the head of the humerus is subjected to a pronounced, inferiorly-directed force, the potential exists for the humeral head to be displaced inferiorly from its articulation with the glenoid cavity. This is the mechanism by which most shoulder dislocations occur. Instances include occasions when a person falls down hard upon an outstretched hand with the arm abducted, extended, and externally rotated at the shoulder joint. The excessive forces exerted across the shoulder joint displace the humeral head anteroinferiorly. The head of the humerus assumes a subglenoid (Fig. 7–21), subcoracoid, or subclavicular position as a result of the impact.

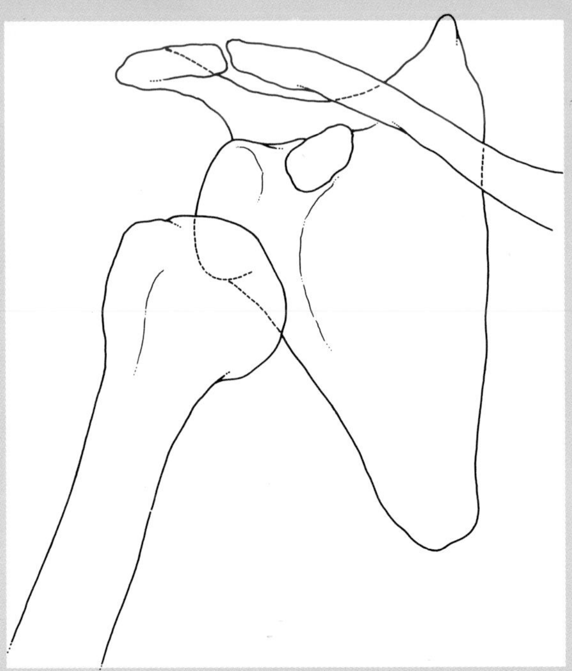

Fig. 7–21: An outline of an AP radiograph of an anterior shoulder dislocation, in which the humeral head rests in a subglenoid position.

Common Signs and Symptoms: The patient typically holds the elbow of the injured upper limb and guards against any attempts to produce active or passive movements at the dislocated joint. The patient presents with a shoulder that has lost its normal rounded contour and acquired the coutour of the acromion.

As is discussed near the end of Chapter 9, an anteroinferior dislocation of the humeral head may damage the axillary nerve along its course beneath the fibrous capsule of the shoulder joint. Damage to the axillary nerve's cutaneous sensory fibers may be tested by assessment of pinprick sensation on the upper lateral aspect of the arm (Figs. 4–12E and 4–12F.) Diminishment or loss of pinprick sensation in this cutaneous area suggests that the axillary nerve's motor fibers may also be damaged, and therefore that the deltoid and/or teres minor muscles may be partially paralyzed.

It originates from the lower lateral border of the scapula and inserts onto the anterior aspect of the upper humerus (Figs. 7–13, 7–16, and 7–17). It can adduct and internally rotate the arm.

Latissimus dorsi is the widest muscle of the back (Fig. 7–13), and it represents most of the muscle mass of the posterior axillary fold (Fig. 7–19). It extends superiorly, laterally and anteriorly from the lower back and the lower thoracic vertebrae (Fig. 7–13) to insert onto the anterior aspect of the upper humerus (Fig. 7–17). It can extend, adduct, and internally rotate the arm.

Latissimus dorsi activity can be assessed by palpation of the lower part of the posterior axillary fold as the patient voluntarily coughs. The cough should elicit a reflexive contraction of latissimus dorsi and thus a distinct increase in the tone of the muscle mass of the lower part of the posterior axillary fold.

CASE **II.1**

The Case of John Robinson

INITIAL PRESENTATION AND APPEARANCE OF THE PATIENT

A young white adult man walks into the emergency center using his left hand to support the right arm close to the side of the body. The patient is perspiring, his hair is tousled, his clothes are grass-stained, and his shoes are muddy.

QUESTIONS ASKED OF THE PATIENT

What is your name? John Robinson.

What is your birth date? [His birth date indicates that he is 20 years old.]

What can I do for you? I would like for you to look at my shoulder. I hurt it pretty badly about an hour ago when I fell down playing touch football. *[The question "What can I do for you?" not only expresses the examiner's desire to help the patient, but also invites the patient to openly and completely discuss his health concerns. In this instance, the examiner quickly learns that the patient sustained a painful shoulder injury about an hour ago. At this point, the examiner tentatively concludes that the patient has sustained one or more of the following injuries: a fractured bone, a partially or completely dislocated joint, a torn ligament, or a torn muscle.]*

Do you recall what happened? Yes. I was running to my left to avoid being tagged, and I thought I could escape from the guy running after me by stopping and cutting to my right. But as he stopped and tried to cut back, he fell on me and slammed me into the ground real hard on my shoulder (the patient gestures that he fell onto his right shoulder).

In what position was your right arm when you fell onto your right shoulder? Uh, it was beside my body. I'm right-handed, so I was holding the football with my right hand close up against my body. *[These last two questions provide the examiner with information about the mechanism of injury.]*

Where does it hurt? Over here [the patient places his left hand over the upper lateral limit of the right shoulder]. *[When pain is a complaint, it is always essential to have the patient mark by word or gesture the exact site(s) at which pain is perceived. In cases of physical injury, pain is generally a reliable indicator of the site(s) of injury.]*

Did you feel anything snap or break when you fell? No. *[This answer indicates that it is the patient's perception that he has not fractured a bone.]*

How would you describe the pain? A bad ache.

Is there anything that makes the pain worse? The pain is worse when I raise my arm.

Is there anything that relieves the pain? Well, the shoulder hurts the least when I keep my arm still. *[In general, these last two questions provide information about the source of pain. In this instance, the finding that the pain is worsened by movement and relieved by rest suggests injury of one or more musculoskeletal structures.]*

Is there anything else that is also bothering you? No, not that I know of. *[This question addresses the examiner's concern that there can be neurovascular complications to the patient's injury or injuries. A complaint of a loss of feeling or an inability to move a part of the upper limb would suggest nerve injury; a complaint of coldness in the distal parts of the upper limb would suggest injury to a major artery.]*

[The examiner finds the patient alert and fully cooperative during the interview.]

PHYSICAL EXAMINATION OF THE PATIENT

Vital signs: Blood pressure
Lying supine: 120/70 left arm and 120/70 right arm
Standing: 120/70 left arm and 120/70 right arm
Pulse: 68
Rhythm: regular
Temperature: 99.1°F
Respiratory rate: 15
Height: 6′0″
Weight: 155 lbs.
HEENT Examination: Normal
Lungs: Normal
Cardiovascular Examination: Normal
Abdomen: Normal

Genitourinary Examination: Normal

Musculoskeletal Examination: Inspection, movement and palpation of the upper and lower limbs are normal except for the following findings for the right upper limb: Inspection reveals a slight upward displacement of the lateral end of the clavicle relative to the acromion. The soft tissues in the region of this step deformity appear swollen. Active and passive abduction of the right arm beyond 90° is painful. Palpation reveals tenderness in the region of the step deformity between the acromion and the lateral end of the clavicle. Downward pressure on the lateral end of the clavicle reduces the step deformity and reveals excessive mobility of the clavicle.

Neurologic Examination: Normal

INITIAL ASSESSMENT OF THE PATIENT'S CONDITION

The patient has sustained an acute injury of the right shoulder region.

ANATOMIC BASIS OF THE PATIENT'S HISTORY AND PHYSICAL EXAMINATION

The patient has sustained an acute injury of musculoskeletal tissues in the right shoulder. The early questioning establishes that it is the patient's perception that he has sustained an acute injury to the right shoulder and that, as a result of this injury, he now has pain upon arm movement. In general, pain upon movement indicates injury or disease of one or more of the bones, joints, and muscles involved in the movement.

The tenderness and edema in the interval between the acromion and the lateral end of the clavicle mark the region as a site of injury. Tenderness is pain elicited by palpation. Tenderness generally marks a region as a site of injury or disease. Edema is excess interstitial fluid in the intercellular spaces of a soft tissue. There are many conditions that produce edema, one being injury to a soft tissue. The palpable tissues in the interval between the acromion and lateral end of the clavicle include (proceeding from the most superficial to the deepest) the skin, the superficial and deep fascia, and the tissues of the AC joint. The tissue or tissues injured in the interval between the acromion and the lateral end of the clavicle appear to be the source of the pain in the upper lateral limit of the shoulder.

The step deformity in the interval between the acromion and the lateral end of the clavicle indicates an altered configuration of the acromioclavicular joint. The AC joint has a subcutaneous position in the upper lateral limit of the shoulder. In general, any significant alteration in the relationship between the articular surfaces of a subcutaneous joint manifests itself as a surface deformity.

The pain upon arm abduction beyond 90° suggests injury of the AC joint. Significant movement between the acromial and clavicular articular surfaces in the AC joint does not occur during arm abduction until about 90° abduction. In general, pain in the region of the AC joint when the arm is abducted beyond 90° points to injury or disease of the AC joint.

INTERMEDIATE EVALUATION OF THE PATIENT'S CONDITION

The history and physical examination shows that the patient has one chief medical problem: injury of the right acromioclavicular joint.

CLINICAL REASONING PROCESS

This case illustrates the general kinds of information an orthopedist seeks from patients who present with musculoskeletal pain. The principal purpose of the clinical examination of musculoskeletal pain is to determine accurately the anatomic site or sites involved. From the history, the examiner seeks information about the location of the pain [Where is the pain? Is its location constant? In cases of joint pain, is more than one joint involved?], the time and mode of onset of the pain [Do you recall when the pain began and what you were doing at that time?], the quality and severity of the pain [How would you describe the pain? How severe is the pain? Has the pain affected your daily or regular activities?], the temporal profile of the pain [Is the intensity of the pain constant, and, if not, how would you describe its duration and variation over a 24-hour period?], aggravating and relieving factors [Is there anything that makes the pain worse? Is there anything that relieves the pain?], and the presence of associated problems [Is there anything else that is bothering you?]. The mnemonic **PQRST** can be used to recall the most important characteristics of pain to ask about: **P**rovocative and palliative circumstances, the **Q**uality of the pain, the precise anatomic **R**egion and referral, the **S**everity of the pain, and its **T**iming.

In the physical exam, the examiner gathers information through an ordered sequence of steps that involves first visual inspection, then testing of movements, and finally palpation. **Visual inspection** can detect surface deformities, areas of swelling and/or redness, and muscle atrophy. **Testing of movements** determines if the patient's pain can be reproduced or

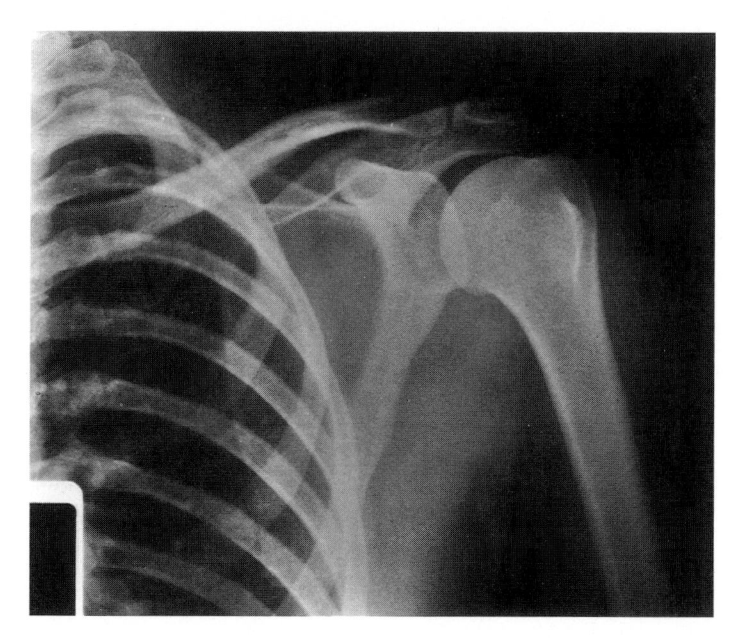

Fig. 7–22: (A) An AP radiograph of the shoulder.

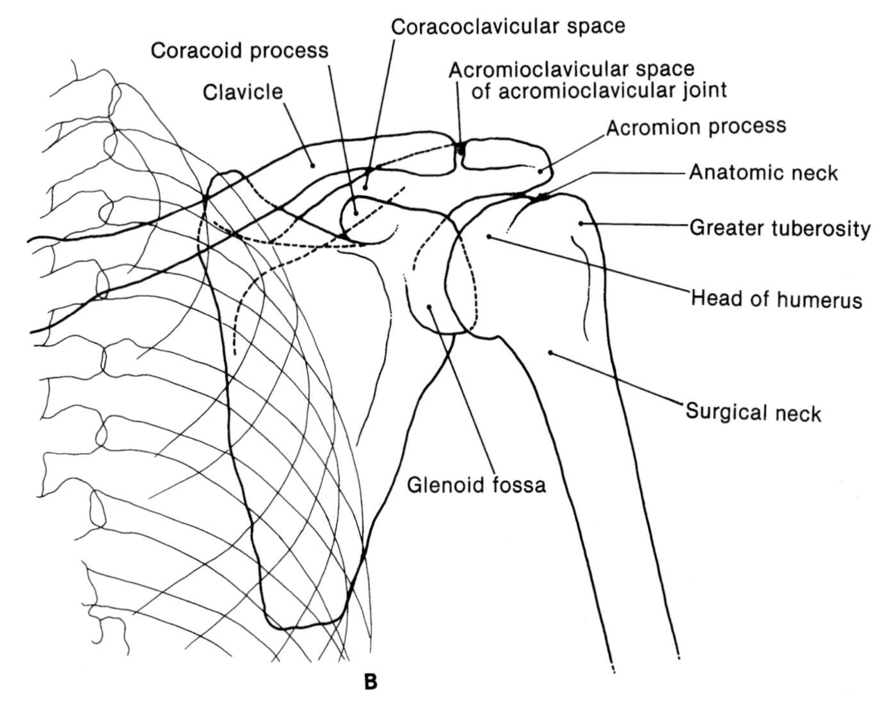

Fig. 7–22: (B) its schematic representation.

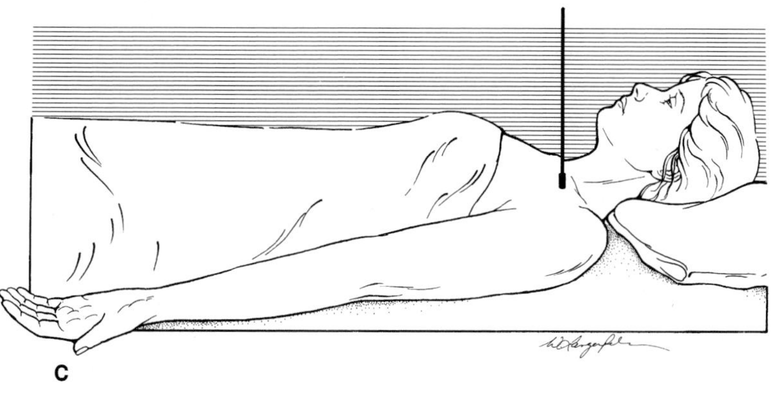

Fig. 7–22: (C) The orientation of a patient's shoulder relative to the X-ray beam and film cassette for an AP radiograph of the shoulder (the film cassette is located beneath the tabletop). In this instance, the arm is externally rotated to the extent that the line between the humeral epicondyles lies in a coronal plane. Such external rotation projects the greater tuberosity in profile, as seen in (A).

intensified by the application of stress to the structures about joints. **Active movements** (that is, movements done under the patient's control) are examined first because they indicate the limits or ranges of motion at which the patient perceives pain or stiffness about a joint and the limits to which the patient tolerates movements about a joint. **Passive movements** (that is, movements performed by the examiner) are conducted second to acquire a feel of the movements at a joint and the extent to which musculotendinous structures passively limit movement at the joint. **Isometric exercises** (that is, procedures in which the examiner restrains movements at a joint by resisting the attempts of the patient to make those movements) are conducted third to test for weakness or loss of muscle action. Finally, areas suspected of injury or disease are **palpated** for tenderness, crepitus (sounds), warmth, and the feel of any swelling.

In this case, the examiner learns from the history that the patient has sustained an injury to the right shoulder within the last hour and now has shoulder pain upon arm movement. Three key findings from the physical exam strongly indicate injury of the AC joint as the cause for the patient's shoulder pain:

(1) inspection shows a surface deformity at the site of the AC joint,
(2) testing of movement yields pain upon arm abduction beyond 90° at the AC joint site, and
(3) palpation reveals tenderness at the AC joint site.

In considering this case, an orthopedist would associate the three key findings from the physical exam with knowledge of the direction of the injurious force (a downward blow on the point of the shoulder) to conclude that the patient has indeed sustained an injury of the AC joint. The principle issue to be resolved is the severity of the injury. AC joint injuries are called **shoulder separations,** and they are graded I to III. A grade I shoulder separation is a simple sprain of the fibrous capsule of the AC joint. A grade II shoulder separation is a subluxation (partial dislocation) of the AC joint; the contact between the acromial and clavicular articular surfaces in the joint is altered because of significant damage to the joint's fibrous caspule. A grade III shoulder separation is a dislocation of the AC joint; the acromial and clavicular articular surfaces of the joint lose all contact because of significant damage to both the joint's fibrous capsule and the coracoclavicular ligament.

RADIOLOGIC EVALUATION AND FINAL RESOLUTION OF THE PATIENT'S INJURY

AP (anteroposterior) radiographs of the patient's shoulders will permit grading of the shoulder separation. An AP radiograph of the shoulder (Fig. 7–22) shows a radiolucent space called the **acromioclavicular space** between the acromion and the lateral end of the clavicle; the acromioclavicular space represents the apposed articular cartilages in the AC joint. The radiograph shows another radiolucent space called the **coracoclavicular space** between the coracoid process and the clavicle; the coracoclavicular space marks the location of the coracoclavicular ligament. An AP radiograph of a grade I shoulder separation shows acromioclavicular and coracoclavicular spaces of normal width. An AP radiograph of a grade II shoulder separation shows a coracoclavicular space of normal width but an acromioclavicular space that is at least 50% wider than that measured in the AP radiograph of the contralateral, uninjured shoulder. An AP radiograph of a grade III shoulder separation shows the acromioclavicular and coracoclavicular spaces are both at least 50% wider than the corresponding spaces in the AP radiograph of the contralateral, uninjured shoulder. In this case, comparison of AP radiographs of the patient's injured and normal shoulders indicated a grade II shoulder separation (Fig. 7–23). The AP radiograph of the right shoulder was also examined for bony fractures, but none were found.

An orthopedist would treat the patient's grade II shoulder separation with rest, support of the right arm in a sling until the pain ceases, and analgesics as required for pain.

THE MECHANISM OF THE PATIENT'S INJURY

The patient suffered a downward blow on the point on the shoulder upon being accidentally slammed onto the ground by an opposing football player. The downward blow on the point of the shoulder (which was, in effect, a downward blow on the acromion) forced the scapula downward from the clavicle. The force was sufficient to tear significantly the fibrous capsule of the AC joint but not the coracoclavicular ligament. Whereas the fibrous capsule of the AC joint attaches the acromion of the scapula to the lateral end of the clavicle, the coracoclavicular ligament suspends the coracoid process of the scapula from the lateral end part of the clavicle. Since the downward blow on the point of the shoulder was sufficient to force the acromion but not the coracoid process downward from the clavicle, the downward blow produced an inferior displacement of the acromion from the lateral end of the clavicle in association with medial rotation of the scapula about its coracoid process (Fig. 7–24).

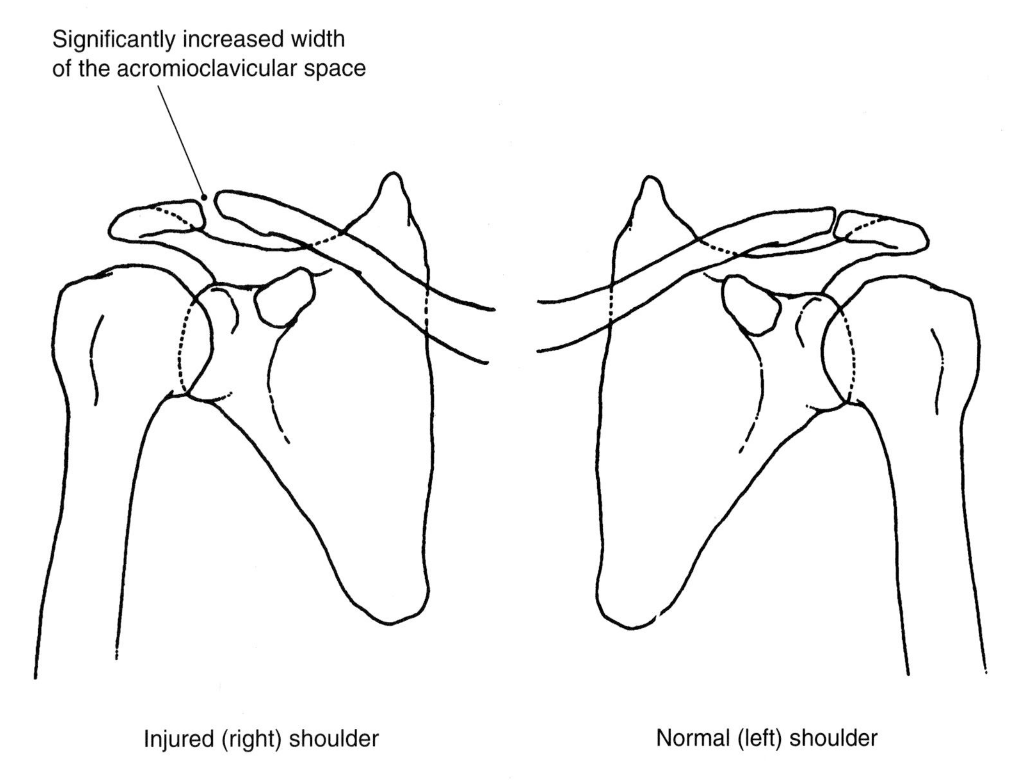

Fig. 7–23: Outlines of the AP radiographs of the normal (left) and injured (right) shoulders of John Robinson.

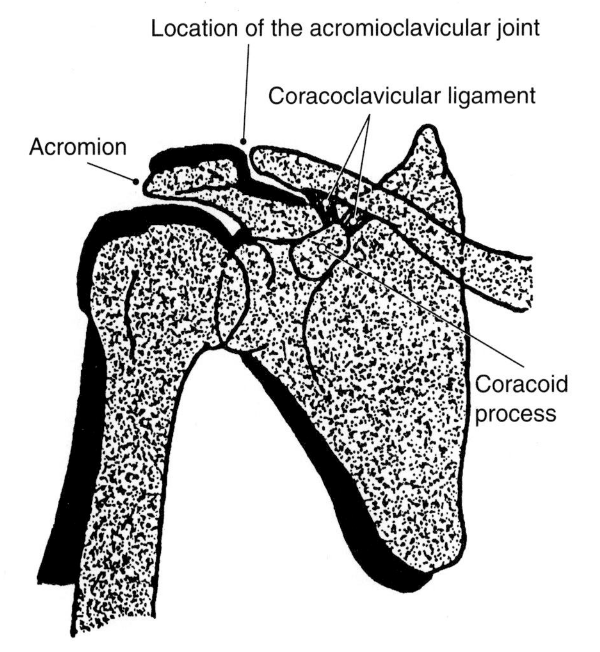

Fig. 7–24: Diagram illustrating the mechanism of a Grade II shoulder separation. The stippled clavicle, scapula, and humerus depict an anterior view of the relationships among these bones immediately following the downward blow on the acromion. The blackened areas indicate the positions of the scapula and humerus immediately prior to the downward blow on the acromion.

The swelling at the site of the AC joint at the time of the physical exam is due to effusion of the AC joint and edema of the soft tissues surrounding the joint (which were colaterally injured during the downward blow on the point of the shoulder). The weight of the upper limb has pulled the acromion slightly downward and laterally from the lateral end of the clavicle, as the fibrous capsule of the AC joint is insufficiently intact to resist this force. The slight downward and lateral displacement of the acromion generates a step deformity in the skin and subcutaneous tissues overlying the subluxed AC joint. The lateral end of the clavicle exhibits greater than normal mobility because it is not securely attached to the acromion by the AC joint's fibrous capsule.

CASE **II.2**

The Case of Joseph Adams

INITIAL PRESENTATION AND APPEARANCE OF THE PATIENT

A 48 year-old black man named Joseph Adams has made an appointment to seek treatment for a painful shoulder. The examiner's partner, who is the patient's primary care physician, cannot examine the patient because of a schedule conflict. Upon entering the examination room, the examiner observes that the patient has a robust physique but keeps his right arm close to the side of the body.

QUESTIONS ASKED OF THE PATIENT

What can I do for you? I would like for you to look at my shoulder; it hurts fairly badly.

Where does it hurt? Over here [the patient places his left hand over and around the point of the right shoulder].

When did the shoulder pain start? About 3 days ago. *[This answer suggests an acute condition or process. It prompts the examiner to inquire about the patient's activities during the week preceding the onset of pain, in order to assess if any of these activities could possibly generate an acute, painful condition or process.]*

Do you recall what you were doing when the pain started? I noticed the pain in my shoulder shortly after waking up in the morning.

Do you recall if you had injured your right shoulder before it became painful? No, I don't think so. Y'see, for the past week I've been painting the basement in my house, and I've been very careful not to fall from a ladder. I painted the entire ceiling of the basement the 2 days just before my shoulder began hurting, and I never once slipped or fell from the ladder.

Do you paint with your right hand? Yes.

Did you find the painting job unusually stressful or tiring? Yes. I'm an insurance broker, and therefore work at a desk most of the time. I found I had to take more breaks from the painting than I thought I would because of muscle cramps and backaches. *[Unusual and/or immoderate physical activity (such as household painting) by a person unaccustomed to such activity can place stress on musculoskeletal structures sufficient to produce an acute, painful condition or process. At this point, the examiner regards the immoderate physical activity associated with the household painting to be the most likely cause of the patient's condition. This working hypothesis implies that the patient is suffering from an acute condition or process in or about the shoulder joint.]*

How would you describe the pain? A dull ache.

Has the nature of the pain or its location changed during the past 3 days? Sometimes it goes down my arm. I haven't slept well the last 2 nights because I've awoken several times with the pain going down my arm.

Can you show me where the pain goes down your arm? Yes, over here [the patient slides his left hand down the lateral aspect of the right upper limb from the point of the shoulder to the region overlying the insertion of deltoid]. *[Shoulder pain that radiates to deltoid's insertion but no further suggests disease or injury of tissues in or about the shoulder joint. The pattern of pain radiation significantly supports the working hypothesis.]*

Has your right shoulder ever been painful like this before? Well, yes. I would say I've had 3 or 4 attacks of shoulder pain during the past few years. However, each time before, I've taken aspirin or Nuprin [ibuprofen] and the pain has gone away after a day or two. But this time the pain is worse, and it hasn't gotten better since it started. *[This answer suggests that the current, acute condition may be an aggravation or complication of a chronic condition.]*

Is there anything that makes the pain worse? The pain is worse when I raise my arm.

Is there anything that relieves the pain? I've been taking Nuprin for the pain. I have the least pain if I don't move my arm. *[In general, these last two questions provide information about the source of pain. In this instance, the finding that the pain is worsened by* *movement and relieved by rest suggests a condition or process involving one or more musculoskeletal structures in or about the shoulder joint.]*

Is there anything else that is also bothering you? No. *[This answer is important because it indicates, in particular, that there are not any neurovascular complications to the patient's condition nor involvement of other joints in the upper and lower limbs.]*

Have you had any recent illnesses or injuries? No.

Are you taking any medications for previous illnesses or conditions? Yes, I am on anti-hypertensive medication.

Is your hypertension being successfully controlled? Somewhat. Actually, I have an appointment with your partner a week from today because the medication has not brought my blood pressure down enough. As your partner has advised, I've gone on a low salt and low saturated fat/cholesterol diet and started walking more for exercise. I've also tried to bring my weight under control, and have been able to lose 10 pounds during the last year; I'm down to 200. I measure my blood pressure almost daily. *[There are complications of arterial hypertension which can produce shoulder pain (such as angina pain resulting from the insufficiency of atherosclerotic coronary arteries to supply the heart's musculature; however, such coronary insufficiency usually produces short-lasting, effort-induced pain). Although the preponderance of evidence points to pain of musculoskeletal origin, assessment of the cardiovascular system during the physical exam now assumes added importance.]*

You mentioned that you're taking Nuprin for your shoulder pain. Are you taking any other over-the-counter drugs? No, except that maybe once a week I'll take an antacid tablet for indigestion. *[The examiner finds the patient alert and fully cooperative during the interview.]*

PHYSICAL EXAMINATION OF THE PATIENT

Vital signs: Blood pressure
Lying supine: 145/95 left arm and 145/95 right arm
Standing: 150/100 left arm and 150/100 right arm
Pulse: 76
Rhythm: regular
Temperature: 99.0°F
Respiratory rate: 18
Height: 6′0″
Weight: 200 lbs.
HEENT Examination: Normal
Lungs: Normal

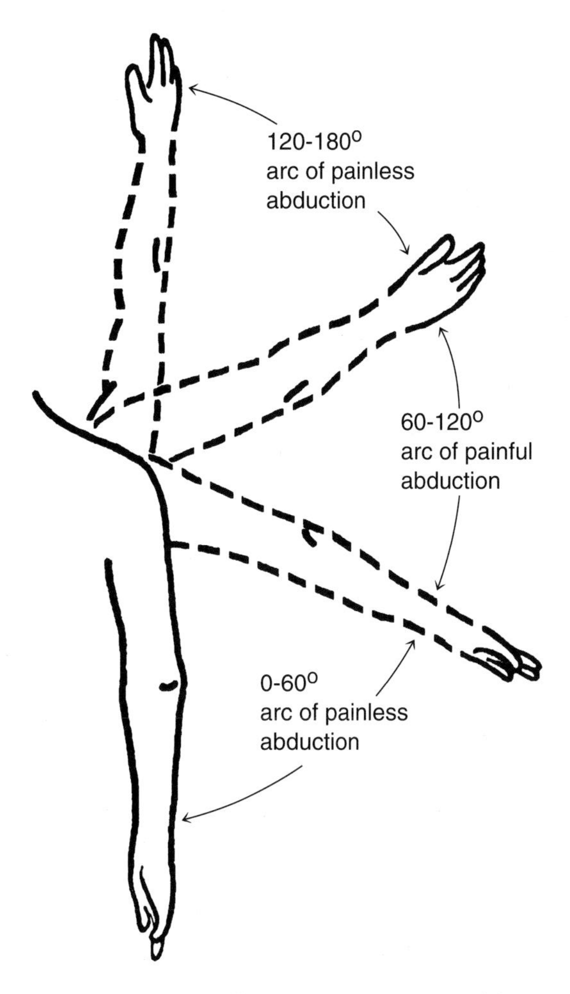

Fig. 7–25: Diagram illustrating a posterior view of the arcs of active or passive abduction in which Joseph Adams experiences painless and painful motion.

Cardiovascular Examination: Examination of the head and neck is normal except for the following findings: In the ocular fundi, the retinal arterioles show evidence of a broadened yellow light reflex. The arteriolar to venous diameter ratio is 1:2. Mild venular depression is noted at two arteriovenous crossings.
Abdomen: Normal
Genitourinary Examination: Normal
Musculoskeletal Examination: Inspection, movement and palpation of the upper and lower limbs are normal except for the following findings for the right upper limb: Active or passive abduction of the right arm from 60° to 120° is painful; active or passive abduction from 0° to 60° and from 120° to 180° is painless (Fig. 7–25). Active or passive adduction is painless from 180° to 120°, painful from 120° to 60° and then painless again from 60° to 0°. The pain produced during the 60° to 120° arc is localized to the region immediately lateral to the acromion; the

severity of the pain disturbs the smoothness of the scapulohumeral rhythm. Abduction of the right arm from the anatomic position against resistance is also painful. Palpation reveals tenderness in the region between the acromion and greater tuberosity.

Neurologic Examination: Normal

INITIAL ASSESSMENT OF THE PATIENT'S CONDITION

The patient appears to be suffering from an acute condition or process in or about the right shoulder.

ANATOMIC BASIS OF THE PATIENT'S HISTORY AND PHYSICAL EXAMINATION

The tenderness in the region between the acromion and greater tuberosity indicates inflammation of one or more tissues in this region. The tissues which can be pressed upon by palpation of this region include (proceeding from the most superficial to the deepest) the skin, the superficial and deep fascia, the middle fibers of deltoid, the subacromion-subdeltoid bursa, the tendinous insertion of supraspinatus, the fibrous capsule and synovial lining of the shoulder joint and the head of the humerus. The inflamed tissue or tissues appear to be the source of the pain felt around the tip of the patient's right shoulder.

The pain upon arm abduction against resistance indicates injury or inflammation of supraspinatus and/or deltoid. When the patient attempts to abduct his arm from the anatomic position against resistance, supraspinatus and deltoid (the muscles chiefly responsible for initiating arm abduction) contract isometrically because the examiner physically prevents any movement at the shoulder joint. It is likely that the pain produced by this procedure involves supraspinatus and/or deltoid, as these two muscles are the only musculotendinous structures in the shoulder significantly stressed by the procedure.

The pain elicited during the 60° to 120° arc of abduction indicates that events which occur during the middle range of arm abduction irritate the inflamed tissue(s) in the region between the acromion and greater tuberosity. The irritating events do not require or involve active concentric contraction of supraspinatus and deltoid, since passive abduction through the 60° to 120° arc is as painful as active abduction. The principal mechanical event that occurs in the vicinity of the tip of the shoulder during the 60° to 120° arc of abduction is that the greater tuberosity of the humerus and the soft tissues about it are drawn proximally into the **subacromial space** (the space underlying the acromion and coracoacromial ligament) (Fig. 7–6). The greater tuberosity and subacromial soft tissues immediately overlying it are drawn into the subacromial space by the upward rotation of the humerus at the shoulder joint. Within the 180° range of arm abduction, the greater tuberosity passes closest to the acromion and coracoacromial ligament during the 60° to 120° arc.

The pain that radiates distally along the upper lateral aspect of the arm appears to be referred pain (inconstant, superficial pain). This presentation suggests injury or inflammation of tissues innervated by C5 and C6 fibers, as the upper lateral aspect of the arm represents parts of the C5 and C6 dermatomes (Figs. 4–12E and 4–12F). The notable tissues in the vicinity about the tip of the shoulder that are innervated by C5 and C6 fibers include the deltoid muscle, all the muscles of the rotator cuff, and the synovial lining and fibrous capsule of the shoulder and AC joints.

The patient's recollection of similar previous but less severe episodes of shoulder pain suggests that the current condition is a complication or aggravation of a chronic condition.

The patient has mild systemic hypertension associated with early stage arteriosclerotic retinopathy. Systemic hypertension is defined as systolic/diastolic blood pressure levels equal to or greater than 140/90 mm Hg. Individuals with a diastolic pressure from 90 to 104 are considered to have mild systemic hypertension. The cause of systemic hypertension in most individuals is not known; individuals with systemic hypertension of unknown etiology are said to have **primary,** or **idiopathic, hypertension.** In this case, the fact that the examiner's partner has advised the patient to modify certain elements of his lifestyle (e.g., reduce sodium, caloric, saturated fat, and cholesterol intake and adopt some program of moderate physical exercise) to help treat his hypertension suggests that the patient has primary hypertension.

The systemic vascular effects of hypertension can be evaluated by ophthalmoscopic examination of the retinal arterioles and venules of the **optic fundus** (the portion of the eye posterior to the lens) (Fig. 7–26). Normal retinal vessels appear as gently curved, reddish bands, each with a narrow, central yellowish stripe about one-fourth the thickness of the band. Each reddish band represents the tubular column of blood within a retinal vessel; the central yellowish stripe represents light reflected from the blood column. The walls of normal retinal vessels are transparent and thus not seen. The arterioles are lighter and narrower than the venules that accompany them; the ratio of arteriolar to venular diameters is approximately

3:4. As an arteriole and its accompanying venule extend toward the periphery, the vessels cross each other, with the arteriole generally passing anterior to the venule. There is no change in the appearance of the vessels at normal arteriovenous crossings.

One of the initial effects of systemic hypertension on the appearance of retinal arterioles is narrowing of the reddish blood column. Generalized arteriolar narrowing to the extent that the ratio of arteriolar to venular diameters becomes 1:2 is a sign of mild hypertension. **Arteriosclerosis** is a major consequence of systemic hypertension; the persistent arterial hypertension promotes proliferation of the smooth muscle and elastic tissue of arterial walls. The sclerotic thickening broadens the central reflex stripe and is associated with changes at arteriovenous crossings. A broadened yellow light reflex and mild venular depression at a few arteriovenous crossings are signs of early stage retinal arteriosclerosis.

INTERMEDIATE EVALUATION OF THE PATIENT'S CONDITION

The history and physical examination indicate a lesion of supraspinatus's tendon of insertion in the right shoulder.

CLINICAL REASONING PROCESS

When considered all together, three key findings from the patient's history and physical exam strongly indicate a lesion of supraspinatus's tendon of insertion as the cause for the patient's shoulder pain.

(1) Testing of movements shows: pain upon either active or passive movement of the arm through the middle range of abduction and
(2) pain upon resisted abduction from the anatomic position;
(3) inspection finds tenderness in the region immediately lateral to the acromion.

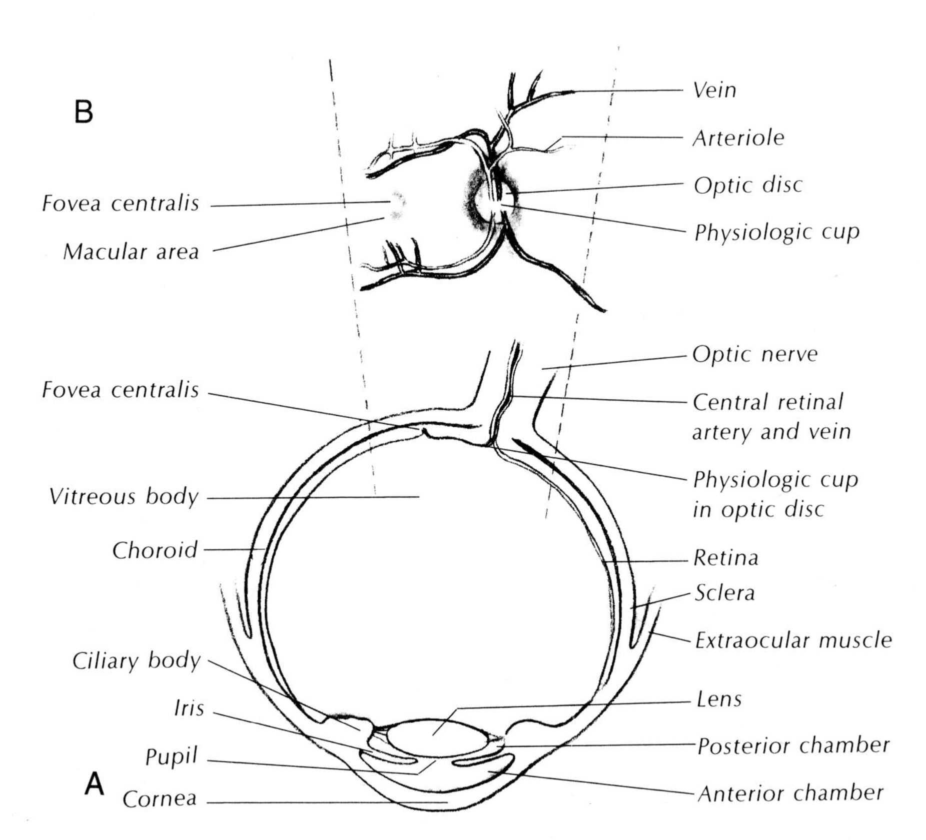

Fig. 7–26: Diagrams illustrating (A) a superior view of a horizontal section of the right eyeball and (B) a portion of the fundus commonly observed with an ophthalmoscope.

The presence of both pain and tenderness in the region immediately lateral to the acromion points to this region as the site of the lesion responsible for the shoulder pain. The pain upon resisted abduction indicates a lesion of supraspinatus and/or deltoid. The pain upon movement through the middle range of abduction points to a lesion of the greater tuberosity or the soft tissues immediately about it, and therefore a supraspinatus lesion, for the following reasons: Within the full 180° range of arm abduction, the greater tuberosity comes into its closest proximity with the overlying acromion and coracoacromial ligament along the 60° to 120° arc. A swollen, tender lesion of or about the greater tuberosity would be most painful during abduction as it is drawn through and maximally compressed in the subacromial space along the 60° to 120° arc.

A major concern for all patients who present with shoulder pain is that the pain may be referred from a disease or disorder in the neck, thorax, or abdomen (the anatomic basis of referred shoulder pain is discussed in the chapters on the neck, thorax, and abdomen). In a patient with a diagnosed systemic cardiovascular disease, there is, in particular, concern that an undiagnosed cardiac condition may be causing referred shoulder pain. However, shoulder pain of cardiac origin usually involves the left shoulder. [If left shoulder pain of cardiac origin is short-lived and effort-induced, it suggests transient insufficiency of coronary arterial blood flow to the heart's musculature. If left shoulder pain of cardiac origin is long-lasting, it indicates infarction (long-term insufficiency of coronary arterial blood flow which results in permanent damage to the heart's musculature).] Furthermore, the findings of pain during active and passive movement through the middle range of abduction and upon isometric testing of supraspinatus and deltoid all point to a musculoskeletal lesion in the shoulder region as the basis of the shoulder pain.

In considering the patient's condition, an orthopedist would recognize the painful midrange of abduction in association with pain-free movement at the limits of abduction as the principal characteristic of the patient's condition. A painful midrange of abduction is encountered so commonly in orthopedics that it is called the **painful arc syndrome.** The syndrome is also called the **subacromial impingement syndrome** because four of the five most common primary causes of the syndrome all involve injury or inflammation of the subdeltoid tissues that become maximally compressed, or impinged, between the greater tuberosity and the overlying acromion during the midrange of abduction. The five most common primary lesions are an incomplete tear of the supraspinatus tendon of insertion, supraspinatus tendinitis, a calcified deposit in the supraspinatus tendon, subacromial bursitis, and a contusion or undisplaced fracture of the greater tuberosity. The painful arc syndrome is almost always a manifestation of one or a combination of these five primary lesions.

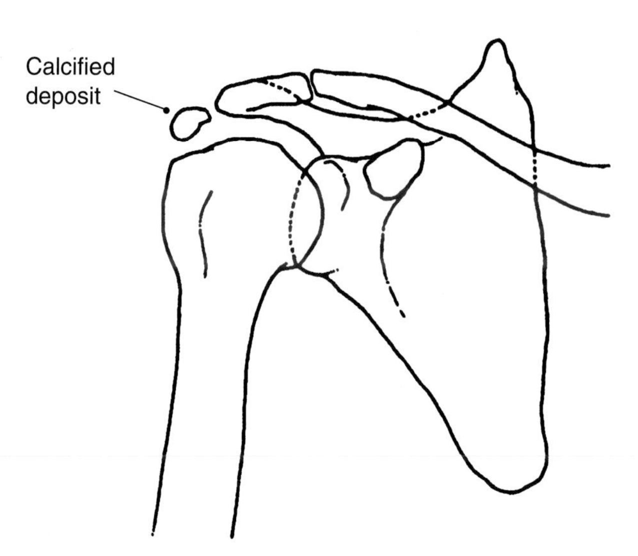

Fig. 7–27: Outline of the AP radiograph of Joseph Adams's right shoulder.

The most likely diagnoses are supraspinatus tendinitis alone or supraspinatus tendinitis in association with a calcified deposit in the supraspinatus tendon and/or subacromial bursitis. This is because the history suggests that the patient's current episode of shoulder pain is a complication or aggravation of a chronic condition. Supraspinatus tendinitis and subacromial bursitis are lesions that frequently exhibit a history of spontaneous exacerbation and abatement over several years. The deposition of a calcified deposit in the supraspinatus tendon is generally a complication of supraspinatus tendinitis. Incomplete tears of the supraspinatus tendon and lesions of the greater tuberosity are generally the product of injury.

RADIOLOGIC EVALUATION AND FINAL RESOLUTION OF THE PATIENT'S CONDITION

An AP radiograph of the patient's shoulder could reveal a calcified deposit in the subacromial soft tissues or an undisplaced fracture of the greater tuberosity. An AP radiograph of the patient's right shoulder showed a calcified deposit immediately superomedial to the greater tuberosity (Fig. 7–27). An orthopedist would conclude that the patient is suffering from supraspinatus tendinitis exacerbated by a calcified deposit in the supraspinatus tendon. Subacromial bursitis may also be present.

Conservative treatment of the patient would involve rest, support of the right arm with a sling, and anti-inflammatory medication for the pain. Recovery would be expected within 4 to 5 days.

THE CHRONOLOGY OF THE PATIENT'S CONDITION

The patient's manual activities during the past several years have led to progressive degeneration of the supraspinatus tendon near its insertion onto the greater tuberosity. The degeneration has occurred in response to the repetitive impingement of this region of the supraspinatus tendon beneath the acromion and the coracoacromial ligament during abduction and flexion of the arm. Episodes of supraspinatus tendinitis have occurred several times during the past few years following periods of excessive manual activity. The most recent period of excessive manual activity (which involved sustained abduction of the arm as the patient painted the basement ceiling) produced not only another episode of supraspinatus tendinitis but also deposition of calcium salts in the inflamed tendon.

RECOMMENDED REFERENCES FOR ADDITIONAL INFORMATION ON THE SHOULDER

Norkin, C.C., and P.K. Levangle, Joint Structure & Function, 2nd ed., F.A. Davis Co., Philadelphia, 1992: *Chapter 7 presents detailed descriptions of the anatomy of the sternoclavicular, acromioclavicular, and shoulder joints and discussions of how motions at these joints are coordinated during various shoulder and arm movements.*

Corrigan, B., and G.D. Maitland, Practical ¸Orthopaedic Medicine, Butterworth & Co., London, 1983: *Chapters 5 and 6 offer descriptions of the clinical examination of the sternoclavicular, acromioclavicular, and shoulder joints and discussions of the signs and symptoms of common lesions at or about these joints.*

Magee, D.J. Orthopaedic Physical Assessment, 2nd ed., W.B. Saunders Co., Philadelphia, 1992: *Chapter 4 provides a comprehensive discussion of the clinical examination of the muscles and joints of the shoulder region.*

Greenspan, A., Orthopedic Radiology, 2nd ed., Gower Medical Publishing, New York, 1992: *This text is a comprehensive source of high quality photographs of orthopedic films. Pages 5.5 through 5.9 in Chapter 5 present numerous radiographic views of the shoulder; each radiographic view is depicted in tandem with a drawing of the patient's orientation to the X-ray beam and film cassette. Fig. 5.20 on page 5.13 summarizes the array of radiologic imaging techniques that may be used to evaluate injuries of the shoulder region. Pages 5.18 through 5.23 present discussions of the radiologic evaluation of patients afflicted with the subacromial impingement syndrome or a shoulder separation.*

DeGowen, R.L., Jochimsen, P.R., and E.O. Theilen, DeGowin & DeGowin's Bedside Diagnostic Examination, 5th ed., Macmillan Publishing Co., New York, 1987: *This text provides exemplary introductions to the clinical analysis of common medical syndromes. Pages 711 through 724 summarize the signs and symptoms of common injuries and diseases of the shoulder region.*

CHAPTER **8**

The Axilla

The axilla is the pyramidal region spanning the space between the lateral surface of the upper chest wall and the medial region of the upper arm. When the arm is partially abducted, it becomes evident that the axilla has muscle-bearing folds of tissue extending between the upper chest wall and upper arm (Fig. 7–19). The anterior fold is called the **anterior axillary fold,** and pectoralis major comprises most of its mass. The posterior fold is called the **posterior axillary fold,** and latissimus dorsi and teres major comprise most of its mass.

The most prominent set of structures coursing through the axilla is a neurovascular bundle comprised of the axillary artery, axillary vein, and the axillary parts of the brachial plexus. The neurovascular bundle begins in the neck, passes over the first rib, extends through the axilla to the lower border of teres major's insertion, and then disperses as it continues into the arm. The axillary parts of the neurovascular bundle consist of only those parts that are distal to the lateral border of the first rib but proximal to the lower border of teres major's insertion onto the humerus. In other words, the cervical (neck) parts of the neurovascular bundle become axillary parts at the lateral border of the first rib, and the axillary parts become brachial (arm) parts at the lower border of teres major.

THE AXILLARY ARTERY

The axillary artery begins at the lateral border of the first rib as a continuation of the subclavian artery and ends at the lower border of the insertion of teres major onto the humerus (Fig. 7–15). At its end, the axillary artery is continuous with the brachial artery. The pulsations of the axillary artery can be palpated in the soft tissues of the upper arm that lie posterior to the uppermost medial border of biceps brachii.

As the axillary artery courses through the axilla, it passes posterior to pectoralis minor (Fig. 7–15). This relationship divides the axillary artery into three parts: the first part is medial to pectoralis minor, the second part is posterior to the muscle, and the third part is lateral to the muscle. The first part of the axillary artery has one branch, the second part two branches, and the third part three branches.

The **superior thoracic artery** is the only branch of the first part of the axillary artery (Fig. 7–15). It supplies pectoralis major and minor.

The **thoracoacromial artery** is the first of the two branches of the second part of the axillary artery (Fig. 7–15). It loops around the medial border of pectoralis minor before dividing into branches that supply deltoid, pectoralis major and minor, subclavius, and the AC joint.

The **lateral thoracic artery** is the second of the two branches of the second part of the axillary artery (Fig. 7–15). It descends along the lateral border of pectoralis minor, giving rise to branches that supply serratus anterior, pectoralis major and minor, subscapularis, and the axillary lymph nodes. It also provides the main blood supply to the mammary glandular tissue in the lateral half of the breast.

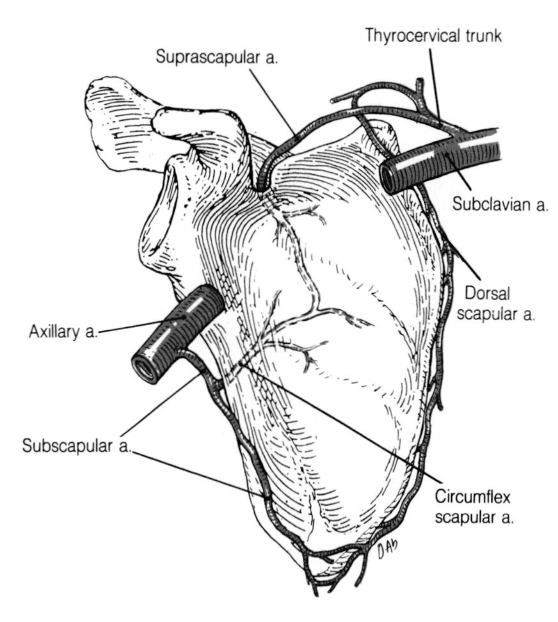

Fig. 8–1: Diagram illustrating the arterial anastomoses around the scapula between branches of the subscapular artery and indirect branches of the thyrocervical trunk.

The **subscapular artery** is the first branch of the third part of the axillary artery, and it is generally the largest branch of the axillary artery (Fig. 7–15). The branches of the subscapular artery form an extensive anastomotic network around the scapula with secondary branches of the subclavian artery (Fig. 8–1). Specifically, the anastomoses occur between branches of the subscapular artery and branches of (a) the **suprascapular artery** and (b) the **transverse cervical artery.** The suprascapular and transverse cervical arteries are branches of the **thyrocervical trunk,** which, in turn, is a branch of the subclavian artery. This anastomotic network around the scapula is important because it provides for collateral circulation across the shoulder joint whenever an arm movement restricts the arterial flow through the primary channelway, the axillary artery. The scapular anastomotic network makes it possible to ligate (under either emergency or surgical conditions) the subclavian-axillary arterial trunk at any point between the origins of the thyrocervical trunk and the subscapular artery.

The **anterior and posterior circumflex humeral arteries** are the last two branches of the third part of the axillary artery (Fig. 7–15). They course around the surgical neck of the humerus and together supply teres major and minor, the long and lateral heads of triceps brachii, the head of the humerus, and the shoulder joint.

THE AXILLARY VEIN

The basilic vein is the major superficial vein on the medial side of the arm (Fig. 8–2). Its tributaries drain superficial tissues on the medial side of the hand, forearm, and arm.

The axillary vein begins at the lower border of the insertion of teres major onto the humerus as the continuation of the basilic vein. Near its origin, the axillary vein receives the venae comitantes of the brachial artery. [It is common for small or medium-sized veins in the limbs to extend as pairs, with the members of each pair having numerous connections between themselves and lying on opposite sides of a similarly sized artery. Since such vascular triads generally serve the same tissues, the pair of veins are called the **venae comitantes** (the accompanying veins) of the artery.] The axillary vein ascends through the axilla beside the axillary artery and ends at the lateral border of the first rib. At its end, the axillary vein is continuous with the subclavian vein.

Near its end, the axillary vein receives its largest tributary, the cephalic vein. The **cephalic vein** is the largest superficial vein on the lateral side of the arm (Fig. 8–2). Its tributaries drain superficial tissues on the lateral side of the hand, forearm, and arm. The cephalic vein passes deeply through the deltopectoral triangle before joining the axillary vein. The **deltopectoral triangle** is the triangular-shaped area in the anterior shoulder region bordered laterally by anterior fibers of deltoid, medially by clavicular fibers of pectoralis major, and superiorly by the clavicle (Fig. 7–20).

THE BRACHIAL PLEXUS

The brachial plexus is so named because it is the plexus, or network, of nerves for the brachium (arm). The brachial plexus forms in the neck, extends into the axilla, and there gives rise to almost all the motor and sensory nerves for the arm, forearm, and hand.

There are four major parts to the brachial plexus: roots, trunks, divisions, and cords (Fig. 8–3). The roots and trunks are located in the neck lateral and superficial to the vertebral column, the divisions extend from the neck into the axilla by passing over the first rib, and the cords are located in the axilla.

The anterior rami of C5, C6, C7, C8, and T1 form the five **roots** of the brachial plexus (Fig. 8–3). Only two nerves arise from the roots. The **dorsal scapular nerve** arises from the C5 root, and the **long thoracic nerve** arises from the C5, C6, and C7 roots (Fig. 8–4). The dorsal scapular nerve innervates levator scapulae, rhomboid major, and rhomboid minor. The long thoracic nerve innervates serratus anterior.

The unions and extension of the roots form the three **trunks** of the brachial plexus (Fig. 8–3). The **upper trunk** is formed from the union of the C5 and C6 roots. The **middle trunk** is an extension of the C7 root. The **lower trunk** is formed from the union of the C8 and T1 roots. The **nerve to subclavius** and the **suprascapular nerve** are the only nerves that arise from the trunks; both arise from the upper trunk (Fig. 8–4). The nerve to subclavius innervates the muscle for which it is named and the sternoclavicular joint. The suprascapular nerve innervates the supraspinatus and infraspinatus muscles and the acromioclavicular and shoulder joints.

The divisions of the trunks form the six **divisions** of the brachial plexus; each trunk divides to form an anterior and a posterior division (Fig. 8–3). The divisions do not give rise to any nerves.

The unions and extension of the divisions form the **cords** of the brachial plexus (Fig. 8–3). Each cord is named for its relationship relative to the second part of the axillary artery.

The **posterior cord** is formed from the union of all three posterior divisions, and it gives rise to the **upper, middle (thoracodorsal), and lower subscapular nerves;** the **axillary nerve;** and the **radial nerve** (Fig. 8–4). Table 8–1 lists the muscles and major joints innervated by the five nerves derived from the posterior cord. Figures 4–12E and 4–12F depict the cuta-

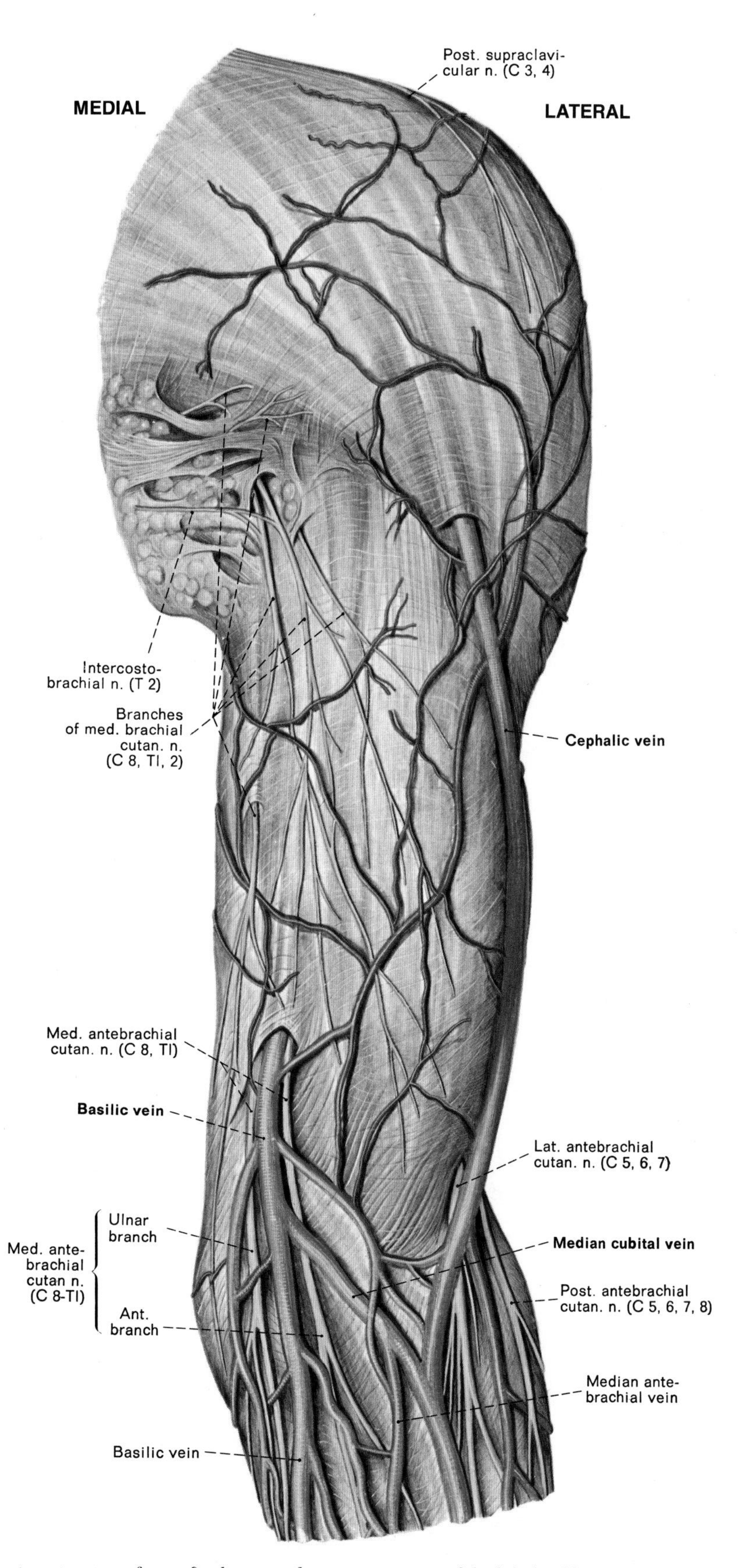

Fig. 8–2: Anterior view of superficial veins and cutaneous nerves of the left shoulder, arm, and upper forearm.

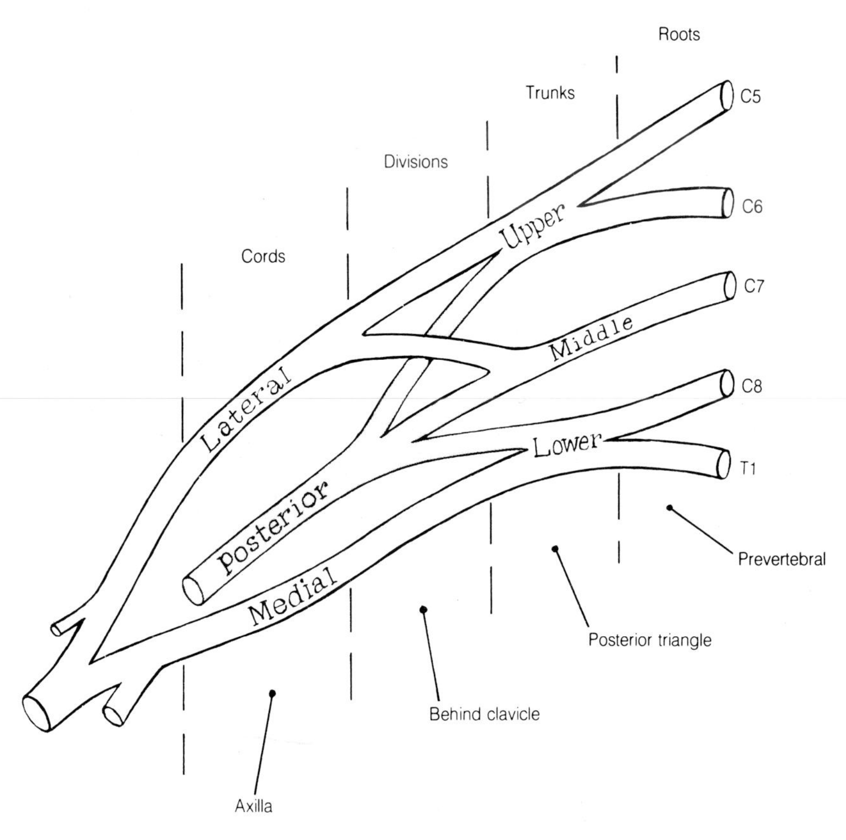

Fig. 8–3: Diagram of the brachial plexus.

neous areas of the upper limb innervated by branches of the axillary and radial nerves.

The **lateral cord** is formed from the union of the anterior divisions from the upper and middle trunks, and it gives rise to the **lateral pectoral nerve,** the **musculocutaneous nerve,** and the **lateral root of the median nerve** (Fig. 8–4). Table 8–2 lists the muscles and major joints innervated by the three nerves derived from the lateral cord. Figures 4–12E and 4–12F depict the cutaneous areas of the upper limb innervated by branches of the musculocutaneous and median nerves.

The **medial cord** is a continuation of the anterior division of the lower trunk, and it gives rise to the **medial pectoral nerve,** the **medial cutaneous nerve of the arm,** the **medial cutaneous nerve of the forearm,** the **ulnar nerve,** and the **medial root of the median nerve** (Fig. 8–4). Table 8–3 lists the muscles and major joints innervated by the five nerves derived from the medial cord. Figures 4–12E and 4–12F depict the cutaneous areas of the upper limb innervated by the medial cutaneous nerves of the arm and forearm and branches of the ulnar nerve.

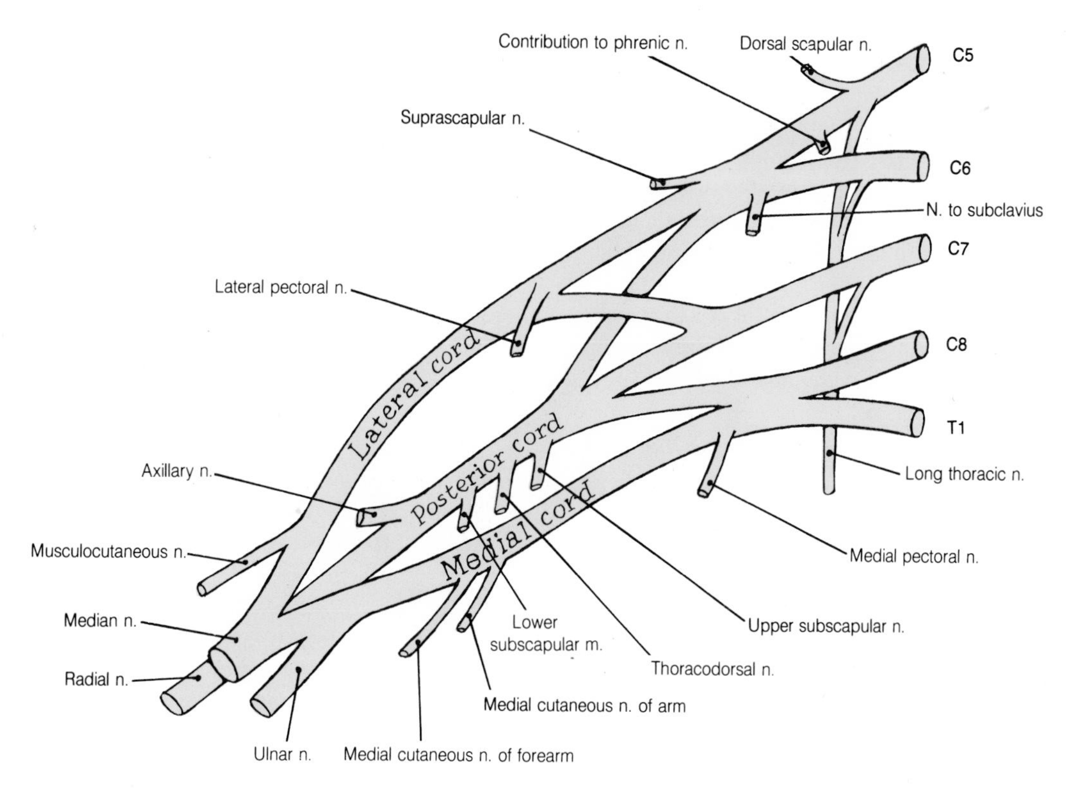

Fig. 8–4: Diagram of the brachial plexus including its branches.

Table 8-1
The Muscles and Major Joints Innervated by the Nerves Derived from the Posterior Cord of the Brachial Plexus

Nerve	Muscles	Major Joints
Upper subscapular	Subscapularis	
Middle subscapular	Latissimus dorsi	
Lower subscapular	Subscapularis Teres major	
Axillary	Deltoid Teres minor	Shoulder
Radial	Triceps and anconeus Brachialis Brachioradialis Extensor carpi radialis longus Extensor carpi radialis brevis Supinator Extensor digitorum Extensor digiti minimi Extensor carpi ulnaris Abductor pollicis longus Extensor pollicis brevis Extensor pollicis longus Extensor indicis	Elbow Proximal radioulnar Distal radioulnar Wrist Carpometacarpal of thumb

CLINICAL PANEL II-3

BRACHIAL PLEXUS INJURIES

Injuries to the brachial plexus can produce a broad spectrum of motor and sensory deficits in the upper limb. Most brachial plexus injuries involve either an injury to the upper parts of the plexus or an injury to the lower parts of the plexus. The following discussion focuses on the principal motor deficits that may result from each type of injury. An understanding of the anatomic basis of the major characteristics of Erb's palsy requires knowledge of the innervation and actions of the shoulder, arm, and forearm muscles. An understanding of the anatomic basis of the major characteristics of Klumpke's palsy requires knowledge of the innervation and actions of the muscles in the hand.

Erb's Palsy

Definition: Erb's palsy refers to the condition in which upper limb muscles are partially or completely paralyzed as a consequence of a lesion involving the C5 and C6 roots, the upper trunk, or the divisions of the upper trunk of the brachial plexus.

Common Mechanism of Injury: Erb's palsy is almost always the result of an injury to the upper parts of the brachial plexus. Excessive traction on the upper parts of the brachial plexus is the most common mechanism of injury. Instances in which such an injury may be sustained include a difficult vaginal delivery of a neonate or accidents in which a cyclist or motorcyclist is thrown from the vehicle and strikes the ground in such a fashion that the shoulder is forcefully depressed at the same time that the head and neck are forcefully flexed to the other side of the body (Fig. 8–5).

Anatomic Basis of the Major Characteristics of the Palsy: The muscles which may be partially or completely paralyzed by excessive traction on the upper parts of the brachial plexus include the upper limb muscles whose sole or major innervation is provided by C5 and/or C6 fibers. If significant damage is sustained by both C5 and C6 fibers, the most important muscular actions compromised or lost by such damage are those of supraspinatus, deltoid, biceps brachii, and supinator (see Tables 7–3, 7–4, 9–2, and 10–4). Significant loss of the actions of these muscles causes the affected upper limb to eventually have the following appearance: The shoulder looses its smooth, rounded contour, and appears to droop somewhat (these changes in appearance are due to the wasting of deltoid). The entire upper limb hangs in a deadened fashion by the side of the body (due to the loss of supraspinatus's and deltoid's abduction actions). The forearm is more pronated than usual as a consequence of the loss of biceps brachii's and supinator's supination actions. In effect, the person's entire upper limb assumes a configuration analo-

Table 8-2
The Muscles and Major Joints Innervated by the Nerves Derived from the Lateral Cord of the Brachial Plexus

Nerve	Muscles	Major Joints
Lateral pectoral	Pectoralis major Pectoralis minor	Acromioclavicular Shoulder
Musculocutaneous	Coracobrachialis Biceps brachii Brachialis	Elbow Proximal radioulnar
Median	Pronator teres Flexor carpi radialis Palmaris longus Fexor digitorum superficialis Lateral half of Flexor digitorum profundus Pronator quadratus Flexor pollicis longus Abductor pollicis brevis Flexor pollicis brevis Opponens pollicis First lumbrical Second lumbrical	Wrist Carpometacarpal of thumb Metacarpophalangeal of all 5 digits Interphalangeal of all 5 digits

CLINICAL PANEL II-3 (continued)

gous to that of the upper limb of a maitre d' when awaiting a gratuity. It is for this reason that Erb's palsy is referred to as the *"waiter's tip palsy."*

Klumpke's Palsy

Definition: Klumpke's palsy is a condition in which upper limb muscles are partially or completely paralyzed as a consequence of a lesion involving the C8 and T1 roots, the lower trunk, or the divisions of the lower trunk of the brachial plexus.

Common Mechanism of Injury: Klumpke's palsy frequently results from injury to the lower parts of the brachial plexus. Excessive traction on these parts is the most common mechanism of injury. Instances in which such an injury may be sustained include accidents in which a person falling from a significant height grasps at some object to break the fall and in the process hyperabducts the arm.

Anatomic Basis of the Major Characteristics of the Palsy: The muscles which may be partially or completely paralyzed by excessive traction on the lower parts of the brachial plexus include the upper limb muscles whose sole or major innervation is provided by C8 and/or T1 fibers. If significant damage is sustained by both C8 and T1 fibers, the most important muscular actions compromised or lost by such damage are those of the intrinsic muscles of the hand (see Tables 11–1 to 11–5). The loss of the actions of these muscles eventually contorts the hand into a clawed configuration distinguished by the hyperextension of the fingers at their MP joints. This hyperextension results from an imbalance between the tonus muscular forces of flexion and extension at the MP joints of the fingers. Reference to Tables 10–3, 10–6, and 11–1 shows that irreversible damage to C8 and T1 motor fibers denervates the MP joint flexors of the fingers (flexor digitorum superficialis, flexor digitorum profundus, and the lumbricals) more adversely than the MP joint extensors of the digits (extensor digitorum, extensor indicis, and extensor digiti minimi). Klumpke's palsy is also characterized by a severe limitation of the thumb's oppositional movement (due to the loss of opponens pollicis's actions) and an inability to abduct the fingers at their MP joints (due to the loss of the actions of the dorsal interossei and abductor digiti minimi).

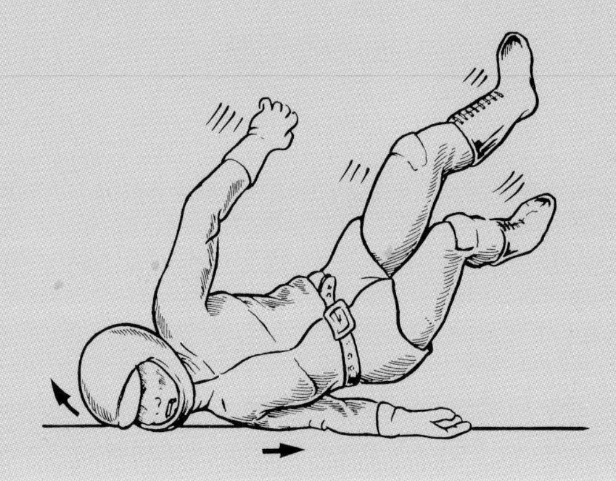

Fig. 8–5: Illustration of the position of a motorcyclist at the moment of impact with the ground in an accident in which the impact produces traction on the upper parts of the brachial plexus as a result of forced shoulder depression combined with forced head and neck flexion. (From Dandy, D. J.: Essential Orthopedics And Trauma, Churchill Livingstone, Edinburgh, 1989, p. 177.)

Table 8-3t
The Muscles and Major Joints Innervated by the Nerves Derived from the Medial Cord of the Brachial Plexus (Excluding the Median Nerve)

Nerve	Muscles	Major Joints
Medial pectoral	Pectoralis major Pectoralis minor	
Ulnar	Flexor carpi ulnaris Medial half of Flexor digitorum profundus Palmaris brevis Abductor digiti minimi Flexor digiti minimi Opponens digiti minimi Third lumbrical Fourth lumbrical All 3 palmar interossei All 4 dorsal interossei Adductor pollicis	Wrist Carpometacarpal of thumb Metacarpophalangeal of all 5 digits Interphalangeal of all 5 digits

CLINICAL PANEL II-4

THORACIC OUTLET SYNDROMES

Upper limb pain can be caused by a number of conditions that intermittently compress the brachial plexus, subclavian artery or vein, or axillary artery or vein in the cervical (neck) and axillary regions between the thoracic outlet and the insertion of pectoralis minor. The different conditions that can produce intermittent upper limb pain by this common mechanism are collectively called the thoracic outlet syndromes. Although symptoms in most cases are neurologic only, they may also be vascular or both vascular and neurologic. Pain and paresthesia (abnormal sensation) are common neurologic symptoms, and may occur in various regions of the upper limb. Common symptoms produced by compression of the subclavian or axillary artery include decrease in skin temperature, easy fatigability, and Raynaud's phenomenon in the upper limb (refer to Chapter 11 for a discussion of Raynaud's phenomenon). Common symptoms produced by compression of the axillary or subclavian vein are edema of the hand, distended superficial veins, and cyanosis.

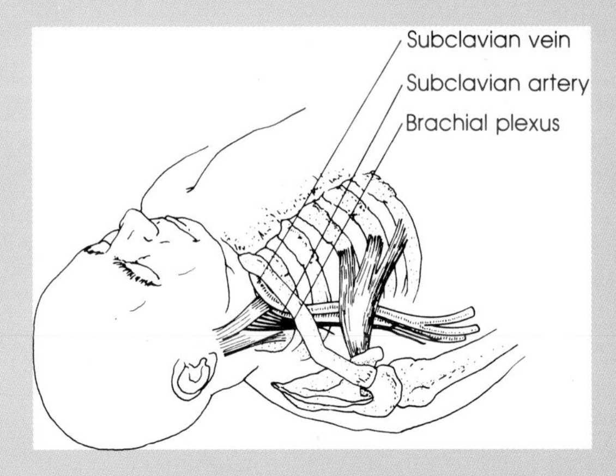

Fig. 8–6: Diagram illustrating the anatomy relevant to the costoclavicular syndrome.

Costoclavicular Syndrome

Definition: This syndrome is the condition in which the subclavian artery, subclavian vein, and/or the divisions of the brachial plexus become compressed between the clavicle above and first rib below (Fig. 8–6). The compression may occur as a consequence of prolonged retraction and depression of the pectoral girdle (such as may occur during the wearing of a heavy coat or backpack or as a result of postural changes associated with advancing age).

Common Symptoms: Common neurologic and vascular symptoms are as described above for thoracic outlet syndromes in general.

Anatomic Basis Of The Examination: As the patient sits erect on an examination table or chair, the examiner palpates the pulsations of the radial artery at the wrist before and after an assistant draws the patient's shoulder downward and backward. The downward and backward movement of the shoulder depresses and retracts the pectoral girdle, and thus draws the clavicle closer to the first rib. This test is positive if the maneuver weakens or abolishes the radial artery's pulsations and elicits the neurologic symptoms experienced by the patient. However, note that this test *can* yield false negative findings. In other words, the patient may be indeed suffering from the costoclavicular syndrome, but the test does not adequately simulate the daily conditions in the patient's life which compress the patient's brachial plexus and/or subclavian vessels between the clavicle and the first rib.

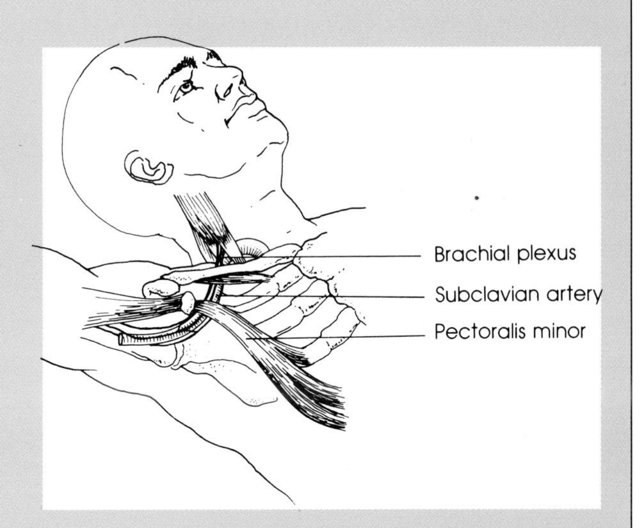

Fig. 8–7: Diagram illustrating the anatomy relevant to the hyperabduction syndrome.

CLINICAL PANEL II-4 (continued)

Hyperabduction Syndrome

Definition: This syndrome is the condition in which the axillary artery, axillary vein, and/or the cords of the brachial plexus become impinged or compressed beneath the coracoid process of the scapula by prolonged hyperabduction of the arm (Fig. 8–7). The compression may occur as a consequence of prolonged hyperabduction during sleep or during the performance of a manual task (such as painting a ceiling).

Common Symptoms: Common neurologic and vascular symptoms are as described above for thoracic outlet syndromes in general.

Anatomic Basis of the Examination: As the patient sits erect on an examination table or chair, the examiner palpates the pulsations of the radial artery at the wrist before and after the patient raises the arm above the head. The test is positive if the maneuver weakens or abolishes the radial artery's pulsations and elicits the neurologic symptoms experienced by the patient.

THE AXILLARY LYMPH NODES

Except for a few nodes located above the elbow, all the lymph nodes of the upper limb are clustered in the axilla. There are six distinct groups of axillary lymph nodes.

The **pectoral (anterior) group** of axillary nodes can be palpated against the posterior aspect of the anterior axillary fold. These nodes drain the superficial tissues of the anterolateral region of the trunk, down to the level of the umbilicus; these nodes also drain the mammary glandular tissue in the lateral half of the breast.

The **subscapular (posterior) group** of axillary nodes can be palpated against the anterior aspect of the posterior axillary fold. These nodes drain the superficial tissues of the posterior region of the trunk, down to the level of the iliac crest.

The **lateral group** of axillary nodes can be palpated against the head and uppermost part of the shaft of the humerus. These nodes drain the deep tissues of the hand, forearm, and arm and the superficial tissues on the medial side of the upper limb.

The **central group** of axillary nodes can be palpated against the chest wall in the center of the axilla. These nodes drain the efferent lymphatics from the anterior, posterior, and lateral groups.

The **deltopectoral (infraclavicular) group** of axillary nodes can be palpated in the deltopectoral triangle. These nodes drain the superficial tissues of the lateral side of the hand, forearm, and arm.

The **apical group** of axillary nodes lies along the axillary vein immediately lateral to the lateral border of the first rib. These nodes drain efferent lymphatics from the other five groups of axillary nodes.

THE MAMMARY GLAND

The mammary gland is discussed in association with the axilla because most of its arterial supply is provided by the lateral thoracic branch of the axillary artery and most of its venous and lymphatic drainage is directed toward the axilla.

The mammary glands are located mainly within the superficial fascia of the breasts. The glands are well-developed in the adult female only. In the adult female, the central region of each gland overlies parts of pectoralis major and serratus anterior, being separated from these muscles by a layer of deep fascia (Figs. 8–8 and 8–9). A prolongation of each gland called the **axillary tail** extends superolaterally along the lower border of pectoralis major into the axilla (not shown in Fig. 8–8).

In the adult female, each gland consists of 15 to 20 well-developed, radially aligned **lobes** of glandular tissue. Each lobe is drained by a **lactiferous duct.** The lactiferous ducts converge upon the **areola** and open onto the **nipple** (the areola is the pigmented, annular area surrounding the nipple) (Fig. 8–9).

Fibrous and adipose tissues occupy the spaces between the lobes of glandular tissue. Strands of connective tissue in the interlobar spaces extend from the deep fascia of the thoracic wall to the dermis of the skin. These strands of connective tissue are called the **suspensory ligaments of the breast** because they help support the mass of the mammary gland (Fig. 8–9).

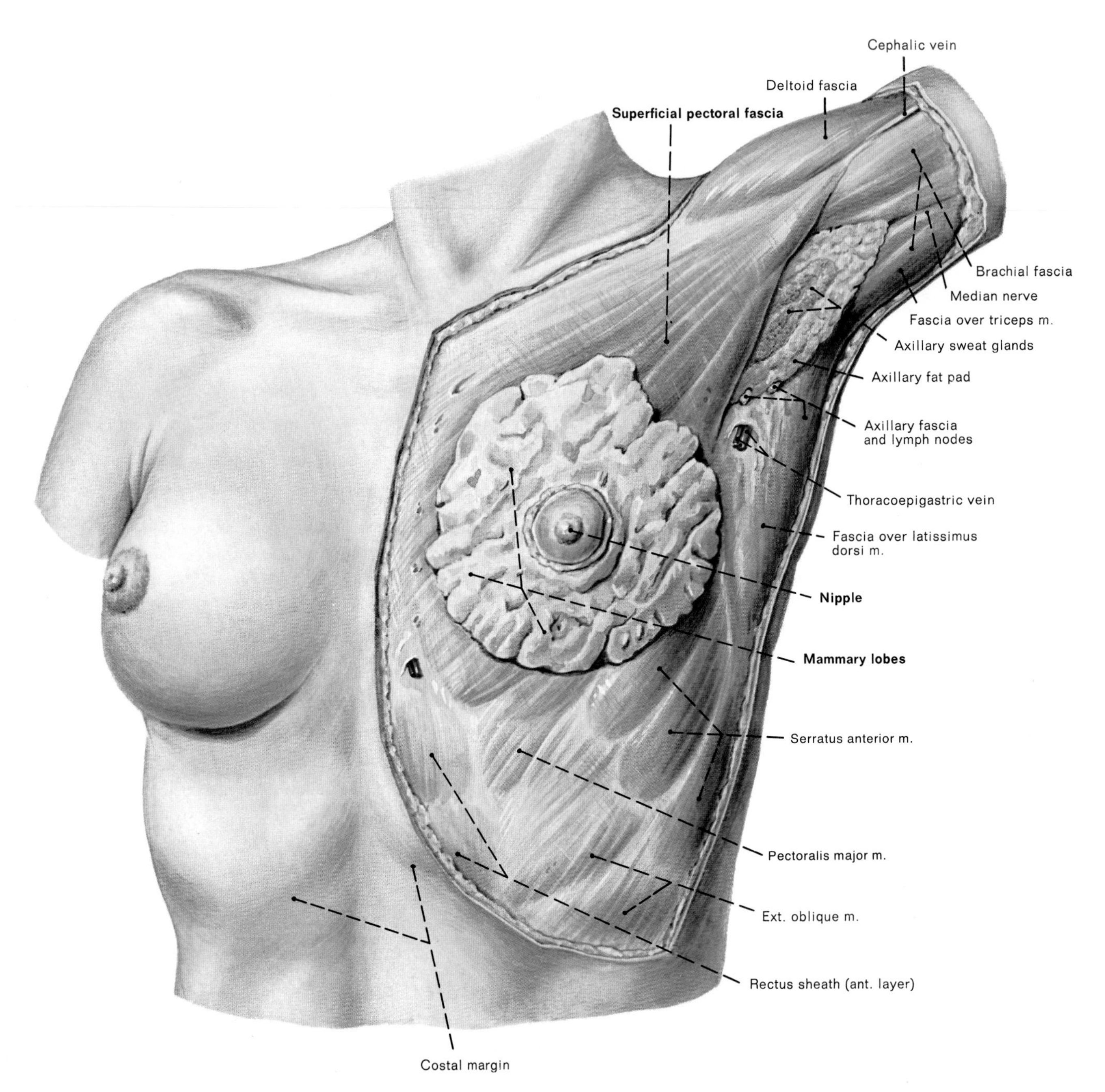

Fig. 8–8: Oblique view of the relationship of the mammary gland in the adult female to the pectoralis major and serratus anterior muscles.

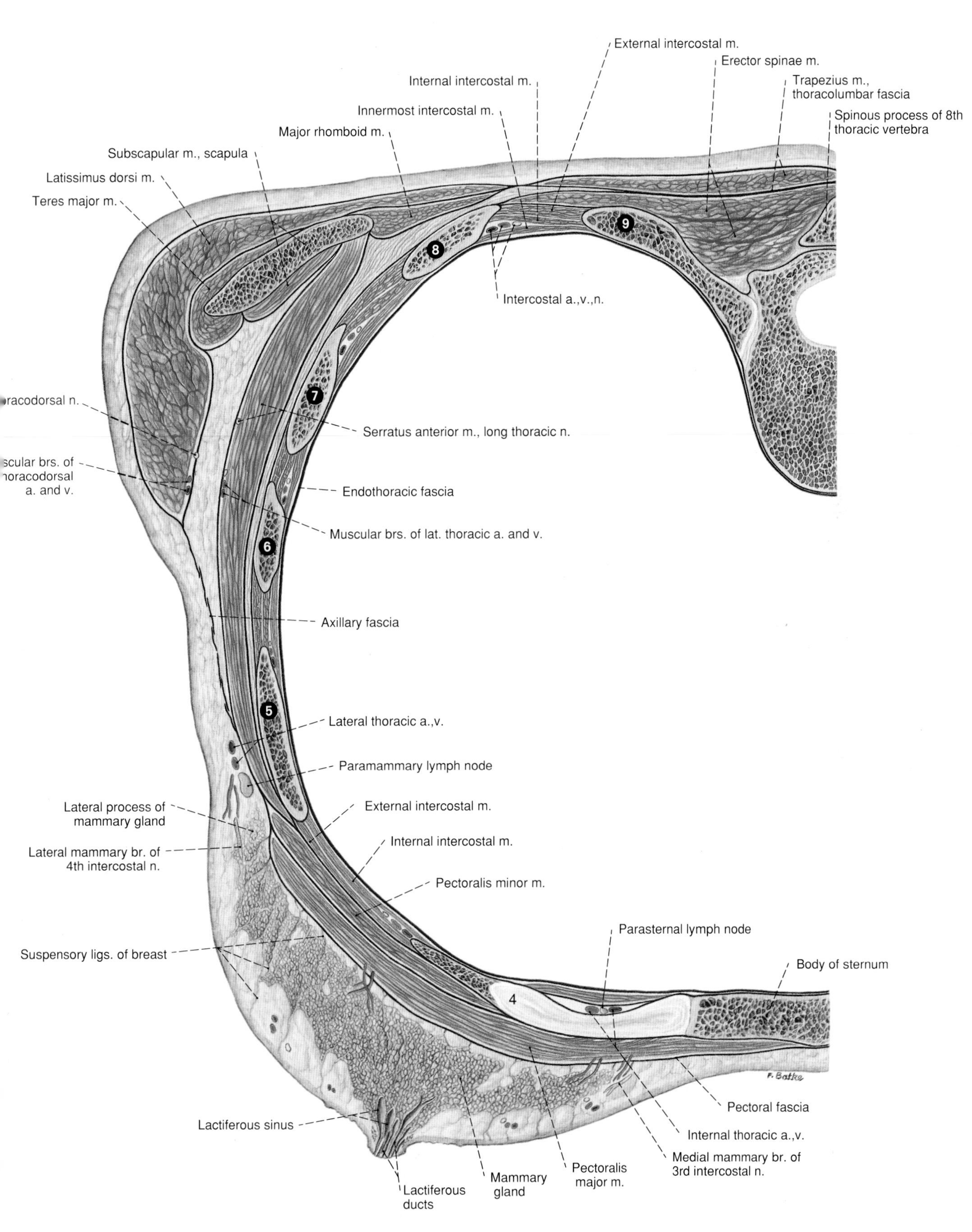

Fig. 8–9: Transverse section through the thoracic wall of an adult female near the inferior angle of the scapula.

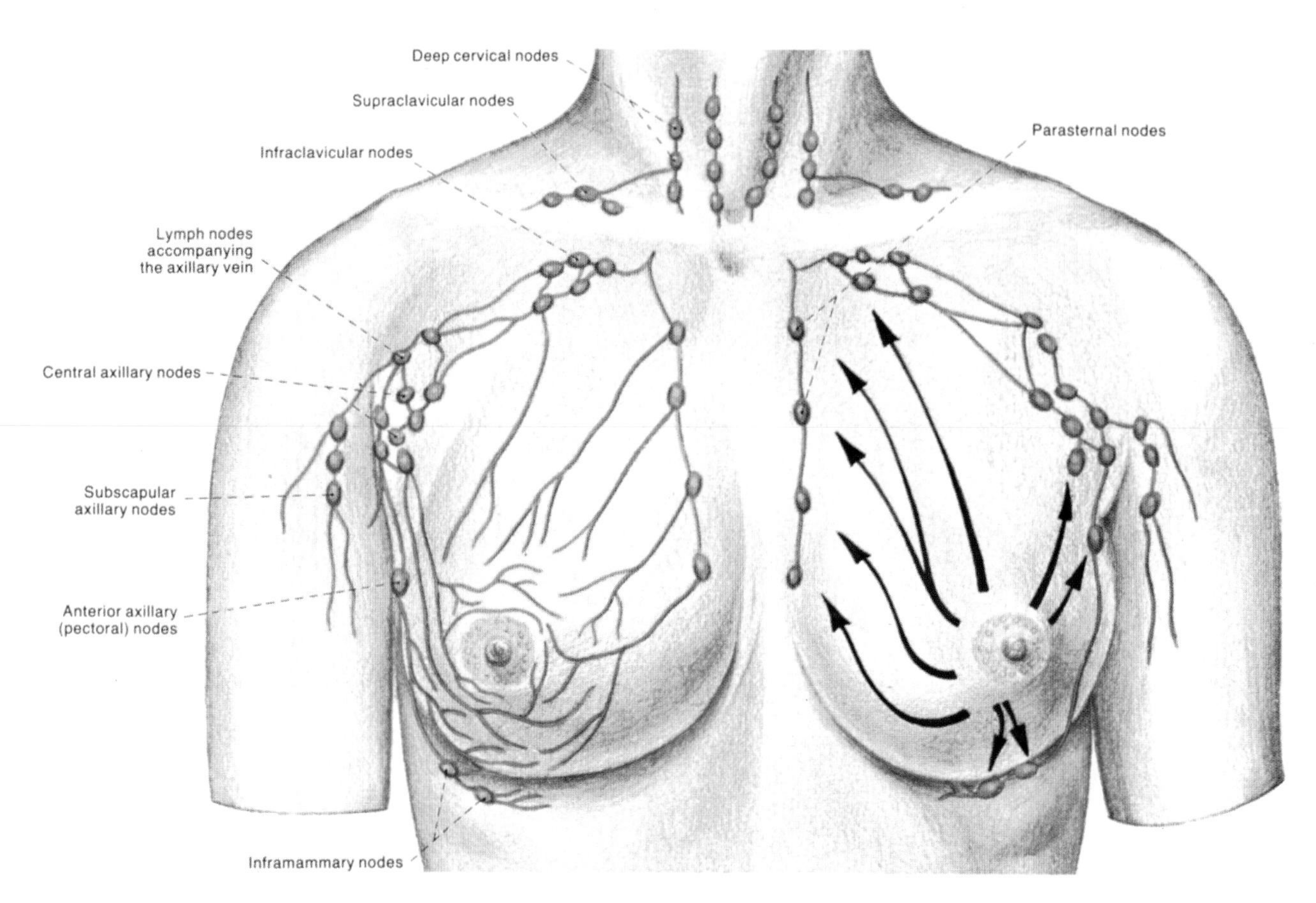

Fig. 8–10: Diagram illustrating lymphatic drainage of the mammary gland in the adult female.

The lateral thoracic artery provides the blood supply to the mammary glandular tissue in the lateral half of the breast. Branches of the **internal thoracic artery** provide the blood supply to the glandular tissue in the medial half of the breast.

The veins that drain the mammary glandular tissue in the lateral half of the breast are tributaries of the axillary vein. The veins that drain the glandular tissue in the medial half of the breast are tributaries of the **internal thoracic veins.**

The lymphatic vessels which drain the mammary glandular tissue in the lateral half of the breast lead primarily into the pectoral group of axillary lymph nodes (Fig. 8–10). This drainage accounts for approximately 75% of the lymph drained from the mammary gland, as about the same percentage of the glandular tissue lies in the lateral half of the breast. The lymphatics that drain the glandular tissue in the medial half of the breast lead primarily into the **internal thoracic (internal mammary) (parasternal) nodes;** these nodes lie beside the internal thoracic veins (Fig. 8–10). Some of the lymphatics that drain the uppermost part of the breast lead into the infraclavicular group of axillary lymph nodes. The **supraclavicular nodes** (which are the lowest members of the deep cervical nodes in the neck) also receive some of the lymph drained from the breast (Fig. 8–10).

CASE **II.3**

The Case of Evelyn Buckings

INITIAL PRESENTATION AND APPEARANCE OF THE PATIENT

A healthy appearing, 31 year-old, white woman named Evelyn Buckings has made an appointment to seek evaluation of a lump in her breast.

QUESTIONS ASKED OF THE PATIENT

What can I do for you? I would like you to examine a lump in my breast.

When did you first notice the lump? About 3 weeks ago during self-examination after a shower. I examine my breasts monthly, and this lump is something I've never felt before. *[The patient's answer suggests that she conscientiously follows a monthly program of breast self-examination. The examiner therefore tentatively assumes at this point that the size of the lump is probably near the limit of palpation (about 1 cm).]*

How would you describe the lump? It feels round and solid. *[The examiner wants the patient to describe in her own words the physical characteristics of the lump, in part to ensure that the examiner will be able to accurately identify and palpate the lump in question.]*

Do you recall bruising or injuring your breasts before you noticed the lump? No. *[This answer suggests that the lump is not a bruise from a recent injury].*

Is the lump painful or tender to the touch? No.

Is this the only lump you have felt in your breasts? Yes.

Have you experienced any recent changes in your breasts other than those you normally experience during the menstrual cycle? No.

Have there been any discharges from the nipples? No. *[The answers to these last two questions could prove useful for the differential diagnosis. For example, some breast lesions are frequently associated with discharges from the nipple.]*

Have you ever had a lump or other problem with your breasts before? No.

Is there any history of breast disease or breast cancer in your family? No, not that I'm aware of.

Do you recall how old you were when you had your first menses? Yes, I was 13.

Have you ever been pregnant? No. *[The last four questions were asked to acquire information about the patient's risk factors for breast cancer. Factors associated with an increased incidence of breast cancer include a prior occurrence of a breast malignancy, a history of breast cancer among close relatives, early menarche (under age 12), late menopause (after age 50), nulliparity, and late age of first pregnancy.]*

What is your marital status at this time? I'm single.

Are you currently sexually active? Yes. I have a long-term relationship with a man I met 2 years ago.

Do you practice contraception? Yes. He uses a condom.

Is it possible that you could be pregnant? No, I don't think so because my last menstrual period ended 4 days ago.

Have you used different contraceptives before? No. *[The last five questions were asked to gather information about the patient's use of contraceptives, the possibility of pregnancy, and her location within the menstrual cycle].*

Is there anything else bothering you? No.

Have you had any recent illnesses or injuries? No.

Are you taking any medications for previous illnesses or conditions? No. *[The examiner finds the patient alert and fully cooperative during the interview.]*

Physical Examination of the Patient

Vital signs: Blood pressure
Lying supine: 120/65 left arm and 120/65 right arm
Standing: 120/65 left arm and 125/65 right arm
Pulse: 70

Rhythm: regular
Temperature: 99.1°F
Respiratory rate: 17
Height: 5′7″
Weight: 130 lbs.
HEENT Examination: Normal
Lungs: Normal
Cardiovascular Examination: Normal
Abdomen: Normal
Genitourinary Examination: Normal
Musculoskeletal Examination: Normal
Neurologic Examination: Normal
Breast Examination: The patient pointed to a spot in the upper lateral quadrant of the left breast as the site of the lump. Inspection and palpation of both breasts are normal except for the following findings: There is a 2 cm × 1 cm × 1 cm discretely bordered, ovoid mass in the left breast at the 2:00 position 4 cm from the nipple. The mass is firm, non-tender, and freely mobile; it does not transilluminate. Its smooth external surface has a slippery quality to it.
Superficial Lymph Node Examination: Normal. In particular, the supraclavicular nodes on both sides and the axillary nodes on both sides are normal in consistency and size.

[Lymphatic dissemination of breast cancer frequently involves the axillary and supraclavicular nodes. Therefore, these nodes are always palpated on both sides in a patient with an undefined breast mass.

The pulp of the fingers is used to press against the axillary walls and in the supraclavicular fossa to assess the size, location, consistency, and mobility of lymph nodes. Normal superficial lymph nodes are frequently difficult to palpate because they are small, non-tender, relatively mobile, and soft. Lymph nodes that are mounting an immunologic response to infectious agents are enlarged but regularly shaped, tender, relatively mobile and firm. Tender lumps can also represent abscesses, however. Lymph nodes involved in carcinamatous metastasis are non-tender but become hard, fixed, and irregularly shaped.]

INITIAL EVALUATION OF THE PATIENT'S CONDITION

The history and physical examination shows that the patient has a dominant mass in the upper lateral quadrant of the left breast.

ANATOMIC BASIS OF THE PATIENT'S HISTORY AND PHYSICAL EXAMINATION

What do we know and what may we deduce from the patient's history and physical examination? The patient has a 2 cm × 1 cm × 1 cm non-tender, mobile mass in the upper lateral quadrant of the left breast. The mass may or may not be malignant, but there are no signs of advanced breast cancer.

CLINICAL REASONING PROCESS

This case illustrates the general kinds of information a gynecologist seeks from patients who present with a breast mass. The principal purpose of the clinical examination of a breast mass is to determine the physical features of the mass and the surrounding breast. From the history, the examiner seeks information about the location of the breast mass and the time of its detection, the presence of pain, the occurrence of discharges from the nipples, recent unusual changes in the breasts, and the presence of factors associated with an increased incidence of breast cancer (history of a previous breast carcinoma, history of breast cancer in primary relatives, and, in premenopausal women, an unusually early age for the onset of menses, nulliparity or a late age for the first pregnancy).

In the physical examination, the examiner gathers information through visual inspection and palpation. Visual inspection is commonly conducted with the patient seated or standing erect with the arms first at rest and then abducted over the head or with the hands pressed on the hips. The breasts are inspected for symmetry, skin abnormalities (such as redness, dimpling, thickening, or retraction), and nipple inversion and discharges. Pressing the hands on the hips (which involves isometric contraction of the pectoralis major muscles) changes the relationships of the glandular, fibrous, and adipose tissues of the breast with the skin and superficial fascia; these changes generally accentuate signs of breast asymmetry and skin dimpling and retraction.

The period of the menstrual cycle that provides optimal conditions for the physical characterization of breast masses are the days immediately following menstruation (this is the period during which there is minimal blood flow and retention of interstitial fluid within the breasts). Palpation of the breasts is best conducted with the patient reclining in a supine position and her hands folded behind her head. The assumption of this position flattens the breasts to improve palpation of any masses. All masses are characterized for location, size,

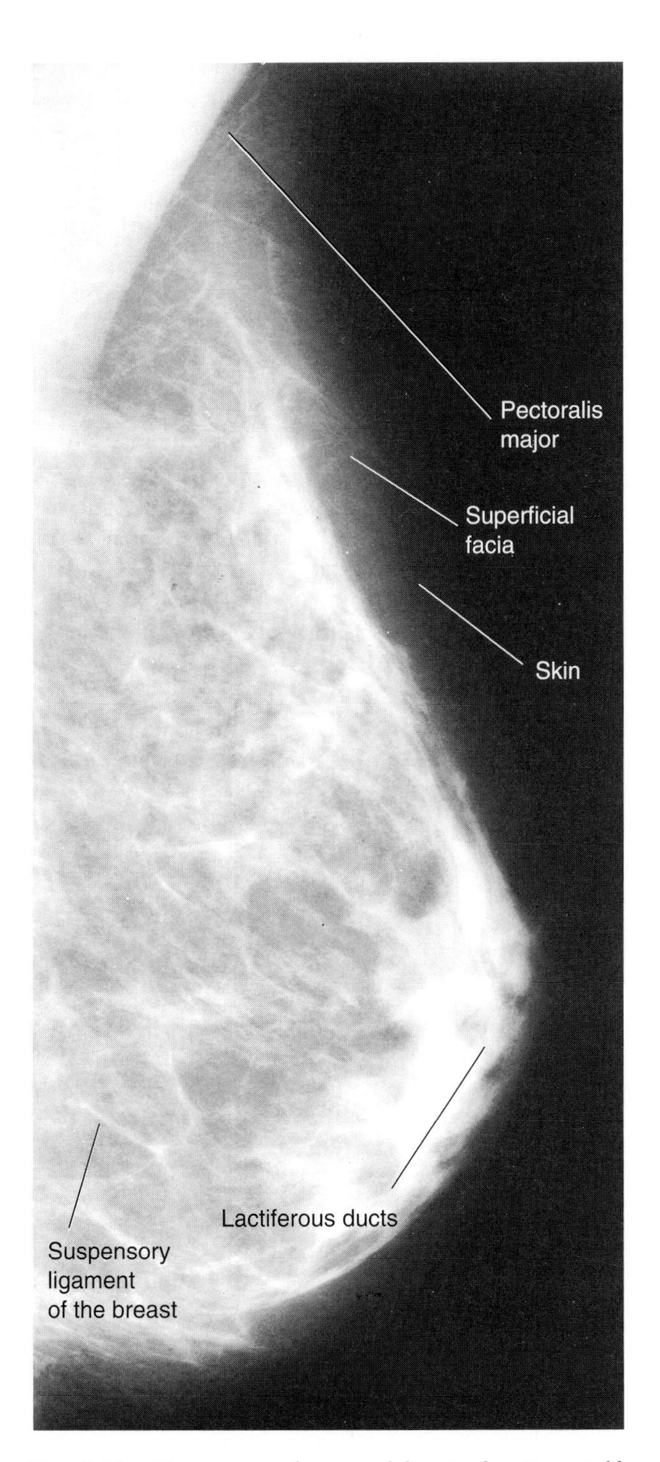

Fig. 8–11: Mammogram of a normal breast of a 41 year-old woman.

shape, consistency, and mobility. In general, a breast mass cannot be discerned by palpation as a discrete entity until it has acquired dimensions of about 1 cm. The average size palpable mass detected by self-examination is 2 to 3 cm; the likelihood of metastasis and axillary or internal mammary node involvement in a woman with a malignant tumor of this size is about 50%.

This case illustrates the limited contribution of the physical examination to the differential diagnosis of breast masses. This is because the physical characteristics typical of benign breast tumors (soft consistency, smooth contours, and relative mobility) can be exhibited by small malignant tumors. The physical characteristics typical of malignant breast tumors (hard consistency; irregular, poorly delineated contours; and immobility) are more consistently expressed in cases of advanced carcinoma. Malignant tumors have poorly delineated borders because of their irregular infiltration of the connective tissue framework of the breast. The fixation of the malignant tumor to the surrounding connective tissue framework progressively diminishes the mobility of the tumor within the breast. Extensions of the tumor can become attached to and exert traction upon the suspensory ligaments of the breast; such retraction is the anatomic basis for the skin or nipple retraction seen in cases of advanced carcinoma. Obstruction of superficial lymphatics and the attendant edema are the bases for skin dimpling encountered in cases of advanced carcinoma.

In consideration of the patient's condition, a gynecologist or oncologist would consider four types of lesions: carcinoma of the breast, a cyst, a fat necrosis, and fibroadenoma of the breast. Cysts are benign, round to ovoid, fluid-filled lesions lined by epithelium; they are believed to be derived from terminal ductal structures of the mammary gland. Cysts of the breast may be slightly compressible and may transilluminate with a bright light. A fat necrosis is a focal mass of calcium and hydrolyzed fat frequently formed after the destruction of adipose tissue from breast trauma; it does not transilluminate. Fibroadenomas are benign lesions of epithelial and connective tissue cells which typically present clinically as well-delineated, round to ovoid masses with a highly smooth surface; they do not transilluminate. Resolution of the nature of the lesion in this case requires mammography and biopsy.

MAMMOGRAPHY AND FINAL RESOLUTION OF THE PATIENT'S CONDITION

Mammography is the radiologic examination of the breast. The most common mammographic method uses X-rays to image the internal anatomy of the breast on X-ray film exposed in combination with sensitive intensifying screens. In film screen mammograms, the glandular and fibrous tissues of the breast cast relatively radiopaque images against the relatively radiolucent backdrop cast by adipose tissue (Fig. 8–11). Although the ductal structures of the glandular tissue do not cast specifically distinguishable linear radiodensities, they do establish a readily identifiable array of linear radiodensities radiating from the nipple. The suspensory ligaments of the breast can be identified as linear radiodensities that extend through the subcutaneous fat to the skin. Normal breasts of equal size generally cast mirror image mammograms.

Film screen mammograms can detect distinct breast lesions as small as 0.5 cm. An 0.5 cm cancer lesion represents a collection of 10^7 to 10^8 malignant cells that has been growing for the past 6 to 7 years and that will not attain a palpable size for another 2 years (these calculations assume an average 100-day linear doubling time). The capacity of film screen mammograms to reveal occult cancers (nonpalpable cancers) accounts for the use of mammography to screen for breast cancer in asymptomatic patients aged 35 and older and to complement the physical examination of most symptomatic patients (patients with one or more breast masses or other suspicious breast change) of all ages.

Film screen mammography provides definition of the macroscopic appearance of breast lesions. However, because there are no macroscopic features exclusively specific to malignant lesions, film screen mammography does not permit unambiguous differentiation between benign and malignant lesions. Moreover, film screen mammograms do not reveal all breast lesions. Accordingly, in most cases involving symptomatic patients, film screen mammography is used to look for occult lesions and to further characterize palpable lesions prior to biopsy of all suspicious lesions.

Because breast cancers arise from the glandular tissue of the breast, malignant lesions have a radiodensity similar to that of the glandular and fibrous tissues of the breast. Mammographic indicators of malignancy include (1) lesions with spiculated, poorly delineated margins, (2) the presence of 5 or more microcalcifications, each less than 1 mm in diameter and all clustered within 1 cm^3 of tissue, and (3) architectual distortion (presence of linear radiodensities that are aligned differently from the normal radial array of linear radiodensities extending from the nipple). Lesions with one or more of these mammographic characteristics need to be biopsied for definitive diagnosis of malignancy.

In this case, film screen mammography revealed the lesion in the upper lateral quadrant of the left breast to be a smoothly marginated oval mass with a radiodensity similar to that of surrounding glandular tissue. The radiologist considered the lesion to be most likely a fibroadenoma or cyst. Ultrasonography revealed that the lesion was not fluid-filled, and therefore not a cyst. Core biopsy (fine needle aspiration cytology) confirmed a diagnosis of fibroadenoma of the left breast. The fibroadenoma was surgically removed.

RECOMMENDED REFERENCES FOR ADDITIONAL INFORMATION ON THE BREAST

Harris, J. R., Hellman, S., Henderson, I. C., and D. W. Kinne, Breast Diseases, J. B. Lippincott Co., Philadelphia, 1991: *Chapter 5 provides a discussion of the roles of physical examination and mammography in the diagnosis of breast disease.*

Mitchell Jr., G. W., and L. W. Bassett, The Female Breast and its Disorders, Williams & Wilkins, Baltimore, 1990: *Chapter 2 presents a discussion of the roles of the history and physical examination in the diagnosis of breast disease.*

CHAPTER 9

The Arm and Elbow

This chapter focuses on the innervation and actions of the arm muscles and clinically important relationships of the major blood vessels and nerves in the arm. Since arm muscles are the prime movers of forearm movements, the discussion begins with an examination of the bones of the arm and forearm and the joints associated with movements of the forearm.

THE HUMERUS: THE BONE OF THE ARM

The shaft of the humerus serves as an attachment site for three of the four muscles of the arm. The distal end of the humerus provides the sites of origin for most of the muscles of the forearm and articulates with the proximal ends of the bones of the forearm.

The most important parts and regions of the shaft and distal end of the humerus include the **spiral groove, medial and lateral epicondyles, trochlea, capitulum, coronoid fossa, and olecranon fossa** (Fig. 9–1). The spiral groove is the groove that spirals along the posterior surface of the midregion of the shaft of the humerus (not shown in Fig. 9–1). The medial and lateral epicondyles are palpable on the sides of the elbow, and the line between them represents the upper border of the **cubital fossa,** the triangular hol-

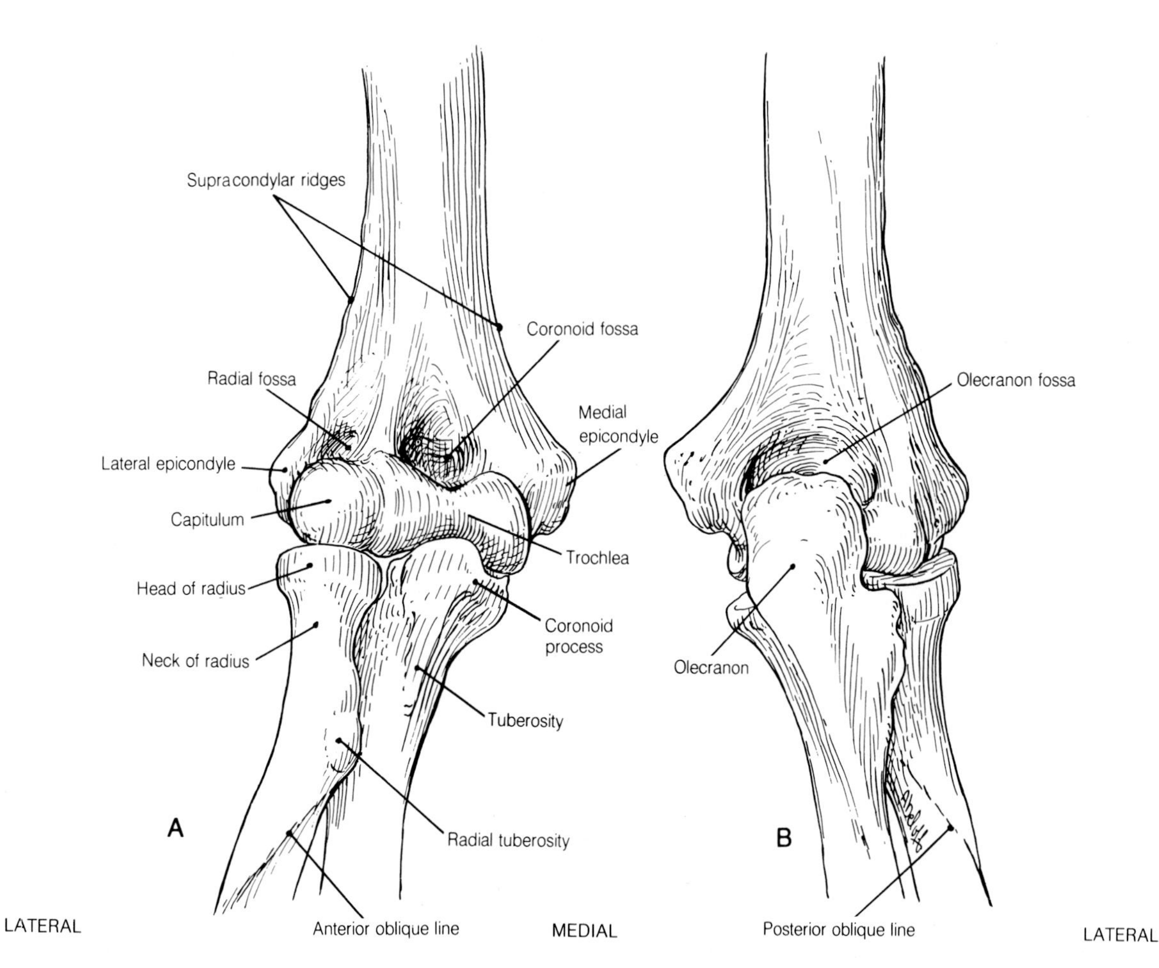

Fig. 9–1: (A) Anterior and (B) posterior views of the distal end of the right humerus and the proximal ends of the right radius and ulna.

low in the front of the elbow. The trochlea is the pulley-shaped surface on the medial side of the distal end of the humerus, and the capitulum is the ball-shaped surface on the lateral side of the distal end. The coronoid and olecranon fossae are the depressions in, respectively, the anterior and posterior surfaces of the humerus between the medial and lateral epicondyles.

The distal end of the humerus exhibits four secondary centers of ossification during childhood and adolescence. There are secondary centers for the capitulum, medial epicondyle, trochlea, and lateral epicondyle. Table 9–1 displays the ages of ossification onset and physeal closure for these centers.

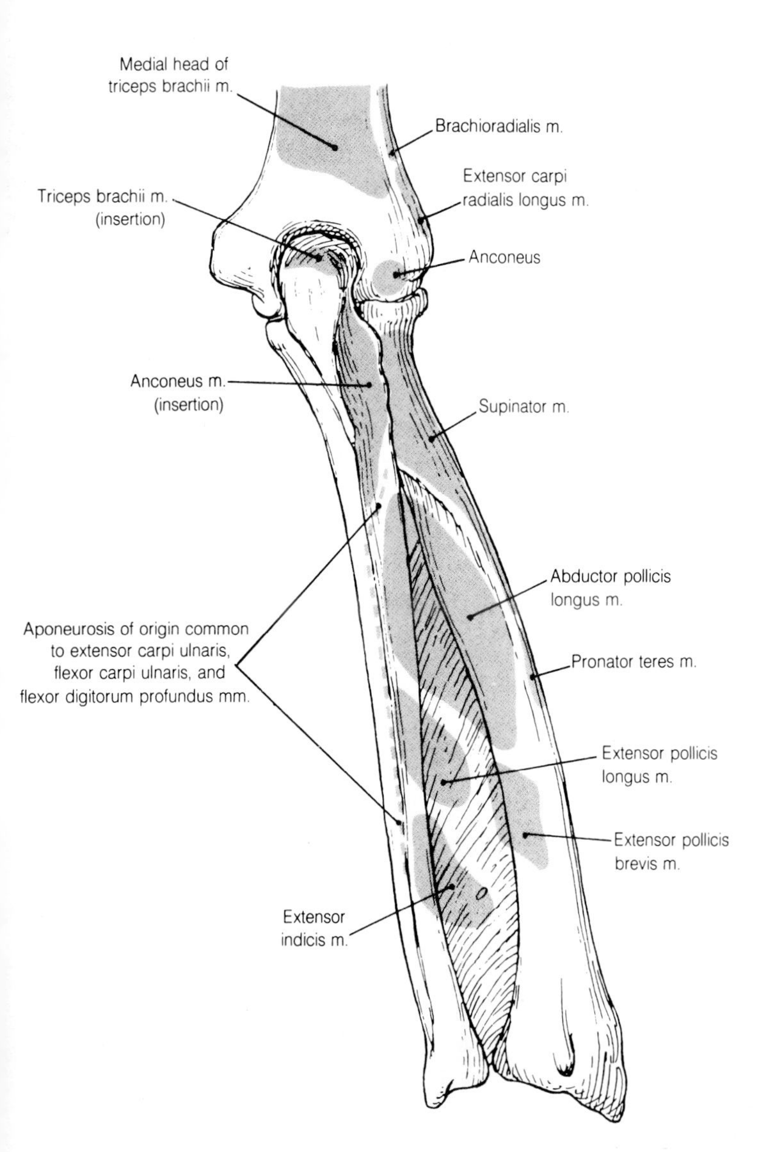

Fig. 9–2: Posterior view of the right radius and ulna and the interosseous membrane joining them.

THE RADIUS AND ULNA: THE BONES OF THE FOREARM

The radius is the lateral bone of the forearm. The **head** of the radius is the disklike, proximal end of the bone; it is palpable on the lateral side of the elbow, immediately distal to the lateral epicondyle of the humerus (Fig. 9–1). The **neck** of the radius is the narrowed segment immediately distal to the head. The prominence that rises from the anteromedial surface of the shaft of the radius (just distal to the neck) is called the **radial tuberosity.** The **shaft** of the radius widens as it extends distally from the neck (Fig. 9–2).

The distal end of the radius has three noteworthy parts (Fig. 9–2). The **styloid process** of the radius is the pointed, pillar-like process that extends distally from the lateral margin of the distal end; it is palpable on the lateral side of the wrist. The **ulnar notch** is the slight depression on the medial surface near the distal end. The **dorsal tubercle** is the small prominence that arises from the posterior surface of the radius near its distal end; it is palpable on the back of the wrist.

The radius has epiphyses at both ends during childhood and adolescence. The epiphysis for the distal end appears at 3 to 18 months of age and fuses with the shaft at 17 to 19 years of age. The epiphysis for the

Table 9–1
Ages of Ossification Onset and Physeal Closure for the Secondary Centers at the Distal End of the Humerus

Center	Ages of Ossification Onset	Ages of Physeal Closure
Capitulum	1-2 months (male) 1-6 months (female)	14-17 years
Medial epicondyle	5-7 years (male) 3-6 years (female)	15-18 years
Trochlea	8-10 years (male) 7-9 years (female)	14-17 years
Lateral epicondyle	12 years (male) 11 years (female)	14-17 years

head appears at 3 to 6 years of age and fuses with the shaft at 14 to 17 years of age.

The ulna is the medial bone of the forearm. The **head** of the ulna is the distal end of the bone. The pointed, pillar-like process that extends distally from the posteromedial margin of the ulna's head is called the **styloid process** of the ulna; it is palpable on the medial side of the wrist (Fig. 9–2). The **shaft** of the ulna widens as it extends toward the proximal end of the bone.

The proximal end of the ulna bears two prominent processes: an anteriorly projecting **coronoid process** and a superiorly projecting **olecranon process** (Figs. 9–1 and 9–3). The olecranon process is palpable in the posterior elbow region. The olecranon and coronoid processes form a large notch called the **trochlear notch** in the anterior surface of the proximal end of the ulna (Fig. 9–3). Immediately distal to the trochlear notch, the lateral surface of the coronoid process bears a slight depression called the **radial notch** (Fig. 9–2).

The ulna has epiphyses at both ends during childhood and adolescence. The epiphysis for the head appears at 4 to 9 years of age and fuses with the shaft at 17 to 19 years of age. The epiphysis for the proximal end appears at 8 to 10 years of age and fuses with the shaft at 14 to 17 years of age.

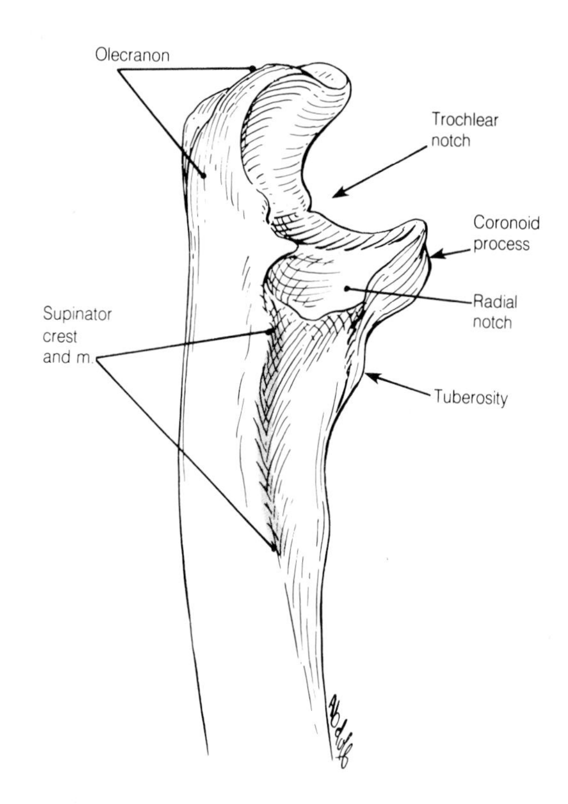

Fig. 9–3: Lateral view of the proximal end of the right ulna.

THE ELBOW JOINT AND THE PROXIMAL, MIDDLE AND DISTAL RADIOULNAR JOINTS: THE JOINTS ASSOCIATED WITH MOVEMENTS OF THE FOREARM

The elbow joint joins the distal end of the humerus with the proximal ends of the radius and ulna (Fig. 9–1). There are two articulations in the joint. In one articulation, the capitulum of the humerus articulates with the head of the radius. In the other, the trochlea of the humerus articulates with the trochlear notch of the ulna.

The elbow joint's fibrous capsule is strengthened on its medial and lateral sides by taut ligaments called, respectively, the **ulnar and radial collateral ligaments** of the elbow joint (Fig. 9–6). By contrast, the anterior and posterior parts of the capsule are comparatively thin and lax, thus allowing flexion and extension of the forearm. Pads of fatty tissue called the **anterior and posterior fat pads** superficially cover, respectively, the anterior and posterior parts of the elbow joint's fibrous capsule.

The proximal radioulnar joint joins the radius and ulna at their proximal ends (Fig. 9–7). The head of the radius articulates in the joint with the radial notch of the ulna. The head of the radius also articulates in the joint with a **C**-shaped ligament called the **annular ligament.** The annular ligament is attached to the anterior and posterior margins of the radial notch of the ulna and to the neck of the radius. Together the annular ligament and the radial notch of the ulna form a part bony-part ligamentous ring around the head of the radius. The synovial cavity of the proximal radioulnar joint is continuous with that of the elbow joint.

The distal radioulnar joint joins the radius and ulna at their distal ends (Fig. 9–8). The head of the ulna articulates in the joint with the ulnar notch of the radius. An integral component of the joint is a triangular-shaped **articular disc** composed of fibrous cartilage. The apex of the articular disc is attached to the styloid process of the ulna, and the base of the disc is attached to the medial margin of the distal end of the radius. The proximal surface of the articular disc faces the synovial cavity of the distal radioulnar joint, and the distal surface of the disc faces the synovial cavity of the wrist joint (Fig. 9–8).

The shafts of the radius and ulna are joined by a thin, fibrous sheet called the **interosseous membrane** of the forearm (Fig. 9–2). Most of the fibers in the interosseous membrane slant distally from their radial attachment to their ulnar attachment. The union of the shafts of the radius and ulna by the interosseous membrane of the forearm is called the middle radioulnar joint.

CLINICAL PANEL II-5

FRACTURES OF THE DISTAL RADIUS IN YOUNG CHILDREN

Common Mechanism of the Injury: Fractures of the distal radius in young children frequently occur as a consequence of an accident in which the child stumbles and falls hard on an outstretched hand.

Anatomic Basis of Torus and Greenstick Fractures: The application of excessive stresses across the long bones of the extremities in an adult tends to produce one or more sudden breaks in the continuity of the bone. When these breaks occur along the shaft of a long bone, they typically extend either transversely or obliquely completely through the shaft. By contrast, the application of excessive stresses across the long bones of the extremities in a young child frequently produces deformations in the structure of the bone. This is because the bones of a young child are relatively supple and easily deformed. Two types of deformation are commonly encountered:

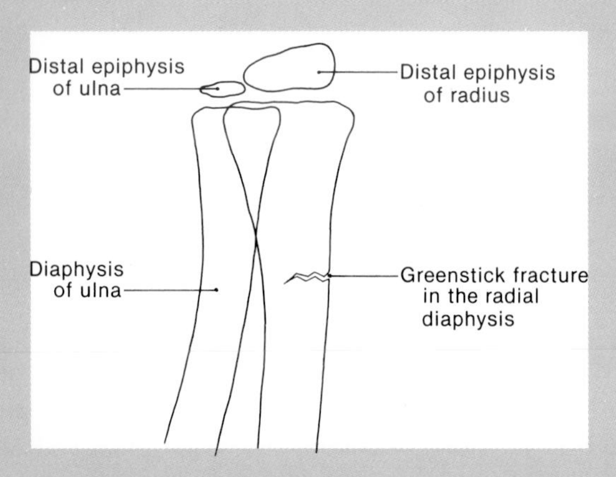

Fig. 9–5: An outline of a radiograph of a greenstick fracture of the distal third of the radial shaft.

Compression failure of the cortical bone in the distal third of a young child's radial shaft can cause the cortical bone to buckle outward. Such a fracture is called a torus fracture because the radiographic outline of the fracture site resembles the outline of an architectural column bearing a torus molding at its base (Fig. 9–4).

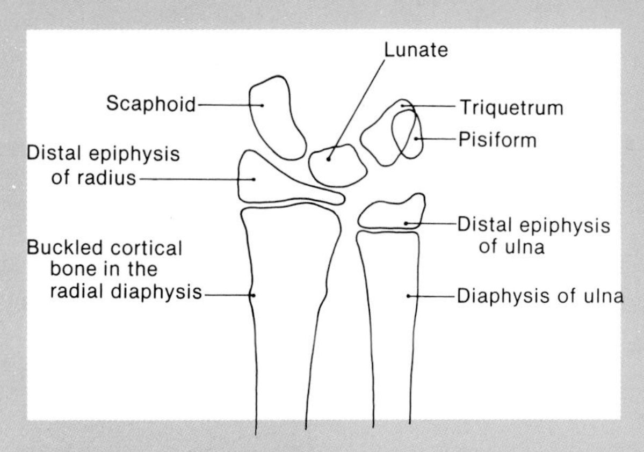

Fig. 9–4: An outline of a PA radiograph of a torus fracture of the distal third of the radial shaft.

Fractures of the distal radius in a young child are frequently incomplete. Radiographs show a discontinuity of the cortical bone on one side of the shaft and a bent cortical bone on the opposite side (Fig. 9–5). These fractures are called greenstick fractures because of their similarity to a break in a supple branch of a tree or bush.

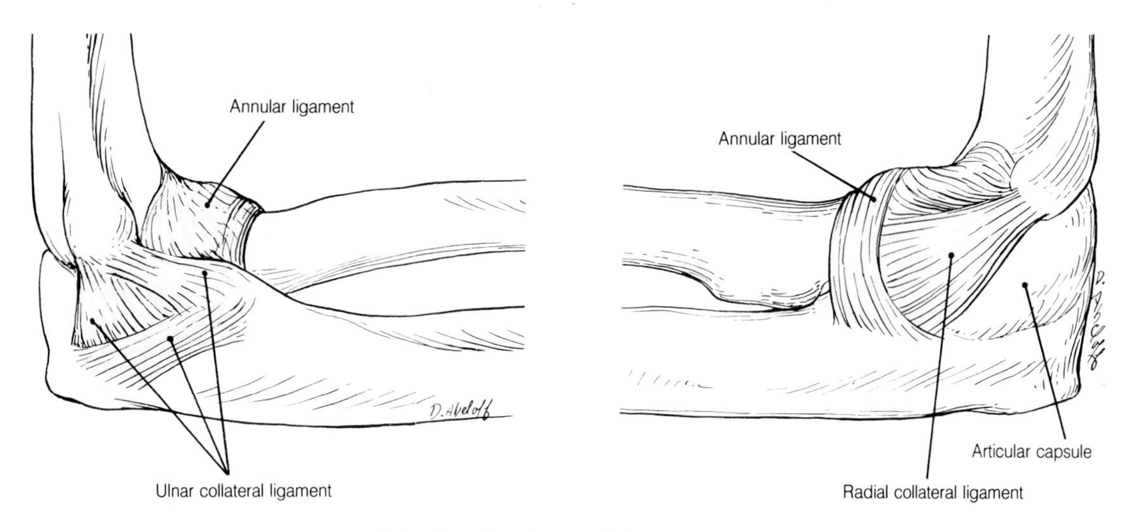

Fig. 9–6: (A) Medial and (B) lateral views of the ligaments of the left elbow joint.

THE MOVEMENTS OF THE FOREARM RELATIVE TO THE ARM

The elbow joint provides for **flexion and extension** of the forearm (Fig. 9–9). During these movements, the head of the radius and the trochlear notch of the ulna rotate about an axis which passes through the humeral epicondyles. Whereas flexion approximates the coronoid process of the ulna with the coronoid fossa of the humerus, extension approximates the olecranon process of the ulna with the olecranon fossa of the humerus.

The proximal, middle and distal radioulnar joints provide for **pronation and supination** of the forearm (Fig. 9–10). During these movements, the radius and ulna rotate about an axis which passes through their heads. The rotary movements that occur at the proximal ends of the radius and ulna differ in their nature from those which occur at the distal ends. At the proximal radioulnar joint, the only pronounced rotary movement is rotation of the head of the radius within its cufflike annular ligament. By contrast, at the distal radioulnar joint, the distal ends of the radius and ulna swing around each other. If the forearm is pronated from the anatomic position, the distal end of the radius moves anteriorly and medially as the head of the ulna moves posteriorly and laterally.

The rotation of the radius and ulna about each other during pronation and supination changes the tension in the interosseous membrane. The membrane is comparatively lax when the forearm is fully supinated or fully pronated. It is taut when the forearm is in the midprone position.

Pronation and supination change the **carrying angle** of the elbow, which is the angle on the lateral side of the elbow between the long axes of the arm and forearm (Fig. 9–10). In the anatomic position (with the forearm fully supinated), the carrying angle averages 169° for males and 167° for females. Pronation to the midprone position brings the long axis of the forearm into co-alignment with that of the arm, and thus increases the carrying angle to 180°.

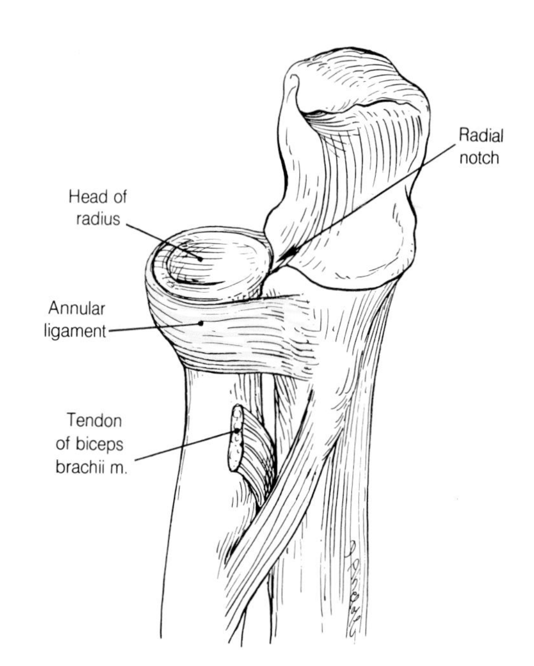

Fig. 9–7: Oblique view of the articulations of the right proximal radioulnar joint.

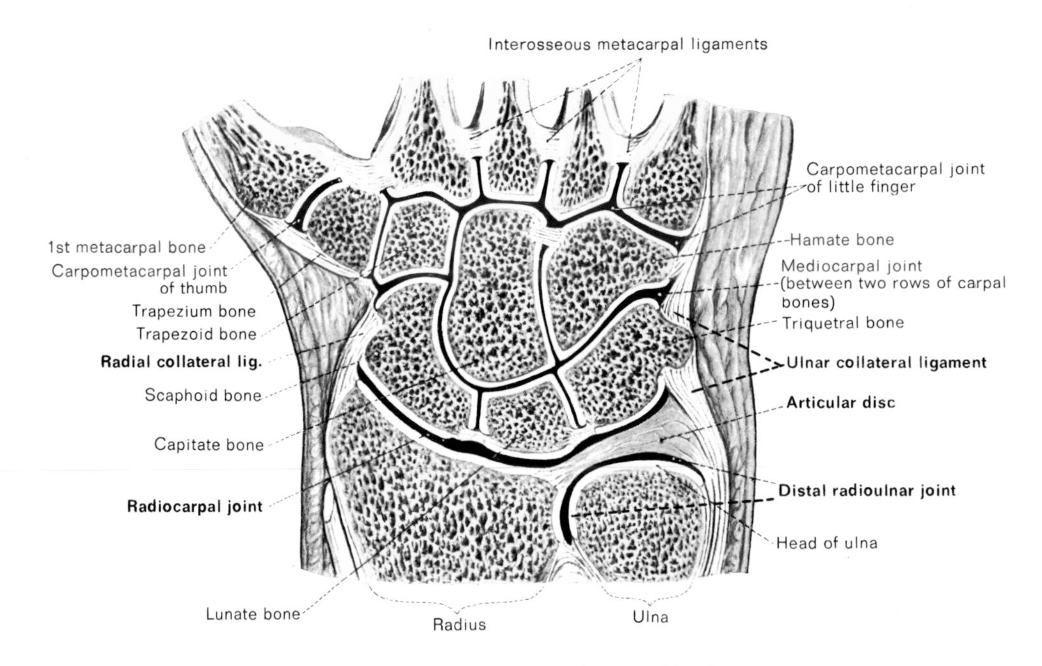

Fig. 9–8: Coronal section of wrist and hand.

THE MUSCLES OF THE ARM

The muscles of the arm are enveloped by a tubular sheath of deep fascia called the **brachial fascia** (Fig. 9–11). On each side of the arm, the brachial fascia gives off a sheet-like layer which extends deeply. The extensions are called the **medial and lateral intermuscular septa,** and each is attached deeply to the shaft of the humerus along a margin extending from the intertubercular groove above to the corresponding epicondyle below. The brachial fascia and its intermuscular septa divide the arm into anterior and posterior muscular compartments.

There are three muscles (coracobrachialis, biceps brachii, and brachialis) in the anterior muscular compartment of the arm, and they are all innervated by the musculocutaneous nerve. Triceps brachii is the only muscle in the posterior muscular compartment of the arm, and it is innervated by the radial nerve. Table 9–2 lists for quick reference the origin, insertion, nerve supply, and actions of the arm muscles.

Coracobrachialis is a comparatively slender muscle which originates from the coracoid process of the scapula and inserts onto the humeral shaft (Fig. 9–12). It can flex and adduct the arm at the shoulder joint.

Biceps brachii is named for being the arm muscle which has two heads of origin: the **long head** originates within the capsule of the shoulder joint from the supraglenoid tubercle of the scapula, and the **short head** originates from the coracoid process of the scapula (Figs. 9–12 and 9–13). The tendon of the long head lies within the intertubercular groove of the humerus as it extends beyond the confines of the shoulder joint's capsule and into the upper part of the arm. The muscle fibers of the two heads merge in the distal half of the arm. Biceps brachii extends distally as a flattened tendon that passes anteriorly to the elbow joint and then inserts onto the radial tuberosity (Fig. 9–12). This tendon also gives rise to a sheet-like extension called the **bicipital aponeurosis,** which blends into the deep fascia on the upper anteromedial side of the forearm (Fig. 9–12). Biceps brachii is the chief supinator and a major flexor of the forearm.

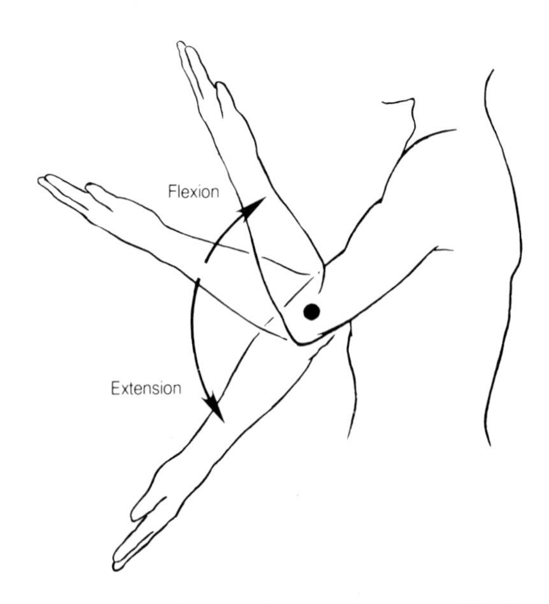

Fig. 9–9: Diagram illustrating flexion and extension of the forearm.

Biceps brachii activity is assessed by the **biceps brachii tendon reflex test.** This test is a simple three-step procedure in which the examiner (1) fully supinates the patient's forearm, then (2) supports the posterior aspect of the patient's partially flexed elbow, and finally (3) uses a reflex hammer to impart a gentle but firm and quick tap upon the biceps brachii's tendon of insertion. The hammer tap, by suddenly stretching the tendon, exacts an abrupt stretching of biceps brachii's muscle fibers, and this should elicit a phasic contraction of the muscle fibers that produces a short, quick flexion of the forearm.

Brachialis originates from the lower half of the anterior surface of the humeral shaft, passes anterior to the elbow joint, and inserts onto the coronoid process of the ulna (Fig. 9–13). It is a chief flexor of the forearm.

Triceps brachii is named for being the muscle of the arm that has three heads of origin: the **long head** originates from the infraglenoid tubercle of the scapula, the **lateral head** originates from that part of the posterior surface of the humeral shaft which is lateral and superior to the spiral groove, and the **medial head** originates from that part of the posterior surface of the humeral shaft which is medial and inferior to the spiral groove (Fig. 9–14). The muscle fibers from all three heads merge in the lower half of the arm and give rise to a common tendon of insertion that passes behind the elbow joint to insert onto the olecranon process of the ulna. Triceps brachii is the chief extensor of the forearm.

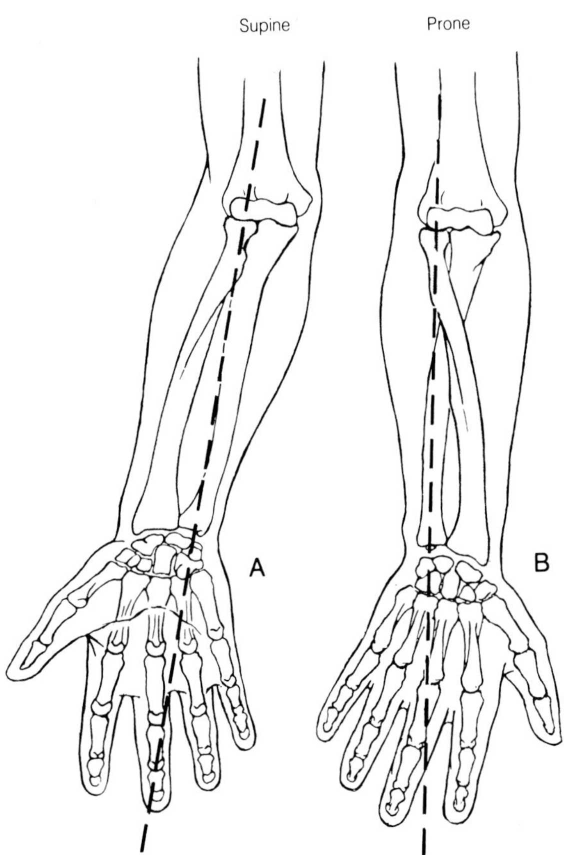

Fig. 9–10: (A) Anterior view of the orientation of the radius and ulna in the fully supinated right forearm. (B) Anterior view of the orientation of the radius and ulna in the fully pronated right forearm. The dashed line in each figure represents the axis about which the radius and ulna rotate during supination or pronation of the forearm.

Table 9–2
The Muscles of the Arm

Muscle	Origin	Insertion	Nerve Supply	Actions
Coracobrachialis	Coracoid process of the scapula	Midregion of the medial aspect of the humeral shaft	Musculocutaneous (C5, **C6,** and C7)	Flexes and adducts the arm
Biceps brachii	Long head from the supraglenoid tubercle of the scapula; short head from the coracoid process of the scapula	Radial tuberosity and deep fascia on the upper anteromedial side of the forearm	Musculocutaneous (C5 and **C6**)	Flexes and supinates the forearm
Brachialis	Lower half of the anterior surface of the humeral shaft	Coronoid process of the ulna	Musculocutaneous (C5 and **C6**) and radial (C7) to a small lateral part	Flexes the forearm
Triceps	Long head from the infraglenoid tubercle of the scapula; lateral head from the posterior surface of the humeral shaft above the spiral groove; medial head from the posterior surface of the humeral shaft below the spiral groove	Olecranon process of the ulna	Radial (C6, **C7,** and **C8**)	Extends the forearm

CLINICAL PANEL II-6

Bicipital Tendinitis

Definition: Bicipital tendinitis is inflammation of the tendon of the long head of biceps brachii. The tendon is generally inflamed along its course within the intertubercular groove (Fig. 7–6) as a result of irritation from repetitive to-and-fro sliding of the groove against the tendon.

Common Symptoms: The patient typically presents with pain in the anterior aspect of the shoulder that may radiate down the anterior aspect of the upper arm. The patient may report that arm movement exacerbates the pain, especially arm movement associated with throwing.

Anatomic Basis of the Examination: As the patient sits erect on an examination table or chair, the examiner begins the test for bicipital tendinitis by grasping the patient's hand and fully extending the patient's arm at the shoulder with the patient's forearm fully extended at the elbow and the forearm fully pronated at the radioulnar joints. This maneuver passively stretches the long head of biceps brachii. The examiner then palpates the intertubercular groove and requests the patient to flex the arm at the shoulder against resistance (the examiner restrains the arm from any flexion). The resulting isometric contraction of biceps brachii exerts added tension upon the long head of the muscle. A positive sign for bicipital tendinitis is production of pain and tenderness in the intertubercular groove upon isometric contraction of the muscle.

Triceps brachii activity is assessed by the **triceps brachii tendon reflex test.** A gentle but firm and quick hammer tap to the muscle's tendon of insertion should elicit a short, quick extension of a partially flexed forearm.

Triceps brachii's long head is the border between two intermuscular spaces in the posterior axillary region. As the long head extends inferiorly from its origin, it lies between the humerus and the lateral border of the scapula (Fig. 7-16). The teres major and minor muscles extend from the lateral border of the scapula to the humerus, with teres minor passing behind the long head and teres major passing in front of the long head. These relationships form two spaces, one medial and the other lateral to the long head. Medially, teres minor, teres major, and the long head of triceps brachii border the **triangular space.** Laterally, teres minor, teres major, the long head of triceps, and the surgical neck of the humerus form the four borders of the **quadrangular space** (Figs. 7–16 and 9–14).

The triangular and quadrangular spaces are notable for the structures that extend posteriorly through them. The axillary nerve and the posterior circumflex humeral artery pass posteriorly through the quadrangular space (Fig. 9–15), and the circumflex scapular artery passes posteriorly through the triangular space. The **circumflex scapular artery,** a branch of the subscapular artery, is one of the arteries that contributes to the anastomoses formed around the scapula by branches of the subscapular artery and thyrocervical trunk.

THE BRACHIAL ARTERY

The brachial artery is the largest artery of the arm. It begins as a continuation of the axillary artery at the lower border of teres major's insertion onto the humerus (Fig. 9–16). It extends distally through the arm on the medial side of the anterior muscular compartment. Finally, it curves laterally as it extends through the cubital fossa into the forearm, passing deep to the bicipital aponeurosis. The brachial artery ends by bifurcating into the ulnar and radial arteries at the level of the neck of the radius.

The **profunda brachii (deep brachial) artery** is the largest branch of the brachial artery (Fig. 9–16). It arises from the brachial artery immediately below the lower border of teres major, passes into the posterior muscular compartment of the arm, and then extends distally through the midregion of the arm by coursing alongside the spiral groove of the humerus (Fig. 9–15). In the distal half of the arm, profunda brachii gives rise to branches that extend across the lateral side of the elbow, both anterior and posterior to the elbow joint (Fig. 9–16). These branches provide for collateral circulation across the lateral side of the elbow.

The **superior and inferior ulnar collateral arteries** arise from the medial side of the brachial artery

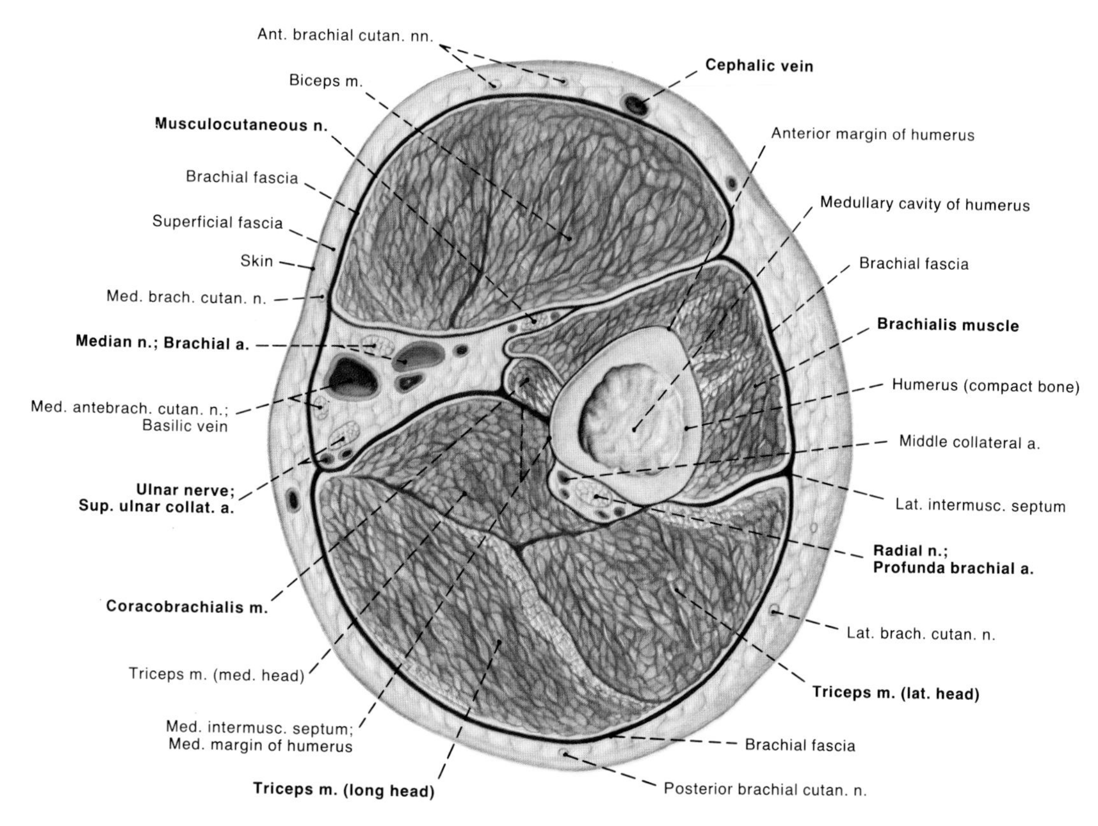

Fig. 9–11: Superior view of a horizontal section of the middle of the right arm.

in the distal half of the arm (Fig. 9–16). They provide collateral circulation across the medial side of the elbow.

The brachial artery is almost always used to measure systemic arterial blood pressure in the office or clinical setting. There are two reasons for its selection:

1. The brachial artery lies approximately at the level of the heart. Two forces contribute to the blood pressure in every vessel of the body: the heart's pumping activity and the earth's gravitational field. The contribution of gravity is proportional to the vertical distance between the vessel in question and the heart. With the arm in the anatomic position (independently of whether the individual is standing, sitting, or reclining), the brachial artery is, for all reasonable purposes of measuring blood pressure in the artery, lying at the level of the heart. Consequently, the blood pressure of the brachial artery under such conditions represents pressure generated almost exclusively by the heart's pumping activity alone.
2. The brachial artery is relatively deep and close to the humeral shaft as it extends distally through the arm, and then assumes a superficial course as it extends through the cubital fossa. The deep course of the artery through the arm makes it possible to stop its blood flow by wrapping an inflatable cuff around the arm and increasing the pressure in the cuff above the systolic pressure. Blood flow in the brachial artery stops because the muscle mass in the arm efficiently transmits the pressure in the cuff and squeezes the brachial artery against the humeral shaft. The superficial course of the artery through the cubital fossa makes it possible to listen to blood flow through the artery.

Determination of brachial artery blood pressure is conducted with the patient's arm in the anatomic position. An inflatable cuff is wrapped securely around the arm, and air is rapidly pumped into the cuff until the cuff's pressure is sufficient to stop brachial pulsations (as palpated, if need be, in the cubital fossa). The diaphragm of a stethoscope is placed over the brachial artery in the cubital fossa, and the pressure within the cuff is slowly dissipated. No blood flow sounds are heard until the cuff pressure falls just below the systolic pressure. At this juncture, blood spurts through the brachial artery only at peak systolic pressure, and the blood spurts produce faint tapping sounds called the **sounds of Korotkoff.** The pressure recorded upon first hearing a tapping sound during each heart-

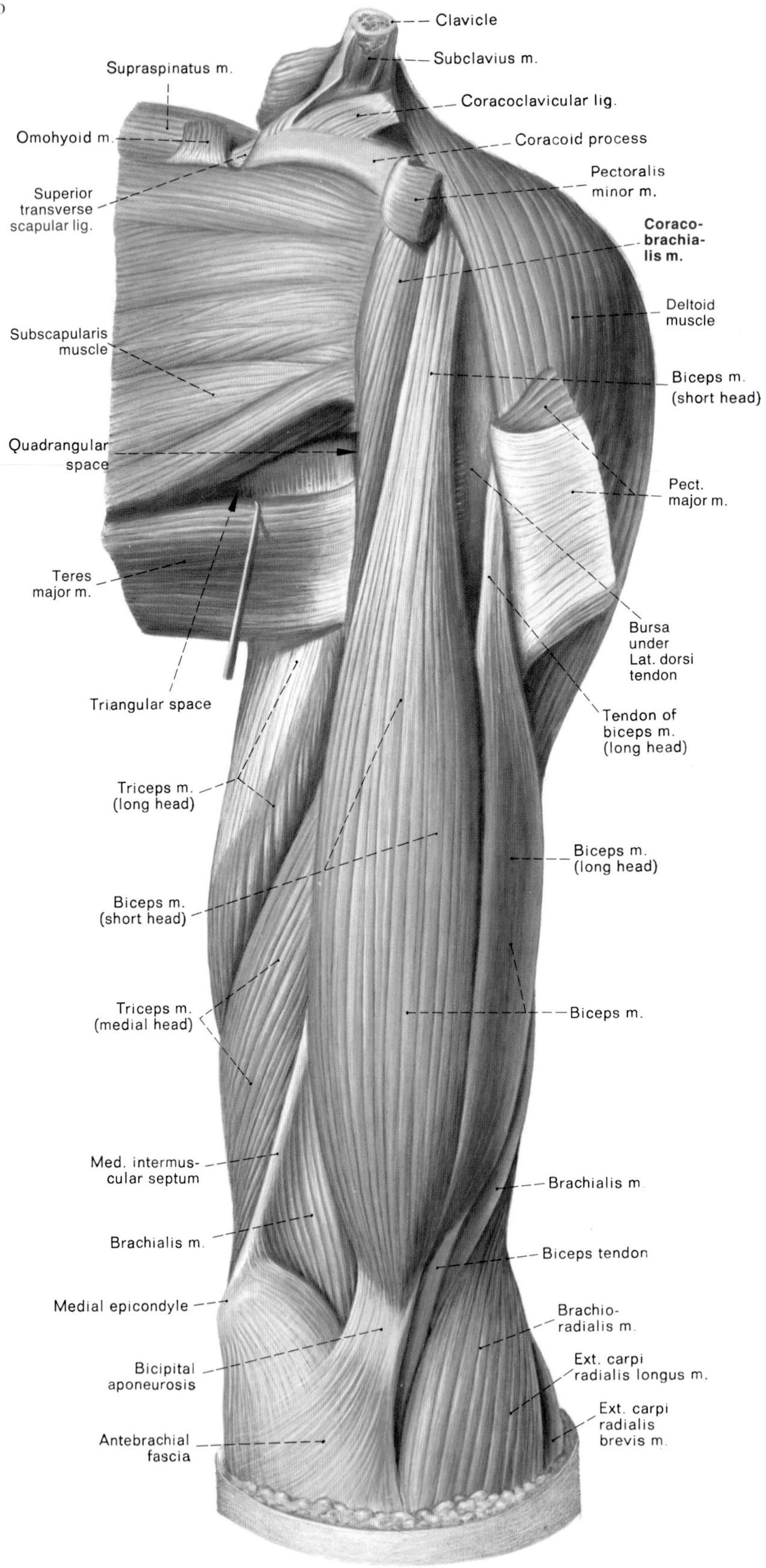

Fig. 9–12: Anterior view of the left arm muscles.

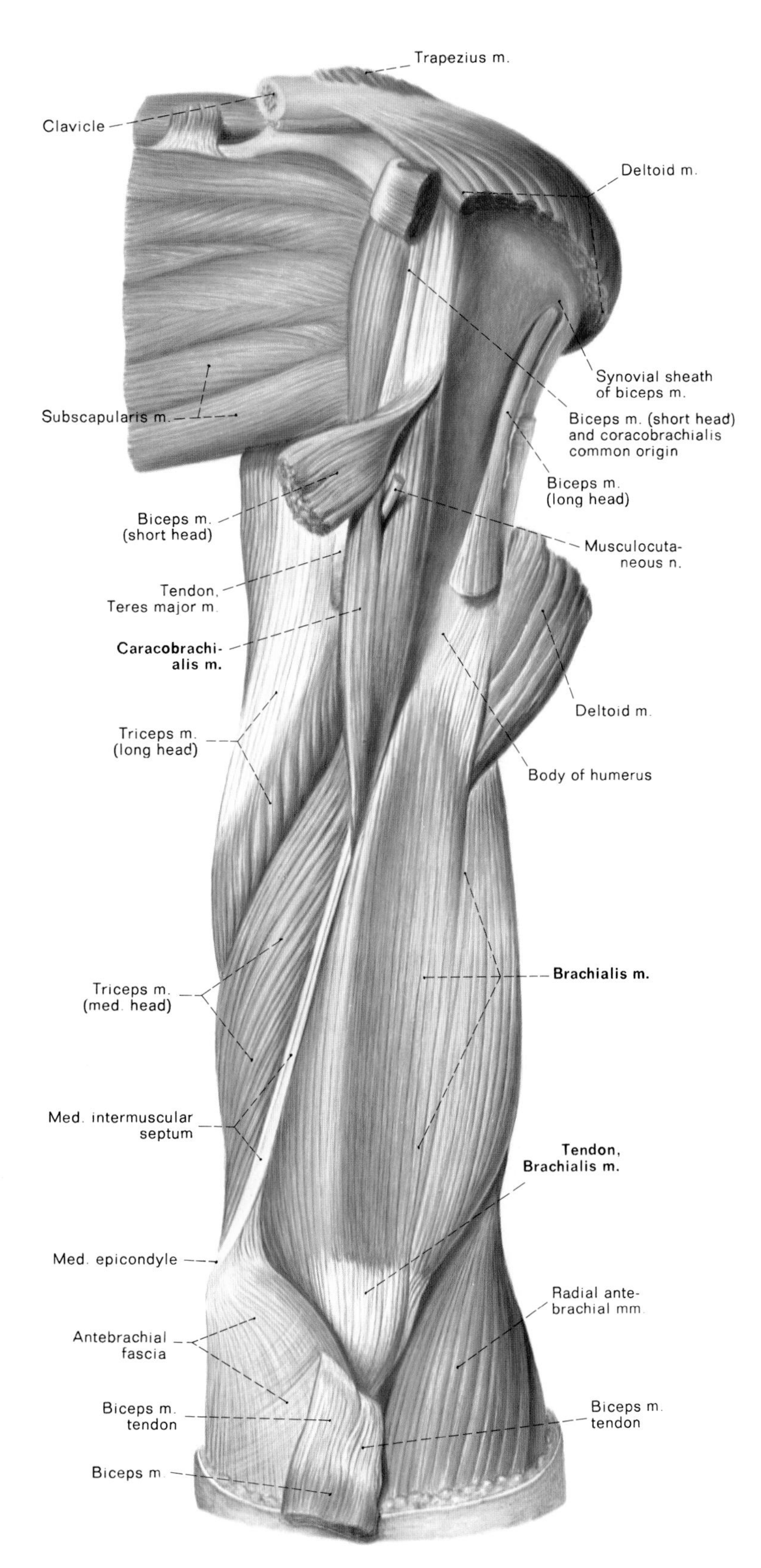

Fig. 9–13: Anterior view of the left arm musculature following reflection of the muscular bellies of the long and short heads of biceps brachii and removal of the midsection of deltoid.

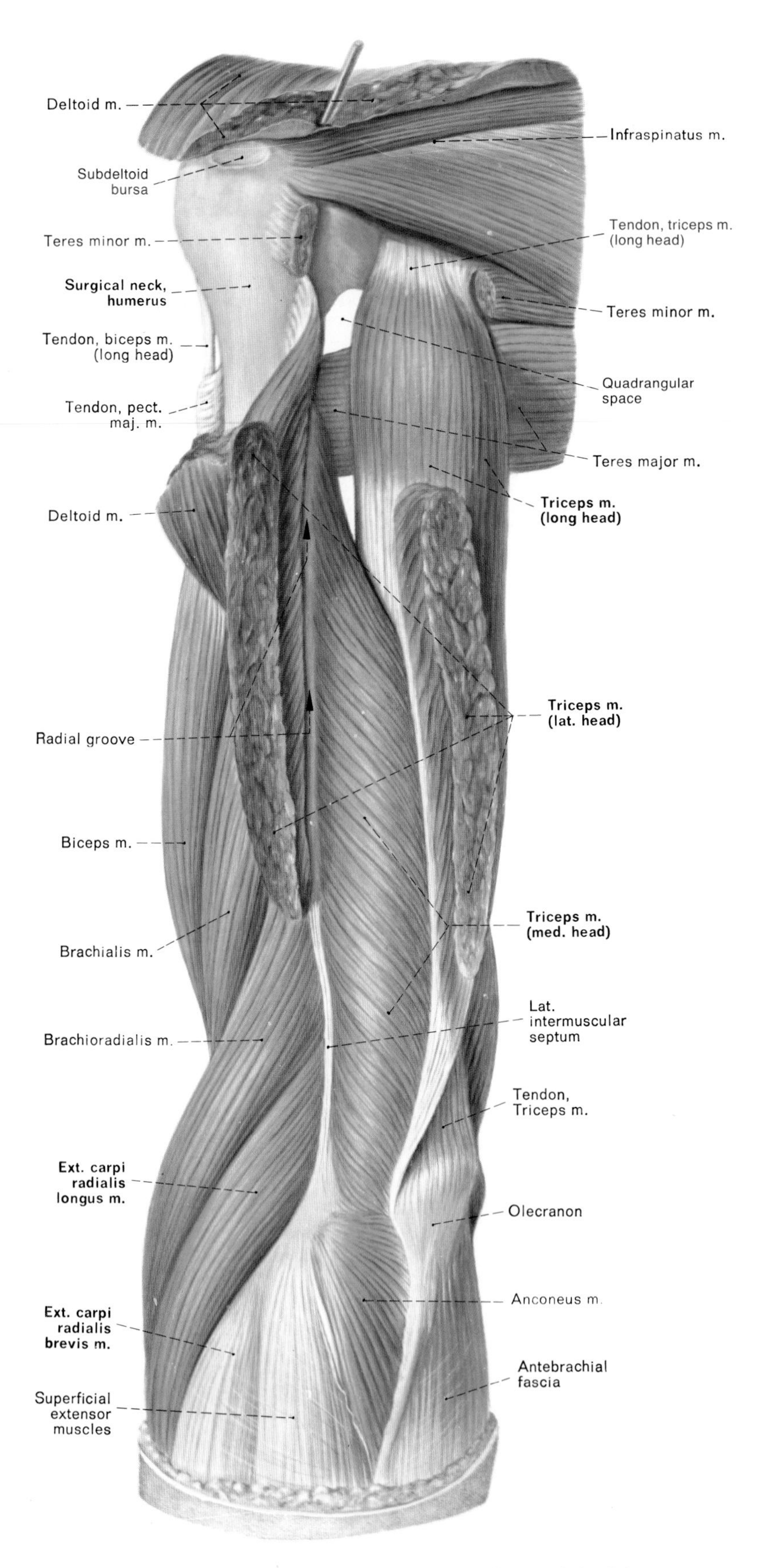

Fig. 9–14: Posterior view of the deep musculature of the left shoulder and arm.

beat marks the systolic pressure. As the cuff pressure is slowly lowered further, the tapping sounds become at first louder (because more blood is spurting through the artery during each heartbeat) and then softer (because the blood flow progressively assumes a more streamlined and silent pattern). The pressure recorded at the resumption of silent streamline flow marks the diastolic pressure. It should be noted, however, that some physicians regard the juncture where the tapping sounds become softer as a more appropriate measure of diastolic pressure. The pressure recorded at the juncture where the tapping sounds become softer is 5 to 10 mm Hg greater than that recorded when silent streamline flow resumes.

THE MAJOR VEINS OF THE ARM

The major veins of the arm are the venae comitantes of the brachial artery, the basilic vein, and the cephalic vein. As the basilic and cephalic veins ascend from the forearm into the arm, they extend through the superficial fascia of the cubital fossa (Fig. 8–2). The two veins are generally connected in the cubital fossa by a short communicating vein called the **median cubital vein.** The superficial location of the median cubital vein and its ready access on the body make it the most preferred vein for drawing samples of venous blood.

THE SUPRATROCHLEAR LYMPH NODES

The supratrochlear lymph nodes can be palpated along the medial side of the basilic vein immediately superior to the medial epicondyle of the humerus. These nodes drain superficial tissues of the medial aspect of the hand and forearm.

RELATIONSHIPS THAT RENDER MAJOR NERVES OF THE ARM SUSCEPTIBLE TO DAMAGE FROM MUSCULOSKELETAL INJURIES

Axillary Nerve: In the upper part of the arm, the axillary nerve lies inferior to the capsule of the shoulder joint and medial to the surgical neck of the humerus as it passes posteriorly through the quadrangular space (Fig. 9–15). These close spatial relationships with the head and surgical neck of the humerus render the axillary nerve vulnerable to injury from shoulder dislocations in which the head of the humerus is displaced inferiorly from its articulation in the shoulder joint and from displaced fractures of the surgical neck.

Radial Nerve: The radial nerve lies along the spiral groove of the humerus in the midregion of the arm (Fig. 9–15). This close relationship renders the radial nerve subject to injury from displaced fractures of the midregion of the humeral shaft.

Ulnar Nerve: The ulnar nerve exits the arm by passing immediately posterior to the medial epicondyle of the arm. This close relationship makes the ulnar nerve susceptible to injury from displaced fractures of the medial epicondyle. This relationship is also the anatomic basis for the colloquial identification of the medial posterior elbow region as the "funny bone." When individuals say that they have hit their "funny bone," they have, in fact, accidentally pressed the ulnar nerve anteriorly against the posterior surface of the medial epicondyle.

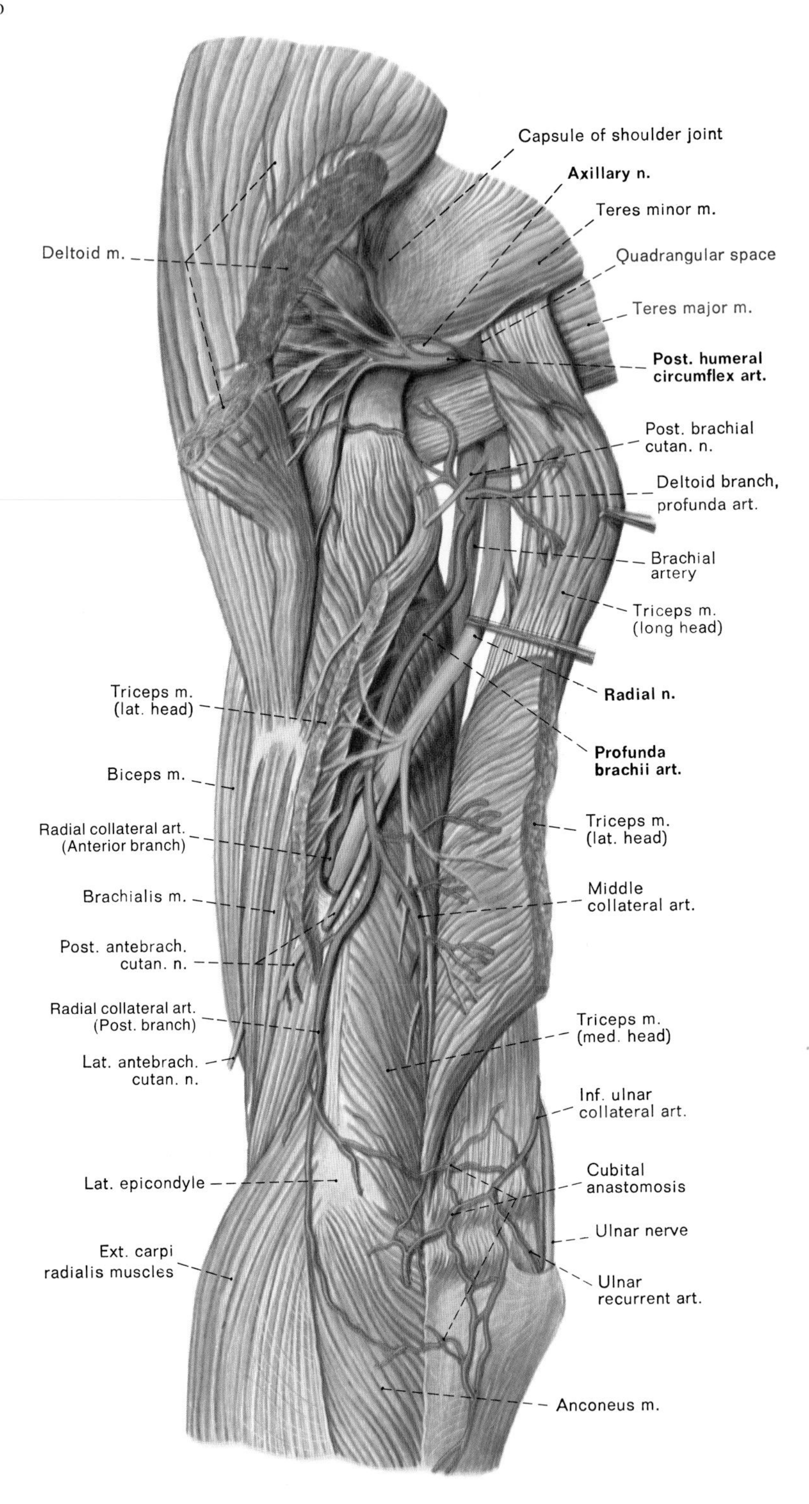

Fig. 9–15: The deep nerves and arteries of the posterior compartment of the left arm.

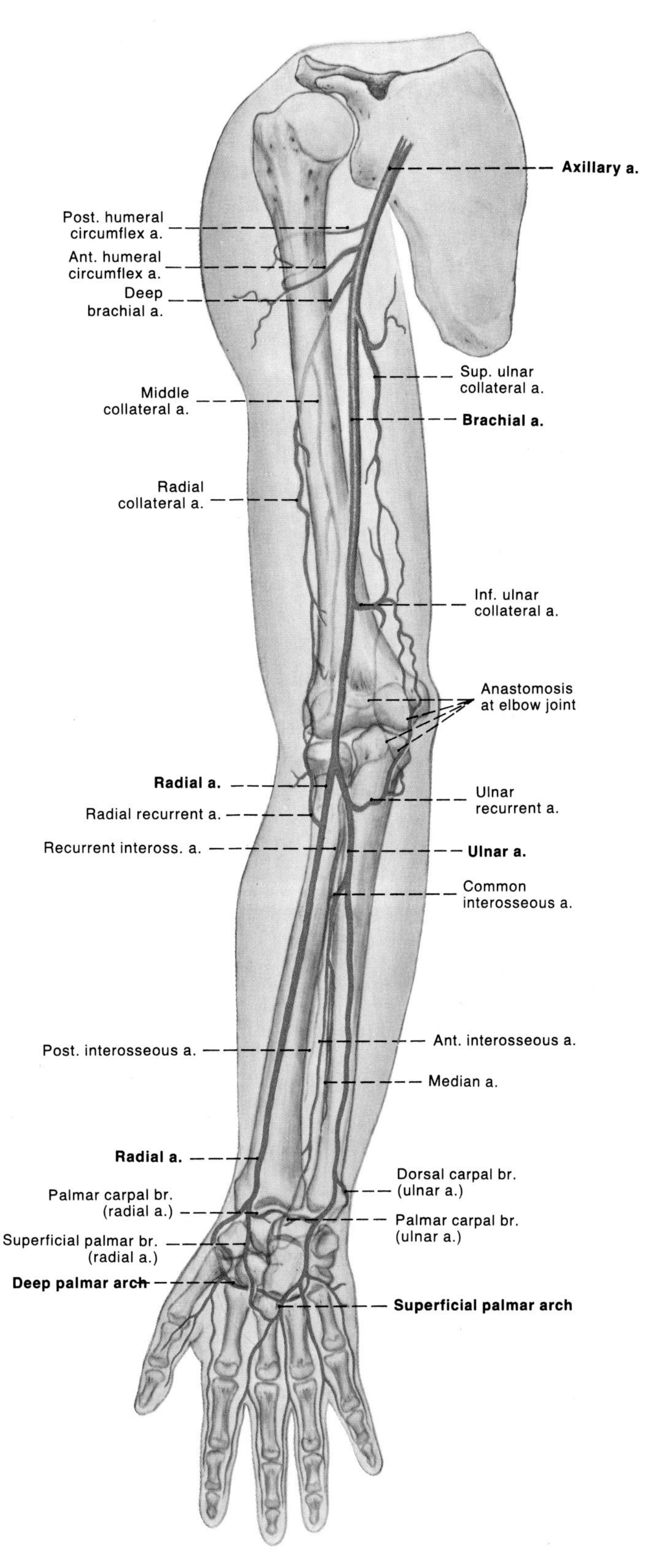

Fig. 9–16: Diagram illustrating the major arteries of the upper limb.

CASE **II.4**

The Case of Rico Alvarez

INITIAL PRESENTATION AND APPEARANCE OF THE PATIENT

Mr. and Mrs. Alvarez, their 7 year-old daughter Maria, and their 3 year-old son Rico walk into the emergency center. Rico is softly crying and holding his right elbow in the palm of his left hand; his right upper limb is held close to the body trunk with the forearm flexed 90° and in the midprone position. The examiner learns from the nurse that the boy's right elbow hurts.

QUESTIONS ASKED OF THE PATIENT AND HIS FAMILY

What happened to your son? [Mrs. Alvarez answers] I don't know. I would like for you to look at his elbow. He says it hurts him very badly.

How long has his elbow hurt him? [Mrs. Alvarez answers] For about the last 2 hours. *[This answer indicates an acute injury or acute inflammation of the elbow.]*

Do you know if Rico injured his elbow? [Mrs. Alvarez answers] No, I don't think so.

Do you know what Rico was doing just before he began to say his elbow hurt? [Mr. Alvarez answers] Yes. We were at the park having a picnic, and Rico and his sister Maria were running around and playing in some bushes nearby. Maria saw a snake and screamed out, and ran over to Rico to pull him away from the snake. I ran over to the kids, and Maria and I both grabbed Rico and ran for safety. With Maria and Rico by my wife, I went back to look for the snake. I found a 2 foot long dead coral snake. When I came back to tell my wife and the kids about the snake, that's when my wife told me that Rico's elbow hurt. *[Mr. Alvarez's account of the incident with the snake suggests that an injury to the patient's elbow occurred in relation to the snake incident. To substantiate this working hypothesis, the examiner now seeks (a) confirmation from the patient that his elbow began to hurt after his father and sister pulled him away from the snake, (b) further information from Mr. Alvarez about how he and his daughter grabbed the patient, and, finally, (c) localization by the patient of the site of pain.]*

Rico, did your elbow begin to hurt after your father and sister took you away from the snake? [Rico nods his head in the affirmative.]

Mr. Alvarez, do you recall how you and Maria grabbed Rico? Uh, let me think; it all happened so fast. I think I grabbed Rico's right hand and Maria grabbed his left hand. We actually lifted him off the ground and ran with him for a few steps until we were far from the area where the snake was. *[Mr. Alvarez's account of grabbing the patient's hand and lifting him off the ground is a textbook description of the history of a pulled elbow.]*

Rico, can you show me where your elbow hurts? [Rico places the fingers of his left hand over the lateral aspect of his right elbow.] *[The site of pain is consistent with a diagnosis of a pulled elbow.]*

Rico, do you hurt anywhere else? [Rico shakes his head in the negative.] *[The examiner finds the patient and his family alert and fully cooperative during the interview.]*

PHYSICAL EXAMINATION OF THE PATIENT

Vital signs: Blood pressure
Seated upright on mother's knee: 110/70 left arm and 110/70 right arm
Pulse: 100
Rhythm: regular
Temperature: 98.6°F
Respiratory rate: 32
Height: 100 cm.
Weight: 16 kg.
[A pediatrician would recognize that the patient's vital signs are within normal limits. The patient's systolic and diastolic blood pressures, weight, and height are in the 90th percentile for boys 3 years of age. The normal range of pulse rate for the patient is 80 to 120. The normal range of respiratory rate for the patient is 20 to 40. The slight trend toward the upper limits of normal for

the pulse and respiratory rate may be secondary to apprehension and/or pain.]

HEENT Examination: Normal
Lungs: Normal
Cardiovascular Examination: Normal
Abdomen: Normal
Genitourinary Examination: Normal
Musculoskeletal Examination: Inspection, movement and palpation of the upper and lower limbs are normal except for the following findings for the right upper limb: The patient will not actively flex, extend, supinate, or pronate the forearm. Passive extension of the forearm is limited to 15° flexion, and passive flexion of the forearm is limited to 120° flexion. The patient strongly resists passive supination of the forearm from the midprone position. Palpation reveals tenderness in the region about the head of the radius.
Neurologic Examination: Normal

INITIAL ASSESSMENT OF THE PATIENT'S CONDITION

The patient appears to have sustained an acute injury of the right elbow.

ANATOMIC BASIS OF THE PATIENT'S HISTORY AND PHYSICAL EXAMINATION

(1) The patient has sustained an acute injury of musculoskeletal tissues in the right elbow. The patient's father describes an incident that suggests he unintentionally injured the patient in the act of pulling him away from a threatening situation. The patient acknowledges that his right elbow became painful after he was forcefully grabbed by his father and sister and pulled away from the snake.

(2) The tenderness in the region about the head of the radius marks the region as a site of injury. Tenderness is pain elicited by palpation. Tenderness generally marks a region as a site of injury or disease. The tissues about the head of the radius that can be pressed upon by palpation include (proceeding from the most superficial to the deepest) the skin, the superficial and deep fascia, three muscles of the forearm (brachioradialis, extensor carpi radialis longus, and supinator), tissues of the lateral aspect of the elbow joint, tissues of the proximal radioulnar joint, and the head of the radius. One or more of these tissues appear to be the source of the pain in the lateral aspect of the elbow.

(3) The elbow injury has restricted supination much more markedly than flexion and extension of the forearm. This difference suggests that the injured elbow tissue or tissues are tissues of the proximal radioulnar joint.

INTERMEDIATE EVALUATION OF THE PATIENT'S CONDITION

The history and physical examination suggest that the patient has sustained an acute injury of the proximal radioulnar joint in the right elbow.

CLINICAL REASONING PROCESS

In consideration of the patient's condition, a pediatrician would recognize five key findings from the patient's history and physical examination as collectively being almost pathognomonic of an injury called a pulled elbow: (1) The patient is between one and five years of age. (2) The patient presents with a partially flexed and pronated forearm supported closely to the trunk of the body. (3) There is a history of the application of sudden distal traction upon the forearm. (4) The patient resists passive supination with the injured forearm. (5) There is no evidence of bony tenderness or soft tissue swelling. This last finding makes a bone fracture very unlikely.

The mechanism by which a pulled elbow injury is sustained involves the application of a comparatively severe distal traction upon the radius with the forearm extended and pronated. The distal traction subluxes the head of the radius from its articulations in both the elbow and proximal radioulnar joints (Fig. 9–17A). Such subluxation may produce a transverse tear in the annular ligament's attachments to the neck of the radius (children five years of age or younger are particularly susceptible to such a tear because their annular ligament's attachments to the neck of the radius are rather thin; the attachments become thicker and more secure after the age of five). A pulled elbow results if the head of the radius then partially protrudes through the transverse tear of the annular ligament and impinges the displaced, proximal part of the ligament between the head of the radius and the capitulum of the humerus when the distal tractive force is released and the head of the radius is drawn back proximally toward the capitulum by the pull of the muscles acting across the elbow joint (Fig. 9–17B).

Reduction of a pulled elbow can be achieved in the office setting by applying posteriorly directed pressure on the head of the radius while slowly, but firmly, supinating and extending the forearm; these manipulations screw the radial head into the annular ligament. Before such reduction is attempted, however, lateral and AP (anteroposterior) radiographs of the elbow should be examined to determine that there are no fractures or dislocations.

RADIOGRAPHIC EVALUATION AND FINAL RESOLUTION OF THE PATIENT'S CONDITION

The ends of the long bones of the limbs in a young child largely consist of hyaline cartilage. These cartilaginous ends, which have a radiographic density of water, are not visualized in radiographs because they have the same radiographic density as that of the tissues (mainly muscles) that surround them. The only structures within the cartilaginous ends that are visualized in radiographs are growing epiphyses (secondary centers of ossification). In elbow radiographs of a 3 year-old boy Figs. 9–18, a pediatric radiologist would expect to find only one or two growing epiphyses: the growing epiphysis for the capitulum of the humerus and possibly the growing epiphysis for the head of the radius.

In pediatric cases in which the cartilaginous ends of the long bones of the limbs may be injured, it is important to examine the cartilaginous ends for epiphyseal plate fractures (epiphyseal plate fractures are classified as one of five types on the basis of the Salter-Harris classification scheme; Figure 9–19 illustrates the five types of Salter-Harris fractures). A Type I fracture, which extends through only the epiphyseal plate, is difficult to detect because the fracture, which has the radiographic density of water, has the same radiographic density as that of the epiphyseal plate and thus cannot be directly visualized in radiographs. Type I and Type II fractures have a more favorable prognosis than do Type III, Type IV, and Type V fractures. The latter group of fractures extend through the epiphyseal side of the epiphyseal plate and thus injure the proliferating chondrocytes of the epiphyseal plate. Injury to these cells increases the likelihood of retarded longitudinal growth at the end of the bone and of premature fusion of the epiphysis with the diaphysis.

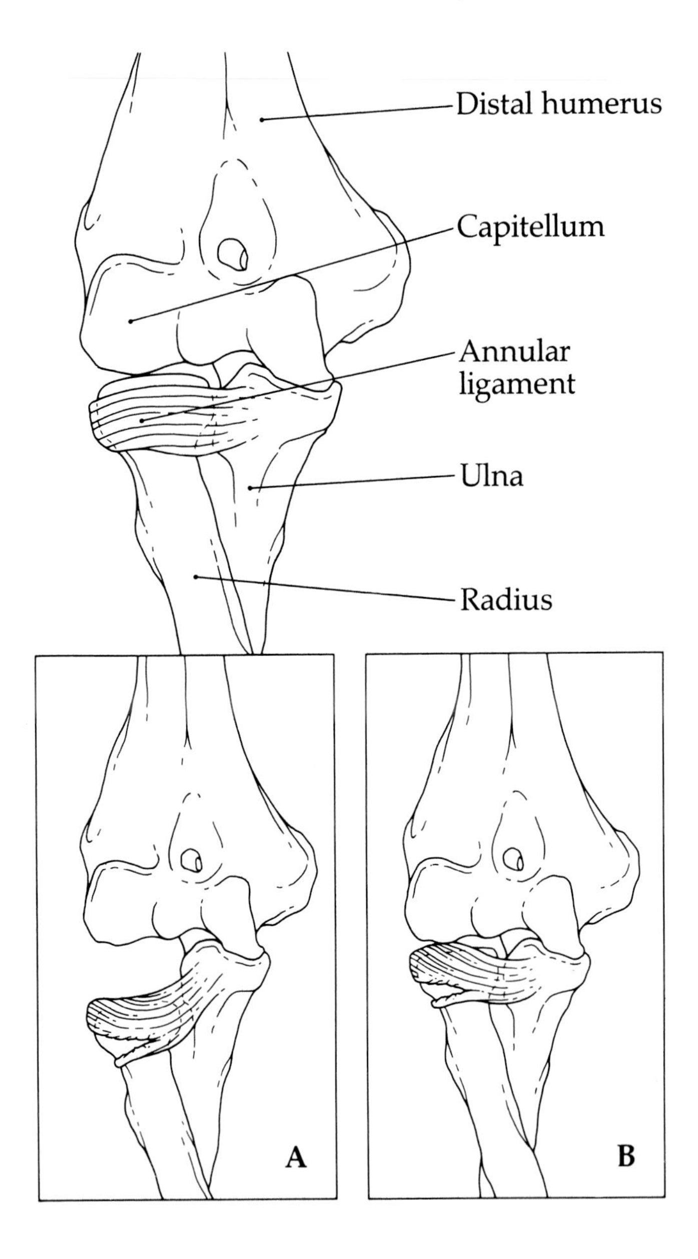

Fig. 9–17: Diagrams illustrating the mechanism of a pulled elbow. (A) Sudden distal traction on the outstretched hand and forearm subluxes the head of the radius and produces a transverse tear through the annular ligament. (B) Release of the distal traction impinges the displaced, proximal fragment of the annular ligament between the head of the radius and the capitulum.

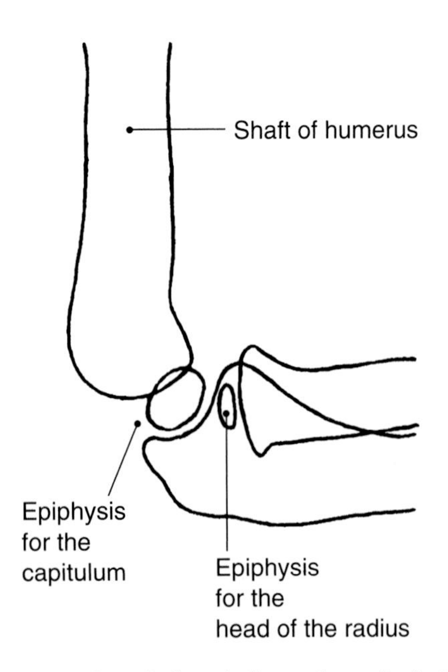

Fig. 9–18: An outline of a lateral elbow radiograph of a 3 year-old boy.

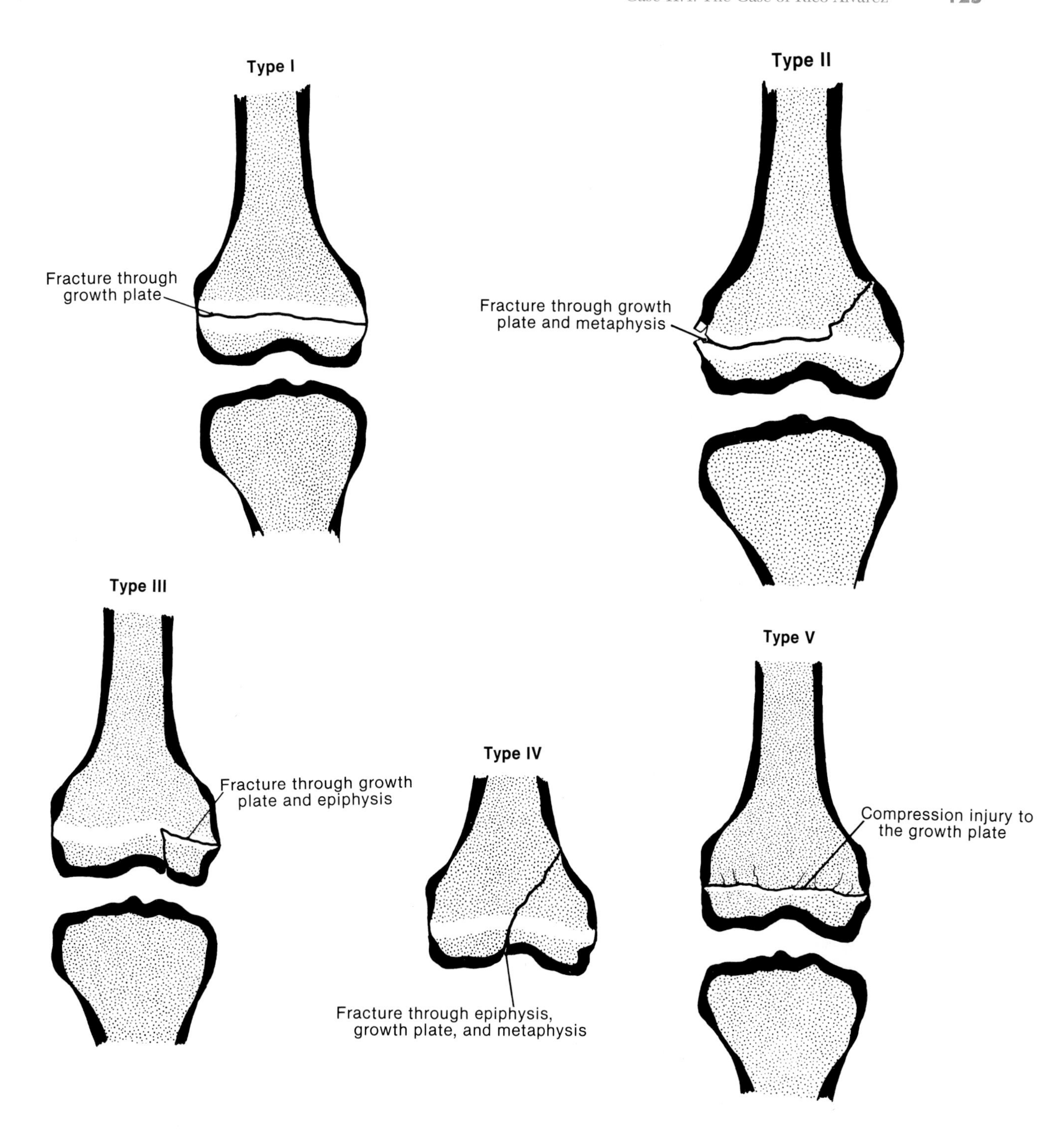

Fig. 9–19: Diagrammatic representations of a (A) Salter-Harris Type I fracture, (B) Salter-Harris Type II fracture, (C) Salter-Harris Type III fracture, (D) Salter-Harris Type IV fracture, and (E) Salter-Harris Type V fracture.

In a lateral elbow radiograph of a young child, the growing epiphysis for the capitulum exhibits three relationships that a pediatric radiologist uses to detect the presence of epiphyseal plate fractures at the distal end of the humerus (Fig. 9–20).

1. The center of the growing epiphysis is colinear with the central axis of the radial shaft. The line defining this colinear relationship is called the radiocapitellar line.
2. The anterior humeral line, which is the line coincident with the anterior margin of the humeral shaft, projects through the middle third of the growing epiphysis.
3. The midhumeral line, which is the line coincident with the central axis of the humeral shaft, projects just posterior to the posterior margin of the growing epiphysis. Any displacement of the growing epiphysis for the capitulum from its normal relationships with the radiocapitellar, anterior humeral, and midhumeral lines may signal a fracture through the adjoining epiphyseal plate.

In this case, lateral and modified AP radiographs of the patient's right elbow did not show any evidence of fractures or dislocations. The lateral elbow radiograph did not show any displacement of the head of the radius because with a pulled elbow it is a part of the annular ligament and not the head of the radius that is displaced at the time of presentation. Upon examination of the radiographs, the examiner reduced the pulled elbow as described previously, bringing almost immediate relief to the patient. The examiner explained to the patient's parents how the injury occurred so as to minimize the likelihood of a recurrent injury.

THE MECHANISM OF THE PATIENT'S INJURY

When Mr. Alvarez grabbed the patient's right hand in the attempt to pull the patient away from the snake, Mr. Alvarez's actions both extended and pronated the patient's right forearm before providing the force that lifted the patient off the ground. The application of this comparatively severe distal traction upon the radius with the forearm extended and pronated produced a pulled elbow as previously described. That part of the annular ligament trapped between the capitulum of the humerus and the head of the radius is the source of the pain in the lateral aspect of the elbow. The tenderness about the head of the radius marks the location of the transverse tear in the annular ligament.

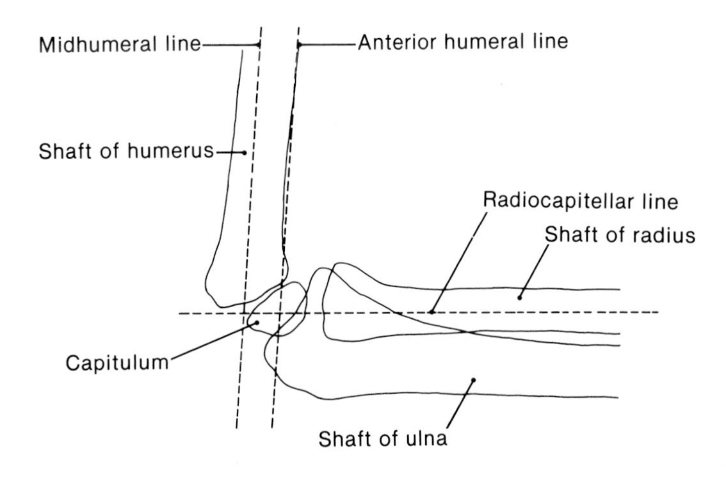

Fig. 9–20: An outline of a lateral radiograph of a child's elbow. Note the relationships of the capitulum to the radiocapitellar, anterior humeral, and midhumeral lines.

CASE **II.5**

The Case of Linda Chin

INITIAL PRESENTATION AND APPEARANCE OF THE PATIENT

Mr. and Mrs. Chin and their 7 year-old daughter Linda walk into the emergency center. Linda is holding her right elbow in the palm of her left hand; her right upper limb is held close to the body trunk with the forearm flexed 70° and in the midprone position. The examiner learns from the nurse that the girl fell and hurt her right elbow.

QUESTIONS ASKED OF THE PATIENT

What happened to your daughter? [Mrs. Chin answers.] She fell down on her elbow at the playground about a half hour ago. She was playing on one of these kiddie merry-go-rounds, you know, where the children grab a bar on the merry-go-round, then push it to get it started, and then jump on for the ride. Well, as she was jumping onto the merry-go-round, she lost her grip and fell down onto the ground, and immediately began crying that her elbow hurt. *[Mrs. Chin's reply indicates that her daughter has suffered an acute injury to her elbow. At this point, the examiner tentatively concludes that the patient has sustained one or more of the following injuries in or about the elbow: a fractured bone, a partially or completely dislocated joint, a torn ligament, or a torn muscle.]*

Linda, what do you remember about how you fell down on your elbow? [Linda answers] I just slipped and fell down. [Mr. Chin answers] I saw what happened, and she was straightening her arm out as she fell on it. *[Mr. Chin's reply suggests that the patient was attempting to stretch out the right upper limb to brace her body against the fall when she hit the ground. The examiner now has a rough idea as to the mechanism of the injury, which is that the elbow injury resulted from a hard fall on the patient's outstretched hand.]*

Where does it hurt? [The patient places her left hand over the medial aspect of her right elbow.] *[When pain is a complaint, it is always essential to have the patient mark by word or gesture the exact site(s) at which pain is perceived. In cases of physical injury, pain is generally a reliable indicator of the site(s) of injury.]*

Do you hurt anywhere else? No. *[This question addresses the examiner's concern that the patient has suffered upper limb injuries other than the major one at the elbow, since stressful forces were probably transmitted along the entire upper limb when the patient fell down.]*

Is there anything else that is also bothering you? [The patient shakes her head in the negative.]

Can you still feel things with your right hand? Yes. *[These last two questions address the examiner's concern that there might be neurovascular complications to the patient's injury or injuries. A complaint of a loss of feeling or an inability to move a part of the upper limb would suggest nerve injury; a complaint of coldness in the distal parts of the upper limb would suggest injury to a major artery. Even though the patient's answers suggest the absence of neurovascular complications at this time, the examiner recognizes that the patient may be too young, too frightened, or too much in pain to notice a symptom such as a slight loss of feeling in the hand.]*

[The examiner finds the patient and her parents alert and fully cooperative during the interview.]

PHYSICAL EXAMINATION OF THE PATIENT

Vital signs: Blood pressure
Lying supine: 110/70 left arm and 110/70 right arm
Standing: 110/70 left arm and 110/70 right arm
Pulse: 105
Rhythm: regular
Temperature: 98.8°F
Respiratory rate: 35
Height: 129 cm.
Weight: 30 kg.
[A pediatrician would recognize that the patient's vital signs are within normal limits. The patient's systolic and diastolic blood pressures, weight, and height are in

the 90th percentile for girls 7 years of age. The normal range of pulse rate for the patient is 70 to 115. The normal range of respiratory rate for the patient is 20 to 40. The slight trend toward the upper limits of normal for the pulse and respiratory rate may be secondary to apprehension and/or pain.]

HEENT Examination: Normal
Lungs: Normal
Cardiovascular Examination: Normal
Abdomen: Normal
Genitourinary Examination: Normal
Musculoskeletal Examination: Inspection, movement and palpation of the upper and lower limbs are normal except for the following findings for the right upper limb: Inspection reveals swelling of the soft tissues overlying the medial epicondyle of the humerus, those of the cubital fossa, and those that surround the olecranon process in the posterior aspect of the elbow. The patient has abrasions on the medial aspect of the palm. The patient will not actively flex, extend, supinate, or pronate the forearm. All passive movements of the forearm are limited by pain. Palpation reveals tenderness in the region about the medial epicondyle.
Neurologic Examination: Normal

INITIAL EVALUATION OF THE PATIENT'S CONDITION

The patient has sustained an acute musculoskeletal injury of the right elbow.

ANATOMIC BASIS OF THE PATIENT'S HISTORY AND PHYSICAL EXAMINATION

(1) The patient has sustained an acute injury of musculoskeletal tissues in the right elbow. The history suggests that the patient fell awkwardly on her outstretched right hand and that the resulting transmission of stressful forces through the right elbow produced a moderately severe injury in the elbow region.

(2) The tenderness in the region about the medial epicondyle marks the region as a site of injury. Tenderness is pain elicited by palpation. The tissues about the medial epicondyle that can be pressed upon by palpation include (proceeding from the most superficial to the deepest) the skin, the superficial and deep fascia, the common tendon of origin of five muscles of the anterior forearm (pronator teres, flexor carpi radialis, palmaris longus, flexor carpi ulnaris, and flexor digitorum superficialis), tissues of the medial aspect of the elbow joint, and the medial epicondyle. One or more of these tissues appear to be the source of the pain in the medial aspect of the elbow.

(3) The swellings anterior and posterior to the elbow joint suggest effusion of the elbow joint and hematoma and/or edema of the soft tissues surrounding the joint. Joint effusion following acute injury to a synovial joint generally indicates hemorrhagic effusion from a moderate-to-severe tearing of the joint's capsule and its associated ligaments, a dislocation (with its attendant capsular and ligamentous tears) and/or an intracapsular fracture (a bone fracture within the fibrous capsule of the joint).

CLINICAL REASONING PROCESS

It is obvious from the history and the patient's initial appearance that the patient has sustained a moderate-to-severe injury of the right elbow. In considering this case upon the completion of the history, a pediatrician or orthopedist would address three major concerns during the physical examination:

1. First, gather as much physical evidence as possible about the musculoskeletal injuries at the elbow.
2. Second, thoroughly examine the remainder of the injured upper limb for physical evidence of other musculoskeletal injuries. In this case, the history suggests that very stressful forces were transmitted along the entirety of the patient's right upper limb, and, therefore, that there may also be injuries proximal and/or distal to the prominent elbow injury.
3. Third, carefully examine the right forearm, wrist, and hand for any signs of neurovascular complications. This is because moderate-to-severe elbow injuries always place the neurovascular supply to the forearm, wrist, and hand at risk. The appropriate physical examinations of neurovascular supply in this case are as follows:
    a. The strength and regularity of the radial and ulnar arterial pulsations should be assessed at both wrists. These pulsations serve as a monitor of the adequacy of blood supply to the wrist and hand.
    b. The speed of capillary refill in the index and little fingers should be examined in both hands as another monitor of the adequacy of blood supply to the hand. The blood in the capillary bed of the soft tissues at the end of the thumb or a finger contributes to the color of the soft tissues as seen through the nail. The color of a patient's nailbeds can be monitored during a simple test to roughly measure the effectiveness by which these soft tis-

sues are being perfused with blood. The test begins by the examiner gently but firmly squeezing (for 2 to 3 seconds) the nailbed of one of the patient's digits (the pressure squeezes most of the blood out of the capillaries in the soft tissues of the nailbed). The examiner then quickly releases the pressure on the nailbed and observes how long it takes for the "ghostly" bed to reassume its normal, pinkish color. It should require no more than 1 to 2 seconds for the capillaries in the soft tissues to become filled with blood (this is because an average capillary has a length of 1 mm, and the average blood velocity in a systemic capillary is 1 mm/sec).

c. Light touch sensation, vibration sensation, and two-point discrimination on the anterior and posterior surfaces of the right forearm and hand should be compared with those sensations on the left forearm and hand. These sensations are sensitive tests of sensory supply to the skin. Light touch sensation may be assessed with a fine wisp of cotton lightly applied to the skin. A relatively low-pitch (128 Hz) tuning fork is commonly used to assess vibration sensation. Two-point discrimination, which is the ability to detect the application of probes at two separate but close points, may be assessed with the ends of an opened paper clip.

d. The strength of numerous hand and digit movements should be assessed in both hands. These movements assess the nerve supply of forearm and hand muscles.

The performance of these examinations did not yield any evidence of neurovascular complications in this case. The examiner also mentally noted that the patient did not give any indication of intense forearm muscle pain, a sign which, in this case, would indicate muscle ischemia. The absence of forearm muscle pain is important because the sequelae of a moderate-to-severe elbow injury (such as the development of a hematoma and/or edema) can significantly compromise blood supply to forearm muscles (sometimes even in the presence of strong radial and ulnar arterial pulses and adequate capillary refill in the fingers). Untreated ischemia of the anterior forearm muscles can lead to **Volkmann's ischemic contracture,** a condition in which the replacement of necrotic muscle tissue by fibrous scar tissue permanently shortens the affected muscles and draws the fingers into marked flexion (Fig. 9–21).

Following completion of the physical examination, a prudent course of action in this case would be to order radiographs of the patient's right upper limb extending from the shoulder to the hand. This is because radiographs of the regions proximal and distal to the elbow may reveal musculoskeletal injuries that were not detected by the physical examination.

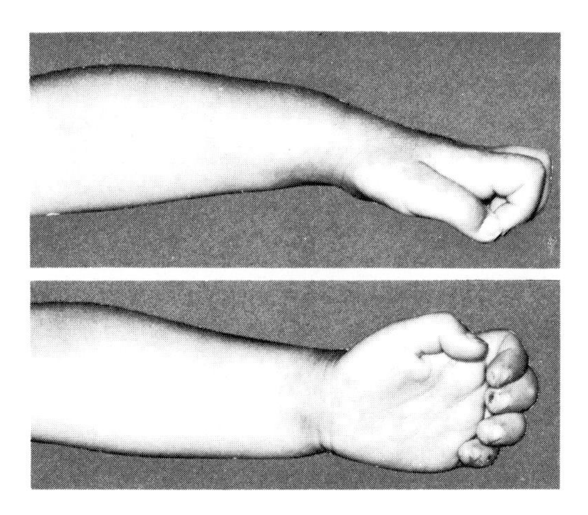

Fig. 9–21: Lateral and anterior views of the deformities of a 6 year-old child's hand and forearm as a result of Volkmann's ischemic contracture of forearm muscles.

RADIOGRAPHIC EVALUATION AND FINAL RESOLUTION OF THE PATIENT'S CONDITION

Lateral and modified AP radiographs of the patient's right elbow revealed a simple, undisplaced, transverse fracture of the medial epicondyle (Figs. 9–22A and 9–22B). The radiographs showed the three growing epiphyses one would expect to find in the elbow of a 7 year-old girl: the growing epiphyses for the capitulum of the humerus, the medial epicondyle of the humerus, and the head of the radius. There was no evidence of displacement of the growing epiphysis for the capitulum from its relationships with the radiocapitellar, anterior humeral, and midhumeral lines.

The lateral elbow radiograph showed both the **sail sign** and the **positive posterior fat pad sign** (Fig. 9–22A). Any significant effusion of the synovial cavity of the elbow joint anteriorly displaces the anterior fat pad of the elbow and posteriorly displaces the posterior fat pad. [When a normal elbow is positioned for a lateral radiograph, the superficial part of the anterior fat pad lies anterior to the bony rim of the coronoid fossa. Consequently, the superficial part of the anterior fat pad appears as a small area of fat radiodensity sandwiched between the bony rim of the coronoid fossa and the water radiodensity shadows cast by brachialis and biceps brachii (Fig. 9–23). By contrast, the posterior fat pad lies deep to the bony rim of the olecranon fossa, and, thus, only the water radiodensity shadow of triceps brachii is seen posterior to the bony rim of the olecranon fossa.] The displacement of the anterior fat pad increases the size of the fat radiodensity area anterior to the coronoid fossa; this feature is called the sail sign because the fat radiodensity area is shaped like a triangular sail. The displacement of the posterior fat pad generates a fat radiodensity area posterior to the olecranon fossa; this feature is called the positive posterior fat pad sign.

Radiographs of the patient's right shoulder, wrist, and hand did not show any evidence of fractures or dislocations.

Since (a) the elbow radiographs showed a simple, undisplaced fracture at the distal end of the humerus proximal to the epiphyseal plate and (b) the physical examination did not reveal any evidence of neurovas-

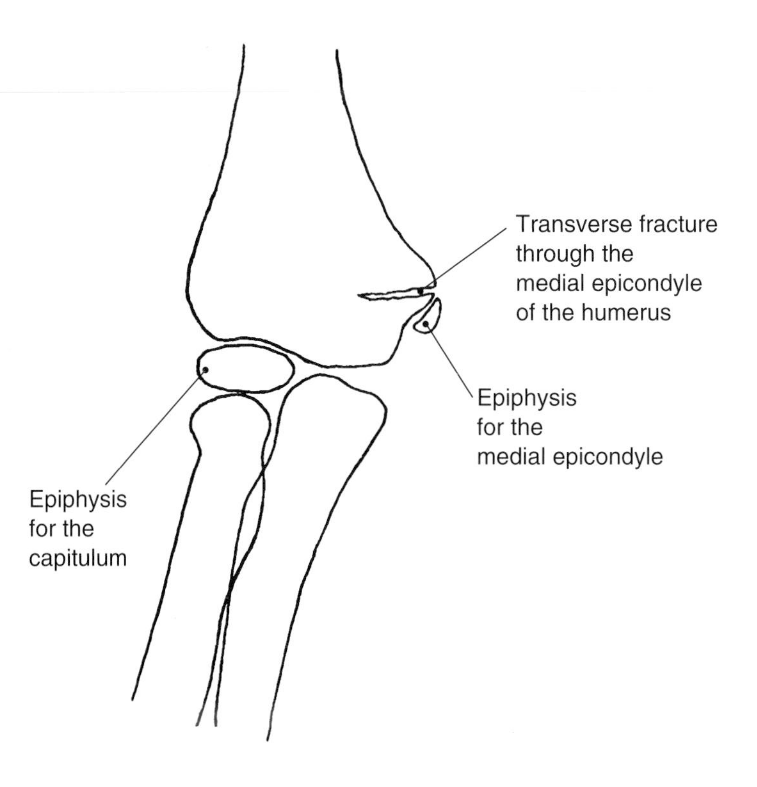

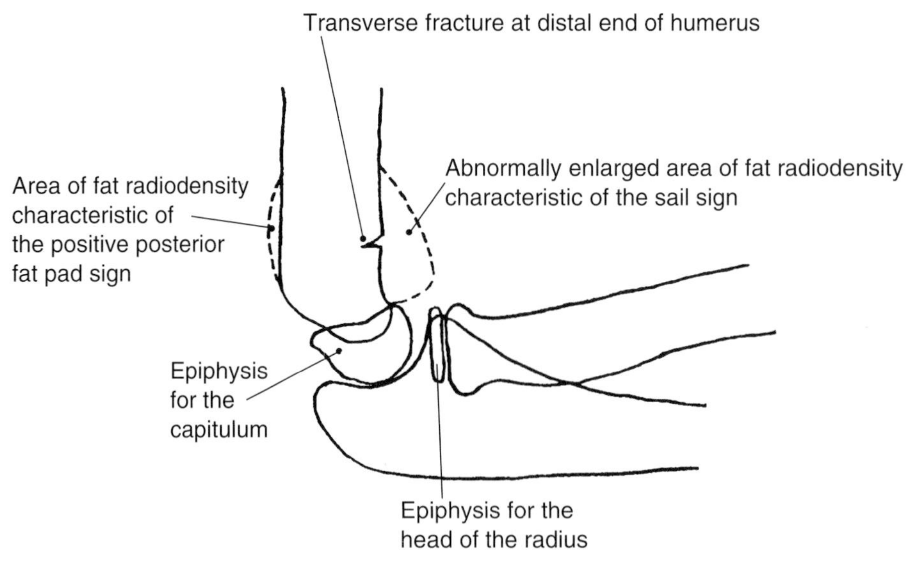

Fig. 9–22: Outlines of the (A) lateral and the (B) AP radiographs of Linda Chin's right elbow. Notice that the AP radiograph does not display the epiphysis for the head of the radius. This is because the relatively thin epiphyseal plate between the shaft of the radius and the epiphysis for the head of the radius was oriented slightly obliquely to the path of the X-ray beam through the patient's elbow.

cular complications distal to the elbow, an orthopedist would plaster cast the upper limb about the elbow for approximately 3 weeks. An orthopedist would also advise the patient and her parents of five signs (pulselessness, pallor, pain, paresthesia, and/or paralysis in the forearm and/or hand) that herald neurovascular complications and thus warrant prompt reevaluation.

THE MECHANISM OF THE PATIENT'S INJURY

The patient suffered an excessive angular force to the medial aspect of the distal end of the humerus when falling down hard onto her outstretched right hand. The right forearm was fully extended and supinated and the right hand fully extended at the moment of impact. The impact of the medial aspect of the palm of the right hand with the ground placed an excessive valgus (laterally directed) stress at the elbow that transversely fractured the medial epicondyle (proximal to the epiphyseal plate at the distal end of the humerus). The intracapsular part of the epicondylar fracture produced a hemorrhagic effusion of the elbow joint. The forearm is held partially flexed at the elbow because flexion minimizes the pressure of the elbow joint effusion and thus the painful tension in the synovial membrane lining of the joint. Due to the limited extent and undisplaced nature of the fracture, the resulting edema of soft tissues in the elbow region was not substantial enough to compromise the neurovascular supply of the more distal parts of the upper limb.

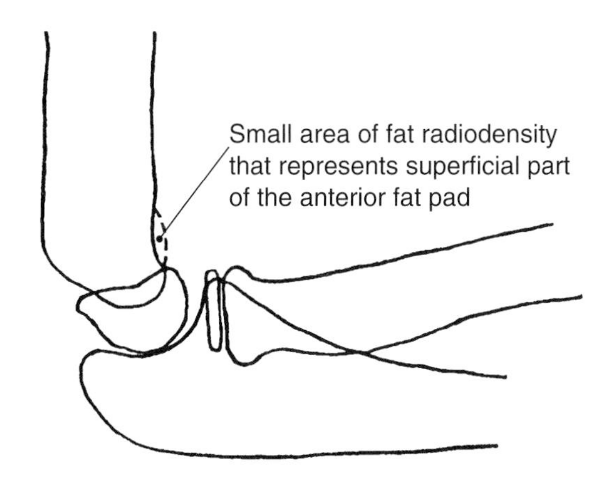

Fig. 9–23: Outline of a normal lateral radiograph of a child's elbow.

RECOMMENDED REFERENCES FOR ADDITIONAL INFORMATION OF THE ELBOW

Corrigan, B., and G. D. Maitland, Practical Orthopaedic Medicine, Butterworth & Co., London, 1983: *Chapter 7 offers a description of the clinical examination of the elbow and discussions of the signs an symptoms of common elbow lesions.*

Magee, D. J., Orthopaedic Physical Assessment, 2nd ed., W. B. Saunders Co., Philadelphia, 1992: *Chapter 5 provides a comprehensive discussion of the clinical examination of the elbow and proximal radioulnar joints.*

Slaby F., and E. R. Jacobs, Radiographic Anatomy, Harwal Publishing Co., Media, PA 1990: *Pages 9 through 14 in Chapter 1 outline the major anatomic features of the most common radiographic views of the elbow.*

Greenspan, A., Orthopedic Radiology, 2nd ed., Gower Medical Publishing New York, 1992: *Fig. 5.56 on page 5.31 summarizes the array of radiologic imaging techniques that may be used to evaluate elbow injuries.*

CHAPTER 10

The Forearm and Wrist

This chapter focuses on the innervation and actions of the forearm muscles and clinically important relationships of major blood vessels and nerves at the wrist. There are two major groups of muscles in the forearm: an anterior group and a posterior group. The anterior forearm muscles lie mostly anterior to the radius and ulna, are innervated by the median and/or ulnar nerve, and, as a group, can pronate the forearm and flex the hand and its digits. The posterior forearm muscles lie mostly posterior to the radius and ulna, are innervated by the radial nerve or one of its branches and, as a group, can supinate the forearm and extend the hand and its digits.

Because all except four of the forearm muscles insert onto bones or deep fascia in the hand, the discussion begins with an examination of the bones, joints, and deep fascial thickenings of the hand.

THE NAMES OF THE DIGITS

In the hand, there are five digits, and they consist of the four fingers plus the thumb (Fig. 10–1). Each digit has two equivalent names:

1st Digit—Thumb
2nd Digit—Index Finger
3rd Digit—Middle Finger
4th Digit—Ring Finger
5th Digit—Little Finger

Note, in particular, that the thumb is a digit but should never be called a finger.

THE CARPALS: THE BONES OF THE WRIST

The eight carpals of the wrist are commonly grouped into a proximal row of four bones and a distal row of four bones (Fig. 10–2). Proceeding from the most lateral to the most medial bone in each row, the bones in each row are arranged as follows:

Proximal Row: **Scaphoid—Lunate—Triquetrum—Pisiform**
Distal Row: **Trapezium—Trapezoid—Capitate—Hamate**

The order of appearance of the carpal ossification centers is quite variable. The ossification centers for the capitate and hamate are generally present at birth. The ossification center for the pisiform is commonly the last to appear.

THE METACARPALS: THE BONES OF THE PALM OF THE HAND

The five metacarpals of the palm are numbered in such a fashion that the first metacarpal is the most lateral metacarpal and the fifth metacarpal is the most medial metacarpal (Fig. 10–2). A metacarpal is associated with each digit:

1st Metacarpal: the metacarpal for the thumb
2nd Metacarpal: the metacarpal for the index finger
3rd Metacarpal: the metacarpal for the middle finger
4th Metacarpal: the metacarpal for the ring finger
5th Metacarpal: the metacarpal for the little finger

Each metacarpal has a **base,** a **shaft,** and a **head.** The base is the proximal end of the metacarpal, and the head is the distal end of the metacarpal. The **knuckles** represent the heads of the second to fifth metacarpals. The **distal palmar crease** (which palm readers refer to as the "love line") overlies the heads of the third to fifth metacarpals (Fig. 10–1).

Each metacarpal bears only a single epiphysis during childhood and adolescence. Whereas the epiphyses of the second to fifth metacarpals reside at the heads of the bones, the epiphysis of the first metacarpal resides at the base of the bone.

THE PHALANGES: THE BONES OF THE DIGITS

The thumb has two phalanges and each finger three phalanges (Fig. 10–2). The phalanges of the thumb are called the proximal phalanx and distal phalanx of the thumb. The phalanges of each finger are called the proximal phalanx, middle phalanx, and distal phalanx of the finger.

Each phalanx has a **base,** a **shaft,** and a **head.** The base is the proximal end of the phalanx, and the head is the distal end of the phalanx. During childhood and adolescence, each phalanx bears only a single epiphysis, residing at the base of the bone.

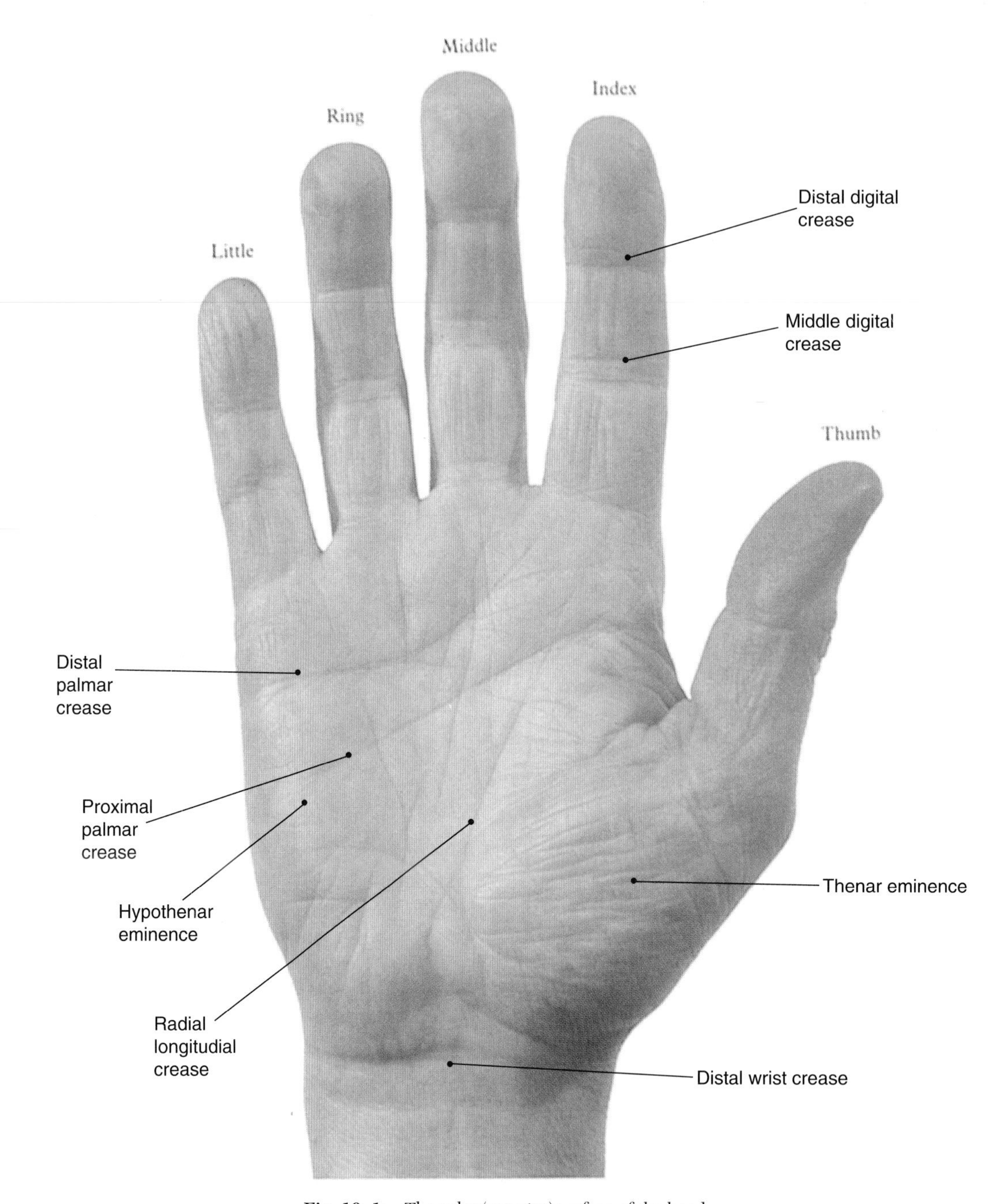

Fig. 10–1: The volar (anterior) surface of the hand.

THE JOINTS OF THE HAND ASSOCIATED WITH ACTIONS OF THE MUSCLES OF THE FOREARM

The **wrist joint** articulates the scaphoid, lunate, and triquetrum with the distal end of the radius and the articular disc distal to the head of the ulna (Fig. 9–8). Note that the ulna does not contribute an articular surface to the wrist joint. The wrist joint provides for **flexion, extension, abduction, and adduction** of the hand (Fig. 10–5). The **distal wrist crease** overlies the wrist joint (Fig. 10–1).

The **midcarpal joint** articulates the distal carpal row with the scaphoid, lunate, and triquetrum (Fig. 9–8). Note that the pisiform does not contribute an articular surface to the midcarpal joint. The midcarpal joint provides for **flexion, extension, abduction, and adduction** of the hand (Fig. 10–5).

The **metacarpophalangeal (MP) joints** each join the head of a metacarpal with the base of its corresponding proximal phalanx. Each MP joint provides for **flexion, extension, abduction, and adduction** of the proximal phalanx of the digit (Figs. 10–6, 10–7 and 10–8).

The **interphalangeal (IP) joints** are the joints between adjacent phalanges in the digits. Whereas the thumb has only one IP joint, each finger has a proximal IP joint and a distal IP joint. The IP joints provide for **flexion and extension** of the more distal phalanx (Fig. 10–9). The **middle and distal digital creases** on the palmar surface of each finger respectively overlie the proximal and distal IP joints (Fig. 10–1).

Normally, when the fingers are flexed at their MP and proximal IP joints so as to form a fist, the index, middle, ring, and little fingers each point to roughly the same spot on the wrist (this spot overlies the proximal pole of the scaphoid). Rotational malalignment of a phalangeal fracture in any of these digits invariably makes that finger point toward a different spot.

THE DEEP FASCIAL THICKENINGS AND SYNOVIAL SHEATHS ASSOCIATED WITH THE TENDONS OF INSERTION OF THE FOREARM MUSCLES

The palmar and dorsal aspects of the wrist and the palmar aspects of the digits are regions in the hand at which thickened bands of deep fascia are attached along their sides to underlying bones (Figs. 10–10 and 10–11). Together the bones and fascial bands form tunnel-shaped spaces through which pass the tendons of certain muscles in route to their insertion sites. These osseofascial (part bony and part fascial) tunnels keep the tendons in constant, close association with the underlying bones during hand movements. The tunnels in the palmar aspect of the wrist transmit tendons of anterior forearm muscles, and the tunnels in the dorsal aspect of the wrist transmit tendons of posterior forearm muscles. The tunnels in the palmar aspects of the digits transmit the tendons of anterior forearm muscles that can flex the digits.

The **flexor retinaculum** forms the fascial roof of two osseofascial tunnels in the palmar aspect of the wrist (Figs. 10–10 and 10–12). The flexor retinaculum is attached medially to the hamate and pisiform and laterally to the scaphoid and trapezium. The **distal wrist crease** overlies the proximal border of the flexor retinaculum (Fig. 10–1).

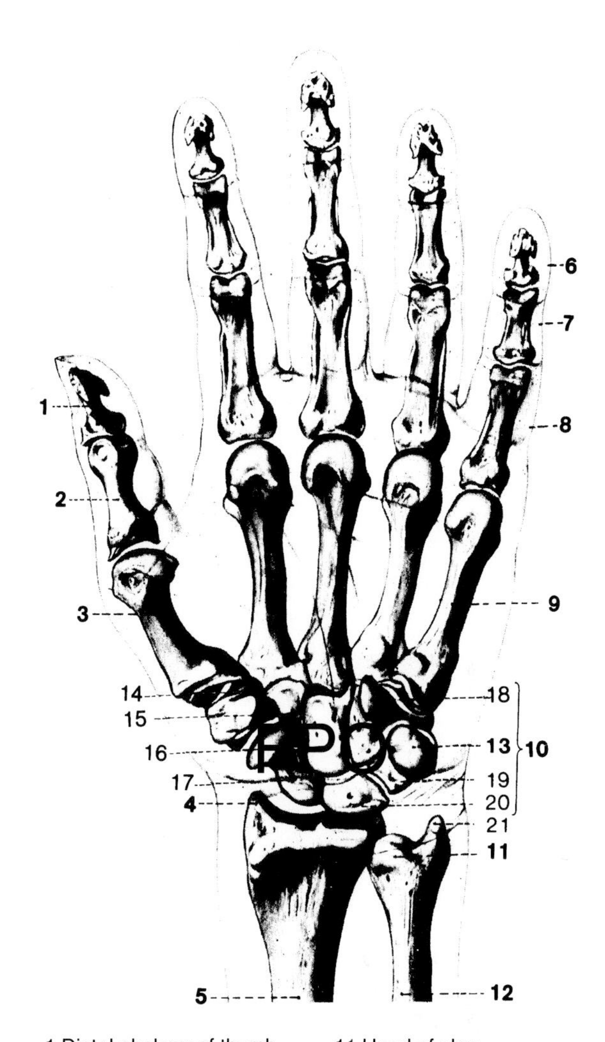

1 Distal phalanx of thumb
2 Proximal phalanx of thumb
3 First metacarpal
4 Styloid process of radius
5 Radius
6 Distal phalanx of little finger
7 Middle phalanx of little finger
8 Proximal phalanx of littlefinger
9 Fifth metacarpal
10 Carpal bones
11 Head of ulna
12 Ulna
13 Pisiform
14 Trapezium
15 Trapezoid
16 Capitate
17 Scaphoid
18 Hook of hamate
19 Triquetrum
20 Lunate
21 Styloid process of ulna

Fig. 10–2: Anterior view of the bones of the hand.

CLINICAL PANEL II-7

Anterior Dislocation of the Lunate

Common Mechanism of the Injury: The lunate is the most commonly dislocated carpal. The dislocation frequently occurs as a consequence of an accident in which a person stumbles and falls hard on an outstretched hand.

Common Symptom: The patient commonly presents with pain in the wrist. An anteriorly dislocated lunate may impinge upon and compress the median nerve in the carpal tunnel (the course of the median nerve through the carpal tunnel is discussed later in this chapter). Such compression may affect the median nerve's sensory and motor supply to the hand (the median nerve's sensory and motor supply to the hand are discussed in Chapter 11).

Radiographic Evaluation of the Injury: When the lunate is dislocated anteriorly, it undergoes an anterior rotation on its convex, proximal surface. This rotation transforms its PA outline from a quadrangular to a triangular shadow (Fig. 10–3). Anterior dislocation of the lunate also displaces the center of the bone from the line in the lateral wrist radiograph that extends from the central axes of the third metacarpal and capitate to the central axis of the radius (Fig. 10–4).

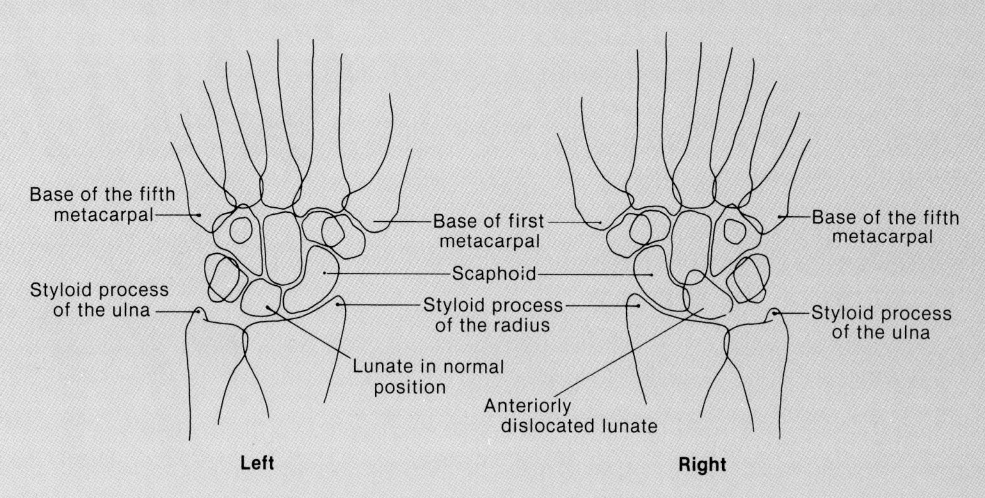

Fig. 10–3: The outlines of the PA radiographs of the normal (left) and injured (right) wrists of a person who has suffered an anteriorly dislocated lunate in the right wrist.

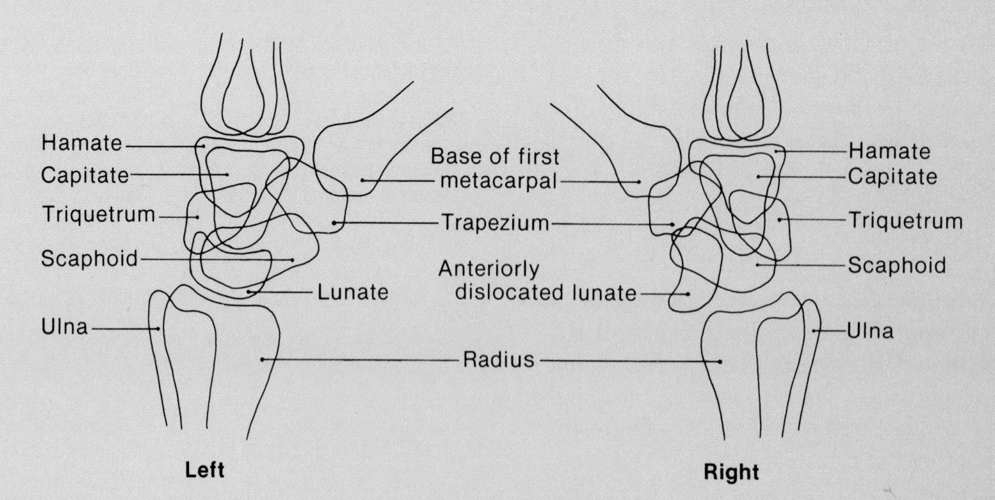

Fig. 10–4: The outlines of the lateral radiographs of the normal (left) and injured (right) wrists of a person who has suffered an anteriorly dislocated lunate in the right wrist. The outlines of the trapezoid and pisiform bones are not shown, for the purpose of clarity.

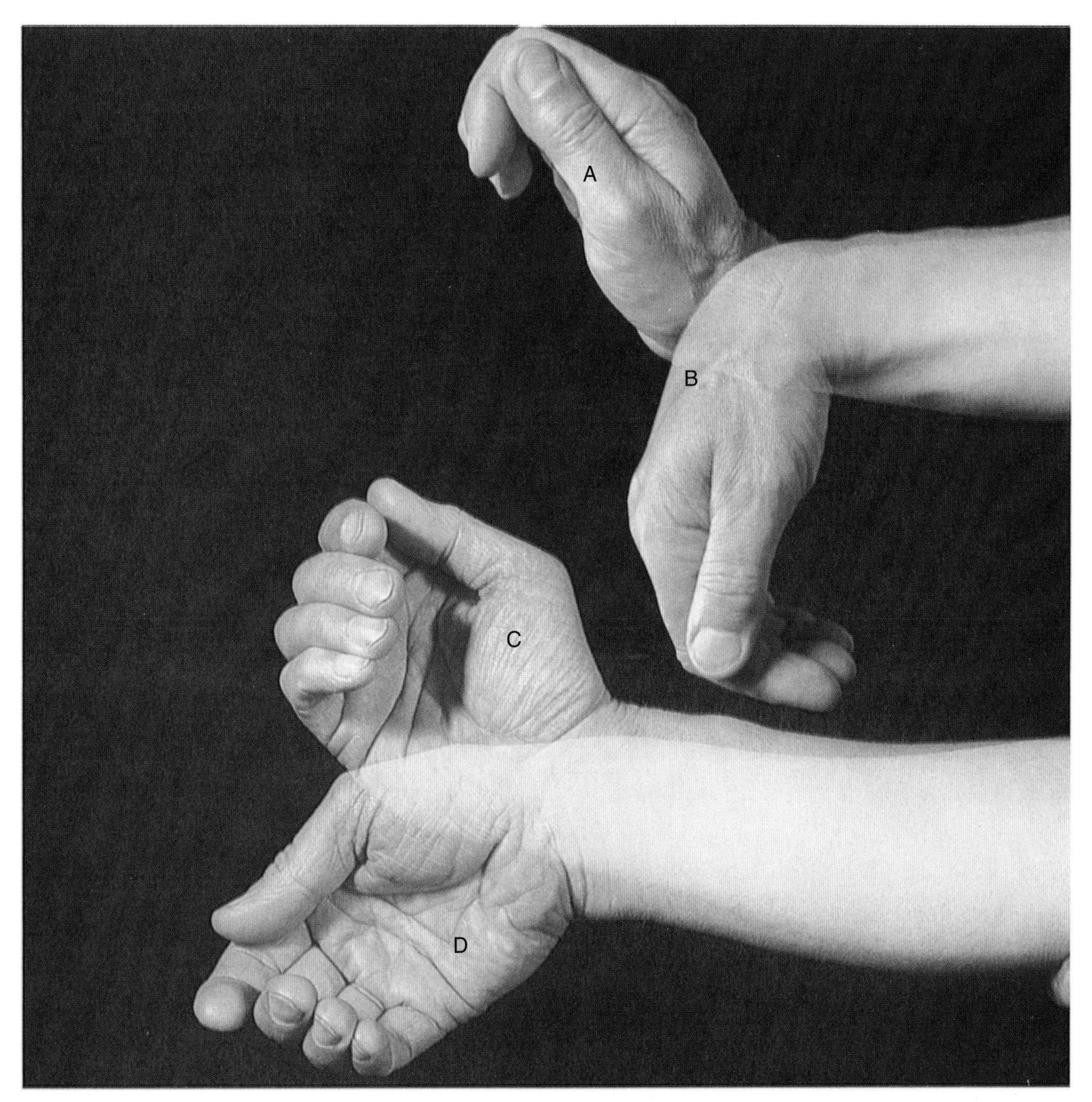

Fig. 10–5: The hand in (A) full extension, (B) full flexion, (C) full abduction, and (D) full adduction at the wrist and midcarpal joints.

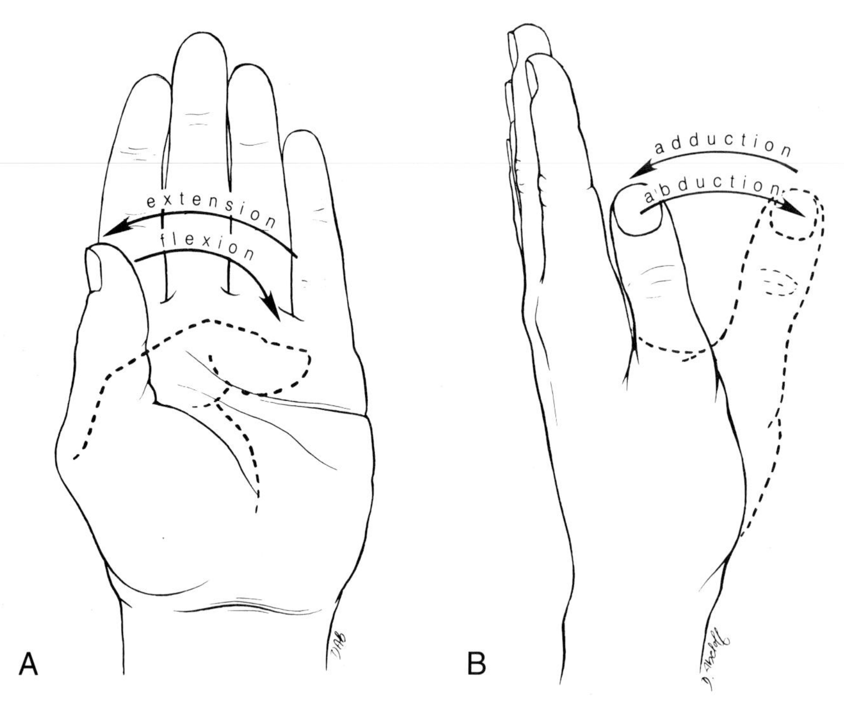

Fig. 10–6: Diagrams illustrating (A) flexion and extension of the thumb at its MP and IP joints and (B) abduction and adduction of the thumb.

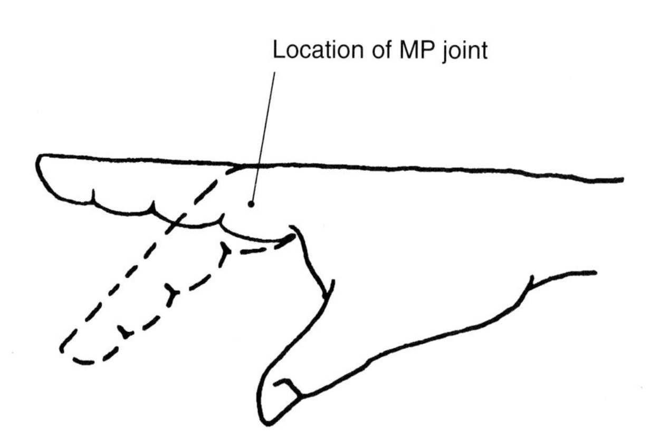

Fig. 10–7: Diagram illustrating flexion of the index finger at its MP joint. The parts of the index finger illustrated with a broken outline represent the index finger partially flexed at its MP joint.

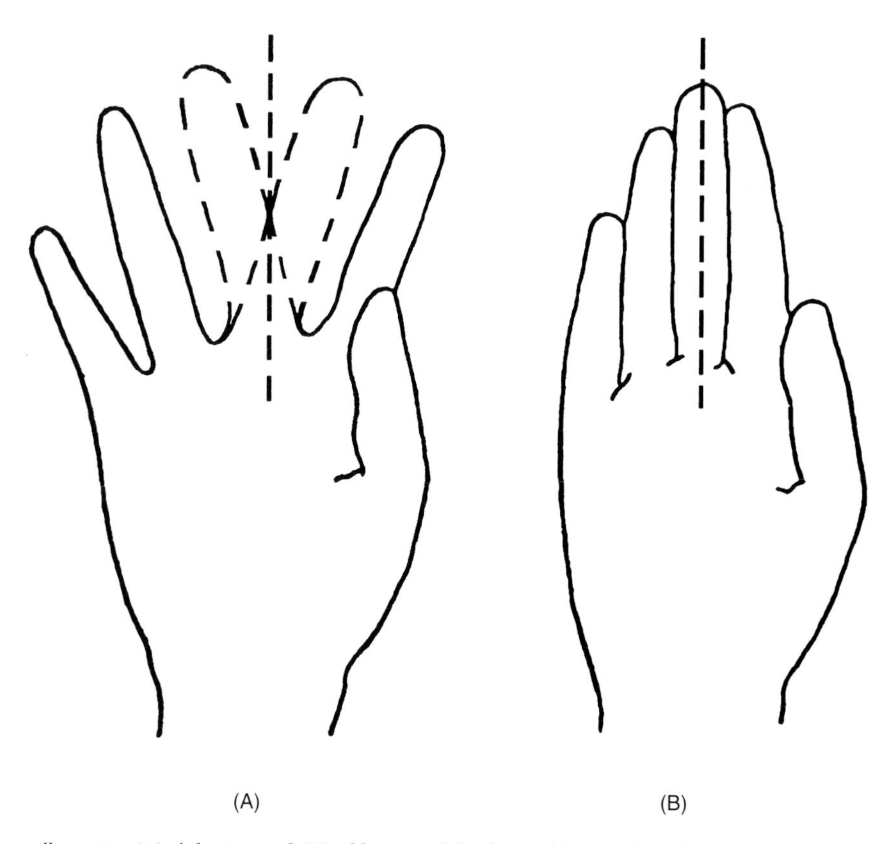

Fig. 10–8: Diagrams illustrating (A) abduction and (B) adduction of the fingers (digits 2-5) at their MP joints. Note that the axial line of the hand extends through the middle finger. Abduction of a finger is movement away from the axial line, and adduction is movement toward the axial line. Observe that the middle finger can be abducted to either side of the axial line, and that adduction restores its central axis coincident with the axial line.

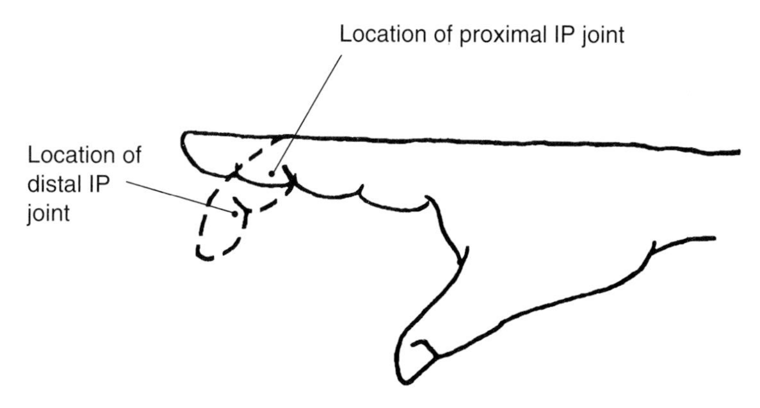

Fig. 10–9: Diagram illustrating flexion of the index finger at its proximal and distal IP joints. The parts of the index finger illustrated with a broken outline represent the index finger partially flexed at its proximal and distal IP joints.

The **carpal tunnel** is the larger of the two osseofascial tunnels in the palmar aspect of the wrist (Fig. 10–12). The carpal tunnel transmits the median nerve and the tendons of three anterior forearm muscles: flexor pollicis longus, flexor digitorum profundus, and flexor digitorum superficialis. The single tendon of flexor pollicis longus inserts onto the distal phalanx of the thumb. Each flexor digitorum muscle gives rise to four tendons, one for each finger. The profundus tendon for each finger inserts onto the distal phalanx of the digit, and the superficialis tendon inserts onto the middle phalanx.

The **extensor retinaculum** forms the fascial roof of six osseofascial tunnels in the dorsal aspect of the wrist (Figs. 10–11 and 10–12). The extensor retinaculum is attached medially to the head of the ulna, the pisiform, and the triquetrum; laterally, it is attached to the distal end of the radius. Fascial septa that extend deeply from the extensor retinaculum to the underlying bones form the walls between the tunnels in the dorsal aspect of the wrist (Fig. 10–12).

The fascial roofs of the osseofascial tunnels in the palmar aspect of the digits are called the **fibrous flexor sheaths** of the digits (Figs. 10–10 and 10–13). Each fibrous flexor sheath extends from the head of the digit's metacarpal to the base of the digit's distal phalanx. The tunnel of the thumb transmits the tendon of flexor pollicis longus. The tunnel of each finger transmits a tendon from flexor digitorum profundus and a tendon from flexor digitorum superficialis (Fig. 10–13).

As the tendons of the forearm muscles extend through the osseofascial tunnels at the wrist and in the digits, the tendons are either singly or collectively wrapped within a synovial sheath (Figs. 10–12 and 10–13). The synovial sheaths permit the tendons to move freely through the tunnels.

In the carpal tunnels of about half the population, the tendon of flexor pollicis longus is enveloped by a synovial sheath separate from the common synovial sheath shared by all the tendons of flexor digitorum profundus and superficialis. In the other half of the population, the common synovial sheath for the flexor digitorum tendons communicates with the synovial sheath enveloping the flexor pollicis longus tendon.

In the osseofascial tunnels of the digits, a single **digital synovial sheath** envelops the one or two tendons extending through the tunnel (Fig. 10–13). The digital synovial sheath enveloping the flexor pollicis longus tendon in the thumb is continuous with the synovial sheath enveloping the tendon in the carpal tunnel; the union of these two sheaths is called the **radial bursa** (Fig. 10–10). The digital synovial sheath enveloping the flexor digitorum tendons in the little finger is continuous with the common synovial sheath for the flexor digitorum tendons in the carpal tunnel; the union of these two sheaths is called the **ulnar bursa** (Fig. 10–10). The digital synovial sheaths enveloping the flexor digitorum tendons in the index, middle, and ring fingers usually are not continuous with the common synovial sheath for the flexor digitorum tendons in the carpal tunnel.

A strong band of deep fascia called the **extensor expansion** extends through the posterior aspect of each finger (Fig. 10–13). Proximally, the extensor expansion forms a fascial hood that covers the posterior and lateral surfaces of the finger's MP joint (Figs. 10–14 and 10–15). From its broad, proximal base, the extensor expansion narrows as it extends distally into the finger. It divides into three slips just proximal to the proximal IP joint. The central slip extends across the proximal IP joint before inserting onto the base of the middle phalanx. The two collateral slips merge as they extend distally, and their union extends across the distal IP joint before inserting onto the base of the distal phalanx.

There are three posterior forearm muscles (extensor digitorum, extensor indicis, and extensor digiti minimi) the tendons of which insert onto the extensor expansions of the fingers (Fig. 10–11). Extensor digitorum gives rise to a tendon for the extensor expansion of each of the four fingers. Extensor indicis gives rise to a tendon for the extensor expansion of the index finger, and extensor digiti minimi gives rise to two tendons for the extensor expansion of the little finger. Whenever a tendon of any of these muscles pulls upon the extensor expansion to which it is attached, it extends the finger at its MP, proximal IP, and distal IP joints.

The **palmar aponeurosis** is a triangular layer of deep fascia in the hand. The apex of the palmar aponeurosis is attached to the distal border of the flexor retinaculum (Fig. 10–16). The aponeurosis widens as it extends distally from its apex, forming a tough, protective fascial layer in the central region of the palm. The aponeurosis is attached superficially to the skin of the palm; this attachment improves the grip of the palmar surface.

THE MUSCLES OF THE FOREARM

The muscles of the forearm are wrapped within an investing layer of deep fascia that extends along the long axis of the forearm (Fig. 10–17). Along the posteromedial side of the forearm, the deep fascia is attached to the margin of the ulna. The deep fascia of the forearm, in association with the radius and ulna and the interosseous membrane between the bones, divides the forearm into two major muscular compartments: a muscular compartment anterior to the radius-interosseous membrane-ulna boundary and a muscular compartment posterior to the osseofascial boundary.

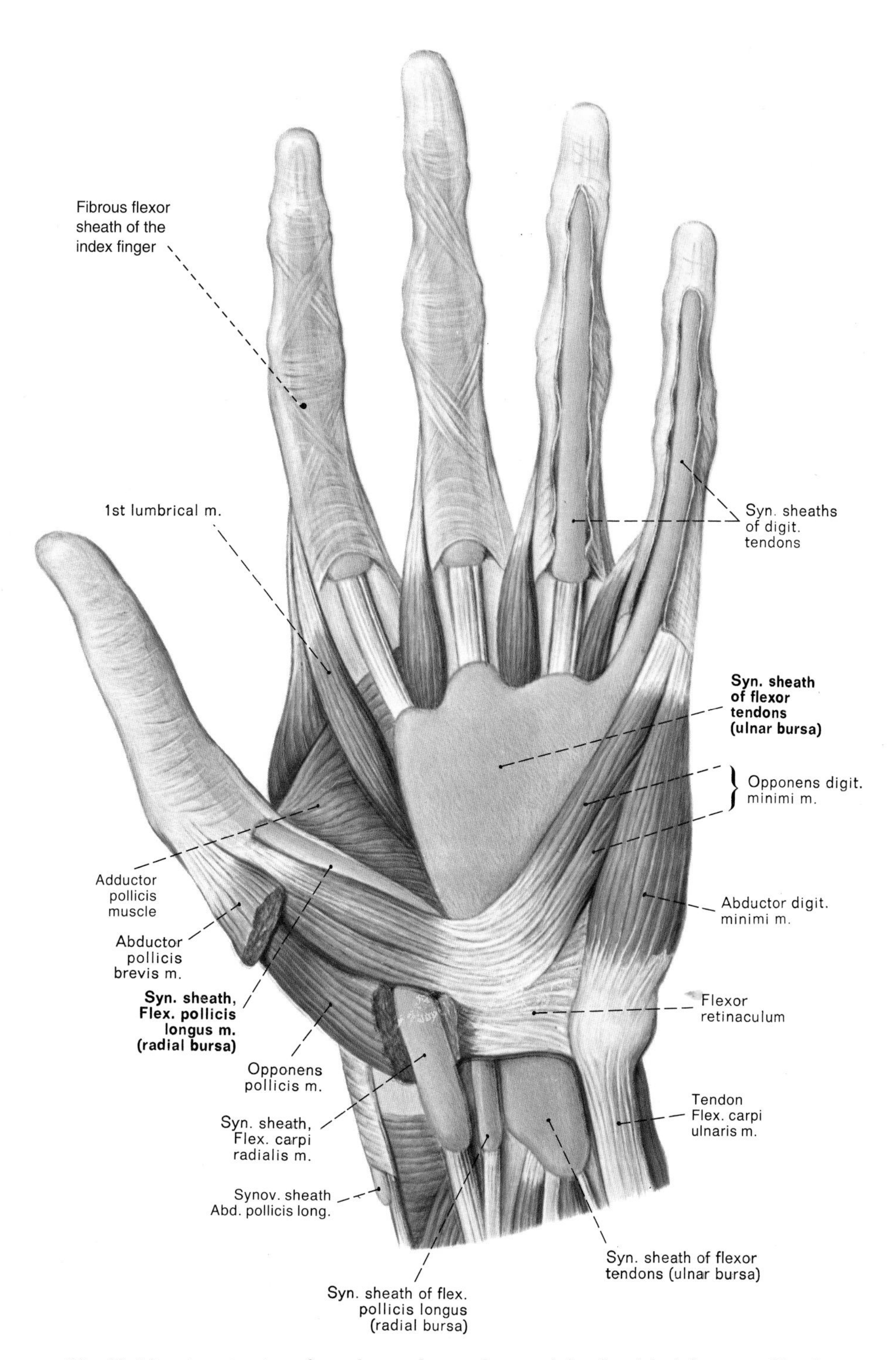

Fig. 10–10: Anterior view of muscles, tendons, and synovial sheaths of the left wrist and hand.

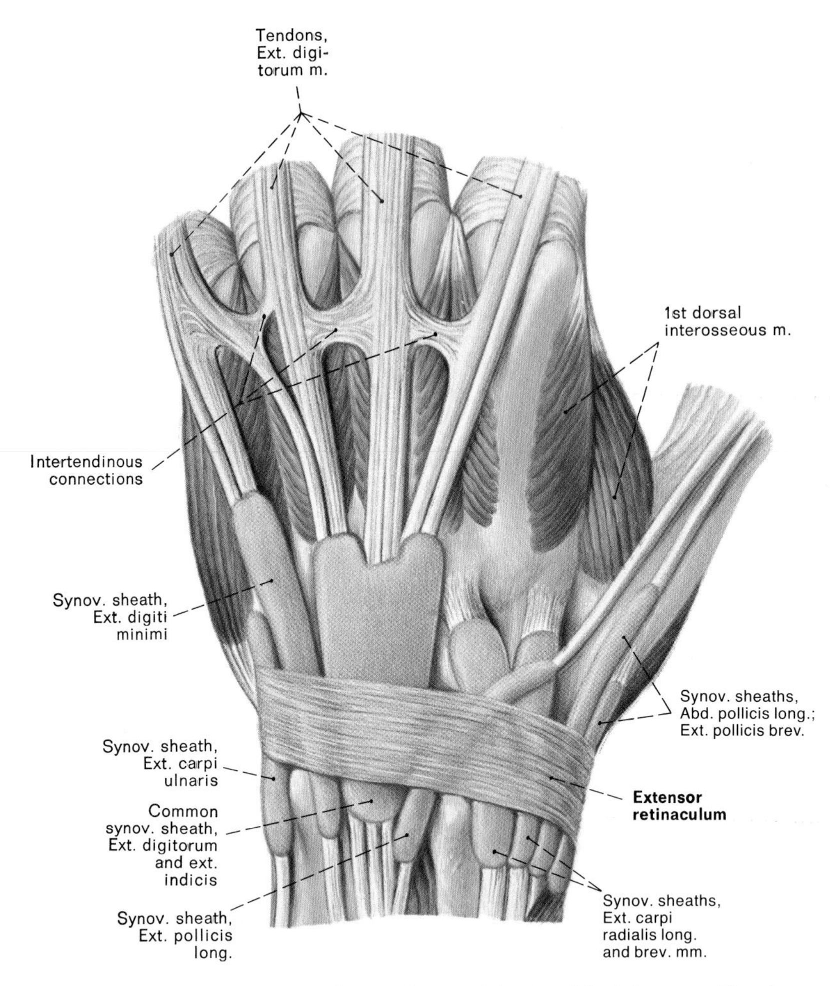

Fig. 10–11: Posterior view of muscles, tendons, and synovial sheaths of the left wrist and hand.

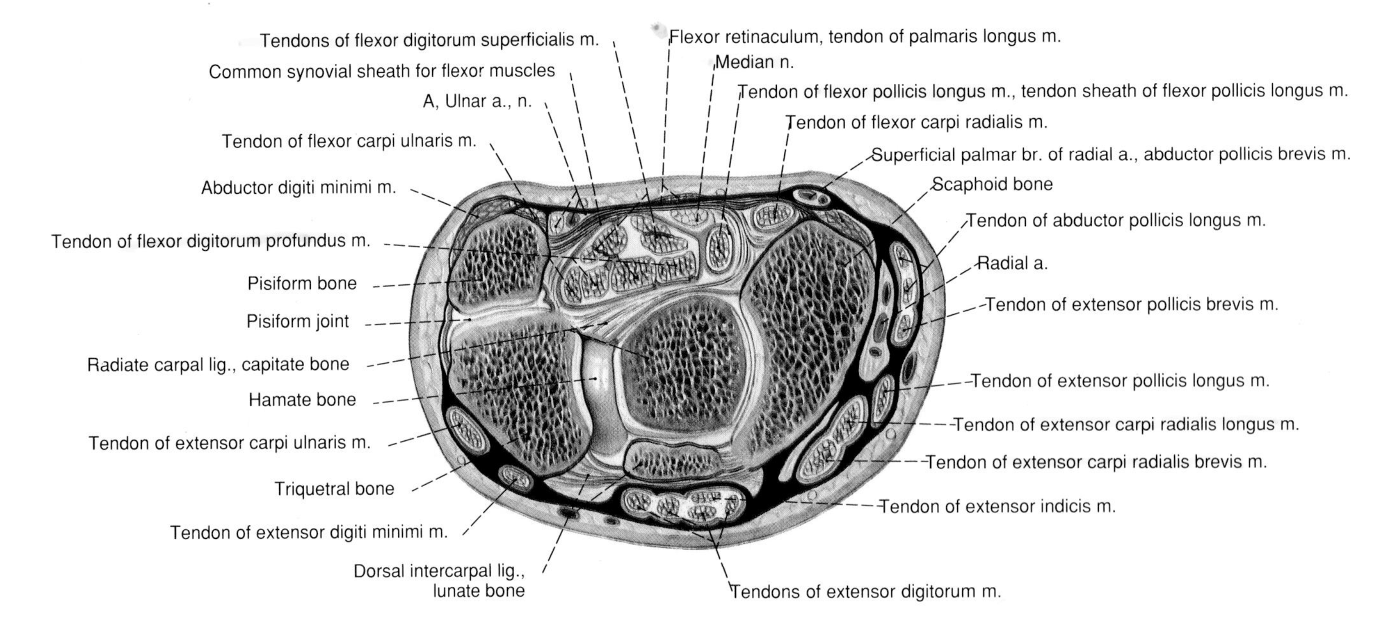

Fig. 10–12: Transverse section of the right wrist through the carpal tunnel. Seen from above. The space transmitting the median nerve and the tendons of flexor pollicis longus, flexor digitorum profundus, and flexor digitiorum superficialis represents the transverse sectional area of the carpal tunnel. The bluish grey bands encircling tendons of anterior and posterior forearm muscles represent synovial sheaths. The black area about the tendons and synovial sheaths of the posterior forearm muscles represents the extensor retinaculum and its septa.

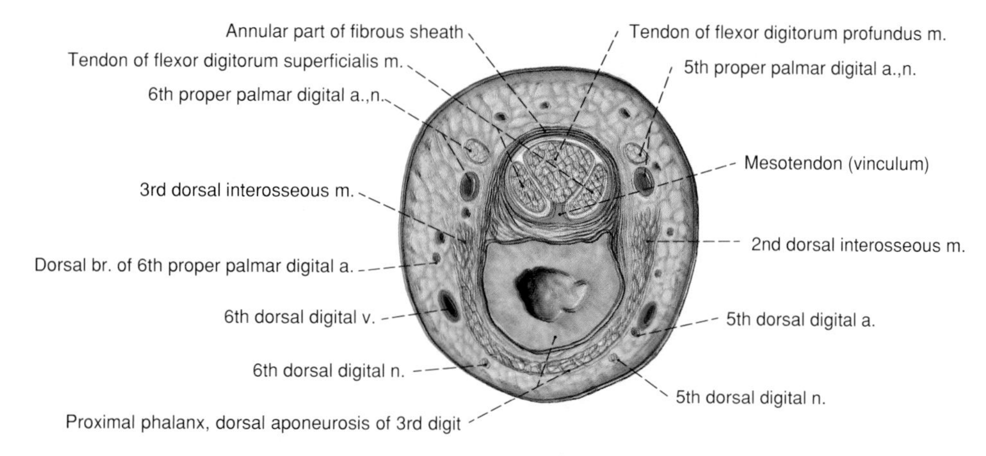

Fig. 10–13: Transverse section of the proximal phalanx of the right middle finger. Seen from above. The bluish-grey band encircling the tendons of flexor digitorum profundus and superficialis represents the digital synovial sheath. The digit's extensor expansion is labelled the dorsal aponeurosis.

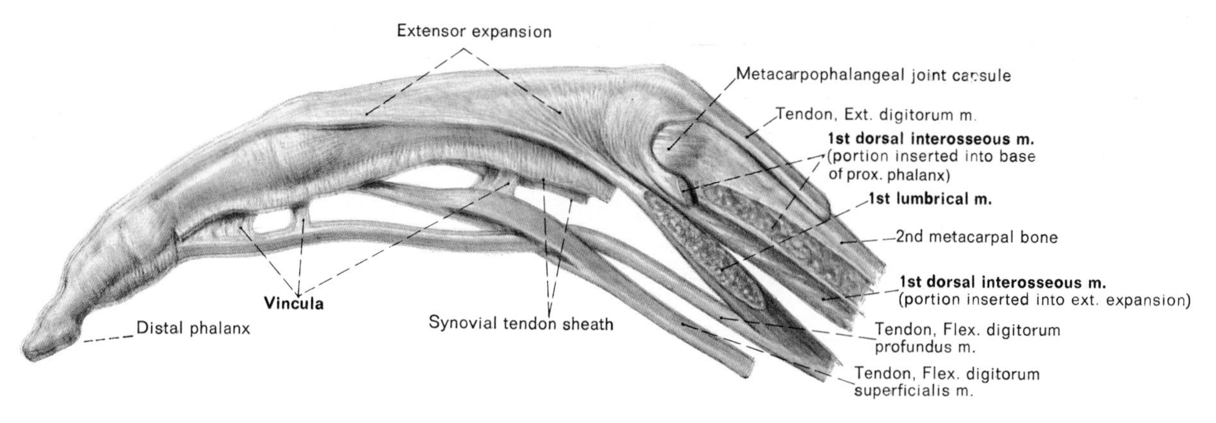

Fig. 10–14: Lateral view of the muscles, tendons, and extensor expansion of the index finger of the right hand.

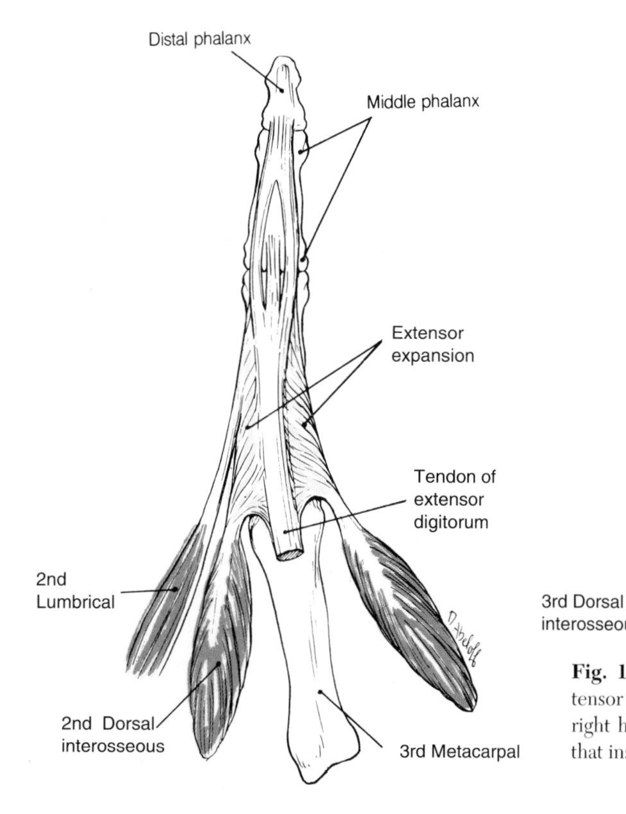

Fig. 10–15: Diagram illustrating the extensor expansion of the middle finger of the right hand and the tendons of the muscles that insert onto the extensor expansion.

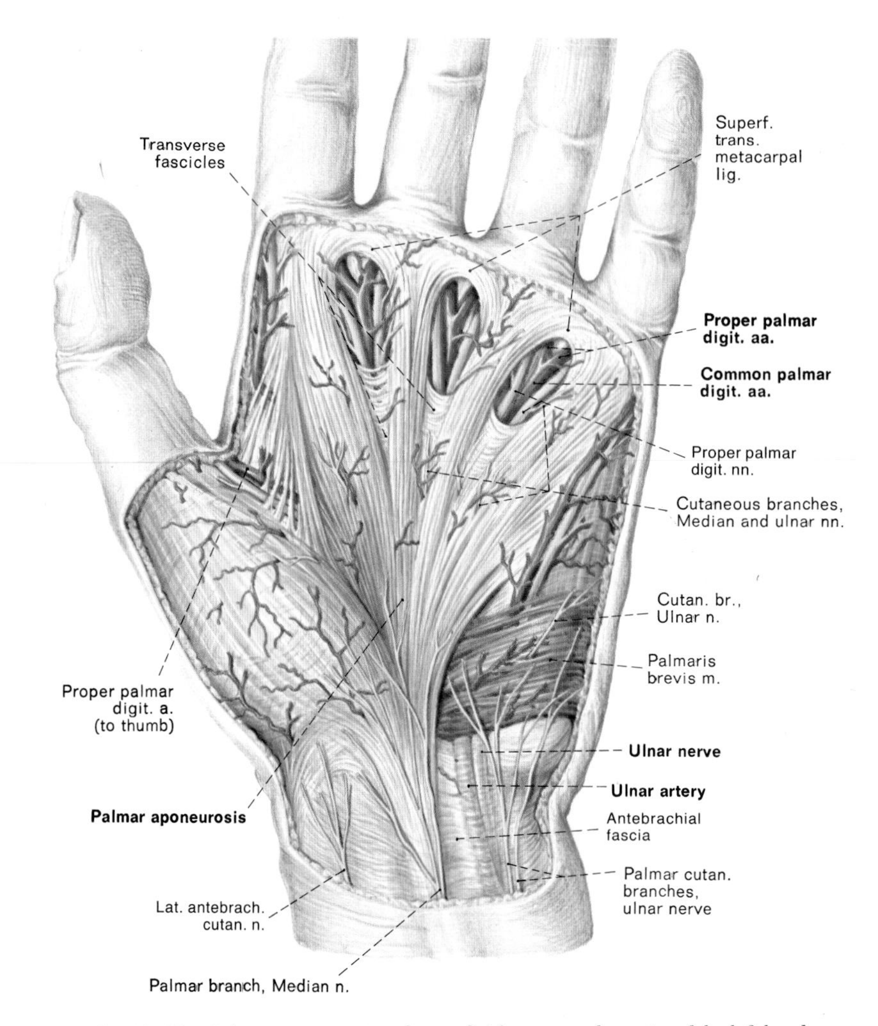

Fig. 10–16: Palmar aponeurosis and superficial nerves and arteries of the left hand.

The Anterior Muscles of the Forearm

There are eight muscles in the anterior muscular compartment of the forearm. All except one and a half of them are innervated by the median nerve or its anterior interosseous branch; the one and a half muscles (flexor carpi ulnaris and the medial half of flexor digitorum profundus) are innervated by the ulnar nerve. The anterior forearm muscles may by classified functionally into three groups: two muscles that pronate the forearm, three muscles that move the hand at the wrist joint, and three muscles that flex the thumb or the fingers. Tables 10–1 to 10–3 list for quick reference the origin, insertion, nerve supply, and actions of the anterior forearm muscles.

Pronator quadratus originates from the distal anterior surface of the ulna, and inserts onto the distal anterior surface of the radius (Fig. 10–18). It is the chief pronator of the forearm.

Pronator teres has two heads of origin: a humeral head that originates from the medial epicondyle of the humerus and an ulnar head that originates from the coronoid process of the ulna. The two heads join to insert onto the shaft of the radius (Fig. 10–19). It can pronate and flex the forearm.

Flexor carpi radialis originates from the medial epicondyle of the humerus (Fig. 10–19). Its tendon passes through an osseofascial tunnel at the wrist (deep to the flexor retinaculum) in route to its insertion onto the second and third metacarpals. It can flex and abduct the hand at the wrist joint.

Palmaris longus originates from the medial epicondyle of the humerus, and inserts onto the flexor retinaculum and the apex of the palmar aponeurosis (Fig. 10–19). Although it can flex the hand at the wrist joint, this action is not especially significant in everyday hand movements. Consequently, its slender tendon of insertion is often chosen to replace damaged tendons of more important muscles or torn ligaments during reconstructive surgery of the limbs.

Table 10–1
The Anterior Muscles of the Forearm Which Pronate the Forearm

Muscle	Origin	Insertion	Nerve Supply	Actions
Pronator quadratus	Distal fourth of the anterior surface of the ulna	Distal fourth of the anterior surface of the radius	Anterior interosseous branch of the median (**C8** and T1)	Pronates the forearm
Pronator teres	Medial epicondyle of the humerus via the common flexor tendon and coronoid process of the ulna	Midregion of the lateral aspect of the radial shaft	Median (C6 and **C7**)	Pronates and flexes the forearm

Table 10–2
The Anterior Muscles of the Forearm Which Move the Hand at the Wrist Joint

Muscle	Origin	Insertion	Nerve Supply	Actions
Flexor carpi radialis	Medial epicondyle of the humerus via the common flexor tendon	Bases of the 2nd and 3rd metacarpals	Median (C6 and **C7**)	Flexes and abducts the hand
Palmaris longus	Medial epicondyle of the humerus via the common flexor tendon	Distal half of the anterior surface of the flexor retinaculum and central part of the palmar aponeurosis	Median (C7 and C8)	Flexes the hand
Flexor carpi ulnaris	Medial epicondyle of the humerus via the common flexor tendon and the olecranon process and posterior border of the ulna	Pisiform, hamate, and 5th metacarpal	Ulnar (C7 and **C8**)	Flexes and adducts the hand

Table 10–3
The Anterior Muscles of the Forearm Which Flex the Thumb or the Fingers

Muscle	Origin	Insertion	Nerve Supply	Actions
Flexor pollicis longus	Anterior surface of the radius and adjoining interosseous membrane	Base of the distal phalanx of the thumb	Anterior interosseous branch of the median (**C8** and T1)	Flexes the phalanges of the thumb
Flexor digitorum superficialis	Medial epicondyle of the humerus via the common flexor tendon, ulnar collateral ligament of the elbow joint, coronoid process of the ulna, and anterior border of the radius	Sides of the middle phalanx of the medial four digits	Median (C7, **C8,** and T1)	Flexes the middle and proximal phalanges of the medial four digits
Flexor digitorum profundus	Proximal three-fourths of the anterior and medial surfaces and posterior border of the ulna and proximal half of the anterior surface of the interosseous membrane	Base of the distal phalanx of the medial four digits	Medial part by the ulnar (**C8** and T1); lateral part by the anterior interosseous branch of the median (**C8** and T1)	Flexes the distal phalanges of the medial four digits

Flexor carpi ulnaris has two heads of origin: a humeral head that originates from the medial epicondyle of the humerus and an ulnar head that originates from the proximal part of the ulna (Fig. 10–19). The two heads join to insert onto the hamate and by means of ligaments to the pisiform and fifth metacarpal. It can flex and adduct the hand at the wrist joint.

Flexor pollicis longus originates from the radius (Fig. 10–20). Its tendon passes through the carpal tunnel and then the osseofascial tunnel of the thumb in route to its insertion onto the distal phalanx of the thumb. It is the flexor of the thumb's distal phalanx.

Flexor digitorum superficialis originates from the medial epicondyle of the humerus and proximal parts of the radius and ulna (Fig. 10–21). The four tendons that arise from the muscle's belly pass through the carpal tunnel and then the osseofascial tunnels of the fingers in route to their insertions onto the middle phalanges of the fingers. The chief action of the muscle is flexion of the middle phalanges of the fingers.

Flexor digitorum profundus originates from the proximal part of the ulna (Fig. 10–20). The four tendons that arise from the muscle's belly pass through the carpal tunnel and the osseofascial tunnels of the fingers in route to their insertions onto the distal phalanges of the fingers. In the carpal tunnel, the tendons of flexor digitorum profundus lie deep to the tendons of flexor digitorum superficialis. In the osseofascial tunnel of each finger, the flexor digitorum profundus tendon extends toward its insertion site by passing through a slit-like opening in the flexor digitorum superficialis tendon (Fig. 10-14). The chief action of flexor digitorum profundus is flexion of the distal phalanges of the fingers.

The two flexor digitorum tendons ensheathed together in the osseofascial tunnel of each finger slide past each other during flexion and extension movements at the finger's MP and IP joints. Specifically, the two tendons slide past each other along the distance from the head of the finger's metacarpal to the base of the finger's middle phalanx (Fig. 10–14). Deep lacerations that cut the flexor digitorum tendons anywhere along this interval may result in significant impairment and disfigurement of the finger, because it requires considerable surgical expertise to repair the injury without post-operatively incurring intertendinous adhesions.

The Posterior Muscles of the Forearm

All the posterior forearm muscles, with the exceptions of anconeus, supinator, and brachioradialis, have tendons with insertion sites in the hand. All these tendons pass through one of the osseofascial tunnels in the dorsal aspect of the wrist (Fig. 10–11). As these tendons extend through the osseofascial tunnels, they are either singly or collectively wrapped within a synovial sheath.

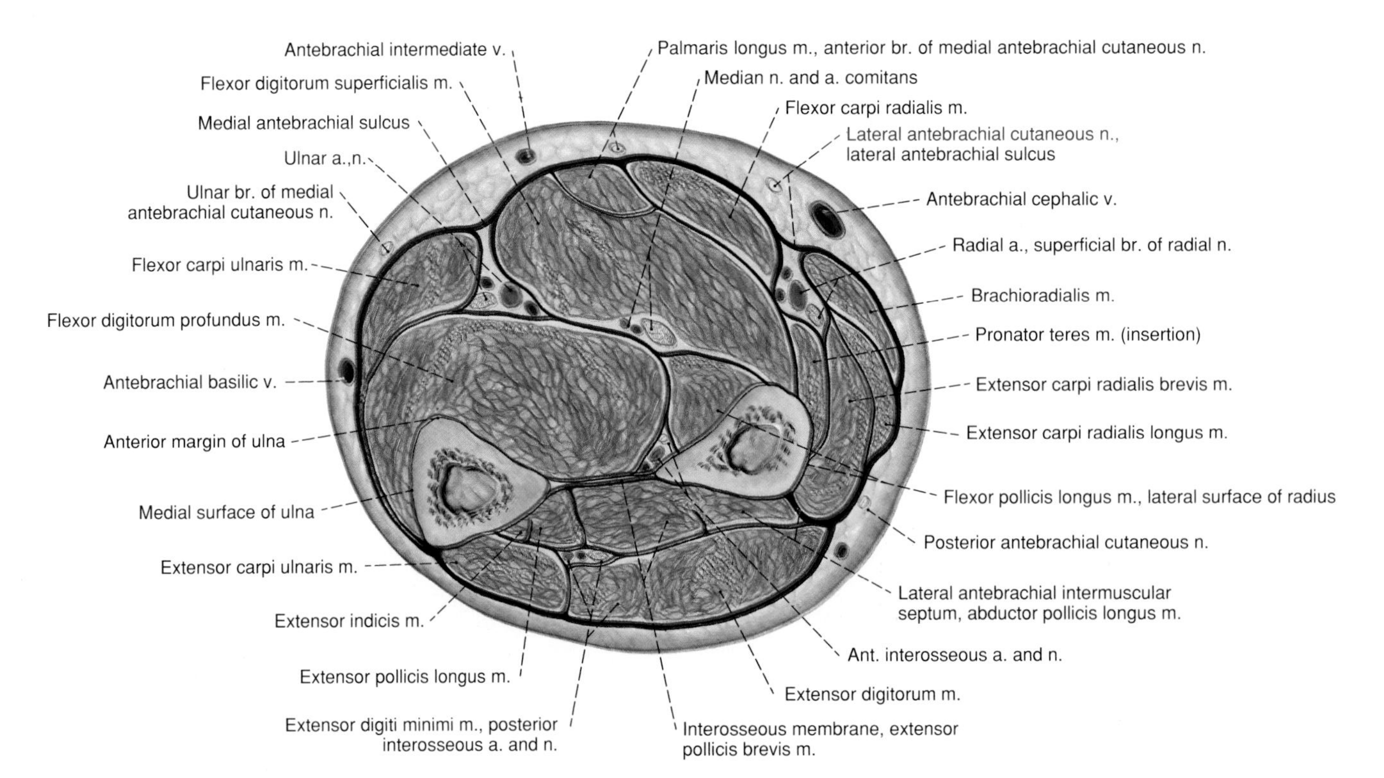

Fig. 10–17: Transverse section through the middle third of the left forearm. Seen from below. The black band directly encircling the forearm muscles represents the investing layer of deep fascia.

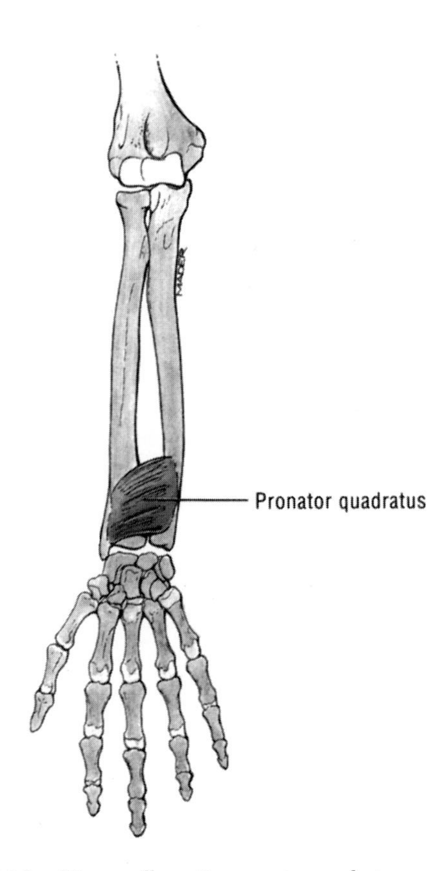

Fig. 10–18: Diagram illustrating pronator quadratus.

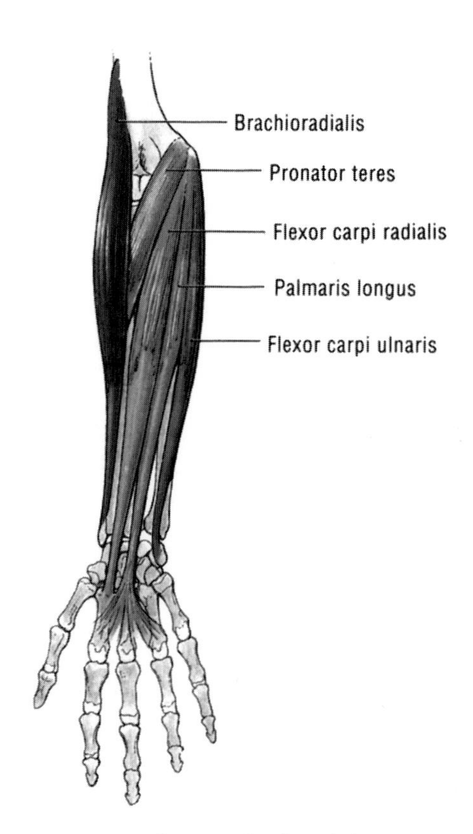

Fig. 10–19: Diagram illustrating brachioradialis, pronator teres, flexor carpi radialis, palmaris longus, and flexor carpi ulnaris.

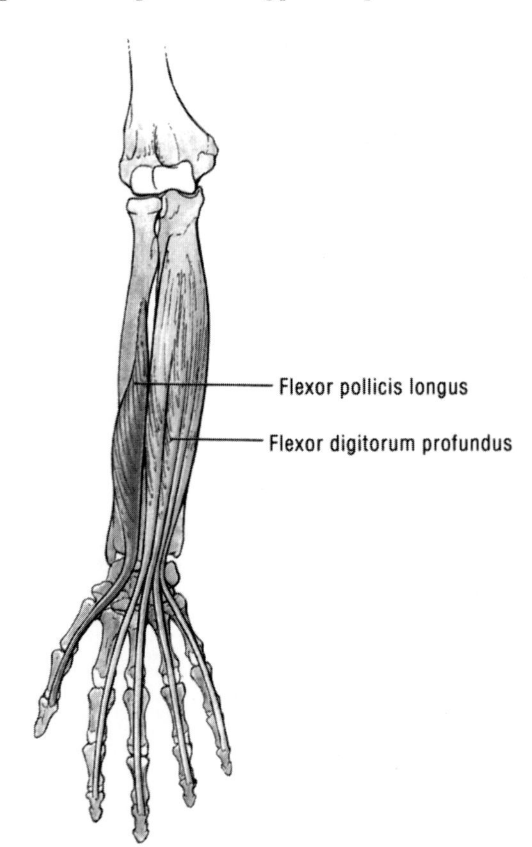

Fig. 10–20: Diagram illustrating flexor pollicis longus and flexor digitorum profundus.

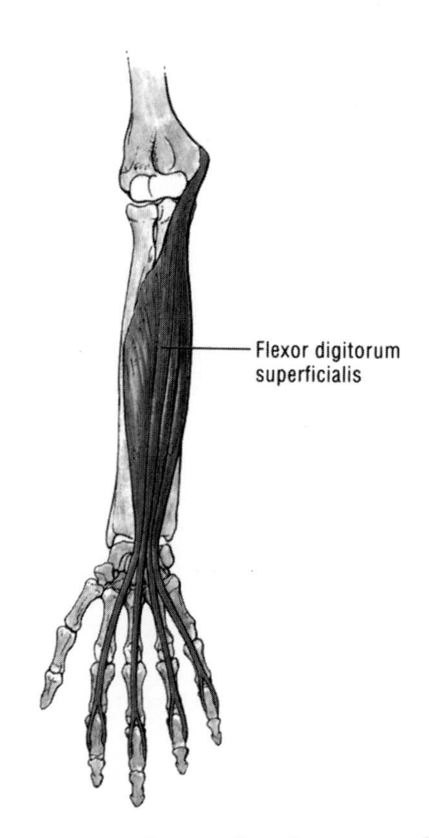

Fig. 10–21: Diagram illustrating flexor digitorum superficialis.

There are twelve muscles in the posterior muscular compartment of the forearm. Nine of them are innervated by the deep branch of the radial nerve or its posterior interosseous branch. The three other muscles (anconeus, brachioradialis, and extensor carpi radialis longus) are innervated by the radial nerve. The posterior forearm muscles may by classified functionally into three groups: three muscles that move the forearm, three muscles that move the hand at the wrist joint, and six muscles that extend the thumb or one or more fingers. Tables 10–4 to 10–6 list for quick reference the origin, insertion, nerve supply, and actions of the posterior forearm muscles.

Anconeus originates from the lateral epicondyle of the humerus, passes posterior to the elbow joint, and inserts onto the olecranon process of the ulna (Fig. 10–22). It assists triceps brachii to extend the forearm.

Brachioradialis originates from the lateral aspect of the distal end of the humerus, passes anterior to the elbow joint, and inserts onto the styloid process of the radius (Fig. 10–23). Its major action is flexion of the forearm. It probably also serves to stabilize the forearm when it is in the midprone position.

Supinator originates from the posterior surface of the lateral epicondyle of the humerus and the proximal part of the ulna (Fig. 10–22). Its fibers spiral around the posterior and lateral sides of the radius to insert onto the anterior, lateral, and posterior sides of the neck and shaft of the radius. Its only action is supination of the forearm.

Extensor carpi radialis longus originates from the lateral epicondyle of the humerus; its tendon passes deep to the extensor retinaculum in route to its insertion onto the second metacarpal (Fig. 10–24). It can extend and abduct the hand at the wrist joint.

Extensor carpi radialis brevis originates from the lateral epicondyle of the humerus; its tendon passes deep to the extensor retinaculum in route to its insertion onto the third metacarpal (Fig. 10–24). It can extend and abduct the hand at the wrist joint.

Extensor carpi ulnaris originates from the lateral epicondyle of the humerus; its tendon passes deep to the extensor retinaculum in route to its insertion onto the fifth metacarpal (Fig. 10–24). It can extend and adduct the hand at the wrist joint.

Table 10–4
The Posterior Muscles of the Forearm Which Move the Forearm

Muscle	Origin	Insertion	Nerve Supply	Actions
Anconeus	Posterior surface of the lateral epicondyle of the humerus	Lateral surface of the olecranon process of the ulna and upper posterior surface of the ulnar shaft	Radial (C7, C8, and T1)	Extends the forearm
Brachioradialis	Proximal two-thirds of the lateral supracondylar ridge of the humerus	Lateral surface of the distal end of the radius	Radial (C5, **C6,** and C7)	Flexes the forearm
Supinator	Lateral epicondyle of the humerus, radial collateral ligament of the elbow joint, annular ligament, and supinator crest of the ulna	Anterior, lateral, and posterior surfaces of the proximal third of the radius	Deep branch of the radial (C5 and **C6**)	Supinates the forearm

Table 10–5
The Posterior Muscles of the Forearm Which Move the Hand at the Wrist Joint

Muscle	Origin	Insertion	Nerve Supply	Actions
Extensor carpi radialis longus	Distal third of the lateral supracondylar ridge of the humerus	Base of the 2nd metacarpal	Radial (C6 and C7)	Extends and abducts the hand
Extensor carpi radialis brevis	Lateral epicondyle of the humerus via the common extensor tendon	Base of the 3rd metacarpal	Deep branch of the radial (**C7** and C8)	Extends and abducts the hand
Extensor carpi ulnaris	Lateral epicondyle of the humerus via the common extensor tendon and the posterior border of the ulna	Base of the 5th metacarpal	Posterior interosseous (C7 and **C8**)	Extends and adducts the hand

Abductor pollicis longus originates from the posterior surfaces of the radius and ulna; its tendon extends through the most lateral osseofascial tunnel under the extensor retinculum (Fig. 10–12) before inserting onto the first metacarpal (Fig. 10–22). It can extend and abduct the thumb at its carpometacarpal joint.

Extensor pollicis brevis originates from the posterior surface of the radius; its tendon extends with the tendon of abductor pollicis longus through the most lateral osseofascial tunnel under the extensor retinaculum (Fig. 10–12) before inserting onto the thumb's proximal phalanx (Fig. 10–22). It can extend the thumb at its MP and carpometacarpal joints.

Extensor pollicis longus originates from the posterior surface of the ulna; its tendon bends around the medial side of the dorsal tubercle of the radius (Fig. 10–22) and then extends through an osseofascial tunnel under the extensor retinaculum (Fig. 10–12) before inserting onto the thumb's distal phalanx (Fig. 10–22). It can extend the thumb at its IP, MP, and carpometacarpal joints.

Extensor indicis originates from the posterior surface of the ulna; its tendon passes deep to the extensor retinaculum before inserting onto the extensor expansion of the index finger (Fig. 10–22). It can extend the index finger at its MP and IP joints.

Extensor digitorum originates from the lateral epicondyle of the humerus. The four tendons that arise from the muscle's belly enter the hand by passing deep to the extensor retinaculum (Fig. 10–12). Each tendon inserts onto the extensor expansion of one of the fingers (Fig. 10–25). It can extend the index, middle, ring, and little fingers at their MP and IP joints.

There are connections among extensor digitorum's tendons as they extend toward their insertion sites (Fig. 10–11). The connections restrict full extension of any finger by the muscle in the absence of some limited extension of the other fingers.

Extensor digiti minimi originates from the lateral epicondyle of the humerus; its two tendons pass deep to the extensor retinaculum before inserting onto the extensor expansion of the little finger (Fig. 10–25). It can extend the little finger at its MP and IP joints.

Table 10–6
The Posterior Muscles of the Forearm Which Extend the Thumb or the Fingers

Muscle	Origin	Insertion	Nerve Supply	Actions
Abductor pollicis longus	Posterior surfaces of the radius and ulna and interosseous membane	Base of the 1st metacarpal	Posterior interosseous (C7 and **C8**)	Abducts the thumb and extends it at the carpometacarpal joint
Extensor pollicis brevis	Posterior surface of the radius and interosseous membrane	Base of the proximal phalanx of the thumb	Posterior interosseous (C7 and **C8**)	Extends the thumb at the metacarpophalangeal and carpometacarpal joints
Extensor pollicis longus	Posterior surface of the ulna and interosseous membrane	Base of the distal phalanx of the thumb	Posterior interosseous (C7 and **C8**)	Extends the thumb at the interphalangeal, metacarpophalangeal, and carpometacarpal joints
Extensor indicis	Posterior surface of the ulna and interosseous membrane	Extensor expansion of the index finger	Posterior interosseous (C7 and **C8**)	Extends the index finger at its MP and IP joints
Extensor digitorum	Lateral epicondyle of the humerus via the common extensor tendon	Extensor expansions of the index, middle, ring, and little fingers	Posterior interosseous (**C7** and C8)	Extends the index, middle, ring, and little fingers at their MP and IP joints and extends the hand
Extensor digiti minimi	Lateral epicondyle of the humerus via the common extensor tendon	Extensor expansion of the little finger	Posterior interosseous (**C7** and C8)	Extends the little finger at its MP and IP joints

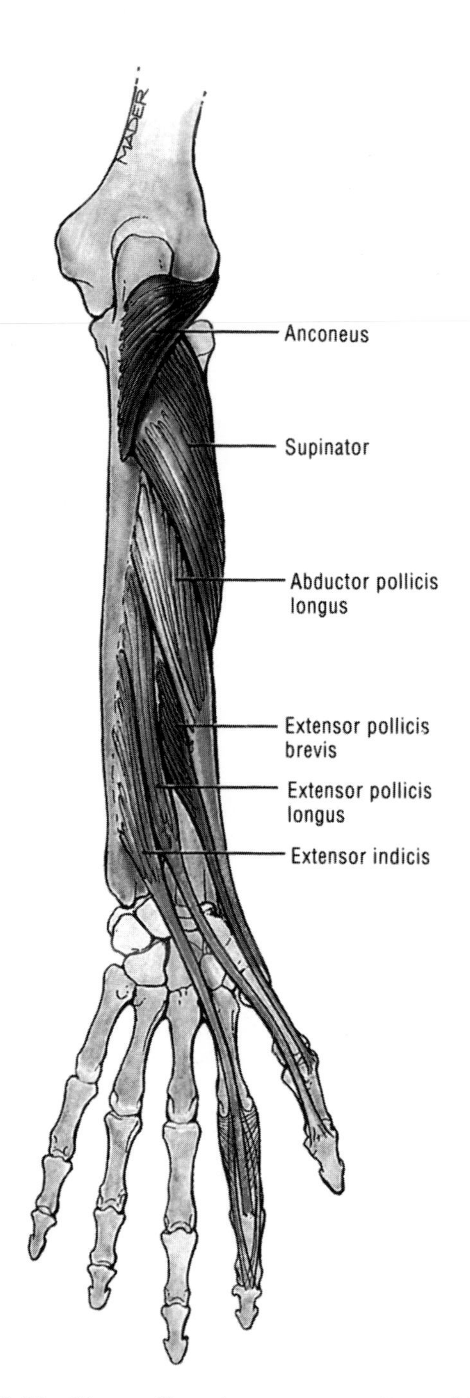

Fig. 10–22: Diagram illustrating anconeus, supinator, abductor pollicis longus, extensor pollicis brevis, extensor pollicis longus, and extensor indicis.

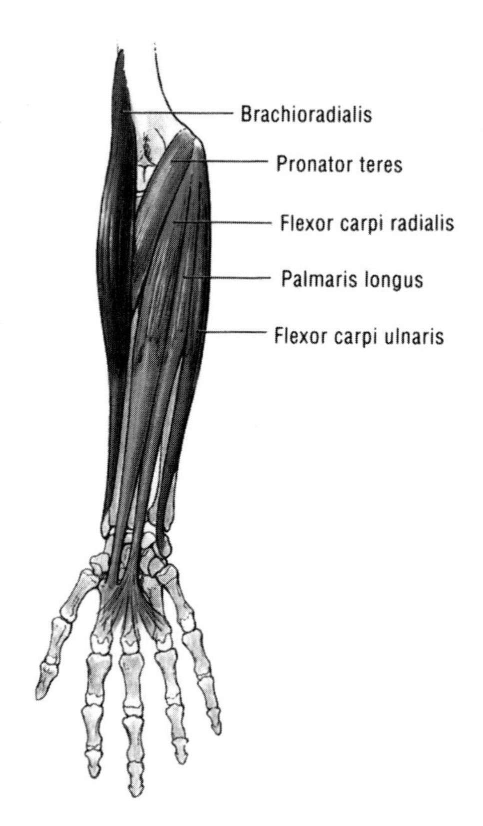

Fig. 10–23: Diagram illustrating brachioradialis, pronator teres, flexor carpi radialis, palmaris longus, and flexor carpi ulnaris.

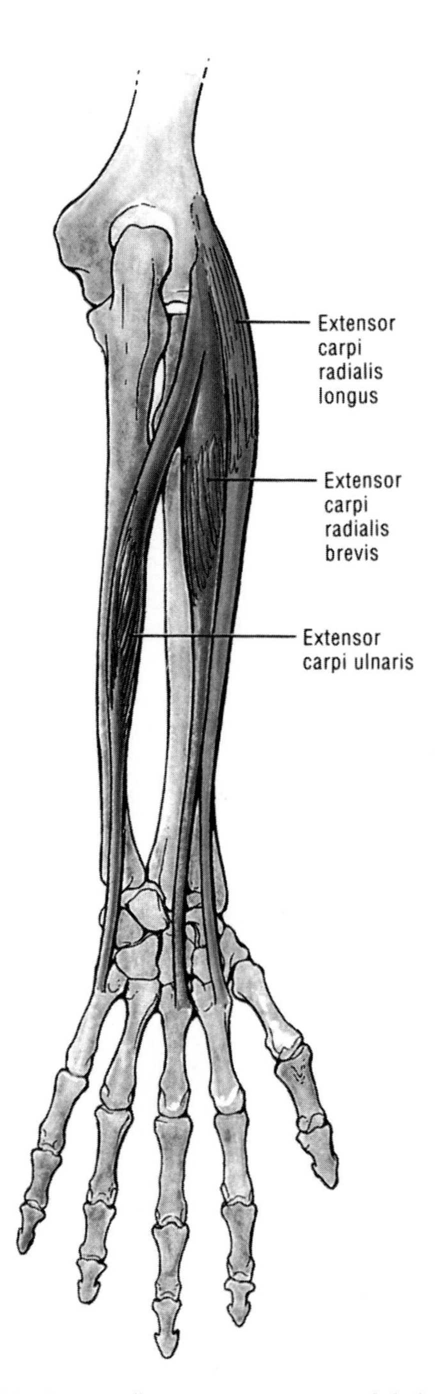

Fig. 10–24: Diagram illustrating extensor carpi radialis longus, extensor carpi radialis brevis, and extensor carpi ulnaris.

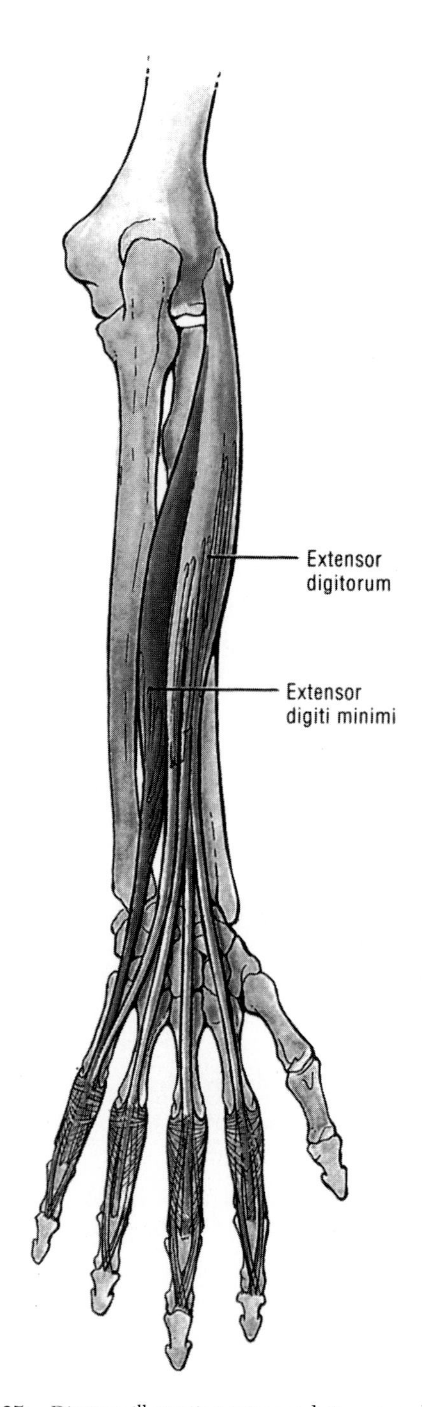

Fig. 10–25: Diagram illustrating extensor digitorum and extensor digiti minimi.

AN OVERVIEW OF THE ACTIONS OF THE ANTERIOR AND POSTERIOR FOREARM MUSCLES

The muscles of the anterior and posterior compartments of the forearm function together to control the movements of the hand. All the carpi muscles of the anterior and posterior forearm (flexor carpi radialis, flexor carpi ulnaris, extensor carpi radialis longus, extensor carpi radialis brevis, and extensor carpi ulnaris) can not only flex or extend the hand, but also abduct or adduct it. The carpi muscles thus function in everyday movements to position and stabilize the hand as a whole.

The carpi muscles also serve as synergists of the forearm muscles that can flex or extend the digits. The extensor carpi muscles, for example, serve as synergists when the only hand movements wanted are flexion of the fingers. Concentric contraction of the flexor digitorum superficialis and profundus muscles will not only flex the fingers at their MP and IP joints, but also flex the hand at the wrist joint. Simultaneous isometric contraction of the extensor carpi muscles negates the unwanted movement produced by the flexor digitorum muscles.

THE COURSE OF THE MEDIAN NERVE THROUGH THE CUBITAL FOSSA, FOREARM AND WRIST

The median nerve accompanies the brachial artery along its descent through the arm and cubital fossa. In the cubital fossa, the median nerve lies deep to the bicipital aponeurosis, medial to the brachial artery, and superficial to brachialis (Fig. 10–26).

The median nerve usually enters the forearm by passing between the two heads of pronator teres (Fig. 10–27). In the interval between the two heads of pronator teres, the median nerve gives rise to its **anterior interosseous branch,** a nerve that extends distally through the anterior compartment of the forearm (anterior to the interosseous membrane of the forearm) and innervates some of the deepest anterior forearm muscles (Fig. 10–28).

Upon passing between the two heads of pronator teres, the median nerve descends through most of the forearm deep to flexor digitorum superficialis (Fig. 10–27). Just proximal to the wrist, it becomes more superficial, and here lies between the tendons of flexor digitorum superficialis and flexor carpi radialis. From here the median nerve extends into the hand by passing through the carpal tunnel (and deep to the flexor retinaculum) (Figs. 10–29 and 10–12).

CLINICAL PANEL II-8

LATERAL EPICONDYLITIS

Definition: This condition is a painful strain of the tendinous origin of the four posterior forearm muscles that originate from the lateral epicondyle of the humerus by the common extensor tendon.

Common Symptoms: The patient presents with pain in the lateral aspect of the elbow that may radiate down the back of the forearm into the dorsum of the hand.

Anatomic Basis of the Examination: Extensor carpi radialis brevis, extensor carpi ulnaris, extensor digitorum, and extensor digiti minimi are the four posterior forearm muscles that originate from the lateral epicondyle of the humerus by the common extensor tendon (see Tables 10–5 and 10–6). In lateral epicondylitis, the part of the common extensor tendon that originates from the anterior aspect of the lateral epicondyle is strained, and thus physical examination reveals that the point of maximum tenderness in the lateral aspect of the elbow region is that which overlies the anterior aspect of the lateral epicondyle.

The pain may be exacerbated by requesting the patient to lie supine with the forearm fully extended (at the elbow joint) and fully pronated (at the radioulnar joints) and then to attempt extension of the hand at the wrist against resistance provided by the examiner. This isometric exercise exacerbates the elbow pain by exerting tension on the anterior part of the common extensor tendon.

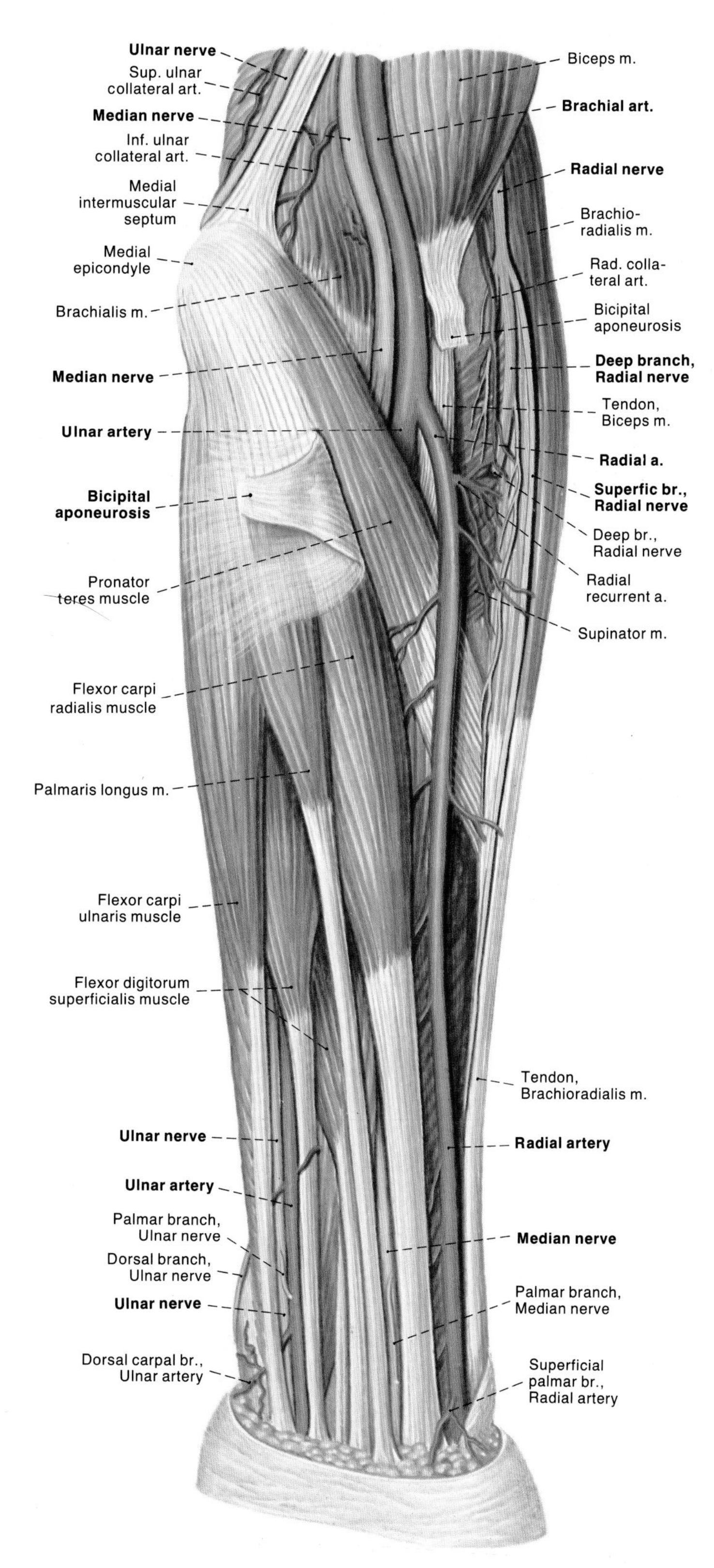

Fig. 10–26: Superficial view of the nerves and arteries of the anterior aspect of the left forearm.

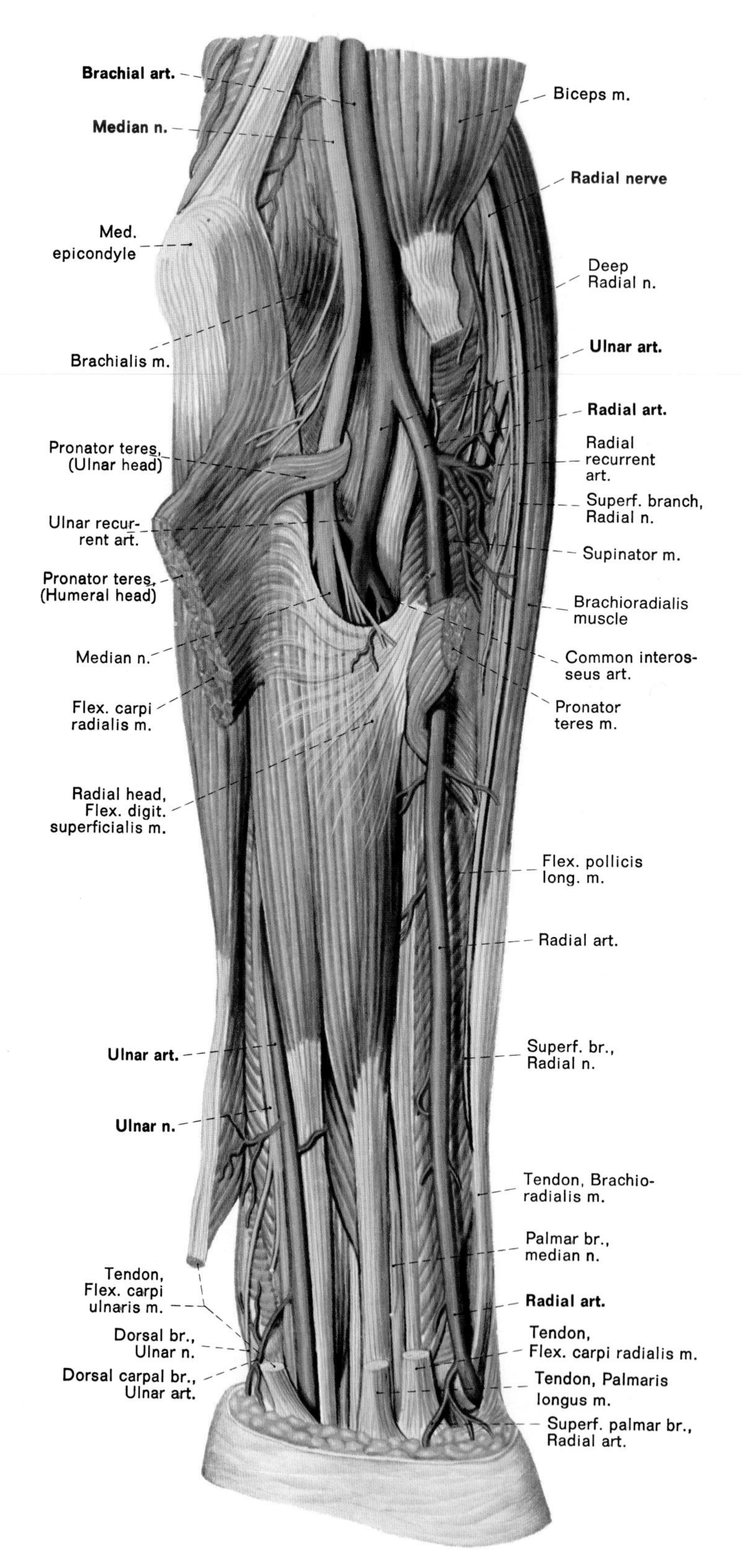

Fig. 10–27: Intermediate view of the nerves and arteries of the anterior aspect of the left forearm.

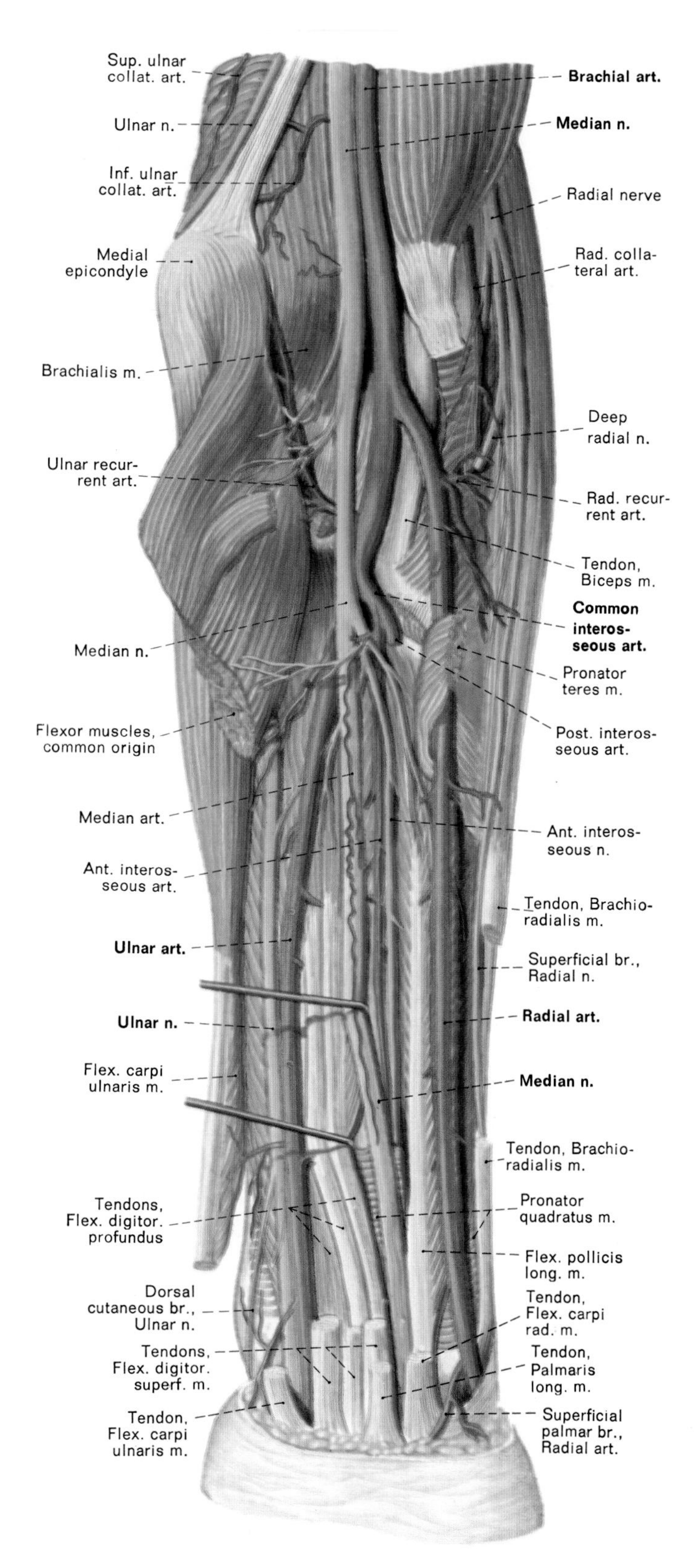

Fig. 10–28: Deep view of the nerves and arteries of the anterior aspect of the left forearm.

THE COURSE OF THE ULNAR NERVE THROUGH THE FOREARM AND WRIST

The ulnar nerve enters the forearm by passing between the two heads of flexor carpi ulnaris. It extends along the medial side of the forearm, lying deep to flexor carpi ulnaris in the proximal half of the forearm but deep to only fascia and skin in the distal half of the forearm (Fig. 10–27). The ulnar nerve gives rise to a **dorsal branch** in the distal half of the forearm; the dorsal branch provides cutaneous innervation for the dorsal (posterior) surface of the medial part of the hand (Figs. 10–27 and 10–28).

The ulnar nerve extends from the forearm into the hand by coursing superficial to the flexor retinaculum. Upon entering the hand, the ulnar nerve divides into two branches: a **superficial branch** and a **deep branch** (Fig. 10–29).

THE COURSE OF RADIAL NERVE BRANCHES IN THE FOREARM

The radial nerve exits the arm by extending distally through the cubital fossa (Fig. 10–26). Upon reaching the level of the lateral epicondyle of the humerus, the radial nerve divides into two branches: a deep branch and a superficial branch.

As the **deep branch of the radial nerve** extends through the cubital fossa, it gives off branches to extensor carpi radialis brevis and supinator. The deep branch exits the cubital fossa by piercing the anterior surface of supinator (Fig. 10–26). The deep branch exhibits a curved course through the substance of the muscle, a course which extends the nerve posteriorly and around the lateral side of the neck of the radius. Upon piercing the posterior surface of supinator, the deep branch of the radial gives rise to its **posterior interosseous branch** (Fig. 10–30), a nerve that extends distally through the posterior compartment of the forearm (posterior to the interosseous membrane of the forearm) and innervates most of the posterior forearm muscles.

The superficial branch of the radial nerve extends through most of the forearm under the cover of brachioradialis (Fig. 10–26). The superficial branch then enters the posterior surface of the hand (Fig. 10–31), where it provides cutaneous innervation for the dorsum of the hand (generally the lateral half) and the posterior surfaces of the lateral two and a half digits (as distally as the proximal IP joints of the index and middle fingers and the IP joint of the thumb).

THE COURSE OF THE ULNAR AND RADIAL ARTERIES THROUGH THE FOREARM AND WRIST

The brachial artery bifurcates into the ulnar and radial arteries at the level of the neck of the radius (Fig. 9–16). The ulnar artery has near its origin a major branch called the common interosseous artery. The **common interosseous artery** bifurcates into the **anterior and posterior interosseous arteries;** these latter two arteries extend through the forearm on opposite sides of the interosseous membrane and supply deep tissues in the anterior and posterior compartments of the forearm.

The courses of the ulnar and radial arteries in the anterior compartment of the forearm become more superficial as the arteries approach the wrist. The ulnar artery accompanies the ulnar nerve along its descent through the distal half of the forearm and extends into the hand by coursing superficial to the flexor retinaculum; here the ulnar artery lies lateral to the ulnar nerve (Figs. 10–12 and 10–29). The pulsations of the ulnar artery can be palpated as it courses superficially to the flexor retinaculum.

At the distal end of the forearm, the radial artery lies superficial to the distal end of the radius (Fig. 9–16). It is here that its pulsations can be palpated. The radial artery courses through the wrist by curving around the lateral sides of the scaphoid and trapezium, after which it enters the hand through its dorsal (posterior) side (Fig. 10–32).

Timing the pulsations of the ulnar and radial arteries measures, in effect, the rate of the heartbeat. The reason is as follows: The high-pressured blood ejected from the left ventricle into the aorta during systole has a pressure greater than that of the blood already present in the aorta. The higher pressure of the ejected blood is great enough to distend (stretch) the region of the aorta immediately adjacent to the heart. Moreover, the higher pressure of the ejected blood also very quickly spreads throughout first all the blood borne by the aorta and its elastic branches and then all the blood borne by the muscular arteries of the systemic circuit. As this pulse of blood pressure-increase spreads in a wave-like fashion through the blood borne by the elastic and muscular arteries of the systemic circuit, so too does the stretching of the arterial walls spread in a wave-like fashion distally from the heart.

The wave of arterial wall stretching that occurs during systole is followed by a smooth, recoil contraction during diastole. The walls of elastic and muscular arteries thus mark the passage of each heartbeat by a cycle

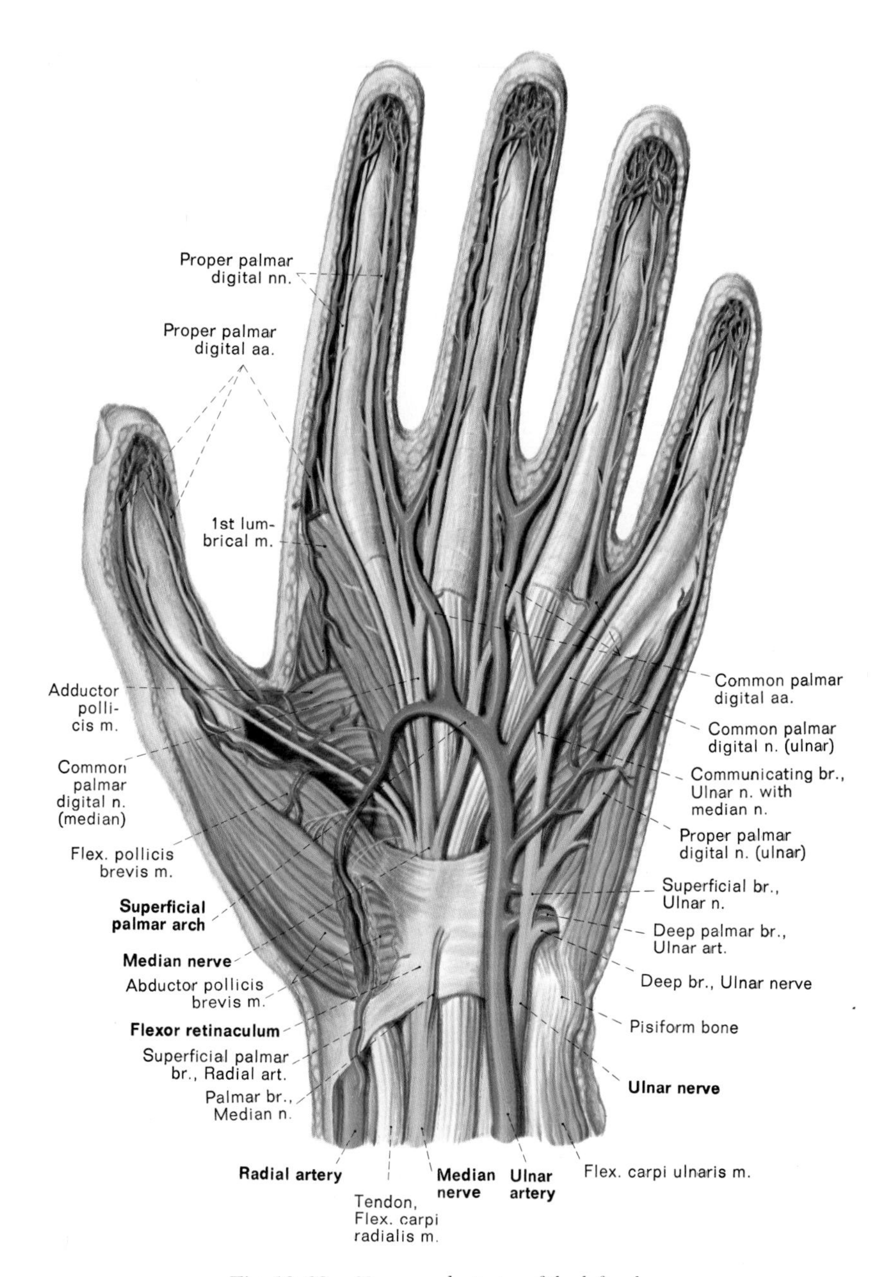

Fig. 10–29: Nerves and arteries of the left palm.

of first distension and then recoil contraction. Because the walls of superficial arteries can be seen and felt to pulse with each heartbeat, the timing of arterial pulsations can be used to measure the rate of the heartbeat.

The pulsations of the ulnar artery can be palpated on the medial side of the anterior forearm about 1 cm proximal to the distal crease of the wrist (Fig. 10–33). The pulsations of the radial artery can be palpated on the lateral side of the anterior forearm about 2 to 3 cm proximal to the distal crease of the wrist (Fig. 10–34). To take the pulse of these arteries, an examiner lightly presses two or three fingers upon the artery (the examiner does not use the thumb, as the pulsations felt may be the pulsations of arteries in the examiner's thumb) (Figs. 10–33 and 10–34).

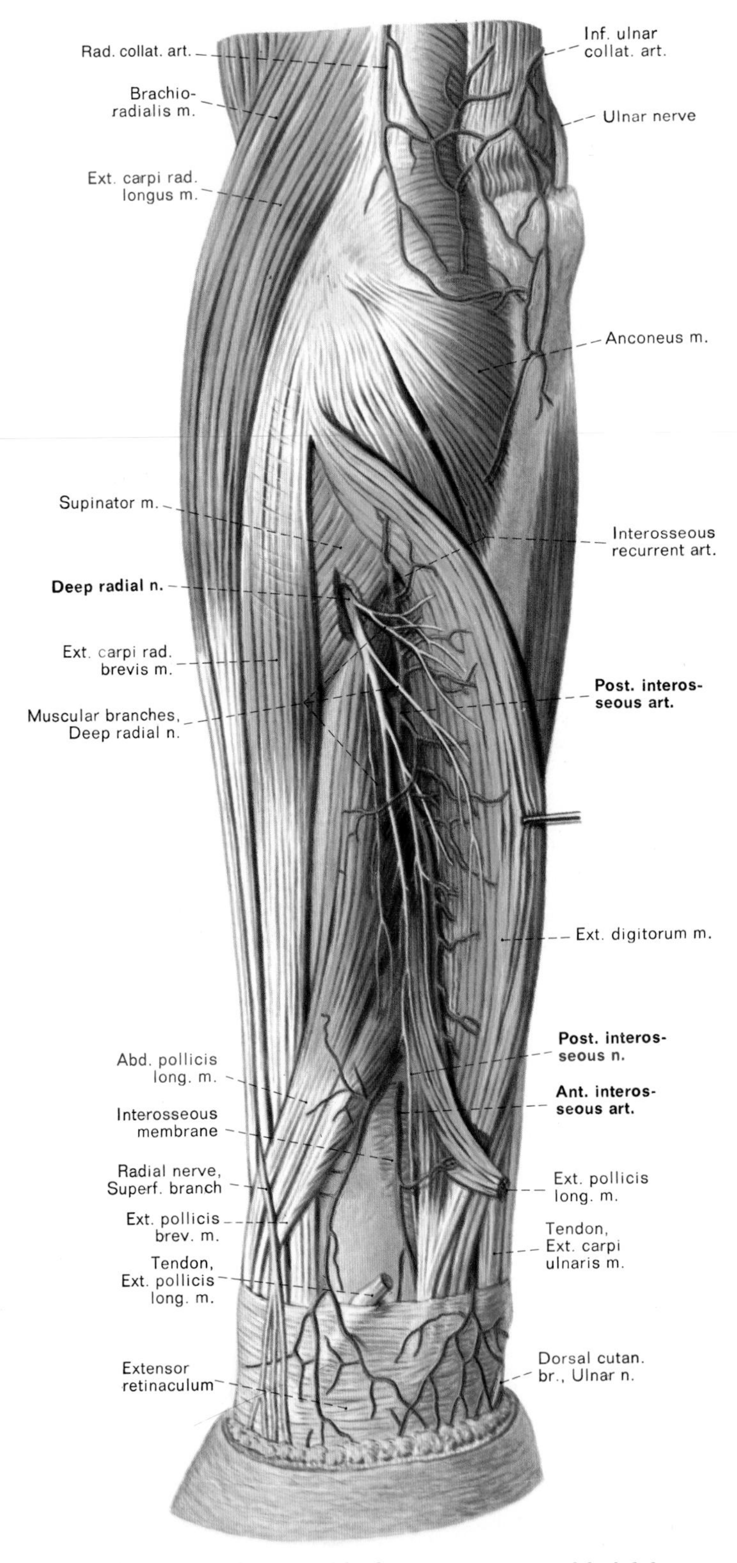

Fig. 10–30: Nerves and arteries of the deep posterior aspect of the left forearm.

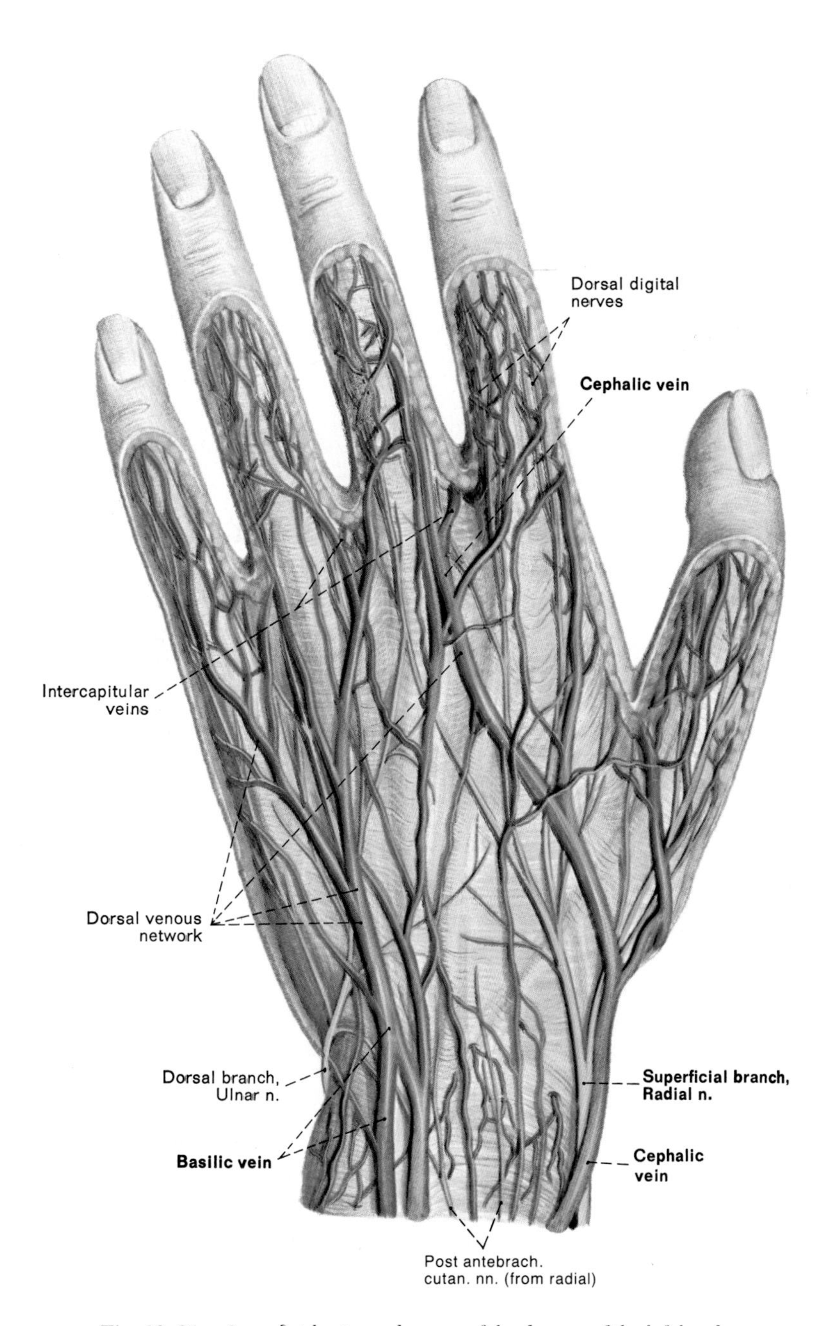

Fig. 10–31: Superficial veins and nerves of the dorsum of the left hand.

THE ANATOMIC SNUFFBOX

This is a surface depression that forms on the lateral side of the wrist when the hand is adducted at the wrist joint and the thumb is extended at its carpometacarpal, MP, and IP joints (Fig. 10–35). The snuffbox has a bony floor composed of the following bony parts: the styloid process of the radius, the scaphoid, the trapezium, and the base of the first metacarpal.

The anatomic snuffbox is bordered medially and laterally by the tendons of the posterior forearm muscles that extend the thumb: the tendons of abductor pollicis longus and extensor pollicis brevis form the lateral border, and the tendon of extensor pollicis longus forms the medial border (Fig. 10–32). As the radial artery curves around the lateral side of the wrist, it courses superficial to the bony floor of the snuffbox.

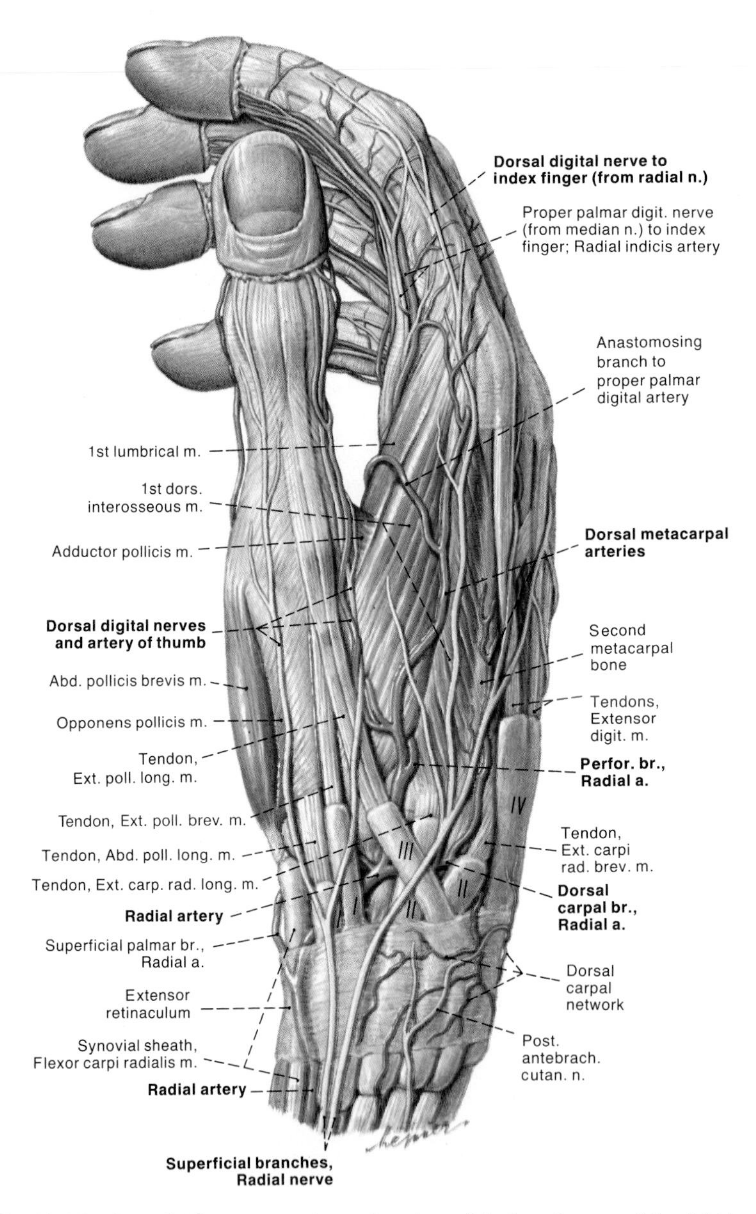

Fig. 10–32: Superficial nerves, arteries, and tendons of the lateral aspect of the right hand.

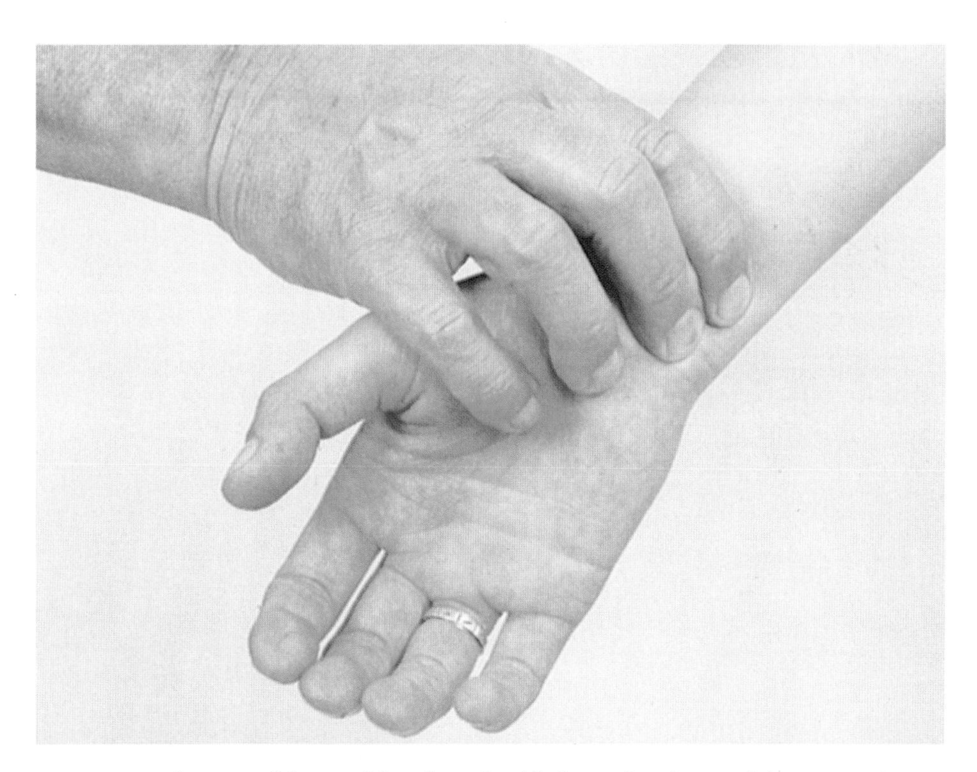

Fig. 10–33: Placement of the tips of the index and middle fingers for palpation of ulnar artery pulsations.

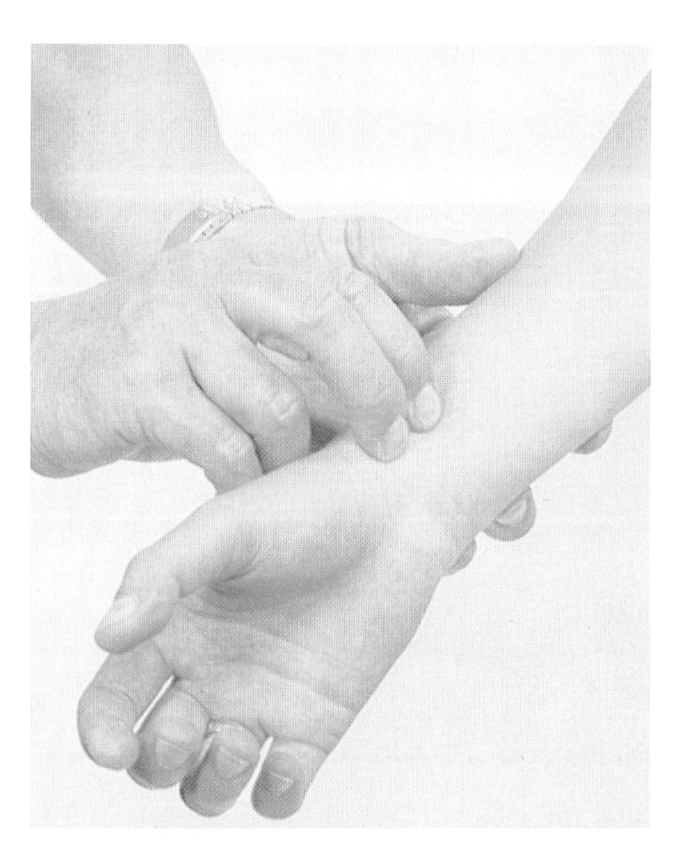

Fig. 10–34: Placement of the tips of the index and middle fingers for palpation of radial artery pulsations.

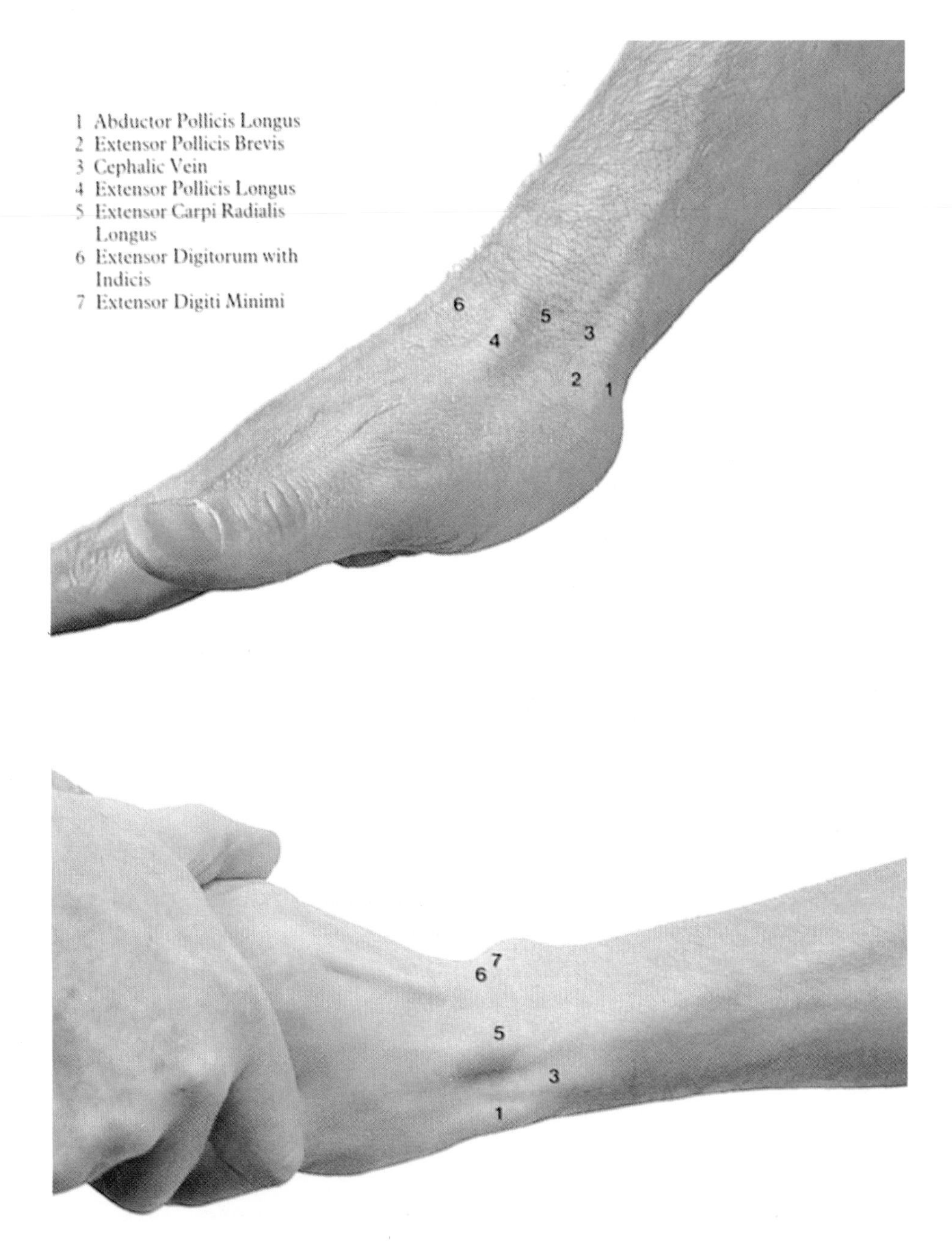

Fig. 10–35: Surface anatomy of the lateral aspect of the right wrist.

CHAPTER 11

The Hand

This chapter addresses the musculature, motor and cutaneous nerve supply, blood supply, and blood and lymphatic drainage of the hand. The chapter begins with a discussion of the actions and innervation of the intrinsic muscles of the hand. These muscles are the 19 muscles of the upper limb located exclusively in the hand.

DIGITAL MOVEMENTS ASSOCIATED SPECIFICALLY WITH THE INTRINSIC MUSCLES OF THE HAND

The intrinsic muscles of the hand include the prime movers of the following digital movements:

Abduction and Adduction of the Fingers at Their MP Joints

The abduction and adduction movements of the fingers are all related to an imaginary axis that extends longitudinally through the middle finger (Fig. 10–8). A finger is abducted when it is moved within the plane of the palm in such a fashion that it moves away from this imaginary axis; a finger is adducted when the opposite movement occurs. Notice that the middle finger can be abducted to either side of this imaginary axis.

Opposition of the Thumb

Opposition of the thumb is basically the movement by which the thumb, starting from the anatomic position, is simultaneously flexed and medially rotated at its carpometacarpal joint. This movement brings the tip of the thumb's palmar surface into opposition with the tips of the palmar surfaces of the fingers (Fig. 11–1). Opposition of the thumb is, in its details, a very complex movement that also involves some abduction of the thumb at its carpometacarpal joint and some flexion at its MP joint.

Reposition of the thumb is the movement opposite to opposition. The prime movers of reposition of the thumb are abductor pollicis longus and extensor pollicis longus.

The opposition and reposition movements of the thumb largely depend upon the freedom of movement in the carpometacarpal joint of the thumb. This synovial joint is an articulation between the base of the first metacarpal and the trapezium (Fig. 10–2). The joint has a saddle-type configuration (the articular surfaces of the two bones are reciprocally concave-convex). Therefore, the thumb can be flexed or extended, abducted or adducted, and medially rotated or laterally rotated at its carpometacarpal joint.

THE INTRINSIC MUSCLES OF THE HAND

The innervation of the 19 intrinsic muscles of the hand may be summarized as follows: The deep branch of the ulnar nerve innervates 13 of the intrinsic muscles, the median nerve five of the intrinsic muscles (abductor pollicis brevis, flexor pollicis brevis, opponens pollicis, and the first and second lumbricals), and the superficial branch of the ulnar nerve one intrinsic muscle (palmaris brevis). All the innervation provided by these nerves to the intrinsic muscles of the hand is derived from the C8 and T1 levels of the spinal cord. Tables 11–1 to 11–5 list for quick reference the origin, insertion, nerve supply, and actions of the intrinsic muscles of the hand.

The Lumbricals

There are four lumbricals, numbered 1 through 4. The lumbricals originate within the palm from the tendons of flexor digitorum profundus (Fig. 11– 2). Each lumbrical extends from its origin by passing anterolaterally to the MP joint of a finger, after which it inserts onto the lateral side of the finger's extensor expansion. The major actions of each lumbrical are flexion of the finger at its MP joint and extension of the finger at its IP joints.

The Dorsal Interossei

There are four dorsal interossei, numbered 1 through 4. Each dorsal interosseous has two heads of origin: the heads originate from the contiguous surfaces of two adjacent metacarpals (Fig. 11– 3). The two heads have a common tendon of insertion that passes anterior to the side of a finger's MP joint before inserting onto the same side of the finger's extensor expansion.

The collective actions of the dorsal interossei can be presented by the acronym **DAB:** starting with the hand in the anatomic position, each **D**orsal interosseous can **AB**duct its finger. Each dorsal interosseous can also flex the finger at its MP joint and extend the finger at its IP joints.

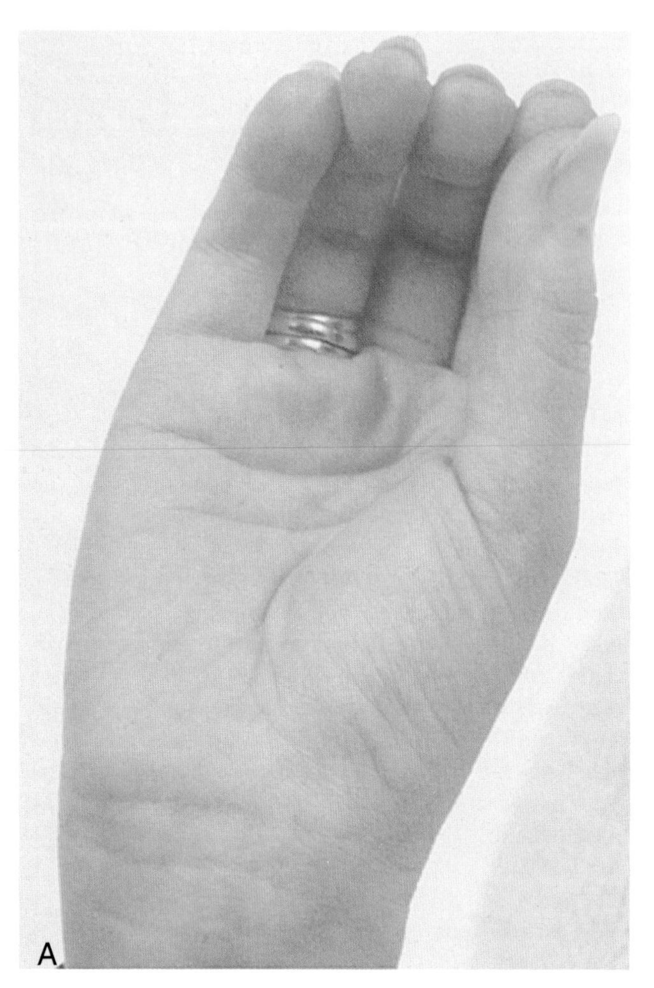

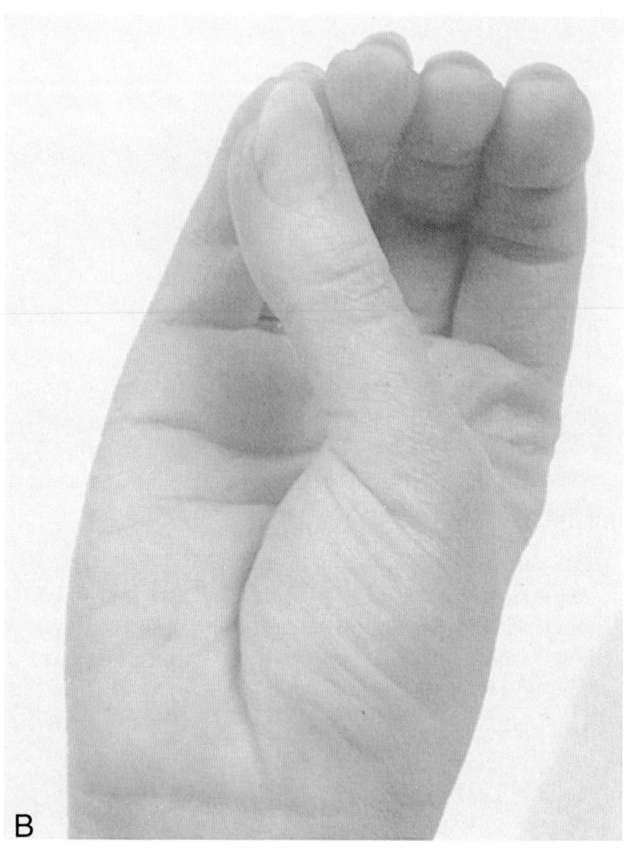

Fig. 11–1: Opposition of the thumb. Opposition of the thumb can bring the tip of the thumb's palmar surface into contact with the tip of any finger's palmar surface. (A) The thumb is opposed to the index finger. (B) The thumb is opposed to the little finger.

Table 11–1
The Lumbricals

Muscle	Origin	Insertion	Nerve Supply	Actions
1st Lumbrical	Tendon of flexor digitorum profundus for the index finger	Lateral side of the extensor expansion of the index finger	Median (C8 and **T1**)	Flexes the index finger at its MP joint and extends it at its IP joints
2nd Lumbrical	Tendon of flexor digitorum profundus for the middle finger	Lateral side of the extensor expansion of the middle finger	Median (C8 and **T1**)	Flexes the middle finger at its MP joint and extends it at its IP joints
3rd Lumbrical	Tendons of flexor digitorum profundus for the middle and ring fingers	Lateral side of the extensor expansion of the ring finger	Deep branch of ulnar (C8 and **T1**)	Flexes the ring finger at its MP joint and extends it at its IP joints
4th Lumbrical	Tendons of flexor digitorum profundus for the ring and little fingers	Lateral side of the extensor expansion of the little finger	Deep branch of ulnar (C8 and **T1**)	Flexes the little finger at its MP joint and extends it at its IP joints

Table 11–2
The Dorsal Interossei

Muscle	Origin	Insertion	Nerve Supply	Actions
1st Dorsal interosseous	Adjacent sides of the 1st and 2nd metacarpals	Lateral sides of the extensor expansion and the base of the proximal phalanx of the index finger	Deep branch of ulnar (C8 and **T1**)	Abducts and flexes the index finger at its MP joint and extends it at its IP joints
2nd Dorsal interosseous	Adjacent sides of the 2nd and 3rd metacarpals	Lateral sides of the extensor expansion and the base of the proximal phalanx of the middle finger	Deep branch of ulnar (C8 and **T1**)	Abducts and flexes the middle finger at its MP joint and extends it at its IP joints
3rd Dorsal interosseous	Adjacent sides of the 3rd and 4th metacarpals	Medial sides of the extensor expansion and the base of the proximal phalanx of the middle finger	Deep branch of ulnar (C8 and **T1**)	Abducts and flexes the middle finger at its MP joint and extends it at its IP joints
4th Dorsal interosseous	Adjacent sides of the 4th and 5th metacarpals	Medial sides of the extensor expansion and the base of the proximal phalanx of the ring finger	Deep branch of ulnar (C8 and **T1**)	Abducts and flexes the ring finger at its MP joint and extends it at its IP joints

Table 11–3
The Palmar Interossei

Muscle	Origin	Insertion	Nerve Supply	Actions
1st Palmar interosseous	Palmar surface of the 2nd metacarpal	Medial side of the extensor expansion of the index finger	Deep branch of ulnar (C8 and **T1**)	Adducts and flexes the index finger at its MP joint and extends it at its IP joints
2nd Palmar interosseous	Palmar surface of the 4th metacarpal	Lateral side of the extensor expansion of the ring finger	Deep branch of ulnar (C8 and **T1**)	Adducts and flexes the ring finger at its MP joint and extends it at its IP joints
3rd Palmar interosseous	Palmar surface of the 5th metcarpal	Lateral side of the extensor expansion of the little finger	Deep branch of ulnar (C8 and **T1**)	Adducts and flexes the little finger at its MP joint and extends it at its IP joints

Some anatomists describe four instead of three palmar interossei. In descriptions of four palmar interossei, the 2nd, 3rd, and 4th palmar interossei respectively correspond to the 1st, 2nd, and 3rd palmar interossei described in this text. When reference is made to four palmar interossei, the first palmar interosseous is described as originating from the palmar surface of the base of the1st metacarpal and inserting onto the medial side of the base of the proximal phalanx of the thumb; its innervation is by the deep branch of the ulnar nerve and its actions are flexion and adduction of the thumb at its MP joint. In this text, this muscle is regarded as part of the adductor pollicis muscle (see Table 11-4).

Table 11–4
The Short Muscles of the Thumb

Muscle	Origin	Insertion	Nerve Supply	Actions
Abductor pollicis brevis	Flexor retinaculum, scaphoid, and trapezium	Lateral side of the base of the proximal phalanx of the thumb	Recurrent branch of median (**C8** and T1)	Abducts the thumb at its carpometacarpal joint
Flexor pollicis brevis	Flexor retinaculum and trapezium	Lateral side of the base of the proximal phalanx of the thumb	Recurrent branch of median (**C8** and T1)	Flexes the thumb at its carpometacarpal and MP joints
Opponens pollicis	Flexor retinaculum and trapezium	Lateral side of the 1st metacarpal	Recurrent branch of median (**C8** and T1)	Flexes and medially rotates the thumb at its carpometacarpal joint
Adductor pollicis	Transverse head from the palmar surface of the shaft of the 3rd metacarpal and oblique head from capitate and the bases of the 2nd and 3rd metacarpals	Medial side of the base of the proximal phalanx of the thumb	Deep branch of ulnar (C8 and **T1**)	Adducts the thumb at its carpometacarpal joint and flexes the thumb at its MP joint

Table 11–5
The Muscles of the Hypothenar Eminence

Muscle	Origin	Insertion	Nerve Supply	Actions
Abductor digiti minimi	Pisiform	Medial side of the base of the proximal phalanx of the little finger	Deep branch of ulnar (C8 and **T1**)	Abducts the little finger at its MP joint
Flexor digiti minimi	Flexor retinaculum and hook of the hamate	Medial side of the base of the proximal phalanx of the little finger	Deep branch of the ulnar (C8 and **T1**)	Flexes the little finger at its MP joint
Opponens digiti minimi	Flexor retinaculum and hook of the hamate	Medial side of the 5th metacarpal	Deep branch of the ulnar (C8 and **T1**)	Flexes and laterally rotates the 5th metacarpal at its carpometacarpal joint
Palmaris brevis	Flexor retinaculum and medial border of the palmar aponeurosis	Dermis on the medial border of the palm	Superficial branch of the ulnar (C8 and **T1**)	Wrinkles the skin on the medial side of the palm

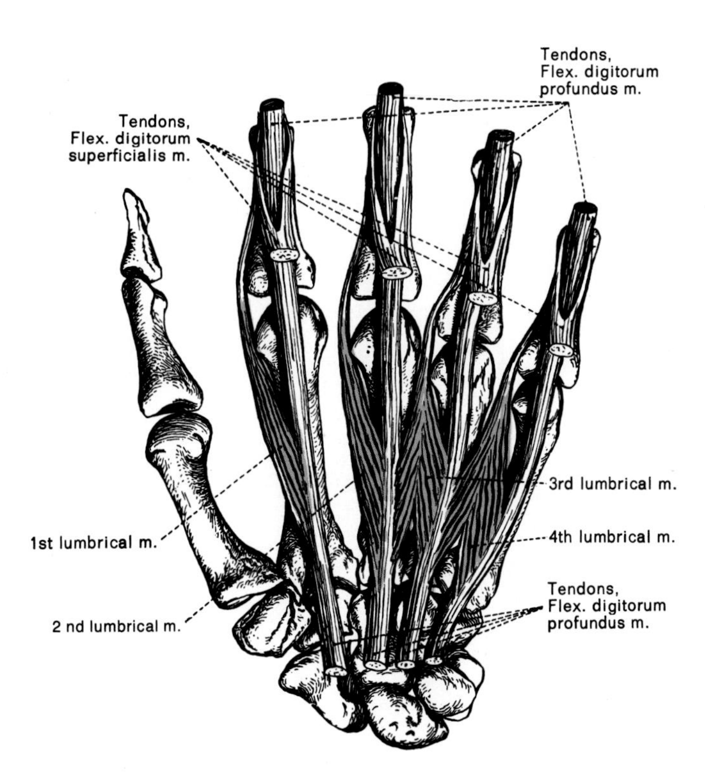

Fig. 11–2: Anterior view of the lumbrical muscles of the left hand.

The Palmar Interossei

There are three palmar interossei, numbered 1 through 3. Each palmar interosseous originates from the metacarpal of a finger (Fig. 11–4). The tendon of insertion passes anterior to the side of a finger's MP joint before inserting into the same side of the finger's extensor expansion.

The collective actions of the palmar interossei can be represented by the acronym **PAD:** if the index, ring, and little fingers are abducted, the **P**almar interossei can **AD**duct these fingers into the anatomic position. Each palmar interosseous can also flex the finger at its MP joint and extend the finger at its IP joints.

The Short Muscles of the Thumb

There are four short muscles of the thumb. Three of these muscles contribute almost all of the mound of soft tissue that covers the anterior surface of the first metacarpal. This mound of soft tissue is called the **thenar eminence** (Fig. 10–1). The **radial longitudinal skin crease of the palm** (which palm readers refer to as the "life line") marks the medial border of the thenar eminence (Fig. 10–1).

The three short muscles of the thumb which form the substance of the thenar eminence are called the **muscles of the thenar eminence.** The muscles of the thenar eminence (abductor pollicis brevis, flexor pollicis brevis, and opponens pollicis) are all innervated by the the **recurrent branch of the median nerve;** this branch is so named because it curves proximally from its origin within the palm of the hand to its muscular innervations.

Abductor pollicis brevis originates from the flexor retinaculum, scaphoid, and trapezium, and inserts onto the lateral side of the base of the thumb's proximal phalanx (Fig. 11–5). It can abduct the thumb at its carpometacarpal and MP joints.

Flexor pollicis brevis originates from the flexor retinaculum, and inserts onto the lateral side of the base of the thumb's proximal phalanx (Fig. 11–5). It can flex the thumb at its carpometacarpal and MP joints.

Opponens pollicis originates from the flexor retinaculum, and inserts onto the lateral side of the first metacarpal (Fig. 11–5). It is the prime mover of opposition of the thumb.

Adductor pollicis has two heads of origin: the transverse head originates from the shaft of the third metacarpal, and the oblique head originates from the bases of the second and third metacarpals. The fibers from both heads converge to insert onto the medial side of the base of the thumb's proximal phalanx (Fig. 11–5). It can adduct the thumb at its carpometacarpal and MP joints and flex the thumb at its MP joint.

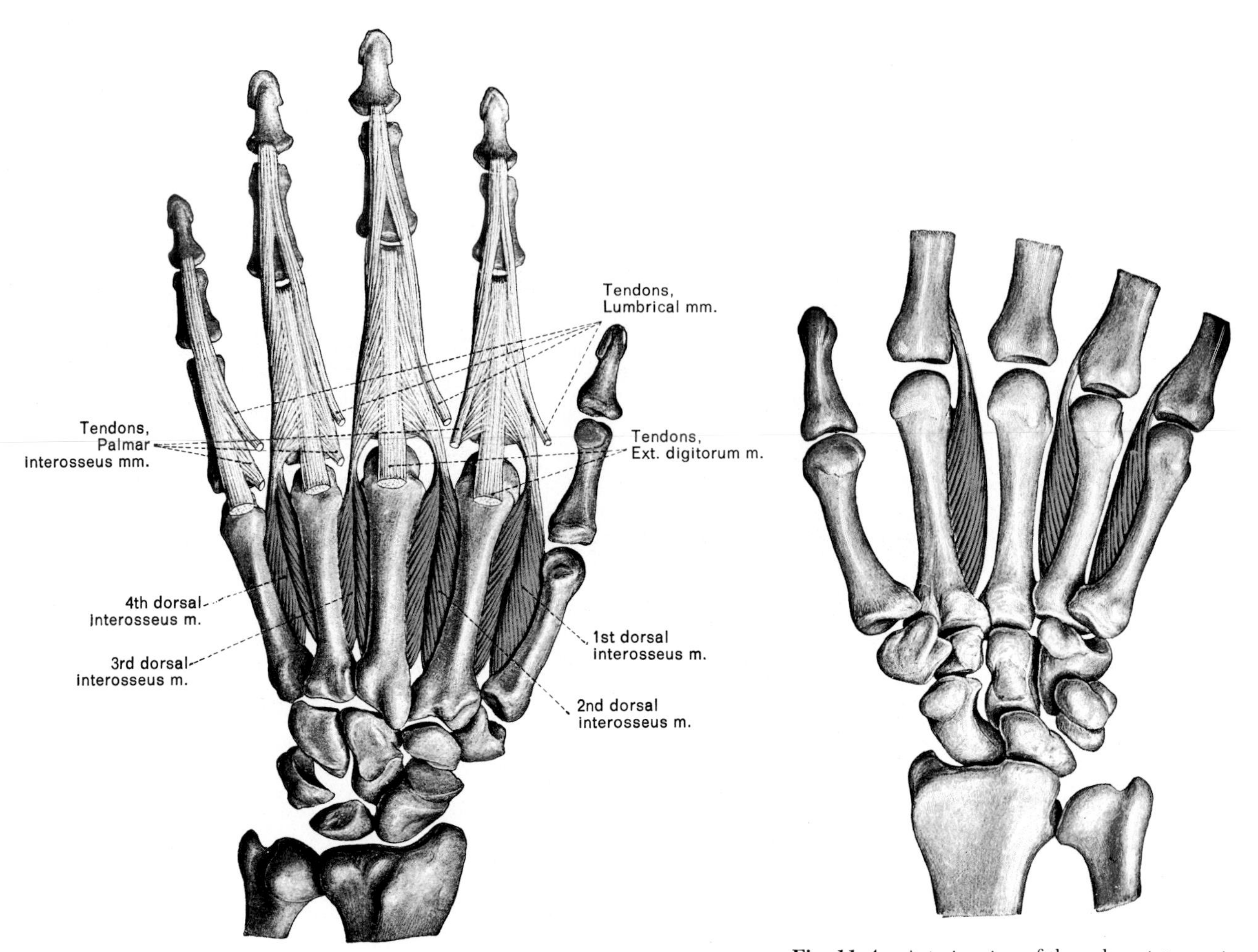

Fig. 11–3: Anterior view of the deep muscles of the right hand.

Fig. 11–4: Anterior view of the palmar interossei of the left hand.

The Muscles of the Hypothenar Eminence

The elongated mound of soft tissue extending along the medial side of the palmar surface of the hand is called the **hypothenar eminence** (Fig. 10–1). The bellies of four muscles form almost all of the hypothenar eminence.

Abductor digiti minimi originates from the pisiform, and inserts onto the medial side of the base of the proximal phalanx of the little finger (Fig. 11–5). It can abduct the little finger at its MP joint.

Flexor digiti minimi originates from the flexor retinaculum, and inserts onto the medial side of the base of the proximal phalanx of the little finger (Fig. 11–5). It can flex the little finger at its MP joint.

Opponens digiti minimi originates from the flexor retinaculum, and inserts onto the shaft of the fifth metacarpal (Fig. 11–5). It can flex and laterally rotate the fifth metacarpal at its carpometacarpal joint; this action deepens the hollow of the palmar surface.

Palmaris brevis extends through the superficial fascia in the palmar side of the hand (Fig. 10–16). It originates from the flexor retinaculum and palmar aponeurosis, and inserts into the dermis on the medial border of the palm. It can wrinkle the skin covering the hypothenar eminence.

THE CUTANEOUS INNERVATION OF THE HAND

The sensory supply of the hand is derived from all three of the major nerves that enter the hand: the median nerve, the ulnar nerve, and the superficial branch of the radial nerve (Fig. 11–6). The specific area of distribution for each nerve is not constant. Nonetheless, there is a common distribution which may be summarized as follows:

On the volar surface of the hand, branches of the ulnar nerve supply the cutaneous innervation for the medial one-third of the palm and the medial one and a half digits (Fig. 11–6). Branches of the median nerve supply the cutaneous innervation for the remainder of the volar surface (the lateral two-thirds of the palm and the lateral three and a half digits). The median and ulnar nerve branches that supply cutaneous innervation for the volar surface of the palm (which are called the **palmar cutaneous branches of the median and ulnar nerves**) arise from their respective nerves proxi-

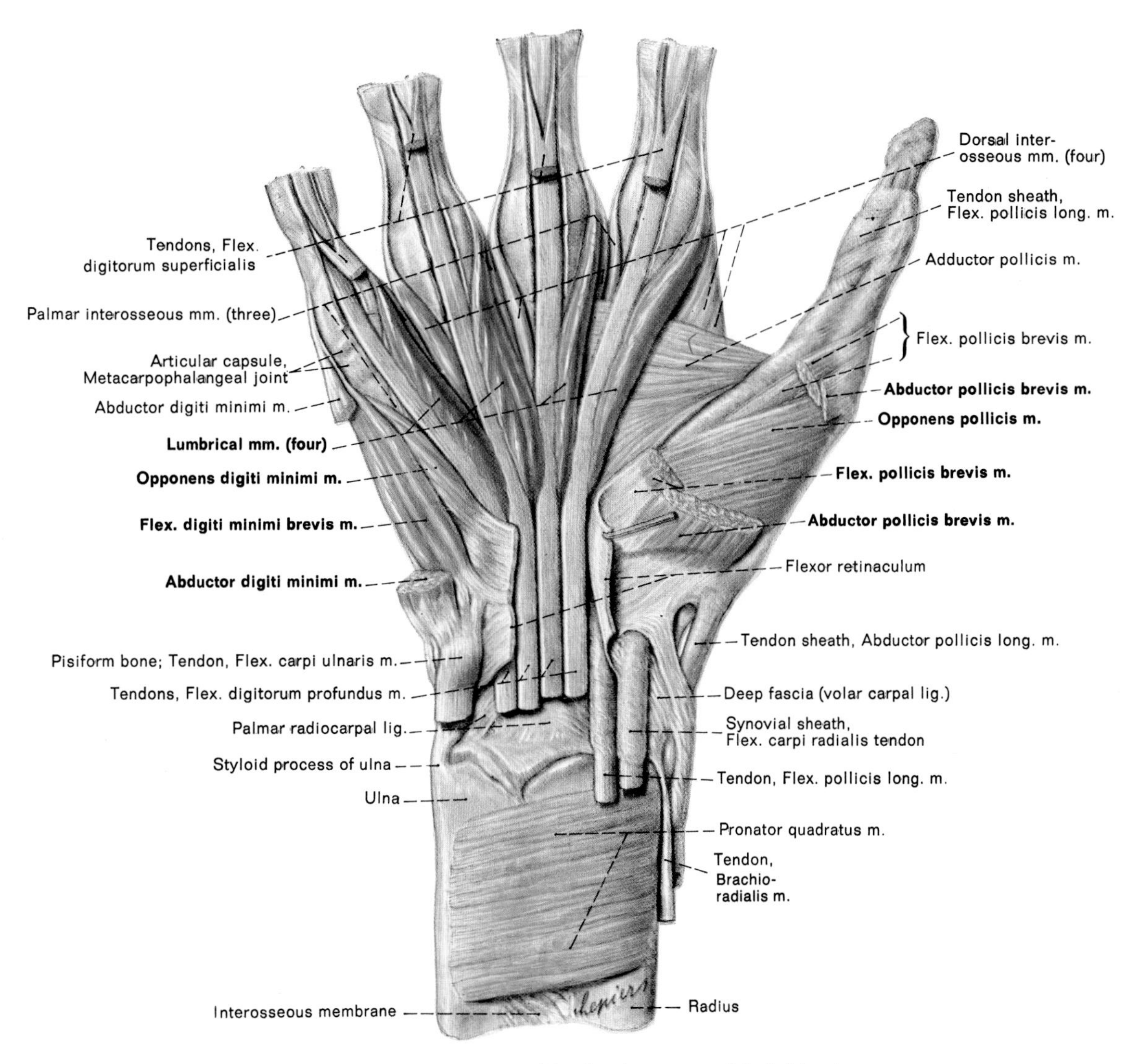

Fig. 11–5: Posterior view of the dorsal interossei of the left hand.

mal to the wrist and cross the wrist anterior to the flexor retinaculum. By contrast, the median and ulnar nerve branches that supply cutaneous innervation for the volar surfaces of the digits (which are called the **palmar digital branches of the median and ulnar nerves**) arise from their respective nerves distal to the flexor retinaculum.

On the dorsal surface of the hand, branches of the ulnar nerve supply the medial third of the dorsum of the hand, and the superficial branch of the radial nerve supplies the lateral two-thirds (Fig. 11–6). The cutaneous innervation for the dorsal surfaces of the digits is subject to much variation; the most common distribution pattern is shown in Figure 11–6.

The hand surfaces receiving sensory innervation from branches of the median nerve are the surfaces by which objects are primarily felt by the hand. It is for this reason that the median nerve is referred to as "the eye of the hand."

THE PALMAR ARTERIAL ARCHES

The blood supply of the hand is mainly provided by two arterial arches which course through the palm of the hand: the superficial palmar arch and the deep palmar arch (Fig. 11–7). Whereas the superficial palmar arch lies anterior to the tendons of flexor digitorum profundus and superficialis, the deep palmar arch lies posterior to these tendons.

The Superficial Palmar Arch

The superficial palmar arch is the direct continuation of the ulnar artery in the hand (Fig. 11–7). The ulnar artery enters the hand by passing superficial to the flexor retinaculum (Figs. 10–29 and 10–12). The ulnar artery gives off a deep branch immediately distal to the flexor retinaculum, and then curves laterally through the palm as the superficial palmar arch. The **proximal palmar crease** (which palm readers refer to as the

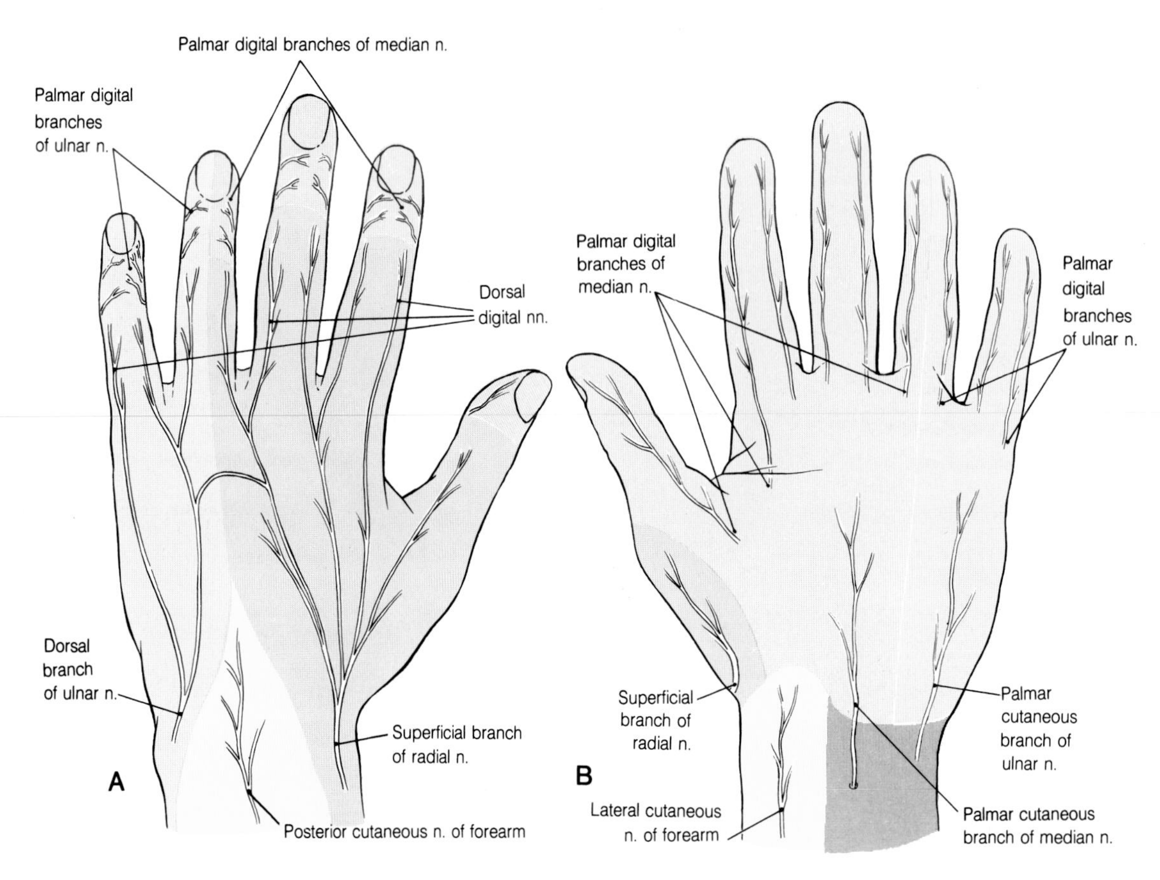

Fig. 11–6: The cutaneous innervation of (A) the dorsal and (B) the volar surfaces of the hand.

"head line") lies just distal to the superficial palmar arch (Fig. 10–1). The superficial palmar arch generally anastomoses in the lateral half of the palm with the superficial branch of the radial artery. The major branches of the superficial palmar arch are the digital arteries that supply the fingers.

The Deep Palmar Arch

The deep palmar arch is the direct continuation of the radial artery in the hand (Fig. 11–7). The radial artery enters the dorsal side of the hand after winding around the lateral side of the wrist (and coursing through the anatomic snuffbox) (Fig. 10–32). The radial artery gives rise to a superficial branch upon entering the hand, and then curves medially through the palm as the deep palmar arch. The deep palmar arch generally anastomoses in the medial half of the palm with the deep branch of the ulnar artery. The major branches of the deep palmar arch are (a) three dorsal metacarpal branches which anastomose with the digital branches of the superficial palmar arch, (b) an artery called **princeps pollicis** which supplies the thumb, and (c) an artery called **radialis indicis** which supplies the index finger.

The Allen Test

The superficial and deep palmar arches each receive blood from both the radial and ulnar arteries. However, there are individuals in whom the palmar arches are effectively supplied by an ulnar artery only or a radial artery only. An example of such an individual can be a hospitalized patient who has recently received fluids and/or medications by an ulnar or radial arterial line. Upon removal of the line, there may be a transient period of poor blood flow through the artery.

The Allen test assesses ulnar and radial arterial blood supply to the hand. The test consists of four simple steps:

1. The patient makes a tight fist with the hand. The making of a tight fist increases pressure on the superficial hand tissues, and empties the capillary beds of these tissues of much of their blood.
2. As soon as the patient has made a tight fist, the examiner presses his thumbs firmly on the ulnar and radial arteries (at the positions selected when taking a pulse). The pressure on the ulnar and radial arteries shuts down most of the blood flow into the palmar arches.

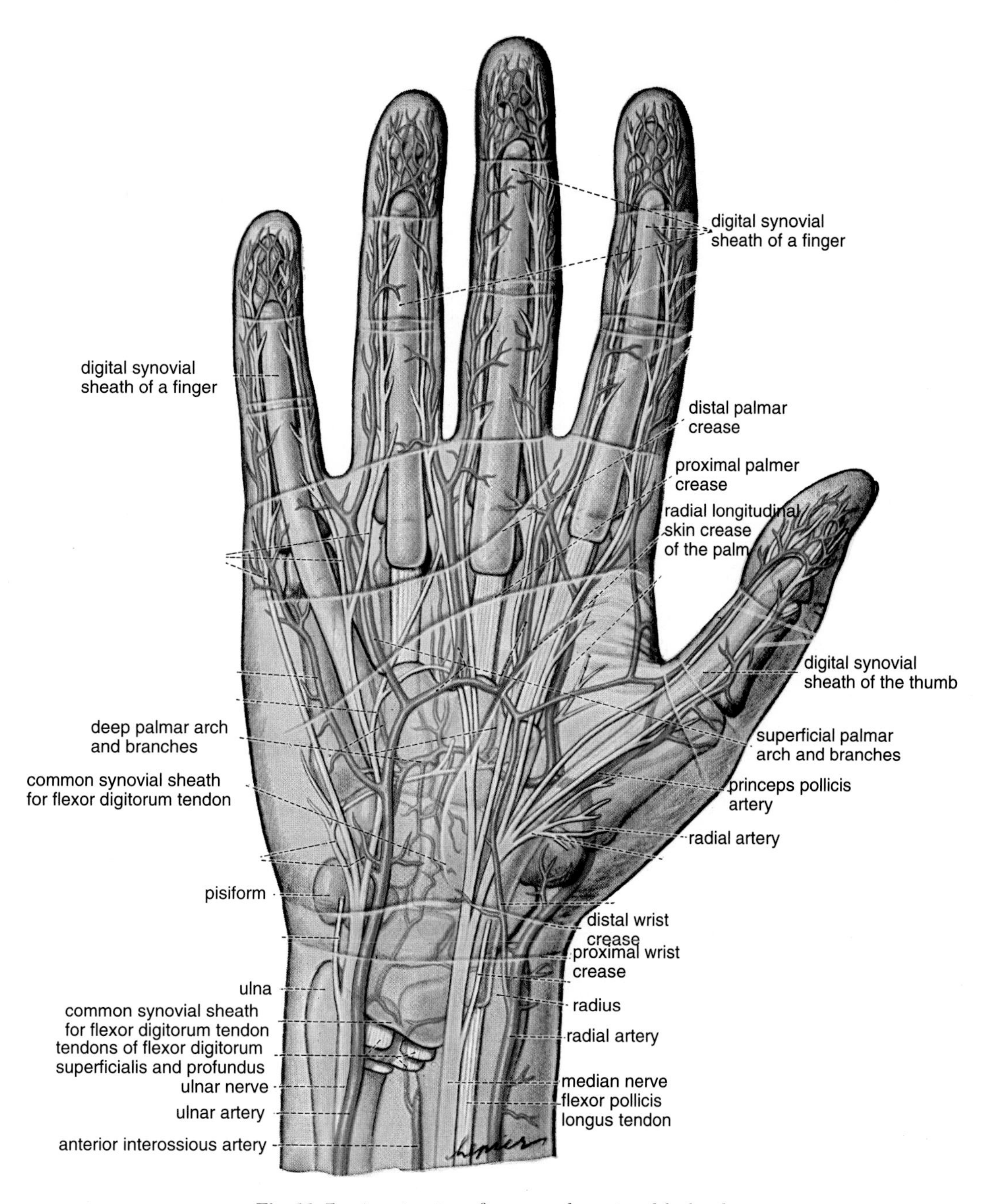

Fig. 11–7: Anterior view of nerves and arteries of the hand.

3. The patient unclenches the hand into a "normal" position. Because the capillary beds of the hand's superficial tissues have a depleted blood level, the color of the palmar surface will be blanched relative to that of the anterior surface of the patient's forearm.
4. The examiner releases his thumb over one of the arteries and notes the time interval required for restoration of hand color (2 to 4 seconds is normally required). The test is repeated with the other artery to ascertain that both arteries are supplying the palmar arches.

VENOUS AND LYMPHATIC DRAINAGE OF THE HAND

The major, net direction of venous flow within the hand is from the palmar side to the dorsal side. Most of this venous flow drains into a superficial venous arch in the dorsum of the hand (Fig. 10–31). The medial end of the venous arch gives rise to the basilic vein, and the lateral end gives rise to the cephalic vein. In the hospital, the venous arch is commonly selected as a site for intravenous administration of drugs and fluids.

The major, net direction of lymph flow within the hand is from the palmar side to the dorsal side. Lymph drained from the superficial tissues in the lateral half of the hand flows primarily into lymphatics that ascend the upper limb along the cephalic vein; these lymphatics drain into the infraclavicular group of axillary nodes. Lymph drained from the rest of the hand (the superficial tissues in the medial half of the hand and the deep tissues of the hand) primarily flows into lymphatics that ascend the upper limb along the basilic vein and deep veins; these lymphatics drain into the lateral group of axillary nodes. Some of the lymphatics draining the medial half of the hand and the middle, ring, and little fingers drain directly into the supratrochlear nodes of the cubital fossa. Finally, a few of the lymphatics draining the middle finger go directly into the supraclavicular nodes.

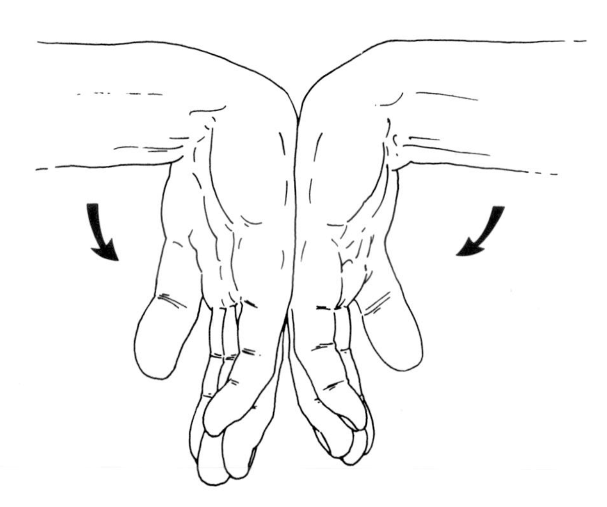

Fig. 11–8: Drawing illustrating the position of the patient's hands during Phalen's test.

CLINICAL PANEL II-9

RAYNAUD'S DISEASE AND RAYNAUD'S PHENOMENON

Definition: Raynaud's disease is an idiopathic vascular disorder in which exposure to cold or experience of stress induces spasmodic constriction of the digital arteries of the limbs. If the symptoms characteristic of Raynaud's disease are produced by an identifiable disorder or condition, such as one of the thoracic outlet syndromes, then the term Raynaud's phenomenon is used to refer to the sequence of vascular changes in the digits.

Common Symptoms: The attacks of Raynaud's disease characteristically occur bilaterally and may involve both the upper and lower limbs. The attacks generally produce, in sequence, pallor, cyanosis, and finally rubor of the affected digits. The intensity of the digital pain associated with the attacks is variable.

Anatomic Basis of Common Symptoms: The skin coloration of the digits becomes first pallid (blanched) because the constriction of the digital arteries depletes the blood level in the capillary beds of the digits. The skin of the digits next acquires a cyanotic (bluish) discoloration because of the deoxygenation of the stagnant blood in the capillary beds. Resolution of the attack by dilation of the digital arteries engorges the capillary beds with oxygenated blood, and the engorgement confers a brilliant ruddy complexion to the skin of the digits.

THE SYMPATHETIC FIBERS CONVEYED BY THE CUTANEOUS NERVES OF THE UPPER LIMB

The nerves of the upper limb that provide sensory fibers to the skin also convey postganglionic sympathetic fibers to the skin. Some of the sympathetic fibers innervate smooth muscle tissue in the walls of the microscopic blood vessels (arterioles and arteriovenous anastomoses) in the skin. Other sympathetic fibers innervate sweat glands and the arrector pili muscles of hair follicles in the skin.

The sympathetic fibers that innervate smooth muscle tissue in the skin's microvasculature help regulate body temperature. The skin of the body (particularly that of the hands and feet) bears an extensive, highly anastomotic, subcutaneous venous plexus; this plexus receives blood directly from the skin's capillaries and arteriovenous anastomoses. The venous plexus has the capacity to hold a comparatively large volume of blood. When the body becomes overheated, the plexus becomes engorged with warm blood, and serves as a radiator of blood-borne heat from the body. By contrast, when the body is exposed to very cold temperatures, the sympathetic fibers that innervate the smooth muscle tissue in the walls of the skin's arterioles and arteriovenous anastomoses stimulate the smooth muscle fibers to shorten. The resulting constriction of the arterioles and arteriovenous anastomoses serves to conserve body heat by greatly diminishing blood flow through the subcutaneous venous plexus.

CASE **II.6**

The Case of Joan Butler

INITIAL APPEARANCE AND PRESENTATION OF THE PATIENT

A 21 year-old white woman, Joan Butler, has made an appointment to seek treatment for a painful wrist. The patient appears to be in good health and not distressed.

QUESTIONS ASKED OF THE PATIENT

How can I help you? I would like for you to look at my wrist [the patient raises her right hand as she refers to her wrist]. I think I hurt it playing basketball two weeks ago, because it's been hurting ever since then. *[The examiner recognizes that the patient is reasonably but not absolutely certain that her wrist pain is a symptom of a sports-related injury. At this point, the examiner tentatively concludes that the patient has sustained one or more of the following injuries: a fractured bone, a partially or completely dislocated joint, a torn ligament, or a torn muscle.]*

Do you recall an injury to your wrist during the basketball game? Yes. I was tripped accidentally by one of the players, and I remember falling down hard on my right hand.

Do you recall how your right hand was positioned when you fell on it? Yes. [The patient demonstrates that she fell down on the palmar surface of her outstretched right hand.] *[These last two questions provide the examiner with information about the mechanism of injury.]*

Where is the pain? Here [the patient gently grasps the lateral aspect of her right wrist between the thumb and fingers of her left hand]. *[When pain is a complaint, it is always essential to have the patient mark by word or gesture the exact site(s) at which pain is perceived. In cases of physical injury, pain is generally a reliable indicator of the site(s) of injury.]*

How would you describe the pain? A dull ache.

Has the pain affected your daily or regular activities? Yes. I'm right-handed, and the thing that's bothered me the most is that the pain will suddenly get worse if I do certain things with my right hand. In particular, I find turning doorknobs painful. So, I've had to change the way I do a lot of things in order to cut down on the use of my right hand.

Is there anything that relieves the pain? Well, I've been taking Advil [ibuprofin] and it helps ease the pain a little. *[In general, these last two questions provide information about the source of the pain. In this instance, the finding that the pain is worsened by movement and minimized with inactivity suggests injury of one or more musculoskeletal structures.]*

Is there anything else that is also bothering you? No. *[As in the cases of John Robinson (Case II-1) and Linda Chin (Case II-5), this question addresses the examiner's concern that there might be neurovascular complications to the patient's upper limb injury or injuries. In this case, a complaint of tingling sensations or a loss of feeling in the hand or a complaint of difficulty in grasping or holding objects would suggest nerve injury. Complaints of pallor or coldness would suggest injury to a major artery.]*

[The examiner finds the patient alert and fully cooperative during the interview.]

PHYSICAL EXAMINATION OF THE PATIENT

Vital signs: Blood pressure
Lying supine: 120/70 left arm and 120/70 right arm
Standing: 120/70 left arm and 120/70 right arm
Pulse: 68
Rhythm: regular
Temperature: 98.5°F
Respiratory rate: 15
Height: 5′11″
Weight: 130 lbs.
HEENT Examination: Normal
Lungs: Normal
Cardiovascular Examination: Normal. Pertinent normal findings in the right upper limb include (1) strong and regular pulsations of the radial and ulnar arteries at the wrist, (2) the occurrence of capillary refill in the index and little fingers within the normal 1 to 2 second interval, and (3) the absence of any signs of vascular disturbance (signs such as pallor, coldness, or cyanosis).

Abdomen: Normal
Genitourinary Examination: Normal
Musculoskeletal Examination: Inspection, movement and palpation of the upper and lower limbs are normal except for the following findings for the right upper limb: Active and passive extension of the hand elicit pain at 70° extension. Supination against resistance (with the examiner holding the patient's right hand in a neutral position) elicits wrist pain. Palpation reveals tenderness in the anatomic snuffbox. Axial compression of the thumb toward the anatomic snuffbox elicits pain.
Neurologic Examination: Normal. Pertinent normal findings include the absence of any sensory or motor loss in the right hand.

INITIAL EVALUATION OF THE PATIENT'S CONDITION

The patient has sustained an injury of the right wrist.

ANATOMIC BASIS OF THE PATIENT'S HISTORY AND PHYSICAL EXAMINATION

1. **The patient has sustained an acute hyperextension injury of musculoskeletal tissues in the right wrist.** The history suggests that forced hyperextension of the right hand produced a musculoskeletal injury in the wrist.
2. **The tenderness in the anatomic snuffbox marks the region as a site of injury.** The musculoskeletal tissues of the anatomic snuffbox consist of four bony components (the styloid process of the radius, scaphoid, trapezium, and the base of the first metacarpal), the ligamentous and capsular tissues that unite these bony components, and the tendons of three posterior forearm muscles (abductor pollicis longus, extensor pollicis brevis, and extensor pollicis longus).

CLINICAL REASONING PROCESS

This case poses the same diagnostic challenges presented by the last case, the case of Linda Chin (Case II.5). Both cases involve a person who has sustained transmission of injurious, stressful forces along the entirety of the upper limb. Consequently, the entire upper limb must be carefully examined for musculoskeletal and neurovascular injuries. Such examination in this case reveals only a wrist injury.

An orthopedist would assume that the patient's wrist injury is most likely the result of forced hyperextension of the hand. Such a mechanism of injury can produce a wide variety of dislocations, fractures, and torn ligaments in the wrist. A **scaphoid fracture** is the most common carpal fracture; the fracture generally occurs in the scaphoid's waist (the narrowed midregion that joins its proximal and distal poles). The carpal dislocations most commonly produced by forced hyperextension of the hand are volar dislocation of the lunate into the carpal tunnel (**anterior lunate dislocation**), dorsal dislocation of all the carpals relative to the lunate (**posterior perilunate dislocation**), and rotational dislocation of the scaphoid (**scapholunate dissociation**).

Orthopedists regard any individual who presents with tenderness in the anatomic snuffbox following forced hyperextension of the hand as having a fractured scaphoid unless proven otherwise. In this case, the examiner establishes tenderness in the anatomic snuffbox by both direct and indirect pressure upon the bony floor of the anatomic snuffbox: direct pressure is applied by palpation, and indirect pressure is applied by axial compression of the thumb.

RADIOGRAPHIC EVALUATION AND FINAL RESOLUTION OF THE PATIENT'S CONDITION

In cases of a suspected scaphoid fracture, four radiographic views of the wrist are routinely obtained: a PA (posteroanterior) view with the hand in the standard (anatomic) position, a PA view with the hand adducted, an oblique view, and a lateral view. Adduction of the hand reduces the volar tilt of the scaphoid in the wrist, and thus the projection of the scaphoid in the adducted PA view is different from that in the standard PA view. Frequently, only one of these four different scaphoid projections reveals a scaphoid fracture during the first 1 to 2 days after the injury; this is because many scaphoid fractures are hairline, undisplaced fractures. In cases in which a scaphoid fracture is strongly suspected but not radiographically confirmed, the wrist is immoblized in a plaster cast (to provide the appropriate conditions for union of the proximal and distal fragments), and radiographic evaluation is repeated 7 to 10 days later. The subsequent bone resorption that occurs at the fracture site renders the site more radiolucent and thus more apparent in the latter set of radiographs.

In this case, all four projections showed a transverse, undisplaced fracture through the waist of the scaphoid (visualization of the fracture was enhanced by the bone resorption that occurred at the fracture site during the past two weeks). Treatment for an undisplaced scaphoid fracture is rigid immobilization of the forearm, wrist, and hand in a plaster cast extending from the proximal forearm to the midshaft of the proximal phalanx of the thumb and the midshafts of the second to fifth metacarpals in the palm of the hand. The only joints in the hand at which the cast permits movement are the MP and IP joints of the fingers and the IP

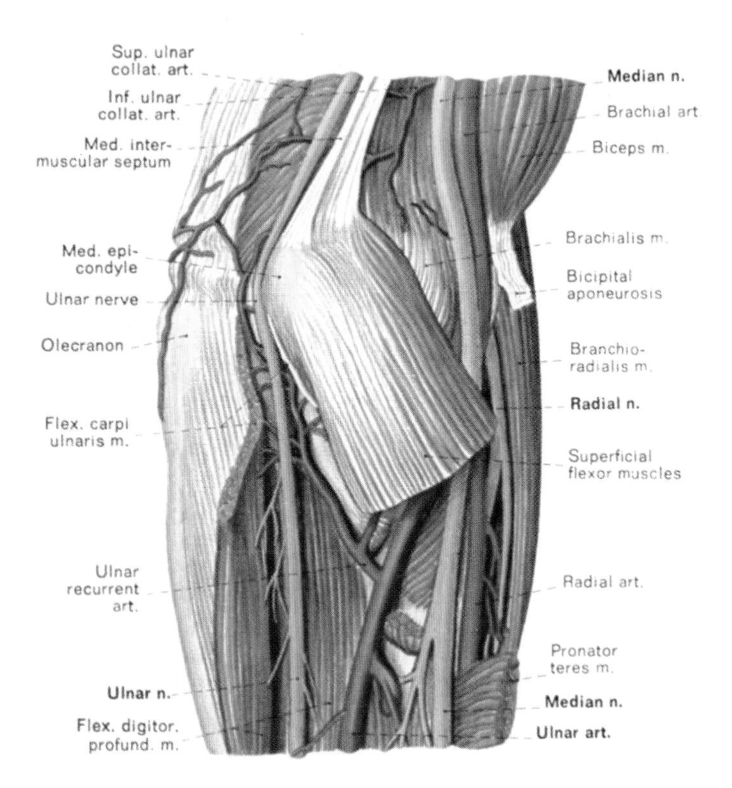

Fig. 11–9: Medial view of the nerves and arteries of the left elbow.

joint of the thumb. Such an extensive plaster cast is necessary because union will not occur unless complete and uninterrupted immobilization is provided for a minimum of six weeks. Union is radiographically assessed by the appearance of trabeculae bridging across the fracture site. In this case, the patient is at risk for nonunion because of the delay in seeking medical attention. Failure of bony union may result in avascular necrosis of the proximal fragment; this is because most or all of the scaphoid's nutrient foramina are located on the surface of the distal half of the bone.

THE MECHANISM OF THE PATIENT'S INJURY

The patient tripped and fell to her right while playing basketball. She used her right upper limb to brace her upper body against the impact with the floor. At the moment of impact, her right arm was partially flexed and abducted at the shoulder, the forearm fully extended at the elbow, and the hand fully extended at the wrist and midcarpal joints. When the hand is fully extended, all the ligaments of the wrist are taut and the carpals are closely packed. Impact with the floor forced the patient's right hand into hyperextension, and the sudden increased transverse load across the waist of the scaphoid produced a fracture at this site.

CASE **II.7**

The Case of Sarah Goldbeck

INITIAL PRESENTATION AND APPEARANCE OF THE PATIENT

A 72 year-old white woman, Sarah Goldbeck, has made an appointment to seek treatment for weakness in the hand. She appears physically fit and robust for a woman of her age.

QUESTIONS ASKED OF THE PATIENT

How can I help you? Doctor, I've lost strength in my right hand, and that's important to me because I'm right-handed. At first, I thought it was just the vagaries of old age, but the problem has persisted and, in fact, I think it has worsened. Starting about six months ago, I noticed I began having difficulty doing a lot of small things with my right hand. One of the first problems I noticed was that it became more difficult for me to hold a pen and write. Also, when I would play bridge with my girl friends, I noticed that I had difficulty in dealing or even holding the cards. However, the problem has become more serious during the past month, and that's why I have come to see you. I now have difficulty in gripping and turning doorknobs to open doors, and, a few days ago, I lost my grip on a cup of hot tea in the morning. *[The patient's chief complaint is muscle weakness, a symptom which signifies neuromuscular dysfunction. Neuromuscular dysfunction may involve disease or injury of (1) upper motor neurons of the cerebral cortex and/or their fibers in the corticospinal tracts of the brainstem and spinal cord, (2) lower motor neurons in the anterior gray columns of the spinal cord and/or their fibers in peripheral nerves, and/or (3) muscle fibers in one or more muscles. Neuromuscular dysfunction thus implies disease or injury of the brain, the spinal cord, the spinal nerves and their peripheral branches, and/or skeletal muscles. In this case, the patient's account of her muscular weakness indicates a chronic condition whose symptoms began insidiously and have progressively worsened. The examiner decides to ask the patient about other changes in her daily activities, changes which would imply neuromuscular dysfunction involving the head and neck, left upper limb, one or both lower limbs, and/or the sphincters that provide control for the timing of urination and defecation. The purpose of this inquiry is to determine the patient's perception of the extent of her muscular weakness.]*

Have you noticed similar weakness in your left hand? No. In fact, I have started using my left hand more. For example, when I have had my morning tea for the past four days, I have used my left hand to hold the cup of hot tea.

Have you noticed any problems getting out of bed in the morning or rising from a chair? No, other than the usual aches and pains.

Have you had any problems standing or walking? No.

Have you had any problems with chewing or swallowing? No. *[Chewing is an activity that involves the prime movers of the lower jaw (the muscles of mastication), the muscles of the tongue, and some of the muscles of facial expression (specifically, those that move the cheeks and lips). Swallowing is an activity that involves the muscles of the soft palate and pharynx.]*

Have you had any problems with bladder or bowel control? No. *[The negative response to each of the last five questions indicates that it is the patient's perception that her muscular weakness is limited to her right upper limb.]*

Correct me if I'm wrong, but I presume you brush your hair with your right hand. Have you had any problems raising and moving your right arm to brush your hair? No, but I have noticed that my grip on the handle of my hair brush is not quite the same way it used to be. *[The patient's negative response to this question indicates that she does not perceive any weakness in her right shoulder muscles. Raising the arm to brush or comb one's hair chiefly involves the abductors of the arm (supraspinatus, deltoid, trapezius, and serratus anterior). Moving an upraised hand forward and backward over the head chiefly involves the arm's medial rotators (pectoralis major, teres major, latissimus dorsi, the anterior fibers of deltoid, and subscapularis) and lateral rotators (posterior fibers of deltoid, infraspinatus, and teres minor).]*

Have you noticed any weakness in carrying heavy things, such as a heavy bag of groceries, in your right hand? No. *[The patient's negative response to this question indicates that she does not perceive any weakness in her right anterior arm and forearm muscles. This is because support of a heavy bag of groceries chiefly involves the flexors of the forearm (biceps brachii and brachialis), the flexors of the hand (flexor carpi radialis, palmaris longus, and flexor carpi ulnaris), and the anterior forearm muscles which can powerfully flex the thumb and fingers (flexor pollicis longus, flexor digitorum superficialis, and flexor digitorum profundus).]*

Have you noticed any weakness in pushing a swinging door forward with your right hand? No. *[The patient's negative response to this question indicates that she does not perceive any weakness in her right posterior arm and forearm muscles. This is because pushing a swinging door forward partly involves the chief extensor of the forearm (triceps brachii), the extensors of the hand (extensor carpi radialis longus, extensor carpi radialis brevis, and extensor carpi ulnaris), and the posterior forearm muscles which can extend the thumb and fingers (abductor pollicis longus, extensor pollicis brevis, extensor pollicis longus, extensor indicis, extensor digitorum, and extensor digiti minimi). [At this point, the examiner concludes that the patient perceives weakness in only one or more of the intrinsic muscles of the right hand.]*

Have you noticed any other problem with your right hand? Yes. I've lost some feeling in my right hand [at this point, the patient awkwardly rubs the tip of the palmar surface of the right thumb back and forth over the tips of the palmar surfaces of the index and middle fingers]. And I'm also frequently woken up at night by tingling or a burning pain in my right hand. I'm able to make the tingling or pain go away by shaking and moving my hand about. *[The response suggests that the neuromuscular dysfunction in the right hand is associated with sensory dysfunction.]*

How long have you noticed a loss of feeling in your right hand? I don't know. I would say it started sometime before I began to lose strength in my hand.

How long have you been bothered at night with the tingling or burning pain in your right hand? I just don't know, but again, I would say it started sometime before I began to lose strength in my hand.

When you're awakened at night with tingling or pain in your right hand, where do you feel the tingling or the pain? Over here [the patient slides the tips of the index and middle fingers of her left hand back and forth over the volar surfaces of the thumb, index finger, and middle finger of her right hand]. *[At this point, the examiner tentatively concludes that the patient is suffering from a chronic condition that has produced three major symptoms, all of which appear to be confined to the right hand: muscle weakness, pain, and hypesthesia.]*

Is there anyting else that is also bothering you? No.

Have you had any recent illnesses or injuries? Well, I fractured my wrist a little over a year ago [the patient indicates with her left hand that she fractured her right wrist].

Did you have any problems with your right wrist or hand following treatment of the fracture? No. The orthopedist who treated me took X-rays after the cast came off and told me that all the bones were in their right places and that the fracture had healed well. I didn't start having problems with my right hand until about nine months after I fractured my wrist.

Are you taking any medications for previous illnesses or conditions? No. *[The examiner finds the patient alert and fully cooperative during the interview.]*

PHYSICAL EXAMINATION OF THE PATIENT

Vital signs: Blood pressure
Lying supine: 130/70 left arm and 130/75 right arm
Standing: 135/70 left arm and 130/75 right arm
Pulse: 75
Rhythm: regular
Temperature: 98.2°F.
Respiratory rate: 17
Height: 5′2″
Weight: 115 lbs.
HEENT Examination: Normal
Lungs: Normal
Cardiovascular Examination: Normal
Abdomen: Normal
Genitourinary Examination: Normal
Musculoskeletal Examination: Inspection, movement, and palpation of the upper and lower limbs are normal except for the following findings for the right upper limb: Isometric exercises reveal a muscle strength of −1 for opposition and abduction of the thumb. Pertinent normal findings include normal strength for flexor pollicis longus, flexor digitorum profundus, flexor digitorum superficialis, extensor digitorum, and the interossei in the right hand.

Many examiners assign limb muscle strength in terms of six grades. Grade 0 represents full and normal strength. Grade –1 represents strength below normal but great enough to move a limb part against appreciable resistance. Grade –2 represents strength sufficient to move a limb part against the force of gravity but not great enough to move the limb part against resistance imposed by the examiner. Grade –3 represents strength just sufficient to move a limb part in the absence of the influence of gravity. Grade –4 represents very severe weakness. Grade –5 represents complete paralysis.

The strength of an individual muscle or muscle group in the right hand is judged relative to the strength of the corresponding muscle(s) in the patient's left hand. Thumb opposition strength is assessed as the patient attempts to keep the thumb in opposition against resistance. Thumb abduction strength is assessed as the patient attempts to abduct the thumb against resistance exerted at the thumb's MP joint. The –1 grade of strength for thumb abduction and opposition in the right hand indicates a slight loss (roughly 10%) of strength.

Flexor pollicis longus strength is assessed as the patient attempts to flex the thumb's distal phalanx against resistance. Flexor digitorum profundus strength is assessed as the patient attempts to flex each finger's distal phalanx against resistance. Flexor digitorum superficialis strength is assessed as the patient attempts to flex each finger's middle phalanx against resistance while not simultaneously flexing the digit's distal phalanx (this latter request minimizes the flexor action of flexor digitorum profundus at each finger's proximal IP joint). Extensor digitorum strength is assessed as the patient attempts to extend the middle and ring fingers at their MP joints against resistance. Palmar and dorsal interossei strength is assessed as the patient attempts to hold a piece of paper between the index and middle fingers against resistance; this procedure specifically tests the strength of the first palmar interosseous and the second dorsal interosseous.

Neurologic Examination: The application of pin pricks reveals hypalgesia on the volar surfaces of the right thumb, index finger, and middle finger (first, second, and third digits). The application of a fine wisp of cotton reveals hypesthesia on the volar surfaces of the right thumb, index finger, and middle finger (first, second, and third digits). Pertinent normal findings include normal pain and light touch sensation on the volar surface of the palm. Tinel's sign is elicited upon percussion applied to the flexor retinaculum of the right wrist; the distribution of pain and tingling matches the volar areas of hypalgesia and hypesthesia. Phalen's test is positive for the right wrist, and produces pain and tingling whose distribution matches the volar areas of hypalgesia and hypesthesia. *[In this case, Tinel's sign is observed upon tapping a reflex hammer on the lateral side of the palmaris longus tendon along the tendon's course over the flexor retinaculum: this site overlies the course of the median nerve through the carpal tunnel (Fig. 10–12). Phalen's test is conducted by requesting the patient to maximally flex both hands at the wrists and to hold this position for 1 minute by pressing the dorsal surfaces of both hands against each other (Fig. 11–8). Elicitation of pain and/or tingling in any part of the volar surfaces of the lateral three and a half digits represents a positive test for both procedures.]*

INITIAL ASSESSMENT OF THE PATIENT'S CONDITION

The patient appears to be suffering from a chronic condition associated with muscle weakness, hypalgesia, hypesthesia and nocturnal pain in the right hand.

ANATOMIC BASIS OF THE PATIENT'S HISTORY AND PHYSICAL EXAMINATION

1. **The patient's description of an insidious onset of muscle weakness in the right hand suggests a chronic disease or condition.** The patient reports progressive loss of muscle strength in the right hand during the past six months.
2. **The cutaneous areas of hypalgesia, hypesthesia, and nocturnal pain in the right hand correspond to areas innervated by the digital cutaneous branches of the median nerve.** This finding points to the right hand as the site of the lesion responsible for the sensory dysfunction in the patient's right hand, because the sensory dysfunction involves only those cutaneous branches of the median nerve that arise distal to the carpal tunnel.
3. **The only movements of the right upper limb whose strength is demonstrably diminished are opposition and abduction of the thumb.** This finding points to the right hand as the site of the lesion responsible for the neuromuscular dysfunction in the patient's right hand. This is because the only weak hand movements objectively detected by physical examination (opposition and abduction of the thumb) are both dependent upon the actions of two of the muscles of the thenar eminence (opponens pollicis and abductor pollicis brevis). Opponens pollicis is the prime mover for opposition of the thumb. Abductor pollicis brevis and abductor pollicis longus are the sole abductors of the thumb. The muscles of the thenar eminence are located exclusively in the hand and are innervated by a branch of the median nerve that arises distal to the carpal tunnel.

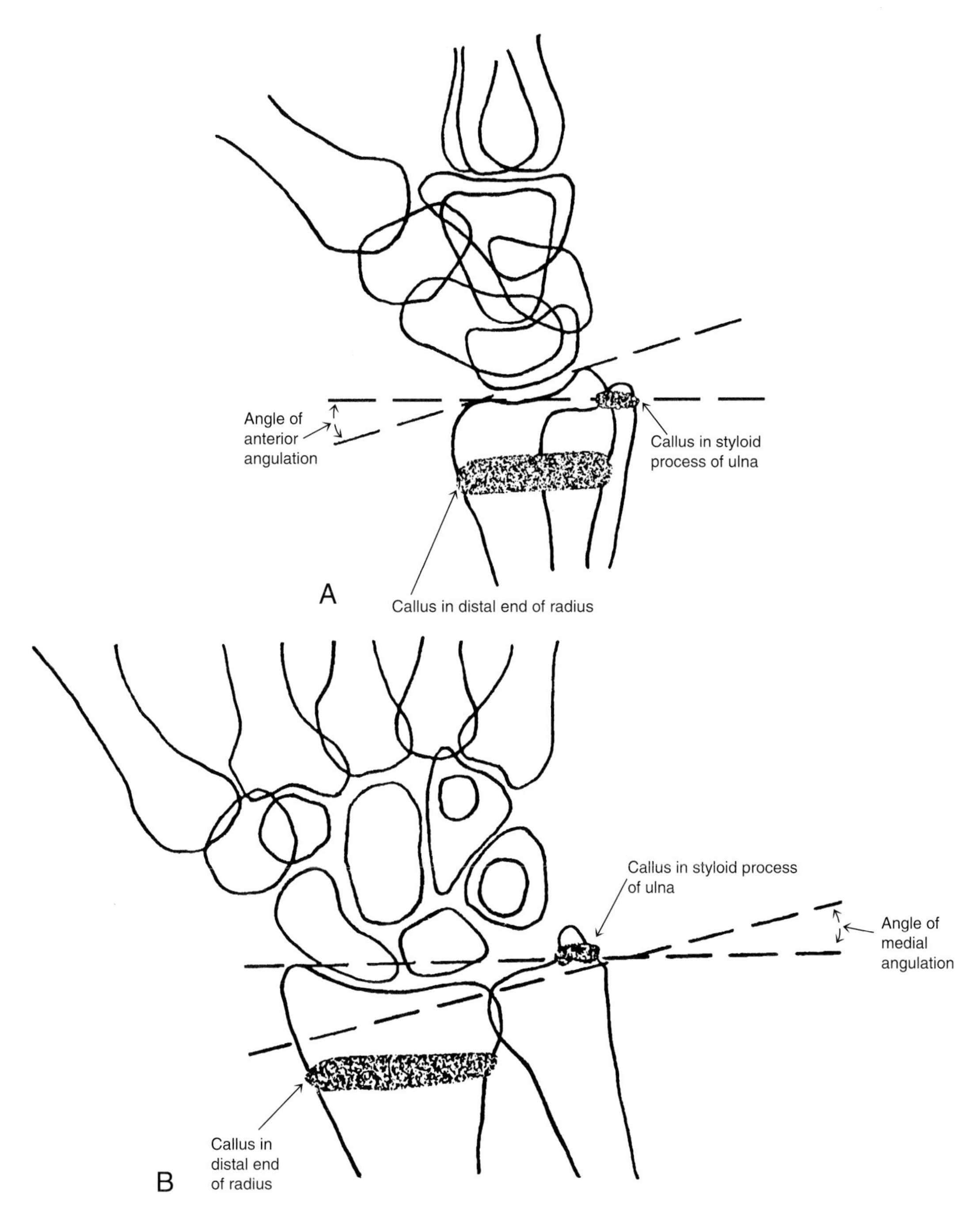

Fig. 11–10: Outlines of the (A) lateral and (B) PA radiographs of Sarah Goldbeck's right wrist.

4. **The elicitation of Tinel's sign upon percussion of the flexor retinaculum and the positive Phalen's test indicate hypersensitivity of the median nerve within the carpal tunnel.** Tinel's sign is pain and/or tingling produced upon percussion of a peripheral nerve of the limbs. The distribution of the pain and/or tingling corresponds to the distribution of the sensory fibers that branch off from the nerve distal to the site of percussion. Tinel's sign indicates hypersensitivity of the nerve's sensory fibers because of a partial lesion or recent regeneration. In this case, the light percussion of a reflex hammer was used to impart a momentary pressure increase on the median nerve along its course through the carpal tunnel, and the percussion produced pain and tingling with a distribution corresponding to the areas innervated by the digital cutaneous branches of the median nerve.

The median nerve extends through the carpal tunnel superficial to the tendons of the flexor digitorum muscles (Fig. 10–12). Flexion of the hand at the wrist and midcarpal joints pulls the flexor digitorum tendons anteriorly within the carpal tunnel; this movement presses the overlying median nerve against the proximal edge of the flexor retinaculum. Phalen's test employs hand flexion to impart a sustained pressure increase on the median nerve at its crossing beneath the proximal edge of the flexor retinaculum. Phalen's test is a procedure alternative to Tinel's test for detecting hypersensitivity of the median nerve along its course through the carpal tunnel. In this case, it produced pain and tingling whose distribution matched that elicited by Tinel's test.

INTERMEDIATE EVALUATION OF THE PATIENT'S CONDITION

The history and physical examination indicate that the patient is suffering from a chronic condition that has impaired the motor and sensory supply of the median nerve to the right hand.

CLINICAL REASONING PROCESS

The patient's initial description of weakness in her right hand indicates that she has been suffering from a chronic condition that has progressively worsened during the past six months. The history indicates that the weakness in the right hand is accompanied by sensory deficits and that the weakness appears to be confined primarily if not exclusively to digital movements.

In evaluation of the patient's condition, an orthopedist, neurologist, or primary care physician would consider diseases or conditions that could produce both sensory and motor disturbances of the upper limb. In this case, the diseases and conditions may be listed according to the lesion site; the following list begins with the most distal sites and proceeds to more proximal or central sites:

I. Peripheral nerve entrapment
  A. Carpal tunnel syndrome
  B. Ulnar tunnel syndrome
  C. Anterior interosseous syndrome
  D. Pronator teres syndrome
  E. Cubital tunnel syndrome
II. Thoracic outlet syndrome (includes scalenus anticus syndrome, Pancoast's syndrome, and costoclavicular syndrome)
III. Cervical radiculopathy
  A. Cervical spondylosis
  B. Herniation of a cervical intervertebral disc
IV. Central nervous system lesions (includes minor stroke, spinal cord tumor, and multiple sclerosis)

Nerve impingement is the mechanism by which most of these conditions produce motor and/or sensory deficits in the upper limb. In the carpal tunnel syndrome, the median nerve becomes entrapped or compressed along its course through the carpal tunnel. In the ulnar tunnel syndrome, the ulnar nerve becomes entrapped or compressed along its course over (superficial to) the flexor retinaculum (where it lies deep to a ligamentous band extending between the pisiform and hamate). In the anterior interosseous syndrome, the anterior interosseous branch of the median nerve becomes entrapped or compressed, frequently near its origin in the proximal forearm. In the pronator teres syndrome, the median nerve becomes entrapped or compressed as it extends into the forearm between the humeral and ulnar heads of pronator teres (Fig. 10–27). In the cubital tunnel syndrome, the ulnar nerve becomes entrapped or compressed along its course behind the medial epicondyle of the humerus (where it lies deep to the tendinous origin of flexor carpi ulnaris [Fig. 11–9]). In the scalenus anticus syndrome, the divisions of the brachial plexus become impinged between the scaleni anterior and medius muscles in the neck. In Pancoast's syndrome, a neoplasm of the upper chest produces motor and/or sensory deficits in the upper limb upon impingement or metastatic invasion of one or more parts of the brachial plexus. In the costoclavicular syndrome, one or more parts of the brachial plexus become compressed between the clavicle and the first rib by prolonged retraction and depression of the pectoral girdle. In cervical spondylosis (osteoarthritis of the cervical part of the vertebral column), cervical spinal nerves become subject to compression within intervertebral foramina narrowed by the thinning of articular cartilage in intervertebral joints and by the formation of periarticular osteophytes. Posterolateral herniations of cervical intervertebral discs have the potential of impinging upon roots of cervical spinal nerves.

Of all the conditions listed, only the carpal and ulnar tunnel syndromes commonly produce motor deficits restricted to the actions of the hand's intrinsic muscles and/or sensory deficits confined to the hand. The motor deficits typical of the ulnar tunnel syndrome include weakness of finger abduction and adduction (due to weakness of the dorsal and palmar interossei); the typical sensory deficits include hypalgesia and hypesthesia of the volar surface of the little finger. The motor deficits typical of the carpal tunnel syndrome include weakness of thumb opposition and abduction (due to weakness of opponens pollicis and abductor pollicis brevis); the typical sensory deficits include hypalgesia and hypesthesia of the volar surfaces of the thumb, index finger, and middle finger. In this case, the patient's chief medical problems match the motor and sensory deficits typical of carpal tunnel syndrome.

Carpal tunnel syndrome generally indicates that the soft tissue contents of the carpal tunnel are being subjected to greater than normal pressures. The signs and symptoms represent median nerve damage from mechanical deformation and/or ischemia. Most of the median nerve lesions are in the form of a neuropraxia (a neuronal lesion in which local demyelination occurs in the absence of axonal disruption) or an axonotmesis (a neuronal lesion in which disruption of axons and their myelin sheaths occurs in the absence of disruption of the endoneurium and other connective tissue sheaths). In cases of insidious onset (such as this one), pain and sensory deficits within the distribution of the digital cutaneous branches of the median nerve are typically the initial neurologic symptoms. Pain and hypalgesia may occur concomitantly as a result of the increased pressure both irritating some pain nerve fibers (thus producing pain) and damaging others (thereby diminishing sensitivity to pain). Motor deficits associated with dysfunction of the muscles of the thenar eminence generally occur subsequent to the onset of the sensory deficits.

A broad spectrum of injuries, diseases, and conditions can produce the carpal tunnel syndrome. Fractures and dislocations at the wrist, chronic inflammatory conditions such as rheumatoid arthritis and gout, and repetitive wrist movements (particularly those involving forceful flexion of the hand) can produce the syndrome by tenosynovitis of the flexor tendons in the carpal tunnel and the attendant thickening of the synovial sheaths. Systemic conditions such as hypothyroidism, diabetes mellitus, acromegaly, and systemic lupus erythematosus can produce the syndrome by retention of interstitial fluid and the attendant swelling of soft tissues. In this case, the most likely basis for the syndrome appears to be the antecedent wrist fracture.

RADIOGRAPHIC EVALUATION AND FINAL RESOLUTION OF THE PATIENT'S CONDITION

Lateral and AP radiographs of the patient's right wrist showed calluses extending transversely through the distal end of the radius and the styloid process of the ulna (Fig. 11–10). The angles of anterior and medial angulation of the distal end of the radius were within normal limits. The radiographic evidence thus indicates that the fracture through the distal end of the radius was properly reduced and that there has been adequate reunion of the fragments. There was no evidence of carpal dislocations or previous carpal fractures.

Because the median nerve damage in this case has progressed to the involvement of the motor fibers innervating the thenar muscles, and because it is likely that synovial thickening subsequent to the wrist fracture is the basis for the increased pressure in the carpal tunnel, an orthopedist would recommend surgical division of the flexor retinaculum to decompress the nerve. An orthopedist would anticipate that surgical release would provide immediate relief from pain and promote partial recovery of both sensory and motor nerve fibers.

THE MECHANISM OF THE PATIENT'S CONDITION

The patient sustained a Colles' fracture of the right wrist fifteen months ago when she tripped and fell on her outstretched right hand. A Colles' fracture is a fracture of the distal radius with posterior displacement and rotation of the distal fragment. More than 50% of Colles' fractures are accompanied by a fracture of the ulnar styloid process. Although the fracture was appropriately reduced and immobilized, posttraumatic thickening of the synovial sheaths about the flexor tendons in the carpal tunnel compressed the median nerve sufficiently to damage its sensory and motor fibers. The difficulties the patient has experienced in many activities (dealing and holding playing cards, writing, turning doorknobs, and supporting a cup of liquid by its handle) are due to a combination of diminished sensation to touch and muscle weakness in her right hand.

RECOMMENDED REFERENCES FOR ADDITIONAL INFORMATION ON THE WRIST AND HAND

Corrigan, B., and G. D. Maitland, Practical Orthopaedic Medicine, Butterworth & Co., London, 1983: *Chapter 8 offers a description of the clinical examination of the wrist and hand and discussions of the signs and symptoms of common musculoskeletal disorders involving the wrist and hand.*

Magee, D. J., Orthopaedic Physical Assessment, 2nd ed., W. B. Saunders Co., Philadelphia, 1992: *Chapter 6 provides a comprehensive discussion of the clinical examination of the forearm, wrist, and hand.*

Slaby, F., and E. R. Jacobs, Radiographic Anatomy, Harwal Publishing Co., Media, PA, 1990: *Pages 21 through 24 in Chapter 1 outline the major anatomic features of the PA radiograph of the hand and the lateral radiograph of the wrist.*

Greenspan, A., Orthopedic Radiology, 2nd ed., Gower Medical Publishing, New York, 1992: *Fig. 6.31 on page 6.16 summarizes wrist and hand injuries. Pages 6.14 through 6.19 discuss the use of trispiral tomography in the radiologic evaluation of scaphoid factures. Trispiral tomography is discussed on page 2.3 in Chapter 2 of the text.*

SECTION III

The Lower Limb

CHAPTER **12**

The Gluteal (Buttock) Region

The discussion of the bones, joints, and muscles of the lower limb in the following chapters will emphasize how these musculoskeletal structures function to bear the weight of the upper body and to provide for the walking gait. The presentation of the lower limb therefore begins with a general description of the walking gait.

THE WALKING GAIT

A complete cycle of the human walking gait is called a **stride** (Fig. 12–1 and Table 12–1). Each lower limb passes through two periods during a stride (Table 12–1): a **stance period,** during which the foot is in contact with the surface below (Fig. 12–1, A-E), and a **swing period,** during which the foot is swung forward above the surface below (Fig. 12–1, E-G).

The basic characteristic of the walking gait is that at least one of the feet is always in contact with the surface below. However, there are two intervals of **double stance** in each cycle of the gait during which both feet are in contact with the surface below. A limb's stance period begins with an interval of double stance (Table 12–1 and Fig. 12–1, A and B) and ends with an interval of double stance (Table 12–1 and Fig. 12–1, D and E). The middle interval of a limb's stance period when it is the sole weight-bearing limb is called its **single stance** interval (Table 12–1).

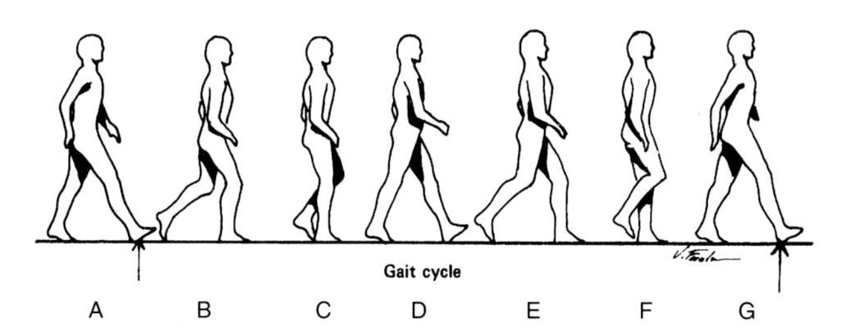

Fig. 12–1: The positions of the lower limbs at various stages during a complete cycle of the human walking gait.

The forward fall of body weight is the principal force that propels the body forward during the walking gait. The lower limb that has just been swung forward and is about to contact the surface below prevents the body from falling down forward at the end of each step. The moment at which initial contact occurs is called **heel strike** because the heel of the foot is the

Table 12–1
The Divisions of the Gait Cycle

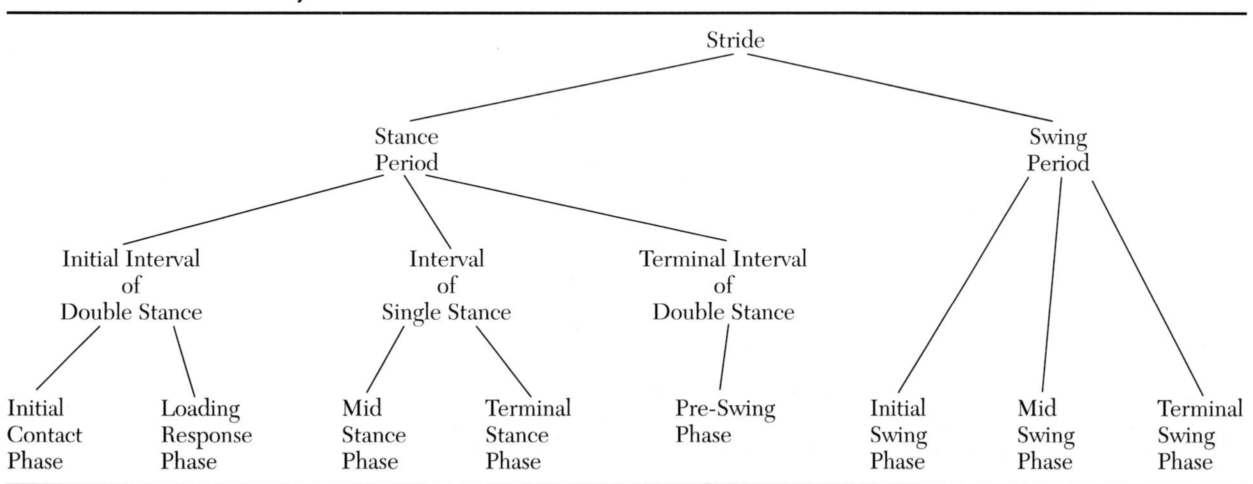

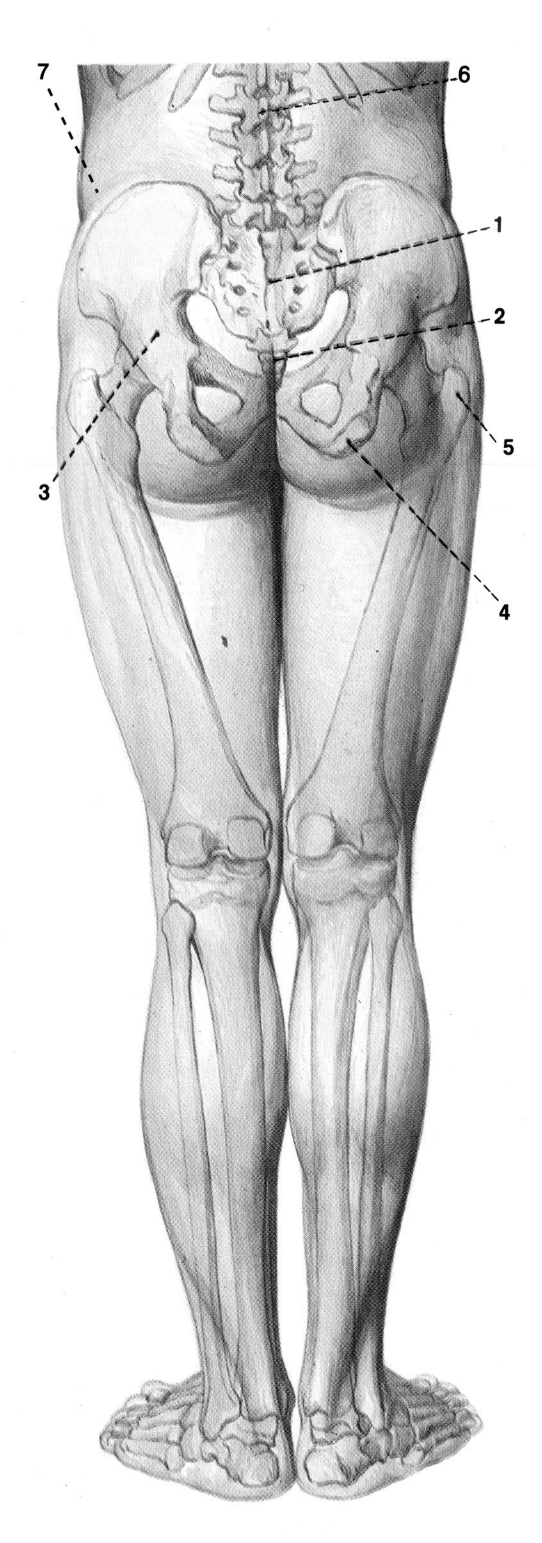

Fig. 12–2: Posterior view of the bones of the lower limb.

first part of the lower limb to strike the surface below (Fig. 12–1A). Heel strike begins the first phase of a stance period, a phase called the **initial contact phase** (Table 12–1). The lower limb begins the initial contact phase flexed approximately 30° at the hip, 0° at the knee, and 0° at the ankle. The lower limb becomes positioned during the initial contact phase to accept support of upper body weight.

The next phase of the stance period is the **loading response phase,** so named because the lower limb responds to the loading of upper body weight upon it. The lower limb rolls forward on its heel during this phase to help sustain the forward momentum of the body. This mechanism by which the body rocks forward over the heel of the newly planted foot is called the **heel rocker** action. The heel rocker action brings the foot into full contact with the surface below, and the moment at which full contact occurs is called **foot flat** (Fig. 12–1B). The initial contact and loading response phases together account for the initial interval of double stance (Table 12–1).

The third phase of the stance period, which is called the **mid stance phase,** begins the middle interval of single stance. An **ankle rocker** action (in which the lower limb rolls forward around the axis of the ankle joint) helps sustain forward body momentum in the mid stance phase. The ankle rocker action helps bring the upper body weight directly over the fully planted foot.

The fourth phase of the stance period is called the **terminal stance phase.** This phase begins with the raising of the heel, called **heel rise (heel off)**. A **forefoot rocker** action (in which the lower limb rolls forward on its forefoot) helps sustain the forward body momentum in the terminal stance phase. The forefoot rocker action helps draw the heel up and advance the body ahead of the sole-supporting foot. The midstance and terminal stance phases together account for the interval of single stance (Table 12–1).

The fifth and final phase of the stance period is called the **pre-swing phase;** it spans the terminal interval of double stance (Table 12–1). Upper body weight is unloaded from the lower limb and transferred to the contralateral lower limb during the pre-swing phase. The pre-swing phase is so named because the lower limb becomes positioned to swing forward rapidly beneath the advancing upper body. The pre-swing phase ends with the forefoot rolling off from the surface below, the final moment of which is called **toe off** because the big toe is the last part of the foot to roll off (Fig. 12–1E). The lower limb ends the pre-swing phase flexed approximately 0° at the hip, 40° at the knee, and 15° at the ankle.

The first phase of the swing period is called the **initial swing phase** (Table 12–1). The foot of the lower limb is lifted above the surface below and the entire limb is accelerated forward. This forward acceleration of the lower limb helps provide the force required to sustain forward body momentum.

The midphase of the swing period is called the **mid swing phase.** During this phase, the lower limb passes beneath the upper body to a position at which the tibia of the leg is vertical.

The third and final phase of the swing period is called the **terminal swing phase.** The lower limb decelerates its forward movement during this phase in preparation for the initial contact phase of the next gait cycle.

THE SKELETAL FRAMEWORK OF THE GLUTEAL REGION

The coxal bone and proximal femur form the skeletal framework of the buttock (Fig. 12–2). The paired coxal bones and the sacrum are all together called the **pelvic girdle** because they form a bony girdle around the pelvis.

The Coxal Bone

The coxal bone during childhood and early adolescence consists of three bones (the **pubis, ischium,** and **ilium**) united by cartilage (Fig. 12–3). The progressive fusion of the pubis, ischium, and ilium produces a single coxal bone in the adult.

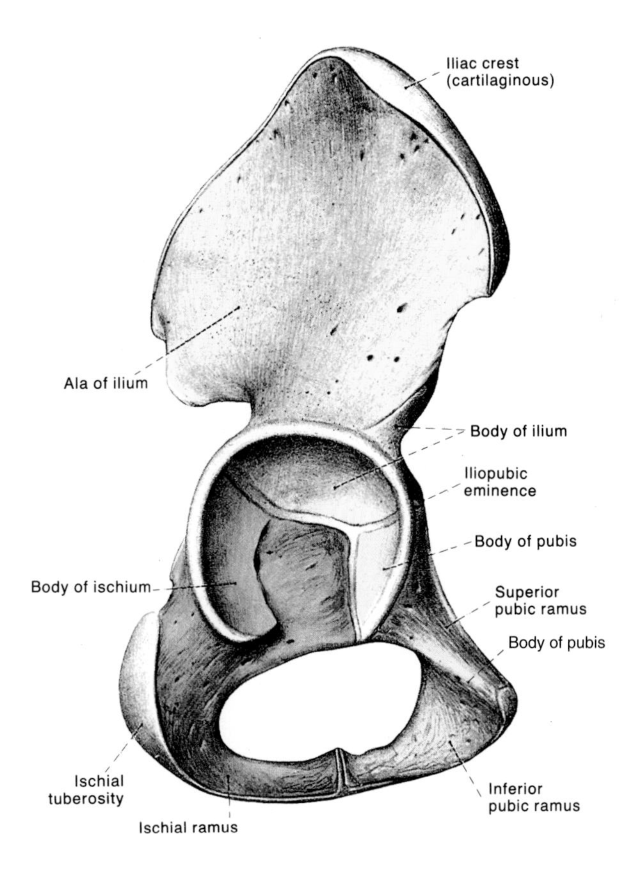

Fig. 12–3: Lateral view of the right coxal bone of a 5 year-old child.

The pubic, ischial, and iliac parts of the coxal bone all contribute to the **acetabulum,** the cup-shaped cavity on the outer surface of the coxal bone (Figs. 12–3 and 12–4). The most distinctive parts of the acetabulum are the **acetabular fossa, lunate surface,** and **acetabular notch.** The acetabular fossa is the central floor region of the acetabular cavity, and the lunate surface is the smooth, horseshoe-shaped surface that almost completely surrounds the acetabular fossa (Fig. 12–4). The acetabular notch is the notch in the lower margin of the rim of the acetabulum. In the body, a ligamentous band called the **transverse acetabular ligament** bridges over the acetabular notch (Fig. 12–11) and converts the acetabular notch into a foramen called the **acetabular foramen.**

The childhood pubis (Fig. 12–3) forms the anteroinferior parts of the adult coxal bone. The most notable pubic parts of the coxal bone are the **body of the pubis, superior and inferior rami of the pubis, pubic crest,** and **pubic tubercle.** The superior and inferior pubic rami are the upper and lower lateral extensions of the body of the pubis (Fig. 12–3). The pubic crest is the superior border of the body of the pubis; the pubic tubercle is the small, round prominence at the lateral end of the pubic crest (Fig. 12–5). The pubic crest and pubic tubercle are palpable near the midline at the lower border of the anterior abdominal wall.

The childhood ischium (Fig. 12–3) forms the posteroinferior parts of the adult coxal bone. The most notable ischial parts of the coxal bone are the **body and ramus of the ischium, ischial tuberosity, ischial spine,** and **lesser sciatic notch.** The body of the ischium forms the posterosuperior part of the ischium, and the ramus of the ischium is the bony beam that extends anteromedially from the lower part of the body (Fig. 12–4). The ischial tuberosity, which is palpable in the buttock, is the large, rough prominence on the posteroinferior surface of the ischium (Figs. 12–2, 12–4, and 12–6); it marks the region of union between the body and ramus of the ischium. The ischial spine is the small, pointed process projecting from the posterior margin of the body of the ischium (Figs. 12–4 and 12–6). The lesser sciatic notch is the curved notch in the posterior margin of the ischium between the ischial tuberosity and the ischial spine (Fig. 12–4).

The paired ischial tuberosities are the parts of the pelvic girdle upon which a person bears weight when seated. In the body, the **ischial bursa** lies directly superficial to each ischial tuberosity. Individuals whose occupation or physical handicap restricts them to a

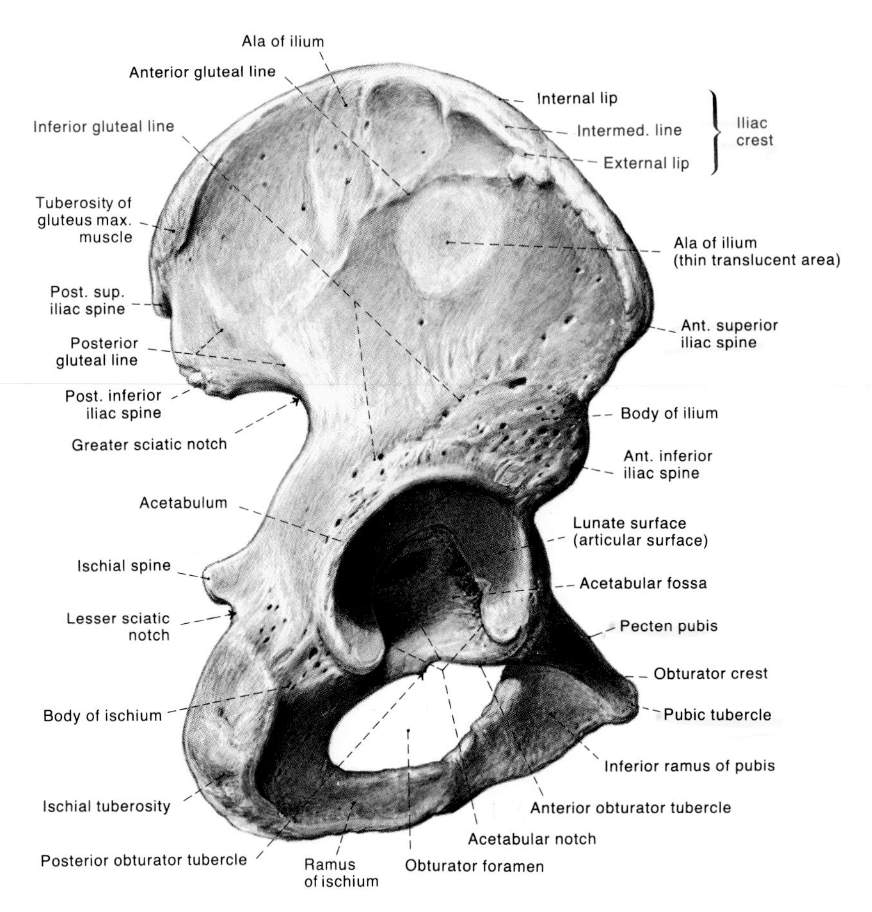

Fig. 12–4: Lateral view of the right coxal bone of an adult.

seated position for long periods experience prolonged irritation of their ischial bursae, and are thus susceptible to **ischial bursitis** (inflammation of the ischial bursa). The ischial bursae may become inflamed because they rest upon only a relatively thin cushion of skin and subcutaneous tissues as they bear the weight of the upper body in a seated individual. Tenderness overlying the ischial tuberosity is a cardinal sign of ischial bursitis. Ischial bursitis is commonly called **Weaver's Bottom** in deference to the Weaver's complaint of this condition in Chaucer's *Canterbury Tales.*

The pubis and ischium together completely encircle the relatively large opening called the **obturator foramen** in the inferior part of the coxal bone (Figs. 12–3, 12–4, and 12–5). The obturator foramen is almost completely enclosed by a thin membrane of fibrous tissue called the **obturator membrane** (Fig. 12–11). The small opening left unfilled by the obturator membrane in the superolateral aspect of the obturator foramen is called the **obturator canal** (Fig. 12–11).

The childhood ilium forms the superior part of the adult coxal bone (Fig. 12–3). The most notable iliac parts of the coxal bone are the **body and wing (ala) of the ilium; iliac fossa and gluteal surface of the wing; anterior superior, anterior inferior, posterior superior, and posterior inferior iliac spines; greater sciatic notch;** and **iliac crest.** The body of the ilium is the part that contributes to the acetabulum; the wing is the fan-shaped, superolateral part of the ilium (Figs. 12–3 and 12–4). The inner and outer surfaces of the wing are respectively called the iliac fossa (Fig. 12–9) and the gluteal surface of the wing.

The anterior margin of the wing of the ilium bears two rounded prominences called the anterior superior and anterior inferior iliac spines (Figs. 12–4 and 12–5). The anterior superior iliac spine is easily palpable near the lateral end of the border region between the front of the thigh and the anterolateral abdominal wall (Fig. 12–7A).

The posterior margin of the wing of the ilium bears two rounded prominences called the posterior superior and posterior inferior iliac spines (Figs. 12–4 and 12–6). The greater sciatic notch is the notch in the posterior margin of the coxal bone between the posterior

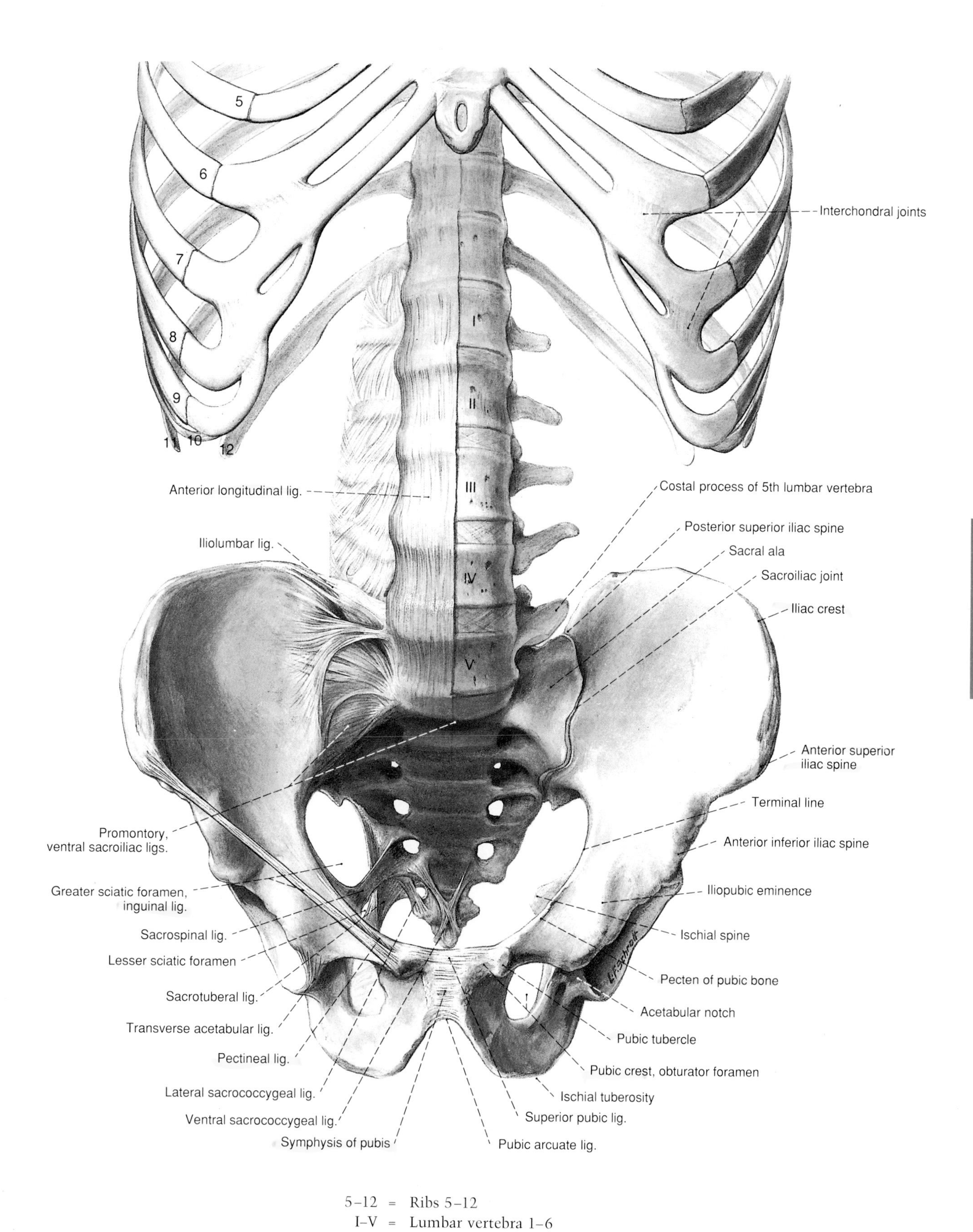

5–12 = Ribs 5–12
I–V = Lumbar vertebra 1–6

Fig. 12–5: Anterior view of the bones and ligaments of the male abdomen and pelvis.

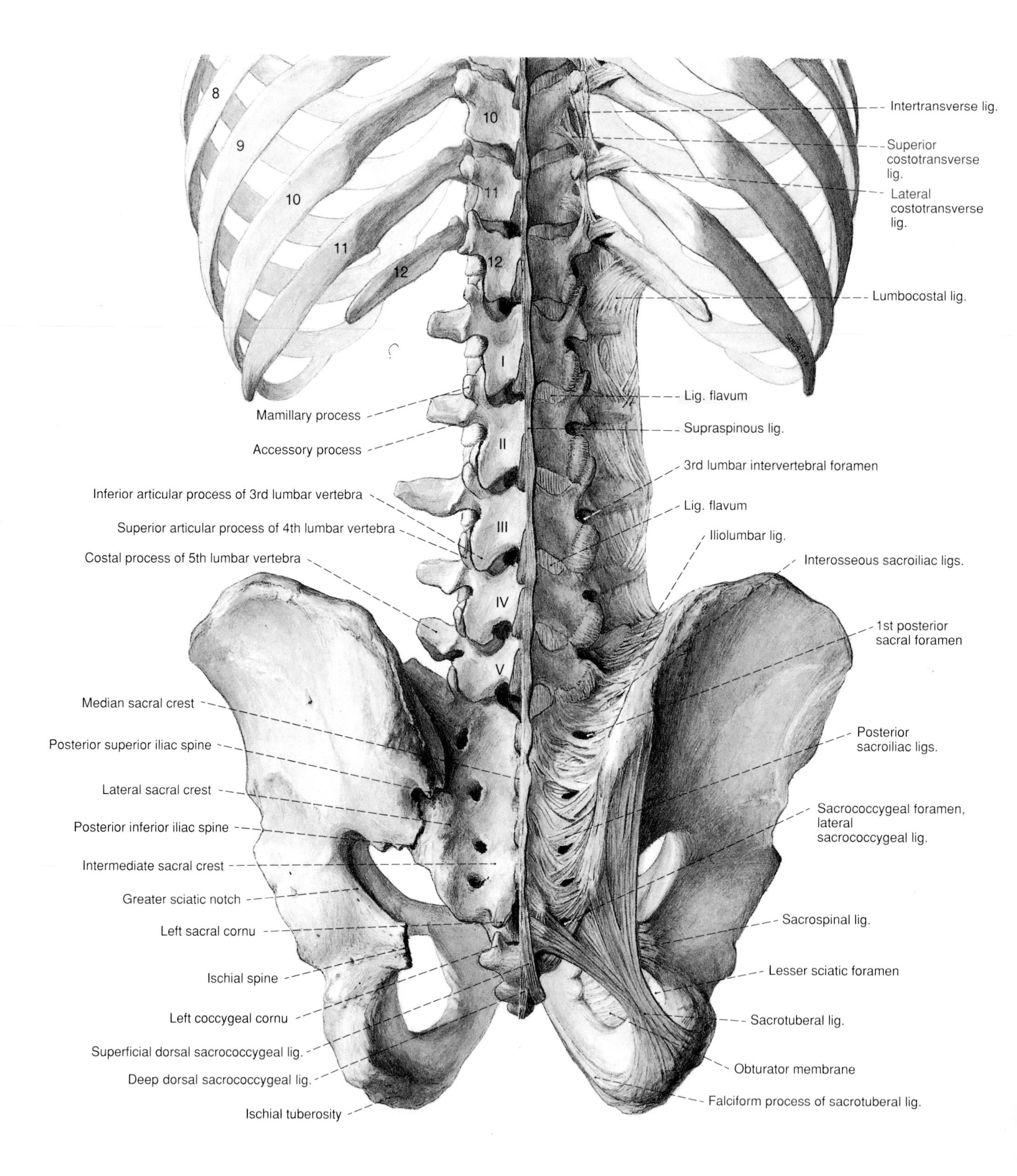

8–12 = Ribs 8–12 and thoracic vertebra10–12
I–V = Lumbar vertebra 1–5

Fig. 12–6: Posterior view of the bones and ligaments of the male abdomen and pelvis.

inferior iliac spine and the ischial spine (Figs. 12–4, and 12–6).

The iliac crest is the gently rounded, upper margin of the wing of the ilium; the iliac crest extends between the anterior superior and posterior superior iliac spines (Figs. 12–4 and 12–5). Most of the iliac crest is palpable along its arching course over the buttock region. Contusions of the muscles attached to the iliac crest and its anterior end, the anterior superior iliac spine, are commonly called **hip pointers.**

The Hip

The femur is the bone of the thigh. The proximal femur (that part extending from the head of the femur to the immediate subtrochanteric region) in clinical practice is called the hip. The shape of the hip is addressed at this point since the hip lies in the gluteal region and contributes to the region's shape. The shaft and distal end of the femur are discussed in the following chapter on the thigh.

The most notable parts of the hip are the **head of the femur** with its **fovea capitis, neck of the femur, greater and lesser trochanters, intertrochanteric line,** and **intertrochanteric crest.** The head of the femur is the spheroidal proximal end of the femur (Fig. 12–8). The fovea capitis is a pit on the inferomedial surface of the head of the femur (fovea capitis literally means "the pit on the head"). The neck of the femur is the narrowed segment immediately distal to the head. The bony prominences called the greater and lesser trochanters mark the upper end of the shaft of the femur (Fig. 12–8). The intertrochanteric line and intertrochanteric crest are the ridges that extend between the trochanters on respectively the anterior and posterior surfaces of the femur.

The greater trochanter of the femur forms a palpable prominence on the uppermost lateral aspect of the thigh (Fig. 12–2). Bedridden individuals who have little ability to alter their position in bed may subject the soft tissues overlying the greater trochanter to lengthy periods of sustained compression. These soft tissues become susceptible to avascular necrosis and ulceration as a result of the sustained pressures on their microvasculature.

The hip has three secondary centers of ossification during childhood and adolescence. The **center for the head** appears during the first 6 months after birth, the **center for the greater trochanter** appears at 4 years of age, and the **center for the lesser trochanter** appears at 12 to 14 years of age.

CLINICAL PANEL III-1

Hip Fractures

Incidence: Fractures of the femoral neck are especially common among the elderly; the contribution of osteoporosis to such fractures accounts for the greater incidence of hip fractures among elderly women than among elderly men. Femoral neck fractures in younger individuals result from severe physical trauma or pathologic bone disorders.

Potential Complication: Most femoral neck fractures in elderly individuals occur immediately distal to the femoral head. Such fractures are called subcapital fractures. The attendant rupture of the retinacular arteries to the head of the femur accounts for the common post-traumatic complication of avascular necrosis of the femoral head (refer to the discussion of the blood supply to the head of the femur in Chapter 13 for a description of the retinacular arteries).

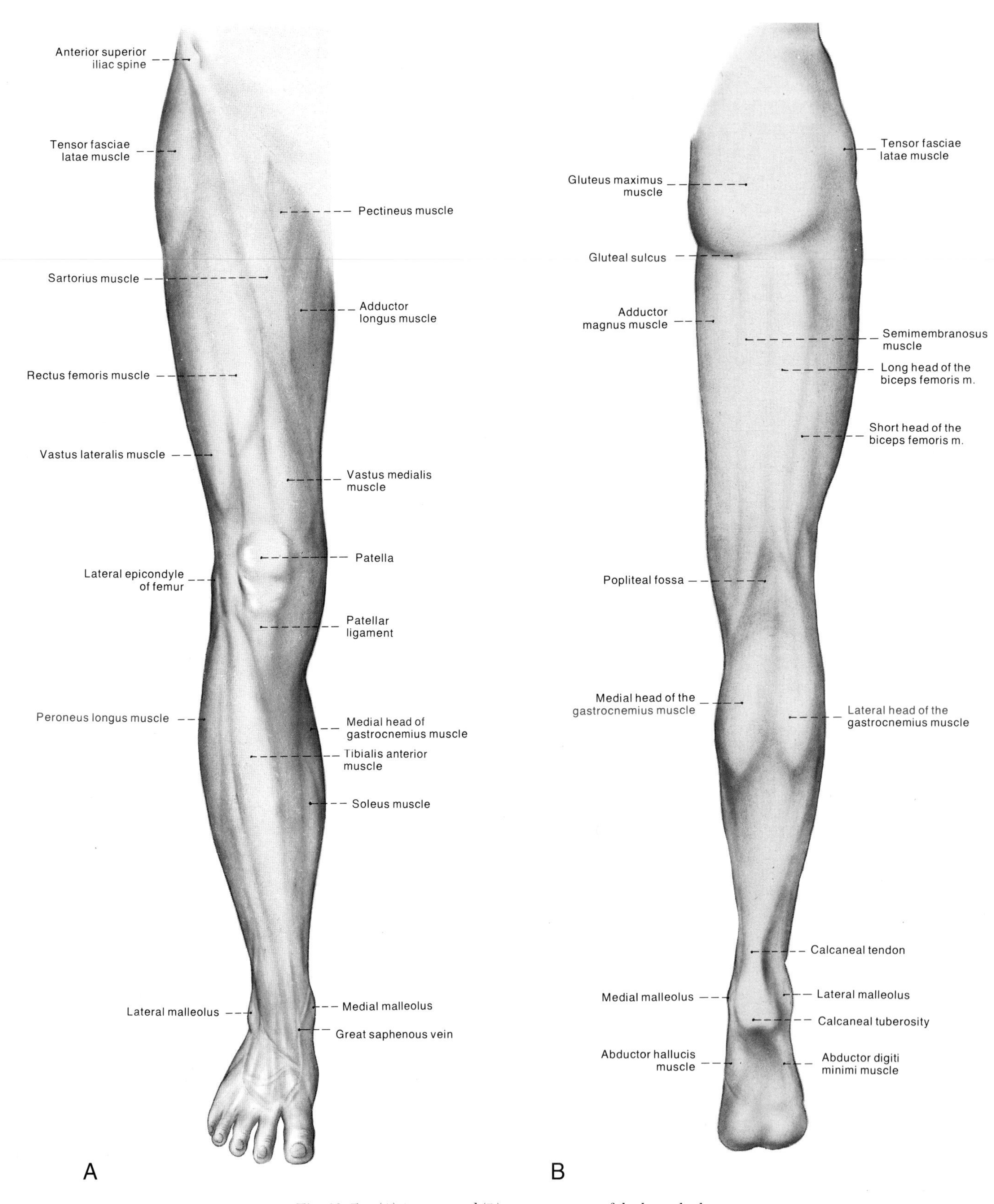

Fig. 12–7: (A) Anterior and (B) posterior views of the lower limb.

THE JOINTS OF THE PELVIC GIRDLE

The Pubic Symphysis and Sacroiliac Joints

Anteriorly in the pelvic girdle, the paired coxal bones articulate with each other by a secondary cartilaginous joint called the pubic symphysis (Fig. 12–5). Posteriorly in the pelvic girdle, the wing of the ilium of each coxal bone articulates with the sacrum by a synovial joint called the sacroiliac joint (Fig. 12–5). The sacroiliac joint on each side is the only joint between a bone of the lower limb and a bone of the axial skeleton. The synovial cavities of the sacroiliac joints gradually become obliterated in late adulthood, and the joints become increasingly fibrosed. Nerves derived from the **superior gluteal nerve,** the **sacral plexus,** and the **dorsal rami of S1 and S2** innervate the sacroiliac joint.

The sacroiliac joints are stabilized by ligaments that bind the coxal bones to the sacrum and coccyx. On each side, there are strong **anterior and posterior sacroiliac ligaments** stretched between the sacrum and the iliac part of the coxal bone (Figs. 12–5 and 12–6). Also on each side, there is a **sacrotuberous ligament** binding the ischial tuberosity to the sacrum and coccyx, and a **sacrospinous ligament** tautly extended from the ischial spine to the sacrum and coccyx (Figs. 12–5, 12–6, and 12–9). The sacrotuberous and sacrospinous ligaments combine to convert the greater and lesser sciatic notches of the coxal bone into osseoligamentous openings called respectively the **greater and lesser sciatic foramina** (Fig. 12–9). In the body, the greater sciatic foramen transmits structures between the pelvis and the buttock, and the lesser sciatic foramen transmits structures between the perineum and the buttock.

The sacroiliac, sacrotuberous, and sacrospinous ligaments all serve to resist the tendency of the upper body weight to rotate the sacrum anteriorly at the sacroiliac joints. When a person is standing erect, the center of gravity of the upper body lies anterior to the sacroiliac joints, and, therefore, the weight of the upper body directs the sacrum to both slide downward and rotate anteriorly at the sacroiliac joints (Fig. 12–10). The wedge-shaped fit of the sacrum between the paired coxal bones limits any downward slide. The sacrotuberous and sacrospinous ligaments, in concert with the most superior and posterior sacroiliac ligaments, markedly limit any anterior rotation of the sacrum (rotation in which the anterosuperior margin of the sacrum moves downward and forward as the tapered lower end of the sacrum moves upward and backward).

The Hip Joint

The hip joint is a synovial joint in which the head of the femur articulates with the lunate surface of the acetabulum of the coxal bone (Fig. 12–11). The femoral head also articulates within the joint with the **acetabular labrum,** a ring of fibrous cartilage that is attached to the bony rim of the acetabulum and its transverse acetabular ligament. The acetabular labrum significantly enhances the stability the joint because it embraces the spheroidal surface of the femoral head just distal to its greatest diameter (Fig. 12–14). The **femoral nerve, obturator nerve, nerve to quadratus femoris, and superior gluteal nerve** innervate the hip joint.

Within the hip joint, the fovea capitis is attached to a ligament called the **ligament to the head of the femur** that arises from the acetabular notch and the transverse acetabular ligament (Fig. 12–11). Although the ligament to the head provides little mechanical support of the joint, it houses an artery (literally called the **artery in the ligament to the head of the femur**) that supplies the head of the femur. The ligament to the head and the fat pad about it occupy the space in the hip joint between the acetabular fossa and the head of the femur (Fig. 12–14).

The joint's fibrous capsule is attached proximally to the rim of the acetabulum; the margin of attachment surrounds the acetabular labrum and the transverse acetabular ligament (Fig. 12–11). The capsule's margin of attachment to the femur extends to the intertrochanteric line, the bases of the trochanters, and the posterior surface of the neck of the femur about a centimeter proximal to the intertrochanteric crest (Figs. 12–12 and 12–13). The distal part of the posterior surface of the neck of the femur thus lies outside the capsule of the hip joint (Fig. 12–13).

The **iliofemoral, pubofemoral, and ischiofemoral ligaments,** which respectively extend from the iliac, pubic, and ischial parts of the coxal bone to the femur, significantly strengthen the fibrous capsule of the hip joint (Figs. 12–12 and 12–13). These ligaments all exhibit a spiralled course around the neck of the femur as they extend from their origin on the coxal bone to blend in with distal parts of the fibrous capsule. On the right side (Figs. 12–12 and 12–13), the ligaments all spiral clockwise as they extend distally; on the left side, the ligaments all spiral counterclockwise as they extend distally. The ligaments on each side thus become more tightly wrapped around the neck of the femur as the thigh becomes increasingly extended.

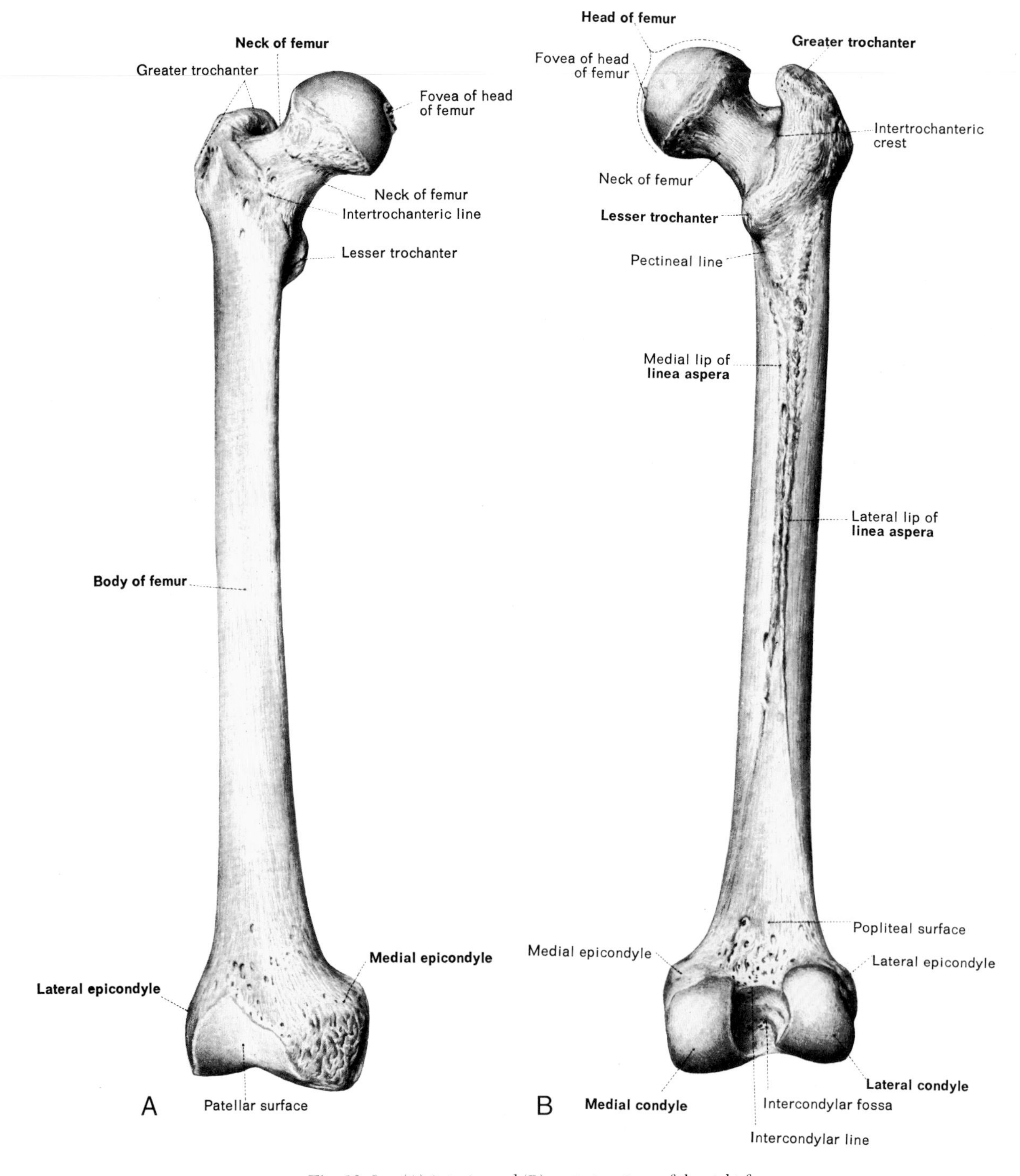

Fig. 12–8: (A) Anterior and (B) posterior views of the right femur.

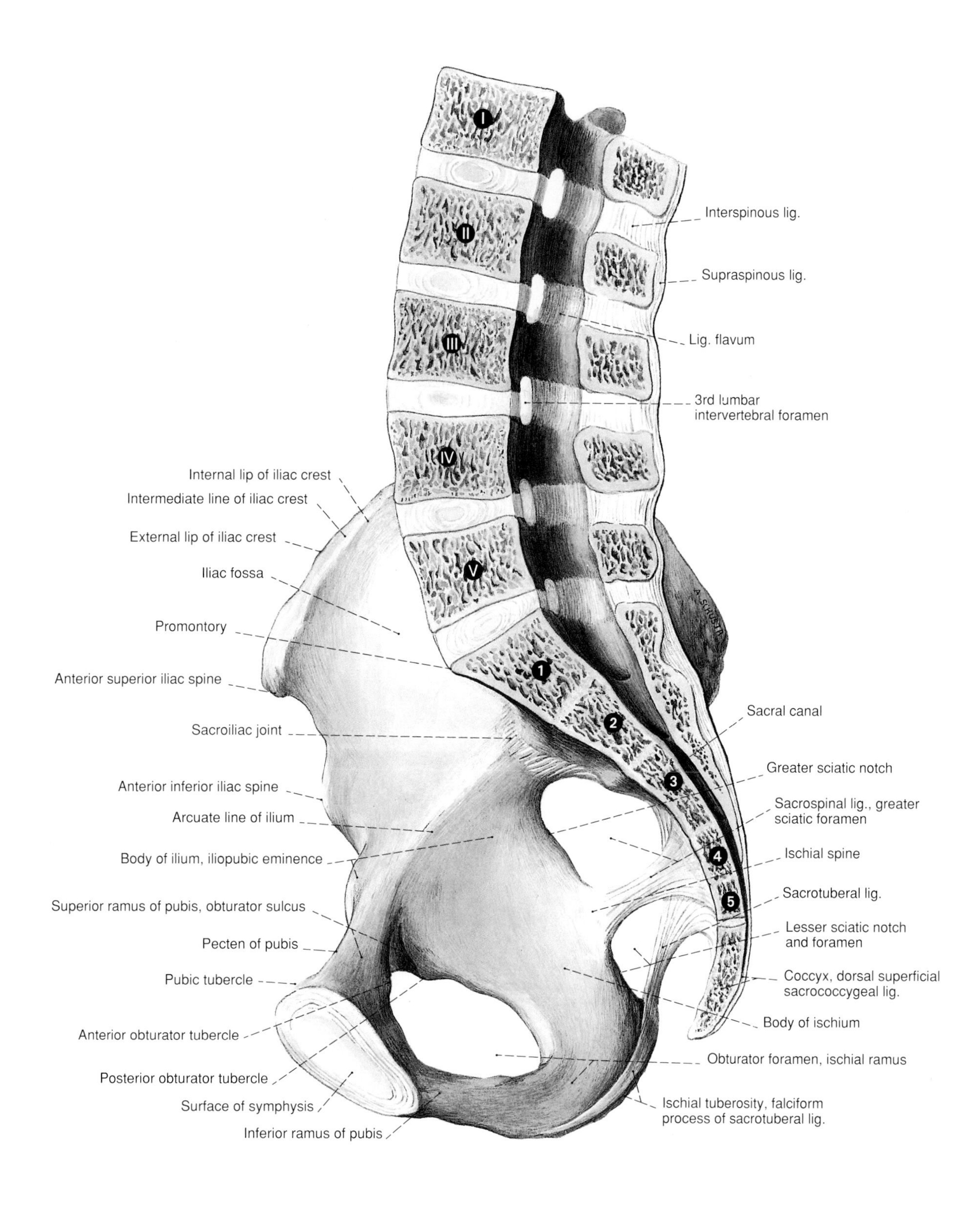

I–V = Lumbar vertebra 1–5
1–5 = Sacral bone, consists of five sacral vertebrae

Fig. 12–9: Median sectional view of the bones and ligaments of the abdomen and pelvis.

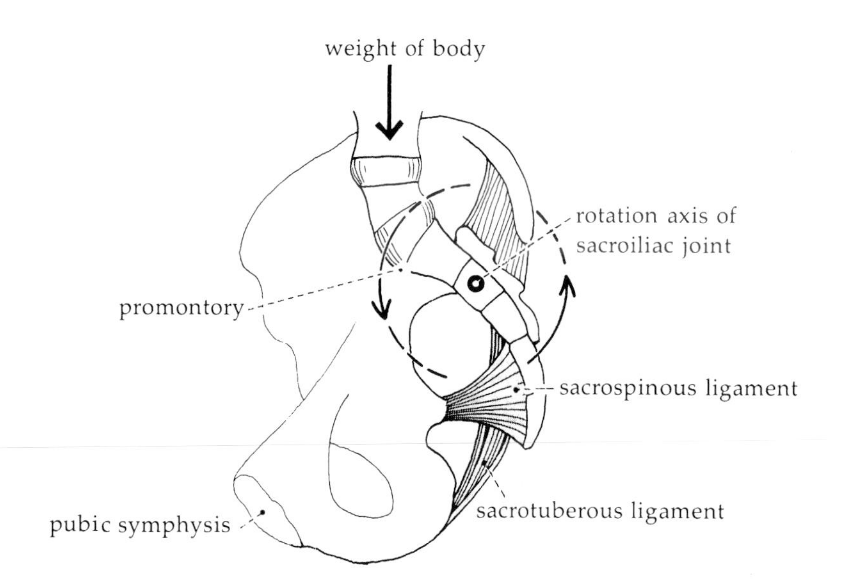

Fig. 12–10: Diagram illustrating the tendency of the weight of the upper body to rotate the sacrum forward at the sacroiliac joints.

The iliofemoral, pubofemoral, and ischiofemoral ligaments all serve to resist the tendency of the upper body weight to tilt the pelvic girdle posteriorly at the hip joints. When a person is standing erect, the center of gravity of the upper body lies just anterior to the body of the second sacral veretebra but posterior to the hip joints, and, therefore, the weight of the upper body directs the pelvic girdle to tilt posteriorly at the hip joints. However, this tendency is resisted by the iliofemoral, pubofemoral, and ischiofemoral ligaments because they are all almost maximally stretched around the neck of the femur when a person is standing erect.

The synovial membrane of the hip joint lines all the intracapsular surfaces except for the intracapsular bony surfaces covered by hyaline cartilage (Fig. 12–14). The synovial membrane thus lines the ligament to the head, the transverse acetabular ligament, the acetabular labrum, the internal surface of the capsule, and the intracapsular surfaces of the neck of the femur. The lunate surface of the acetabulum and the femoral head (but not the fovea capitis) are the only intracapsular surfaces covered by hyaline cartilage. The ligament to the head and the fat pad about it in the acetabular fossa lie outside the synovial cavity.

The ball-and-socket configuration of the hip joint provides for **flexion, extension, abduction, adduction, external (lateral) rotation,** and **internal (medial) rotation** of the thigh (Fig. 12–15). The combination of the flexion, extension, abduction, and adduction movements provides for **circumduction** of the thigh. The narrowness of the neck of the femur enhances the extent to which the femoral head can rotate in any direction on the lunate surface of the acetabulum.

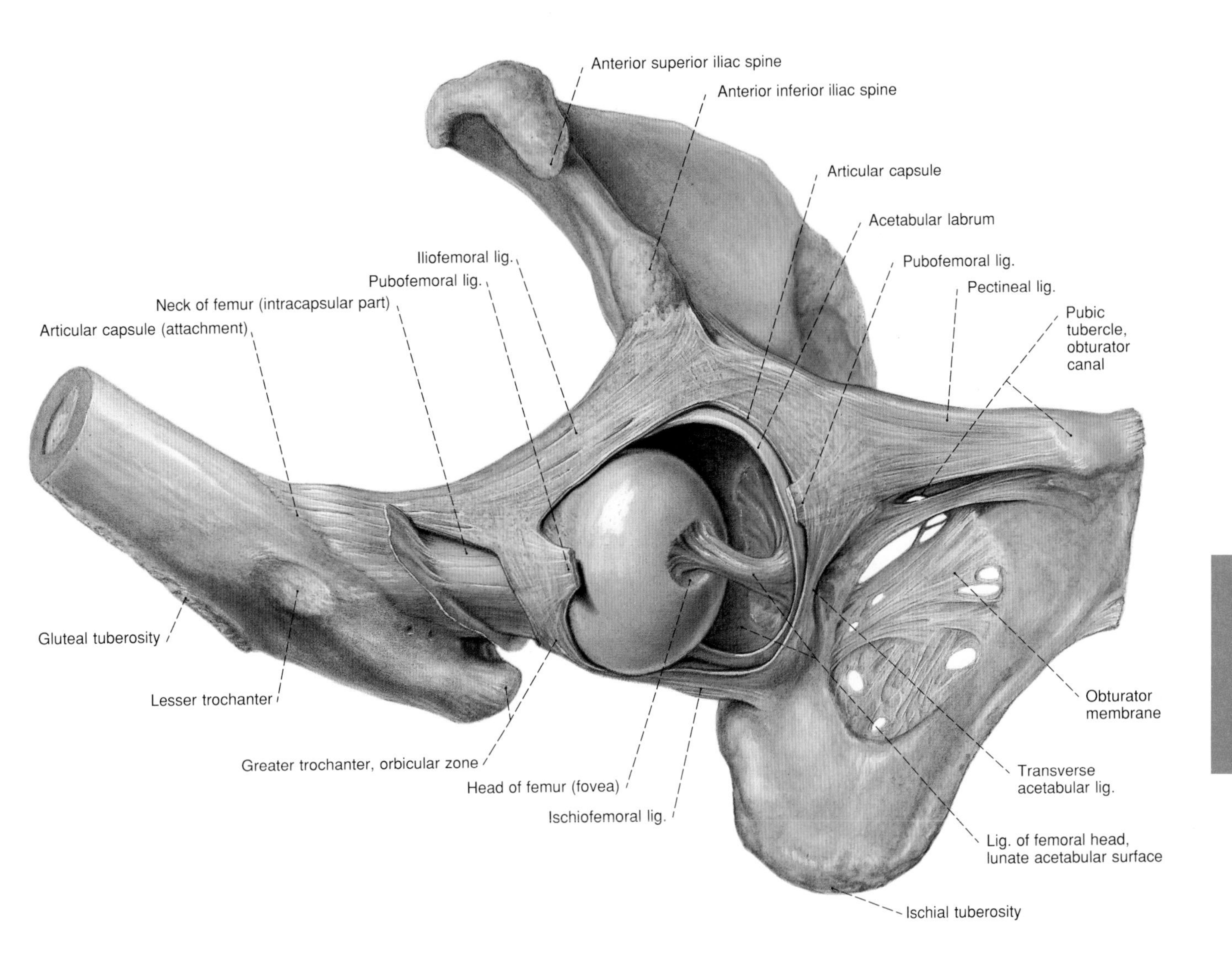

Fig. 12–11: Anteroinferior view of a partially disarticulated right hip joint.

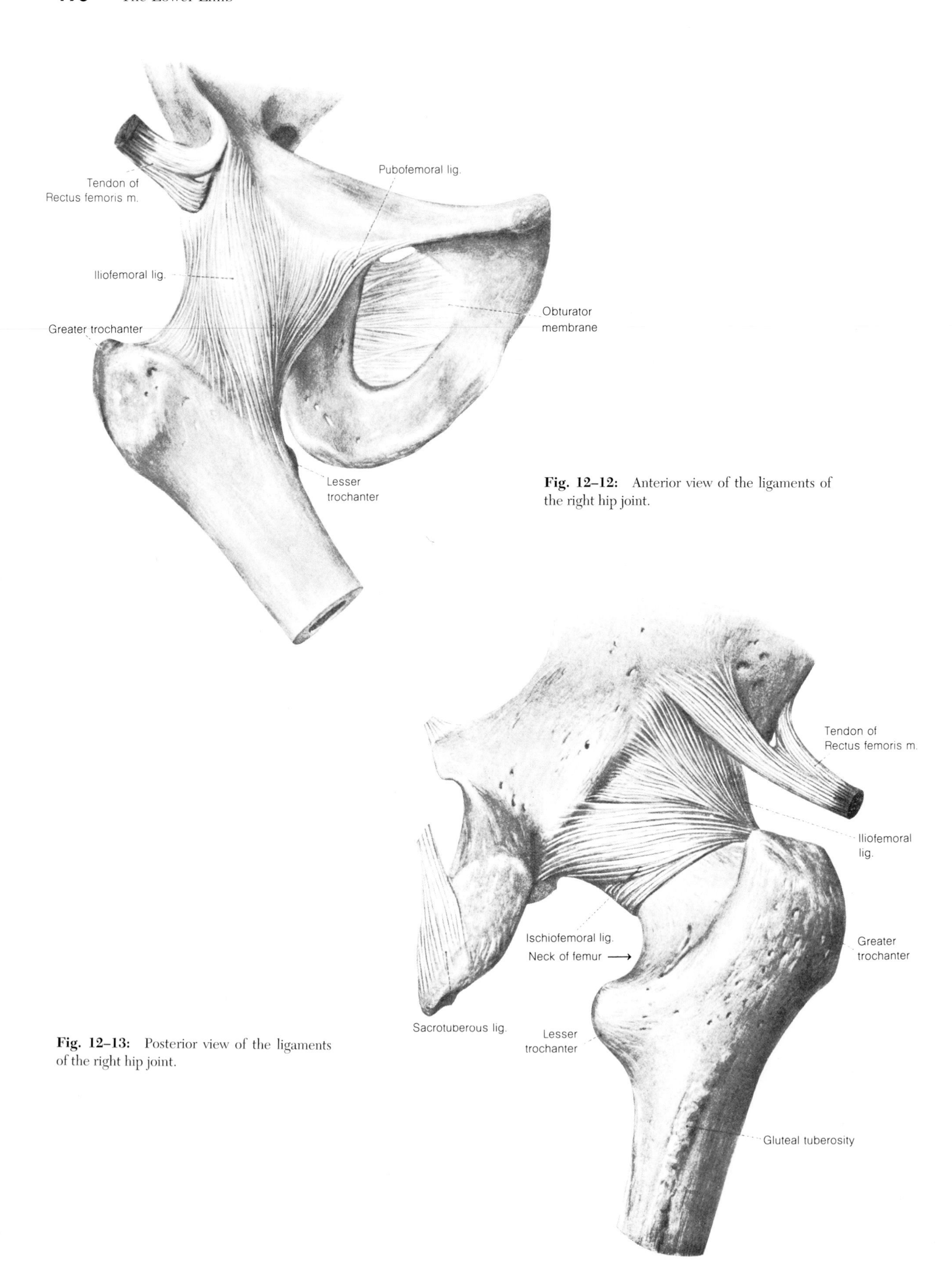

Fig. 12–12: Anterior view of the ligaments of the right hip joint.

Fig. 12–13: Posterior view of the ligaments of the right hip joint.

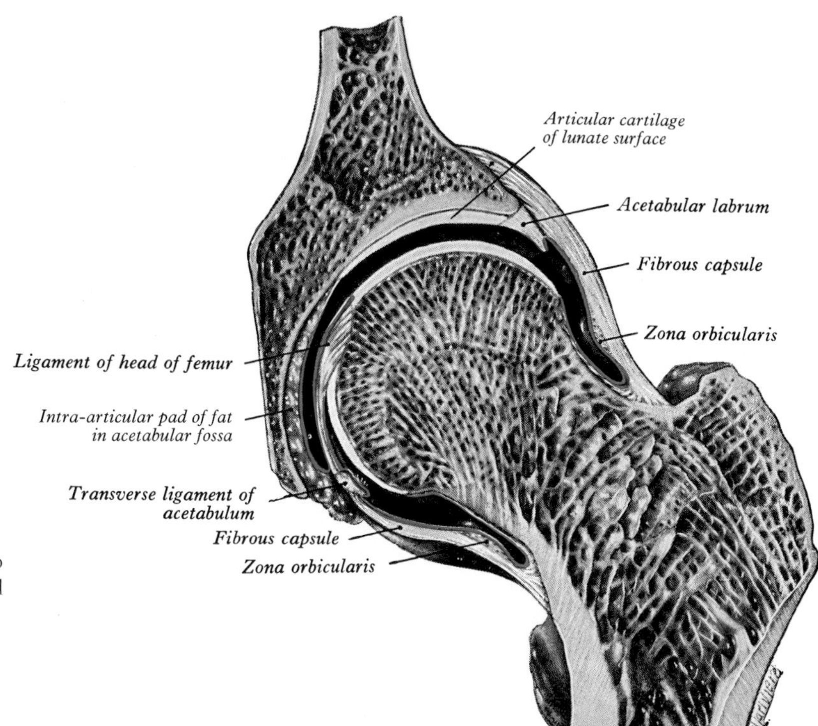

Fig. 12–14: A section through the hip joint. The blue line represents the synovial membrane of the hip joint.

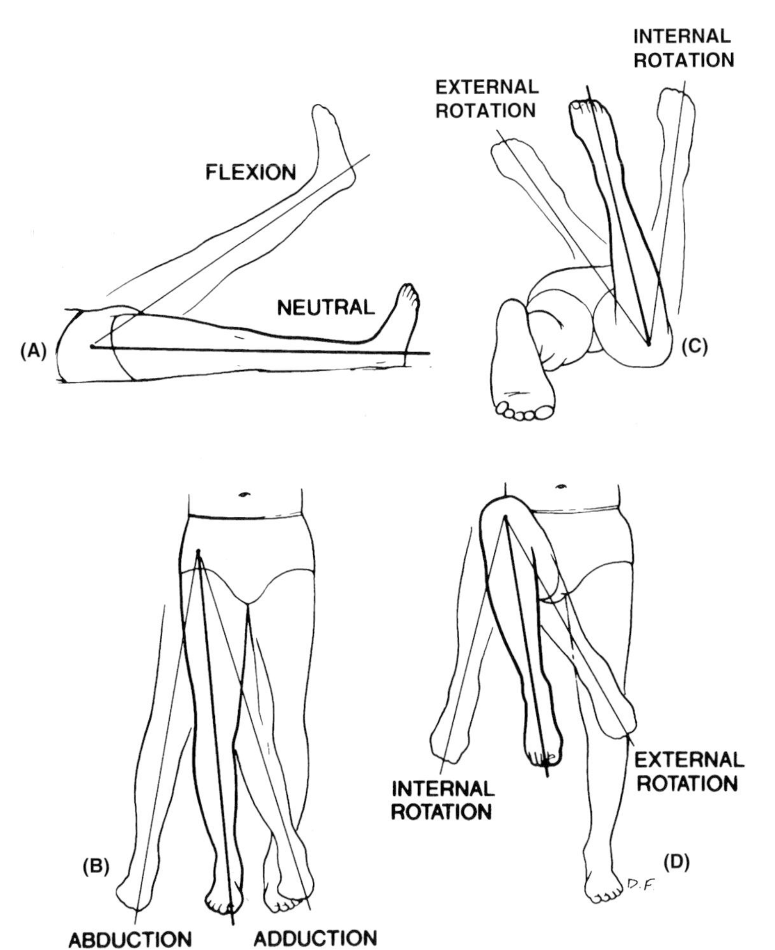

Fig. 12–15: Movements of the thigh at the hip joint: (A) flexion, (B) abduction and adduction, (C) internal and external rotation with the thigh flexed 0°, and (D) internal and external rotation with the thigh flexed 90°.

CLINICAL PANEL III-2

Femoral Antetorsion

Definition: The neck of the femur is normally anteriorly angulated at its union with the shaft of the femur (Fig. 12–16). The term **femoral anteversion** is used to describe this normal anatomic feature. The magnitude of femoral anteversion is defined as the angle between the line coincident with the long axis of the neck of the femur and the line that passes through the posterior surfaces of the condyles at the distal end of the femur (Fig. 12–16). The angle of femoral anterversion ranges from 30 to 50° at birth; it decreases during childhood and adolescence to 10 to 15° by adulthood. The term **femoral antetorsion** is used to decribe the condition in which the angle of femoral anteversion at any age is at least two standard deviations greater than the mean for that age.

Common Symptoms: An absence of or significant delay in the decrease of the angle of femoral anteversion during early childhood results in an intoeing gait [a gait in which the toes of an affected lower limb point inward (markedly medially)]. The child's parents frequently report that the child easily trips while walking or running. They may also note that the child frequently sits on the floor with the legs sticking out (Fig. 12–17A).

Common Signs: The condition is frequently bilateral. In such cases, the child's stance shows both lower limbs turned inward from the hip down (Fig. 12–17B). Range of motion testing shows internal rotation of the thighs markedly greater than external rotation. Maximum external thigh rotation [as measured with the child lying prone and the leg flexed 90° at the knee (as depicted in Fig. 12–15C)] is generally less than 20°.

Common Course of the Condition: In the vast majority of children with femoral antetorsion, the requisite decrease in the angle of femoral anteversion occurs spontaneously and produces normal lower limbs by age 8.

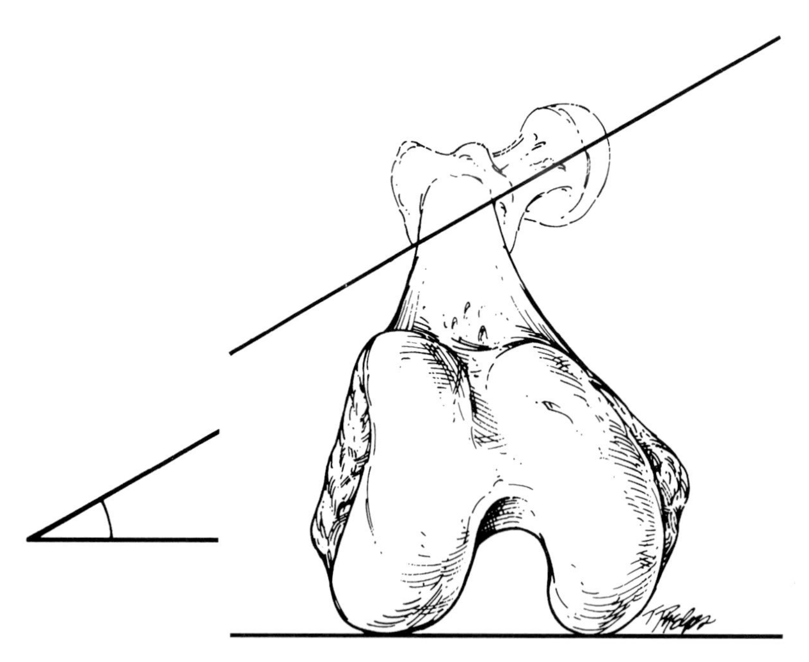

Fig. 12–16: Diagram illustrating the angle of femoral anteversion.

CLINICAL PANEL III-3

Pelvic Fractures

Definition: Pelvic fractures are fractures of the bony pelvis.

Common Features: Since the bony pelvis is structurally a united ring of bones, breaks within the ring generally occur in pairs. The pair of breaks commonly are either a pair of fractures or a fracture accompanied by a joint dislocation. The superior and inferior pubic rami are the most commonly fractured parts of the bony pelvis; dislocation of the pubic symphysis is more common than dislocation of the sacroiliac joints.

There is a high morbidity and mortality associated with pelvic fractures because of attendant hemorrhagic shock and pelvic organ damage. In particular, it must always be assumed with pelvic fractures, until examination proves otherwise, that the bladder and urethra are also damaged.

THE MAJOR MECHANICAL ROLE OF THE BONY PELVIS

The paired coxal bones, sacrum, and coccyx collectively provide a skeletal framework for the walls of the pelvis, and thus are collectively called the **bony pelvis.** In the body, the four bones are maintained in an essentially fixed configuration relative to each other, mainly because the pubic symphysis, the sacroiliac joints, and the sacrococcygeal joint (which is the symphysis between the sacrum and coccyx) all permit very little movement. The weight of the upper body and the pull of the powerful gluteal and thigh muscles can exact only marginal alterations in the bony pelvis's configuration.

The major mechanical role of the bony pelvis is to transmit forces between the upper body and the lower limbs and also between the lower limbs themselves. In this role the bony pelvis serves as the mechanical equivalent of two arches joined at their lateral bases. The bases of these two arches are joined on each side at the acetabulum (Fig. 12–18). The sacrum, the sacroiliac joints and ligaments, and the iliac parts of the coxal bones together form one osseoligamentous arch in the posterosuperior part of the pelvic girdle. This posterosuperior arch transmits the upper body weight onto the lower limbs (Fig. 12–18). The pubic symphysis, the pubic bodies, and the superior pubic rami of the coxal bones together form another osseoligamentous arch in the anteroinferior part of the pelvic girdle. This anteroinferior arch serves not only as a tie beam that resists the tendency of the upper body's weight to thrust the coxal bones apart laterally, but also as a compression strut that resists the tendency of the lower limbs to thrust the coxal bones medially against one another.

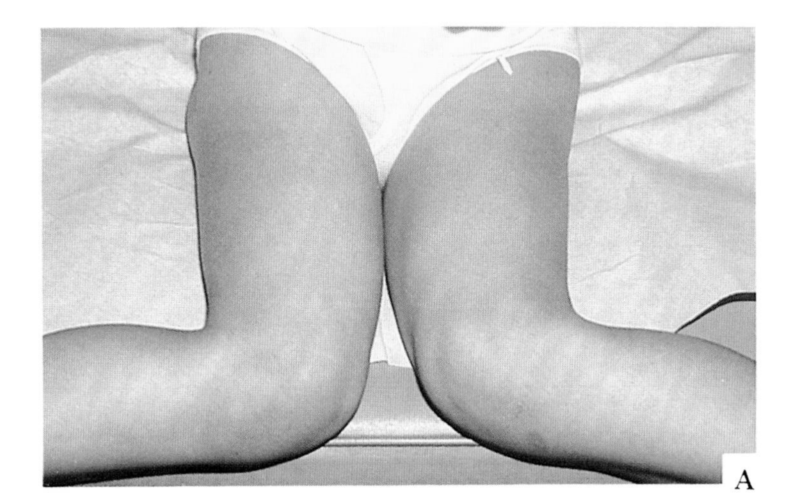

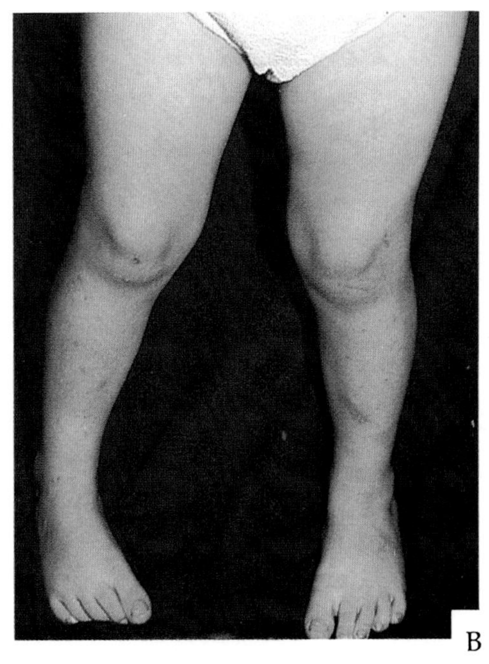

Fig. 12–17: Photographs showing (A) the extent to which a child with bilateral femoral antetorsion can internally rotate the thighs (and thereby stick the feet out) while sitting on a floor and (B) the stance appearance of the lower limbs of a child with bilateral femoral antetorsion.

THE MUSCLES OF THE GLUTEAL REGION

The gluteal muscles may be classified functionally into three groups: one muscle which extends the thigh, three muscles which abduct the thigh, and five muscles which externally rotate the thigh. Tables 12–2 to 12–4 list for quick reference the origin, insertion, nerve supply, and actions of the gluteal muscles.

Gluteus maximus is so named because it is the largest gluteal muscle. It originates from the posterior part of the outer surface of the bony pelvis and extends inferolaterally behind the hip joint and lateral to the greater trochanter (Fig. 12–19). A large bursa called the **trochanteric bursa** intervenes between the greater trochanter and the overlying part of gluteus maximus. Most of its fibers insert into the iliotibial tract of the fascia lata; the remainder of the fibers insert onto the upper posterior surface of the shaft of the femur. The thigh's investing layer of deep fascia is called the **fascia lata,** and the **iliotibial tract** is the vertical tract of the fascia lata that extends from the iliac crest above to the lateral condyle of the tibia below (Fig. 12–20).

The gluteus maximus is the chief extensor of the thigh at the hip joint when the thigh is flexed 90° or more. Therefore, the muscle is the prime mover of extension of the trunk of the body relative to the thigh when someone rises from a seated position.

When walking, gluteus maximus is most active during the time interval that extends from the terminal swing phase of the swing period to the mid stance phase of the subsequent stance period (Fig. 12–21A). During the late part of the terminal swing phase, it contracts eccentrically to restrain the forward movement of the thigh.

During the initial contact and loading response phases of the stance period, gluteus maximus's extensor activity at the hip joint opposes the tendency of the trunk of the body to fall forward about the hip joint.

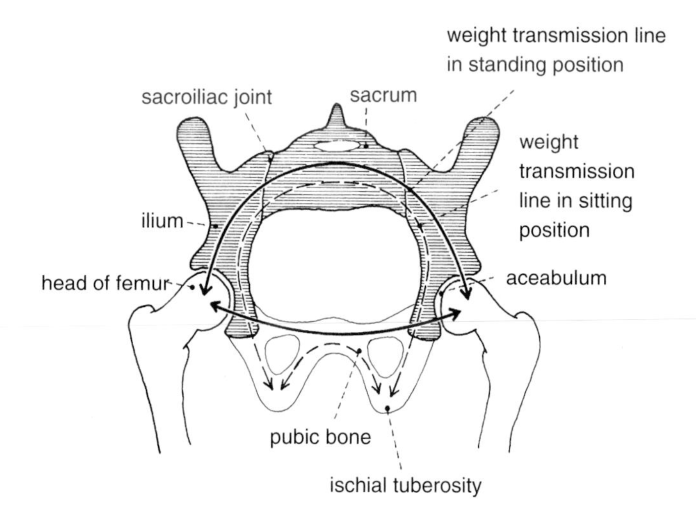

Fig. 12–18: Diagram illustrating the transmission of forces through the bony pelvis.

The body trunk has a tendency during these two phases to fall forward at the hip joint because upper body weight is being loaded upon the limb along a line that projects anterior to the hip joint (Fig. 12–22). The projection of the upper body weight line anterior to the hip joint generates a significant flexion torque at the joint that pulls the body trunk forward at the joint. Individuals who suffer from paralysis of gluteus maximus lean the body trunk backward at heel strike in order to compensate for this loss of gluteus maximus's contribution to the walking gait (Fig. 12–23).

Table 12–2
The Muscle of the Buttock that Extends the Thigh

Muscle	Origin	Insertion	Nerve Supply	Actions
Gluteus maximus	External surface of the wing of the ilium posterior to the posterior gluteal line, posterior surfaces of the sacrum and coccyx, and sacrotuberous ligament	Most fibers insert into the iliotibial tract of the fascia lata; the remainder insert onto gluteal tuberosity	Inferior gluteal (L5, **S1**, and **S2**)	Extends the trunk of the body relative to the thigh when rising from a seated position; extends the thigh

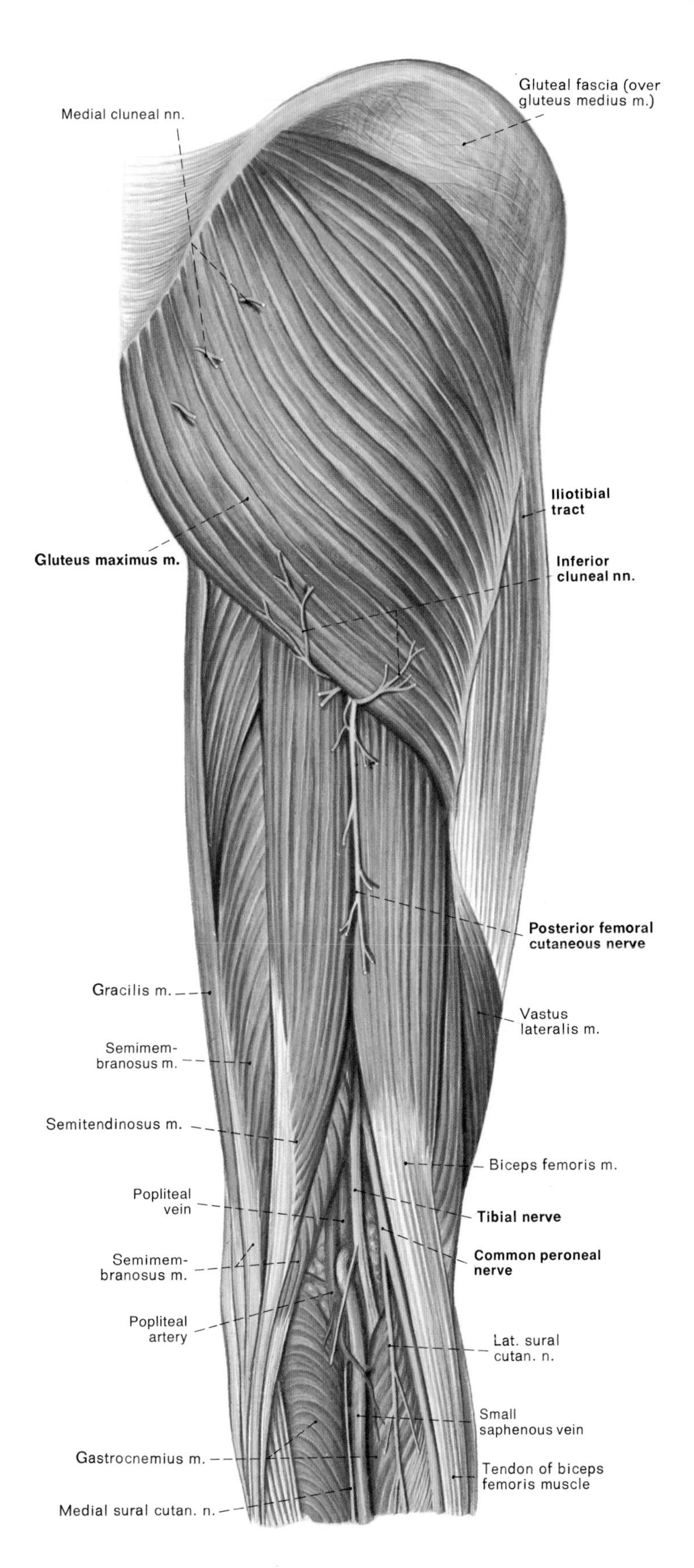

Fig. 12–19: Posterior view of muscles, nerves, and blood vessels of the right thigh.

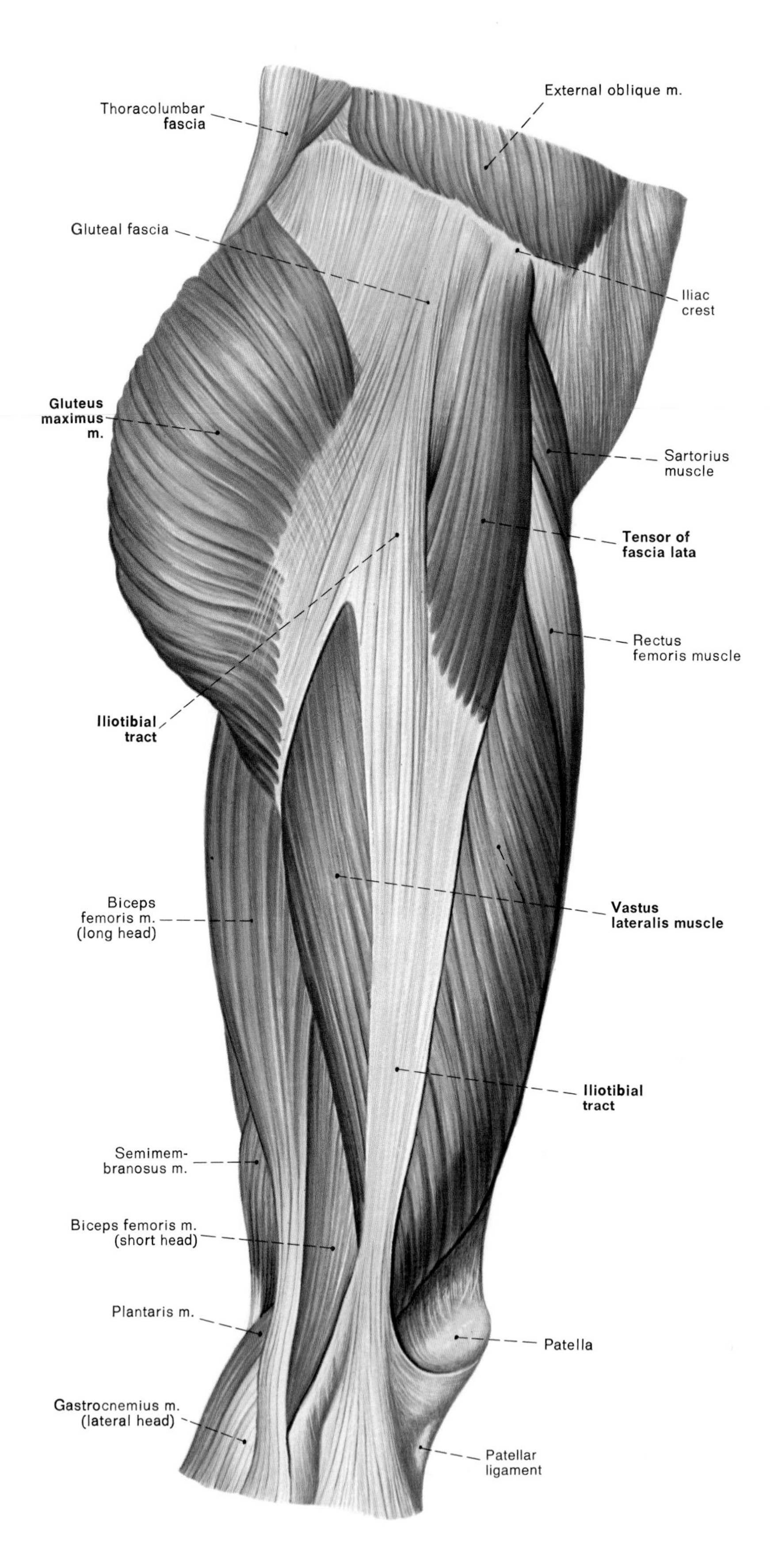

Fig. 12–20: Lateral view of the muscles of the right thigh.

Table 12–3
The Muscles of the Buttock that Abduct the Thigh

Muscle	Origin	Insertion	Nerve Supply	Actions
Gluteus medius	External surface of the ilium between the posterior and anterior gluteal lines	Lateral aspect of the greater trochanter of the femur	Superior gluteal (**L5** and S1)	Abducts and internally rotates the thigh
Gluteus minimus	External surface of the ilium between the anterior and inferior gluteal lines	Anterior aspect of the greater trochanter of the femur	Superior gluteal (**L5** and S1)	Abducts and internally rotates the thigh
Tensor fasciae latae	Anterior superior iliac spine, anterior part of the outer surface of the iliac crest, and deep surface of the uppermost part of the iliotibial tract	Iliotibial tract of the fascia lata	Superior gluteal (L4 and L5)	Tenses the iliotibial tract; abducts, internally rotates, and flexes the thigh

During the initial contact phase, loading response phase, and early part of the mid stance phase, gluteus maximus also exerts abductor activity at the hip joint. The forward placement of the lower limb during these phases aligns many of gluteus maximus's upper muscle fibers lateral to the hip joint; this realignment is responsible for the muscle's capacity to exert an abductor activity at the joint. This abductor activity complements the lateral pelvic tilting action exerted by gluteus medius and gluteus minimus during these three phases.

Although gluteus maximus overlies the ischial tuberosity when a person is standing, the posterior border of the muscle moves anterolaterally from behind the ischial tuberosity when a person reclines to a seated position. It is in this fashion that the ischial tuberosity and its overlying bursa acquire a subcutaneous position when a person is seated.

Gluteus medius and **gluteus minimus** are the chief abductors at the hip joint. Gluteus medius originates from the middle part of the outer surface of the ilium, passes directly over the hip joint, and inserts

Table 12–4
The Muscles of the Buttock that Externally Rotate the Thigh

Muscle	Origin	Insertion	Nerve Supply	Actions
Piriformis	Anterior surface of the sacrum	Greater trochanter of the femur	Branches from the anterior rami of **S1** and S2	Externally rotates the thigh
Superior gemellus	Ischial spine	Greater trochanter of the femur	Nerve to obturator internus (L5 and **S1**)	Externally rotates the thigh
Obturator internus	Pelvic surface of the obturator membrane and the surrounding bony margin	Greater trochanter of the femur	Nerve to obturator internus (L5 and **S1**)	Externally rotates the thigh
Inferior gemellus	Ischial tuberosity	Greater trochanter of the femur	Nerve to quadratus femoris (L5 and S1)	Externally rotates the thigh
Quadratus femoris	Ischial tuberosity	Quadrate tubercle of the femur	Nerve to quadratus femoris (L5 and S1)	Externally rotates the thigh

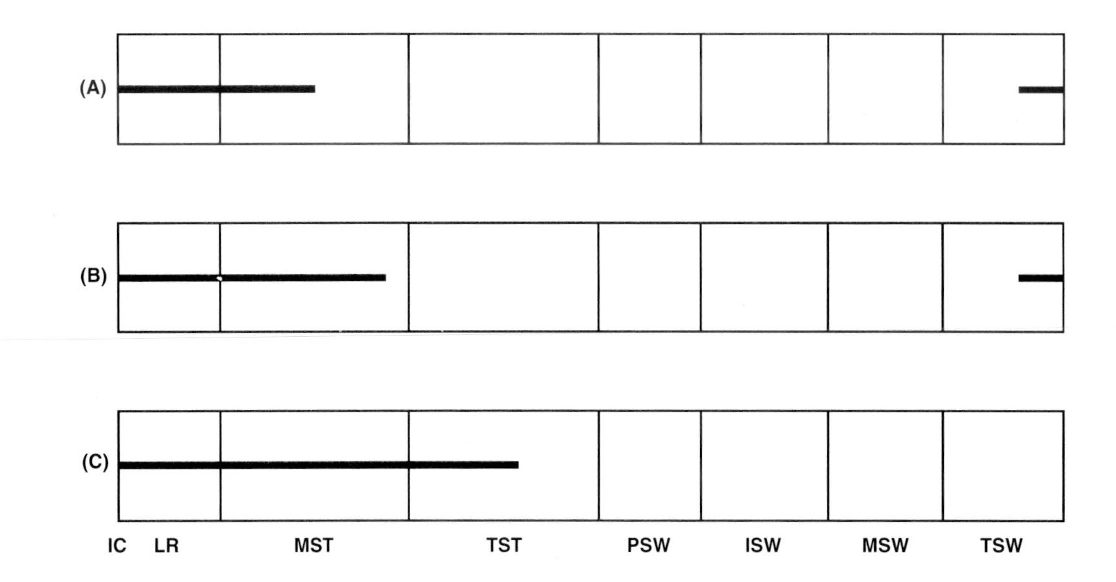

Fig. 12–21: Time frames of consistent significant activity by (A) gluteus maximus, (B) gluteus medius and gluteus minimus, and (C) tensor fasciae latae during a stride of the walking gait. IC = initial contact phase, LR = loading response phase, MST = mid stance phase, TST = terminal stance phase, PSW = pre-swing phase, ISW = initial swing phase, MSW = mid swing phase, and TSW = terminal swing phase.

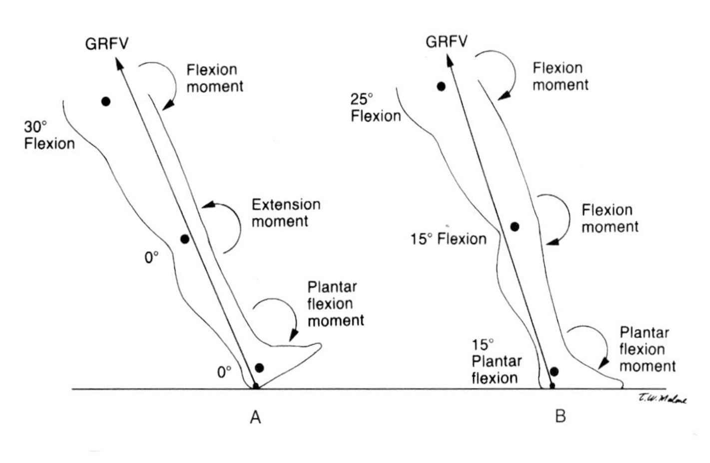

Fig. 12–22: The orientation of the lower limb to the upper body weight line at (A) heel strike and (B) foot flat. The GRFV vector represents the Ground Reaction Force Vector; the ground reaction force is the force that the ground exerts upon the lower limb in reaction to the imposition of upper body weight through the lower limb onto the ground. The GRFV vector is coincident with the upper body weight line.

onto the greater trochanter of the femur (Fig. 12–24). Gluteus minimus originates from the anterior part of the outer surface of the ilium, extends inferolaterally over the hip joint, and inserts onto the greater trochanter.

When walking, gluteus medius and minimus are most active during the interval that extends from the terminal swing phase of the swing period to the mid stance phase of the subsequent stance period (Fig. 12–21B). During the initial contact, loading response, and mid stance phases of the stance period, their abductor activity at the hip joint exerts a **lateral pelvic tilting action** that opposes the tendency of the body trunk during these three phases to fall medially downward at the joint (to fall downward on the medial side of the hip joint). The basis for the tendency of the body trunk to fall downward on the medial side of the hip joint is as follows: The loading of upper body weight upon the reference lower limb during the initial contact and loading response phases occurs in association with the unloading of upper body weight from the contralateral lower limb. The unloading of upper body weight from the contralateral lower limb releases that limb from its support of the pelvis. The loss of contralateral support of the pelvis generates an adduction torque at the hip joint of the reference lower limb (which tilts the pelvic girdle 3° to 4° downward at the hip joint).

Individuals who suffer from paralysis of gluteus medius and minimus compensate for the loss of the lateral pelvic tilting action of these muscles by bending the body trunk laterally over the affected limb during almost all of its stance period in the walking gait (the lateral body trunk bending commences during the loading response phase and extends through the terminal stance phase). The lateral body trunk bending steadies the upper body over the affected limb by placing the center of the gravity of the upper body over the affected limb during all but the initial contact phase of its stance period.

The capacity of gluteus medius and gluteus minimus to adequately exert their lateral pelvic tilting action can be examined by the following test: The patient stands in front of the examiner, with his or her back to the examiner, and alternatively raises each foot off the ground. If gluteus medius and minimus on one side can adequately exert their lateral pelvic tilting action, the contralateral (opposite) side of the pelvis becomes slightly elevated when the contralateral foot is raised off the ground. If the lateral pelvic tilting action of the two gluteal muscles on one side is compromised, however, the contralateral side of the pelvis falls downward when the contralateral foot is raised off the ground. Observation of such an abnormal reaction is reported as a **Trendelenburg sign** for the weight-bearing lower limb. The three most common conditions that produce a Trendelenburg sign are (1) paralysis or weakness of the gluteus medius and minimus muscles, (2) dislocation of the hip joint, and (3) an abnormal angle of inclination between the neck and shaft of the femur.

Tensor fasciae latae is the most superficial muscle in the anterior part of the gluteal region (Fig. 12–20). It originates from the iliac crest and extends inferiorly to insert exclusively into the fascia lata. The muscle is named for its action: it tenses the fascia lata, in particular, the iliotibial tract. The muscle's tonic exertion of tension in the iliotibial tract helps support the hip and knee joints. When walking, tensor fasciae latae exerts a lateral pelvic tilting action through all the phases of the stance period except the pre-swing phase (Fig. 12–21C).

Piriformis, superior gemellus, obturator internus, inferior gemellus, and **quadratus femoris** are all gluteal muscles that can externally rotate the thigh from the anatomic position (Fig. 12–24). They are the deepest muscles of the gluteal region.

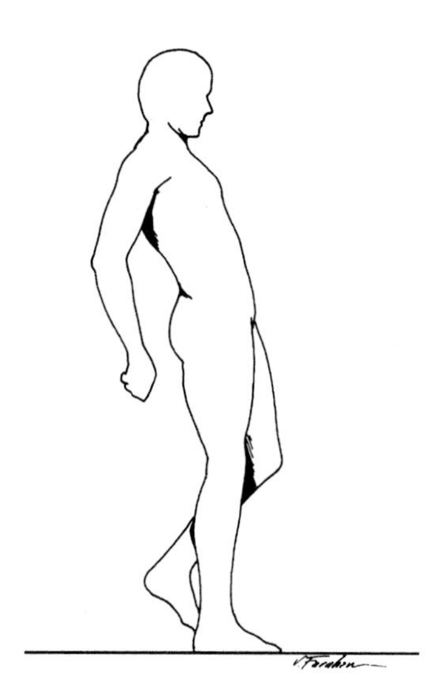

Fig. 12–23: Illustration of the backward lean of the trunk during the walking gait by a person afflicted with paralysis of gluteus maximus.

THE ORIGINS OF THE NERVES THAT INNERVATE THE GLUTEAL MUSCLES

The sacral plexus gives rise to all the nerves that innervate the gluteal muscles. The **sacral plexus** arises from the **lumbosacral trunk** (which is derived from the anterior rami of L4 and L5) and the **anterior rami of S1-S4** (Fig. 12–25). The sacral plexus lies in the lateral wall of the pelvis, where it rests against the pelvic surface of the piriform muscle. Table 12–5 lists the muscles and major joints of the lower limb innervated by the nerves derived from the sacral plexus.

The sacral plexus gives rise to the **sciatic nerve,** the largest nerve in the body. The sciatic nerve is derived from the anterior rami of L4, L5, S1, S2, and S3. The sciatic nerve extends from the sacral plexus into the gluteal region by passing through the greater sciatic foramen. It emerges in the gluteal region inferior to piriformis and deep to gluteus maximus (Fig. 12–26). Its inferolateral course through the lower medial quadrant of the gluteal region ends as it extends inferiorly into the posterior part of the thigh at a point midway between the ischial tuberosity and the greater trochanter of the femur.

Table 12–5
The Muscles and Major Joints of the Lower Limb Innervated by the Nerves Derived from the Sacral Plexus

Nerve	Muscles	Major Joints
Superior gluteal	Gluteus medius Gluteus minimus Tensor fasciae latae	Sacroiliac Hip
Inferior gluteal	Gluteus maximus	
Nerve to obturator internus	Superior gemellus Obturator internus	
Nerve to quadratus femoris	Inferior gemellus Quadratus femoris	Hip
Common peroneal	Short head of biceps femoris Tibialis anterior Extensor hallucis longus Extensor digitorum longus Peroneus tertius Extensor digitorum brevis Peroneus longus Peroneus brevis	Knee Proximal tibiofibular Distal tibiofibular Ankle
Tibial	Semimembranosus Semitendinosus Long head of biceps femoris Gastrocnemius Soleus Plantaris Popliteus Tibialis posterior Flexor hallucis longus Flexor digitorum longus Abductor hallucis Flexor hallucis brevis Adductor hallucis Flexor digitorum brevis Quadratus plantae Lumbricals of the foot Plantar and dorsal interossei of the foot Abductor digiti minimi Flexor digiti minimi brevis	Knee Proximal tibiofibular Distal tibiofibular Ankle

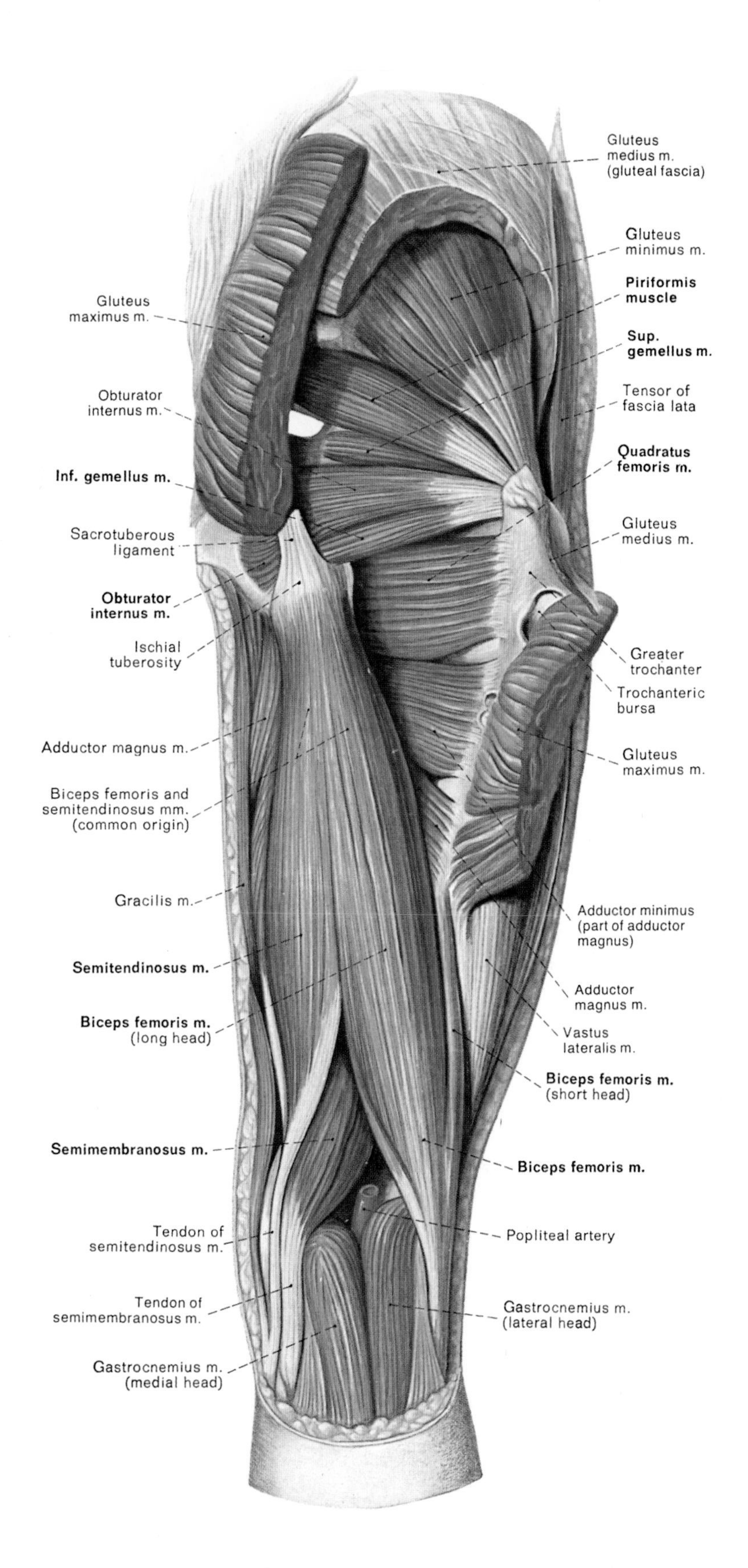

Fig. 12–24: Posterior view of gluteal and posterior thigh muscles of the right thigh.

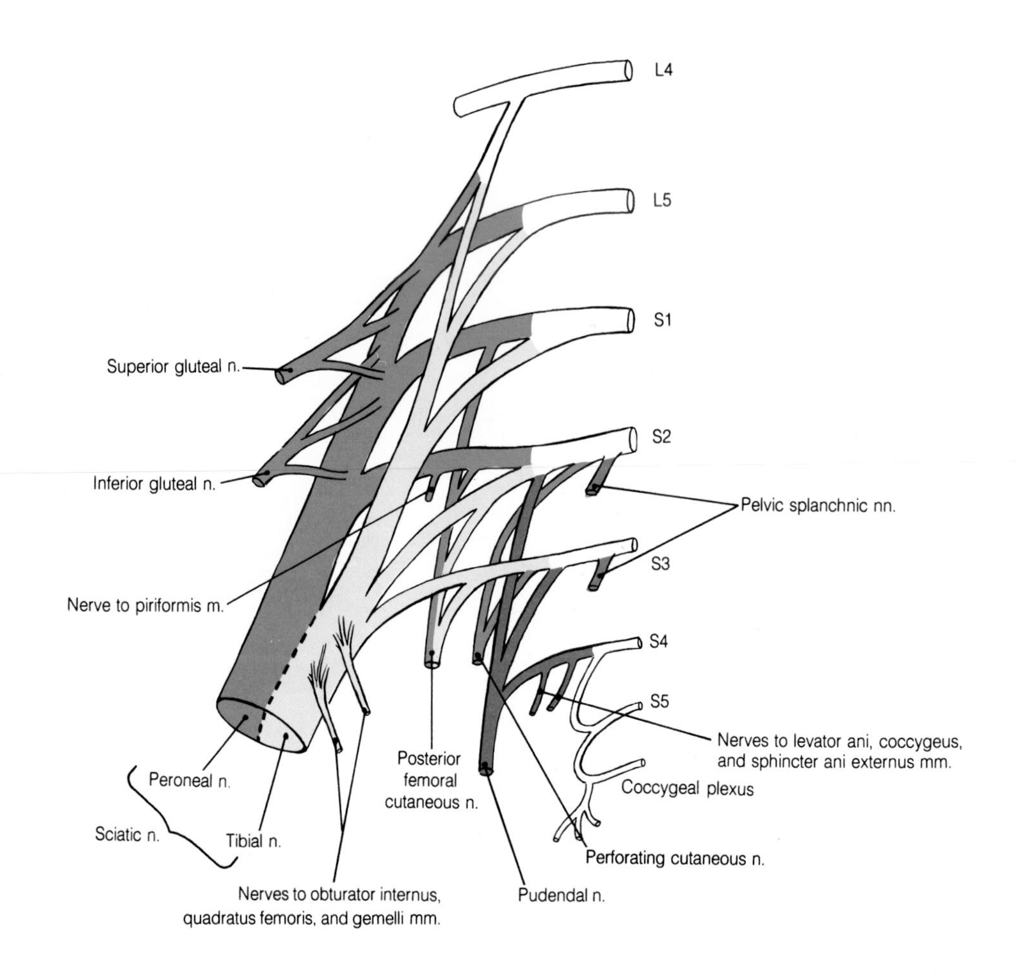

Fig. 12–25: Diagram of the sacral plexus.

The sciatic nerve is the single most important gluteal structure at risk by an intramuscular injection in the buttock. Because the sciatic nerve courses through the lower medial quadrant of the gluteal region, the upper lateral quadrant of the buttock is the safest site for intramuscular injections.

The **posterior cutaneous nerve of the thigh** accompanies the sciatic nerve through the gluteal region (Fig. 12–26). The posterior cutaneous nerve of the thigh is derived from the anterior rami of S1, S2, and S3. It provides sensory innervation for the posterior side of the thigh and the lower medial quadrant of the buttock surface.

THE MAJOR ARTERIES OF THE GLUTEAL REGION

The **superior and inferior gluteal arteries** are the largest arteries of the gluteal region. Both gluteal arteries arise in the pelvis as branches of the internal iliac artery. Both arteries extend from the pelvis into the gluteal region by passing through the greater sciatic foramen. The superior gluteal artery emerges in the gluteal region superior to piriformis (Fig. 12–26) and courses alongside the superior gluteal nerve. The inferior gluteal artery emerges in the gluteal region inferior to piriformis and courses alongside the inferior gluteal nerve.

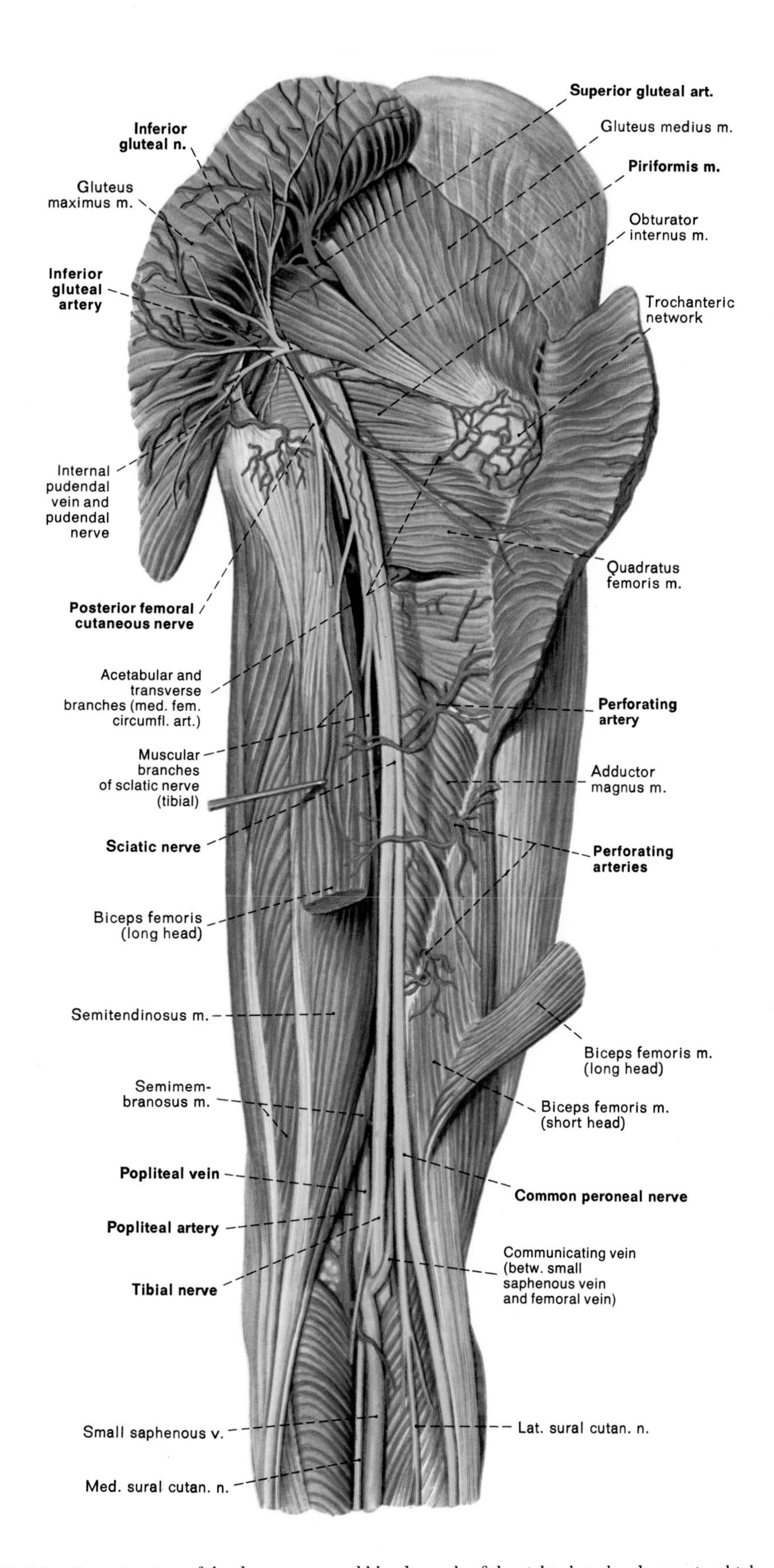

Fig. 12–26: Posterior view of the deep nerves and blood vessels of the right gluteal and posterior thigh regions.

RECOMMENDED REFERENCES FOR ADDITIONAL INFORMATION ON THE WALKING GAIT AND THE HIP

Perry, J. Gait Analysis, McGraw-Hill, Inc., New York, 1992: *Sections 1 and 2 of this text provide a very thorough account of the phases of the walking gait and the major events that occur at the hip, knee, and ankle during each phase.*

Corrigan, B. and G. D. Maitland, Practical Orthopaedic Medicine, Butterworth & Co., London, 1983: *Chapter 9 offers a description of the clinical examination of the hip joint and discussions of the signs and symptoms of common musculoskeletal disorders in and about the hip.*

Magee, D. J., Orthopaedic Physical Assessment, 2nd ed., W. B. Saunders Co., Philadelphia, 1992: *Chapter 10 provides a comprehensive discussion of the clinical examination of the hip joint.*

Greenspan, A., Orthopedic Radiology, 2nd ed., Gower Medical Publishing, New York, 1992: *Chapter 7 presents numerous radiographic views of pelvic girdle trauma. Fig. 7.11 on page 7.8 summarizes the array of radiologic imaging techniques that may be used to evaluate injuries of the pelvic girdle and proximal femur.*

Balderston, R. A., Rothman, R. H., Booth, R. E., and W. J. Hozack, The Hip, Lea & Febiger, Philadelphia, 1992: *This text provides comprehensive discussions of pelvic and hip fractures and some of the most common hip disorders among pediatric patients (congenital dislocations of the hip, Legg-Perthes disease, slipped capital femoral epiphysis and septic arthritis of the hip): Chapter 12 offers a discussion of the clinical presentation of degenerative joint disease, rheumatoid arthritis, and avascular necrosis of the hip.*

Oski, F. A., DeAngelis, C. D., Feigin, R. D., and J. B. Warshaw, Principles and Practice of Pediatrics, J. B. Lippincott Co., Philadelphia, 1990: *Pages 940 through 944 in Chapter 38 present concise descriptions of several common hip disorders among pediatric patients.*

DeGowin, R. L., Jochimsen, P. R., and E. O. Theilen, DeGowin & DeGowin's Bedside Diagnostic Examination, 5th ed., Macmillan Publishing Co., New York, 1987: *This text provides exemplary introductions to the clinical analysis of common medical syndromes. Pages 740 through 752 summarize the signs and symptoms of common injuries, diseases, and conditions involving the buttock, hip, and/or thigh.*

CHAPTER 13

The Thigh and Knee

This chapter focuses upon the innervation and actions of the thigh muscles and the course and distribution of the major arteries, veins, and nerves in the thigh. Since thigh muscles are the prime movers of leg movements, the discussion begins with an examination of the bones of the thigh and leg and the architecture of the knee joint. The head and neck of the femur were discussed in the preceding chapter on the gluteal region.

THE BONES OF THE THIGH AND LEG

The Femur: Its Shaft and Distal End

The shaft of the femur serves as an attachment site for most of the major muscles of the gluteal region and thigh. The distal end of the femur is shaped to articulate with the patella and the proximal end of the tibia. Proceeding from the distal end proximally, the most notable parts of the distal end and shaft of the femur are as follows:

The **medial and lateral condyles** are the pair of rounded, knob-like prominences at the distal end of the femur (Fig. 12–8B). The femoral condyles are the parts of the femur which articulate in the knee joint with the tibia. The **intercondylar fossa (intercondylar notch)** is the deep depression between the condyles in the back of the distal end of the femur (Fig. 12–8B). The **patellar surface** is the smooth depression between the condyles in the front of the distal end of the femur (Fig. 12–8A). The patellar surface articulates in the knee joint with the posterior surface of the patella. The **medial and lateral epicondyles** are the pair of pointed prominences that lie on the side surfaces of the condyles (Fig. 12–8A).

The **adductor tubercle** is a pointed prominence on the medial side of the distal end of the femur. The adductor tubercle marks the uppermost edge of the medial condyle.

The **popliteal surface** is the flat, triangular surface on the posterior aspect of the distal third of the shaft of the femur (Fig. 12-8B). It is the anteror wall of the upper part of the **popliteal fossa,** the space behind the knee joint (Fig. 12–7B). The **medial and lateral supracondylar lines** are respectively the medial and lateral edges of the popliteal surface.

The **linea aspera** is the ridge that rises from the posterior aspect of the middle third of the shaft of the femur (Fig. 12–8B). The medial and lateral edges of the linea aspera are continuous inferiorly with respectively the medial and lateral supracondylar lines.

Longitudinal growth of the femoral shaft during childhood and adolescence occurs more actively at the distal end than at the proximal end. The distal end of the femur bears a single secondary center of ossification during childhood and adolescence. This center, which appears by 36 to 37 weeks of gestation, is the only secondary center of ossification among the long bones of the limbs that always appears before full-term birth. Since this center is always present during the last three or four weeks of a normal gestational period, its presence in the lower limbs of an abandoned and deceased newborn forensically establishes that the child was viable at birth.

The Patella

The patella is the largest sesamoid bone of the body. The patella forms within the quadriceps femoris tendon, which envelops all of the patella except for its posterior surface. Ossification begins at 3 years. The exposed posterior surface of the patella articulates in the knee joint with the patellar surface of the femur.

The Tibia

The tibia is the medial bone of the leg. The expansive, proximal end of the tibia is commonly called the **tibial plateau.** The tibial plateau bears two prominent masses called the **medial and lateral condyles** (Fig. 13–1, A and B); the superior, concave surface of each tibial condyle articulates in the knee joint with the corresponding femoral condyle. The anteroposteriorly-directed prominence that lies between the articular surfaces of the tibial condyles is called the **intercondylar eminence** (Fig. 13–1B). A triangular-shaped prominence called the **tibial tuberosity** projects forward

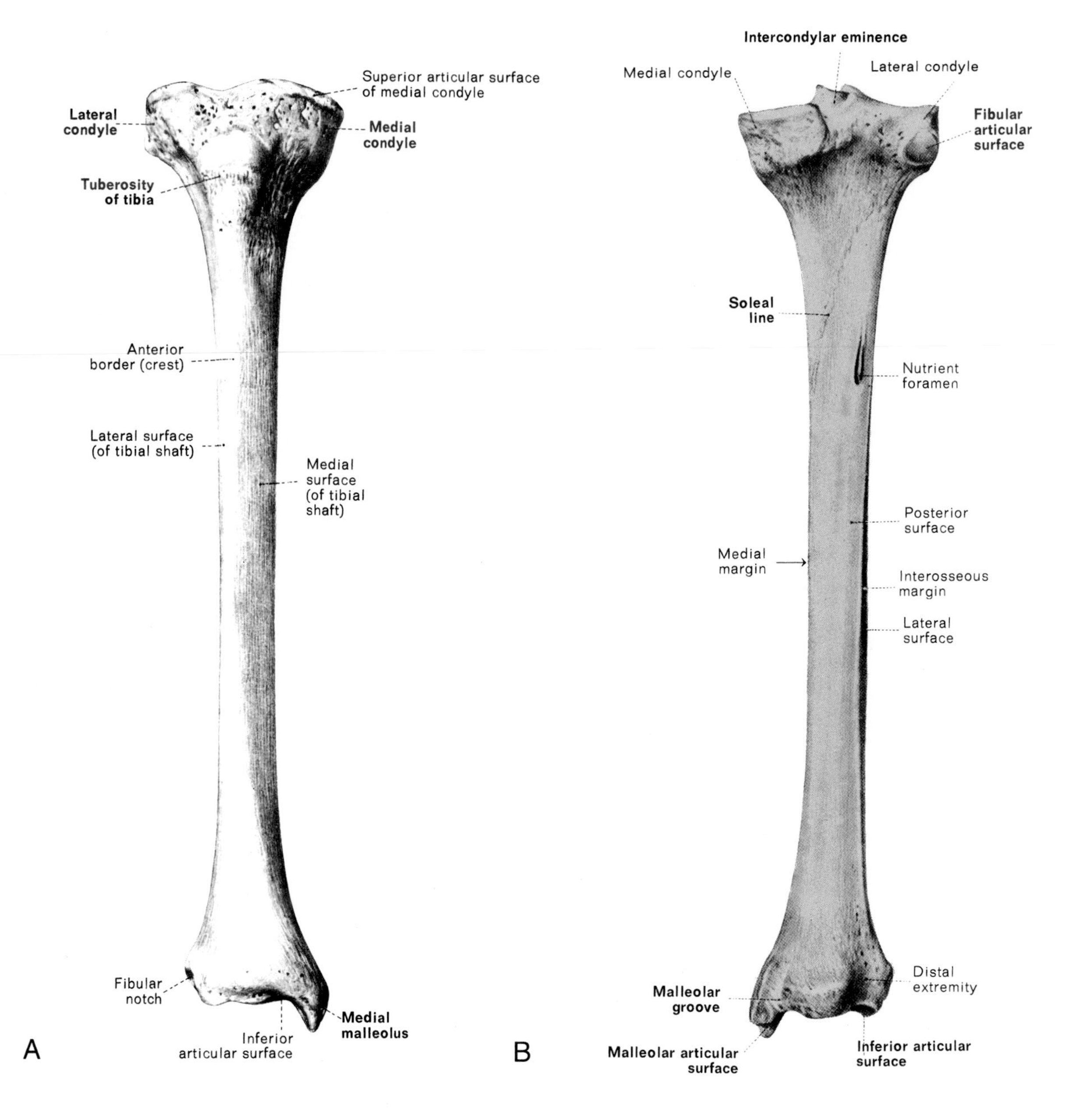

Fig. 13–1: (A) Anterior and (B) posterior views of the right tibia.

from the anterior border of the bone's proximal end (Fig. 13–1A).

The shaft of the tibia has three surfaces: anteromedial, anterolateral, and posterior. The sharply defined, vertical, bony ridge that can be palpated on the front of the leg is the edge between the tibia's anteromedial and anterolateral surfaces; this bony ridge is colloquially referred to as the **shin.** The anteromedial surface of the tibia is subcutaneous along almost all of its length, and is thus responsible for most of the shape of the leg's anteromedial surface.

The stout process that projects inferiorly from the medial side of the distal end of the tibia is called the **medial malleolus** (Fig. 13–1A). The medial malleolus forms the subcutaneous, bony prominence on the medial side of the ankle region (Fig. 12–7, A and B).

The tibia has single secondary centers of ossification at its proximal and distal ends during childhood and adolescence. The center at the proximal end appears at birth, and the center at the distal end appears at 6 months of age.

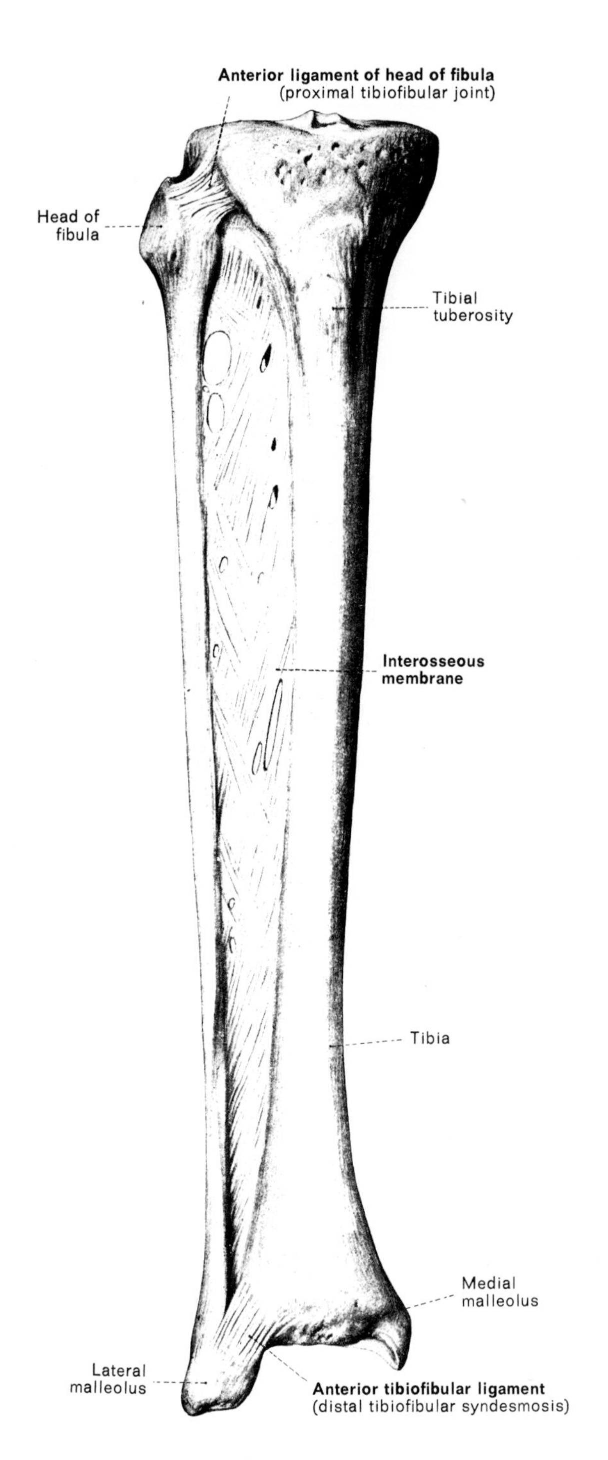

Fig. 13–2: Anterior view of the right tibia, interosseous membrane, and fibula.

CLINICAL PANEL III-4

Acute Hematogenous Osteomyelitis of the Distal Femur or Proximal Tibia

Description: Most cases of bacterial bone infection occur in children and adolescents. In this population, hematogenous spread is the most common mode of acquisition; trauma to a bone region appears to be a major predisposing factor for the seeding of bacteria in the region from a distant site of infection.

Common Symptoms: The most common sites of acute hematogenous osteomyelitis in children and adolescents are the metaphyses of the distal femur and the proximal tibia. When pediatric patients can describe their condition, the complaint is constant and severe pain. Localized warmth and soft tissue swelling may also be present.

Anatomic Basis of the Common Involvement of the Distal Femur and Proximal Tibia: The sluggish and turbulent blood flow through the metaphyses of developing long bones renders these sites susceptible to the deposition and proliferation of blood-borne bacteria, especially following trauma-induced occlusion of the microvasculature at these sites.

Radiographic Features: The initial radiographic signs of acute bacterial bone infection are edema and loss of fascial plane definition of the overlying soft tissues; these signs occur within 24 to 48 hours. [The facial planes that lie between muscles in the limbs have a relatively high fat content and thus a radiographic density of fat. If a fascial plane is parallel to the path of the X-ray beam, it may cast a relatively radiolucent line between the more radiodense shadows cast by the muscles to either side of it (muscles have a radiographic density of water).] If untreated, lytic lesions of the infected metaphysis occur within 7 to 10 days, followed by endosteal sclerosis due to new periosteal bone formation.

The Fibula

The fibula is the lateral bone of the leg. The slightly enlarged, proximal end of the fibula is called the **head** (Fig. 13–2); it can be palpated at the uppermost end of the lateral side of the leg. The blunt process that projects inferiorly from the lateral side of the distal end of the fibula is called the **lateral malleolus** (Fig. 13–2). The lateral malleolus forms the subcutaneous, bony prominence on the lateral side of the ankle region (Fig. 12–7, A and B).

The fibula has single secondary centers of ossification at its proximal and distal ends during childhood and adolescence. The center at the distal end appears at 9 to 12 months of age, and the center at the proximal end appears at 3 to 4 years of age.

CLINICAL PANEL III-5

Bucket-Handle or Corner Fractures of the Long Bones of the Limbs in Infants and Young Children

Definition: A bucket-handle or corner fracture of a long bone of the upper or lower limb is a fracture that extends transversely through the metaphysis (Fig. 13–3, A and C).

Mechanism of Injury: A bucket-handle or corner fracture results from the imposition of very forceful traction upon the involved limb. Infants and young children only rarely encounter circumstances in which they are accidentally subjected to such severe traction. Consequently, a bucket-handle or corner fracture of a long bone of the upper or lower limb in an infant or young child is virtually pathognomonic of non-accidental trauma (physical abuse of the infant or young child). Adults, adolescents, and older children (but not young children) have the strength necessary to impose traction forces of the magnitude that generate bucket-handle or corner fractures.

Bucket-handle or corner fractures in infants and young children are frequently encountered in the distal femur or proximal tibia. They are generally the result of an adult or adolescent grasping and exerting a very forceful pull on the child's lower limb.

Anatomic Basis for the Characterization of Bucket-Handle or Corner Fractures: Identification of bucket-handle or corner fractures requires superior radiographic images. Some metaphyseal fractures of non-accidental origin produce metaphyseal fragments of uniform thickness (Fig. 13–3, A and B). If the fracture plane in such fractures is oriented parallel to the X-ray beam, then the radiograph shows a thin radiolucent line extending transversely through the metaphysis (Fig. 13–3A). However, if the fracture plane is oriented slightly obliquely to the X-ray beam, then the radiograph shows a thin, curved band (with a radiographic density of bone) arching around the metaphysis (Fig. 13–3B). Such radiographic evidence is called a bucket-handle fracture because the thin, curved band of bone density resembles a bucket-handle embracing the metaphysis.

Other metaphyseal fractures of non-accidental origin produce a metaphyseal fragment thicker around its margin than in its center (Fig. 13–3, C and D). If the central region of the fracture plane is oriented parallel to the X-ray beam, then the radiograph shows small but readily identifiable areas of bone density near one or both sides of the metaphysis (Fig. 13–3C). Such radiographic evidence is called a corner fracture because each small area of bone density lies at a "corner" of the metaphysis. If the central region of the fracture plane is oriented slightly obliquely to the X-ray beam, then a bucket-handle fracture appearance results (Fig. 13–3D).

THE KNEE JOINT

The Articulations and Nerve Supply of the Knee Joint

The knee joint is a synovial joint in which the femur articulates with the tibia and the patella. The joint consists of three articulations (Fig. 13–4): (1) The medial femoral condyle articulates with the medial tibial condyle and medial meniscus. (2) The lateral femoral condyle articulates with the lateral tibial condyle and lateral meniscus. (3) The patellar surface of the femur articulates with the posterior surface of the patella.

The **femoral, obturator, tibial,** and **common peroneal nerves** innervate the joint.

The Fibrous Cartilage Discs of the Knee Joint

The **medial and lateral menisci** of the knee joint are crescent-shaped, articular discs of fibrous cartilage (Figs. 13–4 and 13–5). The medial meniscus lies interposed between the medial condyles of the femur and tibia, and the lateral meniscus lies interposed between the lateral condyles of the two bones. Each meniscus is attached by its **horns** (ends) to the tibial condyle below (Fig. 13–5) and by its relatively broad, outer edge to the deep surface of the knee joint's fibrous capsule (Fig. 13–6). The superior and inferior surfaces of each meniscus and its sharp, inner edge are free and unattached, and lie bathed in the joint's synovial fluid (Fig. 13–6).

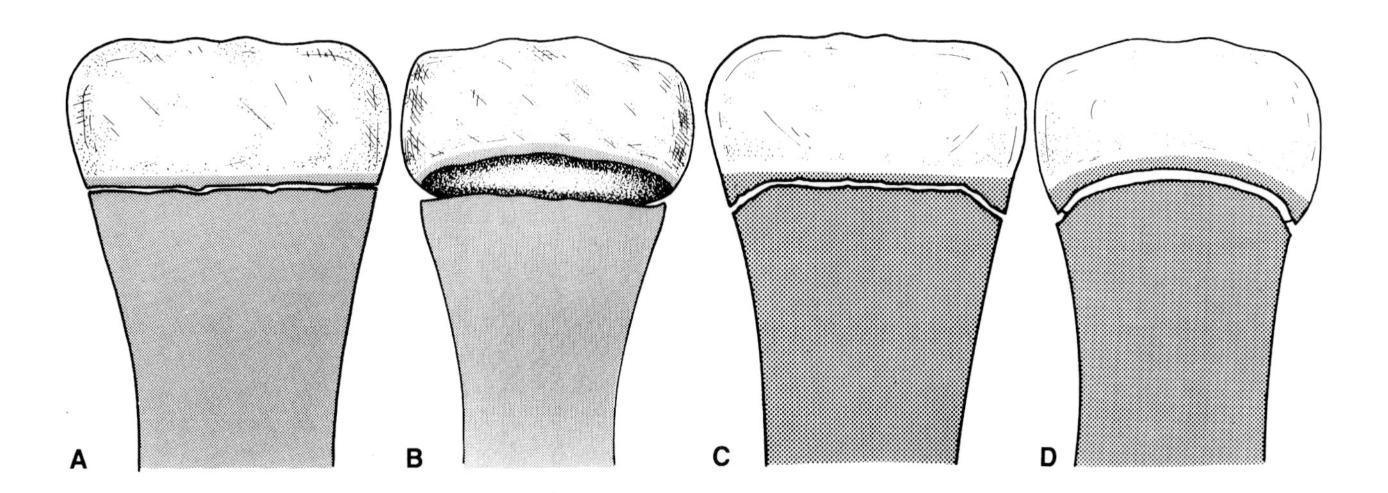

Fig. 13–3: Diagrams illustrating the radiographic appearance of metaphyseal fractures of non-accidental origin. (A) and (B) show two different views of a metaphyseal fracture in which the metaphyseal fragment is of uniform thickness. (A) A thin, straight band of bone density appears at the metaphysis if the fracture plane is oriented parallel to the x-ray beam and (B) a thin, curved band of bone density (resembling a bucket-handle) appears at the metaphysis if the fracture plane is oriented slightly obliquely to the x-ray beam. (C) and (D) show two different views of a metaphyseal fracture in which the metaphyseal fragment is thicker around its margin than in its center. (C) Small areas of bone density appear at the corners of the metaphysis if the central region of the fracture plane is oriented parallel to the x-ray beam and (D) a moderately thick, curved band of bone density (resembling a bucket-handle) appears at the metaphysis if the central region of the fracture plane is oriented slightly obliquely to the x-ray beam.

Each meniscus presents a flat surface to the tibial condyle below and a concave surface to the femoral condyle above (Fig. 13–6). The menisci thus enhance the stability of the knee joint by increasing the concavity of the articular surfaces presented to the femoral condyles. The combined articular surface which each tibial condyle and its meniscus presents to the apposing femoral condyle is referred to as the **meniscotibial surface.**

The Fibrous Capsule of the Knee Joint

The joint's fibrous capsule is attached superiorly just proximal to the articular margins of the femoral condyles medially and laterally and to the upper margin of the intercondylar fossa posteriorly (Fig. 13–7). Inferiorly, the capsule is attached to the articular margins of the tibial condyles; there is no attachment to the fibula. There is a gap in the posterior part of the capsule, which transmits the **popliteus muscle** (Fig. 13–7). Popliteus originates within the knee joint from the lateral side of the lateral femoral condyle and lateral meniscus, and extends inferomedially through the gap in the posterior part of the capsule toward its insertion onto the upper posterior surface of the tibia.

The anterior aspect of the knee joint is covered by the tendon of quadriceps femoris, the patella, and the patellar ligament (Fig. 13–4). The **quadriceps femoris tendon** passes in front of the knee joint to insert onto the tibial tuberosity. The patella lies partially embedded in the tendon at the level where the tendon extends in front of the distal end of the femur. The tendon envelops all but the patella's posterior surface. The **patellar ligament (ligamentum patellae)** is the part of the quadriceps femoris tendon that extends from the inferior margin of the patella to the tibial tuberosity.

The Major Ligaments that Stabilize the Knee Joint

The **tibial** and **fibular collateral ligaments** strengthen the sides of the joint's capsule (Figs. 13–4 and 13–6). The ligaments restrict side-to-side movements of the leg at the knee joint. The tibial collateral ligament (medial collateral ligament) extends from the medial epicondyle of the femur to the medial tibial condyle and the medial surface of the tibial shaft; the ligament is attached to both the joint capsule and the medial meniscus. The fibular collateral ligament (lateral collateral ligament) extends from the lateral epicondyle of the femur to the head of the fibula. Only the superior part of the fibular collateral ligament is attached to the knee joint capsule. The origin of popliteus separates the fibular collateral ligament from the lateral meniscus (Fig. 13–7).

The **anterior** and **posterior cruciate ligaments** directly bond the femur and tibia together within the joint capsule (Fig. 13–4). The anterior cruciate liga-

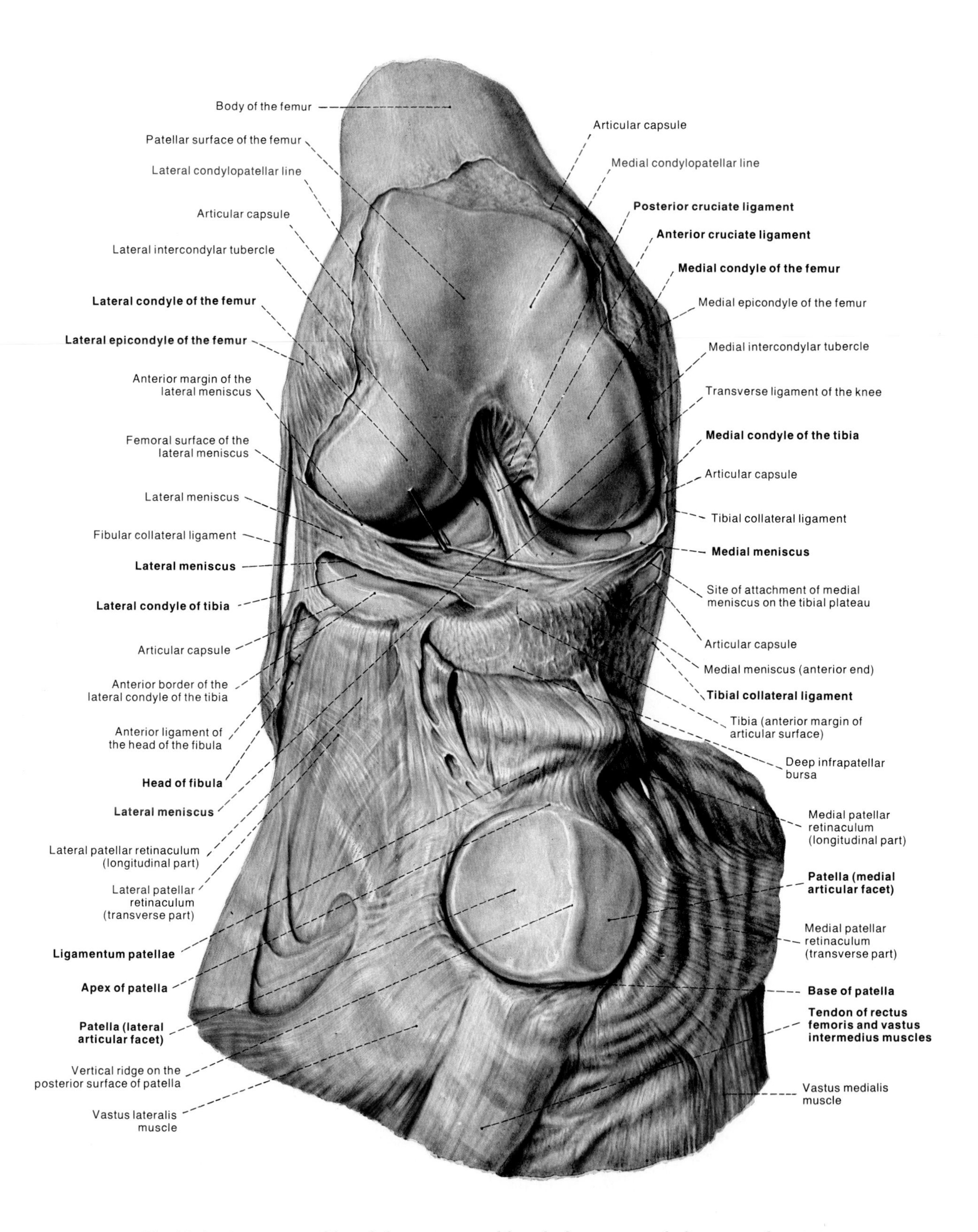

Fig. 13–4: Anterior view of the right knee joint, opened from the front to expose the ligaments and menisci.

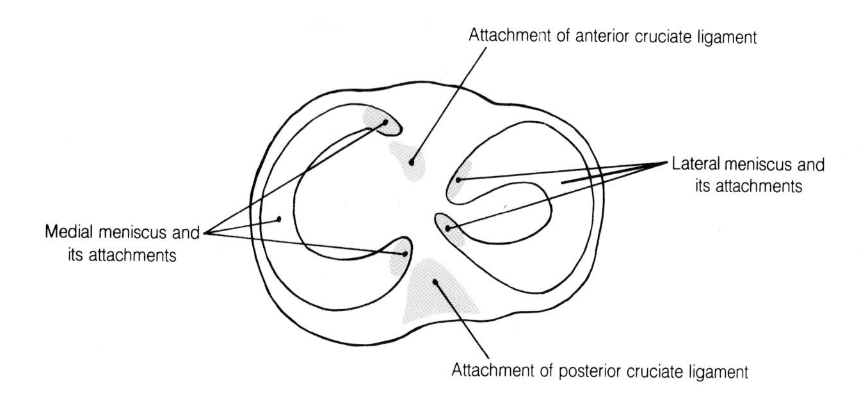

Fig. 13–5: Diagram of the attachments of the cruciate ligaments and menisci to the tibial plateau.

ment extends from the anterior area of the tibial intercondylar surface (Fig. 13–5) to the posterior part of the medial surface of the lateral femoral condyle (Fig. 13–4). The posterior cruciate ligament extends from the posterior area of the tibial intercondylar surface (Fig. 13–5) to the anterior part of the lateral surface of the medial femoral condyle (Fig. 13–4). The ligaments cross each other as they extend from their tibial to femoral attachments (this crossing is the anatomic basis for the naming of the ligaments as cruciate ligaments). The anterior cruciate ligament restricts the tibia from being pulled too far forward during extension of the leg, and the posterior cruciate ligament restricts the tibia from being pulled too far backward during flexion of the leg.

The **oblique popliteal ligament** strengthens the posterior part of the joint capsule. The ligament is an oblique extension of the tendon of insertion of the posterior thigh muscle semimembranous (Fig. 13–8).

The Synovial Membrane and Bursae of the Knee Joint

The synovial membrane lines most of the deep surface of the joint capsule, generally extending to the capsular attachments of the femur and tibia (Fig. 13–9). However, the synovial membrane does not line the deep surface of the middle posterior part of the capsule, since here the synovial membrane is reflected anteriorly around the sides and the anterior aspect of the cruciate ligaments (Fig. 13–10). The cruciate ligaments thus lie outside the knee joint's synovial cavity but inside its fibrous capsule.

The most anterosuperior part of the joint's synovial cavity is a recess that lies between the quadriceps femoris tendon and the distal end of the femur (Fig. 13–9). This synovial membrane-lined recess is called the **suprapatellar bursa** because it extends for about 5 to 7 cm above the upper border of the patella.

In the lower anterior part of the joint, a fat pad intervenes between the patellar ligament and the synovial membrane of the knee joint (Fig. 13–9). This fat pad is called the **infrapatellar fat pad** because it lies below the lower border of the patella.

When fluid accumulates in the knee joint's synovial cavity, the normal hollowed contours around the patella and along the sides of the patellar ligament may swell and the patella may be displaced anteriorly. The contours surrounding the upper margin of the patella may swell because of expansion of the underlying suprapatellar bursa. The hollowed contours along the lower margin of the patella and the sides of the patellar ligament may swell because of expansion of the part of the joint's cavity deep to the patella and the infrapatellar fat pad. The patella may be displaced anteriorly because the patella and the portion of the quadriceps femoris tendon in which it is embedded overly the anterior aspect of the joint cavity.

There are three major bursae anterior to the knee joint which do not communicate with the joint's synovial cavity. They are all related to the patella (Fig. 13–9):

1. The **prepatellar bursa** is a subcutaneous bursa which lies between the skin and (a) the lower half of the patella and (b) the upper half of the patellar ligament. When fluid accumulates in the prepatellar bursa, it produces a swelling over the lower half of the patella and the upper half of the patellar ligament.
2. The **superficial infrapatellar bursa** is a subcutaneous bursa that lies between the skin and the patellar ligament.
3. The **deep infrapatellar bursa** is an intracapsular bursa which lies between the patellar ligament and

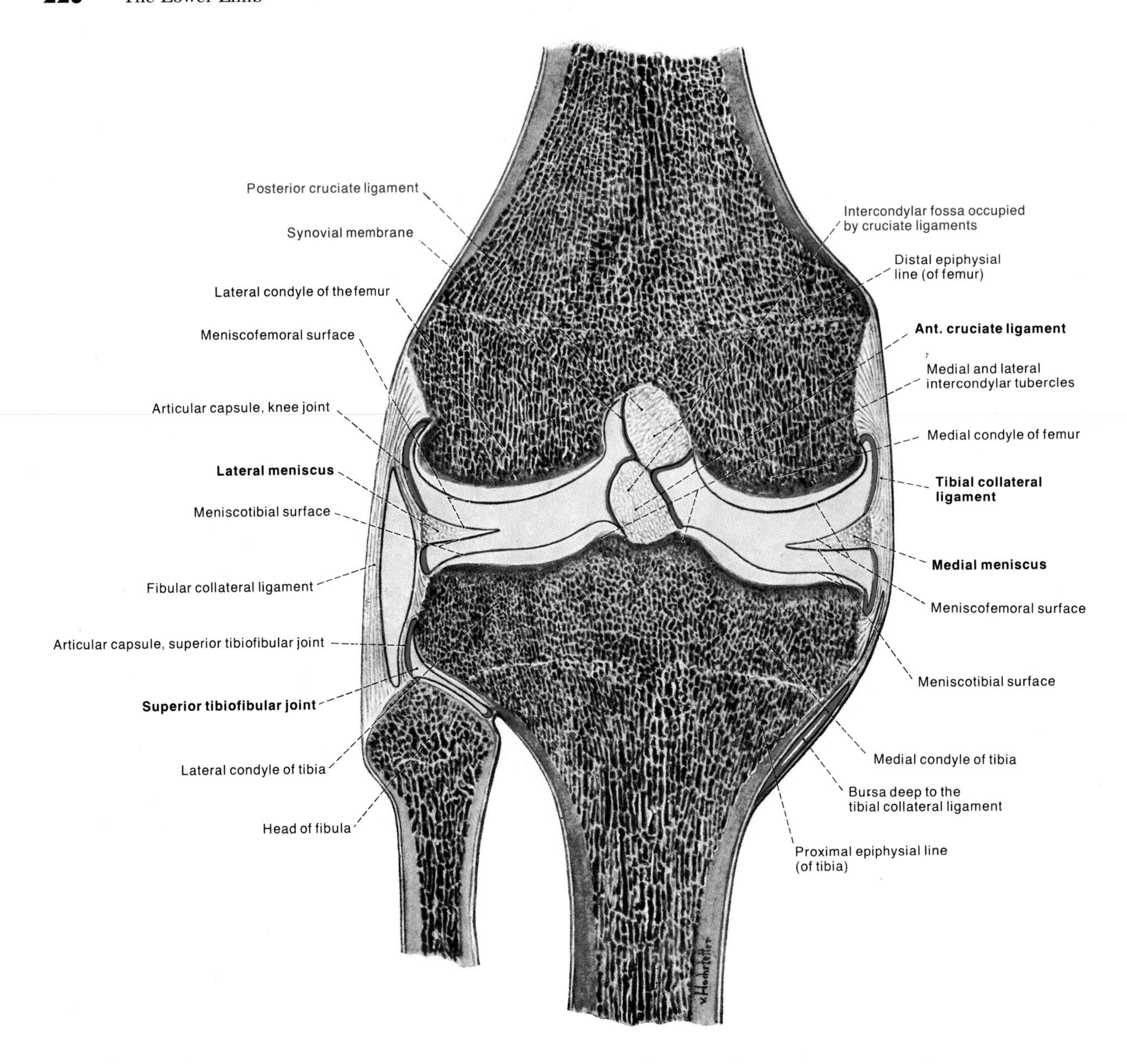

Fig. 13–6: Coronal sectional view of the right knee joint and proximal tibiofibular joint.

the infrapatellar fat pad. When fluid accumulates in the deep infrapatellar bursa, it produces swellings on both sides of the patellar ligament.

There are three major bursae about the knee joint which commonly communicate with the joint's synovial cavity. They are related to muscles about the knee joint:

1. The **popliteus bursa** is associated with the lateral aspect of the joint; it intervenes between popliteus and the lateral femoral condyle (Fig. 13–7).
2. The **semimembranosus bursa** is associated with the medial aspect of the joint; it intervenes between semimembranosus's tendon of insertion and the medial tibial condyle (Fig. 13–7).
3. The **gastrocnemius bursa** is associated with the medial aspect of the joint; it intervenes between the medial head of gastrocnemius and the joint's capsule.

The Movements that Occur at the Knee Joint

The knee joint provides for **flexion, extension, external rotation,** and **internal rotation** of the leg. However, the external and internal rotation movements

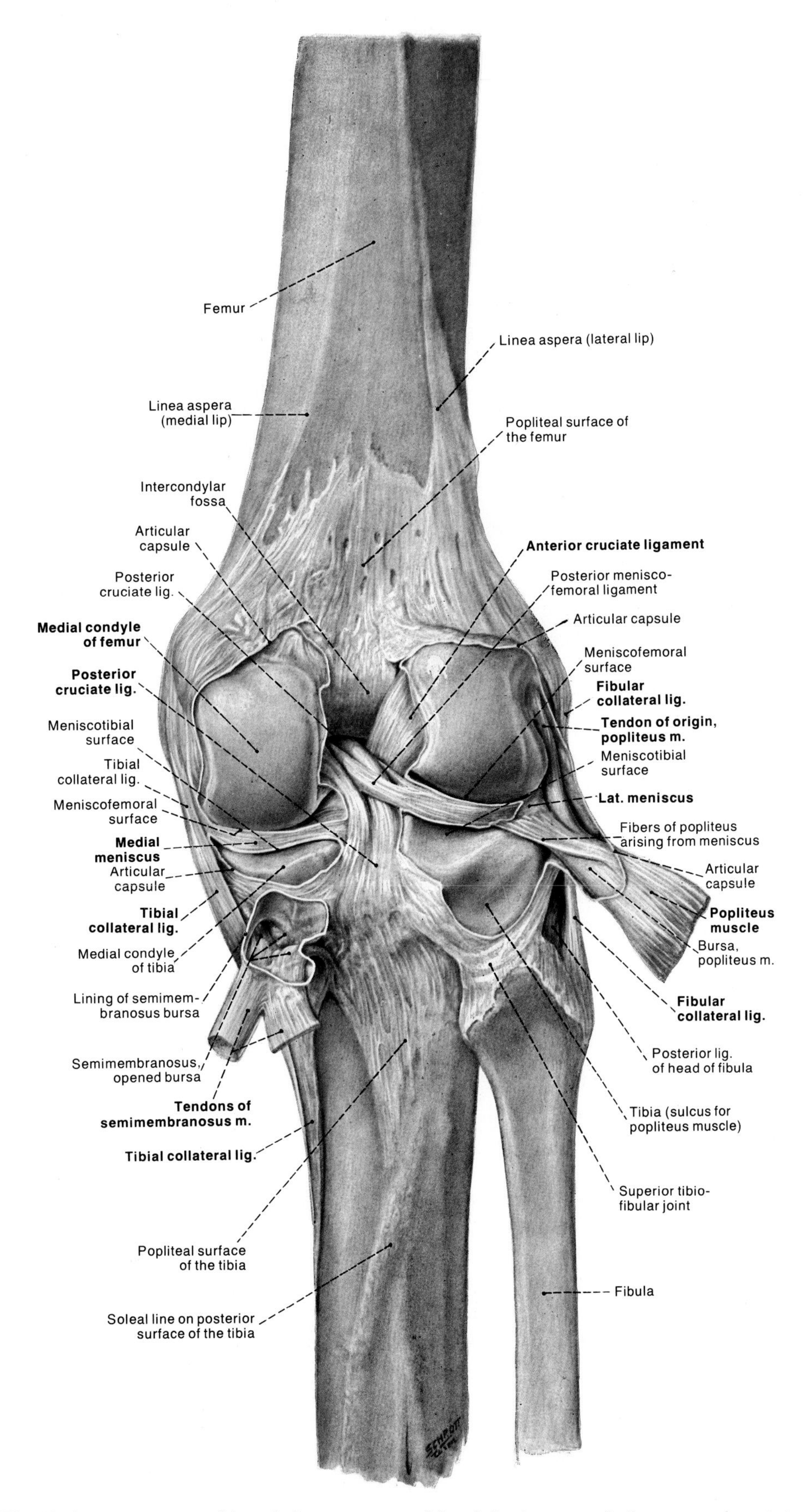

Fig. 13–7: Posterior view of the right knee joint, opened from behind to expose the ligaments and menisci.

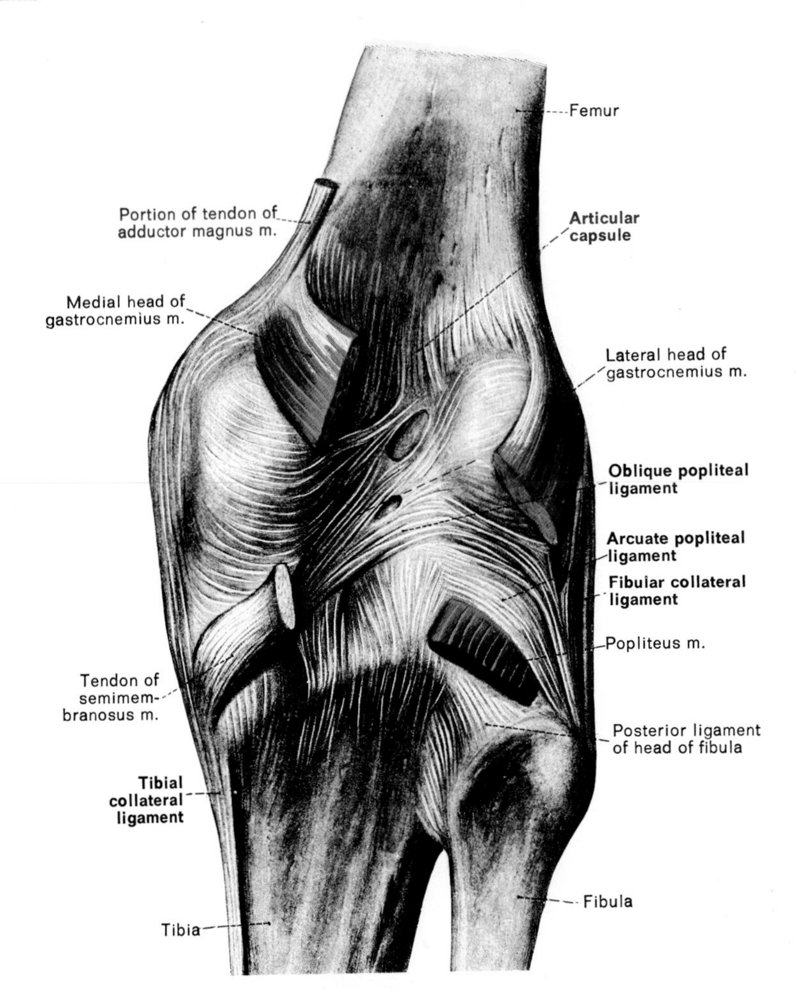

Fig. 13–8: Posterior view of the right knee joint.

cannot always be executed independently of the flexion and extension movements. This is because rotational movements automatically become incorporated into flexion or extension movements when the leg is flexed or extended to any extent between the positions of 0° and 30° flexion.

Extension in the knee joint proceeds mainly by the femoral condyles rolling forward and sliding backward on their meniscotibial surfaces (Fig. 13–11). These anteroposterior rolling and sliding movements generate three major changes in the joint as the leg is extended from the fully flexed to the 30° flexed position:

1. The zones of contact between the femoral and tibial condyles move forward in the joint. When the leg is fully flexed, the femoral condyles are in contact with the posterior parts of the upper surfaces of the tibial condyles. As extension proceeds from this position, the femoral condyles progressively roll forward on the tibial condyles farther than they slide backward. The net effect is that the contact zones between the femoral and tibial condyles move progressively forward during extension from the posterior to the more anterior parts of the upper surfaces of the tibial condyles.
2. The femoral condyles and their meniscotibial surfaces become more congruent. When the leg is fully flexed, the meniscotibial surfaces are in contact with the highly curved, posterior parts of the femoral condyles. As extension proceeds from this position, increasingly flatter parts of the femoral condyles roll and slide into apposition with the meniscotibial surfaces. The less curved, anterior parts of the femoral condyles form a better fit with the shallow, meniscotibial fossae than do the highly curved, posterior parts of the femoral condyles.
3. The menisci are displaced anteriorly and their inner margins are compressed outwardly. The forward movement of the contact zones between the femoral and tibial condyles forces the menisci to pivot about their horns so that the menisci move forward within the joint. At the same time, the rolling into apposition of the flatter, anterior parts of

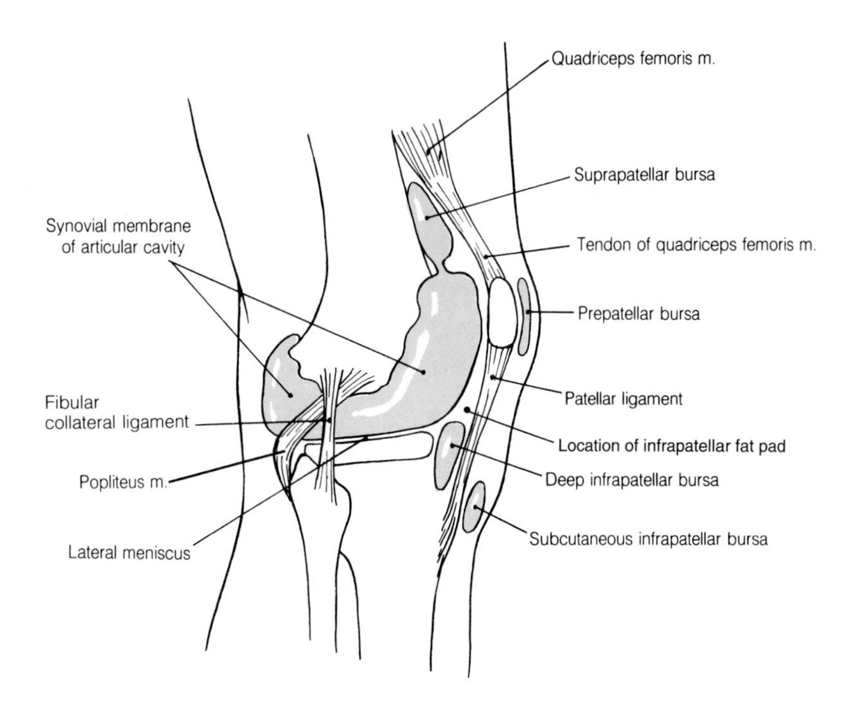

Fig. 13–9: Diagram of the synovial membrane of the knee joint and associated bursae.

the femoral condyles presses the inner margins of the menisci outward.

A fundamental change in the mechanics of extension occurs at the position just 30° short of full extension. When the leg is extended to this semi-flexed position, the lateral femoral condyle suddenly becomes fully congruent with its meniscotibial surface; however, on the medial side of the knee joint, the medial femoral condyle is still incongruent with its meniscotibial surface. Consequently, at this position, only the medial femoral condyle is free to roll forward and slide backward on its meniscotibial surface. Further extension from this position occurs only with conjunctional (integral and concurrent) medial rotation of the femur or lateral rotation of the tibia within the knee joint. Medial rotation of the femur occurs if the foot is planted firmly on the ground and the tibia is thus fixed; lateral rotation of the tibia occurs if the foot is suspended above the ground, and the femur is thus the more fixed of the two bones. In either case, the medial femoral condyle and the medial meniscus move backward relative to the medial tibial condyle as the lateral femoral condyle and lateral meniscus simultaneously move forward relative to the lateral tibial condyle. The slight movement of the anterior end of the lateral meniscus onto the upward sloping surface of the lateral tibial condyle deforms the lateral meniscus to the extent that the meniscotibial surface now becomes incongruent with the lateral femoral condyle. This change frees the lateral femoral condyle to roll forward and slide backward until full congruence with its meniscotibial surface is once again restored. In practice, these rotation and extension movements occur in a concurrent and continuous fashion throughout the last 30° of extension, with most of the rotational changes occurring during the last 5° of extension.

The conjunctional rotation movements that occur within the knee joint during the last 30° of extension not only make the last 30° of extension possible, but

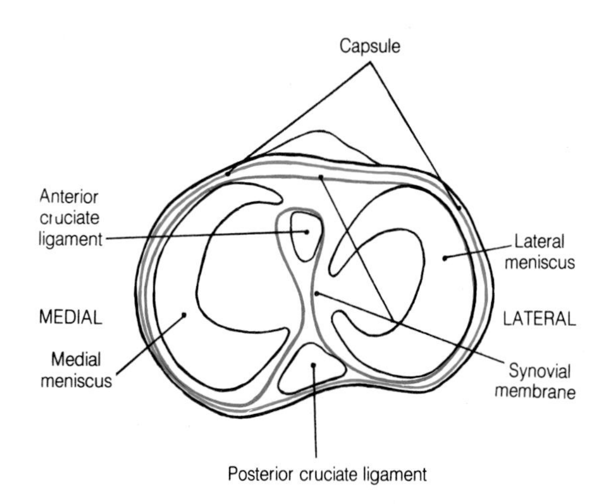

Fig. 13–10: Diagram showing the tibial attachments of the capsule and synovial membrane of the knee joint.

CLINICAL PANEL III-6

PRIMARY OSTEOARTHRITIS OF THE KNEE JOINT

Definition: Primary osteoarthritis is degenerative joint disease. Genetic factors and the aging process are involved in the etiology of the disease. The knee joint is a common site of primary osteoarthritis.

Common Symptoms: Osteoarthritic joints generally become painful when movement occurs within the joints. The pain is exacerbated by cold (and weight-bearing in lower limb joints) and relieved by rest and warmth. Crepitus (joint sounds) may also accompany movements at osteoarthritic joints. Osteoarthritic joints may become stiff at rest. Inflammation is generally minimal and there are no systemic manifestations.

Osteoarthritis of the knee joint more commonly affects the medial femorotibial and patellofemoral articulations than the lateral femorotibial articulation. Marked degeneration of the medial femorotibial articulation can produce a varus deformity at the knee. [The term **varus** indicates that the distal bone points toward the midline of the body; the term **valgus** indicates that the distal bone points away from the midline.]

Radiographic Features: The principal findings are (a) narrowing of involved joint spaces (as a result of degeneration and thinning of the articular cartilage surfaces), (b) subchondral osteosclerosis (increased radiopacity of subchondral bone areas as a result of a net increase of bone tissue deep to the thinned articular cartilage surfaces), and (c) periarticular osteophytes (amorphous outgrowths of newly deposited bone tissue from the margins of subchondral areas).

also "lock" the joint into its most stable and closely-packed configuration. In the anatomic position, the femoral condyles are in maximum congruence and contact with the meniscotibial surfaces. The leg is laterally rotated to the maximal extent within the knee joint. There is maximum compression between the femoral condyles and meniscotibial surfaces because the anterior and posterior cruciate ligaments, the tibial and fibular collateral ligaments, and the oblique popliteal ligament are all somewhat twisted and tightly stretched between the femur and tibia.

All the movements just described occur in reverse order when the leg is flexed from the anatomic to the fully flexed position. The first few degrees of flexion require "unlocking" rotational movements within the knee joint; the "unlocking" of the joint occurs by lateral rotation of the femur or medial rotation of the tibia (as during extension, the less fixed bone is the bone that rotates).

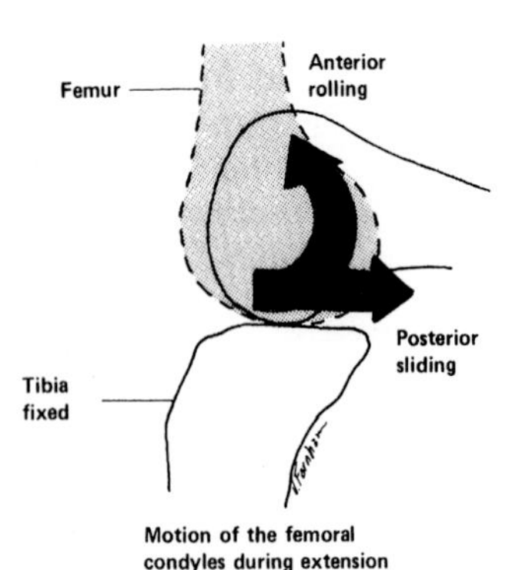

Fig. 13–11: Motions of the femoral condyles during extension of the leg at the knee joint.

During leg extension, the posterior surface of the patella slides upward and forward on the patellar surface of the femur. The patella moves in this fashion as a result of the upward and forward pull of quadriceps femoris. The vastus lateralis and vastus intermedius muscles of the quadriceps group, however, also exert a strong, laterally-directed pull on the patella, which is counteracted in part by a medially-directed pull of the vastus medialis muscle. The laterally-directed pull on the patella is also resisted by the rather prominent, anterior bulge of the lateral femoral condyle. The patella is thus subject to recurrent dislocation during leg extension in individuals who suffer from a markedly weakened vastus medialis muscle or a congenitally underdeveloped lateral femoral condyle.

THE PROXIMAL AND DISTAL TIBIOFIBULAR JOINTS

The tibia and fibula articulate with each other at their proximal ends by the proximal tibiofibular joint (Figs. 13–6 and 13–7) and at their distal ends by the distal tibiofibular joint (Fig. 14–8). The proximal tibiofibular joint is a synovial joint, and the distal tibiofibular joint is a syndesmosis. The tibiofibular joints provide for only a small range of movement.

The shafts of the tibia and fibula are also joined by a thin, fibrous sheet called the **interosseous membrane of the leg** (Fig. 13–2).

THE MUSCLES OF THE THIGH

The muscles of the thigh are enveloped by a tubular sheath of deep fascia called the **fascia lata** (Fig. 13–13). The superior margin of the fascia lata is attached medially to the inferior ramus of the pubis, the ischial ramus and tuberosity, and the sacrotuberous ligament; posteriorly to the sacrum and coccyx; laterally to the iliac crest; and anteriorly to the inguinal ligament. Inferiorly, the fascia lata extends past the knee joint to blend with the deep fascia of the leg and becomes attached to the tibial condyles and the head of the fibula.

The **inguinal ligament** extends between the anterior superior iliac spine and the pubic tubercle, and is the structure in the front of the body which serves as the boundary between the abdomen and the thigh (Fig. 13–13). The inguinal ligament is named for its location in the **inguinal (groin) region** of the body, which is the border region in the front of the body between the abdomen and thigh. [Refer to the discussion of the inguinal canal in Chapter 20 for a description of the inguinal ligament.]

The strongest and most prominent aspect of the fascia lata is the vertical tract called the **iliotibial tract** that extends from the iliac crest to the lateral tibial condyle (Fig. 12–20). All of tensor fasciae latae's fibers and most of gluteus maximus's fibers insert onto the iliotibial tract. The lines of pull which these muscle fibers tonically exert within the iliotibial tract support the lateral side of the knee joint.

In the distal two-thirds of the thigh, the fascia lata gives off on the medial side and on the lateral side sheet-like layers that extend deeply toward the femur. The extensions are called the **medial** and **lateral intermuscular septa,** and they are attached to the shaft of the femur along respectively the medial and lateral edges of the linea aspera and popliteal surface. The fascia lata and its medial and lateral intermuscular septa divide the distal two-thirds of the thigh into two compartments: anterior and posterior.

The muscles of the thigh are generally divided into three groups on the basis of their innervation and general location: a posterior group that is innervated by the sciatic nerve, a medial group which (except for one muscle, the hamstring part of adductor magnus) is innervated by the obturator nerve, and an anterior group which (except for one muscle, psoas major) is innervated by the femoral nerve. In the distal two-thirds of the thigh, the muscles of the anterior group occupy the anterior compartment, and the muscles of the medial and posterior groups occupy the posterior compartment.

CLINICAL PANEL III-7

BAKER'S CYST

Definition: The expression Baker's cyst refers to a fluid-filled herniation of the synovial membrane lining the posterior aspect of the knee joint (Fig. 13–12). Most popliteal cysts represent a distended gastrocnemius or semimembranosus bursa. A Baker's cyst is almost always a complication of chronic knee joint effusion, such as may occur with rheumatoid arthritis of the knee.

Common Symptoms: Most Baker's cysts are asymptomatic. However, rupture of a Baker's cyst and dissection of its fluid contents into either the lower posterior thigh or upper calf may produce acute pain.

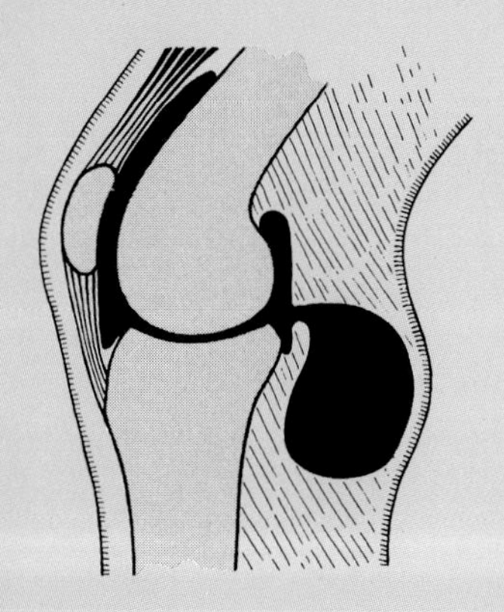

Fig. 13–12: Diagram showing how a Baker's cyst represents a herniation of the synovial membrane lining the posterior aspect of the knee joint.

The Posterior Muscles of the Thigh

The posterior thigh muscles form the surface contour of the back of the thigh (Fig. 12–7B). Table 13–1 lists for quick reference the origin, insertion, nerve supply, and actions of the posterior thigh muscles.

Semimembranosus and **semitendinosus** each have a tendon which extends for about half the length of the muscle (Fig. 12–24). Semimembranosus is named for its thin, membrane-like tendon of origin (not shown in Fig. 12–24), and semitendinosus for its round, cord-like tendon of insertion. Both muscles originate from the ischial tuberosity, extend inferiorly through the posterior compartment of the thigh, pass behind the medial aspect of the knee joint, and insert onto proximal parts of the tibia. Both muscles can ex-

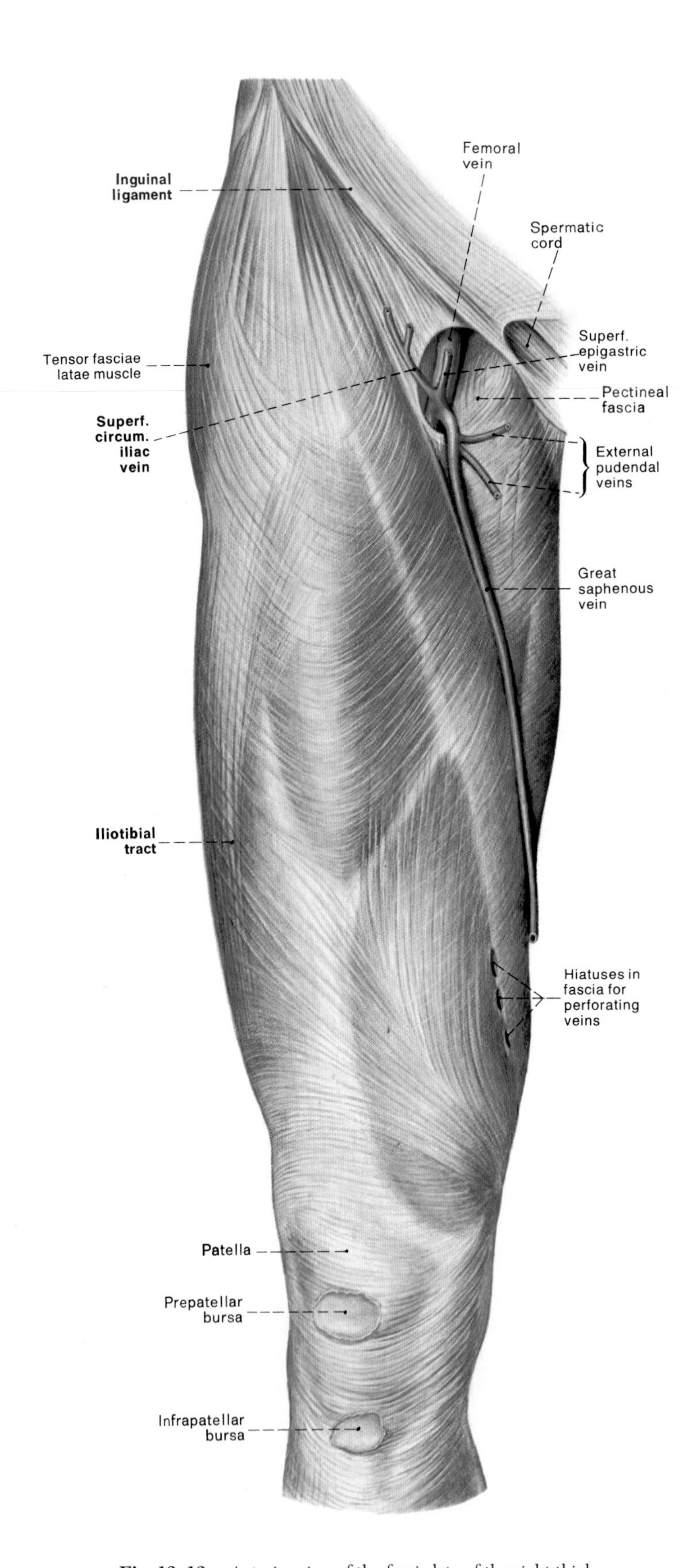

Fig. 13–13: Anterior view of the fascia lata of the right thigh.

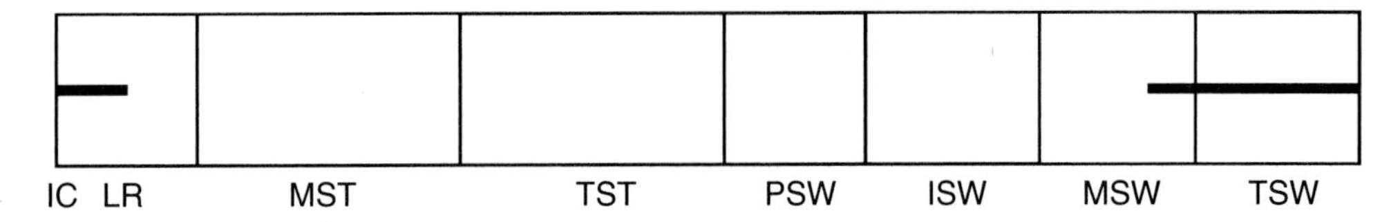

Fig. 13–14: Time frames of consistent significant activity by the hamstring muscles (semimembranosus, semitendinosus, and the long head of biceps femoris) during a stride of the walking gait. Abbreviations: IC = initial contact phase, LR = loading response phase, MST = mid stance phase, TST = terminal stance phase, PSW = pre-swing phase, ISW = initial swing phase, MSW = mid swing phase, and TSW = terminal swing phase.

tend the thigh and flex and internally rotate the leg. The muscles exhibit identical actions because of their nearly identical relationships to the hip and knee joints.

Biceps femoris is named for being the thigh muscle which has two heads of origin: the long head originates from the ischial tuberosity, and the short head originates from the distal part of the femur (Fig. 12–24). The fibers of the two heads merge to give rise to a common tendon of insertion which passes behind the lateral aspect of the knee joint and inserts onto the head of the fibula. Whereas both heads of biceps femoris can flex and externally rotate the leg, only the long head can extend the thigh.

The posterior thigh muscles that act across both the hip and knee joints (namely, semimembranosus, semitendinosus, and the long head of biceps femoris) are commonly called the **hamstring muscles.** This name refers to the historical practice of using the tendons of these thigh muscles as "strings" for the hanging of hams (hams are meat cuts prepared from the buttock and thigh regions of a pig).

When walking, the hamstring muscles are most active during the period that extends from the mid swing phase of the swing period to the loading response phase of the subsequent stance period (Fig. 13–14). The hamstring muscles contract eccentrically during the mid and terminal swing phases of the swing period to control and slow the forward movement of the thigh and leg of the limb in preparation for heel strike (their extensor activity at the hip joint restrains the forward movement of the thigh, and their flexor activity at the knee joint protects against hyperextension of the leg at the knee). Hamstring activity during the initial contact phase and the early part of the loading response phase of the stance period (a) complements gluteus maximus's action of opposing the tendency of the body to fall forward about the limb's hip joint and (b) counteracts the tendency for hyperextension at the knee joint.

The Medial Muscles of the Thigh

The medial thigh muscles form three coronally-oriented muscular planes in the medial aspect of the thigh. These muscular planes intervene between the anterior and posterior thigh muscles. For the purposes of the following discussion, these muscular planes will be called the adductor planes, since each plane bears one of the three major adductors of the thigh. The posterior adductor plane has adductor magnus, the middle adductor plane has adductor brevis, and the anterior adductor plane has adductor longus. Table 13–2 lists for quick reference the origin, insertion, nerve supply, and actions of the medial thigh muscles.

Adductor magnus is so named because it is the largest adductor muscle of the thigh (Fig. 13–15). It lies directly anterior to semimembranosus in the medial aspect of the thigh, forming most of the posterior adductor plane.

Adductor magnus is divisible into two functionally distinct parts: an adductor part which can adduct the thigh and a hamstring part which can extend the thigh. The muscles fibers of the adductor part originate from pubic and ischial parts of the coxal bone, extend inferolaterally in the thigh, and insert onto the linea aspera and upper end of the medial supracondylar line. The muscle fibers of the hamstring part originate from the ischial tuberosity, extend inferiorly in the thigh, and insert onto the lower end of the medial supracondylar line and the adductor tubercle of the femur. The prominent gap between the insertion borders of adductor magnus's two parts is called the **hiatus of adductor magnus.**

When walking, adductor magnus acts primarily as an extensor at the hip joint. It is most active during a time frame that extends from the terminal swing phase of the swing period to the loading response phase of the subsequent stance period (Fig. 13–16A). Its contributions to the walking gait during this time frame are complementary to those of gluteus maximus. Specifically, adductor magnus acts to restrain the forward movement of the thigh during the late part of the terminal swing phase and to oppose the tendency of the body trunk to fall forward about the hip joint during the initial contact and loading response phases.

Gracilis lies in the posterior adductor plane, directly medial to adductor magnus (Fig. 13–15). Gracilis originates from pubic and ischial parts of the coxal

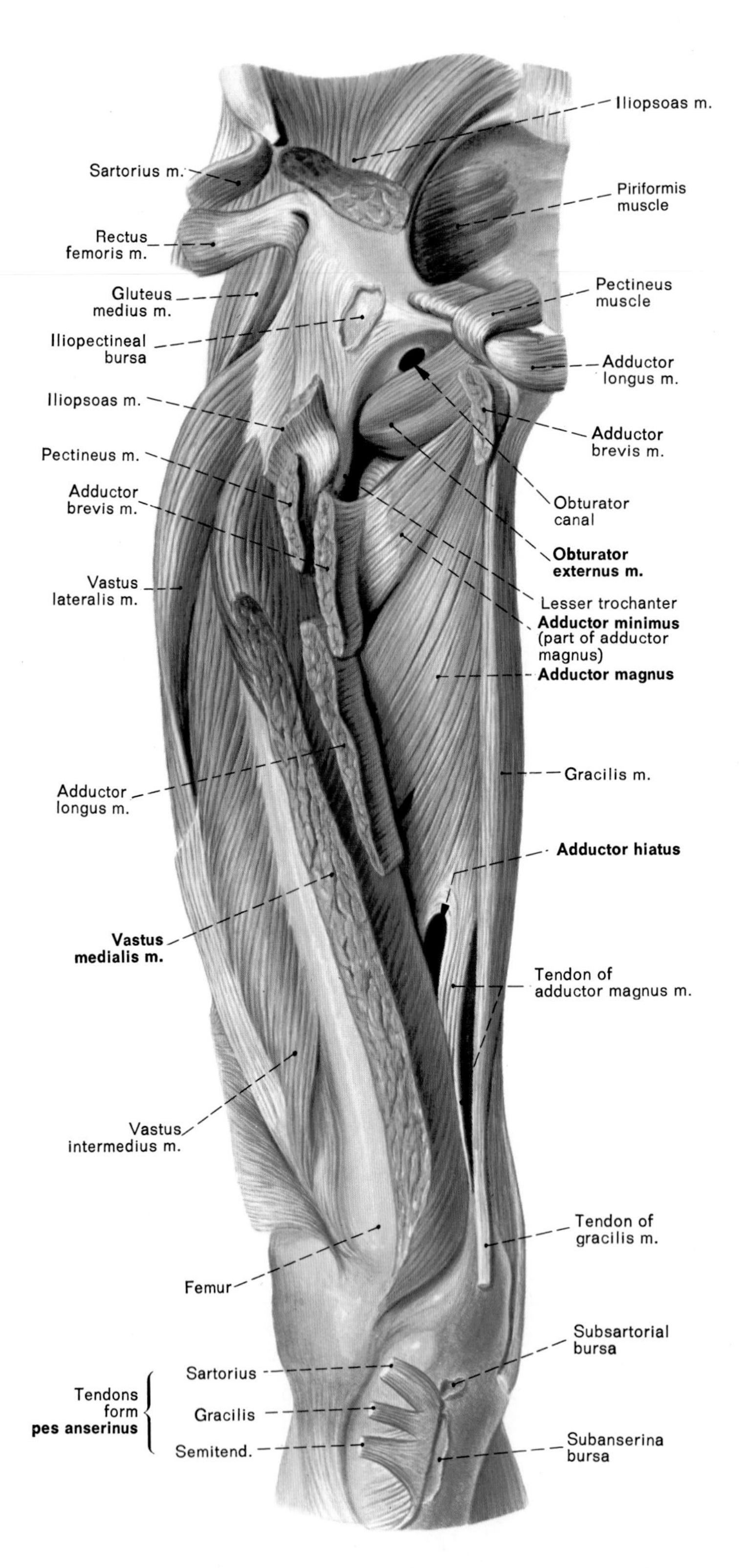

Fig. 13–15: Anterior view of anterior and medial thigh muscles of the right thigh.

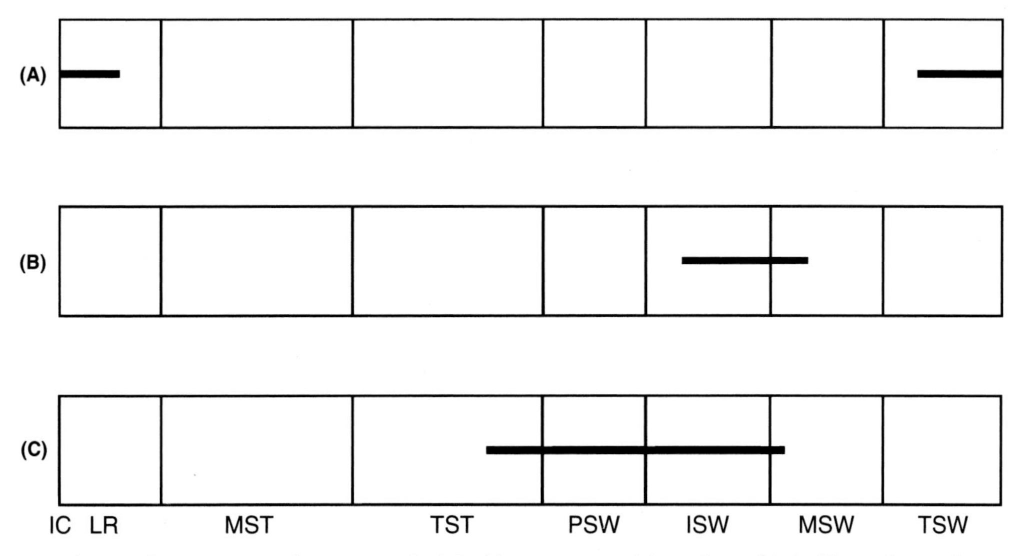

Fig. 13–16: Time frames of consistent significant activity by (A) adductor magnus, (B) gracilis, and (C) adductor longus during a stride of the walking gait. Abbreviations: IC = initial contact phase, LR = loading response phase, MST = mid stance phase, TST = terminal stance phase, PSW = pre-swing phase, ISW = initial swing phase, MSW = mid swing phase, and TSW = terminal swing phase.

bone, extends inferiorly through the medial side of the thigh, passes behind the medial aspect of the knee joint, and inserts onto the upper shaft of the tibia. Gracilis can flex and internally rotate the leg.

When walking, gracilis acts primarily as a flexor and adductor of the thigh during the swing period. It is most active during a brief interval that extends from the initial swing phase to the mid swing phase (Fig. 13–16B).

Adductor brevis forms most of the middle adductor plane; it lies directly anterior to the upper part of adductor magnus (Figs. 13–15 and 13–17). Adductor brevis originates from the pubic part of the coxal bone, and extends inferolaterally in the thigh to insert onto the linea aspera of the femur. Its chief action is adduction of the thigh.

Obturator externus lies in the middle adductor plane, directly superior to adductor brevis (Fig. 13–15).

Table 13–1
The Posterior Muscles of the Thigh

Muscle	Origin	Insertion	Nerve Supply	Actions
Semimembranosus	Ischial tuberosity	Posterior aspect of the medial tibial condyle	Tibial portion of the sciatic (**L5, S1,** and S2)	Extends the thigh and flexes and internally rotates the leg
Semitendinosus	Ischial tuberosity	Upper part of the medial surface of the tibia	Tibial portion of the sciatic (**L5, S1,** and S2)	Extends the thigh and flexes and internally rotates the leg
Biceps femoris	Long head from the ischial tuberosity; short head from the lateral edge of the linea aspera, lateral supracondylar line, and lateral intermuscular septum	Head of the fibula	Tibial portion of the sciatic (L5, **S1,** and S2) for the long head and common peroneal portion of the sciatic (L5, **S1,** and S2) for the short head	Long head extends the thigh; long and short heads flex and externally rotate the leg

Table 13–2
The Medial Muscles of the Thigh

Muscle	Origin	Insertion	Nerve Supply	Actions
Adductor magnus	Adductor part from the inferior ramus of the pubis and ramus of the ischium; hamstring part from the ischial tuberosity	Adductor part onto gluteal tuberosity, linea aspera, and upper part of the medial supracondylar line; hamstring part onto adductor tubercle	Obturator (L2, **L3,** and **L4**) for the adductor part and tibial portion of the sciatic (L4) for the hamstring part	Adductor part adducts the thigh and hamstring part extends the thigh
Gracilis	Body and inferior ramus of the pubis and ramus of the ischium	Upper part of the medial surface of the tibia	Obturator (**L2** and L3)	Adducts the thigh and flexes and internally rotates the leg
Adductor brevis	Body and inferior ramus of the pubis	Upper part of the linea aspera	Obturator (L2, **L3,** and L4)	Adducts the thigh
Obturator externus	Outer surface of the obturator membrane and its bony margin	Trochanteric fossa of the femur	Obturator (L3 and **L4**)	Externally rotates the thigh
Adductor longus	Body of the pubis	Middle part of the linea aspera	Obturator (L2, **L3,** and L4)	Adducts the thigh

It originates from the outer surface of the obturator membrane, and inserts onto the posterior aspect of the femur near the greater trochanter. It acts in association with the gluteal muscles which can externally rotate the thigh.

Adductor longus forms the medial aspect of the anterior adductor plane; it lies directly in front of adductor brevis (Figs. 13–17 and 13–18). Adductor longus, like adductor brevis, originates from the pubic part of the coxal bone, and extends inferolaterally in the thigh to insert onto the linea aspera of the femur. Its chief action is adduction of the thigh.

When walking, adductor longus is most active during a time frame that extends from the terminal stance phase of the stance period to the beginning of the mid swing phase of the subsequent swing period (Fig. 13–16C). Throughout the phases of this time frame, it exerts flexor activity at the hip joint that accelerates forward movement of the thigh. It also exerts adductor activity at the hip joint that helps counterbalance an abduction torque that develops during the pre-swing phase. The generation of this abduction torque during the pre-swing phase occurs as follows: The unloading of upper body weight from the reference lower limb during the pre-swing phase occurs in association with the loading of upper body weight upon the contralateral lower limb. The transfer of support of upper body weight from the reference limb to the contralateral limb is accompanied by a shift of first the pelvic region of the body trunk and then the chest and abdominal regions from the side of the reference limb to that of the contralateral limb. The slight lag in the shift of the chest and abdominal regions generates an abduction torque at the hip joint of the reference limb.

The Anterior Muscles of the Thigh

Table 13–3 lists for quick reference the origin, insertion, nerve supply, and actions of the anterior thigh muscles.

Pectineus is the first of three anterior thigh muscles that extend the muscular plane of adductor longus anterolaterally into the anterior compartment of the thigh (Fig. 13–18). Pectineus originates from the pubic part of the coxal bone, extends past the anteromedial aspect of the hip joint, and inserts onto the upper end of the linea aspera of the femur. It can both flex and adduct the thigh.

Iliacus and **psoas major** are the other two anterior thigh muscles that extend the muscular plane of adductor longus anterolaterally into the anterior compartment of the thigh (Fig. 13–18). Iliacus originates from the iliac fossa, extends inferiorly in front of the hip joint, and inserts onto the lesser trochanter of the femur. Although some fibers insert directly onto the lesser trochanter, most blend into psoas major's tendon of insertion onto the lesser trochanter.

Table 13–3
The Anterior Muscles of the Thigh

Muscle	Origin	Insertion	Nerve Supply	Actions
Pectineus	Pecten pubis	Pectineal line of the femur	Femoral (**L2** and L3) and obturator (L3)	Flexes and adducts the thigh
Iliacus	Iliac fossa and upper lateral part of the sacrum	Insertion tendon of psoas major and lesser trochanter of the femur	Femoral (**L2** and L3)	Flexes the thigh
Psoas major	Bodies and transverse processes of the 12th thoracic and all five lumbar vertebrae and the intervertebral discs between them	Lesser trochanter of the femur	Branches of the anterior rami of **L1, L2,** and L3	Flexes the thigh
Sartorius	Anterior superior iliac spine	Upper part of the medial surface of the tibia	Femoral (L2 and L3)	Abducts, flexes, and externally rotates the thigh and flexes and internally rotates the leg
Rectus femoris	Straight head from the anterior inferior iliac spine; reflected head from a groove above the acetabulum	Tibial tuberosity via the patellar ligament	Femoral (L2, **L3,** and **L4**)	Flexes the thigh and extends the leg
Vastus medialis	Medial edge of the linea aspera and the medial intermuscular septum	Tibial tuberosity via the patellar ligament	Femoral (L2, **L3,** and **L4**)	Extends the leg
Vastus intermedius	Anterior and lateral surfaces of the femoral shaft	Tibial tuberosity via the patellar ligament	Femoral (L2, **L3,** and **L4**)	Extends the leg
Vastus lateralis	Greater trochanter of the femur and the lateral edge of the linea aspera	Tibial tuberosity via the patellar ligament	Femoral (L2, **L3,** and **L4**)	Extends the leg

Psoas major originates mainly from the lumbar part of the vertebral column; its origin thus places the proximal end of the muscle in the posterior abdominal wall. It extends inferolaterally through the lower posterior abdominal wall along the medial border of iliacus, passes in front of the hip joint, and gives rise to its tendon of insertion onto the lesser trochanter.

Since iliacus and psoas major share a common tendon of insertion and usually act together, they are frequently referred to as a single muscular entity called iliopsoas. Iliopsoas is the most powerful flexor of the thigh, and also acts as a flexor of the trunk of the body relative to the thigh when rising from a recumbent to a seated position.

When walking, significant iliopsoas activity does not consistently occur unless the pace is either increased or decreased. Under the conditions of an increasing or decreasing pace, iliopsoas is most active during the ini-

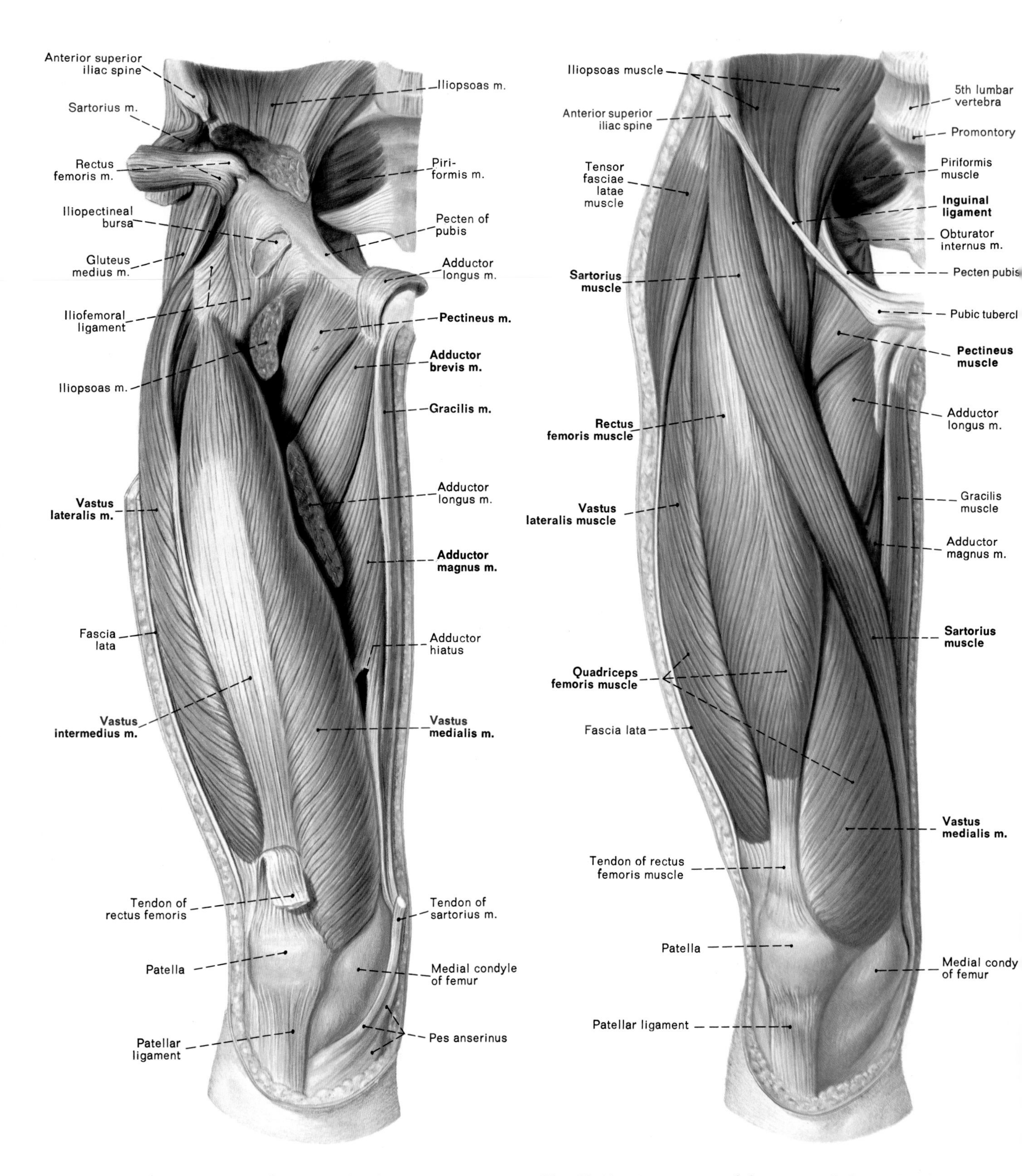

Fig. 13–17: Anterior view of anterior and medial thigh muscles of the right thigh.

Fig. 13–18: Anterior view of the anterior thigh muscles of the right thigh.

tial swing phase of the swing period (Fig. 13–19A).

Sartorius originates from the anterior superior iliac spine, extends inferomedially through the anterior compartment of the thigh, passes behind the medial aspect of the knee joint, and inserts onto the upper part of the tibial shaft (Fig. 13–18). The muscle is named for its use by tailors and seamstresses. When a seated person places the distal end of one leg atop the distal end of the other thigh so as to provide a support for sewing (sartorial) activity, sartorius contributes to most of the wanted movements, namely, flexion, abduction, and exteral rotation of the thigh at the hip joint, and flexion and internal rotation of the leg at the knee joint.

When walking, significant sartorius activity does not consistently occur unless the pace is either increased or decreased. Under the conditions of an increasing or decreasing pace, sartorius is most active during a short time frame that extends from the late part of the pre-swing phase of the stance period to the mid part of the initial swing phase of the swing period (Fig. 13–19B).

Sartorius, iliopsoas, pectineus, and adductor longus together form a triangular, muscular trough in the upper anteromedial region of the thigh called the **femoral triangle** (Fig. 13–18). The femoral triangle is bordered medially by the medial border of adductor longus, laterally by the medial border of sartorius, and superiorly by the inguinal ligament. Adductor longus and the three anterior thigh muscles that anterolaterally extend adductor longus's plane into the anterior compartment of the thigh form the slightly depressed floor of the femoral triangle. The muscles from the most lateral to the most medial are iliacus, psoas major, pectineus, and adductor longus.

Quadriceps femoris is named for being the muscle of the thigh with four heads: **rectus femoris, vastus medialis, vastus intermedius,** and **vastus lateralis.** Rectus femoris itself has two heads of origin: a straight head from the anterior inferior iliac spine and a reflected head from the outer iliac surface (Fig. 13–17). The vastus muscles originate from the femoral shaft and the intermuscular septa of the fascia lata (the vastus muscles are named for their medial-to-lateral relationship with each other) (Fig. 13–17). The four muscles share a common tendon of insertion that partially encloses the patella and passes in front of the knee joint to insert onto the tubal tuberosity (Fig. 13–18).

Rectus femoris and the vastus muscles can all act in concert to extend the leg. Rectus femoris is the sole muscle of the group which has an additional action, namely, flexion of the thigh.

Quadriceps femoris activity is assessed by the **quadriceps femoris tendon reflex (knee-jerk) test.** This test is a simple three-step procedure in which the examiner (1) has the patient sit on an examination table with the legs dangling over the edge, then (2) gently grasps the sides of the thigh just above the knee, and finally (3) uses a reflex hammer to impart a gentle but firm and quick tap upon the patellar ligament. The hammer tap, by suddenly stretching the ligament, exacts an abrupt stretching of quadriceps femoris's muscle fibers, and this should elicit a phasic contraction of the muscle fibers that produces a short, quick extension of the leg.

When walking, significant rectus femoris activity does not consistently occur unless the pace is either increased or decreased. Under the conditions of an increasing or decreasing pace, rectus femoris is most active during a short time frame that extends from the late part of the pre-swing phase of the stance period to the early part of the initial swing phase of the swing period (Fig. 13–19C).

When walking, the vastus muscles are most active as a group during a time frame that extends from the terminal swing phase of the swing period to the mid stance phase of the subsequent stance period (Fig. 13–19D). During the terminal swing phase, the vastus muscles contract concentrically to extend the leg at the knee prior to heel strike. As previously noted, the hamstring muscles act in an antagonistic fashion during this interval to prevent hyperextension at the knee.

During the loading response phase, the extensor activity of the vastus muscles at the knee joint resists the flexion torque that develops there. The generation of this flexion torque during the loading response phase occurs as follows: The loading response phase includes the interval from heel strike to foot flat (Fig. 12–22). During this interval, the anterior leg muscles pull the leg forward at the ankle as the foot rocks forward over the heel. This forward leg movement draws the knee joint anterior to the line of projection of the upper body weight being loaded onto the limb. The projection of the upper body weight line posterior to the knee joint generates a significant flexion torque at the joint.

Articularis genus is a comparatively small muscle which originates from the lower anterior surface of the femoral shaft and inserts onto the most superior aspect of the fibrous lining of the suprapatellar bursa. The muscle proximally retracts the lining when quadriceps femoris acts to extend the leg; this retraction prevents the lining of the suprapatellar bursa from being entrapped during leg extension in a redundantly folded configuration between the femur and the quadriceps femoris tendon.

The Pes Anserinus

One muscle from each of the three major thigh muscle groups inserts onto the upper medial surface of the tib-

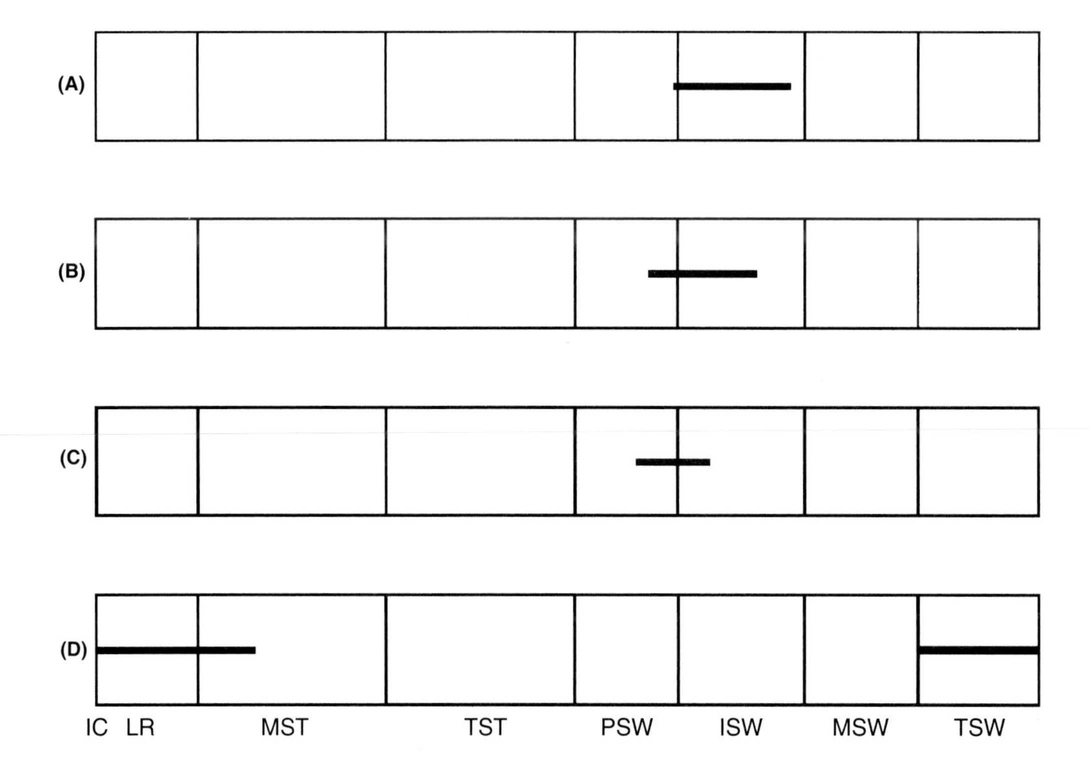

Fig. 13–19: Time frames of consistent significant activity by (A) iliopsoas, (B) sartorius, (C) rectus femoris, and (D) the group of vastus muscles (vastus lateralis, vastus intermedius, and vastus medialis) during a stride of the walking gait. Abbreviations: IC = initial contact phase, LR = loading response phase, MST = mid stance phase, TST = terminal stance phase, PSW = pre-swing phase, ISW = initial swing phase, MSW = mid swing phase, and TSW = terminal swing phase.

ial shaft: semitendinosus from the posterior group, gracilis from the medial group, and sartorius from the anterior group. These three muscles share two actions: flexion and internal rotation of the leg. The tendons of these three muscles overlap each other as they insert onto the tibial shaft (Figs. 13–15 and 13–17). Because their overlapped configuration resembles the outline of a goose foot, the three tendons are commonly referred to by the Latin expression for goose foot: pes anserinus. The **anserine bursa** intervenes between the three tendons and the underlying tibia (Fig. 13–15).

THE COURSE AND DISTRIBUTION OF THE MAJOR NERVES OF THE THIGH

The Femoral Nerve

The femoral nerve is derived from the lumbar plexus, specifically, the anterior rami of L2, L3, and L4. The femoral nerve enters the thigh by passing deep to the inguinal ligament at a point midway between the anterior superior iliac spine and pubic tubercle (Fig. 13–20). As it then extends distally across the floor of the femoral triangle, the femoral nerve ramifies into numerous muscular and cutaneous branches.

The Obturator Nerve

The obturator nerve is derived from the lumbar plexus, specifically, the anterior rami of L2, L3, and L4. The obturator nerve enters the thigh by passing through the obturator canal (Fig. 13–20). As it then extends distally through the medial aspect of the thigh, the obturator nerve ramifies into numerous muscular and cutaneous branches.

The Sciatic Nerve

The sciatic nerve extends from the gluteal region into the posterior part of the thigh at a point midway between the ischial tuberosity and the greater trochanter of the femur (Fig. 12–26). The sciatic nerve has two terminal branches: the **tibial** and **common peroneal nerves.** In most individuals, the division of the sciatic nerve into its two terminal branches occurs in the lower part of the posterior compartment of the thigh. In others, the division occurs as proximally as the gluteal region. All branches of the sciatic nerve are generally identified as being derived from either the **tibial portion or the common peroneal portion of the sciatic nerve.**

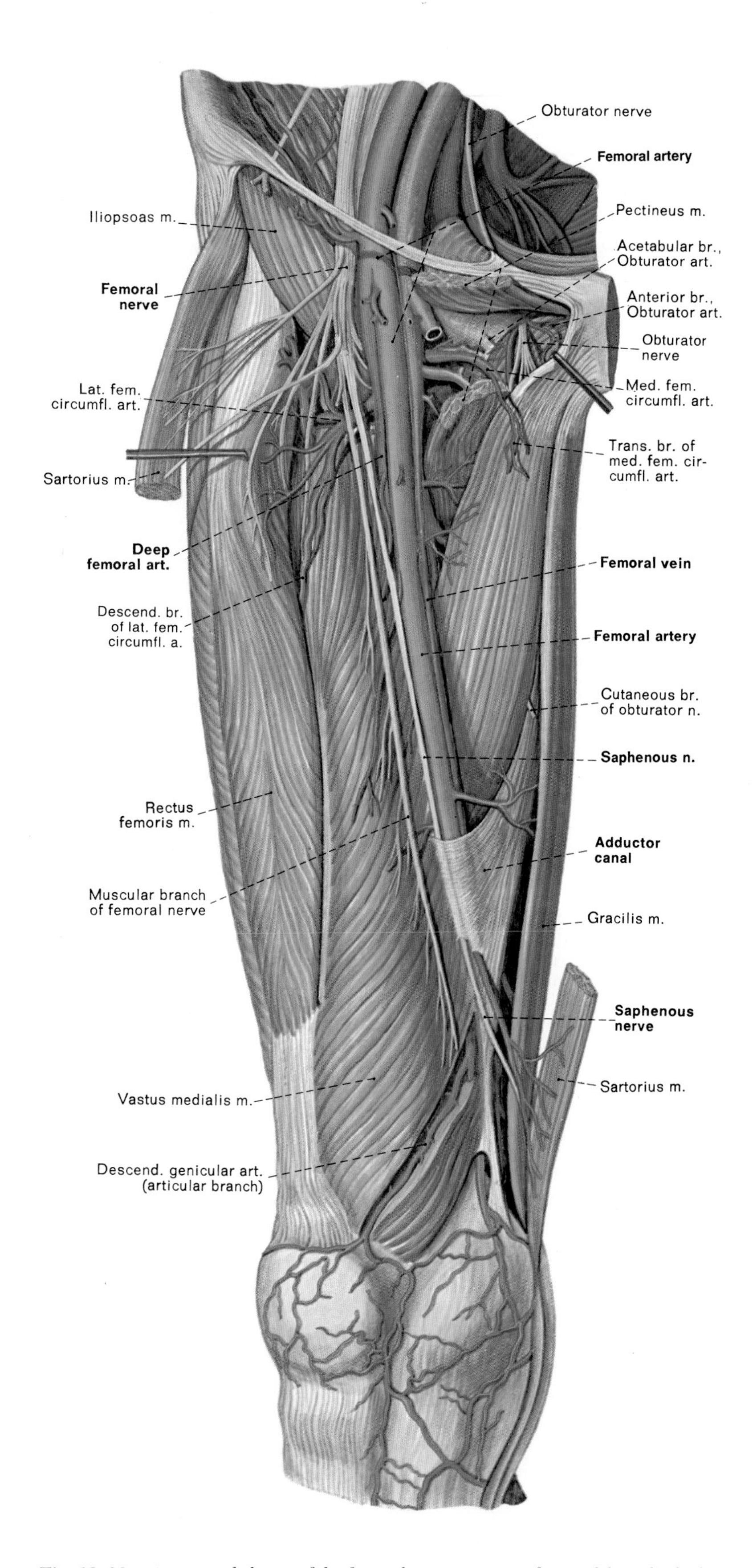

Fig. 13–20: Anteromedial view of the femoral nerve, artery, and vein of the right thigh.

Figs. 4–12G and 4–12H depict the cutaneous areas of the lower limb innervated by branches of the femoral, obturator, tibial, and common peroneal nerves. Some cutaneous areas are innervated by nerves derived directly from the lumbar plexus, sacral plexus, or the posterior rami of L1, L2, L3, S1, S2, and S3.

THE PRIMARY ARTERIAL TRUNK OF THE LOWER LIMB AND ITS COLLATERALS IN THE POSTERIOR COMPARTMENT OF THE THIGH

The primary arterial trunk of the lower limb consists of an artery whose name changes from the external iliac to the femoral and then to the popliteal artery as it extends from the abdomen to the proximal end of the leg. The derivation of this arterial trunk from the common iliac artery and its course through the abdomen and lower limb are as follows:

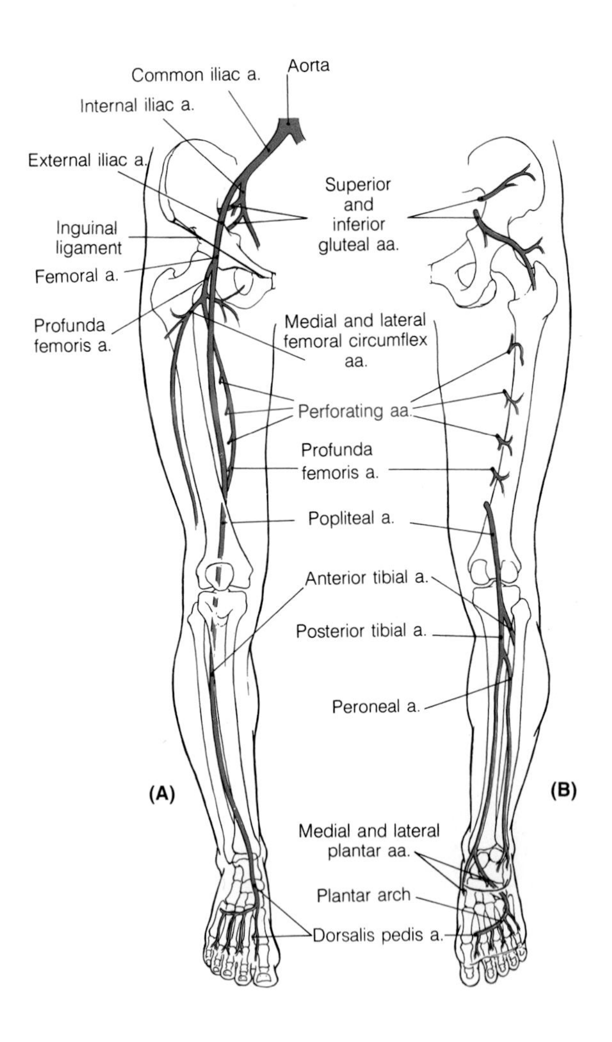

Fig. 13–21: (A) Anterior and (B) posterior views of the major arteries of the right lower limb.

The left and **right common iliac arteries** arise in front of the left side of the body of the fourth lumbar vertebra as the paired terminal branches of the **abdominal aorta** (Fig. 13–21A). The arteries diverge as they extend inferolaterally to opposite sides of the posterior abdominal wall. As each artery passes in front of the sacroiliac joint, it bifurcates into two terminal branches, the **internal** and **external iliac arteries.** The external iliac artery assumes an obliquely downward course in the lateral abdominal wall (beside the medial border of psoas major) which leads the artery to pass deep to the inguinal ligament (Figs. 13–20 and 13–21A). Upon passing deep to the inguinal ligament, the artery enters the thigh, and its name changes to the femoral artery. The **femoral artery** thus begins deep to the inguinal ligament as the continuation of the external iliac artery.

The femoral artery enters the thigh deep to the inguinal ligament at a point midway between the anterior superior iliac spine and pubic symphysis. The femoral artery lies medial to the femoral nerve as both structures extend distally across the floor of the femoral triangle (Fig. 13–20). The first 2 to 3 cm of the femoral artery is the segment of the artery best suited for palpation of the femoral pulse and obstructive compression of the artery. This segment is best suited for obstructive compression because it overlies the extra-acetabular part of the head of the femur in the hip joint and the insertion tendon of psoas major. Therefore, when pressure is applied to this segment, the segment is squeezed against the unyielding femoral head and femoral blood flow can be effectively blocked. Such blockage can prevent catastrophic blood loss when a more distal segment of the femoral artery is severed.

The femoral artery's course through the thigh is a vertical descent that passes it first over the floor of the femoral triangle and then through the **subsartorial canal (adductor canal),** the fascial canal in the anterior compartment of the thigh deep to the middle part of sartorius (Fig. 13–20). At the distal end of the subsartorial canal, the femoral artery extends through the hiatus of adductor magnus (Fig. 13–22) to enter the popliteal fossa, and its name changes to the popliteal artery (Fig. 13–21B). The **popliteal artery** thus begins in the hiatus of adductor magnus as a continuation of the femoral artery.

The popliteal artery extends inferolaterally through the popliteal fossa, descending behind first the popliteal surface of the femur, next the fibrous capsule of the knee joint, and finally popliteus (popliteus is the deepest muscle at the proximal end of the back of the leg). At the lower border of popliteus, the popliteal artery bifurcates into its terminal branches: the **anterior** and **posterior tibial arteries** (Fig. 13–21).

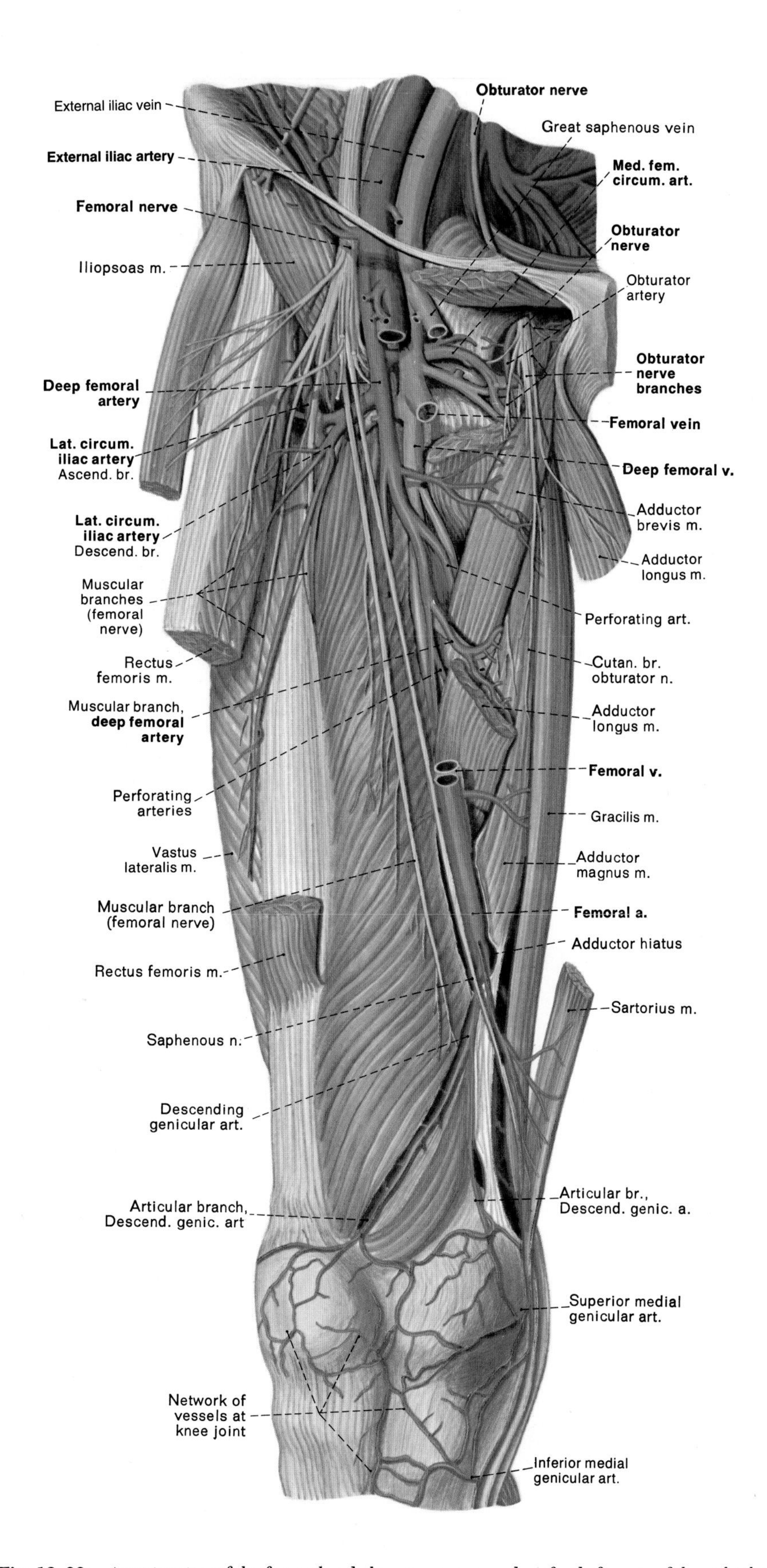

Fig. 13–22: Anterior view of the femoral and obturator nerves and profunda femoris of the right thigh.

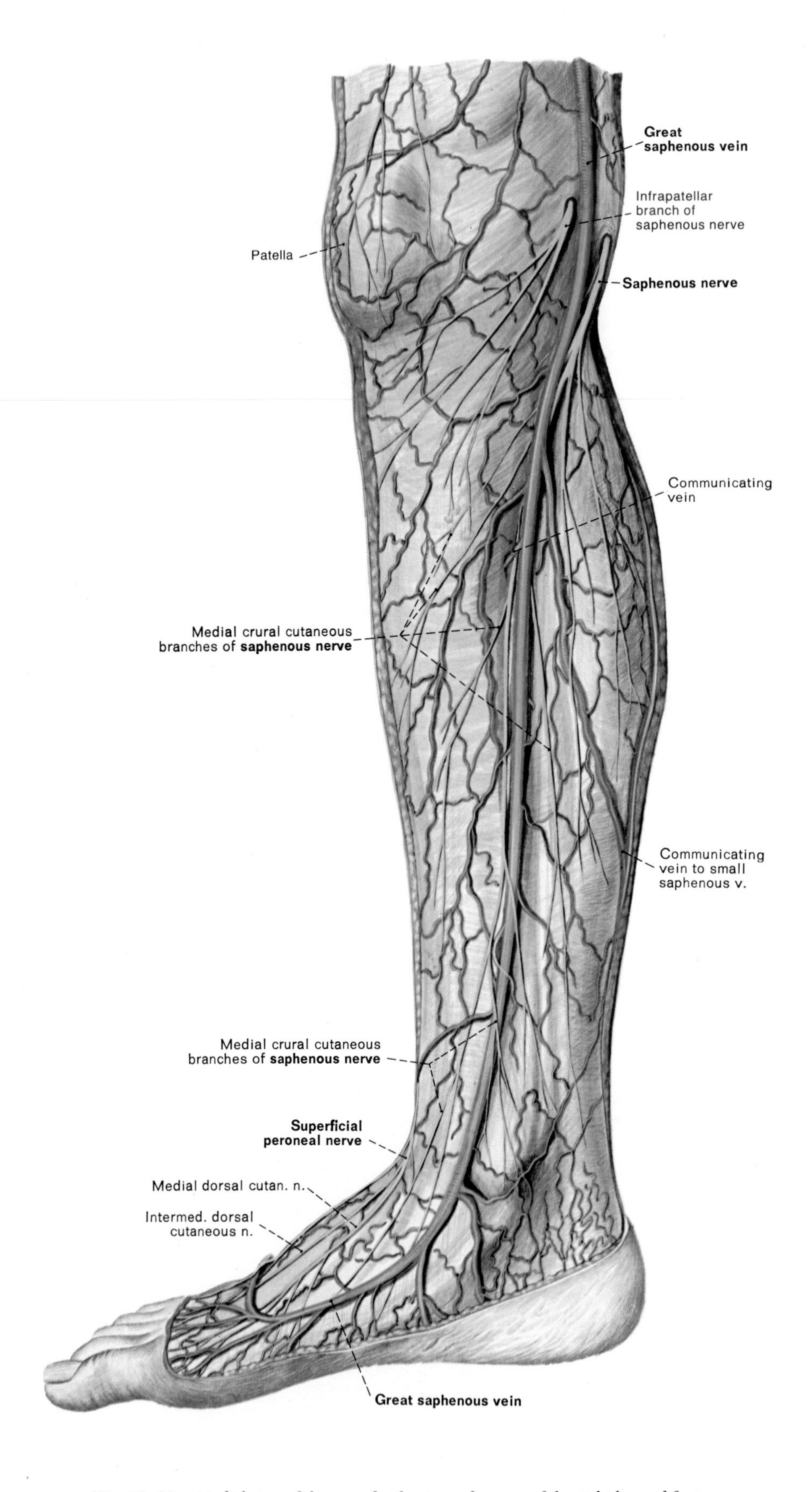

Fig. 13–23: Medial view of the superficial veins and nerves of the right leg and foot.

The pulsations of the popliteal artery can be palpated if the fingers are deeply and firmly, but still gently, pressed into the soft tissues of the popliteal fossa. Deep palpation is required because the popliteal artery lies deeply in the popliteal fossa and is conducted with the leg partially flexed at the knee joint; the partial flexion augments the deep palpation by lessening the tension in the skin and superficial fascia of the popliteal fossa.

The Profunda Femoris (Deep Femoral Artery)

Profunda femoris (the deep femoral artery) is the largest branch of the primary arterial trunk of the lower limb. It arises in the femoral triangle from the posterior aspect of the femoral artery (Figs. 13–20, 13–21A, and 13–22). It extends between pectineus and adductor longus as it begins its vertical descent through the thigh, lying anterior to adductor brevis proximally and adductor magnus distally.

Profunda femoris gives rise to six major branches: the **medial** and **lateral circumflex femoral arteries** and the **first, second, third,** and **fourth perforating arteries** (Fig. 13–21A). The medial and lateral circumflex femoral arteries respectively arise from the medial and lateral aspects of profunda femoris (Fig. 13–22), and then curve around opposite sides of the femur to anastomose with each other in the posterior compartment of the thigh. The four perforating arteries arise from the posterior aspect of profunda femoris as it descends vertically on the anterior surfaces of adductor brevis and magnus (the fourth perforating artery represents the terminal continuation of profunda femoris). The perforating arteries are so named because they extend from the anterior to the posterior compartment of the thigh by perforating (passing through) adductor magnus.

In the buttock and back of the thigh, the superior and inferior gluteal arteries, the medial and lateral circumflex femoral arteries, and the four perforating arteries give rise to ascending and descending branches. The anastomoses among these ascending and descending branches form a vertical chain of arteries in the posterior compartment of the thigh, which extends from branches of the internal iliac artery in the buttock to muscular branches of the popliteal artery in the knee region. Under emergency conditions, the femoral artery can be ligated at any point along its course through the anterior compartment of the thigh without risking total loss of blood supply to the lower limb distal to the site of ligation because the vertical chain of anastomosed arteries in the posterior compartment of the thigh provides collateral circulation to the knee, leg, and foot.

The Blood Supply to the Head of the Femur

Branches of the medial and lateral circumflex femoral arteries form an extracapsular vascular ring around the base of the neck of the femur. Branches of the extracapsular vascular ring called the **retinacular arteries** ascend along the neck of the femur to penetrate the capsule of the hip joint and give rise within the joint to branches that supply the upper end of the femur. The retinacular arteries and their branches are believed to be the chief source of blood suppy to the head of the femur at all ages. At birth, the cartilaginous femoral head is also supplied by arteries entering the head from the femoral shaft. Establishment of an epiphyseal plate between the **capital epiphysis** (the epiphysis for the head of the femur) and the metaphysis by about 4 years of age abolishes all blood supply from the metaphysis. This arrangement persists until about 9 years of age, at which time the **artery in the ligament to the head of the femur** begins to become increasingly prominent in supplying the femoral head. **Medullary arteries of the metaphysis** begin to supply the femoral head upon epiphyseal fusion in late adolescence or early adulthood.

The Obturator Artery

The obturator artery, which is a branch of the internal iliac artery, also supplies a small region of the upper thigh. The obturator artery enters the thigh upon passage through the obturator canal (Fig. 13–22). It gives rise to the artery of the ligament to the head of the femur and to branches which supply the adductor and hamstring muscles.

THE PRIMARY VENOUS TRUNK OF THE LOWER LIMB AND ITS MAJOR TRIBUTARIES

The primary arterial trunk of the lower limb is accompanied throughout its entire length by the primary venous trunk of the lower limb. The segments of the venous trunk that lie beside the external iliac, femoral, and popliteal arteries are respectively called the external iliac, femoral, and popliteal veins. The **popliteal vein** begins near the inferolateral border of the popliteal fossa as the union of the **venae comitantes of the anterior** and **posterior tibial arteries,** and ends in the hiatus of adductor magnus as the origin of the femoral vein. The **femoral vein** ascends the thigh alongside the femoral artery, and ends deep to the inguinal ligament as the origin of the **external iliac vein** (Fig. 13–22). As the femoral vein leaves the thigh, it lies immediately medial to the femoral artery in the femoral triangle.

The largest superficial tributary of the lower limb's primary venous trunk is the **great saphenous vein.**

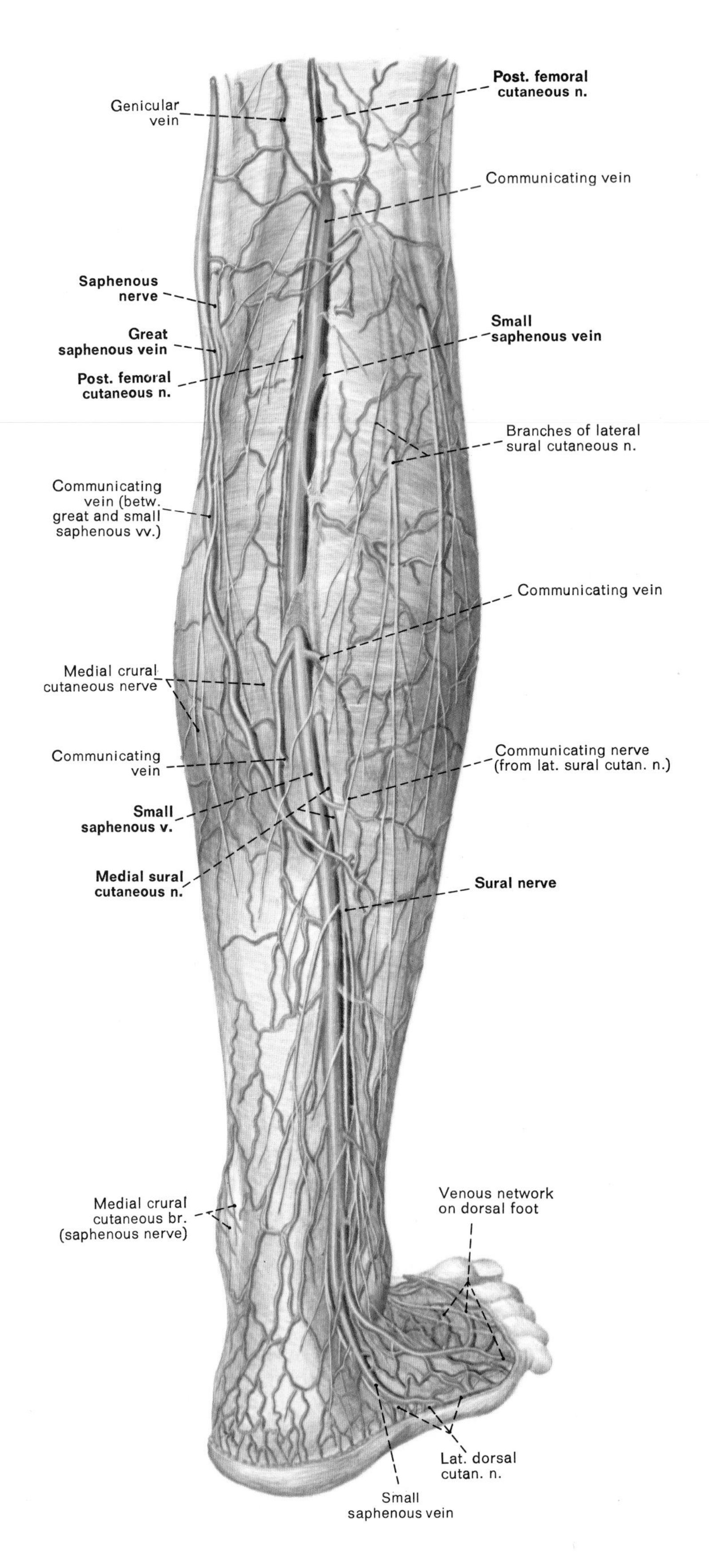

Fig. 13–24: Posterolateral view of the superficial veins and nerves of the right leg and foot.

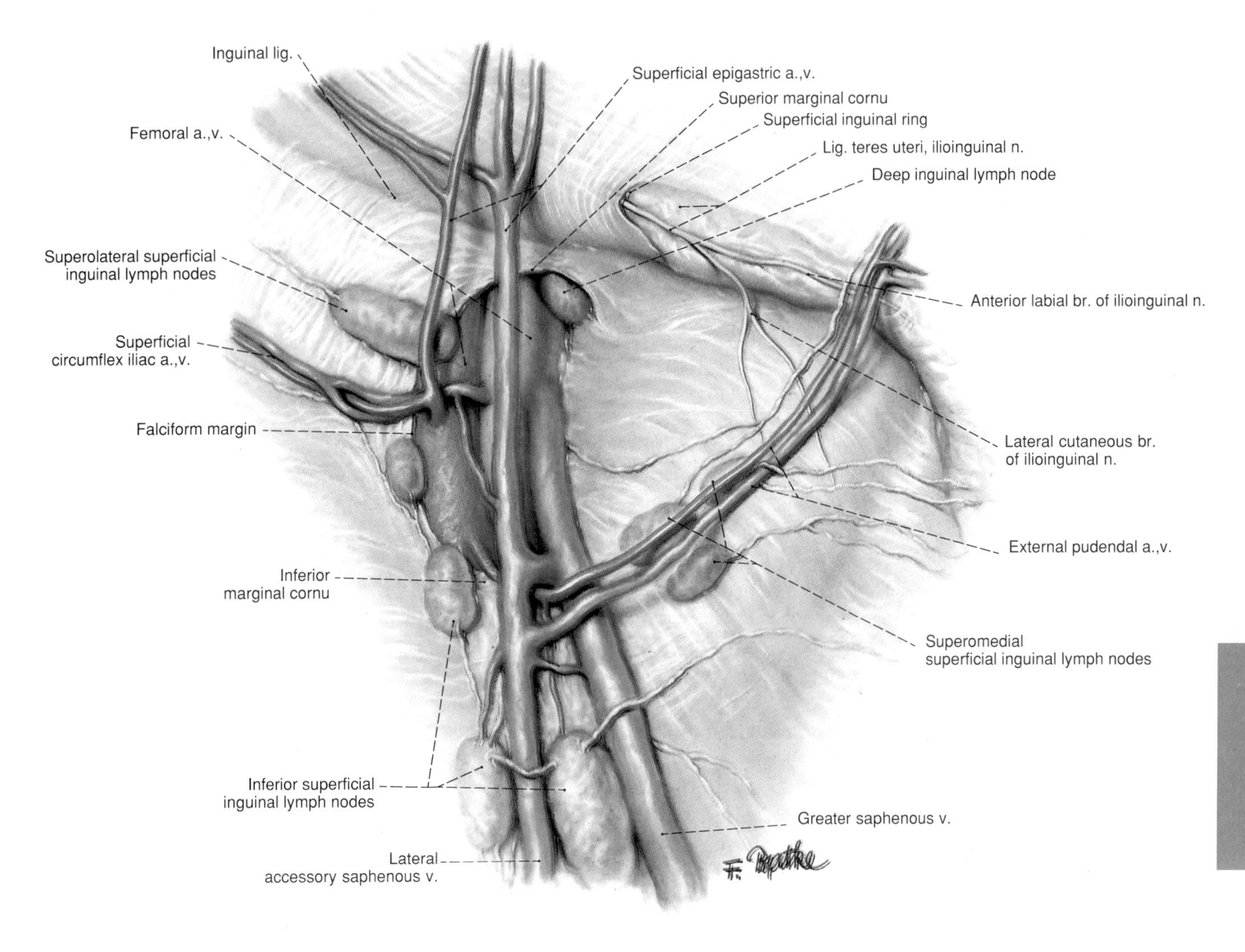

Fig. 13–25: Anterior view of the superficial inguinal lymph nodes. The group of nodes identified as the inferior group of superficial inguinal nodes represents the vertical group of superficial inguinal nodes. The groups of nodes identified as the superolateral and superomedial groups of superficial inguinal nodes respectively represent the lateral and medial members of the horizontal group of superficial inguinal nodes.

The great saphenous vein is the largest superficial vein of the lower limb, and also the longest vein of the body. It begins as the medial extension of the **dorsal venous arch of the foot,** and curves upward into the leg by passing in front of the medial malleolus (Fig. 13–23). The great saphenous vein courses the anteromedial side of the leg, the posteromedial side of the knee, and finally the anteromedial side of the thigh as the vein ascends the lower limb. It ends by passing through an opening called the **saphenous opening** in the fascia lata overlying the femoral triangle and then uniting with the femoral vein (Fig. 13–13). The tributaries of the great saphenous vein drain all the superficial tissues of the lower limb except for those of the lateral side of the foot and the posterolateral aspect of the leg.

The second largest superficial tributary of the lower limb's primary venous trunk is the **small saphenous vein.** The small saphenous vein begins as the lateral extension of the dorsal venous arch of the foot, and curves upward into the leg by passing behind the lateral malleolus (Fig. 13–24). It ascends the posterolateral aspect of the leg and then commonly ends by uniting with the popliteal vein. The tributaries of the small saphenous vein drain the superficial tissues of the lateral side of the foot and the posterolateral aspect of the leg.

THE GENERAL DISTRIBUTION OF VEINS IN THE LOWER LIMB

The general pattern to the distribution of veins in the lower limb features three categories of veins:

1. **superficial veins,**
2. **deep veins,**
3. **perforating (communicating) veins.**

The superficial veins (the largest of which are the great and small saphenous veins) are subcutaneous. They are clinically used as vessels of intravenous injection and venous blood sampling. The superficial veins drain cutaneous tissues, and ultimately empty into deep veins. The deep veins (the largest of which are the external iliac, femoral, and popliteal veins) are more inaccessible. They drain almost all the muscle and bone tissues of the limb. The perforating veins arise from superficial veins, extend into deeper regions of the limb, and end by anastomosis with deep veins. The valves of perforating veins are oriented to ensure net blood flow from superficial to deep veins.

Skeletal muscle action in the lower limbs significantly augments venous return because of (a) the general distribution of the veins and (b) the orientation of the one-way valves in the superficial, perforating, and deep veins. Skeletal muscle activity is a major propulsive force of blood flow in the deep veins. It is common for segments of the deep veins to be aligned parallel to and sandwiched between the bellies of two spindle- or strap-shaped muscles. When the girth of the muscle bellies increases during concentric contraction, the muscle bellies press upon the deep vein segments sandwiched between them, increasing the blood pressure in the segments sufficiently to force open closed valves and propel blood proximally toward the heart. During subsequent muscle relaxation, when gravity generally forces the blood in the deep veins of the lower limb to move downward away from the heart, the closure of the valves limits the extent of such retrograde flow.

Muscle action is equally important in powering blood flow in the superficial and perforating veins, but in these veins the action is indirect. When muscle action empties segments of deep veins of their blood, blood subsequently flows from the superficial and perforating veins into these deep venous segments during periods of muscle relaxation.

Skeletal muscle action in the lower limbs also limits the extent to which gravity can increase hydrostatic blood pressure in the veins of the lower limbs. During periods of limited muscular activity in the lower limbs (e. g., when a military person stands at attention or a salesperson stands in a confined area), there is infrequent muscular compression of the deep veins of the lower limb. The continuous venous drainage of lower limb tissues, in combination with the limited activity of the muscle action pump, causes the veins to become engorged with blood. The veins thereby become lengthy, blood-swollen columns in which gravity generates pronounced hydrostatic pressure. Retrograde transmission of this gravity-generated, hydrostatic pressure increases the blood's hydrostatic pressure in small venules and capillary beds, which, in turn, results in net extravasation (net escape) of fluid from the microvasculature into the interstitial spaces. The tissues of the legs and feet become edematous; they literally swell as their interstitial spaces accommodate the fluid extravasated from the circulating blood. Muscle action in the lower limbs prevents this cascade of events by preventing the formation of the long, venous columns in the thigh and leg.

CLINICAL PANEL III-8

Peripheral Atherosclerotic Occlusive Disease

Definition: Peripheral atherosclerotic occlusive disease is a condition in which medium or larger arteries of the limbs become occluded by plaques.

Common Symptom: The most common initial symptom of the disease in a lower limb is muscular pain or fatigue that occurs with exercise but abates with rest (intermittent claudication).

Anatomic Basis of the Sites of Lower Limb Pain Caused by Intermittent Claudication: Occlusions of the external iliac artery diminish blood supply to almost all of the lower limb's muscles, and thus produce exertion-dependent pain extending distally from the buttock. Occlusions in the femoral artery immediately proximal to the origin of profunda femoris diminish blood supply to thigh and leg muscles, and thus produce exertion-dependent pain extending distally from the thigh. Occlusions in the popliteal artery or femoral artery distal to the origin of profunda femoris produce exertion-dependent pain in the leg muscles. Progressive obstruction in the anterior or posterior tibial arteries does not diminish blood supply to the leg muscles, and thus does not produce intermittent claudication.

A Test for Valve Competence in the Great Saphenous Vein and its Communicating Veins

The competency (functional integrity) of the valves in the great saphenous vein and in the communicating veins that extend from it to deep veins may be tested in the following manner: The great saphenous vein is first drained of most of its luminal blood by laying the patient in a supine position, elevating the entire lower limb above the level of the body trunk, and massaging the vein with long strokes to draw the blood toward the

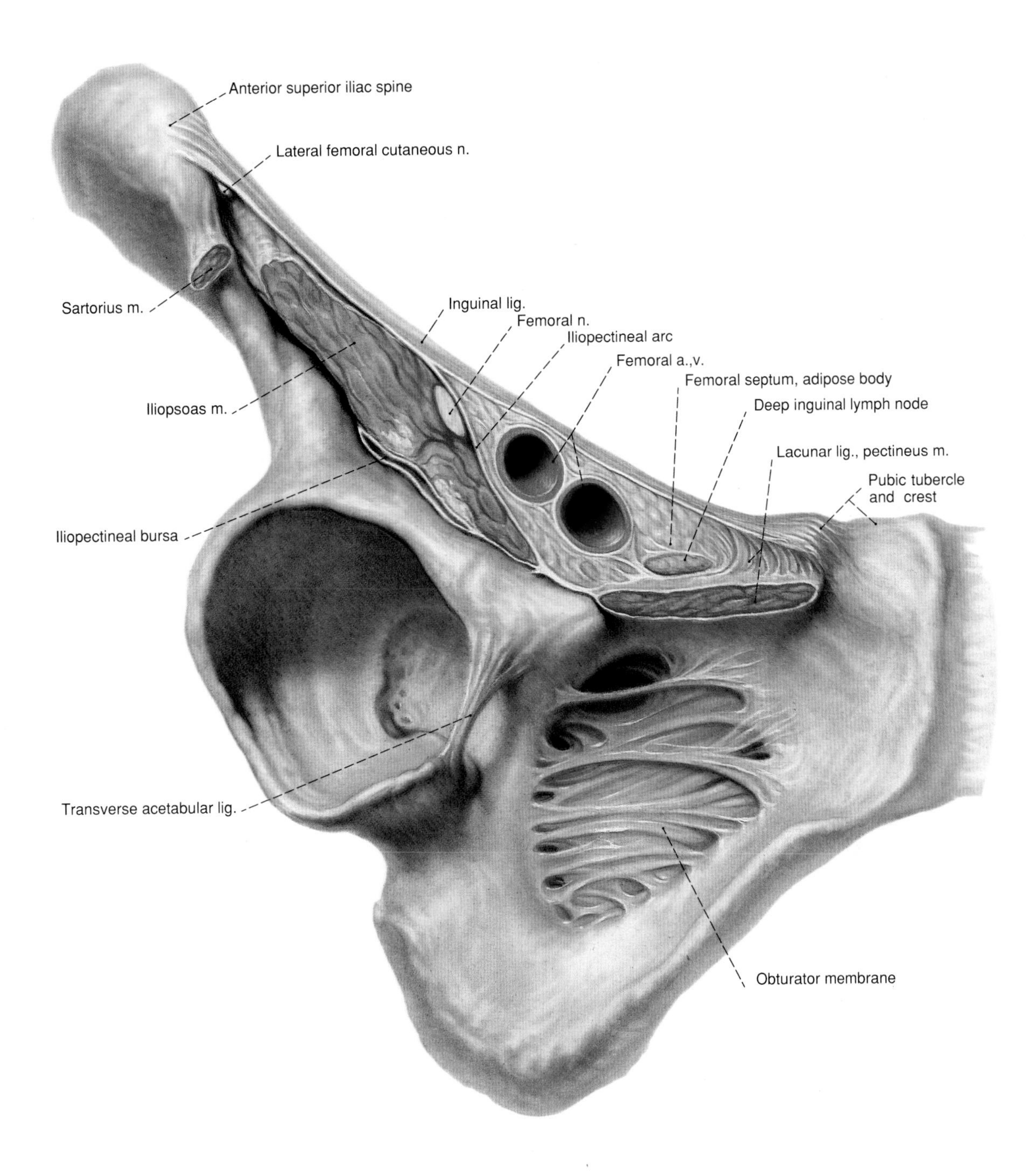

Fig. 13–26: Anteroinferior view of the right femoral sheath. The white margin surrounding the femoral artery, femoral vein, femoral septum, and deep inguinal lymph node represents the femoral sheath. The anterior part of the white margin represents the fascia transversalis, and the posterior part of the white margin represents the iliacus fascia.

vein's union with the femoral vein in the femoral triangle. A tourniquet is applied around the mid-region of the thigh (to constrict the great saphenous vein just below its union with the femoral vein), and the patient is requested to stand. With the tourniquet in place, the examiner observes filling of the great saphenous vein for a period of 60 seconds, and then releases the tourniquet and notes if there is further filling.

The filling that occurs with the tourniquet in place tests the competency of valves in the perforating veins extending from the great saphenous vein. If these valves are all competent, the great saphenous vein receives blood only from its tributaries, and filling of the great saphenous vein near its origin begins within about 35 seconds after application of the tourniquet. Faster and more extensive filling indicates incompetent valves.

The filling that occurs upon release of the tourniquet tests the competency of valves in the great saphenous vein itself. If these valves are all competent, release of the tourniquet will not produce further filling. Rapid filling from above upon release signals incompetent valves.

THE MAJOR LYMPH NODE GROUPS IN THE LOWER LIMB

The **popliteal group of lymph nodes** is the most distal group of lymph nodes in the lower limb. The group lies embedded in the fat of the popliteal fossa. The group drains the lymphatics that accompany the small saphenous vein and its tributaries; these lymphatics drain the superficial tissues of the lateral side of the foot and the posterolateral aspect of the leg. The popliteal nodes also receive lymph from the deep tissues of the leg and foot.

Most of the lymph nodes in the lower limb are clustered in the inguinal region. There are three distinct groups of lymph nodes in the inguinal region:

The **vertical group of superficial inguinal lymph nodes** lies in the superficial fascia of the thigh beside the terminal segment of the great saphenous vein (Fig. 13–25). The group drains the lymphatics that accompany the great saphenous vein and its tributaries. These lymphatics drain all the superficial tissues of the lower limb except those of the lateral side of the foot and the posterolateral aspect of the leg.

The **horizontal group of superficial inguinal lymph nodes** lies in the superficial fascia of the thigh inferior and parallel to the inguinal ligament (Fig. 13–25). The group drains lymph from the urethra, the lower half of the anal canal, and the external genitalia of both sexes (which includes the vagina below the hymen in the female but not the testes in the male). The group also drains lymph from the anterolateral abdominal wall up to the level of the umbilicus and from the superficial tissues of the gluteal region.

The **group of deep inguinal lymph nodes** lies along the medial side of the terminal segment of the femoral vein in the femoral triangle (Fig. 13–25). The group drains lymph from the popliteal nodes, the deep tissues of the thigh, and some of the superficial inguinal nodes.

Most of the efferent lymphatics from the vertical and horizontal groups of superficial inguinal nodes and all of the efferent lymphatics from the deep inguinal nodes drain into the **external iliac nodes.**

THE FEMORAL SHEATH AND ITS CONTENTS

A funnel-shaped fascial sheath (the femoral sheath) envelops the most proximal segments of the femoral artery and vein in the femoral triangle (Fig. 13–26). The femoral sheath is a continuous extension of two deep, abdominal wall fascial layers: the fascia transversalis and the iliacus fascia. The **fascia transversalis** is the fascial layer that underlies the deep surface of transversus abdominis, the deepest muscle in the flank region of the anterolateral abdominal wall. The **iliacus fascia** is the fascial layer that covers the anterior surfaces of iliacus and psoas major in the abdomen. In an adult, the femoral sheath extends for 3 to 4 cm distal to the inguinal ligament, at which point the sheath ends by fusing with the tunica adventitia of the femoral vessels. The great saphenous vein and a number of efferent lymphatics from superficial inguinal lymph nodes pierce the lower medial side of the femoral sheath.

The femoral artery lies lateral to the femoral vein within the femoral sheath (Fig. 13–26). The space medial to the femoral vein within the sheath is a potential space called the **femoral canal.** The femoral canal bears one of the deep inguinal lymph nodes and transmits efferent lymphatics from superficial and deep inguinal lymph nodes toward the external iliac nodes. The proximal (upper) end of the femoral canal is called the **femoral ring,** and it is filled with loose connective tissue called the **femoral septum** (Fig. 13–26). The femoral ring and its septum form a region of the abdominal wall predisposed to herniation. A weakness or opening in the femoral septum, through which a tissue or organ may protrude, is called a **femoral hernia.** [The anatomy of femoral hernias is discussed in conjunction with the anatomy of inguinal hernias in Chapter 20.]

CASE **III.1**

The Case of Michael Kenat

INITIAL PRESENTATION AND APPEARANCE OF THE PATIENT

A senior adult white man has made an appointment to seek treatment for a painful right hip. The examiner observes that the patient stands with the right thigh slightly flexed, adducted, and externally rotated and with a slightly accentuated lumbar lordosis. The patient walks with an abnormal gait in which he (a) leans his body toward the right side during the loading response, mid-stance, and terminal stance phases of the right lower limb's stance period and (b) shortens the stride to minimize the duration of the right lower limb's stance period.

QUESTIONS ASKED OF THE PATIENT

What is your name? Michael Kenat.

What is your birth date? [His birth date indicates that he is 61 years old.]

Please tell me about your problem. Well, I began noticing a little over a year ago that my hip seemed to bother me most of the time [the patient places his right hand over the lateral aspect of his right hip as he refers to his hip problem]. Until recently, I hadn't paid too much attention to it because I thought it was just arthritis or rheumatism that comes with old age, and there's nothing you can really do except take aspirin for the pain. However, after falling down on my hip a few days ago, it's become so painful that I now limp. *[The request "Please tell me about your problem" invites the patient to openly and completely discuss his health concerns. In this instance, the examiner quickly learns that the patient has not only recently sustained a painful injury to the right hip, but also experienced hip pain of over a year's duration. The examiner decides to gather information about the acute injury first, as it is the intensification of the hip pain following the injury that has prompted the patient to seek medical attention.]*

Can you give me more details as to what exactly happened when you fell down? Yes. About 5 days ago, I was walking up the basement stairs and slipped on a piece of paper that was on one of the steps. I slipped to my right and my hip hit the stairwell wall pretty hard. My grip on the handrail saved me from falling down the stairs. My hip has hurt more and more since that happened.

Can you describe or point out the part of your hip that hit the wall? [The patient stands up and places the fingertips of his right hand on the buttock region overlying and immediately posterior to the greater trochanter of the femur.] *[The answers to these last two questions suggest that the fall on the right hip imparted axial and/or transverse forces on the neck of the femur. At this point, the examiner is concerned that the forces may have fractured the neck of the femur. The examiner seeks to obtain information about the exact anatomic site of the hip pain and its features both before and after the hip injury.]*

Can you describe or point out where you have pain? [The patient stands up and places his right hand first over the right midinguinal region. As he adds that it sometimes goes down his leg, he slides his right hand up and down over the upper anteromedial aspect of the right thigh.] *[The patient's description suggests hip joint pain.]*

How would you describe the pain? A deep, dull ache.

Is the intensity of the pain constant? No. My hip doesn't bother me in the morning when I awake, but after standing or walking around for an hour or so, that's when it starts hurting again. *[This answer indicates that the pain emanates from a musculoskeletal structure or structures sensitive to weight-bearing.]*

Has the pain changed in any way since you fell on your hip? No, other than it tends to hurt more and faster when I stand on my leg or walk around. However, I've also started having low back pain if I stand too long. *[In this case, the examiner reasons that there are two likely explanations for the low back pain: either the pa-*

tient injured his lower back when he fell or that the development of an abnormal erect posture (such as the exaggerated lumbar lordosis) is placing an undue strain on lower back muscles.]

You say that the pain becomes worse when you stand or walk around. Have you noticed anything else that makes the pain worse? Yes, before I hurt my hip, I noticed that it hurt more when the weather was cold and damp.

Have you noticed whether lifting heavy objects makes the pain worse? I don't think so. Because, with respect to making the pain worse, it doesn't seem to matter much what I'm doing once I'm up and about. Just being on my feet for about an hour is what seems to bring the pain on. *[The examiner asked this question to explore the possibility that the hip pain is emanating from an injury or lesion of the lower back. Hip pain that emanates from a lower back condition tends to be aggravated by stooping and lifting activities and relieved by walking. The patient's remarks here suggest that it is not likely that a lower back condition is the source of the hip pain.]*

Is there anything that relieves the pain? Yes. Sitting or lying down or taking a hot bath eases the pain. I've also found that aspirin relieves the pain for a while.

Have all of these things relieved the pain both before and after your fall 5 days ago? Yes.

Is there anything else that is bothering you that you haven't talked to me about? Well, I've noticed that although my hip doesn't hurt in the morning when I awake, it's stiff. The stiffness goes away if I walk around a bit, but then the pain starts up if I walk too much. *[At this point, the examiner recognizes that the aggregate characteristics of the hip pain (insidious onset, morning stiffness, aggravation by weight-bearing and cold weather, and relief with rest, warmth, and anti-inflammatory medication) are typical of osteoarthritis of the hip joint. However, none of the patient's remarks have allayed the examiner's concern that the patient may have a femoral neck fracture. Since an arthritic condition appears to be the basis of the chronic hip pain, the examiner decides to ask questions to discriminate among the major categories of arthritis.]*

Have you ever seriously injured your right hip before? No.

Did you have any problems with your right hip when you were a child or adolescent, or is there a family history of hip problems? No. *[The examiner asked these last two questions to explore the possibility that there is a factor in the patient's history that predisposes the hip to osteoarthritis. Injury or disease of a joint predisposes it to osteoarthritis.]*

Are there any other joints that are painful or stiff? No. *[Some arthritic conditions, such as rheumatoid arthritis and systemic lupus erythematosus, frequently involve several joints with a symmetric distribution. The pain of osteoarthritis typically involves a single joint or a small number of joints with an asymmetric distribution.]*

Have you recently had any fevers or chills? No. *[This question was asked to explore the possibility of infectious arthritis. Infectious arthritis is frequently preceded by a fever or flu-like symptoms].*

Have you noticed any recent change in bowel habits? No. *[Hip pain may be referred from pelvic disease. A change in bowel habits may herald disease of the lower gastrointestinal tract.]*

Have you recently lost weight? No, unfortunately, I've put on about 5 pounds in the last year. *[A recent loss of weight may be (but is not necessarily) a sign of metastatic cancer. Cancer in older men frequently originates in certain pelvic organs (prostate, urinary bladder, sigmoid colon, and rectum).]*

Have you had any recent illnesses or injuries, other than your fall on the stairs? No.

Are you taking any medications for previous illnesses or conditions? No. I have only been taking aspirin. *[The examiner finds the patient alert and fully cooperative during the interview.]*

PHYSICAL EXAMINATION OF THE PATIENT

Vital signs:
Blood pressure
Lying supine: 130/75 left arm and 130/70 right arm
Standing: 130/75 left arm and 130/70 right arm
Pulse: 74
Rhythm: regular
Temperature: 98.9°F.
Respiratory rate: 16
Height: 5′9″
Weight: 205 lbs.
HEENT Examination: Normal
Lungs: Normal
Cardiovascular Examination: Normal. A pertinent set of normal findings is the absence of any signs of vascular disease in the lower limbs, signs such as coldness, pallor, cyanosis, weak or absent arterial pulses, and arterial bruits. *[The absence of such signs is significant because stenosis of the external iliac artery may produce exertion-dependent hip pain.]*

Abdomen: Normal
Genitourinary Examination: Normal
Musculoskeletal Examination: Inspection, movement, and palpation of the upper and lower limbs are normal except for the following findings for the right lower limb: Inspection reveals a 5 cm diameter yellow-green bruise overlying and immediately posterior to the greater trochanter. Pain restricts active and passive extension of the thigh to less than 20° flexion, active and passive internal rotation of the thigh greater than 30°, and active and passive abduction greater than 30°. Deep palpation of the midinguinal region at the site immediately lateral to the pulsations of the first 2 to 3 cm of the femoral artery in the femoral triangle elicits tenderness. Palpation of the bruise posterior to the greater trochanter elicits mild tenderness.

A pertinent normal finding is equal lengths of the lower limbs, as measured by the distance from the anterior superior iliac spine to the medial malleolus. Another pertinent normal finding is equal girths of the thighs, as measured with a tape measure.

[The examiner measures the lengths of the lower limbs as the patient lies supine on an examination table with the limbs crossed at the ankles; the crossing at the ankles ensures that the lower limbs have equivalent positions relative to the bony pelvis. The lengths of the lower limbs are measured because the right lower limb appears shorter than the opposite limb when the patient stands. In this case, the right lower limb appears shorter because it is slightly flexed and adducted when the patient stands. Decreased girth of the right thigh would suggest disuse atrophy of the thigh muscles.]

Neurologic Examination: Normal. A pertinent set of normal findings is the absence of any signs of neurologic dysfunction in the lower limbs, signs such as paresthesia, anesthesia, loss of position sensation, and hyperactive, weak, or absent phasic stretch reflexes.

INITIAL ASSESSMENT OF THE PATIENT'S CONDITION

The patient appears to be suffering from both a chronic disorder and an acute injury of the right hip.

ANATOMIC BASIS OF THE PATIENT'S HISTORY AND PHYSICAL EXAMINATION

1. **The patient's description of an insidious onset of right hip pain suggests a chronic disease or condition.** The patient reports progressive worsening of right hip pain during the past 12 to 18 months. His recent trauma on the basement stairs appears to have complicated or aggravated the chronic condition.
2. **The worsening of the hip pain upon weight-bearing by the affected lower limb indicates that one or more musculoskeletal structures that bear weight in the erect position are the source of the hip pain.** The presence of hip pain does not necessarily imply disease or injury of the hip region, because pain can be referred to the hip from disease or injury of the lumbar and sacral parts of the vertebral column, one or more pelvic organs, or the knee.

   In this case, the worsening of the hip pain upon weight-bearing effectively excludes the possibility that the hip pain is pain referred from disease or injury of one or more pelvic organs. The absence of any abnormal findings upon visual inspection, testing of movements, and palpation of the knee joint effectively excludes the possibility that the hip pain is referred from disease or injury of the knee.

   Hip pain referred from disease or injury of the vertebral column is generally exacerbated by stooping movements or the lifting of heavy objects. The absence of such symptoms and signs in this case decreases the likelihood that the hip pain is referred from a disorder of the vertebral column.

   Individuals frequently redistribute weight-bearing forces across one part of the lower limb in order to alleviate disease- or injury-induced pain in other parts of the lower limb. The resulting musculotendinous and/or ligamentous strain in the region of redistributed weight-bearing forces may produce pain in this secondary site. However, in this case, the physical examination did not reveal any evidence of disease or injury in any part of the right lower limb except for the closely related hip and inguinal regions.
3. **The deep tenderness in the midinguinal region marks it as a site of disease or injury.** Tenderness is pain upon palpation and generally marks a region as a site of disease or injury.

   Orthopedists select the site in the front of the thigh immediately lateral to the first 2 to 3 cm of the femoral artery for pressing upon the anterior aspect of the hip joint. The tissues pressed upon at this site include (proceeding from the most superficial to the deepest) the skin, superficial fascia, fascia lata, the common tendon of insertion of the iliacus and psoas major muscles, and tissues of the hip joint. The diseased or injured tissues at this site appear to be the source of the pain in the midinguinal region.
4. **The inconstant, superficial pain that radiates**

distally along the upper anteromedial aspect of the thigh appears to be referred pain. This presentation suggests injury or disease of tissues innervated by L1 and L2 fibers, as the upper anteromedial aspect of the thigh represents parts of the L1 and L2 dermatomes (Fig. 4-12G). The notable tissues in the hip region that are innervated in part by L1 and/or L2 fibers include all the anterior thigh muscles, all the medial thigh muscles except for obturator externus, and the hip joint.

5. **The painful diminishing in the limits of active and passive extension, internal rotation, and abduction of the thigh suggests disease or injury of one or more tissues (which include the bony, cartilaginous, synovial, and capsular tissues) of the hip joint.** In general, pain upon movement indicates injury or disease of one or more of the bones, joints, and muscles involved in the movement. The muscles acting across the hip joint do not appear to be involved because the physical examination does not reveal any weakness or pain upon isometric testing of the thigh flexors, extensors, internal rotators, external rotators, abductors, and adductors.

INTERMEDIATE EVALUATION OF THE PATIENT'S CONDITION

The history and physical examination indicate that the patient is suffering from a chronic condition and an acute injury of the tissues of the right hip joint.

CLINICAL REASONING PROCESS

The patient's initial description of his hip problem indicates he has been suffering from a chronic condition associated with hip pain that was exacerbated or complicated 5 days previously by a hard fall on the hip. The patient is thus suffering from both an acute injury and a chronic condition. The interpretation of the physical findings is thus complicated by the fact that the findings produced by the injury cannot, in general, be distinguished from those resulting from the chronic condition.

In the differential diagnosis of the patient's condition, an orthopedist or primary care physician would consider five major categories of injury and disease:

1. Fracture of the hip and/or bony pelvis
2. Arthritis
   A. Osteoarthritis
   B. Inflammatory arthritis (includes rheumatoid arthritis)
   C. Infectious arthritis
   D. Collagen/vascular disease (includes systemic lupus erythematosus and scleroderma)
   E. Metabolic and endocrine arthritis (includes gout)
3. Avascular necrosis of the head of the femur
4. Paget's disease of bone
5. Tumor (includes benign, primary malignant, and metastatic lesions)

Trauma-related injury heads the list because the intensification of the hip pain following the fall on the basement stairs has prompted the patient to seek medical attention. The patient's account of the fall on the stairs and the location of the bruise on the right buttock indicates that the patient fell hard onto the posterolateral aspect of the greater trochanter. Such an impact raises the possibility that the patient has sustained a nondisplaced fracture of the neck of the femur. It is possible for an individual with such a fracture to be ambulatory, have satisfactory range of motion at the hip joint, and experience comparatively minor pain.

When considered together, four key findings from the patient's history and physical examination strongly indicate that the chronic condition associated with hip pain is a disease of the hip joint: The history establishes that the patient has midinguinal pain associated with occasional upper medial thigh pain. Disease or injury of the hip joint generally presents as midinguinal pain that may be referred to the anterior and/or medial aspects of the thigh as distal as the knee. The history also establishes that the midinguinal pain is exacerbated upon weight-bearing by the lower limb; this is an expected symptom of hip joint disease or injury. Testing of movements shows pain upon either active or passive movement of the thigh to the limits of certain movements but no pain upon isometric testing of the muscles acting across the hip joint. This combination of physical findings is typical of arthritis and capsulitis. Palpation reveals pain upon application of pressure to the anterior aspect of the hip joint and the tissues superficial to it.

The combined characteristics of the patient's hip problem suggest osteoarthritis of the right hip joint. A patient suffering from osteoarthritis of the hip joint typically is 50 years of age or older. The patient commonly gives a history of an insidious onset of midinguinal pain that may be referred to the upper anterior and/or medial aspects of the thigh. The pain is exacerbated by weight-bearing and cold and relieved by rest, warmth, and anti-inflammatory drugs (such as aspirin). Crepitus (a grating or grinding sound or feeling) may accompany movements at the joint. The joint may become stiff upon rest.

RADIOGRAPHIC EVALUATION AND FINAL RESOLUTION OF THE PATIENT'S CONDITION

The examiner ordered an anteroposterior (AP) radiograph of the pelvis and groin and lateral radiographs of both hips. The radiologist reported that the radiographs did not reveal any evidence of fractures in the proximal femurs or the bony pelvis. Nondisplaced or impacted fractures of the neck of the femur are difficult to detect in conventional radiographs. In this case, a technetium bone scan would have been warranted if the radiologist had determined that there was ambiguous evidence in the conventional radiographs for a nondisplaced femoral neck fracture. A technetium bone scan is a nuclear medicine imaging procedure in which a patient's body or a body region is scanned after intravenous injection of radiolabeled technetium phosphates. The radiolabeled technetium phosphates are deposited with other inorganic salts at sites of new bone formation. Accordingly, bone fractures appear as sites of increased radioactivity because they are sites of increased bone formation.

The AP radiograph of the pelvis confirmed the presence of osteoarthritic changes in both hip joints. Osteoarthritis is a degenerative condition involving articular cartilage. Primary osteoarthritis is the idiopathic form of the condition, and its onset is believed to be related to senescence (the aging process) and genetic factors. Secondary osteoarthritis is the form of the condition that appears to occur in response to previous physical or infectious injury of a joint. Both forms of osteoarthritis involve focal degeneration of articular cartilage surfaces and a chronic shedding of cartilaginous fragments from degenerate foci. In a synovial joint, the cartilaginous fragments are shed into the joint cavity and absorbed by the synovial membrane lining. The inflammatory response by the synovial membrane to this chronic absorption of cartilaginous particles leads to hyperplasia and increasing fibrosis of the lining. The thinning of articular cartilage surfaces at the sites of focal degeneration leads to the formation of new subchondral bone at these sites and also at the margins of articular surfaces. These processes produce characteristic features in the radiographic image of an osteoarthritic joint:

1. One or more joint space regions are narrowed in width; these regions mark sites of focal degeneration and thinning of articular cartilage.
2. One or more subchondral bone regions are osteosclerotic; these regions mark sites of new subchondral bone formation.
3. One or more articular margins bear osteophytes, bony outgrowths which are markers of irregular deposition of new bone.

In this case, the AP radiograph of the pelvis showed, in both hip joints, a narrowing of the width of the superior joint space and osteosclerosis of the subchondral bone areas of the femoral head and acetabulum that are apposed to each other across the superior joint space. The AP radiograph also displayed in both hip joints osteophytes along the acetabular rim and the margin of the femoral head with the neck of the femur. The degenerative changes appeared more marked in the right hip joint. The extent of radiographic changes in an osteoarthritic joint does not, however, always correspond with the severity of the pain and loss of function in the joint.

Given the body of evidence from the history, physical exam, and radiographs in support of a diagnosis of osteoarthritis and the lack of evidence of other disease or injury, an orthopedist or primary care physician would decide to treat the patient for osteoarthritis of the hip joints. Conservative treatment would begin with complete bed rest for a few days to provide time for reduction of the acute inflammation that resulted from the fall on the hip. The physician would recommend a program for weight loss, avoidance of weight-bearing during the day when pain becomes a problem, and analgesics as required for pain.

THE CHRONOLOGY OF THE PATIENT'S CONDITION

The patient has primary osteoarthritis of several joints, with the most marked degenerative changes having occurred in the hip and knee joints (the degenerative changes in the knee joints and the left hip joint are asymptomatic). Primary osteoarthritis has been the basis of the patient's chronic right hip pain.

The hard fall on the greater trochanter of the right hip strained the fibrous capsule and produced minute fractures through the hyaline cartilage and subchondral bone of both the acetabular and femoral head surfaces that are apposed to each other across the superior joint space of the right hip joint. The minute trabecular fractures in the subchondral bone of the acetabulum and femoral head have intensified the right hip pain to the extent that the patient has adopted an altered gait in order to minimize weight-bearing by the right hip joint (the minute fractures in the articular cartilage are painless because the articular cartilage is devoid of neural elements). In particular, the patient leans his upper body toward the right side during the loading response, mid stance, and terminal stance phases of the right lower limb's stance period. This measure reduces the total compressive force on the right hip joint during the right lower limb's stance period. The reason is as follows: When walking, the total compressive force acting across the hip joint increases severalfold during the loading response, mid stance, and terminal stance phases of the stance period (relative to the total compressive force acting during the swing period). The hip joint is compressed during these phases not only by the burden of bearing all the gravitational weight of the upper body but also by the lateral pelvic tilting action exerted by the chief thigh abductors (gluteus medius and minimus) to support and steady the upper body over the lower limb. Leaning the upper body toward the side of the lower limb reduces the activity of gluteus medius and minimus, and thus reduces the compressive force exacted across the hip joint by the chief thigh abductors.

The minute fractures in the articular cartilage of the acetabulum and femoral head accelerated the extrusion of degenerated cartilage fragments into the synovial cavity. The fragments have collected (through the force of gravity) mainly in the lower synovial recesses of the joint cavity. The absorption of these cartilaginous fragments by the synovial membrane lining the lower recesses of the joint cavity has intensified the lining's chronic hyperplasia and progressive fibrosis. The intensification of this inflammatory response has led to synovial adhesions and a loss of laxity in the most inferior part of the joint capsule. Attempts by the patient to stand produce a painful stretching of the taut inferior part of the right hip joint capsule, which, in turn, elicits a reflexive contraction of medial thigh muscles (because the obturator nerve supplies sensory nerve fibers to the inferior part of the hip joint capsule). The reflexive contraction of the medial thigh muscles causes the patient to stand with the right thigh slightly flexed, adducted, and externally rotated at the hip joint. The patient compensates for the thigh flexion by an increased lumbar lordosis which produces a strain on the lumbar part of the vertebral column, and this strain is the cause of the low back pain.

CASE **III.2**

The Case of Jason Rudenko

INITIAL PRESENTATION AND APPEARANCE OF THE PATIENT

Mr. and Mrs. Rudenko have made an appointment for their son Jason, a 9 year-old white male with a painful left knee. Jason is an alert, well-nourished young boy who appears mildly apprehensive as he sits on the examination table.

QUESTIONS ASKED OF THE PATIENT

What seems to be the problem? My knee hurts [the patient points to the anteromedial aspect of the left knee].

When did your knee start to hurt? Three days ago. *[This answer suggests an acute condition or process. It prompts the examiner to inquire about the patient's activities during the week preceding the onset of pain, in order to assess if any of these activities could possibly generate an acute, painful condition or process.]*

Do you recall what you were doing when the pain started? Nothin' that I can remember. I was at school, and my knee started hurting around noon.

Do you recall if you had injured your left knee before it became painful? No, I don't think so. *[It is the patient's perception that the knee pain is not a consequence of an injury or immoderate physical activity.]*

How would you describe the pain? I'm not sure what you mean. My knee just hurts. *[Although information about the quality and severity of the knee pain would be helpful to the examiner, the examiner recognizes that children frequently do not know the vocabulary appropriate for the description of pain.]*

Has your feeling of the pain or its location changed during the past 3 days? Ah, let me think. I'd say the pain feels pretty much the same all the time, and is in about the same place all the time.

Has your left or right knee ever been painful like this before? No.

Is there anything that makes the pain worse? My knee hurts more when I walk. Even sometimes just standing around hurts.

Is there anything that relieves the pain? I find that my knee feels better when I'm sitting down, like right now. *[The answers to the last two questions indicate that the pain emanates from a musculoskeletal structure or structures sensitive to weight-bearing.]*

Is there anything else that is also bothering you? No.

Do you have any pain in your left hip, ankle, or foot? No.

Do you have any pain in your right hip, knee, ankle, or foot? No.

Do you have any pain in your shoulders, elbows, wrists, or hands? No. *[These answers are important because they indicate, in particular, that there are not any neurovascular complications to the patient's condition nor pain in other joints of the upper and lower limbs.]*

Have you had any recent illnesses or injuries? No.

Are you taking any medicine for previous illnesses or injuries? No.

Have you taken any medicine for your knee pain? No. *[The examiner finds the patient alert and fully cooperative during the interview.]*

PHYSICAL EXAMINATION OF THE PATIENT

Vital signs:
Blood pressure: 117/74
Pulse: 105
Rhythm: regular
Temperature: 99.5°F.
Respiratory rate: 21
Height: 4′7″ (139.7 cm.)
Weight: 78 lbs. (35.5 kg.)

[A pediatrician would recognize that the patient's vital signs are within normal limits. The patient's systolic and diastolic blood pressures, weight, and height are in

the 90th percentile for boys 9 years of age. The normal range of pulse rate for the patient is 70 to 115. The normal range of respiratory rate for the patient is 15 to 25. The slight trend toward the upper limits of normal for the pulse and respiratory rates may be secondary to apprehension and/or pain.]

HEENT Examination: Normal

Lungs: Normal

Cardiovascular Examination: Normal

Abdomen: Normal

Genitourinary Examination: Normal

Musculoskeletal Examination: Inspection, movement, and palpation of the upper and lower limbs are normal except for the following findings for the left lower limb: Left knee pain restricts active and passive external rotation of the thigh at the hip greater than 35°, active and passive internal rotation greater than 20°, active and passive abduction greater than 30°, and active and passive adduction greater than 20°. Passive internal rotation to the 20° limit elicits thigh muscle spasm. Full extension of the thigh (0° flexion) is painful. Deep palpation of the midinguinal region at the site immediately lateral to the pulsations of the first 2 to 3 cm of the femoral artery in the femoral triangle elicits tenderness.

The patient walks with a gait in which he (a) leans his body toward the left side during the loading response, mid stance, and terminal stance phases of the left lower limb's stance period and (b) shortens the stride to minimize the duration of the left lower limb's stance period. The initial contact phase of the left lower limb's stance period begins with forefoot strike (the forefoot is the first part of the foot to strike the surface below). Flexion of the left leg at the knee is greater than normal at forefoot strike and throughout the stance period. Heel rise occurs earlier than normal, and thigh extension is less than normal during the terminal stance phase of the stance period. External rotation of the left thigh at the hip is greater than normal throughout the walking gait.

Pertinent normal findings include full active and passive ranges of motion at the left knee and absence of any tenderness about the left knee. Isometric testing of muscle actions at the left hip and knee do not exacerbate left knee pain.

Neurologic Examination: Normal. A pertinent set of normal findings is the absence of any signs of neurologic dysfunction in the lower limbs, signs such as paresthesia, anesthesia, loss of position sensation, and hyperactive, weak, or absent phasic stretch reflexes.

INITIAL ASSESSMENT OF THE PATIENT'S CONDITION

The patient's chief medical problem appears to be painful motion at the left hip joint.

ANATOMIC BASIS OF THE PATIENT'S HISTORY AND PHYSICAL EXAMINATION

1. **The history suggests that an acute disease or condition is responsible for the patient's left knee pain.** The left knee pain is of 3 days duration and does not appear to have been preceded by injury or immoderate physical activity.
2. **The worsening of the knee pain upon weight-bearing by the affected lower limb indicates that one or more musculoskeletal structures that bear weight in the erect position are the source of the knee pain.** The presence of knee pain does not necessarily imply disease or injury of the knee region. This is because pain can be referred to the knee from disease or injury of the hip.

   In this case, the left knee is remarkable for the absence of any signs of disease or a pathologic condition other than the presence of apparently constant pain. Visual inspection does not reveal any surface deformities, areas of swelling and/or redness, or muscle atrophy about the left knee. Active and passive movements at the left knee are full in range. Isometric exercises at the left knee do not reveal any weakness or loss of muscle action and do not exacerbate knee pain. Palpation does not detect tenderness, crepitus, warmth, or the feel of any swelling.
3. **The deep tenderness in the midinguinal region marks the region as a site of disease or injury.** Tenderness is pain upon palpation. Tenderness generally marks a region as a site of disease or injury.

   Orthopedists select the site in the front of the thigh immediately lateral to the first 2 to 3 cm of the femoral artery for pressing upon the anterior aspect of the hip joint. The tissues pressed upon at this site include (proceeding from the most superficial to the deepest) the skin, superficial fascia, fascia lata, the common tendon of insertion of the iliacus and psoas major muscles, and tissues of the hip joint.
4. **The painful diminishing in the limits of active and passive external rotation, internal rotation, abduction, and adduction of the thigh suggests disease or a pathologic condition of one or more tissues (which includes the bony, cartilaginous, synovial, and capsular tissues) of the hip joint.** In general, pain upon movement indicates injury or disease of one or more of the bones, joints,

and muscles involved in the movement. The muscles acting across the hip joint do not appear to be involved because the physical examination does not reveal any weakness or pain upon isometric testing of the thigh flexors, extensors, internal rotators, external rotators, abductors, and adductors.

5. **The constant pain along the anteromedial aspect of the left knee appears to be pain referred from the hip.** This presentation suggests disease or a pathologic condition of tissues innervated by L3 fibers, as a part of the L3 dermatome covers most of the anteromedial aspect of the knee (Fig. 4–12G). The notable tissues in the hip region that are innervated in part by L3 fibers include all the medial and anterior thigh muscles and the hip joint.

INTERMEDIATE EVALUATION OF THE PATIENT'S CONDITION

The patient appears to be suffering from an acute disease or condition involving the left hip joint.

CLINICAL REASONING PROCESS

In resolving this case further, a pediatrician or orthopedist would know that three types of pathologic processes are responsible for most non-traumatic hip problems in school-age children: disorders of endochondral ossification, infection, and collagen/vascular disease. A **slipped capital femoral epiphysis** [a Salter-Harris Type I fracture of the epiphyseal plate for the head of the femur (refer to the case of Rico Alvarez, Case II-4, for a description of Salter-Harris type fractures)] and **Legg-Perthes disease** (idiopathic avascular necrosis of the head of the femur) are the endochondral ossification disorders that must be considered. The most common infectious processes are **osteomyelitis** (infection of one or more bones about the hip joint), **septic arthritis** (infection of the hip joint's synovial cavity), and **transient synovitis of the hip** (a nonspecific inflammation of the hip joint's synovial cavity). **Juvenile rheumatoid arthritis** (the pediatric form of this rheumatic disease) is the most likely collagen/vascular disorder. A physician would primarily consider pathologic processes that typically involve only one joint because the history and physical examination did not reveal any evidence for a polyarticular problem.

Slipped capital femoral epiphysis is a disorder which occurs most commonly during late childhood and adolescence. The epiphyseal growth plate for the femoral head changes its orientation during the early years of adolescence, placing the plane of the growth plate in greater alignment with the direction of weight-bearing forces through the hip joint. This change in orientation subjects the growth plate to greater disruptive shear forces during the course of daily activities. The increased mechanical stress on the growth plate in combination with certain hormonal imbalances renders young adolescents vulnerable to a Salter-Harris Type I fracture in which the metaphysis becomes displaced anteriorly, laterally, and superiorly.

Legg-Perthes disease is a disorder believed to be related to the developmental changes in the blood supply to the hip during childhood (for a discussion of these changes, refer to the section on the blood supply to the head of the femur). The childhood period from 4 to 9 years of age, when the retinacular arteries are the sole significant source of blood supply to the capital epiphysis, spans the years of the highest incidence of the disease. Limitations in the sources of blood supply apparently increase the vulnerability of the femoral head to avascular necrosis as a result of mechanical or infectious insults.

Most cases of osteomyelitis occur in children and adolescents. In this population, hematogenous spread is the most common mode of acquisition; trauma to a bone region appears to be a major predisposing factor for the seeding of bacteria in the region from a distant site of infection. The most common sites of acute hematogenous osteomyelitis in children and adolescents are the metaphyses of the long bones of the limbs. The sluggish and turbulent blood flow through the metaphyses of developing long bones renders these sites susceptible to the deposition and proliferation of blood-borne bacteria, especially following trauma-induced occlusion of the microvasculature at these sites.

Septic arthritis in children frequently occurs as a result of penetrating injuries which expose joints directly to the external environment or as a consequence of hematogenous dissemination of bacteria from a distant site of infection. Septic arthritis of the hip joint may also occur from direct extension of a bacterial infection of the metaphysis at the proximal end of the femur. Direct extension is possible because the metaphysis is intra-articular (the synovial membrane of the hip joint extends over the cortical margin of the metaphysis).

The etiology of transient synovitis of the hip is unknown. The condition produces a mild inflammation of the hip joint whose duration is self-limited.

Juvenile rheumatoid arthritis is a chronic inflammatory disease that occurs in a variety of forms. The disease always involves at least one joint. The systemic form of the disease can cause rash, generalized lymphadenopathy, and splenomegaly.

In this case, a pediatrician would judge that the patient is most likely suffering from Legg-Perthes disease or transient synovitis of the hip. The reasons for this judgment are that the patient exhibits the signs and symptoms typical of transient synovitis of the hip and the initial stage of Legg-Perthes diseases (an antalgic limp associated with hip and/or knee pain, marked limitations to internal rotation and abduction of the thigh, and a normal or slightly elevated temperature). Moreover, these two conditions are the most common causes of hip disease in children of the patient's age. A slipped femoral capital epiphysis is not considered likely because it generally occurs in older children or adolescents. Osteomyelitis and septis arthritis are not considered likely because children of the patient's age suffering from such infections frequently appear ill, are febrile, and complain of severe pain at the site of infection (because of the intensity of the pain, patients may even resist all movement at the involved joint). There is also no evidence in the history of a recent infection (such as a cold or the flu) which could serve as a source of blood-borne bacteria.

RADIOGRAPHIC EVALUATION AND FINAL RESOLUTION OF THE PATIENT'S CONDITION

AP and frog-leg lateral radiographs of the patient's hips showed inferolateral displacement of the head of the left femur from the acetabular cavity. There was no evidence on the left side of changes in the architecture or the extent of mineralization in the head of the femur or the parts of the coxal bone that form the acetabulum. There was also no evidence of widening or irregularities of the epiphyseal plate in the left femur (which are early radiographic signs of a slipped femoral capital epiphysis).

The displacement of the head of the left femur from the acetabulum indicates effusion of the hip joint. The displacement occurs in part because of the swelling of the soft tissues in the hip joint between the acetabular fossa and the head of the femur (these soft tissues include the ligament to the head, a fat pad, and the synovial membrane surrounding the ligament and fat pad). This radiographic finding is important but non-specific; it indicates the presence of an inflammatory process in the synovial cavity of the left hip joint but does not point to the cause of the inflammation.

A pediatrician or orthopedist would conclude that the patient is most likely suffering from transient synovitis of the hip or early stage Legg-Perthes disease (because early in the initial phase of Legg-Perthes disease, hip radiographs are generally normal). The physician would recommend complete bed rest and analgesics as needed and a follow-up examination a week from the initial visit. If the patient is suffering from transient synovitis of the hip, a week's bed rest should result (at a minimum) in significant improvement of the patient's condition. If pain and limping persist but do not become significantly worse, further physical reevaluation and radiologic imaging may be necessary to identify the underlying pathology.

The patient's parents would also be advised to contact the physician immediately if the patient begins to develop a fever, complain of greater pain, or show a decline in general health (as any of these changes would suggest the presence of osteomyelitis or septic arthritis). Clinical judgment is required to decide whether to conduct arthrocentesis at this time to collect a sample of joint fluid aspirate for microbial analysis, bearing in mind that such an invasive procedure always carries a small risk of inadvertent bacterial inoculation of a sterile joint cavity. In this case, arthrocentesis was deferred because there were no findings in the history, physical exam, or radiographs that strongly suggested, far less indicated, septic arthritis.

EXPLANATION OF THE PATIENT'S CONDITION

The patient is suffering from transient synovitis of the left hip joint, which began 3 days before. The inflammatory process has increased the permeability of the microvasculature serving the synovial membrane of the hip joint, and this, in turn, has resulted in an effusion of the hip joint's synovial cavity. The effusion has produced a painful distention of the synovial membrane lining. In this patient, the stimulation of pain fibers, which extend from the synovial membrane lining to the left dorsal root of L3, produce pain that is perceived not in the hip but instead referred to the anteromedial aspect of the left knee (the cutaneous nerve supply of which is provided by anterior femoral cutaneous branches of the femoral nerve and branches of the saphenous nerve) (Fig. 4–12G). The irritability of the synovial membrane and the effusion itself in the left hip joint have reduced the ranges of motion at the joint.

A seated position relieves the patient's left knee pain because the position in which the thigh is flexed about 90° and slightly externally rotated and abducted at the hip joint is the position that maximizes the encapsulation of the femoral head by the acetabular cavity and labrum. Maximization of the encapsulation reduces the tension and thus the stimulation of pain fibers in the synovial membrane to a minimum.

Walking exacerbates the patient's left knee pain because walking imposes forces and extremes of range of motion that unduly stretch certain parts of the already distended synovial membrane and fibrous capsule of the hip joint. First, walking increases the compressive forces exerted across the hip joint. These forces intensify the pain emanating from the hip by compressing the already inflamed soft tissues that occupy the space in the hip joint between the acetabular fossa and the head of the femur. The patient reduces the imposition of these compressive forces by leaning his body toward the left side during the loading response, mid stance, and terminal stance phases of the left lower limb's stance period and by flexing the leg more than normal during the left lower limb's stance period. The manner in which the leftward leaning of the body reduces compressive force across the hip joint is explained in the case of Michael Kenat, Case III.1. Leg flexion permits the muscles acting across the knee joint to counteract some of the compressive forces normally exerted at the hip joint.

Second, a normal walking gait requires about 10° hyperextension of the thigh at the hip during the terminal stance phase of the stance period. Pain has rendered the patient reluctant to hyperextend the thigh. Hyperextension is painful because the fibrous capsule of the left hip joint is distended by the effusion within the joint, and thus the iliofemoral, pubofemoral, and ischiofemoral ligaments all become maximally stretched (and painful) upon full thigh extension. The iliofemoral, pubofemoral, and ischiofemoral ligaments of the hip joint become more tightly wrapped around the neck of the femur as the thigh becomes increasingly extended; in a normal hip joint, these ligaments are all almost maximally stretched at full thigh extension.

The reluctance by the patient to hyperextend the left thigh during the terminal stance phase and the greater than normal leg flexion throughout the stance period combine to shorten the effective length of the left lower limb prior to heel rise. The patient compensates for this shortening by plantarflexing the foot at the ankle joint. The foot plantarflexion accounts for the early heel rise during the left lower limb's stance period.

Third, and last, a normal walking gait requires internal rotation of the thigh (from the neutral position) during a time frame that extends from the mid swing phase of the swing period to the early part of the terminal stance phase of the subsequent stance period. In the hip joint, the extent of encapsulation of the head of the femur by the acetabular cavity and labrum progressively decreases as the extent of internal rotation of the femur from the neutral position increases. This relationship explains why in patients suffering from painful hip effusion (with no other complications) internal rotation of the thigh is markedly limited and the thigh is excessively externally rotated throughout the walking gait.

CASE **III.3**

The Case of Thomas Mayhew

INITIAL PRESENTATION AND APPEARANCE OF THE PATIENT

Thomas Mayhew, a 22 year-old Native American man, has made an appointment to seek treatment for an unstable right knee. As the patient walks to the examination table, the examiner observes that the patient exhibits a stiff-legged gait in which the right leg is fully extended at the knee during the terminal swing phase of the swing period and throughout the first half of the stance period.

QUESTIONS ASKED OF THE PATIENT

Please tell me about your problem. I was playing touch football 6 days ago when a guy slammed into my knee while I was running with the football. [The patient uses the palm of his right hand to demonstrate that the force of the blow was directed against the posterolateral aspect of the distal end of the femur.] I thought I heard something pop in my knee when I was hit. The knee began to swell up and hurt so much that I stopped playing. My knee was so swollen for the next 2 days that I had trouble bending it. I guess I should have gone to a doctor, but I hoped that if I just rested the knee for a few more days, I'd be all right. Well, most of the swelling's gone down, but my knee sometimes gives out when I walk.

Can you describe more specifically what you mean when you say your knee gives out? Well, it's like the bones in my knee slide past each other. [The patient demonstrates the sensation by making fists with both hands, placing the left fist on top of the right fist, and then simultaneously sliding the left fist backward and the right fist forward.] *[The examiner's invitation to the patient for an open and thorough discussion of his injury has revealed much information. First, the mechanism of injury was a severe blow to the posterolateral aspect of the distal end of the femur. Second, the patient's recollection of a popping sound upon impact suggests sudden rupture of a ligament, tear of a meniscus, and/or partial or complete dislocation of an articulation within the knee joint. Third, the rapid, marked swelling of the right knee joint indicates that the impact produced significant injury to one or more musculoskeletal structures. Fourth, the patient's report that the knee intermittently gives out when walking indicates dynamic instability in the right knee.]*

Do you recall any more details about how your knee was hit? Well, I was running to my right, and I thought I could escape from the guy running after me by stopping and cutting to my left. But as he tried to cut back to follow me, he fell down, and his shoulder slammed into my knee just as I was pushing off from my right foot. *[This answer indicates that the patient's right lower limb was firmly planted on the ground and the right knee partially flexed at the moment the knee sustained a severe blow to its posterolateral aspect.]*

Does your knee hurt now? A little bit.

Where does it hurt? On the inside of my knee. *[The patient uses the fingertips of his left hand to demonstrate the presence of pain in the medial aspect of the right knee.]*

[When pain is a complaint, it is always essential to have the patient mark by word or gesture the exact site(s) at which pain is perceived. In cases of physical injury, pain is generally a reliable indicator of the site(s) of injury.]

Has the location of the knee pain changed during the past 6 days? Well, right after I got hit, my whole knee hurt real bad as it swelled up. As the swelling's gone down, most of the pain has also gone away except for on the inside of the knee. *[The patient's description of severe pain in the swollen knee soon after the injury suggests hemarthrosis (a bloody effusion of the joint). Extravasated blood (blood that has escaped from the vasculature into a body cavity or the interstitial spaces of a tissue) is a profound irritant to the serous membranes that line the serous cavities of the body (such as the cavities of the synovial joints and the pleural, pericardial, and peritoneal cavities). A serous (non-bloody) effusion of a joint can produce pain, but the pain is* *comparatively moderate.]*

How would you describe the pain? It varies from an ache to a sharp pain.

Is there anything that makes the pain worse? If I'm walking and I put a twist on my knee, it hurts real bad.

Is there anything that makes the pain better? Just staying off my feet and sitting or lying down is when I'm most comfortable. *[In general, these last two questions provide information about the source of pain. In this instance, the finding that the pain is worsened by movement and relieved by rest suggests injury of one or more musculoskeletal structures.]*

Is there anything else that is also bothering you? Not now, but there was something that bothered me up until this morning. For the past 2 days I wasn't able to straighten my knee all the way out. But this morning when my knee gave out, something happened, I don't know what, and all of a sudden I could straighten out my knee. *[In this case, an inability to fully extend the leg suggests a torn meniscus, the presence of a loose body of tissue in the knee joint, an injury to the quadriceps femoris tendon, or spasm of the hamstring muscles. A meniscal fragment or loose tissue body restricts full extension by becoming entrapped between one of the femoral condyles and its apposing tibial condyle.]*

Do you have any loss of feeling in your right leg or foot? No.

Have you noticed any weakness in moving your right foot? No.

Have you felt any coldness in your right leg or foot? No. *[When the examiner asks if there are other things bothering the patient in addition to the instability and pain in the right knee, the examiner is addressing a concern that there might be neurovascular complications to the patient's knee injury. Although the patient does not mention any symptoms that would suggest neurovascular complications (such as the symptoms described in the last three questions), it is nonetheless still possible that he has one or more of these symptoms and has not mentioned them because he judges them to be inconsequential. To exclude the presence of such symptoms, the examiner elects to pose direct questions to obtain this critical information.*

In taking a history, it is prudent initially to ask open-ended questions, questions that invite the patient to discuss openly and completely his/her medical problems and related concerns. This approach is designed to give the examiner as much information as possible about the patient's problems in the patient's own words. However, after a certain amount of information has been learned, an examiner may begin to construct a differential diagnosis and therefore pose direct questions to obtain specific positives and negatives. Alternatively, an examiner may pose direct questions to learn information that a patient may regard as inconsequential or may be reluctant to provide unless given a specific request.]

Have you previously injured your right knee? No.

Have you had any recent illnesses or injuries? No.

Are you taking any medications for previous illnesses or conditions? No. *[The examiner finds the patient alert and fully cooperative during the interview.]*

PHYSICAL EXAMINATION OF THE PATIENT

Vital signs:
Blood pressure
Lying supine: 120/75 left arm and 120/75 right arm
Standing: 120/75 left arm and 120/70 right arm
Pulse: 68
Rhythm: regular
Temperature: 98.5°F.
Respiratory rate: 12
Height: 6′3″
Weight: 160 lbs.
HEENT Examination: Normal
Lungs: Normal
Cardiovascular Examination: Normal. A pertinent set of normal findings is the absence of any signs of vascular disease in the right lower limb, signs such as coldness, pallor, cyanosis, weak or absent arterial pulses, and arterial bruits.
Abdomen: Normal
Genitourinary Examination: Normal
Musculoskeletal Examination: Inspection, movement, and palpation of the upper and lower limbs are normal except for the following findings for the right lower limb: Inspection reveals an 8 cm diameter greenish bruise overlying the medial aspect of the knee, swollen soft tissues on both sides of the patellar ligament, and slightly diminished muscle mass in the thigh region immediately superomedial to the patella.
Valgus stress applied at the knee with the knee at 30° flexion produces valgus subluxation of the tibia on the femur. Valgus stress applied at the knee with the knee fully extended also produces valgus subluxation of the tibia on the femur. Varus stress applied at the knee with the knee at 30° flexion does not produce a

varus subluxation of the tibia on the femur. The Lachman test with the right knee at 30° flexion is positive (the same test on the left knee is negative). The posterior drawer test with the knee at 90° flexion produces about 0.5 cm posterior movement of the tibia on the femur (the same test on the left knee yields the same result). McMurray's test with the leg externally rotated produces a painful, audible click upon leg extension.

[As used here, the term valgus indicates a laterally-directed stress applied to the lower end of the leg in conjunction with a medially-directed stress applied at the knee, and the opposite term varus indicates a medially-directed stress applied to the lower end of the leg in conjunction with a laterally-directed stress applied at the knee. A valgus subluxation of the tibia on the femur is an abnormal widening of the medial joint space in the knee joint (the joint space between the medial tibial and femoral condyles). A varus subluxation of the tibia on the femur is an abnormal widening of the lateral joint space in the knee joint.]

The knee is positive for the bulge sign. Palpation reveals tenderness directly below the medial epicondyle of the femur and along the posterior half of the joint space between the medial tibial and femoral condyles. Measurement of the girths of the thighs shows the girth of the right thigh to be 5% less than that of the left thigh.

Neurologic Examination: Normal. A pertinent set of normal findings is the absence of any signs of neurological dysfunction in the right lower limb, signs such as paresthesia, anesthesia, loss of position sensation, and hyperactive, weak, or absent phasic stretch reflexes.

INITIAL ASSESSMENT OF THE PATIENT'S CONDITION

The patient's chief medical problem appears to be dynamic instability of the right knee following traumatic injury.

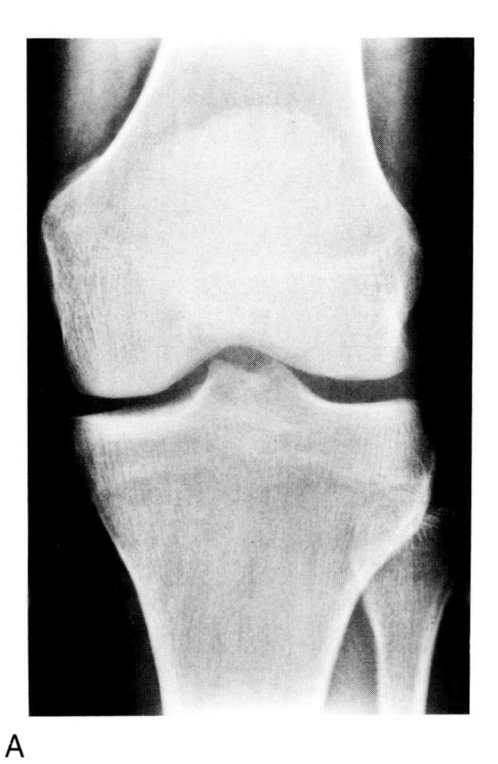

Fig. 13–27: (A) An AP radiograph of the knee, (B) its schematic representation, and (C) the orientation of a patient's knee relative to the x-ray beam and film cassette for the radiograph.

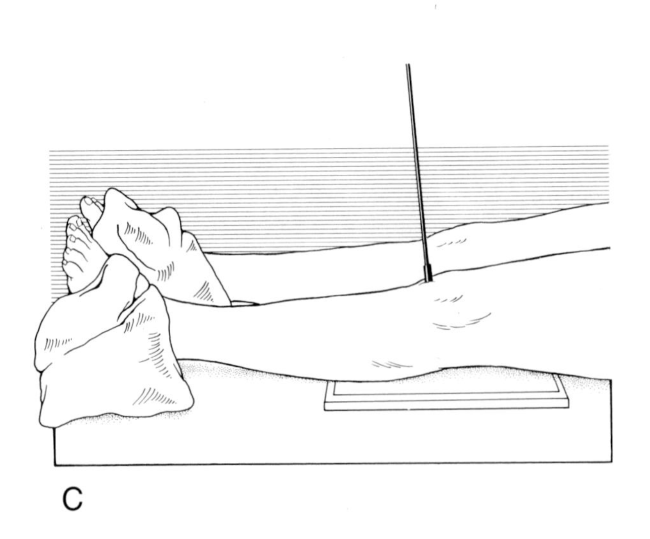

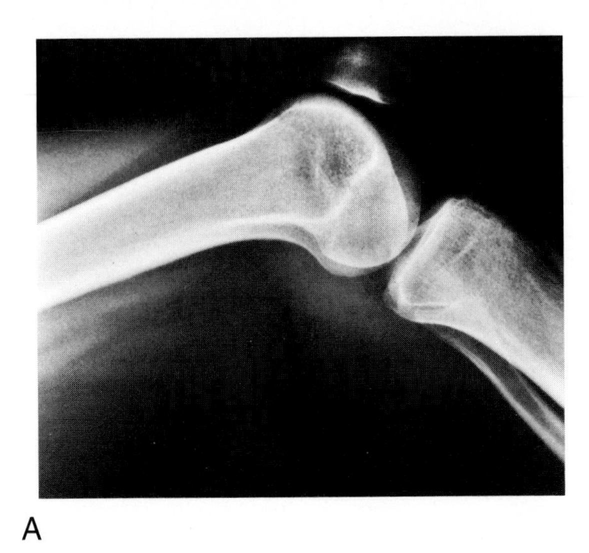

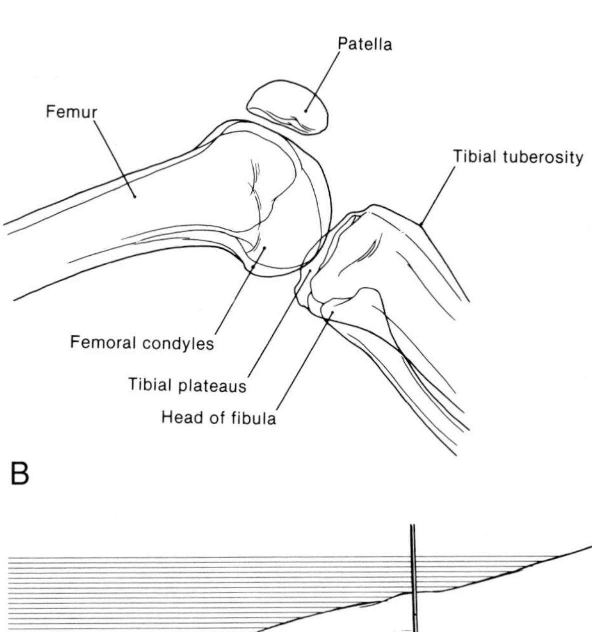

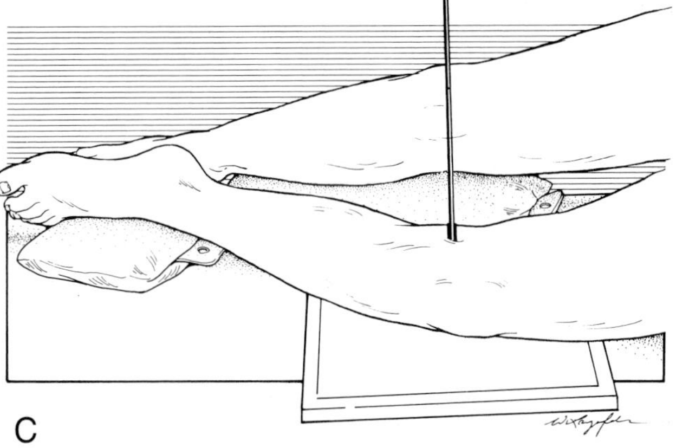

Fig. 13–28: (A) A lateral radiograph of the knee, (B) its schematic representation, and (C) the orientation of a patient's knee relative to the x-ray beam and film cassette for the radiograph.

ANATOMIC BASIS OF THE PATIENT'S HISTORY AND PHYSICAL EXAMINATION

1. **The patient has sustained a severe injury of musculoskeletal tissues in the right knee.** The production of a popping sound at impact, the rapid onset of marked swelling in association with severe pain immediately after impact, the temporary limitation of leg extension, and the intermittent, posttraumatic episodes of dynamic instability all point to significant injury of one or more musculoskeletal tissues.

2. **The results of the valgus and varus stress tests, the Lachman test, and posterior drawer test collectively indicate significant damage to the tibial collateral and anterior cruciate ligaments.** Valgus and varus stress tests at the knee assess the integrity of the structures that respectively provide medial and lateral stability of the knee. When the knee is at 30° flexion, the anterior and posterior cruciate ligaments are comparatively lax and the tibial and fibular collateral ligaments are the most effective ligamentous supports, respectively, of medial and lateral stability of the knee joint. When the knee is fully extended, the anterior and posterior cruciate ligaments are taut and can by themselves provide medial and lateral stability in the absence of collateral ligament support.

   The negative finding obtained upon applying a varus stress with the knee at 30° flexion indicates the absence of significant damage to the fibular collateral ligament. The subluxation obtained upon applying a valgus stress with the knee at 30° flexion suggests significant damage to the tibial collateral ligament. The subluxation obtained upon applying a valgus stress with the knee fully extended suggests significant damage to at least one of the following structures: anterior cruciate ligament, posterior cruciate ligament, or posteromedial aspect of the fibrous capsule.

   The Lachman test assesses primarily the integrity of the anterior cruciate ligament. This test is conducted as the patient lies supine on an examination table with the examiner standing beside the involved knee. The examiner grasps the distal end of the thigh with one hand and the proximal end of the leg with the other hand, holds the knee in a position of 30° flexion, and then firmly pulls the leg forward. A "mushy" or "soft" feel to the end point of the anterior movement of the tibia beneath the femur is a positive test; a "hard" feel to the end point is a negative test. The positive Lachman test

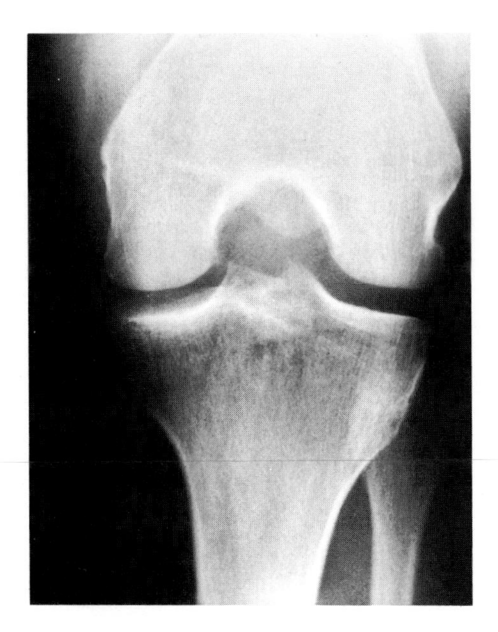

A

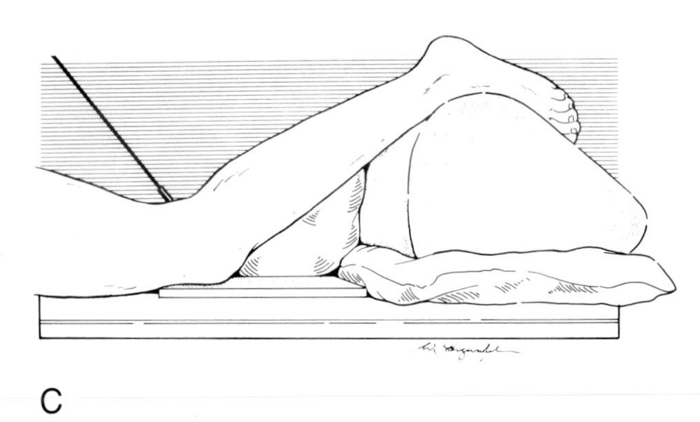

C

Fig. 13–29: (A) A tunnel radiograph of the knee, (B) its schematic representation, and (C) the orientation of a patient's knee relative to the x-ray beam and film cassette for the radiograph.

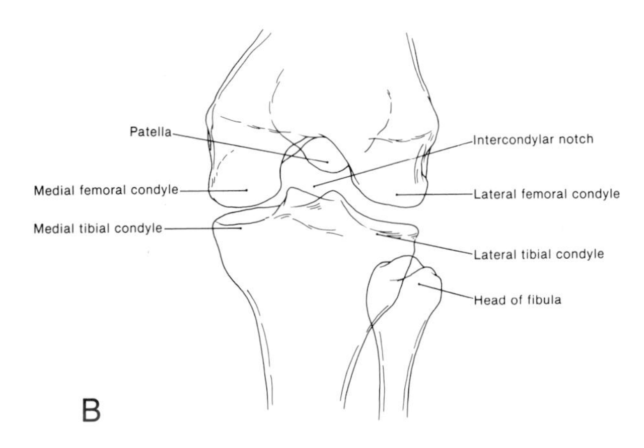

B

in this case suggests significant damage to the anterior cruciate ligament.

The posterior drawer test assesses primarily the integrity of the posterior cruciate ligament. Test results of about 1.0 cm or less of posterior movement of the tibia beneath the femur are considered to be negative findings. However, test results on the injured knee should always be compared with those on the uninjured knee. The test is conducted as the patient lies supine on an examination table with the injured lower limb flexed 90° at the knee and the foot resting on the table. The examiner grasps the upper end of the leg with both hands and firmly forces it backward. The negative test result in this case suggests the absence of significant damage to the posterior cruciate ligament.

3. **The painful, audible click produced in McMurray's test upon extension of an externally rotated leg indicates a tear in the middle and/or posterior thirds of the medial meniscus.** McMurray's test is used to detect tears of the menisci. As the patient lies supine with the thigh and leg flexed, the examiner grasps and presses on the medial and lateral aspects of the knee with one hand and holds the plantar surface of the foot with the other hand. The examiner flexes the leg until the heel almost touches the buttock. The examiner then rotates the leg externally if the patient is suspected of having a tear of the medial meniscus or internally if the patient is suspected of having a tear of the lateral meniscus. External rotation of the leg draws the medial meniscus toward the center of the joint, and internal rotation draws the lateral meniscus toward the center of the joint. While forcefully maintaining the leg in either rotated configuration, the examiner extends the leg to 90° flexion at the knee. The test is positive for a meniscal tear if the leg extension maneuver produces a palpable or audible click in association with pain. It is believed that the click occurs when the femoral condyle rolls over the meniscal tear. McMurray's test may reveal tears in the posterior and middle thirds of the meniscus but not in the anterior third.
4. **The bulge sign indicates a small effusion of the right knee joint.** When a small amount of excess fluid is in the synovial cavity of the knee joint, the only visual evidence commonly consists of swollen soft tissues on the medial and/or lateral sides of the patellar ligament. To distinguish these swellings from swellings of the prepatellar, superficial infrapatellar, and deep infrapatellar bursae (bursae that do not communicate with the knee joint's synovial cavity), the examiner massages the swollen soft tissues on the medial and/or lateral sides of the patellar ligament as the patient lies

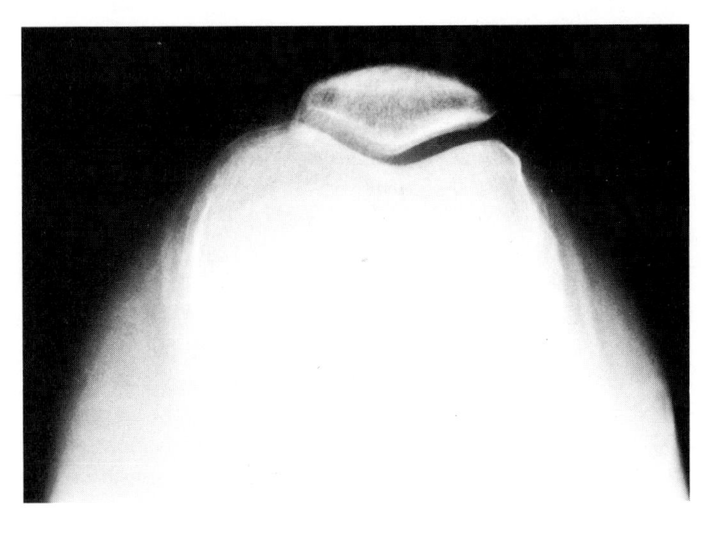

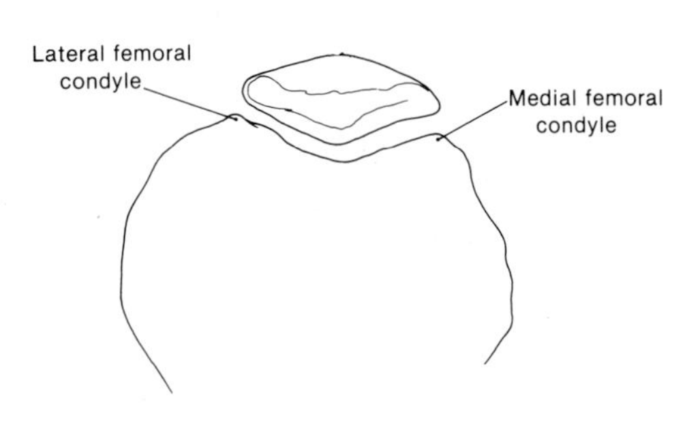

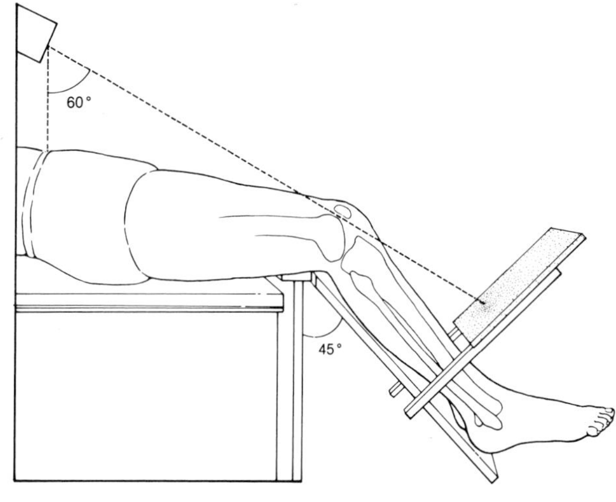

Fig. 13–30: (A) An axial radiograph of the knee, (B) its schematic representation, and (C) a commonly selected orientation of the patient's knee relative to the x-ray beam and film cassette for the radiograph. (p. 65.)

supine on an examination table. If most or all of the swellings on the sides of the patellar ligament represent swellings of soft tissues or bursae exterior to the knee joint's synovial cavity, then the massage will not significantly diminish the swellings. If most or all of the swellings represent a knee joint effusion, however, then the massage will displace the excess fluid superiorly into the suprapatellar bursa and the swellings will be markedly diminished upon withdrawal of the examiner's hand. But within about a minute the soft tissues alongside the patellar ligament will swell, or bulge, again (the bulge sign) as the excess fluid redistributes itself back into the anteroinferior parts of the synovial cavity alongside the patellar ligament. The bulge sign is the most sensitive physical test for the presence of a small amount of excess fluid in the synovial cavity of the knee joint.

5. **The tenderness directly below the medial epicondyle of the femur and the tenderness along the posterior half of the medial joint space mark these regions as sites of injury.** Tenderness is pain elicited by palpation. Tenderness generally marks a region as a site of injury or disease. The palpable tissues directly below the medial epicondyle of the femur include (proceeding from the most superficial to the deepest) the skin, the superficial and deep fascia, the tendon of sartorius, the tibial collateral ligament, the fibrous capsule of the knee joint, and the medial femoral condyle. The palpable tissues along the posterior half of the medial joint space include (proceeding from the most superficial to the deepest) the skin, the superficial and deep fascia, the tendons of sartorius, gracilis, semitendinosus, and semimembranosus, the medial head of gastrocnemius, the tibial collateral ligament, the fibrous capsule of the knee joint, and the medial meniscus.
6. **The diminished muscle mass in the thigh region immediately superomedial to the patella indicates atrophy of vastus medialis.** The lowest part of the muscular belly of vastus medialis accounts for all the muscle mass in the thigh region immediately superomedial to the patella.
7. **The decreased girth of the right thigh suggests atrophy of one or more thigh muscles.** Skeletal muscles suffer a loss of bulk mass when the general level of their activity significantly decreases. A change in thigh girth generally indicates a change in

the collective bulk mass of one or more of the three major thigh muscle groups (the anterior, medial, and posterior thigh muscles). Other observations, measurements, and tests are needed to determine which thigh muscle groups have suffered a loss of tissue mass. In this case, the finding of vastus medialis atrophy suggests that the other muscles of the quadriceps femoris group have also undergone some atrophy, and, therefore, that slight atrophy of the quadriceps femoris muscles probably accounts for the minimal (5%) decrease in right thigh girth.

INTERMEDIATE EVALUATION OF THE PATIENT'S CONDITION

The patient appears to have suffered significant damage to the tibial collateral and anterior cruciate ligaments and a medial meniscal tear in the right knee.

CLINICAL REASONING PROCESS

It is obvious from the history that the patient has sustained a moderate-to-severe injury of the right knee. In considering this case upon the completion of the history, a primary care physician or orthopedist would address three major concerns during the physical exam:

1. Gather as much physical evidence as possible about the musculoskeletal injuries at the knee.
2. Thoroughly examine the remainder of the injured lower limb for physical evidence of other musculoskeletal injuries. In this case, the history suggests that very stressful forces were applied mainly perpendicularly to the long axis of the lower limb, and, therefore, that it is not likely (although still possible) that there are injuries proximal and/or distal to the prominent knee injury.
3. Carefully examine the right leg, ankle, and foot for any signs of neurovascular complications. This is because moderate-to-severe knee injuries always place the neurovascular supply to the leg, ankle, and foot at risk. The appropriate physical examinations of neurovascular supply in this case are as follows:
   i. The strength and regularity of the dorsalis pedis and posterior tibial arterial pulsations should be assessed at both ankles. These pulsations serve as one monitor of the adequacy of blood supply to the ankle and foot. However, it is important to recognize that some individuals have a rather small or absent dorsalis pedis and thus weak or absent arterial pulsations on the dorsum of the foot just immediately lateral to the tendon of extensor hallucis longus (refer to the discussion of the major arteries of the leg in Chapter 14).
   ii. The speed of capillary refill in the big toe should be examined in both feet as another monitor of the adequacy of blood supply to the foot. The test is conducted in a manner similar to that described in the case of Linda Chin (Case II-5) for assessing the speed of capillary refill in the fingers.
   iii. Light touch sensation, vibration sensation, and two-point discrimination on the anterior and posterior surfaces of the right leg and the dorsal and plantar surfaces of the right foot should be compared with those sensations on the corresponding surfaces of the left leg and foot. These sensations are sensitive tests of sensory supply to the skin. Light touch sensation may be assessed with a fine wisp of cotton lightly applied to the skin. A relatively low-pitch (128 Hz) tuning fork is commonly used to assess vibration sensation. Two-point discrimination, which is the ability to detect the application of probes at two separate but close points, may be assessed with the ends of an opened paper clip.
   iv. The strength of numerous foot and toe movements should be assessed in both feet. These movements assess the nerve supply of leg and foot muscles.

The performance of these examinations in this case did not yield any evidence of neurovascular complications. The examiner also mentally noted that the patient did not give any indication of intense leg pain, a sign which, in this case, might indicate muscle ischemia.

In considering this case, a primary care physician or orthopedist would appreciate that the set of circumstances that led to the patient's injury are fairly common in the games of rugby and football and that the patient has apparently sustained one of the more severe set of injuries that can be produced under these circumstances. If a rugby or football player is hit on the lateral aspect of the knee (with the lower limb being in the position as described in the history) with sufficient force to produce a complete tear of one or more ligaments about the knee, it is relatively common for the player to sustain a triad of injuries: complete rupture of the tibial collateral ligament, complete rupture of the anterior cruciate ligament, and a tear of the medial meniscus. This triad of injuries matches those indicated by the physical examination.

RADIOGRAPHIC EVALUATION AND FINAL RESOLUTION OF THE PATIENT'S CONDITION

Upon completion of the physical exam, an orthopedist

would order a set of radiographs of the patient's right knee to determine if there any bony fractures or bodies of bony tissue in the knee joint's synovial cavity. Figures 13–27, 13–28, 13–29, and 13–30 respectively show the normal appearance of AP, lateral, tunnel, and axial radiographs of the knee. The radiographs of the patient's right knee did not show any evidence of bone fractures. The tunnel radiograph (Fig. 13–29) is so named because it displays the relatively deep, concave profile of the intercondylar notch at the distal end of the femur. A tunnel radiograph frequently provides the best evidence of a loose body of bony tissue in the knee joint's synovial cavity; examination of this patient's radiographs did not reveal the presence of any loose osseous bodies. The axial radiograph (Fig. 13–30) provides a view that best displays the patellofemoral articulation and evidence of patellar fractures.

An orthopedist would next use arthroscopy to examine the menisci and cruciate ligaments of the right knee. Arthroscopy is the procedure by which the interior of a synovial joint is examined with a small, flexible tubular device with a fiberoptic interior. The device, an arthroscope, is inserted under local anesthesia through a small incision in the capsule of the synovial joint. In this case, arthroscopic examination revealed tears in the medial aspect of the capsule, a complete, shredded tear through the midportion of the anterior cruciate ligament, and a bucket-handle tear of the medial meniscus (Fig. 13–31C). The central portion of the inner fragment of the medial meniscus was found displaced into the intercondylar area of the tibial plateau.

Many orthopedists would advise the patient that his injuries can be treated either conservatively (which involves essentially rest and immobilization of the knee for a few weeks) or surgically. Surgery is the more appropriate option if the patient is very active in physical or sports activities and wants the right knee capable of withstanding the stresses imposed by such activities. Conservative treatment may be the more appropriate option if the patient is willing to tolerate the limitations imposed by a right knee which, after it has healed, is capable of withstanding the stresses of light-to-moderate but not strenous physical activity.

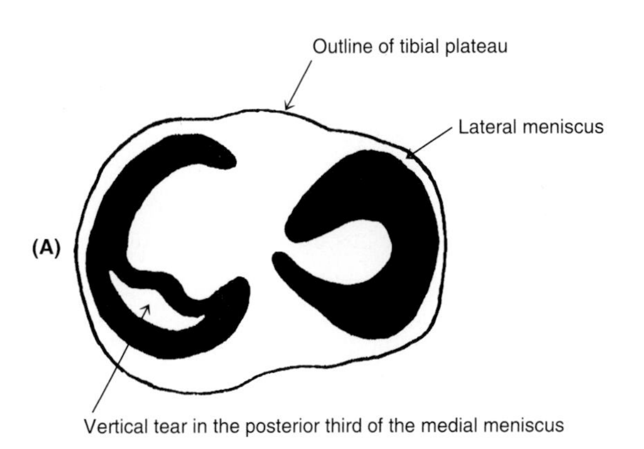

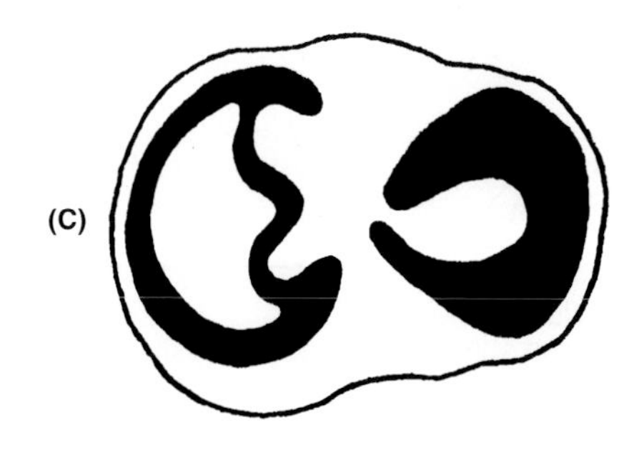

Fig. 13–31: Drawings of the superior view of the medial and lateral menisci in Thomas Mayhew's right knee (a) at the time of injury, (B) following extension of the medial meniscal tear into the anterior third of the meniscus, and (C) at the time of arthroscopic examination.

MECHANISM OF THE PATIENT'S INJURY AND AN ACCOUNT OF HIS CONDITION

At the moment of impact, the patient's right foot was firmly planted on the ground below, the right leg flexed 40° at the knee, and the right thigh maximally internally rotated at the knee (the right thigh was maximally internally rotated at the knee because the patient had just begun his movement to the left). The severe impact upon the posterolateral aspect of the distal end of the femur exerted a marked valgus force on the right knee. This valgus force produced a marked abduction of the right leg at the knee, which, in turn, significantly widened the medial joint space and stretched the medial part of the fibrous capsule beyond its limit of extension. The excessive extension of the medial part of the fibrous capsule caused this part of the fibrous capsule and the deep part of the tibial collateral ligament (the part attached to the fibrous capsule) to tear, followed in quick succession by the tearing of the superficial part of the tibial collateral ligament (near its attachment to the medial epicondyle of the femur). Continued abduction of the right leg at the knee now stretched the anterior cruciate ligament so tightly over the medial aspect of the lateral femoral condyle that the midportion of the ligament suffered an abrupt rupture, generating the popping sound heard by the patient.

The marked widening of the medial joint space had permitted the medial meniscus to move centrally within the joint. As the valgus force on the right knee dissipated (subsequent to the tearing of the above ligamentous structures), muscle spasms of the muscles acting across the knee joint quickly entrapped and pressed the medial meniscus between the medial tibial and femoral condyles, in the process producing a vertical tear in the posterior third of the meniscus and some tearing of the posterior part of the external border of the mensicus from its attachment to the fibrous capsule (Fig. 13–31A). The retinacular blood vessels torn in association with the ligamentous and capsular tears rapidly filled the knee joint with a bloody effusion, producing a stiff and painful knee.

By 96 hours after the injury, the patient's knee effusion had diminished to the extent that the patient became aware that the right leg could not be extended at the knee to less than 30° flexion. The entrapment of the inner fragment of the medial meniscus between the apposing surfaces of the medial tibial and femoral condyles was responsible for this limitation in leg extension. As the patient began to walk around on the injured knee, however, repetitive flexion and extension movements began to extend the anterior limit of the medial meniscal tear until it extended into the anterior third of the meniscus (Fig. 13–31B). When the patient's knee gave out the morning of the visit, the conditions responsible for this instability forced the central portion of the inner fragment from between the apposing surfaces of the medial tibial and femoral condyles into the intercondylar area of the tibial plateau (Fig. 13–31C). Because the central portion of the inner fragment was no longer interposed between contact points of the medial tibial and femoral condyles, the patient suddenly discovered that he had regained full leg extension at the right knee.

The episodes of instability in the patient's right knee have occurred when the patient unwittingly recapitulates some of the movements that occurred in the right knee immediately prior to impact on the football field. The episodes of instability have occurred when, as the right foot is firmly planted on the surface below and the right leg partially flexed at the knee, the patient suddenly turns to his left, and thus simultaneously imposes a valgus force on the knee and an internal rotatory force on the femur at the knee. The combination of these forces produces an anterior subluxation of the medial tibial condyle beneath the medial femoral condyle (the medial tibial condyle partially dislocates anteriorly relative to the medial femoral condyle). The patient describes this anterior subluxation as a "giving out" of the knee. As the patient reflexively attempts to shift all the upper body weight onto only the left lower limb (because he is apprehensive about bearing weight on the right knee), he flexes the right thigh at the hip and the right leg at the knee to raise the right foot off the ground. As the right leg is flexed, the sum of the flexion forces exerted across the knee [in particular, those exerted by the medial rotators of the leg (sartorius, gracilis, semitendinosus, and semimembranosus)] suddenly reduce the anterior tibial subluxation with a thud, and the patient finds the right knee temporarily stable again. The patient walks with a stiff-legged gait in which the right leg is fully extended at the knee during the terminal swing phase of the swing period and throughout the first half of the stance period so as to reduce the incidence of anterior tibial subluxation.

The tenderness immediately below the medial epicondyle of the femur marks the location at which there occurred a complete tear of the tibial collateral ligament. The tenderness along the posterior half of the joint space between the medial tibial and femoral condyles marks the sites at which there occurred tearing of the posterior part of the external border of the mensicus from its attachment to the fibrous capsule. The 5% decrease in the girth of the right thigh and the diminishment of the vastus medialis muscle mass are due to disuse atrophy of the quadriceps femoris muscles.

RECOMMENDED REFERENCES FOR ADDITIONAL INFORMATION ON THE KNEE

Corrigan, B., and G. D. Maitland, Practical Orthopaedic Medicine, Butterworth & Co., London, 1983: *Chapter 10 offers a description of the clinical examination of the knee and discussions of the signs and symptoms of common knee lesions.*

Magee, D. J., Orthopaedic Physical Assessment, 2nd ed., W. B. Saunders Co., Philadelphia, 1992: *Chapter 11 provides a comprehensive discussion of the clinical examination of the hip joint.*

Slaby, F., and E. R. Jacobs, Radiographic Anatomy, Harwal Publishing Co., Media, PA, 1990: *Pages 52 through 58 in Chapter 2 outline the major anatomic features of the AP, tunnel, lateral, and axial radiographs of the knee. Pages 58 through 68 outline the major anatomic features of T1-weighted magnetic reasonance images of coronal and sagittal sections of the knee.*

Greenspan, A., Orthopedic Radiology, 2nd ed., Gower Medical Publishing, New York, 1992: *Fig. 8.20 on page 8.13 summarizes the array of radiologic imaging techniques that may be used to evaluate knee injuries. Pages 8.26 through 8.34 present discussions of the radiologic evaluation of patients with soft tissue injuries in and about the knee.*

CHAPTER **14**

The Leg and Ankle

This chapter focuses upon the innervation and actions of the leg muscles. There are three major groups of muscles in the leg: anterior, posterior, and lateral. The anterior leg muscles lie mostly anterior to the tibia and fibula, are innervated by the deep peroneal nerve, and, as a group, can invert and dorsiflex the foot and extend the toes. The posterior leg muscles lie mostly posterior to the tibia and fibula, are innervated by the tibial nerve, and, as a group, can invert and plantarflex the foot and flex the toes. The lateral leg muscles lie lateral to the fibula, are innervated by the superficial peroneal nerve, and, as a group, can evert and plantarflex the foot.

Because all except one of the leg muscles insert onto bones in the foot, the chapter begins with an examination of the bones and joints of the foot.

THE BONES OF THE FOOT

The bones of the foot can be divided into three groups: the bones of the hindfoot, the bones of the midfoot, and the bones of the forefoot.

The Tarsals: The Bones of the Hindfoot and Midfoot

There are seven tarsal bones (Fig. 14–1). The two most proximal tarsals, the **talus** and **calcaneus,** comprise the bones of the hindfoot. In the hindfoot, the talus rests atop the calcaneus. The calcaneus is palpable in the heel.

The five distal tarsals comprise the bones of the midfoot. Four of these tarsals are aligned in a row that extends transversely across the breadth of the midfoot. Proceeding from the most medial to the most lateral of these four tarsals, they are aligned as follows:

Medial Cuneiform–Intermediate Cuneiform–Lateral Cuneiform–Cuboid

The fifth tarsal of the midfoot, the navicular, articulates anteriorly with the three cuneiforms and laterally with the cuboid.

The ossification centers for the calcaneus and talus are present at birth. The lateral cuneiform begins to ossify in the first year, the medial cuneiform in the second year, and the intermediate cuneiform in the third year.

The Metatarsals and Phalanges: The Bones of the Forefoot

The five metatarsals of the foot are numbered in such a fashion that the first metatarsal is the most medial metatarsal and the fifth metatarsal is the most lateral metatarsal (Fig. 14–1). A metatarsal is associated with each toe:

1st metatarsal: the metatarsal for the big toe
2nd metatarsal: the metatarsal for the second toe
3rd metatarsal: the metatarsal for the third toe
4th metatarsal: the metatarsal for the fourth toe
5th metatarsal: the metatarsal for the little toe

Each metatarsal has a **base,** a **shaft,** and a **head.** The base is the proximal end of the metatarsal, and the head is the distal end of the metatarsal. The heads of the metatarsals lie embedded in the foot pad at the bases of the toes.

Each metatarsal generally bears only a single epiphysis during childhood and adolescence. Whereas the epiphysis of the first metatarsal resides at the bone's base, the epiphyses of the second to fifth metatarsals reside at the heads of the bones.

The phalanges are the bones of the toes. The big toe has two phalanges and each other toe three phalanges (Fig. 14–1). The phalanges of the big toe are called the

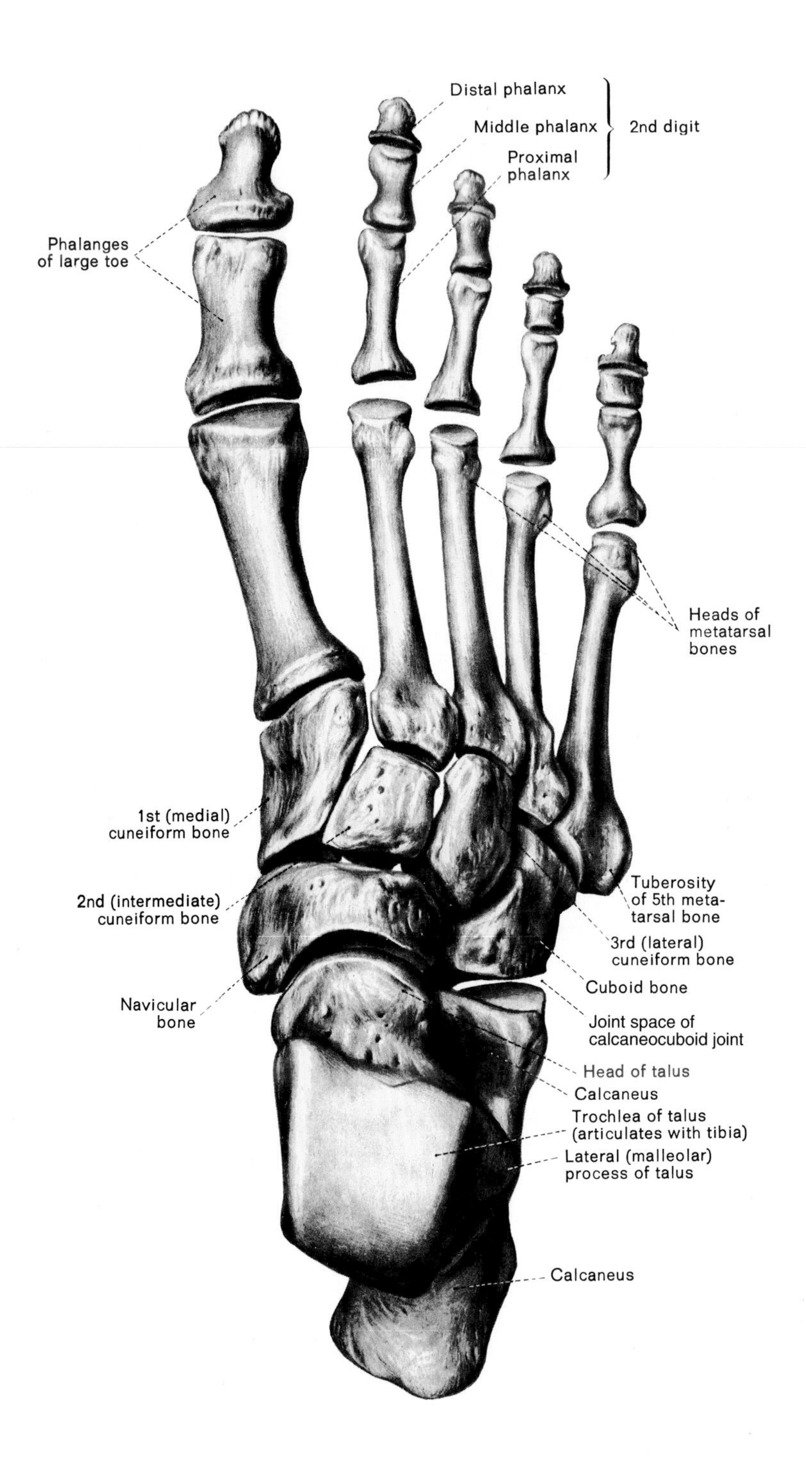

Fig. 14–1: Dorsal view of the bones of the right foot.

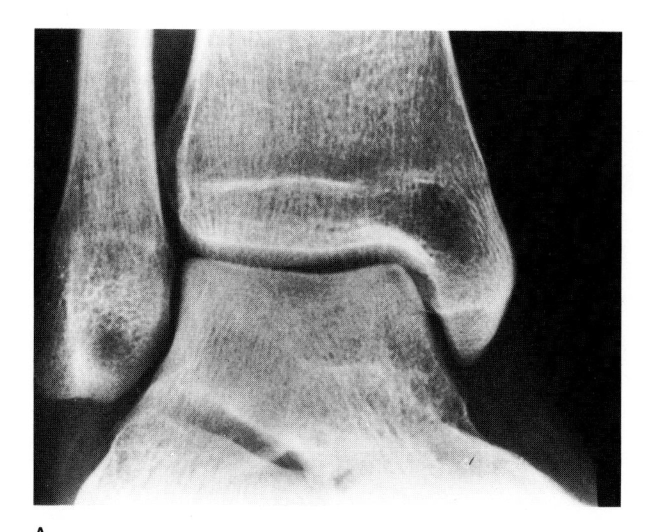
A

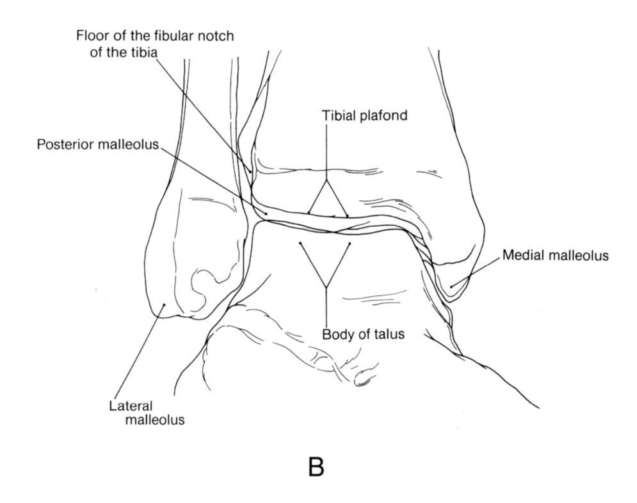

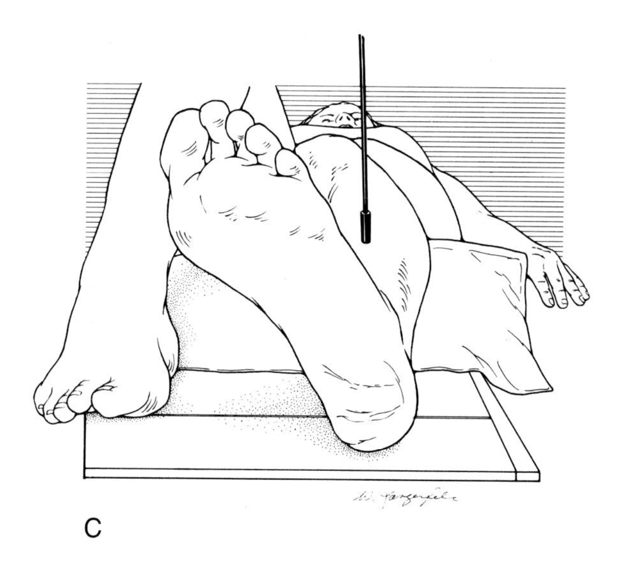

Fig. 14–2: (A) A mortise radiograph of the ankle, (B) its schematic representation, and (C) the orientation of a patient's foot relative to the X-ray beam and film cassette for the radiograph. Note that the foot is internally rotated to render the line between the medial and lateral malleoli perpendicular to the path of the X-ray beam.

proximal phalanx and distal phalanx of the big toe. The phalanges of the other toes are called the proximal phalanx, middle phalanx, and distal phalanx of the toe.

Each phalanx has a **base,** a **shaft,** and a **head.** The base is the proximal end of the phalanx, and the head is the distal end of the phalanx. During childhood and adolescence, each phalanx bears only a single epiphysis, residing at the base of the bone.

THE MAJOR JOINTS OF THE FOOT ASSOCIATED WITH ACTIONS OF THE MUSCLES OF THE LEG

The Ankle Joint

The ankle joint is a synovial joint in which the distal ends of the tibia and fibula articulate with the talus (Fig. 14–2). The joint has a tenon-and-mortise configuration [the terms tenon and mortise are commonly used to describe woodwork joints in which a projection (the tenon) of one piece of wood fits into the groove or slot (the mortise) of another piece]. In the ankle joint, the body of the talus forms the tenon, and the horizontal surface at the distal end of the tibia, in combination with the inferiorly projecting medial and lateral malleoli, form a groove-like mortise. The horizontal articular surface at the distal end of the tibia is called the **tibial plafond** because it represents the plafond (ceiling) of the ankle joint.

The curved articular surface that the body of the talus provides in the ankle joint is called the **trochlea,** and it is shaped similar to a cone truncated (cut off) at its apex (Fig. 14–3). The trochlea is oriented in the ankle joint such that its conical surface (the convex surface between the basal and truncated apical ends) faces the tibial plafond above. The truncated apical end surface is slightly concave and faces the medial malleolus. The basal end surface is slightly convex and faces the lateral malleolus. The size and radius of curvature of the basal end surface are greater than those of the apical end surface.

Movement in the ankle joint consists basically of the cone-shaped tenon (that is formed by the trochlea of the talus) rotating backward or forward within the groove-like mortise (that is formed by the medial malleolus, tibial plafond, and lateral malleolus). The axis of rotation passes laterally through the tip of the lateral malleolus and medially through a point near the tip of the medial malleolus (Fig. 14–4). The backward and forward rotating motions of the trochlea respectively provide for **dorsiflexion and plantarflexion of the foot** (Fig. 14–5). The forefoot is drawn upward during dorsiflexion and pulled downward during plantarflexion.

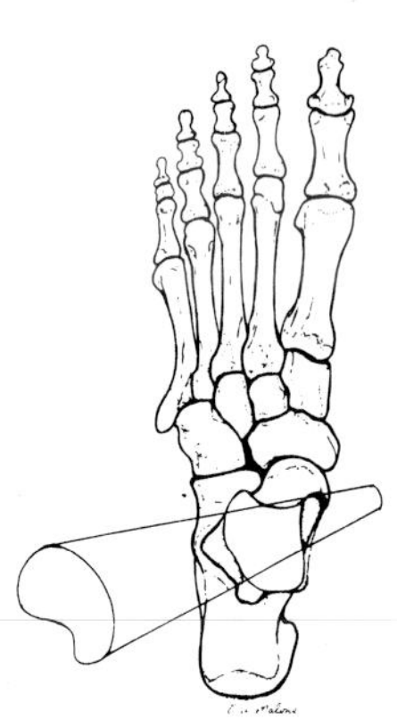

Fig. 14–3: Diagram illustrating the similarity in shape between the trochlea of the talus and a truncated cone.

CLINICAL PANEL III-9

Sprain of the Anterior Talofibular ligament

Definition: First- and second-degree sprains of ligaments are partial tears of ligaments. The limited partial tear of a first-degree sprain does not affect the stability of the joint that the ligament protects. The more extensive partial tear of a second-degree sprain diminishes the ligament's contribution to the stability of its joint. Third-degree sprains of ligaments are complete tears of ligaments. A third-degree sprain completely abolishes a ligament's contribution to the stability of its joint. Sprain of the anterior talofibular ligament is one of the most common ligamentous injuries of the ankle region.

Common Mechanism of Injury: A common (but not universal) mechanism of injury is the sudden application of a marked inversion force on a plantarflexed foot. [The following section discusses inversion and eversion of the foot.]

Common Symptoms: The anterolateral aspect of the ankle region becomes painful and swollen within a few hours following the injury.

Common Sign: Since the anterior talofibular ligament is most commonly torn near its fibular attachment, point tenderness is generally greatest anteroinferior to the lateral malleolus.

When the foot is dorsiflexed or plantarflexed in the ankle joint, however, the trochlea of the talus and the malleoli also rotate relative to each other around a vertical axis. If the foot is non-weight-bearing, and thus the talus is freer to rotate vertically in the ankle joint than the malleoli of the tibia and fibula, dorsiflexion of the foot is accompanied by a slight external (lateral) rotation of the talus relative to the malleoli. In other words, the foot rotates slightly externally to the leg; this movement is called **abduction of the foot.** If the foot is weight-bearing, and thus the malleoli are freer to rotate vertically in the ankle joint than the talus, dorsiflexion of the foot is accompanied by a slight internal (medial) rotation of the malleoli relative to the talus. In other words, the leg rotates slightly internally relative to the foot.

Opposite rotary movements between the leg and foot occur when the foot is plantarflexed in the ankle joint. If the foot is non-weight-bearing, plantarflexion of the foot is accompanied by a slight internal rotation of the foot relative to the leg; this movement is called **adduction of the foot.** If the foot is weight-bearing, plantarflexion of the foot in the ankle joint is accompanied by a slight external rotation of the leg relative to the foot.

The ankle joint is strengthened on its medial side by a taut ligament named for its triangular (deltoid) shape: the **deltoid ligament** (Fig. 14–6). The deltoid ligament spreads out like a fan as it extends from the medial malleolus to the navicular bone, the plantar calcaneonavicular ligament, the sustentaculum tali of the calcaneus, and the talus.

Three ligaments strengthen the lateral side of the ankle joint: the **anterior talofibular, calcaneofibular,** and **posterior talofibular ligaments** (Fig. 14–7). Each of these ligaments is named for its proximal and distal attachments. The three ligaments may be collec-

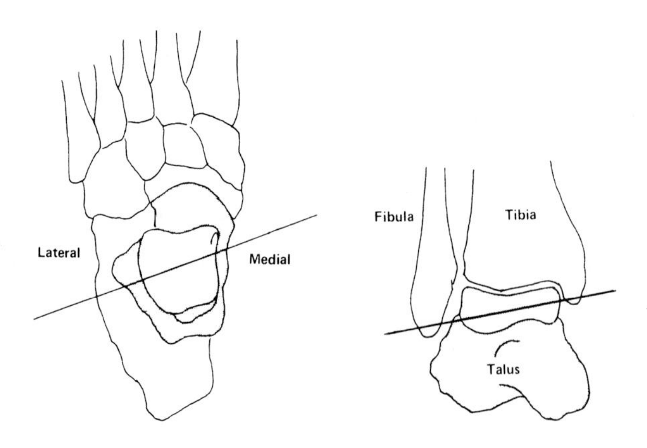

Fig. 14–4: (A) Superior and (B) anterior views of the axis of rotation of the ankle joint.

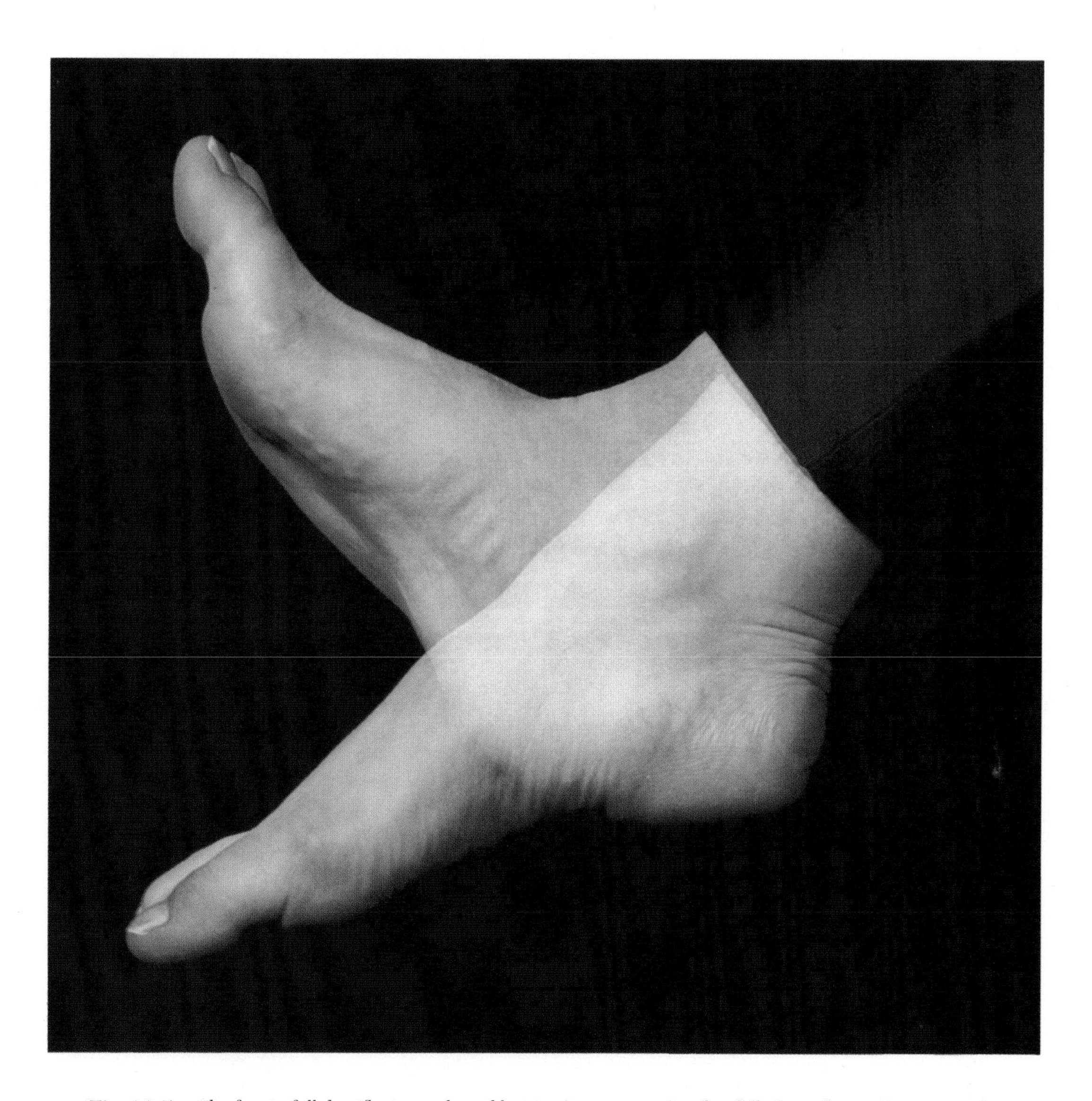

Fig. 14–5: The foot in full dorsiflexion at the ankle joint (upper image) and in full plantarflexion (lower image).

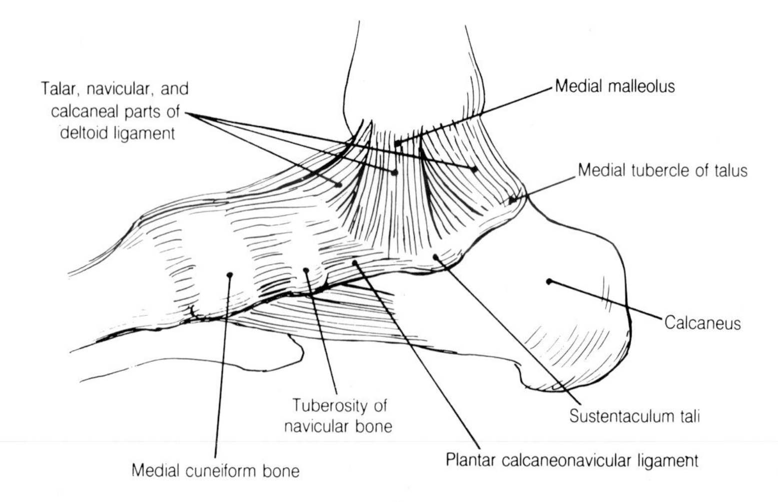

Fig. 14–6: The deltoid ligament of the ankle joint.

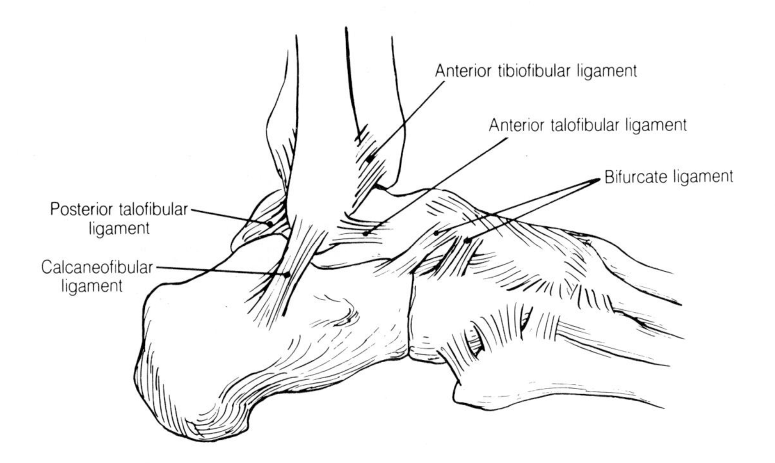

Fig. 14–7: The lateral collateral ligament of the ankle joint.

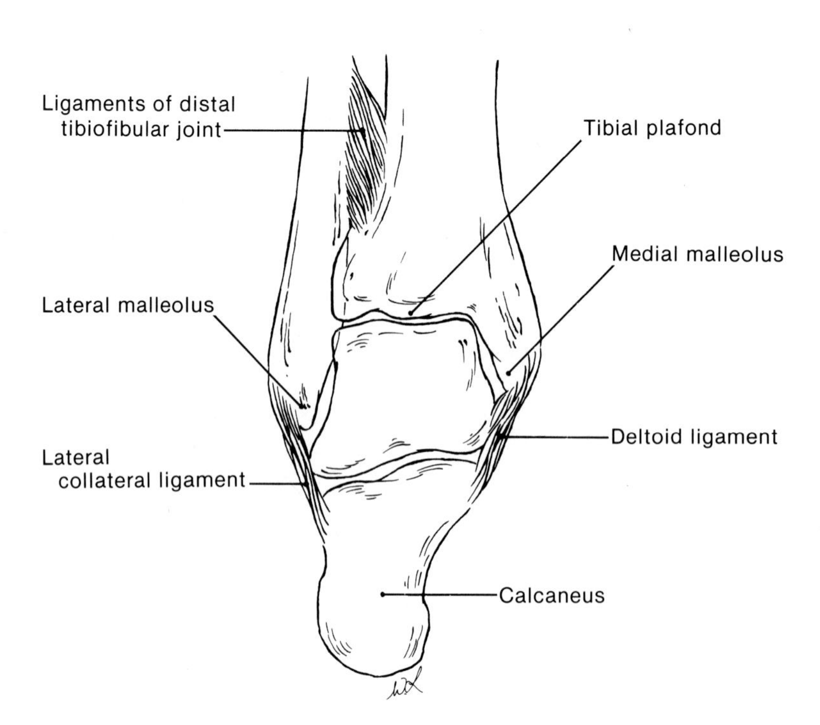

Fig. 14–8: A diagram of the ring of bones and ligaments that encircles the talus in the ankle region.

tively referred to as the **lateral collateral ligament of the ankle joint.**

When considering the mechanisms that commonly produce injuries about the ankle joint, it is helpful to appreciate that the ankle region should be regarded structurally as a united ring of bones and ligaments that encircles the talus (Fig. 14–8). This osseoligamentous ring consists of

a. the fibrous capsule and associated ligaments of the distal tibiofibular joint,
b. the tibial plafond and medial malleolus,
c. the deltoid ligament,
d. the calcaneus,
e. the lateral collateral ligament of the ankle joint,
f. the lateral malleolus.

Breaks within this osseoligamentous ring generally occur in pairs. In most instances, a pair of breaks consists of a fracture of either the medial or lateral malleolus plus a ruptured ligament. Forces that produce a break in the medial aspect of the osseoligamentous ring around the talus more commonly produce avulsion of the medial malleolus rather than rupture of the deltoid ligament, because the deltoid ligament can bear greater tensile forces than the medial malleolus. Forces that produce a break in the lateral aspect of the osseoligamentous ring more commonly produce a rupture of one of the component ligaments of the lateral collateral ligament of the ankle joint than a fracture of the lateral malleolus.

The Joints of Foot Supination and Pronation

Pronation and supination of the foot are complex foot movements. Each movement may be described as the summation of three component conjunctional movements (the term conjunctional signifies that the movements cannot occur independently). Supination of the non-weight-bearing foot is a movement in which approximately equal parts of foot adduction and foot inversion are accompanied by a slight degree of plantarflexion of the foot (Fig. 14–9). **Inversion of the foot** is the movement by which, starting from the anatomic postion, the medial edge of the foot is elevated and the lateral edge lowered so that the sole faces slightly medially. Pronation of the non-weight-bearing foot is a movement in which approximately equal parts of foot abduction and foot eversion are accompanied by a slight degree of dorsiflexion of the foot (Fig. 14–9). **Eversion of the foot** is the movement by which, starting from the anatomic position, the lateral edge of the foot is elevated and the medial edge lowered so that the sole faces slightly laterally.

Supination and pronation of the foot occur primarily at the group of articulations that join the four most proximal tarsals: the talus, calcaneus, navicular, and cuboid. These articulations reside within the confines of three separate fibrous capsules, and thus three anatomically separate joints: the anatomic subtalar, talocalcaneonavicular, and calcaneocuboid joints.

The talus and calcaneus articulate via anterior and posterior articulations (Fig. 14–10). The synovial cavity of the posterior talocalcaneal articulation and the fibrous capsule about it represent the **anatomic subtalar joint.** The anterior talocalcaneal articulation (Fig. 14–10) is anatomically part of the **talocalcaneonavicular (TCN) joint** (Fig. 14–11). The TCN joint has a ball-and-socket configuration in which the ball-shaped head of the talus (Fig. 14–10) and the facets on the undersurface of the talus for the anterior talocalcaneal articulation articulate with a socket (Fig. 14–11) formed by the posterior surface of the navicular, the superior surface of the plantar calcaneonavicular ligament, and the facets on the upper surface of the talus for the anterior talocalcaneal articulation. The **calcaneocuboid joint** articulates the anterior surface of the calcaneus with the posterior surface of the cuboid (Fig. 14–1).

The articulations among the anatomic subtalar, TCN, and calcaneocuboid joints may be alternatively grouped between two functional joints: the functional subtalar and transverse tarsal joints. This is because supination and pronation of the foot may be also described as the summation of principally two interdependent motions: (1) a screwlike motion in which the talus and calcaneus twist about each other and (2) a motion in which the talus and calcaneus glide and rotate relative to the navicular and cuboid. The screwlike motion between the talus and calcaneus involves all of the subtalar articulations between the two bones; these subtalar articulations, which consist of the anterior and posterior talocalcaneal articulations (Fig. 14–10), comprise the **functional subtalar joint.** The motion in which the talus and calcaneus move relative to the navicular and cuboid involve the calcaneocuboid joint and the talonavicular articulation in the TCN joint; these two articulations together are called the **transverse tarsal joint** because they extend transversely across the foot.

The movements which occur among the tibia, fibula, talus, calcaneus, navicular, and cuboid during supination and pronation of the non-weight-bearing foot are different from those that occur in the weight-bearing foot. In the non-weight-bearing foot, the calcaneus is free to accomplish all the conjunctional movements of supination (adduction, inversion, and plantarflexion) or pronation (abduction, eversion, and dorsiflexion) at the functional subtalar joint. Supination and pronation of the calcaneus at the functional subtalar joint respectively produce, in effect, supination and pronation of the hindfoot. The articulations of the

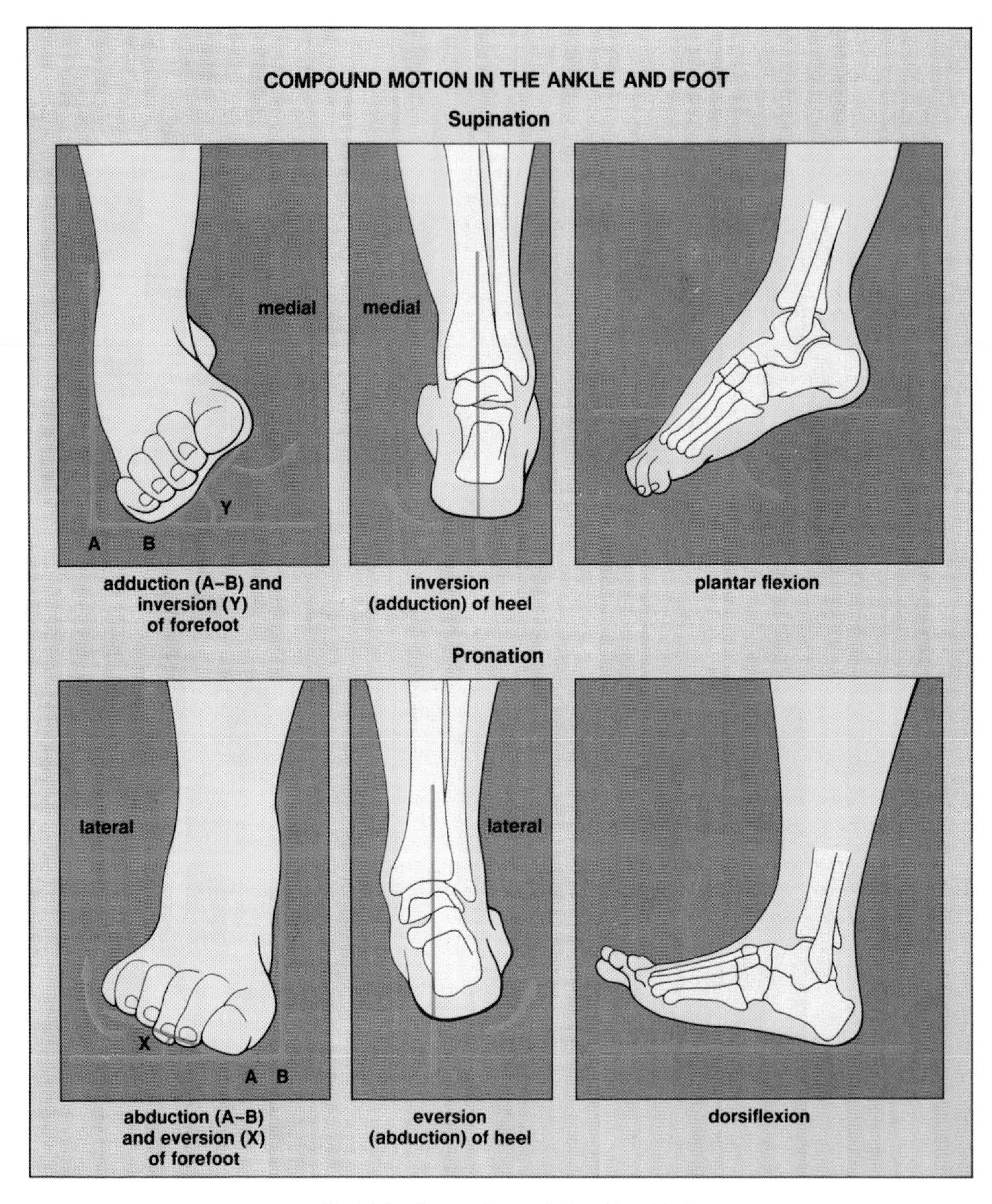

Fig. 14–9: Compound motion in the ankle and foot.

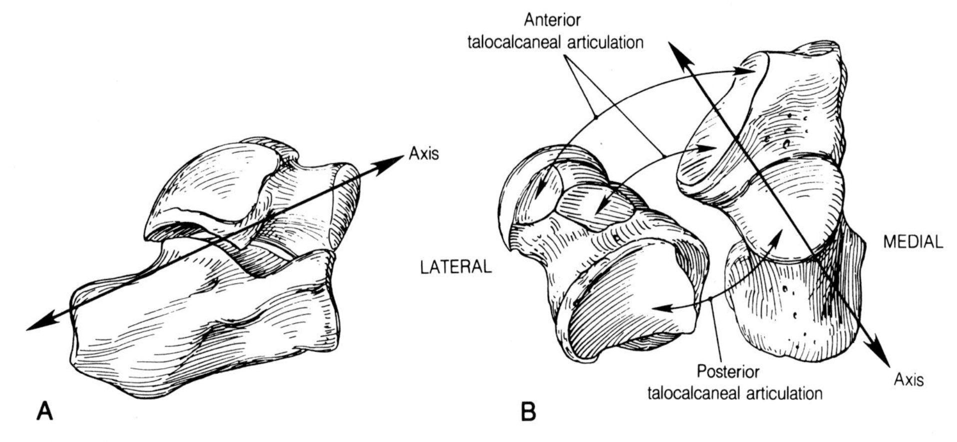

Fig. 14–10: Diagrams illustrating (A) lateral and (B) superior views of the axis and articular surfaces of the functional subtalar joint. In (B), the talus has been turned upside-down to show the articular surfaces for its anterior and posterior articulations with the calcaneus. The synovial cavity of the posterior talocalcaneal articulation and the fibrous capsule about it represent the anatomic subtalar joint.

transverse tarsal joint transmit any supination or pronation twist of the calcaneus to the bones of the midfoot and forefoot. Full supination of the non-weight-bearing foot renders the bones in the functional subtalar and transverse tarsal joints relatively immobile and tightly packed. By contrast, full pronation renders the bones in these joints mobile and relatively loosely packed.

When the foot is weight-bearing, however, the conjunctional movements of hindfoot supination or pronation are distributed in the following fashion: hindfoot inversion and eversion stresses are accomplished principally by movement of the calcaneus at the functional subtalar joint, plantarflexion and dorsiflexion stresses are addressed by movement of the talus at the ankle joint, and foot adduction and abduction stresses are addressed by rotational movements of the tibia and fibula in the leg. A supination stress upon the weight-bearing foot produces hindfoot inversion, dorsiflexion, and external rotation of the leg. Hindfoot inversion occurs because the calcaneus is free to respond to a supination stress by inversion in the functional subtalar joint. The supination stress also attempts either to plantarflex the calcaneus relative to the talus or to dorsiflex the talus relative to calcaneus; the latter action occurs because the talus is freer to move in this regard than the calcaneus. The dorsiflexion stress on the talus in the functional subtalar joint produces dorsiflexion at the ankle joint. Finally, the supination stress also attempts to either adduct the calcaneus or abduct the talus at the functional subtalar joint. Neither movement is as free as internal/external rotation of the tibia and fibula in the leg. The abduction stress on the talus thus produces external rotation of the leg. By analogy, it should be evident that a pronation stress upon the weight-bearing foot produces hindfoot eversion, plantarflexion, and internal rotation of the leg.

The Metatarsophalangeal Joints

The metatarsophalangeal joints each join the head of a metatarsal with the base of its corresponding proximal phalanx. Each metatarsophalangeal joint provides for **flexion, extension, abduction,** and **adduction** of the proximal phalanx of its toe.

The reference axial line for the abduction and adduction of toes extends through the second toe. Abduction of a toe is movement away from this axial line, and adduction is movement toward the axial line.

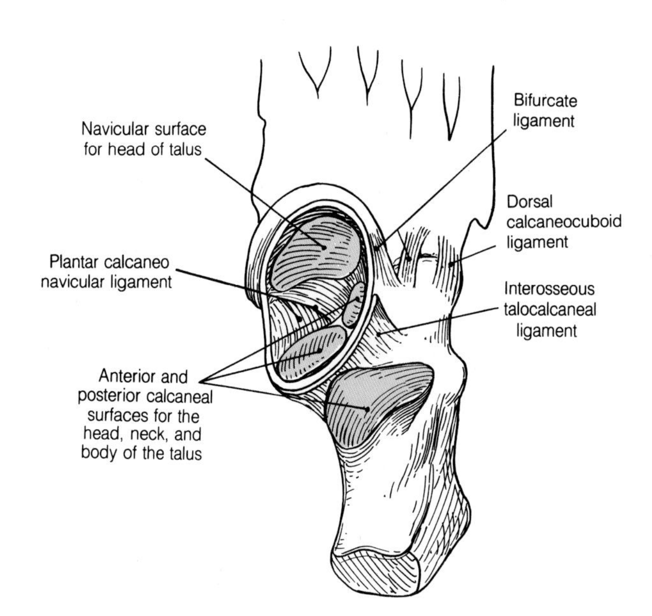

Fig. 14–11: Diagram illustrating the socket shape of the distal articular surfaces of the talocalcaneonavicular (TCN) joint.

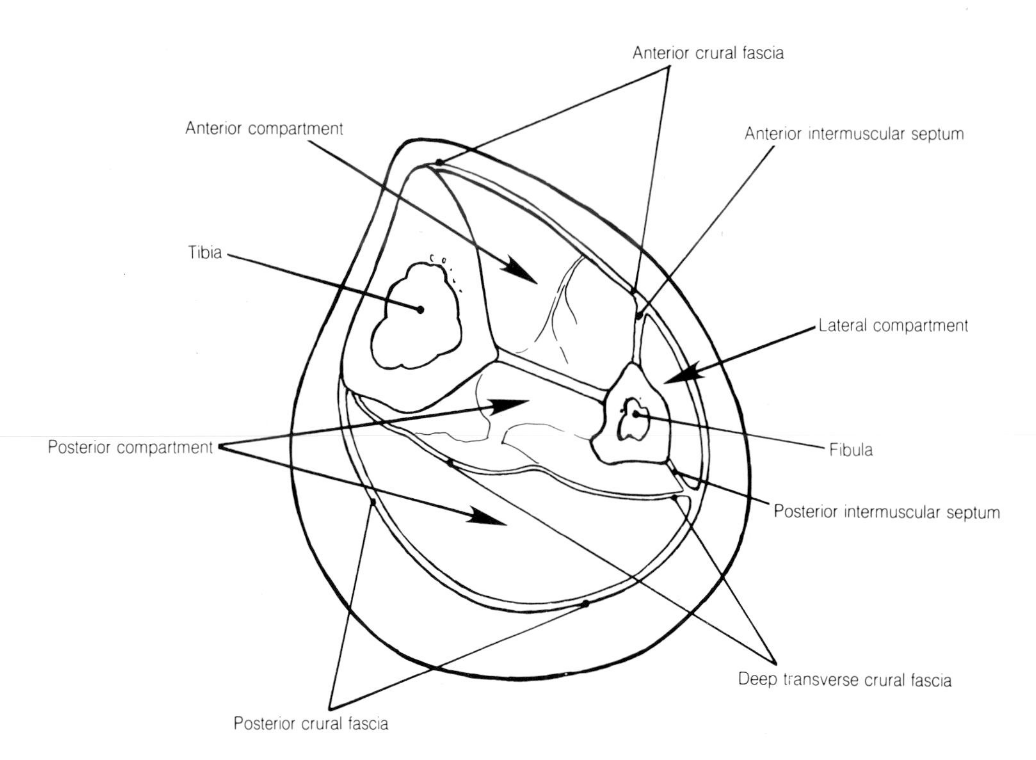

Fig. 14–12: Diagram illustrating a superior view of the four muscle compartments of the right leg.

The Interphalangeal Joints

The interphalangeal joints are the joints between adjacent phalanges in the toes. Whereas the the big toe has only one interphalangeal joint, each of the four lateral toes has a proximal interphalangeal joint and a distal interphalangeal joint. The interphalangeal joints provide for **flexion and extension** of the more distal phalanx.

THE MUSCLES OF THE LEG

The muscles of the leg are enveloped by a tubular sheath of deep fascia called the **crural fascia** (Fig. 14–12). The crural fascia laterally gives off two extensions called the **anterior** and **posterior crural intermuscular septa** that are attached deeply to the shaft of the fibula. The crural fascia posteriorly gives off an extension called the **deep transverse crural intermuscular septum** which extends transversely through the posterior part of the leg. The crural fascia and its intermuscular septa divide the leg into four compartments: anterior, lateral, superficial posterior, and deep posterior.

The Anterior Muscles of the Leg

The tendons of insertion of the anterior leg muscles pass in front of the ankle joint as they extend from the anterior compartment of the leg onto the dorsum of the foot. All the anterior leg muscles can thus dorsiflex the foot. The tendons of insertion are tethered in position in the lower anterior part of the leg and in front of the ankle joint by two thickenings of the crural fascia called the **superior** and **inferior extensor retinacula** (Fig. 14–13). The tendons are either singly or collectively wrapped within synovial sheaths as they extend through or beneath the inferior extensor retinaculum. Table 14–1 lists for quick reference the origin, insertion, nerve supply, and actions of the anterior leg muscles.

Tibialis anterior originates from the upper anterolateral surface of the tibial shaft and inserts onto the medial cuneiform and the base of the 1st metatarsal (Figs. 14–14 and 14–15). It is a major dorsiflexor and invertor of the foot.

Extensor hallucis longus originates from the fibula and inserts onto the base of the distal phalanx of the big toe (Figs. 14–14 and 14–16). It can extend the phalanges of the big toe and contributes to dorsiflexion of the foot.

Extensor digitorum longus originates from the tibia and fibula (Fig. 14–14). The four tendons which arise from its belly insert onto the extensor expansions of the lateral four toes (Fig. 14–16). The anatomic relationships of these extensor expansions to the bones and joints of the lateral four toes are identical to the corresponding relationships of the extensor expansions of the fingers in the hand. The muscle's chief action is extension of the phalanges of the lateral four toes at the metatarsophalangeal and interphalangeal joints.

Peroneus tertius originates from the fibula and inserts onto the base of the 5th metatarsal (Figs. 14–14 and 14–16). It contributes to eversion of the foot.

When walking, the chief dorsiflexors of the foot (tibialis anterior, extensor hallucis longus, and extensor digitorum longus) are most active during a time frame that extends from the pre-swing phase of the stance period to the loading response phase of the subsequent stance period (Fig. 14–17). The chief dorsiflexors contract concentrically during the first half of the swing period to dorsiflex the foot from its plantarflexed position at toe off to an almost neutral position by the mid swing phase; this activity helps clear the toes from the ground immediately following toe off. Their activity during the remainder of the swing period maintains the ankle in an almost neutral position until heel strike.

The chief dorsiflexors act during the loading response phase of the stance period to resist the plantarflexion torque at the ankle joint (which is a consequence of the projection of the upper body weight line posterior to the ankle joint) (Fig. 12–22). Their resistance (a) controls and slows the plantarflexion that brings the sole of the foot into full contact with the surface below and (b) pulls the leg forward at the ankle as the foot is plantarflexed. These two effects help define the heel rocker action that occurs during the loading response phase.

Table 14–1
The Anterior Muscles of the Leg

Muscle	Origin	Insertion	Nerve Supply	Actions
Tibialis anterior	Lateral tibial condyle, upper half of the lateral surface of the tibia, and adjoining part of the anterior surface of the interosseous membrane	Medial cuneiform and base of the 1st metatarsal	Deep peroneal (**L4** and L5)	Dorsiflexes and inverts the foot
Extensor hallucis longus	Middle two-fourths of the medial surface of the fibula and adjoining part of the anterior surface of the interosseous membrane	Base of the distal phalanx of the big toe	Deep peroneal (**L5** and S1)	Extends the big toe and dorsiflexes the foot
Extensor digitorum longus	Lateral tibial condyle, upper three-fourths of the medial surface of the fibula, and upper part of the anterior surface of the interosseous membrane	Extensor expansions of the lateral four toes	Deep peroneal (L5 and S1)	Extends the lateral four toes and dorsiflexes the foot
Peroneus tertius	Lower third of the medial surface of the fibula and adjoining part of the anterior surface of the interosseous membrane	Base of the 5th metatarsal	Deep peroneal (L5 and S1)	Dorsiflexes and everts the foot

The Lateral Muscles of the Leg

Table 14–2 lists for quick reference the origin, insertion, nerve supply, and actions of the lateral leg muscles. **Proneus longus and brevis** respectively originate from the upper two-thirds and lower two-thirds of the lateral surface of the fibula (Fig. 14–18). Each muscle's tendon of insertion passes behind the lateral malleolus (and thus behind the lateral side of the ankle joint) to enter the sole of the foot on its lateral side. Both tendons are tethered in position on the lateral side of the foot by two retinacula called the **superior** and **inferior peroneal retinacula.** The synovial sheaths around the two tendons are continuous with each other deep to the superior peroneal retinaculum.

Whereas peroneus brevis's tendon remains on the lateral side of the foot and inserts onto the base of the fifth metatarsal (Fig. 14–16), peroneus longus's tendon curves around the lateral surface of the cuboid and then extends obliquely through the sole to the medial side of the foot, where it inserts onto the base of the 1st metatarsal (Figs. 14–15 and 14–19).

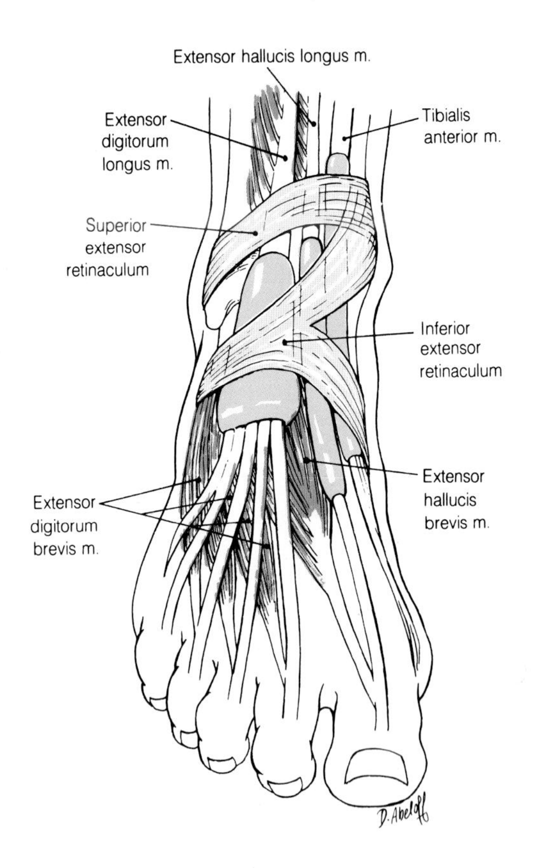

Fig. 14–13: The extensor retinacula of the ankle.

The two muscles are the chief evertors and pronators of the foot. Both muscles can also plantarflex the foot. When walking, peroneus longus and brevis are most active primarily during the limb's stance period when it is the sole weight-bearing limb (Fig. 14–20). These muscles' primary role during this interval is to serve as pronators of the foot.

The Superficial Posterior Muscles of the Leg

Table 14–3 lists for quick reference the origin, insertion, nerve supply, and actions of the superficial posterior leg muscles. **Gastrocnemius** has two heads of origin: the medial head originates from the medial femoral condyle, and the lateral head originates from the lateral femoral condyle (Fig. 14–21). **Soleus** originates from the upper, posterior surfaces of the tibia and fibula. **Plantaris** originates from the lateral supracondylar line of the femur. The tendons of insertion of the three muscles pass behind the ankle joint and fuse with each other to form the tendon called the **Achilles tendon (tendo calcaneus),** which inserts onto the posterior surface of the calcaneus.

The principal action of all three muscles is plantarflexion of the foot. Whereas gastrocnemius and soleus are the major plantarflexors of the foot, plantaris is a relatively weak plantarflexor. Gastrocnemius can also assist in flexion of the leg at the knee joint.

Gastrocnemius and soleus activity is assessed by the **Achilles tendon reflex test.** This test is a simple three-step procedure in which the examiner (1) has the patient sit on an examination table with the legs dangling over the edge, then (2) tenses the Achilles tendon by gently grasping the foot and dorsiflexing it, and finally (3) uses a reflex hammer to impart a gentle but firm, quick tap upon the Achilles tendon. The hammer tap, by suddenly stretching the tendon even further, exacts an abrupt stretching of gactrocnemius and soleus's muscle fibers, and this should elicit a phasic contraction of the muscle fibers that procuce a short, quick plantarflexion of the foot.

When walking, gastrocnemius and soleus are most active during a time frame that extends from the loading response phase to the terminal stance phase of the stance period (Fig. 14–22). Throughout this interval, the two muscles act to resist a dorsiflexion torque at the ankle joint that pulls the leg forward about the ankle joint axis. This dorsiflexion torque is generated near the beginning of the mid stance phase, when the upper body weight line changes from an alignment in which it projects posterior to the ankle to an alignment in which it projects anterior the ankle joint (Fig. 14–23). The dorsiflexion torque progressively increases throughout the mid stance phase and most of the terminal stance phase as the upper body moves progressively forward first over and then beyond the supporting foot.

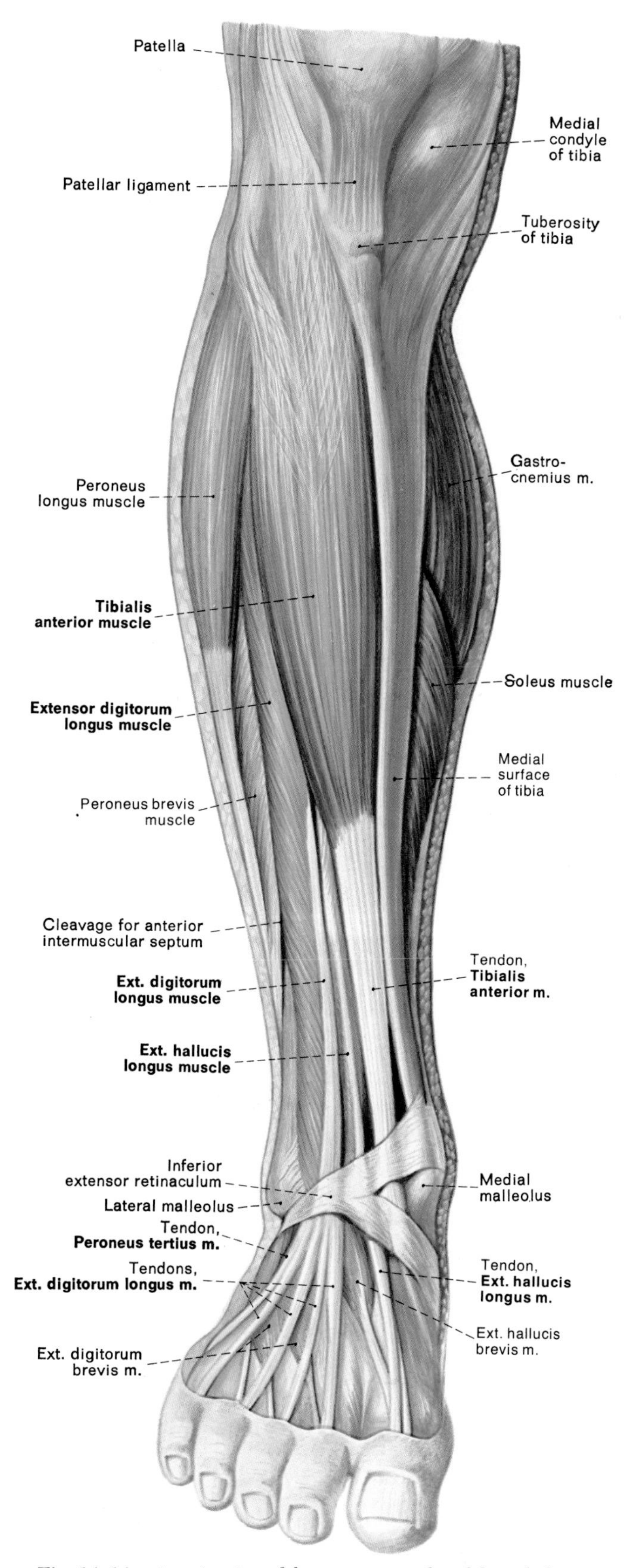

Fig. 14–14: Anterior view of the anterior muscles of the right leg.

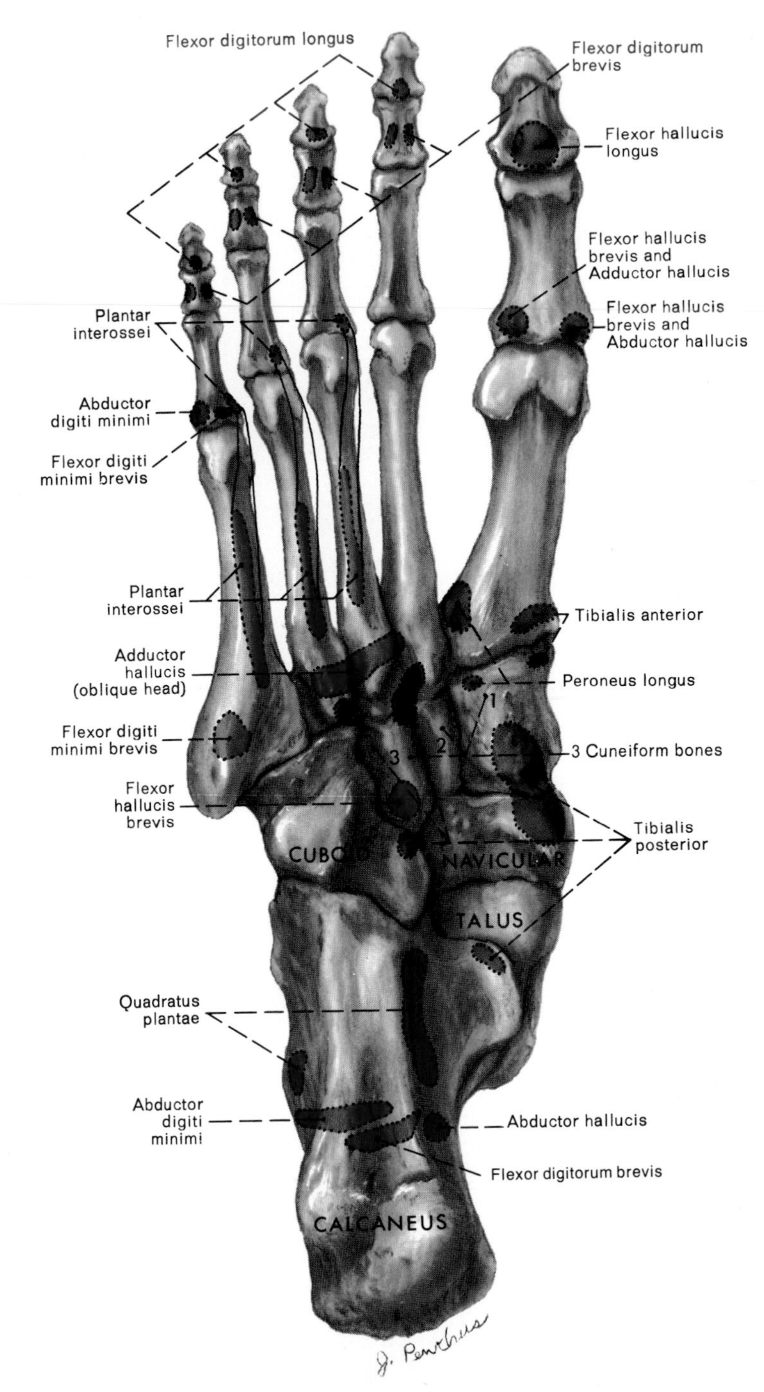

Fig. 14–15: Plantar view of the bones of the right foot and attachment sites of muscles.

The dorsiflexion resistance provided by gastrocnemius and soleus during the mid stance phase controls and slows the forward movement of the leg around the axis of the ankle joint, and thus helps define the ankle rocker action that characterizes the mid stance phase. Near the end of the mid stance phase, the dorsiflexion resistance provided by two muscles begins to significantly restrain further dorsiflexion at the ankle joint. This restraint of dorsiflexion at the ankle joint in combination with the continued forward pull exerted on the leg by the advancing body leads to heel rise, a movement in which the hindfoot and midfoot roll upward and forward around the metatarsophalangeal joints (Fig. 14–24).

Heel rise marks the beginning of the terminal stance phase and initiates the forefoot rocker action in which the lower limb rolls forward on the forefoot. Gastrocnemius and soleus help define the forefoot rocker action during the terminal stance phase through their restraint of dorsiflexion at the ankle joint. The forward movement of the upper body beyond the supporting foot and the roll of the body around the fore-

Table 14–2
The Lateral Muscles of the Leg

Muscle	Origin	Insertion	Nerve Supply	Actions
Peroneus longus	Upper two-thirds of the lateral surface of the fibula	Medial cuneiform and base of the 1st metatarsal	Superficial peroneal (**L5, S1,** and S2)	Everts and plantarflexes the foot
Peroneus brevis	Lower two-thirds of the lateral surface of the fibula	Base of the 5th metatarsal	Superficial peroneal (**L5, S1,** and S2)	Everts and plantarflexes the foot

Table 14–3
The Superficial Posterior Muscles of the Leg

Muscle	Origin	Insertion	Nerve Supply	Actions
Gastrocnemius	Medial head from the politeal surface of the femur and the upper posterior part of the medial tibial condyle; lateral head from the lateral surface of the lateral femoral condyle	Posterior surface of the calcaneus	Tibial (S1 and **S2**)	Plantarflexes the foot and flexes the leg
Soleus	Posterior surface of of the head and upper one-fourth of the shaft of the fibula, soleal line of the tibia, and the oblique popliteal ligament	Posterior surface of the calcaneus	Tibial (S1 and **S2**)	Plantarflexes the foot
Plantaris	Lower end of the lateral supra-condylar line of the femur and the oblique popliteal ligament	Posterior surface of the clacaneus	Tibial (S1 and **S2**)	Weakly plantarflexes the foot and weakly flexes the leg

foot rocker during the terminal stance phase combine to force the body to fall freely forward. This free forward fall is the principal force that propels the body forward during the walking gait.

The Deep Posterior Muscles of the Leg

Table 14–4 lists for quick reference the origin, insertion, nerve supply, and actions of the deep posterior leg muscles.

Popliteus originates from the lateral femoral condyle and the lateral meniscus of the knee joint, extends inferomedially behind the knee joint, and inserts onto the upper posterior surface of the tibial shaft (Fig. 14–25). If the foot of the lower limb is planted on the surface below, concentric contraction of popliteus externally rotates the thigh relative to the leg. If the foot is suspended above the surface below, concentric contraction of popliteus internally rotates the leg relative to the thigh. When walking, popliteus is active throughout the stride except for the initial swing phase of the swing period (Fig. 14–26A).

The tendons of insertion of all the other deep posterior leg muscles pass behind the medial malleolus (and thus behind the medial side of the ankle joint) to enter the sole of the foot on its medial side. The tendons are tethered in position on the medial side of the foot by a single retinaculum called the **flexor retinaculum** (Fig. 14–27). The tendons are each individually wrapped within a synovial sheath as they extend deep to the flexor retinaculum.

Tibialis posterior originates from the posterior surfaces of the tibia and fibula (Fig. 14–25) and inserts principally onto the navicular bone (Fig. 14–15). It is a chief invertor and supinator of the foot, and it can also plantarflex the foot.

When walking, tibialis posterior is consistently most active during two time frames: the first interval extends from the initial contact phase to the early part of the

Table 14–4
The Deep Posterior Muscles of the Leg

Muscle	Origin	Insertion	Nerve Supply	Actions
Popliteus	Lateral surface of the lateral tibial condyle and the lateral meniscus	Upper posterior surface of the tibia	Tibial (**L4, L5,** and S1)	Internally rotates the leg (in the non-weight-bearing limb) and flexes the leg
Tibialis posterior	Midregions of the posterior surfaces of the shafts of the tibia and fibula and the posterior surface of the intervening interosseous membrane	Tuberosity of the navicular, medial cuneiform, cuboid, sustenaculum tali, and the bases of the 2nd, 3rd, and 4th metatarsals	Tibial (L4 and L5)	Inverts and plantarflexes the foot
Flexor hallucis longus	Lower two-thirds of the posterior surface of the shaft of the fibula and the lower part of the posterior surface of the interosseous membrane	Base of the distal phalanx of the big toe	Tibial (**S2** and S3)	Flexes the big toe at its metatarsophalangeal and interphalangeal joints and plantarflexes the foot
Flexor digitorum longus	Posteromedial surface of the tibia below the soleal line	Bases of the distal phalanges of the lateral four toes	Tibial (**S2** and S3)	Flexes the lateral four toes and plantarflexes the foot

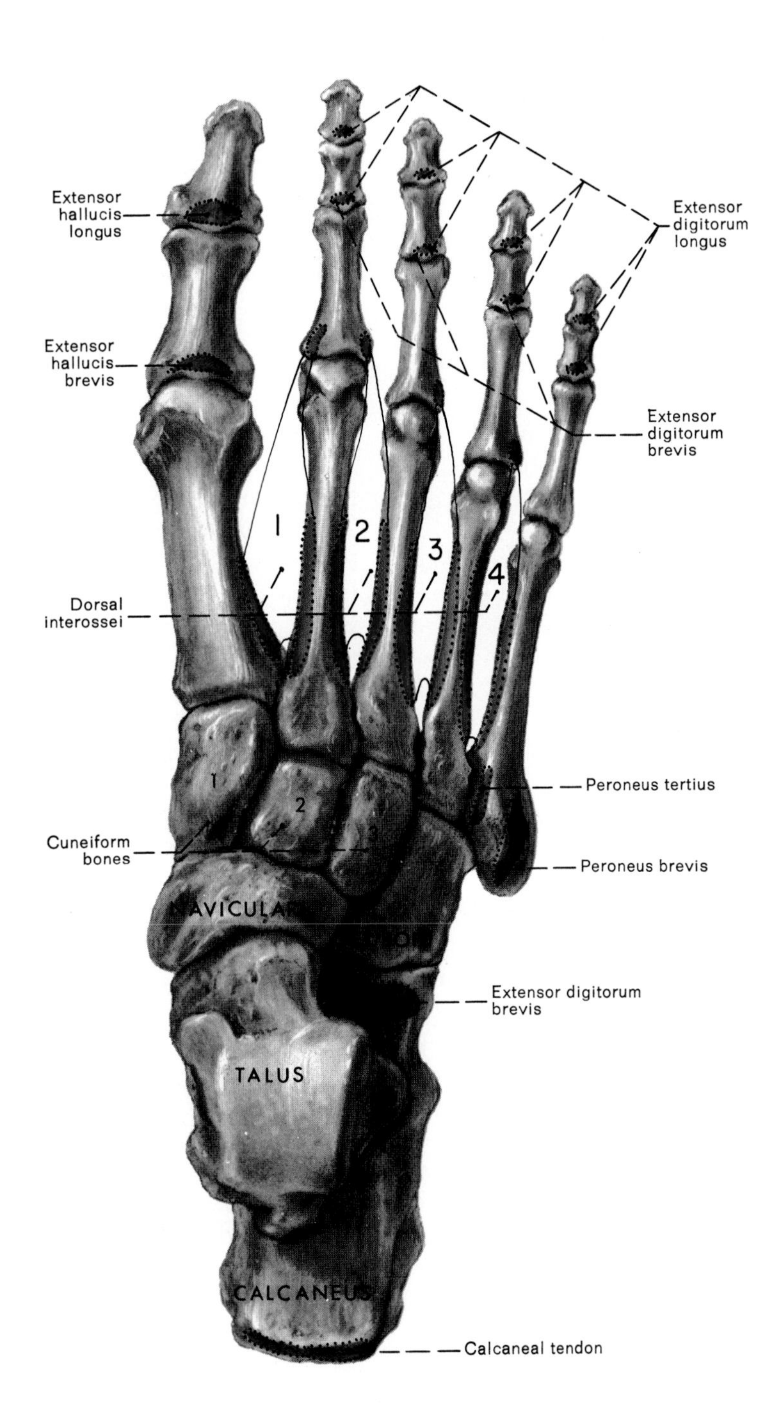

Fig. 14–16: Dorsal view of the bones of the right foot and attachment sites of muscles.

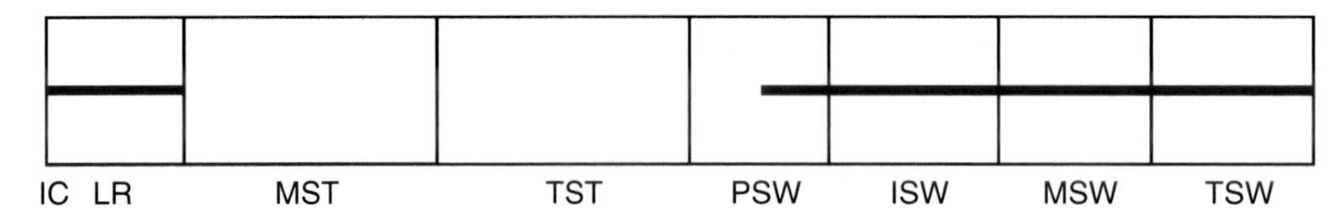

Fig. 14–17: Time frame of consistent significant activity by the chief dorsiflexors of the foot (tibialis anterior, extensor hallucis longus, and extensor digitorum longus) during a stride of the walking gait. Abbreviations: IC = initial contact phase, LR = loading response phase, MST = mid stance phase, TST = terminal stance phase, PSW = pre-swing phase, ISW = initial swing phase, MSW = mid swing phase, and TSW = terminal swing phase.

mid stance phase, and the second interval extends from the late part of the mid stance phase to almost the end of the terminal stance phase (Fig. 14–26B). Its primary role during these intervals is to serve as a supinator of the foot.

Flexor hallucis longus originates from most of the posterior surface of the fibula (Fig. 14–25) and inserts onto the base of the distal phalanx of the big toe (Fig. 14–15). Its chief actions are flexion of the phalanges of the big toe; it can also plantarflex the foot.

When walking, flexor hallucis longus is most active during a time frame that extends from the mid stance phase to the pre-swing phase of the stance period (Fig. 14–26C). It acts during this interval to firmly press the tissue pad at the distal end of the big toe against the surface below. This action augments forefoot rocker action during the terminal stance phase in three ways: First, the action reduces the force per unit area borne by the forefoot rocker by increasing the forefoot surface area in contact with the surface below. Second, the action lengthens the stride by increasing the extent to which the lower limb can roll forward on its forefoot. Third, the action controls and slows the lower limb's forward roll on the forefoot by resisting the dorsiflexion torque at the metatarsophalangeal joints. This dorsiflexion torque is generated during the terminal stance phase when the upper body weight line changes from an alignment in which it projects posterior to the metatarsophalangeal joints to an alignment in which it projects anterior to the joints (Fig. 14–24).

Flexor digitorum longus originates from most of the posterior surface of the tibia (Fig. 14–25). Its four tendons of insertion insert onto the bases of the distal phalanges of the lateral four toes (Fig. 14–15). Its chief actions are flexion of the phalanges of the lateral four toes; it can also plantarflex the foot.

When walking, flexor digitorum longus is most active during a time frame that extends primarily through the mid stance and terminal stance phases of the stance period (Fig. 14–24D). It acts during this interval to firmly press the tissue pads at the distal ends of the lateral four toes against the surface below. This action augments forefoot rocker action during the terminal stance phase by mechanisms similar to those just described for flexor hallucis longus.

THE NERVES OF THE LEG

Almost all of the major nerves of the leg are derived from the sciatic nerve. The sciatic nerve commonly divides into its two terminal branches, the tibial and common peroneal nerves, as it descends through the popliteal fossa.

The Tibial Nerve

As the tibial nerve descends through the popliteal fossa, it lies superficial to the popliteal artery and vein (Fig. 12–26). As the tibial nerve descends through the leg, it lies adjacent to the deep transverse crural intermuscular septum, and thus between the superficial and deep posterior compartments of the leg. The tibial nerve enters the sole of the foot by passing behind the medial malleolus deep to the flexor retinaculum (where it lies between the posterior tibial vessels and

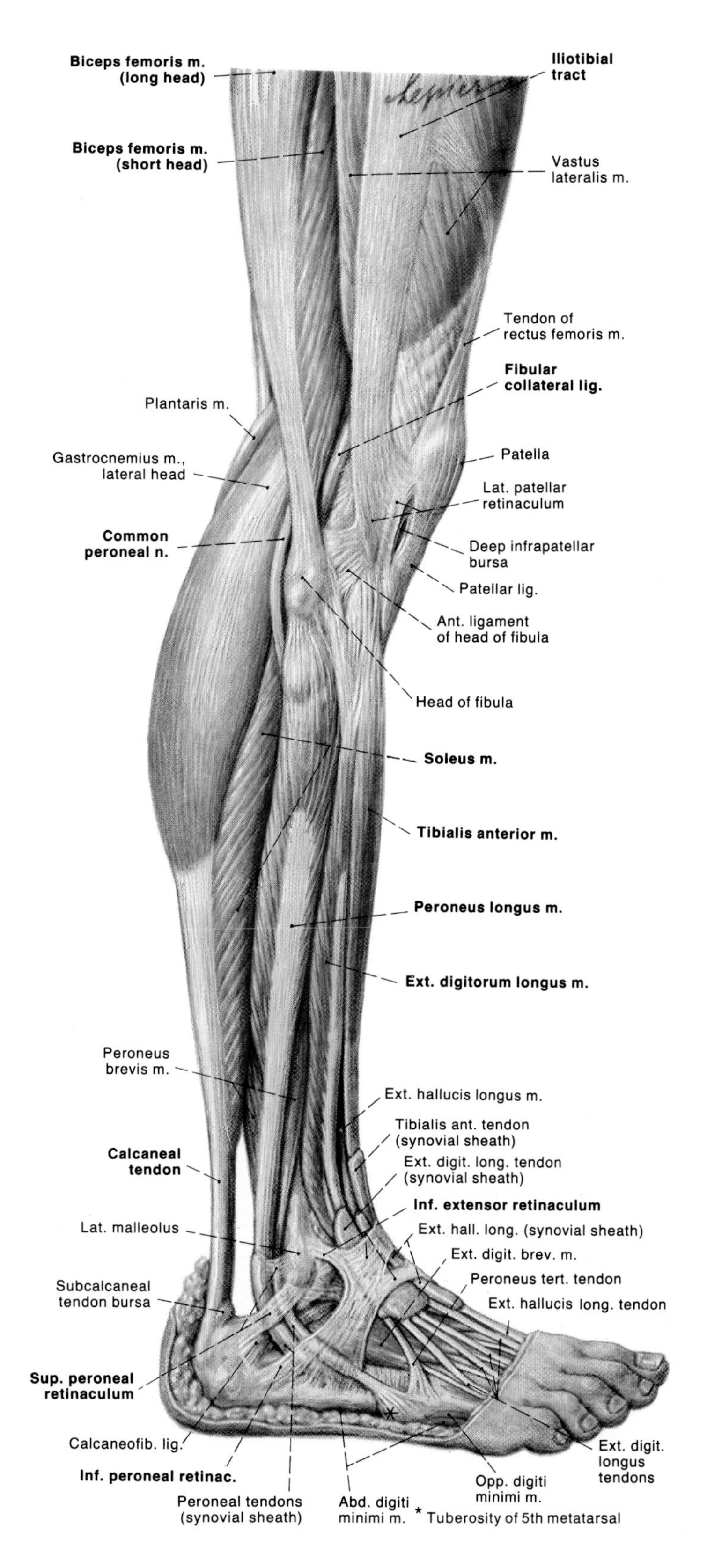

Fig. 14–18: Lateral view of the superficial muscles and tendons of the right lower thigh and leg.

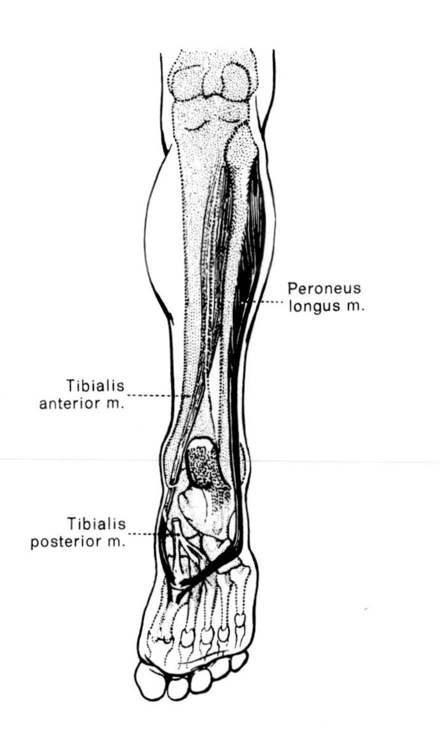

Fig. 14–19: Diagram illustrating the tendons of insertion of tibialis anterior and peroneus longus in the plantar aspect of the foot.

the tendon of flexor hallucis longus) (Fig. 14–28). The tibial nerve divides in the sole of the foot into its two terminal branches: the **medial** and **lateral plantar nerves.**

The tibial nerve innervates all the superficial and deep posterior leg muscles. The medial and lateral plantar nerves innervate all the plantar muscles of the foot.

The Common Peroneal Nerve

As the common peroneal nerve descends through the popliteal fossa, it parallels the medial border of biceps femoris's tendon of insertion (Fig. 12-26). The common peroneal nerve curves around the lateral side of the neck of the fibula as it enters the leg (Fig. 14–29). It ends in this vicinity by dividing into its two terminal branches, the superficial and deep peroneal nerves.

The Superficial Peroneal Nerve

The superficial peroneal nerve begins deep to peroneus longus, pierces peroneus longus and brevis as it descends through the leg, and then descends within the subcutaneous tissues on the lateral side of the distal third of the leg (Fig. 14–29). The superficial peroneal nerve innervates the lateral leg muscles.

The Deep Peroneal Nerve

The deep peroneal nerve begins deep to peroneus longus, and descends through much of its course in the leg anterior to the interosseous membrane (Fig. 14–29). Upon passing deep to the superior extensor retinaculum, it exits the leg by passing in front of the

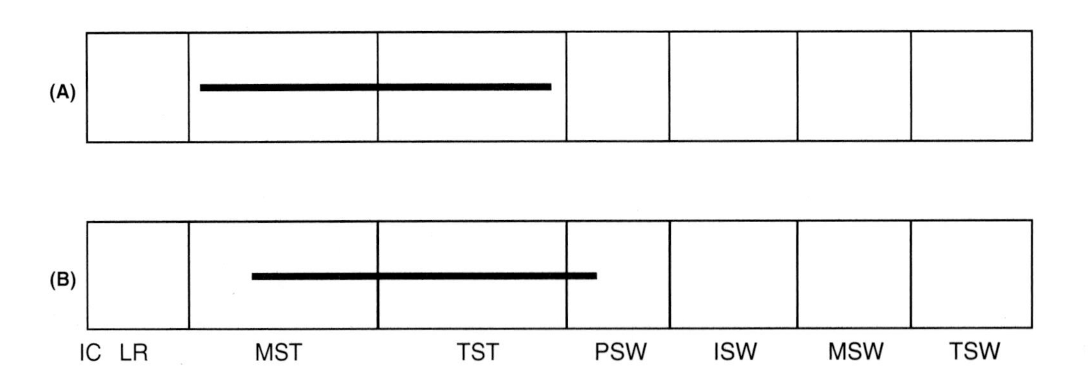

Fig. 14–20: Time frames of consistent significant activity by (A) peroneus longus and (B) peroneus brevis during a stride of the walking gait. Abbreviations: IC = initial contact phase, LR = loading response phase, MST = mid stance phase, TST = terminal stance phase, PSW = preswing phase, ISW = initial swing phase, MSW = mid swing phase, and TSW = terminal swing phase.

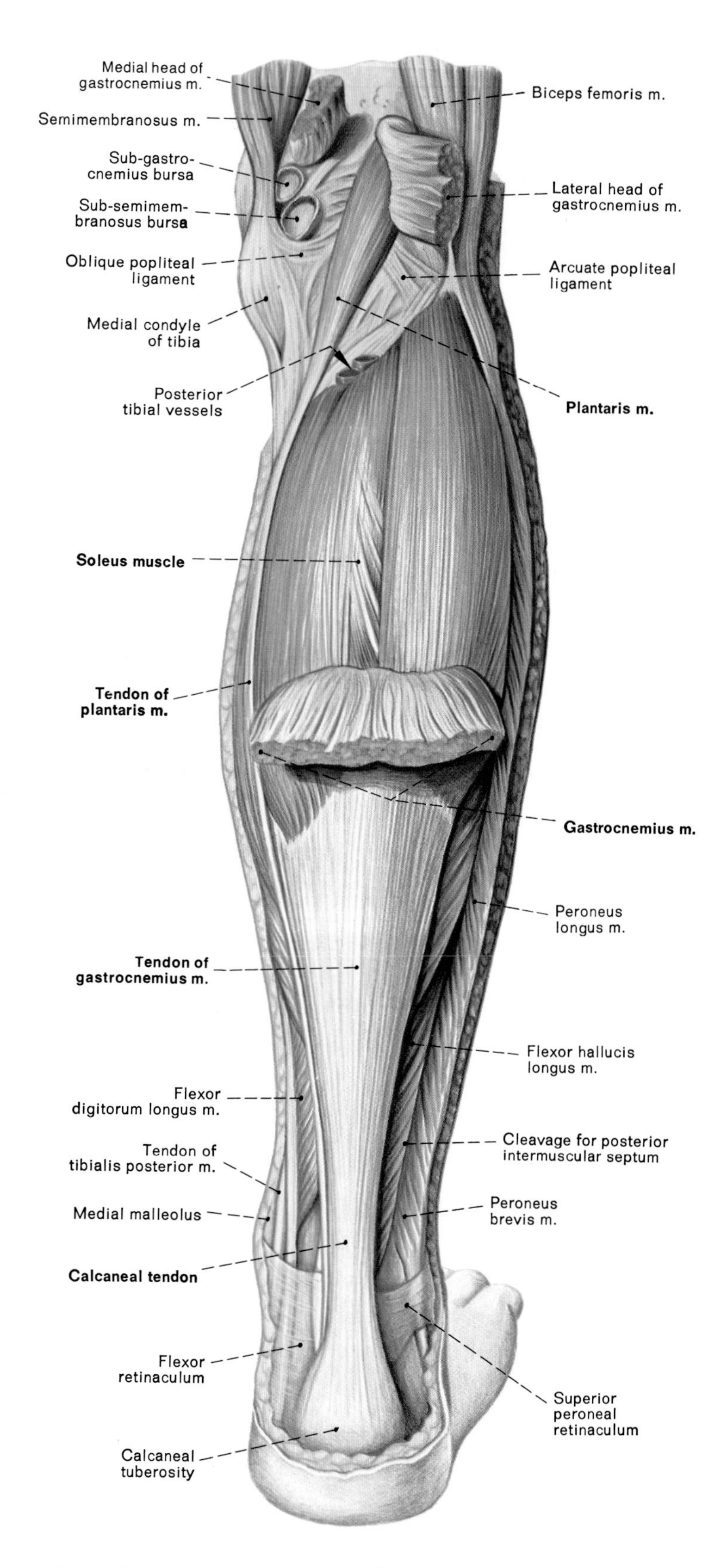

Fig. 14–21: Posterior view of the superficial posterior muscles of the right leg.

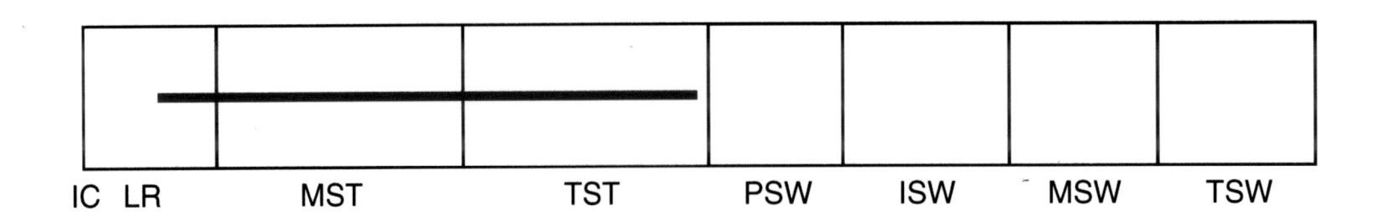

Fig. 14–22: Time frame of consistent significant activity by gastrocnemius and soleus during a stride of the walking gait. Abbreviations: IC = initial contact phase, LR = loading response phase, MST = mid stance phase, TST = terminal stance phase, PSW = pre-swing phase, ISW = initial swing phase, MSW = mid swing phase, and TSW = terminal swing phase.

ankle joint (where it lies between the anterior tibial vessels and the tendon of extensor hallucis longus). The deep peroneal nerve innervates the anterior leg muscles and extensor digitorum brevis.

THE MAJOR ARTERIES OF THE LEG

In the lower part of the popliteal fossa, the popliteal artery divides into its two terminal branches: the anterior and posterior tibial arteries (Fig. 13–21).

The Anterior Tibial Artery

The anterior tibial artery lies through most of its descent in the leg anterior to the interosseous membrane. Upon passing deep to the inferior extensor retinaculum, it ends in front of the ankle joint (and midway between the malleoli) as the origin as the **dorsalis pedis** (the "artery of the dorsum of the foot") (Fig. 14–29). The pulsations of the dorsalis pedis can be palpated on the dorsum of the foot immediately lateral to the tendon of extensor hallucis longus.

The Posterior Tibial Artery

The posterior tibial artery descends through the leg and then behind the medial malleolus in company with the tibal nerve (Fig. 14–28). The pulsations of the posterior tibial artery can be palpated posteroinferiorly to the medial malleolus (the pulsations are most easily palpated if the foot is both dorsiflexed and inverted). The posterior tibial artery ends deep to the flexor retinaculum upon division into its two terminal branches: the **medial** and **lateral plantar arteries.**

The **peroneal artery** is the largest branch of the posterior tibial artery (Fig. 13–21B). The peroneal artery begins in the upper posterior part of the leg. It supplies muscular branches to posterior and lateral leg muscles and a nutrient artery to the fibula.

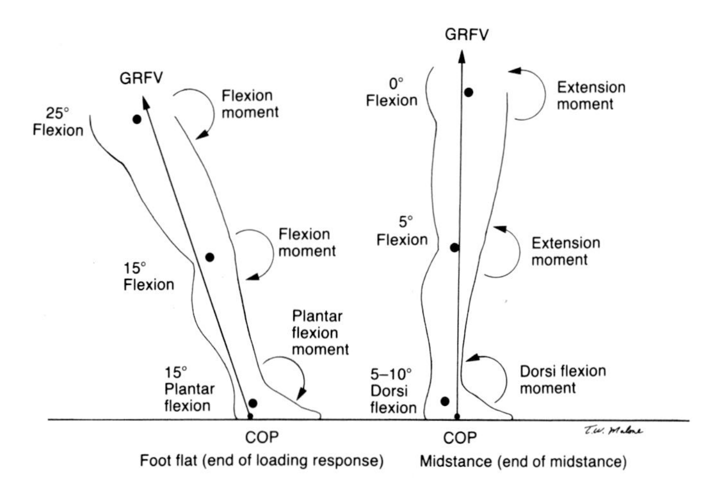

Fig. 14–23: The orientation of the lower limb to the upper body weight line (A) at foot flat and (B) during the mid stance phase when the GRFV is vertical. The GRFV vector represents the Ground Reaction Force Vector; the ground reaction force is the force that the ground exerts upon the lower limb in reaction to the imposition of upper body weight through the lower limb onto the ground. The GRFV vector is coincident with the upper body weight line.

Gait

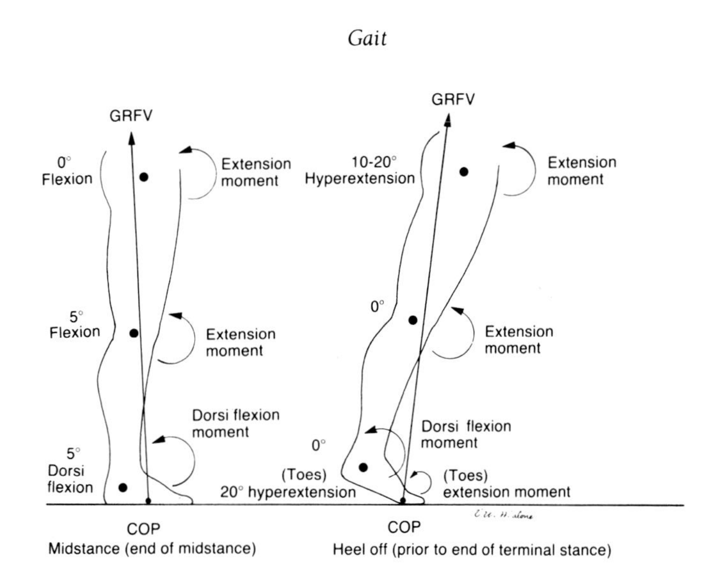

Fig. 14–24: The orientation of the lower limb to the upper body weight line (A) during the midstance phase when the GRFV is vertical and (B) during the terminal stance phase shortly after heel rise. The GRFV vector represents the Ground Reaction Force Vector; the ground reaction force is the force that the ground exerts upon the lower limb in reaction to the imposition of upper body weight through the lower limb onto the ground. The GRFV vector is coincident with the upper body weight line.

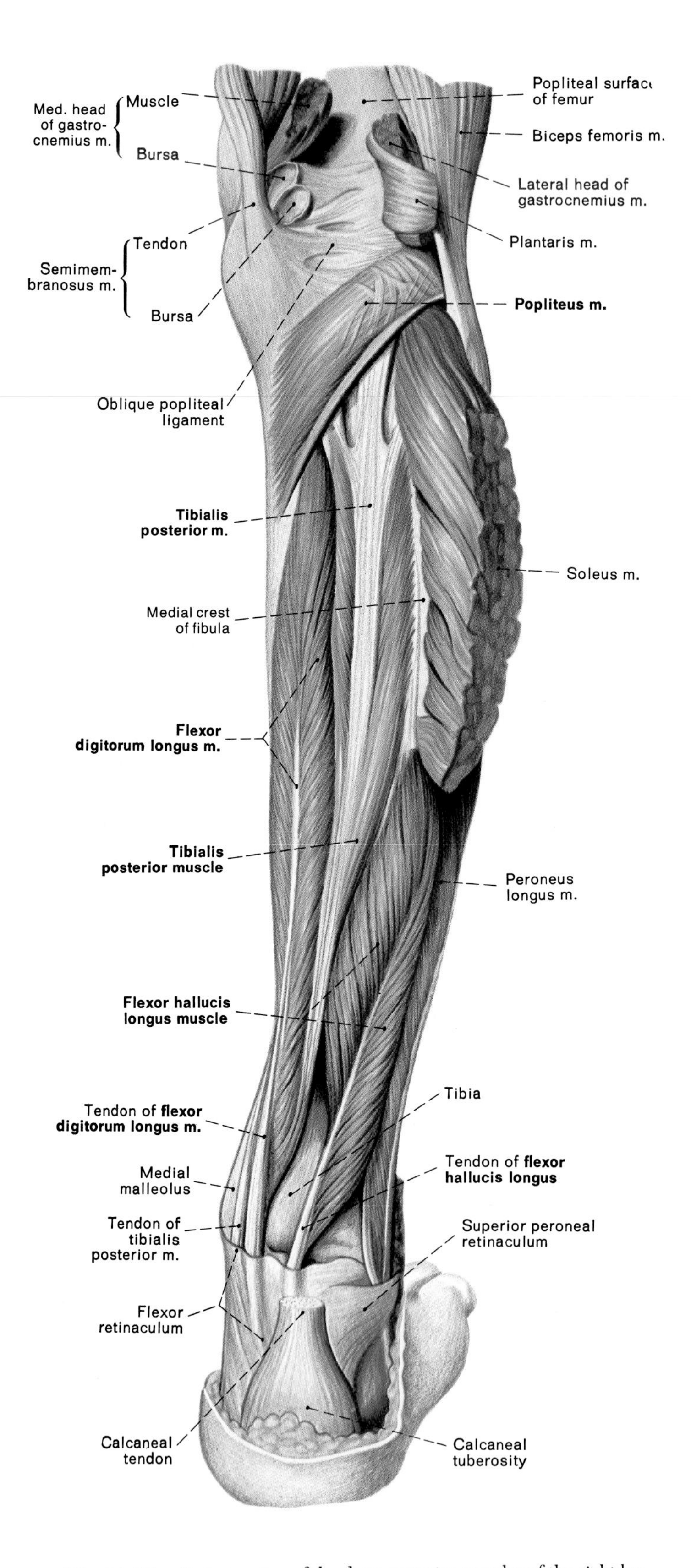

Fig. 14–25: Posterior view of the deep posterior muscles of the right leg.

CLINICAL PANEL III-10

Acute Anterior Tibial Syndrome

Definition: The acute anterior tibial syndrome refers to any condition in which there is an acute increase in pressure in the anterior compartment of the leg. Such a pressure increase can result from muscular edema following strenuous physical exercise or from generalized swelling following a tibial and/or fibular fracture.

Common Symptoms: The patient presents with pain in the anterior compartment of the leg. The skin overlying the anterior leg compartment may become reddened, warm, and edematous.

Anatomic Basis of a Common Sign: An attempt by the patient to dorsiflex the foot against resistance (provided by the examiner) exacerbates the anterior compartment pain. This isometric exercise intensifies the anterior compartment pain because the maneuver increases pressure in the anterior leg compartment through isometric contraction of the compartment's muscles. These muscles include all of the major dorsiflexors of the foot: tibialis anterior, extensor hallucis longus, and extensor digitorum longus.

If increased pressure in the anterior leg compartment is not relieved, the result can be compression neuropathy of the deep peroneal nerve and ischemic necrosis of the compartment's muscles. Loss of the actions of the anterior leg muscles results in an abnormal gait called **foot drop.** In foot drop, the patient is unable either to dorsiflex the foot during the first half of the swing period or to maintain the foot in an almost neutral position during the latter half of the swing period in preparation for heel strike. The consequence of these losses is that the toes of the affected foot are the first part of the foot to strike the ground during the initial contact phase. The patient compensates for the loss of action of the foot's dorsiflexors by raising the foot higher than normal (through increased flexion of the thigh at the hip) during the swing phase. This alteration in the gait permits almost the entire plantar surface of the affected foot to strike the ground at the initial contact phase.

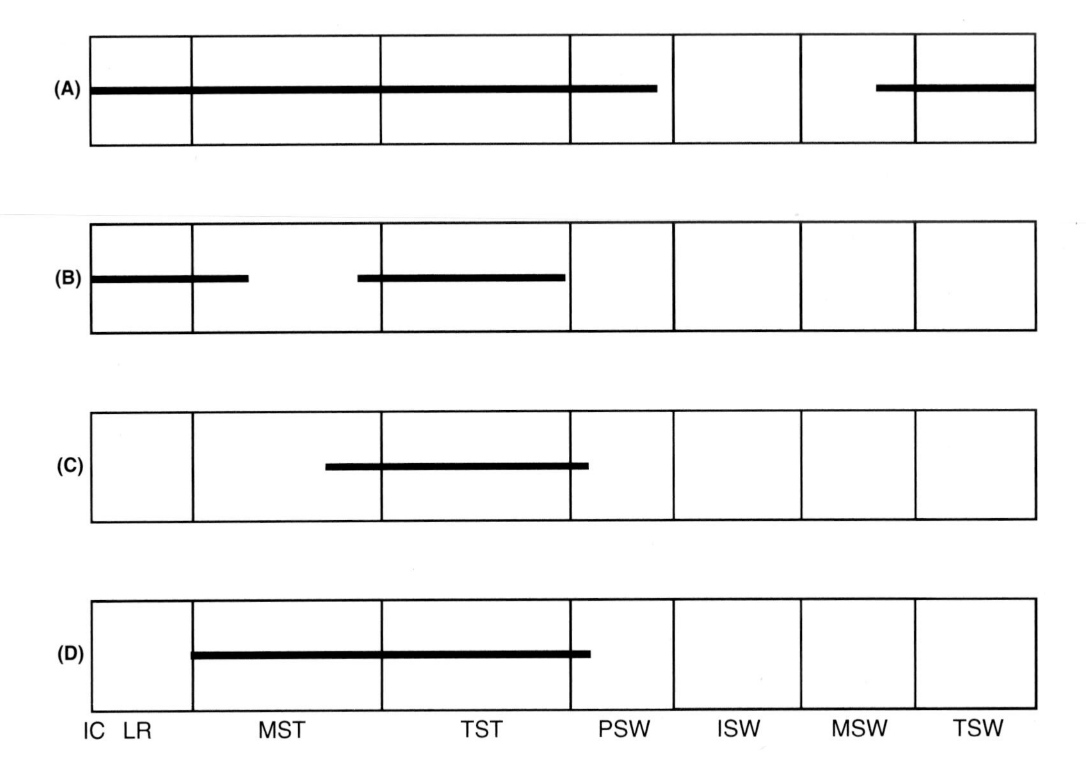

Fig. 14–26: Time frames of consistent significant activity by (A) popliteus, (B) tibialis posterior, (C) flexor hallucis longus, and (D) flexor digitorum longus during a stride of the walking gait. Abbreviations: IC = initial contact phase, LR = loading response phase, MST = mid stance phase, TST = terminal stance phase, PSW = pre-swing phase, ISW = initial swing phase, MSW = mid swing phase, and TSW = terminal swing phase.

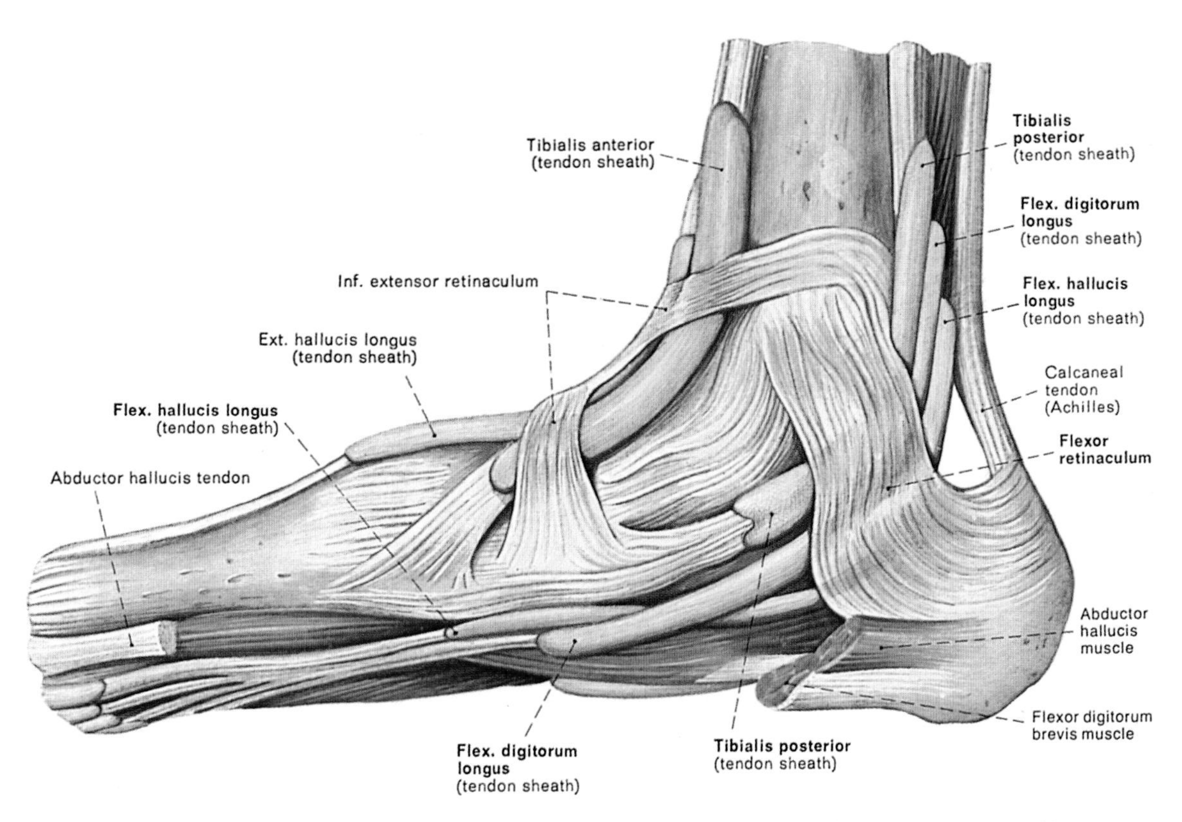

Fig. 14–27: Medial view of the retinacula, tendons, and synovial sheaths in the medial aspect of the right ankle.

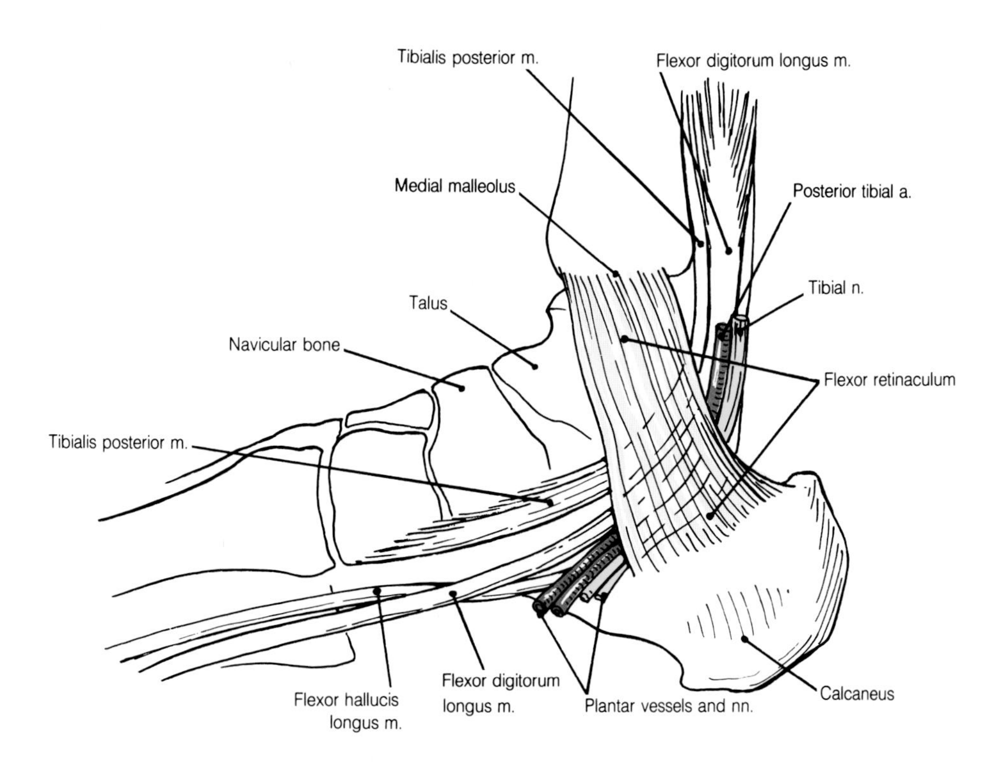

Fig. 14–28: Diagram depicting structures deep to the flexor retinaculum at the right ankle.

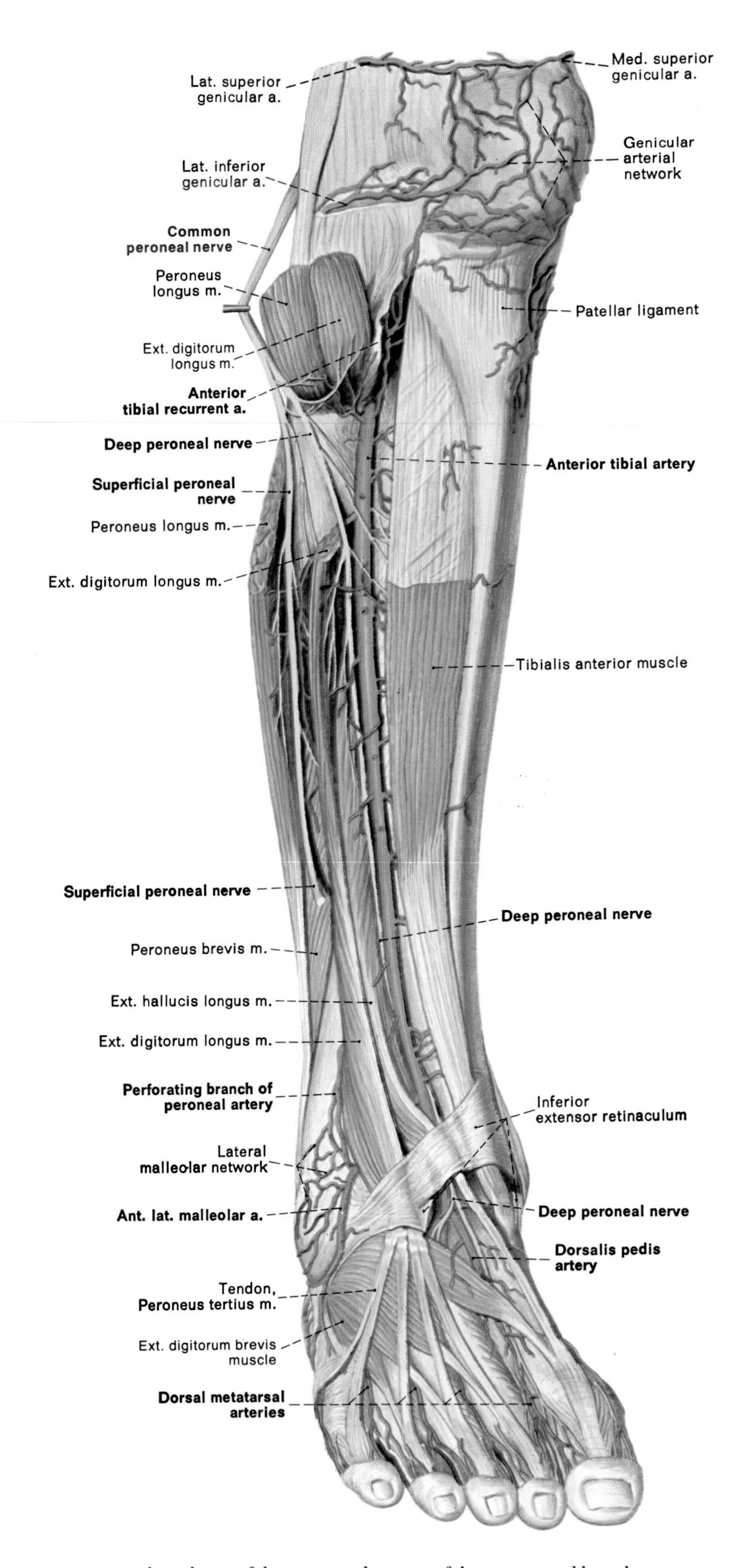

Fig. 14–29: Anterolateral view of the nerves and arteries of the anterior and lateral compartments of the right leg.

CHAPTER **15**

The Foot

This chapter considers the mechanical role of the arches of the foot during the stance period of the walking gait. This discussion begins with a description of the arches of the foot and their structural and dynamic supports.

THE ARCHES OF THE FOOT

The bones of the foot are arched longitudinally on each side and transversely in the middle of the foot. The **medial longitudinal arch** of the foot consists of the calcaneus, talus, navicular, the cuneiforms, and the three most medial metatarsals (Fig. 15–1). The **lateral longitudinal arch** consists of the calcaneus, cuboid, and the two most lateral metatarsals. The difference in height between the medial and lateral longitudinal arches forms a **transverse arch** consisting of the cuboid, the cuneiforms, and the bases of the metatarsals (Fig. 15–1).

The anterior and posterior ends of the longitudinal arches form the skeletal framework of the weight-bearing regions of the feet. The tuberosity of the calcaneus represents the common posterior end of the medial and lateral longitudinal arches, and the heads of the metatarsals form the anterior ends of the arches. The regions of the sole of the foot formed by the tuberosity of the calcaneus and the heads of the metatarsals are respectively called the **heel** and the **ball of the foot.** If a person distributes body weight equally between both lower limbs while standing in sand at a beach, the heels and balls of the feet form the deepest recesses of the impressions cast in the sand by the foot, as these are the weight-bearing regions of the feet (Fig. 15–2).

When a person distributes body weight equally between both lower limbs while standing, the tibia and fibula in each leg transmit half the body weight across the ankle joint to the underlying talus. The talus, in turn, transmits half of this weight posteriorly through the anatomic subtalar joint to the calcaneus and the other half anteriorly through the TCN and calcaneocuboid joints to the heads of the metatarsals. The weight borne by the heads of the metatarsals is distributed in a 2:1:1:1:1 ratio, proceeding from the head of the first metatarsal to the head of the fifth metatarsal.

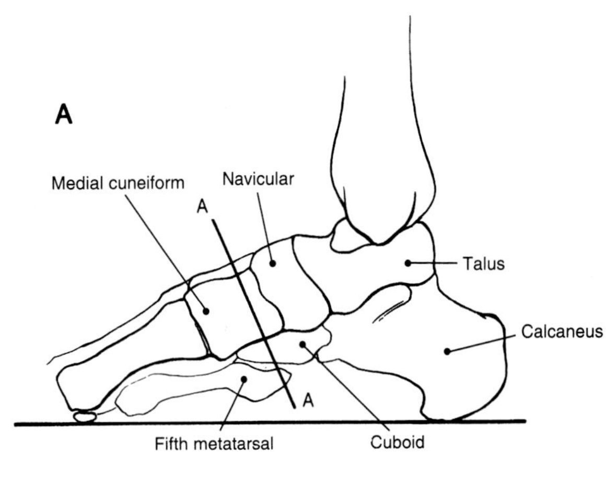

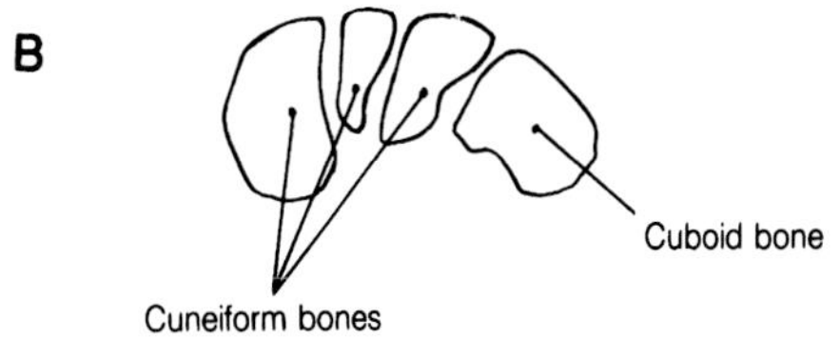

Fig. 15–1: Diagrams illustrating the (A) longitudinal and (B) transverse arches of the foot.

The Structural and Dynamic Supports of the Arches

There are three major sets of factors that generate and stabilize the arches of the foot:

1. The shapes of the bones of the foot and the orientations of their articulations are the foremost structural factors responsible for the arches. The structural component of particular importance to the medial and lateral longitudinal arches is the emplacement

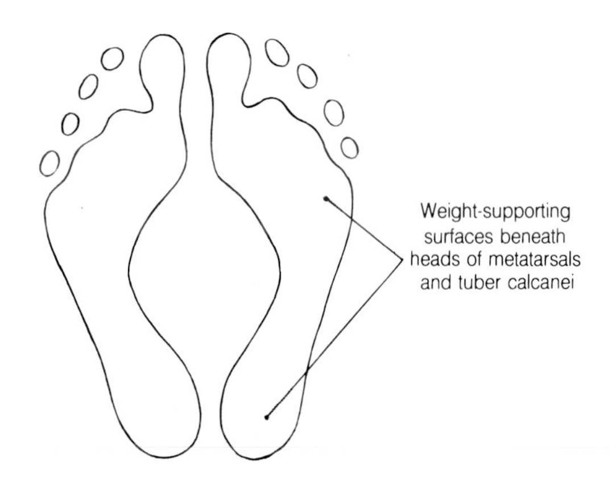

Fig. 15–2: Illustration depicting the appearance of footprints in sand at a beach.

in each arch of a keystone (a wedge-shaped bone at the crown that locks the other bones in place). The talus is the keystone for the medial arch, and the cuboid is the keystone for the lateral arch.

2. The interosseous ligaments that bind together the plantar aspects of the bones are the structures which provide the greatest static support of the arches. Three ligaments and one deep fascial layer are, in particular, the major fibrous supports:

   The **plantar calcaneonavicular ligament** is the principal fibrous support of the medial longitudinal arch (Fig. 15–3). The ligament not only serves as a tie beam between the calcaneus and navicular, but also articulates with the head of the talus in the TCN joint. Although the ligament provides some resiliency to the medial longitudinal arch, the resiliency is not sufficient to justify the common reference to the ligament as the foot's spring ligament.

   The **long plantar ligament** is the principal fibrous support of the lateral longitudinal arch (Fig. 15–3). It serves not only as a tie beam between the calcaneus posteriorly and the cuboid and the bases of the second to fourth metatarsals anteriorly, but also converts the groove in the plantar surface of the cuboid into an osseofascial tunnel for the tendon of insertion of peroneus longus.

   The **plantar calcaneocuboid ligament (short plantar ligament)** is the second most important fibrous support of the lateral longitudinal arch (Fig. 15–3). Lying deep to the long plantar ligament, the short plantar ligament extends from the calcaneus to the cuboid.

   The **plantar aponeurosis** is the tough, inelastic, layer of deep fascia that extends through the sole of the foot from the calcaneus to the digital fibrous sheaths of the toes (Fig. 15–4). It supports both the medial and lateral arches. The lateral arch is especially supported by the part of the plantar aponeurosis called the calcaneometatarsal ligament which extends from the calcaneus to the base of the fifth metatarsal.

3. The muscles of the leg and foot support the arches during ambulation. However, there is uncertainty as to the manner in which the muscles provide this support. It is currently believed that when the foot is planted on the surface below during walking or running movements, the muscles support the arches more through their capacity to distribute stresses through the foot than through their capacity to eccentrically contract and thus act as elastic tie beams or suspensory bands. The long muscles of the leg which are regarded as extensile supports of the arches during locomotory activities include the following: Peroneus longus and tibialis anterior both insert onto the medial cuneiform and the base of the first metatarsal. Together the two muscles form a musculotendinous sling under the foot which supports the transverse arch (Fig. 14–19). Flexor hallucis longus and tibialis posterior both help support the medial longitudinal arch. Flexor digitorum longus helps support both the medial and lateral arches.

The Function of the Arches during the Walking Gait

During the walking gait, the arches of the foot and the interosseous ligaments that support them function as a twisted osseoligamentous plate (Fig. 15–5). This twisted plate confers both stability and flexibility to the foot. When the heel of the bare foot strikes the surface below at the beginning of a stance phase, the center of pressure (COP) in the foot lies at the posterior end of the lateral longitudinal arch (the COP represents the central point in the foot at which the upper body weight is acting) (Fig. 15–6). From heel strike to foot flat, the COP moves anteriorly directly under the lateral arch. This loading of body weight under the lateral aspect of the hindfoot imparts a pronation stress on the hindfoot. This pronation stress produces movements in the articulations of the functional subtalar and transverse tarsal joints that render the talus, calcaneus, navicular, and cuboid relatively mobile and loosely packed at these articulations. In other words, the pronation stress enhances the flexibility of the osseoligamentous plate. Specifically, the pronation stress produces hind-

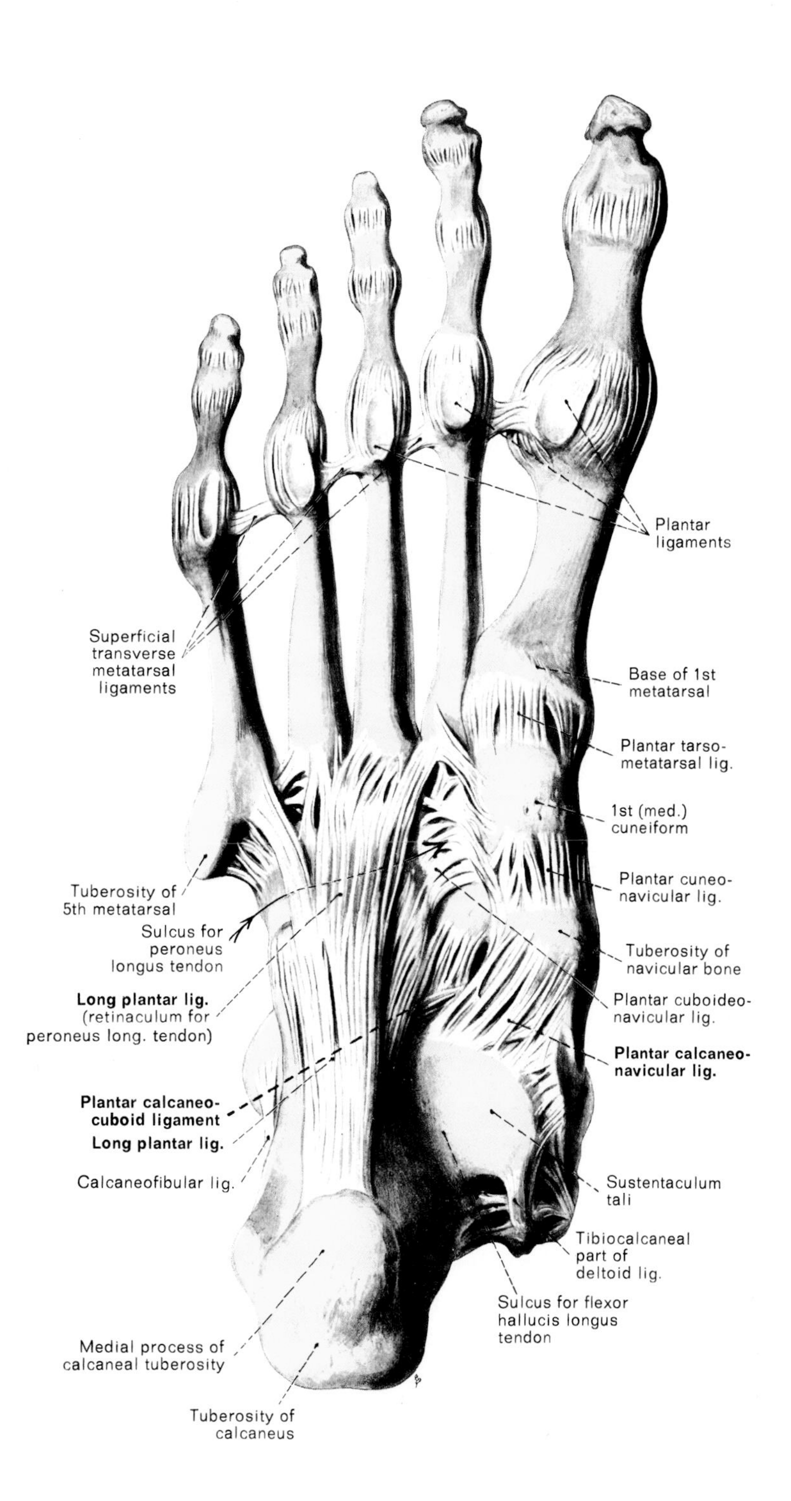

Fig. 15–3: Ligaments in the plantar aspect of the right foot.

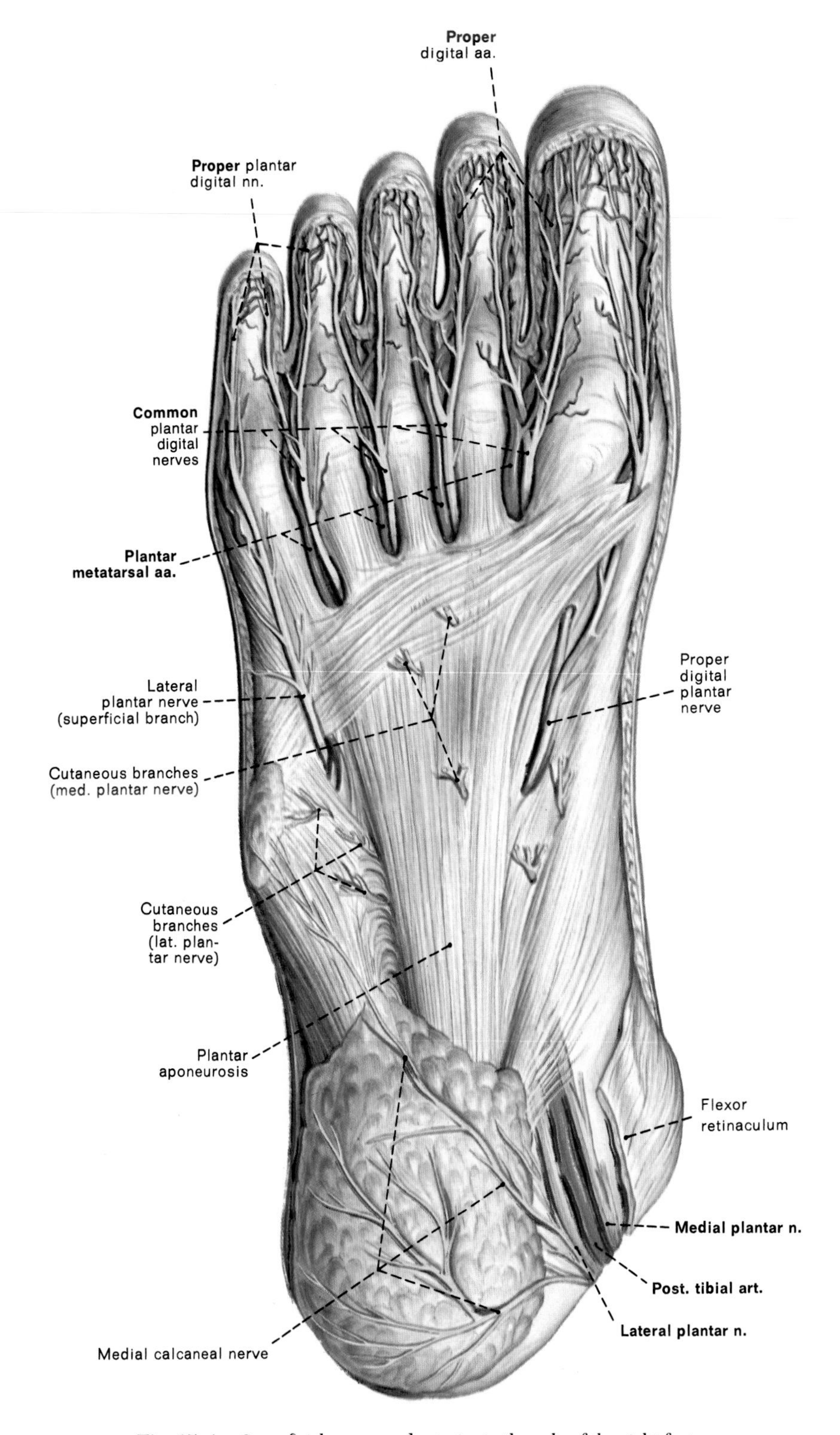

Fig. 15–4: Superficial nerves and arteries in the sole of the right foot.

foot eversion and plantarflexion of the talus in the ankle joint; the talar plantarflexion, in turn, moves the navicular anteroinferiorly. The overall effect of these movements is a slight flattening of the medial and lateral longitudinal arches and a slight stretching of the plantar interosseous ligaments that support them. The osseoligamentous plate thus serves from heel strike to foot flat as a flexible twisted plate that absorbs some of the force of upper body weight through a slight untwisting and the concurrent stretching of its ligamentous tie beams (the plantar calcaneonavicular and the long and short plantar ligaments).

From foot flat to heel rise, the COP moves anteromedially from the lateral edge of the midfoot to a point between the bases of the first and second metatarsals (Fig. 15–6). During this interval of the stance phase, soleus and tibialis posterior impose a supination stress on the hindfoot and midfoot. This supination stress produces movements in the articulations of the functional subtalar and transverse tarsal joints that render the talus, calcaneus, navicular, and cuboid relatively immobile and tightly packed at these articulations. In effect, these movements elevate the arches and convert the osseoligamentous plate into a rigid lever extending from the heel (calcaneus) posteriorly to the ball of the foot (heads of the metatarsals) anteriorly.

From heel rise to toe off, the COP moves anteriorly from a point between the bases of the first and second metatarsals to a point midway between the big and second toes (Fig. 15–6). The rigid osseoligamentous plate is pulled upward and forward around the ball of the foot (Fig. 15–7). The extension of the toes at the metatarsophalangeal joints converts the heads of the metatarsals into a pulley under which the plantar aponeurosis is tightened. The tightened plantar aponeurosis binds the bones of the foot into a more rigid configuration, and this greater rigidity enhances the stability of the foot. The pads at the distal ends of the toes become the last plantar surfaces from which the foot rolls off from the surface below.

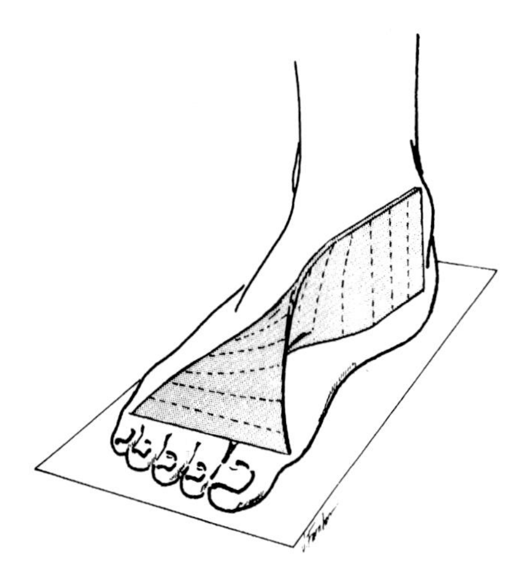

Fig. 15–5: Diagram illustrating the analogy of the arches of the foot and the ligaments that support them to a twisted osseoligamentous plate.

THE INTRINSIC MUSCLES OF THE FOOT

Extensor digitorum brevis is the only intrinsic foot muscle on the dorsum of the foot (Fig. 14–14). It originates from the calcaneus and the inferior extensor retinaculum (Fig. 14–16). The most medial of its four tendons of insertion inserts onto the base of the proximal phalanx of the big toe; the other three tendons insert onto the extensor expansions of the second, third, and fourth toes. It assists extensor hallucis longus in extending the proximal phalanx of the big toe and extensor digitorum longus in extending the phalanges of the second, third, and fourth toes. It is innervated by the deep peroneal nerve.

Abductor hallucis originates from the calcaneus and flexor retinaculum and inserts onto the medial side of the base of the proximal phalanx of the big toe (Fig. 15–8). It can abduct and flex the big toe. It is innervated by the medial plantar nerve.

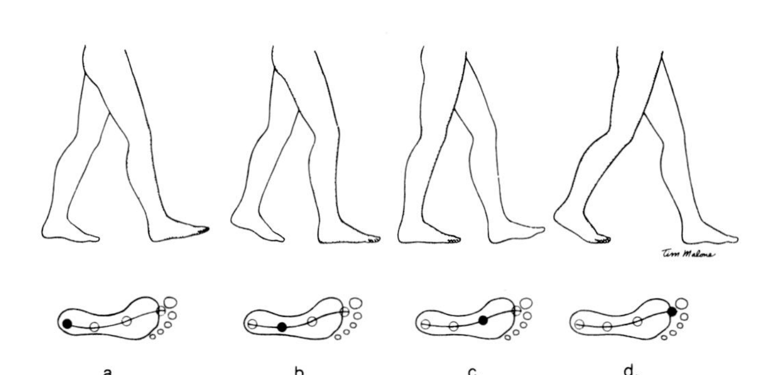

Fig. 15–6: Diagrams illustrating a center of pressure (COP) pathway in the foot (a) at heel strike, (b) at foot flat, (c) immediately prior to heel rise, and (d) at toe off.

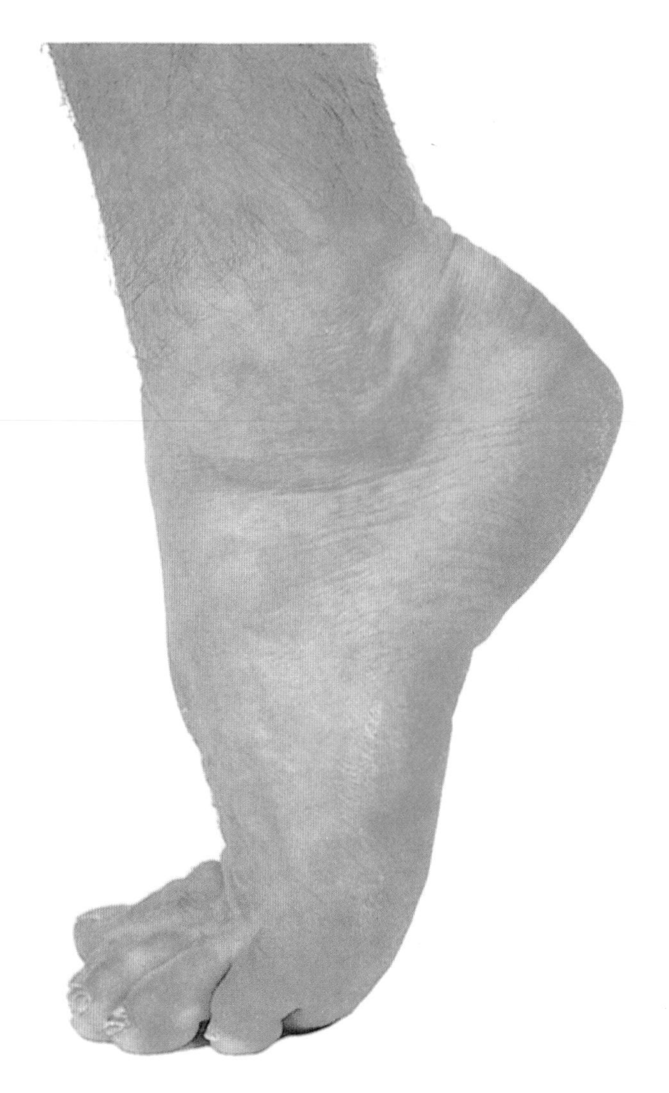

Fig. 15–7: A foot position in which the hindfoot and midfoot have been pulled upward and forward around the ball of the foot.

Flexor hallucis brevis originates from the cuboid and lateral cuneiform and inserts by a pair of tendons onto the medial and lateral sides of the base of the proximal phalanx of the big toe (Fig. 15–9). Each tendon has a sesamoid bone. When the hindfoot and midfoot are rolled upward and forward around the metatarsophalangeal joints during heel rise, the sesamoid bones in flexor hallucis brevis's tendons act as roller bearings about which the plantar aponeurosis is stretched taut. Flexor hallucis brevis can flex the big toe at its metatarsophalangeal joint. It is innervated by the medial plantar nerve.

Adductor hallucis originates by two heads of origin from the metatarsals, plantar ligaments, and metatarsophalangeal joints (Fig. 15–10). It inserts onto the lateral side of the base of the proximal phalanx of the big toe. It can flex and adduct the big toe. It is innervated by the lateral plantar nerve.

Flexor digitorum brevis originates from the calcaneus (Fig. 15–8). The anatomic relationships of its tendons of insertion to flexor digitorum longus's tendons of insertion and to the phalanges of the lateral four toes are identical to the corresponding relationships in the hand of flexor digitorum superficialis's tendons of insertion to flexor digitorum profundus's tendons of insertion and to the phalanges of the fingers (Fig. 15–9). It assists flexor digitorum longus in flexing the proximal and middle phalanges of the lateral four toes. It is innervated by the medial plantar nerve.

Quadratus plantae originates by two heads from the sides of the calcaneus, and inserts onto the posterolateral margin of flexor digitorum longus's tendon in the sole of the foot (Fig. 15–9). It assists the actions of

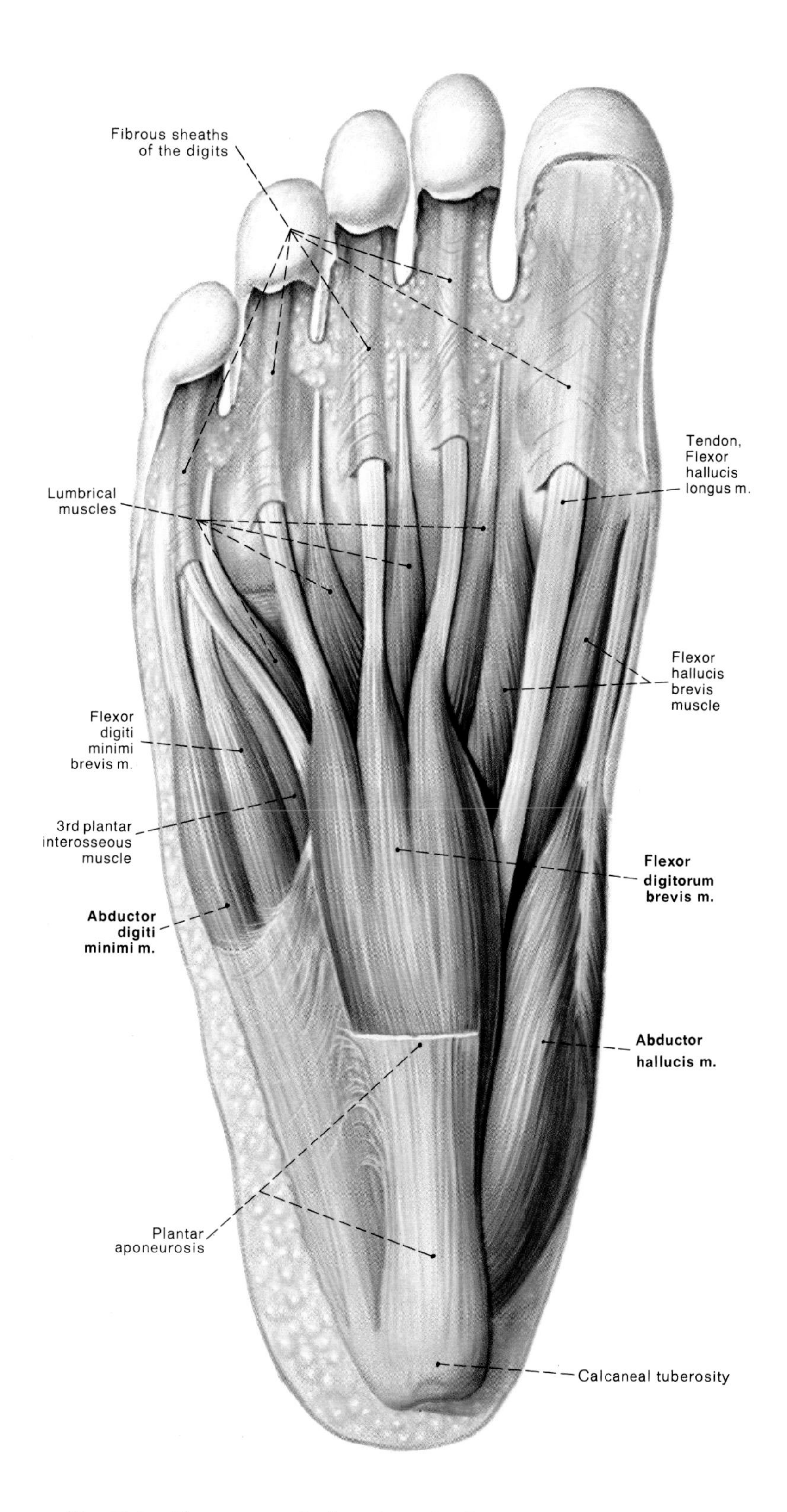

Fig. 15–8: The most superficial muscles in the plantar aspect of the right foot.

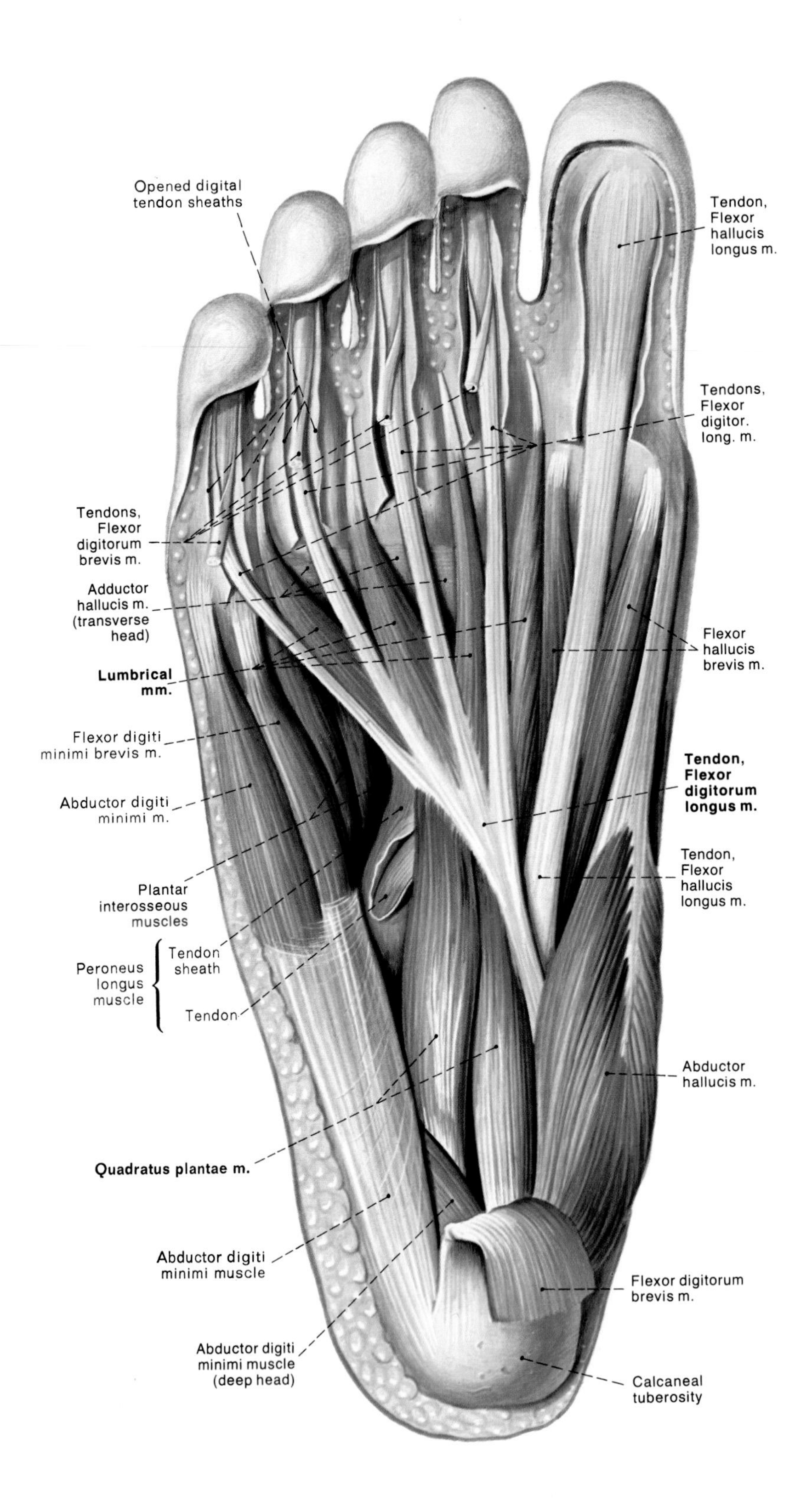

Fig. 15–9: Illustration depicting chiefly the muscles in the plantar aspect of the right foot immediately deep to the most superficial muscles.

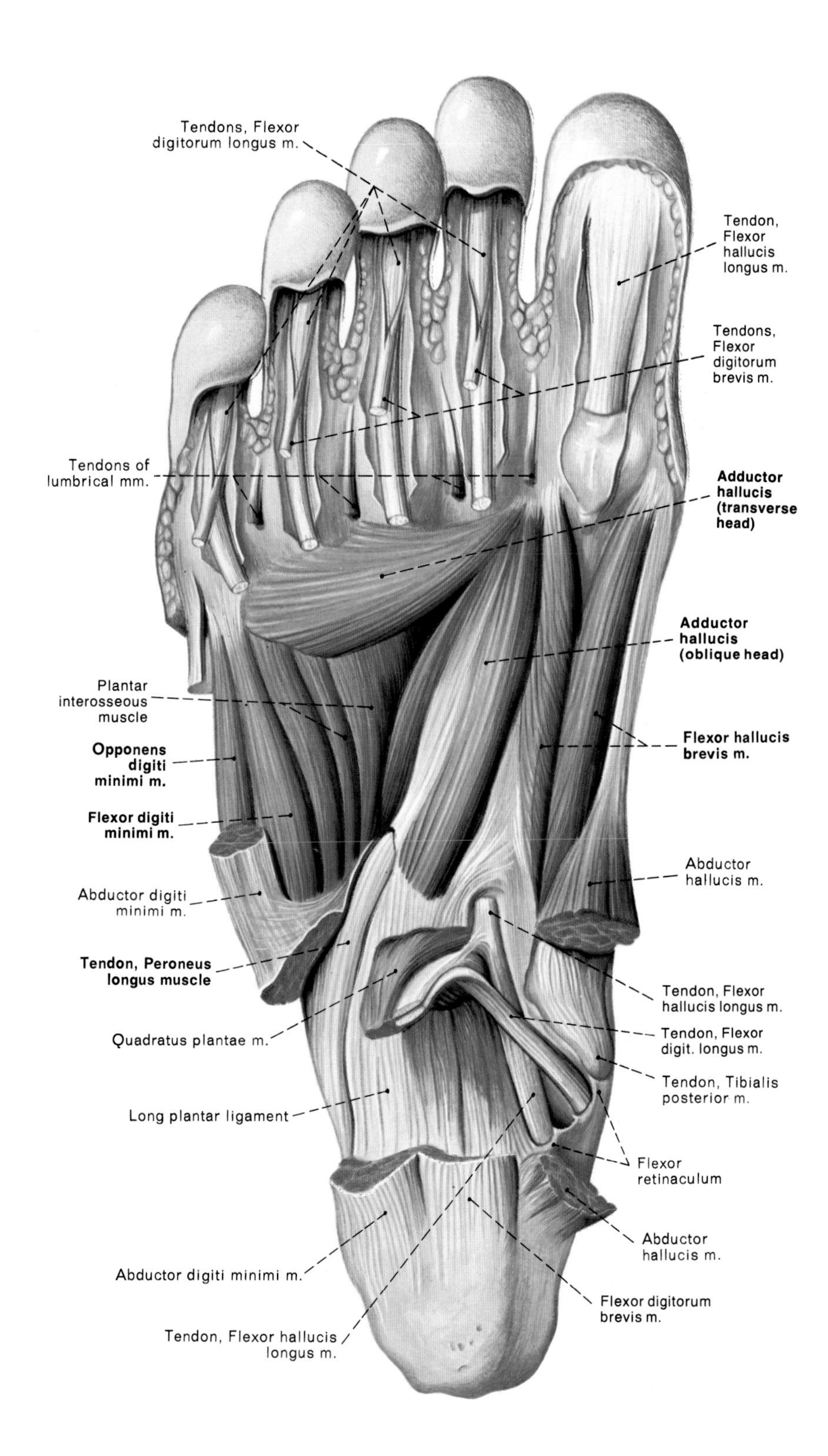

Fig. 15–10: Deep muscles in the plantar aspect of the right foot.

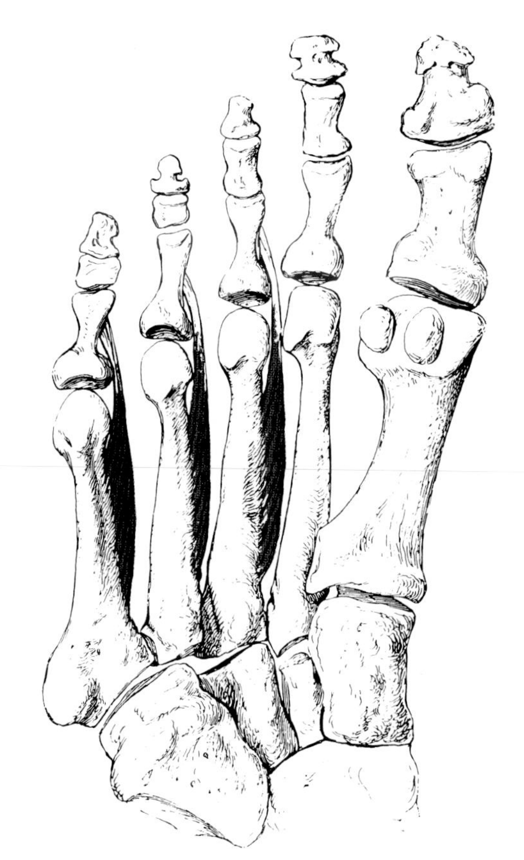

Fig. 15–11: Plantar view of the plantar interossei of the right foot.

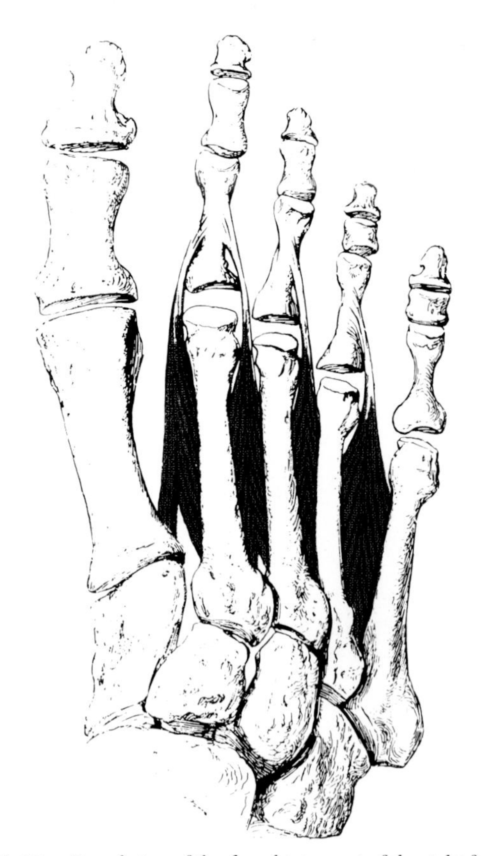

Fig. 15–12: Dorsal view of the dorsal interossei of the right foot.

flexor digitorum longus. It is innervated by the lateral plantar nerve.

The **four lumbrical muscles of the foot** originate from the tendons of insertion of flexor digitorum longus and insert onto the medial sides of the bases of the proximal phalanges and the extensor expansions of the lateral four toes (Fig. 15–9). The lumbricals can flex the metatarsophalangeal joints and extend the interphalangeal joints of the lateral four toes. The extensor action across the interphalangeal joints contributes to the walking gait by preventing the lateral four toes from being buckled under the foot when they are flexed by flexor digitorum longus and brevis during the mid stance and terminal stance phases of the stance period. The most medial lumbrical is innervated by the medial plantar nerve, and the three most lateral lumbricals are innervated by the lateral plantar nerve.

There are **three plantar and four dorsal interosseous muscles of the foot** (Figs. 15–11 and 15–12). The plantar and dorsal interossei all originate from the metatarsals. Each interosseous muscle inserts onto a side of the proximal phalanx and the extensor expansion of one of the lateral four toes. The muscles are innervated by the lateral plantar nerve.

Abductor digit minimi originates from the calcaneus and inserts onto the lateral side of the base of the proximal phalanx of the little toe (Fig. 15–9). It can abduct and flex the little toe. It is innervated by the lateral plantar nerve.

Flexor digiti minimi brevis originates from the base of the fifth metatarsal and inserts onto the base of the proximal phalanx of the little toe (Figs. 15–9 and 15–10). It can flex the little toe. It is innervated by the lateral plantar nerve.

THE PLANTAR ARTERIAL ARCH

The **lateral plantar artery,** which is the larger of the two terminal branches of the posterior tibial artery, begins deep to the flexor retinaculum (Fig. 15–13). It extends from its origin obliquely through the sole of the foot to the base of the fifth metatarsal, at which point it sharply curves medially to begin the major arterial arch of the foot: the plantar arterial arch. The arch is joined medially by the **deep plantar artery,** a branch of the dorsalis pedis, which descends into the sole of the foot by passing between the first and second metatarsals.

THE SENSORY INNERVATION OF THE FOOT

Figs. 4–12G and 4–12H depict the cutaneous areas of the dorsum and sole of the foot innervated by the superficial and deep peroneal, saphenous, sural, tibial, and medial and lateral plantar nerves.

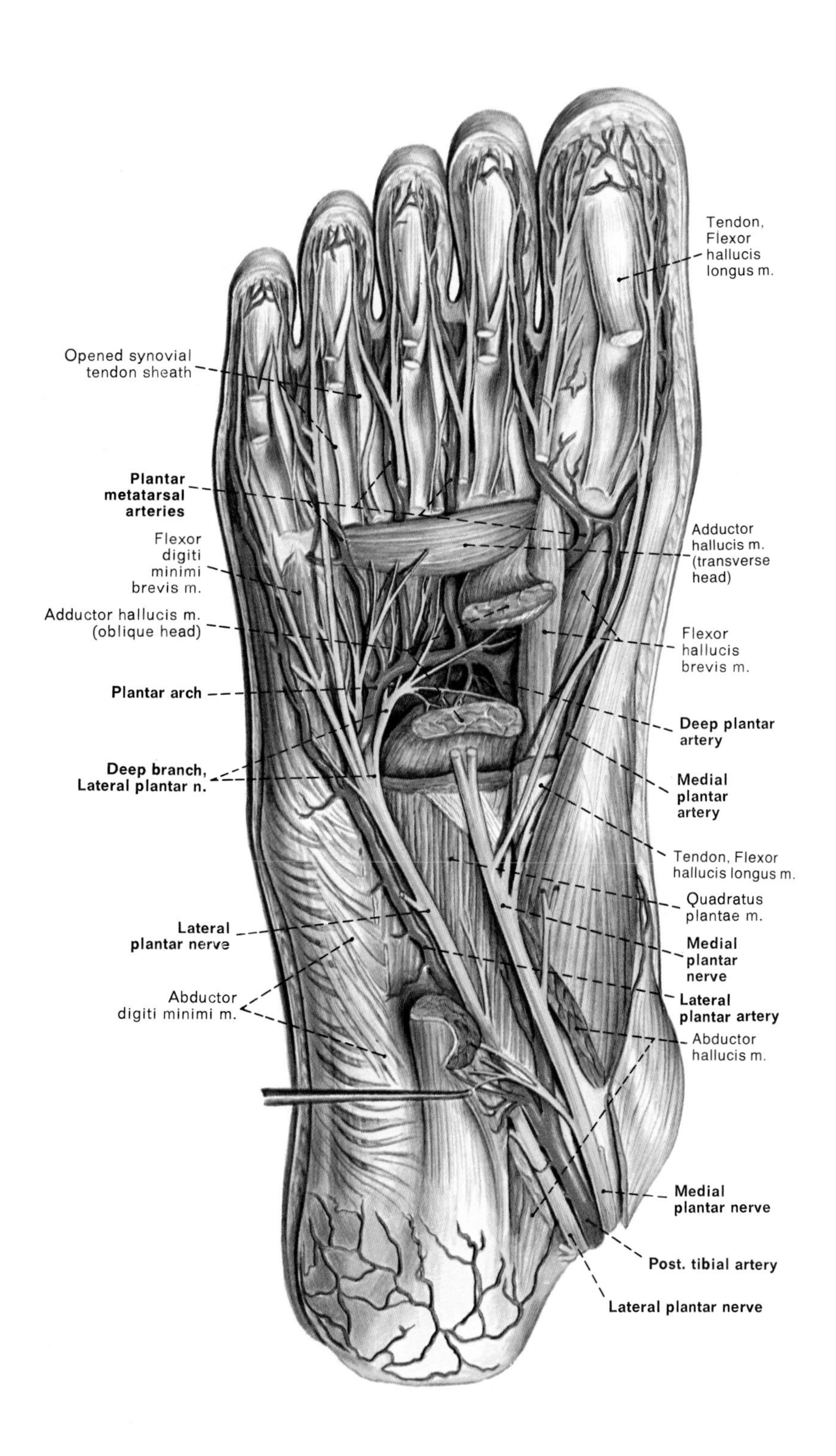

Fig. 15–13: The plantar arch and deep nerves and blood vessels in the plantar aspect of the right foot.

CASE **III.4**

The Case of Bill Jackson

INITIAL PRESENTATION AND APPEARANCE OF THE PATIENT

Bill Jackson, a 28 year-old black man, appears in the ambulatory clinic. The patient walks into the examination room with an abnormal gait in which he (a) leans his body toward the right side during the single stance interval of the right lower limb's stance period and (b) shortens the stride to minimize the duration of the right lower limb's stance period. The patient appears extremely uncomfortable as he slowly sits down.

QUESTIONS ASKED OF THE PATIENT

Please tell me about your problem. I have a very bad pain in my leg.

Can you point to where you have pain? Yes. It goes down my leg [the patient demonstrates the extent of the pain by sliding his right hand over the right buttock,down the posterolateral aspect of the right thigh, and then down over the lateral aspect of the right leg].

How would you describe the pain? It sometimes feels sharp like a knife. At other times, it has a burning quality to it.

When did the pain start? Three days ago, while playing squash. I had been playing about an hour when I went for a low fast return. As soon as I went for the ball, I knew something was wrong, because I felt this sharp pain in my lower back. During the last two days, the pain has moved from the lower back into my leg. *[The answer to this last question indicates that it is the patient's perception that the lower limb pain is a consequence of an acute injury. The extension of the pain from the lower back to the lower limb suggests the progression of a pathologic condition or process subsequent to the acute injury.]*

Is there anything that relieves the pain? Well, I've been taking Advil [ibuprofen] for the past three days, and that helps for a while. But I find that just lying in bed is what eases the pain the most. *[This answer indicates that the pain emanates from a musculoskeletal structure or structures sensitive to weight-bearing. Since ibuprofen is a non-steroidal anti-inflammatory drug, it appears that an inflammatory condition is involved in the production of much of the patient's pain.]*

Is there anything that makes the pain worse? Yes. The more I'm up and about, the worse the pain gradually becomes. However, once I am up and about, I prefer standing and walking around to sitting. For example, I feel more pain now just sitting here than I did walking into this room. *[This answer indicates that the musculoskeletal structure or structures sensitive to weight-bearing are part of or related to the vertebral column and/or the paired coxal bones. This is because they are the only musculoskeletal structures that bear weight in both the standing and seated positions.*

At this point, the examiner decides to interrupt the asking of open-ended questions with two questions designed to elicit simple yes-or-no answers about three specific activities (coughing, sneezing, and bearing down during bowel movement). The reasons are as follows:

The examiner recognizes that lower limb pain associated with acute injury of one or more musculoskeletal structures in the vertebral column can be referred and/or radicular pain. ***Referred pain*** *is pain that a diseased or injured structure refers to a cutaneous region of the body. The dermatomes of the painful cutaneous region are innervated by the same spinal cord segments that innervate the diseased or injured structure. The painful cutaneous region is generally anatomically distant from the site of the diseased or injured structure.* ***Radicular pain*** *is pain produced by compression and/or inflammation of the roots of one or more spinal nerves. Radicular pain is frequently perceived as radiating (extending) along the courses of those nerves transmitting nerve fibers to and from the involved spinal nerves.*

Coughing, sneezing, and bearing down during bowel movement are all activities that require the performance of the Valsalva maneuver (refer to the section on the interior of the larynx in Chapter 32 for a discus-

*sion of the Valsalva maneuver). The Valsalva maneuver compresses the roots of the spinal nerves by an increase in intrathecal pressure (**intrathecal pressure** is an expression used to refer to the pressure of the cerebrospinal fluid in the subarachnoid space of the cranial and spinal meninges). The examiner appreciates that if activities such as coughing, sneezing, and bearing down during bowel movement intensify the patient's lower limb pain, then it is likely that radicular pain is at least a major component of the lower limb pain. Such information would imply that the patient is suffering from some compression and/or irritation of the roots of one or more spinal nerves.*

The Valsalva maneuver increases intrathecal pressure in the following manner: The intrathoracic pressure (the pressure within the thoracic cavity of the chest) increases during the performance of the Valsalva maneuver. The increase in intrathoracic pressure decreases the return of venous blood into the right atrium from the superior and inferior venae cavae. This decrease in the flow of blood through the superior and inferior venae cavae results in an increase in systemic venous pressures, which are transmitted in a retrograde fashion to the vertebral venous plexuses and the dural venous sinuses of the cranial cavity. It is the increased blood pressure in the vertebral venous plexuses and the dural venous sinuses which finally imposes an increase in cerebrospinal fluid, or intrathecal, pressure.]

This may seem like an odd question, but have you noticed if coughing or sneezing worsens the hip, thigh and leg pain? Well, now that you mention it, yes, I have found that coughing and sneezing make the pain worse.

How about when you bear down during a bowel movement? Does that also make the pain worse? Yes, but not as much as when I sneeze or cough.

Is there anything else that is also bothering you, or that you think may be associated with your hip, thigh, and leg pain? Yes, possibly. The skin on the top of my foot seems numb some of the time. *[This answer suggests that the acute injury responsible for the patient's lower limb pain has also resulted in an area of hypesthesia or anesthesia.]*

Did you have any pain or numbness in the lower part of your body before you got this pain playing squash 3 days ago? Yes. The right side of my lower back has been hurting me for months. It's odd, but ever since this leg pain came on, I no longer have much pain in my lower back. *[This response indicates that the patient has been suffering for several months duration from a chronic condition associated with low back pain.]*

Can you show me where you had the low back pain? [The patient slides his right hand up and down over the right paraspinal region of the lower back.]

Have you had any other recent injuries or illnesses? No.

Are you taking any medications for previous injuries or illnesses? No. *[The examiner finds the patient alert and fully cooperative during the interview.]*

PHYSICAL EXAMINATION OF THE PATIENT

Vital signs:
Blood pressure
Lying supine: 110/65 left arm and 110/65 right arm
Standing: 110/70 left arm and 110/65 right arm
Pulse: 68
Rhythm: regular
Temperature: 98.7°F.
Respiratory rate: 16
Height: 6′2″
Weight: 155 lbs.
HEENT Examination: Normal
Lungs: Normal
Cardiovascular Examination: Normal
Abdomen: Normal
Genitourinary Examination: Normal
Musculoskeletal Examination: Inspection, movement, and palpation of the lower back and the upper and lower limbs are normal except for the following findings: Inspection of the lower back shows a flattening of the normal lumbar lordosis and a listing of the lumbar spine to the right.

Pain restricts active flexion of the lumbosacral spine to less than 20°. Extension of the lumbosacral spine to 20° is relatively painless. Lateral flexion of the lumbosacral spine to the right does not intensify the pain; pain restricts lateral flexion of the lumbosacral spine to the left.

The straight leg raising test exacerbates right lower limb pain at 45° elevation; lowering the leg to the highest level at which pain is relieved and then dorsiflexing the foot reintensifies the pain. The crossed straight leg raising test exacerbates the right lower limb pain at 40° elevation.

Palpation of the lower back reveals bilateral paraspinal muscle spasm in the lumbar region.

Neurologic Examination: The patient reports diminished pin and cotton touch sensation on the medial aspect of the dorsum of the right foot and the contiguous areas of the big and second toes. Extension

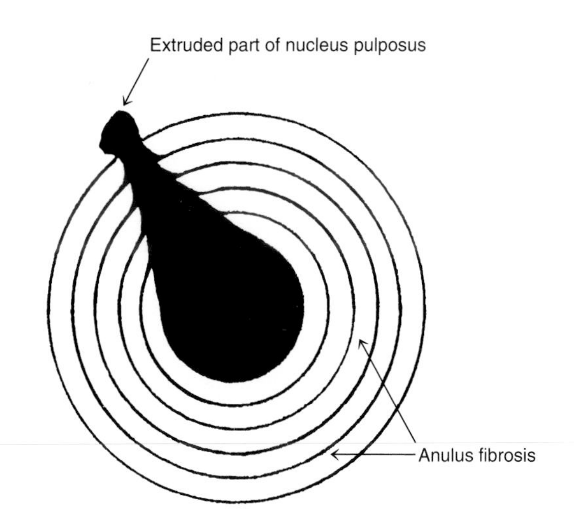

Fig. 15–14: Diagram illustrating the extrusion of nucleus pulposus material through the anulus fibrosis in a herniated intervertebral disc.

of the big toe in the right foot has a strength of grade −1. Pertinent normal findings include bilaterally symmetrical knee jerk and Achilles tendon reflex tests.

INITIAL ASSESSMENT OF THE PATIENT'S CONDITION

The patient appears to be suffering from an acute injury that has resulted in (a) right hip, thigh, and leg pain, (b) muscle weakness of an anterior leg muscle in the right leg, and (c) hypesthesia of part of the right foot. Prior to the acute injury, the patient has experienced chronic low back pain.

ANATOMIC BASIS OF THE PATIENT'S HISTORY AND PHYSICAL EXAMINATION

1. **The patient's description of the acute onset of right low back pain immediately following a stressful physical movement indicates an acute injury. The subsequent extension of the low back pain down into the right lower limb suggests the development and/or progression of a pathologic condition or process subsequent to the acute injury.**
2. **The alleviation of right lower limb pain in the recumbent position and its intensification in both the standing and seated positions indicate injury of one or more musculoskeletal structures that bear weight in the vertebral column.** The seated position effectively isolates the lower limbs from the hip joints downward from any major role in supporting upper body weight.
3. **The intensification of right lower limb pain by coughing, sneezing, and bearing down during bowel movement suggests compression and/or inflammation of the roots of one or more spinal nerves.**
4. **The results of the straight leg raising (SLR) test suggest compression and/or inflammation of the dural coverings of the roots of the L5 and/or S1 spinal nerves.** This test is frequently conducted to explore the anatomic basis of low back and/or lower limb pain. The test is conducted with the lower limb on the painful side of the body as the patient lies supine on an examination table. The test has two parts. In the first part, the examiner flexes the thigh at the hip with the leg extended at the knee until low back and/or lower limb pain is elicited or intensified. In the second part, the examiner lowers the lower limb to the highest level at which pain is relieved and then dorsiflexes the foot.

The anatomic basis of the SLR test is as follows: In the supine position, the sciatic nerve and its roots exhibit slack as they extend from the vertebral column through the buttock into the thigh. Flexion of the thigh at the hip from 0° to 35° flexion (with the leg fully extended at the knee and the foot in a neutral position at the ankle) stretches the sciatic nerve primarily along its course through the buttock; flexion to 35° takes up almost all the slack in the sciatic nerve along its course in the buttock and thigh. Flexion from 35° to 70° places increased tension on the sciatic nerve, but now almost all of the increased tension is exacted in the vertebral column upon the roots of the spinal nerves (L4, L5, S1, S2, and S3) that contribute to the sciatic nerve. The greatest traction is exerted upon the L5 and S1 spinal nerves along their passage through their respective intervertebral foramina and the roots of these two spinal nerves along their descent in the cauda equina of the vertebral canal (the roots of L4, S2, and S3 are only slightly stretched). Thigh flexion greater than 70° continues to increase tension in the sciatic nerve by exacting tension on the nerve and its branches along their descent through the lower limb.

If the first part of the SLR test elicits or intensifies low back and/or lower limb pain (at the sites identified by the patient during the history) within the 35° to 70° arc of thigh flexion, then the finding **suggests** that traction upon the dural coverings of the L5 and S1 spinal nerves and/or their roots represents at least part of the mechanical basis for the patient's low back and/or lower limb pain. The dural coverings are implicated because receptors sensitive

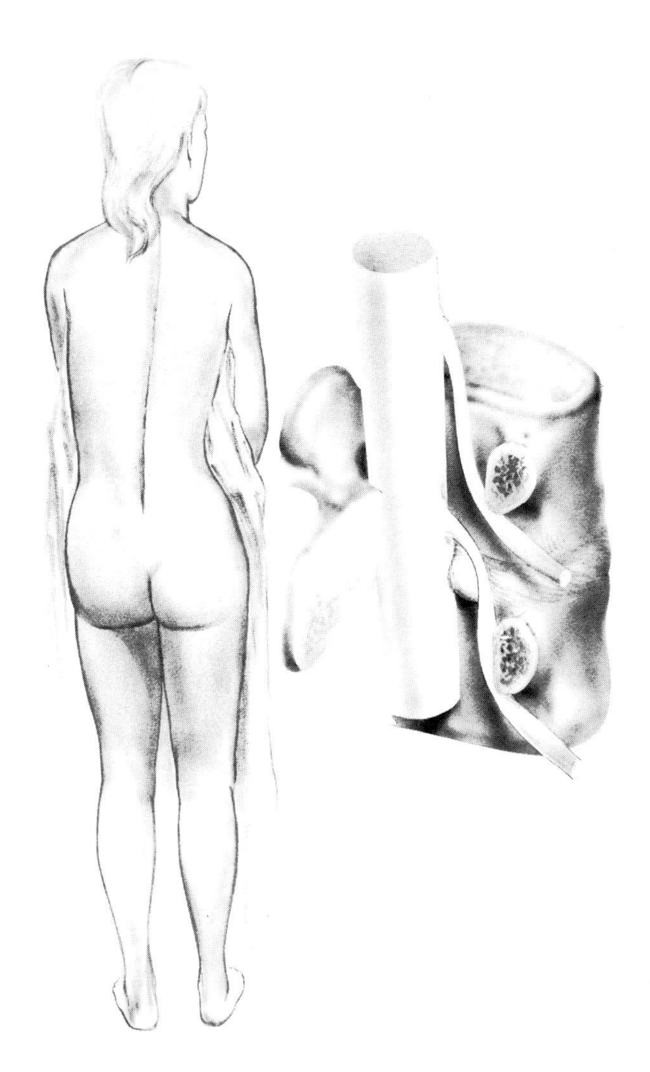

Fig. 15–15: (A) Drawing of the extruded nuclear material of a herniated intervertebral disc in which the extruded material lies medial to the descending roots of a right lumbar spinal nerve. (B) Drawing of the patient's lumbar spine list to the right; the list to the right avoids drawing the inflamed roots of the right lumbar spinal nerve to the left and into contact with the disc protrusion.

to the traction reside in the dural coverings but not the spinal nerves or their roots themselves. There are mechanisms (such as the painful stretching of hamstring muscles), however, which can intensify lower limb pain within the 35° to 70° arc of thigh flexion other than increased traction upon the dural coverings of the L5 and S1 spinal nerves. Almost all of these other mechanisms require movement in joints of the lumbosacral spine, the sacroiliac joint, or the hip joint in order to intensify pain within the 35° to 70° arc of thigh flexion. The second part of the SLR test obviates consideration of almost all these other mechanisms because dorsiflexion of the foot at the ankle does not evoke movement in joints of the lumbosacral spine, the sacroiliac joint, or the hip joint. The foot dorsiflexion exerts traction upon the dural coverings of the L5 and S1 spinal nerves by stretching of the tibial nerve along its passage behind the medial malleolus and deep to the flexor retinaculum. Consequently, if the second part of the SLR test elicits or intensifies low back and/or lower limb pain, then the finding **indicates** that traction upon the dural coverings of the L5 and S1 spinal nerves and/or their roots represents at least part of the mechanical basis for the patient's pain.

5. **The result of the crossed straight leg raising test indicates compression of the dural coverings of the roots of the L5 or S1 spinal nerve by a herniated intervertebral disc.** A herniated intervertebral disc is a disc in which the gelatinous material of the nucleus pulposus has herniated through the anulus fibrosis and extruded radially beyond the confines of the anulus fibrosis (Fig. 15–14).

The crossed SLR test is another test frequently conducted to explore the anatomic basis of low back and/or lower limb pain. The test is identical to the first part of the SLR test except that (a) the test is conducted on the painless side of the body and (b) the painless lower limb is raised until pain is elicited or intensified on the painful side of the body. If the crossed SLR test elicits or intensifies low back and/or lower limb pain (on the painful side) within the 35° to 70° arc of thigh flexion on the painless side, then the finding is **almost pathognomonic** of the presence of a herniated intervertebral disc.

The anatomic basis of the crossed SLR test is as follows: When the thigh of the painless lower limb is flexed from 35° to 70° at the hip, the traction on the roots of the L5 and S1 spinal nerves (on the painless side of the body) pulls the dural sac about the cauda equina toward the painless side. This shift of the dural sac toward the painless side also draws the dural sac about the L5 and S1 roots on the painful side toward the painless side. If the extruded nuclear material of a herniated disc lies immediately medial to the descending L5 or S1 roots, the drawing of the dural sac about these roots against the extruded nuclear material will impose traction on the dural sac and thereby elicit or intensify pain (Fig. 15–15).

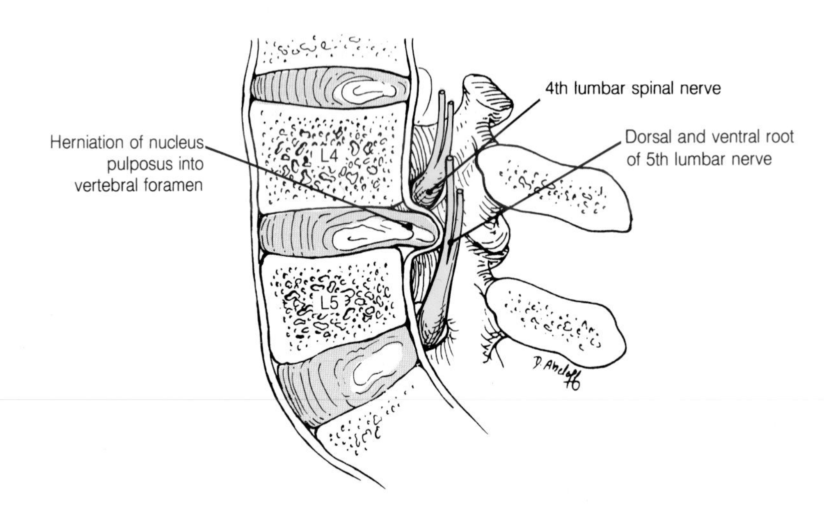

Fig. 15–16: Diagram illustrating impingement of the roots of the fifth lumbar spinal nerve by a posterolateral herniation of the L4/L5 intervertebral disc.

6. **The cutaneous areas of hypesthesia in the right foot correspond to part of the L5 dermatome** (Fig. 4–12G).
7. **The diminished strength of big toe extension indicates neuromuscular dysfunction of extensor hallucis longus.** The fifth lumbar spinal nerve provides extensor hallucis longus with most of its motor nerve fibers (Table 14–1).

INTERMEDIATE EVALUATION OF THE PATIENT'S CONDITION

The patient appears to be suffering from an acute herniation of an intervertebral disc in the lumbosacral spine (the lumbosacral region of the vertebral column).

CLINICAL REASONING PROCESS

In considering this case, an orthopedist or primary care physician would recognize the patient's description of radicular pain in the right lower limb as typical of sciatica. **Sciatica** is the term given to the syndrome in which pain occurs anywhere along the course of the sciatic nerve and its branches. Given the acute onset of the lower limb pain following a stressful physical movement, a physican would regard three key findings from the patient's history and physical exam as collectively being almost pathognomonic of an acutely herniated intervertebral disc:

1. Coughing, sneezing, and bearing down during bowel movement intensify the radicular pain.
2. The straight leg raising (SLR) test is positive.
3. The crossed SLR test is positive.

The principal issues to be resolved are identification of the involved spinal nerve roots and intervertebral disc. The area of hypesthesia on the dorsum of the foot is the key finding in the identification of the involved spinal nerve roots: the L5 roots. Compression and/or irritation of the L5 roots suggests that the herniation has occurred in the L4/L5 disc. The reasoning is as follows:

The posterolateral margins of the intervertebral discs in the lumbosacral spine are the most common sites of herniation. As the roots of the L5 and S1 spinal nerves end their descent within the vertebral canal, the L5 roots are the spinal nerve roots that pass most directly behind the posterolateral margin of the L4/L5 disc (Fig. 15–16), and the S1 roots are the spinal nerve roots that pass most directly behind the posterolateral margin of the L5/S1 disc. Therefore, compression and/or irritation of the L5 roots by a herniated intervertebral disc is most likely to involve a posterolateral herniation of the L4/L5 disc.

RADIOLOGIC EVALUATION AND FINAL RESOLUTION OF THE PATIENT'S INJURY

Magnetic resonance imaging (MRI) is the preferred method for many physicians in the radiologic evaluation of a presumed herniated intervertebral disc. This is because MRI provides superb visualization of the integrity of the bony, cartilaginous, ligamentous, and neural tissues of the vertebral column. In this case, MRI revealed a posterolateral herniation of the L4/L5 intervertebral disc on the right side. The extruded material lies for the most part medial to the descending course of the L5 roots.

Many orthopedists would advise the patient that his herniated disc can be treated either conservatively (essentially bedrest and traction for a few weeks) or surgically. Surgery frequently involves limited laminectomy and partial or complete discectomy.

MECHANISM OF THE PATIENT'S INJURY AND AN ACCOUNT OF HIS CONDITION

The patient's chronic low back pain on the right side was due to repetitive episodes of severe compression on the right side of the L4/L5 disc (the episodes occurred during the playing of squash). Extrusion of nucleus pulposus material into the inner part of the anulus fibrosis increased tension on the outer posterolateral margin of the anulus and the overlying posterior longitudinal ligament. Pain fibers innervate the ligaments of the vertebral column and probably also the outermost lamella (layer) of the anulus fibrosis of the intervertebral discs. The sensory fibers innervating a segment of the posterior longitudinal ligament are provided not only by the spinal nerves at the level of the segment but also by spinal nerves one to two levels above and below the level of the segment. Consequently, stimulation of pain fibers in the posterior longitudinal ligament produced pain in the right paraspinal region that was referred one to two levels above and below the level of the L4/L5 disc.

While the patient was playing squash three days before, he suddenly and very forcefully laterally flexed the lumbar spine to the right. The resulting compression of the nucleus pulposus in the L4/L5 disc pressed the nuclear material radially outward through the already weakened posterolateral margin of the anulus fibrosis. The marked tension on the outermost lamella of the anulus and the overlying posterior longitudinal ligament prior to the renting of the anulus produced the severe low back pain. The tear in the outermost lamella of the anulus released the tension in this lamella, and thus relieved much of the low back pain previously experienced by the patient.

The irritation imposed by the protruding disc lesion on the roots of the right fifth lumbar spinal nerve has initiated an inflammatory response in the spinal nerve. Any movement of the lumbar spine (such as anterior flexion or lateral flexion to the left) that stretches the inflamed L5 roots over the disc protrusion intensifies pain along the course of the sciatic nerve. Reflexive spasm of the paraspinal muscles has flattened the normal lumbar lordosis. The patient lists the lumbar spine to the right to avoid drawing the inflamed L5 roots to the left and thus into contact with the disc protrusion.

RECOMMENDED REFERENCES FOR ADDITIONAL INFORMATION ON THE ANKLE AND FOOT AND THE CASE OF BILL JACKSON

Norkin, C. C., and P. K. Levangle, Joint Structure & Function, 2nd ed., F. A. Davis Co., Philadelphia, 1992: *Chapter 12 presents detailed descriptions of the motions that occur within the ankle, functional subtalar, and transverse tarsal joints during the walking gait.*

Corrigan, B., and G. D. Maitland, Practical Orthopaedic Medicine, Butterworth & Co., London, 1983: *Chapter 12 offers a description of the clinical examination of the ankle and discussions of the signs and symptoms of common ankle lesions.*

Magee, D. J., Orthopaedic Physical Assessment, 2nd ed., W. B. Saunders Co., Philadelphia, 1992: *Chapter 12 provides a comprehensive discussion of the clinical examination of the leg, ankle, and foot.*

Greenspan, A., Orthopedic Radiology, 2nd ed., Gower Medical Publishing, New York, 1992: *Fig. 9.27 on page 9.16 summarizes the array of radiologic imaging techniques that may be used to evaluate ankle and foot injuries. Pages 9.25 through 9.31 present discussions of the radiologic evaluation of patients with soft tissue injuries about the ankle.*

Hardy, Jr., R. W., Lumbar Disc Disease, 2nd ed., Raven Press, New York, 1993: *Chapter 4, 5, and 6 present discussions of the clinical diagnosis of a herniated lumbar disc, nondiscogenic causes of low back pain and sciatica, and radiologic imaging of the lumbar spine.*

Cailliet, R., Low Back Pain Syndrome, 4th ed., F. A. Davis Co., Philadelphia, 1988: *Chapter 5 provides a discussion of the tissue sites of low back pain. Chapter 7 presents a discussion of the clinical diagnosis of low back pain. Chapter 11 covers various aspects of disc disease.*

SECTION **IV**

The Thorax

CHAPTER 16

The Thoracic Wall

The thorax is the region in the trunk of the body that lies between the neck and abdomen. The thorax has an inner **thoracic cavity** surrounded by a **thoracic wall** (Fig. 16–1). The thoracic cavity houses the thymus, heart, lungs, lower part of the trachea, and most of the esophagus.

The thoracic wall consists mainly of the bony and muscular tissues that enclose the organs and great blood vessels of the thorax. Every breath an individual takes requires movement by the thoracic wall. Each breath begins with outward movements of the thoracic wall, movements which, by increasing the dimensions of the thorax, serve to expand the lungs and thereby bring air into the lungs. Each breath then ends with inward movements of the thoracic wall, movements which, by restoring the thorax to its initial, lesser dimensions, permit recoil retraction of the lungs and the concurrent expulsion of air from them.

The skeletal framework of the thoracic wall is the rib cage. The rib cage consists of the sternum in front, the ribs on the sides, and the thoracic vertebrae posteriorly. The superior opening into the space surrounded by the rib cage is called the **thoracic inlet** (Fig. 16–1). The thoracic inlet is roofed over on each side by the very tenuous muscle scalenus minimus and its fascia-like tendon of insertion, the suprapleural membrane. The inferior opening out from the space surrounded by the rib cage is called the **thoracic outlet.** At the thoracic outlet, the **diaphragm** separates the thoracic cavity from the abdominal cavity (Fig. 16–1).

THE SKELETAL AND CARTILAGINOUS FRAMEWORK

The Thoracic Vertebrae

The thoracic vertebrae have the structure of typical vertebrae (Fig. 16–2). The thoracic vertebrae have the most inferiorly-directed spinous processes in the vertebral column. The tip of the spinous process of each thoracic vertebra lies approximately at the level of the body of the vertebra below it. The tips of the spinous processes of the thoracic vertebrae are all prominent and palpable in the midline of the back. The spinous process of the first thoracic vertebra is the second highest prominent spinous process in the vertebral column (the highest prominent spinous process is that of the seventh cervical vertebra; the spinous processes of the second through sixth cervical vertebrae are all palpable but not prominent in the midline of the back).

With the exception of the two or three lowest thoracic vertebrae, all the thoracic vertebrae have six small surfaces called **costal facets (rib facets)** for articulation with ribs (Fig. 16–2). There is a pair of superior costal facets on the upper sides of the body, a pair of inferior costal facets of the lower sides of the body, and a third pair of costal facets on the transverse processes.

The Sternum

The sternum consists of three flat bones (Fig. 16–3): the uppermost is called the **manubrium,** the middle, and largest, bone is called the **body,** and the lowest, arrowhead-shaped bone is termed the **xiphoid process.**

The articulation between the manubrium and the body of the sternum is a symphysis called the **manubriosternal joint.** The manubriosternal joint can be easily palpated as a rather prominent, transverse ridge in the midline of the anterior chest wall. The joint's prominence is due in large part to the fact that the posterior angle between the manubrium and the body is less than 180°. This angle is called the **sternal angle** or **angle of Louis** (Fig. 16–3). When reference is made in clinical notes to the location of the manubriosternal joint in a patient, it is common practice to identify the location as the sternal angle itself.

The sternal angle is the single most important landmark of the anterior chest wall. This is because, on each side, the costal cartilage of the second rib almost invariably articulates with the sternum at the level of the sternal angle (Figs. 16–3 and 16–4). Accordingly, palpation of the sternal angle permits unambiguous identification of not only the second ribs at the sides of the sternal angle, but also all the lower ribs (by simply

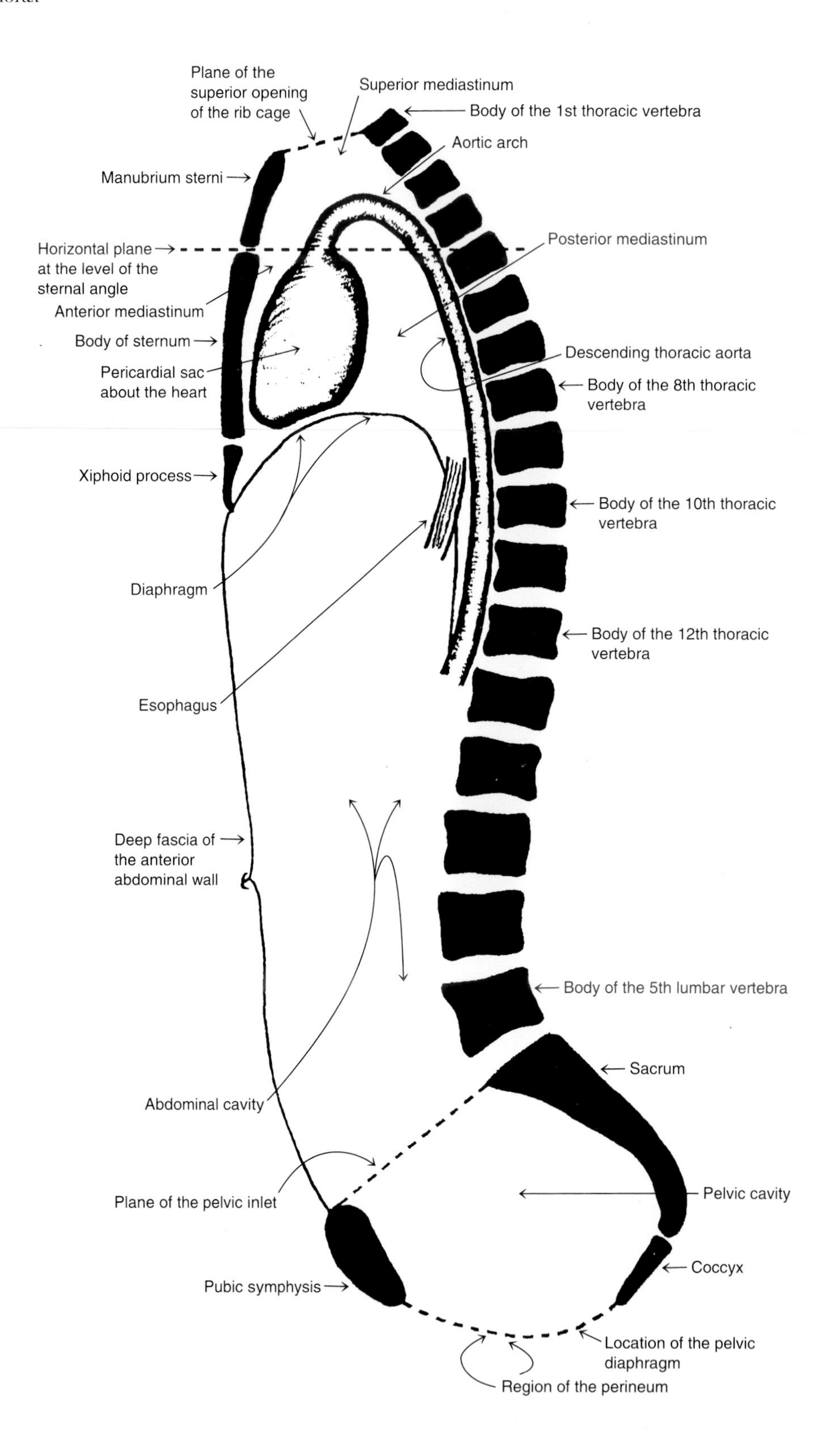

Fig. 16–1: Diagram of a median section of the thoracic, abdominal and pelvic cavities. The superior opening of the rib cage is called the thoracic inlet by anatomists but the thoracic outlet by physicians. The opening is bordered anteriorly by the superior margin of the manubrium sterni, laterally by the first ribs, and posteriorly by the body of the 1st thoracic vertebra. The thoracic cavity is the space bounded superiorly by the superior opening of the rib cage, anteriorly by the sternum, inferiorly by the diaphragm, and posteriorly by the bodies of the 12 thoracic vertebrae.

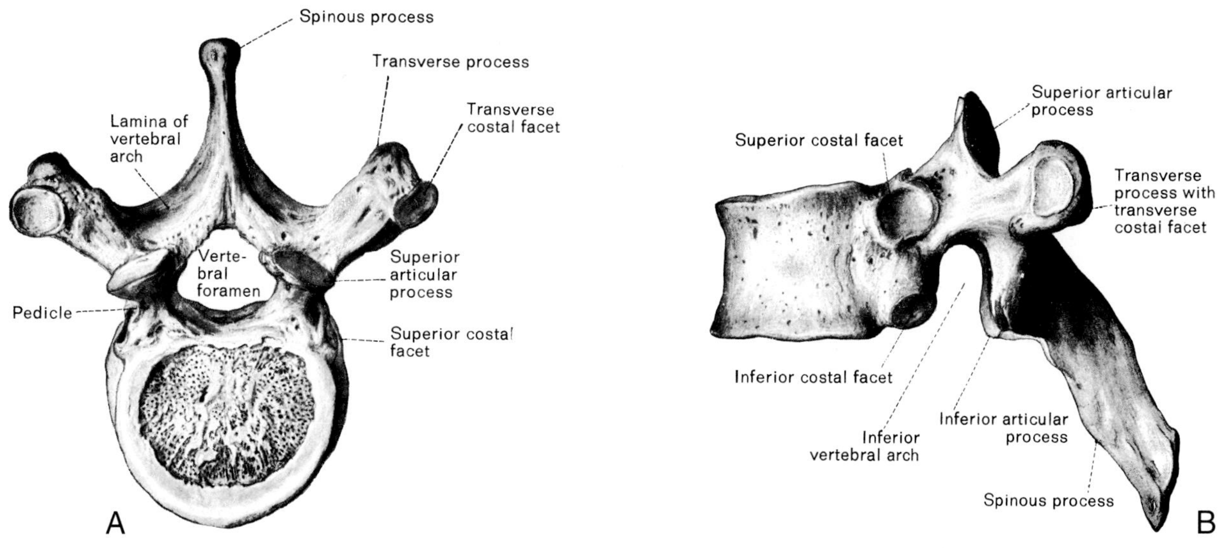

Fig. 16–2: (A) Superior and (B) left lateral views of the sixth thoracic vertebra.

counting them in their sequential palpation from the second rib downward).

The body of the sternum is commonly selected for needle biopsy of hematopoietic (blood-forming) bone marrow in adult patients. Hematopoietic (red) bone marrow occurs in only certain bones in the adult, among which are the bones of the rib cage. The body of the sternum is commonly selected for bone marrow biopsy in adults because of its breadth and subcutaneous position in the chest wall.

The Costal Cartilages

The costal cartilages are bars of hyaline cartilage that cap the anterior tips of the ribs (Fig. 16–4). The junction between the anterior tip of each rib and its costal cartilage is called a **costochondral junction.**

The costal cartilage of the first rib forms a synchondrosis with the manubrium. This joint (at which no movement is possible) is the only one of its type in the adult skeleton.

The costal cartilage of the second rib forms a synovial joint with the sternum at the sternal angle (Figs. 16–3 and 16–4). The costal cartilages of the third through seventh ribs form synovial joints with the body of the sternum (Fig. 16–4).

The costal cartilages of the eighth, ninth, and tenth ribs all form synovial joints with the costal cartilages of the rib above (Fig. 16–4). These articulations among the costal cartilages of the seventh, eighth, ninth, and tenth ribs form, on each side, a scalloped, cartilaginous margin extending from the tip of the tenth rib to the sternum. These two margins are called the **left and right costal margins.** The paired costal margins and the intervening xiphoid process form the surface demarcation between the thorax and abdomen in the front of the body.

The costal cartilages of the eleventh and twelfth ribs do not form any articulations.

Because the costal cartilages are avascular tissues, they have a low reparative capacity. It is for this reason that physically traumatized costal cartilages heal relatively slowly, and that a focal infection of either costal margin requires aggressive treatment to prevent dissemination of the pathogens throughout both margins.

The Ribs

There are 12 pairs of ribs. All the rib pairs bear a cancellous interior of hematopoietic marrow throughout adulthood. Each rib pair is identified by its number, with the first pair being the most superior pair.

The spaces between adjacent ribs are called **intercostal spaces.** Each intercostal space is also identified by its number; the number corresponds to that of the rib bordering the upper margin of the space. For example, the intercostal space between the fifth and sixth ribs on the right side is called the right fifth intercostal space.

Each rib pair is distinguishable from all the other pairs. However, the second through ninth rib pairs have many specific features in common, and hence are regarded as typical ribs.

A typical rib has three segments, called the head, neck and body (Fig. 16–5). The **head** is the segment that forms the slightly enlarged posterior end of the rib. The **neck** is a relatively short segment that extends from the head to the point where a small prominence, called the **tubercle** of the rib, projects from the rib surface. The **body** is the longest segment of a typical rib; it extends from the tubercle to the anterior tip of

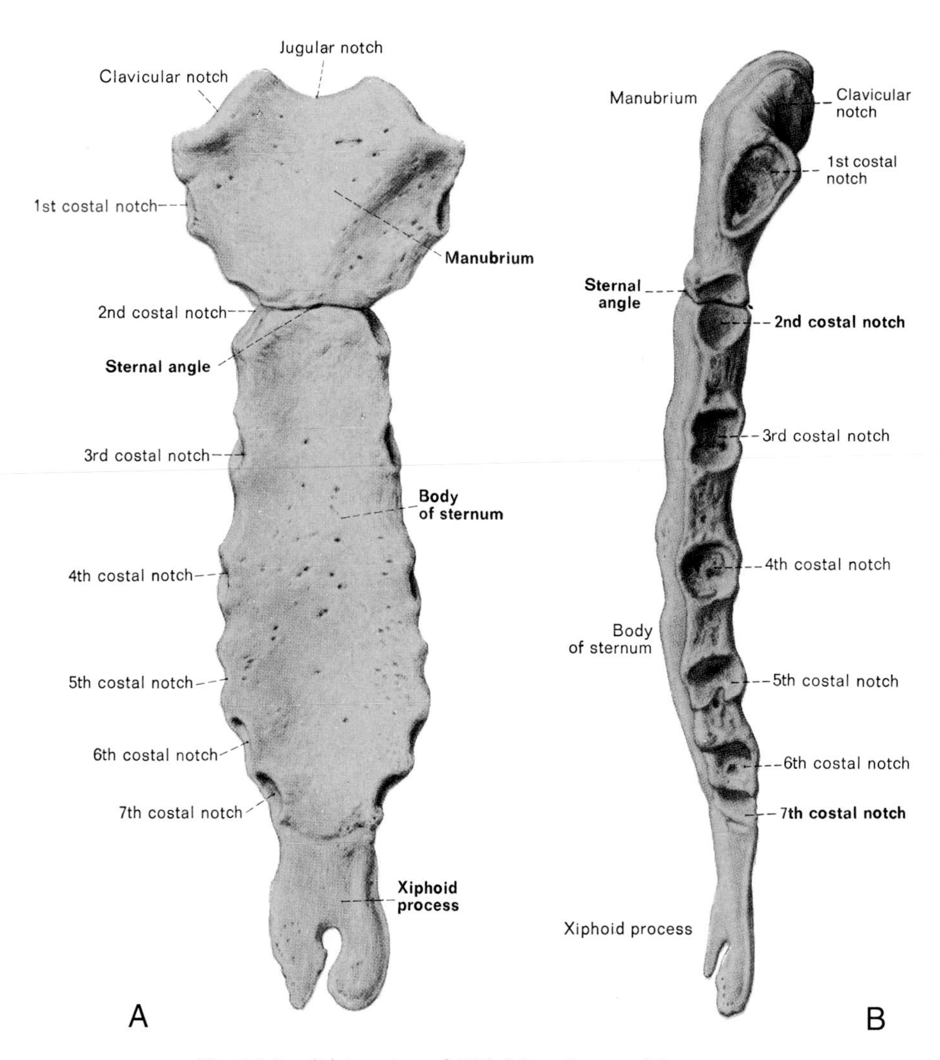

Fig. 16–3: (A) Anterior and (B) left lateral views of the sternum.

the rib. The body initially extends backward and laterally before curving sharply forward and sweeping around to the front. The region where the body turns sharply forward is called the **angle** of the rib.

A cross sectional view of the body of a typical rib shows that whereas the upper border is gently curved, the lower border is internally grooved (Fig. 16–6). The shallow groove on the lower, inner surface of a rib is called the **costal groove.**

Each typical rib forms two synovial joints with thoracic vertebrae. In one of the joints, the head of the rib articulates with both the superior costal facet on the body of the thoracic vertebra at its own level and the inferior costal facet on the body of the vertebra above. This joint is called the **costovertebral joint** of the rib. In the other joint, the tubercle of the rib articulates with the costal facet on the transverse process of the vertebra at its own level. The joint is called the **costotransverse joint** of the rib. To be specific, consider the synovial joints which the fourth rib forms with the vertebral column. In the fourth rib's costovertebral joint, the head of the rib articulates with the superior costal facet on the body of the fourth thoracic vertebra and the inferior costal facet on the body of the third thoracic vertebra. In the fourth rib's costotransverse joint, the tubercle of the rib articulates with the costal facet on the transverse process of the fourth thoracic vertebra.

THE INTERCOSTAL MUSCLES

The intercostal spaces of the chest wall each contain three overlapping layers of muscles. The muscles of each intercostal space are innervated by the intercostal nerve which courses through the space. The intercostal muscles are all muscles of respiration (muscles of breathing).

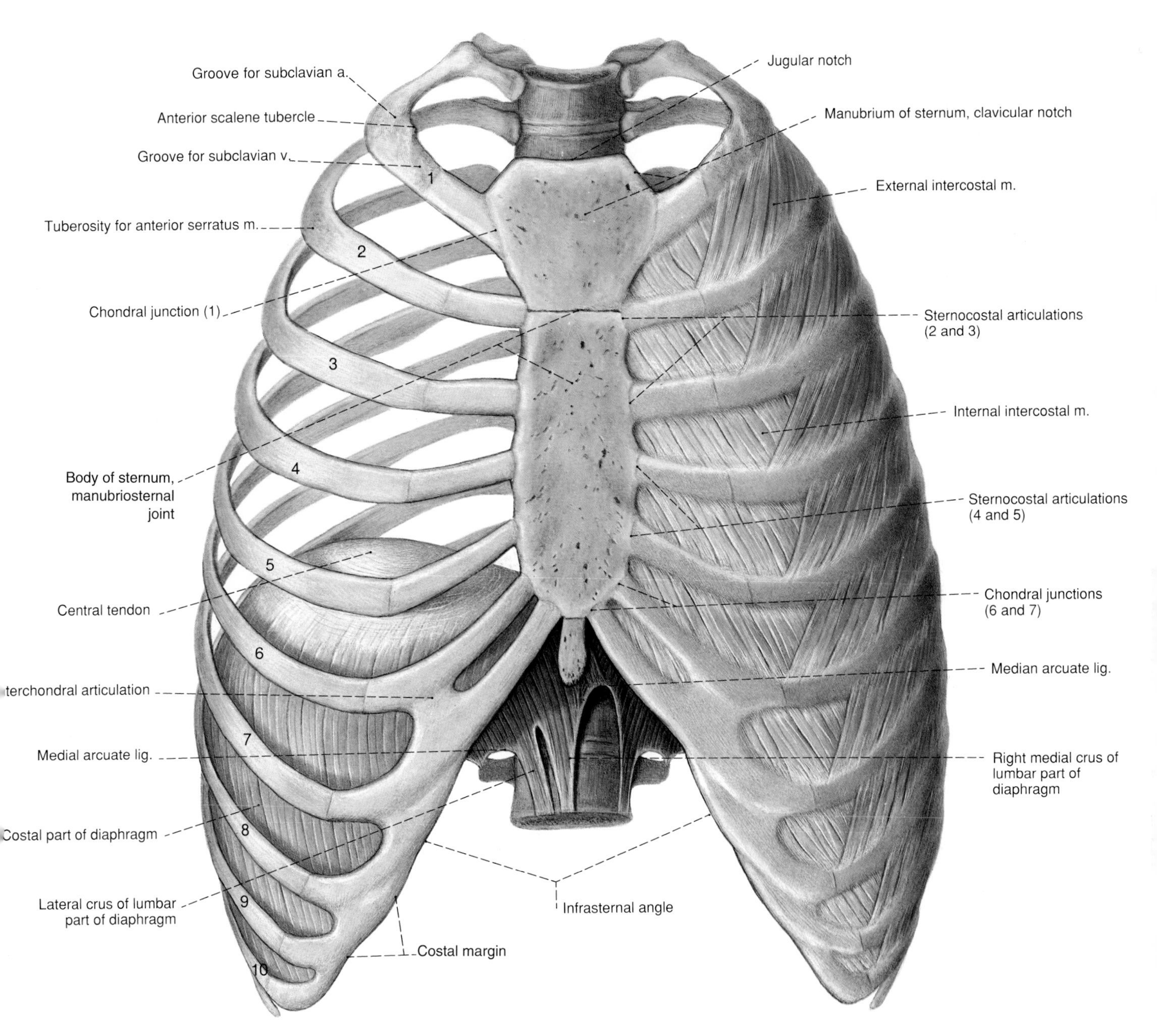

Fig. 16–4: Anterior view of the rib cage.

The External Intercostal Muscles

The most superficial muscle in an intercostal space is the external intercostal muscle (Figs. 16–4, 16–6, and 16–7). Each external intercostal muscle originates from the lower border of the body of the rib above, and inserts onto the upper border of the body of the rib below. The muscle fibers extend downward and forward from their origin. The external intercostal muscles extend posteriorly as far as the tubercle of the ribs and anteriorly as far as the anterior tips of the ribs. In the interchondral interval of each intercostal space (the interval which is bordered by the costal cartilages), the external intercostal muscle is replaced by a thin, fibrous membrane called the **external intercostal membrane.**

The Internal Intercostal Muscles

The middle muscle layer of each intercostal space is the internal intercostal muscle (Figs. 16–4, 16–6, and 16–7). Each internal intercostal muscle originates from the upper border of the lower rib, and inserts onto the floor of the costal groove of the upper rib. The muscle fibers extend upward and forward from their origin.

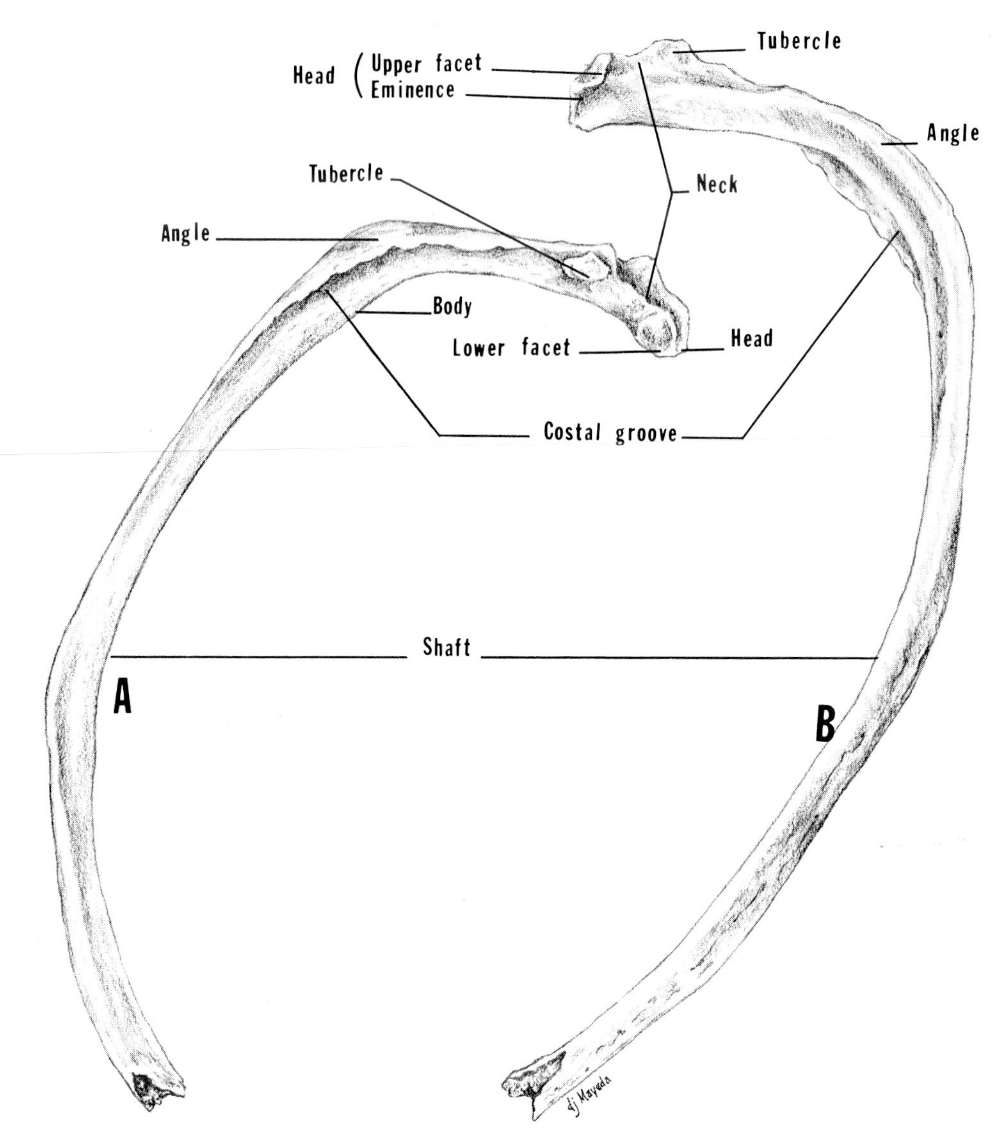

Fig. 16–5: (A) Inferior and (B) superior views of a typical left rib.

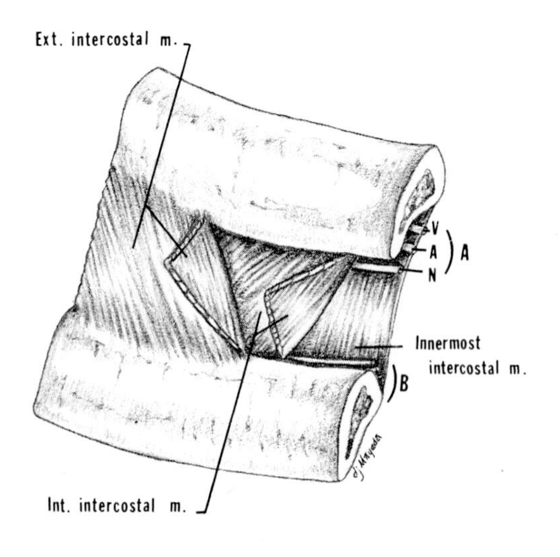

Fig. 16–6: Diagram of a lateral view of the structures in an intercostal space. The major neurovascular bundle in an intercostal space is shown at A; collateral blood vessels are shown at B. In the acronym **VAN,** the letter **V** represents an intercostal vein, **A** an intercostal artery, and **N** an intercostal nerve.

The internal intercostal muscles extend anteriorly as far as the lateral border of either the sternum or cóstal margin and posteriorly as far as the angles of the ribs. In the most posterior interval of each intercostal space (the interval extending from the rib angles to the vertebral column), the internal intercostal muscle is replaced by a thin, fibrous membrane called the **internal intercostal membrane** (Fig. 24–1).

The Deepest Intercostal Muscle Layer

The deepest muscle layer of the intercostal spaces is formed by thin, vestigial muscles. Their occurrence and location are thus variable. In general, the muscles occur in three groups: an anterior group called the **transversus thoracis** (Fig. 16–8), a lateral group called the **innermost intercostal muscles** (Fig. 16–7), and a posterior group called the **subcostal muscles.**

THE NEUROVASCULAR ELEMENTS

The nerve and blood supply of the chest wall consists largely of the neurovascular elements which course through the intercostal spaces. The major elements in each space consist of an intercostal vein, an intercostal artery, and an intercostal nerve (Fig. 16–6). The vein, artery, and nerve extend as a neurovascular bundle which lies (a) partially under the cover of the costal groove of the upper rib and (b) between the two deepest intercostal muscle layers.

The major neurovascular elements in an intercostal space are thus partially covered superficially by the lower edge of the upper rib (Fig. 16–6). The clinical importance of this anatomic relationship relates to the many instances in which diagnosis or medical treatment involves inserting a needle completely through the tissues of an intercostal space. When such a procedure is performed, incidental damage to the major neurovascular bundle in the intercostal space is mini-

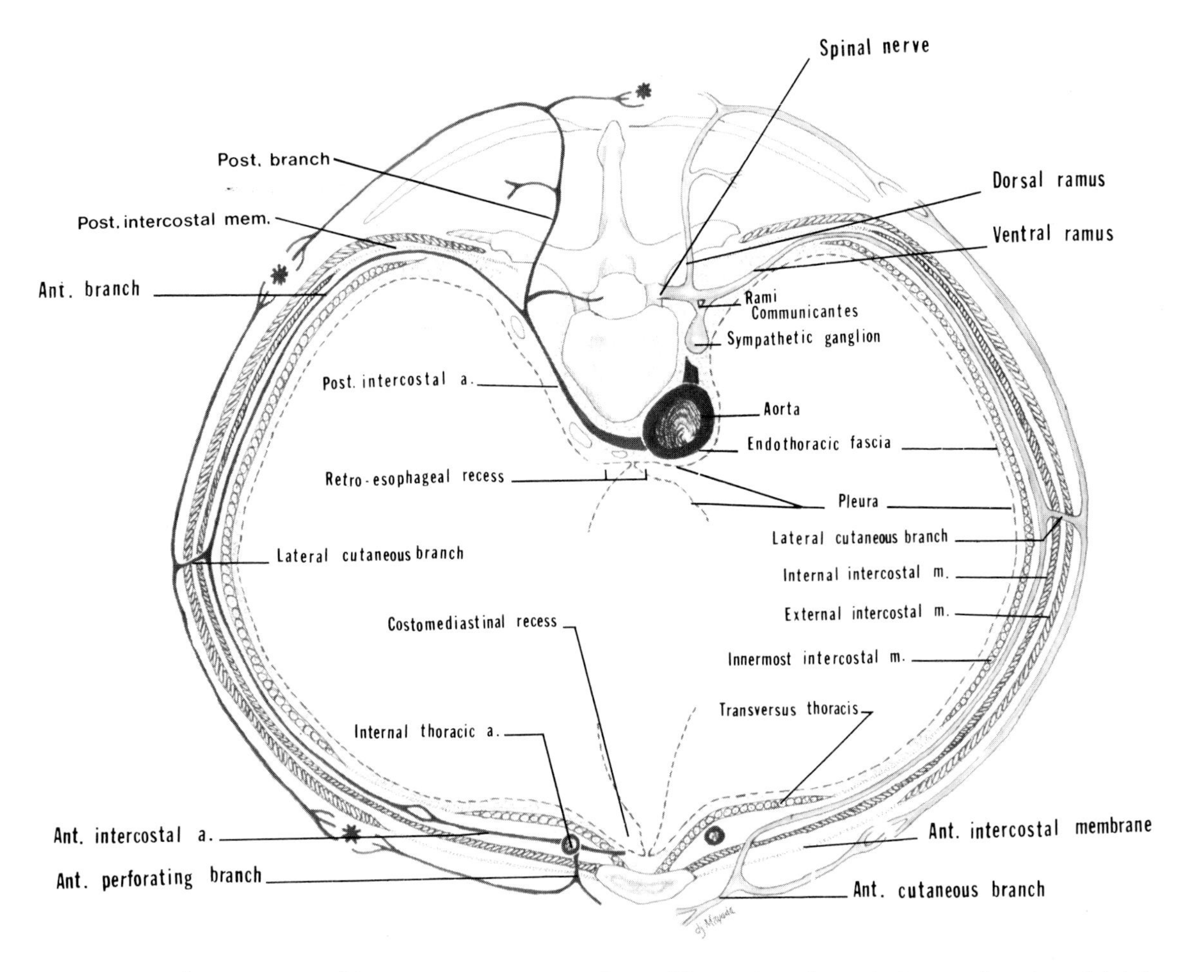

Fig. 16–7: Diagram of a superior view of the structures in an intercostal space. The anterior and posterior intercostal arteries and their branches are shown on the left, and the branches of the intercostal nerve as shown on the right.

mized by inserting the needle just above the lower rib bordering the space.

The major neurovascular bundle in each intercostal space provides a route by which pathogens can readily spread. The pathogens commonly spread anteriorly along the bundle from the initial point of infection, simply because the bundle follows a downgrade course as it extends from the posterior (vertebral) end to the anterior end of the intercostal space. If an infection tracks down to the anterior (costal margin) end of one of the seventh through tenth intercostal spaces and then produces an abscess, the abscess presents generally as a palpable bulge near the costal margin. By contrast, if an infection tracks down to the anterior (sternal) end of one of the six uppermost intercostal spaces, physical evidence of the abscess is readily concealed by the mass of the overlying pectoralis major muscle. If the infection continues to spread from this point, it generally remains deep to pectoralis major and spreads laterally toward the axilla. The sternocostal origin of pectoralis major hinders the infection from spreading medially over the sternum. Consequently, infections of the six uppermost intercostal spaces only rarely produce an abscess over the midline of the sternum. This observation is diagnostically important because it signifies that a palpable swelling over the midline of the sternum is only rarely an abscess of intercostal origin.

The Nerve Supply

The nerve supply of the chest wall is provided by **eleven pairs of intercostal nerves** (Fig. 16–9). These intercostal nerves are derived from the anterior rami of the first eleven thoracic spinal nerves, and numbered accordingly (Fig. 16–7). For example, the right fifth intercostal nerve is derived from the anterior ramus of the right fifth thoracic spinal nerve, and courses through the right fifth intercostal space under the partial cover of the costal groove of the right fifth rib.

Each intercostal nerve, with the exception of the first one, begins as almost the entire anterior ramus of the corresponding thoracic spinal nerve. The first intercostal nerve differs from this general pattern in that it begins as merely a minor branch of the anterior ramus of T1. Most of the anterior ramus of T1 forms the T1 root of the brachial plexus.

The course of the six uppermost intercostal nerves is more limited than that of the five lowest intercostal nerves. The six uppermost intercostal nerves course through their respective intercostal spaces only. By contrast, the five lowest intercostal nerves extend beyond their course through the intercostal spaces. After coursing deep to the costal cartilage of the anterior tip of their respective rib, the five lowest intercostal nerves all assume a medial or inferomedial course within the anterolateral abdominal wall.

The anterior rami of T12 are commonly called the **subcostal nerves** (Fig. 16–9). The subcostal nerves course within the posterior and anterolateral abdominal walls.

In many diseases affecting just the origins of the intercostal nerves, pain is often referred to the anterolateral cutaneous innervations of the nerves. For example, patients with tuberculous osteomyelitis of the lower thoracic vertebrae may experience pain in those regions of the anterolateral abdominal wall innervated by the lower intercostal nerves.

The Arterial Supply

Each of the nine uppermost intercostal spaces is supplied by a **pair of anterior intercostal arteries** and a single **posterior intercostal artery.** The posterior intercostal artery and its collateral branch each anastomose with one of the anterior intercostal arteries. The posterior intercostal artery and the anterior intercostal artery with which it forms anastomoses are both part of the major neurovascular bundle which lies under partial cover of the costal groove of the rib above the space. The two lowest intercostal spaces are supplied by only a posterior intercostal artery.

The anterior intercostal arteries are direct or indirect branches of the internal thoracic artery (Figs. 16–7 and 16–8). On each side of the thorax, the **subclavian artery** gives rise to a major branch called the **internal thoracic artery.** The internal thoracic artery extends inferiorly alongside the deep surface of the chest wall (just lateral to the sternum). At the level of the sixth intercostal space, it bifurcates into two major arteries, the **musculophrenic** and **superior epigastric arteries.** On each side, the anterior intercostal arteries of the five uppermost intercostal spaces are all branches of the internal thoracic artery, and those of the next four lower intercostal spaces are all branches of the musculophrenic artery.

The posterior intercostal arteries for the nine lowest intercostal spaces are all branches of the descending thoracic aorta (Figs. 16–7 and 16–9). The posterior intercostal arteries for the two uppermost intercostal spaces are all branches of the superior intercostal artery. The **superior intercostal artery (supreme intercostal artery)** is a branch of the costocervical trunk, which, itself, is a branch of the subclavian artery (Fig. 16–9).

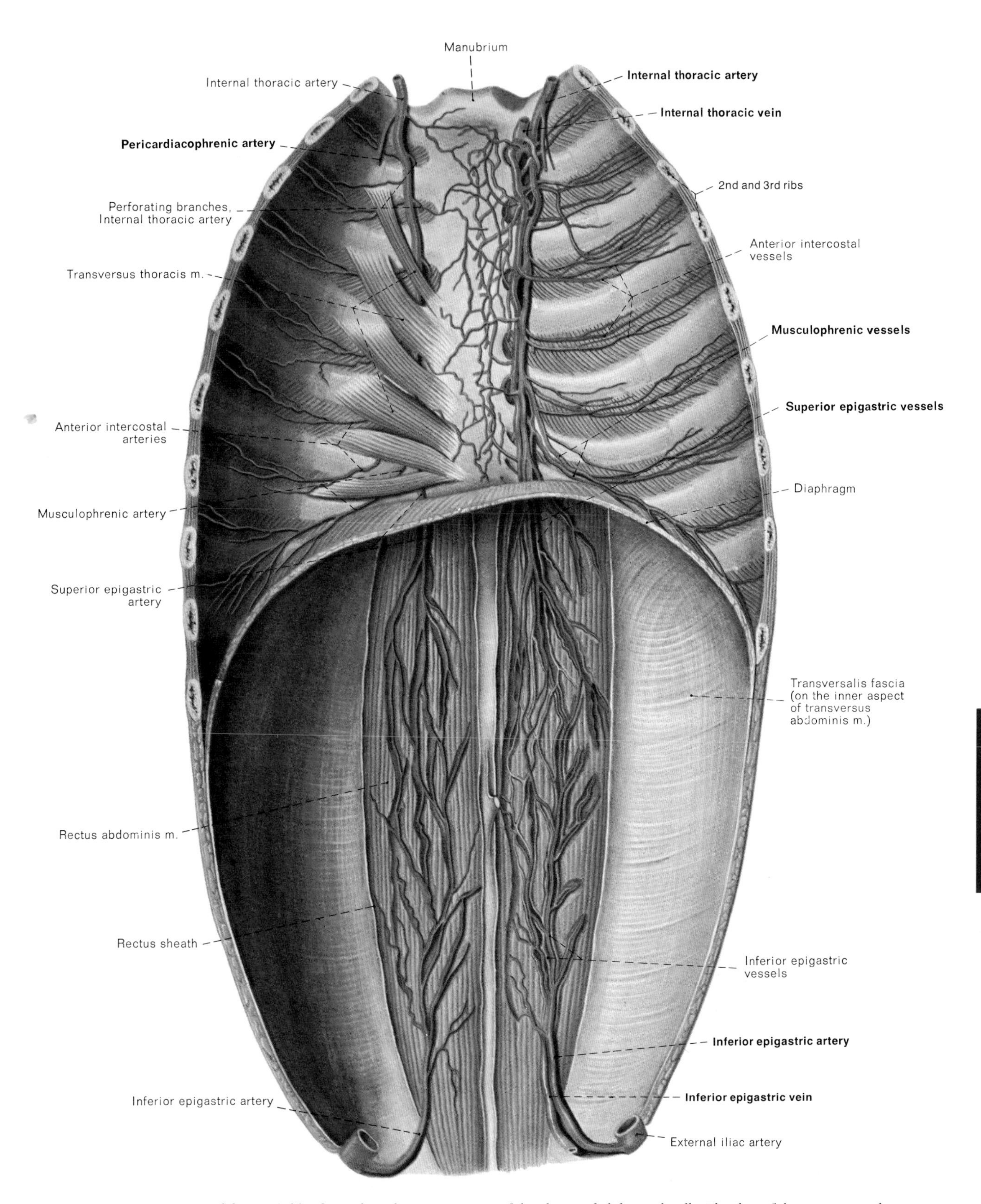

Fig. 16–8: Posterior view of the major blood vessels in the anterior parts of the chest and abdominal walls. The slips of the transversus thoracic muscle have been removed on the right side. The posterior part of the rectus sheath has been removed on both sides.

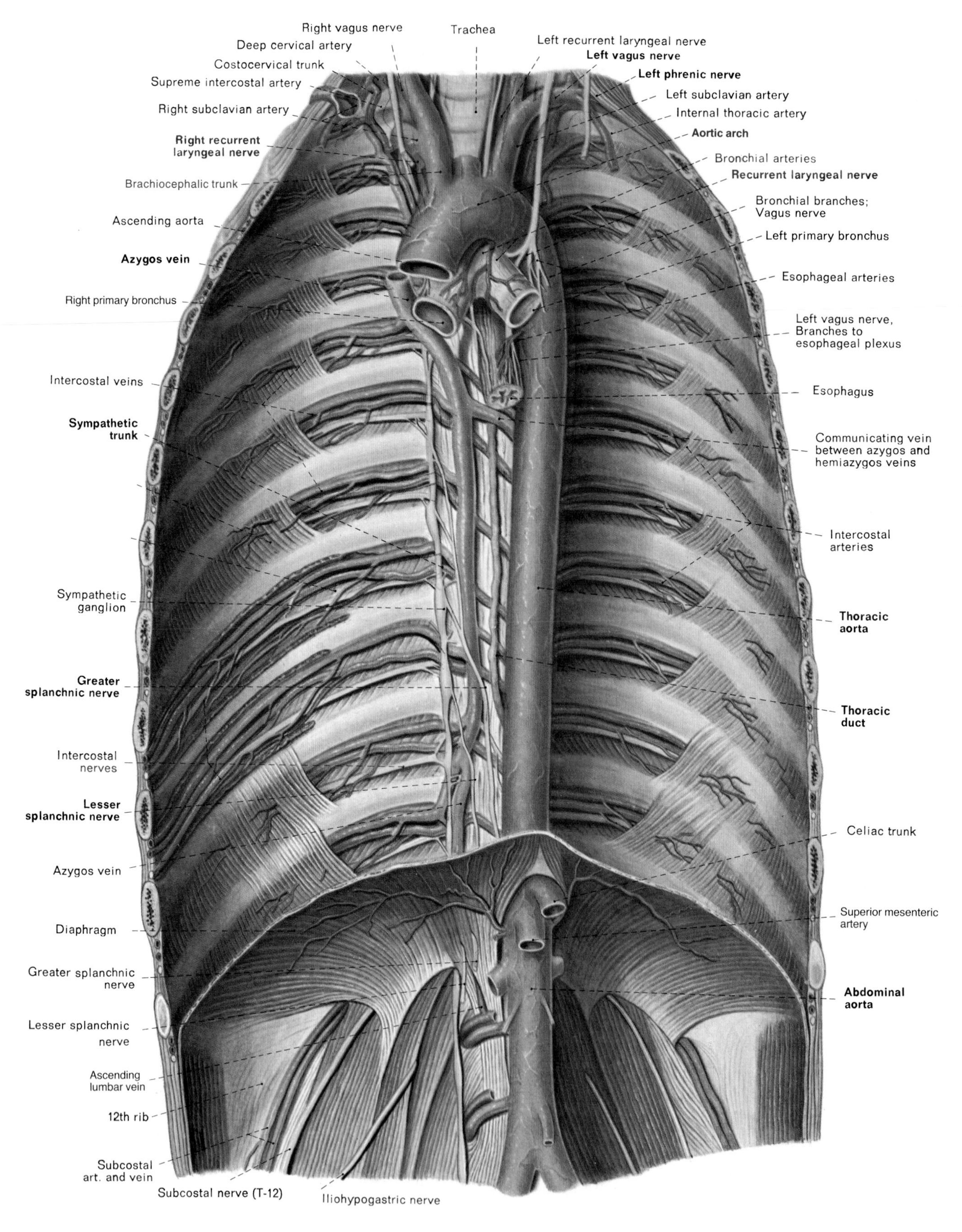

Fig. 16–9: Anterior view of the nerves and major blood vessels in the posterior part of the chest wall. The innermost intercostal muscles have been removed from the right eighth and ninth intercostal spaces.

CLINICAL PANEL IV-1

CONGENITAL COARCTATION OF THE THORACIC AORTA

Description: Congenital coarctation (constriction) of the aorta occurs in one of every 2000 births. The coarctation frequently occurs in the segment of the aorta immediately distal to the origin of the left subclavian artery. In order to simplify the discussion of the clinical features of congenital aortic coarctation, the following discussion is limited only to cases in which (a) the coarctation occurs immediately distal to the origin of the left subclavian artery (Fig. 16–10) and (b) there are no other congenital cardiovascular anomalies.

Common Symptoms: Congenital aortic coarctation is generally asymptomatic if the diameter of the aorta has not been effectively reduced by more than 50%. If the coarctation is physiologically significant, health-related effects, such as decreased exercise tolerance and fatigability, may initially appear at any time from infancy to early childhood.

Common Physical Signs: A physiologically significant aortic coarctation typically delays the timing of the pulse and diminishes the pulse volume of those arteries that arise from the aorta distal to the coarctation relative to those arteries that arise proximal to the coarctation. Accordingly, simultaneous palpation of radial and posterior tibial pulsations typically reveals the posterior tibial pulses to be delayed and of lesser pulse volume than the radial pulses.

Characteristic Radiographic Signs: For young children of school age with a physiologically significant aortic coarctation, PA chest films may show the outline of the number "3" in the upper part of the left border of the cardiovascular shadow (Fig. 16–11B). The anatomic basis of this abnormal radiographic finding is as follows:

In PA chest films, the heart and the major blood vessels attached to it cast almost all of the central radiodense shadow in the film (compare Figs. 6–2 and 16–11A). Therefore, the central radiodense shadow can be called the cardiovascular shadow. The radiolucent lung fields border the sides of the cardiovascular shadow (Fig. 6–2). The aortic segment immediately distal to the origin of the left subclavian artery casts part of the upper left border of the cardiovascular shadow; the rounded outline of this aortic segment is called the aortic knob (compare Figs. 16–11A and 17–13). A coarctation along this aortic segment can produce dilatations of the aorta in the regions immediately proximal and distal to the coarctation. The separation of these dilatations by the narrow coarctation can transform the shape of the upper left border of the cardiovascular shadow into an outline of the number "3" (Fig. 16–11B). The upper and lower rounded parts of the number "3" respectively represent the dilatations immediately above and below the coarctation. The indentation in the number "3" outline marks the location of the coarctation.

For young children of school age with a physiologically significant aortic coarctation, PA chest films may also reveal notches in the lower borders of the posterior segments of any of the third to seventh ribs (Fig. 16–12). The anatomic basis of the rib notches is as follows:

Congenital aortic coarctation impedes aortic blood flow to the lower limbs. A collateral circulation develops in afflicted individuals, which compensates for this decrease in aortic flow. The collateral circulation takes advantage of the anastomoses between the superior and inferior epigastric arteries in the anterior abdominal wall (Figs. 16–8 and 16–10). The superior epigastric arteries are branches of the internal thoracic arteries, which, in turn, are branches of the subclavian arteries. The inferior epigastric arteries are branches of the external iliac arteries. The anastomoses between the superior and inferior epigastric arteries thereby provide an arterial route from the aortic arch to the femoral arteries which is collateral to that provided by the descending thoracic and abdominal aorta and the latter's common iliac branches.

The manner in which the intercostal arteries contribute to this collateral arterial route is as follows: The subclavian and axillary arteries give rise to branches which anastomose with each other around the medial and lateral borders of the scapula (Figs. 8–1 and 16–10). The arterial anastomoses extending along the medial border of the scapula anastomose, in turn, with branches of upper posterior intercostal arteries. In a normal individual, these latter anastomoses are poorly developed and do not conduct much blood. In a person afflicted with aortic coarctation, however, the anastomoses between the arteries bordering the scapula and the posterior intercostal arteries become very well developed. A prominent blood flow is soon established which travels from the subclavian and axillary arteries to the internal thoracic arteries (and also the descending thoracic aorta) by way of the intercostal arteries.

CLINICAL PANEL IV-1, continued

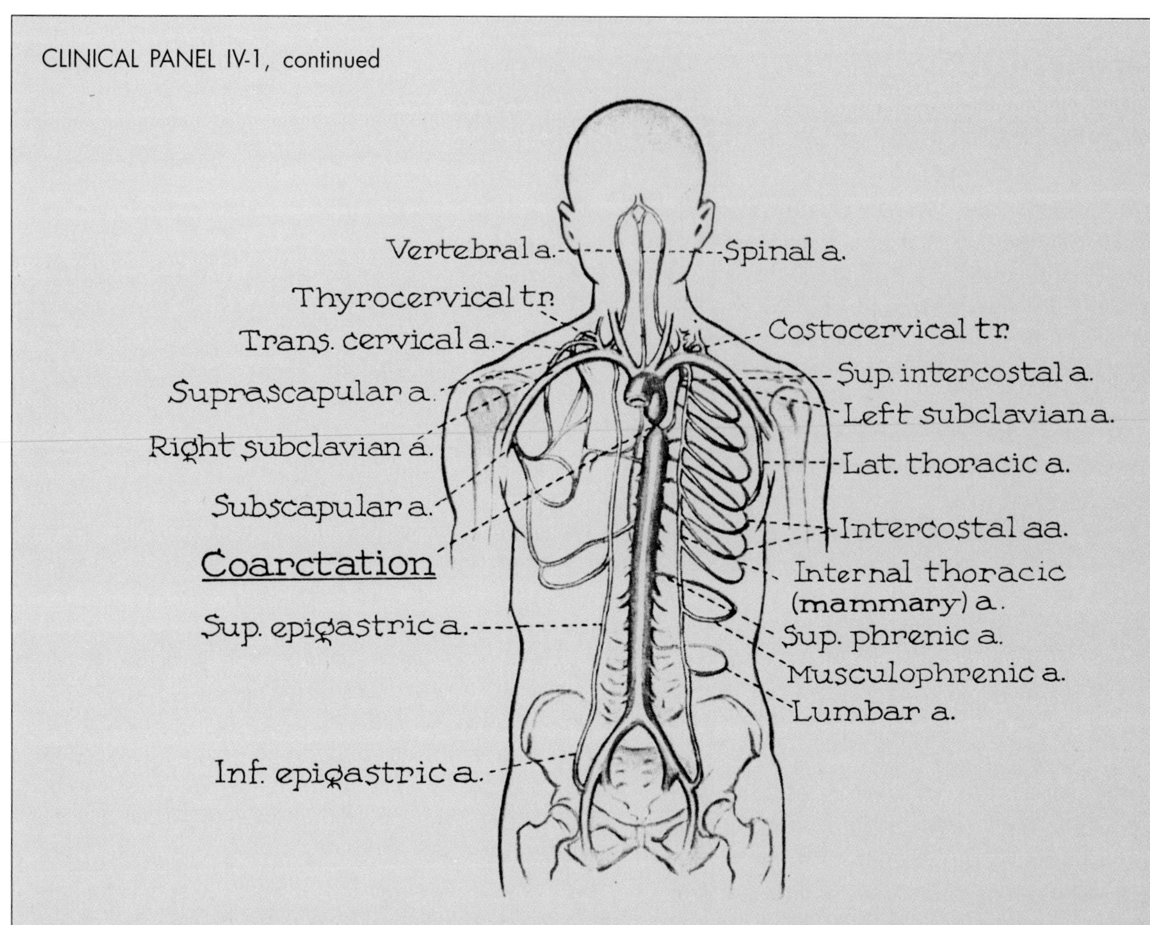

Fig. 16–10: Diagram of the routes of collateral circulation to the lower limbs in an individual with an aortic coarctation immediately distal to the origin of the left subclavian artery.

When intercostal arteries conduct a relatively high blood flow in infants and children with aortic coarctation, the walls of the arteries develop dilatations. These dilatations ultimately develop, in turn, notches in the ribs of young, school-age children. Accordingly, radiographic evidence of prominent notching of the lower border of the upper ribs is a frequent finding in children with congenital aortic coarctation. Such notching is commonly most pronounced along the posterior segments of the ribs. This is because the neurovascular bundle which courses through each intercostal space lies closest to the overlying costal groove of the rib above along the posterior interval of the space.

CLINICAL PANEL IV-1, continued

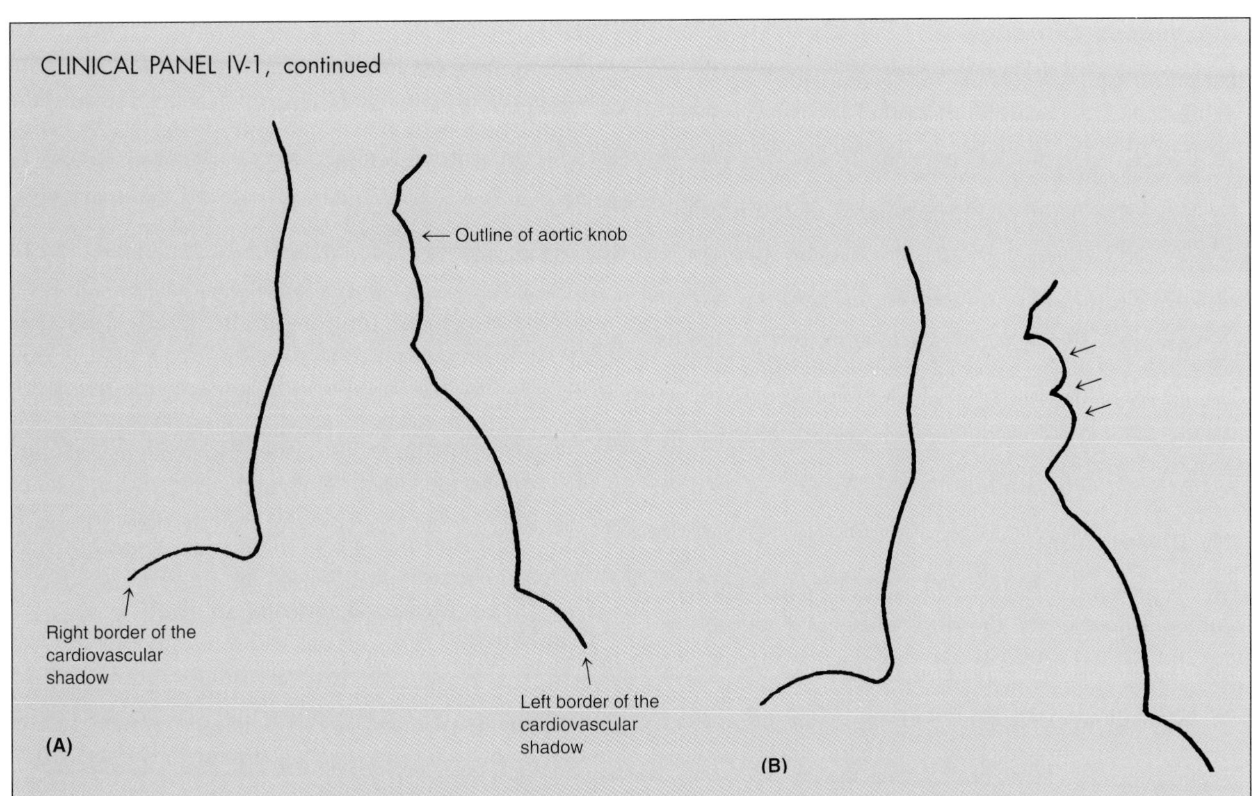

Fig. 16–11: (A) A schematic drawing of the left and right borders of the cardiovascular shadow of a normal PA chest film. The left and right borders of the cardiovascular shadow respectively represent the radiographic outlines of the left and right sides of the patient's heart and the major blood vessels attached to it. (B) A schematic drawing of the borders of the cardiovascular shadow of an individual with dilatations immediately above and below a coarctation in the thoracic aortic segment immediately distal to the origin of the left subclavian artery. The dilatations and the intervening coarctation cast an outline in the shape of the number "3" (arrows).

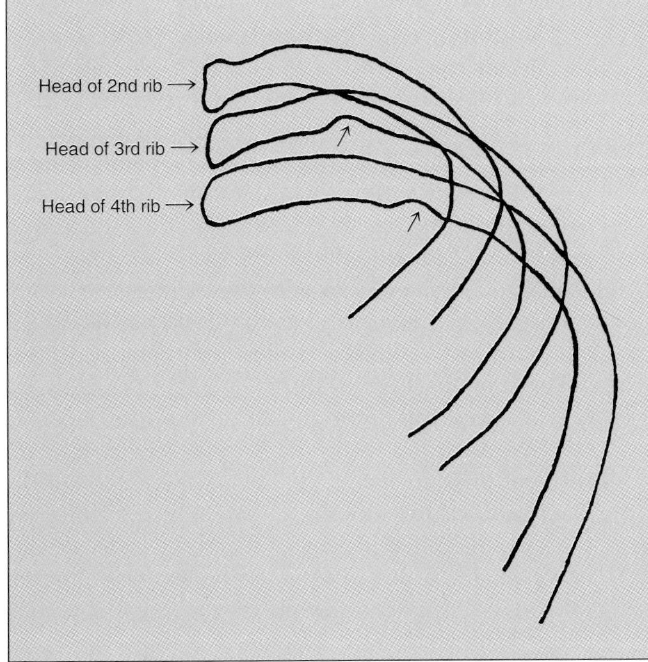

Fig. 16–12: A schematic drawing of the shadows cast by the second, third and fourth left ribs in a PA chest film. The drawing shows the typical locations of notches (arrows) in the lower borders of the third and fourth ribs that may appear in individuals suffering from a coarctation in the thoracic aortic segment immediately distal to the origin of the left subclavian artery.

The Venous Drainage

The intercostal arteries are all accompanied by venae comitantes. The venae comitantes of the anterior intercostal arteries ae tributaries of the venae comitantes of the musculophrenic and internal thoracic arteries (Fig. 16–8), which, in turn, are tributaries of the brachiocephalic veins.

The venae comitantes of the nine lowest posterior intercostal arteries are tributaries of the azygos system of veins (Fig. 19–5): the azygos system of veins drains ultimately into the superior vena cava. The venae comitantes of the two uppermost posterior intercostal arteries are ultimately drained by the brachiocephalic veins.

THE DIAPHRAGM

The diaphragm is the partially muscular and partially tendinous floor of the thorax. It presents a convex surface to the thoracic cavity above and a concave surface to the abdominal cavity below.

The diaphragm presents not only an overall convex surface to the thoracic cavity above, but also two localized convexities, one on each side of the body (Fig. 22–4). These smaller, upward bulges in the diaphragm's contour are called the **domes (cupulae)** of the diaphragm. The right dome is generally higher than the left dome. The right dome represents the region of the diaphragm which rests directly atop the most superior surface of the liver. The left dome represents a region of the diaphragm which commonly rests atop the fundus of the stomach.

The diaphragm's muscle fibers originate from three distinct sets of structures which border the periphery of the diaphragm (Fig. 16–13).

1. Anteriorly in the midline, muscle fibers originate from the deep surface of the xiphoid process.
2. Anterolaterally, muscle fibers originate from the deep surfaces of the six lowest pairs of costal cartilages.
3. Posteriorly, muscle fibers originate from upper lumbar vertebrae and deep fascial thickenings bordering the upper margins of psoas major and quadratus lumborum (Fig. 24–1).

The thickened fascial layers bordering the upper margins of psoas major and quadratus lumborum are respectively called the **medial** and **lateral arcuate ligaments.**

On each side, the muscle fibers that originate from the upper lumbar vertebrae extend superomedially as a muscular pillar (Fig. 24–1). The muscular pillar on the left is called the **left crus,** and its muscle fibers originate from the bodies of the two uppermost lumbar vertebrae. The pillar on the right is called the **right crus,** and its muscle fibers originate from the bodies of the three uppermost lumbar vertebrae. The left and right crura cross into one another as they extend toward the central region of the diaphragm. A thickened fascial layer called the **median arcuate ligament** covers the undersurface of the region where the two crura cross into each other (Fig. 17–4A).

The muscle fibers of the diaphragm extend centrally to insert into its central tendinous region, which is called the **central tendon of the diaphragm** (Fig. 16–13). The central tendon of the diaphragm is thus the structure which the diaphragm's muscle fibers primarily pull upon when they concentrically contract.

The diaphragm is innervated by the left and right phrenic nerves. The phrenic nerve arises from the cervical plexus. The C3, C4, and C5 spinal cord levels commonly contribute fibers to the phrenic nerve.

The Three Major Openings in the Diaphragm

There are three major openings in the diaphragm for transmission of large structures between the thorax and abdomen. Each opening is named for the largest structure which passes through it.

The **aortic opening (aortic hiatus)** is the lowest lying opening (Fig. 24–1); it lies at the level of the body of the twelfth thoracic vertebra (Fig. 16–1). Strictly speaking, the aortic opening is not an opening through the diaphragm, but rather a large indentation in its posterior margin. The left and right crura form the left and right borders of the aortic opening (Fig. 24–1). The aortic opening transmits the aorta, azygos vein, and thoracic duct (Fig. 16–13 does not depict the transmission of the thoracic duct through the aortic opening).

The **esophageal opening (esophageal hiatus)** is the intermediate-level opening (Fig. 24–1); it lies at the level of the body of the tenth thoracic vertebra (Fig. 16–1). The esophageal opening usually splits the muscle fibers of the right crus (Fig. 24–1). Some of these fibers decussate with (cross over) each other above and to the left of the opening; they form, in effect, a muscular sling around the esophageal opening. The esophageal opening transmits the esophagus, the left and right vagus nerves, the esophageal branches of the left gastric artery, and the esophageal tributaries of the left gastric vein (Fig. 16–13).

The **caval opening** is the highest lying opening (Fig. 24–1); it lies at the level of the eighth thoracic vertebra. It is an opening in the central tendon of the diaphragm just right of the midline. The caval opening transmits the inferior vena cava and the right phrenic nerve (Fig. 16–13). The left phrenic nerve passes through the diaphragm by piercing the central tendon of the diaphragm on the left side.

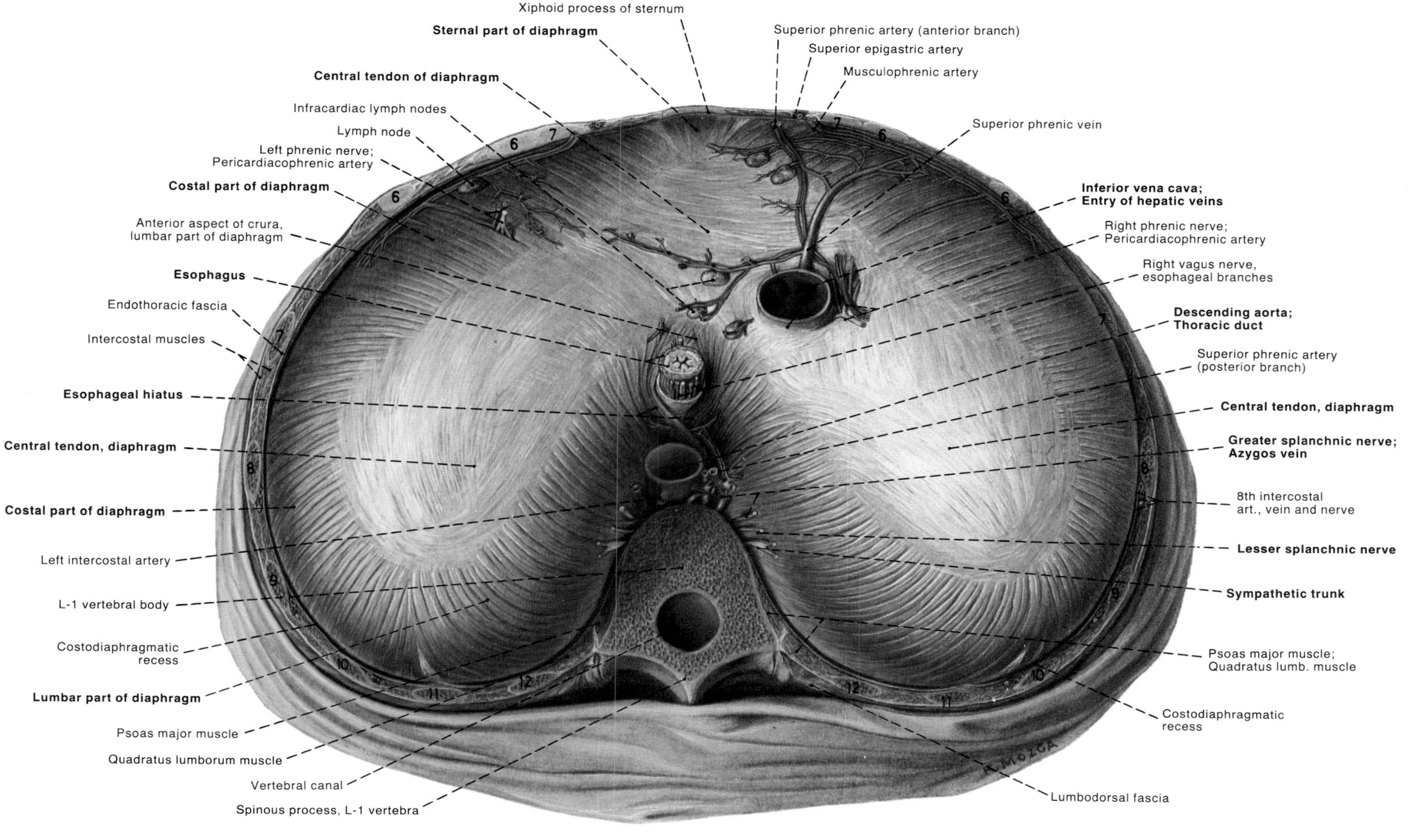

Fig. 16–13: Superior view of the diaphragm, its openings, blood vessels, and nerves.

THE SUPRAPLEURAL MEMBRANE

Each side of the thoracic inlet is roofed over by a distinct fascial layer called the suprapleural membrane (Sibson's fascia) (Fig. 16–14). A minor muscle called scalenus minimus, which originates from the transverse process of the seventh cervical vertebra, inserts its muscle fibers into the suprapleural membrane. Accordingly, the suprapleural membrane can be regarded as the aponeurotic tendon of insertion of scalenus minimus.

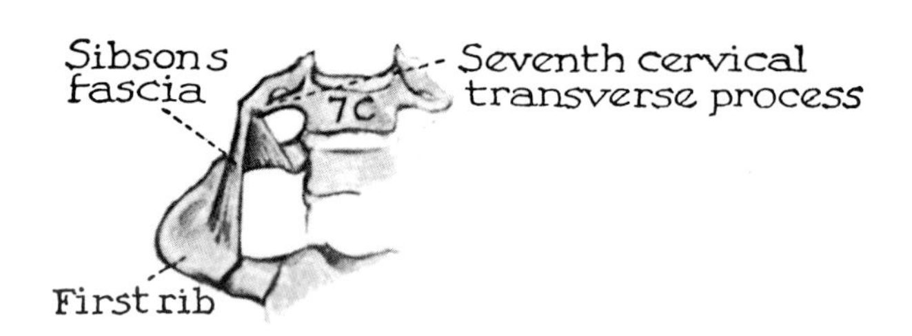

Fig. 16–14: Diagram of an anterior view of the suprapleural membrane (Sibson's fascia).

CASE **IV.1**

The Case of Sean O'Brian

INITIAL PRESENTATION AND APPEARANCE OF THE PATIENT

A 19 year-old, white man named Sean O'Brian walks into the student infirmary. The patient is perspiring, his hair is tousled, his clothes are grass-stained, and his shoes are covered with dirt.

QUESTIONS ASKED OF THE PATIENT

What can I do for you? I'd like for you to look at my chest. I hurt it pretty badly about an hour ago playing football. *[The question "What can I do for you" not only expresses the examiner's desire to help the patient, but also invites the patient to openly and completely discuss his health concerns. In this instance, the examiner quickly learns that the patient sustained a painful chest injury about an hour ago. At this point, the examiner tentatively concludes that the patient has sustained an injury to one or more of the following: the chest wall, a thoracic viscus, or an abdominal viscus. An abdominal viscus may be injured because the lower ribs of the rib cage lie at the level of not only the outer lower margins of the lungs but also the viscera in the upper third of the abdomen.]*

Do you recall what happened? Yes. I was running after the guy with the football when I was knocked down by another player. I fell pretty hard on the left side of my chest when I hit the ground. *[This question provides the examiner with information about the mechanism of injury.]*

Where does it hurt? Over here [the patient puts his right hand behind the back and places the dorsum of the right hand over the posterior aspect of the lower part of the rib cage on the left side; the patient winces as he makes the necessary moves]. *[When pain is a complaint, it is always essential to have the patient mark by word or gesture the exact site(s) at which pain is perceived. In cases of physical injury, pain is generally a reliable indicator of the site(s) of injury.]*

How would you describe the pain? It's a very sharp pain. *[The presence of sharp pain suggests injury to the chest wall, as it is innervated by sensory fibers which can produce the sensation of sharp pain.]*

Did you feel anything snap or break when you hit the ground? No, but I felt the pain right after hitting the ground.

I noticed that you winced when you tried to show me where your chest hurts. Did putting your hand behind your back make the pain worse? Yes.

Is there anything else that makes the pain worse? Yes. It hurts for me to take a deep breath.

Is there anything that relieves the pain? Well, my chest hurts the least when I keep still. *[The last three questions provide information about the source of pain. In this instance, the finding that the pain is worsened by physical and respiratory movements of the rib cage and relieved by rest suggests injury of one or more musculoskeletal structures of the chest wall.]*

Do you have trouble catching your breath or feel short of breath? No. *[The examiner asks this question because injury to the chest wall can result in partial or complete collapse of a lung. Such a collapse would cause the patient to be dsypneic (short of breath). The patient's negative response indicates to the examiner that the patient's lungs are not collapsed or that, at worst, one of the lungs may be minimally collapsed.]*

Is there anything else that is also bothering you? No. *[The examiner finds the patient alert and fully cooperative during the interview.]*

PHYSICAL EXAMINATION OF THE PATIENT

Vital signs: Blood pressure
Lying supine: 120/70 left arm and 120/70 right arm
Standing: 120/70 left arm and 120/70 right arm
Pulse: 65
Rhythm: regular
Temperature: 98.3°F.
Respiratory rate: 14
Height: 6′1″
Weight: 155 lbs.
HEENT Examination: Normal
Lungs: Inspection, palpation, auscultation, and percussion of the lungs are normal except for the following findings on the left side of the chest: Palpation lo-

cates the site of pain to a region overlying the posterolateral segment of the left tenth rib. The site is tender to palpation. Compression to the rib cage intensifies pain at the site. Pertinent normal findings include normal respiratory and voice sounds and normal percussion around the site of pain.

Cardiovascular Examination: Normal

Abdomen: Normal. Pertinent normal findings include the absence of LUQ pain and any abdominal tenderness. (The acronym LUQ refers to the left upper quadrant of the abdomen.)

Genitourinary Examination: Normal

Musculoskeletal Examination: Normal

Neurologic Examination: Normal

INITIAL ASSESSMENT OF THE PATIENT'S CONDITION

The patient has sustained an acute injury of the posterolateral aspect of the left side of the chest wall.

ANATOMIC BASIS OF THE PATIENT'S HISTORY AND PHYSICAL EXAMINATION

1. **The patient has sustained an acute injury of musculoskeletal tissues on the left side of the chest wall.** The questioning establishes that it is the patient's perception that he has sustained an acute injury to the left side of the chest and that, as a result of the injury, he now has pain upon movement. In general, pain upon movement indicates injury or disease of one or more of the bones, joints, and muscles involved in the movement.
2. **The combination of pain and tenderness over the posterolateral aspect of the left tenth rib marks the region as a site of injury.** Tenderness is pain elicited by palpation. Tenderness generally marks a region as a site of injury or disease. The palpable tissues overlying the posterolateral aspect of the left tenth rib include (proceeding from the most superficial to the deepest) the skin, the superficial and deep fascia, latissimus dorsi, and the rib itself.
3. **The intensification of the left tenth rib pain upon compression of the rib cage indicates fracture of the rib.** An examiner can apply compression to a patient's rib cage by placing one hand against the patient's sternum, the other hand against the patient's upper thoracic vertebrae, and then pressing the hands toward each other. Intensification of chest wall pain indicates a rib fracture.

FINAL RESOLUTION OF THE PATIENT'S CONDITION

The history and physical examination indicate that the patient has sustained a fracture of the posterolateral segment of the left tenth rib.

CLINICAL REASONING PROCESS

This case illustrates the general kinds of information an examiner seeks from patients who present with chest pain. The principal purpose of the clinical examination of chest pain is to determine accurately the anatomic site or sites involved. From the history, the examiner seeks information about the location of the pain [Where is the pain? Is its location constant?], the time and mode of onset of the pain [Do you recall when the pain began and what you were doing at that time?], the quality and severity of the pain [How would you describe the pain? How severe is the pain? Has the pain affected your daily or regular activities?], the temporal profile of the pain [Is the intensity of the pain constant, and, if not, how would you describe its duration and variation over a 24-hour period?], aggravating and relieving factors [Is there anything that makes the pain worse? Is there anything that relieves the pain?], and the presence of associated problems [Is there anything else that is bothering you?].

In the physical exam of the lungs, the examiner gathers information through an ordered sequence of steps that involves first visual inspection, then palpation, third, percussion, and finally, auscultation of the chest wall. **Visual inspection** provides information about the rate, rhythm, depth, symmetry, and effort of respiratory movements and any deformity of the chest wall. **Palpation** reveals sites of tenderness, provides physical assessment of respiratory movements (as determined by the extent to which respiratory movements move the examiner's hands when the hands are laid on the patient's chest wall), and monitors tactile fremitus (the detection of palpable chest wall vibrations produced by the passage of the patient's spoken voice sounds through the lungs to the overlying chest wall regions). **Percussion** provides information about the average density of peripheral lung regions subjacent to the chest wall. **Auscultation** assesses the nature of respiratory and voice sounds and detects adventitious sounds (unusual or abnormal sounds).

In this case, the history of recent trauma to the chest, the co-location of pain and tenderness over a rib segment, and the intensification of the pain by chest compression are collectively diagnostic of a rib fracture. The absence of any palpable physical discontinuity over the site of the rib fracture suggests an undisplaced fracture. Physicians would differ in their assessment of this case with regard to the necessity of obtaining chest or rib films to document radiographically the presence of a rib fracture, because undisplaced rib fractures are frequently difficult to detect radiographically. In most instances, an emergency or primary care physician would advise rest and light physical activity for two weeks to promote healing of the fracture and analgesics to provide less painful respiration.

The posterolateral segments of the ninth, tenth and eleventh ribs on the left side overlie the posterolateral base of the left lung and the pleural space about it (Fig. 22–10). Immediately deep to this part of the left lung lie the diaphragm and the spleen. Therefore, fractures of these ribs along their posterolateral segments may be associated with injury to the left pleural space, the left lung, and the spleen. Injury to the pleural space and/or lung could result in a pneumothorax or a hemothorax. Injury to the spleen could result in a ruptured spleen or intracapsular hemorrhage of the spleen. These considerations explain the pertinence in this case of the normal findings upon auscultation and percussion of the posterolateral base of the left lung and the absence of abdominal tenderness or LUQ abdominal pain. A physican would advise the patient to seek immediate medical attention if the patient suddenly found it difficult to breathe or began to experience abdominal pain.

RECOMMENDED REFERENCES FOR ADDITIONAL INFORMATION ON THE THORACIC WALL

Bates, B., A *Guide to Physical Examination and History Taking*, J. B. Lippincott Co., Philadelphia, 1991: *Pages 241 through 255 in Chapter 8 present a discussion of the visual inspection, palpation, percussion, and auscultation of the chest.*

DeGowin, R. L., Jochimsen, P. R., and E. O. Theilen, *DeGowin & DeGowin's Bedside Diagnostic Examination, 5th ed.*, Macmillan Publishing Co., New York, 1987: *Pages 227 through 231 summarize the signs and symptoms of a number of injuries and diseases of the chest wall.*

CHAPTER 17

The Lungs and the Mechanics of Respiration

The lungs occupy the left and right lateral regions of the thorax. A serous membrane-lined space, called a **pleural space (pleural cavity),** envelops each lung (Fig. 17–1A). The pleural space provides a lubricated, microscopically-thin, free space between the lung and the thoracic wall regions which the lung faces. The pleural space also partially separates the lung from the thoracic viscera which lie between the lungs in the mid-region of the thorax.

The central mid-region of the thorax is called the **mediastinum.** The most prominent mediastinal viscera are the heart and the great veins and arteries extending to and from it, the trachea, the esophagus, and the thymus.

THE MAJOR SURFACE FEATURES OF THE LUNGS

The Apex and Surfaces of Each Lung

Each lung is basically pyramidal, and has an apex and three external surfaces. The **apex** projects into the base of the neck (Fig. 22–4), and is roofed over by the suprapleural membrane. The three surfaces are identically named in both lungs (Figs. 17–2 and 17–3): each lung has a **mediastinal surface** which faces the mediastinal viscera, a **diaphragmatic surface** which lies atop the diaphragm, and a **costal surface** which is apposed to the inner surface of the rib cage.

The Root and Hilum of Each Lung

Each lung is attached to viscera in the mediastinum by a bundle of structures called the root of the lung (Fig. 17–1A). The root of each lung consists of all those structures which enter or exit the lung; the most prominent structures are the lung's main stem bronchus, pulmonary artery, and pulmonary veins (Figs. 17–2B and 17–3B). The hilum of each lung is that region of the lung's mediastinal surface at which structures are transmitted between the root of the lung and the parenchyma of the lung.

The Fissures and Lobes of Each Lung

Each lung bears a deep fissure, called the **oblique fissure,** which runs obliquely through the lung (Figs. 17–2A and 17–3A). The right lung only has an additional fissure, called the **horizontal fissure,** which extends horizontally through it (Fig. 17–3A).

The two fissures of the right lung divide it into three lobes: an **upper lobe,** a **middle lobe,** and a **lower lobe** (Fig. 17–3A). The upper lobe of the right lung is the lobe which lies above both fissures, the middle fissure is the lobe which lies between the two fissures, and the lower lobe is the lobe which lies below the oblique fissure.

The single fissure of the left lung divides it into two lobes: an **upper lobe** and a **lower lobe** (Fig. 17–2A). The anteroinferior region of the upper lobe is called the **lingula** (Fig. 17–4A); the lingula is considered to be the left lung's homologue of the right lung's middle lobe.

THE SURFACE PROJECTIONS OF THE LUNGS

When conducting a physical examination, it is important to know the bony and cartilaginous structures which mark the surface projections of the lungs and their lobes. There are four notable surface relationships:

1. The apex of each lung projects above the medial third of the clavicle (Fig. 22–4). In an adult, the apex of each lung projects about an inch above the medial third of the clavicle. The obliquity of the thoracic inlet accounts for this surface relationship. The oblique lie of the thoracic inlet within the body places all but the inlet's most anterior region above the clavicles.

   The apex of the lung and the outer lining of the pleural space above the apex are always in jeopardy when a penetrating wound is suffered above the medial third of the clavicle. If the outer pleural lining is penetrated, air may enter the pleural space (producing a condition known as a **pneumothorax**) or blood may enter the space (producing a condition known as a **hemothorax**).

Thorax: Pleural Cavities and Pleurae

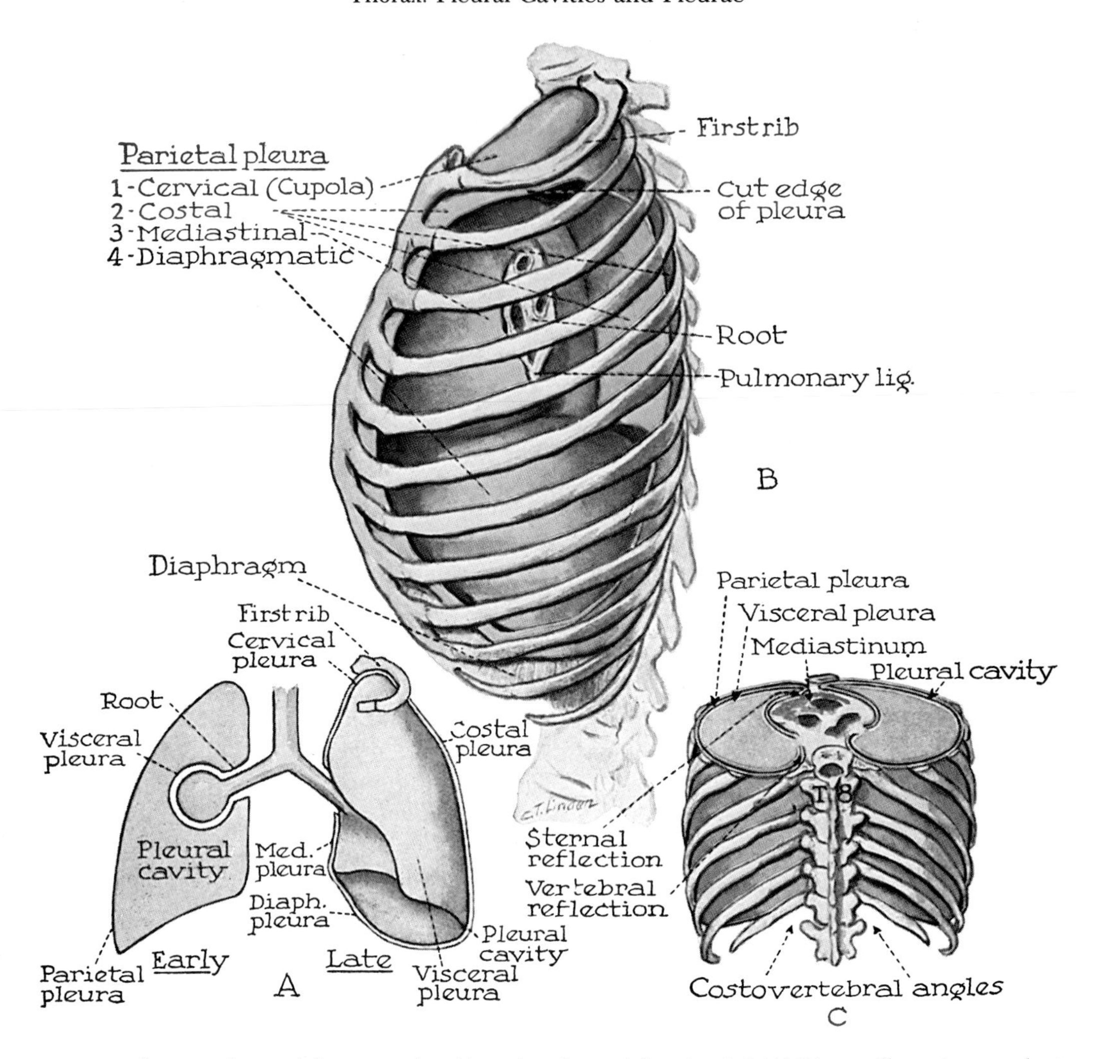

Fig. 17–1: Diagrams illustrating how each lung is enveloped by a pleural space (pleural cavity). (A) Diagram illustrating an early stage of pleural space development and the final relations between the visceral pleura of the lung and the parietal pleura of the chest wall. (B) Diagram of the lateral view of the left pleural space. The lung and most of the costal pleura have been removed. (C) Oblique superior view of the anterior (sternal) and posterior (vertebral) reflections between the visceral and parietal pleura.

2. When a person is breathing quietly, the inferior margin of each lung lies at the level of the sixth rib at the midclavicular line, the level of the eighth rib at the midaxillary line, and the level of the tenth rib at the lateral border of the vertebral column (Fig. 17–4). The peripheral part of each lung's inferior margin is called the **costodiaphragmatic margin** of the lung, as it is the margin between the costal and diaphragmatic surfaces of the lung. This margin marks the lowest border of the lung within the pleural space. It is not constant in position, however; it moves down when a person breathes in and up when a person breathes out. During quiet breathing, its level remains approximately constant, and can be described by its relationships to the sixth, eighth, and tenth ribs.

3. For both lungs, the surface projection of the oblique fissure is an arc that begins posteriorly between the tips of the spinous processes of the second to fourth thoracic vertebrae (Fig. 17–4B) and then follows a curved course around the chest wall that ends near the sixth costochondral junction (Fig. 17–4A). If a person abducts both arms and places both hands behind the head, the medial borders of the laterally rotated scapulae approximate the surface projections of the oblique fissures to the back. For the right lung, the surface projection of the horizontal fissure is an arc which courses approximately along the length of the fourth rib from the midaxillary line to the lateral border of the sternum (Fig. 17–4A).

Because of the relatively low surface projection of each lung's oblique fissure to the front of the chest, each lung's upper lobe is the lobe which casts the largest surface projection to the front of the chest (Fig. 17–4A). Because of the relatively high surface projection of each lung's oblique fissure to the back of the chest wall, each lung's lower lobe is the lobe which casts the largest surface projection to the back of the chest (Fig. 17–4B). The middle lobe of the right lung does not project to the back of the chest.

4. The anterior costomediastinal margin of the left lung bears a laterally-deviated notch called the cardiac notch (Figs. 17–4A). The anterior costomediastinal margin of the right lung gently curves first medially and then laterally as it extends from the level of the second to that of the sixth costal cartilage. The anterior costomediastinal margin of the left lung follows a similar course except for a laterally-deviated notch, called the **cardiac notch,** between the fourth and sixth costal cartilages. The lingula borders the lower part of the cardiac notch. The left lung's cardiac notch is caused by the heart's asymmetric position in the thorax (two-thirds of the heart's mass lies to the left of the midline).

THE PLEURAL SPACE ABOUT EACH LUNG

The external surfaces of each lung (including the surfaces within the fissures) are completely covered by a serous membrane called **visceral pleura** (Fig. 17–1, A and C). The mediastinal viscera and thoracic wall regions apposed to the lung surfaces are also lined by a serous membrane called the **parietal pleura.** The visceral and parietal pleura are continuous with each other at the root of the lung. The region of continuity is marked by a double layer of pleura, called the **pulmonary ligament,** which hangs down from the root of each lung (Figs. 17–1B, 17–2B, and 17–3B).

Each lung is thus enveloped by a separate, membrane-lined, fluid-filled space called a pleural space (Fig. 17–1, A and C). Each pleural space is an airtight space in which a microscopically-thin film of fluid separates the lung's visceral surface from the apposing, parietal pleural lining.

The parietal pleura of each pleural space is commonly divided for descriptive purposes into four regions. Each region is referred to by the name of the external lung surface which it faces. Accordingly, the **mediastinal, diaphragmatic, costal,** and **apical pleura** are the regions of parietal pleura which respectively face the mediastinal, diaphragmatic, costal, and apical lung surfaces (Fig. 17–1B).

The mediastinal and diaphragmatic pleura in each pleural space form a closed, continuous boundary with the costal pleura. This boundary projects to the chest surface as a closed, continuous line. These lines on the left and right sides of the chest wall are called the **lines of pleural reflection.**

The highest point of each line of pleural reflection is located about an inch above the medial third of the clavicle (Fig. 22–4). Each line curves inferomedially from this highest point to a point on the anterior chest wall almost in the midline at the level of the sternal angle (Fig. 17–4A). From this point, each line extends almost straight downward to the level of the fourth costal cartilage. The subsequent courses of the lines differ as they descend from the level of the fourth costal cartilage to that of the xiphisternal joint (which is the articulation between the xiphoid process and the body of the sternum). On the right side, the line exhibits a gently curved inferolateral course as it descends to the right lateral endpoint of the xiphisternal joint. On the left side, the line exhibits an abrupt lateral deviation to just beyond the left lateral border of the body of the sternum before descending to the level of the xiphisternal joint. The left line of pleural reflection has this laterally-directed deviation in its course because of the heart's asymmetric position in the thorax.

The further course of both lines is nearly identical. Each line curves inferolaterally from its point at the level of the xiphisternal joint, passing through the eighth rib at the midclavicular line (Fig. 17–4A), the tenth rib at the midaxillary line, and the twelfth rib at the lateral border of the vertebral column (Fig. 17–4B). From this last intersection, each line extends straight upward to rejoin its highest point.

The outer margins of the lungs underlie, for the most part, the lines of pleural reflection. However, there is one major recess in each pleural space into which the lung does not extend during quiet breathing. This recess, a space two ribs high, appears in each pleural space between the costodiaphragmatic margin of the lung and the costodiaphragmatic margin of the parietal pleura (Fig. 17–4). This recess is called the **costodiaphragmatic recess.** During quiet breathing, the angle between the costal and diaphragmatic pleura at the pleural costodiaphragmatic margin is so acute that the costal and diaphragmatic pleura near the margin face directly opposite each other, separated by only a microscopically-thin film of pleural fluid. This microscopically-thin space represents the costodiaphragmatic recess. The costodiaphragmatic margin of the lung descends completely into the recess only during a full and deep inward breath.

There is another major recess in the left pleural space into which the lung does not extend during quiet breathing. This recess, which is called the **costomediastinal recess,** is a microscopically-thin space bordering the anterior costomediastinal margin of the left pleural space between the levels of the fourth and sixth costal cartilages (Fig. 17–4A). The margin of the left lung that forms the cardiac notch extends completely into the costomediastinal recess only during a full and deep inward breath.

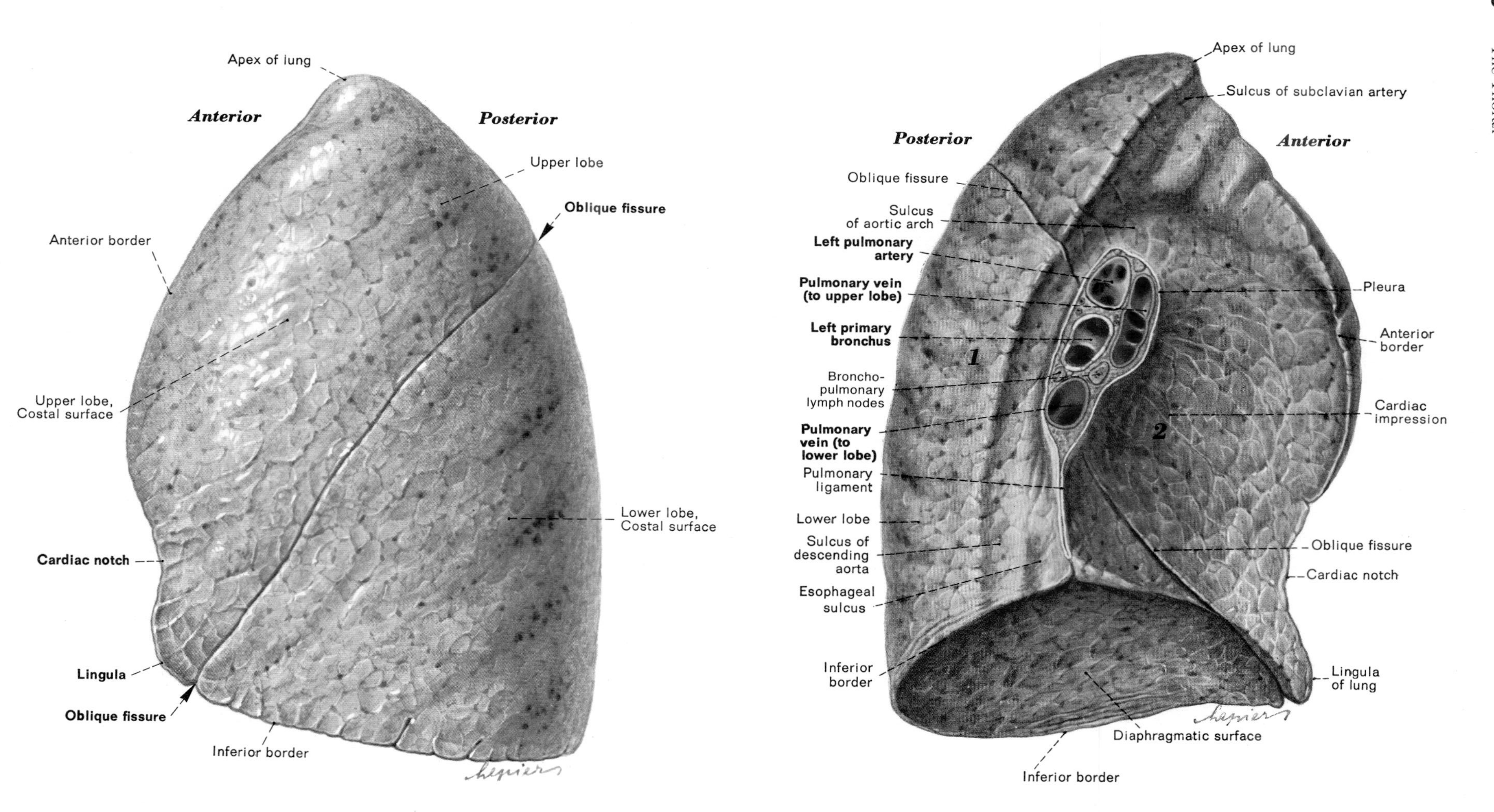

Fig. 17–2: (A) Lateral and (B) medial views of the left lung.

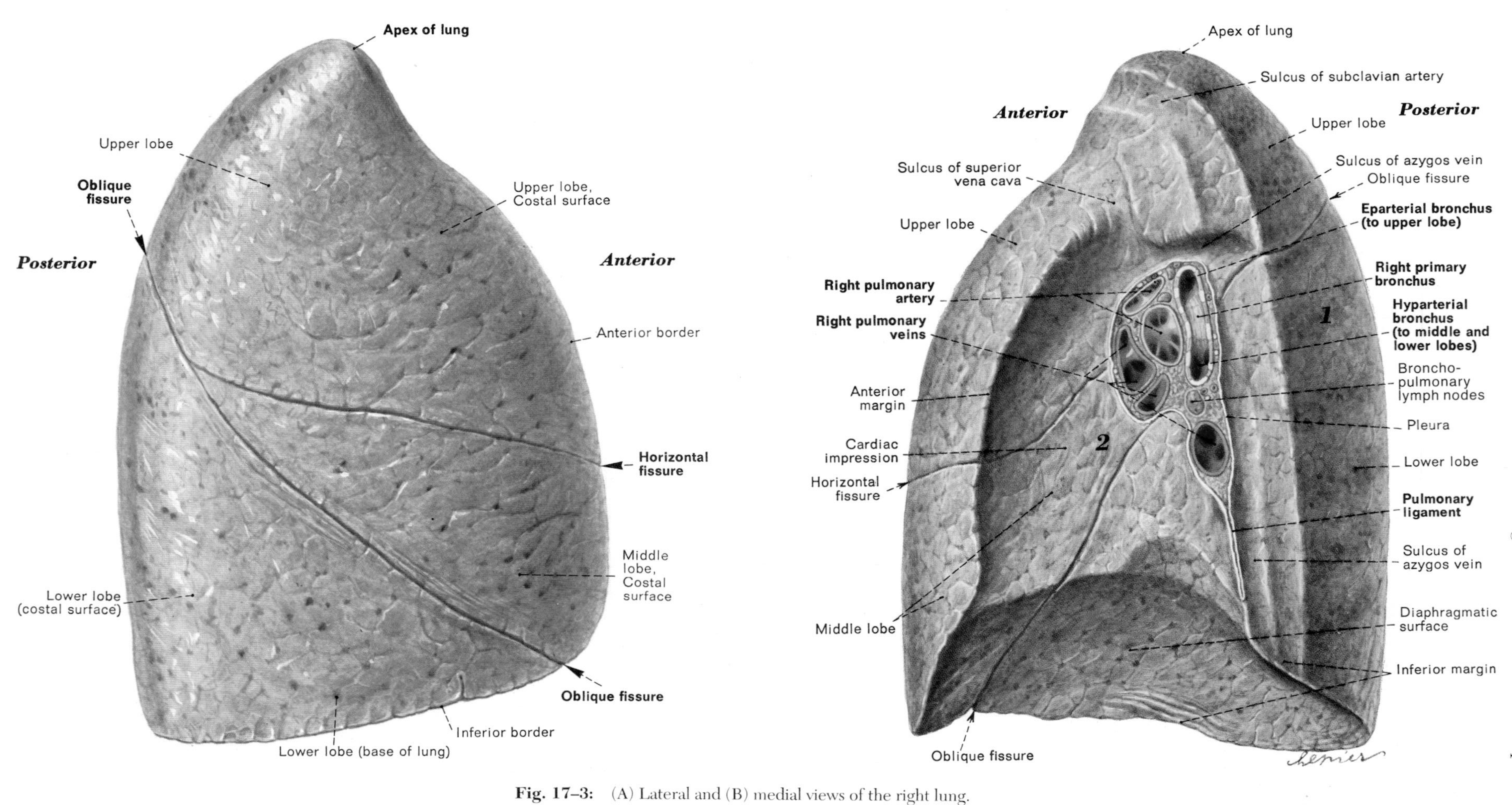

Fig. 17–3: (A) Lateral and (B) medial views of the right lung.

The Sensory Innervation of the Visceral and Parietal Pleura

The visceral afferents of the lung also supply the sensory innervation of the visceral pleura. Consequently, stimuli that evoke pain from the visceral pleura include ischemia and sudden increases in tension, and the pain evoked is commonly vague and diffuse.

The intercostal nerves and the phrenic nerves supply the sensory innervation of the parietal pleura. Whereas the intercostal nerves supply the costal pleura and the circumferential region of the diaphragmatic pleura, the phrenic nerves supply the mediastinal pleura and the central region of the diaphragmatic pleura.

The sensory fibers of the parietal pleura are sensitive to touch, pain, and temperature. Pain evoked by these fibers is generally sharp and superficial, and aggravated when air is breathed in. The pain may also be referred. If the costal pleura or the circumferential region of the diaphragmatic pleura is irritated, pain may be referred to the thoracic and abdominal walls along the course of the irritated intercostal nerves. If the mediastinal pleura or the central region of the diaphragmatic pleura is irritated, pain may be referred to the lower part of the neck and over the point of the shoulder [the cutaneous innervation of these regions is provided by the supraclavicular nerves; the sensory fibers of the supraclavicular nerves enter the spinal cord at the same levels (C3 and C4) that receive sensory fibers from the phrenic nerves].

The Production and Absorption of Pleural Fluid

Pleural fluid is essentially interstitial fluid that has entered the pleural space. Under normal physiologic conditions, it is believed that pleural fluid is produced primarily by the flow of interstitial fluid from the systemic capillary beds of the parietal pleura to the pleural space. It is also believed that pleural fluid is absorbed primarily through the lymphatics of the parietal pleura.

THE INTERNAL ANATOMY OF THE LUNGS

The airways of each lung exhibit an arborescent (branching) pattern. The total aggregate of each lung's airways is known as its **bronchial tree.**

There are two basic types of airways in the lungs, those of the respiratory type and those of the conduction type. The respiratory airways are those in which there occurs diffusional exchange of gases between (a) the air breathed into the lungs and (b) the blood in the pulmonary circulation. The conduction airways serve only as conduits for the mass flow of air between the respiratory airways and the trachea.

The Conduction Airways

The bronchial tree of each lung begins in the root of the lung with the lung's **main stem (primary) bronchus.** The left and right main stem bronchi are the terminal branches of the trachea (Fig. 17–5).

As each main stem bronchus enters the hilum of its lung, it gives rise to a lobar bronchus for each of the lung's lobes (Figs. 17–5 and 17–6). Each lobar bronchus is named for the lobe it serves.

In the right lung, the right main stem bronchus bifurcates into the **upper lobar bronchus** and the intermediate bronchus; the **intermediate bronchus** subsequently bifurcates into the **middle** and **lower lobar bronchi** (Fig. 17–5). As the right pulmonary artery enters the hilum of its lung, it passes beneath the origin of the upper lobar bronchus but above the origins of the middle and lower lobar bronchi. These anatomic relationships are the basis for the common reference to the right upper lobar bronchus as the **eparterial bronchus** and the middle and lower lobar bronchi as the **hyparterial bronchi** of the right lung (the prefixes ep- and hyp- respectively mean above and below).

In the left lung, the left main stem bronchus bifurcates into **upper** and **lower lobar bronchi** (Fig. 17–5). As the left pulmonary artery enters the hilum of its lung, it passes above the division of the left primary bronchus into its two lobar bronchi. The two lobar bronchi of the left lung are thus both hyparterial bronchi.

The lobar bronchi in each lung give rise to segmental bronchi (Figs. 17–5 and 17–6). The **segmental bronchi** serve separate, pyramidal segments of the lung called **bronchopulmonary segments.** These segments are the basic, macroscopic respiratory units of the lung. They are also the smallest respiratory units of a lung that can be identified and excised individually by a surgeon.

Each segment is enveloped by a connective tissue sheath that keeps air breathed into the segment contained within the segment. Each segment is supplied by a single segmental bronchus and a single pulmonary arterial branch. The segmental bronchus conducts almost all the air which enters and exits the segment; the pulmonary arterial branch conducts all the blood pumped to the segment for oxygenation. The segmental bronchus and pulmonary arterial branch enter the segment at its apex. The apices of each lung's segments all project toward the hilum of the lung; the bases of the segments all together form the costal and diaphragmatic surfaces of the lung.

There are ten bronchopulmonary segments in each lung. In the left lung, the upper and lower lobes each consist of five segments. In the right lung, three segments form the upper lobe, two segments form the middle lobe, and five segments form the lower lobe.

Each lung segment is both numbered and named. In both lungs, the most superior segment is numbered 1, and the most posteroinferior segment is numbered 10. With the exception of the upper lobe of the left lung, the segments in each lobe are named according to their relative positions within the lobe.

The three segments of the upper lobe of the right lung are numbered and named **(1) apical, (2) posterior,** and **(3) anterior** (Figs. 17–5 and 17–6). The three uppermost segments of the upper lobe of the left lung are collectively homologous to the upper lobe of the right lung, and thus numbered and named identically.

The two lowest segments in the upper lobe of the left lung form the lingula; these two segments are numbered and named **(4) superior lingular** and **(5) inferior lingular** (Figs. 17–5 and 17–6). The combined positions of the superior and inferior lingular segments in the left lung are homologous to the combined positions occupied in the right lung by the two segments of its middle lobe. This homology is the anatomic basis for the characterization of the lingula as being the left lung's homologue of the right lung's middle lobe. The two segments of the right lung's middle lobe are numbered and named **(4) lateral** and **(5) medial.**

The five segments of the lower lobe of each lung are numbered and named **(6) apical, (7) medial basal, (8) anterior basal, (9) lateral basal,** and **(10) posterior basal** (Figs. 17–5 and 17–6).

The segmental bronchus for the apical segment of each lung's lower lobe is the segmental bronchus which arises most directly and posteriorly from the main stem bronchus. Consequently, when a supine person accidentally aspirates a small object into either lung, the object will most likely lodge in the lower lobe's apical segment.

Each segmental bronchus gives rise to approximately ten more generations of airways within its bronchopulmonary segment, all of which may be referred to simply as **bronchi** (Fig. 17–6). The smallest bronchi (which are the smallest conducting airways to have glands and cartilage in their walls) give rise to the next two or three generations of airways, airways which are called **bronchioles**. The last generation of bronchioles are called **terminal bronchioles** (Fig. 17–7), and they are the smallest conduction airways in the bronchial tree.

The branching pattern of the conduction airways is such that their diameters become reduced by a constant factor of the cube root of 0.5 at each branching. Such a branching pattern minimizes the magnitude of kinetic energy lost by the air as it moves via mass flow through progressively smaller conduction airways.

$\sqrt[3]{.5}$ $\sqrt[3]{\frac{1}{2}}$ $(2^{-1})^{\frac{1}{3}}$

The Respiratory Airways

Each terminal bronchiole serves an arborescent collection of respiratory airways. Each terminal bronchiole and the respiratory airways it serves is called an **acinus.** Acini represent the basic, microscopic respiratory units of the lungs.

Terminal bronchioles give rise to the largest respiratory airways, the **respiratory bronchioles** (Fig. 17–7). All of the smaller airways in the respiratory portion of the bronchial tree [**alveolar ducts, alveolar sacs (alveolar atria), and alveoli**] stem from the respiratory bronchioles. The air contained within the respiratory airways of the lung accounts (on the average) for 80% of the air in the bronchial tree.

The diameters of alveolar ducts do not significantly decrease at each branching. This kind of branching pattern optimizes the diffusional flow of gases which occurs along alveolar ducts.

There are various microscopic connections that provide collateral ventilation of respiratory airways. The smallest communications are the 2 to 13 micrometer diameter openings in alveolar walls called **pores of Kohn** that connect adjacent alveoli directly. The next largest connections are 30 micrometer diameter airways called **Lambert's canals** that connect terminal and respiratory bronchioles with adjacent alveoli. There are 80 to 150 micrometer diameter airways that connect respiratory airways in adjacent bronchopulmonary segments. The largest known collateral connections are 200 micrometer diameter interacinar airways.

It is believed that these various microscopic collateral connections do not significantly affect gas flow into and out from respiratory airways during normal respiration. However, these collateral connections do perform an essential role in support of the cough mechanism's capacity to remove obstructive material from conducting airways. This role occurs at the acinar level in the following manner: When the primary conducting airways serving an acinus become obstructed, the microscopic collateral connections to the acinus help maintain inflation of its alveoli via collateral ventilation. When the individual coughs in an attempt to clear the obstruction (in cases in which the obstructive material consists of a movable mass such as a viscous plug of mucus), the chest wall and diaphragm are pressed inward on the lungs in an attempt to suddenly and significantly increase intra-acinar air pressure. The sudden increase in intra-acinar air pressure generates a pressure gradient across the length of the mucus plug, which, if sufficient in magnitude, forces the mucus plug centrally through the bronchial tree toward the trachea for final expulsion. If the various microscopic collateral connections among acini did not exist, obstruction of the conducting airways to a group of acini

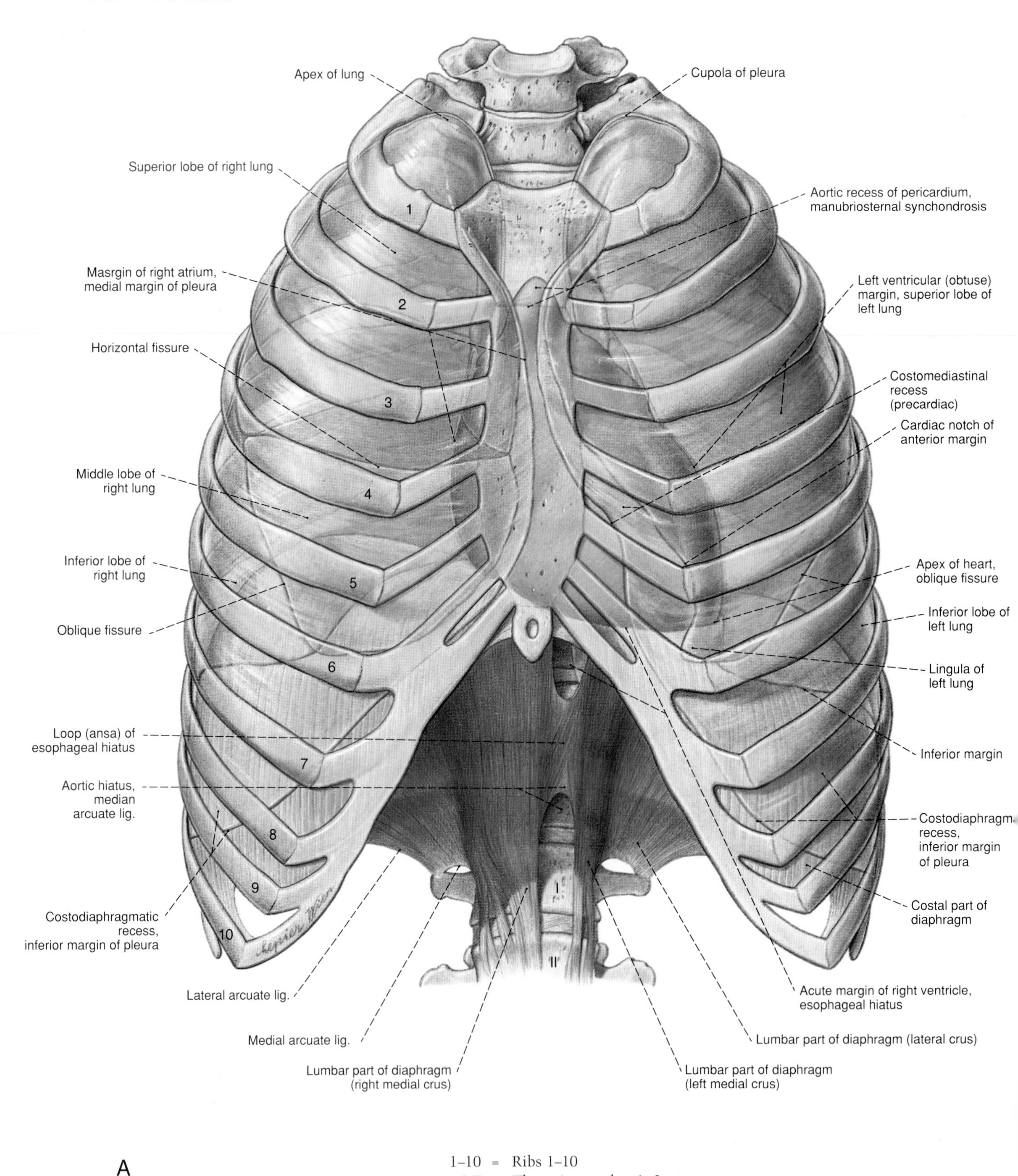

1–10 = Ribs 1–10
I,II = Thoracic vertebra 1–2

Fig. 17–4: (A) Anterior view of the lungs and pericardial sac as seen through a translucent pleura. (B) Posterior view of the lungs and pericardial sac as seen through a translucent pleura.

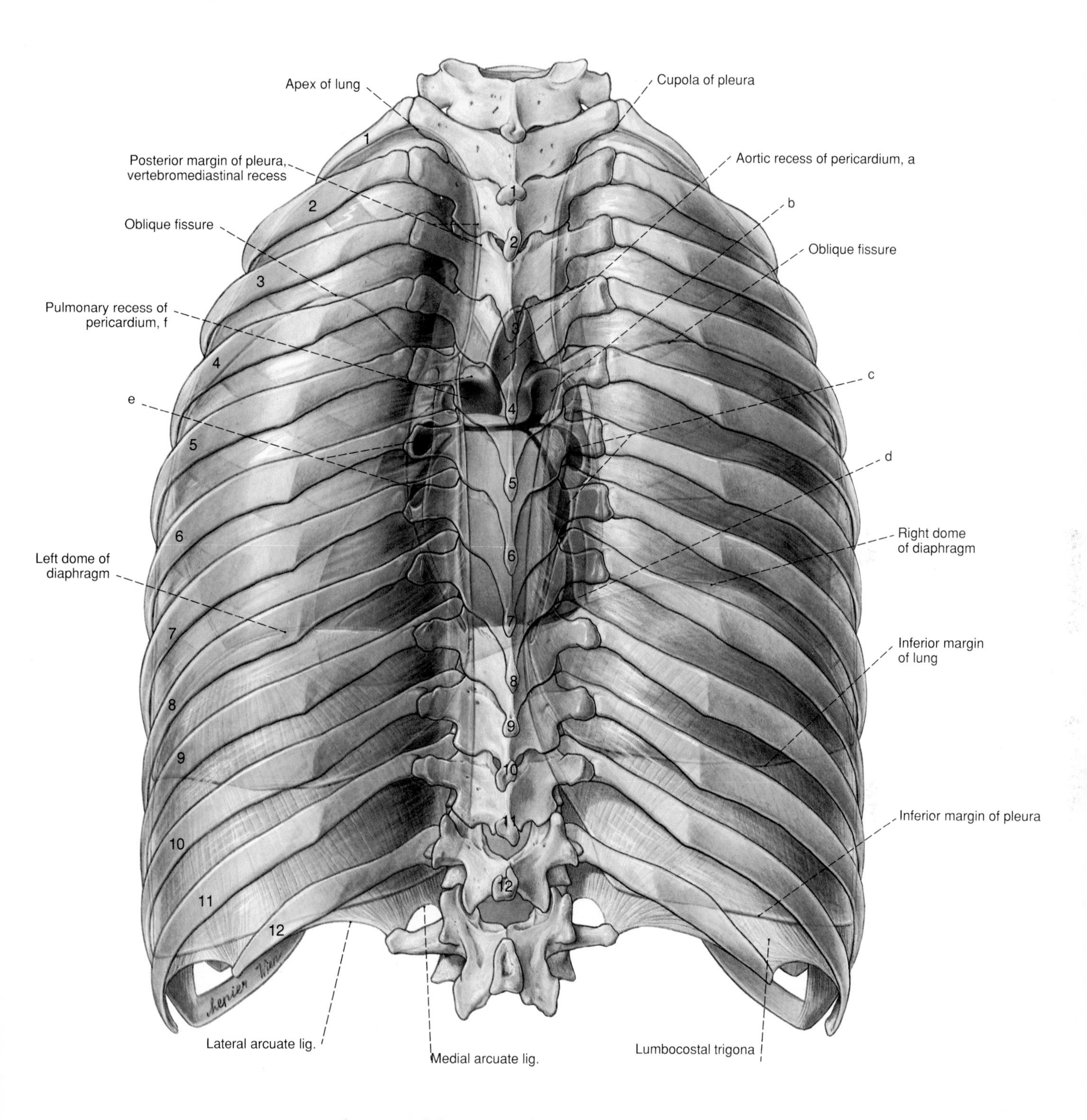

1–12 = Ribs 1–12 and thoracic vertebra 1–12 (spinous processes)

B

a = Exit of ascending aorta
b = Entry of superior vena cava
c = Entry of right pulmonary veins
d = Entry of inferior vena cava
e = Entry of left pulmonary veins
f = Exit of pulmonary trunk

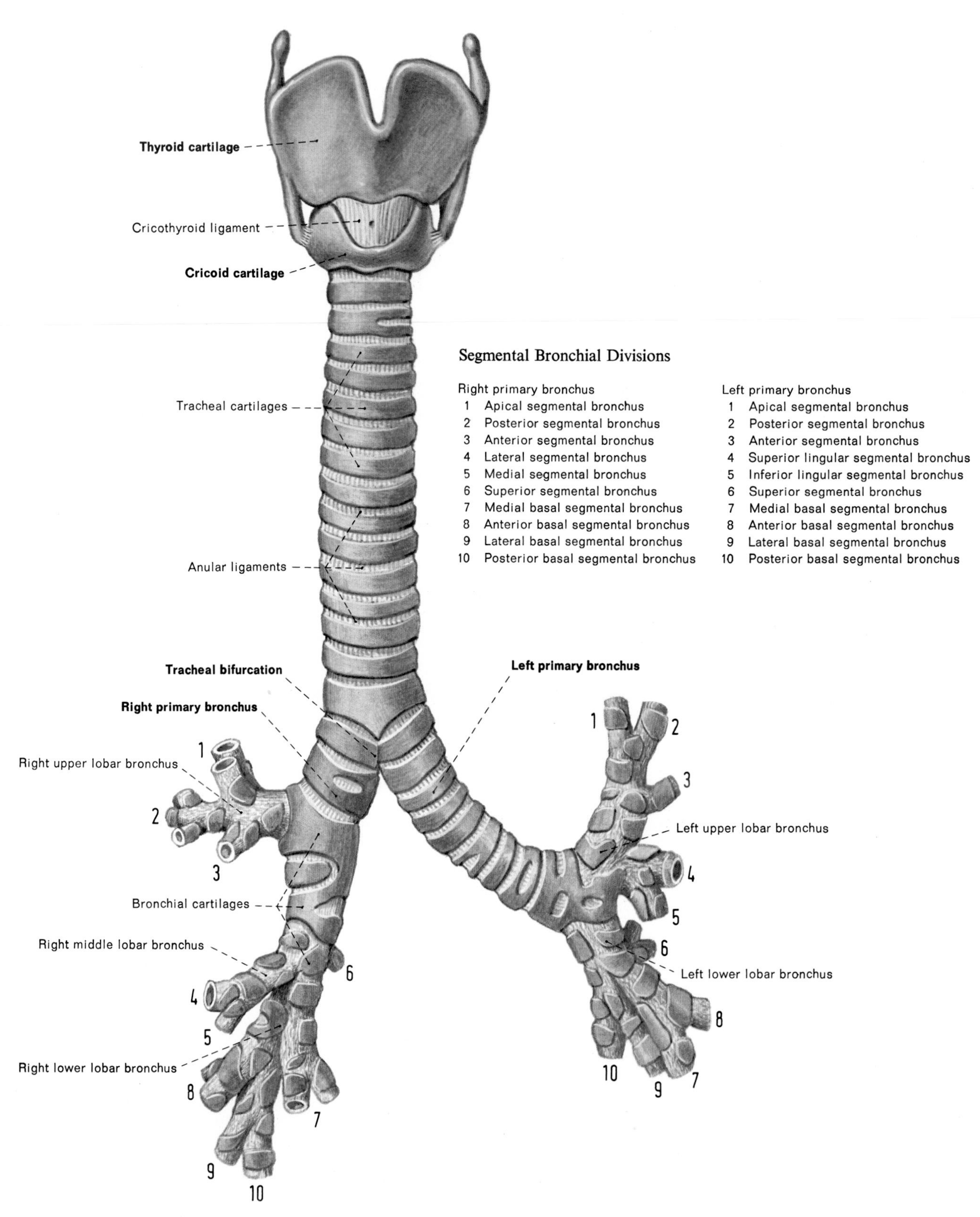

Fig. 17–5: Anterior view of the larynx, trachea, and the primary, lobar, and segmental bronchi of the lungs. The names of the numbered segmental bronchi are given in the text.

would almost always lead to their atelectasis (collapse), and thus a marked reduction in the efficacy of the cough mechanism to expulse obstructive material.

The Arterial Supply

Each lung has two distinct arterial supplies, one emanating from one or two bronchial arteries and the other emanating from a pulmonary artery.

The **one or two bronchial arteries** of each lung provide for the nutrition of the conduction airways in its bronchial tree. Bronchial arteries arise from either the descending thoracic aorta or the upper posterior intercostal arteries.

The **pulmonary artery** of each lung conducts oxygen-poor blood to it for gaseous exchange. The **left** and **right pulmonary arteries** are the terminal branches of the **pulmonary trunk.** During the systolic phase of every heartbeat, the pulmonary trunk receives oxygen-poor blood ejected from the right ventricle.

Each pulmonary artery generates an arterial vascular tree in the lung whose branching pattern matches, in general, that of the bronchial tree. In other words, all the lobar, segmental, and smaller bronchi of a lung are generally accompanied by a corresponding pulmonary arterial branch. These branches, after several generations of further branching, conduct oxygen-poor blood into the capillary networks which envelop the alveolar sacs, and it is here that the blood becomes re-oxygenated (Fig. 17–7). These capillary networks also receive some of the blood supplied to the lungs by the bronchial arteries; this blood is also re-oxygenated as it passes through the capillary networks.

The Venous Drainage

Each lung has two venous drainage systems. The tributaries of one system empty into a bronchial vein and those of the other system empty into two pulmonary veins.

The single **bronchial vein** which emanates from the hilum of each lung conducts oxygen-poor blood into the azygos system of veins. Whereas the bronchial vein from the right lung empties into the azygos vein, the bronchial vein from the left lung empties into either the superior hemiazygos vein or a left upper posterior intercostal vein.

The tributaries of the bronchial veins drain the capillary beds of the conduction airways in the bronchial tree. However, some of the veins that drain these capillary beds empty into tributaries of the pulmonary veins. The capillary beds of the conduction airways are thus drained by both the bronchial and pulmonary veins.

The **pulmonary veins** conduct oxygen-rich blood from the lungs directly to the left atrium. There are commonly two such veins from each lung; one is called the **superior pulmonary vein** and the other the **inferior pulmonary vein.** Whereas the superior pulmonary vein of the left lung drains the lung's upper lobe, the superior pulmonary vein of the right lung drains both the upper and middle lobes of that lung. The inferior pulmonary vein of both lungs drains oxygen-rich blood from the lower lobe of the lung.

The tributaries of the pulmonary veins drain the capillary networks that envelop the respiratory airways of the bronchial tree. The blood drained from these capillary networks in each bronchopulmonary segment empties partly into intrasegmental veins and partly into intersegmental veins. **Intrasegmental veins** extend along a segment's airways; by contrast, **intersegmental veins** course in the connective tissue septa separating adjacent segments. Whereas intrasegmental veins drain oxygen-rich blood from only their own segment, intersegmental veins receive oxygen-rich blood from at least two adjacent segments.

Therefore, there is a marked difference between the segmental distribution of the pulmonary arterial branches and the segmental distribution of the tributaries of the pulmonary veins in each lung. Pulmonary arterial branches form just one arborescent vascular system in a bronchopulmonary segment, one which parallels that of the airways. By contrast, pulmonary venous tributaries form two arborescent vascular systems among the segments, one which parallels that of the airways and another which courses within the connective tissue septa separating the segments.

The Lymphatic Drainage

Lymphatics commence in the bronchopulmonary tree near the centers of acini. The macrophages concentrated near the commencement of these lymphatics are commonly carbon- and dust-laden in adult individuals. In adult lungs, the collection of macrophages near an acinar center is frequently sufficiently carbon- and dust-laden to form a dark speck visible to the unaided eye.

Each lung is drained by a deep plexus and a superficial plexus of lymphatics. The lymphatic vessels of the deep plexus, which drain tissues in the inner two-thirds of the bronchopulmonary segments, extend centrally alongside the airways of each segment to empty into nodes, called **pulmonary nodes,** lying near the origins of the segmental bronchi. The efferents from the pulmonary nodes empty into nodes, called **bronchopulmonary nodes,** which lie in the root of the lung. The efferents from the bronchopulmonary nodes empty

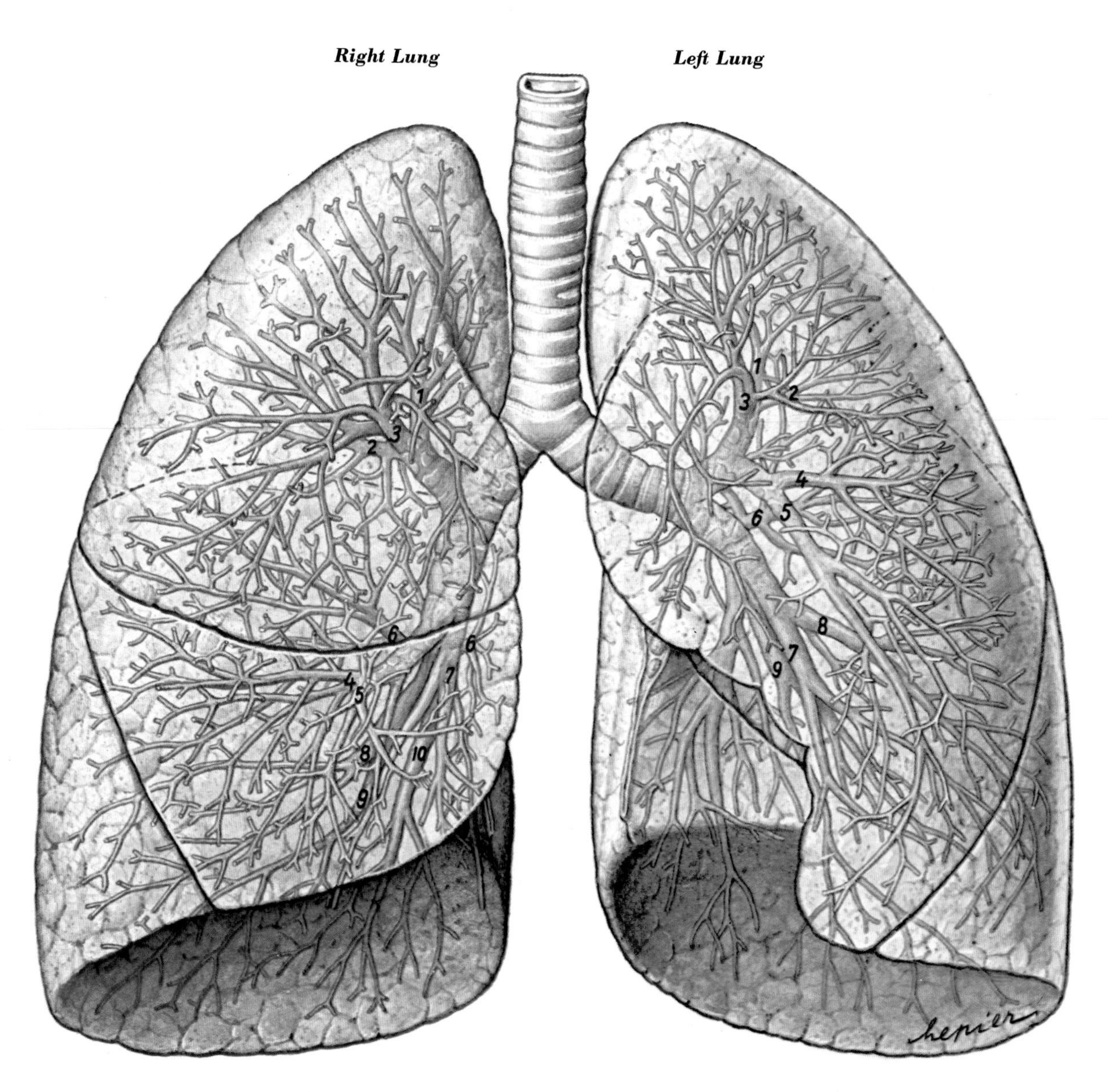

Fig. 17–6: Anterior view of the bronchial tree and its lobar and bronchopulmonary divisions in each lung. The names of the numbered segmental bronchi are given in the text.

into the **tracheobronchial nodes;** these nodes lie about the tracheal termination and in the angle between the origins of the two main stem bronchi. In adults, the pulmonary, bronchopulmonary, and tracheobronchial nodes are all internally speckled with darkened regions, each representing a collection of carbon- and dust-laden macrophages.

The lymphatic vessels of the superficial plexus, which drain tissues in the outer third of the bronchopulmonary segments, extend first outward along connective tissue septa to the visceral pleura and then inward along the outer surfaces and fissures of the lung to empty into bronchopulmonary nodes. In adult lungs, collections of carbon- and dust-laden macrophages render many of the lymphatics in the connective tissue septa and visceral pleura visible as thin, dark streaks. Extensive anastomoses between the deep and superficial lymphatics occur in the hilum. There are also some interconnecting vessels between the deep and superficial lymphatics in the peripheral regions of each lung.

The Nerve Supply

Each lung bears an extensive autonomic plexus, called the **pulmonary plexus,** in the hilum. The pulmonary plexus receives both parasympathetic and sympathetic innervation.

On each side, the cranial origins of the vagus nerve bear the cell bodies of the preganglionic parasympathetic neurons for the lung's autonomic innervation. The fibers that emanate from these neurons are trans-

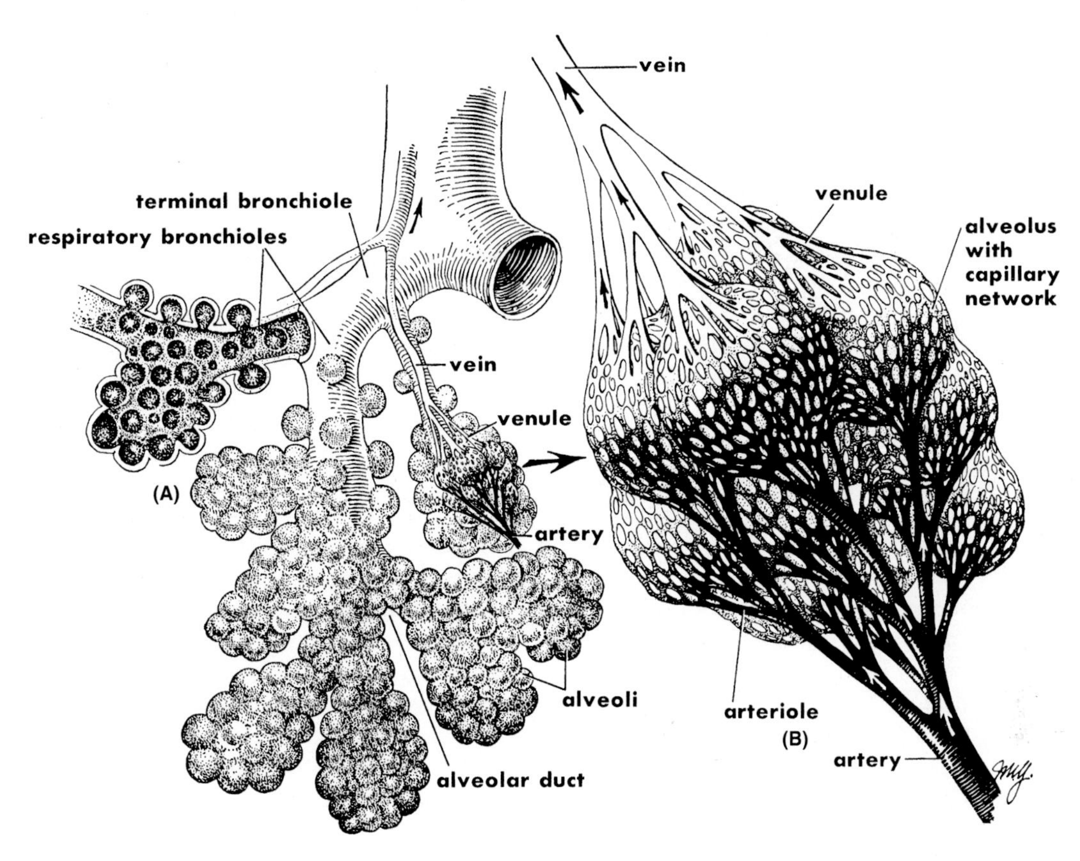

Fig. 17–7: Diagrams illustrating (A) the respiratory airways that stem from terminal bronchioles and (B) the pulmonary capillary networks that envelop the respiratory airways of pulmonary acini.

mitted to the pulmonary plexus mainly by the direct branches of the vagus nerve and branches of the vagus nerve's cardiac branches (the cardiac branches of each vagus nerve are the branches directed toward the cardiac plexus).

On each side, the lateral grey columns of the T2-T5 spinal cord segments bear the cell bodies of the preganglionic sympathetic neurons for the lung's autonomic innervation. The fibers that emanate from these neurons synapse with postganglionic sympathetic neurons in all three cervical ganglia and in the T2-T5 ganglia of the sympathetic chains. Plexus branches of these paravertebral ganglia transmit postganglionic sympathetic fibers to the pulmonary plexus.

Parasympathetic stimulation of the smooth muscle tissue in the walls of the conduction airways causes bronchial constriction; the sympathetic supply is inhibitory to this effect.

THE MECHANICS OF RESPIRATION

Respiration consists of two phases: **inspiration** and **expiration.** Air flows into the lungs when they are expanded, and then flows out when they retract. When a lung is expanded, the aggregate volume of its bronchial tree increases, and thus the average air pressure within its airways decreases. When this average airway pressure becomes less than atmospheric pressure, air is inspired (it flows into the lungs). When the lung subsequently retracts, the average airway pressure increases as the aggregate volume of the airways decreases. When the average airway pressure exceeds that of atmospheric pressure, air is expired (it flows out from the lungs).

Respiration can be classified as quiet or forced. The relaxed breathing which occurs when resting or sleeping is called **quiet respiration.** The strenuous breathing which occurs when exercising vigorously is called **forced respiration.**

The Mechanics of Inspiration

The lungs are expanded during inspiration as a consequence of increases in the dimensions of the thorax. When an adult or adolescent inspires, the dimensions of the thorax increase along three axes: vertical, anteroposterior, and transverse.

The vertical dimension of the thorax increases during inspiration due to the concentric contraction of the diaphragm (Fig. 17–8). When the diaphragm's muscle fibers concentrically contract, they act from their relatively secure origins to pull upon, and thus pull down, the central tendon of the diaphragm. This pull lowers the overall level of the diaphragm in the trunk of the body, and, in so doing, increases the vertical dimension of the thorax. It is by this action that the diaphragm serves as the chief muscle of inspiration.

The anteroposterior and transverse dimensions of the thorax of an adult or adolescent increase during inspiration because of the movements of the ten uppermost rib pairs. The inspiratory movements of these rib pairs increase the cross-sectional dimensions of the thorax in the following manner: In the adult or adolescent chest wall, these ribs all have a marked oblique lie. The body of each rib slopes downward as it sweeps forward from the costotransverse joint to its costochondral junction. The bodies of all these ribs are pulled upward during inspiration in such a manner that each rib pivots or rotates about an axis which extends through its costotransverse joint. The orientation of this axis of rotation determines whether the upward movement of the rib's downward-sloping body causes either (1) a pronounced upward and forward movement of the rib's anterior end or (2) a marked upward and lateral movement of the rib's side. In the first instance, the upward movement of the rib's body contributes primarily to an increase in the anteroposterior dimension of the thorax. In the second instance, the upward movement of the side of the rib's body contributes principally to an increase in the transverse dimension of the thorax.

In the costotransverse joints of the six uppermost rib pairs, the rib's tubercle has a convex articular facet which is reciprocal to an anteriorly-facing, concave costal facet on the transverse process of the vertebra at the same level. The curvature and orientation of these facets cause each of these ribs to pivot primarily during respiration about an axis which runs from its costotransverse joint to its costovertebral joint. Because rotation of just a few degrees about this axis produces pronounced up-and-down movements of the anterior ends of these ribs, the movements of the first through sixth rib pairs during respiration are generally likened to the movements of a water pump handle (Fig. 17–9). The combined superior and anterior displacement of the anterior ends of these ribs during inspiration accounts for the upward and forward movement of the sternum during inspiration. This is the mechanism by which the movements of the six uppermost rib pairs generate most of the increase in the anteroposterior dimension of the thorax during inspiration.

In the costotransverse joints of the seventh through tenth rib pairs, the tubercle of each rib has a flat articular surface that faces a flat costal facet on the trans-

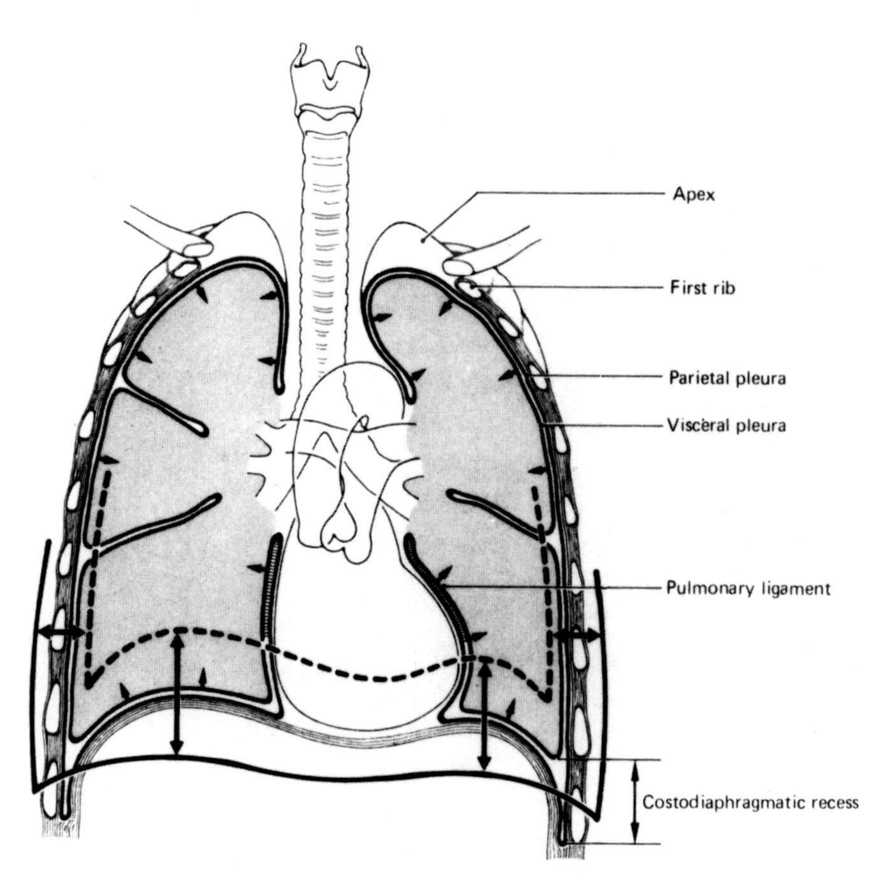

Fig. 17–8: Diagram of a coronal section of the chest illustrating the increase in the volume of the thoracic cavity during inspiration produced by descent of the diaphragm and upward and lateral movements of the sides of the ribs. The single arrows represent the retractive forces of the lungs, and the double arrows show the excursion of the lung bases and periphery between deep inspiration and expiration.

verse process. The flatness and orientation of these facets restrict each of these ribs to pivot primarily during respiration about an axis which runs from its costotransverse joint to the synovial joint at the tip of its costal cartilage. Since rotation of just a few degrees about this axis produces pronounced up-and-down movements of the sides of the ribs, the movements of the seventh through tenth rib pairs during respiration are commonly likened to the movements of a water bucket handle (Fig. 17–10). The combined superior and lateral displacement of the sides of these ribs during inspiration produces much of the increase in the thorax's transverse dimension (Fig. 17–8).

In an adult or adolescent, there are thus three major thoracic wall movements during inspiration: (1) the diaphragm moves inferiorly, (2) the upper anterior chest wall moves anteriorly, and (3) the lower lateral chest wall regions move laterally (Fig. 17–8). These thoracic wall movements are accompanied by almost identical movements of the apposed lung surfaces, and it is in this fashion that the lungs are expanded.

In a child, however, there is only one major thoracic wall movement during inspiration: the diaphragm moves inferiorly. A child's breathing is therefore essentially just diaphragmatic breathing. Rib movements do not contribute significantly to inspiration in a child simply because the ribs lie almost horizontally in the chest wall. Consequently, when the bodies of a child's ten uppermost rib pairs rotate upward about axes through their costotransverse joints, there occurs only a slight anterior or lateral displacement of the ribs.

The lung surfaces apposed to thoracic wall regions always move with the thoracic wall regions. Each lung exerts within itself retractive forces which tend to pull it away from the chest wall and diaphragm. The capacity of these retractive forces to collapse the lung can be counteracted by an average pleural fluid pressure of 4 mm Hg below atmospheric pressure. Under normal physiologic conditions, pleural fluid exhibits a pressure which averages 8 to 10 mm Hg below atmospheric pressure. The average magnitude of the pleural fluid's subatmospheric pressure thus always ensures that the costal and diaphragmatic surfaces of the lung are closely apposed to their respective thoracic wall regions.

CLINICAL PANEL IV-2

TENSION PNEUMOTHORAX

Definition: A tension pneumothorax is generally the product of an injury which produces a tissue flap that opens a pleural space to air intake during inspiration but prevents air escape during expiration. This situation quickly leads to total atelectasis (maximum retraction and collapse) of the lung and the establishment of an air pressure within the pleural space that is greater than atmospheric pressure. The relatively high air pressure in the affected pleural space presses the mediastinal viscera of the chest (in particular, the heart) toward the contralateral (opposite) side of the chest, in the process markedly compressing the only functional lung that the injured person has. Consequently, a tension pneumothorax is a medical emergency that requires prompt evacuation of air from the affected pleural space.

Common Symptoms: The patient is generally in a state of intense anxiety because of severe dyspnea (difficulty in breathing). The dyspnea is the result of the loss of the function of one lung and the severe compromise of the function of the other lung. The air in the affected pleural space irritates the parietal pleura and thus is frequently responsible for sharp pain on the affected side of the chest.

Common Signs: The patient is tachypneic (has an elevated respiratory rate) and tachycardic (has an elevated heartbeat rate). The trachea (as visualized in the anterior part of the neck) is frequently deviated to the side opposite that of the tension pneumothorax.

The lips and nailbeds of the patient generally exhibit a purplish-to-bluish discoloration. Such discoloration is a sign of central cyanosis (decreased oxygenation of blood in both central and peripheral tissues). The colors of the oral mucosa (or lips) and nailbeds respectively provide a rough index of the oxygenation of blood in central and peripheral tissues of the body.

Auscultation reveals the absence of breath sounds on the affected side of the chest. Breath sounds are absent because the lung is markedly retracted from the chest wall and not being filled with air during inspiration. Percussion finds a note of tympany over the affected side of the chest wall. The tympanic note is due to the replacement of normal peripheral lung tissue with air.

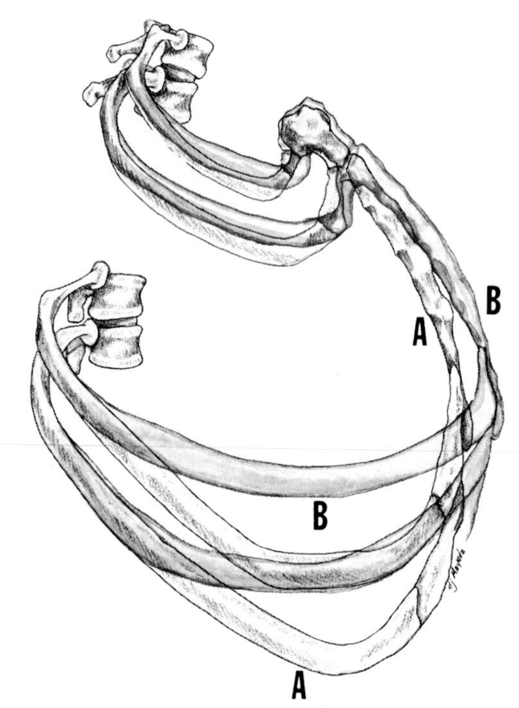

Fig. 17–9: Diagram illustrating the water pump handle movement of the ribs during respiration. Compare the positions of the sternum and ribs at (A) the beginning and (B) the end of inspiration. Note the increase in the anteroposterior dimension of the rib cage.

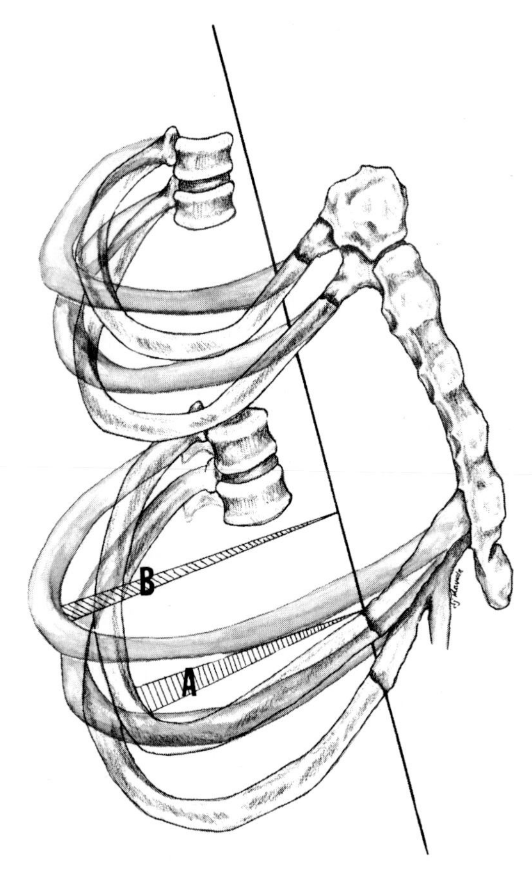

Fig. 17–10: Diagram illustrating the water bucket handle movement of the ribs during respiration. Compare the distances of the ribs from the central axis of the thorax at (A) the beginning and (B) the end of inspiration. Note the increase in the lateral dimension of the rib cage.

The Mechanics of Expiration

The lungs retract during expiration primarily as a result of the exertion of their retractive forces. When the inspiratory muscles relax during expiration, the lung's retractive forces effect a smooth recoil contraction of the lungs and the thoracic wall regions to which they are apposed. There are two sources of the lung's retractive forces:

1. A retractive tension is exerted throughout expiration by the lung's elastic connective tissue fibers; these fibers were stretched during inspiration.
2. Most of the lung's volumetric increase during inspiration occurs through alveolar expansion. The expansion of the alveoli increases the surface tension of the surfactant lining the alveoli. Surfactant is a lipoprotein mixture secreted by type II alveolar epithelial cells. It forms a multilayered film at the interface between the fluid lining the luminal surface of the alveoli and the air in the alveoli. Surfactant is required for normal alveolar expansion and retraction during respiration. The surface tension exerted by surfactant during expiration accounts for about two-thirds of the lung's retractive forces.

The Effects of Anterolateral Abdominal Wall Movement

The movements of the thoracic wall during respiration affect, and can be affected by, movements of the anterolateral abdominal wall. The downward thrust of the diaphragm during inspiration increases pressure in the abdominopelvic cavity, and thus tends to outwardly displace the anterolateral abdominal wall. In converse fashion, concentric contraction of the muscles of the anterolateral abdominal wall during expiration compresses the abdomen, and thus tends to thrust the diaphragm upward. The reciprocal effects of rib cage and anterolateral abdominal wall movements on each other vary with the depth of respiration.

THE ACTIONS OF THE MUSCLES OF RESPIRATION

The specific actions and activity of the respiratory muscles vary with body posture and position and the depth of breathing. In general, the respiratory muscles are the most effective in supporting quiet respiration when a person is seated upright. This relationship explains why a person suffering from dyspnea breathes with minimum effort when seated upright.

The Inspiratory Actions of the Diaphragm

The diaphragm is the chief muscle of inspiration. It can account for about two-thirds of the lungs' volumetric increase during inspiration. It is the most effective inspiratory muscle when a person is supine.

The inspiratory actions of the diaphragm change with the depth of the inspiration. During quiet respiration, the principal action of the diaphragm is to increase the vertical dimension of the thorax. During forced respiration, however, the diaphragm serves to increase both the vertical and lateral dimensions of the thorax. The diaphragm's capacity to increase the vertical dimension is limited by the extent to which it can pull down upon the abdominal viscera (especially the liver). Once it attains this limit during forced inspiration, the central tendon of the diaphragm becomes a stabilized platform from which the diaphragm's muscle fibers can now pull upward upon their sternal and costal origins (in other words, the diaphragm contributes at this stage to the inspiratory movements of the seventh through tenth rib pairs).

The diaphragm exerts different effects during inspiration upon the major structures which pass through it. At the caval opening, the central tendon of the diaphragm pulls outward upon the wall of the inferior vena cava. The lumen of the inferior vena cava is thus stretched open as the diaphragm exerts its "thoracic pump" role of augmenting venous return to the heart. At the esophageal opening, the muscle fibers of the diaphragm's right crus pull the esophagus downward and to the right during inspiration. This action helps prevent regurgitation of stomach contents as the fundus of the stomach is compresssed from above. At the aortic opening, the diaphragm does not exert any effect upon the aorta during inspiration, as the aorta passes through an indentation in the posterior margin of the diaphragm.

The average level of the diaphragm in the trunk of the body and the extent of its respiratory excursions (movements) during normal breathing vary with posture. The diaphragm rests at its highest levels and exhibits its greatest excursions when the body is supine. The diaphragm rests at its lowest levels and exhibits its smallest excursions during normal breathing when a person is seated upright.

The descent of the diaphragm during inspiration increases the pressure on the abdominopelvic cavity by decreasing the vertical dimension of the abdomen. Such an increase in abdominopelvic cavity pressure is used for postural stabilization of the trunk of the body and to help void waste products from pelvic viscera (to help evacuate feces from the rectum and urine from the urinary bladder).

The Respiratory Actions of the External and Internal Intercostal Muscles

The respiratory functions of the intercostal muscles are in dispute. Almost all investigators agree that the tonicity of the intercostal muscles serves to prevent the intercostal spaces from bellowing inward during inspiration and bulging outward during expiration.

Some investigators maintain that the external intercostal muscles and the interchondral parts (the parts extending between the costal cartilages) of the internal intercostal muscles act during inspiration to help rotate the ribs upward at their costotransverse joints. The interosseous parts (the parts extending between the ribs) of the internal intercostal muscles are believed to act during expiration to help rotate the ribs downward at their costotransverse joints.

Figure 17–11 depicts the mechanical basis for the presumed respiratory actions of the external and internal intercostal muscles. The ribs and costal cartilages bordering an intercostal space may be likened to two curved but parallel levers (L1 and L2), with the ribs hinged posteriorly to an unmovable, vertical column VC (the vertebral column) and the costal cartilages hinged anteriorly to a movable, vertical column SC (the sternum and its costal margins). The intercostal space follows a downward course from the costotrans-

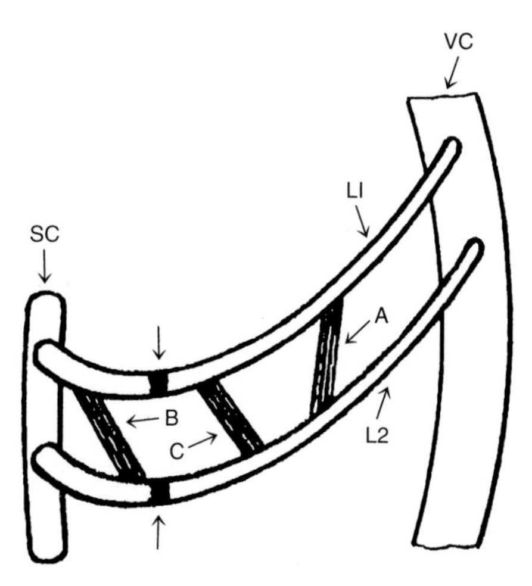

Fig. 17–11: Diagram illustrating the actions of intercostal muscles. The arrows point to the costochondral junctions between the ribs and their costal cartilages. See text for details.

verse joint to the costochondral junction, and then an upward course from the costochondral junction to the sternum or costal margin. If the fibers of an intercostal muscle in a given interval of the intercostal space are roughly aligned with the slope of the interval (fiber groups A and B), then concentric contraction of those fibers will rotate the ribs upward at their costotransverse joints (the fibers of group A represent external intercostal muscle fibers, and the fibers of group B represent internal intercostal muscle fibers in the interchondral part of the intercostal space). If the fibers of an intercostal muscle in a given interval of the intercostal space are roughly perpendicular to the slope of the interval (fiber group C), then concentric contraction of those fibers will rotate the ribs downward at their costotransverse joints (the fibers of group C represent internal intercostal muscle fibers in the interosseous part of the intercostal space).

The Inspiratory Actions of Scalenus Anterior and Scalenus Medius

Scalenus anterior and scalenus medius are neck muscles which extend inferolaterally from cervical vertebrae to insert upon the upper surface of the first rib. These scaleni muscles can be employed to assist directly in the inspiratory movements of the first rib and indirectly in the inspiratory movements of lower ribs.

The Respiratory Actions of Scalenus Minimus

It is believed that scalenus minimus acts during respiration to resist the bellowing inward of the suprapleural membrane during inspiration and its bulging outward during expiration.

The Inspiratory Action of Sternocleidomastoid

Sternocleidomastoid is one of the most prominent muscles of the neck. It extends superolaterally from its attachments to the manubrium of the sternum and the clavicle to its insertion onto the occipital bone and the mastoid process of the temporal bone of the skull. It can assist inspiration by pulling upward directly on the sternum and indirectly on the upper ribs.

The Inspiratory Actions of Pectoralis Major and Pectoralis Minor

If the arms are abducted at the shoulder joint, concentric contraction of the pectoralis muscles can assist inspiration by pulling upward on the sternum and the upper ribs.

The Respiratory Actions of the Muscles of the Anterolateral Abdominal Wall

The rectus abdominis muscles, the external and internal obliques, and the transversus abdominis muscles are relatively inactive during quiet respiration if a person is supine. If a person assumes a sitting or standing posture, however, these muscles become tonically active. Tonic activity of the muscles during the inspiratory phase limits the extent to which the downward thrust of the diaphragm outwardly displaces the anterolateral abdominal wall.

During forced respiration, the anterolateral abdominal wall muscles concentrically contract strongly during the expiratory phase. This expiratory activity compresses the abdomen, and thus pushes the relaxed diaphragm upward. The anterolateral abdominal wall muscles thus greatly increase the magnitude of the forces which serve to retract the lungs during the expiratory phase of forced respiration.

CASE **IV.2**

The Case of Marsha Hadley

INITIAL PRESENTATION AND APPEARANCE OF THE PATIENT

A 19 year-old, white woman named Marsha Hadley walks into the student infirmary at 9:00 AM. She appears apprehensive and uncomfortable. She winces and splints her chest to the right when she coughs.

QUESTIONS ASKED OF THE PATIENT

What is it that has brought you to the infirmary? I've had a bad chest cold for the past two days, and this morning I noticed blood in the mucus I've been coughing up from my lungs. *[The patient's answer indicates that she has been suffering during the past 48 hours from a disease or condition whose symptoms she associates with lower tract respiratory infections (infectious processes involving primarily the lungs). The examiner decides to ask the patient first about those symptoms that patients frequently associate with lower respiratory tract diseases: chest pain, dyspnea (difficulty with breathing), cough, sputum expectoration, and hemoptysis (the act of coughing up blood). The patient's initial remarks indicate that she has at least three of these symptoms.]*

How would you describe the blood in your mucus? There's what looks to be like streaks of bright-red blood in the mucus. *[Blood in the sputum suggests severe inflammation of the trachea or the bronchopulmonary tree of one or both lungs or necrosis infarction of these same tissues.]*

How would you describe the color and thickness of the mucus that you cough up from from your lungs? It's yellowish and very thick. I have a lot of difficulty in being able to clear my lungs because of how thick the mucus is. *[Sputum production is an indicator of increased bronchial secretions. Sputum that is thick and yellowish (or greenish) is called mucopurulent sputum. The production of mucopurulent sputum suggests that an infectious inflammatory process is the basis for the increased bronchial secretions.]*

How long have you had your cough? It started 2 days ago when I felt this chest cold coming on. *[Cough receptors (receptors whose stimulation initiates the desire to cough) are located in the larynx, trachea, and the main stem bronchi of the lungs. The presence of a cough frequently indicates an irritative process or irritative substances at one or more sites in the upper respiratory tract.]*

I notice that when you cough, you bend your chest to the right and wince. Is that because it hurts you to cough? Yes, very much so. The pain is really sharp. *[Coughing commonly involves exaggerated respiratory movements. Deep inspiration, which first draws a relatively large volume of air into the lungs, is quickly followed by marked concentric contraction of the anterolateral abdominal wall muscles (coincident with relaxation of the diaphragm) to forcefully expel the air from the lungs through the larynx, pharynx, and oral cavity.]*

Does it also hurt you to take a deep breath? Yes.

Where does it hurt? Here on the right side of my chest [the patient abducts the right arm and places the palm of the left hand over the lateral aspect of the right breast and the fingers over the medial wall of the right axilla]. *[Sharp chest pain associated with breathing movements suggests inflammation of parietal pleura on the affected side.]*

Do you have any trouble breathing? Yes. I started having trouble breathing yesterday, and it became worse last night. *[Dyspnea is an indicator of inadequate ventilation or perfusion of the lungs.]*

Have you felt feverish recently? Yes. I have felt very feverish for the past 2 days. I have found that Tylenol helps reduce the fever. However, I haven't taken any Tylenol since last night.

Do you have a sore throat? No. *[A sore throat indicates inflammation of the pharynx]*

Is there anything else that is also bothering you that I haven't asked about? No.

Do you know any individuals around you, such as friends or classmates, who have also recently suffered from a similar illness? Yes, some of my friends have recently had colds or the flu, but not as bad as my chest cold.

Do you have or have you ever had any problems with your lungs? No.

Do you have allergies? No.

Do you smoke? No.

Have you recently been exposed to any noxious gases or fumes? No.

Do you use any illegal drugs? No.

Have you had any recent illnesses or injuries prior to this illness? No.

Are you taking any medication for previous illnesses? No.

Have you tried any over-the-counter medications except Tylenol? No. *[The examiner finds the patient alert and fully cooperative during the interview.]*

PHYSICAL EXAMINATION OF THE PATIENT

Vital signs:
Blood pressure
Lying supine: 125/75 left arm and 125/75 right arm
Standing: 125/75 left arm and 125/75 right arm
Pulse: 125
Rhythm: regular
Temperature: 104.2°F.
Respiratory rate: 23
Height: 5′10″
Weight: 150 lbs.

[The examiner recognizes that the patient's body temperature, pulse, and respiratory rate are not within normal limits. 99.0°F. may be regarded as the normal upper limit of body temperature as recorded by an oral thermometer. Normal adult heartbeat rates vary from 60 to 100 beats per minute. Normal adult respiratory rates vary from 8 to 16 cycles per minute.]

HEENT Examination: Normal

Lungs: Inspection, palpation, percussion, and auscultation of the lungs are normal except for the following findings on the right lung: With the patient seated upright, auscultation reveals the following findings at the second and third intercostal spaces along the right midaxillary line: (a) diminished vesicular breath sounds, (b) enhanced bronchial breath sounds, (c) increased intensity and clarity of spoken voice and whispered sounds, (d) egophony, and (e) fine rales near the end of deep inspiration. Percussion dullness is found at the same sites.

Cardiovascular Examination: The cardiovascular examination is normal except for the tachycardia.

Abdomen: Normal

Genitourinary Examination: Normal

Musculoskeletal Examination: Normal

Neurologic Examination: Normal

INITIAL ASSESSMENT OF THE PATIENT'S CONDITION

The patient appears to be suffering from an acute lower respiratory tract disease.

ANATOMIC BASIS OF THE PATIENT'S HISTORY AND PHYSICAL EXAMINATION

1. **The patient's chest pain appears to be somatic pain originating from the parietal pleura of the right pleural space (a sharp pain worsened by respiratory movements).** The movements of the chest wall and diaphragm during inspiration stretch the parietal pleura of the pleural spaces. The increased tension in the parietal pleura elicits a sharp, knife-like pain if the parietal pleura is inflamed. Sharp chest pain produced by the stretching of inflamed parietal pleura is called **pleuritic pain.**
2. **The diminished vesicular breath sounds, enhanced bronchial breath sounds, increased intensity and clarity of spoken and whispered voice sounds, and egophony heard over the upper right lateral chest wall collectively indicate consolidation of the upper lobe of the right lung.** The regions of the second and third intercostal spaces along the right midaxillary line overlie the upper lobe of the right lung (Fig. 17–4B).

   Auscultation of the lung fields provides assessment of the intensity and pitch of the breath sounds being transmitted through the lungs. Vesicular breath sounds (the soft, low-pitched sounds produced by the swirling of air within respiratory airways) are normally heard over all regions of the lung fields except the manubrium sterni and the upper interscapular region. Bronchovesicular breath sounds (vesicular breath sounds mixed with the harsh, high-pitched, bronchial breath sounds produced by the passage of air through large conducting airways) are normally heard over the manubrium sterni and the upper interscapular region.

   Auscultation of the lung fields also provides assessment of the intensity and clarity of spoken voice and whispered sounds being transmitted through the airways of the lungs to the chest wall. Spoken voice sounds are sounds that emanate mainly from the vocal cords. Whispered sounds are sounds generated by the passage of air through the mouth and the lips; no sounds emanate from the vocal cords. Spoken voice and whispered sounds heard in any region of a normal lung field are indistinct (individual words are not discernable) and not as loud as when heard directly. The intensity of spoken voice sounds is greatest over the upper interscapular region and the anterior ends of the first and second intercostal spaces and least over the bases of the lungs. The intensity of whispered sounds is greatest in the lung field regions where bronchovesicular breath sounds are normally heard.

Acute diminishment of vesicular breath sounds over a peripheral lung region indicates diminished air flow into and out from the lung region and/or the dampening of the sounds by excess pleural fluid or pleural thickening. There are four general pathologic mechanisms which diminish air flow into and out from a lung region:

a. obstruction of one or more of the conducting airways bringing air to the lung region,
b. loss of the elasticity of the interstitial tissues about the respiratory airways (the normal contraction of these interstitial tissues during expiration forces air out from the respiratory airways),
c. atelectasis (collapse of the respiratory airways), and
d. consolidation (accumulation of fluid within the interstitial tissues and/or airways of the lung).

An acute enhancement of bronchial breath sounds or an acute increased intensity and clarity of spoken voice and whispered sounds over a peripheral lung region suggest consolidation of the peripheral lung region. To be more specific, each of these findings suggests the presence of a lung region of increased density that extends from an open (patent) bronchus to the surface of the lung. The reasoning is as follows: The peripheral lung regions are largely air-filled because they consist principally of acini (the basic, microscopic respiratory units of the lung). The only relatively solid tissues in peripheral lung regions are the interstitial and vascular tissues that support the respiratory airways of the acini. This air-filled, honeycombed structure to peripheral lung regions makes these regions poor transmitters of audible sounds. The poor transmission of audible sounds accounts for the absence of bronchial breath sounds and the indistinct nature of voice sounds in peripheral lung regions. Any uniform or approximately uniform accumulation of fluid within the interstitial tissues and/or the respiratory airways of a peripheral lung region make the region a better transmitter of audible vibrations. If the fluid accumulation extends continuously from the conducting airways [the bronchi that conduct air (and also audible sounds) into the peripheral lung region] to the costal surface of the peripheral lung region, then the improved transmission of audible sounds through the consolidated lung region will reach the chest wall and be detected upon auscultation.

Egophony is the auscultation of spoken voice sounds which have a nasal or bleating quality. An examiner generally tests for egophony by requesting the patient to voice an extended phonation of the vowel "e." Egophony is said to exist if the extended phonation sounds like the vowel "a" over a peripheral lung region. The physical basis for egophony is not understood. Egophony is most commonly heard over a region of consolidation or the upper level of a pleural effusion.

3. **The fine rales heard over the right upper lateral chest wall near the end of inspiration indicate inflammation and/or congestion of acini in the upper lobe of the right lung.** Rales are an abnormal respiratory sound. Fine rales sound like the sharp crackling sounds heard when a lock of hair is held close to the ear and rubbed between the thumb and index finger. Fine rales heard near the end of inspiration are caused by air suddenly rushing into and reinflating acinar airways that have become congested with fluid.
4. **The percussion dullness over the right upper lateral chest wall supports the ausculatory evidence indicating consolidation of the upper lobe of the right lung.**

Percussion is the physical examination technique in which various regions of a patient's body are tapped to produced audible sounds. The walls of the chest and abdomen are the regions most commonly percussed during a physical exam. The quality of the audible sounds produced provides assessment of the average density of tissues lying to a depth of 5 to 7 cm beneath the wall area being percussed.

Percussion of a chest or abdominal wall area that overlies a pocket of air or gas (such as a gas-filled bowel loop) emits a percussion note called **tympany** that is relatively loud and has a musical timbre (similar to that produced by the beating of a drum). Percussion of the chest wall areas which overlie tissues whose average density approximates that of peripheral lung tissue emit a percussion note called **resonance** that is of relatively loud intensity, low pitch, and long duration. Percussion of a chest or abdominal wall area that overlies tissues whose average density is greater that of lung tissue but less than that of liver tissue emits a percussion note called **dullness** that is of relatively medium intensity, medium pitch, and medium duration. Percussion of a chest or abdominal wall area that overlies tissues whose average density approximates that of the liver emits a percussion note called **flatness** that is of relatively soft intensity, high pitch, and short duration. It should be noted that in practice examiners encounter a spectrum of percussion notes that range from the muscial timbre of tympany to the unresonanting sound of flatness. However, in characterizing the percussion notes, examiners generally classify a note as being one of the four basic categories just described.

In this case, the finding of percussion dullness in place of percussion resonance over a peripheral lung region is an important finding: it indicates the presence of (a) a peripheral lung region of increased density, (b) pleural fluid, or (c) pleural thickening.

5. **The patient's dyspnea indicates inadequate ventilation or perfusion of the lungs.** The patient's tachypnea and tachycardia are probably due to her dyspnea.
6. **The fever of 104.2°F. suggests the presence of a major inflammatory process.**

INTERMEDIATE EVALUATION OF THE PATIENT'S CONDITION

The patient appears to be suffering from an acute disease involving chiefly the upper lobe of the right lung.

CLINICAL REASONING PROCESS

In this case, all of the major signs and symptoms point to an acute disease that involves inflammation of the upper lobe of the right lung and the parietal pleura of the right pleural space. The inflammatory process in the upper lobe of the right lung has produced consolidation within the lobe; the detection of physical signs of consolidation indicate that the consolidation process is relatively extensive.

An emergency or primary care physician would recognize that the patient's signs and symptoms are typical of lobar pneumonia (pneumonia confined to the lobe of a lung). Pneumonia is infectious inflammation of the acinar respiratory units in one or more lobes of a lung. Pneumonia may be caused by bacteria, viruses, or Mycoplasma. The infectious agents can enter the lung by direct inhalation, aspiration of oral and/or nasopharyngeal secretions, or hematogenous dissemination from a distant site of infection in the body.

The physician would appreciate that the history and physical findings are typical of bacterial pneumonia, specifically, Streptococcus pneumoniae (pneumococcal) pneumonia. In pneumococcal pneumonia, the inflammation of the respiratory airways produces an exudate that accumulates within the airways. The exudate disperses in a rather arbitrary fashion to neighboring acini by the pores of Kohn. If this pathologic process is not abated, it produces a number of ill-defined and irregularly shaped regions of consolidation which coalesce into progressively larger regions of consolidation within the infected lobe. The visceral pleura and connective tissue sheath enveloping the lobe retard dispersion of the exudate to a neighboring lobe.

RADIOGRAPHIC EVALUATION AND FINAL RESOLUTION OF THE PATIENT'S CONDITION

In cases of suspected pneumonia, PA and lateral chest films confirm the presence of inflammation of lung parenchyma and define the site(s) of inflammation. Before reviewing the anatomic basis of the radiographic findings in this case, a discussion of the normal radiographic appearance of thoracic viscera in chest films is in order.

Chest films are taken as a patient holds his/her breath after taking a deep inward breath of air. The deep inspiration maximally fills the lungs with air (thereby increasing their radiolucency) and thrusts the costodiaphragmatic margin of each lung into the costodiaphragmatic recess of its pleural space.

The normal PA chest film is a composite of the images cast by the bones and soft tissues of the chest wall, diaphragm, the viscera of the upper abdomen, the viscera of the mediastinum, and the major blood vessels and airways of the lungs. (Fig. 17–12). The large radiopaque shadow in the center of the film can be called the **cardiovascular shadow,** since the heart and the major vessels attached to it cast almost all of the shadow (the vertebral column and sternum also contribute to the shadow). The curved water-density outlines of four sets of structures define the right border of the cardiovascular shadow (a water-density structure is a structure with a radiographic density similar to that of water) (compare Figs. 17–12A and 17–13). Proceeding from the uppermost to the lowermost, these structures are:

a. the brachiocephalic artery and right brachiocephalic vein together
b. the superior vena cava and ascending aorta together
c. the right atrium
d. the inferior vena cava.

The curved water-density outlines of five sets of structures define the left border of the cardiovascular shadow (compare Figs. 17–12A and 17–13). Proceeding from the uppermost to the lowermost, these structures are:

a. the left subclavian artery and left brachiocephalic vein together
b. the terminal part of the aortic arch (the prominent, rounded outline cast by the terminal part of the aortic arch is called the **aortic knob**)
c. the pulmonary trunk
d. the auricle of the left atrium
e. the left ventricle.

The lung fields in a normal PA chest film appear as radiolucent fields whose only prominent markings are water-density bands which branch and radiate into the lung fields from about the midregion of the left and

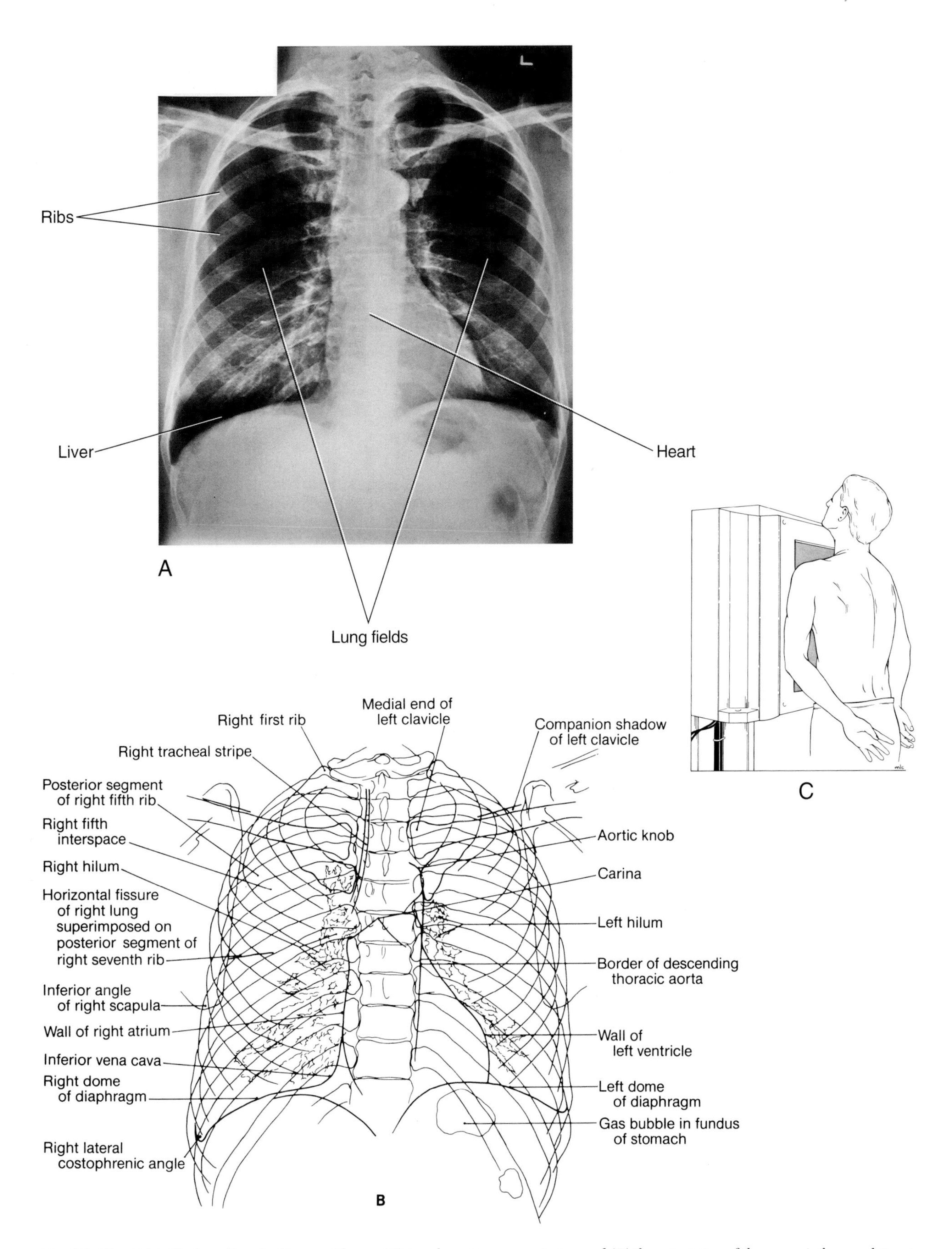

Fig. 17–12: (A) A PA chest film of a 54 year-old man, (B) its schematic representation, and (C) the orientation of the person's thorax relative to the X-ray beam and film cassette for the radiograph.

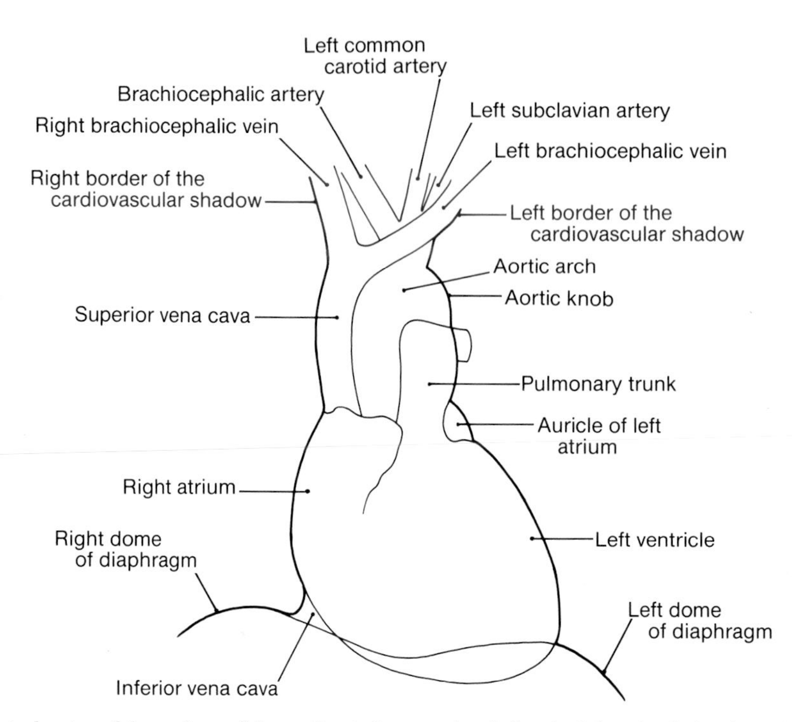

Fig. 17–13: A schematic drawing of the outlines of the mediastinal viscera that define the left and right borders of the cardiovascular shadow of a PA chest film.

right borders of the cardiovascular shadow. The central area on each side is referred to as the **hilum of the lung field.** [Note that the definition of the hilum of a lung field in a chest film is different from the anatomic definition of the hilum of a lung.] The pulmonary artery and its major branches and the pulmonary veins and their major tributaries in each lung cast the water-density bands of each hilum. The walls of the lobar and segmental bronchi are too obliquely oriented to cast distinct shadows in the lung fields. The pulmonary and bronchopulmonary lymph nodes are too small to cast distinct shadows in the lung fields. The domes of the diaphragm cast water-density shadows at the bases of the lung fields. The inferolateral corner of each lung field is called the **lateral costophrenic angle;** this sharply acute angle represents the projection of the lateral aspect of the lung's costodiaphragmatic margin into the costodiaphragmatic recess.

Lateral chest films are commonly taken to allow better visualization of a lesion or abnormality confined to one side of the thorax (Fig. 17–14). In a normal lateral chest film, the heart casts a bulbous shadow overlying the anterosuperior margins of the domes of the diaphragm. The tracheal lumen appears as an almost vertical radiolucent band which ends immediately behind the superior limit of the posterior border of the cardiac shadow. The aortic arch casts a water-density shadow that arches over the cardiac shadow and crosses the tracheal lumen.

The most radiolucent lung field areas in a lateral chest film are the retrosternal and retrocardiac areas (Fig. 17–14). The retrosternal area is bounded by the sternum anteriorly and the cardiac shadow and ascending aorta posteriorly. The retrosternal area shows the superimposed radiodensities of principally the anterior bronchopulmonary segments of the upper lobes of both lungs. The retrocardiac area is the radiolucent area directly posterior to the lower part of the cardiac shadow. The retrocardiac area shows the superimposed radiodensities of the basal bronchopulmonary segments of the lower lobes of both lungs. The posteroinferior corner of the retrocardiac area is called the **posterior costophrenic angle,** or **posterior sulcus.** This sharply acute angle represents the superimposed projections of the posterior aspect of each lung's costodiaphragmatic margin into its costodiaphragmatic recess.

In a normal lateral chest film, the radiodensities of the lungs' hila are superimposed in the region directly posterior to the upper part of the cardiac shadow. In this region of the superimposed hila, the walls of a few lobar or segmental bronchi frequently are sufficiently parallel to the path of the X-ray beam for each to cast a fine water-density circle around an air-density interior.

In this case, PA and lateral chest films of the patient showed consolidation and an air bronchogram of the upper region of the right lung field. Consolidation was indicated by multiple, irregularly shaped, water-density areas evenly distributed throughout the upper region

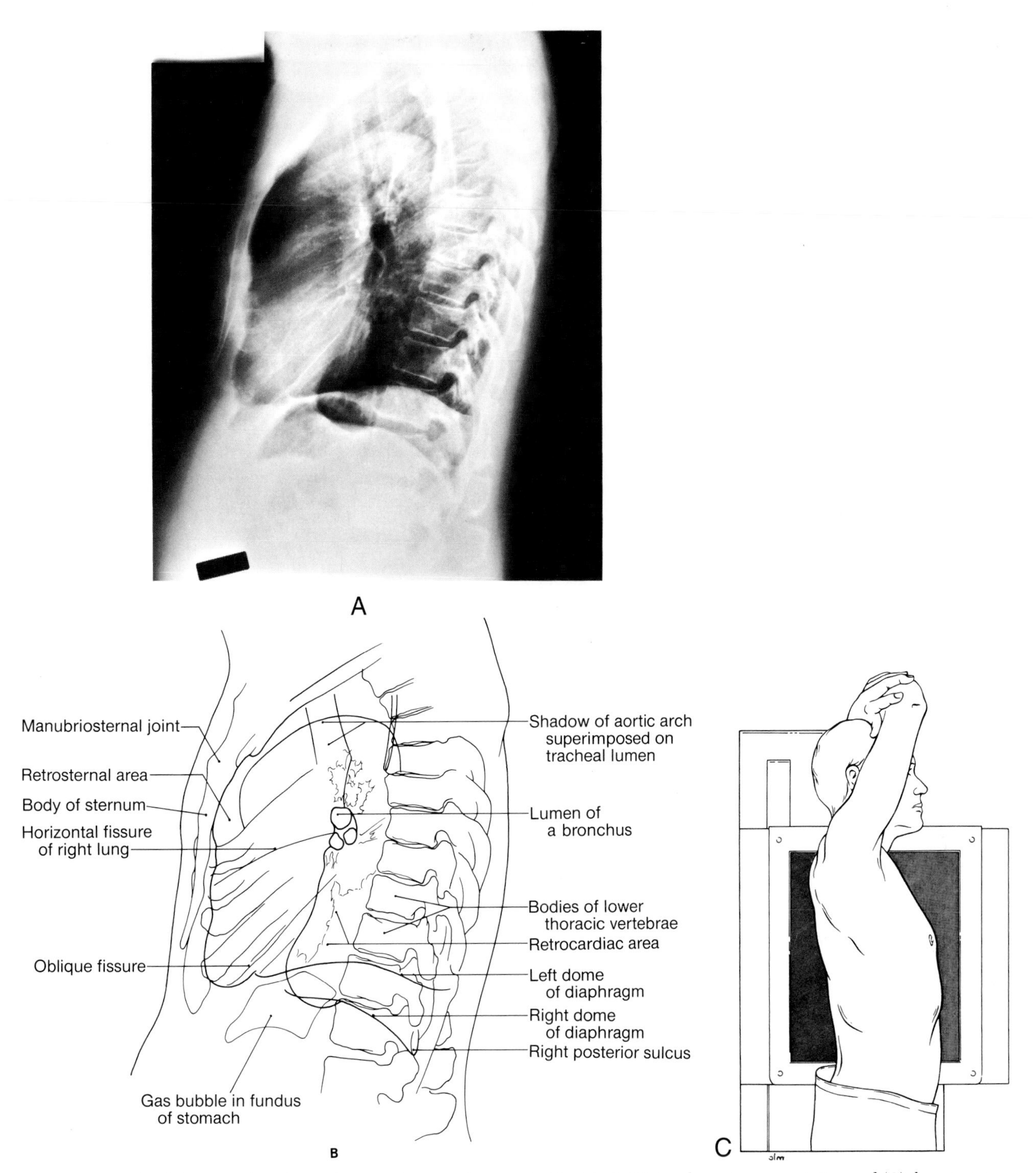

Fig. 17–14: (A) A left lateral chest film of the same 54 year-old man as in Fig. 17–12, (B) its schematic representation, and (C) the orientation of the patient's thorax relative to the x-ray beam and film cassette for the radiograph.

of the right lung field. The PA chest film revealed obscuration of the silhouettes of the superior vena cava, the ascending aorta, and the upper end of the right atrium along the right border of the cardiovascular shadow.

An **air bronchogram** is a radiographic sign in which the lumina of air-filled bronchi appear in chest films as branching, radiolucent bands within a relatively radiopaque consolidated lung region. The sign is observed when the consolidation process in a lung region does not replace the air in the larger conducting bronchial airways.

The obscuration of the silhouettes of the superior vena cava, the ascending aorta, and the upper end of the right atrium along the right border of the cardiovascular shadow in the PA chest film is a silhouette sign. Any loss of the silhouettes of the mediastinal viscera that define the borders of the cardiovascular shadow along the medial margins of the lung fields in a PA chest film, or any loss of the silhouettes of the domes of the diaphragm along the inferior margins of the lung fields, is called a **silhouette sign.** Silhouette signs are frequently produced by consolidation of certain bronchopulmonary segments, specifically, those segments in each lung that directly face the mediastinal viscera or the diaphragmatic domes. If any of these bronchopulmonary segments acquire a radiographic density equal to that of the mediastinal viscera or diaphragmatic domes, a silhouette sign is produced because part or all of the border between the radiopaque segments and the mediastinal viscera or diaphragmatic domes cannot be discerned. The anterior segment of the right lung's upper lobe directly faces segments of the superior vena cava and ascending aorta and the upper end of the right atrium. The obscuration of the silhouettes of these three mediastinal viscera demonstrates that the consolidation process in the right lung's upper lobe involves in particular the anterior segment of the lobe.

A radiologist would recognize that the features of consolidation in the upper region of the right lung field suggest pneumonia of bacterial etiology. Further evaluation of the patient would entail sputum collection for microscopic examination and culture of infectious agents. Treatment would involve administration of antibiotics, fluids, and antipyretic drugs. If effective antibiotics are administered and the patient's condition remains uncomplicated, the patient's condition should resolve within 1 to 2 weeks.

EXPLANATION OF THE PATIENT'S CONDITION

The patient has pneumococcal pneumonia in the upper lobe of the right lung. The exudate produced in the respiratory airways dispersed in the fashion described above to produce extensive consolidation of the entire upper lobe. Extension of the inflammatory process to the parietal pleura accounts for the pleuritic pain. The patient splints (bends) the chest to the right when coughing so as to minimize the inspiratory excursions of the chest wall on the right side, and thereby minimize the exacerbation of pleuritic pain. The pneumonia has also increased bronchial secretions, which, in turn, have prompted coughing and the expectoration of mucopurulent sputum. The blood streaks in the sputum represent blood extravasated from the rupture of small blood vessels in bronchial walls.

CASE **IV.3**

The Case of John Nguyen

INITIAL PRESENTATION AND APPEARANCE OF THE PATIENT

A 20 year-old Asian man named John Nguyen walks into the student infirmary at 10:00 AM. He appears apprehensive and uncomfortable and is coughing.

QUESTIONS ASKED OF THE PATIENT

What is it that has brought you to the infirmary? I have had a cough and felt flu-like for the past week. I don't seem to be getting any better. *[Cough receptors (receptors whose stimulation initiates the desire to cough) are located in the larynx, trachea, and the main stem bronchi of the lungs. The presence of a cough frequently indicates an irritative process or irritative substances at one or more sites in the upper respiratory tract.]*

Do you bring up any sputum when you cough? Occasionally, but not usually.

How would you describe the color and thickness of the sputum that you occasionally can bring up? It's yellowish and a bit thick. *[Sputum production is an indicator of increased bronchial secretions. Sputum that is thick and yellowish (or greenish) is called mucopurulent sputum. The production of mucopurulent sputum suggests that an infectious inflammatory process is the basis for the increased bronchial secretions.]*

Have you seen any streaks of blood in your sputum? No.

Do you have any chest pains? No, not really.

Do you have any trouble breathing? No.

Can you explain what you mean when you say you feel flu-like? Yes. I feel feverish and tired, and I've had headaches on and off. I just don't seem to have much energy.

Do you have a sore throat? No, not now. But before I got this cough and flu, I did have a sore throat for a few days last week. *[A sore throat indicates inflammation of the pharynx.]*

Is there anything else that is also bothering you that I haven't asked about? No.

Do you know any individuals around you, such as friends or classmates, who have also recently suffered from a similar illness? Yes, some of my friends have recently had colds or the flu.

Do you have or have you ever had any problems with your lungs? No.

Do you have allergies? No.

Do you smoke? No.

Have you recently been exposed to any noxious gases or fumes? No.

Do you use any illegal drugs? No.

Have you had any recent illnesses or injuries prior to this illness? No.

Are you taking any medication for previous illnesses? No.

Have you tried any over-the-counter medications for your cough and flu-like symptoms? Yes, throat lozenges. *[The examiner finds the patient alert and fully cooperative during the interview.]*

PHYSICAL EXAMINATION OF THE PATIENT

Vital signs: Blood pressure
Lying supine: 125/75 left arm and 125/75 right arm
Standing: 125/75 left arm and 125/75 right arm
Pulse: 115
Rhythm: regular
Temperature: 102.4°F.
Respiratory rate: 15
Height: 5′10″
Weight: 150 lbs.

[The examiner recognizes that the patient's body temperature and pulse are markedly elevated. 99.0°F. may be regarded as the normal upper limit of body temperature as recorded by an oral thermometer. Normal adult heartbeat rates vary from 60 to 100 beats per minute.]

HEENT Examination: Normal
Lungs: Inspection, palpation, percussion, and auscultation of the lungs are normal except for the following finding on the left lung: With the patient sitting upright, auscultation reveals fine rales near the end of deep inspiration along the base of the lung.
Cardiovascular Examination: The cardiovascular examination is normal except for the tachycardia.
Abdomen: Normal
Genitourinary Examination: Normal
Musculoskeletal Examination: Normal
Neurologic Examination: Normal

INITIAL ASSESSMENT OF THE PATIENT'S CONDITION

The patient appears to be suffering from an acute lower respiratory tract disease.

ANATOMIC BASIS OF THE PATIENT'S HISTORY AND PHYSICAL EXAMINATION

1. **The fine rales heard over the base of the left lung near the end of inspiration indicate inflammation and/or congestion of acini in the lower lobe of the left lung.** Rales are an abnormal respiratory sound. Fine rales sound like the sharp crackling sounds heard when a lock of hair is held close to the ear and rubbed between the thumb and index finger. Fine rales heard near the end of inspiration are due to air suddenly rushing into and reinflating acinar airways that have become congested with fluid.
2. **The fever of 102.4°F. suggests the presence of a major inflammatory process.**

INTERMEDIATE EVALUATION OF THE PATIENT'S CONDITION

The patient appears to be suffering from an acute disease involving in part the lower lobe of the left lung.

CLINICAL REASONING PROCESS

In this case, almost all the signs and symptoms (nonproductive cough, fever, malaise, intermittent headache of one week's duration) are non-specific. The only significant localizing physical sign (rales) points to lower respiratory tract disease. The history of an antecedent sore throat (which signifies inflammation of the pharynx) suggests an acute infectious disease of the respiratory tract.

A primary or emergency care physician would consider pneumonia as the most likely diagnosis. Pneumonia is infectious inflammation of the acinar respiratory units in one or more lobes of a lung. Pneumonia may be caused by bacteria, viruses, or Mycoplasma. The infectious agents can enter the lung via direct inhalation, aspiration of oral and/or nasopharyngeal secretions, or hematogenous dissemination from a distant site of infection in the body.

The physician would appreciate that the history and physical findings are typical of viral or mycoplasmal pneumonia. In these types of pneumonia, the inflammation typically involves chiefly the tissues in the interalveolar septa of the respiratory airways and results in the accumulation of fluid within these tissues.

RADIOGRAPHIC EVALUATION AND FINAL RESOLUTION OF THE PATIENT'S CONDITION

In cases of suspected pneumonia, PA and lateral chest films confirm the presence of inflammation of lung parenchyma and define the site(s) of inflammation. In this case, PA and lateral chest films showed the presence of a network of linear radiopacities in the basal regions of both lung fields; the density of the network was greater in the base of the left lung field. A radiologist would interpret such findings as indicative of a pulmonary disease or condition that is producing a thickening of interstitial pulmonary tissues.

A radiologist would regard the radiographic findings supportive of a diagnosis of pneumonia of viral or mycoplasmal etiology. A radiologist would consider pneumonia of viral or mycoplasmal etiology to be more likely than pneumonia of bacterial infection because whereas bacterial pneumonia typically produces consolidation of pulmonary airways, viral or mycoplasmal pneumonia characteristically produces a thickening of interstitial pulmonary tissues. Further evaluation of the patient would entail sputum collection (if possible) for microscopic examination and culture of infectious agents. Treatment would involve administration of antibiotics, fluids, and antipyretic drugs. If effective antibiotics are administered and the patient's condition remains uncomplicated, the patient's condition should resolve within 1 to 2 weeks.

EXPLANATION OF THE PATIENT'S CONDITION

The patient has mycoplasmal pneumonia in the lower lobe of both lungs. The inflammatory process has thickened interalveolar septa (via edematous swelling). The thickened septa account for the reticular pattern of linear radiopacities in the basal regions of both lung fields. The pneumonia has also marginally increased bronchial secretions, which, in turn, have prompted coughing and infrequent expectoration of mucopurulent sputum.

CASE **IV.4**

The Case of Anne Nokomura

INITIAL PRESENTATION AND APPEARANCE OF THE PATIENT

A 57 year-old, Asian woman named Anne Nokomura walks into the ambulatory clinic at 9:00 AM. She appears apprehensive, uncomfortable, and short of breath.

QUESTIONS ASKED OF THE PATIENT

What is it that has brought you to the clinic? I have trouble catching my breath. *[The symptom of breathlessness or difficulty in breathing is called dyspnea. Dyspnea is an indicator of inadequate ventilation or perfusion of the lungs. The term ventilation refers to the inhalation and exhalation of air from the respiratory airways. The term perfusion refers to the circulation of blood through the pulmonary vasculature.]*

When did you first notice that you were having trouble catching your breath? Last night, around 8:00 PM, just after I came home.

Do you recall specifically what you were doing when you first noticed your breathlessness? Let me think. Yes, I was hanging up some clothes when I first noticed I was having trouble catching my breath.

Do you have trouble catching your breath all the time now or just some of the time? Well, when I sit or lie down for a few minutes, I still have a little trouble catching my breath, but it doesn't bother me that much. But if I get up and even just walk around, that's when I begin to really notice that it's difficult for me to catch my breath. *[The patient's answers to the last three questions suggests that the dyspnea is the consequence of an acute disease or condition and becomes manifest with physical activity.]*

Does it hurt for you to breathe? No, not at all.

Have you felt faint since last evening? No, not at all. *[The symptom of feeling faint is called syncope. In this case, syncope would suggest dyspnea severe enough that there is an inadequate supply of oxygen to the brain.]*

Have you had a cough recently or since last evening? No.

Is there anything else that is also bothering you, or that you think may be associated with your difficulty in catching your breath? Well, I don't know, but I have also been bothered since last evening by a feeling of tightness right here in my chest [the patient places her left hand over her sternum]. And this morning, I noticed that my right leg and foot seem swollen. My right shoe feels tight, and the stocking around my right leg feels tighter than that around my left leg. *[A feeling of tightness in the chest may represent stimulation of visceral sensory fibers in the heart and/or lungs. The swelling of the right leg indicates edema.]*

Have you ever had a previous episode of breathlessness like this one? No.

Do you smoke? No.

Have you taken any medications recently? No.

Have you had any recent illnesses or injuries? No.

Do you have any heart or lung problems? No, not that I know of. I've been fairly healthy all my life.

Have any members of your family had heart or lung problems? No, not that I know of.

Have you recently traveled? Yes, I was on a nonstop, 6-hour flight yesterday. That's why I didn't get home until 8 in the evening. *[The examiner finds the patient alert and fully cooperative during the interview.]*

PHYSICAL EXAMINATION OF THE PATIENT

Vital signs: Blood pressure
Lying supine: 130/75 left arm and 130/75 right arm
Standing: 125/75 left arm and 130/70 right arm
Pulse: 95
Rhythm: regular
Temperature: 98.8°F.
Respiratory rate: 28
Height: 5′4″
Weight: 125 lbs.

[The examiner recognizes that the patient has tachypnea (an elevated rate of respiration) and a heartbeat rate near the upper limit of the normal range. Normal adult respiratory rates vary from 8 to 16 cycles per

minute. In adults, the normal range of the heartbeat rate is 60 to 100 beats per minute.]

HEENT Examination: The head, eyes, ears, nose, and throat exam is normal except for a purplish cast to the oral mucosa.
Lungs: Normal
Cardiovascular Examination: The cardiovascular exam is normal except for a purplish cast to the nailbeds of the fingers and toes and non-pitting edema of the right leg and foot.

[The examiner inspects the patient's oral mucosa and the nailbeds of the upper and lower limbs because the colors of these tissues are monitors of the level of oxygenation of arterial blood, and thus monitors of the severity of the patient's respiratory distress.]

Abdomen: Normal
Genitourinary Examination: Normal
Musculoskeletal Examination: Normal
Neurologic Examination: Normal

INITIAL ASSESSMENT OF THE PATIENT'S CONDITION

The patient appears to be suffering from an acute disease or condition that has rendered her short of breath.

ANATOMIC BASIS OF THE PATIENT'S HISTORY AND PHYSICAL EXAMINATION

1. **The history suggests that an acute event has made the patient dyspneic.** The tachypnea and high normal heartbeat rate appear to be autonomic responses associated with the patient's dyspnea.
2. **The mild central cyanosis indicates insufficent oxygenation of arterial blood.** The normal reddish color of the oral mucosa and nailbeds comes from the oxygenated blood circulating through the dense capillary network of these tissues. A bluish cast is added to the color of these tissues (and they are said to be cyanotic) when the capillary concentration of reduced hemoglobin is 5 gms percent or greater. The colors of the oral mucosa and nailbeds thus respectively provide a rough index of the oxygenation of central and peripheral tissues of the body. Bluish discoloration of both the oral mucosa and nailbeds indicates central cyanosis.
3. **Edema of the right leg and foot indicate localized inflammation, venous obstruction, and/or lymphatic obstruction.** Edema is the condition in which there is an increase in the volume of a soft tissue's interstitial fluid.

Under normal physiologic conditions, interstitial fluid is formed at the arteriolar origins of a tissue's blood capillary beds and absorbed at its venular ends. The forces which govern the flow of fluid across a capillary wall are the hydrostatic and oncotic osmotic pressures of (a) the blood plasma within the capillary and (b) the interstitial fluid surrounding the capillary (oncotic osmotic pressure is the partial osmotic pressure generated in each body fluid by its constituent proteins). The blood's hydrostatic pressure is, on the average, about 30 mm Hg at the arteriolar origins of systemic capillaries and 10 mm Hg at their venular ends. Interstitial fluid hydrostatic pressure is rather uniform throughout a tissue, and is roughly −5 mm Hg. The oncotic osmotic pressure of blood plasma, with its 7% protein concentration, is 28 mm Hg, and that of interstitial fluid, with its 1% protein concentration, is 6 mm Hg.

At the arteriolar origins of systemic capillaries, the 30 mm Hg hydrostatic pressure of the blood and the −5 mm Hg hydrostatic pressure of interstitial fluid combine to force fluid out from the capillaries with a net force of 35 mm Hg. By contrast, the 22 mm Hg oncotic osmotic pressure difference between the blood plasma and interstitial fluid is directed to forcing fluid into the capillaries. Consequently, at the arteriolar origins of systemic capillaries, there is a net 13 mm Hg pressure difference directed to forcing fluid out from the blood plasma; this bulk fluid flow out from the blood plasma produces interstitial fluid. Similar calculations show that at the venular ends of systemic capillaries, there is a 7 mm Hg pressure difference directed to forcing fluid into the blood plasma; this bulk fluid flow contributes to interstitial fluid absorption.

Under normal physiologic conditions, interstitial fluid production at the arteriolar origins of systemic capillaries is greater than the interstitial fluid absorption that occurs at the venular ends. Lymphatic capillaries draw off the excess interstitial fluid. The venous and lymphatic drainage of a tissue are thus both important in maintaining a constant level of interstitial fluid.

Edema can be either generalized or localized. **Generalized edema** is edema in which most of an individual's subcutaneous soft tissues become bilaterally swollen. Disorders which can produce generalized edema include right-sided congestive heart failure and hypoproteinemia. In the former disorder, failure by the right atrium and right ventricle of the heart to pump into the lungs the blood which is drained by the right atrium during each cardiac cycle leads to an increase in the hydrostatic pressure of systemic capillaries throughout the body. This increase results in an increase in interstitial fluid production. In cases of hypoproteinemia, a marked decrease in blood plasma protein concentration decreases its oncotic osmotic pressure. Again, the result is an increase in interstitial fluid production.

Localized edema is edema confined to one or a

few related body regions, generally localized to one side of the body. The causes of localized edema include increased permeability of blood capillary beds (as a result of the action of chemical or bacterial agents, heating, or physical trauma), venous obstruction, and lymphatic obstruction.

The expression pitting edema can apply to either generalized or localized edema. **Pitting edema** is edema severe enough that the application of fingertip pressure for 5 seconds upon the edematous subcutaneous tissues in question leaves a temporary impression, or pit, in the tissues upon removal of the finger (the fingertip pressure forces some of the excess interstitial fluid into deeper tissues). The severity of pitting edema can be assessed by measuring in millimeters the depth of the impression.

INTERMEDIATE EVALUATION OF THE PATIENT'S CONDITION

This patient appears to be suffering from an acute disease or condition that has resulted in a significant deficiency in the capacity of the patient's lungs to oxygenate the blood.

CLINICAL REASONING PROCESS

The patient is suffering from a disease or condition which has acutely impaired ventilation and/or perfusion of the lungs. The two most common mechanisms by which ventilation may be acutely impaired are aspiration of a foreign body into a bronchus and pneumothorax. A pneumothorax is the condition in which the lung retracts from the chest wall as a result of the introduction of air into a pleural space. The patient's history precludes aspiration of a foreign body. A large pneumothorax would result in the replacement of percussion resonance with percussion tympany over a significant area of one of the lung fields. Although a small pneumothorax might not manifest any physical signs, it is unlikely that it would result in mild central cyanosis.

The two most common mechanisms by which perfusion may be acutely impaired are sudden cardiac dysfunction and pulmonary embolism. The history and physical examination do not reveal any evidence of cardiac disease or dysfunction. A pulmonary embolism is the partial or complete obstruction of a pulmonary artery by a foreign body. In about 90% of all cases, a dislodged thrombus (dislodged blood clot) forms the embolus. Other less commonly encountered emboli are fat masses (which may be mobilized upon fracture of bones with a fat-filled cancellous interior) and air bubbles (which may be drawn into the circulation from exposure of cut veins to the exterior). In this case, a primary care physician would tentatively judge pulmonary embolism to be the most likely basis of the patient's dyspnea and mild central cyanosis, and would regard a venous thrombus to be the most likely source of the embolus. This is because when venous return is impaired for a long period (such as may occur in an individual's lower limbs during a lengthy flight), the venous stasis may lead to a hypercoagulable state and subsequent thrombosis.

Most cases of pulmonary emboli do not result in pulmonary infarction. [The term infarction refers to the necrosis of tissue resulting from acute insufficiency of arterial supply and/or venous drainage.] This is because the respiratory airways and the connective tissues supporting them in each bronchopulmonary segment have a dual arterial supply: one that emanates from a pulmonary artery and another from one or two bronchial arteries. The blood supply provided by the bronchial arterial branches generally provides enough blood-borne nutrients to sustain the respiratory airways and the surrounding tissues in the absence of any pulmonary arterial supply. Moreover, the cells that comprise the walls of the respiratory airways can also acquire oxygen and release carbon dioxide through the lung's respiratory tree.

When a pulmonary embolism does result in pulmonary infarction, extravasation of blood from the necrotic tissue can seep into respiratory airways and result in hemoptysis (coughing up of blood) and/or the pleural space and inflame the parietal pleura. Pleuritic chest pain and hemoptysis in cases of suspected pulmonary embolism are thus strongly suggestive of pulmonary infarction.

RADIOGRAPHIC EVALUATION AND FINAL RESOLUTION OF THE PATIENT'S CONDITION

PA and lateral chest films would effectively screen for other possible causes of the patient's dyspnea, such as pneumonia or interstitial lung disease. Chest films would not provide information highly specific for pulmonary embolism. They might show abnormalities that can be produced by pulmonary embolism: decreased vascularity in a limited region of the lung field, abrupt cutoff or rapid tapering of an occluded pulmonary arterial branch, or a small infiltrate representing a region in which infarction or atelectasis has occurred. They might also show no abnormalities. In this case, PA and lateral chest films did not show any abnormalities.

Perfusion and ventilation lung scans would provide valuable information for the evaluation of a suspected pulmonary embolism. A **perfusion scan** is a nuclear medicine technique that images the pulmonary vascular distribution of intravenously injected, radiolabelled microspheres or macroaggregates of albumin; the radiolabelled particles become entrapped in the pulmonary capillary bed because their size exceeds that of the average diameter of the capillaries. A **ventilation scan** is a nuclear medicine technique which images the distribution of inhaled, radiolabelled inert gas throughout the bronchial tree. Pulmonary embolism is indicated if a region segmental or larger in size shows a perfusion defect but normal ventilation. In this case, perfusion and ventilation lung scans showed normal ventilation in both lungs but a segmental-sized perfusion defect in the right lung's lower lobe.

A pulmonary angiogram would provide the most specific evidence for a pulmonary embolism. Pulmonary angiography images the distribution of a radiopaque medium through the pulmonary vasculature upon release of the medium into the right ventricle. However, because pulmonary angiography requires catheterization of the heart and the use of a potentially irritating radiopaque medium, the procedure is generally not employed in cases of suspected pulmonary embolism if the results of the ventilation and perfusion scans provide adequate evidence of a pulmonary embolism (as in this case). A pulmonary angiogram of the patient would show a filling defect in a segmental branch of the right lung's lower lobar artery.

CHRONOLOGY OF THE PATIENT'S CONDITION

A thrombus developed in the patient's right popliteal vein during her lengthy jet plane trip. Most of the venous thrombus embolized shortly after her return home, and (after passing through the right femoral vein, right external iliac vein, right common iliac vein, inferior vena cava, right atrium, right ventricle, pulmonary trunk, and right pulmonary artery) became lodged near the origin of the segmental arterial branch to the lateral basal segment of the right lung's lower lobe. The embolus markedly diminished pulmonary arterial blood flow to the segment; the acute stretching of the segmental arterial branch is the basis of the patient's substernal tightness. The embolus has markedly diminished the capacity of the segment to provide for gas exchange in its respiratory airways; the deficiency is sufficient to have produced arterial hypoxemia, which is the basis for the patient's mild central cyanosis, dyspnea, and tachypnea. The part of the thrombus still retained in the patient's right popliteal vein is partially obstructing venous drainage from the right leg and foot, and thus responsible for the slight edema of the right leg and foot.

CASE **IV.5**

The Case of Mark Mehlman

INITIAL PRESENTATION AND APPEARANCE OF THE PATIENT

A 21 year-old, white man named Mark Mehlman walks into the emergency center at 9:00 PM. The patient appears anxious, apprehensive, and short of breath. He splints the chest to the right when he takes a deep breath.

QUESTIONS ASKED OF THE PATIENT

How can I help you? My chest hurts real bad and I'm having a real problem catching my breath. I was sitting at home working with the computer when all of a sudden I got this bad pain in my chest and found it difficult to catch my breath.

How long has it been since the chest pain and difficulty in catching your breath began? It started within the last hour. *[The symptom of breathlessness or difficulty in breathing is called dyspnea. Dyspnea is an indicator of inadequate ventilation or perfusion of the lungs. The term ventilation refers to the inhalation and exhalation of air from the respiratory airways. The term perfusion refers here to the circulation of blood through the pulmonary vasculature.]*

Have you felt faint or dizzy? No. *[The symptom of feeling faint is called syncope. Syncope is an indicator of inadequate perfusion of the brain or insufficient concentrations of oxygen and/or glucose in the blood circulating through the cerebral vasculature. The term perfusion refers here to the circulation of blood through the cerebral vasculature.]*

Where does your chest hurt? All along here [the patient slides his left hand up and down along the right side of the chest wall].

How would you describe the pain? It's real sharp.

Has the nature of the pain or its location changed during the past hour? No.

Does it hurt to breathe? Yes, especially when I try to breathe air in.

Does it hurt to bend or twist your rib cage? A little bit. *[Sharp chest pain associated with breathing movements suggests inflammation of parietal pleura on the affected side.]*

Is there anything else that is also bothering you? No.

Have you ever had a previous episode of breathlessness and chest pain like this one? No.

Did you injure yourself in any way today before you became breathless? No.

Have you had any previous diseases or problems with your lungs? No.

Have you had any previous diseases or problems with your heart? No.

Have you had any recent illnesses or injuries? No.

Have you recently been exposed to any noxious gases or fumes? No.

Do you smoke? Yes, about a pack a day.

How long have you been smoking a pack of cigarettes a day? About 3 years.

Now, this next question I'm about to ask, I'm asking it because I want to have as much correct information as possible about your difficulty with breathing. Have you taken any illegal drugs? No. *[The examiner inquires about illegal drug use because (a) the patient is a young adult (and therefore in one of the age groups most commonly involved with illegal drug use) and (b) certain illegal drugs, such as cocaine, can produce respiratory disorders.]*

Are you taking any medications for past illnesses or injuries? No.

Did you take any over-the-counter medications today? No.

Have you recently travelled? No. *[The examiner finds the patient alert and fully cooperative during the interview.]*

PHYSICAL EXAMINATION OF THE PATIENT

Vital signs: Blood pressure
Lying supine: 120/65 left arm and 120/65 right arm
Standing: 120/65 left arm and 120/65 right arm
Pulse: 105
Rhythm: regular
Temperature: 98.9°F.
Respiratory rate: 30

Height: 6′4″
Weight: 145 lbs.

[The examiner recognizes that the patient exhibits tachycardia (an elevated heartbeat rate) and tachypnea (an elevated respiratory rate).]

HEENT Examination: The head, eyes, ears, nose, and throat exam is normal except for the following findings: (a) The trachea exhibits pendular deviations during respiration, swinging to the right during inspiration and to the left during expiration. (b) The oral mucosa has a purplish cast.

Lungs: Inspection, palpation, percussion, and auscultation of the lungs are normal except for the following findings on the right side of the chest: Inspection shows that the intercostal spaces on the right side of the chest are wider than those on the left sid. With the patient seated upright, auscultation at the second to seventh intercostal spaces along the right midaxillary line reveals very faint vesicular breath sounds and a marked decrease in the intensity and clarity of spoken voice and whispered sounds. Percussion tympany is found at the same sites.

Cardiovascular Examination: The cardiovascular examination is normal except for the tachycardia and a purplish cast to the oral mucosa and the nailbeds of the fingers and toes.

Abdomen: Normal

Genitourinary Examination: Normal

Musculoskeletal Examination: Normal

Neurologic Examination: Normal

INITIAL ASSESSMENT OF THE PATIENT'S CONDITION

The patient is suffering from an acute condition or disease, the major symptoms of which are chest pain and dyspnea.

ANATOMIC BASIS OF THE PATIENT'S HISTORY AND PHYSICAL EXAMINATION

1. **The qualities of the patient's chest pain (a sharp pain worsened by respiratory movements) indicates inflammation of the parietal pleura in the right pleural space.** The movements of the chest wall and diaphragm during inspiration stretch the parietal pleura of the pleural spaces. The increased tension in the parietal pleura elicits a sharp, knife-like pain if the parietal pleura is inflamed. Sharp chest pain produced by the stretching of inflamed parietal pleura is called pleuritic pain.
2. **The patient's dyspnea indicates inadequate ventilation or perfusion of one or both lungs.**
3. **The mild central cyanosis indicates insufficent oxygenation of arterial blood.** The normal reddish color of the oral mucosa and nailbeds comes from the oxygenated blood circulating through the dense capillary network of these tissues. A bluish cast is added to the color of these tissues (and they are said to be cyanotic) when the capillary concentration of reduced hemoglobin is 5 gms percent or greater. The colors of the oral mucosa and nailbeds thus respectively provide a rough index of the oxygenation of central and peripheral tissues of the body. Bluish discoloration of both the oral mucosa and nailbeds indicates central cyanosis.
4. **Acute diminution of vesicular breath sounds over a peripheral lung region indicates diminished air flow into and out from the lung region and/or the dampening of the sounds by excess pleural fluid or pleural thickening.** There are four general pathologic mechanisms which diminish air flow into and out from a lung region:
   a. obstruction of one or more of the conducting airways bringing air to the lung region,
   b. loss of the elasticity of the interstitial tissues about the respiratory airways (the normal contraction of these interstitial tissues during expiration forces air out from the respiratory airways),
   c. atelectasis (collapse of the respiratory airways), and
   d. consolidation (accumulation of fluid within the interstitial tissues and/or airways of the lung).
5. **Acute diminution of the intensity and clarity of spoken voice and whispered sounds over a peripheral lung region indicates interference in the normal conduction of audible vibrations from the central conducting airways of the lung's hilum to the chest wall overlying the peripheral lung region.** Conditions which create such interference include complete bronchial obstruction, pleural effusion, pleural thickening, pneumothorax, and hyperinflation of lung tissue.
6. **The percussion tympany over the right lung field suggests percussion of primarily air beneath the chest wall on the right side.** Percussion of the lung fields provides assessment of the average density of tissues lying to a depth of 5 to 7 cm beneath the chest wall areas being percussed. The finding of percussion tympany in place of percussion resonance over a peripheral lung region is an important and specific finding: it indicates a replacement of air-filled peripheral lung tissue with either air or hyperinflated peripheral lung tissue. The presence of air would signify a pneumothorax, a condition in which air enters the pleural space and the lung's re-

tractive forces pull the lung away from the chest wall.

7. **The widened intercostal spaces on the right side of the rib cage indicate replacement of air-filled peripheral lung tissue with either air or hyperinflated peripheral lung tissue.** As just noted, the presence of air would signify a pneumothorax. A pneumothorax produces widening of intercostal spaces because the overlying chest wall is no longer subjected to the recoil contractive forces of the now retracted lung. Hyperinflation of peripheral lung tissue produces widening of intercostal spaces because of the abnormal distension of the lung parenchyma.

INTERMEDIATE EVALUATION OF THE PATIENT'S CONDITION

The patient has suffered a pneumothorax on the right side.

CLINICAL REASONING PROCESS

In this case, the findings of the history and physical exam are diagnostic. The patient's account of a sudden onset of sharp chest pain and severe dyspnea are complemented by the findings of a marked decrease of breath and voice sounds and the appearance of percussion tympany and widened intercostal spaces over a vast extent of the right lung field. This combination of findings is highly specific for a large pneumothorax.

The only other condition which manifests similar physical findings (a decrease of breath and voice sounds, widened intercostal spaces and percussion tympany over a vast extent of a lung field) is emphysema. However, the pathophysiologic changes of emphysema typically progress in both lungs over several years to produce an insidious onset of dyspnea. Emphysema is a condition in which the respiratory airways become permanently distended (hyperinflated) as a result of the loss of elasticity in their walls. Since both lungs are typically equally affected, the widening of intercostal spaces which occurs upon hyperinflation of the lungs is generally symmetric on both sides of the chest wall. The trachea generally retains its midline position throughout the inspiratory and expiratory phases of respiration. Moreover, the hyperinflation of an entire lung which is encountered in advanced cases of emphysema also produces a lowered diaphragmatic dome, which is not observed in this case.

When an individual who has apparently been in good health and injury-free reports a sudden onset of severe dyspnea and sharp chest pain, the most likely diagnoses are idiopathic spontaneous pneumothorax and pulmonary embolism with pulmonary infarct (refer to the case of Anne Nokomura, Case IV-4, for discussions of pulmonary embolism and pulmonary infarct). Acute myocardial infarction with congestive failure would also be considered if the individual is elderly; however, the chest pain is typically dull and associated with a feeling of substernal tightness. Upon completing the history but before conducting the physical exam, a primary or emergency care physician would regard idiopathic spontaneous pneumothorax as more likely than pulmonary embolism because the patient is of the sex and an age (male and between the ages of 20 and 30) and has the habit (the cigarette smoking habit) and body habitus (tall and lean) that are relatively common to those individuals who suffer a spontaneous pneumothorax. The rupture of a pleural bleb in the apex of the lung is the mechanism responsible for many cases of idiopathic spontaneous pneumothorax. The etiology of the formation of these apical pleural blebs is not known.

RADIOGRAPHIC EVALUATION AND FINAL RESOLUTION OF THE PATIENT'S CONDITION

A PA chest film can provide evidence of a small pneumothorax or a rough estimate of the extent of a large pneumothorax. In this case, a PA chest film revealed an outer border (of 4.0 cm average width) in the right lung field devoid of lung vascular markings and bordered medially by a thin line with the radiographic density of water (this water-density line represents the visceral pleura lining the partially retracted right lung). The radiologist judged the right lung to have retracted to approximately 50% its normal size.

On the basis of the history and the physical exam and radiographic findings, a physician would make a diagnosis of idiopathic spontaneous pneumothorax. The patient would be hospitalized for placement of chest tube (tube thoracostomy) and evacuation of the pleural space. In the absence of any complications, the patient's condition should resolve within 2 to 4 days. The attending physician would inform the patient that there is a 25 to 50% probability that the patient will suffer another spontaneous pneumothorax. The physician would also advise the patient to stop smoking because of the numerous, deleterious effects of smoking on the cardiovascular and respiratory systems.

EXPLANATION OF THE PATIENT'S CONDITION

The patient suffered a spontaneous 50% pneumothorax as a consequence of rupture of an apical pleural bleb of the right lung. The rupturing of the apical pleural bleb resulted in a temporary escape of air from the right lung into the surrounding cavity. The intrapleural air intervened between the chest wall and the lung's costal surface, and thus permitted separation and a partial retraction of the lung from the chest wall.

The inadequate ventilation provided by the partially retracted right lung has forced the patient to respire at a greater than normal rate in order to achieve adequate oxygenation of the blood supply. The patient's perception of the need to breathe more rapidly and more deeply in order to achieve adequate oxygenation of the blood supply is the basis of the complaint of a difficulty in catching his breath. The tachypnea and tachycardia are autonomic responses associated with the patient's dyspnea.

The tearing of small blood vessels coincident with the rupture of the pleural bleb resulted in the extravasation of a small amount of blood into the right pleural space. The irritation of the parietal pleura upon contact with the extravasated blood is the basis for the sharp chest pain on the right side. The patient splints (bends) the chest wall to the right during inspiration so as to minimize the inspiratory excursions of the chest wall on the right side, and thereby minimize the exacerbation of pleuritic pain.

The anatomic basis for the pendulant motion of the trachea during the inspiratory and expiratory phases of respiration is as follows: The patient's pneumothorax at the time of the physical exam is a closed pneumothorax: the torn apical pleural bleb has become sealed, entrapping the air in the right pleural space. When the patient inspires (attempts to breathe air into the lungs), the expansion of the vertical, anteroposterior, and lateral dimensions of the right pleural space increases the volume of the air entrapped in the space, and thus decreases its pressure. When the pressure of the air entrapped within the right pleural space decreases to a subatmospheric level, the partially retracted lung and the trachea to which it is attached are drawn to the right. When the patient next expires (attempts to breath air out from the lungs), the retraction movements of the rib cage and diaphragm decrease the volume of air entrapped in the right pleural space, and thus increase its pressure. When the pressure of the air entrapped within the right pleural space increases to a supra-atmospheric level, the partially retracted lung and the trachea to which it is attached are pushed to the left.

RECOMMENDED REFERENCES FOR ADDITIONAL INFORMATION ON THE LUNGS

DeGowin, R. L., Jochimsen, P. R., and E. O. Theilen, *DeGowin & DeGowin's Bedside Diagnostic Examination*, 5th ed., Macmillan Publishing Co., New York, 1987: *Pages 295 through 315 present a discussion of the physical assessment of the lungs and pleural spaces.*

Wilson, J. D., Braunwald, E., Isselbacher, K. J., Petersdorf, R. G., Martin, J. B., Fauci, A. S., and R. K. Root, *Harrison's Principles Of Internal Medicine*, 12th ed., McGraw-Hill, New York, 1991: *This text is one of the most authoritative and comprehensive texts on internal medicine. In Part 7 on disorders of the respiratory system, Chapter 200 provides a discussion of the common signs and symptoms of respiratory disease, and Chapter 203 presents descriptions of the noninvasive and invasive procedures commonly employed to investigate respiratory disorders.*

Kelley, W., N., DeVita, Jr., V. T., Dupont, H. L., Harris, Jr., E. D., Hazzard, W. R., Holmes, E. W., Hudson, L. D., Humes, H. D., Paty, D. W., Watanabe, A. M., and T. Yamada, *Textbook Of Internal Medicine*, J. B. Lippincott Co., Philadelphia, 1992: *This text is one of the most authoritative and comprehensive texts on internal medicine. In Part VIII on pulmonary and critical care medicine, Chapter 379 provides a discussion of the significance of chest pain, cough, sputum production, hemoptysis, dyspnea, wheezing, stridor, and cyanosis in the analysis of respiratory disorders.*

Slaby, F., and E. R. Jacobs, *Radiographic Anatomy*, Harwal Publishing Co., Malvern, PA, 1990: *Pages 100 through 107 in Chapter 3 outline the major anatomic features of PA and lateral chest films. These outlines describe chest films in terms of the images cast by four major sets of anatomically distinct structures: (1) the bones and soft tissues of the chest wall, (2) the diaphragm and underlying abdominal viscera, (3) the mediastinal viscera, and (4) the major blood vessels and airways of the lungs.*

Blank, N., *Chest Radiographic Analysis*, Churchill Livingstone, Edinburgh, 1989: *Chapter 1 provides an extensive discussion of the analysis of PA and lateral chest films.*

Sutton, D., *A Textbook Of Radiology And Imaging*, Churchill Livingstone, Edinburgh, 1993: *Chapter 11 presents a discussion of the roles of plain film radiography, CT scanning, radionuclide studies, and pulmonary angiography in the analysis of respiratory disorders.*

Light, R. W., *Pleural Diseases*, Lea & Febiger, Philadelphia, 1990: *Chapter 1 presents a description of the anatomy of the visceral and parietal pleura, and Chapter 2 provides a discussion of the physiology of the pleural space.*

CHAPTER **18**

The Heart

The heart lies in the middle mediastinum completely enclosed within a tissue sac called the pericardium (Fig18–1). The space inside the pericardium that surrounds the heart is called the pericardial cavity. The pericardial cavity separates the heart from other mediastinal viscera.

THE PERICARDIUM AND PERICARDIAL CAVITY

The pericardium consists of two tissue layers. The outer tissue layer is composed of fibrous connective tissue, and is called the fibrous pericardium. The fibrous pericardium is attached to all the large veins and arteries extending to and from the heart. Figure 18–2 shows the attachment of the fibrous pericardium to the superior vena cava, ascending aorta, and pulmonary trunk. It spreads out from these vascular attachments to form a fibrous sac in which the heart is completely sealed. It is in this fashion that the fibrous pericardium physically separates the heart from the lungs and all the mediastinal viscera. The fibrous pericardium is also adherent to the posterior surface of the sternum and the superior surface of the central tendon of the diaphragm (Fig. 18–2).

The inner layer of the pericardium is composed of a serous membrane, and thus is called the **serous pericardium.** The serous pericardium has two parts. One part covers the inner surface of the fibrous pericardium, and is called the **parietal part of the serous pericardium.** The other part covers the outer surface of the heart and the proximal segments of the arteries and veins attached to the heart, and is called the **visceral part of the serous pericardium.** Since the visceral part of the serous pericardium is the most superficial tissue layer of the heart, it is sometimes called the **epicardium.** The parietal and visceral layers of the serous pericardium are continuous with each other at the sites where the fibrous pericardium is attached to the large veins and arteries extending to and from the heart (the site labelled transition of parietal and visceral pericardium in Figure 18–3 is one of these sites of continuity).

The parietal and visceral layers of the serous pericardium thus form a closed cavity around the heart called the **pericardial cavity** (Fig. 18–3). The pericardial cavity in an adult contains 20 to 30 ml of pericardial fluid. The pericardial cavity can be accessed for paracentesis (aspiration of pericardial fluid) by needle puncture in the left infrasternal angle (which is the inferior angle between the lower margin of the left seventh costal cartilage and the left margin of the xiphoid process) (Fig. 18–4). The needle is directed posterosuperiorly and to the left to pierce the pericardial sac just above the margin of fusion with the central tendon of the diaphragm.

There are only two continuous margins, or lines of reflection, within the pericardial sac where the visceral part of the serous pericardium becomes continuous with the parietal part of the serous pericardium. One encircles the great arteries (the ascending aorta and pulmonary trunk), and the other insinuates the great veins (the superior and inferior venae cavae and pulmonary veins). The line of reflection about the great arteries is called the **arterial mesocardium,** and that about the large veins is called the **venous mesocardium** (Fig. 18–6). The venous mesocardium is roughly in the shape of the Greek capital letter Γ (pronounced "gamma").

There are two spaces about the heart partially bordered by the arterial and venous mesocardia. The more superior of these spaces extends transversely above the heart, and is bordered by the arterial mesocardium anteriorly and by the horizontal part of the venous mesocardium posteriorly. This space is a subregion of the pericardial cavity, and is called the **transverse pericardial sinus** (as it extends between the extreme left and right sides of the pericardial cavity) (Figs. 18–2, 18–3, and 18–6). The more inferior of the two spaces lies behind the heart (specifically, the left atrium), and is bordered above by the horizontal part of the venous mesocardium and on the right side by the vertical part of the venous mesocardium. This subregion of the pericardial cavity is called the **oblique pericardial sinus** (Figs. 18–3 and 18–6).

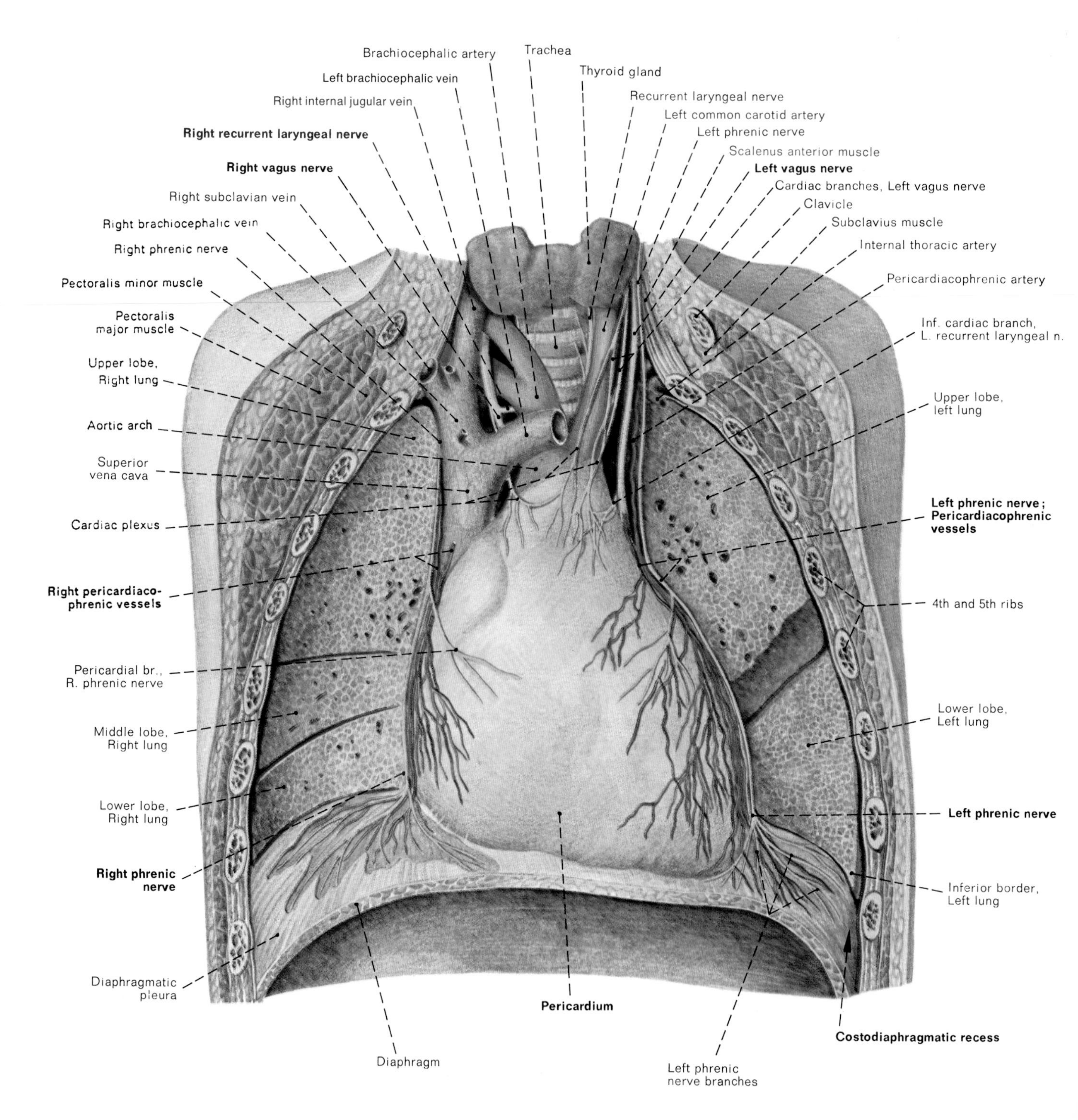

Fig. 18–1: Anterior view of the pericardial sac about the heart. In this coronal section through the thorax, the anterior part of the chest wall and the anterior parts of the lungs and diaphragm have been removed to display the pericardium and the vessels and nerves of the superior mediastinum.

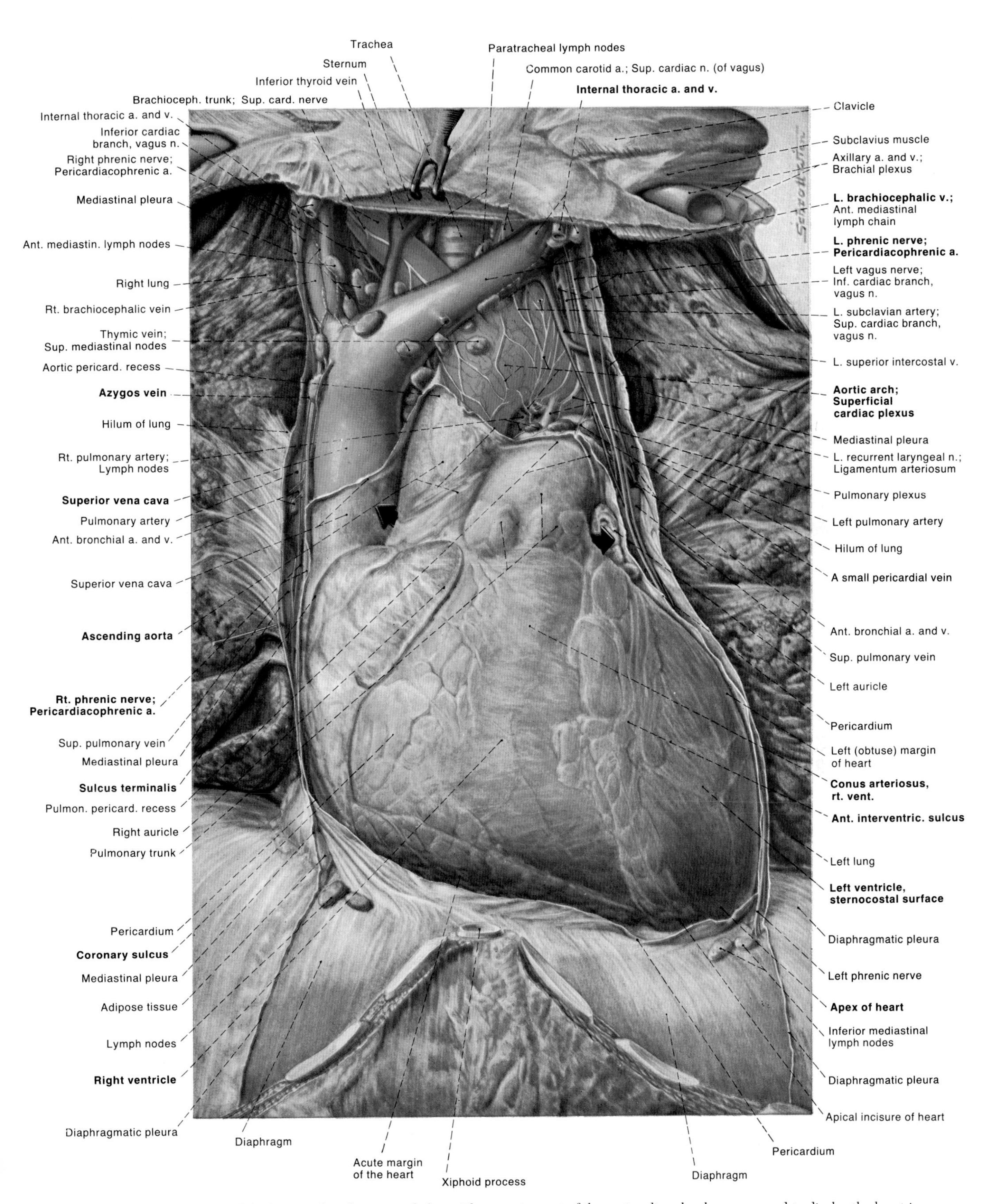

Fig. 18–2: Anterior view of the heart within the pericardial sac. The anterior part of the pericardium has been removed to display the heart in its normal position within the middle mediastinum. The arrow indicates the location of the transverse pericardial sinus.

THE DISTINCTION BETWEEN THE FUNCTIONAL SIDES OF THE HEART AND THE ANATOMIC SURFACES OF THE HEART

The heart is an organ whose functional sides correspond only approximately to its anatomic sides. Accordingly, one set of conventional expressions has been adopted for referring to the functional sides of the heart and another set of expressions for referring to the heart's anatomic surfaces. The expressions "left side of the heart" and "right side of the heart" refer to the functional sides of the heart (to the left atrium and ventricle distinct from the right atrium and ventricle). Consequently, the expression "right-sided heart failure" denotes failure by the right atrium and the right ventricle to adequately propel onward into the pulmonary trunk the blood that is drained by the right atrium during each cardiac cycle.

By contrast, the anatomic sides of the heart's exterior consist of its apex, three surfaces, and four borders. These terms relate to the heart's shape and orientation within the thorax. The shape of the heart is roughly that of a cone with a truncated apex, a flat base, and two curved surfaces that diverge as they extend from the apex toward the base. In the thorax, this cone of myocardial tissue is oriented such that one of its curved surfaces lies atop the diaphragm and the other behind the anterior chest wall. The truncated apex points anteriorly, downward, and to the left, and the flat base faces posteriorly, upward, and to the right. Consequently, the bluntly pointed region of the heart at its lower left margin is called the **apex** of the heart (Figs. 18–2 and 18–5), and the flat heart surface facing diametrically opposite from the apex is called the **base** of the heart. The heart's base consists primarily of that part of the heart's posterior surface formed by the left atrium (Fig. 18–6).

The region of the heart's exterior which can be seen from an anterior viewpoint is one of the heart's curved surfaces, and is called the **sternocostal surface** (Figs. 18–2 and 18–5). The sternocostal surface is named for its placement directly behind the sternum and the costal cartilages of the five uppermost ribs. An outline of the sternocostal surface forms a rough trapezoid, and it is the borders of this trapezoid that form the four borders of the heart. The borders of the heart's sternocostal surface have simple, direct names: **superior, left, inferior,** and **right.**

The remainder of the heart's exterior forms its other curved surface: **the diaphragmatic surface.** The diaphragmatic surface consists of that region of the heart's exterior, exclusive of the base, which can be seen from a posteroinferior viewpoint (Fig. 18–6). The surface is named for its placement directly above that portion of the fibrous pericardium which is fused with the central tendon of the diaphragm. Note that the heart, as it lies within the thorax, rests upon its diaphragmatic surface but not its base.

THE HEART'S SURFACE GROOVES

There are two major grooves located on the heart's sternocostal and diaphragmatic surfaces (Fig. 18–7). One of these grooves encircles the base of the heart. This groove is called the **atrioventricular groove** because both atria lie posterior and to the right of it and both ventricles lie anterior and to the left of it. The atrioventricular groove is thus a surface marker of the internal structures that separate the heart's atrial chambers from its ventricular chambers.

The other major groove forms an arch over the heart's apex. It marks on the heart's exterior the internal location of the **interventricular septum,** the wall of myocardial tissue that separates the left and right ventricles. Accordingly, this groove is called the **interventricular groove.** More specifically, the part of the groove which courses on the sternocostal surface is called the **anterior interventricular groove,** and the part which courses on the diaphragmatic surface is called the **inferior interventricular groove** (Fig. 18–7). The point where the inferior interventricular groove meets the atrioventricular groove is called the **crus** (cross) of the heart. The combined tracings of the atrioventricular and interventricular grooves delineate the location of the left and right ventricles.

THE HEART'S CHAMBERS

The Right Atrium

The right atrium occupies the right most part of the heart as it lies within the thorax (Fig. 18–5). The wall musculature of the right atrium forms the right third of the heart's sternocostal surface and all of its right border. The anterior roof of the right atrium projects upward and to the left to form a small, conical pouch called the **right auricle.**

The right atrium is divisible into two parts, an anterior part and a posterior part. These two parts differ in their embryologic derivation. Whereas the walls of the anterior part develop from the primitive atrium, the walls of the posterior part develop from an embryonic venous sinus attached to the primitive atrium. Evidence of this difference in embryologic origins is usually indicated by a vertically-oriented sulcus (groove) on the outer surface of the lateral wall of the right atrium. This sulcus is called the **sulcus terminalis** because it marks the junction on the lateral wall where the wall musculature which developed from the

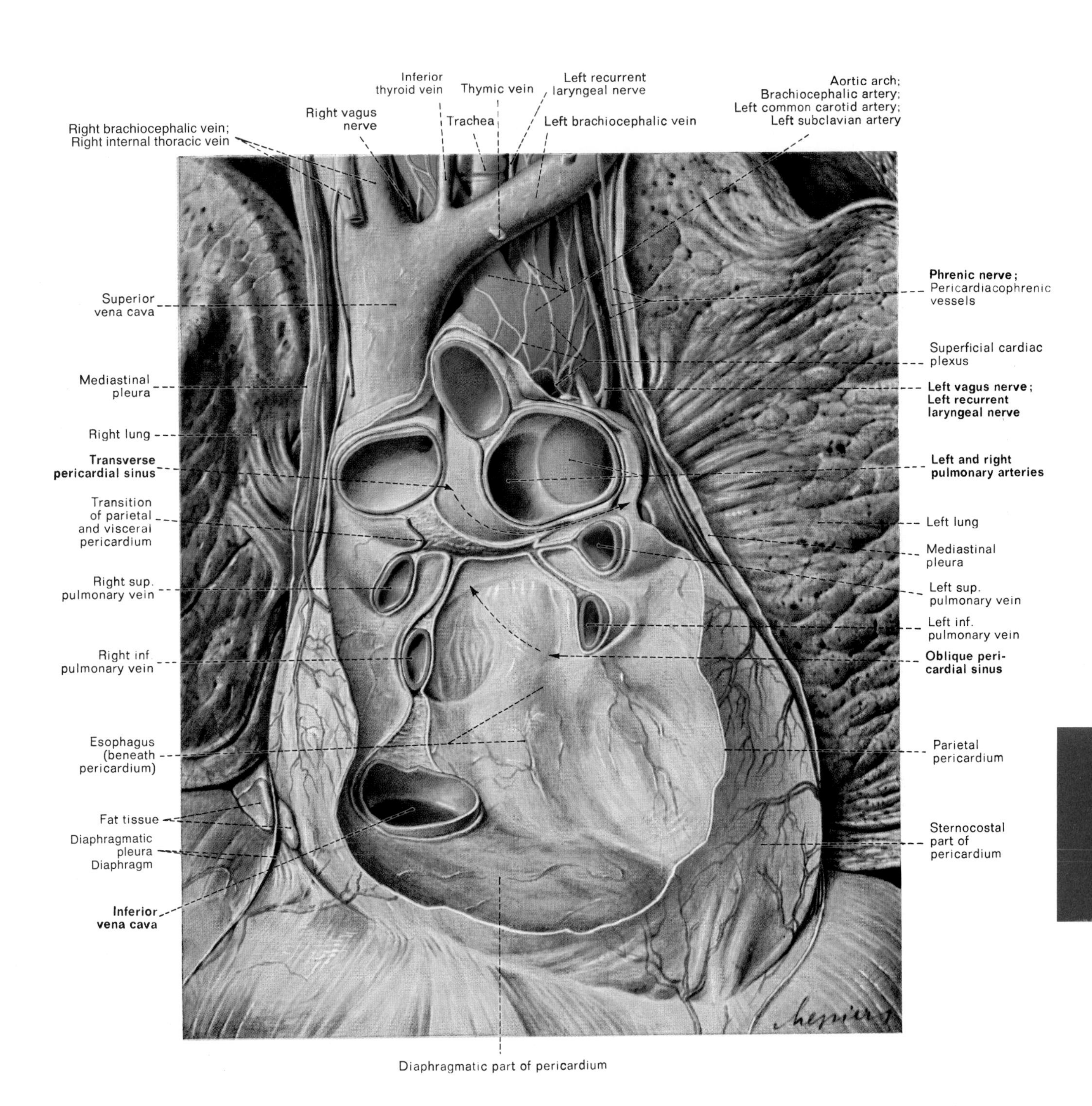

Fig. 18–3: Anterior view of the posterior part of the pericardial cavity. The anterior part of the pericardium and the heart have been removed to reveal the region of the pericardial cavity surrounded by the posterior part of the pericardium. The illustration depicts the 8 large vessels attached to the heart within the pericardial cavity: the superior vena cava, inferior vena cava, ascending aorta, pulmonary trunk, and four pulmonary veins.

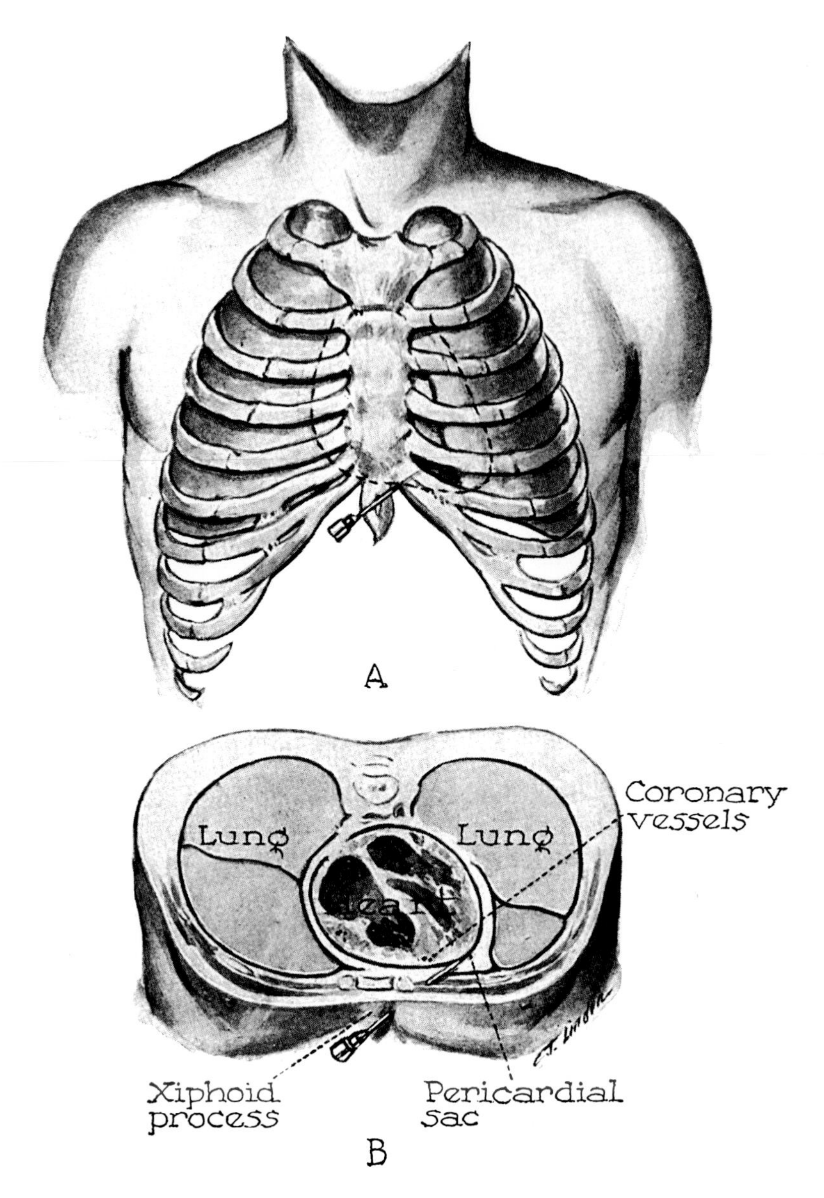

Fig. 18–4: Diagrams illustrating (A) anterior and (B) superior oblique views of needle placement for pericardiocentesis.

primitive atrium terminates, and is replaced by wall musculature which developed from the embryonic venous sinus (the sulcus terminalis is not depicted in Figure 18–5).

The walls of the posterior part of the right atrium are comparatively thin and have a smooth, inner surface. These walls contain all three major venous openings into the right atrium (Fig. 18–8). The openings are those of the superior and inferior venae cavae and the coronary sinus. The openings of the venae cavae into the posterior part of the right atrium are the anatomic basis for the naming of this part of the right atrium the **sinus venarum** (sinus of the venae cavae).

The venae cavae approach the atrial chamber from opposite directions, with the superior vena cava opening into the posterior roof region and the inferior vena cava opening into the posterior floor region of the chamber. The coronary sinus opens into the posterior floor region of the right atrium to the left of the inferior vena cava opening (Fig. 18–8). The openings of the inferior vena cava and the coronary sinus are each guarded by a valve consisting of a small ridge of tissue rising from the floor region of the chamber. These valves do not have any postnatal function.

Most of the posteromedial wall of the right atrium is also the anteromedial wall of the left atrium. The posteromedial wall of the right atrium is thus called the **interatrial septum.** The surface of the interatrial septum is postnatally marked by a thin, oval depression called the fossa ovalis (Fig. 18–8). The **fossa ovalis** marks the location of the **foramen ovale** in the fetal heart.

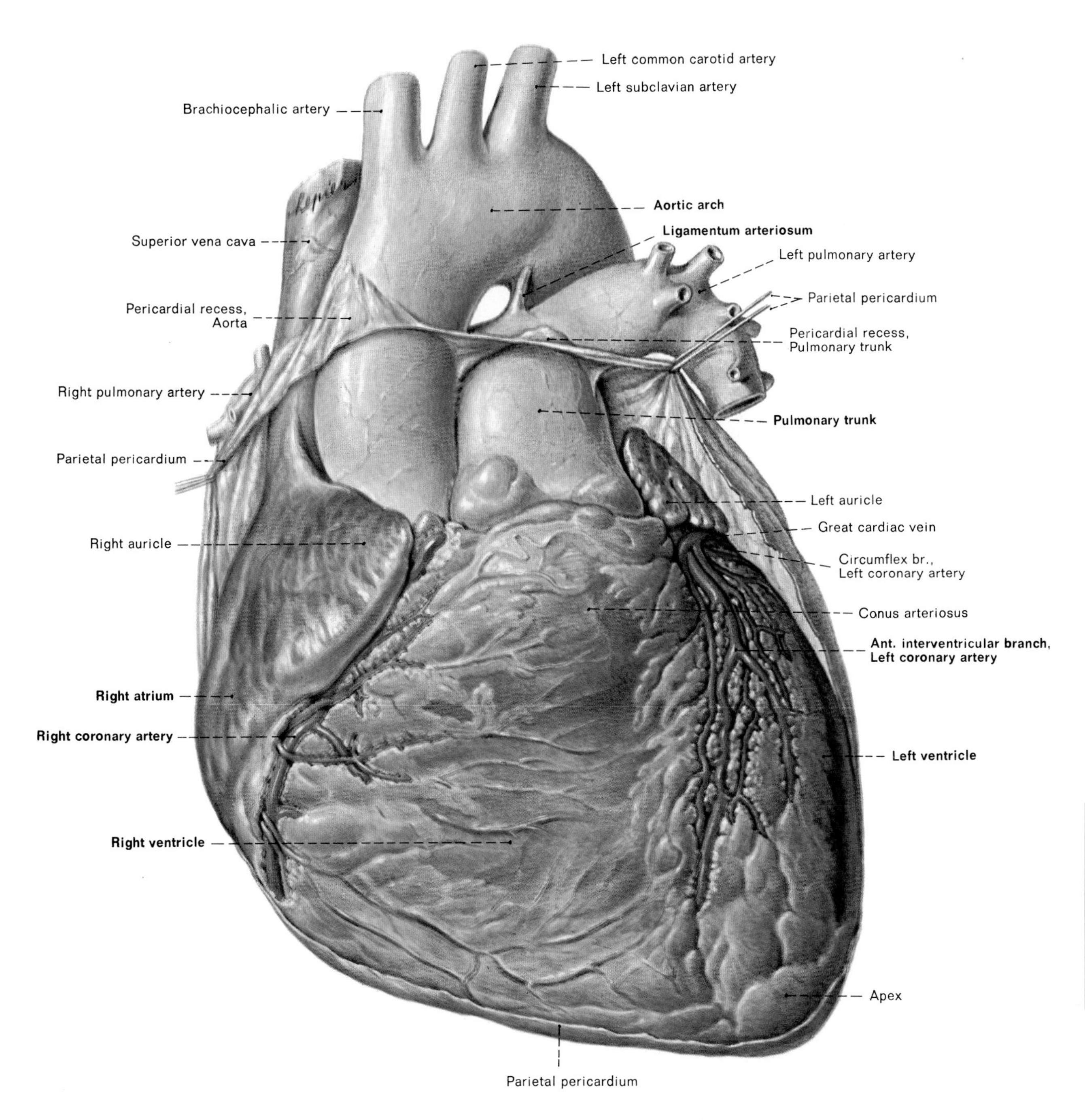

Fig. 18–5: Anterior view of the sternocostal surface of the heart.

A small part of the posteromedial wall of the right atrium separates the right atrium from the left ventricle. This small region of the posteromedial wall of the right atrium is called the **atrioventricular septum** (Fig. 18–8).

The walls of the anterior part of the right atrium have muscular ridges on their inner surfaces. The largest of these ridges is called the **crista terminalis** (terminal crest) (Fig. 18–8). The crista terminalis extends vertically within the lateral wall of the right atrium from its roof to its floor. The position of the crista terminalis on the inner surface of the lateral wall indicates that of the sulcus terminalis on the outer surface.

Smaller muscular ridges, collectively called **musculi pectinati (pectinate muscles),** extend anteriorly from the crista terminalis (like the teeth of a comb) into the anterior wall of the right atrium (the pectinate muscles extending anteriorly from the crista terminalis are not depicted in Figure 18–8). In the wall of the right auricle, the pectinate muscles form an interlaced, sponge-like network of muscular ridges (Fig. 18–8).

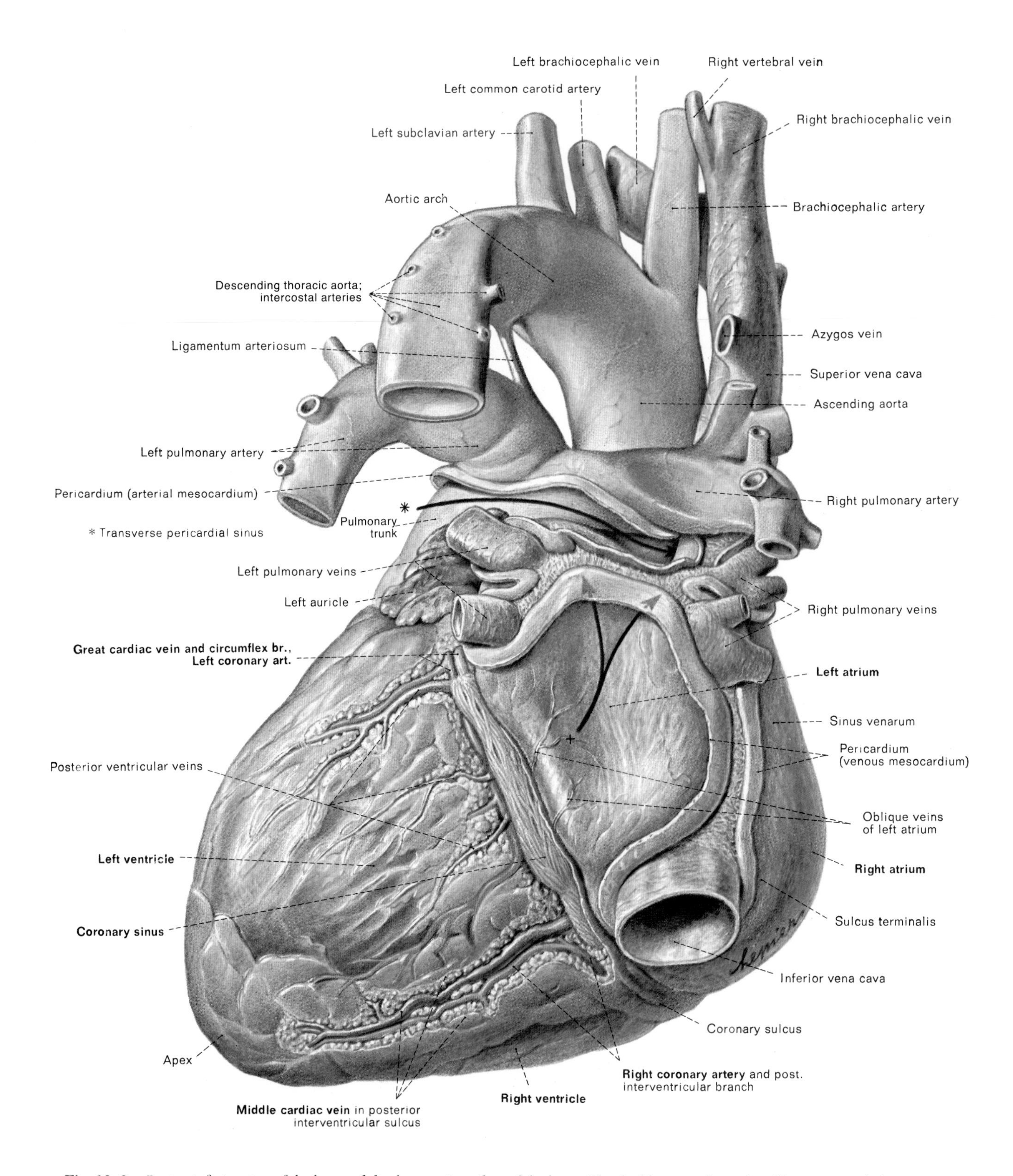

Fig. 18–6: Posteroinferior view of the base and diaphragmatic surface of the heart. The double arrows lie in the oblique pericardial sinus.

The opening out from the right atrium, the right atrioventricular orifice, is located in the anteromedial floor of the chamber (Fig. 18–8). The orifice leads into the right ventricle, and is guarded by a valve known equivalently as the **right atrioventricular valve** or the **tricuspid valve** (Figs. 18–8 and 18–9). All of the heart's valves serve to control blood flow through the orifice which each guards.

The Right Ventricle

The right ventricle projects to the left of the right atrium, and occupies the left anterior region of the heart (Fig. 18–5). The approximate location of the ventricle's internal borders are indicated on the heart's exterior by the atrioventricular and interventricular grooves.

The most posterior wall of the right ventricle forms the interventricular septum (Fig. 18–9). By contrast, the ventricle's anterior wall contributes to the heart's exterior. The anterior wall accounts for almost all of the heart's sternocostal surface to the left of the right atrium (Fig. 18–5).

When blood begins to flow from the right atrium into the right ventricle during the early moments of diastole, most of the blood flow occurs along a tract within the right ventricle appropriately called the ventricle's inflow tract (Figs. 18–8 and 18–10). This inflow tract extends from the atrioventricular orifice toward the left lower corner of the right ventricular chamber. The walls lining the inflow tract are essentially the walls of the lower right half of the chamber, and they are distinguished on their inner surfaces by the presence of thick, muscular ridges and columns. These internal elevations of the ventricular myocardium are collectively called **trabeculae carneae** ("little beams of flesh") (Figs. 18–8 and 18–9).

When blood is ejected from the right ventricle into the pulmonary trunk during systole, the blood is propelled through a channelway in the chamber called the ventricle's outflow tract (Figs. 18–9 and 18–10). The outflow tract extends upward from the left lower corner of the right ventricular chamber to the root of the pulmonary trunk. The walls lining the outflow tract are mostly the walls of the upper left half of the chamber. This region of the ventricle has the shape of an inverted cone or funnel, and is therefore referred to as either the **conus arteriosus** (arterial cone) or the **infundibulum** (funnel) (Fig. 18–9). The walls of the infundibulum have the smoothest inner surfaces in the right ventricle. The infundibular walls become progressively more elastic and fibrous and less muscular as they approach the root of the pulmonary trunk.

The infundibular walls also become thinner as they narrow into the root of the pulmonary trunk. This thinning is especially noticeable in the infundibular wall formed by the interventricular septum. The interventricular septum region at the root of the pulmonary trunk is called the **membranous part of the interventricular septum** because of its extreme thinness (Figs. 18–8 and 18–12). The remainder of the septum (that portion extending toward the apex) is thickened with myocardium, and thus is called the **muscular part of the interventricular septum.**

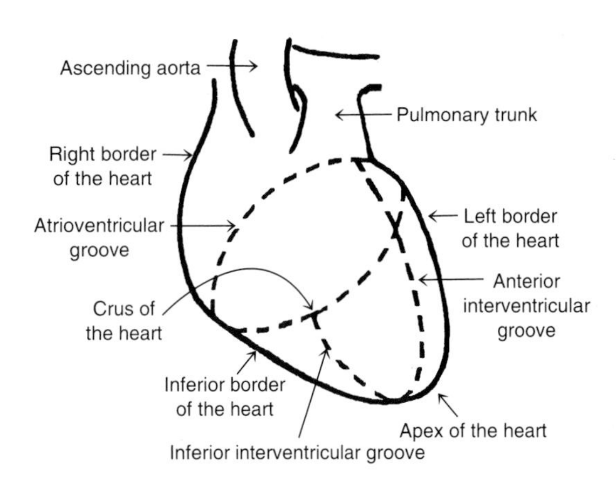

Fig. 18–7: Diagram illustrating an anterosuperior view of the atrioventricular and interventricular grooves on the surface of the heart.

The origin of the pulmonary trunk marks the opening out from the right ventricle. The opening is guarded by a valve known as the **pulmonary valve** (Fig. 18–9).

The Left Atrium

The base of the heart is formed largely by the left atrium (Fig. 18–6). The left atrium lies behind and to the left of the right atrium, and behind, above, and to the right of the left ventricle. A small, conical, muscular pouch, called the **left auricle,** projects forward and to the left from the anterior roof of the left atrium. The left auricle forms the uppermost part of the heart's left border (Fig. 18–5).

The walls of the posterior part of the left atrium are similar to those of the right atrium in that they are thin, have smooth inner surfaces, and receive the openings of all the large veins draining into the atrium (Fig. 18–11). There are generally four large veins that open into the posterior wall of the left atrium: the pair of pulmonary veins from the left lung and the pair of pulmonary veins from the right lung.

Most of the walls of the anterior part of the left atrium differ from those of the right atrium in that they do not bear musculi pectinati on their inner surfaces. The only exceptions are the walls of the left auricle (Fig. 18–11).

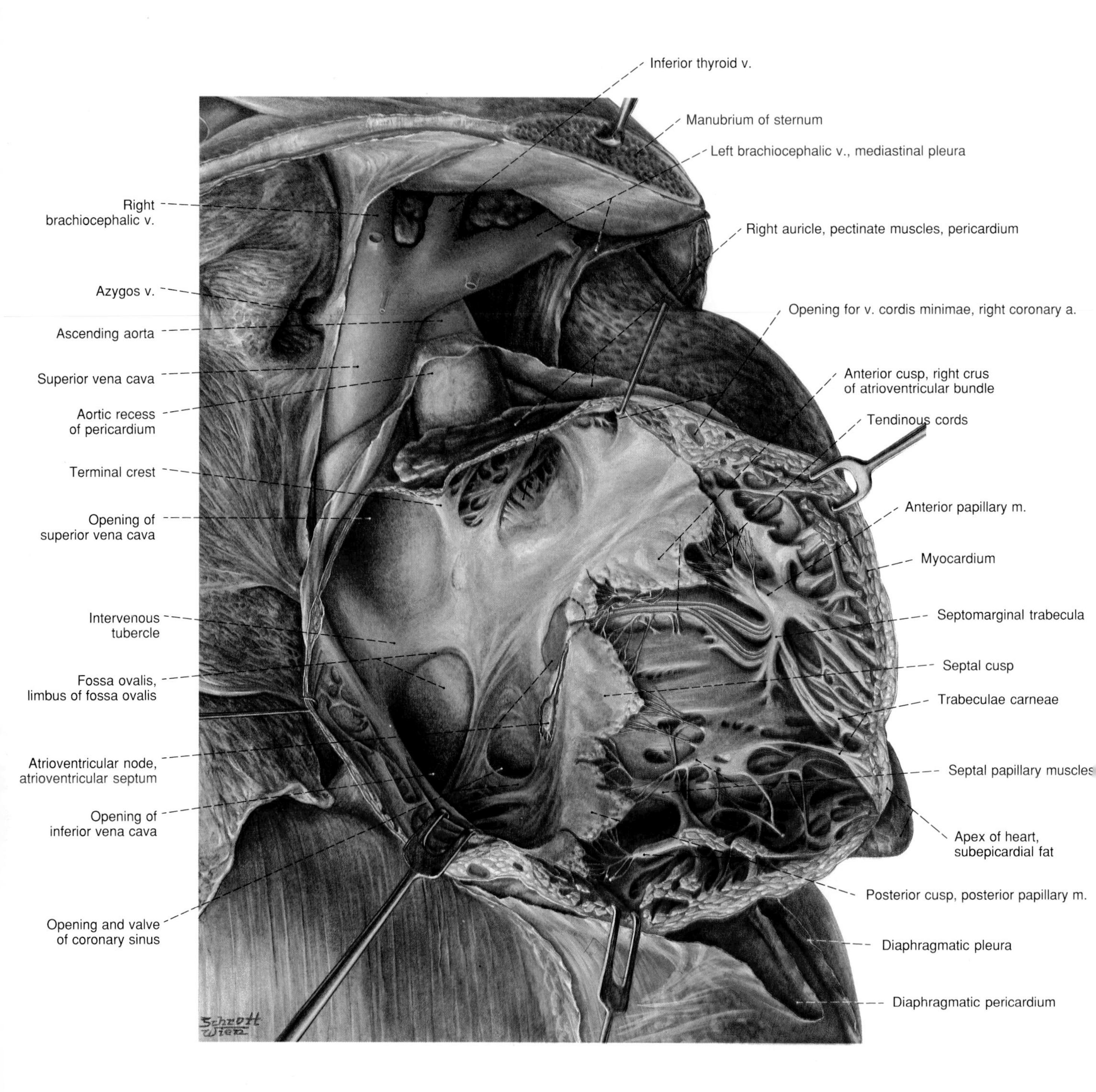

Fig. 18–8: Anterior view of the interior of the right atrium and the inflow tract of the right ventricle. The anterior walls of the right atrium and right ventricle have been cut and retracted to provide the view. Parts of the heart's conducting system are schematically displayed in yellow. The atrioventricular septum is colored blue, and the dashed lines demarcate the membranous part of the interventricular septum.

The exit from the left atrium, the left atrioventricular orifice, is located in the left anteroinferior wall of the chamber. The orifice leads into the left ventricle, and is guarded by a valve with three synonymous names: **left atrioventricular valve, bicuspid valve,** or **mitral valve** (Figs. 18–11 and 18–12).

The Left Ventricle

The left ventricle projects anteriorly and inferiorly and to the left from the left atrium (Fig. 18–6). The left ventricle forms the heart's apex and almost all of the heart's left border (Fig. 18–5). The left ventricle also forms the part of the heart's sternocostal surface to the left of the anterior interventricular groove (Fig. 18–5) and the part of its diaphragmatic surface to the left of the inferior interventricular groove (Fig. 18–6).

The walls of the left ventricle are the thickest walls of myocardium in the heart. In an adult, the left ventricle's walls near the apex are about three times thicker than the anterior wall of the right ventricle (compare the thickness of the anterior wall of the right ventricle as depicted in Figs. 18–8 and 18–9 with that of the left ventricle's wall as depicted in Figs. 18–11 and 18–12). This difference in wall thickness enables the musculature of the left ventricle to generate a 5 to 6 times higher systolic pressure than the musculature of the right ventricle. The left ventricle must generate a severalfold greater systolic pressure than that produced by the right ventricle because of the greater resistance to blood flow offered by the systemic circuit as compared to that offered by the pulmonary circuit. At the time of birth, however, the walls of the left ventricle are as thick as those of the right ventricle, since both ventricles (up to that instant) have been pumping blood through circuits of comparable resistance. As the resistance of the systemic circuit becomes more pronounced postnatally, the walls of the left ventricle become progressively thicker than that of the anterior wall of the right ventricle.

When blood flows from the left atrium into the left ventricle during diastole, it is directed to flow along a tract within the ventricle called its inflow tract (Figs. 18–10 and 18–11). The inflow tract extends from the mitral valve orifice to the apex. The inflow tract is lined, in part, by the walls of the left posterior half of the left ventricle. These walls bear an intricate network of trabecular carneae on their inner surfaces (Fig. 18–11). These trabeculae carneae are finer than those found in the walls lining the inflow tract of the right ventricle (compare Figs. 18–8 and 18–11).

When blood is ejected from the left ventricle into the aorta during systole, the blood is propelled through a channelway in the chamber called its outflow tract (Figs. 18–10 and 18–12). The outflow tract extends upward from the apex region of the chamber to the root of the aorta, and is lined, in part, by the interventricular septum.

The origin of the aorta marks the opening out from the left ventricle. This opening is guarded by a valve known as the **aortic valve** (Fig. 18–12).

THE HEART'S VALVES

The Atrioventricular Valves

Each of the atrioventricular valves consists of four kinds of structures:

1. an **anulus** (ring) of fibrous connective tissue which serves as a foundation for the valve (Fig. 18–13),
2. thin leaflets of connective tissue called **cusps** which together can completely drape over the orifice (Fig. 18–13),
3. slender cords of connective tissue called **chordae tendinae** that are tethered to the ventricular margins of the cusps (Figs. 18–8, 18–11, and 18–12), and
4. stout pillars of ventricular myocardium called **papillary muscles** that project from the walls of the ventricle and give rise to the chordae tendinae (Figs. 18–8, 18–11, and 18–12).

The anulus of each atrioventricular valve is an irregularly-shaped ring of fibrous connective tissue that lies embedded in the musculature encircling the atrioventricular orifice (Fig. 18–13). The anulus marks the boundary between the musculature of the atrium and the ventricle on each side of the heart. The anulus secures the basal margins of the cusps of each valve.

The cusps of each valve are, in fact, divisions of a single, tubular curtain of tissue that extends from the atrioventricular orifice into the ventricular chamber (Fig. 18–13). The indentations in the ventricular margin of the tissue curtain divide the curtain into the parts known as cusps (Figs. 18–8 and 18–11).

The major difference between the left and right atrioventricular valves is that whereas the right valve has three cusps, the left valve has only two cusps (Figs. 18–8, 18–11, and 18–13). This difference is the basis for the right and left atrioventricular valves being also respectively known as the tricuspid and bicuspid valves. The bicuspid valve is also commonly called the mitral valve because when the edges of the two cusps appose each other to close the left atrioventricular orifice, their closed configuration resembles the shape of a miter (a pointed hat worn on ceremonial occasions by Catholic or Episcopal bishops).

The cusps of each atrioventricular valve are individually named according to the marginal sector of the

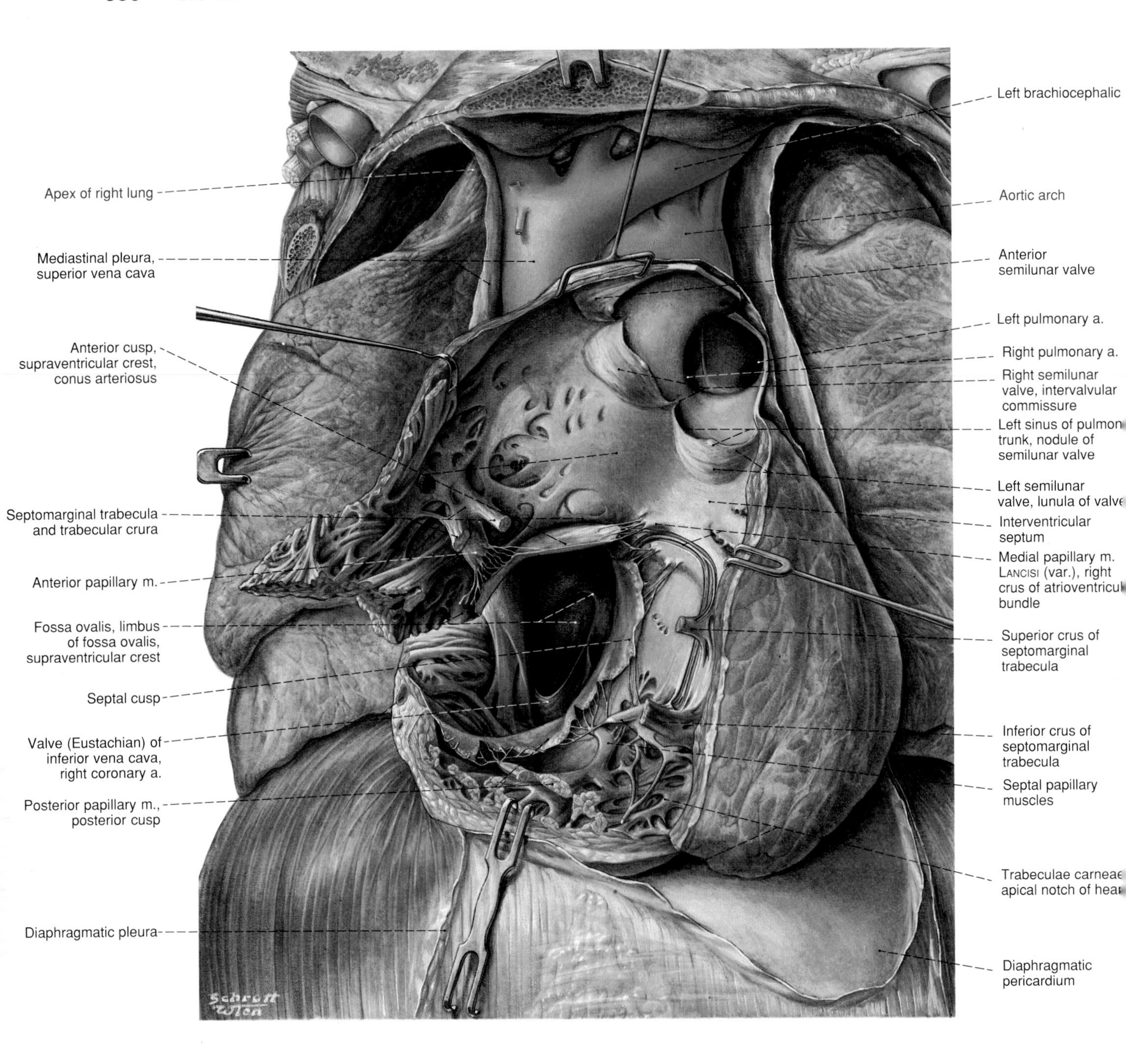

Fig. 18–9: Anterior view of the outflow tract of the right ventricle and the origin of the pulmonary trunk. The anterior wall of the right ventricle has been cut horizontally along a margin immediately above and parallel to the inferior border of the heart and vertically along a margin just to the right of the anterior interventricular groove. The right bundle branch of the heart's conducting system is schematically displayed in yellow.

CLINICAL PANEL IV-3

MITRAL VALVE PROLAPSE

Definition: Mitral valve prolapse is the condition in which there is excessive excursion of the valve's leaflets through the valve's anulus and into the left atrium during systole. The structural basis for the excessive excursion is commonly excessive redundant leaflet tissue; the chordae tendinae may also be elongated and redundant.

Common Symptoms: Mitral valve prolapse in the vast majority of affected individuals is asymptomatic.

Common Signs: An understanding of the anatomic basis of the common signs of mitral valve prolapse requires knowledge of the mechanism of production of the heart's normal sounds S1 and S2.

The characteristic auscultatory findings associated with mitral valve prolapse are a midsystolic click followed by a late systolic crescendo-decrescendo murmur (the murmur is generally best heard at the site of the apex beat). [A crescendo-decrescendo murmur is a murmur that begins softly, becomes progressively louder, and then progressively softer.] It is believed that the midsystolic click emanates from the prolapsed mitral valve leaflets or their chordae tendinae when they are acutely tensed at the moment of maximum excursion. The late systolic murmur indicates some regurgitation of blood from the left ventricle into the left atrium through the valve's partially unapposed leaflets.

Procedures which decrease systemic venous return to the heart (administration of a vasodilator, moving from a supine to a standing position, or performance of the Valsalva maneuver) decrease left ventricular end-diastolic volume and thus move the click and murmur toward S1. [Left ventricular end-diastolic volume is the volume of blood in the left ventricle at the end of diastole.] By contrast, procedures that increase systemic venous return to the heart (squatting or isometric exercises) increase left ventricular end-diastolic volume and thus delay the onset of the click and murmur.

orifice from which they arise. Accordingly, the tricuspid valve has an **anterior (ventral) cusp,** a **posterior (dorsal) cusp,** and a **septal cusp** (the last being the cusp that arises from the sector nearest the interventricular septum). The bicuspid valve has **anterior** and **posterior cusps.**

The anterior cusp of the mitral valve is considerably larger than the posterior cusp. The anterior cusp plays a major role in the definition of both the inflow and outflow tracts of the left ventricle.

Because the papillary muscles of the atrioventricular valves consist of ventricular myocardium, they contract during systole. The pull they exert upon the chordae tendinae (which represent their tendons of insertion) prevents eversion of the atrioventricular cusps into the respective atria during systole.

The Aortic and Pulmonary Valves

The aortic and pulmonary valves each consists of two kinds of structures: (1) an **anulus** of fibrous connective tissue which serves as a foundation for the valve, and (2) three thin leaflets of connective tissue called **semilunar cusps** which, by apposition of their free margins, can completely close shut the opening at the root of the arterial trunk (Fig. 18–13). The semilunar cusps each have a fibrous nodule at the midpoint of the free margin; the three nodules collectively plug the central hole remaining when the free margins are apposed (the nodules at the midpoints of the free margins of the semilunar cusps are depicted in Figs. 18–9 and 18–12 but not in Fig. 18–13).

The anuli of the aortic and pulmonary valves are each in the shape of a three-pointed crown. The points of the crown project away from the heart within the wall of the arterial trunk.

The connective tissue leaflets of the aortic and pulmonary valves are known as semilunar cusps since each has a half moon-shaped outline in cross section. Each semilunar cusp has an attached and a free margin. The attached margin is secured to the anulus, along an arc extending between two of its points. The free margin extends between these two points into the lumen of the arterial trunk. The semilunar cusps thus establish three cup-shaped spaces at the root of the arterial trunk.

The semilunar cusps of the aortic and pulmonary valves are individually named (Figs. 18–9, 18–12, and 18–13). The aortic valve has a **posterior cusp,** a **left cusp,** and a **right cusp.** The posterior cusp is the most posterior of the three cusps; the left and right cusps lie on opposite sides of the posterior cusp. The pulmonary valve has an **anterior cusp,** a **left cusp,** and a **right cusp.** The anterior cusp is the most anterior of the three cusps; the left and right cusps lie on opposite sides of the anterior cusp.

The regions of the arterial trunk opposite the semilunar cusps are relatively thin, and bulge outward at all times. The dilatations in the trunk wall formed by the bulges are called **sinuses** (Figs. 18–9 and 18–12).

The sinuses opposite the left and right cusps of the aortic valve each has an opening that marks the origin of one of the heart's coronary arteries (Figs. 18–12 and 18–13). The opening in the sinus opposite the left cusp is the origin of the left coronary artery, and the opening in the sinus opposite the right cusp is the origin of the right coronary artery. These relationships provide an alternate scheme for naming the cusps of the aortic valve: The left cusp of the aortic valve may be alternatively called the **left coronary cusp,** the right cusp the **right coronary cusp,** and the posterior cusp the **non-coronary cusp.**

The Fibrous Skeleton of the Heart

The fibrous anuli of the tricuspid, mitral, pulmonary, and aortic valves are collectively referred to as the fibrous skeleton of the heart. They not only provide a structural framework for the attachment of the heart's musculature and the cusps of its valves, but also serve the heart's conducting system by insulating the electrical activity of the atrial musculature from that of the ventricular musculature.

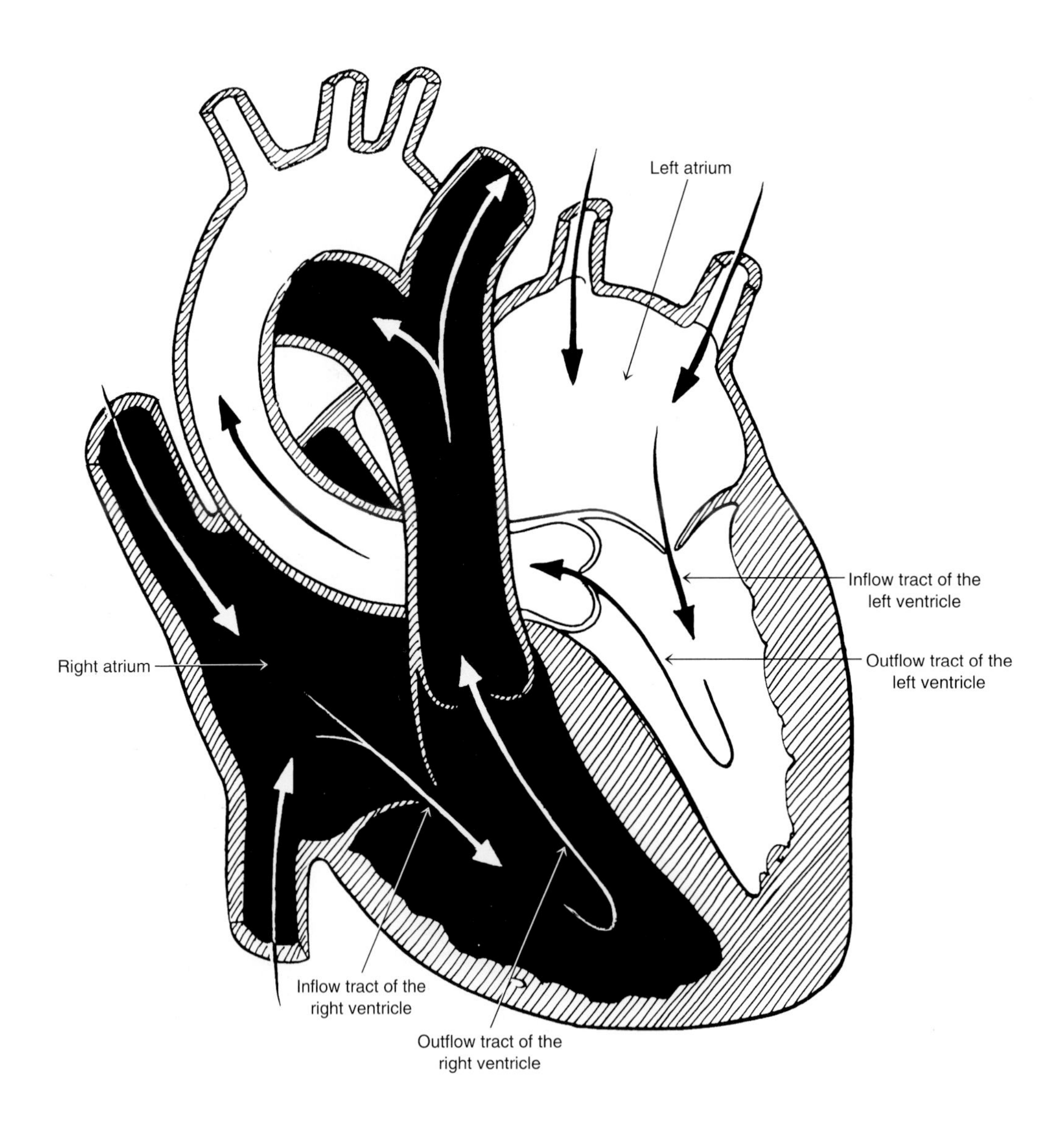

Fig. 18–10: Diagram of blood flow through the heart's chambers.

THE HEART'S VASCULATURE

The heart's major arteries and veins all lie on the heart's exterior, mostly extending within the atrioventricular and interventricular grooves. Early anatomists were impressed with the observation that the vessels coursing through the atrioventricular groove appear to crown the heart. Accordingly, the largest vessels found in the atrioventicular groove were termed coronary (crown-like) vessels. This explains why the heart's two greatest arteries are called the left and right coronary arteries and its largest venous channelway is termed the coronary sinus. Each of these vessels or one of its major branches extends for almost all of its course within the atrioventricular groove.

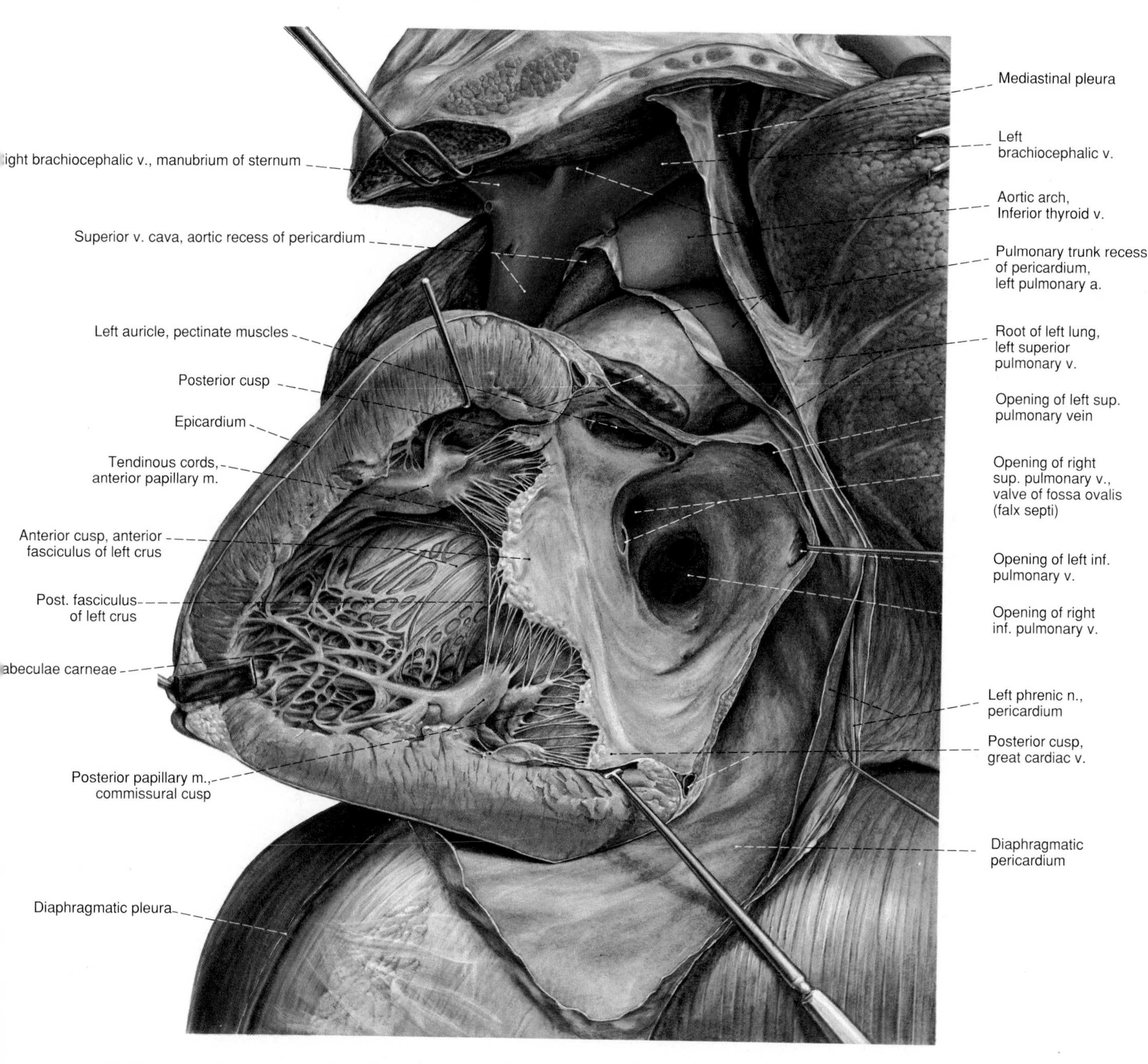

Fig. 18–11: View of the interior of the left atrium and the inflow tract of the left ventricle. To secure this view, the heart as a whole was displaced from its anatomic position by a probe which pulled the apex of the heart forward and to the right. The outer walls of the left ventricle and left atrium were next cut along a margin parallel to the left border of the heart and retracted. Branches of the left bundle branch of the heart's conducting system are schematically displayed in yellow.

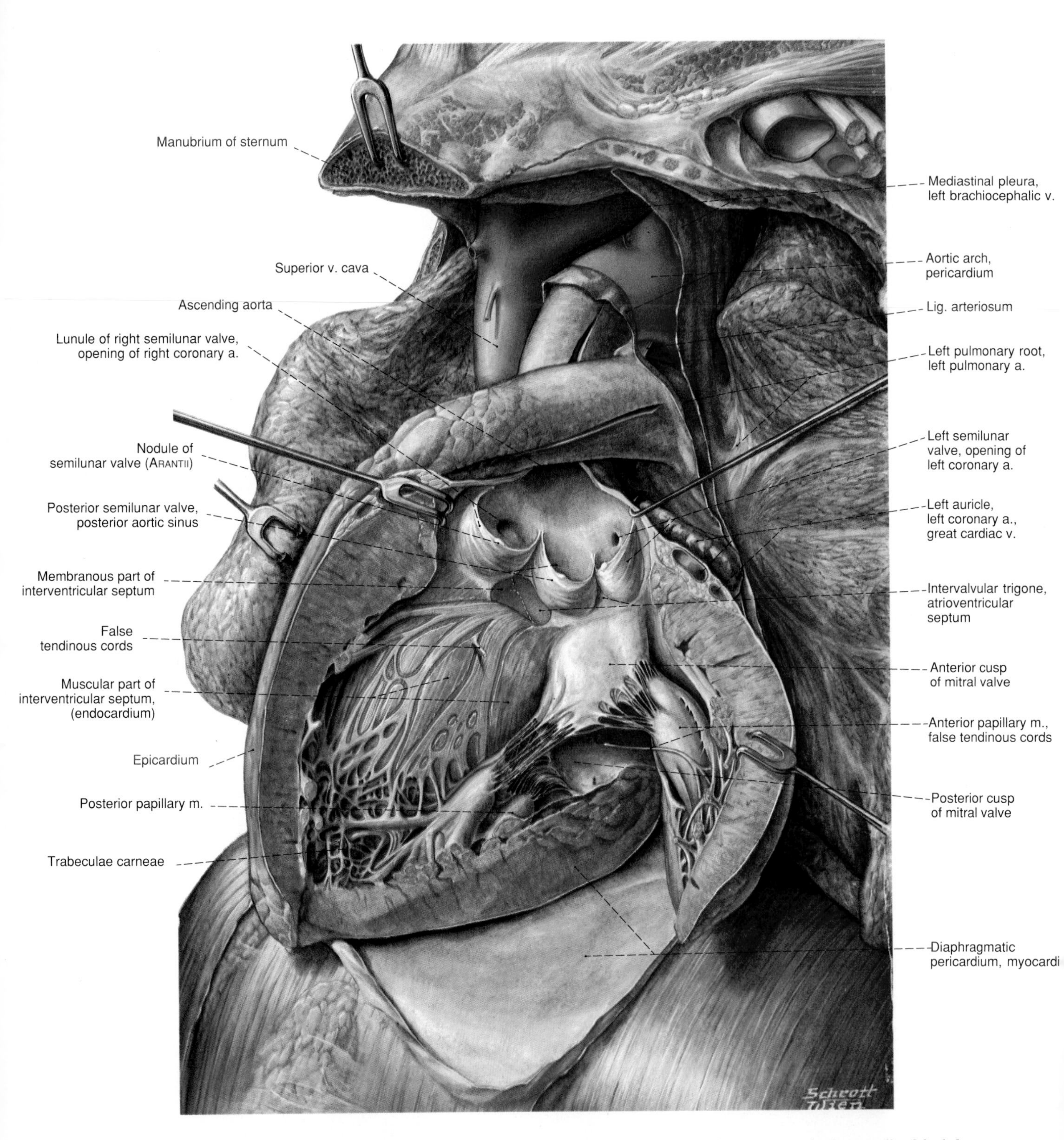

Fig. 18–12: View of the outflow tract of the left ventricle and the origin of the ascending aorta. The anterior and inferior walls of the left ventricle have been cut along margins immediately to the left of the interventricular septum. The left bundle branch of the heart's conducting system and its branches are schematically displayed in yellow. The atrioventricular septum and the membranous part of the interventricular septum are colored blue, and the boundary between indicated by a broken line.

The Coronary Arteries

The heart is supplied by a left coronary artery and a right coronary artery. Blood flow through the coronary arteries and their branches is greatest during diastole and least during systole. This is because the contraction of the ventricular musculature during systole compresses the intramyocardial and subendocardial coronary arterial branches of the ventricular walls, and this compression diminishes blood flow (intramyocardial branches are branches that extend from the external to the internal surface of the heart; subendocardial branches are branches that course directly deep to the endocardium). The systolic decrease in coronary arterial blood flow is more marked in the intramyocardial and subendocardial branches of the left ventricle than that in the corresponding branches of the right ventricle. This is because whereas the left ventricular musculature generates intramyocardial and intraventricular pressures in the range of 120 mm Hg during systole, the right ventricular musculature generates intramyocardial and intraventricular pressures in the range of only 25 mm Hg during systole.

Both coronary arteries are branches of the ascending aorta; each arises from one of the aortic sinuses (Figs. 18–12 and 18–13). The left coronary artery emerges from the sinus of the left cusp, and the right coronary artery emerges from the sinus of the right cusp.

The **right coronary artery** descends on the heart's sternocostal surface in the atrioventricular groove (Figs. 18–5 and 18–14). It gives rise near the inferior border of the heart to a branch called the **marginal branch** that extends to the left along the heart's inferior border (Fig. 18–14). The right coronary artery continues its course in the atrioventricular groove as it extends on the heart's diaphragmatic surface (Fig. 18–6). At the crus of the heart, the artery enters the inferior interventricular groove, and here is called either the **posterior interventricular artery** or the **posterior descending artery** (Figs. 18–6 and 18–14). The right coronary artery supplies all of the right atrium, much of the right ventricle, and variable parts of the left atrium and left ventricle.

The **left coronary artery** and its branches supply in all individuals the heart regions complementary to those supplied by the right coronary artery and its branches. The left coronary artery emerges on the heart's surface by extending between the left auricle and the root of the pulmonary trunk (Fig. 18–14). Upon reaching the atrioventricular groove, it divides into its two major branches. One major branch descends on the heart's sternocostal surface in the anterior interventricular groove. This branch is called the **anterior interventricular artery** or the **left anterior descending (LAD) artery** (Figs. 18–5 and 18–14). It gives rise to branches, called **diagonal branches of the LAD artery,** which extend to the left over the sternocostal surface area formed by the left ventricle (Fig. 18–14). The LAD artery arches over the apex to meet the posterior interventricular artery in the inferior interventricular groove (Fig. 18–14). The LAD artery supplies both ventricles; it is the major source of arterial supply to the interventricular septum.

The other major division of the left coronary artery is called the **circumflex artery** (Figs. 18–5 and 18–14). It extends within the atrioventricular groove as it curves around the left border of the heart and onto the diaphragmatic surface (Fig. 18–6). Near the left border of the heart, the circumflex artery gives rise to a **lateral branch.** On the heart's diaphragmatic surface, it gives rise to **one or more posterolateral branches** (the lateral and posterolateral branches of the circumflex artery are not illustrated in Fig. 18–14). The circumflex artery commonly meets a branch of the right coronary artery in the atrioventricular groove on the diaphragmatic surface. The circumflex artery supplies much of the left atrium and left ventricle.

In many individuals, the left coronary artery divides into three major branches upon reaching the atrioventricular groove. The third branch is called the **left diagonal artery,** and it descends on the sternocostal surface of the left ventricle between the LAD artery and the lateral branch of the circumflex artery (a left diagonal artery is not depicted in Fig. 18–14). A left diagonal artery supplies part of the left ventricle only.

There is much variability in the extent to which each coronary artery supplies the individual atria and ventricles and their parts. The left coronary artery, however, always supplies the greater mass of heart tissue, as its branches supply most of the left ventricle.

The expressions **"left dominance"** and **"right dominance"** refer to whether the left versus the right coronary artery gives rise to the posterior interventricular artery. Ninety percent of all hearts are right dominant. This 90% incidence explains why the posterior interventricular artery has been described as a branch of the right coronary artery.

Clearly, there are intercoronary anastomoses in normal hearts. The anastomoses generally consist of arterioles (with diameters of about 40 microns) located in contiguous regions of left and right coronary arterial supply. However, there is uncertainty as to the magnitude of blood flow through these anastomoses during normal conditions or following occlusion of some arteries which channel blood flow through them. Accordingly, the coronary arteries are generally classified as functional end arteries.

Atherosclerotic lesions of coronary arterial vessels occur in epicardial but not intramyocardial vascular segments (epicardial vascular segments are the segments that course over the external surface of the

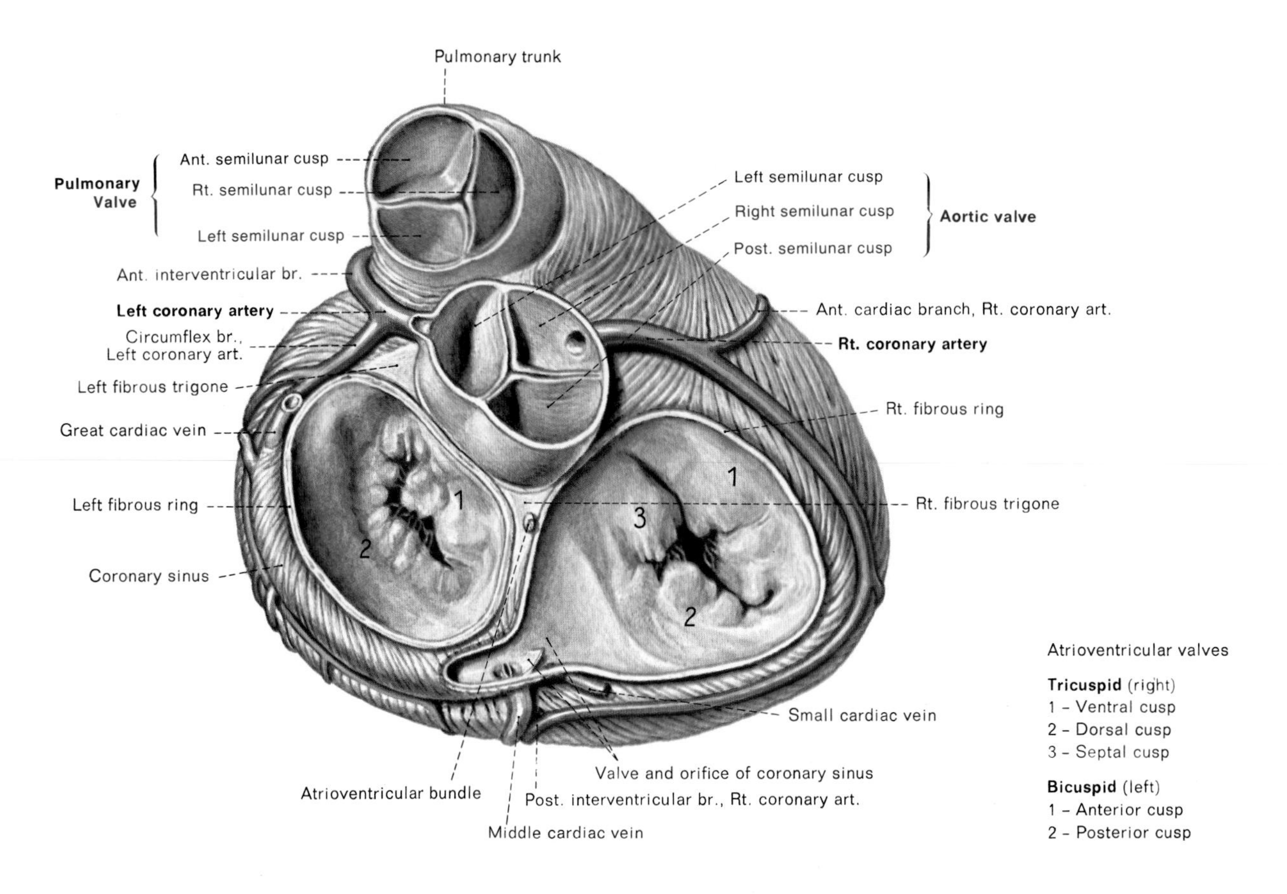

Fig. 18–13: Superior view of the heart's valves. The pulmonary trunk has been transected immediately distal to the pulmonary valve, the aorta has been transected immediately distal to the aortic valve, and the walls of the left and right atria have been removed.

heart). Of the heart's major arteries, the three vascular segments most commonly occluded by atheromatous plaques (in the order of decreasing incidence) are:

1. the portion of the right coronary artery immediately proximal to the crus of the heart
2. the proximal half of the LAD artery
3. the portion of the right coronary artery that courses over the sternocostal surface of the heart.

The Cardiac Veins

The **coronary sinus** is the largest of the cardiac veins (Fig. 18–6). It receives about 60% of the heart's venous drainage. It lies throughout its length within the atrioventricular groove on the diaphragmatic surface, extending from the heart's left border to almost its inferior border. The coronary sinus terminates at its opening into the posterior floor region of the right atrium (Fig. 18–8).

The **great cardiac vein** is the second largest surface vein of the heart (Fig. 18–5). It arises near the apex and ascends on the sternocostal surface within the anterior interventricular groove. Upon reaching the atrioventricular groove, it turns and extends to the left within this groove until it gives rise to the coronary sinus at the heart's left border (Fig. 18–6). The great cardiac vein drains parts of the left atrium and both ventricles.

The **middle cardiac vein** arises near the apex and ascends on the diaphragmatic surface within the inferior interventricular groove (Fig. 18–6). It is a major tributary of the coronary sinus.

The **anterior cardiac veins** drain the anterior wall of the right ventricle. They extend to the right on the sternocostal surface of the heart, passing over the atrioventricular groove. Upon reaching the anterior wall of the right atrium, the veins penetrate the atrial musculature to open directly into the right atrium.

The **venae cordis minimae** are named for what they are: the smallest veins of the heart. Because they were first described by Adam Christian Thebesius in 1708, they are also known as the **Thebesian veins.** These veins are found within the walls of all the heart's chambers (most predominantly in the right atrium), and they drain directly into the chambers.

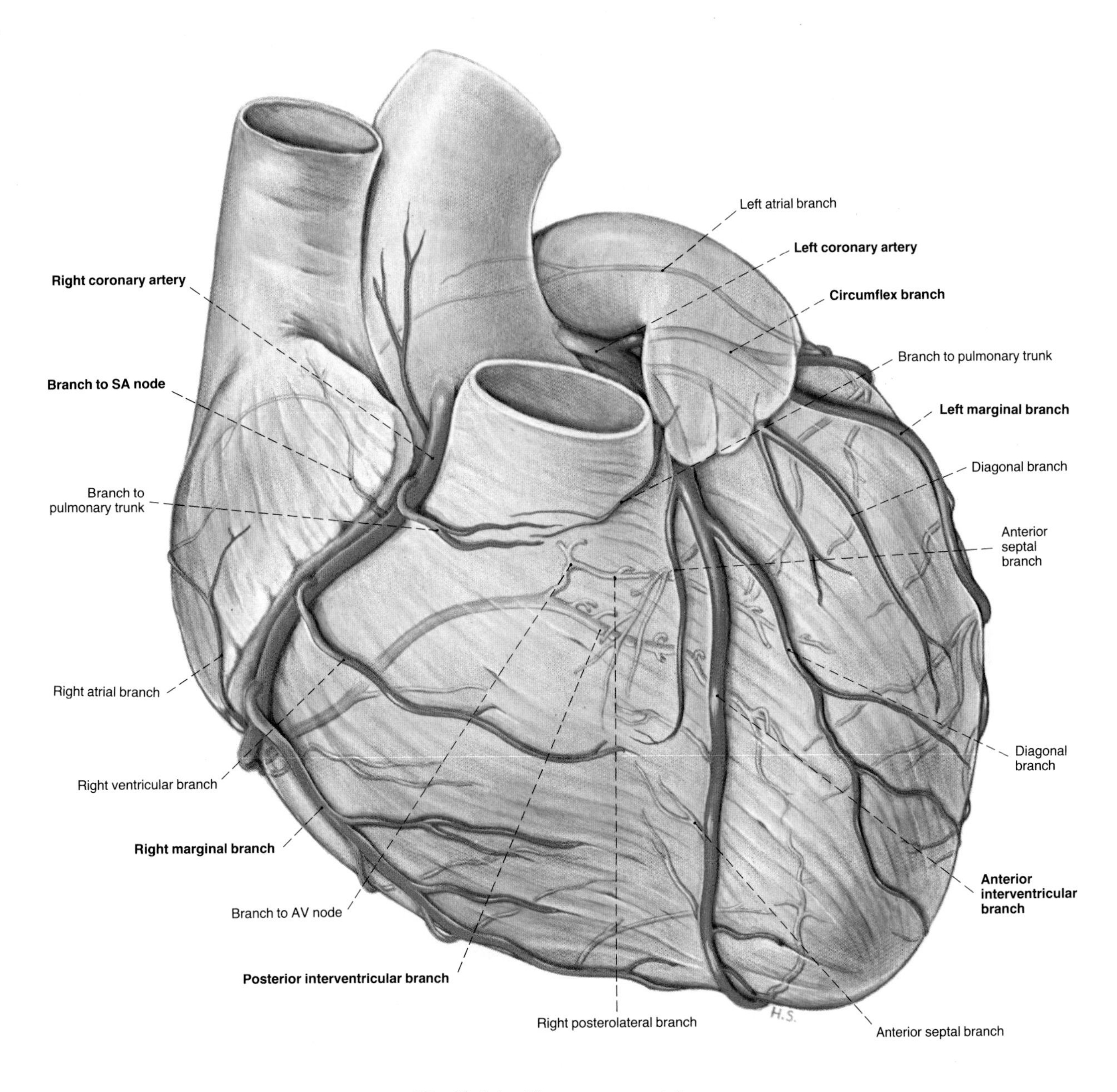

Fig. 18–14: The coronary arterial system.

THE HEART'S CONDUCTING SYSTEM

On each functional side of the heart, the atrium and ventricle work together as one pump. Because the pumping activities of the two atrioventricular pumps are synchronous, they can be described by a single cycle of events. This cycle of events is called the **cardiac cycle.**

The contraction and relaxation of the heart's musculature is highly coordinated during the cardiac cycle. This coordination is controlled by a system of highly specialized, myocardial cells called the conducting system of the heart. The heart's conducting system provides a nodular cluster of cells which initiates the signal for myocardial contraction during each cardiac cycle. The system also supplies cellular tracts to propagate this signal in a highly concerted fashion through first the atrial and then the ventricular musculature.

Three parts of the conducting system are in the walls of the right atrium. These parts are the sinoatrial node, the internodal tracts, and the atrioventricular node.

The **sinoatrial (SA) node** is a crescent-shaped cluster of specialized myocardial cells which extends from the superior vena caval opening in the right atrium into the upper end of the terminal crest (Fig. 18–15). Since the myocardial cells of the SA node have the fastest spontaneous rate of depolarization of all the heart's myocardium, the cells of the SA node are the first myocardial cells to spontaneously depolarize during each cardiac cycle. The SA node thus generates the depolarization signal which begins electrical activity in the heart during each cardiac cycle.

The contraction signal generated by the SA node spreads in a wave-like fashion throughout the musculature of both atria, evoking in its wake a wave of contraction which spreads throughout the musculature of both atria. The signal generated by the SA node also spreads to the internodal tracts, which, in turn, rapidly transmit the signal from the SA to the atrioventricular node. The **internodal tracts** extend from the SA node through the walls of the right atrium, and terminate in the atrioventricular node. The specific location of the tracts and their cellular composition are disputed.

The **atrioventricular (AV) node** is an oval-shaped cluster of specialized myocardial cells that is located in the posteroinferior region of the interatrial septum, close to the orifice of the coronary sinus (Figs. 18–8 and 18–15). The AV node lies immediately adjacent to the fibrous anulus of the tricuspid valve. At the appropriate moment during the cardiac cycle, the AV node relays the contraction signal it receives (from the SA

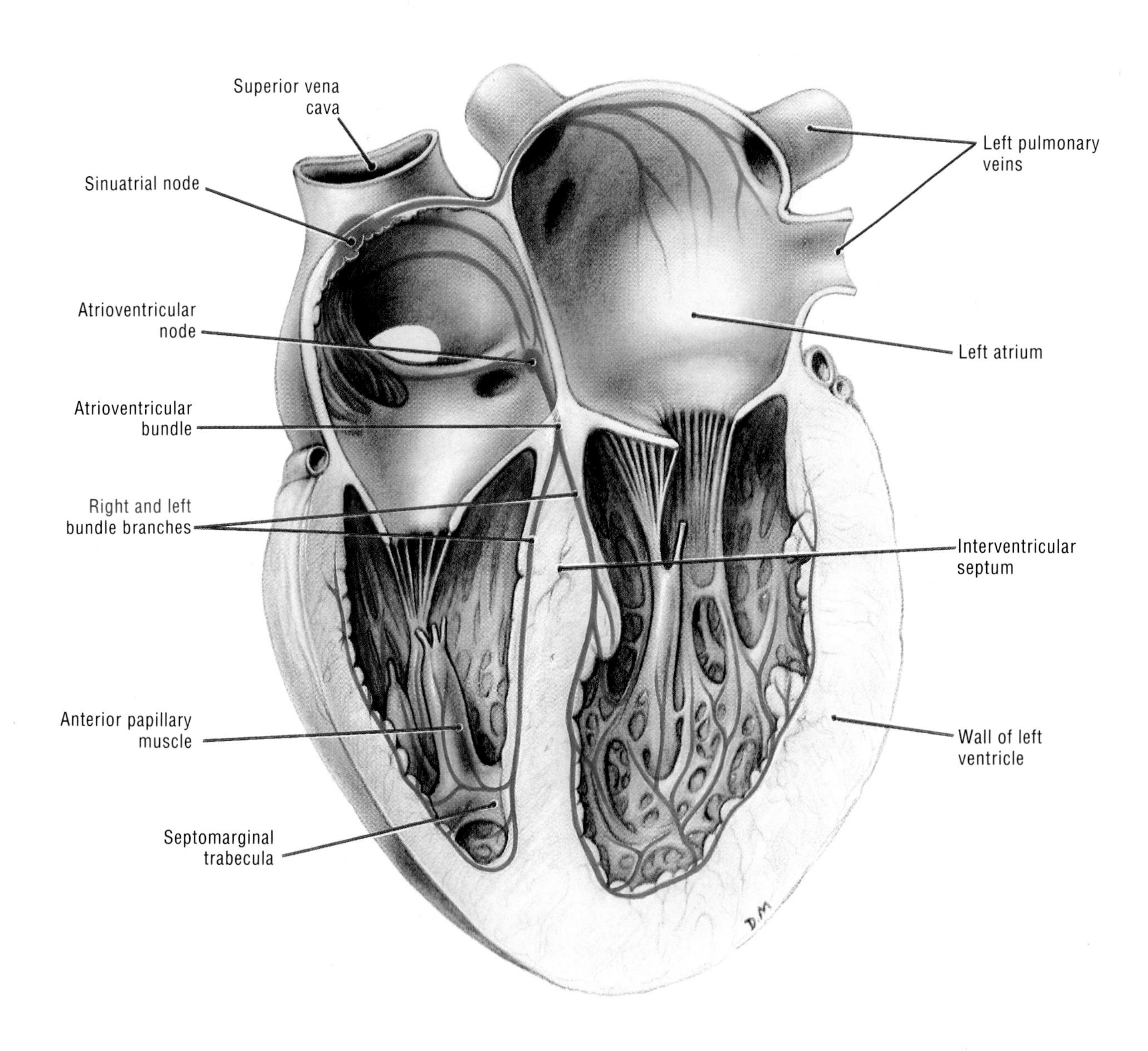

Fig. 18–15: The conducting system of the heart.

nodes by the internodal tracts) to those parts of the conducting system that transmit it to the ventricular musculature.

These latter parts of the conducting system all consist of tracts of specialized myocardial cells located within the walls of the ventricles. The larger tracts function during each cardiac cycle to rapidly transmit the contraction signal from the AV node through the ventricular walls. The finer tracts disperse the signal to a multitude of points within the ventricular musculature.

The large tracts which rapidly transmit the contraction signal through the ventricular walls include the atrioventricular bundle and the left and right bundle branches. These tracts each consist of a large bundle of specialized myocardial cells wrapped in a connective tissue sheath. The sheath separates the specialized cells both physically and electrically from the great mass of ventricular, contractile myocardial cells. By contrast, the fine tracts which disperse the signal throughout the ventricular musculature consist of individual, specialized cells called **Purkinje fibers.** The Purkinje fibers have direct electrical contact with neighboring, contractile myocardial cells.

The **atrioventricular (AV) bundle (bundle of His)** is a tract of cells which originates from the AV node, passes through an opening in the fibrous anulus of the tricuspid valve, and then extends through the membranous part of the interventricular septum (Fig. 18–15). At the border between the membranous and muscular parts of the interventricular septum, the AV bundle divides into two major branches, the left and right bundle branches (Fig. 18–15). Because the fibrous anuli of the tricuspid and mitral valves electrically insulate the atrial musculature from the ventricular musculature, the AV bundle serves as the only major tract of myocardial cells that can conduct the contraction signal from the atria to the ventricles. The left and right bundle branches then conduct the signal further through the respective walls of the left and right ventricles.

The **right bundle branch** initially extends for a distance on the right side of the muscular part of the interventricular septum (Fig. 18–15). It then veers off into a muscular beam in the right ventricle which extends from the interventricular septum to the lower margin of the ventricle's anterior wall. This myocardial bridge is called the **septomarginal trabecula** (Figs. 18–8 and 18–15). [Early anatomists called this bridge the moderator band because they believed that its contraction during systole moderated outward bulging of the ventricle's anterior wall. There is no evidence, however, to support this hypothesis.] Having passed through the septomarginal trabecula and into the anterior wall of the right ventricle, the right bundle branch ramifies into many smaller tracts.

The **left bundle branch** differs in two important respects from the right bundle branch. First, the left bundle branch ramifies near its origin into about a half dozen smaller tracts (Fig. 18–12). The left branch thus does not extend through much of its course as a solitary tract, as the right branch does. Second, the several ramifications of the left branch spread throughout the musculature of the left ventricle by extending apically on the left side of the muscular part of the interventricular septum (Figs. 18–12 and 18–15), and then assume a recurrent course through the other walls of the chamber (Fig. 18–15).

The **Purkinje fibers** are the terminal cells of the conducting system in both ventricles. They are very large diameter cells whose terminal processes end in the subendocardial myocardium. During the cardiac cycle, the Purkinje fibers transmit the signal for contraction directly to the subendocardial myocardium of the ventricular walls.

The nutrient arteries for the SA and AV nodes are each a branch (generally an indirect branch) of either the right coronary artery or the circumflex artery. The nutrient artery for the SA node is a branch of the right coronary artery in about 55% of the population and a branch of the circumflex artery in the rest of the population. In the case of the nutrient artery for the AV node, there is a high degree of correlation between its origin and the origin of the posterior interventricular artery. In other words, the AV node is almost always supplied by a branch of the dominant coronary artery. This high degree of correlation explains why 90% of all hearts are right dominant and have their AV node supplied by a branch of the right coronary artery.

The electrical activity of the heart's conducting system is modulated by an extensive autonomic plexus, called the **cardiac plexus,** located at the base of the heart. The cardiac plexus receives both sympathetic and parasympathetic innervation. Sympathetic stimulation increases the force and rate of the heartbeat, and results in dilatation of the coronary arteries; parasympathetic stimulation produces the opposite effects.

The cell bodies of the preganglionic sympathetic neurons for the heart's autonomic innervation are located in the upper five thoracic spinal cord segments. The fibers that emanate from these neurons synapse with postganglionic neurons in all three cervical ganglia and in the upper five thoracic ganglia of the sympathetic chains. Cardiac branches of these paravertebral ganglia transmit postganglionic sympathetic fibers to the cardiac plexus. The cardiac branches which emanate from the three cervical ganglia are commonly called the **superior, middle, and inferior cardiac nerves.** The middle and inferior cardiac nerves and the cardiac branches of the upper five thoracic ganglia also transmit sensory fibers from the heart and aorta which are sensitive to ischemia. These sensory fibers mediate the visceral pain associated with angina pectoris and myocardial infarctions. Such myocardial ischemic pain is commonly referred to regions of the T1-T5 dermatomes simply because the visceral sensory fibers enter the spinal cord at the levels of the segments for the upper five thoracic spinal nerves.

The cranial origins of the vagus nerves bear the cell bodies of the preganglionic parasympathetic neurons for the heart's autonomic innervation. On each side, the fibers that emanate from these neurons are transmitted to the cardiac plexus by the **superior** and **inferior cardiac branches of the vagus nerve** and **cardiac branches of the vagus nerve's recurrent laryngeal branch.** These cardiac branches also transmit sensory fibers from baroreceptors and chemoreceptors that reside in and about the walls of the venae cavae, aorta, and pulmonary trunk. The baroreceptors monitor hydrostatic pressure and the chemoreceptors monitor oxygen tension of the blood borne by these vascular trunks. The sensory information provided by these baroreceptors and chemoreceptors initiate reflexive changes in cardiac output (the expression cardiac output refers to the rate at which the heart pumps blood into the aorta).

THE APEX BEAT AND THE HEART SOUNDS

When ventricular blood is ejected posterosuperiorly into the roots of the aorta and pulmonary trunk during the systolic period of every cardiac cycle, an equal and opposite force propels the heart anteroinferiorly. This force causes the heart's apex to literally beat against the left side of the anterior chest wall. The distance of the palpable apex beat from the midline is a rough measure of heart size. In a normal adult, the apex beat is located in the left fifth intercostal space, 8 to 9 cm from the midline.

The heart produces two sounds during each cardiac cycle. The sounds can be described by the expression "lub-dup." The first heart sound (the lub) is of longer duration and of a lower frequency than the second heart sound (the dup). The first heart sound is labelled S1 and the second heart sound S2. S1 occurs as a consequence of the near-simultaneous closure of the mitral and tricuspid valves (the mitral valve is the first of the two valves to close). S2 occurs as a consequence of the near-simultaneous closure of the aortic and pulmonary valves (the aortic valve is the first of the two valves to close).

Each heart sound thus has two components to it. In the case of S1, one component results from the closure of the mitral valve, and the other component results from the closure of the tricuspid valve. The mitral component is generally best heard on the chest wall region overlying the apex beat (Fig. 18–16). The tricuspid component is generally best heard over the left lower end of the body of the sternum [Figure 18–16 illustrates an alternative site (the right lower end of the body of the sternum) for auscultation of the tricuspid component].

In the case of S2, one component results from the closure of the aortic valve, and the other component results from the closure of the pulmonary valve. The aortic component is generally best heard over the anterior end of the right second intercostal space (Fig. 18–16). The pulmonary component is generally best heard over the anterior end of the left second intercostal space.

The time interval between the closure of the aortic and pulmonary valves at the end of systole is generally long enough to be detected. The expression **S2 splitting** refers to the common, normal detection of the separate aortic and pulmonary components of S2. Deep inspiration increases the time interval between the closure of the aortic and pulmonary valves at the end of systole. The mechanism of this normal phenomenon, which is called **physiologic splitting of S2,** is as follows: Deep inspiration decreases intrathoracic pressure, which, in turn, increases systemic venous return to the right atrium and ventricle. The resulting increase in right ventricular end-diastolic volume lengthens the period of ejection of blood from the right ventricle during systole, which, in turn, delays closure of the pulmonary valve.

Deep inspiration not only delays the closure of the pulmonary valve but also hastens the closure of the aortic valve. The basis of this latter effect is not known. Some investigators believe that deep inspiration increases the capacity of the pulmonary vasculature to an extent greater than the inspiratory increase of right ventricular output. If these changes do indeed occur, it follows that there is pooling of blood in the pulmonary vasculature and a consequent decrease of pulmonary venous return to the left atrium and ventricle. The resulting decrease in left ventricular end-diastolic volume shortens the period of ejection of blood from the left ventricle during systole, which, in turn, hastens the closure of the aortic valve.

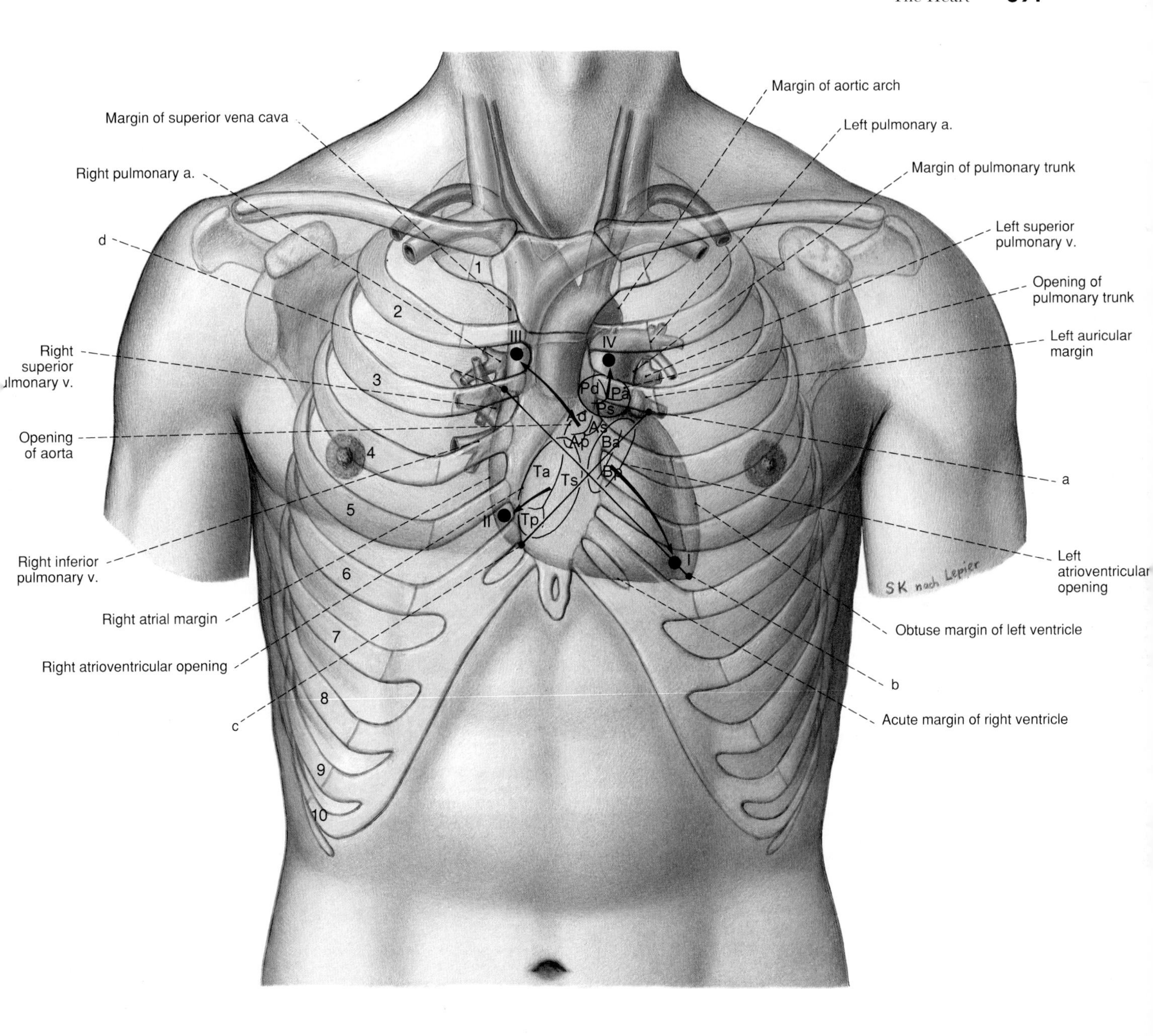

1–10 = Ribs 1–10

Aortic valve
- Ad = Right semilunar valve
- Ap = Posterior semilunar valve
- As = Left semilunar valve

Valve of pulmonary trunk
- Pa = Anterior semilunar valve
- Pd = Right semilunar valve
- Ps = Left semilunar valve

Bicuspid (mitral) valve
- Ba = Anterior cusp
- Bp = Posterior cusp

Tricuspid valve
- Ta = Anterior cusp
- Tp = Posterior cusp
- Ts = Septal cusp

Auscultation points
- I = Mitral valve
- II = Tricuspid valve
- III = Aortic valve
- IV = Pulmonary valve

- a–c = Transverse diameter of ventricular base
- b–d = Cardiac axis
- a = Site of base of left ventricle
- b = Apex of heart
- c = Site of base of right ventricle
- d = Site of right atrium

Fig. 18–16: Diagram illustrating the anterior surface projections of the heart's valves and the sites on the anterior chest wall over which the sounds of valve closure can best be heard.

CASE **IV.6**

The Case of Maria Velasquez

INITIAL PRESENTATION AND APPEARANCE OF THE PATIENT

A 42 year-old, Hispanic woman named Maria Velasquez has made an appointment to seek treatment for attacks of chest tightness. The examiner's partner, who is the patient's primary care physician, cannot examine the patient because the partner is involved in the management of an emergency case. The patient appears somewhat apprehensive.

QUESTIONS ASKED OF THE PATIENT

What can I do for you? I've been having attacks of tightness in my chest for about the past month, and I'm afraid I'm going to have a heart attack.

Where do you feel the tightness in your chest? Right here, deep in the center of my chest [the patient marks the location by making a tight fist with her right hand and placing the fist over the sternum]. *[When any chest discomfort or pain is a complaint, it is always essential to have the patient mark by word or gesture the exact site(s) at which the discomfort or pain is perceived. The examiner recognizes that stimulation of visceral pain fibers in thoracic viscera can result in a variety of uncomfortable sensations, including feelings of tightness, heaviness, or pressure, choking or smothering sensations, and dull, deep aches. At this point, the examiner believes that the patient's attacks of chest tightness most likely represent attacks of stimulation of visceral pain fibers in one or more thoracic viscera.]*

Do you always feel the tightness in just the center of your chest? No. Sometimes the tightness extends up the left side of my chest to the front of my shoulder. If the tightness goes up to the front of my shoulder, I also sometimes feel like numbness down the inside of my left arm [the patient marks the location of the left arm numbness by rubbing the dorsal surfaces of the fingers of her right hand against the medial aspect of her left arm]. *[At this point, the examiner tentatively regards the left shoulder and arm discomfort as referred pain.]*

How often do you have these attacks of chest tightness? About once a day or every other day.

You said that the attacks started about a month ago. As the month has progressed, have you noticed any change in how often you have the attacks? No.

How would you describe a typical attack of chest tightness? Well, once I start to feel some tightness coming on, the tightening will increase for about a half a minute or so, and generally grip me so bad that it stops me from doing whatever I'm doing at that time. The attack will then gradually subside and be over 5 to 10 minutes later.

Have you noticed if there is anything that seems to bring on the attacks? Yes, and this is bothering me, because I've noticed that most of the attacks have occurred at work when I become stressed. You see, I'm a news editor at one of the local television stations, and I'm frequently under pressure to get the news copy and videotape footage put together properly for the evening news program.

Have you found anything that relieves the chest tightness? Yes, I've found that if I divert my attention for a few minutes and try to relax myself, then the tightness goes away in about 5 to 10 minutes.

Have you noticed if physical activity, such as running, brings on an attack of chest tightness? Yes. On the few occasions that I've tried to jog this past month, I've had tightness develop in my chest. *[The examiner recognizes that the patient's attacks of chest tightness appear to be provoked by emotional or physical stress and relieved by rest and relaxation.]*

Do you ever get an attack if you're resting or in a relaxed frame of mind? No.

Have you noticed if sudden exposure to cold weather brings on an attack of chest tightness? No, I haven't experienced that.

Do you ever get an attack while eating or after eating? No. Oh, I take that back. Two weeks ago I went out to dinner with a representative of one of our biggest commercial clients. I ate more that evening than I usually do, and I did get an attack while driving home after the dinner.

Have you noticed any changes in the way your heart beats, such as that it seems to beat too fast or to skip a beat at times? No. *[The examiner poses this question to ascertain if the patient's attacks of chest tightness are associated with palpitations. The term palpitations refers to the conscious sensation of an unusual heartbeat, generally one that has increased in frequency and/or strength or is irregular in frequency.]*

Do you have any breathing problems, such as difficulty in catching your breath? No, except for when I have an attack, and then I do feel like I can't catch my breath.

Have you had any fainting spells? No. *[The examiner poses this question to ascertain if the patient's attacks of chest tightness are associated with syncope. Syncope is an indicator of inadequate perfusion of the brain or insufficient concentrations of oxygen and/or glucose in the blood circulating through the cerebral vasculature.]*

Is there anything else that is also bothering you? No.

Have you ever had any problems with your heart? No, not that I know of. But your partner told me 2 years ago that I have hypertension and that hypertension can lead to heart problems. That's why I'm concerned that my attacks of chest tightness are a prelude to a heart attack.

Are you taking any medications for hypertension? Yes, your partner has put me on medications to control the hypertension. I've been on the medication for the past 2 years.

Is there any history of hypertension or heart problems in your family? Yes. My father had hypertension and he died of a heart attack when he was 65. I have a 45 year-old brother, and he's also being treated for hypertension.

Have you noticed any changes or problems with urination? No.

Is there any history of kidney problems in your family? No.

Is there any history of diabetes in your family? No.

Do you smoke? Unfortunately, yes. I smoke about a pack a day. Your partner has repeatedly tried to help me to kick the habit, but I just can't seem to do it. I've been able to follow up on all the other things your partner advised me to do to control my hypertension except stop smoking.

What sort of lifestyle changes have you made to better control your hypertension? I don't cook with salt and I'm very conscious of the caloric, fat, and cholesterol contents of the food I eat. I had wanted to start a jogging program this month, but, as I've said, on the occasions when I've tried to jog, I've always had an attack of tightness in my chest.

How long have you been smoking about a pack of cigarettes a day? Ah, I guess about 9 years now.

Do you drink alcoholic beverages? No.

Have you had any recent illnesses or injuries? No.

Are you taking any medications for previous illnesses or injuries other than your anti-hypertensive medication? No.

When was your last period? I had my last period 5 days ago.

Have your last several menstrual periods been normal in their timing and amount of discharge? Yes. *[The examiner asks the last two questions in order to inquire if the patient is pregnant or experiencing hormonal changes associated with menopause.]*

[The examiner finds the patient alert and fully cooperative during the interview.]

PHYSICAL EXAMINATION OF THE PATIENT

Vital signs: Blood pressure
Lying supine: 145/95 left arm and 145/95 right arm
Standing: 150/100 left arm and 150/100 right arm
Pulse: 75
Rhythm: regular
Temperature: 98.6°F
Respiratory rate: 15
Height: 5′ 10″
Weight: 145 lbs.
HEENT Examination: Normal
Lungs: Normal
Cardiovascular Examination: Examination of the head and neck is normal except for the following findings: In the ocular fundi, the retinal arterioles show evidence of a broadened yellow light reflex. The arteriolar to venous diameter ratio is 1:2. Mild venular depression is noted at two arteriovenous crossings.

Abdomen: Normal
Genitourinary Examination: Normal
Musculoskeletal Examination: Normal
Neurologic Examination: Normal

INITIAL ASSESSMENT OF THE PATIENT'S CONDITION

The patient suffers from brief attacks of chest tightness during periods of emotional and physical stress.

ANATOMIC BASIS OF THE PATIENT'S HISTORY AND PHYSICAL EXAMINATION

1. **The patient's attacks of substernal tightness suggest disease of one or more thoracic viscera.** Substernal tightness is a common indicator of stimulation of visceral pain fibers in thoracic viscera.
2. **The patient's left anterior shoulder tightness and left medial arm numbness appear to be referred pain (inconstant, superficial discomfort).** The anterior aspect of the shoulder represents parts of the C4 and C5 dermatomes (Fig. 4–12, A and E). The medial aspect of the arm represents part of the T2 dermatome (Fig. 4–12, E and F).
3. **The patient has mild systemic hypertension associated with early stage arteriosclerotic retinopathy.** Systemic hypertension is defined as systolic/diastolic blood pressure levels equal to or greater than 140/90 mm Hg. Individuals with a diastolic pressure from 90 to 104 are considered to have mild systemic hypertension. The cause of systemic hypertension in most individuals is not known; individuals with systemic hypertension of unknown etiology are said to have **primary,** or **idiopathic, hypertension.** In this case, the examiner's partner has advised the patient to modify certain elements of her lifestyle (e.g., reduce sodium, calories, saturated fat, and cholesterol intake and adopt some program of moderate physical exercise) to help treat her hypertension. This suggests that the patient has primary hypertension.

   The systemic vascular effects of hypertension can be evaluated by ophthalmoscopic examination of the retinal arterioles and venules of the **optic fundus** (the portion of the eye posterior to the lens) (Fig. 7–26). Normal retinal vessels appear as gently curved, reddish bands, each with a narrow, central yellowish stripe about one-fourth the thickness of the band. Each reddish band represents the tubular column of blood within a retinal vessel; the central yellowish stripe represents light reflected from the blood column. The walls of normal retinal vessels are transparent and thus not seen. The arterioles are lighter and narrower than the venules that accompany them; the ratio of arteriolar to venular diameters is approximately 3:4. As an arteriole and its accompanying venule extend toward the periphery, the vessels cross each other, with the arteriole generally passing anterior to the venule. There is no change in the appearance of the vessels at normal arteriovenous crossings.

   One of the initial effects of systemic hypertension on the appearance of retinal arterioles is narrowing of the reddish blood column. Generalized arteriolar narrowing to the extent that the ratio of arteriolar to venular diameters becomes 1:2 is a sign of mild hypertension. **Arteriosclerosis** is a major consequence of systemic hypertension; the persistent arterial hypertension promotes proliferation of the smooth muscle and elastic tissue of arterial walls. The sclerotic thickening broadens the central reflex stripe and is associated with changes at arteriovenous crossings. A broadened yellow light reflex and mild venular depression at a few arteriovenous crossings are signs of early stage retinal arteriosclerosis.

INTERMEDIATE EVALUATION OF THE PATIENT'S CONDITION

The patient appears to be suffering from angina pectoris in association with mild systemic hypertension and atherosclerosis.

CLINICAL REASONING PROCESS

Central chest (substernal) discomfort indicates a disorder, disease, or injury of one or more thoracic and/or upper abdominal viscera. The thoracic viscera most commonly involved are the esophagus, the heart, the trachea and larger bronchi, the pericardium, the pulmonary trunk and arteries, and the thoracic aorta (the viscera have been listed in the order of decreasing frequency in which they are involved). The upper abdominal viscera occasionally involved are the stomach, duodenum, liver, gallbladder, and pancreas.

Central chest discomfort is, in fact, visceral pain. Although the discomfort may indeed be perceived as pain (in the form of a dull ache), it is more commonly perceived as tightness, pressure, a squeezing sensation, or a smothering sensation. Independent of its manifestation, the discomfort is almost always deep and diffuse in location. The surface body regions to which central chest discomfort is most commonly referred are those of the T1-T6 dermatomes. Pain may also be referred to the neck, lower jaw, or upper aspect of the shoulder.

The best strategy for discerning the visceral origin of central chest discomfort is to seek the information symbolized by the mnemonic **PQRST:** the **P**rovocative and palliative circumstances of the discomfort, the **Q**uality of the discomfort, its precise anatomic **R**egion and referral, the **S**everity of the discomfort, and its **T**iming. The correlation of this information with the findings of the physical exam generally identifies the diseased or injured thoracic or abdominal viscus.

A primary care physician would attribute the patient's central chest discomfort to angina pectoris (the expression is generally shortened to simply angina). The term **angina** refers to the type of central chest discomfort produced by **transient myocardial ischemia.** The expression transient myocardial ischemia refers, in turn, to the syndrome of conditions in which the heart's coronary arteries are insufficient in their capacity to oxygenate the myocardium during periods of increased metabolic demand. Characterization of a patient's central chest discomfort as angina therefore identifies coronary insufficiency as the process responsible for the patient's discomfort but not the disorder responsible for the coronary insufficiency.

A primary care physician would tentatively characterize the patient's angina as **stable angina,** primarly because the attacks predictably occur when certain physical activities are attempted (jogging) or emotional stress of sufficient intensity (completing a task at work under the pressure of a deadline) is encountered. The PQRST aspects of the patient's angina represent the most common characteristics of angina: the attacks

1. are provoked by physical or emotional stress and relieved by rest
2. feel like tightness or pressure
3. are located deep in the substernal region and radiate to the anterior aspect of the left shoulder and medial aspect of the left arm
4. are generally severe enough to interrupt the patient's activities at the time, and
5. reach maximum intensity within one minute and subside within 5 to 10 minutes if rest and relaxation are immediately sought.

In general, angina may be provoked by any condition which stimulates cardiac output. Typical conditions include physical exercise, emotional stress, sudden exposure to the cold, and the initiation of digestive processes after a heavy meal. It is also important to know that although angina most commonly radiates to the left side, it can also radiate to the right shoulder and arm.

The results of the cardiovascular exam are of primary importance in the identification of the disorder responsible for a patient's transient myocardial ischemia. The cardiovascular exam employs every region of the body to gather direct and indirect evidence of cardiovascular disease: (a) The fundi of the eyes are examined for pathologic changes in retinal blood vessels. (b) The neck is examined for the strength, shape and timing of carotid pulses, the height to which blood accumulates in the jugular veins, and the nature of jugular pulsations. The characteristics of the carotid pulses provide an assessment of left atrioventricular function. The height to which blood accumulates in the jugular veins is a measure of central venous pressure. Marked tricuspid regurgitation manifests characteristic changes in the jugular pulsations. (c) The chest is examined for the site and force of the apex beat, the rhythm of the heartbeat, and the characteristics of the normal heart sounds (S1 and S2) and (if present) abnormal heart sounds (such as extra heart sounds, murmurs, and clicks). The site of the apex beat provides a rough measure of heart size. Certain combinations of normal and abnormal heart sounds point to specific cardiac disorders. (d) The abdomen is examined for arterial bruits (abnormal sounds of turbulent arterial blood flow). A bruit commonly signals the presence of an aneurysm. (e) The extremities are examined for the strength, shape, and timing of the radial and ulnar pulses in the upper limb and the femoral, dorsalis pedis, and posterior tibial pulses in the lower limb. The extremities are also examined for abnormalities such as varicose veins, edema, cyanosis, and pallor.

The finding of mild systemic hypertension and early stage arteriosclerotic retinopathy but no other signs of cardiovascular disease suggest in this case that coronary artery disease (coronary atherosclerosis) is the anatomic basis of the patient's angina. In other words, the evidence suggests that atherosclerotic changes in the coronary arteries and their branches have narrowed the lumen of one or more arteries to the extent that a portion of the myocardium cannot be adequately perfused during intervals of increased metabolic demand.

RADIOGRAPHIC EVALUATION AND FINAL RESOLUTION OF THE PATIENT'S CONDITION

PA and lateral chest films of the patient were negative. Although chest films cannot provide direct evidence of coronary artery disease, they were ordered in this case to look for radiographic signs of those cardiovascular and pulmonary diseases that can present as angina.

In this case, the patient needs to undergo further testing to confirm the diagnosis of angina due to coronary artery disease. An electrocardiographic recording of the electrical activity in the patient's heart before, during, and after monitored physical exercise can provide evidence of myocardial ischemia and grade its severity. Analysis of blood and urine samples could provide evidence of non-cardiac disorders (such as diabetes mellitus, kidney disease, and thyroid disease) which may be contributing to the patient's myocardial ischemia. If such testing indicates coronary artery disease severe enough to warrant coronary bypass surgery, coronary angiography is employed to assess the extent of occlusion of the coronary arteries and their branches. Coronary angiography is the radiographic procedure which provides direct visualization of the patency of the coronary arteries and their branches by individual catheterization of the left and right coronary arteries followed by injection of a radiodense contrast agent. Coronary bypass surgery is a surgical procedure in which segments of a patient's great saphenous vein or internal thoracic artery are used to construct a vascular bypass around a site of coronary arterial occlusion. The great saphenous vein is commonly used for coronary bypass surgery because (a) it can be readily resected from the leg, (b) its diameter closely approximates that of the coronary arteries and their major branches, and (c) its wall bears a comparatively high content of elastic tissue.

Last, the patient would be counselled to continue treatment of her hypertension and that her continued cigarette smoking is significantly contributing to the worsening of her coronary heart disease.

EXPLANATION OF THE PATIENT'S CONDITION

The patient suffers from angina due to atherosclerotic occlusion of the coronary arteries and their branches. In particular, atherosclerotic plaques near the origin of the patient's LAD artery have reduced its effective diameter by 75%, and it is the insufficiency of this artery which is chiefly responsible for the angina upon emotional and physical stress.

CASE **IV.7**

The Case of Dorothy Simpson

INITIAL PRESENTATION AND APPEARANCE OF THE PATIENT

A 32 year-old, white woman named Dorothy Simpson has made an appointment to seek treatment for episodes of chest tightness. The patient is a thin woman who appears apprehensive and fatigued.

QUESTIONS ASKED OF THE PATIENT

What can I do for you? I feel a severe tightness in my chest whenever I exert myself too much.

Where do you feel the tightness in your chest? Right here, deep in the center of my chest [the patient marks the location by making a tight fist with her right hand and placing the fist over the sternum]. *[When any chest discomfort or pain is a complaint, it is always essential to have the patient mark by word or gesture the exact site(s) at which the discomfort or pain is perceived. The examiner recognizes that stimulation of visceral pain fibers in thoracic viscera can result in a variety of uncomfortable sensations, including feelings of tightness, heaviness, or pressure, choking or smothering sensations, and dull, deep aches. At this point, the examiner believes that the patient's attacks of chest tightness most likely represent attacks of stimulation of visceral pain fibers in one or more thoracic viscera.]*

Do you always feel the tightness in just the center of your chest? No. Sometimes the tightness extends up the right side of my chest to the front of my shoulder. If the tightness goes up to the front of my shoulder, I also sometimes feel like numbness down the inside of my right arm down to the elbow [the patient marks the location of the right arm numbness by rubbing the dorsal surfaces of the fingers of her left hand against the medial aspect of her right arm]. *[At this point, the examiner tentatively regards the right shoulder and arm discomfort as referred pain.]*

Do you recall when you first began to experience tightness in your chest? Yes. The first time I was struck with an attack of chest tightness was about 2 weeks ago at work, when the elevators in our building were out-of-order for a day. The three times I had to climb 2 flights of stairs that day, I was attacked each time with chest tightness about halfway up the second flight.

How often do you now have these attacks of chest tightness? Maybe once a day, and the attacks are always about the same.

How would you describe a typical attack of chest tightness? Well, once I start to feel some tightness coming on, the tightening will increase for about a half a minute or so, and generally grip me so bad that it stops me from doing whatever I'm doing at that time. The attack will then gradually subside and be over 5 to 10 minutes later.

Have you noticed if there is anything else in addition to physical exertion that seems to bring on the attacks? Yes. My chest becomes tight if I become a bit overstressed at work.

Have you found anything that relieves the chest tightness? Yes, I've found that if I rest for a few minutes and try to relax myself, then the tightness goes away in about 5 to 10 minutes. *[The examiner recognizes that the patient's attacks of chest tightness appear to be provoked by emotional or physical stress and relieved by rest and relaxation.]*

Have you noticed if sudden exposure to cold weather brings on an attack of chest tightness? No, I've not experienced that.

Do you ever get an attack if you're resting or in a relaxed frame of mind? No.

Do you ever get an attack while eating or after eating? No.

Have you noticed any changes in the way your heart beats, such as that it seems to beat too fast or to skip a beat at times? No. *[The examiner poses this question to ascertain if the patient's attacks of chest tightness are associated with palpitations. The term palpitations refers to the conscious sensation of an unusual heartbeat, generally one that has increased in frequency and/or strength or is irregular in frequency.]*

Do you have any breathing problems, such as difficulty in catching your breath? Yes. Starting about 6 months ago, I began noticing that I would get short of breath whenever I exert myself. My shortness of breath has been getting worse over the past few months. A few times, I've become so short of breath that I have felt light-headed. At first, I didn't think that anything was wrong with my health because I've been under a lot of stress at work, and there's been times when the stress has made me feel tired and short of breath, and so I've sort of felt that these problems were all work-related. Now, however, beginning about 2 weeks ago, I have these attacks of chest tightness soon after I become short of breath. *[The patient describes an insidious onset of dyspnea (shortness of breath) which began about 6 months ago and has progressively worsened to the extent that it produces light-headedness (near syncope). Dyspnea is an indicator of inadequate ventilation or perfusion of the lungs.]*

You just mentioned that your stress at work has made you feel tired. Do you feel, in general, that you fatigue more easily than in the past? Yes. Actually, I began noticing about a year ago that I fatigue rather easily.

Have you had any nausea or vomiting? No.

Is there anything else that is also bothering you? No.

Have you ever had any problems with your heart? No, not that I know of.

Is there any history of hypertension or heart problems in your family? No.

Have you noticed any changes or problems with urination? No.

Is there any history of kidney problems in your family? No.

Is there any history of diabetes in your family? No.

Do you smoke? No.

Do you drink alcoholic beverages? No.

Have you had any recent illnesses or injuries? No.

Are you taking any medications for previous illnesses or injuries? No.

When was your last period? I had my last period 5 days ago.

Do you practice birth control? Yes. My husband uses a condom. *[The examiner finds the patient alert and fully cooperative during the interview.]*

PHYSICAL EXAMINATION OF THE PATIENT

Vital signs: Blood pressure
Lying supine: 120/75 left arm and 120/75 right arm
Standing: 120/75 left arm and 120/75 right arm
Pulse: 95
Rhythm: regular
Temperature: 99.0°F.
Respiratory rate: 18
Height: 5′ 9″
Weight: 120 lbs.

[The examiner recognizes that the patient has a heartbeat rate near the upper limit of the normal range. In adults, the normal range of the heartbeat is 60 to 100 beats per minute.]

HEENT Examination: Normal
Lungs: Normal
Cardiovascular Examination: The cardiovascular examination is notable for the following findings: The carotid, radial, and posterior tibial pulses are readily palpable bilaterally but reduced in volume. The feet and ankles show evidence of minimal edema and there is a faint purplish cast to the nailbeds of the toes. Pertinent normal findings are a reddish cast to the oral mucosa and the nailbeds of the fingers.

The right internal jugular venous (IJV) pulse is elevated to 5 cm above the sternal angle with prominent *a* and *v* components. The right IJV venous pulse is abolished by minimal pressure, but the pulsations do not disappear upon inspiration. Sustained pressure over the right upper quadrant of the abdomen causes a 1 cm rise in right IJV pressure.

There is a palpable sustained elevation of the left border of the sternum during systole. Auscultation finds that the S2 splitting on deep inspiration is narrow. Auscultation at the anterior end of the left 2nd intercostal space reveals accentuated sounds emanating from the closure of the pulmonary valve; the accentuation is palpable at the site of auscultation. Auscultation finds a grade 2/6 diastolic murmur beginning upon closure of the pulmonary valve; the murmur is loudest at the anterior end of the left second intercostal space.

Abdomen: Normal
Genitourinary Examination: Normal
Musculoskeletal Examination: Normal
Neurologic Examination: Normal

INITIAL ASSESSMENT OF THE PATIENT'S CONDITION

The patient suffers from fatigue of 12 months' duration, stress-induced dyspnea of 6 months' duration, and stress-induced angina of 2 weeks' duration.

ANATOMIC BASIS OF THE PATIENT'S HISTORY AND PHYSICAL EXAMINATION

1. **The patient's history suggests that the patient is suffering from a chronic condition which began insidiously 6 to 12 months ago and has progressively worsened.**
2. **Transient myocardial ischemia appears to be the basis of the patient's attacks of chest tightness.** Four features of the attacks indicate that the chest tightness is anginal in character: the chest tightness (i) has a deep, substernal location, (ii) is provoked by physical or emotional stress, (iii) is palliated by rest, and (iv) lasts no longer than 5 to 10 minutes when provocative factors are removed.
3. **The patient's dyspnea indicates inadequate ventilation or perfusion of one or both lungs.** On occasion, the dyspnea is severe enough to produce near syncope.
4. **The height of the patient's right IJV pulse indicates that the right atrial pressure is 10 cm water, which is about 25% greater than the normal 7 to 8 cm water pressure.** [The reader is referred to the discussion of the internal and external jugular veins in Chapter 27 for an explanation of how the elevation of the right IJV pulse to a height 5 cm above the sternal angle accounts for the estimation of the right atrial pressure to be 10 cm water.] Right atrial pressure is frequently called central venous pressure, since the blood pressure of the right atrium approximates that of the large systemic veins converging upon the right atrium.
5. **The palpable sustained elevation (lifting) of the left border of the sternum during systole indicates right ventricular hypertrophy (abnormally thickened walls of the right ventricle).**

   During each heartbeat, each ventricle attempts to eject during systole all of the blood that it receives during diastole. If an abnormality arises in which one or both ventricles are chronically subjected to an excessive hemodynamic demand, the heart responds to the increased workload by ventricular hypertrophy. One of two mechanisms is commonly the basis for the increased workload: (1) the ventricle is volumetrically overloaded with blood during diastole (a mechanism known as diastolic volume overload) or (2) the ventricle needs to generate an excessive intraventricular pressure during systole in order to maintain its output (a mechanism known as systolic pressure overload). Valvular disorders are a frequent cause of both types of overload. Aortic and pulmonary valve regurgitation respectively generate diastolic volume overload in the left and right ventricles. Aortic valve stenosis and systemic hypertension generate systolic pressure overload in the left ventricle, and pulmonary valve stenosis and pulmonary hypertension generate systolic pressure overload in the right ventricle. If a condition that promotes ventricular hypertrophy is not corrected, it ultimately produces ventricular dilatation and dysfunction.

   Elevation of the left border of the sternum during systole by a heart beating within the normal rate limits (60 to 100 beats per minute) suggests a markedly hypertrophic right ventricle. The greater than normal mass of the right ventricle and its increased workload combine to increase the force with which the heart's sternocostal surface beats against the anterior chest wall during systole. Lifting (anterior displacement) of the anterior chest wall is greatest in the chest wall region overlying the anterior wall of the right ventricle; the lower left border of the sternum and the anterior ends of the left third–sixth intercostal spaces overlie most of the anterior wall of the right ventricle (Fig. 22–4). If the left sternal lift is sustained, it is likely that the cause of the right ventricular hypertrophy has been a systolic pressure overload.

6. **The characteristics of the grade 2/6 diastolic murmur indicate pulmonary valve regurgitation during diastole.**

   When blood flow into or out from a heart chamber becomes turbulent, the turbulence generates audible sounds that are termed heart murmurs. Many heart murmurs arise at or near one of the heart's valves. Common mechanisms that produce turbulent blood flow at or near a heart valve include

   1. blood flow through an abnormally narrowed valve,
   2. blood flow through an irregularly structured valve,
   3. excessive blood flow through a normal valve,
   4. retrograde blood flow through an incompetent or defective valve, and
   5. blood flow through the aortic valve into an abnormally dilated ascending aorta or through the pulmonary valve into an abnormally dilated pulmonary trunk.

   Heart murmurs can also be produced by the turbulent shunting of blood through a patent ductus arteriosus or defects in the interventricular and interatrial septa. It should be noted that heart murmurs are occasionally encountered in individuals who are in good health and have a normal heart; such murmurs are called innocent murmurs because their presence cannot be associated with any disease or disorder.

The loudness of heart murmurs is graded on a scale of 1 to 6. A grade 1 murmur is barely audible; frankly, only cardiologists can consistently detect grade 1 murmurs, and they generally cannot distinguish such murmurs until they have listened to a few heartbeats after the stethoscope has been applied to the patient's chest. Grade 2 murmurs are faint but distinct murmurs that a cardiologist can detect during the first few heartbeats after applying the stethoscope. A grade 3 murmur is moderately loud; its sounds can be readily distinguished as being additional to those of S1 and S2. A grade 4 murmur is a murmur whose intensity is so great that it generates palpable vibrations in an area of the patient's chest wall; such palpable vibrations are termed a **thrill.** All murmurs of grade 4 and above generate thrills. Grade 5 murmurs are the loudest murmurs whose detection still requires application of a stethoscope to a patient's chest. A grade 6 murmur is a murmur whose intensity is so great that it produces audible sounds that can be heard without the application of a stethoscope. In this case, the designation of the murmur as a grade 2/6 murmur means that the murmur has a grade of 2 on a scale from 1 to 6.

The timing of a murmur during the cardiac cycle is an important distinguishing characteristic. **Systolic murmurs** occur during systole: they begin at or after S1 and end before or at S2. **Diastolic murmurs** occur during diastole: they begin at or after S2 and end before or at S1.

The most common causes of a diastolic murmur are semilunar valve regurgitation, turbulent blood flow through a stenosed atrioventricular valve, and increased blood flow through an atrioventricular valve. In this case, the site of loudest auscultation and the timing of the diastolic murmur indicate that the murmur is a product of pulmonary valve regurgitation: (i) The murmur is heard loudest over the anterior end of the left second intercostal space. This is also the site for optimal auscultation of the pulmonary component of S2. (ii) The murmur begins immediately after the production of the pulmonary component of S2. Diastolic murmurs produced by turbulent blood flow through a stenosed atrioventricular valve or increased blood flow through an atrioventricular valve occur only when the atrioventricular valve is open. Since the mitral and tricuspid valves do not normally open immediately after closure of the semilunar valves, turbulent blood flow through either atrioventricular valve is not a likely explanation of the patient's diastolic murmur.

7. **The increased intensity of the sounds emanating upon closure of the pulmonary valve suggests either increased compliance of the valve's cusps or an increased rate of change of the pressure differences across the valve near the end of systole.**

Toward the end of systole, after the ventricles have ejected most of their blood, there begins a retrograde (backward) flow of blood from the aorta and pulmonary trunk into the ventricles (as a consequence of the blood pressure in each arterial trunk being greater than that in the ventricle). The retrograde blood flow in the aorta snaps the semilunar cusps of the aortic valve shut and then stretches the cusps back toward the left ventricle. The retrograde blood flow in the pulmonary trunk almost concurrently snaps the semilunar cusps of the pulmonary valve shut and then stretches its cusps back toward the right ventricle. The closure and elastic distention of the cusps of each valve forces the blood in the corresponding arterial trunk to subsequently recoil back into the arterial trunk. The elastic distention of the cusps of each valve and the subsequent recoil of the blood into the downstream arterial trunk establish vibrations in the blood borne by the arterial trunk and in the wall of the trunk itself. When these vibrations reach the chest wall and set its tissues into audible vibratory motions, the audible vibrations collectively produce S2.

There are thus two major factors that determine the intensity of the vibrations produced upon the closure of either the aortic and pulmonary valve: (1) the *compliance* of the valve's cusps (compliance is a measure of the ease by which the cusps can be elastically stretched) and (2) the *rate of change* of the pressure differences across the valve near the end of systole (which ultimately leads to the blood pressure in the arterial trunk being greater than that in the ventricle). An increase in the compliance of either valve's cusps (in the absence of any other changes) enhances the intensity of the vibrations produced after the valve's closure. This is because the more compliant (the less stiff and unyielding) the cusps (to a limited extent), the lesser the dampening of the vibrations produced in the blood borne by the downstream arterial trunk. Similarly, an increase in the rate of change of the pressure differences across either valve near the end of systole (in the absence of any other changes) enhances the intensity of the vibrations produced after the valve's closure. This is because the greater the rate of change, the greater the force of the retrograde blood flow that snaps the valve shut and stretches its cusps.

8. **The reduced volume of the carotid, radial, and posterior tibial pulses suggests decreased cardiac output.**
9. **The purplish cast to the nailbeds of the toes indicates peripheral cyanosis.**

INTERMEDIATE EVALUATION OF THE PATIENT'S CONDITION

The patient is suffering from a disorder or disease associated with elevated central venous pressure, right ventricular hypertrophy, pulmonary valve regurgitation, and decreased cardiac output.

CLINICAL REASONING PROCESS

A primary care physician would tentatively conclude that the patient is suffering from chronic pulmonary hypertension (chronically elevated blood pressure in the pulmonary circuit). The two key findings are (a) the sustained lift of the left border of the sternum during systole (which suggests right ventricular hypertrophy that has occurred in response to systolic pressure overload) and (b) the accentuated, palpable pulmonary component of S2 (which suggests elevated pulmonary arterial pressure). A primary care physician would recognize that chronic pulmonary hypertension is the most common mechanism responsible for the combination of these two key physical findings.

The next step in the consideration of the patient's condition is to determine the pathophysiologic basis of the presumed chronic pulmonary hypertension. The most common disorders include primary pulmonary hypertension, pulmonary thromboembolic disease, pulmonary veno-occlusive disease, obstructive or interstitial lung disease, left-sided valvular heart disease, left ventricular dysfunction, and collagen vascular disease. Analysis of these diagnostic possibilities generally requires utilization of some combination of the following procedures: radiology of the chest, electrocardiography, pulmonary function tests, analysis of arterial blood gases, perfusion and ventilation lung scans, Doppler echocardiography, and pulmonary angiography.

RADIOLOGIC EVALUATION AND FINAL RESOLUTION OF THE PATIENT'S CONDITION

PA and lateral chest films of the patient showed enlarged pulmonary arteries in the hila of the lung fields, evidence supportive of pulmonary hypertension. An echocardiogram revealed enlargement of the right ventricle and distortion of the interventricular septum. The radiologist noted that the nature of the interventricular septal distortion was typical of that produced by right ventricular systolic pressure overload. Doppler studies confirmed pulmonary valve regurgitation during diastole. Perfusion and ventilation lung scans were normal.

These radiologic procedures were conducted in association with electrocardiography, arterial blood gas analysis, and pulmonary function tests. Analysis of the collective findings did not provide any evidence of pulmonary thromboembolic disease, pulmonary veno-occlusive disease, obstructive or interstitial lung disease, left-sided valvular heart disease, left ventricular dysfunction, or collagen vascular disease.

Having excluded other cardiovascular and pulmonary causes of the patient's pulmonary hypertension, a primary care physician would diagnose the patient's condition as primary pulmonary hypertension. This is a pulmonary disorder of unknown etiology in which resistance to pulmonary blood flow increases as a consequence of smooth muscle hypertrophy and intimal proliferation of the pulmonary arteries and arterioles. Current medical treatment of primary pulmonary hypertension offers afflicted individuals some relief but does not effectively reduce pulmonary artery pressure. Patients generally survive 2 to 3 years from the time of diagnosis, with right ventricular failure being a common cause of death.

CHRONOLOGY OF THE PATIENT'S CONDITION

Smooth muscle hypertrophy and intimal proliferation of the patient's pulmonary arteries and arterioles in the initial stage of the patient's disease increased resistance to pulmonary blood flow. Right ventricular output was maintained in the presence of this initially progressively increasing resistance by right ventricular hypertrophy and increase of pulmonary arterial pressure. During the past 12 months pulmonary vascular resistance has reached a plateau and cardiac output (left ventricular output) has progressively declined. The significant elevation of pulmonary arterial pressure has enlarged the pulmonary arteries and causes a more forceful closure of the pulmonary valve at the end of systole (and thereby produces the accentuated, palpable pulmonary component of S2). The reduced cardiac output is the basis for the current reduced volume of the patient's carotid, radial, and posterior tibial pulses.

The patient's current stress-induced dyspnea, near syncope, and fatigue are manifestations of the inability of the pulmonary vasculature to support a significant increase in cardiac output. Increased metabolic demands on the patient's hypertrophied right ventricular myocardium have (within the last 2 weeks) produced episodes of transient myocardial ischemia of the right ventricular myocardium and the concomitant attacks of angina. Incipient right ventricular dysfunction has led to damming of blood in the right atrium and a consequent increase in central venous pressure (which is responsible for the heightened right IJV pulse and the pedal edema).

The 1 cm rise in the height of the right IJV pressure upon application of sustained pressure over the right upper quadrant of the abdomen represents a positive hepatojugular reflex test. The anatomic basis of this test is as follows: The liver occupies much of the right upper quadrant of the abdomen. If the right upper quadrant of a patient's abdomen is subjected to sustained pressure as the patient lies in a supine position and maintains normal quiet respiration, the abdominal compression (a) increases hepatic blood flow into the inferior vena cava and (b) indirectly compresses the right atrium through elevation of the diaphragm. These two changes combine to transiently increase right atrial (central venous) pressure. The right atrioventricular pump of a normal heart beating at a basal rate can accommodate the transient diastolic volume overload from the liver so quickly that the height of the right IJV pulse remains unchanged. If the sustained abdominal compression produces a sustained increase in the height of the right IJV pulse, then this positive finding indicates that the right atrioventricular pump is impaired in its capacity to propel onward into the pulmonary trunk the blood which is drained by the atrium during each cardiac cycle.

SUGGESTED REFERENCES FOR ADDITIONAL INFORMATION ON THE HEART

Bates, B., *A Guide to Physical Examination and History Taking*, J. B. Lippincott Co., Philadelphia, 1991: *Pages 288 through 303 in Chapter 9 present a discussion of the physical assessment of the heart.*

Kelley, W. N., DeVita, Jr., V. T., Dupont, H. L., Harris, Jr., E. D., Hazzard, W. R., Holmes, E. W., Hudson, L. D., Humes, H. D., Paty, D. W., Watanabe, A. M., and T. Yamada, *Textbook Of Internal Medicine*, J. B. Lippincott Co., Philadelphia, 1992: *This text is one of the most authoritative and comprehensive texts on internal medicine. In Part II on cardiology, Chapter 50 provides a discussion of the significance of dyspnea, fatigue, edema, chest pain, palpitations, syncope, and murmurs in the analysis of cardiovascular disorders.*

Wilson, J. D., Braunwald, E., Isselbacher, K. J., Petersdorf, R. G., Martin, J. B., Fauci, A. S., and R. K. Root, *Harrison's Principles Of Internal Medicine*, 12th ed., McGraw-Hill, New York, 1991: *This text is one of the most authoritative and comprehensive texts on internal medicine. In Part 6 on disorders of the cardiovascular system, Chapter 175 offers discussions on arterial pulse pressure, the jugular venous pulse, precordial palpation, and cardiac auscultation.*

Lilly, L. S., *Pathophysiology Of Heart Disease*, Lea & Febiger, Philadelphia, 1993: *Chapter 2 offers a discussion of heart sounds and murmurs, and Chapter 3 presents a discussion of the roles of plain film radiography, echocardiography, cardiac catheterization, nuclear imaging, CT scanning, and magnetic resonance imaging in the analysis of cardiac disorders.*

Marriott, H. J. L., *Bedside Cardiac Diagnosis*, J. B. Lippincott Co., Philadelphia, 1992: *Chapters 1 through 18 present concise discussions of normal and abnormal findings in the history and physical examination of the cardiovascular system.*

Chesler, E., *Clinical Cardiology*, Springer-Verlag, New York, 1993: *Chapter 2 presents a discussion of the physical assessment of the cardiovascular system.*

CHAPTER **19**

The Mediastinum: Its Divisions and Viscera

The mediastinum is the median region of the thoracic cavity (Fig. 16–1). It is bounded anteriorly by the sternum, posteriorly by the thoracic vertebrae, and, on each side, by mediastinal pleura. The mediastinum is bounded above by the thoracic inlet and below by the diaphragm.

THE DIVISIONS OF THE MEDIASTINUM

The Superior Mediastinum

The superior mediastinum is the part of the mediastinum which lies above the level of the sternal angle (Fig. 16–1). This level in the thorax passes through the manubriosternal joint in the front of the body and the intervertebral disc between the bodies of the fourth and fifth thoracic vertebrae in the back.

The Inferior Mediastinum

The inferior mediastinum is that part of the mediastinum which lies below the level of the sternal angle. The inferior mediastinum is divisible into three regions: the **anterior mediastinum,** which is the region of the inferior mediastinum between the sternum and the pericardial sac, the **middle mediastinum,** which is the mediastinal region consisting of the pericardial sac and its contents, and the **posterior mediastinum,** which is the region of the inferior mediastinum between the pericardial sac and the vertebral column (Fig. 16–1).

THE VISCERA OF THE MEDIASTINUM (EXCLUDING THE HEART)

The Trachea

The trachea begins in the neck as an airway continuous with the lower border of the cricoid cartilage of the larynx (Fig. 17–5). It occupies a midline position in the body as it passes through the thoracic inlet. The trachea descends completely through the superior mediastinum and terminates in the posterior mediastinum. The lower third of the trachea may deviate slightly to the right. The trachea divides at its termination into the left and right main stem bronchi.

The trachea is always patent (open) because of the 16 to 20 **C**-shaped cartilaginous rings distributed along its length (Fig. 19–1). The rings stretch the anterior and lateral walls of the trachea into a permanent archway; the ends of the rings stretch the posterior wall of the trachea into a flat surface.

The trachea moves downward during inspiration and upward during expiration. Upon deep inspiration, the tracheal bifurcation can be displaced inferiorly as far as the level of the body of the seventh thoracic ver-

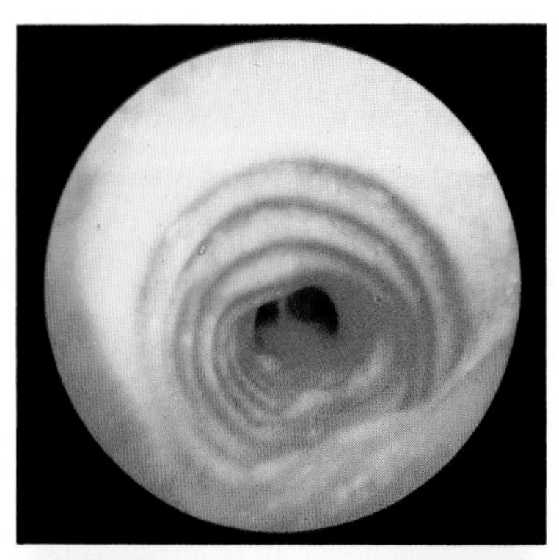

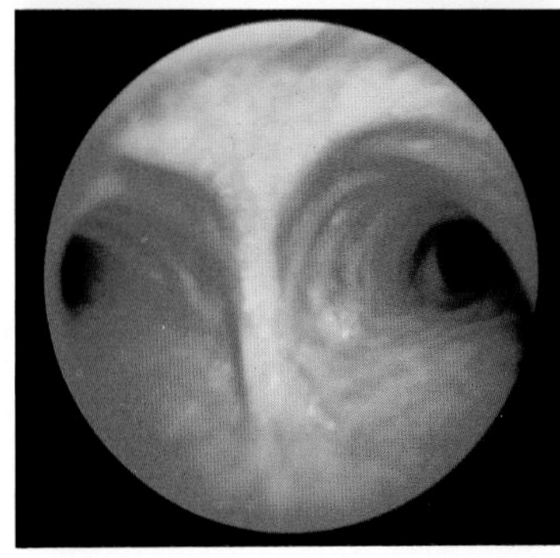

Fig. 19–1: Bronchoscopic views of the trachea (upper photograph) and the carina (lower photograph). The curved ridges in the upper photograph represent the outlines of the inner ridges of the C-shaped cartilages of the trachea.

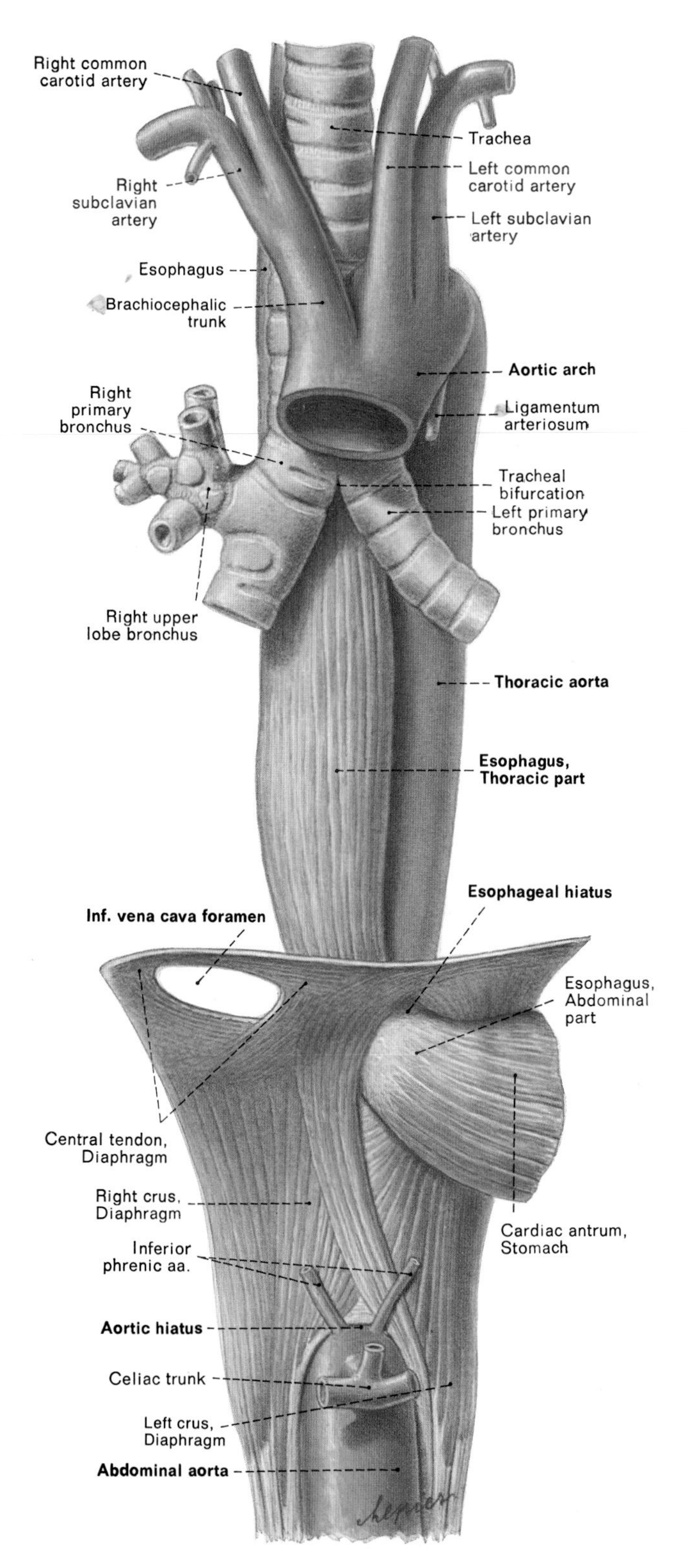

Fig. 19–2: Anterior view of the aorta and lower esophagus at the tracheal bifurcation and the diaphragm.

tebra. At the conclusion of expiration, it can rebound superiorly to just above the level of the sternal angle.

In the superior mediastinum, the trachea lies behind the arch of the aorta (Fig. 19–2). In the posterior mediastinum, the trachea lies behind the pericardal sac, extending inferiorly to the level of the base of the heart. The trachea lies throughout its course in the mediastinum directly in front of and slightly to the right of the esophagus.

The main stem bronchi lie exclusively in the posterior mediastinum as they extend from the tracheal bifurcation to the roots of the lungs. The right main stem bronchus is more vertical, larger, and shorter than the left main stem bronchus (Fig. 17–5). Consequently, if a small object is accidentally aspirated into the trachea when a person is standing or seated upright, it is more likely that the object will pass into the right main stem bronchus than the left one.

A prominent ridge, called the **carina,** marks the internal margin between the origins of the left and right main stem bronchi (Fig. 19–1). This anteroposterior ridge is called the carina because it resembles the keel of a boat when examined with a bronchoscope (carina is a term derived from the Latin word for the keel of a boat).

The space immediately inferior to the tracheal bifurcation is called the **subcarinal space.** The left and right main stem bronchi flank the sides of the subcarinal space. The space contains fat, connective tissue, and a few bronchomediastinal lymph nodes.

The Esophagus

The esophagus begins in the neck as a muscular tube continuous with the laryngopharynx (Fig. 19–3). It descends through the superior and then posterior mediastinum (compare Figs. 16–1 and 19–3), pierces the diaphragm (Fig. 19–2), and finally ends at the cardiac orifice of the stomach. In the superior mediastinum, the esophagus lies directly behind the trachea (Fig. 19–3), directly in front of the upper thoracic vertebrae, and slightly to the left of the midline. In the posterior mediastinum, it lies directly behind the pericardial sac (Fig. 19–3). Whereas the muscularis externa of the cervical segment of the esophagus bears primarily striated muscle fibers, the muscularis externa of the lower segments bear almost exclusively smooth muscle fibers.

When the esophagus is empty, it has a slit-like lumen, and its mucosal lining is gathered in long longitudinal folds. When food boli descend the esophagus, the boli expand the esophageal lumen and stretch out the longitudinal folds of its mucosal lining. The expansions elicit reflexive peristalsis in the lower two-thirds of the esophagus.

Three segments of the esophagus are physically restricted from fully expanding during peristaltic trans-

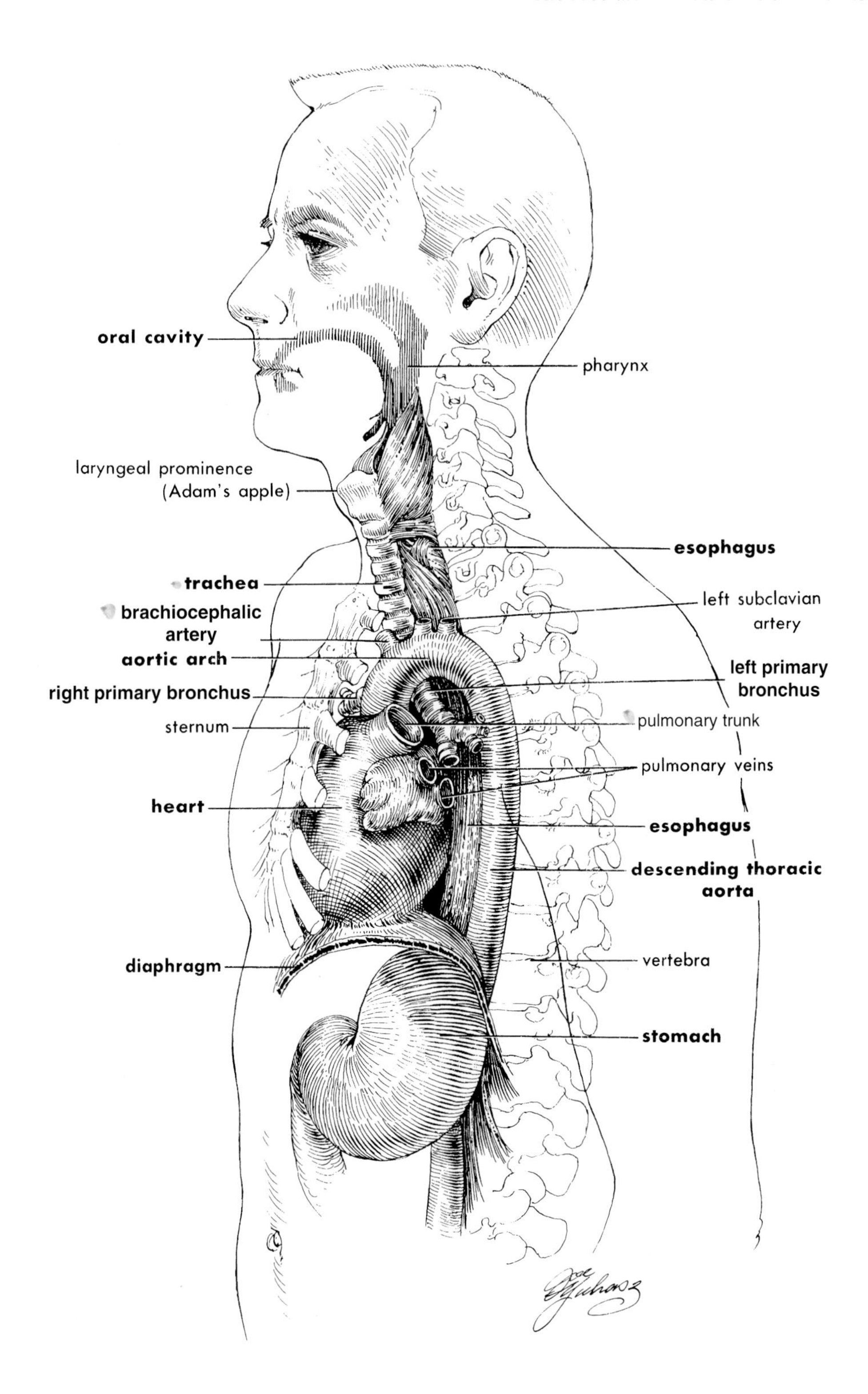

Fig. 19–3: Diagram of a lateral view of the upper body trunk to show relations among the visera of the mediastinum.

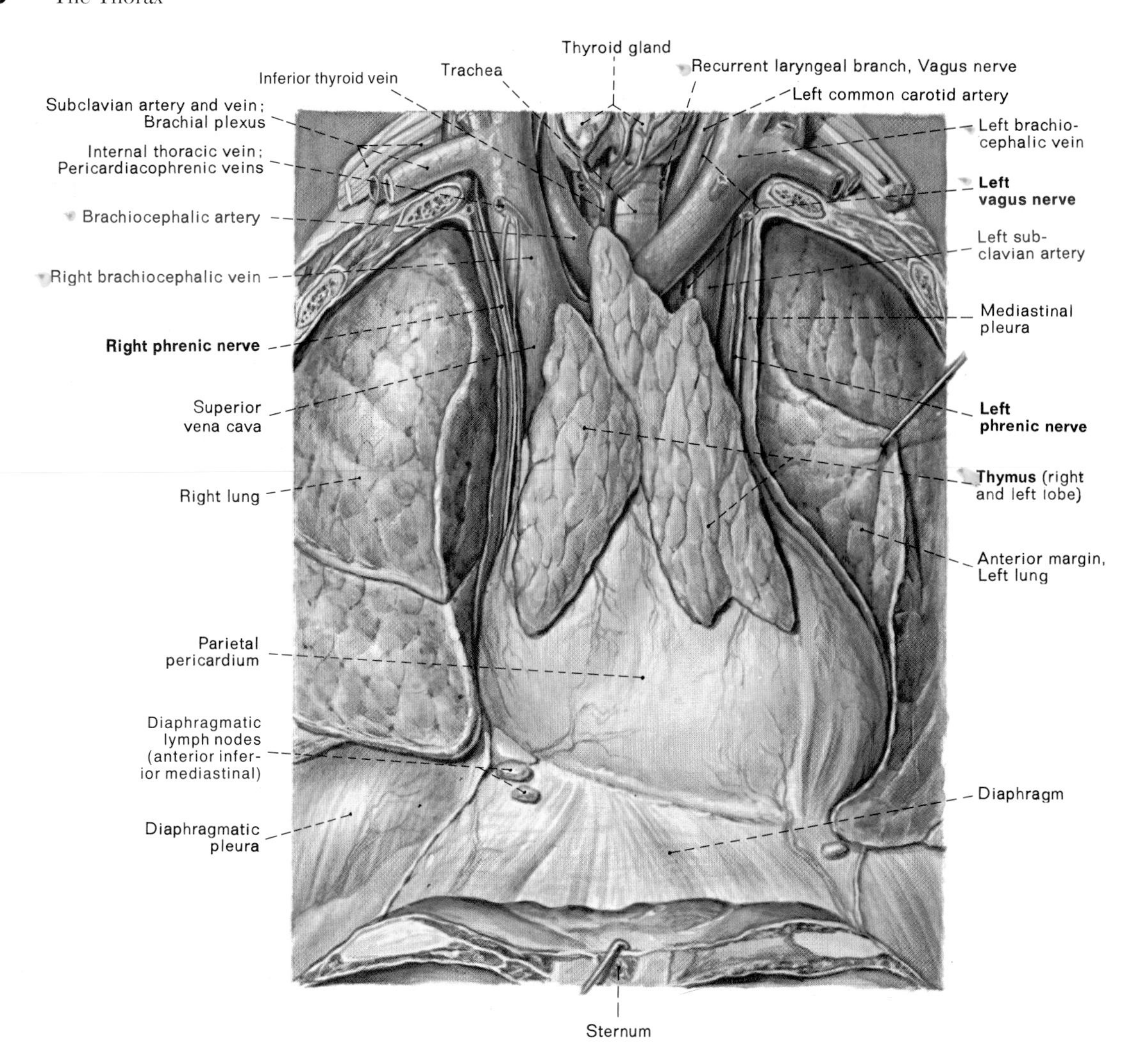

Fig. 19–4: Anterior view of the thymus in the mediastinum of a young boy.

port of food boli. Each of these esophageal segments is physically restricted because it is in contact with a structure which resists displacement by the passage of food boli down the esophagus. The three segments are the most common sites of stricture following swallowing of caustic agents, and also the most common primary sites of esophageal carcinoma. The aortic arch, left main stem bronchus, and muscular border of the esophageal opening of the diaphragm are the structures which respectively restrict the uppermost, middle, and lowest segments.

The upper third of the esophagus is supplied by the inferior thyroid arteries and is drained by the inferior thyroid veins. The middle third of the esophagus is supplied by branches of the descending thoracic aorta and is drained by tributaries of the azygos system of veins. The lower third of the esophagus is supplied by branches of the left gastric artery and is drained by tributaries of the left gastric vein.

The border region between the middle and lower thirds of the esophagus is one of the body regions drained by tributaries of both the portal vein and the inferior or superior vena cava. These regions are known as regions of portal-systemic anastomoses, since some of the veins which drain toward the portal vein form anastomoses with veins draining toward the inferior or superior vena cava (which are the largest systemic veins).

There are numerous pathologic processes which produce portal hypertension (elevated blood pressure in the portal vein). Since all the tributaries of the portal vein are valveless, any increase in portal venous pressure is transmitted retrogradely throughout all of the portal vein's tributaries. When such elevated pressures are transmitted to regions of portal-systemic anastomoses, significant differences in blood pressure are generated along many of the venous anastomoses. These pronounced pressure differences produce dilatations in the walls of the veins, which, in time, may rupture with marked loss of blood. Therefore, it is not uncommon for persons afflicted with long-term portal hypertension to have esophageal varices (pronounced venous dilatations in the border region between the middle and lower thirds of the esophagus). The varices raise tortuous swellings in the esophageal mucosa. Rupture of esophageal varices can result in a fatal loss of blood.

The Thymus

The thymus is a bilobar, lymphoid organ which degenerates by late adolescence. It is mostly located in the anterior part of the superior mediastinum (Fig. 19–4). The inferior poles of its lobes extend into the anterior mediastinum. It is sufficiently prominent in children to cast its own shadow in PA (posteroanterior) chest films.

THE MAJOR BLOOD AND LYMPHATIC VESSELS OF THE MEDIASTINUM

The Aorta

The aorta is divided into three segments as it courses through the mediastinum: the ascending aorta, the aortic arch, and the descending thoracic aorta.

The **ascending aorta** is the segment which is enveloped within the fibrous pericardium, and hence lies in the middle mediastinum (Fig. 18–2). The ascending aorta extends superiorly and to the right from its origin at the base of the left ventricle, and ends at the level of the sternal angle. The ascending aorta has two branches: the left and right coronary arteries.

The **aortic arch (arch of the aorta)** is the segment which lies in the superior mediastinum (Fig. 16–1). The aortic arch begins as a continuation of the ascending aorta (Fig. 18–2), lying directly in front of the right pulmonary artery. As it arches upward over the right pulmonary artery, it is also extending posteriorly and to the left. At its superior limit, it lies immediately above the bifurcation of the pulmonary trunk into the left and right pulmonary arteries (Fig. 18–5). The space here between the curved undersurface of the aortic arch and the upper surface of the T-shaped, pulmonary trunk bifurcation is called the **aortic-pulmonary window.** The aortic-pulmonary window is crossed by a cord of fibrous connective tissue called the ligamentum arteriosum, which is attached above to the curved undersurface of the aortic arch and below to the upper surface of the pulmonary trunk birfurcation (the **ligamentum arteriosum** is the postnatal remnant of the **ductus arteriosus**) (Figs. 18–2 and 18–5). After the aortic arch curves down from its upper limit, it lies directly behind the left pulmonary artery and directly to the left of the trachea and esophagus. The aortic arch ends at the level of the sternal angle.

The aortic arch has three branches: the brachiocephalic artery, the left common carotid artery, and the left subclavian artery (Fig. 18–5). The **brachiocephalic artery** is the first branch of the aortic arch; it extends superiorly and to the right to its ending deep to the right sternoclavicular joint, at which point it bifurcates into the **right common carotid** and **right subclavian arteries.** The **left common carotid** and **left subclavian arteries** are respectively the second and third branches of the aortic arch.

The descending thoracic aorta is the segment which lies in the posterior mediastinum (Fig. 16–1). It ends as it enters the abdomen through the aortic opening of the diaphragm. It lies throughout much of its course directly anterior to the vertebral column, just left of the midline. The descending thoracic aorta has numerous small branches, including eleven pairs of posterior intercostal arteries, one or two bronchial arteries, and esophageal arteries to the middle third of the esophagus.

The Superior Vena Cava and its Major Tributaries

On each side of the body, the **internal jugular vein** and the **subclavian vein** unite in the root of the neck to form a brachiocephalic vein. At this union on the right, it is common for three large lymphatic vessels (the right jugular trunk, the right subclavian trunk, and the right bronchomediastinal trunk) to directly join one of the veins. At this union on the left, it is common for the thoracic duct and the left bronchomediastinal trunk to directly join one of the veins.

From its origin, the right brachiocephalic vein extends almost directly inferior through the superior mediastinum (Fig. 18–2). By contrast, the left brachiocephalic vein assumes a course that extends it both downward and to the right, directly in front of the aortic arch. At the level just above that of the sternal angle, the left and right brachiocephalic veins unite to form the superior vena cava. The superior vena cava descends vertically to its opening into the posterior roof region of the right atrium.

In addition to draining the venous systems of the head, neck, and upper limbs, the superior vena cava also drains the **azygos system of veins** (Fig. 19–5). There are three major veins in the azygos system: the **azygos vein,** the **hemiazygos vein,** and the **accessory hemiazygos vein.**

On the right side of the body, the azygos vein begins in the abdomen as a continuation of the union of the right lumbar veins; on the left side, the union of the left lumbar veins gives rise to the hemiazygos vein. The azygos vein ascends into the thorax by passing through the aortic opening of the diaphragm; the hemiazygos vein ascends into the thorax by passing through the left crus of the diaphragm. The azygos vein generally drains all the right posterior intercostal veins (except for the first one); the hemiazygos vein drains a variable

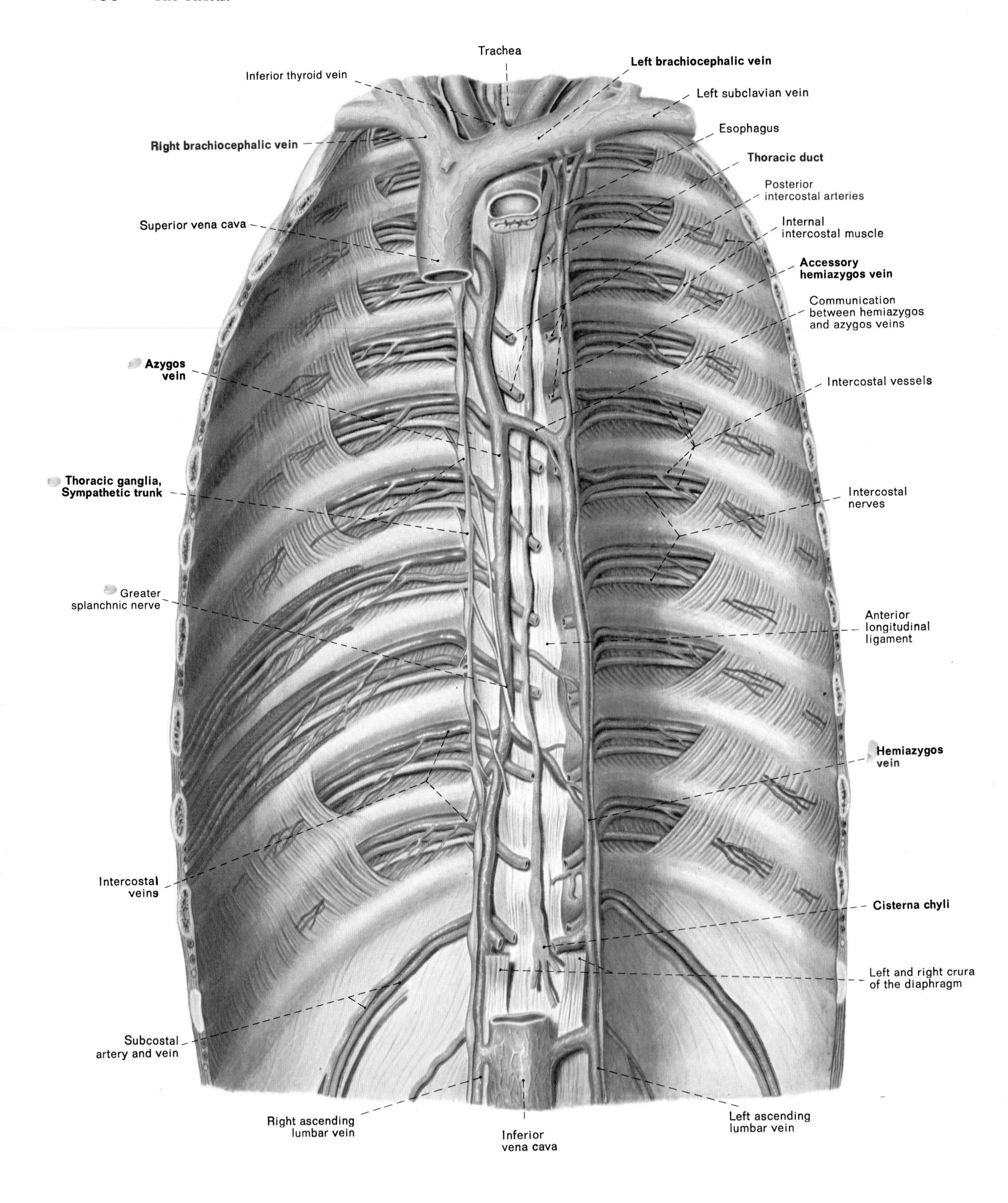

Fig. 19–5: Anterior view of the relationships of the azygos system of veins and the thoracic duct to the posterior chest wall.

number of the lower left posterior intercostal veins before curving to the right, arching over the vertebral column, and finally emptying into the azygos vein. The accessory hemiazygos vein arises from an upper left posterior intercostal vein, and drains a variable number of upper left posterior intercostal veins before it also empties into the azygos vein. The uppermost left posterior intercostal veins generally empty into the left brachiocephalic vein.

The azygos vein terminates by passing behind the root of the right lung and then arching forward over the root to empty into the superior vena cava.

The Inferior Vena Cava

The inferior vena cava enters the thorax by passing through the caval opening of the diaphragm (Fig. 16–13). It then almost immediately opens into the posterior floor region of the right atrium (Fig. 18–8).

The Thoracic Duct

The thoracic duct begins in the abdomen as a continuation of the cisterna chyli (which is the bulbous lymphatic sac formed by the union of the lymphatics draining the intestines) (Fig. 19–5). The thoracic duct enters the thorax by passing through the aortic opening of the diaphragm. As it ascends within the posterior mediastinum, its course takes it from behind and to the right of the esophagus to a position directly to the left of the esophagus at the level of the sternal angle. As it then ascends within the superior mediastinum, it courses along the left side of the esophagus to the root of the neck, where it finally curves laterally and empties into the origin of the left brachiocephalic vein.

THE MAJOR NERVES OF THE MEDIASTINUM

The Phrenic Nerves

The phrenic nerves arise in the neck from the cervical plexus (generally receiving fibers from C3, C4, and C5), descend through the superior and then middle mediastinum, and finally end upon passing through the diaphragm. In their descent through the middle mediastinum, the phrenic nerves course alongside the pericardial sac and pass in front of the roots of the lungs (Figs. 18–2 and 18–3). The right phrenic nerve exits the middle mediastinum through the caval opening of the diaphragm; the left phrenic nerve exits the middle mediastinum by piercing the left dome of the diaphragm (Fig. 16–13). The phrenic nerves provide motor innervation to the diaphragm; they also provide sensory supply to the mediastinal pleura, the diaphragmatic pleura overlying the central tendon of the diaphragm, and the peritoneum underlying the central tendon.

The Vagus Nerves

The vagus nerves are the tenth pair of cranial nerves. The vagus nerves are the cranial nerves with the greatest field of distribution: they innervate structures in the head, neck, thorax, and abdomen. In the thorax, the vagus nerves provide all the parasympathetic innervation of the viscera; specifically, they provide the preganglionic parasympathetic fibers which course through the autonomic plexuses of the thoracic viscera.

In passing from the neck to the abdomen, the vagus nerves descend through the superior and then posterior mediastinum. In their descent through the superior mediastinum, the vagus nerves assume opposite anteroposterior relations with the aortic arch: whereas the right vagus nerve passes behind the beginning of the aortic arch, the left vagus nerve passes in front of the terminal part of the aortic arch (Fig. 18–2 illustrates the descent of the left vagus nerve in front of the terminal part of the aortic arch). As the left vagus nerve descends in front of the terminal part of the aortic arch, it gives rise to its recurrent laryngeal branch, the left recurrent laryngeal nerve (Fig. 18–2). The left recurrent laryngeal nerve curves sharply from its origin, first to the right and then upward; this course extends it first behind the ligamentum arteriosum and then loops it underneath the aortic arch. Upon looping through the aortic-pulmonary window in this fashion, the left recurrent laryngeal nerve ascends into the neck, lying between the trachea and esophagus on the left side. [The right recurrent laryngeal nerve arises from the right vagus nerve in the root of the neck. The paired recurrent laryngeal nerves innervate all of the intrinsic muscles of the larynx except for one pair (the cricothyroids). They also provide sensory innervation to the membrane lining the lumen of the larynx below the vocal folds.]

Each vagus nerve begins its descent through the posterior mediastinum by passing behind the root of the lung. Whereas the right vagus nerve then assumes a position on the posterior surface of the esophagus, the left vagus nerve assumes its position on the anterior surface of the esophagus. From this level downward in the posterior mediastinum, the vagus nerves progressively give rise to a plexus of nerve fibers that envelops the esophagus. It is in this dispersed configuration that the vagus nerves exit the thorax by passing through the esophageal opening of the diaphragm. Just above this opening, nerve fibers from both vagus nerves coalesce on the anterior surface and also the posterior surface of the esophagus to form nerve trunks called, respectively, the **anterior** and **posterior vagal trunks.**

CASE **IV.8**

The Case of Tony Chiancone

INITIAL PRESENTATION AND APPEARANCE OF THE PATIENT

A 67 year-old, white man named Tony Chiancone has made an appointment to seek treatment for chest pain. The patient appears somewhat apprehensive.

QUESTIONS ASKED OF THE PATIENT

What can I do for you? I've been having chest pains for the past 3 months, and lately I've been getting them more frequently.

Do you have chest pain now? No.

How would you describe the chest pains? They're hard to describe. I guess the best way to describe them is that they feel like both a tightness and an ache deep in my chest.

Can you show me or tell me where the chest pains are located? Right here, deep below the breastplate [the patient indicates the location of the chest pains by making a fist with his right hand and placing it over the sternum]. *[When pain is a complaint, it is always essential to have the patient mark by word or gesture the exact site(s) at which pain is perceived.]*

Has the nature of the pain or its location changed during the past 3 months? No. *[The substernal location and quality (a deep tightness and ache) of the patient's chest pains suggest visceral pain of thoracic origin.]*

How often do you have chest pains? Now, maybe once or twice a day. When I first started getting the chest pains about 3 months ago, I got them maybe once every 2 or 3 days. *[The patient's answer suggests that the pathology of the underlying disease or disorder is progressively worsening.]*

Have you recognized if there's anything that seems to cause the chest pains? Yes, physical activity. You see, I'm an avid gardener, and ever since I retired 2 years ago, I've been able to spend a lot of time fixing up the large area behind our house. Well, starting about 3 months ago, I noticed that if I did something strenuous, like hauling a load of mulch in a wheelbarrow, that's when I would get the chest pains. I found that if I broke my work down into tasks that were easier to do physically, I could avoid getting chest pains. However, in the past month, even some of the easier tasks have brought on chest pains.

What happens when you get chest pains? What do you do? Well, they come on pretty fast, and sometimes grip me so hard that I have to stop and sit down. I've found out that if I just stop what I'm doing and sit or lie down for 5 to 10 minutes, the chest pains will go away.

Have you found any other things that bring on the chest pains? Uh, yes, sex. The last time my wife and I made love, I developed chest pains, and I had to stop. *[The patient's description of strenuous physical activity and intense emotion as provocative factors and rest as a palliative factor suggest that the chest pains represent angina pectoris (visceral pain occurring as a result of transient myocardial ischemia). The 5-10 minute duration of the chest pain attacks is also consistent with this hypothesis.]*

Is there anything else that is also bothering you? Well, during the past year, I've noticed that I get winded pretty easily and tire rather quickly when I do physical work. And when I get the chest pains, I've also noticed that I get light-headed. *[This reply identifies two additional important symptoms: exertional dyspnea (difficulty with breathing upon exertion) and exertional presyncope (a feeling of lightheadedness upon exertion). Whereas the exertional dyspnea began about 9 months prior to the onset of chest pains, the exertional presyncope appears to occur only in concert with the onset of chest pains. Dyspnea is an indicator of inadequate ventilation or perfusion of the lungs. Presyncope and syncope are indicators of inadequate perfusion of the brain or insufficient concentrations of oxygen and/or glucose in the blood circulating through the cerebral vasculature.]*

Does the feeling of being light-headed come on when you get the chest pains and then go away when the chest pains stop? Yes.

Have you had any palpitations? No. *[The term palpitations refers to the conscious sensation of an unusual heartbeat, generally one that has increased in frequency and/or strength or is irregular in frequency.[*

Have you ever had any heart or lung problems? No, not that I know of.

Have any members of your family had heart or lung problems? No.

Have you ever had rheumatic fever? No, not that I know of.

Have you had any recent diseases or injuries? No.

Are you taking any medications for previous diseases or injuries? No. *[The examiner poses this question because many medications have the potential of producing cardiac and/or pulmonary disorders.]*

Have you taken any medications for your chest pains? No.

Do you take any over-the-counter medications on a regular basis? Yes, I take aspirin every now and then for arthritis in my hands. *[The examiner finds the patient alert and fully cooperative during the interview.]*

PHYSICAL EXAMINATION OF THE PATIENT

Vital signs: Blood pressure
Lying supine: 130/80 left arm and 125/80 right arm
Standing: 130/80 left arm and 125/75 right arm
Pulse: 75
Rhythm: regular
Temperature: 99.0°F.
Respiratory rate: 17
Height: 5′10″
Weight: 150 lbs.
HEENT Examination: Normal
Lungs: Normal
Cardiovascular Examination: The cardiovascular exam is normal except for the following findings: Palpation finds an accentuated and sustained apical beat (located in the left fifth intercostal space 10 cm from the midline) with the patient seated upright. A systolic thrill is palpable in the suprasternal area and along the common carotid arteries. The carotid pulses are small in amplitude and slow to rise and fall. The radial and ulnar pulses at the wrists and the dorsal pedis and posterior tibial pulses about the ankles are all small in amplitude.

Auscultation at the anterior end of the right second intercostal space reveals diminished sounds emanating from the closure of the aortic valve. Auscultation also reveals a click during early systole and a grade 4/6 crescendo-decrescendo murmur that originates after S1 and terminates before S2. The murmur is heard loudest at the anterior end of the right second intercostal space and radiates to the common carotid arteries. The loudness of the murmur increases with squatting and decreases during the Valsalva maneuver.

Abdomen: Normal
Genitourinary Examination: Normal
Musculoskeletal Examination: Normal
Neurologic Examination: Normal

INITIAL ASSESSMENT OF THE PATIENT'S CONDITION

The patient's chest pains, exertional dyspnea, and exertional presyncope appear to be the result of a cardiac disorder.

ANATOMIC BASIS OF THE PATIENT'S HISTORY AND PHYSICAL EXAMINATION

1. **Transient myocardial ischemia appears to be the basis of the patient's chest pains.** Five features of the chest pains indicate that the pain is anginal in character: the pains (i) have a substernal location, (ii) convey a feeling of tightness deep in the chest, (iii) are provoked by physical or emotional stress, (iv) are palliated by rest, and (v) last no longer than 5-10 minutes when provocative factors are removed.
2. **The exertional dyspnea indicates inadequate ventilation or perfusion of the lungs during periods of physical exercise.**
3. **The exertional presyncope indicates inadequate perfusion of the brain or insufficient concentrations of oxygen in the blood circulating through the cerebral vasculature during periods of physical exercise.**
4. **The accentuated apical beat indicates left ventricular hypertrophy (abnormally thickened walls of the left ventricle).** During each heartbeat, each ventricle attempts to eject during systole all of the blood that it receives during diastole. If an abnormality arises in which one or both ventricles are chronically subjected to an excessive hemodynamic demand, the heart responds to the increased workload by ventricular hypertrophy. One of two mechanisms is commonly the basis for the increased workload: (1) the ventricle is volumetrically over-

loaded with blood during diastole (a mechanism known as diastolic volume overload) or (2) the ventricle needs to generate an excessive intraventricular pressure during systole in order to maintain its output (a mechanism known as systolic pressure overload). Valvular disorders are a frequent cause of both types of overload. Aortic and pulmonary valve regurgitation respectively generate diastolic volume overload in the left and right ventricles. Aortic valve stenosis and systemic hypertension generate systolic pressure overload in the left ventricle, and pulmonary valve stenosis and pulmonary hypertension generate systolic pressure overload in the right ventricle. If a condition which promotes ventricular hypertrophy is not corrected, it ultimately produces ventricular dilatation and dysfunction.

An apical beat of increased intensity by a heart beating within the normal rate limits (60 to 100 beats per minute) suggests a markedly hypertrophic left ventricle. The greater than normal mass of the left ventricle and its increased workload combine to increase the intensity of the apical beat. If the intensified apical beat is sustained to the extent that it approaches the time of S2, it is likely that the cause of the left ventricular hypertrophy has been a systolic pressure overload.

5. **The characteristics of the grade 4/6 crescendo-decrescendo murmur indicate turbulent blood flow in the left ventricle, at the aortic valve, or in the ascending aorta during systole.**

When blood flow into or out from a heart chamber becomes turbulent, the turbulence generates audible sounds that are termed heart murmurs. Many heart murmurs arise at or near one of the heart's valves. Common mechanisms which produce turbulent blood flow at or near a heart valve include

1. blood flow through an abnormally narrowed valve,
2. blood flow through an irregularly structured valve,
3. excessive blood flow through a normal valve,
4. retrograde blood flow through an incompetent or defective valve, and
5. blood flow through the aortic valve into an abnormally dilated ascending aorta or through the pulmonary valve into an abnormally dilated pulmonary trunk.

Heart murmurs can also be produced by the turbulent shunting of blood through a patent ductus arteriosus or defects in the interventricular and interatrial septa. It should be noted that heart murmurs are occasionally encountered in individuals who are in good health and have a normal heart; such murmurs are called innocent murmurs because their presence cannot be associated with any disease or disorder.

The loudness of heart murmurs is graded on a scale of 1 to 6. A grade 1 murmur is barely audible; frankly, only cardiologists can consistently detect grade 1 murmurs, and they generally cannot distinguish such murmurs until they have listened to a few heartbeats after the stethoscope has been applied to the patient's chest. Grade 2 murmurs are faint but distinct murmurs that a cardiologist can detect during the first few heartbeats after applying the stethoscope. A grade 3 murmur is moderately loud; its sounds can be readily distinguished as being additional to those of S1 and S2. A grade 4 murmur is a murmur whose intensity is so great that it generates palpable vibrations in an area of the patient's chest wall; such palpable vibrations are termed a **thrill.** All murmurs of grade 4 and above generate thrills. Grade 5 murmurs are the loudest murmurs whose detection still requires application of a stethoscope to a patient's chest. A grade 6 murmur is a murmur whose intensity is so great that it produces audible sounds that can be heard without the application of a stethoscope. In this case, the designation of the murmur as a grade 4/6 murmur means that the murmur has a grade of 4 on a scale from 1 to 6.

The timing of a murmur during the cardiac cycle is an important distinguishing characteristic. **Systolic murmurs** occur during systole: they begin at or after S1 and end before or at S2. **Diastolic murmurs** occur during diastole: they begin at or after S2 and end before or at S1.

The shape of a murmur is another important distinguishing characteristic. The term shape refers to the shape of a plot of the murmur's intensity (loudness) as a function of time during the cardiac cycle. Most systolic murmurs exhibit one of two distinctive shapes: a crescendo-decrescendo shape or a plateau shape (Fig. 19–6).

A crescendo-decrescendo murmur is a murmur which begins softly, becomes progressively louder, and then becomes progressively softer until it ends (Fig. 19–6A). Crescendo-decrescendo systolic murmurs which begin shortly after S1 and end shortly before S2 are called **midsystolic murmurs** because their intensity is greatest during the midsystolic period. They begin shortly after S1 and end shortly before S2 because they emanate from turbulence that is generated as blood is ejected from the left ventricle through the aortic valve into the ascending aorta or from the right ventricle through the pulmonary valve into the pulmonary trunk during systole. It is for this reason that midsystolic murmurs are also called **ejection murmurs.** There are many anatomic and functional abnormalities that can produce ejection murmurs.

A plateau systolic murmur is a murmur whose intensity is fairly constant from its beginning at S1 to its end at S2 (Fig. 19–6B). Plateau systolic murmurs are called **pansystolic (holosystolic) murmurs** because they extend throughout systole. The most common causes of pansystolic murmurs are mitral regurgitation, tricuspid regurgitation, and ventricular septal defect. In mitral and tricuspid regurgitation, the involved atrioventricular valve is incompetent or defective in its capacity to restrict retrograde blood flow during systole from the contracting, high-pressured ventricle into the relaxed, low-pressured atrium. Ventricular septal defects permit shunting of blood during systole from the high-pressured left ventricle into the lower-pressured right ventricle.

In this case, the grade 4/6 systolic murmur is a midsystolic (ejection) murmur. There are a number of findings which collectively indicate that the murmur is a product of turbulent blood flow that originates within the left ventricle, at the aortic valve, or within the ascending aorta: (i) The murmur is heard loudest over the anterior end of the right second intercostal space and radiates to the common carotid arteries. This intercostal area overlies the ascending aorta, and is the site that generally affords optimal auscultation of the aortic component of S2. (ii) The palpable vibrations coincident with the murmur are most easily palpated in the suprasternal area and along the common carotid arteries. Since the suprasternal area overlies the aortic arch, systolic thrills in this area indicate palpable vibrations in the blood column of the aortic arch. (iii) The rise and fall of blood pressure within the common carotid arteries (which are among the most proximal branches of the thoracic aorta) during systole is abnormal.

6. **The decreased intensity of the sounds emanating upon closure of the aortic valve suggests either decreased compliance of the valve's cusps or a decreased rate of change of the pressure differences across the valve near the end of systole.**

Toward the end of systole, after the ventricles have ejected most of their blood, there begins a retrograde (backward) flow of blood from the aorta and pulmonary trunk into the ventricles (as a consequence of the blood pressure in each arterial trunk being greater than that in the ventricle). The retro-

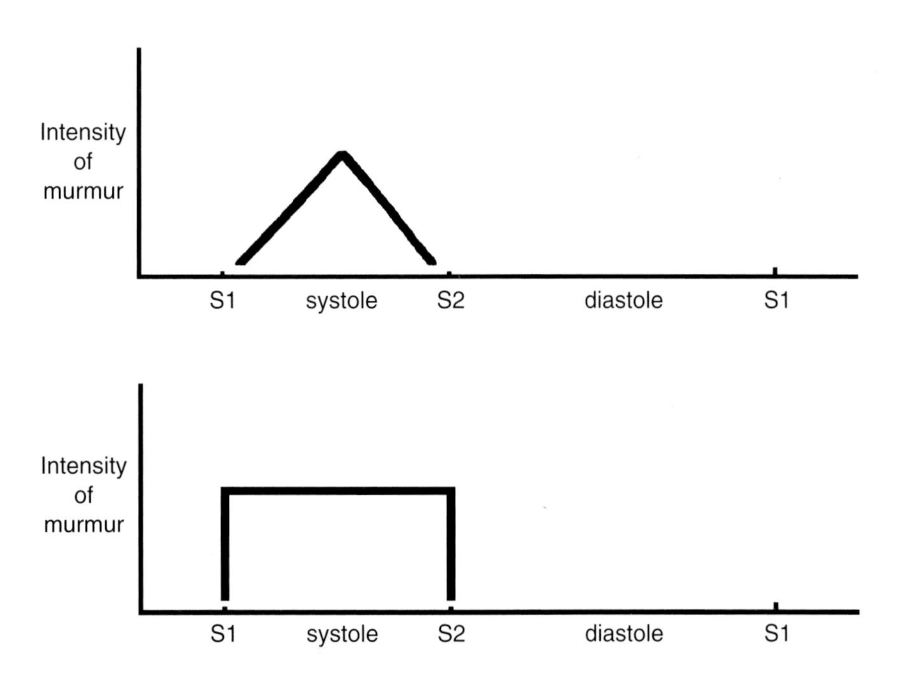

Fig. 19–6. The shape of (A) a crescendo-decrescendo midsystolic murmur and (B) a plateau holosystolic murmur. The shape of a murmur refers a plot of its intensity as a function of time during the cardiac cycle.

grade blood flow in the aorta snaps the semilunar cusps of the aortic valve shut and then stretches the cusps back toward the left ventricle. The retrograde blood flow in the pulmonary trunk almost concurrently snaps the semilunar cusps of the pulmonary valve shut and then stretches its cusps back toward the right ventricle. The closure and elastic distention of the cusps of each valve forces the blood in the corresponding arterial trunk to subsequently recoil back into the arterial trunk. The elastic distention of the cusps of each valve and the subsequent recoil of the blood into the downstream arterial trunk establish vibrations in the blood borne by the arterial trunk and in the wall of the trunk itself. When these vibrations reach the chest wall and set its tissues into audible vibratory motions, the audible vibrations collectively produce S2.

There are thus two major factors that determine the intensity of the vibrations produced upon the closure of either the aortic and pulmonary valve: (1) the *compliance* of the valve's cusps (compliance is a measure of the ease by which the cusps can be elastically stretched) and (2) the *rate of change* of the pressure differences across the valve near the end of systole (which ultimately leads to the blood pressure in the arterial trunk being greater than that in the ventricle). A decrease in the compliance of either valve's cusps (in the absence of any other changes) diminishes the intensity of the vibrations produced after the valve's closure. This is because the less compliant (the more stiff and unyielding) the cusps, the greater the dampening of the vibrations produced in the blood borne by the downstream arterial trunk. Similarly, a decrease in the rate of change of the pressure differences across either valve near the end of systole (in the absence of any other changes) diminishes the intensity of the vibrations produced after the valve's closure. This is because the lesser the rate of change, the lesser the force of the retrograde blood flow that snaps the valve shut and stretches its cusps.

INTERMEDIATE EVALUATION OF THE PATIENT'S CONDITION

The patient appears to be suffering from a chronic disease or disorder that produced an anatomic or functional abnormality of the left ventricle, the aortic valve, and/or the ascending aorta.

CLINICAL REASONING PROCESS

A primary care physician or cardiologist would recognize that the patient's symptoms and physical signs are collectively characteristic of **left ventricular outflow obstruction** (obstruction of blood flow out from the left ventricle during systole). Pre-eminent among the findings are the accentuated and sustained apical beat. An accentuated apical beat indicates left ventricular hypertrophy. The sustainment of the accentuated apical beat during systole suggests that the left ventricular hypertrophy has occurred in response to systolic pressure overload. The most common causes of left ventricular systolic pressure overload are [proceeding from the most proximal to the most distal site]:

a. **subvalvular stenosis** (obstruction of the left ventricular outflow tract below the aortic valve),
b. **aortic stenosis** (narrowing of the aortic valve), and
c. **supravalvular stenosis** (stenosis of the ascending aorta).

Secondly, the patient's report of fatigue, exertional dyspnea, exertional presyncope, and angina represent a constellation of symptoms characteristic of severe left ventricular outflow obstruction.

A primary care physician or cardiologist would also include in his/her considerations a fourth cardiac disorder known as hypertrophic cardiomyophathy. **Hypertrophic cardiomyopathy** is an idiopathic and hereditary cardiac disorder in which the left and/or right ventricle undergoes hypertrophy. A common pattern of hypertrophy is one in which there occurs heterogeneous hypertrophy of the walls of the left ventricle, with the most disproportionate hypertrophy occurring in the upper portion of the interventricular septum. In this common pattern of hypertrophic cardiomyopathy, left ventricular outflow obstruction (and its attendant systolic pressure overload) frequently occur **after** the myocardial hypertrophy. When left ventricular outflow obstruction does occur, it is generally due to anterior motion of the anterior cusp of the mitral valve during systole. The anterior motion of the mitral valve's anterior cusp brings the cusp into apposition with the hypertrophied, upper portion of the interventricular septum, thereby narrowing the subaortic area of the left ventricular outflow tract and producing a dynamic form of left ventricular outflow obstruction. The presenting symptoms and signs of an individual afflicted with this common pattern of hypertrophic cardiomyopathy are occasionally indistinguishable from the symptoms and signs of an individual in which left ventricular hypertrophy has occurred in response to systolic pressure overload.

In this case, the examiner knew that the effects of two dynamic maneuvers (squatting and the Valsalva maneuver) on the intensity (loudness) of the ejection murmur might help identify the patient's disorder. The examiner found that squatting intensified the murmur and that the Valsalva maneuver softened the murmur.

As is now explained, these findings diminish the likelihood that hypertrophic cardiomyopathy is the basis of the patient's cardiac problems.

Squatting from a standing position increases systemic venous return to the right atrium and increases peripheral arterial resistance. These two changes act in concert to increase the end diastolic volume of the left ventricle, which, in turn, increases the volume of blood ejected by the left ventricle during systole (the **stroke volume**). Increase in the stroke volume generally increases the loudness of the murmur of aortic stenosis but generally decreases the loudness of the murmur of hypertrophic cardiomyopathy (the loudness of the murmur of hypertrophic cardiomyopathy decreases because the increased stroke volume decreases the extent of the dynamic outflow obstruction that is typical of this disorder).

The Valsalva maneuver subjects the viscera of the thoracic cavity (in particular, the heart and the lungs) to sustained elevated pressure. [Refer to Chapter 32 for a discussion of the Valsalva maneuver and the role of the vestibular and vocal folds of the larynx in the performance of the maneuver.] Initiation of the manuever prompts a sustained decrease in systemic venous return to the right atrium but a transient increase in pulmonary venous return to the left atrium (this latter change results in a transient increase in the left ventricular stroke volume and the systemic blood pressure). However, after hemodynamic equilibrium is established within 5-10 seconds, the left ventricular stroke volume and the systemic blood pressure decrease as a consequence of the sustained decrease in systemic venous return to the right atrium. This new state of hemodynamic equilibrium persists as long as the elevated intrathoracic pressure is sustained. The decrease in the stroke volume generally decreases the loudness of the murmur of subvalvular, aortic, and supravalvular stenosis but generally increases the loudness of the murmur of hypertrophic cardiomyopathy (the loudness of the murmur of hypertrophic cardiomyopathy increases because the decreased stroke volume increases the extent of the dynamic outflow obstruction that is typical of this disorder).

RADIOLOGIC EVALUATION AND FINAL RESOLUTION OF THE PATIENT'S CONDITION

A lateral chest film revealed curved bands of calcium density inferior to the emergence of the ascending aorta shadow from the heart's shadow. The rapid motion of these curved, radiodense bands on fluoroscopy were judged by the radiologist to indicate that the bands represent areas of calcification of the aortic valve. The lateral chest film also indicated dilatation of the ascending aorta. The results of echocardiography documented left ventricular hypertrophy and suggested the presence of calcified aortic cusps of reduced mobility.

On the basis of the collective physical and radiologic findings, a cardiologist would make a diagnosis of aortic stenosis. This judgment would be based, in part, upon the knowledge that aortic valve calcification and dilatation of the ascending aorta are common findings associated with aortic stenosis in elderly individuals but rarely encountered in cases of subvalvular or supravalvular stenosis.

Aortic stenosis does not generally manifest the symptoms of fatigue, exertional dyspnea, exertional presyncope or syncope, and angina until the effective diameter of the aortic valve has been reduced by approximately two-thirds. Aortic stenosis of such severity cannot be effectively managed medically. In this case, the patient would be advised that surgical replacement of the aortic valve is the most appropriate treatment. Coronary angiography would also be necessary to evaluate for the presence of coronary artery disease.

EXPLANATION OF THE PATIENT'S CONDITION

The patient was born with a congenitally abnormal, bicuspid aortic valve (his aortic valve consists of two instead of three semilunar cusps). [A congenital bicuspid aortic valve, rheumatic heart disease, and degenerative calcification are the most common etiologies of aortic stenosis among adults.] The abnormal configuration of the aortic valve's cusps has produced lifelong turbulent blood flow about the cusps. The continuous exposure of the valve's cusps to abnormal hemodynamic stresses prompted over the years first fibrotic thickening and then calcification of the cusps. The increasing rigidity of the cusps produced in effect a progressive narrowing of the aortic valve opening.

The insidious, progressive narrowing of the aortic valve opening was the basis for the progressive obstruction of the patient's left ventricular outflow tract. As the aortic valve opening progressively narrowed, left ventricular output was maintained through establishment of systolic hypertension of the left ventricular chamber and a systolic pressure gradient between the left ventricle and the aorta. The left intraventricular systolic hypertension imposed an increased stress on the chamber's walls, which was reduced to near-normal levels by progressive, concentric hypertrophy of the left ventricular musculature. [Ventricular wall stress is proportional to intraventricular blood pressure but inversely proportional to wall thickness.]

The progressive left ventricular hypertrophy adequately maintained the patient's cardiac output for several years in the presence of increasing left ventricular outflow obstruction. However, this maintenance was sustained at the expense of decreasing left ventricular compliance. [In this instance, left ventricular compliance refers to the distensibility of the left ventricle, as measured in terms of units of volume increase per unit of intraventricular pressure increase.] The reduced compliance of the left ventricle (that is, its reduced capacity to accommodate end diastolic volumes at low pressure) had decreased about a year ago to the extent that the end diastolic pressures of the left ventricle and left atrium would become significantly increased during periods of physical exercise. During such periods, the pulmonary venous vasculature would retrogradely transmit the increased end diastolic pressures to the pulmonary capillary beds of the lungs. The elevation of blood pressure in the pulmonary capillary beds of the lungs would result in transudation of fluid into pulmonary interstitial spaces and a consequent reduction of pulmonary compliance. [In this instance, pulmonary compliance refers to the expansibility of the lungs, as measured in terms of units of volume increase per unit of increase in intra-respiratory airway pressure.] The reduced compliance of the lungs would increase the work of respiration and thus give the patient a feeling of shortness of breath. It is in this fashion that the reduced left ventricular compliance began to manifest itself about a year ago via the onset of exertional dyspnea.

The recent onset of exertional presyncope is a consequence of a significant decrease in the maximum output of the left ventricle. During periods of physical exercise, there occurs a marked reduction in the vascular resistance of the active muscles. In a healthy individual, this vascular dilatation does not produce a drop in systemic arterial pressure because there occurs a rapid compensatory increase in cardiac output. However, in an individual with severe aortic stenosis, the maximum cardiac output is insufficient to maintain systemic arterial pressure. The exercise-induced drop in systemic arterial pressure results in reduced cerebral perfusion and attendant presyncope.

The recent onset of exercise-induced angina is a consequence of the combined effects of two factors. The first major factor is that the left ventricular musculature has hypertrophied to such an extent that it can no longer be adequately supplied by the coronary arteries during periods of exercise-induced tachycardia. It is important to appreciate that the patient's coronary arteries are only minimally occluded (as a consequence of age) and that they would be able to adequately supply the left ventricular musculature during exercise-induced tachycardia if the musculature were of normal mass. Secondly, the marked thickening and calcification of left and right coronary cusps of the aortic valve have reduced the effective diameters of the origins of the coronary arteries and thus blood flow into these arteries. The combined effects of these two factors accounts for the patient's exertional angina.

The systolic thrill in the suprasternal area are palpable vibrations emanating from the aortic arch during the time frame of the grade 4/6 ejection murmur. The early ejection click is regarded by many cardiologists as an audible marker of the opening of the aortic valve. The presence of an early ejection click in cases of aortic stenosis is a reliable indicator that the ascending aorta is abnormally dilated.

CASE **IV.9**

The Case of Bruce Hicks

INITIAL PRESENTATION AND APPEARANCE OF THE PATIENT

A 25 year-old, black man named Bruce Hicks has made an emergency appointment to seek treatment for chest pain. The patient appears apprehensive.

QUESTIONS ASKED OF THE PATIENT

What can I do for you? I've been having sharp chest pains since yesterday evening. *[The patient's characterization of the pains as sharp in nature indicates somatic pain.]*

Do you have chest pains now? Yes, but not all the time.

Can you show me or tell me where the chest pains are located? Right here, deep in the center of my chest [the patient indicates the location of the chest pains by placing his left index and middle fingertips over the sternum]. *[When pain is a complaint, it is always essential to have the patient mark by word or gesture the exact site(s) at which pain is perceived.]*

Has the nature of the pain or its location changed since yesterday evening? No.

Have you recognized if there's anything that seems to cause the chest pains? Well, I guess the only thing I've found is that lying down consistently makes the pains worse.

Have you found any other things that bring on the chest pains? Yes, a number of things. Sometimes if I move a certain way, take a deep breath, or cough, the pains will get worse. *[The last two answers strengthen the hypothesis that much or all of the patient's chest pains are somatic in nature. Some or all of the pain may be described as pleuritic in nature, since pain intensity is affected by respiratory movements.]*

Have you found anything that relieves the chest pains? Not really. I find that sitting quietly and leaning forward, just like I'm doing right now, seems to cut down the pains the most.

Is there anything else that is also bothering you? Yes, but I don't know if it's important. I have felt feverish and tired for the past 3 days. I just haven't had a lot of energy.

Have you had any palpitations? No. *[The term palpitations refers to the conscious sensation of an unusual heartbeat, generally one that has increased in frequency and/or strength or is irregular in frequency.]*

Do you have trouble catching your breath? No. *[Difficulty with breathing is called dyspnea. Dyspnea is an indicator of inadequate ventilation or perfusion of the lungs.]*

Have you felt light-headed or faint? No. *[Syncope is a feeling of faintness. A less severe form of this symptom, a feeling of light-headedness, may be called presyncope (near syncope). Presyncope and syncope are indicators of inadequate perfusion of the brain or insufficient concentrations of oxygen and/or glucose in the blood circulating through the cerebral vasculature.]*

Have you ever had any heart or lung problems? No, not that I know of.

Have any members of your family had heart or lung problems? No.

Have you had any recent diseases or injuries? No.

Are you taking any medications for previous diseases or injuries? No. *[The examiner poses this question because many medications have the potential for producing cardiac and/or pulmonary disorders.]*

Have you taken any medications for your chest pains? No.

Do you take any over-the-counter medications on a regular basis? No. *[The examiner finds the patient alert and fully cooperative during the interview.]*

PHYSICAL EXAMINATION OF THE PATIENT

Vital signs: Blood pressure
Lying supine: 115/70 left arm and 115/70 right arm
Standing: 115/70 left arm and 115/70 right arm
Pulse: 72

Rhythm: regular
Temperature: 103.6°F.
Respiratory rate: 15
Height: 6′0″
Weight: 150 lbs.

[The only abnormal vital sign is marked elevation of the body temperature.]

HEENT Examination: Normal
Lungs: Normal
Cardiovascular Examination: The cardiovascular examination is normal except for the following findings: Auscultation reveals a scratchy rub in the precordial area (the chest wall region overlying the sternocostal surface of the heart). There are three components to the rub during each cardiac cycle: the first occurs during early diastole, the second at the end of diastole, and the third during systole.
Abdomen: Normal
Genitourinary Examination: Normal
Musculoskeletal Examination: Normal
Neurologic Examination: Normal

INITIAL ASSESSMENT OF THE PATIENT'S CONDITION

The patient is suffering from acute chest pains and is febrile.

ANATOMIC BASIS OF THE PATIENT'S HISTORY AND PHYSICAL EXAMINATION

1. **Much or all of the patient's substernal chest pain appears to be somatic in nature.** The chest pains are sharp in nature and assumption of a supine position, deep inspiration, and the act of coughing all intensify the pains.
2. **The three-component precordial rub is virtually pathognomonic of pericarditis (inflammation of the serous pericardium of the pericardial sac).** A precordial friction rub is a common physical sign of pericarditis. There may be only one or two components to the rub per cardiac cycle. However, auscultation of three components (the first in early diastole, the second at the end of diastole, and the last during systole) are virtually diagnostic of pericarditis. The components of the rub are commonly attributed to the rubbing together of the parietal and visceral parts of the serous pericardium during the cardiac cycle. It is also known, however, that a friction rub is frequently heard in the presence of substantial pericardial effusion (substantive effusion of the pericardial cavity).

CLINICAL REASONING PROCESS

An acute febrile illness antecedent to acute onset of sharp, pleuritic pains suggests infectious inflammation of the parietal pleura of one or both pleural spaces and/or the parietal part of the serous pericardium of the pericardial cavity. In this case, the central, substernal location of the chest pains suggests pericardial involvement.

In cases in which acute pericarditis appears to be the most likely diagnosis, it is important to exclude the possibility of acute myocardial infarction. An acute myocardial infarction (a "heart attack") is an acute necrosis of myocardial tissue resulting from insufficiency of arterial supply; the insufficiency can be the result of an acute thrombosis, coronary spasm, or systemic hypotension. The pain of acute myocardial infarction is frequently anginal in location and quality (a deep, retrosternal ache) but generally of greater severity and longer duration than typical anginal pain. Electrocardiography and measurement of cardiac enzyme levels in the blood provide important information regarding the possible presence of an acute myocardial infarction.

An initial concern in the treatment of acute pericarditis is to explore the possiblity of a bacterial infection using blood cultures and a complete blood count. If the evidence points to idiopathic or viral acute pericarditis, treatment typically involves administration of analgesics and anti-inflammatory medication and frequent observation for the development of pericardial effusion (effusion of excess fluid in the pericardial cavity). Progressive pericardial effusion can result in cardiac tamponade, the life-threatening condition in which intrapericardial cavity pressure increases to the extent that it severely restricts venous return to both sides of the heart.

RADIOLOGIC EVALUATION AND FINAL RESOLUTION OF THE PATIENT'S CONDITION

Echocardiography did not show any evidence of pericardial effusion or cardiac abnormality. On the basis of the findings from the history, physical exam, and echocardiographic exam, a primary or emergency care physican would make a diagnosis of acute pericarditis.

EXPLANATION OF THE PATIENT'S CONDITION

The patient is suffering from a viral infection of 3 days duration that has inflamed the visceral and parietal parts of the serous pericardium.

CASE **IV.10**
The Case of Darrell Miller

INITIAL PRESENTATION AND APPEARANCE OF THE PATIENT

A 39 year-old, black man named Darrell Miller has made an appointment to seek treatment for chest pains. The patient appears healthy and relaxed but overweight.

QUESTIONS ASKED OF THE PATIENT

What can I do for you? I've been having bad chest pains for the past 3 months.

Do you have chest pains now? No.

Where do you feel the chest pains? Right here, deep in the center of my chest [the patient indicates the location of the chest pains by moving the fingertips of his right hand up and down over the lower part of the sternum]. *[When pain is a complaint, it is always essential to have the patient mark by word or gesture the exact site(s) at which pain is perceived.]*

How would you describe the chest pains? It feels like an intense burning sensation, sometimes going up to my throat. *[The patient's description of the chest pains suggests that they present as visceral pain that occasionally radiates superiorly to the pharynx.]*

Has the nature of the chest pains or their location changed during the past few months? No, not really.

Have you recognized if there's anything that seems to cause the chest pains? No, I can't say that I have.

Have you recognized if there's any time during the day when you are more likely to have the chest pains? Yes. Most of the attacks occur in the evening after dinner or later when I go to bed.

When you get the chest pains in the evening, is there anything that seems to worsen the pains? Yes. I've noticed that if I try to do some work around the house, the pains become worse. I've found that activities like lifting something heavy or bending over to pick up something off the floor will make the pains worse.

Have you found anything that relieves the pains? Well, when I get the chest pains in bed, the pains build up to the point that I have to sit up in bed or get up and walk around. Sitting or standing up does seem to help a bit. I have also found out that if I take an antacid and stay up for a while watching TV or reading the newspaper, then the pains generally go away and I can return to bed to sleep.

Do you ever get chest pains after other meals, such as breakfast or lunch? Sometimes, but then it's usually after a big meal. I'm a foreman at construction work sites, and I generally don't have the time to have a leisurely breakfast or lunch. So the big meal of the day for me is usually dinner.

Do you ever get chest pains if you're doing some heavy work before a meal? Rarely.

Is there anything else that is also bothering you? No.

Have you had any problems swallowing food or liquids? No. *[Difficulty with swallowing is called dysphagia.]*

Have you had any nausea or vomiting? No.

Has there been any change in your bowel habits or in the appearance of your stool? No.

Have you ever had any problems with your digestive system? No.

Have any members of your family ever had an ulcer or problems with their digestive system? No.

Have you had any trouble with shortness of breath? No.

Have you ever had any heart or lung problems? No, not that I know of.

Have any members of your family had heart or lung problems? No.

Do you smoke? Yes, about a pack of cigarettes a day.

How long have you been smoking about a pack a day? Oh, about 10 years.

Do you drink alcoholic beverages? No.

Have you had any recent diseases or injuries? No.

Are you taking any medications for previous diseases or injuries? No.

Do you take any over-the-counter medications on a regular basis other than antacids? Yes, I take aspirin for headaches. *[The examiner finds the patient alert and fully cooperative during the interview.]*

PHYSICAL EXAMINATION OF THE PATIENT

Vital signs: Blood pressure
Lying supine: 135/80 left arm and 135/80 right arm
Standing: 135/80 left arm and 135/80 right arm
Pulse: 76
Rhythm: regular
Temperature: 98.3°F.
Respiratory rate: 18
Height: 5′10″
Weight: 235 lbs.
HEENT Examination: Normal
Lungs: Normal
Cardiovascular Examination: Normal
Abdomen: Normal
Genitourinary Examination: Normal
Musculoskeletal Examination: Normal
Neurologic Examination: Normal

INITIAL ASSESSMENT OF THE PATIENT'S CONDITION

The patient suffers from nocturnal chest pains of 3 months' duration.

ANATOMIC BASIS OF THE PATIENT'S HISTORY AND PHYSICAL EXAMINATION

1. **The patient's low substernal pain appears to be visceral pain of thoracic and/or abdominal origin.** The diffuse burning quality of the patient's chest pains is one of the more common presentations of thoracic visceral pain. The low substernal location of the visceral pain suggests disease of thoracic and/or upper abdominal viscera.
2. **Relief of pain by taking an antacid suggests that abnormal exposure of one or more segments of the digestive tract to gastric chyme is the basis for the nocturnal chest pains.** The digestive tract segments immediately proximal and distal to the stomach (respectively, the esophagus and duodenum) are subject to abnormal exposure to gastric contents. Antacids frequently relieve the pain of inflammation of esophageal or duodenal mucosa that has been abnormally exposed to gastric contents.

INTERMEDIATE EVALUATION OF THE PATIENT'S CONDITION

The patient appears to be suffering from a disease or disorder that involves abnormal exposure of the upper digestive tract to gastric chyme.

CLINICAL REASONING PROCESS AND FINAL RESOLUTION OF THE PATIENT'S CONDITION

A primary care physician or gastroenterologist would recognize the patient's history as a typical record of the symptoms of mild reflux esophagitis. Reflux esophagitis is the condition in which esophageal inflammation occurs as a consequence of regurgitation of gastric contents into the esophagus. The pain and discomfort of reflux esophagitis is frequently indistinguishable from the pain and discomfort of cardiac disease. This is because most of the visceral pain fibers of the esophagus enter the spinal cord at the same spinal cord segments (primarily T1 through T4) that receive most of the visceral pain fibers of the heart.

The circumferentially-oriented smooth muscle fibers in the lower end of the wall of the esophagus form a physiologic sphincter that normally minimizes reflux of gastric contents into the esophagus. Esophageal reflux is minimized during periods of increased intra-abdominal pressure by a reflex pathway that increases the tone of the physiologic sphincter. Most patients suffering from reflex esophagitis are afflicted with some aspect of neuromuscular dysfunction of the physiologic sphincter. [The difference between an anatomic and a physiologic sphincter is as follows: All sphincters consist of a collection of muscle fibers circumferentially-oriented around a visceral tube; at appropriate times, the muscle fibers concentrically contract either to constrict the lumen of the visceral tube or to control its size. An anatomic sphincter is a sphincter in which the collection of the circumferentially-oriented muscle fibers is so distinct and separate from adjacent tissues that it can be identified visually at the microscopic or gross level of resolution. The pylorus of the stomach and the sphincter urethrae around the urethra are examples of anatomic sphincters. A physiologic sphincter is a sphincter in which the collection of the circumferentially-oriented muscle fibers cannot be identified visually but can be demonstrated physiologically.]

The common presentation of reflux esophagitis is substernal pain that is worse after meals, aggravated by recumbency or any activity that increases intra-abdominal pressure, and relieved by antacids. Recumbency brings the gastric contents into contact with the lower end of the esophagus. Activities that increase intra-abdominal pressure (such as bending over forward or lifting heavy objects) increase pressure within the lumen of stomach, which, in turn, exerts greater pressure on the lumen at the lower end of the esophagus. In these situations, the incompetent physiologic sphincter is deficient in its capacity to prevent reflux of the gastric contents. [The act of lifting heavy objects requires performance of the Valsalva maneuver. This dynamic maneuver involves, in part, measures which increase intra-abdominal pressure. Refer to Chapter 32 for a discussion of the Valsalva maneuver and the roles of the vestibular and vocal folds of the larynx, the diaphragm, and the muscles of the anterolateral abdominal wall in the performance of the maneuver.]

In this case, which involves a patient with mild symptoms common of reflux esophagitis, a primary care physician or gastroenterologist would recommend dietary modifications [particularly lowering caloric intake to reduce the patient's weight, lowering fat intake (because dietary fat delays gastric emptying and decreases the pressure exerted by the physiologic sphincter at the lower end of the esophagus), decreasing the size of the patient's dinner meals, increasing (if possible) the time between dinner and retirement to bed, and avoiding coffee, citrus juices, and tomato products (as these latter food substances exacerbate esophageal inflammation)], cessation of smoking, sleeping with the head and trunk of the body elevated, and a prescribed regimen of antacids and histamine receptor blocking agents.

Patients suffering from reflux esophagitis may present with more atypical symptoms or symptoms suggestive of disease of the stomach and/or duodenum, angina, or even acute myocardial infarction. In these cases, other investigative measures, such upper GI contrast radiographic studies, upper GI endoscopy, or electrocardiography, are required for diagnosis.

SUGGESTED REFERENCES FOR ADDITIONAL INFORMATION ON THE MEDIASTINUM

Slaby, F., and E. R. Jacobs, *Radiographic Anatomy*, Harwal Publishing Co., Media, PA, 1990: *Pages 107 through 114 in Chapter 3 describe the principal relations among mediastinal viscera in CT scans of the superior mediastinum and the upper part of the inferior mediastinum.*

Sutton, D., A *Textbook Of Radiology And Imaging*, Churchill Livingstone, Edinburgh, 1993: *Chapter 12 provides a discussion of the radiologic analysis of mediastinal lesions.*

SECTION **V**

The Abdomen, Pelvis, and Perineum

The abdomen is the region in the trunk of the body that lies between the thorax and pelvis. The abdomen has an inner **abdominal cavity** (Fig. 16–1) surrounded by outer **abdominal walls.** The abdominal cavity houses the adrenal glands, kidneys, liver, gallbladder, pancreas, spleen, stomach, small intestine, and most of the large intestine.

The pelvis is the region in the trunk of the body that lies between the abdomen and perineum. The pelvis has as inner **pelvic cavity** (Fig. 16–1) surrounded by outer **pelvic walls.** The pelvic cavity houses the terminal segments of the large intestine, the urinary bladder, and reproductive viscera in both sexes.

The abdominal and pelvic cavities are continuous with each other, and the cavity that they form together is called the **abdominopelvic cavity.** The **diaphragm,** which serves as the floor of the thoracic cavity, also serves as the roof of the abdominopelvic cavity (Fig. 16–1). The floor of the abdominopelvic cavity is called the **pelvic diaphragm** (Fig. 16–1), and it consists primarily of the **left and right levator ani muscles.**

The perineum is the most inferior region in the trunk of the body (Fig. 16–1). The perineum in both sexes contains the external genitalia and the anal canal (the external genitalia are the external sex organs, and the anal canal is the terminal segment of the digestive tract).

CHAPTER **20**

The Anterolateral Abdominal Wall

The outer wall of soft tissue layers in the anterior and flank regions of the abdomen is called the anterolateral abdominal wall. The anterolateral abdominal wall consists of five major soft tissue layers, which, proceeding from the most superficial to the deepest, are:

(1) the **skin,**
(2) the **subcutaneous fascia,**
(3) a **layer of overlapping muscles and their sheet-like tendons,**
(4) a layer of fascia called the **transversalis fascia,** and
(5) a serous membrane called the **peritoneum.**

Above the level of the umbilicus, the subcutaneous fascia consists only of a single layer of connective tissue bearing a variable amount of fat. Below the level of the umbilicus, however, the subcutaneous fascia is divided into two layers, a fatty superficial layer known as **Camper's fascia** and a membranous deep layer known as **Scarpa's fascia.** The peritoneum is the deepest tissue layer of all the abdominal and pelvic walls.

THE BORDERS OF THE ANTEROLATERAL ABDOMINAL WALL

The **xiphoid process** and the **costal margins of the rib cage** mark the upper border of the anterolateral abdominal wall (Fig. 12–5). The **iliac crests, anterior superior iliac spines, pubic tubercles,** and **pubic crests of the coxal bones** together with the **inguinal ligaments** and **pubic symphysis** mark the lower border of the anterolateral abdominal wall (Fig. 12–5). The **anterior axillary lines** approximate the lateral borders of the anterolateral abdominal wall; the anterior axillary line is the vertical line on each side that passes through the anterior axillary fold.

THE MUSCLES OF THE ANTEROLATERAL ABDOMINAL WALL

The **left and right rectus abdominis muscles** comprise most of the muscle mass in the median region of the anterolateral abdominal wall (Fig. 20–1C). On each side, the lengthy, straplike muscle extends vertically from its origin on the coxal bone to its insertion onto the rib cage.

A tendinous sheath called the **rectus sheath** envelops each rectus abdominis. The lateral margin of each rectus sheath is called its **linea semilunaris** (Fig. 20–1A). The medial margins of the left and right rectus sheaths are united along the midline to form a connective tissue seam called the **linea alba.** Whereas the anterior part of each rectus sheath completely covers the anterior surface of rectus abdominis (Fig. 20–1A), the posterior part of the sheath underlies only the upper three-quarters of the muscle, thereby leaving a lower, unattached border to the posterior part of the sheath, called the **arcuate line** (Fig. 20–1D).

The **pyramidalis muscle** is inconstantly found in the rectus sheath; when present, it lies anterior to the most inferior part of rectus abdominis (Fig. 26–8) and extends from the pubic crest of the coxal bone to the linea alba.

Three overlapping muscles **(external oblique, internal oblique,** and **transverse abdominis)** form the muscle mass in each flank region of the anterolateral abdominal wall (Fig. 20–1, A, B, and C). Each of the three muscles originates as a muscular belly that extends medially within the anterolateral abdominal wall; just lateral to the lateral border of rectus abdominis, the muscular belly gives rise to a sheet-like tendon of insertion called the **aponeurosis** of the muscle. The aponeuroses of the three muscles fuse to form the rectus sheath on each side. Below the level of the arcuate line, the aponeuroses of all three muscles form the anterior part of the rectus sheath (Fig. 20–2B). Above the level of the arcuate line, external oblique's aponeurosis and a superficial layer of internal oblique's aponeurosis form the anterior part of the rectus sheath, and a deep layer of internal oblique's aponeurosis and transversus abdominis's aponeurosis form the posterior part of the sheath (Fig. 20–2A).

Actions of the Anterolateral Abdominal Wall Musculature

Rectus abdominis, external oblique, internal oblique, and transversus abdominis can flex the lumbar part of the vertebral column. They can also aid respiration, micturition, defecation, and parturition through their capacity to compress the abdominopelvic cavity. Their muscle tone decreases during inspiration and increases during expiration, micturition, defecation, and parturition. Pyramidalis chiefly acts to exert tension on the linea alba.

Innervation of the Anterolateral Abdominal Wall Musculature

Rectus abdominis and external oblique are each innervated by the anterior rami of spinal nerves T7 through T12. Internal oblique and transversus abdominis are each innervated by the anterior rami of T7 through T12 and L1. Pyramidalis is innervated by the anterior ramus of T12.

Reflex Tests of the Anterolateral Abdominal Wall Musculature

The collective activities of rectus abdominis, external oblique, internal oblique, and transversus abdominis may be assessed by a phasic stretch reflex test. The test is conducted in three parts; each part begins with the patient lying supine on an examination table. In the first part of the test, the examiner uses a reflex hammer to impart a gentle but firm and quick tap on the anterolateral abdominal wall at a point along the midclavicular line and immediately below the costal margin. In

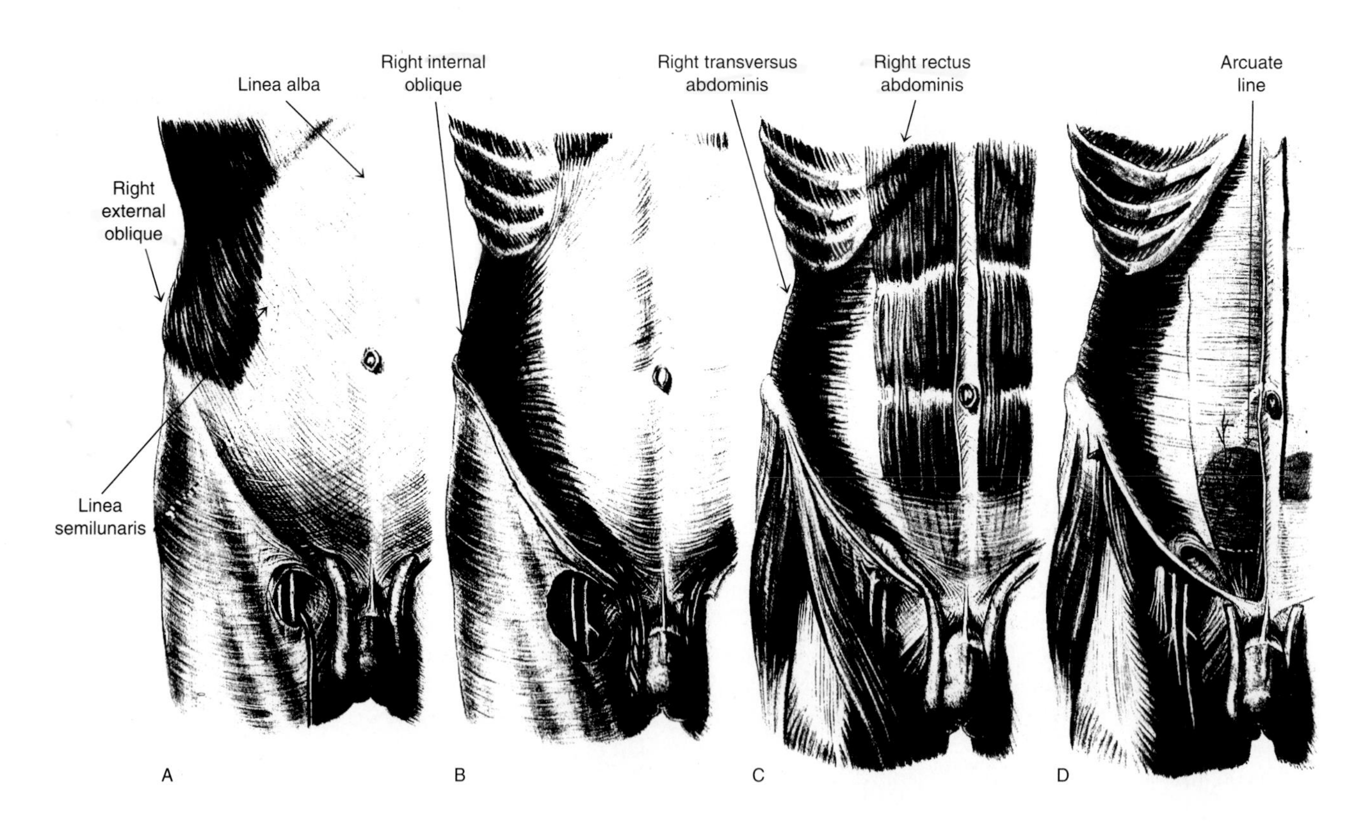

Fig. 20–1: Anterior views of the anterolateral abdominal wall upon removal of first (A) the skin and subcutaneous fascia, then (B) the external oblique and its aponeurosis, (C) the internal oblique and its aponeurosis, and finally (D) the rectus abdominis.

the second part of the test, the examiner uses the reflex hammer to impart a gentle but firm and quick tap on the anterolateral abdominal wall at a point immediately lateral to the umbilicus. In the third part of the test, the examiner uses the reflex hammer to impart a gentle but firm and quick tap on the anterolateral abdominal wall at a point along the midclavicular line and immediately above the inguinal ligament. The hammer taps suddenly stretch the underlying muscles and/or their tendons, and this should elicit a phasic contraction of the muscle fibers that produces a quick retraction of the underlying anterolateral abdominal wall region. The first, second, and third parts of the test, respectively, assess spinal cord reflex activity at the T7-T9, T9-T11, and T11-L1 levels of the spinal cord.

The collective activities of rectus abdominis, external oblique, internal oblique, and transversus abdominis may also be assessed by a superficial reflex test. A **superficial reflex test** is a spinal cord reflex test in which stimulation of one or more cutaneous sensory nerves elicits a reflexive contraction of one or more muscles. The superficial reflex test for the anterolateral abdominal wall muscles is conducted in three parts; each part begins with the patient lying supine on an examination table. In the first part of the test, the examiner gently draws a blunt wooden blade medially across the anterolateral abdominal wall along a course parallel to and about one inch below the costal margin. In the second part of the test, the examiner gently draws the blunt wooden blade medially across the anterolateral abdominal wall at the level of the umbilicus. In the third part of the test, the examiner gently draws the blunt wooden blade medially across the anterolateral abdominal wall along a course parallel to and about one inch above the inguinal ligament. The cutaneous stimulation should elicit a quick retraction of the underlying anterolateral abdominal wall region. The first, second, and third parts of the test, respectively, assess spinal cord reflex activity at the T7–T9, T9–T11, and T11–L1 levels of the spinal cord.

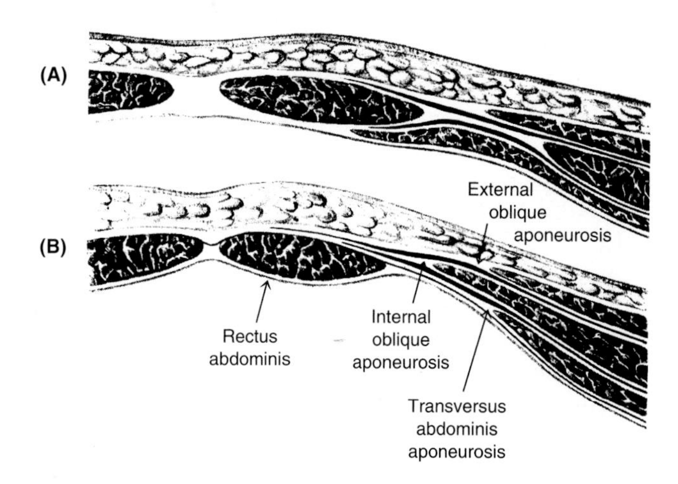

Fig. 20–2: Horizontal cross-sectional views of the the rectus sheath (A) above the arcuate line and (B) below the arcuate line.

THE MAJOR NEUROVASCULAR ELEMENTS OF THE ANTEROLATERAL ABDOMINAL WALL

On each side, the **anterior rami of spinal nerves T7 through T12** and **L1** course slightly downward as they extend anteriorly from the lateral border of the anterolateral abdominal wall toward the linea alba. These nerves provide the sensory supply of the anterolateral abdominal wall's skin and peritoneal lining. L1 provides its sensory supply through two major terminal branches: the **iliohypogastric nerve** and the (more inferior) **ilioinguinal nerve.** The skin over the xiphoid process is part of the T7 dermatome on the body surface, the skin around the umbilicus is part of the T10 dermatome, and the skin immediately above the pubic crests is part of the L1 dermatome (Fig. 4–12C).

The chief arteries of the anterolateral abdominal wall are the **lumbar branches of the abdominal aorta,** the **superior** and **inferior epigastric arteries,** and the **deep circumflex iliac artery.** The superior epigastric artery, which is one of the two terminal branches of the internal thoracic artery, enters the rectus sheath at the upper border of the posterior part of the sheath (Fig. 16–8). The inferior epigastric artery, which is a branch of the external iliac artery, enters the rectus sheath at the arcuate line (the lower border of the posterior part of the sheath). Each epigastric artery extends vertically within the rectus sheath between rectus abdominis and the posterior part of the sheath, finally anastomosing with each other in the umbilical region (Fig. 16–8). The deep circumflex iliac artery is a branch of the external iliac artery.

The major veins draining the anterolateral abdominal wall are the venae comitantes of its chief arteries. The subcutaneous fascia bears the **lateral thoracic vein,** a tributary of the axillary vein, and the **superficial epigastric vein,** a tributary of the femoral vein. The lateral thoracic and superficial epigastric veins anastomose at about the level of the umbilicus.

The para-umbilical region of the anterolateral abdominal wall has several relatively small veins called the **para-umbilical veins,** so named because they drain the region around the umbilicus and extend from the perimeter of this region towards the umbilicus.

Blood conducted by the para-umbilical veins is ultimately drained by the portal vein. In instances of severe portal hypertension, the para-umbilical veins may become markedly distended, forming a radiating pattern of varicose veins about the umbilicus. In remembrance of Medusa, a Greek mythological character whose head was covered with a multitude of snakes, the pattern of varicose, para-umbilical veins is termed a **caput medusae** (the head of Medusa).

Lymphatic drainage from the deeper tissues of the anterolateral abdominal wall proceeds to the external iliac, common iliac, and lumbar lymph nodes. Lymphatic drainage from the skin and subcutaneous fascia of the anterolateral abdominal wall above the level of the umbilicus proceeds to the anterior group of axillary lymph nodes; below the umbilicus, drainage proceeds to the horizontal group of superficial inguinal lymph nodes.

THE INGUINAL CANAL

The **inguinal region** is the anterolateral abdominal wall region that immediately borders the anterior thigh. The **inguinal canal** is a passageway in the inguinal region through which the **spermatic cord in the male** and the **round ligament of the uterus in the female** traverse the abdominal wall.

The Development of the Inguinal Canal: The Events Common to Both Sexes

The formation of the inguinal canal spans a time period in the fetal life of both sexes extending from the fifth to the thirty-second week. The development which occurs up to the twenty-eighth week is very similar in both sexes.

The formation of the gonads (the testes in the male and the ovaries in the female) begins in both sexes during the fifth week of fetal development. Each gonad begins its development in the lower region of the posterior abdominal wall, in a layer of connective tissue sandwiched between the peritoneum and the deep fascia (Fig. 20–3A). Concurring with the development of the gonad is the formation of a cord of connective tissue (the **gubernaculum**) which extends from the gonad to a swelling in the genital area destined to become the scrotum in a male and the labium majus in a female. The gubernaculum lies throughout its length superficial to the peritoneum as it extends downward from the gonad and then curves forward to pass into the anterolateral abdominal wall, after which it continues into the genital swelling. The course of the gubernaculum through the anterolateral abdominal wall marks the site of the future inguinal canal. Moreover, in a male, when the testis descends from its original

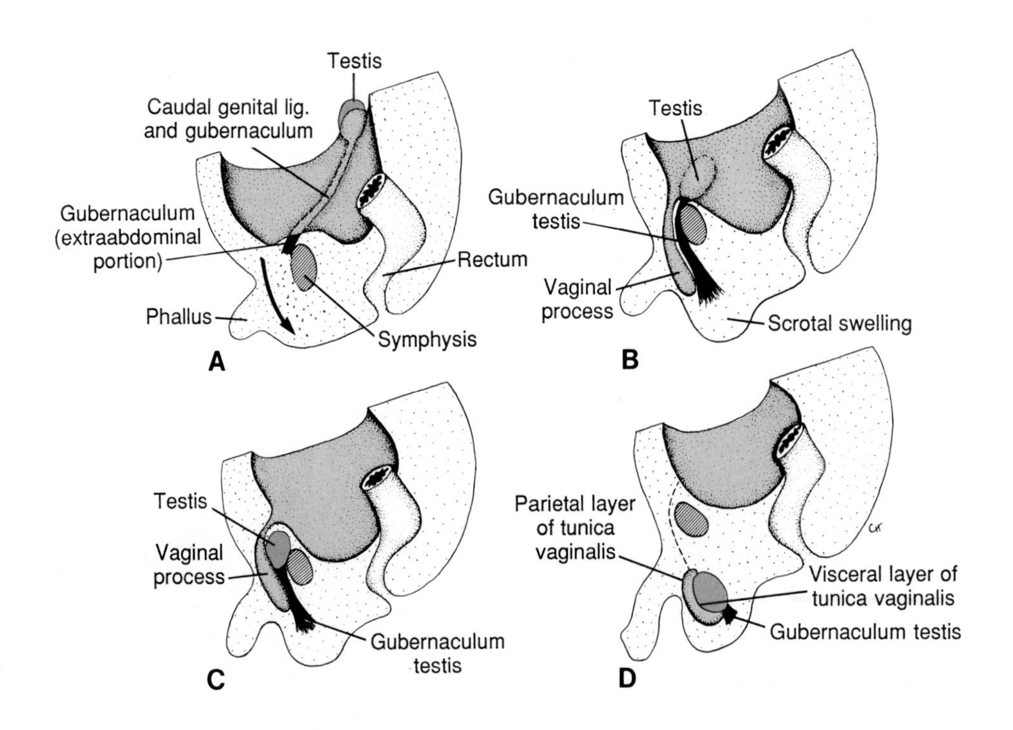

Fig. 20–3: Schematic diagrams of the descent of the testis during (A) the second month of gestation, (B) the middle of the third month, (C) the seventh month, and (D) shortly after birth.

position in the posterior abdominal wall to its final position in the scrotum, the testis's passage follows the course of the gubernaculum. The specific role that the gubernaculum plays in the descent of the testis, however, is unknown.

The formation of the inguinal canal begins with the forward protrusion of a peritoneal evagination called the **processus vaginalis** (**vaginal process**) (Fig. 20–3B). As the processus vaginalis pushes forward into the transversalis fascia, it acquires a tubular fascial covering called the **internal spermatic fascia.** The rounded margin where the internal spermatic fascia is continuous with the transversalis fascia is called the **deep inguinal ring.** The deep inguinal ring is the lateral, or deep, end of the inguinal canal, and it lies midway between the anterior superior iliac spine and the pubic symphysis, immediately lateral to the origin of the inferior epigastric artery from the external iliac artery.

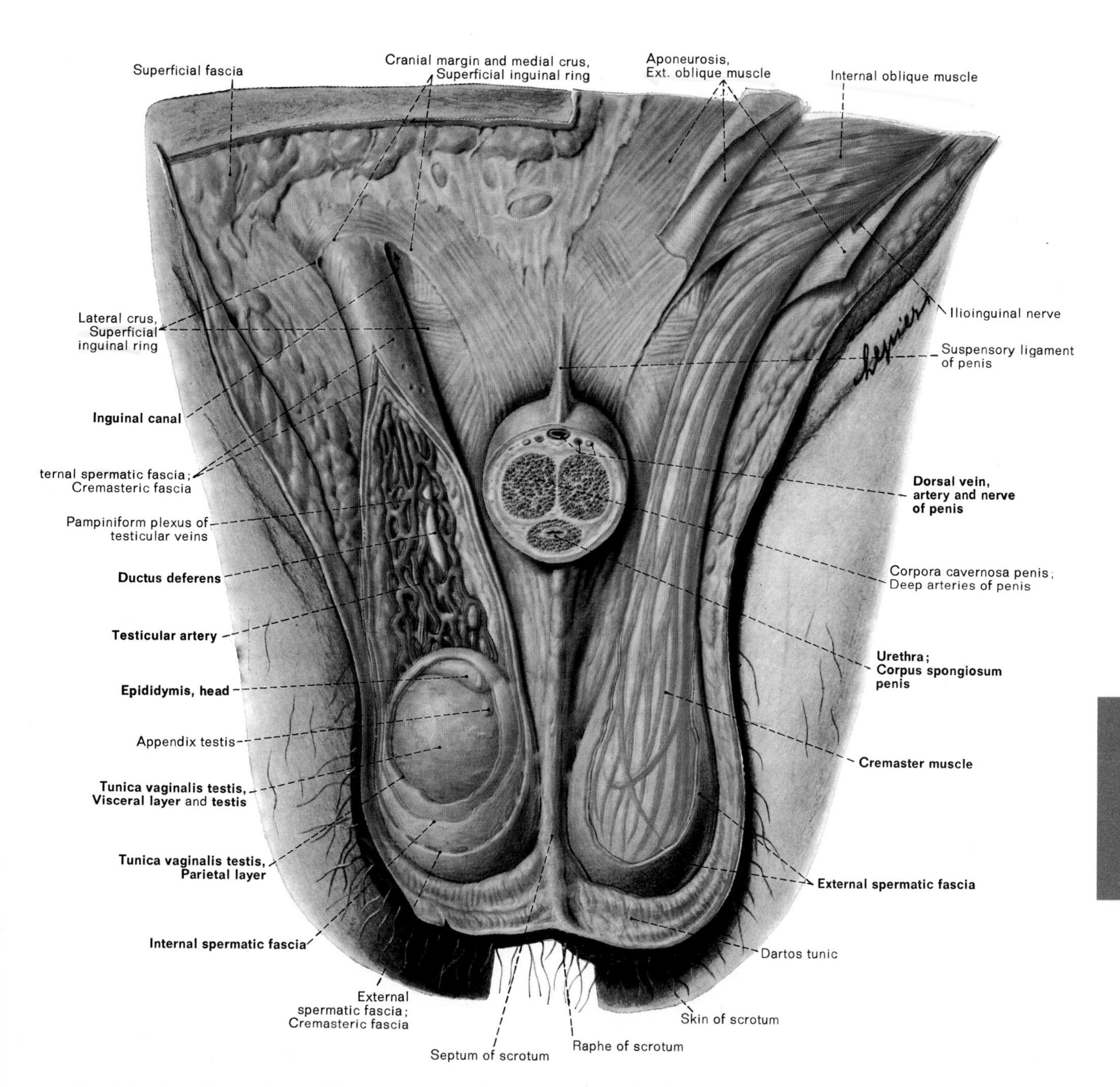

Fig. 20–4: Coronal sectional view of the penis, scrotum and spermatic cord. On the left side, the skin, superficial fascia, and external spermatic fascia of the scrotal sac have been removed to show the cremaster muscle as a continuation of the internal oblique muscle. On the right side, the contents of the spermatic cord are exposed.

The processus vaginalis (with its internal spermatic fascia covering) generates the inguinal canal by pushing forward and somewhat medially, first passing beneath the arching fibers of transversus abdominis and then pushing forward into the internal oblique. As the processus vaginalis pushes forward into the internal oblique, it acquires a second tubular fascial covering called the **cremasteric fascia.**

The formation of the inguinal canal ends as the processus vaginalis (with its internal spermatic and cremasteric fascia coverings) pushes forward into external oblique's aponeurosis to acquire a third tubular fascial covering called the **external spermatic fascia.** The triangular margin where the external spermatic fascia is continuous with external oblique's aponeurosis is called the **superficial inguinal ring.** The superficial inguinal ring is the medial, or superficial, end of the inguinal canal, and it lies immediately above and medial to the pubic tubercle (Fig. 20–4).

The inguinal canal thus marks the route by which the processus vaginalis protrudes through the fascial and muscle layers of the anterolateral abdominal wall during fetal development. In both sexes, this route follows, in part, the course of the ilioinguinal nerve through the anterolateral abdominal wall. The ilioinguinal nerve lies between the internal and external obliques as it courses through the inguinal canal, and emerges from the inguinal canal at the superficial inguinal ring.

The anterolateral abdominal wall structures that border the inguinal canal in both sexes form a roof, an anterior wall, a posterior wall, and a floor for the inguinal canal. The lower free borders of internal oblique and transversus abdominis form the canal's roof. External oblique's aponeurosis forms most of the canal's anterior wall. Transversalis fascia forms most of the canal's posterior wall. The inguinal ligament forms the canal's floor. The **inguinal ligament** is the curved, lower free border of external oblique's aponeurosis; it extends between the pubic tubercle and the anterior superior iliac spine (Fig. 12–5).

The Development of the Inguinal Canal: The Events Specific to the Male (Descent of the Testis)

During the period when the inguinal canal is being formed in a male, the testis migrates progressively closer to the deep inguinal ring (relative to the initial location of the gonad in the posterior abdominal wall) (Fig. 20–3, A and B). This apparent descent of the testis is actually caused by both migration of the testis within the abdominal walls and unequal rates of growth of related structures within these walls.

Normally, the testis begins to pass through the deep inguinal ring and into the inguinal canal during the twenty-eighth week of fetal development. It descends through the inguinal canal along a pathway that is directly behind, but topologically superficial to, the processus vaginalis (Fig. 20–3C); this pathway is enclosed within the fascial coverings that the processus vaginalis acquired as it formed the inguinal canal. Accordingly, by the time that the testis passes through the superficial inguinal ring and into the developing scrotum (an event which commonly occurs during the thirty-second week of fetal age), it too is ensheathed within the same fascial coverings (the internal spermatic, cremasteric, and external spermatic fasciae) that ensheath the processus vaginalis.

As the testis descends from the posterior abdominal wall, through the inguinal canal, and into the scrotum, it extends along with itself the nerves and blood and lymphatic vessels that initially served it when it was forming in the posterior abdominal wall. It also extends along with itself the **vas deferens,** the tubelike vessel which, during sexual climax, transmits sperm from the epididymis to the ejaculatory duct. This collection of structures (the vas deferens, nerves, and blood and lymphatic vessels) that extends from the testis is referred to as the **spermatic cord.** It should be apparent that, within the inguinal canal and the scrotum, the spermatic cord is ensheathed by the same fascial coverings that ensheath the processus vaginalis.

Once the testis has descended into the scrotum, the formation of the inguinal canal and its contents in the male is virtually finished. The only development of importance still to occur is the sealing-up of the distal part of the processus vaginalis (the end part in front of the testis) and the degeneration of the remaining, proximal part during the first few weeks following birth (Fig. 20–3D). The only postnatal, gross remnant of the processus vaginalis is thus a closed sac called the **tunica vaginalis** which covers all but the posterior surface of the testis. Accordingly, the only structures generally remaining in the inguinal canal of the male a few weeks after birth are the spermatic cord (with its fascial coverings) and the ilioinguinal nerve.

The Development of the Inguinal Canal: The Events Specific to the Female

The only significant differences between the sexes in inguinal canal development occur after the formation of the canal itself. In the female, the ovary migrates from its original position in the posterior abdominal wall, and, like the testis, follows the course of the gubernaculum. In the female fetus, however, the gubernaculum, at a point between its attachment to the ovary and its entrance into the anterolateral abdominal wall, becomes attached to the developing uterus. Consequently, the ovary's migration leads it to a position lateral to the uterus within the pelvic cavity. That portion of the gubernaculum that extends from the uterus toward and through the anterolateral abdominal wall into the labial swelling remains after inguinal canal formation as a cordlike structure called the **round ligament of the uterus.** Because that portion of the gubernaculum destined to become the round ligament of the uterus exists prior to the formation of the inguinal canal, the round ligament of the uterus is not ensheathed by the same fascial coverings that ensheath the processus vaginalis as it forms the inguinal canal in the female.

The processus vaginalis completely degenerates in the female during the first few postnatal weeks. There is no postnatal, gross remnant of the processus vaginalis in the labium majus analogous to the tunica vaginalis in the male. Accordingly, the only major structures remaining in the inguinal canal of the female a few weeks after birth are the round ligament of the uterus and the ilioinguinal nerve.

ABDOMINAL WALL HERNIAS

Definition of an Abdominal Wall Hernia

A hernia is most commonly defined as a protrusion of a tissue or organ through a wall by which it is normally contained. However, in order to avoid the semantic argument of whether the term hernia refers to the protrusion versus the wall opening itself, it is best to recognize that the common denominator to all hernia definitions is the anatomic defect in the supporting structures. Bearing this in mind, a hernia of an abdominal wall is probably best defined as a weakness or opening in the muscular and fascial layers through which a contained tissue or organ may protrude.

The Parts of an Abdominal Wall Hernia

An abdominal wall hernia almost always consists of three parts, which, proceeding from the innermost to

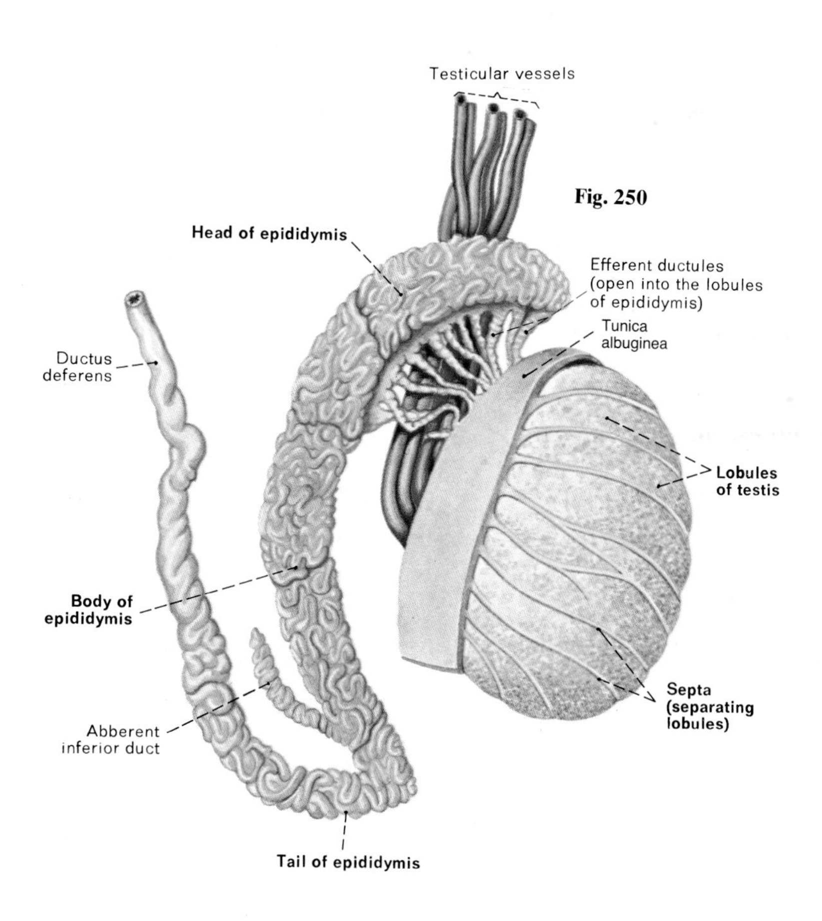

Fig. 20–5: Testis, epididymis, and the beginning of the vas deferens. Most of the tunica albuginea has been removed.

the most superficial part, are the contents of the hernia, its sac, and the sac's coverings. The **contents of a hernia** may include any tissue or organ in the abdominopelvic cavity, commonly a segment of the small or large intestine. The **sac of a hernia** consists of an outpouching of parietal peritoneum; this parietal diverticulum envelops the contents of the hernia as they protrude through the abdominal wall region. The **neck of the sac of a hernia** is the tapered, proximal end of the sac (it is the end at which the sac is continuous with the parietal peritoneum). The **sac coverings of a hernia** are the tissue layers of the abdominal wall region (superficial to the peritoneum) through which the contents of the hernia and its sac protrude.

Inguinal Hernias

The inguinal region of the anterolateral abdominal wall is predisposed to hernias because it bears a preformed passageway through its muscular and fascial layers: the inguinal canal. Herniations of abdominopelvic viscera into the inguinal canal are defined by whether they are indirect or direct.

An **indirect inguinal hernia** is a hernia whose contents protrude into the inguinal canal through the deep inguinal ring; the neck of the sac of an indirect inguinal hernia thus always lies lateral to the inferior epigastric artery. The contents may ultimately traverse the inguinal canal and extend through the superficial inguinal ring into the scrotum or labium majus. An indirect inguinal hernia has an embryologic derivation because it almost always occurs in an individual with a **patent processus vaginalis** (a person in whom the processus vaginalis persists after birth); accordingly, the sac of an indirect inguinal hernia generally consists of a patent processus vaginalis.

A **direct inguinal hernia** is a hernia whose contents protrude into the inguinal canal through a distention of its posterior wall. The neck of the sac of a direct inguinal hernia always lies medial to the origin of the inferior epigastric artery and within the **inguinal,** or **Hesselbach's, triangle.** The inguinal triangle is the area within the inguinal region bounded by the inferior epigastric artery laterally, the lateral edge of rectus abdominis medially, and the inguinal ligament inferiorly. A direct inguinal hernia is acquired.

Femoral Hernias

A **femoral hernia** is a hernia whose contents protrude through the femoral ring into the femoral canal. The neck of a femoral hernia always lies immediately lateral and inferior to the pubic tubercle, and extends inferior to the inguinal ligament. A femoral hernia is acquired.

Abdominal wall hernias occur about six times more frequently in males than in females. In males, about 97% of abdominal wall hernias are inguinal, 2% femoral, and 1% umbilical. In females, about 50% of abdominal wall hernias are inguinal, 34% femoral, and 16% umbilical.

THE SCROTUM

The contents of the scrotum are enveloped on each side by a sac of five tissue layers, which, proceeding from the most superficial to the deepest, are:

(1) the **skin,**
(2) the **superficial fascia,**
(3) the **external spermatic fascia,**
(4) the **cremasteric fascia,** and
(5) the **internal spermatic fascia** (Fig. 20–4).

The superficial fascia contains smooth muscle fibers collectively called the **dartos muscle.** The tonicity of the dartos muscle is responsible for the rugosity of the skin of the scrotal sac.

The cremasteric fascia contains skeletal muscle fibers collectively called the **cremaster muscle** (Fig. 20–4). The cremaster muscle acts to raise the testis. The cremaster muscle is innervated by the genital branch of the genitofemoral nerve. The **genitofemoral nerve,** which is derived from the L1 and L2 roots of the lumbar plexus, has two major terminal branches: the genital branch that provides motor supply to the cremaster muscle and a femoral branch that provides sensory supply for the upper medial surface of the thigh. If a fingernail is quickly stroked along the upper medial surface of the thigh of a male, such stimulation of the femoral branch of the genitofemoral nerve will elicit a superficial reflex (the **cremasteric reflex**) which (by stimulating the genital branch of the genitofemoral nerve and thereby concentrically contracting the cremaster muscle) suddenly raises the testis. The presence of this superficial withdrawal reflex in males is tested in a neurologic examination of trauma or disease involving the L1 and L2 spinal cord segments.

The contents of the scrotum on each side include the testis, epididymis, and the spermatic cord (Fig. 20–4). The **testis** functions as an exocrine gland that produces the male sperm cells (spermatozoa) and as an endocrine gland that produces the male sex hormone, testosterone. The oval-shaped testis is surrounded by a white, fibrous capsule called the **tunica albuginea** (Fig. 20–5). The **tunica vaginalis,** which is the sealed distal end of the processus vaginalis, lies closely apposed to the front and sides of the testis (Fig. 20–4). The spermatozoa produced within the testis exit the gonad at its superior pole to enter the head of the epididymis.

The comma-shaped **epididymis** is generally subdivided for descriptive purposes into a head, body, and tail (Fig. 20–5). The head of the epididymis lies atop the superior pole of the testis; the body and tail of the epididymis extend inferiorly from the head, lying adjacent to the posterior surface of the testis.

The **vas deferens** (**ductus deferens**) is the most prominent, solitary structure of the spermatic cord. The vas deferens begins as a direct continuation of the tail of the epididymis (Fig. 20–5), and ends in the pelvis at its union with the duct of the seminal vesicle to form the ejaculatory duct. The vas deferens is basically a thick-walled tube rich in smooth muscle fibers responsive to sympathetic stimulation. Such stimulation during sexual orgasm produces peristaltic contractions that propel mature spermatozoa through the lumen of the vas deferens.

The spermatic cord has three major arteries:

(1) The **testicular artery** is a branch of the abdominal aorta, and it is the principal artery supplying the testis and epididymis.
(2) The **cremasteric artery** is a branch of the inferior epigastric artery, and it supplies the cremaster muscle.
(3) The **artery to the vas deferens** is a branch of the inferior vesical artery.

The veins of the spermatic cord comprise much of the cord's bulk; they arise from the testis and form a highly anastomotic plexus called the **pampiniform plexus** (Fig. 20–4). As the pampiniform plexus extends within the inguinal canal, it gives rise to a small number of veins which ultimately coalesce to form the **testicular vein.** The right testicular vein drains into the inferior vena cava; the left testicular vein drains into the left renal vein.

The lymphatics which drain the testis are afferent to aortic, or lumbar, lymph nodes. The lymphatics which drain the skin of the scrotal sac are afferent to the horizontal group of superficial inguinal lymph nodes.

CASE **V.1**

The Case of Jonathan Anderson

INITIAL PRESENTATION AND APPEARANCE OF THE PATIENT

Samantha Anderson calls at 10:30 AM to arrange an emergency appointment for her infant son, Jonathan. Jonathan, a two month-old white male, is John and Samantha Anderson's first child; the examiner has been Jonathan's pediatrician since the child was born. Although Jonathan was born at 36 weeks' gestation (weighing 2700 gms and measuring 47 cm crown to heel), his birth occurred without incident and his first check-up was normal. Samantha Anderson enters the reception room with Jonathan at 11:30 AM; Jonathan is crying.

QUESTIONS ASKED OF THE PATIENT'S MOTHER

Why have you brought Jonathan in? He vomited this morning and since then has been crying almost continuously. *[Vomiting is an autonomic response that may be elicited by numerous diseases and conditions involving the abdomen, thorax, or central nervous system. Vomiting thus is not a specific symptom (it does not point to a specific class of pathologic conditions nor a specific body region).]*

How many times has he vomited this morning? Just once.

Do you know the time at which he vomited? Around 8:30.

Had he been fed this morning before he vomited? Yes, around 7 o'clock.

Do you recall what the vomit looked like? Yes, and it was the color of the vomit that really concerned me; his vomit was greenish in color. *[The green emesis indicates the eructation of bile-containing intestinal chyme, and thus implies that the patient's vomiting episode involved the eructation of material from primarily the upper gastrointestinal tract.]*

What happened after he vomited? Well, he started to cry, and since then has been crying almost continuously. I have tried walking with him and rocking him, and three or four times, he's stopped crying for about five minutes, but then he starts all over again.

Has he been fed since he vomited? No.

When did Jonathan awaken this morning? Well, his crying woke John and me up at 5 o'clock this morning. When we checked on him, we changed his diaper because we found a bowel movement. He's always fussy when his diaper is soiled or wet. He quickly went back to sleep after the diaper change, and didn't waken until around 6:30, when he was crying again. When I picked him up, he seemed unusually fussy, but quieted down a bit after I rocked him. He looked okay, and his diaper was not soiled. I thought he was hungry, so that's when I gave him a feeding of formula around 7 o'clock.

Have you noticed anything else unusual about Jonathan this morning? Yes. As I dressed him to come over to your office, I noticed that his belly seems a bit swollen. *[Acute abdominal distention indicates an acute accumulation of gas and/or fluid in the abdomen. This sign is the first evidence acquired from the history which points to the abdomen as the site of Jonathan's condition. Acute abdominal distention is an ominous sign because it is frequently caused by a disease, injury, or condition which (a) will not resolve in the absence of medical and/or surgical intervention and (b) will shortly worsen into a condition of significantly increased morbidity and mortality.]*

How has his health been in the last few days? Fine. He had some colic two nights ago, but that kept him awake for only about ten minutes. *[This answer indicates that it is the mother's perception that Jonathan's condition has developed within the last 24 hours.]*

[The examiner finds Samantha Anderson to be alert and fully cooperative during the interview.]

PHYSICAL EXAMINATION OF THE PATIENT

Vital signs:
Blood pressure
Lying supine: 80/40
Pulse: 130
Rhythm: regular

Temperature: 99.8°F. per axilla
Respiratory rate: 32
Height: 54 cm crown-heel
Weight: 4200 gms

[A pediatrician would recognize that, except for the axillary temperature, the patient's vital signs are within normal limits.]

HEENT Examination: Normal
Lungs: Normal
Cardiovascular Examination: Normal
Abdomen: Inspection, auscultation, percussion, and palpation of the abdomen are normal except for the following findings: Inspection reveals moderate, generalized abdominal distention. Auscultation finds loud, high-pitched bowel sounds; rushes of these sounds are coincident with enhanced crying.
Genitourinary Examination: Inspection reveals slightly erythematous skin overlying a left inguinal swelling. Light palpation of the left inguinal swelling reveals a mass extending from above the midregion of the inguinal ligament to a point above and medial to the pubic tubercle, and from there, into the upper part of the scrotal sac; the overlying skin is warmer than that of the rest of the anterior abdominal wall. Light palpation of both the left inguinal mass and the left testis elicit enhanced crying.
Musculoskeletal Examination: Normal
Neurologic Examination: Normal

INITIAL ASSESSMENT OF THE PATIENT'S CONDITION

The history and physical examination shows that the patient has two chief medical problems: acute abdominal distention and an inflammatory left inguinal lesion.

ANATOMIC BASIS OF THE PATIENT'S HISTORY AND PHYSICAL EXAMINATION

(1) **Acute abdominal distention signals that a disease or an event has generated a sudden accumulation of gas and/or fluid within the abdominopelvic cavity.** Samantha Anderson's account established that Jonathan's abdominal distention is of less than 12 hours duration.

(2) **The rushes of loud, high-pitched bowel sounds coincident with enhanced crying represent the painful passage of very forceful peristaltic contractions through distended bowel segments.** Bowel sounds are a monitor of peristaltic activity in the bowel; they are produced by the gurgling of gas and fluid within a narrowed bowel segment during the passage of a peristaltic wave. The greater the strength of the peristaltic contractions, the greater the intensity, or loudness, of bowel sounds; the greater the bowel wall tension, the higher the pitch of the sounds. The loud, high-pitched bowel sounds suggest that the patient's acute abdominal distention may be due to acute accumulation of gas and/or fluid within the small and/or large bowel; such an accumulation of gas and/or fluid would distend (and make highly tense) the affected bowel segments and generate very forceful peristaltic contractions in an attempt to move the gas and/or fluid through the partially or totally obstructed bowel.

Each wave of very forceful peristaltic contractions produces a sudden increase in intraluminal pressure in the bowel segment immediately proximal to the obstruction. This sudden increase in intraluminal pressure even further stretches the distended wall of the bowel segment, thereby eliciting a short interval of intense abdominal pain (which is visceral in nature). The term **intestinal colic** is used to refer to the sporadic series of short intervals of intense abdominal pain that occur in this fashion following bowel obstruction.

(3) **The patient's green emesis indicates the eructation of bile-containing intestinal chyme.** Near the beginning of the small intestine, there is the opening of a duct (the common bile duct) that conducts a fluid called bile from the liver and gallbladder to the lumen of the small intestine. Bile contains organic acids which aid in the emulsification of dietary fats and their absorption following digestion. Bile is variably colored yellow, brown, or olive green. The addition of olive green bile to chyme in the proximal part of the small intestine will color the chyme green (chyme is the mash of partly digested foodstuffs that passes from the stomach to the small intestine).

Vomiting is a common autonomic response to abdominal disease, and thus is not a specific finding (it does not point to a specific disease or event). Moreover, vomiting does not necessarily suggest abdominal disease, because it may also occur in response to diverse mechanisms, such as physical or chemical stimulation of the vomiting center in the brain.

(4) **Inspection and physical examination of the left inguinal swelling indicate the presence of an indirect inguinal hernia whose visceral contents are inflamed.** The surface landmarks of the left inguinal swelling correspond to the anatomic location of an indirect inguinal hernia: the lateral end of the swelling overlies the location of the deep

inguinal ring; the most medial part of the swelling extends into the upper left part of the scrotal sac from a point overlying the superficial inguinal ring. The warm, erythematous skin overlying the swelling and the tenderness of the swelling to gentle palpation suggest inflammation of the hernia's contents.

(5) **A tender left testis indicates inflammation of the gonad.** Inflammation of the testis in this setting suggests that the pressure-generating, inflamed mass in the inguinal canal is markedly impinging upon the spermatic cord somewhere along its course through the inguinal canal.

(6) **The patient has a minimally elevated body temperature.** A safe and easy way to assess an infant's body temperature is to place the bulb of a thermometer against the undersurface of the axilla, adduct the infant's arm against the body, and wait for about a minute in order to insure a proper reading. Normal axillary temperature for an infant ranges from 97.7° to 99.3°F.

INTERMEDIATE EVALUATION OF THE PATIENT'S CONDITION

The patient appears to be suffering from complications of a left indirect inguinal hernia.

CLINICAL REASONING PROCESS

When considered all together, the presence of an inflamed indirect inguinal hernia in association with acute abdominal distention, hyperactive bowel sounds, and bilious vomiting, strongly suggests the hypothesis that a loop of small or large bowel is entrapped within the left inguinal canal and has become inflamed due to ischemia (loss of blood supply). A pediatrician would tentatively diagnose a strangulated indirect inguinal hernia (an incarcerated, or non-reducible, indirect inguinal hernia whose visceral contents have become ischemic). This condition is a surgical emergency. The pediatrician would presume that a loop of small or large bowel represents the contents of the hernia, because the infant shows a constellation of physical signs indicative of acute bowel obstruction (acute abdominal pain and distention, hyperactive bowel bounds, and vomiting).

RADIOLOGIC EVALUATION AND FINAL RESOLUTION OF THE PATIENT'S CONDITION

Abdominal plain films would provide evidence of acute bowel obstruction. In this case, plain films of the patient's abdomen showed evidence of small bowel distention: A supine abdominal plain film showed centrally located, distended, gas-filled bowel loops traversed by thin, closely spaced, uninterrupted water-density lines (the central location of the distended bowel loops suggests that they are small bowel loops; the uninterrupted water-density lines traversing the loops represent the submucosal circular folds of the small bowel). An erect abdominal plain film showed air-fluid levels arranged in a step-ladder pattern among a number of the centrally located, distended bowel loops.

CHRONOLOGY OF THE PATIENT'S CONDITION

Normal testicular passage through the inguinal canals and descent into the scrotal sac occurred within the patient during the twenty-eighth through thirty-second weeks in utero. His premature birth interrupted normal closure of the processus vaginalis on the left side, thereby establishing the potential for the development of an indirect inguinal hernia. By the time the patient was put to bed the night before the condition presented itself, a loop of mid-jejunum had already entered the left deep inguinal ring and migrated part way through the inguinal canal; the sharply curved leading edge of the bowel loop produced a partial obstruction of the jejunum. The pressure increases within the abdominopelvic cavity that aided the early morning bowel movement also pushed the herniated jejunal loop through the superficial inguinal ring into the scrotal sac, and thus produced an incarcerated indirect inguinal hernia (a hernia whose contents could not be manually manipulated backward through the inguinal canal). At this time, the constriction of the jejunal loop by the superficial and deep inguinal rings not only produced a complete mechanical obstruction of the jejunum, but also strangulated (rendered ischemic) the incarcerated jejunal loop. The ischemia evoked intense visceral pain, which awoke the patient at 6:30 AM and provoked his crying.

The acute complete obstruction of the jejunum prompted progressive accumulation of intestinal gas and fluid proximal to the site of obstruction. The patient's intestinal gas was being significantly increased by air swallowed as he cried, and his 7:00 AM feeding introduced a significant amount of fluid. Other mechanisms also contributed to gas and fluid accumulation. Small bowel obstruction typically stimulates the small bowel mucosa to secrete fluid and electrolytes into the intraintestinal lumen. In addition, following obstruction, the normally low numbers of predominantly gram-positive bacterial flora in the small bowel changes to large numbers of gram-negative and anaerobic organisms more like those found in the large bowel. In this case, the nutrients entering the stagnant jejunal segment are metabolized by these bacteria partly into gas.

The entrapped intestinal gas and fluid distended the proximal jejunum, which, in turn, distended the contours of his abdomen. The bilious vomiting at 8:30 AM partially relieved the intraluminal pressures being borne by distended segments of the proximal jejunum. The passage of very forceful peristaltic contractions through these distended bowel segments generated loud, high-pitched bowel sounds accompanied by a short period of intense visceral pain (which a young child or older individual would recognize as severe intestinal cramps).

Strangulation of the jejunal loop incarcerated within the left inguinal canal and the upper part of the scrotal sac promptly evoked an inflammatory response that distended the jejunal loop and rendered it tender to gentle palpation. The inflammation quickly extended to the overlying superficial tissues, as evidenced by the overlying warm and erythematous skin. The distended jejunal loop impinged upon the segment of the spermatic cord traversing the inguinal canal to the extent that venous and lymphatic drainage of the testis and epididymis was markedly reduced, thereby distending the testis and rendering it tender to gentle palpation. The ischemia of the jejunal loop will soon result in a disruption of its mucosal barrier, leading to translocation of bacteria across the jejunal wall, bacteremia, and sepsis. Prompt recognition of this emergency can reduce the risk of serious and life-threatening complications.

RECOMMENDED REFERENCES FOR ADDITIONAL INFORMATION ON THE ANTEROLATERAL ABDOMINAL WALL

Bates, B., *A Guide to Physical Examination and History Taking,* J. B. Lippincott Co., Philadelphia, 1991: *Pages 343 through 361 in Chapter 11 present a discussion of the visual inspection, auscultation, percussion, and palpation of the abdomen.*

DeGowin, R. L., Jochimsen, P. R., and E. O. Theilen, *DeGowin & DeGowin's Bedside Diagnostic Examination,* 5th ed., Macmillan Publishing Co., New York, 1987: *Pages 489 through 496 offer a discussion of the physical examination of abdominal hernias.*

Silen, W., *Cope's Early Diagnosis of the acute Abdomen,* 18th ed., Oxford University Press, New York, 1991: *Chapter 15 provides a discussion of the early diagnosis of strangulated and obstructed hernias.*

CHAPTER **21**

A General Survey of the Abdominal and Pelvic Organs

This chapter presents an overview of the spatial relationships among the organs of the abdomen and pelvis. This discussion focuses on the relationship of each organ to the peritoneum, the deepest tissue layer of the walls of the abdominopelivic cavity.

THE REGIONS OF THE ABDOMINAL CAVITY

The intersection of the **median sagittal plane** with the **horizontal plane which passes through the umbilicus** divides the abdominal cavity into four quadrants: the **RUQ,** or right upper quadrant; the **LUQ,** or left upper quadrant; the **LLQ,** or left lower quadrant; and the **RLQ,** or right lower quadrant.

The intersections of the **midclavicular planes** with the **transpyloric and intertubercular planes** divide the abdominal cavity into nine regions (Fig. 21–1): the **right hypochondriac, epigastric, left hypochondriac, right lumbar, umbilical, left lumbar, right inguinal, hypogastric,** and **left inguinal regions.** The transpyloric plane, which lies at the level of the intervertebral disc between the first and second lumbar vertebrae, is named for its intersection with the pylorus, the end of the stomach which joins the duodenum. The intertubercular plane, which lies at the level of the body of the fifth lumbar vertebra, is named for its intersection with the iliac tubercles of the coxal bones. The iliac tubercle is a tubercle on the outer margin of the iliac crest (Fig. 12–4 does not depict the iliac tubercle). In the adult coxal bone, the iliac tubercle is about 5 cm posterosuperior to the anterior superior iliac spine.

THE PERITONEAL CAVITY AND INTRAPERITONEAL AND RETROPERITONEAL VISCERA

The **peritoneal cavity** is the cavity within the abdominopelvic cavity that is lined by the peritoneum (Fig. 21–2). The organs of the abdomen and pelvis all lie within the abdominopelvic cavity but superficial to the peritoneal cavity; in other words; the organs all lie deep to the abdominopelvic walls but superficial to the peritoneum which lines those walls. Almost all of the abdominopelvic organs exhibit one of two relationships with the peritoneum: Viscera that are almost completely covered by an extended invagination of the peritoneal lining, such as the jejunum and ileum, are called **intraperitoneal viscera** (Fig. 21–2). Viscera that lie sandwiched between the deep fascial lining of the posterior abdominal wall and its peritoneal lining, such as the kidneys and the abdominal aorta, are called **retroperitoneal viscera** (Fig. 21–2). Organs that were initially intraperitoneal during fetal development but became retroperitoneal by the time of birth are called **secondarily retroperitoneal viscera.** Table 21–1 summarizes the peritoneal relationships of the major abdominal and pelvic viscera.

PARIETAL AND VISCERAL PERITONEUM

Parietal peritoneum is peritoneum that lines an abdominal or pelvic wall. Parietal peritoneum receives the same nerve and blood supply as that of the wall region it lines. Visceral peritoneum is peritoneum that envelops or covers the surface of an abdominal or pelvic organ. Visceral peritoneum receives the same nerve and blood supply as that of the organ it covers. A **peritoneal ligament** is a double layer of peritoneum extending between two organs or an organ and an abdominopelvic wall region. Peritoneal ligaments transmit the vascular elements and nerves that supply intraperitoneal organs (Fig. 21–2).

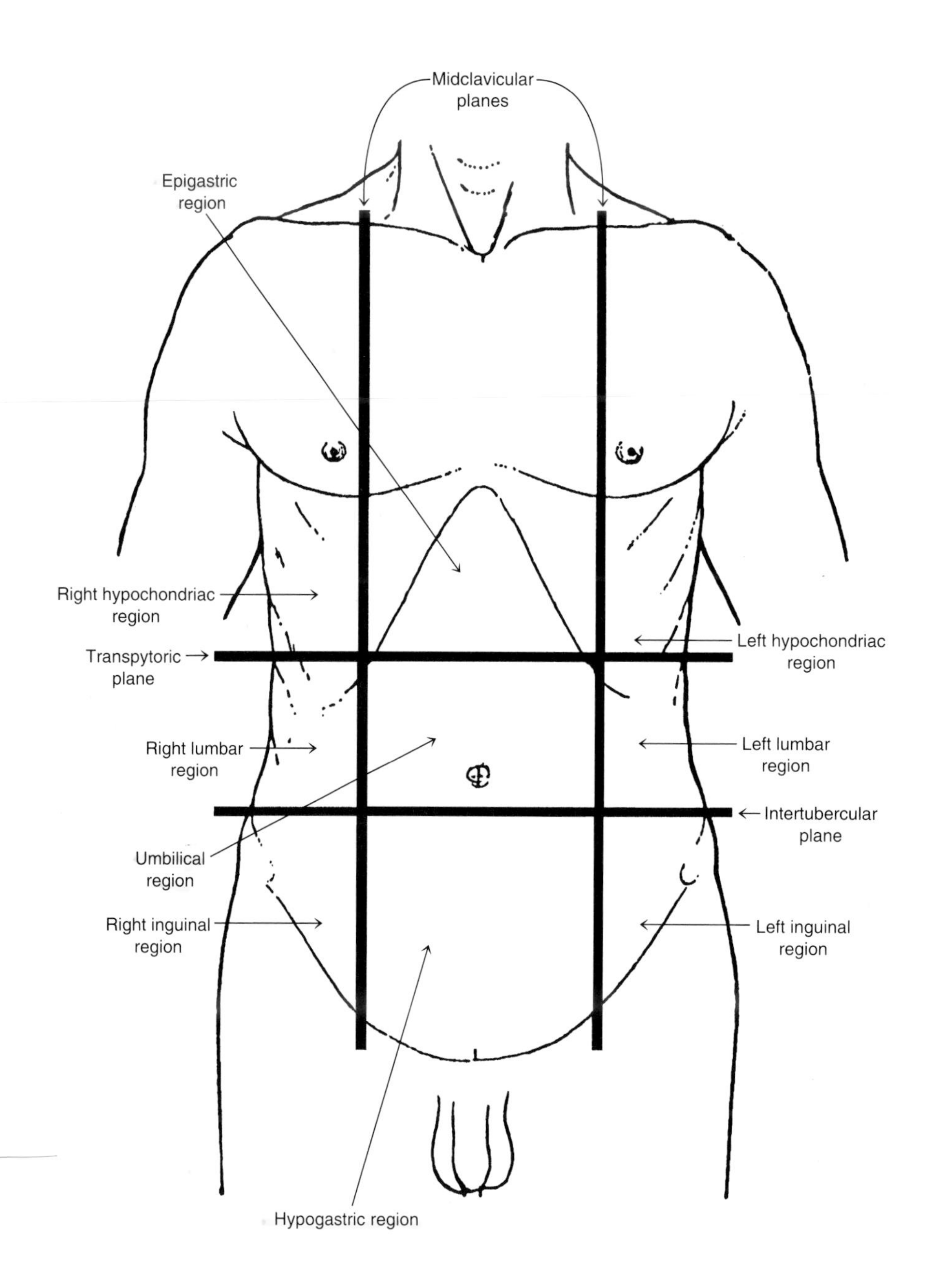

Fig. 21–1: Diagram illustrating the division of the abdominal cavity into nine regions by the transpyloric and intertubercular planes and the paired midclavicular planes.

Table 21–1
The Peritoneal Relationships of the Major Abdominal and Pelvic Viscera

The Major Retroperitoneal Viscera of the Abdominopelvic Cavity

Abdominal aorta

Inferior vena cava

Kidneys

Adrenal glands

Ureters

The Peritoneal Relationships of the Abdominal and Pelvic Segments of the Digestive Tract

Stomach and proximal half of the first part of the duodenum – Intraperitoneal

Duodenum distal to the proximal half of the first part – Secondarily retroperitoneal

Jejunum and ileum – Intraperitoneal

Cecum and appendix – Intraperitoneal

Ascending colon – Secondarily retroperitoneal

Transverse colon – Intraperitoneal

Descending colon – Secondarily retroperitoneal

Sigmoid colon – Intraperitoneal

Upper third of the rectum – Anterior and lateral surfaces are covered with peritoneum

Middle third of the rectum – Anterior surface only is covered with peritoneum

Lower third of the rectum – Lies beneath the peritoneum lining the floor of the pelvis

The Peritoneal Relationships of Remaining Major Abdominal and Pelvic Organs

Spleen – Intraperitoneal

Head, neck, and body of pancreas – Secondarily retroperitoneal

Tail of pancreas – Intraperitoneal

Liver – Most of the liver's exterior is covered with peritoneum; its bare area lies directly against the undersurface of the diaphragm.

Gallbladder – In most instances, only the fundus and the posteroinferior surface of the body and neck of the gallbladder are covered with peritoneum; the remaining gallbladder surfaces lie directly against the liver's visceral surface.

Urinary bladder – Superior surface is covered with peritoneum

Body of the uterus, supravaginal part of the cervix of the uterus, and the uterine tubes –Covered with peritoneum

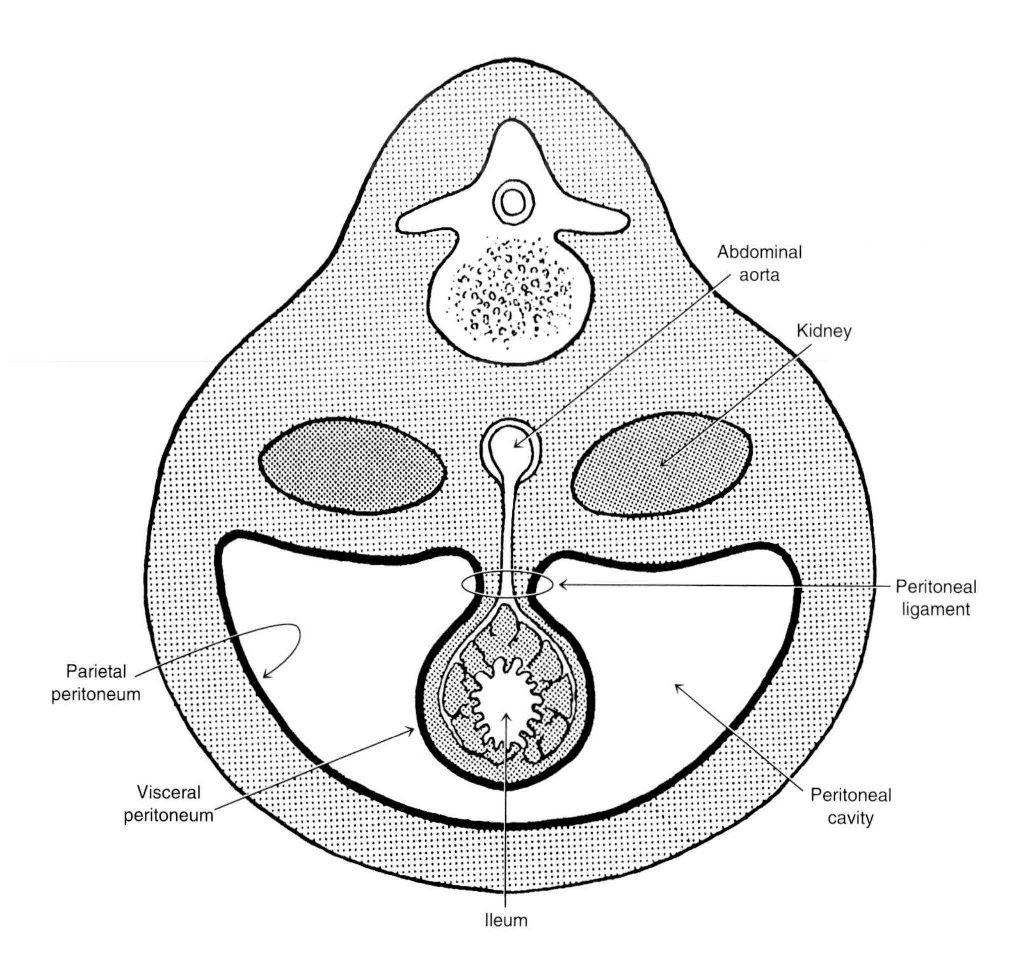

Fig. 21–2: Schematic drawing of the peritoneal relationships of the abdominal aorta, kidneys, and ileum during fetal development. These relationships persist throughout postnatal life.

THE ORGANIZATION OF THE ABDOMEN AND PELVIS

A helpful way to picture in your mind the organization of the abdomen and pelvis is to visualize in the order shown the placement of the following five groups of organs:

(1) the retroperitoneal viscera of the abdomen,
(2) the secondarily retroperitoneal viscera of the abdomen,
(3) the intraperitoneal segments of the small and large intestines in the mid and lower abdomen,
(4) the liver and gallbladder and the intraperitoneal viscera in the upper abdomen, and
(5) the pelvic viscera.

The Retroperitoneal Viscera of the Abdomen: the Abdominal Aorta, Inferior Vena Cava, Kidneys, Adrenal Glands, and Ureters

All of these viscera lie directly against regions of the posterior abdominal wall (Fig. 21–3). The abdominal aorta descends immediately in front and slightly to the left of the vertebral column, extending from the aortic opening of the diaphragm to the level of the fourth lumbar vertebra. The inferior vena cava ascends immediately to the right of the abdominal aorta, extending from the level of the fourth lumbar vertebra to the caval opening of the diaphragm.

The kidneys lie nestled in the paravertebral gutters within the upper quadrants of the abdominopelvic cavity. The adrenal glands rest atop the superior poles of the kidneys. The ureters emerge from the hila of the

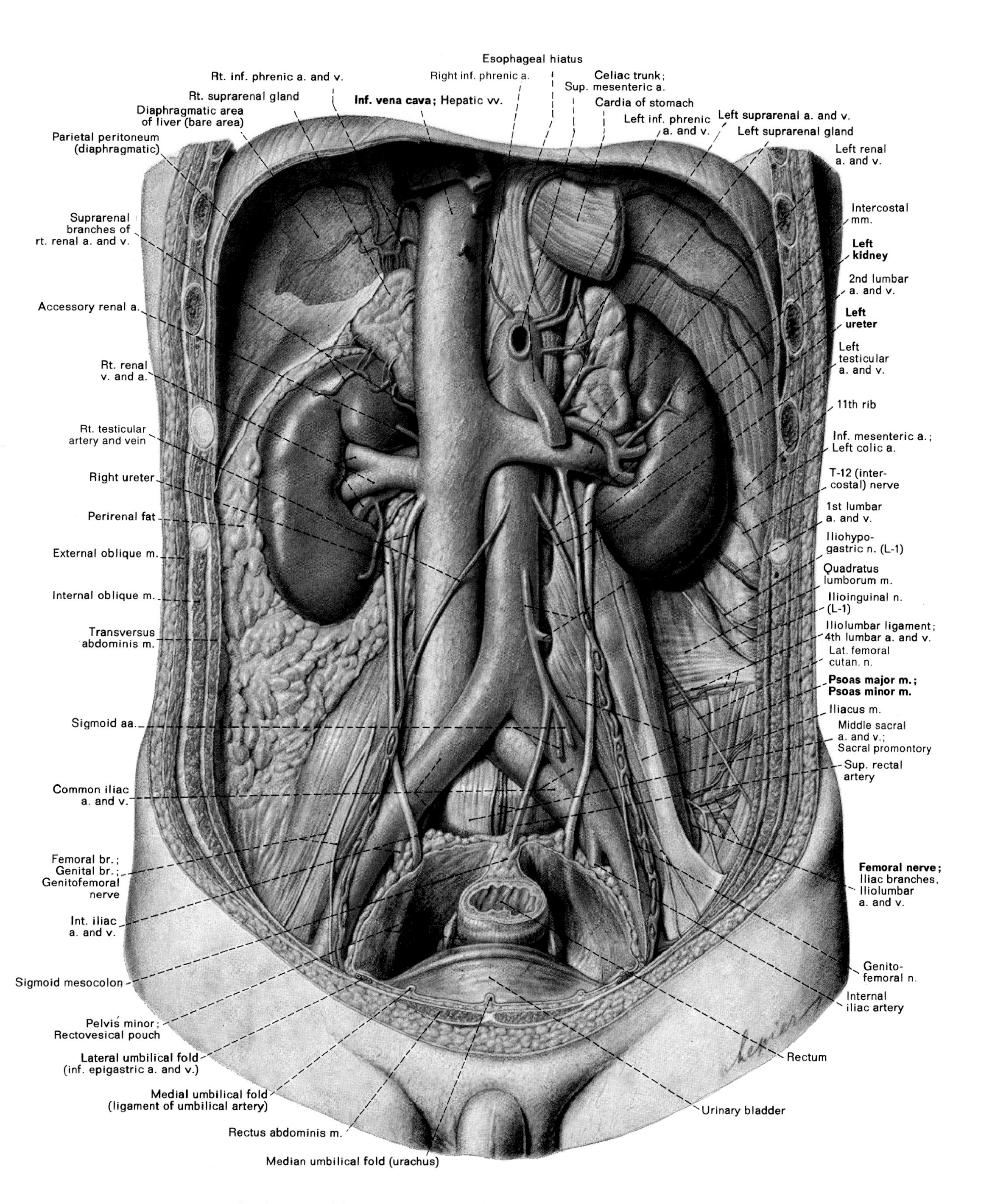

Fig. 21–3: The placement of the major retroperitoneal viscera against the posterior abdominal wall in a male.

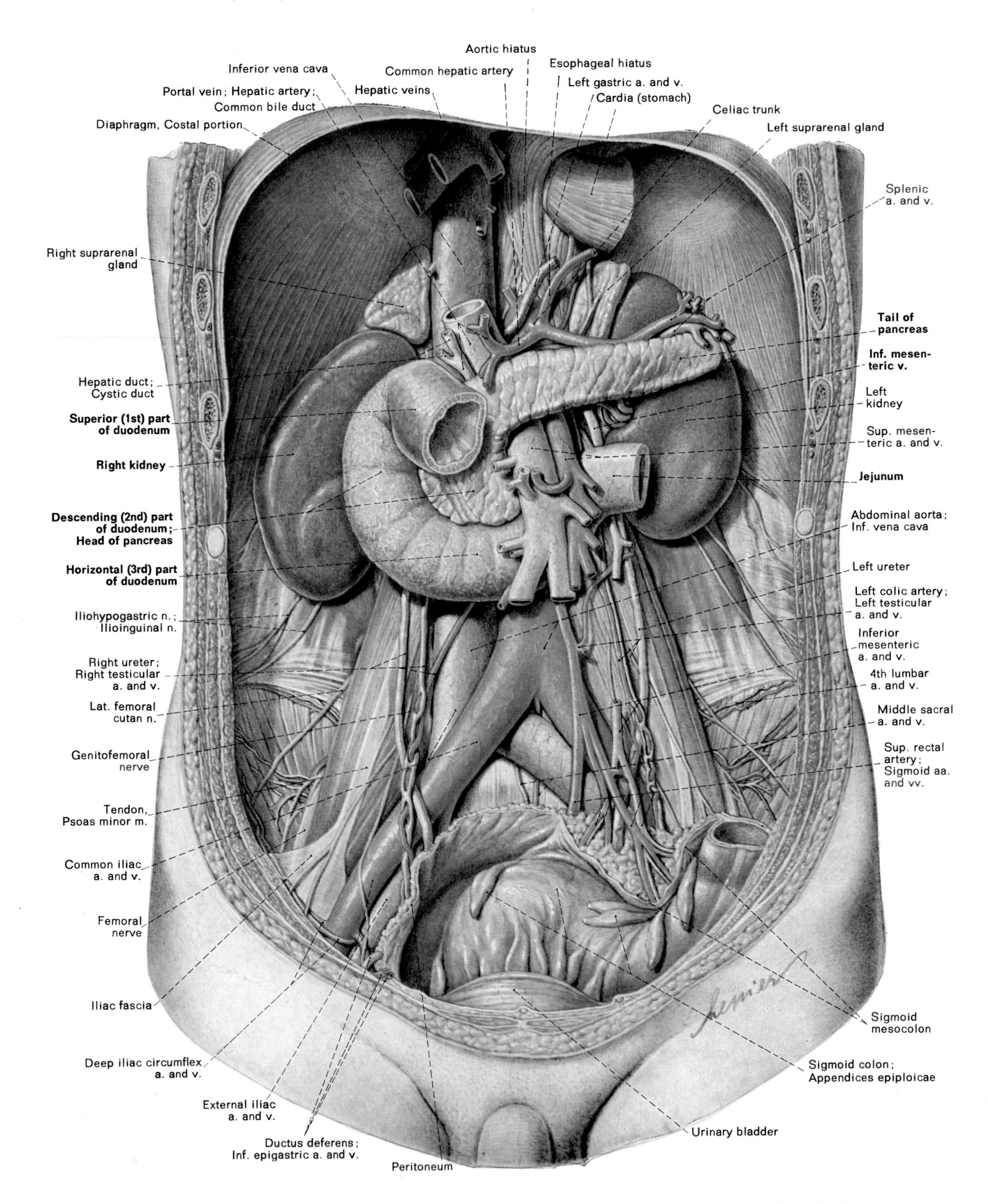

Fig. 21–4: The placement of the pancreas, duodenum, and major retroperitoneal viscera against the posterior abdominal wall in a male.

kidneys, descend in front of the psoas major muscles, and finally traverse inferiorly along the lateral walls of the pelvis before piercing the upper lateral corners of the posterior surface of the urinary bladder.

The Secondarily Retroperitoneal Viscera of the Abdomen: the Head, Neck and Body of the Pancreas; the Distal Half of the First Part of the Duodenum and the Second, Third, and Fourth Parts of the Duodenum; the Ascending Colon with the Hepatic Flexure at its Upper End; and the Descending Colon with the Splenic Flexure at Its Upper End

All of these viscera lie in part directly against regions of the posterior abdominal wall and in part directly against the anterior surfaces of one or more retroperitoneal viscera. The head, neck, and body of the pancreas extend across most of the breadth of the posterior abdominal wall in the upper quadrants of the abdominopelvic cavity (Fig. 21–4). As these regions of the pancreas extend from the right to the left, they lie directly against first the inferior vena cava, second, the portal vein, third, the abdominal aorta, and, finally, the perirenal fascia about the left kidney.

The distal half of the first part of the duodenum and the second, third, and fourth parts of the duodenum follow a course around the head of the pancreas and then along the lower border of the neck and body of the pancreas (Fig. 21–4). In following this course, the second part of the duodenum comes into direct contact with the perirenal fascia about the right kidney, the third part of the duodenum comes into direct contact with the inferior vena cava, and the third and fourth parts at their border come into direct contact with the abdominal aorta.

In the RLQ and RUQ, the ascending colon ascends from approximately the lower border of the right lumbar region to the upper border of the right lumbar region (Fig. 21–5). The hepatic flexure at the upper end of the ascending colon lies directly against the perirenal fascia about the right kidney. In the LUQ and LLQ, the descending colon descends from the left hypochondriac region to the left inguinal region. The splenic flexure at the upper end of the descending colon lies directly against the perirenal fascia about the left kidney.

The Intraperitoneal Segments of the Small and Large Intestines in the Mid and Lower Abdomen: the Cecum, Transverse Colon, Sigmoid Colon, Jejunum, Ileum, and Appendix

Each of these intestinal segments, except for the cecum, lies suspended in the abdominopelvic cavity by a broad peritoneal ligament called a **mesentery.** In the right inguinal region, the cecum lies closely apposed to the posterior abdominal wall (Fig. 21–5). Across the breadth of the midregion of the abdomen, the transverse colon lies suspended from the posterior abdominal wall by a mesentery called the **transverse mesocolon;** the root of the transverse mesocolon extends from the hepatic flexure to the splenic flexure (Fig. 21–6). In the left inguinal region, the sigmoid colon lies suspended from the posterior abdominal and pelvic walls by a fan-shaped mesentery called the **sigmoid mesocolon** (Fig. 21–5).

In and about the umbilical region and encircled by the ascending, transverse, descending, and sigmoid segments of the large intestine, the coils of the jejunum and ileum lie suspended from the posterior abdominal wall by a mesentery called the **mesentery of the small intestine** (Fig. 21–5); the root of the mesentery of the small intestine extends from the duodenojejunal junction to the ileocecal junction (Fig. 21–6). In the right inguinal region, the appendix lies suspended from the mesentery of the small intestine by a mesentery called the **mesoappendix.**

The Liver and Gallbladder and the Intraperitoneal Viscera of the Upper Abdomen: the Tail of the Pancreas, Spleen, Stomach, and the Proximal Half of the First Part of the Duodenum

The liver lies mainly in the right hypochondriac and epigastric regions, suspended from the diaphragm and anterolateral abdominal wall by peritoneal ligaments (Fig. 21–5). Much of the gallbladder lies directly apposed to the visceral surface of the liver.

In the left hypochondriac region, a peritoneal ligament called the **splenorenal (leinorenal) ligament** extends from the anterior surface of the perirenal fascia about the left kidney to the hilum of the spleen (Fig. 21–7). The splenorenal ligament transmits the tail of the pancreas and suspends the spleen from the posterior abdominal wall.

Between the liver on the right and the spleen on the left lie the stomach and the proximal half of the first part of the duodenum (Fig. 21–8). A broad peritoneal ligament called the **lesser omentum** extends from the visceral surface of the liver to the lesser curvature of the stomach and the upper border of the proximal half of the first part of the duodenum. Another broad peritoneal ligament called the **greater omentum** arises from the greater curvature of the stomach and the lower border of the proximal half of the first part of the duodenum. The greater omentum is divisible into left, middle, and right parts. The leftmost part of the greater omentum is called the **gastrophrenic liga-**

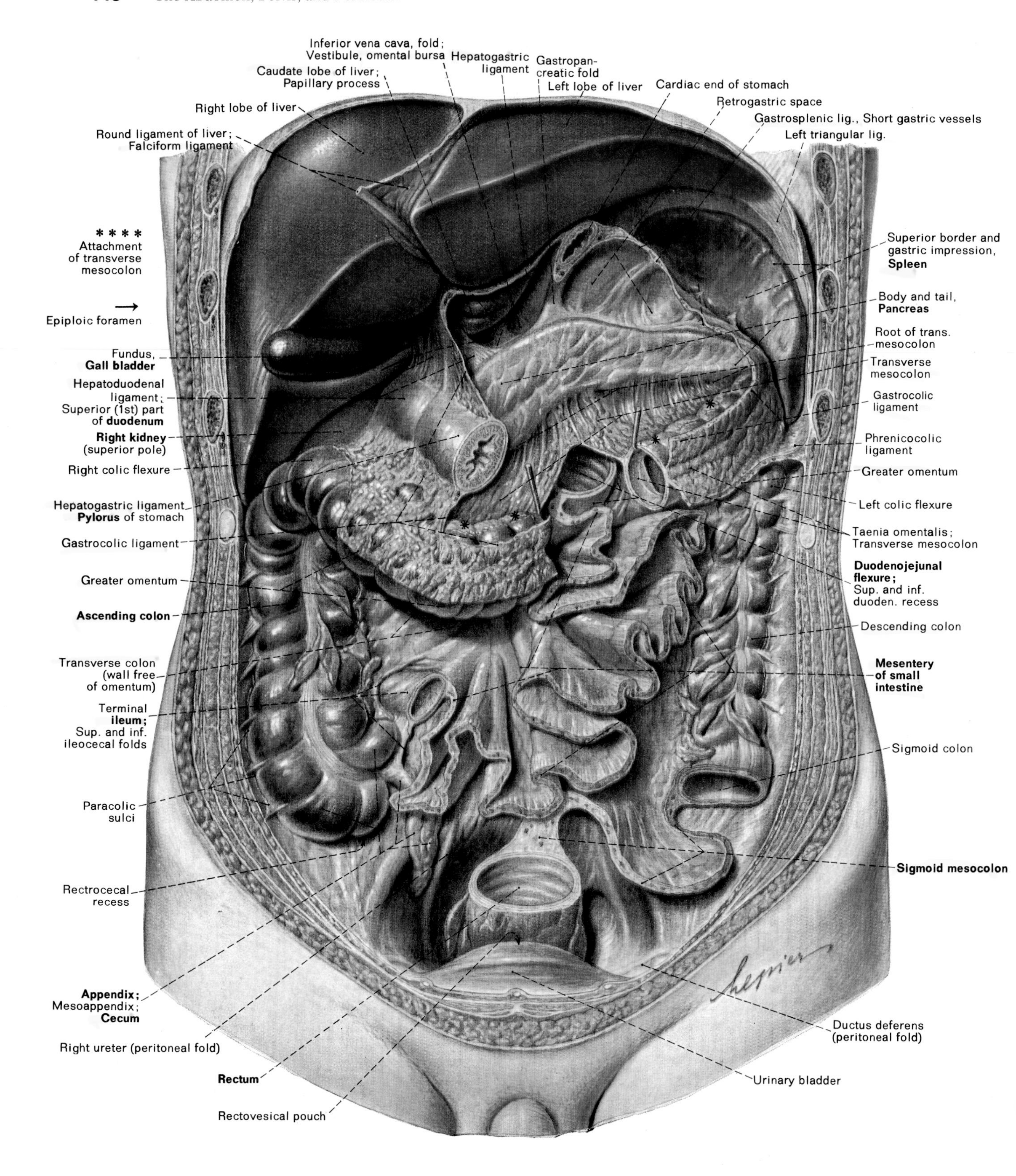

Fig. 21–5: Anterior view of the abdominal cavity in a male following excision of selected viscera. The stomach has been excised following transection at the cardiac opening and just proximal to the pylorus; the stomach's attachments to the greater and lesser omenta have also been severed. The jejunum and ileum have been excised following transection at the duodenojejunal junction and at the distal ileum and severance from its mesentery. A portion of the transverse colon and the greater omentum attached to it have been excised. The sigmoid colon has been excised following transection at its union with the descending colon and rectum and severance from its sigmoid mesocolon.

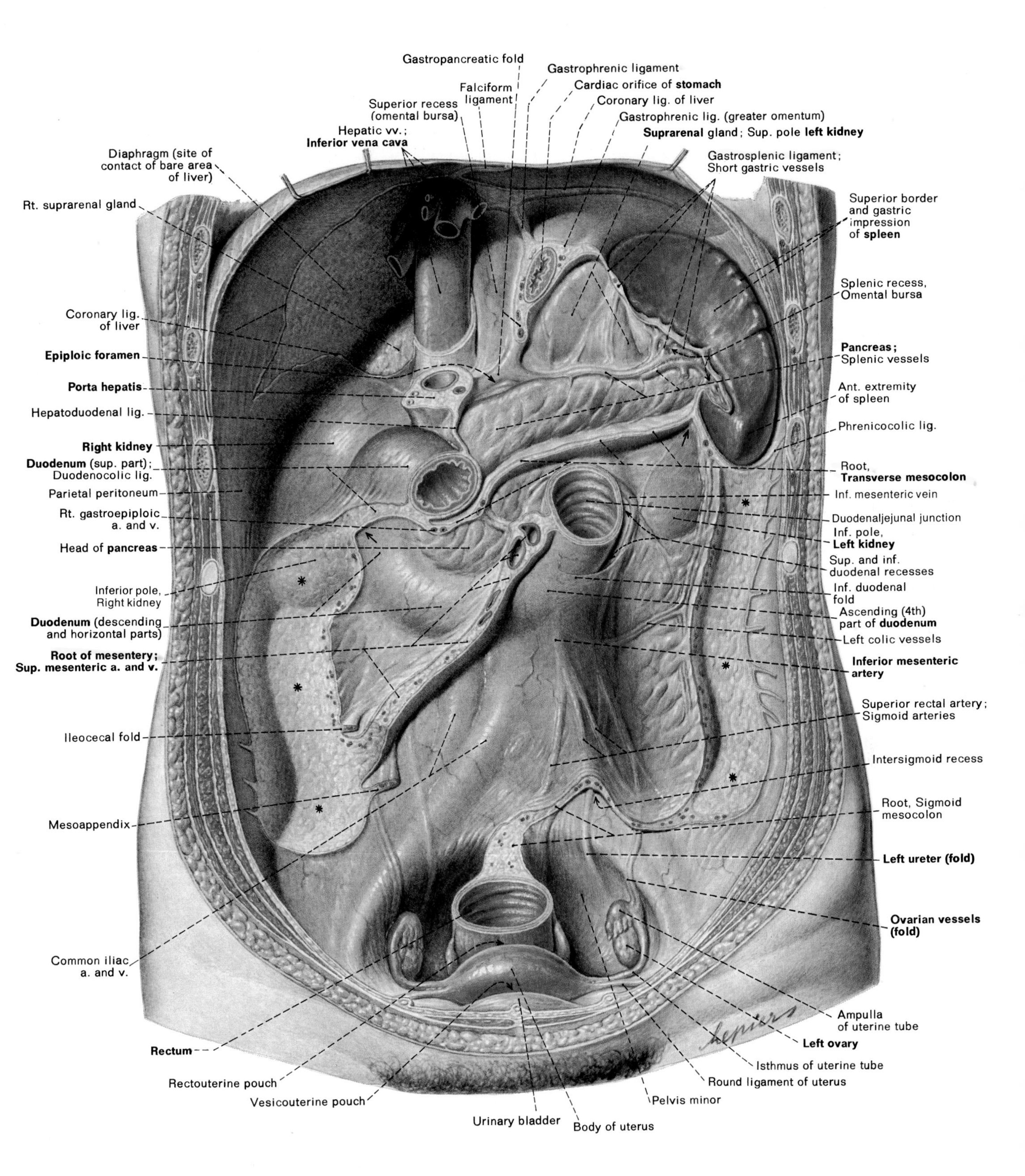

Fig. 21–6: Anterior view of the abdominal cavity in a female following excision of the liver, gallbladder, stomach, jejunum, ileum, appendix, cecum, ascending colon, transverse colon, descending colon, and sigmoid colon.

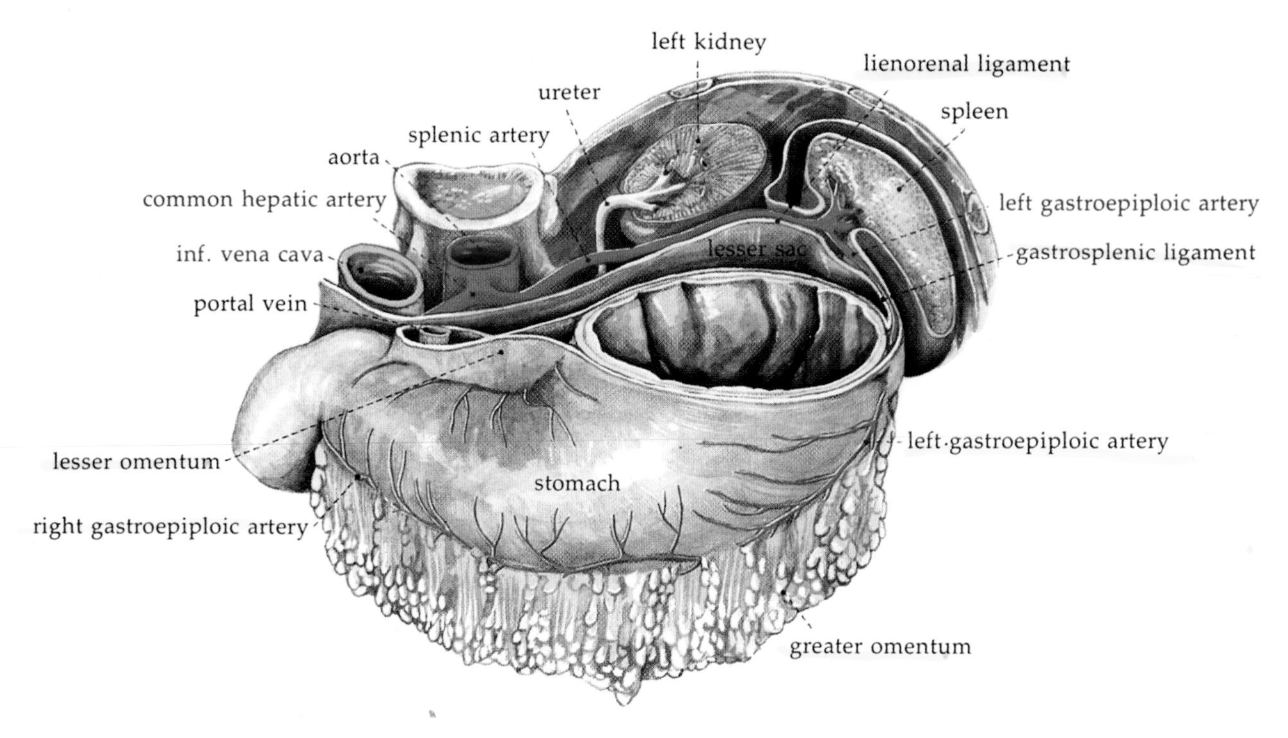

Fig. 21–7: Illustration of the spleen and its relationships to upper abdominal viscera.

ment because it reflects onto the undersurface of the diaphragm (Fig. 21–6). The middle part is called the **gastrosplenic ligament** because it reflects onto the hilum of the spleen (Figs. 21–7 and 21–8). The rightmost part is called the **gastrocolic ligament** because it extends to the transverse colon (Fig. 21–9). The gastrocolic ligament is relatively lengthy; it descends from its gastroduodenal margin into the lower abdomen before it folds back on itself to ascend back up to its colic margin. In most adults, the anterior descending fold of the gastrocolic ligament is fused with the posterior ascending fold (Fig. 21–9).

The greater and lesser omenta together with the stomach, transverse colon, diaphragm, and upper posterior abdominal wall serve to wall off a space within the peritoneal cavity called the **lesser sac** (**omental bursa**) (Figs. 21–7 and 21–9). The remaining space within the peritoneal cavity is called the **greater sac** because its volume is greater than that of the lesser sac (Fig. 21–9).

The lesser sac is sealed off superiorly by the diaphragm (Fig. 21–9), anteriorly by the lesser omentum, stomach and gastrocolic ligament (Fig. 21–9), inferiorly by the transverse colon and transverse mesocolon (Fig. 21–9), posteriorly by the upper posterior abdominal wall (Fig. 21–9), and on the left by the gastrophrenic, gastrosplenic, and splenorenal ligaments (Figs. 21–6 and 21–7). On the right, the lesser sac communicates with the greater sac through a passageway called the **epiploic foramen** (Fig. 21–8). The epiploic foramen is, in fact, the only passageway between the greater and lesser sacs. The epiploic foramen has four borders: The caudate lobe of the liver borders the foramen superiorly. The free right margin of the lesser omentum borders the foramen anteriorly; this margin transmits the portal vein, hepatic artery proper, and bile duct. The proximal half of the first part of the duodenum borders the foramen inferiorly. The inferior vena cava borders the foramen posteriorly.

The Pelvic Viscera: the Rectum, Urinary Bladder, and Uterus and Uterine Tubes in the Female

In both sexes, as the peritoneum descends from the posterior pelvic wall onto the floor of the pelvis, it covers in the midregion of the pelvis first the anterior and lateral surfaces of the upper third of the rectum and then the anterior surface of the middle third of the rectum (Fig. 21–9). In the female, the peritoneum continues anteriorly to next cover reproductive organs (specifically, the supravaginal part of the cervix of the uterus, the body of the uterus, and the uterine tubes) and finally the superior surface of the urinary bladder before reflecting onto the anterolateral abdominal wall (Fig. 21–9). In the male, the peritoneum extends anteriorly from the anterior surface of the middle third of

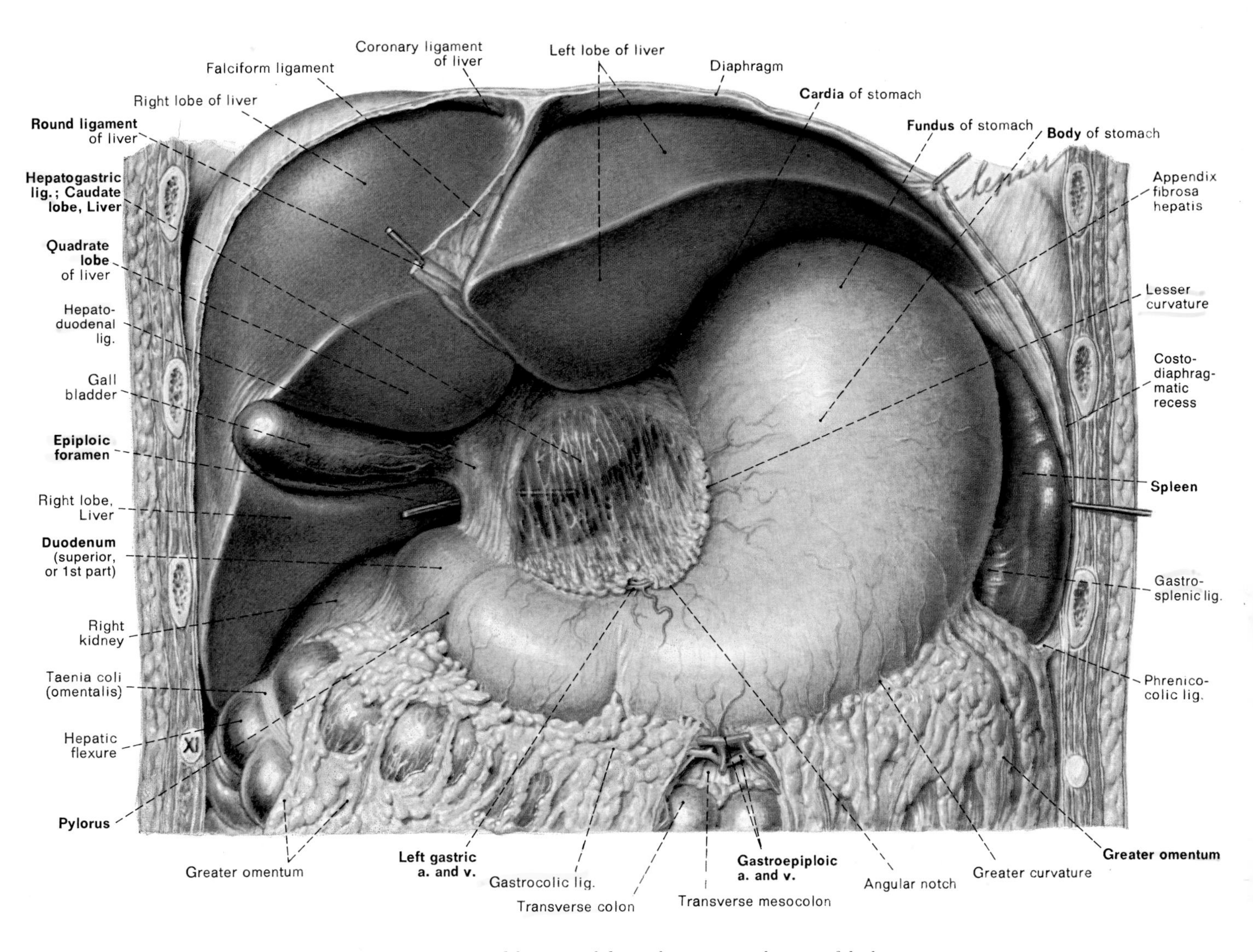

Fig. 21–8: Anterior view of the upper abdominal cavity upon elevation of the liver.

the rectum directly onto the superior surface of the urinary bladder and then onto the anterolateral abdominal wall.

THE TYPES OF PAIN ASSOCIATED WITH DISEASE AND INJURY OF ABDOMINAL AND PELVIC VISCERA

Disease or injury of an abdominal or pelvic viscus can produce three types of pain:

(1) **Visceral pain** is a dull, sickening pain which is poorly localized to one of the midline regions (epigastric, umbilical, or hypogastric) of the abdomen. Visceral pain is produced by the stimulation of visceral pain fibers. Visceral pain fibers are sensitive to acute stretching and anoxia. Visceral pain fiber endings are located in the muscular walls of hollow abdominal and pelvic viscera (such as the walls of the small and large bowel, the extrahepatic biliary ducts, the ureters, and the bladder) and in the fibrous capsules of solid abdominal viscera (such as the capsules of the liver, spleen, and kidneys).

(2) **Referred pain** is pain that a diseased or injured viscus refers to a cutaneous region of the body. The dermatomes of the painful cutaneous region are innervated by the same spinal cord segments that innervate the diseased or injured viscus. The painful cutaneous region is generally anatomically distant from the site of the diseased or injured viscus.

(3) **Somatic pain** is produced by the stimulation of parietal peritoneum pain fibers. Inflammation of the parietal peritoneum of the anterolateral abdominal wall produces somatic pain that is sharp and limited to the region of inflammation. By contrast, inflammation of the parietal peritoneum of the posterior abdominal wall produces somatic pain that is

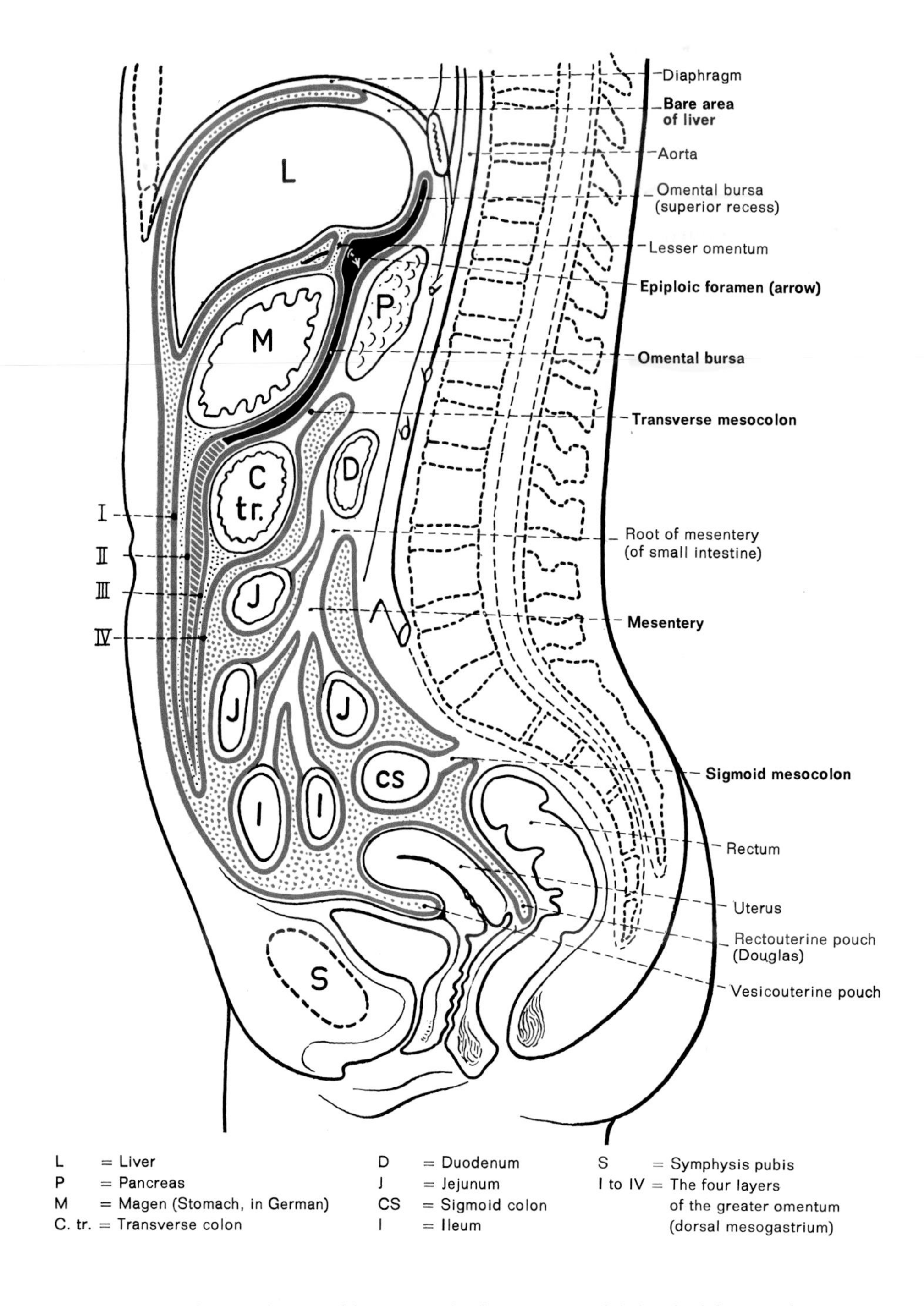

L = Liver
P = Pancreas
M = Magen (Stomach, in German)
C. tr. = Transverse colon
D = Duodenum
J = Jejunum
CS = Sigmoid colon
I = Ileum
S = Symphysis pubis
I to IV = The four layers of the greater omentum (dorsal mesogastrium)

Fig. 21–9: Schematic drawing of the peritoneal reflections in an adult female abdominopelvic cavity.

diffuse and localized to the midline. A diseased abdominal or pelvic viscus can produce somatic pain when its inflammatory process extends to the parietal peritoneum of those abdominopelvic wall regions with which the viscus contacts.

Stretching exacerbates the somatic pain of inflamed parietal peritoneum. Minimally inflamed parietal peritoneum in an anterolateral abdominal wall region causes an individual to voluntarily **guard,** or **splint** (stiffen), the inflamed region against unwanted movement. Worsening of the inflammation ultimately elicits reflex contraction and rigidity of the abdominal wall muscles overlying the inflamed region.

The sensitivity of inflamed parietal peritoneum to sudden stretching can be detected by pressing the approximated fingers of one hand into a region of the anterolateral abdominal wall and then suddenly withdrawing the fingers. The wave of sudden rebound movement that spreads throughout the abdominal walls elicits tenderness called **rebound tenderness** in those wall regions lined by inflamed parietal peritoneum. It is preferable to apply the finger pressure at a site distant from suspected sites of inflammation. An examiner must use discretion in applying appropriate pressure during testing for rebound tenderness, as the pain can be quite severe.

CHAPTER **22**

The Viscera Supplied by the Celiac Artery

The celiac artery and its branches supply the following viscera: the liver, gallbladder, extrahepatic biliary ducts, pancreas, spleen, lower third of the esophagus, stomach, and duodenum.

THE LIVER

The liver is a wedge-shaped organ that lies mainly in the right hypochondriac and epigastric regions (Fig. 21–5). Most of the liver's dome-shaped, upper surface lies immediately beneath the right dome of the diaphragm. The liver's lower surface faces both posteriorly and inferiorly; it is called the liver's visceral surface because it faces a variety of viscera: specifically, the inferior vena cava, gallbladder, extrahepatic biliary ducts, stomach, duodenum, hepatic flexure, right adrenal gland, and right kidney.

The inferior vena cava, gallbladder, ligamentum venosum, and ligamentum teres imprint an **H**-shaped group of grooves and fissures on the liver's visceral surface (the **ligamentum venosum** is the postnatal, cordlike remnant of the **fetal ductus venosus,** and the **ligamentum teres** (**round ligament**) is the postnatal, cordlike remnant of the **fetal umbilical vein**) (Fig. 22–1). The inferior vena cava and gallbladder impress grooves that form, respectively, the upper and lower parts of the right vertical limb of the **H.** The ligamentum venosum and ligamentum teres furrow fissures that form, respectively, the upper and lower parts of the left vertical limb of the **H.** The fissure that represents the crossbar of the **H** is called the **porta hepatis,** as it is the portal through which the **portal vein** and **hepatic artery proper** enter the liver and the **left** and **right hepatic ducts** exit the liver.

Most of the liver's exterior is covered with peritoneum. The **bare area of the liver** is the small, posterosuperior region of the liver's exterior which is not covered with peritoneum; it lies directly against the undersurface of the diaphragm (Figs. 22–1 and 22–2).

Three peritoneal ligaments (the **coronary ligament** and the **left** and **right triangular ligaments**) suspend the liver from the diaphragm. The coronary ligament is so named because it crowns the breadth of the upper surface of the liver (Fig. 22–2). The coronary ligament consists of anterior and posterior layers of peritoneum. The anterior and posterior layers of the coronary ligament form most of the peritoneal reflections that border the bare area of the liver. The triangular ligaments each consist of two triangular-shaped layers of peritoneum. The left and right triangular ligaments represent, respectively, the leftmost and rightmost limits of the anterior and posterior layers of the coronary ligament (Figs. 22–2 and 22–3).

The **falciform ligament** consists of two layers of peritoneum that extend between the midline of the anterolateral abdominal wall and the anterior surface of the liver (Fig. 22–3). The ligament is named for its sharp, sickle-shaped, lower border (falx is the Latin term for scythe, or sickle). The falciform ligament transmits the ligamentum teres (round ligament) along its lower border and the **para-umbilical veins** from the umbilical region toward the portal vein. The left and right layers of the falciform ligament diverge as they approach the superior surface of the liver and become continuous with the anterior layer of the coronary ligament.

The falciform ligament's margin of attachment to the liver's anterior surface demarcates the border between the liver's **left** and **right lobes** (Fig. 22–3). The **H**-shaped group of grooves and fissures on the liver's visceral surface further subdivides the right lobe into the **quadrate** and **caudate lobes** (Fig. 22–1). The crossbar and the upper parts of the vertical limbs of the **H** mark the borders of the caudate lobe; the crossbar and the lower parts of the vertical limbs of the **H** mark the borders of the quadrate lobe.

Percussion can be used during a physical examination to assess the size of the liver. As the patient lies

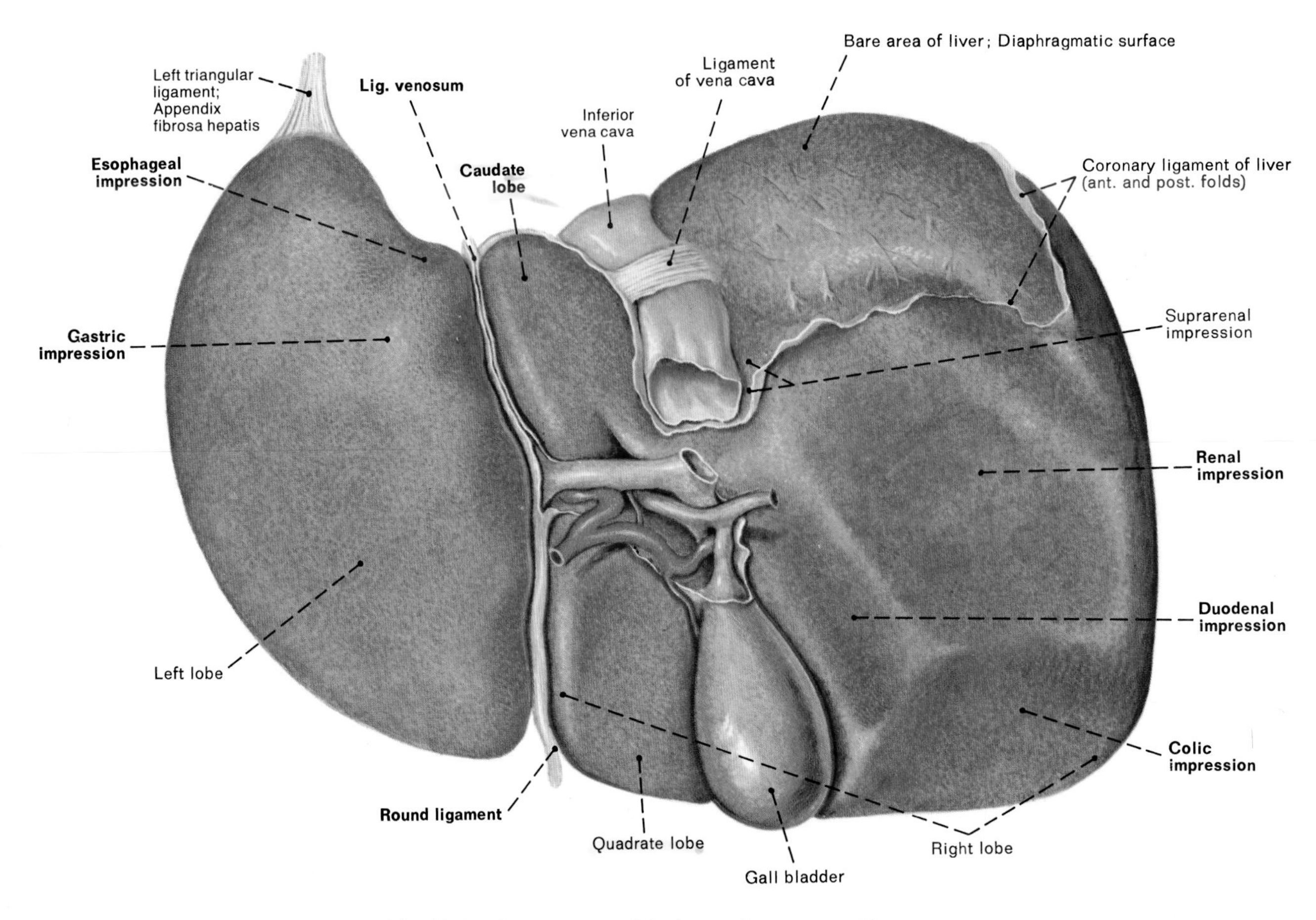

Fig. 22–1: Posterior view of the liver and its peritoneal ligaments.

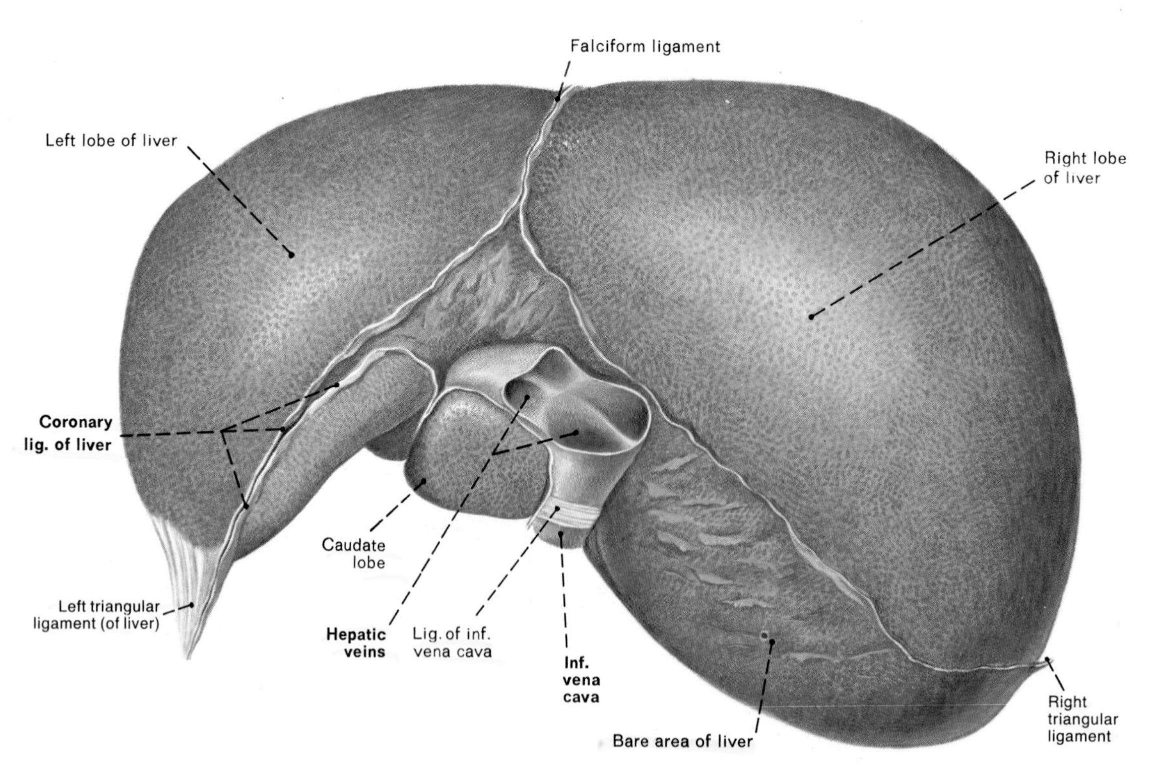

Fig. 22–2: Superior view of the liver and its peritoneal ligaments.

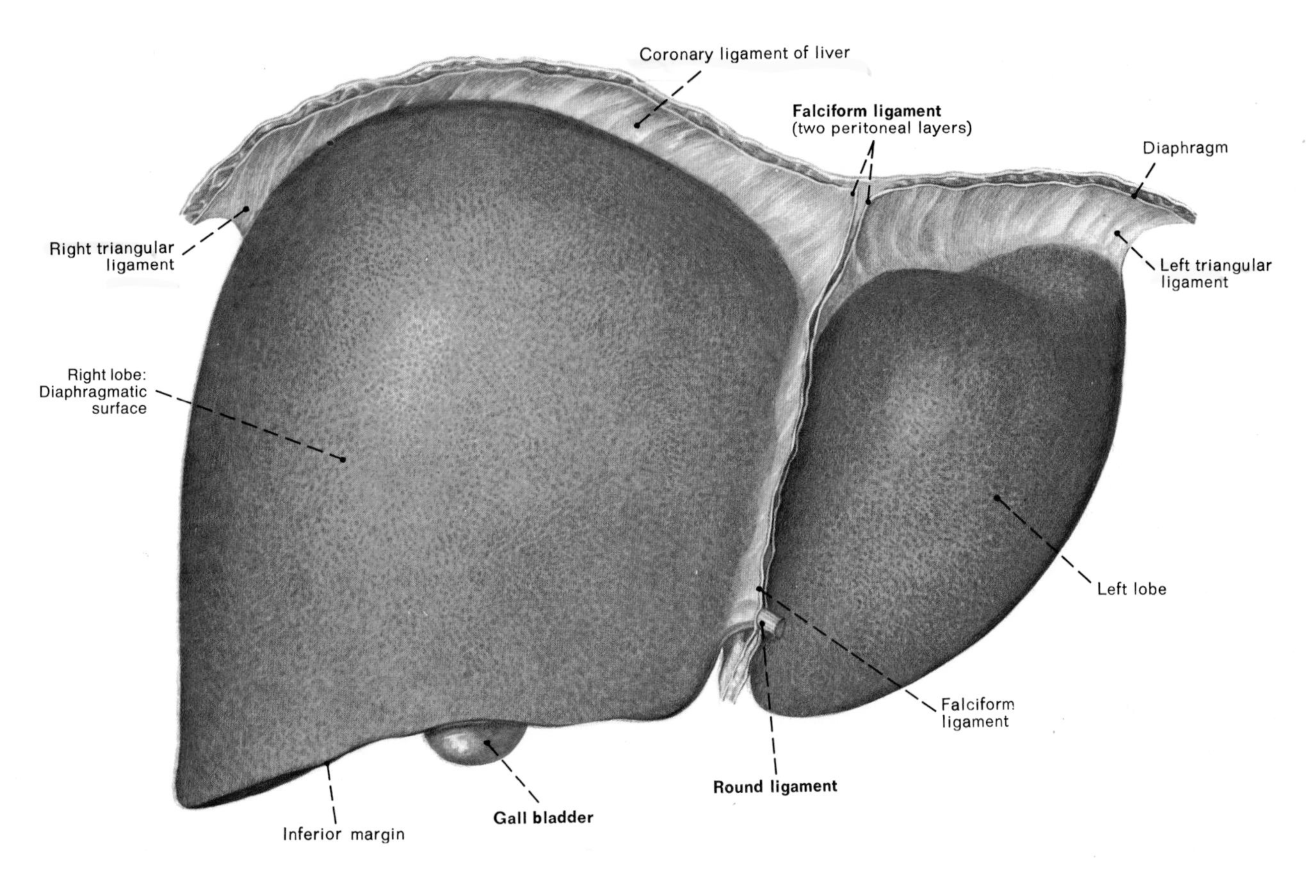

Fig. 22–3: Anterior view of the liver and its peritoneal ligaments.

supine and holds the breath at full expiration, the examiner percusses the anterior chest wall and anterolateral abdominal wall from the right second intercostal space downward along the right midclavicular line. In a normal adult, **percussion resonance** (due to percussion of the right lung) is encountered down to the highest level (which is typically that of the fourth intercostal space) at which the liver crosses the right midclavicular line (Fig. 22–4). A thin wedge of the right lung's lower lobe anteriorly covers the diaphragm and the underlying liver down to the level of the sixth rib; percussion of both the right lung and the liver between the levels of the fourth intercostal space and the sixth rib produces a zone of **percussion dullness** called **hepatic dullness.** A zone of **percussion flatness** called **hepatic flatness** is generally encountered from the level of the sixth rib down to that of the liver's anteroinferior margin. Percussion of bowel segments inferior to the liver produce percussion dullness, resonance, or tympany below the zone of hepatic flatness. The combined heights of the zones of hepatic dullness and hepatic flatness are a measure of the size of the liver; the normal range for the combined heights in an adult is 6 to 12 cm. It should be noted, however, that the presence of gas-filled bowel segments immediately posterior to the lower part of the liver's visceral surface can obscure determination of the lower limit of the zone of hepatic flatness, and thus lead to a faulty underestimate of liver size.

Deep palpation of the anterolateral abdominal wall generally permits palpation of the liver's anteroinferior margin. A normal liver's margin is soft, smooth, and non-tender; its course generally parallels the right costal margin of the rib cage.

THE GALLBLADDER

The gallbladder is a pear-shaped sac that lies suspended from the lower part of the liver's visceral surface (Fig. 21–5). The expanded, blind end of the gallbladder is called its **fundus,** the tapered, open end its **neck,** and the intervening region its **body;** the duct that extends from the gallbladder's neck is called the **cystic duct** (Fig. 22–5).

Occasionally, the entire gallbladder is covered with peritoneum, and lies suspended from the liver's visceral surface by a very short peritoneal ligament. In most instances, however, the anterosuperior surfaces of the body and neck of the gallbladder lie directly against the lower part of the liver's visceral surface, leaving only the fundus and the posteroinferior surfaces of the body and neck covered with peritoneum (at the borders of

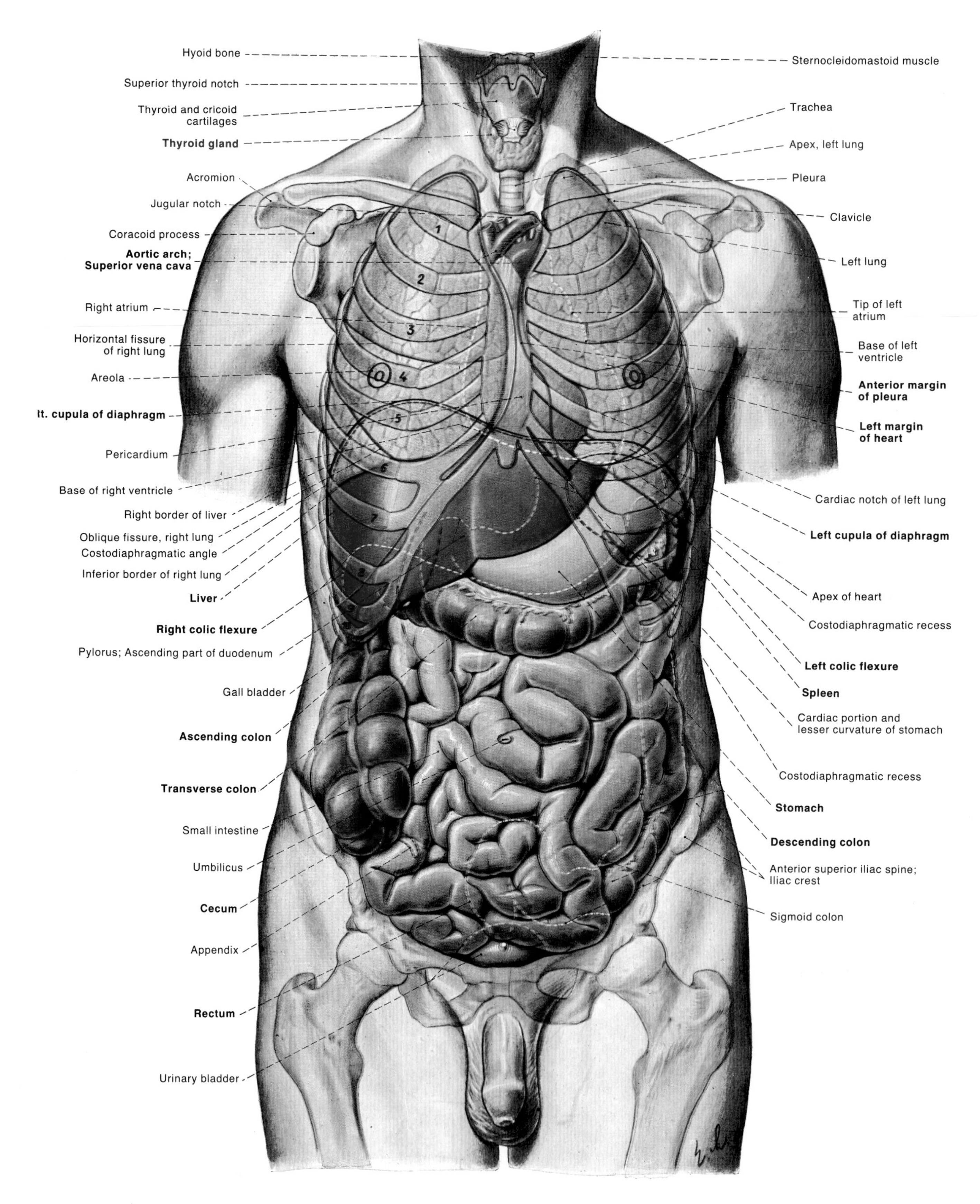

Fig. 22–4: Anterior view of the thoracic and abdominal viscera and their relationships to the anterior aspect of the rib cage.

these posteroinferior surfaces, the peritoneum extends onto the liver's visceral surface). The fundus projects below the liver's anteroinferior margin, and thus can come into contact with the anterior abdominal wall.

THE EXTRAHEPATIC BILIARY DUCTS

The extrahepatic biliary ducts conduct bile from the liver and gallbladder to the duodenum. The extrahepatic biliary apparatus begins with the emergence of the left and right hepatic ducts from the liver parenchyma at the porta hepatis.

The **left** and **right hepatic ducts** unite within the lesser omentum to form the **common hepatic duct** (Fig. 22–5). The common hepatic duct unites, in turn, with the cystic duct to form the **(common) bile duct.** The common bile duct descends within the lesser omentum immediately to the right of the hepatic artery proper and anterior to and to the right of the portal vein.

Upon leaving the lesser omentum, the common bile duct passes behind the first part of the duodenum and the head of the pancreas (Fig. 22–6) before traversing the posteromedial wall of the second part of the duodenum along the **main pancreatic duct** (Figs. 22–5 and 22–7). The two ducts unite to form a dilated saccule called the **hepatopancreatic ampulla (of Vater),** which opens into the lumen of the duodenum at the summit of the **major duodenal papilla** (**greater duodenal papilla**). The hepatopancreatic ampulla is encircled by a smooth muscle sphincter called the **hepatopancreatic sphincter (of Oddi).**

THE PANCREAS

The pancreas is a gland that extends across most of the breadth of the posterior abdominal wall at approximately the level of the transpyloric plane (Fig. 21–4). The expanded, rounded right end of the gland is called its **head** (Figs. 21–4, 22–7, and 22–8); the head is joined with the **body** of the gland (Fig. 22–8) by a constricted part called the **neck.** The tapered, left end of the gland is called its **tail** (Figs. 21–4, 22–7, and 22–8). The **uncinate process** is the hooked part that extends inferomedially from the lower left margin of the head of the gland (Figs. 22–6 and 22–8).

Proceeding from right to left, the most important relations of the pancreas are as follows: The first, second, and third parts of the duodenum encircle the head of the pancreas (Figs. 21–4 and 22–6). The head of the pancreas lies posteriorly directly against segments of the inferior vena cava (Fig. 21–4) and common bile duct (Fig. 22–6). The **superior mesenteric and splenic veins** join to form the **portal vein** behind the neck of the pancreas (Fig. 22–6). As the body of the pancreas extends to the left from the neck of the gland toward the tail, it comes into direct contact posteriorly with the abdominal aorta (at the level of the origin of the superior mesenteric artery) and the perirenal fascia about the left kidney (Fig. 21–4). As the splenic artery extends leftward from the celiac artery toward the spleen, it runs along the upper border of the body and tail of the pancreas (Fig. 22–8). As the splenic vein extends to the right from the spleen toward the origin of the portal vein, it runs first near the upper border of the tail and then behind the body and neck of the pancreas (Fig. 22–9). The **inferior mesenteric vein** joins the splenic vein behind the body of the pancreas (Fig. 22–9). The tail of the pancreas extends through the splenorenal ligament to

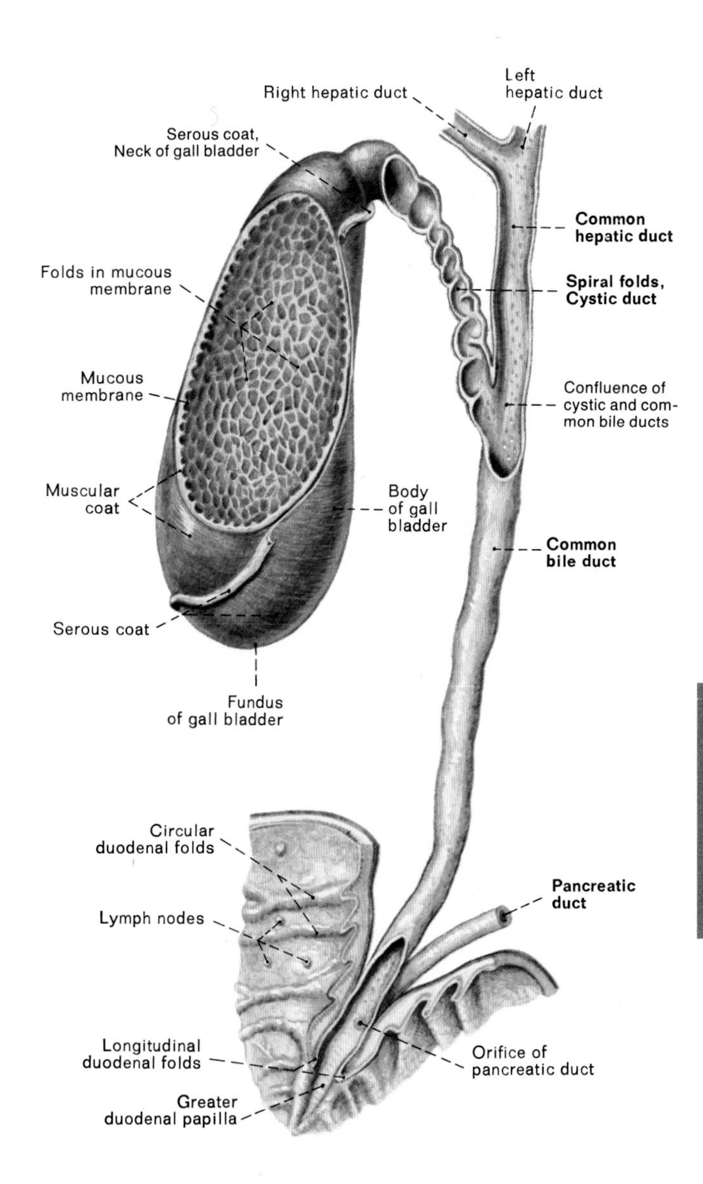

Fig. 22–5: Anterior view of the gallbladder and the extrahepatic biliary ducts.

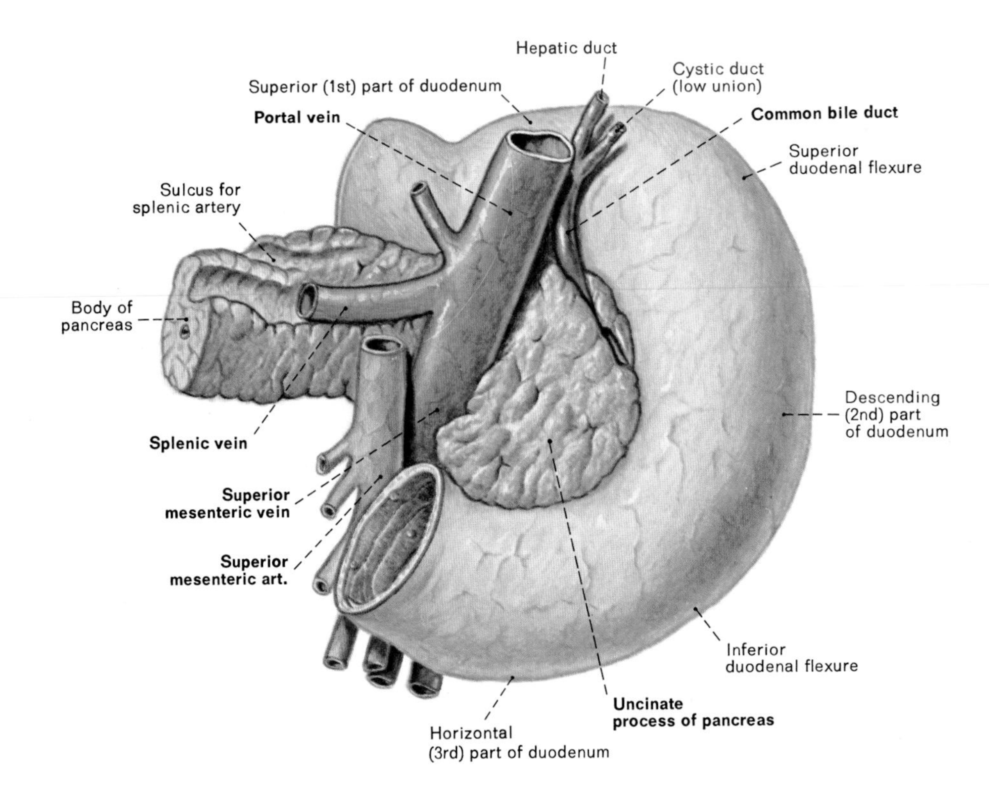

Fig. 22–6: Posterior view of the relations of the superior mesenteric, splenic, and portal veins to the head, neck, and body of the pancreas and the duodenum.

come into direct contact with the hilum of the spleen.

The main pancreatic duct begins in the tail of the pancreas, and then runs to the right through the body, neck, and head of the gland (Fig. 22–7). An **accessory pancreatic duct** may originate from the main pancreatic duct within the head of the pancreas; this accessory pancreatic duct generally opens into the lumen of the second part of the duodenum at the summit of a minor (lesser) duodenal papilla (Fig. 22–7).

THE SPLEEN

The spleen is the largest lymphatic organ of the body. It lies mainly in the left hypochondriac region, pressed upward, backward, and to the left against the diaphragm (Fig. 21–5). It lies above and to the left of the fundus of the stomach and immediately above the **phrenicocolic ligament** (the phrenicocolic ligament is a peritoneal ligament that secures the splenic flexure to the left flank region of the anterolateral abdominal wall) (Fig. 21–5).

The spleen's surface projection to the left side of the body trunk spans the levels of the ninth, tenth, and eleventh ribs; the long axis of the organ overlaps the downward course of the posterolateral part of the left tenth rib (Fig. 22–10). The spleen's anterosuperior border is firm and notched; its palpation below the left costal margin in an adult indicates splenomegaly.

THE STOMACH

The stomach openings at the esophageal and duodenal junctures are called, respectively, the **cardiac and pyloric openings** (Fig. 22–11). The stomach has an anterior surface, a posterior surface, a small, upper curved border to the right called the **lesser curvature,** and a large, lower curved border to the left called the **greater curvature.** The lesser curvature is indented along its lower margin by a sharply-defined notch called the **angular notch (incisura angularis)** (not shown in Fig. 22–11).

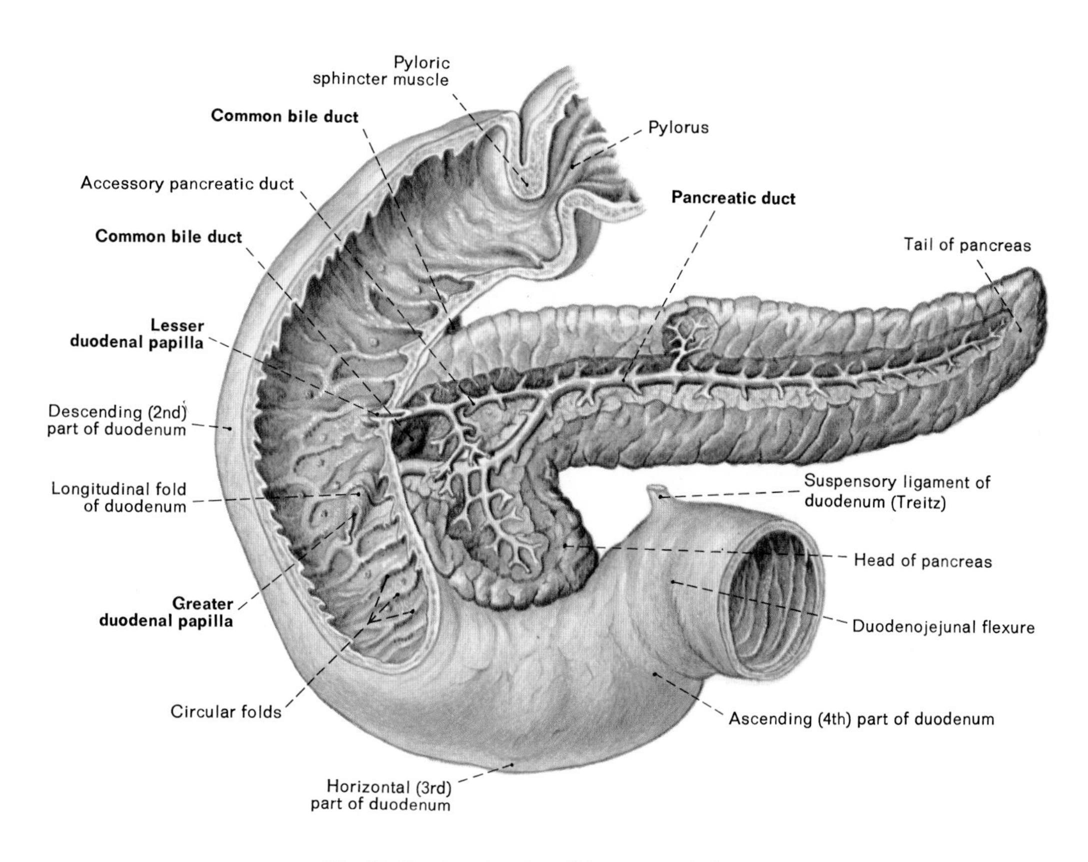

Fig. 22–7: Anterior view of the pancreatic duct system.

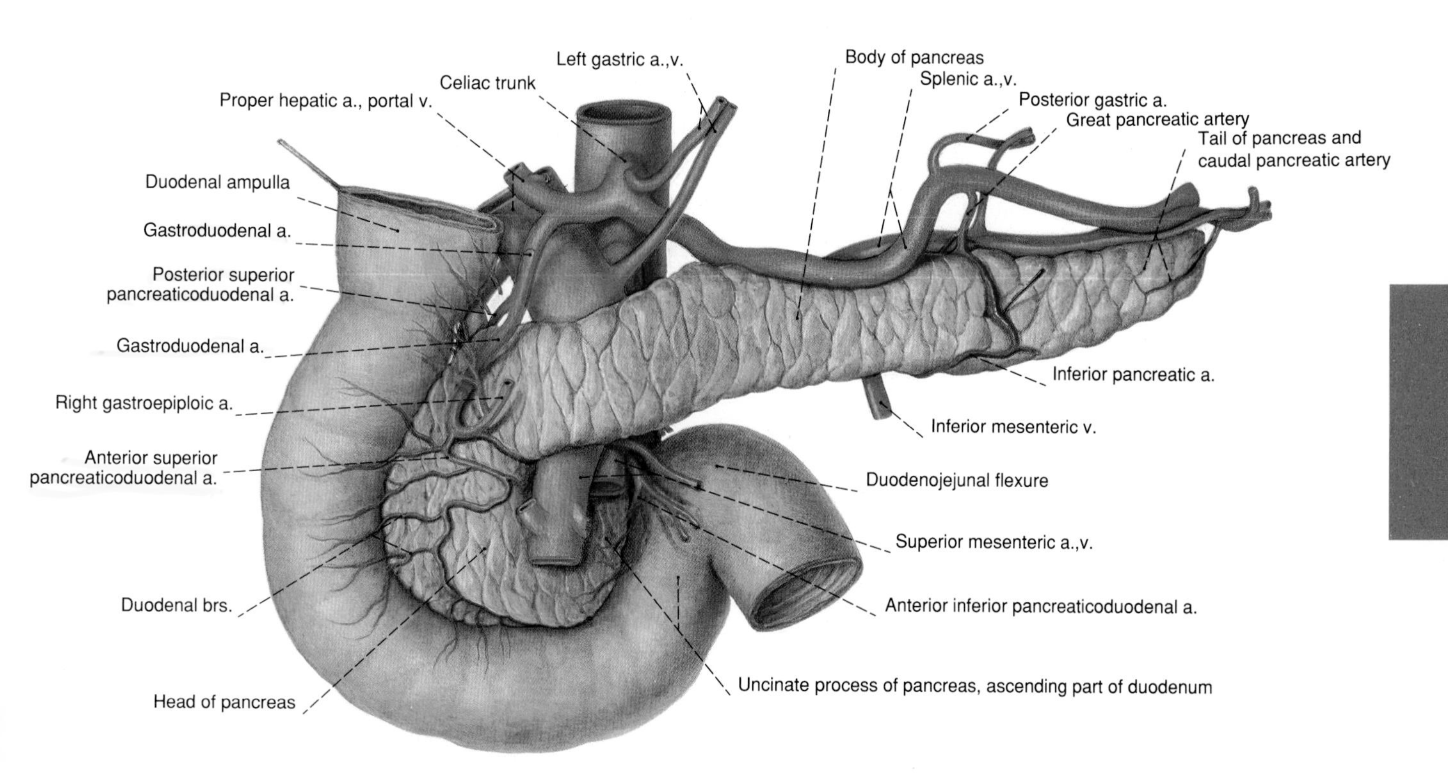

Fig. 22–8: Anterior view of the relations of branches of the celiac arterial tree and tributaries of the portal vein to the duodenum and pancreas.

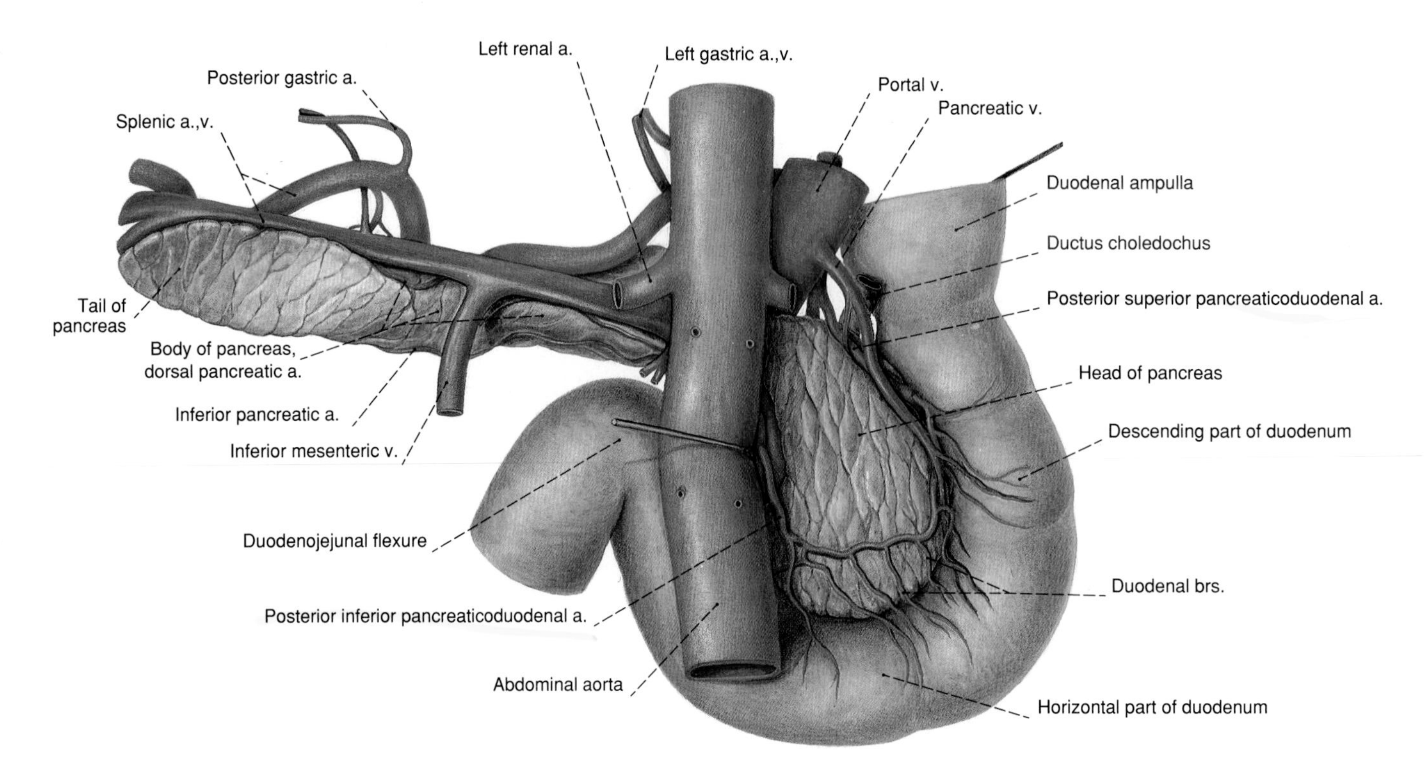

Fig. 22–9: Posterior view of the relations of branches of the celiac arterial tree and tributaries of the portal vein to the duodenum and pancreas.

The stomach has three parts: The **fundus** is the part that lies above the cardiac opening, the **pyloric part** is the part that extends from the angular notch to the pyloric opening, and the **body** is the part that intervenes between the fundus and pyloric part (Fig. 22–11). The pyloric part begins as a comparatively wide chamber, called the **pyloric antrum,** which suddenly narrows into a small caliber chamber called the **pyloric canal.** The region of thickened, stomach-wall musculature, called the **pyloric sphincter,** or **pylorus,** marks the end of the stomach and its junction with the duodenum.

The position and shape of the stomach are not fixed; they vary with the volume of stomach contents, the position of the body, and the phases of respiration. The stomach is commonly found mainly in the left upper quadrant.

THE DUODENUM

The duodenum is the shortest and widest segment of the small intestine. As it extends from the pylorus to the **duodenojejunal flexure** (which is the curved junction between the duodenum and jejunum) (Figs. 22–8 and 22–9), the duodenum traces a C-shaped course around the head, neck and body of the pancreas (Figs. 21–4 and 22–8). The duodenum is divided into four parts: the **first part,** or **superior part;** the **second part,** or **descending part;** the **third part,** or **horizontal part;** and the **fourth part,** or **ascending part.** The duodenum is the most fixed segment of the small intestine, and lies mainly in the umbilical region.

THE BLOOD SUPPLY TO THE LIVER, GALLBLADDER, PANCREAS, SPLEEN, STOMACH, AND DUODENUM

The **celiac artery** arises from the abdominal aorta just below the aortic opening of the diaphragm (Fig. 21–4). The celiac artery extends for only a very short distance before dividing into its three direct branches: the left gastric, splenic, and hepatic arteries.

The **left gastric artery** is the smallest branch of the celiac artery (Fig. 22–12). The left gastric artery gives rise to **esophageal branches** which supply the lower third of the esophagus and **gastric branches** which supply the stomach along the upper part of its lesser curvature.

CLINICAL PANEL V-1

DUODENAL ULCER DISEASE

Description: Duodenal ulcers are inflammatory erosions of the duodenal wall that extend through the mucosa into the submucosa and even into the muscularis mucosa. More than 95% of duodenal ulcers occur in the first part of the duodenum.

Common Symptoms: The pain of a duodenal ulcer is characteristically a sharp or burning pain in the epigastric region that typically appears within 1 to 3 hours after a meal. Sufferers generally find that certain foods and antacids generally provide temporary relief from pain.

Common Sign: The most common physical sign of a duodenal ulcer is epigastric tenderness. The tenderness is visceral pain elicited upon compression of the ulcer.

Radiographic Evaluation: An upper gastrointestinal (GI) contrast radiographic examination is generally the first radiologic procedure used to evaluate a patient suspected of having a duodenal ulcer. Examination generally begins with the patient swallowing a suspension of powdered barium sulfate. The finely particulate barium sulfate is swallowed in order to increase the radiographic density of the lumina of the segments of the upper GI tract (the lumina of the esophagus, stomach, and duodenum). Barium sulfate has a very high radiographic density (it greatly attenuates X-rays) because of the relatively high atomic number (56) of barium. Barium sulfate's chemical inertness and virtual insolubility in all the luminal environments of the GI tract render it the contrast medium of choice for examining the lumina and mucosa of the GI tract from the esophagus to the rectum.

When a segment of the GI tract is filled with a barium sulfate suspension, the suspension casts in radiographs and CT scans a radiopaque area whose border faithfully outlines that of the segment's lumen. The passage of the barium sulfate suspension through the segment is monitored by fluoroscopy so as to distinguish permanent strictures in the caliber of the segment from those temporarily produced by peristaltic activity. Some of the particulate barium sulfate remains temporarily adherent to the segment's mucosa after the bulk of the suspension has passed through the segment; this thin, adherent coat of particulate barium sulfate casts in radiographs and CT scans fine radiopaque shadows that detail surface features of the mucosal surface.

Resolution of a barium sulfate-coated, mucosal surface is generally enhanced if the GI tract segment is distended with gas. Such a radiographic procedure is called a double-contrast study because media of extremely different radiodensity (barium sulfate versus air) are used to image the wall of the segment. Effervescent powder is generally swallowed to generate gas in the stomach and duodenum. The sharp contrast between a thin, barium sulfate coat on the mucosal lining of the tract and the air in the lumen of the tract enhances resolution of the mucosal surface features.

Double-contrast studies permit identification of more than 90% of the duodenal ulcers confirmed upon fiberoptic endoscopic examination. A duodenal ulcer generally appears as a distinct outpouching of the duodenal wall.

The **splenic artery** is the largest branch of the celiac artery (Fig. 22–12). It presents a highly tortuous configuration as it extends along the superior border of the pancreas (Fig. 22–8), giving rise to branches [the **great pancreatic artery** (Fig. 22–8) and **dorsal pancreatic artery** (Fig. 22–9)] which supply the body of the pancreas. It then extends within the splenorenal ligament [where it gives rise to a branch, the **caudal pancreatic artery** (Fig. 22–8), which supplies the tail of the pancreas] to attain the hilum of the spleen, where it gives rise to the short gastric arteries, the left gastroepiploic artery, and many splenic branches. The **short gastric arteries** (Fig. 22–12) extend upward within the gastrosplenic ligament to supply the fundus of the stomach and anastomose with branches of the left gastric artery. The **left gastroepiploic artery** extends downward within the gastrosplenic ligament to supply the stomach and greater omentum along the upper part of the stomach's greater curvature. The **splenic branches** supply the spleen.

The **hepatic artery** follows an almost straight course as it extends from its origin to the upper margin of the first part of the duodenum (Fig. 22–12). In the vicinity of the upper margin of the first part of the duodenum, the hepatic artery first gives rise to the **gastroduodenal** and **right gastric arteries,** and then ascends within the lesser omentum (near the lesser omentum's free right border) to reach the porta he-

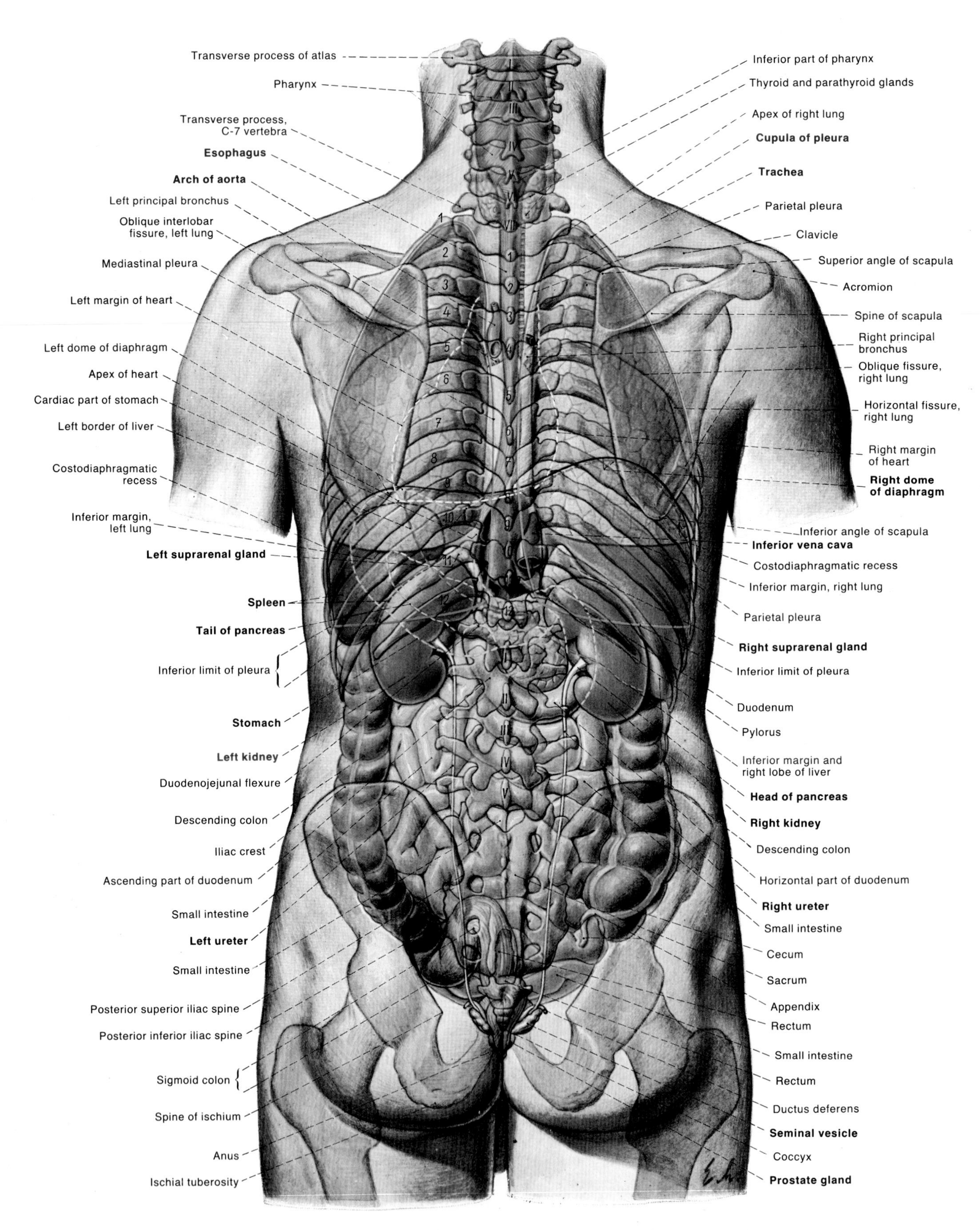

Fig. 22–10: Posterior view of the thoracic and abdominal viscera and their relationships to the scapulae and the posterior aspect of the rib cage.

patis, where it divides into its two terminal branches, the **left** and **right hepatic arteries.** The segment of the hepatic artery that extends from the celiac artery to the origin of the gastroduodenal artery is called the **common hepatic artery.** The segment of the hepatic artery that extends from the origin of the gastroduodenal artery to the origins of the left and right hepatic arteries is called the **hepatic artery proper.**

The distribution of the hepatic artery's branches is as follows: The right gastric artery supplies the stomach along the lower part of its lesser curvature; its terminal branches anastomose with gastric branches of the left gastric artery (Fig. 22–12). The gastroduodenal artery passes behind the first part of the duodenum before giving rise to the right gastroepiploic and superior pancreaticoduodenal arteries. The **right gastroepiploic artery** supplies the stomach and greater omentum along the lower part of the stomach's greater curvature; its terminal branches anastomose with terminal branches of the left gastroepiploic artery. There are generally two superior pancreaticoduodenal arteries, an **anterior superior pancreaticoduodenal artery** that lies anterior to the head of the pancreas (Fig. 22–8) and a **posterior superior pancreaticoduodenal artery** that lies posterior to the head of the pancreas (Fig. 22–9). Both arteries supply the duodenum and the head and neck of the pancreas.

The following section on the venous drainage of the liver, gallbladder, pancreas, spleen, stomach, and duodenum describes the intrahepatic distribution of the blood supplied by the left and right hepatic arteries. The right hepatic artery commonly gives rise to the cystic artery. The **cystic artery** (which is not shown in Fig. 22–12) supplies the gallbladder and the cystic duct.

THE VENOUS DRAINAGE OF THE LIVER, GALLBLADDER, PANCREAS, SPLEEN, STOMACH, AND DUODENUM

The **portal vein** is the recipient of the venous drainage from the gallbladder, pancreas, spleen, stomach, and duodenum (Fig. 22–13). Major portal tributaries from these viscera include:

(1) the short gastric veins, which drain the fundus of the stomach
(2) the left gastric vein, which drains the lower third of the esophagus and the stomach along the upper part of its lesser curvature
(3) the right gastric vein, which drains the stomach along the lower part of its lesser curvature
(4) the left gastroepiploic vein, which drains the stomach along the upper part of its greater curvature
(5) the right gastroepiploic vein, which drains the stomach along the lower part of its greater curvature
(6) the splenic vein, which drains the spleen and the body and tail of the pancreas
(7) the superior pancreaticoduodenal veins, which drain the duodenum and the head and neck of the pancreas.

The portal vein is formed behind the neck of the pancreas by the union of the splenic and superior mesenteric veins (Fig. 22–6). It ascends within the lesser omentum to reach the porta hepatis; in the lesser omentum, it lies posterior to and between the hepatic artery proper and common bile duct. At the porta hepatis, the portal vein divides into its two terminal branches, the **left** and **right branches of the portal vein.**

Blood flow within the portal vein is laminar; in other words, the blood flowing within the "left half" of the portal vein, which comes mainly from the splenic vein, does not mix with the blood flowing within the "right half" of the portal vein, which comes mainly from the superior mesenteric vein. The left and right branches of the portal vein thus receive blood that was primarily conducted into the portal vein by, respectively, the splenic and superior mesenteric veins.

The liver thus receives blood from two sources: the portal vein and the hepatic artery proper. The portal vein and hepatic artery proper conduct, respectively, about 70% and 30% of the blood entering the liver.

At the porta hepatis, the left and right branches of the portal vein and the left and right hepatic arteries each give rise to a vascular tree (a tree-like network of vessels whose branches become progressively smaller as they extend throughout the parenchyma of the liver). The vascular tree that emanates from the left branch of the portal vein parallels the vascular tree that stems from the left hepatic artery, and the vascular tree that emanates from the right branch of the portal vein parallels the vascular tree that stems from the right hepatic artery. There is, therefore, in each side of the liver, a tree of portal venous vessels and a tree of hepatic arterial vessels that supply identical sectors of the liver in a parallel fashion.

There is also in each side of the liver a tree-like network of ducts that conducts bile toward the porta hepatis. Moreover, the tree of biliary ducts parallels the portal venous and hepatic arterial trees in the same side of the liver. The tree-like networks of biliary ducts in the left and right sides of the liver respectively conduct bile towards the left and right hepatic ducts.

The left and right trees of portal venous vessels, hepatic arterial vessels, and biliary ducts thus functionally divide the liver into left and right sides. The left and right sides of the liver do not have anatomic names and do not directly correspond to the left and right lobes of the liver. The left side of the liver consists of the left lobe, the quadrate lobe, and part of the caudate lobe. The remainder of the liver represents its right side.

The lobar distribution of the liver's blood supply and the lobar pattern of its biliary drainage may be summarized as follows: The liver's left lobe and quadrate lobe are supplied by the left branch of the portal vein and the left hepatic artery, and their bile is drained by the left hepatic duct. The liver's caudate lobe is supplied by both the left and right branches of the portal vein and by both the left and right hepatic arteries, and its bile drains into both the left and right hepatic ducts. The remainder of the liver (which is the part of the liver's right lobe exclusive of the quadrate and caudate lobes) is supplied by the right branch of the portal vein and the right hepatic artery, and its bile is drained by the right hepatic duct.

The main branches of the left and right trees of portal venous vessels, hepatic arterial vessels, and biliary

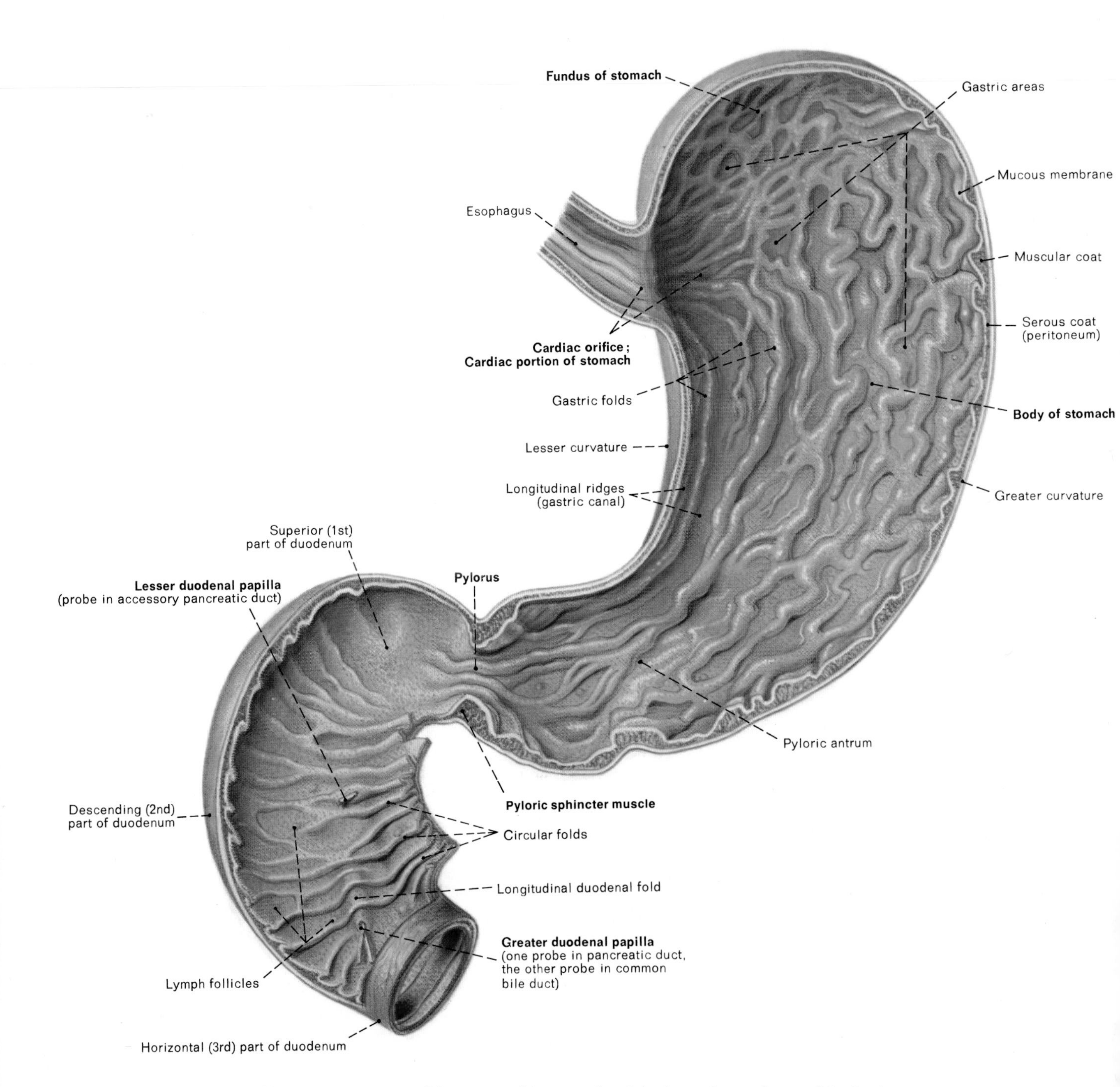

Fig. 22–11: Anterior view of the interior of the stomach and the first and second parts of the duodenum.

ducts also functionally divide the liver into eight **segments.** The reader should refer to surgical texts for discussions of the segmental anatomy of the liver. The segments of the liver are the smallest sectors of the liver that can be individually excised without compromising blood supply to or biliary drainage from other sectors.

In each side of the liver, blood is drained by small veins called **central veins,** that ultimately give rise (by repetitive anastomosis within the liver) to three large veins called the **left, middle,** and **right hepatic veins.** Each of these large hepatic veins drains blood from two or more segments of the liver. The hepatic veins emerge from the bare area of the liver to unite with the inferior vena cava immediately below its entrance into the caval opening in the diaphragm. In most individuals, the left and middle hepatic veins do not directly unite with the inferior vana cava, but, instead, join each other to form a common trunk that directly unites with the inferior vena cava.

THE CELIAC PLEXUS

The **celiac plexus** is the plexus of the autonomic nervous system that surrounds the celiac artery; it contains two large, sympathetic ganglia called the **left** and **right celiac ganglia.** The **left** and **right greater splanchnic nerves** transmit almost all the preganglionic sympathetic fibers to the celiac ganglia. The postganglionic sympathetic fibers that exit the celiac ganglia and plexus innervate target tissues in the viscera supplied by the celiac artery; almost all these postganglionic sympathetic fibers attain their target tissues by coursing along the branches of the arterial tree that sprouts from the celiac artery.

The dorsal root ganglia of spinal nerves T5-T9 contain the cell bodies of almost all the sensory nerves that innervate the viscera supplied by the celiac artery. The fibers of these sensory nerves extend from their target tissues first to the celiac plexus, by coursing in a retrograde fashion along the branches of the celiac arterial

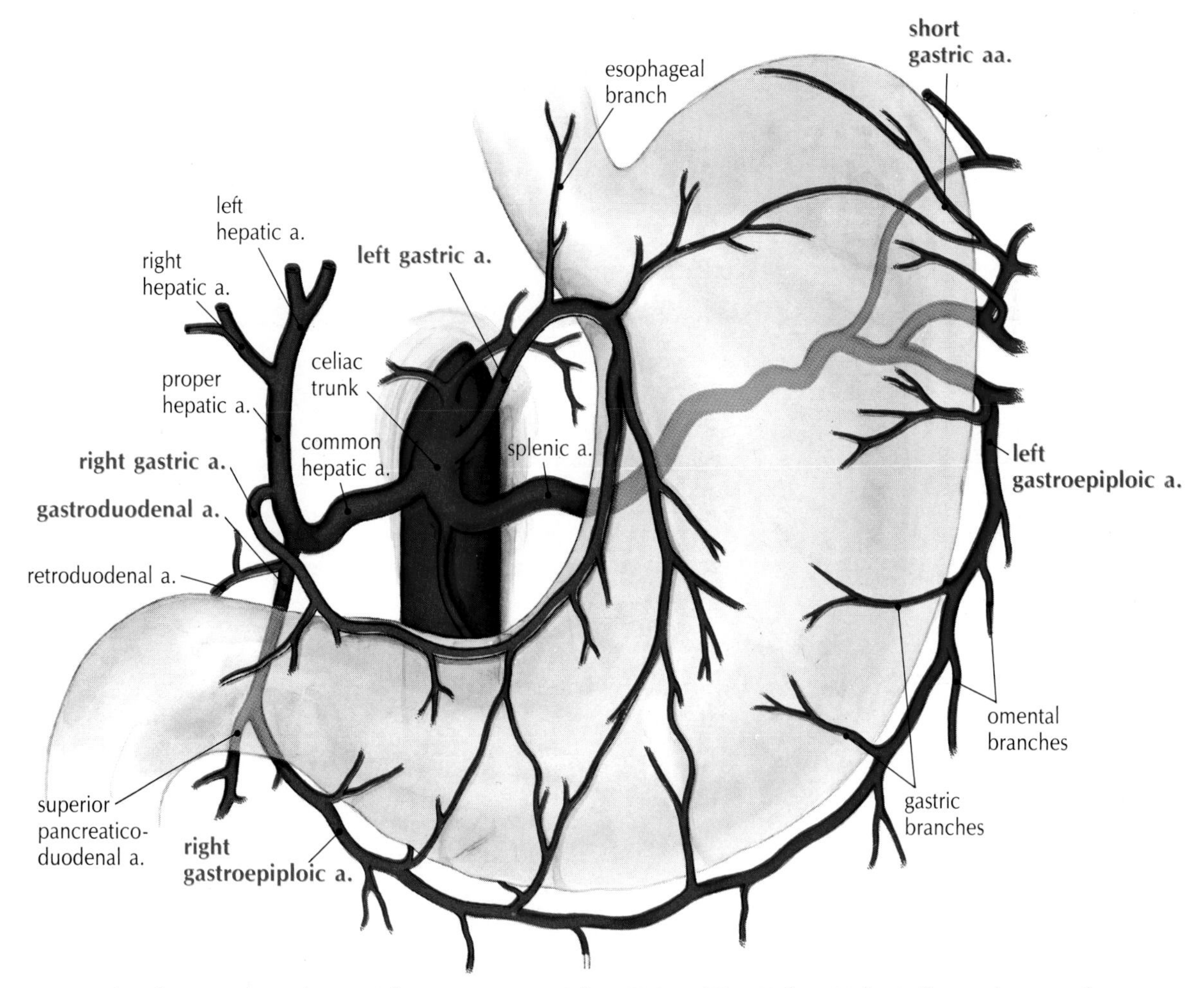

Fig. 22–12: The celiac arterial tree. (From Melloni, J. L., Dox, I., Melloni, H. P., and B. J. Melloni: Melloni's Illustrated Review of Human Anatomy, J. B. Lippincott Co., Philadelphia, 1988, p. 19.)

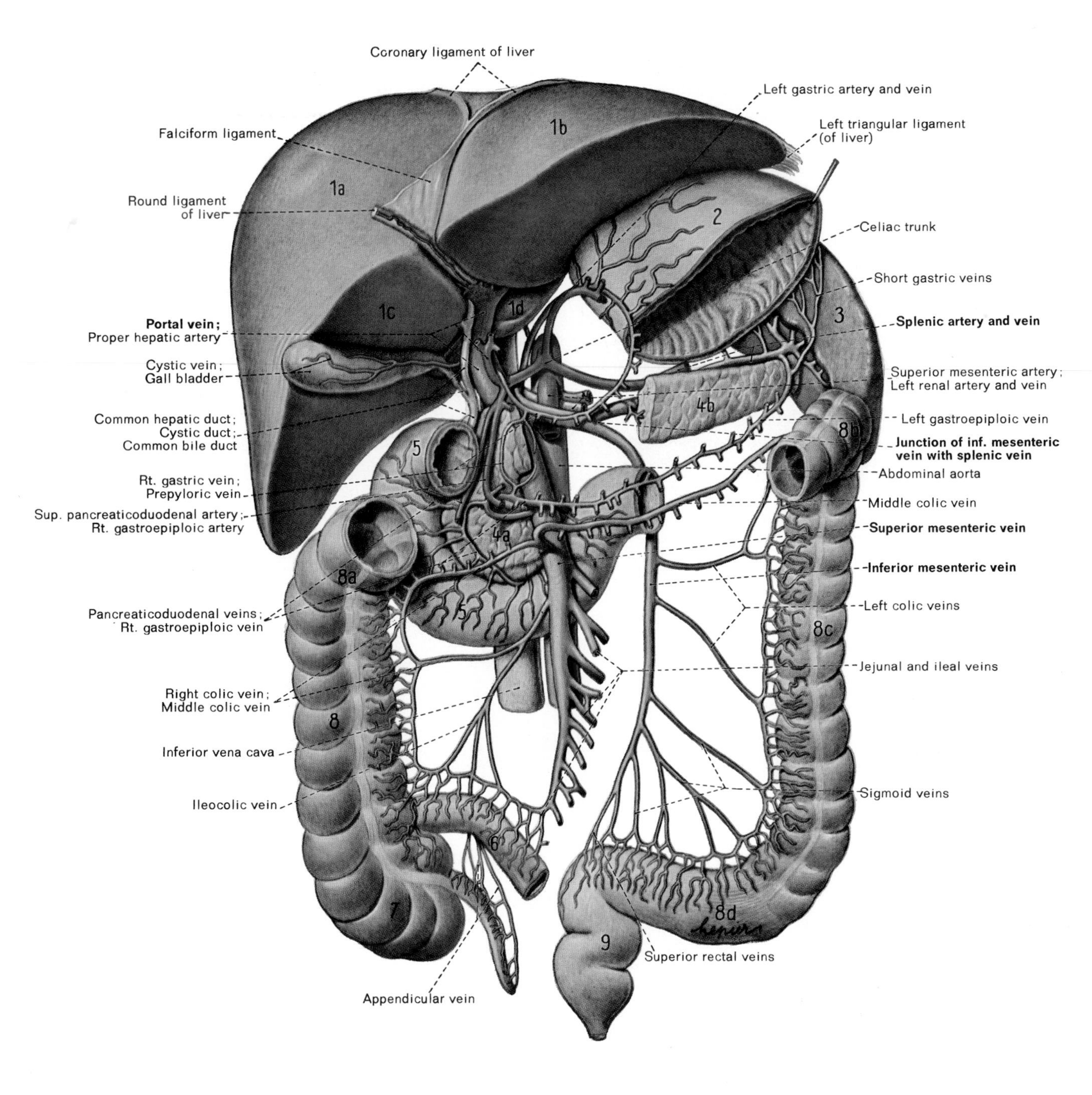

1a Right lobe of liver
1b Left lobe of liver
1c Quadrate lobe of liver
1d Caudate lobe of liver
2 Stomach
3 Spleen
4a Head of pancreas
4b Tail of pancreas
5 Duodenum
6 Ileum
7 Cecum
8 Ascending colon
8a Right colic flexure
8b Left colic flexure
8c Descending colon
8d Sigmoid colon
9 Rectum
↑ = Junction of sup. mesenteric v. and splenic vein to form portal vein

Fig. 22–13: The abdominal portal system of veins.

tree. From the celiac plexus, the sensory fibers accompany the left and right greater splanchnic nerves to the sympathetic trunk ganglia associated with the T5-T9 spinal cord segments; the white rami communicantes of these ganglia transmit the fibers to the appropriate dorsal roots and their ganglia. Almost all the sensory fibers that innervate the viscera supplied by the celiac trunk thus enter the spinal cord at the spinal cord segments that give rise to the preganglionic sympathetic fibers for the greater splanchnic nerves. The only exceptions are some sensory fibers from the liver, gallbladder, and the extrahepatic biliary ducts that enter the spinal cord at the C3-C5 segments; the **right phrenic nerve** transmits these fibers to their cell bodies in the dorsal root ganglia of spinal nerves C3, C4, and C5.

Disease or injury of the viscera supplied by the celiac artery produces visceral pain that is (most commonly) poorly localized to the epigastric region. Disease or injury of any of the viscera may refer pain to parts of the T5-T9 dermatomes. Disease or injury of the liver, gallbladder, and extrahepatic ducts may also refer pain to the right shoulder (as it represents parts of the C3-C5 dermatomes on the right side).

THE ANTERIOR AND POSTERIOR VAGAL TRUNKS

The **left and right vagus nerves** provide parasympathetic innervation to all the viscera supplied by the celiac artery. The preganglionic parasympathetic fibers provided by the vagus nerves enter the abdominal cavity distributed within the **anterior and posterior vagal trunks,** which are principally formed from, respectively, the left and right vagus nerves. The anterior vagal trunk gives rise to gastric, hepatic, and duodenal branches; the posterior vagal trunk gives rise to gastric and celiac branches (the latter are branches to the celiac plexus).

THE CELIAC LYMPH NODES

The **celiac lymph nodes** are the cluster of pre-aortic nodes around the celiac artery. They are the terminal group of nodes to receive lymph from the liver, gallbladder, pancreas, spleen, stomach, and duodenum. Their afferent lymphatics extend from groups of nodes strung out along the branches of the celiac artery.

CASE **V.2**

The Case of Joseph Bruich

INITIAL PRESENTATION AND APPEARANCE OF THE PATIENT

A 51 year-old, white man named Joseph Bruich walks into the emergency center at 11:00 AM. He appears apprehensive and uncomfortable; he bends his body forward as he walks around within the examination room.

QUESTIONS ASKED OF THE PATIENT

What is it that has brought you to the emergency center? I have a severe pain in my abdomen.

Where does it hurt? Right across here [The patient sweeps his left hand from the epigastric region through the left hypochondriac region all the way over to the left flank].

When did the pain start? Yesterday.

Do you recall what you were doing when the pain started? Yes, I was sitting down by my desk to begin work after lunch when the pain struck.

I notice that you just used the word "struck" to describe how the pain started. Do you recall how long it took for the pain to build up to its greatest intensity from the moment you were first aware of it? Oh, I'd say it gradually increased over an hour or so. *[In general, the striking of sudden, intense abdominal pain suggests a catastrophic event or a rapidly developing pathologic condition.]*

Has the intensity, or severity, of the pain changed since it built up to its greatest intensity? No, it's been pretty much the same.

How would you describe the pain? It's like a deep ache in my abdomen that bores all the way through to my back. *[Deep abdominal pain that bores all the way through the back suggests a disease or condition involving one or more retroperitoneal and/or secondarily retroperitoneal viscera. The patient's description of a constant, deep back pain suggests that all or at least a major component of the pain is visceral pain.]*

Has the nature of the pain or its location changed since it struck? No, not really.

Have you ever had pains like this before? No.

Is there anything that makes the pain worse? Yes, lying down. The pain bothered me so much last night that I didn't sleep well. The pain is also worse when I take a deep breath.

Is there anything that relieves the pain? Yes, I've found that if I bend over like this [the patient flexes his abdomen anteriorly], the pain is less intense. The only rest and sleep I got last night was when I bent myself over like this and also brought my legs up [the patient flexes his thighs].

Is there anything else that has also started bothering you since yesterday, or that you think may be associated with the pain in your abdomen? Well, I vomited right after the pain struck me yesterday, and then retched twice after I went home. I'm still a bit nauseous even now. I don't know if this is important or not, but I first noticed this morning that I'm having difficulty catching my breath, and it hurts if I take a deep breath. I've also been coughing, but I don't bring anything up.

Before we get to your difficulty with breathing, I'd like to first ask about your nausea and vomiting. Do you recall what your vomit looked like when you vomited right after the pain struck? Sort of. As I remember it, it looked pretty much like what I had eaten for lunch.

Now, when did you first notice that you have difficulty catching your breath? I noticed the difficulty this morning when I first went up the stairs at home. Every time I had to go up the stairs, I noticed I was really breathing hard by the time I got to the top step.

You just mentioned that it hurts to take a deep breath. Can you show me where it hurts? It hurts me deep in the left side of my back. [As the examiner touches several points on the left side of the patient's back in order to more specifically locate the pain, the patient locates the pain at the levels of the posterior segments of the left tenth, eleventh, and twelfth ribs.]

When you take a deep breath and it hurts, what does the pain feel like? It feels like a sharp knife cutting through me. *[A sharp, knife-like pain upon inspiration generally indicates inflammation of parietal pleura on the painful side of the chest. The stretching of inflamed parietal peritoneum in the upper abdomen can also be the basis of sharp pain upon deep inspiration.]*

I'd like to get back now to your abdominal pain. How has the abdominal pain affected your diet? I have not felt like eating much. I tried eating a little of the dinner that my wife prepared last night, but that only made the pain worse for a time, so much so that I vomited most of my dinner.

Has there been any change in your bowel habits or bowel movement? Not really.

Do you feel any abdominal fullness? Yes. Now that you mention it, I have felt bloated since this morning.

Have you had any recent illnesses prior to this attack of pain in your abdomen? No.

Have you had a problem with indigestion, either recently or in the past? No, not really. You know, occasionally, I'll eat too much and get a lot of gas in my stomach, but Pepto-Bismol [an antacid] helps that problem right away.

Do you know any individuals around you, such as members of your family or people you work with, who have also recently suffered from abdominal pain? No.

Is there a family history of your abdominal pain? No.

Are you taking any medication for previous illnesses? No.

Have you tried any over-the-counter medications for the pain in your abdomen? Yes. I've tried Pepto-Bismol, but it hasn't helped at all.

Do you drink alcoholic beverages, and if so, how much each day? I may have two or three beers, maybe more, in the evening.

How long have you had two or more alcoholic drinks every day? A long time, I guess ever since I was young.

Have you ever felt you should cut down on your drinking? Yes. My wife and brother have told me to cut down on my drinking, but I find that I need two or three beers just to relax at the end of the day.

Did you drink any beer last night? Yes, I had two beers.

What effect, if any, has beer had on your abdominal pain? Not much. *[The patient's answers to the last five questions suggest that he may be suffering from alcoholism. The patient indicates that his daily intake of alcohol during the past 30 years has been two or more drinks, that he has been advised by family members to cut down on his drinking, and that even he has thought that he may be drinking too much. The answer to the last two questions indicates that the acute onset of rather severe abdominal pain yesterday did not restrain him from alcohol. The examiner also recognizes that since many individuals do not admit to their actual daily intake of alcohol, the amount admitted to should be regarded as a minimum.] [The examiner finds the patient alert and fully cooperative during the interview.]*

PHYSICAL EXAMINATION OF THE PATIENT

Vital signs: Blood pressure
Lying supine: 130/75 left arm and 130/80 right arm
Standing: 120/65 left arm and 120/70 right arm
Pulse: 110
Rhythm: regular
Temperature: 100.5°F.
Respiratory rate: 22
Height: 5′10″
Weight: 210 lbs.

[The examiner recognizes that the patient has mild tachycardia (a slightly elevated heartbeat rate), tachypnea (an elevated respiratory rate), and orthostatic (body-position-dependent) changes in blood pressure. In adults, the normal range of the heartbeat rate is 60 to100 beats per minute and the normal range of the respiratory rate is 8 to16 breaths per minute.]

HEENT Examination: Normal

Lungs: Inspection, palpation, percussion, and auscultation of the lungs are normal except for the following findings on the left lung: With the patient sitting upright, auscultation reveals diminished vesicular breath sounds and decreased intensity and clarity of spoken voice and whispered sounds along the posterolateral part of the base of the left lung. Percussion flatness is also found along the posterolateral part of the base of the left lung. Auscultation reveals a pleural friction rub above the zone of percussion flatness along the posterolateral part of the base of the left lung.

Cardiovascular Examination: The cardiovascular examination is normal except for the mild tachycardia and the orthostatic changes in blood pressure.

Abdomen: Inspection, auscultation, percussion, and palpation of the abdomen are normal except for the following findings: Auscultation reveals hypoactive bowel sounds. There is guarding of the anterolateral abdominal wall (but no rebound tenderness). The epigastric and left hypochondriac regions are tender to deep palpation.
Genitourinary Examination: Normal
Musculoskeletal Examination: Normal
Neurologic Examination: Normal

INITIAL ASSESSMENT OF THE PATIENT'S CONDITION

The patient appears to be suffering an acute condition involving thoracic and/or upper abdominal viscera.

ANATOMIC BASIS OF THE PATIENT'S HISTORY AND PHYSICAL EXAMINATION

(1) **The rapid onset of the patient's upper abdominal pain suggests a catastrophic event or a rapidly developing pathologic condition involving one or more thoracic and/or upper abdominal viscera.** The sudden onset of the patient's problem(s) is indicated by the patient's ability to specifically recall what he was doing and where he was when the pain first struck and that his upper abdominal pain achieved maximum intensity within 1 to 2 hours of its occurrence.

(2) **The patient's upper abdominal pain appears to be epigastric visceral pain (a deep upper abdominal ache) associated with LUQ somatic pain (position-dependent pain). The presentation suggests inflammation of the parietal peritoneum lining the posterior abdominal wall of the LUQ [the somatic pain is exacerbated when lying supine (a position that stretches the parietal peritoneum lining the posterior abdominal wall) but minimized when lying in a fetal position (a position that minimizes tension within the parietal peritoneum lining the posterior abdominal wall)].** There is no indication of inflammation of the parietal peritoneum lining the anterolateral abdominal wall.

(3) **The epigastric and left hypochondriac tenderness suggests inflammation and/or acute distention of one or more LUQ abdominal viscera and/or inflammation of LUQ parietal peritoneum.**

(4) **The sharp pain on the left side upon deep inspiration is an indicator of inflammation of the parietal pleura covering the diaphragm's upper surface on the left side and/or inflammation of the parietal peritoneum lining the diaphragm's undersurface on the left side.** The descent of the diaphragm upon deep inspiration exerts tension on the diaphragm's parietal pleural and peritoneal linings. If one or both linings are inflamed, the increased tension will elicit a sharp, knife-like pain.

(5) **The diminished vesicular breath sounds and decreased intensity and clarity of spoken and whispered voice sounds along the posterolateral part of the base of the left lung suggest diminished air flow into and out from this part of the left lung, a pleural effusion in the costodiaphragmatic recess of the left pleural space, and/or parietal pleural thickening in the costodiaphragmatic recess of the left pleural space.** Auscultation of the lung fields provides assessment of the intensity and pitch of the breath sounds being transmitted through the lungs. Vesicular breath sounds (the soft, low-pitched sounds produced by the swirling of air within respiratory airways) are normally heard over all regions of the lung fields except the manubrium sterni and the upper interscapular region. Bronchovesicular breath sounds (vesicular breath sounds mixed with the harsh, high-pitched, bronchial breath sounds produced by the passage of air through large conducting airways) are normally heard over the manubrium sterni and the upper interscapular region.

Auscultation of the lung fields also provides assessment of the intensity and clarity of spoken voice and whispered sounds being transmitted through the airways of the lungs to the chest wall. Spoken voice sounds are sounds that emanate mainly from the vocal cords. Whispered sounds are sounds generated by the passage of air through the mouth and the lips; no sounds emanate from the vocal cords. Spoken voice and whispered sounds heard in any region of a normal lung field are indistinct (individual words are not discernable) and not as loud as when heard directly. The intensity of spoken voice sounds is greatest over the upper interscapular region and the anterior ends of the first and second intercostal spaces and least over the bases of the lungs. The intensity of whispered sounds is greatest in the lung field regions where bronchovesicular breath sounds are normally heard.

Acute diminishment of vesicular breath sounds and decreased intensity and clarity of spoken voice and whispered sounds over a peripheral lung region are important but non-specific findings: they indicate diminished air flow into and out from the lung region and/or the dampening of the sounds by excess pleural fluid or pleural thickening. There are four general pathologic mechanisms that diminish air flow into and out from a lung region:

(a) obstruction of one or more of the conducting airways bringing air to the lung region,
(b) loss of the elasticity of the interstitial tissues about the respiratory airways (the normal contraction of these interstitial tissues during expiration forces air out from the respiratory airways),
(c) atelectasis (collapse of the respiratory airways), and
(d) consolidation (accumulation of fluid within the interstitial tissues and/or airways of the lung).

(6) **The percussion flatness along the posterolateral part of the base of the left lung suggests a pleural effusion in the costodiaphragmatic recess of the left pleural space.** Percussion of the lung fields provides assessment of the average density of tissues lying to a depth of 5 to 7 cm beneath the chest wall areas being percussed. The finding of percussion flatness in place of percussion resonance over a peripheral lung region is an important and fairly specific finding: it indicates a replacement of air-filled peripheral lung tissue with pleural fluid.

(7) **The pleural friction rub above the zone of percussion flatness along the posterolateral part of the base of the left lung indicates inflammation of parietal and visceral pleura of the left pleural space.** A pleural friction rub sounds like the sounds that are generated if the palm of one hand is pressed over the ear and the dorsum of that hand is then slowly and lightly rubbed by the fingers of the other hand. Pleural friction rubs occur when the roughened surfaces of inflamed parietal and visceral pleura rub against each other during respiratory movements.

(8) **The patient's dyspnea indicates inadequate ventilation or perfusion of the lungs.** The patient's tachypnea is probably due to his dyspnea.

(9) **The patient's nausea and vomiting are common autonomic responses to abdominal (and sometimes even thoracic) disease.**

(10) **The mild tachycardia and orthostatic changes in blood pressure suggest a shift in the equilibrium distribution of water from the patient's bloodstream to his interstitial spaces.** Extensive inflammation involving the left lung and its pleural space and one or more LUQ abdominal viscera could account for such a shift.

INTERMEDIATE EVALUATION OF THE PATIENT'S CONDITION

The patient appears to be suffering from an acute disease involving the left lung and its pleural space and one or more LUQ abdominal viscera.

CLINICAL REASONING PROCESS

This case illustrates the general kinds of information an examiner seeks from patients who present with abdominal pain. The principal purpose of the clinical examination of abdominal pain is to determine accurately the anatomic site or sites involved. From the history, the examiner seeks information about the location of the pain [Where is the pain? Is its location constant?], the time and mode of onset of the pain [Do you recall when the pain began and what you were doing at that time?], the quality and severity of the pain [How would you describe the pain? How severe is the pain? Has the pain affected your daily or regular activities?], the temporal profile of the pain [Is the intensity of the pain constant, and, if not, how would you describe its duration and variation over a 24-hour period?], aggravating and relieving factors [Is there anything that makes the pain worse? Is there anything that relieves the pain?], and the presence of associated problems [Is there anything else that is bothering you?].

In the physical examination, the examiner gathers information through an ordered sequence of steps that involve first, visual inspection, then, auscultation, third, percussion, and finally, palpation. **Visual inspection** detects skin lesions and general distention; it may also reveal local swellings, pulsations, and peristalsis. **Auscultation** detects bowel sounds, friction rubs, and arterial bruits. **Percussion** provides information about the size of organs and masses and the location of gas and fluid. **Palpation** reveals tenderness and the feel of organs and masses.

In this case, all of the major physical signs and symptoms point to an acute disease that involves inflammation of (a) the parietal pleura of the left pleural space (and possibly also the left lung) and (b) parietal peritoneum and one or more abdominal viscera of the epigastric and left hypochondriac regions. The inflammation appears to be confined to the left side of the body.

An emergency care physician would recognize that the patient's signs and symptoms are not characteristic of the more common acute diseases and conditions of the aorta and esophagus (the aorta, inferior vena cava, and esophagus are the major viscera that extend between the thoracic and abdominal cavities). An individual with an aortic dissection (an acute tearing of the aortic wall) or an aortic aneurysm (dilatation of the aortic wall) typically presents with notable cardiovascular signs (increased pulse and markedly elevated systemic systolic and diastolic pressures in cases of aortic dissection and a pulsatile abdominal mass in cases of a large aortic aneurysm). A person suffering from reflux esophagitis (pain from reflux of gastric contents into the esophagus) usually has central chest pain that is readily relieved by antacids. An individual suffering esophageal spasms typically has pain associated with swallowing. A person with an esophageal rupture (an acute tearing of the esophagus) commonly presents with severe central chest pain and is in cardiovascular shock.

An emergency care physician would also appreciate that the patient's signs and symptoms are not characteristic of the more common acute diseases and conditions of the lungs, kidneys, stomach, small bowel, large bowel, and spleen (the spleen, left kidney, and parts of the stomach, small bowel, large bowel, and pancreas are the major viscera located in the LUQ). In general, acute pulmonary diseases and conditions do not evoke marked inflammation of the upper abdomen. A person suffering from acute pyelonephritis (acute inflammation of the kidney, calyces, and ureteral pelvis) typically presents with fever and chills and flank pain and exhibits tenderness to percussion of the back in the region immediately overlying the inflamed kidney. Acute diseases and conditions involving the stomach, small bowel, and large bowel do not commonly create conditions that lead to pleural effusion. An individual with a ruptured spleen usually has a history of recent trauma.

An emergency care physician would tentatively diagnose acute pancreatitis complicated by a left pleural effusion. The diagnosis is based on three major considerations: (1) The patient exhibits signs and symptoms typical of acute pancreatitis: abrupt onset of epigastric pain extending to the back (the patient describes the pain as a deep ache that bores all the way to the back), nausea, and vomiting. (2) The patient gives a history of excessive alcohol drinking. Alcohol abuse and obstructive biliary tract disease are the two most common etiologic causes of pancreatitis. (3) A left pleural effusion is a frequent complication of an attack of pancreatitis.

RADIOGRAPHIC EVALUATION AND FINAL RESOLUTION OF THE PATIENT'S CONDITION

Evidence provided by PA and lateral chest films would help differentiate among the possible mechanisms responsible for the abnormal physical findings relative to the left lung field. In this case, the PA chest film of the patient showed an elevated left dome of the diaphragm, and the lateral chest film showed an elevated left dome of the diaphragm and evidence of a left pleural effusion (an elevated, blunted posterior costophrenic angle for the left lung overlying a water-density region bearing a curved upper margin). The PA chest film did not show the left pleural effusion because only 250 ml of fluid has accumulated in the left subpulmonic pleural space. In this patient, the upper level of pleural fluid would not attain the level of the lateral costophrenic angle until 350 ml of fluid have accumulated in the left subpulmonic space.

An AP erect abdominal plain film would provide general anatomic information about the upper abdomen and determine whether there is a pneumoperitoneum from a ruptured hollow viscus. In this case, the AP erect abdominal plain film showed a distended, partially fluid-filled, loop of small bowel overlying the bodies of the second and third lumbar vertebrae. The distended loop of small bowel represents a segment of the distal duodenum afflicted with paralytic ileus because of its contact with inflamed pancreatic tissue. Such a distended bowel loop is called a sentinel loop because it marks the site of localized inflammation.

CT (computed tomography) is the best radiologic method to visualize the general anatomy of the entire pancreas. CT scans can define the anatomic extent and severity of the inflammatory process of acute pancreatitis. A CT scan of the patient's abdomen showed an enlarged pancreas whose radiographic attenuation is +15 Hounsfield units (down from a normal magnitude of +50 Hounsfield units). The enlargement is the result of edema. Edema also accounts for the diminished radiographic attenuation of the parenchyma of the pancreas.

Abdominal ultrasonography is frequently ordered early in radiologic evaluation of suspected pancreatitis. The procedure is able to detect pancreatic enlargement, pseudocysts, gallbladder stones, and common bile duct enlargement unless there is significant gas in adjacent bowel segments. Bowel gas interferes with the reflected signal.

Acute pancreatitis of the severity encountered in this patient requires hospitalization and supportive treatment: analgesics for pain; intravenous administration of fluids, electrolytes, and colloids; and possibly nasogastric suction to empty the stomach of its contents if vomiting persists. Controlled studies have shown that nasogastric suction is of unproven benefit in shortening the course of pancreatitis. It is of use if paralytic ileus persists and the patient continues to vomit. The disease is often self-limited and resolves in 1 to 2 weeks, but serious life-threatening complications can result.

CHRONOLOGY OF THE PATIENT'S CONDITION

The patient was struck with an attack of acute pancreatitis at 1:15 PM the previous day. The acute inflammation evoked sudden, severe, epigastric visceral pain, which attained its maximum intensity within two hours. The inflammatory process occurred throughout the entire gland. The retroperitoneal location of the head, neck, and body of the pancreas accounts for the perception of pain boring all the way through the back.

Within the next few hours, the inflammatory process began to extend into the tissues about the gland. Anterior extension has involved the parietal peritoneum of the posterior abdominal wall that overlies the sweep of the pancreas from the epigastric region through the left hypochondrium to the left flank; inflammation of this stretch of posterior abdominal wall peritoneum accounts for the LUQ somatic pain and the tenderness of the epigastric and left hypochondriac regions to deep palpation. The guarding of the anterolateral abdominal wall is a conscious attempt by the patient to resist pressure on the epigastric and left hypochondriac regions. Posterior extension has involved the posterior part of the diaphragm on the left side, and this involvement is responsible for the left pleural effusion, pleural friction rub, pain upon deep inspiration, and elevation of the left dome of the diaphragm. The left pleural effusion has compressed and raised the posterolateral part of the costodiaphragmatic margin of the left lung; the compression is the basis for the patient's dyspnea and non-productive cough. The left pleural effusion accounts for the percussion dullness along the posterolateral part of the base of the left lung. The left pleural effusion and the compressed posterolateral part of the costodiaphragmatic margin of the left lung account for the diminished vesicular breath sounds and decreased intensity and clarity of spoken voice and whispered sounds along the posterolateral part of the base of the left lung. Rightward and downward extension has involved the duodenum; this involvement has produced paralytic ileus, which is the basis for the vomiting, hypoactive bowel sounds and the distended, partially fluid-filled, small bowel loop in the AP erect abdominal plain film.

The orthostatic changes in the patient's blood pressure are a measure of the net amount of fluid transferred from the systemic circulation into the interstitial spaces of the pancreas and the retroperitoneal connective tissues about it plus the left subpulmonic pleural space.

CASE **V.3**

The Case of Frank Dobrovsky

INITIAL PRESENTATION AND APPEARANCE OF THE PATIENT

A 62 year-old white man named Frank Dobrovsky walks into the emergency center at 5:00 PM. He appears apprehensive and uncomfortable; his body is bent forward.

QUESTIONS ASKED OF THE PATIENT

What is it that has brought you to the emergency center? I have a severe pain in my abdomen.

Where does it hurt? Right across here [the patient sweeps his right hand from the epigastric region through the left hypochondriac region all the way over to the left flank].

When did the pain start? Today.

Do you recall what you were doing when the pain started? Yes, I was sitting down by my desk to begin work after lunch when the pain struck.

I notice that you have just used the word "struck" to describe how the pain started. Do you recall how long it took for the pain to build up to its greatest intensity from the moment you were first aware of it? Oh, I'd say within 10 to 15 seconds. *[In general, the striking of sudden, intense abdominal pain suggests a catastrophic event or a rapidly developing pathologic condition.]*

Has the intensity, or severity, of the pain changed since it struck? Yes, the pain now is not as severe as it was when it first struck.

How would you describe the pain? It's like a deep ache in my abdomen that bores all the way through to my back.

Has the nature of the pain or its location changed since it struck? Yes. When the pain first struck, it felt like someone was stabbing me with a knife right here [the patient points to the epigastric region], and the pain was really intense. In the last few hours, the pain has become more of an ache and has spread to the left [the patient again sweeps his right hand from the epigastric region through the left hypochondriac region all the way over to the left flank].

Have you ever had pains like this before? No, it's the worst pain I've ever had.

Is there anything that makes the pain worse? Yes, lying down. I tried to lie down after the pain, but that just made the pain much worse. The pain is also worse when I take a deep breath or cough. Any motion makes it worse. The ride to the hospital was really bad, especially when we went over bumps. *[Abdominal pain exacerbated by motion or movement is an indicator of inflamed parietal peritoneum.]*

Is there anything that relieves the pain? Yes, lying very still. I've also found that if I bend over like this [the patient flexes his abdomen anteriorly], the pain is less intense.

Is there anything else that is also bothering you, or that you think may be associated with the pain in your abdomen? Well, I vomited right after the pain struck me, and I'm still a bit nauseous even now. I don't know if this is important or not, but I noticed about two hours ago that my left shoulder also hurts.

Before we get to the pain in your left shoulder, I'd like to first ask about your nausea and vomiting. Do you recall what your vomit looked like? Yes, it's interesting that you should ask, because it looked like I had vomited coffee grounds.

You're right, that is interesting. Have you noticed any change in the nature or color of your bowel moment, either today or during the past few weeks? No.

Now, how would you describe the pain in your left shoulder? It's like a deep ache.

Can you point to where the pain is in your left shoulder? Uh, no. It hurts all up in here [the patient rubs his right hand over the top of his left shoulder, from the base of the neck to the tip of the shoulder].

Does the pain in your left shoulder change in intensity when you move your left arm? No. [The patient partially extends, abducts, adducts, and then flexes his arm before answering the question.]

Have you tried to drink or eat anything since the pain struck? No. I took an antacid tablet after I vomited, but it didn't do anything.

Do you feel any abdominal fullness? No, not really.

Have you had any recent illnesses prior to this attack of pain in your abdomen? No.

Have you had a problem with indigestion, either recently or in the past? Occasionally I'll get some burning in my belly one or two hours after I eat, and sometimes I'll wake up in the middle of the night with burning and have to take an antacid tablet. You know, I have been taking a lot of antacid tablets lately.

Do you know any individuals around you, such as members of your family or people you work with, who have also recently suffered from abdominal pain? No.

Is there a family history of your abdominal pain? No.

Are you taking any medication for previous illnesses? No.

Do you drink alcoholic beverages, and if so, how much each day? I may have a beer in the evening two or three times a week. I really like beer, especially dark beers, but I have to watch my weight. *[The examiner finds the patient alert and fully cooperative during the interview.]*

PHYSICAL EXAMINATION OF THE PATIENT

Vital signs: Blood pressure
Lying supine: 135/80 left arm and 135/75 right arm
Standing: 130/75 left arm and 130/70 right arm
Pulse: 95
Rhythm: regular
Temperature: 100.5°F.
Respiratory rate: 22
Height: 5′10″
Weight: 210 lbs.

[The examiner recognizes that the patient has a heartbeat rate near the upper limit of the normal range and tachypnea (an elevated respiratory rate). In adults, the normal range of the heartbeat rate is 60 to 100 beats per minute and the normal range of the respiratory rate is 8 to 16 breaths per minute.]

HEENT Examination: Normal
Lungs: Normal
Cardiovascular Examination: Normal
Abdomen: Inspection, auscultation, percussion, and palpation of the abdomen are normal except for the following findings: Auscultation reveals hypoactive bowel sounds. There is guarding of the anterolateral abdominal wall (but no rebound tenderness). The epigastric and left hypochondriac regions are tender to deep palpation.
Genitourinary Examination: Normal
Musculoskeletal Examination: Inspection and palpation of the left shoulder region reveal the following pertinent normal findings: Active and passive movements of the left arm and isometric tests of left arm movements involving the shoulder, acromioclavicular, and sternoclavicular joints are all normal and do not produce any change in left shoulder pain.
Neurologic Examination: Normal

INITIAL ASSESSMENT OF THE PATIENT'S CONDITION

The patient appears to be suffering from an acute condition involving upper abdominal viscera.

ANATOMIC BASIS OF THE PATIENT'S HISTORY AND PHYSICAL EXAMINATION

(1) **The sudden onset of the patient's upper abdominal pain suggests a catastrophic event or rapidly developing pathologic condition involving one or more thoracic and/or upper abdominal viscera.** The sudden onset of the patient's problem(s) is indicated by the patient's ability to specifically recall what he was doing and where he was when the pain first struck and that his upper abdominal pain achieved maximum intensity within seconds of its occurrence.

(2) **The coffee grounds emesis indicates hemorrhage of the upper gastrointestinal tract (hemorrhage from the esophagus, stomach, or duodenum).** The 'coffee grounds' in the vomitus represent precipitated blood clots rendered black by the action of gastric acid.

(3) **The patient's upper abdominal pain appears to be epigastric visceral pain (a deep upper abdominal ache) associated with upper abdominal somatic pain (position-dependent pain). The presentation suggests inflammation of the parietal peritoneum lining the upper part of the posterior abdominal wall [the somatic pain is exacerbated when lying supine (a position that stretches the parietal peritoneum lining the posterior abdominal wall) but minimized when flexing the body forward (a position that minimizes tension within the parietal peritoneum lining the posterior abdominal wall)].** There is no indication of inflammation of the parietal peritoneum lining the anterolateral abdominal wall.

(4) The epigastric and left hypochondriac tenderness suggest inflammation and/or distention of one or more LUQ abdominal viscera and/or inflammation of LUQ parietal peritoneum.

(5) The left shoulder pain appears to be referred pain (superficial pain independent of the position or movement of the left arm). This presentation suggests inflammation of the parietal pleura covering the central region of the diaphragm's upper surface on the left side and/or inflammation of the parietal peritoneum lining the central region of the diaphragm's undersurface on the left side. The skin covering the shoulder from the base of the neck to the tip of the shoulder receives its sensory innervation from the supraclavicular nerves, and thus represents parts of the C3 and C4 dermatomes of the body.

(6) The patient's nausea and vomiting are common autonomic responses to abdominal (and sometimes even thoracic) disease.

(7) The high normal heartbeat rate and tachypnea represent psychological responses to the severe upper abdominal pain.

INTERMEDIATE EVALUATION OF THE PATIENT'S CONDITION

The patient has suffered an acute catastrophic event originating in the upper abdomen causing peritoneal inflammation and hemorrhage from the upper gastrointestinal tract.

CLINICAL REASONING PROCESS

When considered together, the aggregate characteristics of the patient's hematemesis and abdominal pain and tenderness suggest an acute hemorrhage and perforation of the upper gastrointestinal tract. Perforation is suspected because the parietal peritoneum lining the central undersurface of the diaphragm on the left side and the posterior abdominal wall in the upper abdomen appears to be inflamed. The other findings from the history and physical examination are non-specific.

Acute hemorrhage and perforation of the upper gastrointestinal tract is a surgical emergency.

RADIOGRAPHIC EVALUATION AND FINAL RESOLUTION OF THE PATIENT'S CONDITION

PA and lateral chest films would show whether there is a pneumoperitoneum from a ruptured hollow viscus and provide general anatomic information about the thorax. In this case, PA and lateral chest films showed a pneumoperitoneum (gas within the peritoneal cavity) underlying the left dome of the diaphragm.

An AP erect abdominal plain film would show whether there is a pneumoperitoneum from a ruptured hollow viscus and provide general anatomic information about the upper abdomen. In this case, the AP erect abdominal plain film showed a pneumoperitoneum underlying the left dome of the diaphragm.

CHRONOLOGY OF THE PATIENT'S CONDITION

An ulcer in the posterior wall of the patient's gastric antrum perforated into the lesser sac at 1:15 PM this afternoon. The perforation was accompanied by hemorrhaging at the site of the perforation, soilage of the peritoneal lining of the lesser sac by gastric contents, and extravasation of gas into the lesser sac. These concurrent catastrophic events evoked sudden, severe, epigastric visceral pain and vomiting; the hemorrhage at the site of the perforation was the basis for the hematemesis. The free gas and acidic gastric contents in the lesser sac soon collected beneath the left side of the dome of the diaphragm; the resulting irritation of the parietal peritoneum lining the central region of the diaphragm's undersurface on the left side has evoked referred pain in the left shoulder through stimulation of sensory fibers in the left phrenic nerve. Extravasation of gastric contents into the peritoneal cavity has been limited to the lesser sac.

Inflammation of the parietal peritoneum lining the lesser sac accounts for the LUQ somatic pain and the tenderness of the epigastric and left hypochondriac regions to deep palpation. The inflammation of retroperitoneal tissues accounts for the perception of pain boring all the way through the back. The inflammation of the peritoneal lining of the lesser sac has involved the duodenum; this involvement has produced paralytic ileus, that is the basis for the hypoactive bowel sounds.

CASE **V.4**

The Case of Nancy Adams

INITIAL PRESENTATION AND APPEARANCE OF THE PATIENT

A 45 year-old, black woman named Nancy Adams walks into the emergency center. She appears apprehensive and very uncomfortable; her body is bent slightly forward and to the right. The sclera of her eyes are moderately yellowish.

QUESTIONS ASKED OF THE PATIENT

What is it that has brought you to the emergency center? I have this pain in my belly that has been getting worse.

Where does it hurt? Over here on the right [the patient places her hand over the RUQ of her abdomen].

When did the pain start? About 5 days ago.

Do you recall what you were doing when the pain started? No, not really, but I believe that the pain started in the morning shortly after breakfast.

How would you describe the pain? I'm not sure what you mean; I feel a deep, steady ache in my belly.

Has the nature of the pain or its location changed during the past 5 days? Yes. When the pain first appeared, it was more in the middle [the patient's hand circumscribes the epigastric region], and it didn't hurt as much or as sharply as it does now. The pain has been getting worse each day. And sometimes I also feel pain in my back, up between my shoulders. [When the examiner touches several points on the upper part of the patient's back in order to more specifically locate her back pain, she locates the pain in the lower right part of the interscapular region.]

Have you ever had pains like this before? Well, when the pain in my belly first started about 5 days ago, I wasn't too worried because I had had that kind of pain on and off for the past several months. The attacks of pain would come on an hour or so after a meal, last for one or two hours and then gradually fade away. Several times I also had the upper back pain at the same time that I'd have an attack of belly pain. But this time, the belly pain hasn't gone away; instead, it has gotten worse.

Is there anything that makes the pain worse? The pain is worse when I take a deep breath.

Is there anything that relieves the pain? No.

Is there anything else that is also bothering you? Beginning yesterday, I have felt feverish and had chills at times; I actually shake when I get the chills. I have also been nauseous the past three days, and vomited twice last night. *[A fever accompanied by chills and shivering indicates a spiking fever (an abrupt increase in body temperature). The spiking fever is generated by the combined actions of (a) pyrogens, (b) forceful, heat-producing, muscular contractions (called the rigors), (c) erection of body hair (the formation of "goosebumps"), and (d) cutaneous vasoconstriction. The rigors produce the shivering, and the formation of the "goosebumps" on cold skin give the feeling of chills. A fever accompanied by rigors and chills suggests an infectious disease.]*

Have you noticed that the white parts of your eyes are somewhat yellowish? No, but my husband said that very same thing to me two days ago.

Did your husband mention if the day before yesterday was the first day that he observed your eyes to be yellowish? Yes.

Is the color of your urine normal? No, it's been darker, looking like Coke, for the last four days.

Is it possible that you may be pregnant? No, I don't think so, because my husband and I practice birth control.

What sort of birth control do you and your husband use? Condoms.

When was your last menstrual period? My last period was 15 days ago; I'm very regular. *[The possibility* *of pregnancy always has to be investigated in cases of fertile women presenting with acute abdominal pain. This is because, first of all, a hemorrhaging ectopic pregnancy (which is a surgical emergency) is one of the more common causes of acute abdominal pain in women of child-bearing age. Second, the presence of pregnancy significantly affects the choice of radiologic and medical procedures that can be used to further resolve the patient's condition.]*

Have you had any recent illnesses prior to this pain in your abdomen? No.

Are you taking any medication for previous illnesses? No.

Have you tried any over-the-counter drugs for the pain in your abdomen? Well, I normally take Advil [ibuprofen] for aches and pain, but the Advil has not helped the pain in my abdomen.

Do you drink alcoholic beverages, and if so, how much each day? I may have a beer with my husband on the weekends. *[The examiner finds the patient alert and fully cooperative during the interview.]*

PHYSICAL EXAMINATION OF THE PATIENT

Vital signs: Blood pressure
Lying supine: 120/65 left arm and 125/70 right arm
Standing: 105/55 left arm and 110/60 right arm
Pulse: 115
Rhythm: regular
Temperature: 103.8°F.
Respiratory rate: 22
Height: 5′8″
Weight: 180 lbs.

HEENT Examination: Inspection of the oral cavity shows that the mucous membrane of the hard palate is yellowish.
Lungs: Normal
Cardiovascular Examination: The cardiovascular exam is normal except for the tachycardia and the orthostatic changes in blood pressure.
Abdomen: Inspection, auscultation, percussion, and palpation of the abdomen are normal except for the following findings: Percussion shows that the combined heights of the zones of hepatic dullness and hepatic flatness span a distance of 14 cm along the right midclavicular line. Deep palpation reveals a firm, regular, smooth but tender liver edge palpable below the right costal margin on inspiration. Fist percussion applied to the lower ribs above the right costal margin elicits an increase in upper abdominal pain. The gallbladder is not palpable, but Murphy's sign (interruption of inhaled breath when the patient is palpated underneath the right costal margin) is observed. Light and deep palpation elicit localized tenderness in the RUQ only.
Genitourinary Examination: Normal
Musculoskeletal Examination: Normal
Neurologic Examination: Normal

INITIAL ASSESSMENT OF THE PATIENT'S CONDITION

The patient appears to be suffering from progressive worsening of an acute disease or condition in the upper abdomen.

ANATOMIC BASIS OF THE PATIENT'S HISTORY AND PHYSICAL EXAMINATION

(1) **The patient's initial abdominal pain presented as epigastric visceral pain (a deep epigastric ache). This presentation suggests disease of one or more thoracic and/or upper abdominal viscera,** because the viscera that most commonly produce visceral pain poorly localized to the epigastric region are the abdominal viscera supplied by the celiac artery and thoracic viscera.

(2) **The patient's interscapular pain appears to be referred pain (inconstant, non-pleuritic, superficial pain). This presentation suggests disease of one or more abdominal viscera supplied by the celiac artery,** as the lower right part of the interscapular region of the back represents the posterior parts of the T5-T9 dermatomes on the right side of the body.

(3) **The patient's RUQ pain is partly or entirely somatic (the anterolateral abdominal wall of the RUQ exhibits rebound tenderness). This presentation indicates inflammation of the RUQ parietal peritoneum.** The inflammation of the RUQ parietal peritoneum could have been acquired from direct contact with inflamed RUQ abdominal viscera or from migration of inflammatory agents from the thorax, the other abdominal quadrants, or the pelvis. The absence of RUQ abdominal wall rigidity suggests that the parietal peritoneum inflammation is not marked at this time.

(4) **The patient's liver is enlarged (as assessed by percussion) and tender (as demonstrated by deep palpation).** The finding of a tender liver explains the exacerbation of upper abdominal pain upon deep inspiration or fist percussion to the right costal margin. This is because any force (such as that delivered from fist percussion to the right costal margin or from the downward thrust of the diaphragm during deep inspiration) that puts in-

creased pressure upon a tender liver will produce upper abdominal pain.

(5) **Murphy's sign demonstrates tenderness of the gallbladder and/or liver; Murphy's sign is generally regarded as an indication of gallbladder inflammation.** The test for Murphy's sign is for the patient to take a deep breath as the examiner presses his fingers deeply underneath the patient's right costal margin. Sudden inspiratory arrest (Murphy's sign) occurs when the downward thrust of the diaphragm impinges a tender gallbladder or liver upon the examiner's fingertips. The inability to palpate the patient's gallbladder is not particularly significant, since a tender, distended gallbladder is not always palpable.

(6) **The patient's history suggests prior episodes of biliary colic (gallstone disease).** The evidence for biliary colic comes from the patient's description of previous bouts of epigastric visceral pain associated with interscapular referred pain. Biliary colic is the pain produced upon distention of the cystic or common bile duct from the attempted passage of a gallstone. Biliary colic consists of epigastric visceral pain that is commonly associated with RUQ, interscapular, or right subscapular referred pain. The pain commonly appears suddenly and is moderately severe.

(7) **The patient's yellowed sclera and oral cavity mucosa and darkened urine indicate jaundice.** Jaundice (icterus) is the yellow staining of the skin, the sclera of the eyes, and the mucous membrane of the oral cavity that occurs upon the deposition of bile pigments (the yellowing of the patient's skin was not observed because of its natural pigmentation). Bile pigments become deposited in these tissues as a consequence of hyperbilirubinemia (elevated serum levels of unconjugated and/or conjugated bilirubin). Conjugated hyperbilirubinemia produces conjugated hyperbilirubinuria (elevated urine levels of conjugated bilirubin); the relatively high concentration of conjugated bilirubin darkens the urine. Although there are disorders other than hepatobiliary disease that can produce jaundice, the preponderance of historical and physical evidence in this case points to hepatobiliary disease as the basis of the patient's jaundice.

(8) **The patient's markedly elevated temperature, orthostatic changes in blood pressure, tachycardia, and tachypnea collectively indicate septicemia (infection of the circulating blood with pathogenic microoganisms).** It is logical to assume in this case that the patient's hepatobiliary disease is the source for the septicemia.

(9) **The patient's nausea and vomiting are not specific findings, as both are common autonomic responses to abdominal disease.**

INTERMEDIATE EVALUATION OF THE PATIENT'S CONDITION

The patient is suffering from hepatobiliary disease (disease of the liver, gallbladder and/or extrahepatic biliary ducts).

CLINICAL REASONING PROCESS

A gastroenterologist would presume that this patient's condition is most likely a complication of her gallstone disease. The physician would recognize that in all of the patient's prior episodes of biliary colic, a gallstone was only temporarily entrapped in the cystic or common bile duct, as the episodes resolved spontaneously. In each instance, the temporarily entrapped gallstone either fell back into the gallbladder lumen from the cystic duct or passed from the common bile duct into the second part of the duodenum.

The latest episode of biliary colic (which occurred five days ago) appears to be complicated by the unrelieved entrapment of a gallstone in a biliary duct. If the gallstone were entrapped within the cystic duct, it would obstruct bile flow from the gallbladder but not the liver, and the patient (by this time) would be suffering from only acute cholecystitis (acute inflammation of the gallbladder). If the gallstone were entrapped within the common bile duct, it would obstruct bile flow from both the gallbladder and the liver, and the patient (by this time) would be suffering from acute inflammation of both the gallbladder and liver. Because this latter condition more closely matches the aggregate characteristics of the patient's history and physical findings, a gastroenterologist would tentatively diagnose cholecystitis complicated by a common bile duct obstruction from a gallstone and ascending cholangitis (ascending sepsis of the extrahepatic and intrahepatic biliary ducts and the gallbladder).

RADIOLOGIC EVALUATION AND FINAL RESOLUTION OF THE PATIENT'S CONDITION

Ultrasound examination of the liver, gallbladder, and extrahepatic biliary ducts would provide highly specific and accurate information. Ultrasonography can reveal dilated intrahepatic biliary ducts (the presence of that indicate biliary tract obstruction) and detect 95% of all gallstones. Ultrasound examination of the patient's biliary tract showed gallstones in the lumen of the gallbladder, dilated gallbladder with thickened walls, dilated biliary ducts, and a gallstone lodged in the lower part of the common bile duct. The radiologist attributed the thickening of the gallbladder wall to edema.

In this case, the patient would be advised that surgery is necessary for a cholecystectomy (excision of the gallbladder) with choledocholithotomy (removal of the entrapped gallstone).

CHRONOLOGY OF THE PATIENT'S CONDITION

The patient has intermittently suffered from cholelithiasis (the presence of gallstones in the gallbladder and biliary ducts) for the past year. Shortly after breakfast five days ago, a gallstone in the patient's gallbladder entered and obstructed the cystic duct, initiating an episode of biliary colic (consisting of epigastric visceral pain and interscapular referred pain). Repetitive peristaltic contractions moved the gallstone into the common bile duct, where it became entrapped in the lower part of the duct (choledocholithiasis) and began to significantly obstruct bile flow. The patient's biliary pain persisted as contractions failed to move the gallstone any further.

During the next few days, intraluminal biliary tract pressures progressively increased (and dilated biliary passageways) as the gallbladder was able to store and concentrate only a portion of the bile excreted by the liver into the gallbladder and extrahepatic biliary ducts. The biliary stasis (stagnation of bile flow) has lead to an increase in the serum level of conjugated bilirubin (deposition of blood-borne conjugated bilirubin in the sclera of the patient's eyes and the mucosa of her oral cavity has yellowed these tissues) and a corresponding increase in the urinary excretion of conjugated bilirubin (that has darkened the patient's urine).

Ascending sepsis of the patient's biliary passageways began about 24 hours ago from a point of origin immediately above the site of the entrapped gallstone in the common bile duct. The ascending sepsis has produced acute inflammation and distention of the gallbladder and liver that account for the worsened epigastric visceral pain and the tenderness of the two organs to palpation. Extension of the liver and gallbladder's inflammatory process to the surrounding parietal peritoneum is responsible for the RUQ parietal pain and rebound tenderness.

Within the last few hours, the microorganisms responsible for the hepatobiliary sepsis have begun to enter the bloodstream; the patient's septicemia is responsible for the elevated body temperature, orthostatic changes in blood pressure, tachycardia, and tachypnea.

RECOMMENDED REFERENCES FOR ADDITIONAL INFORMATION ON THE VISCERA SUPPLIED BY THE CELIAC ARTERY

Kelley, W., N., DeVita, Jr., V. T., Dupont, H. L., Harris, Jr., E. D., Hazzard, W. R., Holmes, E. W., Hudson, L. D., Humes, H. D., Paty, D. W., Watanabe, A. M., and T. Yamada, *Textbook of Internal Medicine*, J. B. Lippincott Co., Philadelphia, 1992: *This text is one of the most authoritative and comprehensive texts on internal medicine. In Part III, on gastroenterology, Chapter 95 provides a brief discussion of the approach to the patient with abdominal pain. Chapter 96 offers an introductory discussion on the approach of choice for the patient with an acute abdomen (an abdominal condition requiring immediate or prompt surgical intervention). Chapters 97, 100, and 101 respectively present discussions of the approaches to patients with nausea and vomiting, gastrointestinal bleeding, and jaundice.*

Wilson, J. D., Braunwald, E., Isselbacher, K. J., Petersdorf, R. G., Martin, J. B., Fauci, A. S., and R. K. Root, *Harrison's Principles of Internal Medicine*, 12th ed., McGraw-Hill, New York, 1991: *This text is one of the most authoritative and comprehensive texts on internal medicine. Part 9 covers disorders of the gastrointestinal system: Chapter 235 provides an introductory discussion of the approach to the patient with gastrointestinal disease.*

DeGowin, R. L., Jochimsen, P. R., and E. O. Theilen, *DeGowin & DeGowin's Bedside Diagnostic Examination*, 5th ed., Macmillan Publishing Co., New York, 1987: *Pages 540 through 551 offer a discussion of common diseases and conditions which produce acute upper abdominal pain. Many of these diseases and conditions involve the viscera supplied by the celiac artery.*

Silen, W., *Cope's Early Diagnosis of the Acute Abdomen*, 18th ed., Oxford University Press, New York, 1991: *Chapters 2 and 3 respectively provide discussions of pertinent aspects of the history and physical examination in the diagnosis of acute abdominal pain. Chapter 4 offers a discussion of the manner in which signs and symptoms may be grouped in the formulation of a differential diagnosis in cases of acute abdominal pain. Chapter 8 presents a discussion of the perforation of gastric and duodenal ulcers. Chapter 9 provides a discussion of the signs and symptoms of acute pancreatitis. Chapter 10 presents a discussion of cholecystitis and other causes of acute RUQ pain.*

de Dombal, F. T., *Diagnosis of Acute Abdominal Pain*, 2nd ed., Churchill Livingstone, Edinburgh, 1991: *Chapters 4 and 5 respectively provide discussions of pertinent aspects of the history and physical examination in the diagnosis of acute abdominal pain. Chapter 8 presents a discussion of the differential diagnosis of acute cholecystitis. Chapter 9 offers discussions of the differential diagnosis of perforated peptic ulcer and acute pancreatitis.*

Slaby, F., and E. R. Jacobs, *Radiographic Anatomy*, Harwal Publishing Co., Malvern, PA, 1990: *Pages 144 through 148 in Chapter 4 outline the major anatomic features of the abdominal plain film. The outline describes the abdominal plain film in terms of the images cast by three major sets of anatomically distinct structures: (1) the bones of the lower chest wall, the lumbar vertebrae, sacrum, coccyx, and coxal bones, (2) the soft tissues of the abdominal wall, and (3) abdominopelvic viscera.*

CHAPTER **23**

The Viscera Supplied by the Superior and Inferior Mesenteric Arteries

The superior mesenteric artery and its branches supply the pancreas, duodenum, jejunum, ileum, cecum, vermiform appendix, ascending colon, and transverse colon. The inferior mesenteric artery and its branches supply the transverse colon, descending colon, sigmoid colon, rectum, and upper half of the anal canal.

THE JEJUNUM AND ILEUM

The jejunum and ileum form the small bowel coils of the abdominopelvic cavity. The jejunum begins at the **duodenojejunal flexure** (Figs. 22–7, 22–8, and 22–9), and the ileum ends at the **ileocecal junction** (which is the union of the ileum with the large intestine at the junction of the cecum and ascending colon). The jejunum represents roughly the proximal two-fifths of the small bowel coils. The jejunal and ileal coils generally lie bracketed to the right by the cecum and ascending colon, above by the transverse colon, to the left by the descending colon, and below by the sigmoid colon (Fig. 23–1).

The major distinctions between the jejunum and ileum are as follows: (1) The **jejunal branches of the superior mesenteric artery** form fewer tiers of arterial arches (arterial anastomoses) in the mesentery of the small intestine than do the **ileal branches. Straight arteries** (**vasa recta**) arise from the most distal tiers to extend straight toward the small bowel segment they supply. (2) The vascularity in the jejunal wall is denser than that in the ileal wall. (3) The **plicae circulares** of the jejunal mucosa are thicker and more numerous than those of the ileal mucosa. This difference contributes to the greater thickness of the jejunal wall; palpation of wall thickness can be used during abdominal surgery to distinguish upper jejunal coils from lower ileal coils. (4) **Peyer's patches** are less numerous in the jejunum than in the ileum. Peyer's patches are aggregates of lymphoid tissue distributed along the antimesenteric border of the jejunum and ileum. (5) The mesentery to the jejunum bears less fat than the mesentery to the ileum.

THE ABDOMINAL SEGMENTS OF THE LARGE INTESTINE

The cecum, vermiform appendix, ascending colon, transverse colon, descending colon, and sigmoid colon comprise the abdominal segments of the large bowel (Fig. 23–1).

The cecum is the first segment of the large intestine; it is continuous superiorly with the ascending colon (Fig. 22–4). Most of the cecum lies in the right inguinal region.

The vermiform appendix forms a blind passage at the lower end of the cecum (Fig. 22–10). It is not fixed in position; it most commonly lies either behind the cecum or draped over the pelvic brim.

The ascending colon ascends in the right lumbar region up to the **hepatic flexure** (**right colic flexure**), where it becomes continuous with the transverse colon (Fig. 22–4). The transverse colon extends leftward across the breadth of the abdominal cavity to the **splenic flexure** (**left colic flexure**), where it is continuous with the descending colon (Fig. 22–4). The descending colon descends from the left hypochondriac region to the left inguinal region, where it becomes continuous with the sigmoid colon. The sigmoid colon lies mainly in the LLQ; at its end, it is continuous with the rectum.

Small masses of fat called **appendices epiploicae** are scattered over the exposed surfaces of the ascending, transverse, descending, and sigmoid colon (Fig. 23–2). Each appendix epiploica is covered with peritoneum.

Along the length of most of the large bowel, the layer of longitudinally-oriented, smooth muscle fibers in the muscularis externa is thickened or gathered into three prominent, longitudinal bands called the **teniae coli** (Fig. 23–2). The three teniae coli are distributed equidistantly around the circumference of the colon. The three teniae coli of the cecum converge at the base of the vermiform appendix (Fig. 22–10).

The wall of most of the large intestine is intermittently gathered circumferentially; the sacculations produced by these gatherings are called **haustra** (Fig. 23–2). The haustra are inconstant in position and presence.

The sides of the ascending and descending colons bulge backward against the peritoneal lining of the posterior abdominal wall to enclose narrow spaces along the lengths of the colons. These gutter-shaped spaces within the peritoneal cavity are called the paracolic gutters. The paracolic gutters on the medial and lateral sides of the ascending colon are called, respectively, the **right medial** and **right lateral paracolic gutters.** The paracolic gutters on the medial and lateral sides of the descending colon are called, respectively, the **left medial** and **left lateral paracolic gutters.**

The upper end of the right lateral paracolic gutter is continuous superiorly with a recess within the peritoneal cavity called the **hepatorenal recess, or Morrison's pouch.** The hepatorenal recess is bordered anteriorly by the visceral surface of the right lobe of the liver, posteriorly by the right kidney, and superiorly by the posterior layer of the coronary ligament. The hepatorenal recess is the region of the greater sac first encountered upon rightward passage through the epiploic foramen.

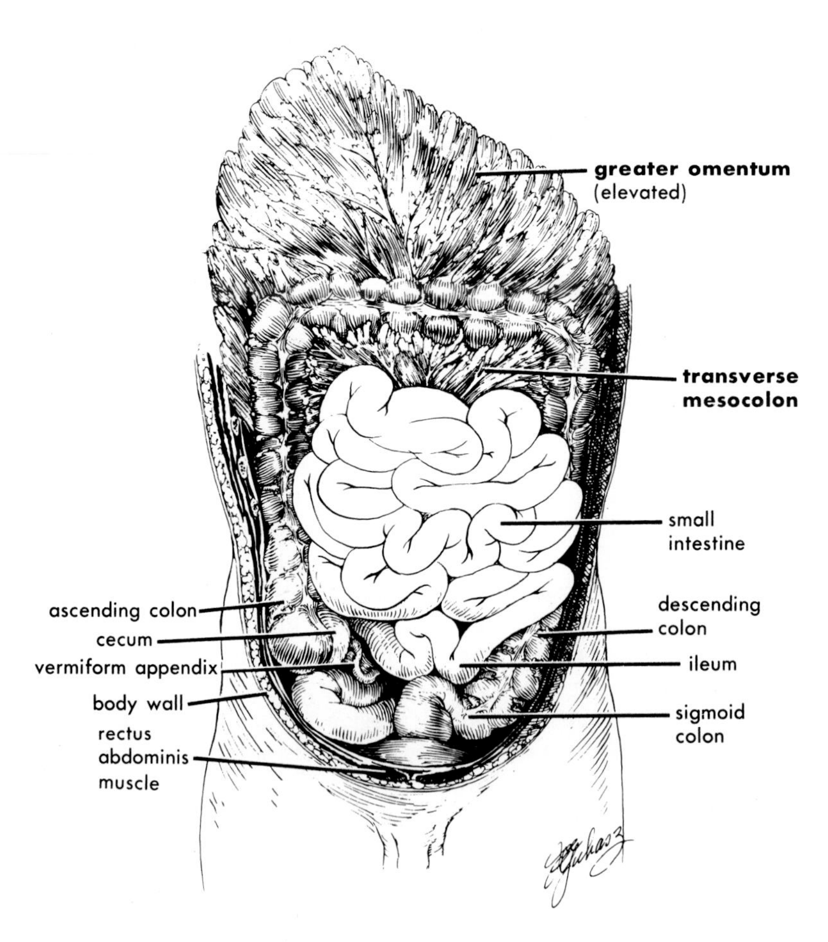

Fig. 23–1: Diagram illustrating the relations of the coils of the small intestine to the segments of the large intestine. The greater omentum and transverse colon have been reflected upward to reveal the underlying intestines.

THE BLOOD SUPPLY TO THE JEJUNUM, ILEUM, AND THE ABDOMINAL SEGMENTS OF THE LARGE INTESTINE

The Superior Mesenteric Artery and Its Branches

The **superior mesenteric artery** arises from the abdominal aorta directly behind the body of the pancreas (Figs. 21–4 and 22–8). The artery passes in front of the left renal vein (Fig. 21–3), uncinate process of the pancreas (Fig. 22–8), and third and/or fourth part of the duodenum (Fig. 21–4) before entering the root of the mesentery of the small intestine (Fig. 21–6) and descending obliquely toward the right inguinal region (Fig. 23–3).

The **inferior pancreaticoduodenal artery** is among the first branches of the superior mesenteric artery (Fig. 23–3); it supplies the pancreas and duodenum. The inferior pancreaticoduodenal artery bifurcates into anterior (Fig. 22–8) and posterior (Fig. 22–9) branches. The anterior and posterior inferior pancreaticoduodenal arteries respectively anastomose with the anterior and posterior superior pancreaticoduodenal arteries (Figs. 22–8 and 22–9).

The superior mesenteric artery gives rise from its left side to 12 to 20 branches to the jejunum and ileum (Fig. 23–3). These **jejunal** and **ileal branches** parallel each other as they extend from their origins (Figure 23–4 depicts the ileal but not the jejunal branches). Each jejunal and ileal branch ends by dividing into two branches; these terminal bifurcations anastomose with adjacent terminal branches to form a tier of **arterial arches** (**arterial arcades**). Most of the branches that emanate from this first tier of arterial arches form in a similar fashion a second tier of arterial arches. The number of tiers of arterial arches within the mesentery of the small intestine progressively increases as the mesentery extends from the duodenojejunal flexure to the ileocecal junction. Whereas there is generally only one tier of arterial arches among the arteries supplying the proximal end of the jejunum, there may be as many

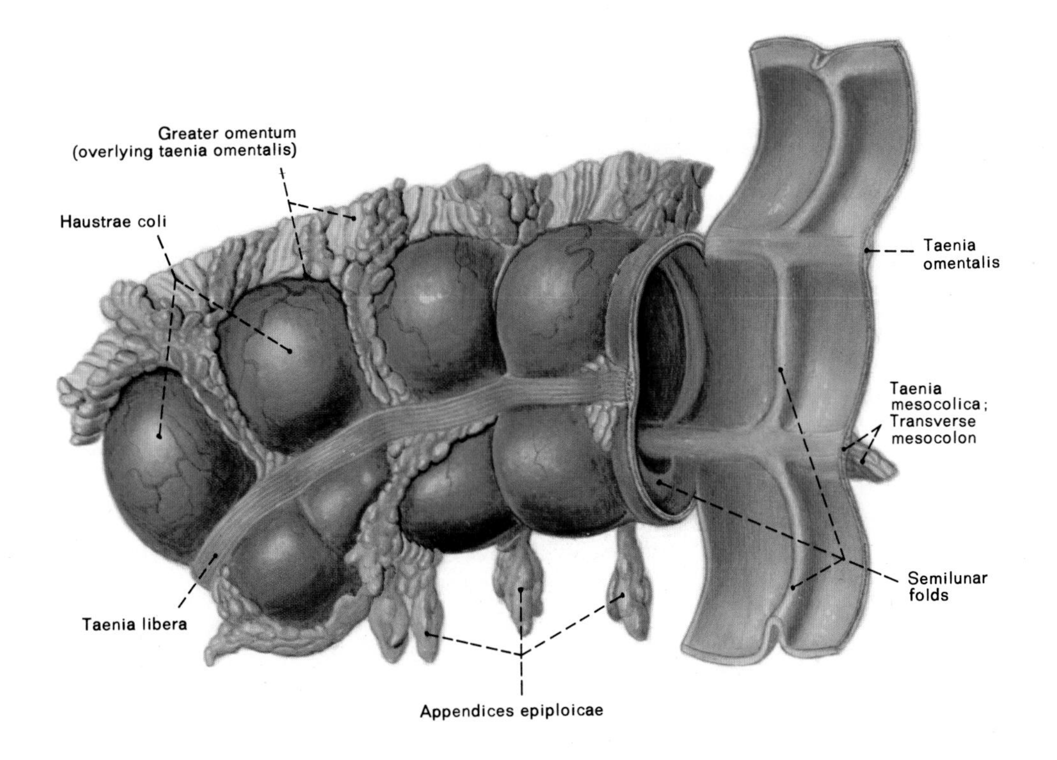

Fig. 23–2: A segment of transverse colon.

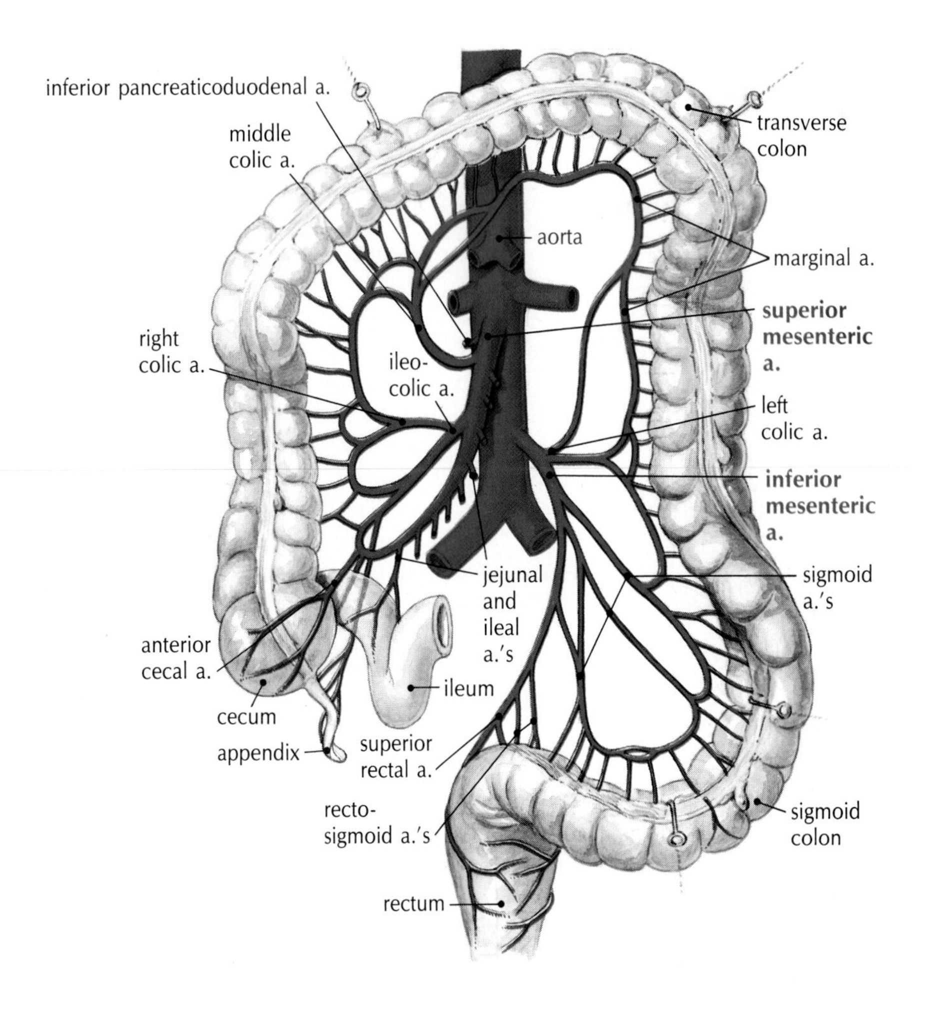

Fig. 23–3: Diagram of the branches of distribution of the superior and inferior mesenteric arteries. (From Melloni, J. L., Dox, I., Melloni, H. P., and B. J. Melloni: Melloni's Illustrated Review of Human Anatomy, J. B. Lippincott Co., Philadelphia, 1988, p. 29.)

as five to six tiers of arterial arches among the arteries supplying the distal end of the ileum (Figure 23–4 depicts only one or two of the tiers of arterial arches among the arteries supplying the ileum). The branches from the terminal tier of arterial arches in each segment of the mesentery are called the **straight arteries** (**vasa recta**) because they follow a straight course from their origins to either the jejunum or ileum (Fig. 23–4).

The **middle colic artery** originates from the right side of the superior mesenteric artery; it is the first colic branch of the superior mesenteric artery (Fig. 23–3). The middle colic artery extends through the extraperitoneal space of the transverse mesocolon to reach the margin of the transverse colon; there it bifurcates into a left branch that courses leftward along the transverse colon toward the splenic flexure and a right branch that courses rightward toward the hepatic flexure. The middle colic artery supplies mainly the transverse colon.

The **right colic artery** is the second colic branch of the superior mesenteric artery (Figure 23–3 depicts a variation in the origins of the right colic and ileocolic arteries in which the arteries arise in common from a colic branch of the superior mesenteric artery). The right colic artery follows a retroperitoneal course to the margin of the ascending colon; there it bifurcates into an ascending branch that anastomoses near the hepatic flexure with the right branch of the middle colic artery and a descending branch that courses inferiorly toward the ileocecal junction.

The **ileocolic artery** is the third and final colic branch of the superior mesenteric artery. It courses near the root of the mesentery of the small intestine as it extends toward the ileocecal junction, at which point it gives rise in a variable fashion to several branches. The **colic branch of the ileocolic artery** ascends along the margin of the ascending colon and anastomoses with the descending branch of the right colic

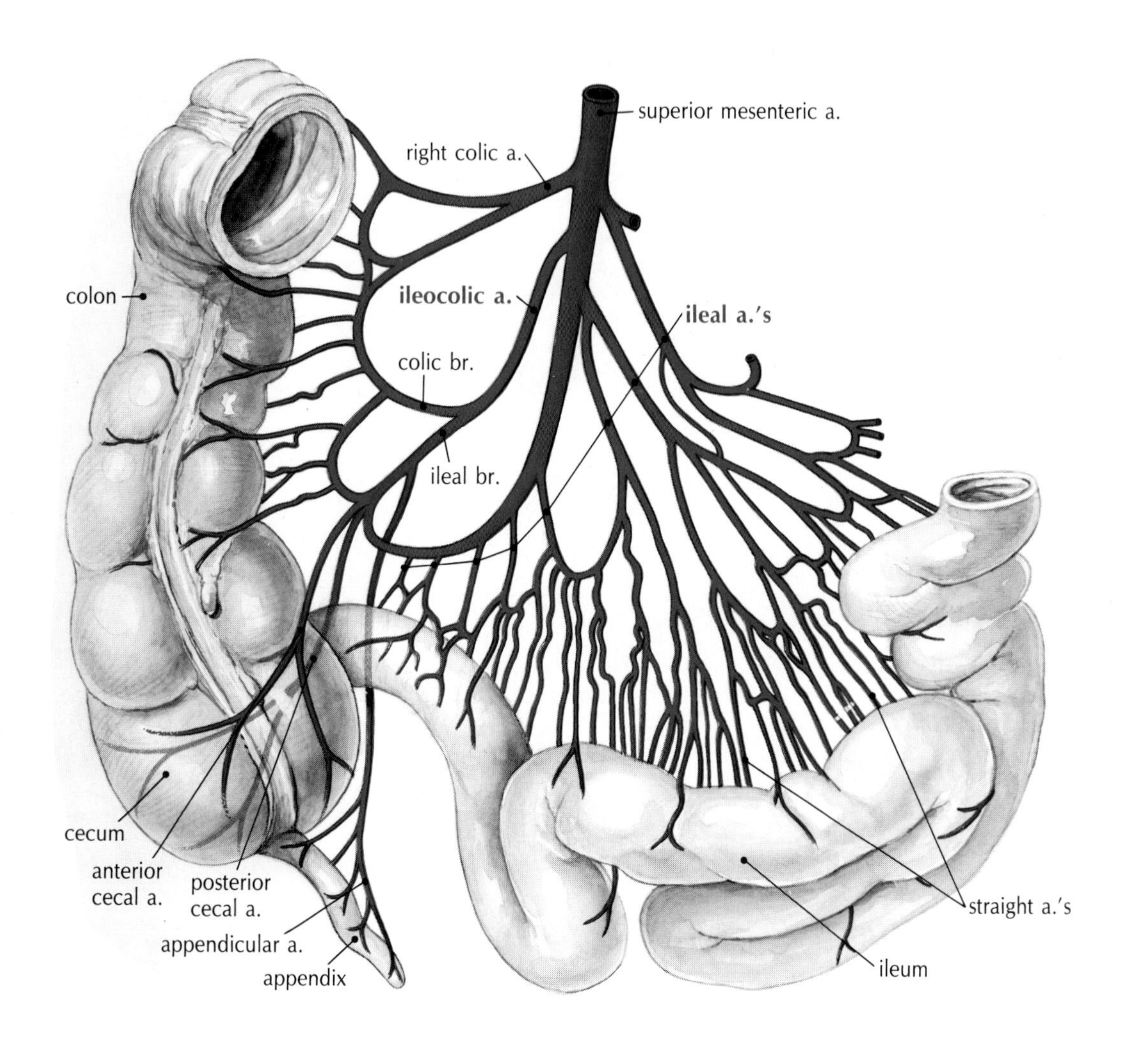

Fig. 23–4: Diagram of some of the branches of distribution of the superior mesenteric artery. (From Melloni, J. L., Dox, I., Melloni, H. P., and B. J. Melloni: Melloni's Illustrated Review of Human Anatomy, J. B. Lippincott Co., Philadelphia, 1988, p. 23.)

artery (Fig. 23–4). The **ileal branch of the ileocolic artery** anastomoses near the ileocecal junction with the terminal end of the superior mesenteric artery (Fig. 23–4). The **anterior and posterior cecal arteries** are distributed, respectively, to the anterior and posterior sides of the cecum, and supply this segment of the large intestine. The **appendicular artery** is transmitted by the mesoappendix to the vermiform appendix.

The Inferior Mesenteric Artery and Its Branches

The **inferior mesenteric artery** arises from the abdominal aorta (Figs. 21–3 and 23–3). It descends on the posterior abdominal wall, extending first to the left of the abdominal aorta, then crossing in front of the left common iliac artery, and finally entering the sigmoid mesocolon.

The **left colic artery** originates from the left side of the inferior mesenteric artery; it is the first colic branch of the inferior mesenteric artery (Fig. 23–3). It follows a retroperitoneal course to the margin of the descending colon; there it bifurcates into an ascending branch that anastomoses near the splenic flexure with the left branch of the middle colic artery and a descending branch that courses inferiorly toward the origin of the sigmoid colon to anastomose with the ascending branch of the first sigmoid artery.

The second and subsequent colic branches of the inferior mesenteric artery are called the **sigmoid arteries** because they supply the sigmoid colon (Fig. 23–3). The terminal branches of the sigmoid arteries anastomose with each other along the mesenteric margin of the sigmoid colon.

The **superior rectal artery** is the terminal extension of the inferior mesenteric artery (Fig. 23–3). The superior rectal artery supplies the rectum and the

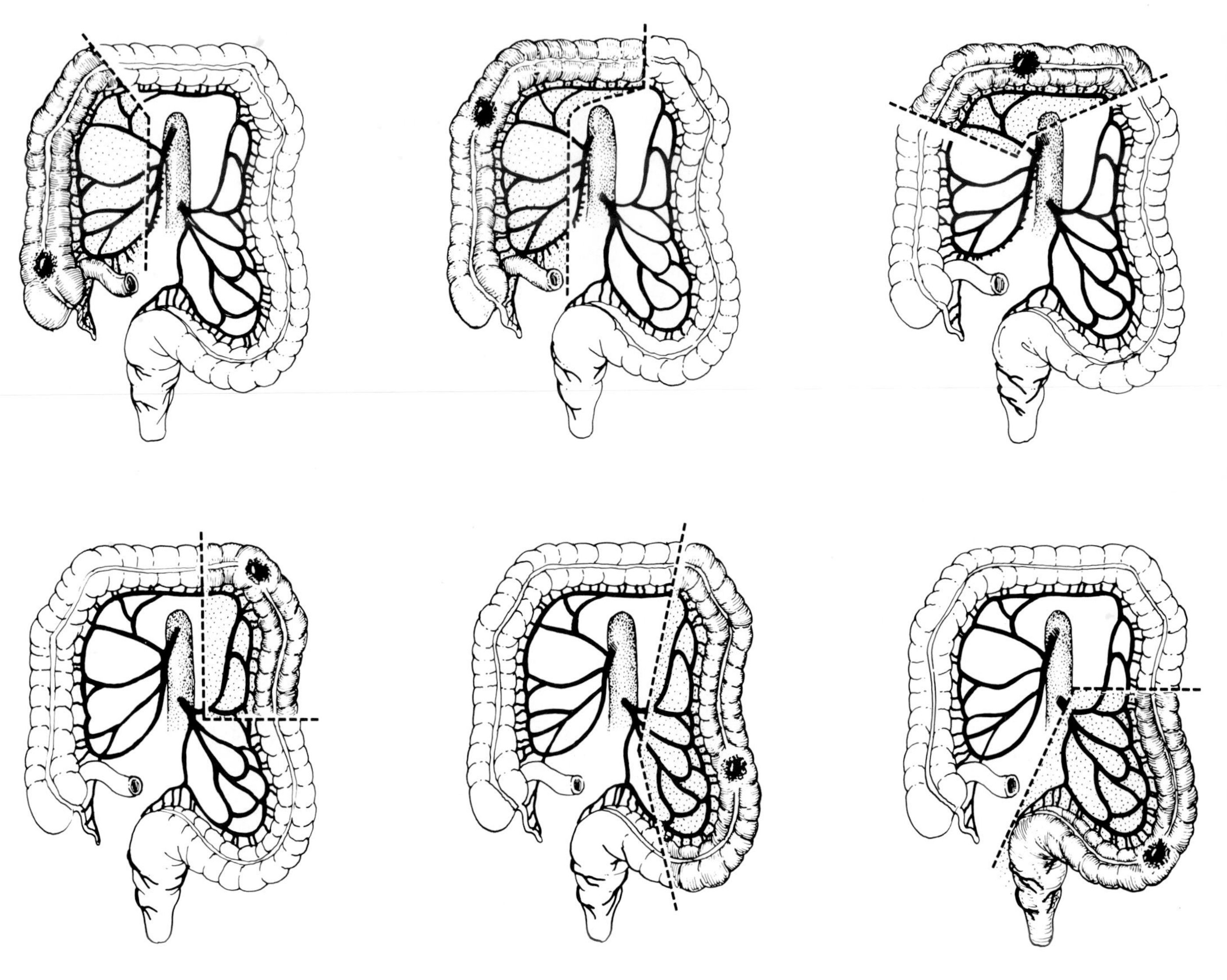

Fig. 23–5: Diagrams illustrating the ileal and large bowel segments that are resected for colon cancer at different sites.

upper half of the anal canal; it is almost the sole arterial supply of the mucosal lining of these digestive tract segments.

The anastomoses among the terminal branches of the colic branches of the superior and inferior mesenteric arteries form, in effect, an artery that runs along the medial margins of the ascending and descending colons and the mesenteric margins of the transverse and sigmoid colons. This artery is called the **marginal artery (artery of Drummond)** (Fig. 23–3). Functionally inadequate anastomoses are encountered most often in one or both of two locations: (1) the anastomoses near the ileocecal junction between the descending branch of the right colic artery and the colic branch of the ileocolic artery, and (2) the anastomoses near the splenic flexure between the left branch of the middle colic artery and the ascending branch of the left colic artery. The anastomoses between the terminal branches of the last sigmoid artery and the superior rectal artery vary among individuals in their functional adequacy.

THE VENOUS DRAINAGE OF THE JEJUNUM, ILEUM, AND THE ABDOMINAL SEGMENTS OF THE LARGE INTESTINE

The portal vein is the recipient of the venous drainage from the jejunum, ileum, and the abdominal segments of the large intestine (Fig. 22–13). Tributaries of the **superior mesenteric vein** drain the jejunum, ileum, and the large bowel segments supplied by the superior mesenteric artery. Tributaries of the **inferior mesenteric vein** drain the large bowel segments supplied by the inferior mesenteric artery.

THE SUPERIOR MESENTERIC PLEXUS

The **superior mesenteric plexus** is the plexus of the autonomic nervous system that surrounds the origin of the superior mesenteric artery; this plexus bears a large, sympathetic ganglion called the **superior mesenteric ganglion.**

The **greater, lesser,** and **least splanchnic nerves** provide the preganglionic sympathetic innervation for the viscera supplied by the superior mesenteric artery. The superior mesenteric plexus transmits postganglionic sympathetic fibers to these viscera; these postganglionic sympathetic fibers attain their target tissues in these viscera by running along the branches of the arterial tree that sprouts from the superior mesenteric artery.

The sensory fibers that innervate the viscera supplied by the superior mesenteric artery enter the spinal cord at those spinal cord segments that provide preganglionic sympathetic innervation for the viscera (specifically, the sensory fibers enter at the T8-T12 levels). Disease or injury of the viscera supplied by the superior mesenteric artery produces visceral pain that is (most commonly) poorly localized to the umbilical region. Disease or injury of any of the viscera may refer pain to parts of the T8-T12 dermatomes.

The **vagus nerves** provide the preganglionic parasympathetic innervation for the viscera supplied by the superior mesenteric artery. Preganglionic parasympathetic fibers from the vagus nerves pass through the superior mesenteric plexus and then run along branches of the superior mesenteric arterial tree to reach minute parasympathetic ganglia strung along the jejunum, ileum, appendix, cecum, ascending colon, and proximal two-thirds of the transverse colon; postganglionic parasympathetic fibers emanate from these ganglia to innervate nearby target tissues.

THE INFERIOR MESENTERIC PLEXUS AND THE PELVIC SPLANCHNIC NERVES

The **inferior mesenteric plexus** is the plexus of the autonomic nervous system that surrounds the origin of the inferior mesenteric artery; the plexus bears a large, sympathetic ganglion called the **inferior mesenteric ganglion.**

The **first** and **second lumbar splanchnic nerves** provide the preganglionic sympathetic innervation for the viscera supplied by the inferior mesenteric artery. The inferior mesenteric plexus transmits postganglionic sympathetic fibers to these viscera; these postganglionic sympathetic fibers attain their target tissues in these viscera by running along the branches of the arterial tree that sprouts from the inferior mesenteric artery.

The pelvic splanchnic nerves provide the preganglionic parasympathetic innervation for the viscera supplied by the inferior mesenteric artery. The **pelvic splanchnic nerves** are small branches of the anterior rami of S2-S4 that transmit preganglionic parasympathetic fibers which arise from neurons residing in the second, third, and fourth sacral segments of the spinal cord. Many of these preganglionic parasympathetic fibers extend directly toward minute parasympathetic ganglia strung out along the large bowel from the distal third of the transverse colon to the anal canal; postganglionic parasympathetic fibers emanate from these ganglia to innervate nearby target tissues.

The sensory fibers that innervate the viscera supplied by the inferior mesenteric artery enter the spinal cord at those spinal cord segments that provide preganglionic sympathetic and preganglionic parasympathetic innervation for the viscera (in other words, the sensory fibers enter at the L1, L2, S2, S3, and S4 levels). Disease or injury of the viscera supplied by the inferior mesenteric artery produces visceral pain that is (most commonly) poorly localized to the hypogastric region. Disease or injury of the viscera may refer pain to parts of the L1, L2, S2, S3, and S4 dermatomes.

THE SUPERIOR AND INFERIOR MESENTERIC LYMPH NODES

The **superior mesenteric nodes** are the cluster of pre-aortic nodes around the origin of the superior mesenteric artery. These nodes receive lymph from all the tissues supplied by the superior mesenteric artery.

The **inferior mesenteric nodes** are the cluster of pre-aortic nodes around the origin of the inferior mesenteric artery. These nodes receive lymph from all the tissues supplied by the inferior mesenteric artery.

Figure 23–5 illustrates the ileal and large bowel segments that are resected for colon cancer at different sites. The boundaries of resection lie just beyond (a) the most proximal and most distal arteries that supply the site in question and (b) the most proximal and most distal lymphatics that drain lymph from the site in question.

CASE **V.5**

The Case of Anna Quintinelle

INITIAL PRESENTATION AND APPEARANCE OF THE PATIENT

A 16 year-old, Hispanic girl named Anna Quintinelle walks into the ambulatory clinic at 10:30 AM. She appears uncomfortable. She is accompanied by her mother.

QUESTIONS ASKED OF THE PATIENT

What is it that has brought you to the clinic? I have a pain in my belly.

Where does it hurt? Over here [the patient places her right hand over the umbilical region of her abdomen].

When did the pain start? Late yesterday afternoon, after school.

Do you recall what you were doing when the pain started? No, not really. I think I may have been listening to some music.

How would you describe the pain? It's like a deep ache that never goes away.

Has the nature of the pain or its location changed since yesterday? No.

Have you ever had a pain like this before? Yes, when I've had the flu, but it's never been as constant as this pain.

Is there anything that makes the pain worse? No.

Is there anything that relieves the pain? No.

Is there anything else that is also bothering you, or that you think may be associated with your abdominal pain? I don't know. I haven't felt like eating since last evening, and, off and on, I have felt nauseous. I haven't vomited though.

Do you know any individuals around you, such as members of your family, friends, classmates, or teachers, who have also recently suffered from abdominal pain? No. *[The examiner asks this question in order to learn if several people whom the patient comes into contact with on a daily basis are presently suffering from GI tract dysfunction (as indicated by abdominal pain, nausea, vomiting, and/or diarrhea). A positive answer would indicate that viral gastroenteritis (intestinal flu) should be considered as a possible cause of the patient's condition.]*

Are you taking any medications for previous illnesses? No.

Are you sexually active? No.

When was your last period? About two weeks ago.

How frequent are your periods? Every 27 to 28 days.

How long do they last? Two to three days.

Has there been any recent discharge from your vagina? No. *[The possibility of pregnancy always has to be investigated in cases of fertile women presenting with acute abdominal pain. This is because, first of all, a hemorrhaging ectopic pregnancy (which is a surgical emergency) is one of the more common causes of acute abdominal pain in women of child-bearing age. Second, the presence of pregnancy significantly affects the choice of radiologic and medical procedures that can be used to further resolve the patient's condition.]*

[The examiner finds the patient alert and fully cooperative during the interview.]

PHYSICAL EXAMINATION OF THE PATIENT

Vital signs: Blood pressure
Lying supine: 115/70 left arm and 110/75 right arm
Standing: 110/70 left arm and 110/70 right arm
Pulse: 70
Rhythm: regular
Temperature: 101.1°F.
Respiratory rate: 18
Height: 5′5″
Weight: 115 lbs.

[The examiner recognizes that the patient is febrile (has an elevated temperature, as recorded by an oral thermometer). The normal range of oral temperatures is from 96.4°F (35.8°C) to 99.1°F (37.3°C).]

HEENT Examination: Normal

Lungs: Normal

Cardiovascular Examination: Normal

Abdomen: Inspection, auscultation, percussion, and palpation of the abdomen are normal except for the following finding: The region of the RLQ midway between the anterior superior iliac spine and the umbilicus is tender to deep palpation.

Genitourinary Examination: Normal

Musculoskeletal Examination: Pertinent normal findings are that (a) flexion of the right thigh against resistance does not produce pain and (b) passive external and internal rotation of the right thigh does not produce pain.

Neurologic Examination: Normal

INITIAL ASSESSMENT OF THE PATIENT'S CONDITION

The patient appears to be suffering from an acute abdominal disease or condition.

ANATOMIC BASIS OF THE PATIENT'S HISTORY AND PHYSICAL EXAMINATION

(1) The patient's midabdominal pain appears to be umbilical visceral pain (a deep, constant midabdominal ache). This presentation suggests disease of one or more viscera supplied by the superior mesenteric artery.

(2) The RLQ tenderness to deep palpation suggests inflammation and/or acute distention of one or more RLQ abdominal viscera and/or inflammation of the parietal peritoneum lining the posterior abdominal wall of the RLQ (as there is no rebound tenderness in the anterolateral abdominal wall of the RLQ). The presentation suggests inflammation in or about the base of the appendix, as the RLQ tenderness can be localized to the midregion between the anterior superior iliac spine and the umbilicus.

(3) The patient's moderately elevated temperature indicates the presence of an inflammatory process.

(4) The patient's anorexia and nausea are not specific findings, as both are common autonomic responses to abdominal disease.

INTERMEDIATE EVALUATION OF THE PATIENT'S CONDITION

The patient appears to be suffering from inflammation of the appendix or one or more viscera about the base of the appendix.

CLINICAL REASONING PROCESS

An emergency care physician would consider that the most likely causes of the patient's condition are acute appendicitis (acute inflammation of the appendix), acute salpingitis (acute inflammation of the right uterine tube), or a ruptured ectopic pregnancy (rupture of an extra-uterine implantation). Because the patient presents with a "classic" pattern of signs and symptoms of an early stage of acute appendicitis (as will be shortly discussed), an emergency care physician would admit the patient to the hospital for observation and preparation for surgery.

When an adolescent girl presents with acute abdominal pain and RLQ tenderness, the examiner has to decide on an individual basis the necessity of conducting a bimanual pelvic examination. In a bimanual pelvic examination, the examiner inserts the digits of one hand into the female patient's vagina, anal canal, and rectum and uses the digits of the other hand to press deeply into the lower anterolateral abdominal wall in an attempt to palpate the various parts of the patient's pelvic reproductive viscera (specifically the cervix and body of the uterus, the uterine tubes, and the ovaries). Such an examination is physically unpleasant, even when there are no abnormal findings. If, under circumstances such as this case, the examiner wants to conduct a bimanual pelvic examination on an adolescent girl, it is incumbent upon the examiner to thoroughly explain the nature of the examination and to discuss its significance with regards to the health and well-being of the patient. The paramount factor in determining the necessity of a bimanual pelvic examination is the examiner's index of suspicion of the likelihood of acute salpingitis or ruptured ectopic pregnancy.

In this case, the examiner discussed with the patient the three most likely causes of her acute abdominal pain and RLQ tenderness and how a bimanual pelvic examination may provide important diagnostic evidence. The patient consented to the examination, the results of which were normal. The patient was also informed of the importance of a urine pregnancy test, the results of that were also negative. In this case, although the examiner believed the patient's denial of sexual activity, the examiner also recognized that adolescent boys and girls frequently have incorrect information regarding the kinds of sexual activities that can lead to pregnancy and may not be as forthcoming as adults about the nature of their sexual activities.

Acute inflammation of the appendix is thought to occur as a consequence of either luminal obstruction or ulceration of its mucosa. Deposition of a fecalith may be the cause of luminal obstruction.

Appendicitis commonly presents as a recognizable sequence of symptoms as the condition proceeds from initial inflammation to gangrene and perforation of the appendix. A common sequence of symptoms prior to perforation is as follows:

(1) Dull, diffuse pain in the umbilical or epigastric region is the common initial symptom. The pain in either region is visceral and/or referred pain.

Visceral pain in the umbilical region is most likely the product of stimulation of sensory fibers which innervate the appendix. These sensory fibers enter the spinal cord at the levels of the segments for the T10-T12 spinal nerves. Stimulation of these sensory fibers is the basis of pain referred to the part of the anterolateral abdominal wall overlying the umbilical region (Fig. 4–12C).

Stimulation of the visceral sensory fibers of an inflamed appendix may also elicit a reflexive spasm of the pyloric sphincter. The spasmodic contractions can stimulate sensory fibers that innervate the stomach, and it is the stimulation of these sensory fibers which generally accounts for visceral pain in the epigastric region. The sensory fibers that innervate the stomach enter the spinal cord at the levels of the segments for the T6-T9 spinal nerves. Stimulation of these sensory fibers is the basis of the pain referred to the part of the anterolateral abdominal wall overlying the epigastric region (Fig. 4–12C).

(2) Acute loss of appetite is generally the second symptom to appear. The anorexia may be accompanied by nausea and/or vomiting.

(3) Localized deep tenderness is commonly the third symptom to appear in the sequence. The deep tenderness is either wholly or in part visceral pain that emanates from the appendix itself upon deep palpation, and thus the site of deep tenderness always corresponds to the location of the appendix. Irritated parietal peritoneum bordering the inflamed appendix may also contribute to the pain perceived upon deep palpation.

Because the base, or proximal end, of the appendix (that is, the part of the appendix most constant in position) commonly underlies the midpoint between the umbilicus and the right anterior superior iliac spine, deep palpation of this specific point will generally (but not always) detect a tender appendix.

Some region of the RLQ will commonly prove tender to deep palpation of the overlying anterolateral abdominal wall if the inflamed appendix lies (i) within the right lateral paracolic gutter (along the cecum), (ii) on the parietal peritoneum overlying the iliacus muscle in the iliac fossa, or (iii) anterior to the terminal ileum. Tenderness upon deep palpation of the RLQ may be minimal or absent if the inflamed appendix lies (i) posterior to the cecum (in a retrocecal position), (ii) posterior to the terminal ileum, or (iii) draped over the right lateral wall of the pelvis (specifically, the lateral wall region formed by the obturator internus muscle). An inflamed appendix lying draped over the right lateral wall of the pelvis commonly elicits tenderness upon rectal examination.

(4) Fever is commonly the fourth symptom to appear in the sequence, generally appearing during the first 24 hours and usually with a temperature no more elevated than 102°F.

(5) Leukocytosis [an increase in the white blood cell (WBC) count to the range of 11,000 to 18,000 cells/mm^3 and a concurrent appearance of immature cells] is generally the fifth symptom to appear in the sequence prior to perforation.

(6) A region of hyperesthesia (increased sensitivity to irritation, such as that produced by gentle pinpricks) in the anterolateral abdominal wall overlying the RLQ is an inconstant finding prior to perforation.

Although perforation rarely occurs during the first 24 hours, it does generally occur during the next 24 hour period. Perforation should be suspected when one or more more of the following signs and symptoms are detected: (1) a WBC count of 20,000 cells/mm^3 or greater, (2) production of a fever of 104°F., and (3) signs of local or diffuse peritonitis. The inflammatory fluid released upon perforation tends to produce an abscess and local peritonitis if confined by surrounding viscera to a small region of the abdominopelvic cavity. For example, an abscess and local peritonitis are likely to occur if the ruptured appendix is retrocecal or lies posterior to the terminal ileum. By contrast, if the ruptured appendix lies anterior to the terminal ileum or in the right lateral paracolic gutter, the inflammatory fluid tends to quickly spread and produce diffuse peritonitis.

Signs of peritonitis include rebound tenderness, voluntary guarding of anterolateral abdominal wall muscles, or reflexive contraction and rigidity of musculature in inflamed abdominal and/or pelvic wall regions. There are three examples of reflexive muscular contraction of diagnostic significance:

(i) Unabated inflammation of the anterolateral abdominal wall ultimately produces a persistent, boardlike rigidity of the wall's musculature, a rigidity that resists respiratory movements.

(ii) Inflammation of the right iliopsoas may occur if a ruptured appendix directly overlies the iliacus and/or psoas major in the lower part of the abdominal cavity. Such inflammation may force the individual to hold the right thigh in either fixed flexion or extension. Inflammation of the right iliopsoas can be tested as the patient lies in a supine position on an examination table and attempts to flex the right thigh against resistance. Pain upon active flexion is a positive sign for iliopsoas inflammation. Alternatively, right iliopsoas inflammation can be tested if the examiner attempts to hyperextend the right thigh as the patient lies in the left lateral decubitus position. Pain upon passive hyperextension is a positive sign.

(iii) Inflammation of the right obturator internus may occur if a ruptured appendix lies draped over the right lateral wall of the pelvis. Such inflammation can be tested as the patient lies in a supine position (with the right thigh flexed 90° at the hip joint and the right leg flexed 90° at the knee joint) and the examiner alternatively internally and externally rotates the thigh. Pain upon passive rotation is a positive sign for obturator internus inflammation.

In this case, the tests for iliopsoas and obturator internus inflammation were conducted during the physical examination. The pertinence of the negative test results was that they suggested that if the patient were suffering from acute appendicitis of a retrocecal or pelvic appendix, then it is likely that the appendix has not yet perforated.

RADIOGRAPHIC EVALUATION AND FINAL RESOLUTION OF THE PATIENT'S CONDITION

An AP erect abdominal film of the patient did not reveal any abnormalities. The abdominal film was taken in preparation of the patient for surgery. It is important to appreciate that there are no radiographic signs that either unambiguously indicate or exclude acute appendicitis.

Laparoscopic examination of the patient's RLQ revealed an inflamed appendix in the right lateral paracolic gutter. The inflamed appendix was excised and the patient discharged after an uneventful recovery.

THE CHRONOLOGY OF THE PATIENT'S CONDITION

Entrapment of a fecalith in the patient's appendix resulted in its luminal obstruction and consequent inflammation. The patient's history indicated a "classic" sequence of symptoms: first visceral pain in the umbilical region followed by anorexia and vomiting. At the time of examination, the patient had a low grade fever and deep tenderness in the RLQ. Blood analysis revealed leukocytosis (16,000 cells/mm^3).

RECOMMENDED REFERENCES FOR ADDITIONAL INFORMATION ON THE VISCERA SUPPLIED BY THE SUPERIOR AND INFERIOR MESENTERIC ARTERIES

Silen, W., *Cope's Early Diagnosis of the Acute Abdomen,* 18th ed., Oxford University Press, New York, 1991: *Chapter 7 presents a discussion of the differential diagnosis of appendicitis.*

de Dombal, F. T., *Diagnosis of Acute Abdominal Pain,* 2nd ed., Churchill Livingstone, Edinburgh, 1991: *Chapter 7 presents a discussion of the differential diagnosis of appendicitis.*

CHAPTER **24**

The Posterior Abdominal Wall and the Viscera Supplied by the Renal Arteries

The renal arteries and their branches supply the adrenal glands, the kidneys, and the uppermost parts of the ureters. Ureteral branches of the abdominal aorta, gonadal arteries, and common iliac arteries supply lower parts of the ureters.

THE POSTERIOR ABDOMINAL WALL

The **body of the twelfth thoracic vertebra** and the **twelfth ribs** mark the upper border of the posterior abdominal wall (Fig. 24–1). The **body of the fifth lumbar vertebra** and the **portions of the iliacus muscles that overlie the iliac fossae of the bony pelvis** form the lowest regions of the wall.

Quadratus lumborum, psoas major, and **iliacus** comprise much of the muscle mass of the posterior abdominal wall (Fig. 24–1). Quadratus lumborum is a muscle of respiration, which, by pulling down upon the twelfth rib during inspiration, aids the inspiratory excursions of the diaphragm. Quadratus lumborum can also laterally flex the body trunk in the lumbar region. Quadratus lumborum is innervated by the anterior ramus of T12 and the anterior rami of L1 to L3 (or L4).

THE LUMBAR PLEXUS

The lumbar plexus arises from the anterior rami of L1, L2, and L3 and a part of the anterior ramus of L4 (Fig. 24–2). On each side of the posterior abdominal wall, the lumbar plexus lies embedded within the muscular mass of psoas major. The major branches of the lumbar plexus and their segmental derivation are as follows: **iliohypogastric nerve (L1), ilioinguinal nerve (L1), genitofemoral nerve (L1 & L2), lateral cutaneous nerve of the thigh (L2 & L3), femoral nerve (L2, L3 & L4), and obturator nerve (L2, L3 & L4).**

THE ADRENAL GLANDS

The adrenal glands lie atop the **superior poles of the kidneys** (Fig. 21–3). Each gland has an outer cortex and an inner medulla.

Each adrenal gland is supplied by three sets of arteries: the **superior adrenal arteries** are branches of the inferior phrenic artery (the paired inferior phrenic arteries arise from the abdominal aorta, near the aortic opening of the diaphragm), the **middle adrenal arteries** are branches of the abdominal aorta, and the **inferior adrenal arteries** are branches of the renal artery. The adrenal gland may also be supplied by branches of the gonadal artery.

Each adrenal gland is drained by just a single vein. The **left adrenal vein** ends by union with the left renal vein, and the **right adrenal vein** ends by union with the inferior vena cava.

THE KIDNEYS AND THE URETERS

The kidneys lie nestled in the paravertebral gutters within the upper quadrants of the abdominopelvic cavity (Fig. 21–3). Each kidney rests posteriorly against the quadratus lumborum and psoas major muscles, tilted in such a fashion that its superior pole lies closer to the vertebral column than its inferior pole (the poles of each kidney are its rounded upper and lower ends).

The left kidney commonly lies at a level slightly higher than that of the right kidney. In a supine indi-

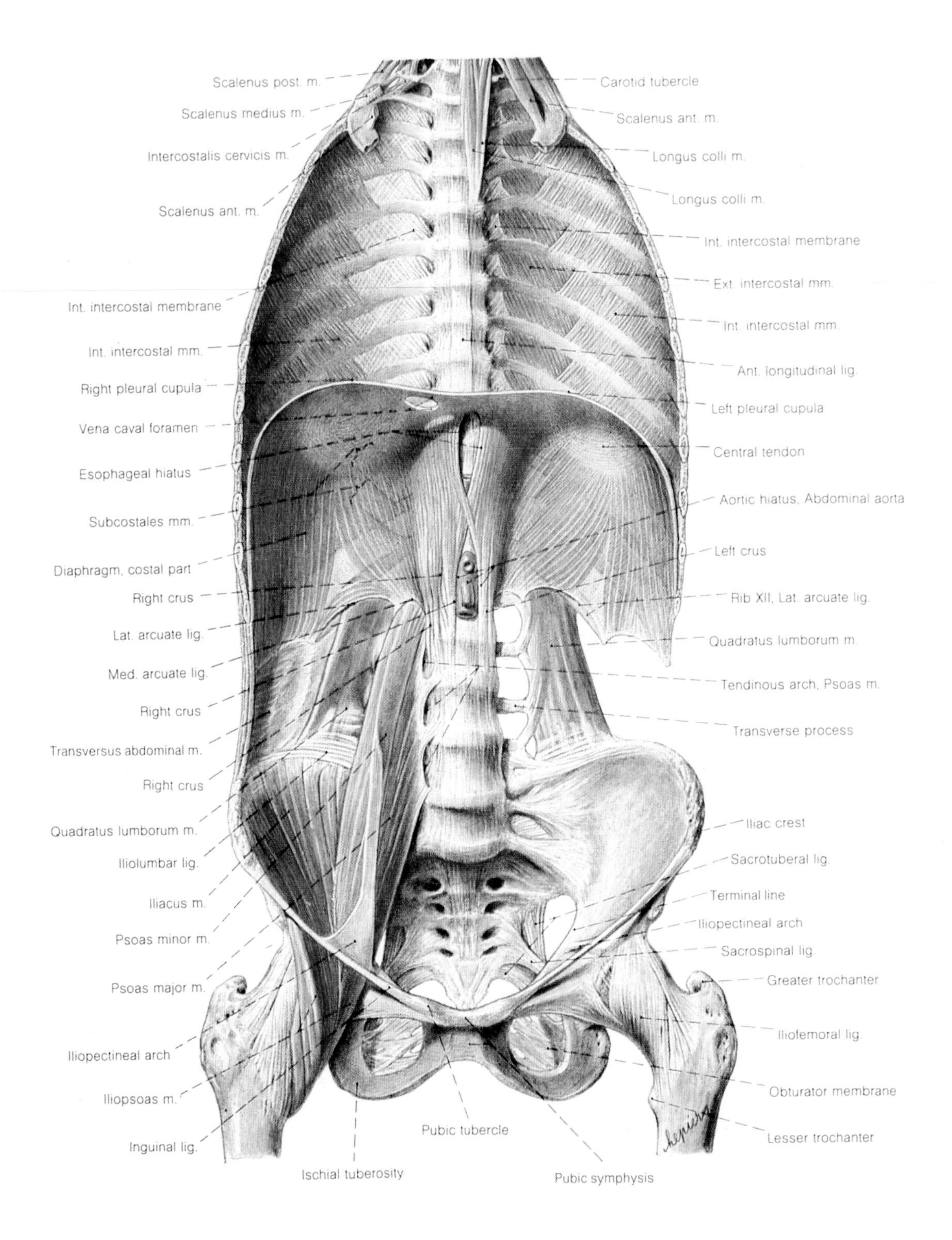

Fig. 24–1: Anterior view of the posterior abdominal wall and posterior aspect of the chest wall.

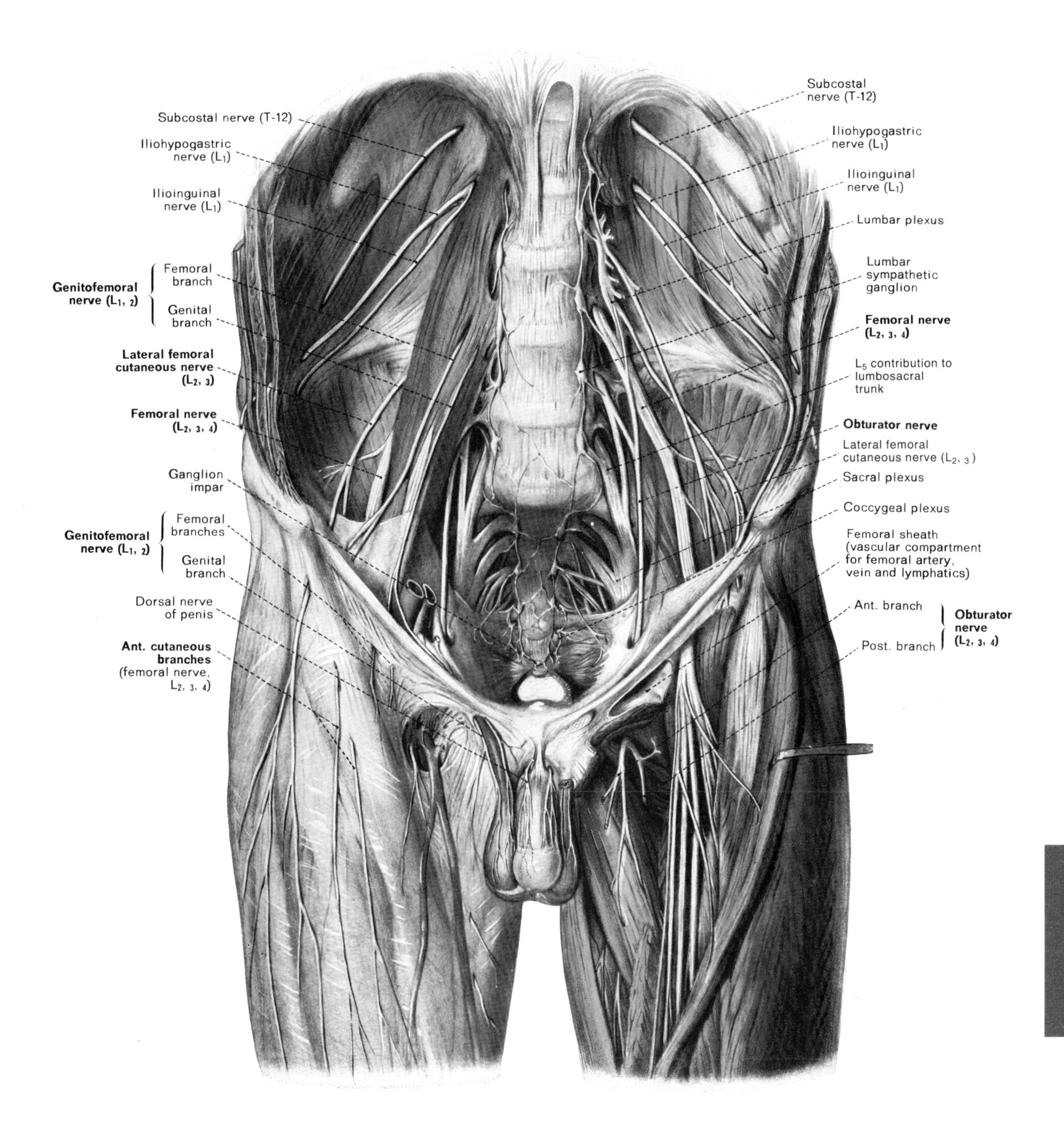

Fig. 24–2: Anterior view of the nerves of the posterior abdominal wall. On the left side, the psoas major and minor muscles and the genitofemoral nerve have been removed to reveal the lumbar plexus and the major nerves (except for the genitofemoral nerve) arising from it.

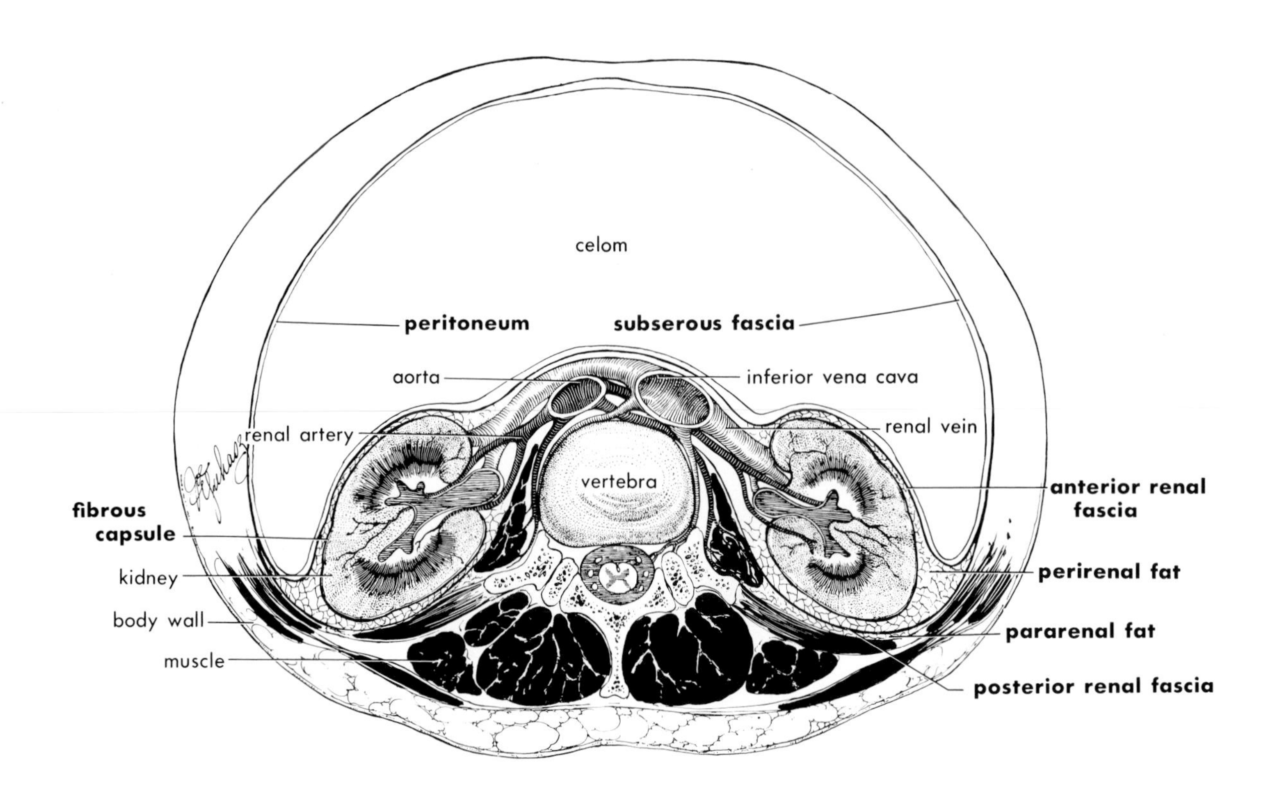

Fig. 24–3: Diagram of a superior view of a horizontal section of the kidneys and the surrounding layers of perirenal fascia.

vidual, the left kidney's superior pole extends above the posterior segment of the left eleventh rib, and the right kidney's superior pole extends above the posterior segment of the right twelfth rib (Fig. 22–10). The transpyloric plane passes through the lower part of the left kidney's hilum and the upper part of the right kidney's hilum (the **hilum** is the vertical fissure in the kidney's medial margin). The length of a kidney (the distance between its poles) averages 3.7 times the thickness of the body of the second lumbar vertebra.

Each kidney is surrounded by three distinct layers of perirenal fascia: a fatty inner layer called the perirenal fat, a membranous intermediate layer called the renal fascia, and a fatty outer layer called the pararenal fat (Fig. 24–3). The pararenal fat is simply that part of the fatty, extraperitoneal connective tissue of the posterior abdominal wall cavity that encases the kidney. The liquified constituency of the fat in the surrounding layers of perirenal fascia at body temperature render the kidneys very mobile. This mobility is the basis for the inferior displacement of the kidneys during inspiration and assumption of an upright position.

The kidneys are retroperitoneal organs. The anterior surface of the perirenal fascia about them, however, is not completely covered with peritoneum because each kidney's perirenal fascia is in direct contact anteriorly with other retroperitoneal viscera. The perirenal fascia about the right kidney is in direct contact anteriorly with the right adrenal gland, the second part of the duodenum, and the hepatic flexure; the perirenal fascia about the left kidney is in direct contact anteriorly with the left adrenal gland, the body of the pancreas, and the splenic flexure (Fig. 24–4).

Each kidney presents anterior and posterior surfaces, superior and inferior poles, a convex outer margin, and an indented inner margin (Fig. 21–3). The **renal artery** enters and the **renal vein** and **renal pelvis** exit the kidney through its hilum (Figs. 21–3 and 24–4). The hilum leads into a central, blind, fat-filled space within the kidney called the **renal sinus.**

The kidney has an inner medulla and an outer cortex (Fig. 24–5). The medulla consists of 5 to 18 conical regions into which are densely packed the ascending and descending limbs of the renal tubules and the collecting tubules. The conical regions are called the **renal pyramids,** and their apices, which are called **renal papillae,** all project toward the renal sinus (Fig. 24–5). Small, chalice-shaped vessels called **minor calyces** envelop the renal papillae, and collect the urine released from the collecting tubules. The minor calyces coalesce into two or three larger, chalice-shaped

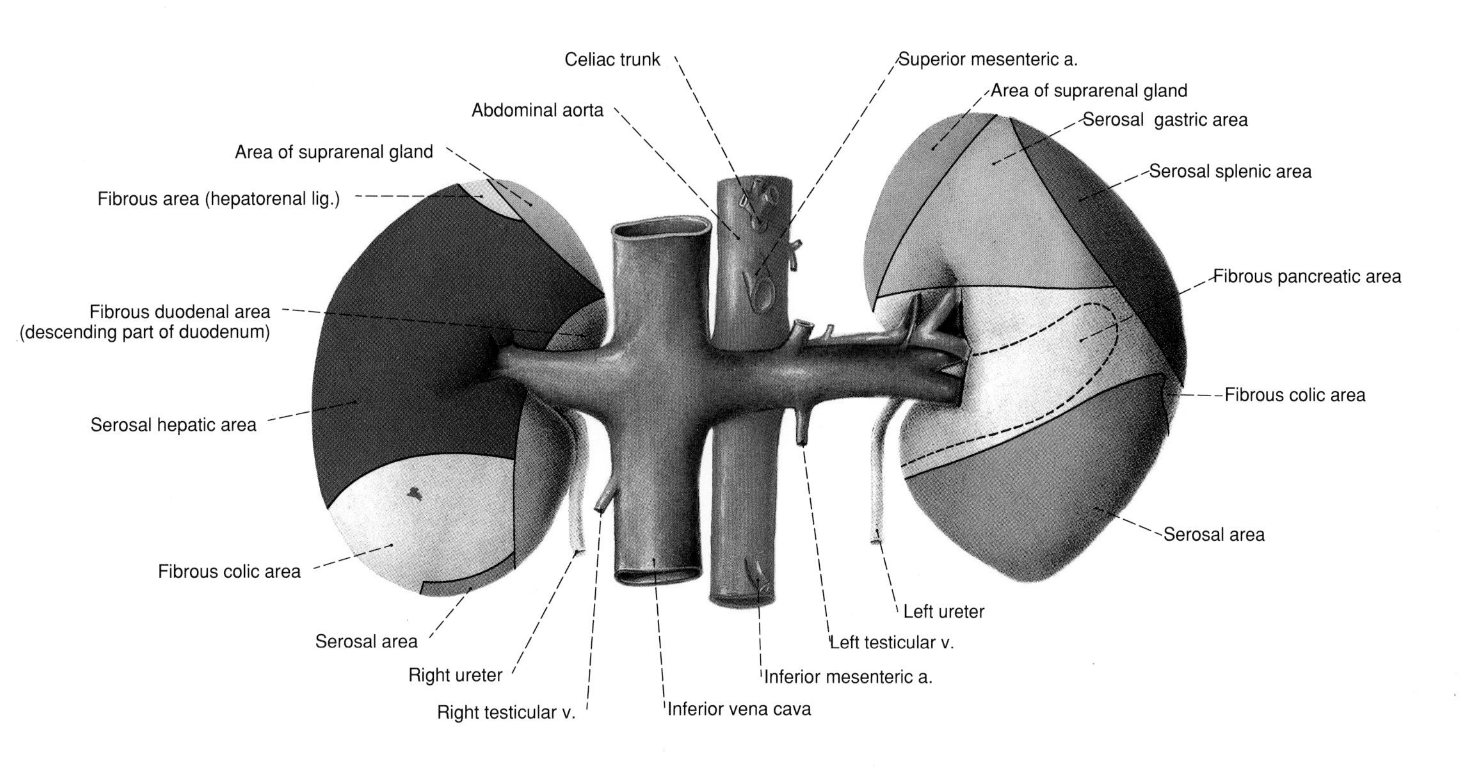

Fig. 24–4: Diagram of the anterior aspects of the kidneys and their relations. The serosal areas are the areas in which visceral peritoneum-covered abdominal organs come into contact with peritoneum-covered parts of the anterior aspects of the kidneys.

vessels called **major calyces,** which, in turn, coalesce to form the **renal pelvis,** the upper, funnel-shaped end of the ureter (Fig. 24–5). The renal pelvis arches downward and narrows into the ureter in the renal hilum.

The cortex consists of the regions into which are densely packed the glomeruli and the convoluted portions of the renal tubules. The cortical regions that extend between the renal pyramids toward the renal sinus are called the **renal columns** (Fig. 24–5). Each renal pyramid and the cortical region that lies at its base represents a **lobe;** each kidney is thus divisible into 5 to 18 lobes.

The renal arteries arise from the abdominal aorta at the level of the first or second lumbar vertebra. Each **renal artery** divides into **anterior and posterior rami** as it extends through the renal hilum (Fig. 24–6). Each ramus gives rise to one or more **segmental arteries;** each segmental artery supplies a separate vascular segment of the kidney. The segmental branches of the renal arteries are anatomic end arteries.

The **left** and **right renal veins** join the inferior vena cava at approximately the same level (Fig. 21–3). The left renal vein crosses in front of the abdominal aorta and behind the superior mesenteric artery as it extends toward the inferior vena cava.

The ureters conduct urine from the kidneys to the urinary bladder by peristalsis. The ureters descend through the abdominal cavity lying directly against the psoas major muscles (Fig. 21–3). Each ureter enters the pelvic cavity near the point where it passes in front of the bifurcation of the common iliac artery into the external and internal iliac arteries.

THE RENAL PLEXUS

The renal plexus is the plexus of the autonomic nervous system that surrounds the renal artery; it contains a large, sympathetic ganglion (the **aorticorenal ganglion**) near the origin of the renal artery and a small, sympathetic ganglion (the **renal ganglion**) along the midregion of the renal artery. The renal plexus transmits postganglionic sympathetic fibers to the adrenal gland, kidney, and the upper part of the ureter. The renal plexus also transmits preganglionic sympathetic fibers from the lesser, least, and first lumbar splanchnic nerves that terminate on the chromaffin cells of the medulla of the adrenal gland; these chromaffin cells are considered to be homologous to postganglionic sympathetic neurons, and they secrete norepinephrine and epinephrine into the venous sinusoids of the adrenal gland.

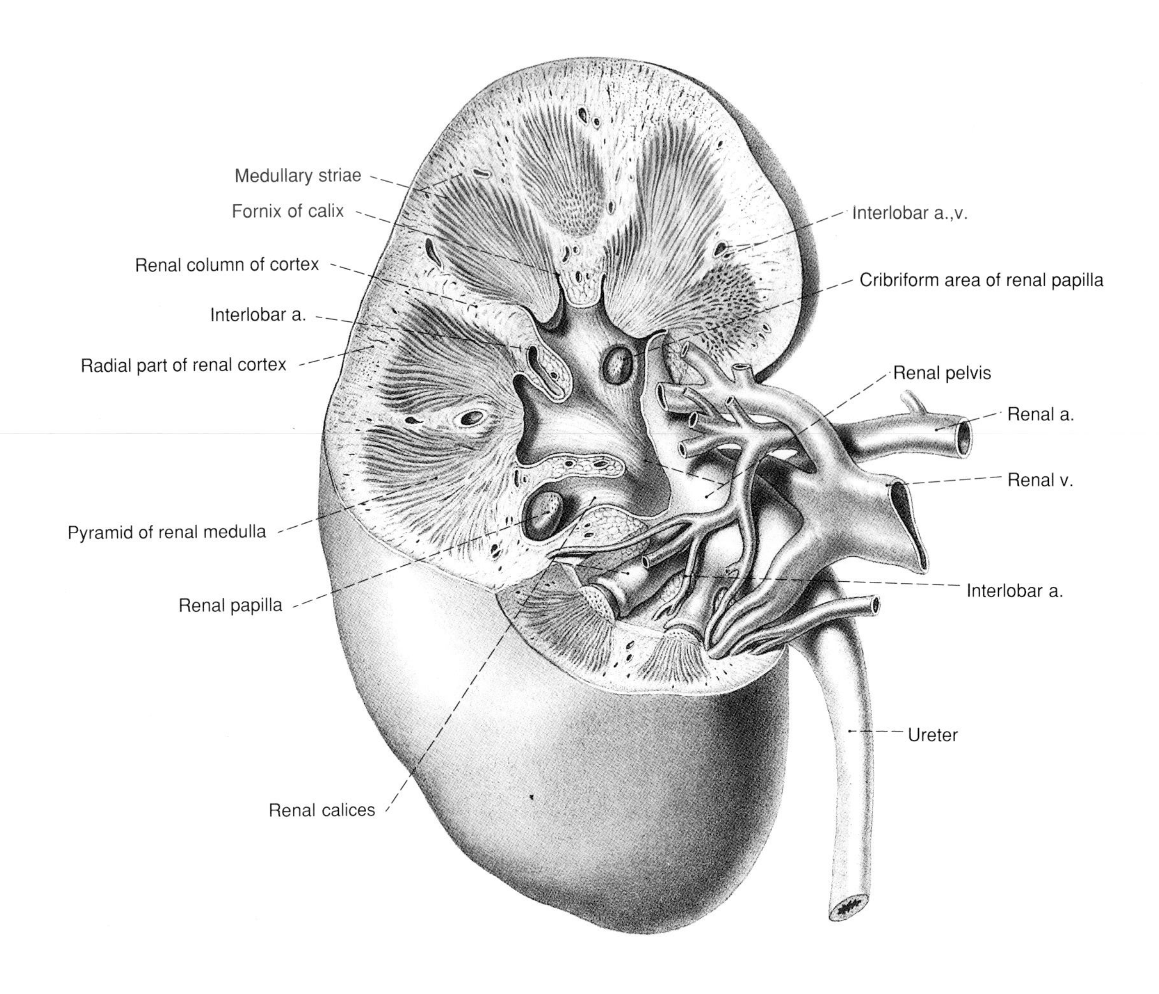

Fig. 24–5: Diagram of the anterior view of a partial coronal section of the right kidney.

The **lesser** and **least splanchnic nerves** and the **first lumbar splanchnic nerve** provide preganglionic sympathetic innervation for the kidney. The sensory fibers innervating the kidney enter the spinal cord at those spinal cord segments that provide preganglionic sympathetic innervation for the kidney (in other words, the sensory fibers enter at the T10-L1 levels). Disease of or injury to the kidney produces visceral pain that is most commonly localized to the costovertebral and posterior subcostal regions. Disease or injury of the kidney may refer pain to parts of the T10-L1 dermatomes.

The lesser and least splanchnic nerves and the first and second lumbar splanchnic nerves provide preganglionic sympathetic innervation for the upper part of the ureter. The sensory fibers innervating the upper part of the ureter enter the spinal cord at those spinal cord segments that provide preganglionic sympathetic innervation for the upper part of the ureter (in other words, the sensory fibers enter at the T10-L2 levels). Disease or injury of the upper part of the ureter produces visceral pain that is (most commonly) localized to the costovertebral angle. Disease or injury of the upper part of the ureter may refer pain to parts of the T10-L2 dermatomes.

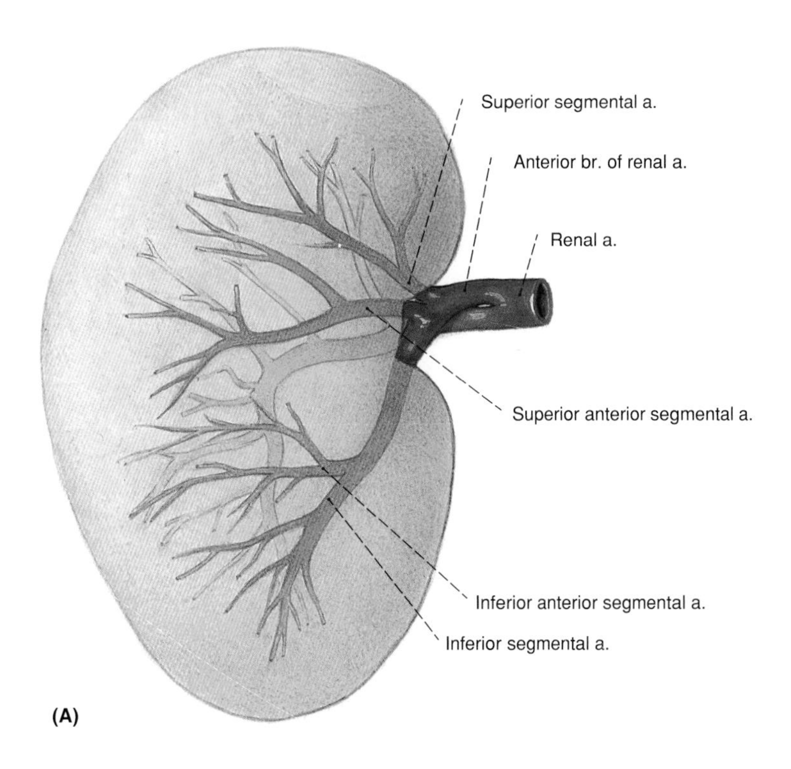

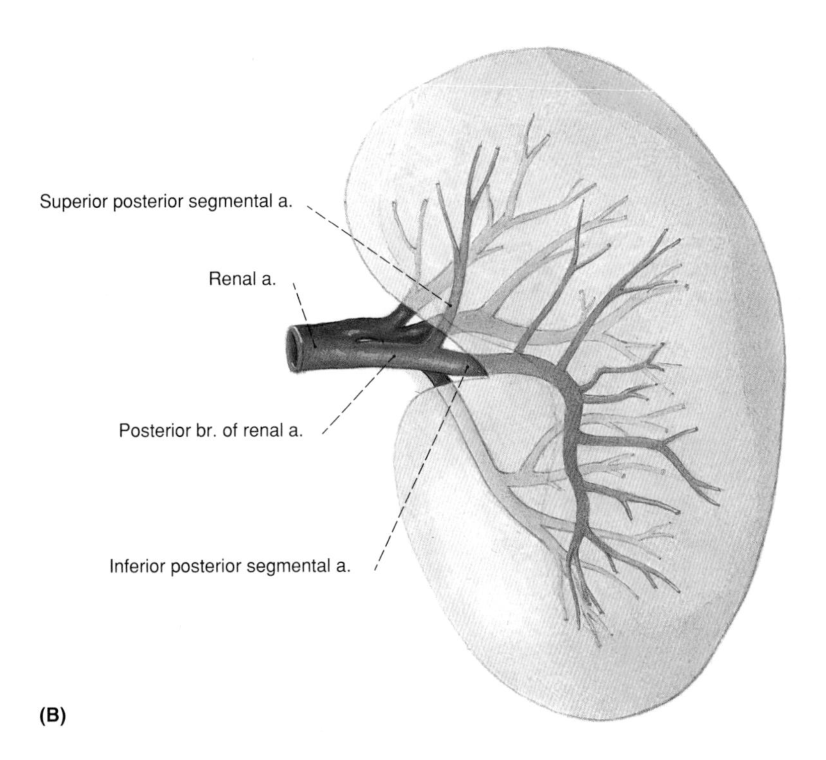

Fig. 24–6: Diagrams of (A) an anterior view and (B) a posterior view of the segmental arteries of the right kidney.

CASE **V.6**

The Case of Albert Skrinjar

INITIAL PRESENTATION AND APPEARANCE OF THE PATIENT

A 41 year-old, white man named Albert Skrinjar walks into the emergency center at 8:00 AM to seek treatment for a painful abdomen. The examiner observes that the patient is in continuous motion: he either quickly paces the floor or, if seated, frequently changes his body position.

QUESTIONS ASKED OF THE PATIENT

Please tell me about your problem. Well, I was struck with a very bad pain on the left side of my back shortly after dinner last night. The pain built up in about five minutes to a severity I've never experienced before. The pain was so bad that I became nauseous and vomited. After I vomited, it seemed like some of the pain had gone away, and so I thought that maybe I had a virus or something that upset my stomach. But the pain kept me awake almost all night, and it hasn't got any better. *[The request "Please tell me about your problem" invites the patient to openly and completely discuss his health concerns. In this instance, the examiner quickly learns that within the last 12 hours the patient has experienced rather severe left back pain associated with nausea and vomiting.]*

Can you show me or explain to me where you felt pain when the pain first struck last evening? Here, on the left side of my back. [As the examiner touches several points on the patient's back in order to more specifically locate the patient's back pain, the patient locates the pain in a broad expanse of the left side of the back that extends superiorly to about the posterior segment of left eleventh rib and inferiorly to the level of the spinous process of the second lumbar vertebra.]

Can you describe the pain that first struck last evening? It was like an excruciating deep ache.

Has the nature of the pain or its location changed at any time during the past 12 hours? Yes, most of the pain now is here on the left side of my body [the patient indicates the present location of pain by sliding his left hand up and down the left flank region.] The pain now is not as bad as it was when it struck last evening, but it's there all the time. And sometimes the pain shoots down the left side of my body into my left testicle [the patient indicates the path of the extended pain by sliding his left hand down from the left flank region over the left inguinal region to the left side of the scrotum]. *[The examiner recognizes the present pain pattern as "loin-to-groin" pain [pain that extends from a loin (flank) region of the abdomen to the ipsilateral groin]; the pain may extend further to the scrotal region in a male or the labial region in a female. Loin-to-groin pain suggests acute distention of the ureter.]*

Is there anything that relieves the pain? No.

Is there anything that worsens the pain? No. *[The patient's inability to find a body position or movement that either relieves or intensifies the pain indicates that somatic pain is not a major component of the patient's left flank pain.]*

Have you ever had a pain like this before? No.

Is there anything else that is also bothering you in addition to your abdominal pain and nausea? No.

Have you had any recent problems with urination? No. But when I urinated before coming here, I noticed that my urine was darker than normal. *[The sudden production of dark urine in this patient suggests blood-stained urine, and thus disease or injury of the urinary tract.]*

Have you had any recent illnesses or injuries? No.

Are you taking any medications for previous illnesses or conditions? No. *[The examiner finds the patient alert and fully cooperative during the interview.]*

PHYSICAL EXAMINATION OF THE PATIENT

Vital signs: Blood pressure
Lying supine: 125/75 left arm and 125/70 right arm
Standing: 125/75 left arm and 125/70 right arm
Pulse: 80
Rhythm: regular
Temperature: 98.9°F.
Respiratory rate: 16
Height: 5′9″
Weight: 170 lbs.
HEENT Examination: Normal
Lungs: Normal
Cardiovascular Examination: Normal
Abdomen: Inspection, auscultation, percussion, and palpation of the abdomen are normal except for tenderness in the left costovertebral region.
Genitourinary Examination: Normal
Musculoskeletal Examination: Normal
Neurologic Examination: Normal

INITIAL ASSESSMENT OF THE PATIENT'S CONDITION

The patient appears to be suffering from an acute disease or condition of the urinary tract system.

ANATOMIC BASIS OF THE PATIENT'S HISTORY AND PHYSICAL EXAMINATION

(1) **The abrupt onset of the patient's left subcostal back pain appears to be visceral pain (a deep ache) evoked by a catastrophic event or rapidly developing pathologic condition involving the left kidney and/or upper part of the left ureter.** The sudden onset of the patient's problem(s) is indicated by the patient's ability to recall where he was when the pain first struck and that his left subcostal back pain initially achieved maximum intensity within minutes of its occurrence. Visceral pain localized to the costovertebral and/or posterior subcostal regions suggests disease or injury of the kidney and/or upper part of the ureter.

(2) **The left costovertebral tenderness suggests inflammation and/or distention of the left kidney and/or upper part of the left ureter.** In cases of acute inflammation and/or distention of the kidney and/or upper part of the ureter, it is also common for gentle fist percussion to elicit tenderness in the costovertebral region.

(3) **The pain in the left flank region of the abdomen that radiates through the left groin into the left side of the scrotum suggests acute inflammation and/or distention of the ureter.** The pain frequently elicited by acute distention of the ureter is a combination of visceral and referred pain extending from the costovertebral and subcostal regions of the back along the loin and groin regions of the abdomen into the ipsilateral scrotal (or labial) region. Although such a distribution of pain is called **renal colic,** the term is a misnomer because the intensity of the pain is typically constant or slowly increasing (the term colic connotes pain that repetitively waxes and wanes). Visceral pain accounts for most of the costovertebral and posterior subcostal pain. The pain radiating through the loin and groin regions into the ipsilateral scrotal region mostly represents referred pain associated with the T12-L2 dermatomes (Fig. 4–12C). The referred pain commonly occurs in cutaneous areas supplied by the subcostal, iliohypogastric, ilioinguinal, and genitofemoral nerves.

(4) **The passage of dark urine suggests blood-stained urine, and thus disease or injury of the urinary tract.** The disease or injury may involve the upper urinary tract (kidney and ureter) and/or the lower urinary tract (urinary bladder and urethra).

(5) **The patient's nausea and vomiting are important but non-specific symptoms.** In this case, they appear to represent autonomic responses to disease or injury of one or more abdominal and/or pelvic viscera.

INTERMEDIATE EVALUATION OF THE PATIENT'S CONDITION

The patient appears to be suffering from acute left upper urinary tract inflammation and/or distention.

CLINICAL REASONING PROCESS

A primary care or emergency medicine physician would consider four major findings from the patient's history and physical examination as collectively indicating acute disease or disorder of the left kidney and/or left ureter: (1) The initial attack of pain occurred in the left subcostal region of the back. (2) The patient has left costovertebral tenderness. (3) The patient has loin-to-groin pain; the loin pain is fairly constant. (4) There is an acute onset of the passage of dark urine.

The key finding in the consideration of this case is the abrupt onset of excruciating upper urinary tract pain. When a part of the collecting system of the upper urinary tract becomes distended, the intensity of the visceral pain that is evoked by the distention is primarily a measure of the rate of distention (the collecting system of the upper urinary tract consists of the minor and major calyces, the pelvis of the ureter, and the remainder of the ureter). Therefore, an abrupt onset of excruciating pain is highly suggestive of an acute high rate of distention of some part of the collecting system of the upper urinary tract.

The most common causes of an acute high rate of distention of some part of the collecting system of the upper urinary tract are a renal calculcus (a "kidney stone") or a blood clot. The entrapment of a calculus or blood clot at a particular level of the collecting system leads to partial or complete obstruction of the flow of urine (obstructive uropathy), and the continuous accumulation of urine proximal to the level of entrapment leads to the painful acute distention of the more proximal parts of the collecting system.

RADIOGRAPHIC EVALUATION AND FINAL RESOLUTION OF THE PATIENT'S CONDITION

An AP abdominal plain film did not show any distinct abnormalities except for an 9 × 10 mm polygonal radiopaque density to the left of the tip of the transverse process of the first lumbar vertebra. An intravenous pyelogram (urogram) revealed retarded passage of contrast material through the calyces and ureter of the left kidney. The calyces and pelvis of the left ureter were distended relative to the calyces and pelvis of the right ureter, but the left kidney did not appear distended relative to the right kidney. The intravenous pyelogram revealed that the upper portion of the ureter was only partially obstructed by the 9 × 10 mm polygonal radiopaque mass (because urine bearing contrast material was evident along the entire course of the left ureter distal to the site of the polygonal mass).

Intravenous urography is a radiologic procedure that uses a radiopaque substance (contrast material) that is filtered from the circulating blood by the kidneys and concentrated before excretion into the calyces. Under normal conditions, most of the radiopaque substance is filtered from the circulating blood and concentrated in the renal tubules two minutes after intravenous injection of the radiopaque substance. Radiopaque substance appears within the calyces four minutes after intravenous injection. Films taken at appropriate intervals provide both anatomic and physiologic information concerning the kidneys. The films provide anatomic information about the size of a kidney, the thickness of its parenchyma, the size and shape of renal papillae, minor and major calyces, and the pelvis of the ureter, and the location of any sites of ureteral obstruction. The films provide physiologic information about the onset and distribution of renal parenchymal opacification and the onset and concentration of contrast material within the calyces.

In this case, a urologist might recommend extracorporeal lithotripsy to reduce the entrapped ureteral stone to fragments small enough to pass through the ureter into the urinary bladder and discharge through the urethra. Extracorporeal lithotripsy is a procedure in which the patient is submerged in a tank of water and exposed to shock waves directed at the region of the back overlying the entrapped ureteral stone. The shock waves disrupt the calculus into fragments that can pass through the ureter and the lower urinary tract.

THE CHRONOLOGY OF THE PATIENT'S CONDITION

A renal calculus in the patient's left kidney was dislodged last evening from its attachment to a renal papilla, passed into the ureter, and became entrapped immediately below the ureteropelvic junction. Its entrapment initially produced complete obstruction, resulting in the attack of excruciating left posterior subcostal pain. A small sharply-edged fragment of the calculus broke off and descended the ureter during the subsequent period in which the patient was vomiting. This release of this fragment rendered the obstruction incomplete, and thereby allowed urine to slowly pass by the entrapped calculus. As the fragment descended the ureter, the fragment cut the ureteral epithelial lining at several sites. These minute cuts are the sites at which blood has extravasated into the patient's urine.

The vain attempt by the patient in the examination room, to find a body position that will alleviate the pain is typical of individuals afflicted with acute obstructive uropathy. In general, this behavior is typical of any individual suffering from severe visceral abdominal pain in the absence of somatic abdominal pain.

RECOMMENDED REFERENCES FOR ADDITIONAL INFORMATION ON THE VISCERA SUPPLIED BY THE RENAL ARTERIES

Kelley, W., N., DeVita, Jr., V. T., Dupont, H. L., Harris, Jr., E. D., Hazzard, W. R., Holmes, E. W., Hudson, L. D., Humes, H. D., Paty, D. W., Watanabe, A. M., and T. Yamada, *Textbook of Internal Medicine,* J. B. Lippincott Co., Philadelphia, 1992: *This text is one of the most authoritative and comprehensive texts on internal medicine. In the nephrology section of Chapter 134 provides a discussion of the approach to the patient with urinary retention and obstruction. Chapter 135 offers a discussion of the approach to the patient with renal lithiasis. Chapter 128 presents a discussion of the approach to the patient with hematuria.*

Wilson, J. D., Braunwald, E., Isselbacher, K. J., Petersdorf, R. G., Martin, J. B., Fauci, A. S., and R. K. Root, *Harrison's Principles of Internal Medicine,* 12th ed., McGraw-Hill, New York, 1991: *This text is one of the most authoritative and comprehensive texts on internal medicine. In Part 8 on disorders of the kidney and urinary tract, Chapter 221 provides an introductory discussion of the approach to the patient with disease of the kidneys and urinary tract.*

Silen, W., *Cope's Early Diagnosis of the Acute Abdomen,* 18th ed., Oxford University Press, New York, 1991: *Chapter 12 presents a discussion of the various types of colics: intestinal, biliary, renal, gastric, and pancreatic.*

de Dombal, F. T., *Diagnosis of Acute Abdominal Pain,* 2nd ed., Churchill Livingstone, Edinburgh, 1991: *Chapter 15 presents a discussion of the signs and symptoms of urinary tract problems.*

CHAPTER 25

The Pelvic Walls

The walls of the pelvis are so pronouncedly basin-shaped that they are often collectively referred to as the pelvic basin. The pelvic walls support and protect the pelvic viscera.

THE BONY PELVIS

The skeletal framework of the pelvic basin is the bony pelvis. In an adult, the bony pelvis consists of the **left and right coxal bones,** the **sacrum,** and the **coccyx** (Fig. 12–5). The major joints of the bony pelvis are the **pubic symphysis** (which joins the pubic bodies of the coxal bones) and the **paired sacroiliac joints** (which together join the iliac parts of the coxal bones with the sacrum).

The major ligaments of the bony pelvis bind the coxal bones to the sacrum and coccyx. On either side, there are **sacroiliac ligaments** stretched between the sacrum and the iliac part of the coxal bone; the most superior and posterior of these ligaments are among the strongest ligaments in the body (Fig. 12–6). Also on each side, there is a **sacrotuberous ligament** binding the ischial tuberosity to the sacrum and coccyx (Fig. 12–6), and a **sacrospinous ligament** tautly extended from the ischial spine to the sacrum and coccyx (Fig. 12–5) (the sacrotuberous and sacrospinous ligaments are respectively labelled the sacrotuberal and sacrospinal ligaments in Figs. 12–5 and 12–6).

The bony pelvis and its sacrotuberous ligaments provide the landmarks that define the superior and inferior openings of the pelvic basin. The superior opening of the pelvis is called the **pelvic inlet.** The border of the pelvic inlet is called the **pelvic brim,** and it is formed anteriorly by the pubic crests of the coxal bones, laterally by the pectineal and arcuate lines of the coxal bones, and posteriorly by the sacral promontory (the bony rim labelled the terminal line in Figure 12–5 represents the pectineal and arcuate lines of the coxal bone). The **sacral promontory** is the anterosuperior edge of the body of the first sacral vertebra (Fig. 12–9). The inferior opening of the pelvis is called the **pelvic outlet,** and it is bordered anteriorly by the pubic arch, laterally by the ischial tuberosities, and posteriorly by the sacrotuberous ligaments and the tip of the coccyx (Fig. 12–6). The **pubic arch** is the midline arch of the bony pelvis bordered above by the pubic symphysis and on the sides by the inferior pubic rami.

When a person stands erect, the bony pelvis is tilted forward (because, in part, of the lordotic curvature of the lumbar portion of the vertebral column) to the extent that the anterior superior iliac spines and the upper margin of the pubic symphysis all lie in a vertical plane (Fig. 12–9). In this orientation, the pelvic brim and the plane of the pelvic outlet lie tilted forward at, respectively, 50 to 60° and 15° angles with the horizontal plane. This forward tilt of the pelvic basin places the pubic parts of the coxal bones partly inferior to the bladder, thereby providing bony support for this viscus.

There are significant differences between the average dimensions of the pelvic cavity of the male versus female bony pelvis: (1) The pelvic basin in a female is shallower and more cylindrical than in a male. (2) The subpubic angle (the angle below the pubic symphysis and between the inferior pubic rami) is more rounded and 20 to 30° greater in a female. (3) The pelvic outlet has a greater circumference in a female, because the ischial tuberosities are more everted.

THE MUSCLES OF THE PELVIC WALLS

The bony pelvis furnishes most of the sites of origin for the four pairs of muscles that line the pelvic walls: the levator ani and coccygeus muscles form the floor of the pelvis, the obturator internus muscles line the lateral walls, and the piriformis muscles fashion much of the posterior wall. The term **pelvic diaphragm** refers to the two pairs of muscles that form the pelvic floor (the levator ani pair and the coccygeus pair).

The **left** and **right levator ani** form most of the muscular floor of the pelvis. Levator ani has a gently curved origin which extends between the pelvic surface of the body of the pubis and the ischial spine (Fig. 25–1). The origin adheres to the medial surface of the obturator internus through a fascial thickening called the **tendinous arch** or **tendinous arc** (Fig. 25–2). The fibers within each muscle extend backward and medially from their origin (Fig. 25–1). Levator ani is

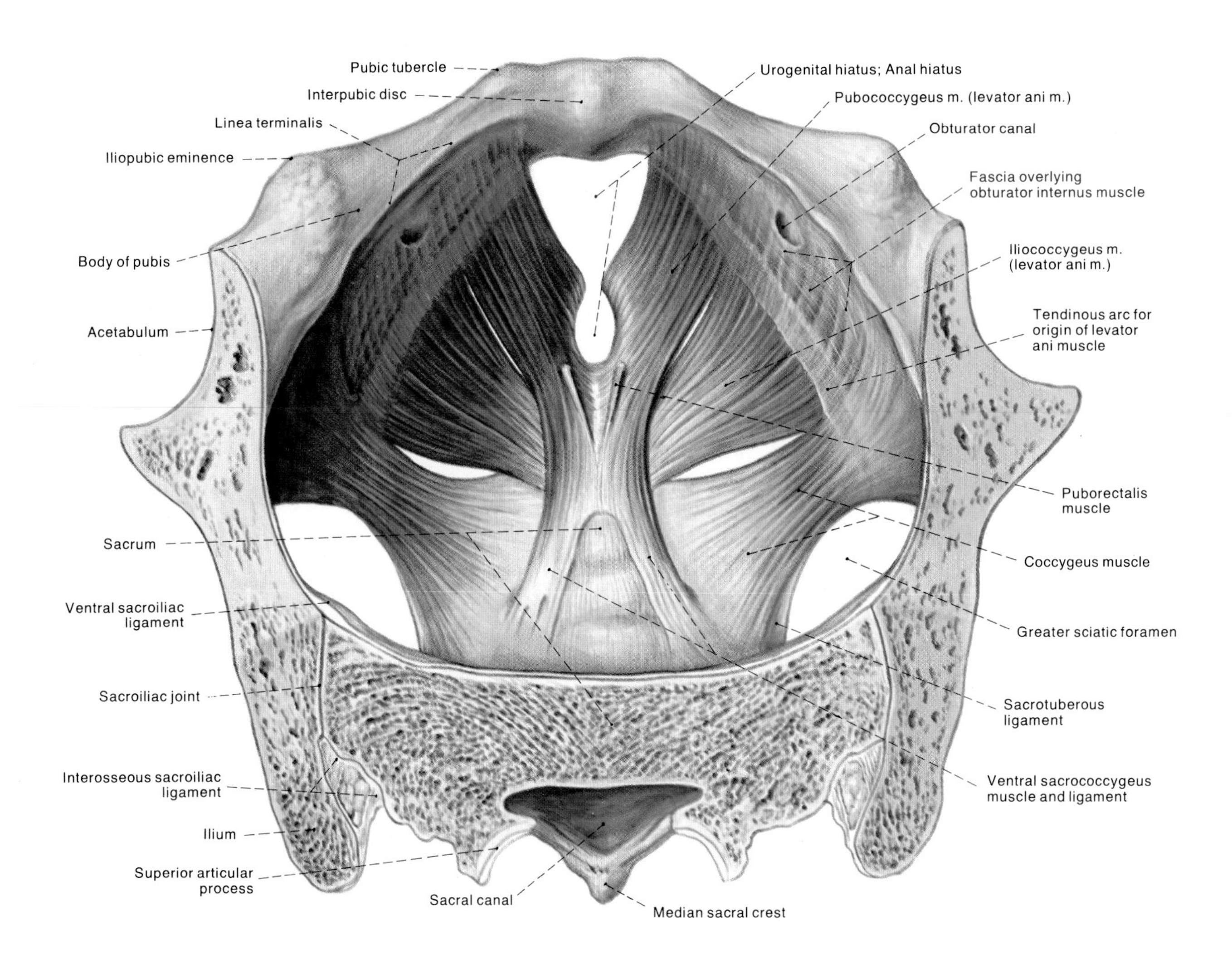

Fig. 25–1: Superior view of the pelvic diaphragm (muscular floor) of the female pelvis. Pubococcygeus and iliococcygeus are the names of respectively the pubic and ischial parts of levator ani.

innervated by nerve fibers that arise from the anterior ramus of S4 and from the pudendal nerve.

The left and right levator ani unite with each other along the midline, except for (a) anteriorly where there is an opening called the **urogenital hiatus** (**genital hiatus**) that transmits the urethra (and also the vagina in the female) and (b) posteriorly where there is an opening called the anal hiatus that allows the passage of the anorectal junction (the junction between the rectum and the anal canal) (Fig. 25–1). This union of the left and right levator ani forms a gutter-shaped floor for the pelvis that slopes gently upward as it extends backward (Fig. 25–2).

In both sexes, the most anterior fibers of each levator ani extend backward from their pubic origin to insert onto a fibromuscular body (the **central tendon of the perineum**) located in front of the anorectal junction (Fig. 25–2). In the female, these muscle fibers course beside the vagina before inserting onto the central tendon of the perineum; accordingly, these fibers are called the **pubovaginalis part of levator ani** (in Figure 25–1, the levator ani muscle fibers that border the sides of the urogenital hiatus represent the pubovaginalis parts of the left and right levator ani muscles). In the male, the fibers extend beneath the prostate before inserting onto the central tendon of the perineum, and so are called the **levator prostatae part of levator ani.**

In both sexes, a prominent group of fibers in the levator ani form a U-shaped muscular sling around the posterior and lateral surfaces of the anorectal junction. This group of fibers in each levator ani represents the **puborectalis part of levator ani** (Fig. 25–1). Tonic contraction of the puborectalis part of levator ani draws the anorectal junction forward and thus increases the angle between the rectum and the anal canal; this action helps prevent unwanted passage of feces from the rectum into the anal canal.

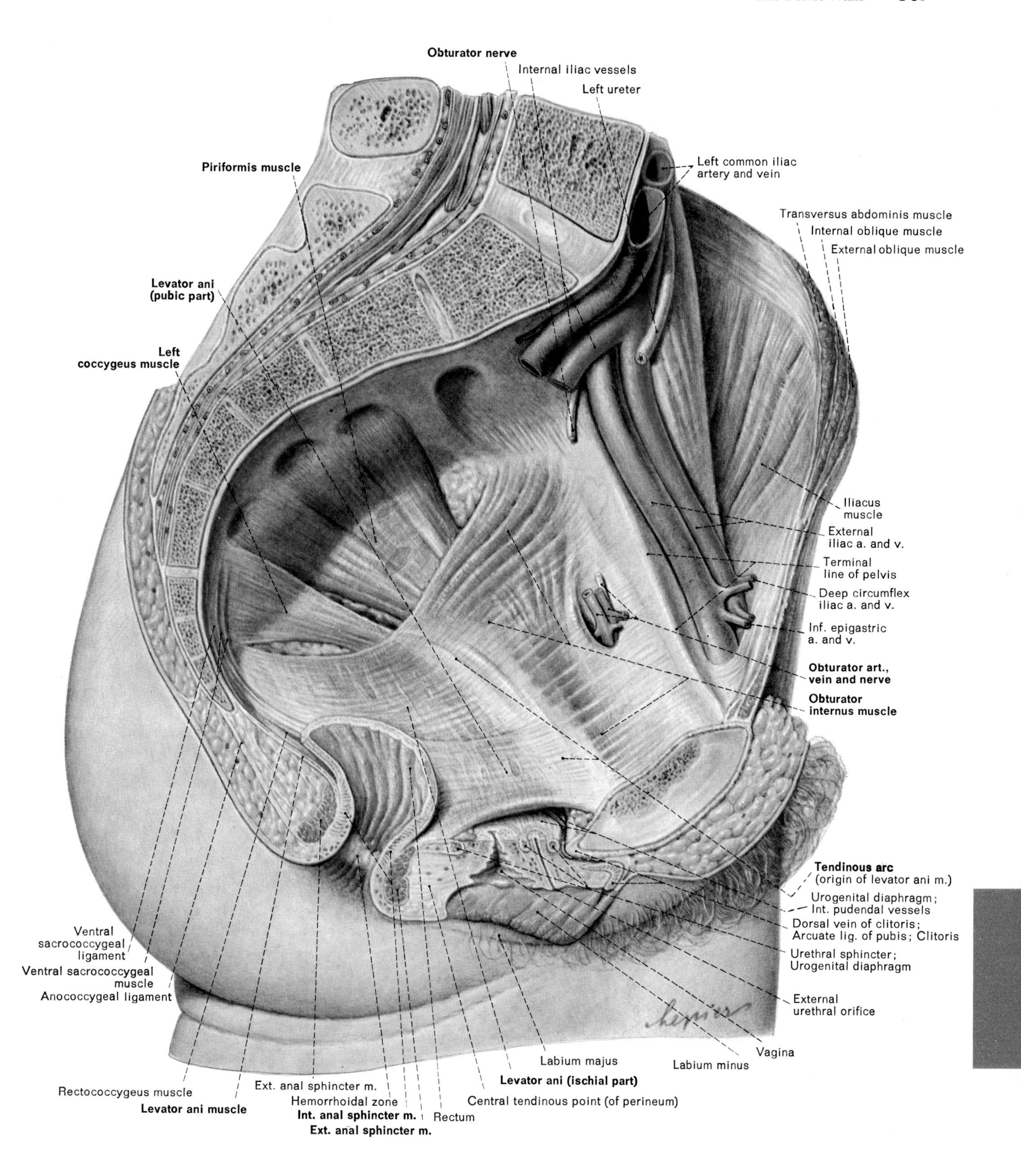

Fig. 25–2: Medial view of the muscles of the left lateral wall and left side of the floor of the female pelvis.

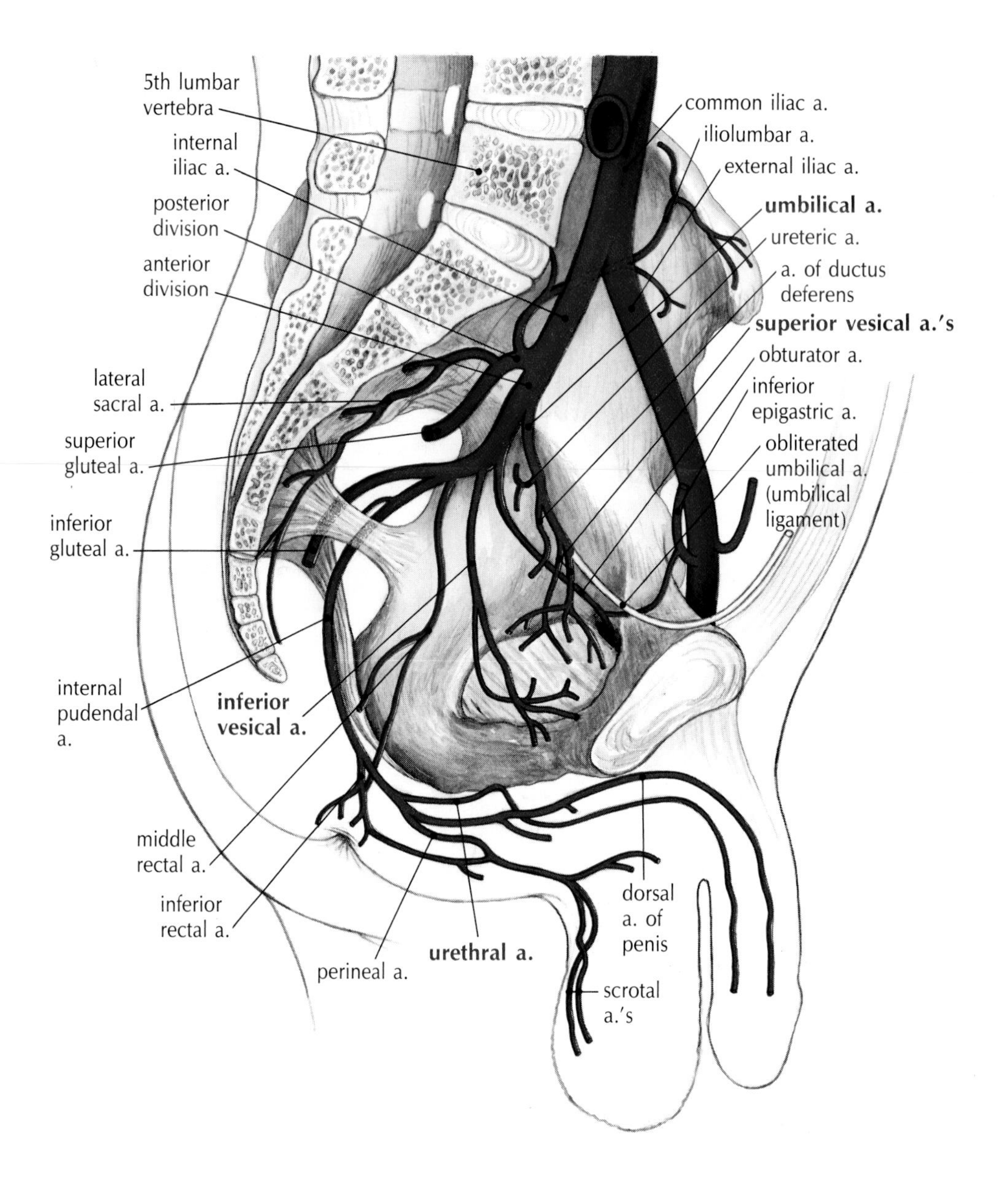

Fig. 25–3: Diagram of the branches of the internal iliac artery.

The **left and right coccygeus muscles** form the most posterior part of the pelvic floor (Fig. 25–1). Coccygeus originates from the pelvic sides of the ischial spine and the sacrospinous ligament, and inserts mainly onto the coccyx. Coccygeus is innervated by fibers that arise from S4 and S5.

Obturator internus originates from the pelvic surface of the obturator membrane and the bony margin surrounding the membrane. The anterosuperior half of the muscle's origin lines the lateral pelvic wall (Fig. 25–2); the posteroinferior half of the muscle's origin lies below the pelvic floor. A fascial thickening, called the tendinous arch of the obturator internus, extends across the medial surface of the muscle along the boundary between the anterosuperior and posteroinferior halves of its origin. This is the fascial thickening from which most of levator ani's fibers originate.

The part of **piriformis** near its origin from the second, third, and fourth sacral vertebrae lines the posterior pelvic wall (Fig. 25–2).

THE SACRAL PLEXUS

The sacral plexus arises from the lumbosacral trunk (which is derived from the anterior rami of L4 and L5), the anterior rami of S1, S2, and S3, and a part of the anterior ramus of S4 (Fig. 12–25). On either side of the pelvis, the sacral plexus rests against the pelvic surface of piriformis. The sacral plexus provides, in combination with nerves from the lumbar plexus, almost all the sensory and motor innervation of the lower limb. The sacral plexus gives rise to almost all the nerves that innervate the pelvic wall muscles and the muscles of the perineum. The plexus also provides cutaneous sensory innervation of the perineum.

The **pudendal nerve** arises from the sacral plexus, receiving fibers from S2, S3, and S4 (Fig. 12–25). Its course from the pelvis, through the gluteal region, and finally into the perineum parallels the course of the internal pudendal artery (Fig. 25–3). The pudendal nerve extends between piriformis and coccygeus to exit the pelvis through the greater sciatic foramen. It extends inferiorly through the gluteal region by coursing past the lateral side of the sacrospinous ligament (near its attachment to the ischial spine). The pudendal nerve then enters the ischiorectal fossa of the perineum by passing through the lesser sciatic foramen. The pudendal nerve's branches which innervate levator ani can arise in either the pelvis or perineum. The pudendal nerve's perineal branches innervate all the muscles of the perineum and supply cutaneous innervation to the external genitalia.

THE INTERNAL ILIAC ARTERY

On each side, the common iliac artery bifurcates into its terminal branches, the external and internal iliac arteries, in front of the sacroiliac joint. The internal iliac artery descends along the lateral pelvic wall for a relatively short distance before commonly dividing into two branches called the **anterior** and **posterior divisions,** or **trunks, of the internal iliac artery** (Fig. 25–3).

The anterior division of the internal iliac artery gives rise to branches that supply pelvic and perineal viscera. The anterior division also gives rise to two branches (the obturator and inferior gluteal arteries) which supply the lower limb. The posterior division generally gives rise to branches that supply the sacral vertebrae and muscles of the posterior abdominal wall and gluteal region.

CHAPTER **26**

The Viscera of the Pelvis and Perineum

The perineum is named from the ancient Greek term for "the space around the natal area." It is the lowermost region of the body trunk, lying between the thighs. The perineum is bordered superiorly by the pelvic diaphragm (Fig. 16–1); it extends inferiorly to the limits of the external genitalia and the buttocks.

This chapter begins with a discussion of the anterior, or urogenital, region of the perineum and its related pelvic viscera and ends with a discussion of the posterior, or anal, region of the perineum and its related pelvic viscera.

THE UROGENITAL REGION OF THE PERINEUM

When the perineum is directly viewed from the lithotomy position (the position in which a person lies on the back with the legs flexed at the knee joints and the thighs flexed and abducted at the hip joints), the perineum exhibits a diamond-shaped boundary whose vertices are formed by the pubic symphysis anteriorly, the ischial tuberosities laterally, and the tip of the coccyx posteriorly (Fig. 26–1). The diamond-shaped perineum is divided for descriptive purposes into two triangular regions by a line drawn through the ischial tuberosities: the anterior and posterior triangular regions are called, respectively, the urogenital and anal regions.

In both sexes, the deep tissues of the urogenital region are referred to as the tissues of the **urogenital (UG) diaphragm.** The UG diaphragm consists basically of a muscle layer sandwiched between two layers of fascia (Fig. 26–2, A and B). The two fascial layers are called the **superior and inferior layers of fascia of the UG diaphragm,** and the tissue-filled space sandwiched between them is called the **deep perineal space.** The UG diaphragm lies directly below the urogenital hiatus [the anterior midline defect in the pelvic diaphragm (the floor of the pelvis) which transmits the urethra in both sexes and also the vagina in the female] (compare Fig. 25–1, which provides a superior view of the urogenital hiatus in the pelvic diaphragm of the female pelvis, with Fig. 26–3, which depicts an inferior view of the urogenital diaphragm, in its location directly inferior to the urogenital hiatus in the pelvic diaphragm of the female pelvis).

In both sexes, the superficial tissues of the urogenital region form the external genitalia. Most of the external genitalia tissues are located in the tissue-filled space of the urogenital region called the **superficial perineal space.** The superficial perineal space lies bracketed between two layers of fascia: the deep fascial layer is the **inferior layer of fascia of the urogenital diaphragm** (also called the **perineal membrane**), and the superficial fascial layer is the **superficial perineal fascia (Colles' fascia)** (Fig. 26–2B). Colles' fascia represents the membranous layer of the superficial fascia in the perineum; it is continuous superiorly with Scarpa's fascia, the membranous layer of the superficial fascia in the anterolateral abdominal wall.

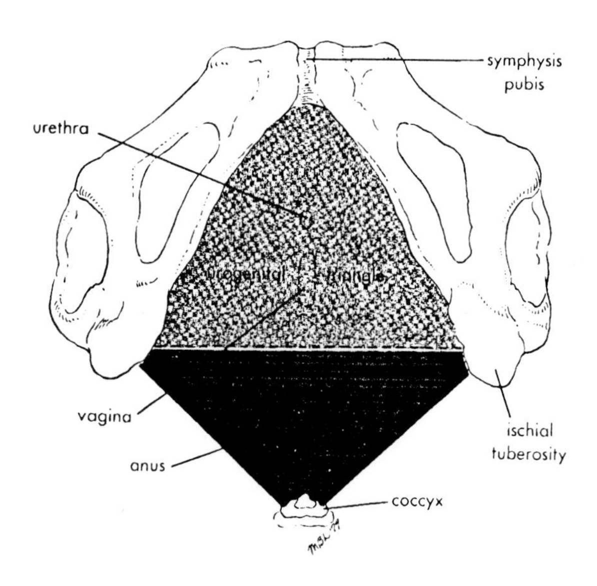

Fig. 26–1: Diagram of the inferior view of the diamond-shaped boundary of the female perineum.

The Superficial Tissues of the Female Urogenital Region

The external genitalia of the female are referred to as the vulva; the **vulva** represents the superficial tissues of the female urogenital region (Fig. 26–4). The surface features of the vulva include the mons pubis, labia majora, labia minora, clitoris, vestibule, external urethral orifice, and vaginal orifice.

The **mons pubis** is the body of fatty tissue that overlies the pubic symphysis and pubic crests. It is covered by hair after puberty.

The **labia majora** are the major paired folds of skin of the vulva. The labia majora extend posteriorly from the mons pubis and unite with each other posterior to the vaginal orifice. The round ligaments of the uterus end within the labia majora.

The **labia minora** are the minor paired folds of skin of the vulva. The labia minora lie medial to the labia majora. Posteriorly, the labia minora unite with each other posterior to the vaginal orifice. Anteriorly, they form a prepuce (hood) in front of the clitoris and a frenulum (fold) behind the clitoris.

The **clitoris** is an erectile organ composed of four bodies of erectile tissue: a **pair of vestibular bulbs** and a **pair of corpora cavernosa** (Fig. 26–5). The vestibular bulbs are attached to the undersurface (that is, the superficial surface) of the perineal membrane (Figs. 26–2, and 26–5). The expanded, posterior portions of the vestibular bulbs lie lateral to the vaginal and urethral orifices; the tapered, anterior portions of the vestibular bulbs unite with each other in the midline anterior to the urethral orifice. The proximal portions of the corpora cavernosa, which are called the **left** and **right crura of the clitoris,** are attached to the sides of the pubic arch (Fig. 26–5); the distal portions of the corpora cavernosa unite to form most of the pendant part of the clitoris.

A **pair of greater vestibular glands** (**Bartholin's glands**) lie immediately posterior to the vestibular bulbs (Figs. 26–2A and 26–5). The duct of each

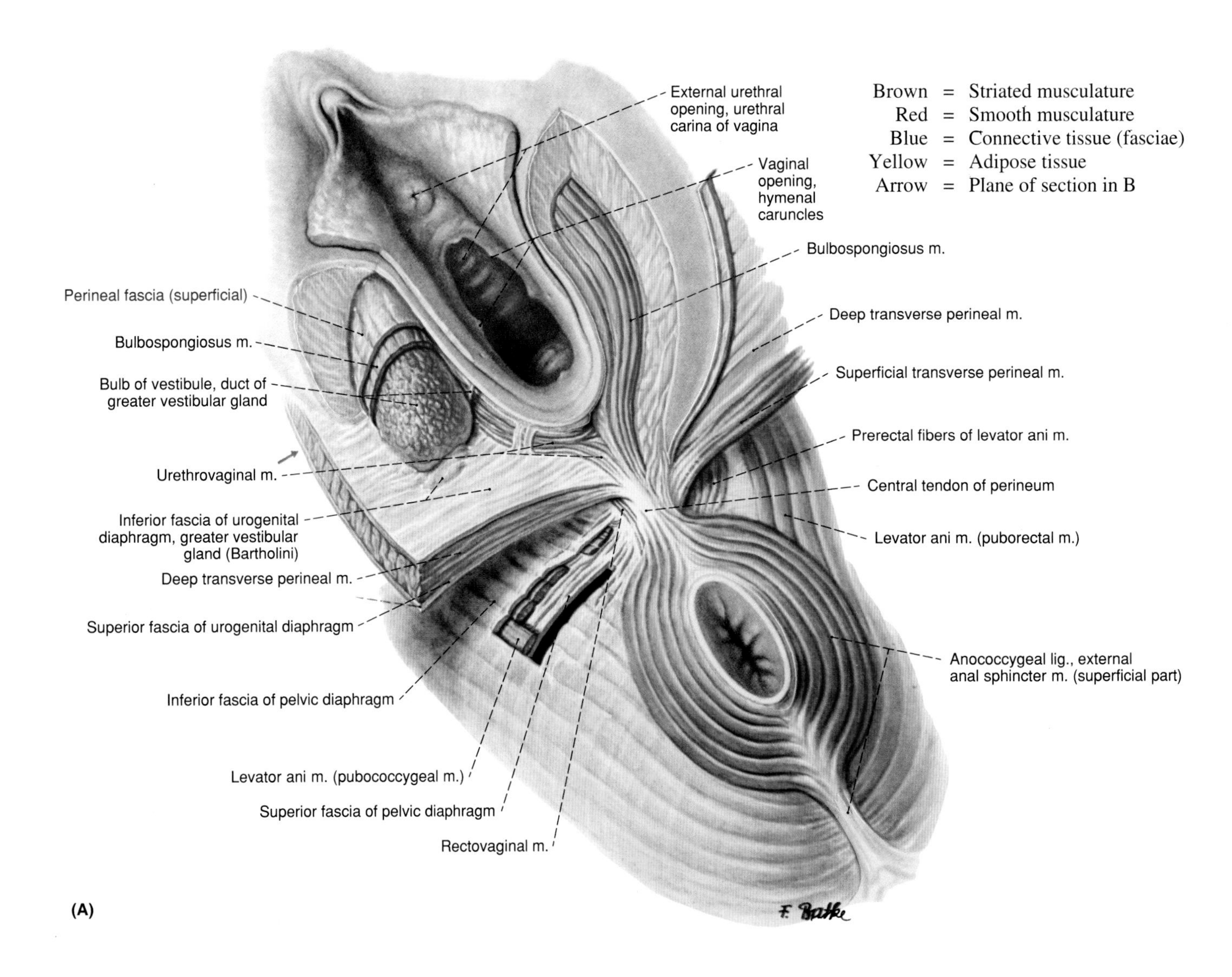

Bartholin's gland opens into the vaginal orifice lateral to the hymen; the secretions of the glands contribute to vaginal lubrication during sexual arousal.

The muscles of the vulva consist of three pairs of small skeletal muscles: (1) A **pair of bulbospongiosus muscles** superficially envelop the vestibular bulbs and the greater vestibular glands (Figs. 26–2 and 26–5). (2) A **pair of ischiocavernosus muscles** superficially envelop the crura of the clitoris (Figs. 26–2B and 26–5). (3) A **pair of superficial transverse perineal muscles** originate from the ischial tuberosities (Fig. 26–5), extend medially behind the vagina (Figs. 26–2A and 26–5), and insert onto the central tendon of the perineum (Fig. 26–2A). This muscle pair serves to stabilize the position of the central tendon of the perineum.

The **vestibule** is the region of the vulva between the labia minora (Fig. 26–4). The urethral and vaginal orifices open into the vestibule.

The Superficial Tissues of the Male Urogenital Region

The penis and scrotum form the external genitalia of the male, and they represent the superficial tissues of the male urogenital region.

The **penis** is an erectile organ composed of an unpaired and two paired bodies of erectile tissue. The unpaired body of erectile tissue is called the **corpus spongiosum** (Fig. 26–6). The corpus spongiosum transmits the terminal portion of the urethra (the **penile urethra**) in a male. The corpus spongiosum is enlarged at its proximal and distal ends. The proximal enlargement of the corpus spongiosum is called the **bulb of the penis,** and it is attached to the undersurface of the perineal membrane. The distal enlargement of the corpus spongiosum is called the **glans penis;** the glans penis forms most of the distal end of the penis. Although the corpus spongiosum becomes turgid during sexual arousal, its turgidity does not significantly contribute to penile erection.

The paired bodies of erectile tissue in the penis are called the **corpora cavernosa** (Fig. 26–6). The proximal portions of the corpora cavernosa, which are called the **left and right crura of the penis,** are attached to the sides of the pubic arch; the distal portions of the corpora cavernosa unite to form most of the pendant part of the penis. The enlargement and rigidity of the corpora cavernosa during sexual excitement give rise to penile erection.

The superficial muscles of the male urogenital region consist of three pairs of small skeletal muscles (Fig. 26–7): (1) A **pair of bulbospongiosus muscles**

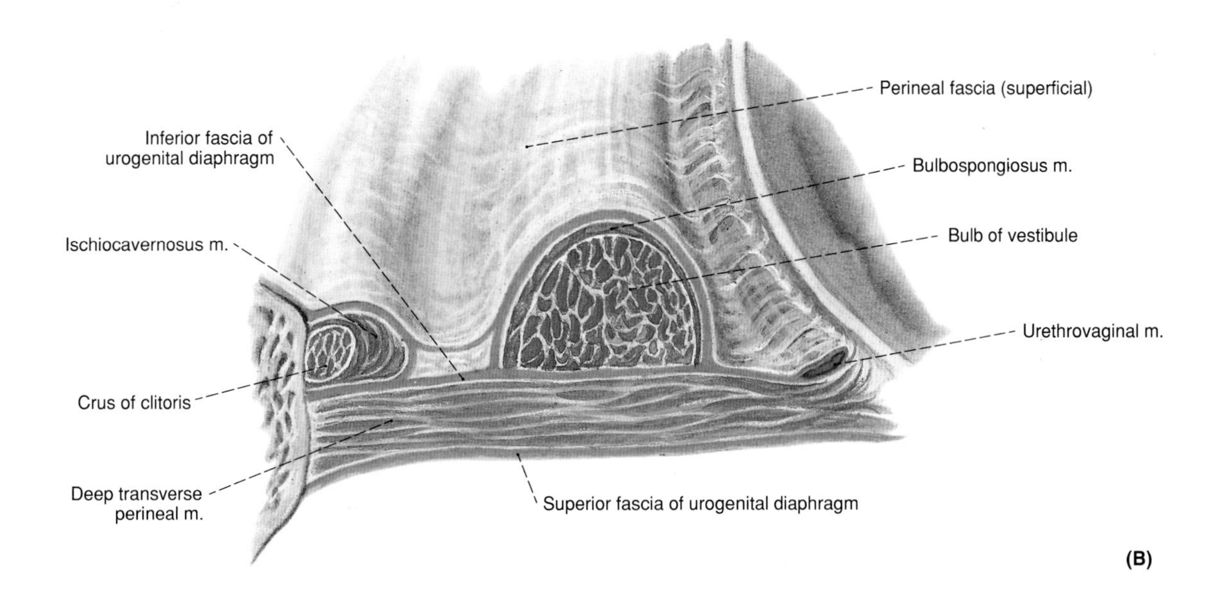

Fig. 26–2: (A) Schematic illustration of an inferior view of some of the tissues of the superficial and deep perineal spaces of the female perineum. Immediately to the right of the central tendon of the perineum, a rectangular area of the inferior fascia of the pelvic diaphragm (which is the layer of fascia immediately superficial to the pelvic diaphragm) has been removed to reveal part of the levator ani muscle deep to the fascia. A small part of the levator ani muscle has been removed to reveal the superior fascia of the pelvic diaphragm (which is the layer of fascia immediately deep to the pelvic diaphragm). Most of the right bulbospongiosus muscle and all of the right superficial transverse perineal muscle have been removed. (B) Schematic illustration of a cross sectional view of the urogenital diaphragm in the plane indicated by an arrow in A.

superficially envelop the bulb of the penis. This muscle pair contracts at the termination of micturition to empty the penile urethra of urine. The rhythmic contractions of the bulbospongiosus muscles during ejaculation are responsible for the propulsion of semen through the penile urethra. (2) A **pair of ischiocavernosus muscles** superficially envelop the crura of the penis. Contraction of the ischiocavernosus muscles during sexual excitement contributes to penile erection by diminishing venous outflow from the corpora cavernosa. (3) A **pair of superficial transverse perineal muscles** originate from the ischial tuberosities, extend medially behind the bulb of the penis, and insert onto the central tendon of the perineum; this muscle pair serves to stabilize the position of the central tendon of the perineum.

The terminal branches of the internal pudendal arteries provide the main arterial supply of the erectile tissues of the penis (the internal pudendal artery is a branch of the anterior division of the internal iliac artery) (Fig. 25–3). On either side, these arteries are the **deep artery of the penis** (which extends along the central axis of the corpus cavernosum) and the **dorsal artery of the penis** (which extends forward through the penis near its dorsal surface) (Fig. 26–8). Increased parasympathetic activity during sexual arousal increases blood flow through these arteries more than ten times. Penile erection occurs as a consequence of the resulting engorgement of the cavernous sinuses in the corpora cavernosa.

The pudendal nerve provides sensory supply to the penis by the **dorsal nerve of the penis** (Fig. 26–8). On each side, the dorsal nerve extends through the penis alongside the dorsal artery.

The Deep Tissues of the Urogenital Region

As previously noted, the tissues of the UG diaphragm represent the deep tissues of the urogenital region, and the deep perineal space is the tissue-filled space between the superior and inferior layers of fascia of the UG diaphragm.

The muscles of the UG diaphragm lie in the deep perineal space. **Sphincter urethrae** is a voluntary muscle which encircles the urethra as it passes through

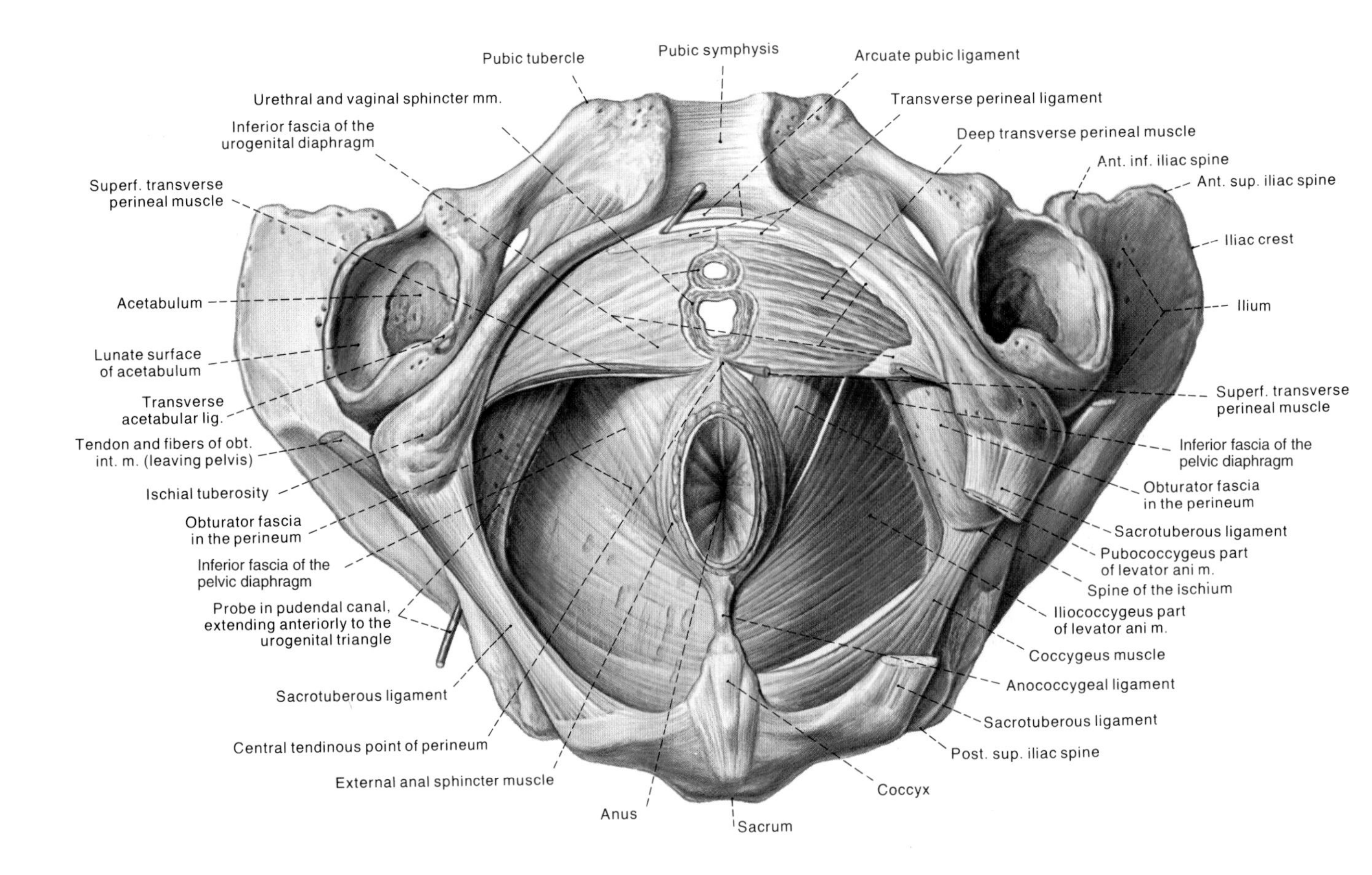

Fig. 26–3: Inferior view of the muscles of the deep perineal space and floor of the pelvis in a female.

the deep perineal space (Fig. 26–3); sphincter urethrae's name describes its action. The **left** and **right deep transverse perineal muscles** originate from the ischial rami and extend medially to insert onto the central tendon of the perineum (Fig. 26–3); they serve to stabilize the position of the central tendon of the perineum.

In the male, the deep perineal space also contains on each side a **bulbourethral gland** (Fig. 26–7). These small glands, which open into the urethra upon its entrance into the superficial perineal space, contribute secretions to semen during ejaculation.

THE URINARY BLADDER AND THE URETHRA

In both sexes, the urinary bladder is one of the major pelvic viscera lying over (superior to) the urogenital region of the perineum.

The urinary bladder is the reservoir viscus of urine. When empty, the bladder assumes the shape of an inverted tetrahedron which has a superior surface cov-

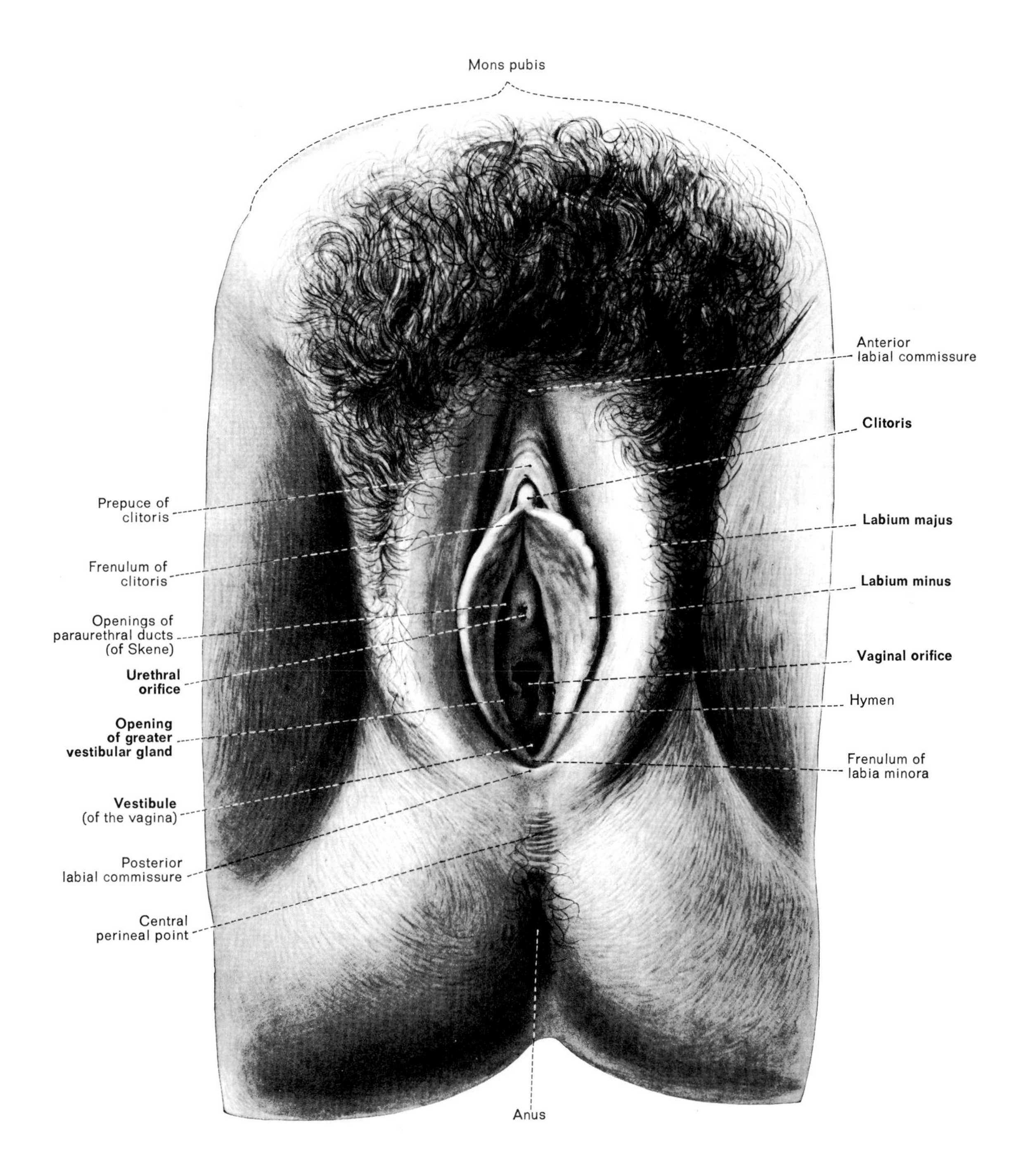

Fig. 26–4: Drawing of the vulva of an adolescent virgin.

ered with peritoneum, a posterior surface called the **base** of the bladder, and two inferolateral surfaces resting atop the narrowed, anterior region of the pelvic floor (Fig. 26–6). The **apex of the bladder** is the most anterior, midline point of its superior surface.

In a newborn child, most of the bladder lies in the abdominal cavity, immediately behind the lowest part of the anterolateral abdominal wall. The bladder progressively assumes a greater lie in the pelvic cavity as the bony pelvis enlarges during childhood. After puberty, the bladder (when empty) lies completely in the pelvic cavity, immediately posterior to the pubic parts of the coxal bones (Fig. 26–9).

The bladder accommodates increases in urine volume through primarily distention of its superior surface. As urine accumulates in the bladder, the bladder's superior surface bulges upward and becomes more rounded. In an adult, this upward bulging commonly elevates the bladder's superior surface above the pelvic inlet. Continuing expansion of the bladder's superior surface now brings it into direct contact with the posterior surface of the anterolateral abdominal wall; concurrently, peritoneum is lifted from the posterior surface of the anterolateral abdominal wall onto the expanding superior surface of the bladder. Under such normal conditions, the bladder and anterolateral abdominal wall come into direct contact with each other above the pubic symphysis (Fig. 26–10). This anatomic relationship makes it possible to access (with a needle) the interior of the bladder from the suprapubic anterolateral abdominal wall without simultaneously traversing the peritoneal cavity.

The ureters pierce the bladder at the upper lateral corners of its posterior surface (Fig. 26–6). The ureters extend obliquely through the posterior wall of the bladder before opening into the interior of the bladder.

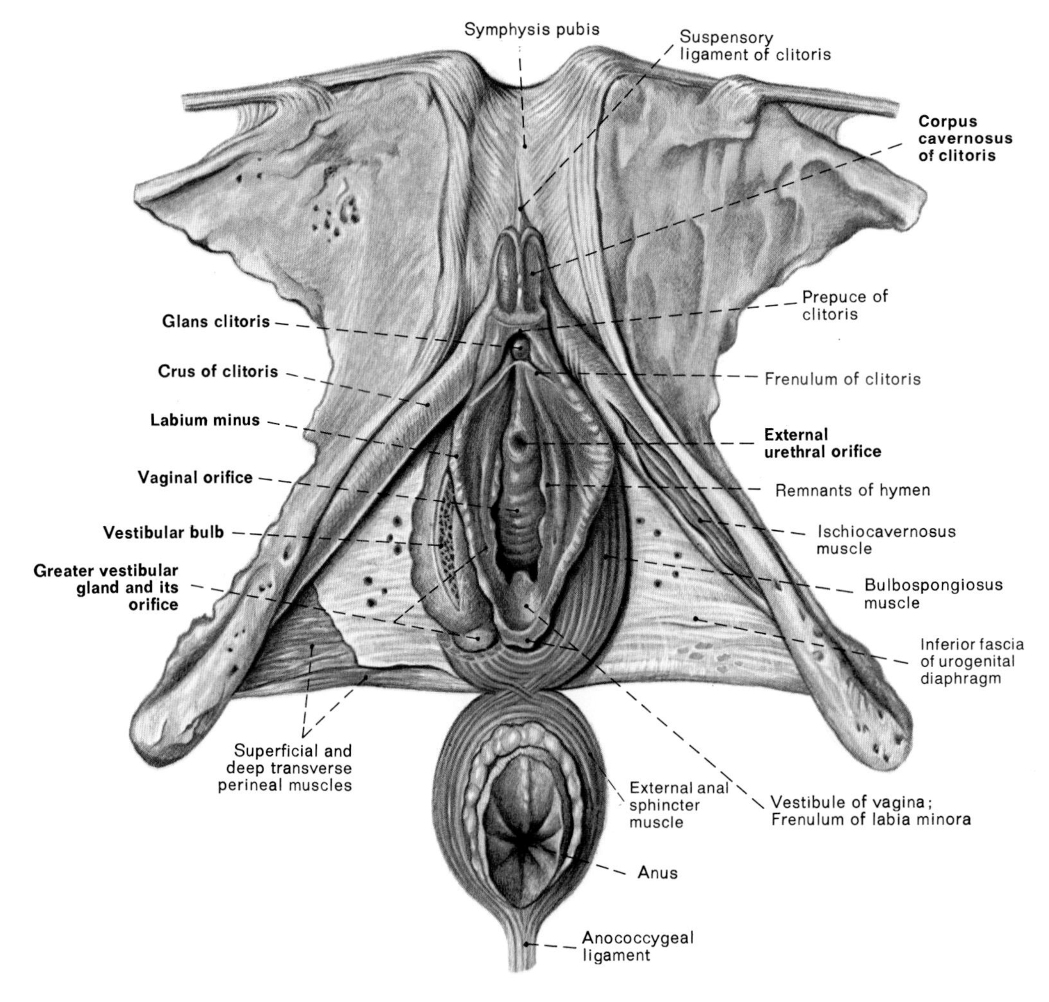

Fig. 26–5: Dissection of the female external genitalia.

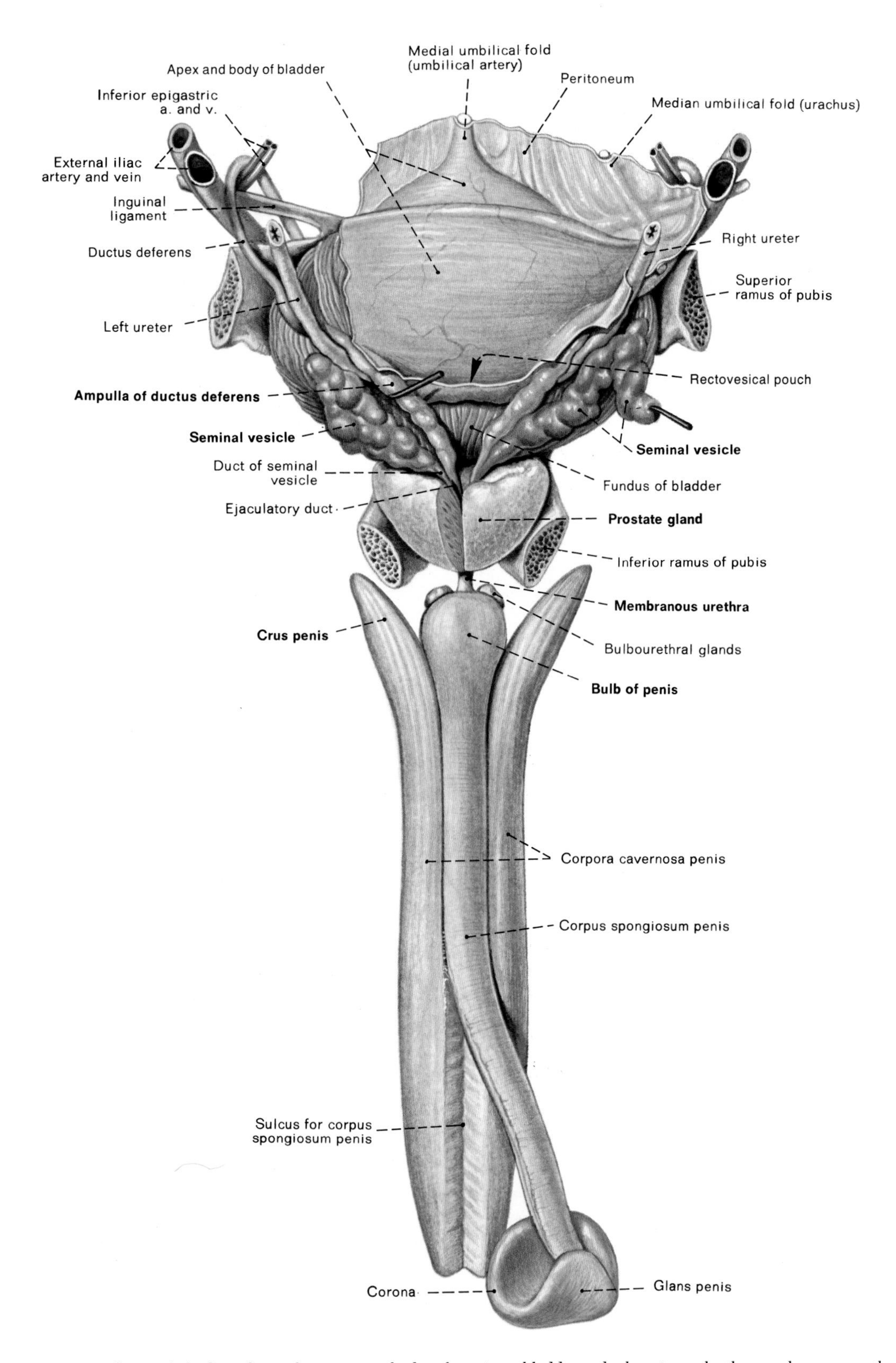

Fig. 26–6: The penile bodies of erectile tissue attached to the urinary bladder and other viscera by the membranous urethra.

When the bladder is filling with urine, the hydrostatic pressure of the urine flattens shut the ureteric segments coursing through the posterior wall of the bladder; these segments are only temporarily forced open when the passage of peristaltic waves down the ureters propels more urine into the bladder.

When the bladder is empty, all the inner surfaces of the bladder (except for a triangular-shaped, posterior area) are thrown into folds; these folds flatten out as urine fills the bladder. The triangular, posterior area of the inner surface of the bladder (which is always fold-free) is called the **trigone** (Fig. 26–11). The entrance of the ureters into the bladder mark the two superolateral corners of the trigone; the neck of the bladder marks the midline, inferior corner of the trigone.

The **neck of the bladder** is the lowest region of the bladder. The neck of the bladder gives rise to the urethra. The neck has a high concentration of circumferentially-oriented connective tissue fibers which serve as a passive sphincter.

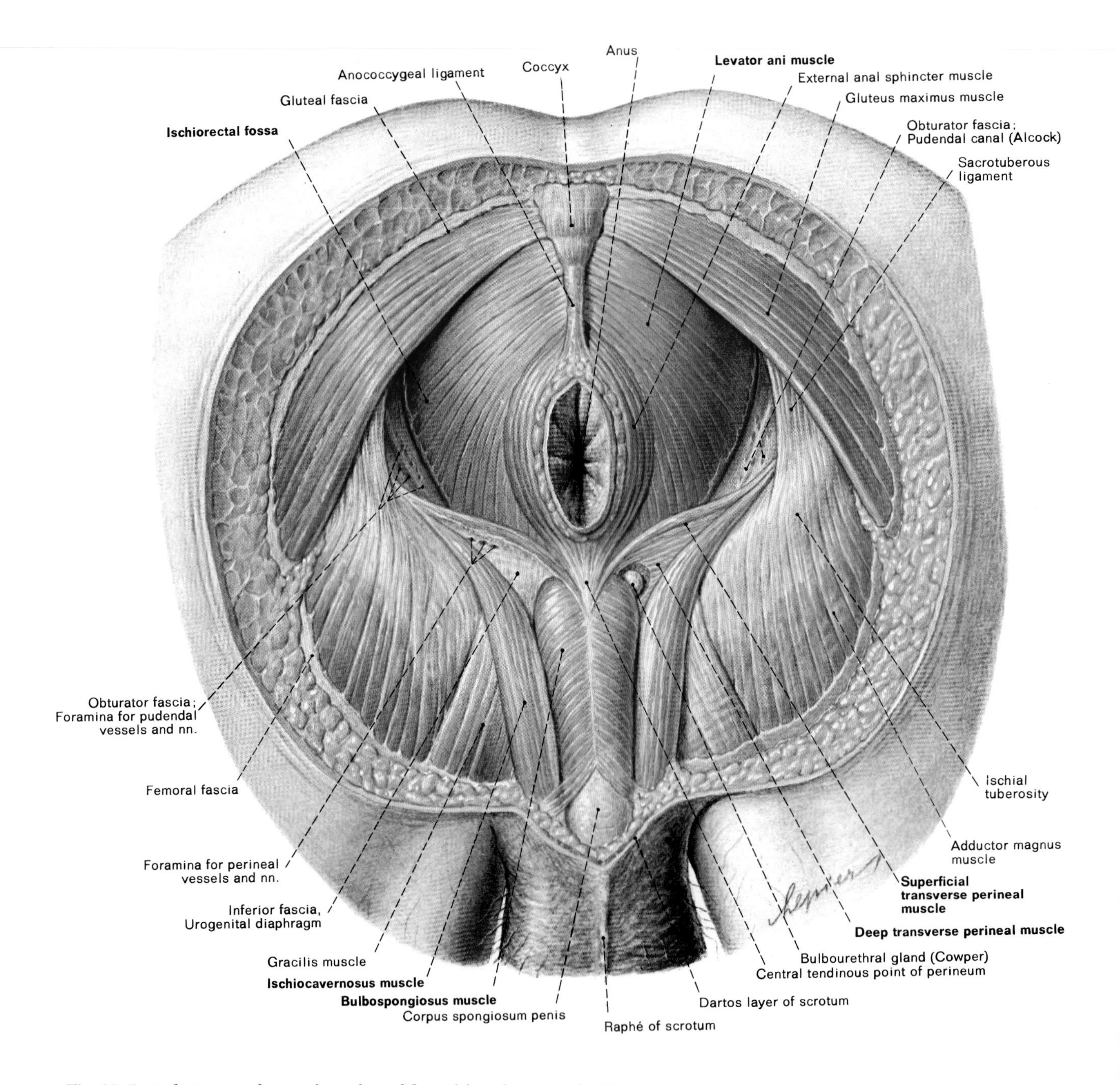

Fig. 26–7: Inferior view of perineal muscles and floor of the pelvis in a male. The right deep transverse perineal muscle and right bulbourethral gland have been revealed by removal of the inferior layer of fascia of the urogenital diaphragm.

The smooth muscle tissue of the bladder wall is called the **detrusor.** The detrusor is innervated by parasympathetic fibers provided by the pelvic splanchnic nerves. In the neck of the bladder, smooth muscle fibers are oriented both circumferentially and longitudinally. It is likely that the circumferentially-oriented fibers in the neck of the bladder are innervated by sympathetic fibers, and, therefore, that these fibers represent a physiologic sphincter (which is called sphincter vesicae). These fibers contract in a male during ejaculation, thus preventing the introduction of semen into the bladder.

In both sexes, the **superior vesical artery,** which is a branch of the anterior division of the internal iliac artery, supplies the upper part of the bladder and the terminal segment of the ureter (Fig. 25–3). In a male, another branch of the anterior division of the internal iliac artery called the **inferior vesical artery** supplies the posteroinferior part of the bladder (in addition to the prostate and the seminal vesicles) (Fig. 25–3). In a female, branches of the **vaginal artery** provide the major arterial supply to the posteroinferior part of the bladder.

The urethra conducts urine from the bladder to the exterior. In a female, the urethra, upon emerging from the neck of the bladder, passes immediately through the deep perineal space of the urogenital diaphragm, where it is encircled by the sphincter urethra (Fig. 26–3). It then traverses the superficial perineal space and opens into the vestibule at the external urethral orifice.

In a male, the urethra, upon emerging from the neck of the bladder, immediately traverses the prostate (Figs. 26–9 and 26–11). This segment of the urethra is called the **prostatic urethra,** and it is the widest and most dilatable segment of the entire urethra in the male. Upon emerging from the prostate, the urethra then traverses the deep perineal space, where it is encircled by the sphincter urethrae. This segment is called the **membranous urethra,** and it is the least dilatable segment of the urethra in the male (Figs. 26–9 and 26–11). Upon entering the superficial perineal space, the urethra traverses the corpus spongiosum of the penis (Figs. 26–9 and 26–11). This terminal segment of the urethra is called the **penile urethra.** The external urethral orifice is the narrowest part of the urethra in the male.

CLINICAL PANEL V-2

Acute Cystitis

Description: Acute cystitis is acute inflammation of the urinary bladder. Infection is a common cause of the acute inflammation. The female is more susceptible to acute cystitis from infection than the male. Acute cystitis from infection is commonly accompanied by proximal urethritis (inflammation of the proximal urethra).

Common Symptoms: Common symptoms of combined acute cystitis and proximal urethritis from infection include dysuria (difficulty or pain upon urination), increased urgency and frequency of urination, suprapubic pain or discomfort, and the passage of cloudy urine. The pain upon urination is frequently a burning sensation resulting from the stimulation of thermal receptors in the inflamed urethra during the passage of urine. Stimulation of proprioceptive and sensory fibers in the bladder wall and urethra accounts for the increased urgency and frequency of urination. Spasmodic contraction of the bladder wall's detrusor muscle can elicit suprapubic pain or discomfort.

Common Sign: A common physical sign is suprapubic tenderness. The suprapubic tenderness is visceral pain elicited upon compression of the inflamed urinary bladder.

The Innervation of the Bladder and Urethra

The **left and right pelvic plexuses** (which are also known as the left and right inferior hypogastric plexuses) are the chief autonomic plexuses of the pelvis. The pelvic plexus on each side is formed on the posterior pelvic wall by sympathetic fibers from the superior hypogastric plexus (the autonomic plexus immediately below the aortic bifurcation) and from the pelvic part of the sympathetic trunk and by parasympathetic fibers from the pelvic splanchnic nerves (Fig. 26–12).

The first and second lumbar splanchnic nerves provide preganglionic sympathetic innervation for the bladder and the lower part of the ureter, and the pelvic plexuses transmit the postganglionic sympathetic fibers to these viscera.

The pelvic splanchnic nerves (Fig. 26–12) provide the preganglionic parasympathetic innervation for the bladder and the lower part of the ureter. The pelvic plexuses transmit the preganglionic parasympathetic fibers of the pelvic splanchnic nerves directly toward minute parasympathetic ganglia embedded within the wall of the bladder and the wall of the lower part of the ureter; postganglionic parasympathetic fibers emanate from these ganglia to innervate nearby target tissues.

The sensory fibers that innervate the bladder and the lower part of the ureter enter the spinal cord at those spinal cord segments that provide preganglionic sympathetic and preganglionic parasympathetic innervation for these viscera (in other words, the sensory

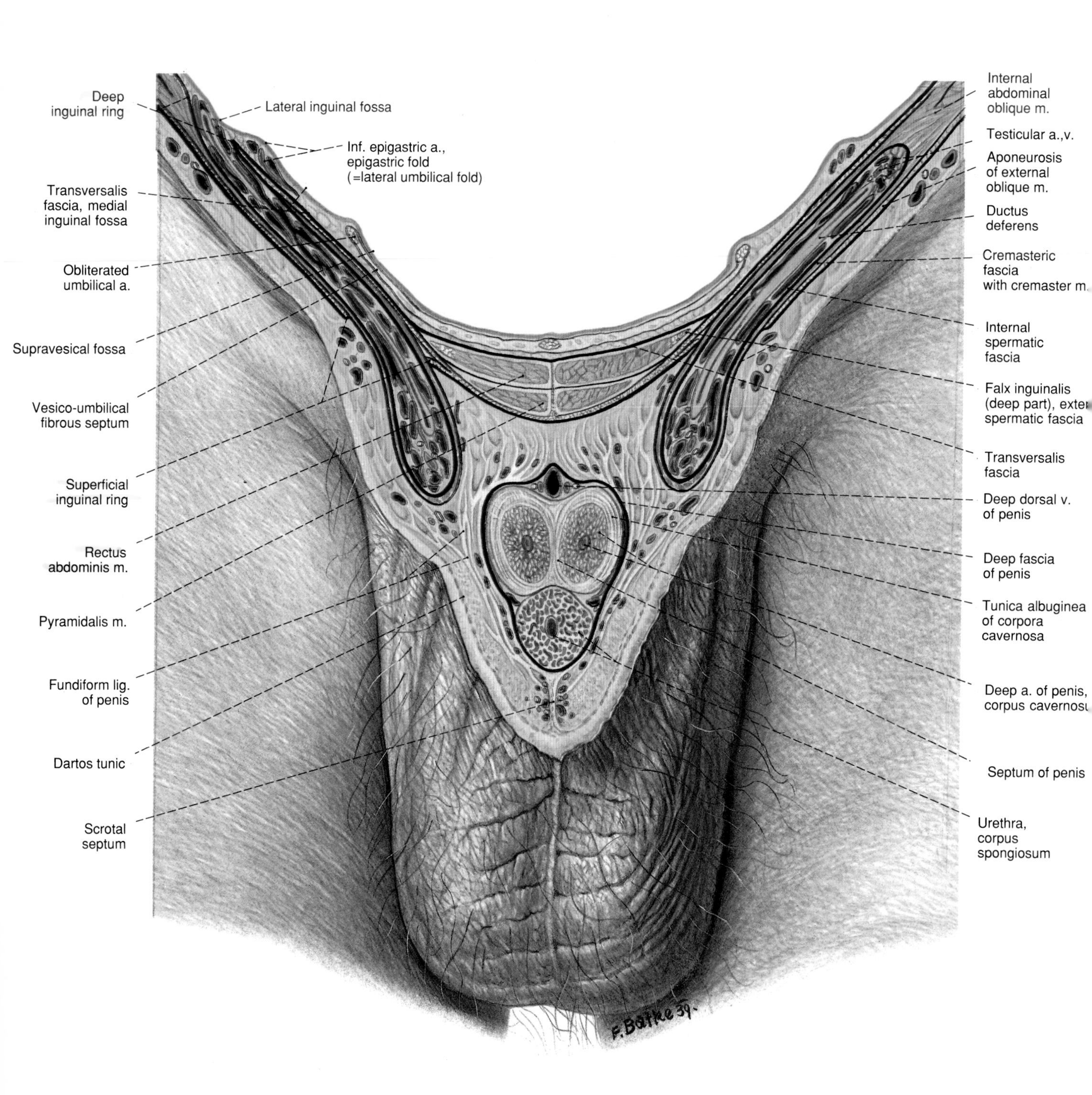

Fig. 26–8: Oblique sectional view of the inguinal regions of the anterolateral abdominal wall, the upper part of the scrotal sac, and the pendant part of the penis. The dorsal artery (colored red) and the dorsal nerve (colored yellow) of the penis are illustrated but not labelled on each side of the deep dorsal vein of the penis.

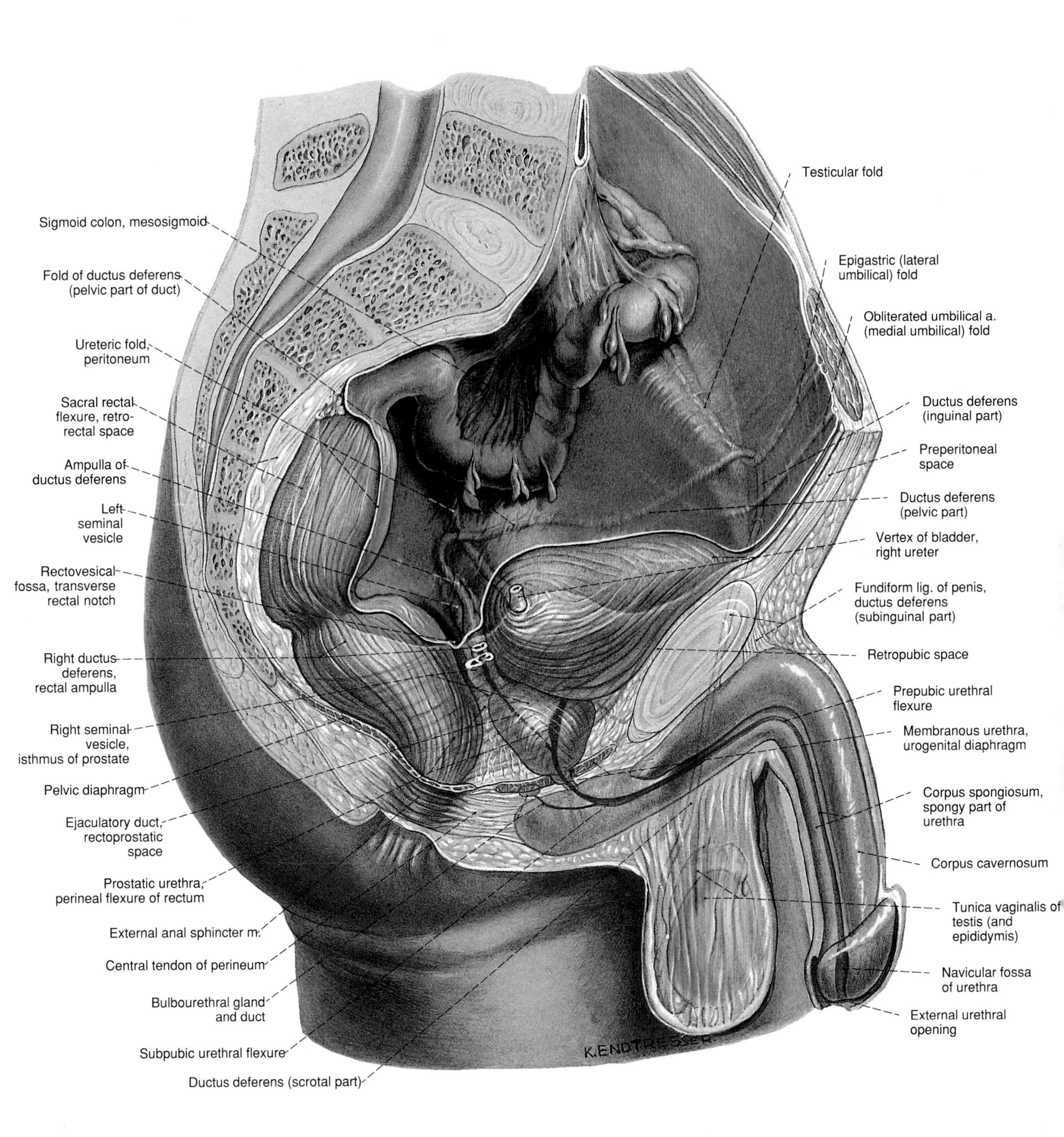

Fig. 26–9: Schematic illustration of the male pelvic viscera projected onto a sagittal plane. All spaces lined by peritoneum are colored green including those spaces on the sides of the urinary bladder and rectum which can be seen through the viscera. The entire course of the ductus deferens and urethra are depicted.

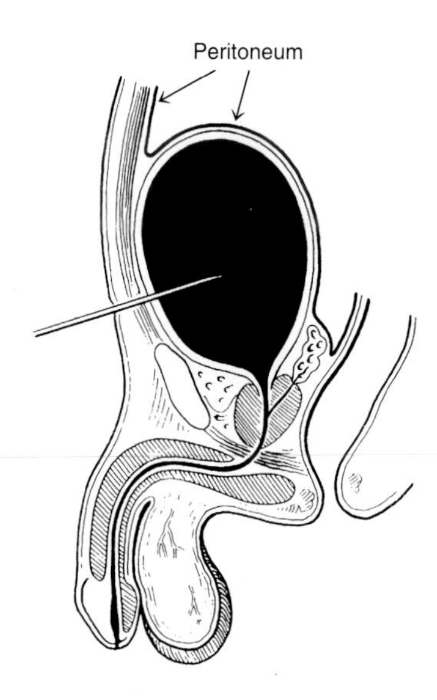

Fig. 26–10: Illustration of how it is possible to access (with a needle) the interior of a bladder markedly distended with urine without simultaneously traversing the peritoneal cavity. The needle is inserted through the immediate suprapubic region of the anterolateral abdominal wall. This procedure may be used to drain the bladder if there is an obstruction of the urethra.

fibers enter at the L1, L2, S2, S3, and S4 levels). Disease or injury of the bladder and lower part of the ureter produces visceral pain that is usually poorly localized to the suprapubic region. Disease or injury of these viscera may refer pain to parts of the L1, L2, S2, S3, and S4 dermatomes.

The left and right pudendal nerves provide sensory innervation for the urethra. The urethra has pressure and thermal receptors.

Micturition

The bladder accommodates increases in urine volume primarily through distention of its superior surface. As urine accumulates in the bladder, the bladder's superior surface bulges upward and becomes more rounded. The accumulation of urine does not significantly increase intravesical pressure until the bladder is almost fully distended. Sensory fibers from the bladder (which are sensitive to both distension and pain) excite the urge to void. The act of micturition requires voluntary relaxation of the paired levator ani muscles and the sphincter urethrae. The relaxation of the levator ani not only pulls the neck of the bladder downward (and thus decreases resistance in the neck), but also reflexively stimulates contraction of the detrusor. Abdominal wall musculature is contracted if additional external pressure is required.

THE REPRODUCTIVE VISCERA OF THE FEMALE PELVIS

The **vagina** is a fibromuscular tube that extends posterosuperiorly through the vulva and UG diaphragm into the pelvis (Fig. 26–13). Anteriorly, the vagina is in direct contact with the urethra and the posterior surface of the bladder. Posteriorly, the vagina is in direct contact with the central tendon of the perineum and the lower third of the rectum. The upper quarter of the posterior surface of the vagina is covered with peritoneum (Fig. 26–13).

The **uterus** is a hollow, pear-shaped organ which is descriptively divided into a cervix and a body (Fig. 26–14). The **cervix of the uterus** protrudes through the uppermost anterior wall of the vagina (Fig. 26–15); the cervix thus has an upper, supravaginal part and a lower, intravaginal part. The narrow cylindrical recess in the uppermost part of the vaginal vault which encircles the intravaginal part of the cervix is divided into anterior, posterior, and lateral spaces called **fornices** (the vaginal fornix displayed in Figure 26–13 is specifically the posterior fornix).

The **body of the uterus** comprises the upper two-thirds of the viscus (Figs. 26–14 and 26–15). The **fundus of the uterus** is that region of the body which lies above the level marking the points of entrance of the uterine tubes into the uterus (Fig. 26–15). On each side, the **round ligament of the uterus** extends from the side of the uterus in route to its entrance into the inguinal canal via the deep inguinal ring (Fig. 26–16).

When the bladder is empty, it is common for the uterus to be both anteverted and anteflexed (Fig. 26–13). An **anteverted uterus** has its longitudinal axis bent forward (generally at an approximate 90° angle) relative to the longitudinal axis of the vagina (the uterus is said to be **retroverted** if its longitudinal axis is bent backward relative to the longitudinal axis of the vagina). An **anteflexed uterus** has its body bent forward relative to the cervix (the uterus is said to be **retroflexed** if its body is bent backward relative to the cervix).

The **paired uterine tubes** enter the sides of the body of the uterus immediately below the level of the fundus of the uterus (Fig. 26–15). Each uterine tube is divided for descriptive purposes into four segments (Fig. 26–15): The **uterine** (**intramural**) part of the uterine tube is the segment that traverses the wall of the uterus. The next most lateral part is the narrowest segment of the uterine tube, and thus is called the **isthmus.** The isthmus is continuous laterally with the widest and longest segment of the uterus, the **ampulla.** The most lateral segment of the uterine tube is funnel-shaped, and thus is called the **infundibulum.** The end of the infundibulum bears a number of finger-like processes called **fimbriae.** The fimbriae surround the opening of

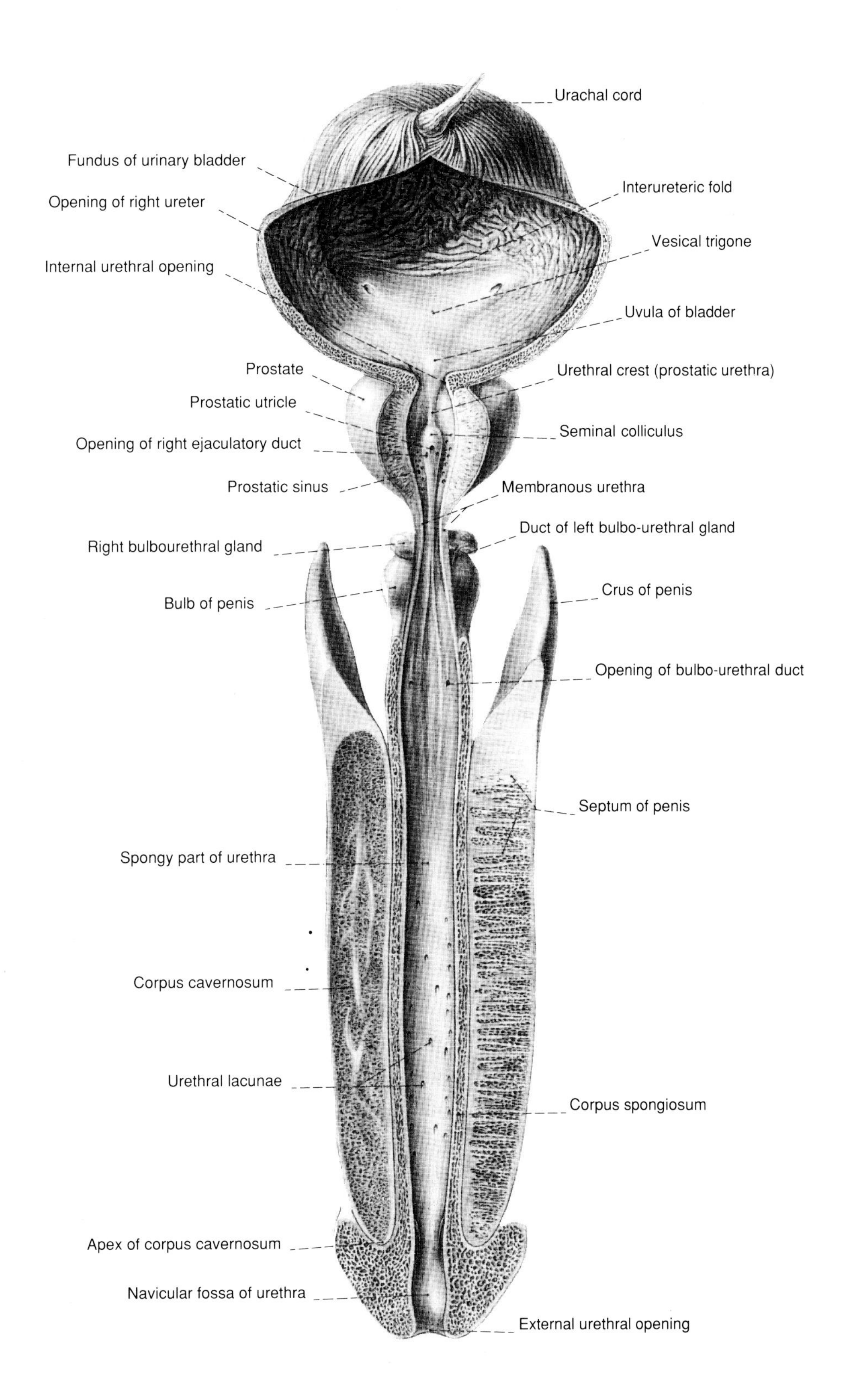

Fig. 26–11: Anterior view of the urinary bladder, prostate, and penis. The urinary bladder, prostate, and penile bodies of erectile tissue have been incised in the midline and opened.

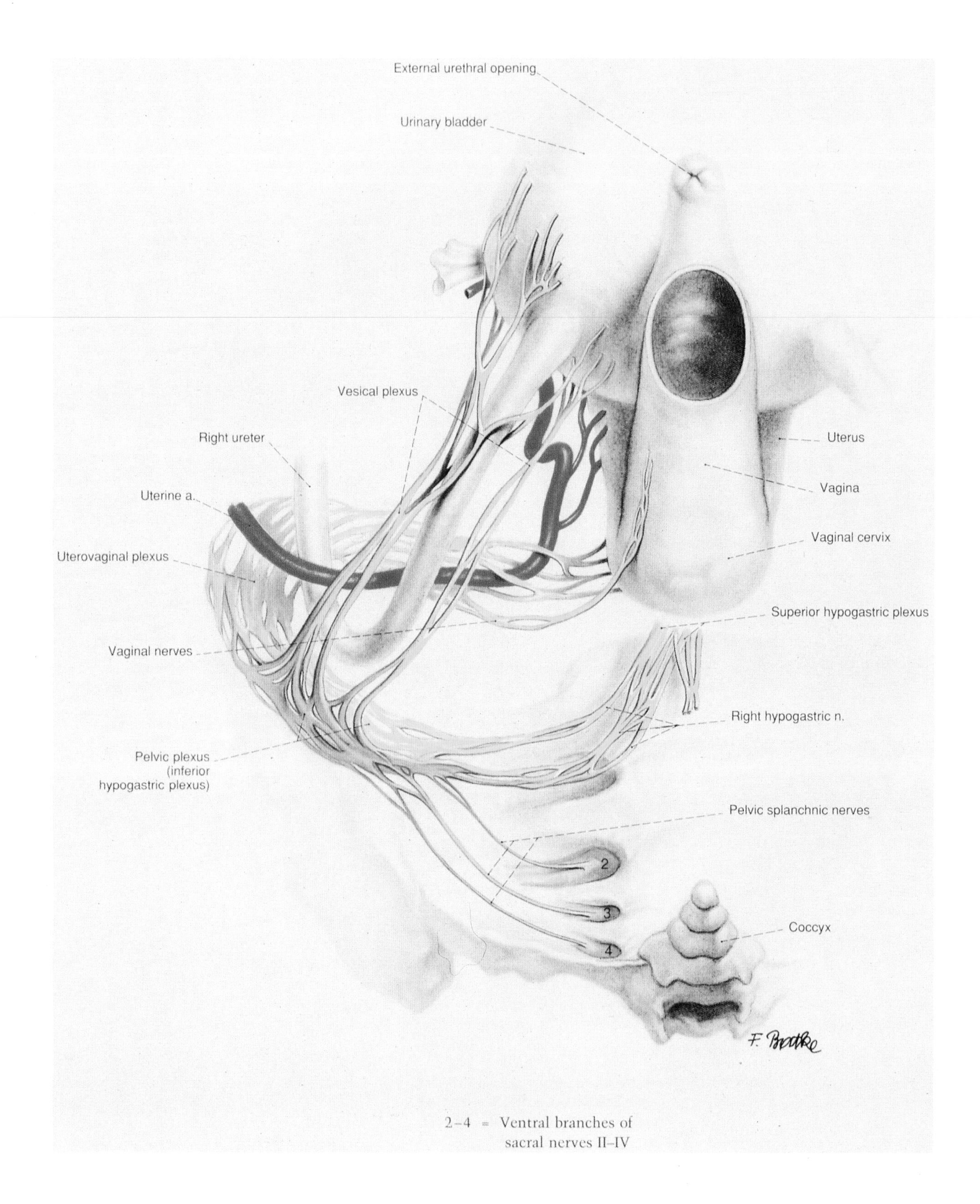

Fig. 26–12: Schematic illustration of an anteroinferior view of the sensory innervation of the urinary bladder, uterus, and vagina.

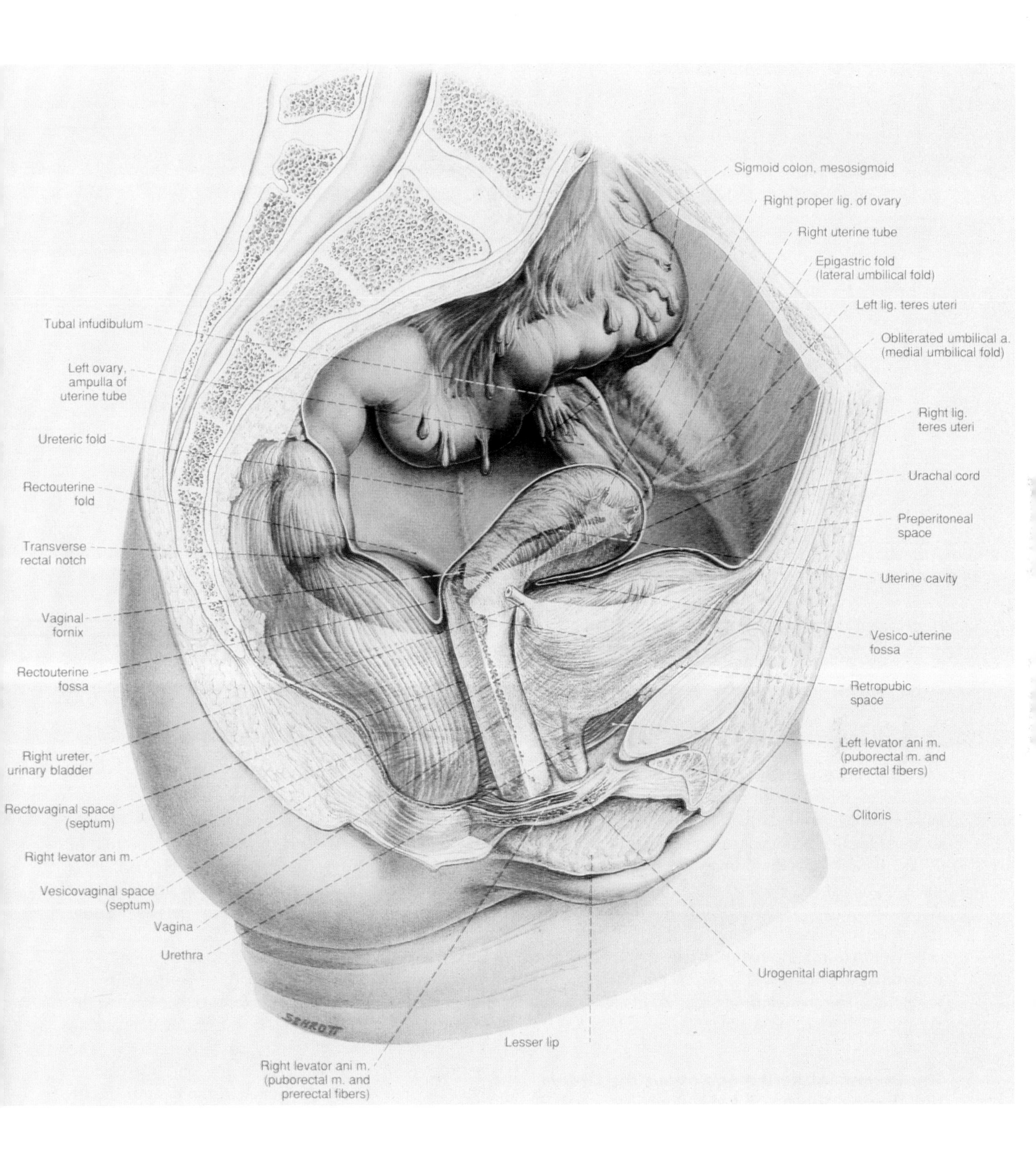

Fig. 26–13: Schematic illustration of the female pelvic viscera projected onto a sagittal plane. All spaces lined by peritoneum are colored green including those spaces on the sides of the urinary bladder and rectum which can be seen through the viscera. The lumina of the vagina and uterus are depicted.

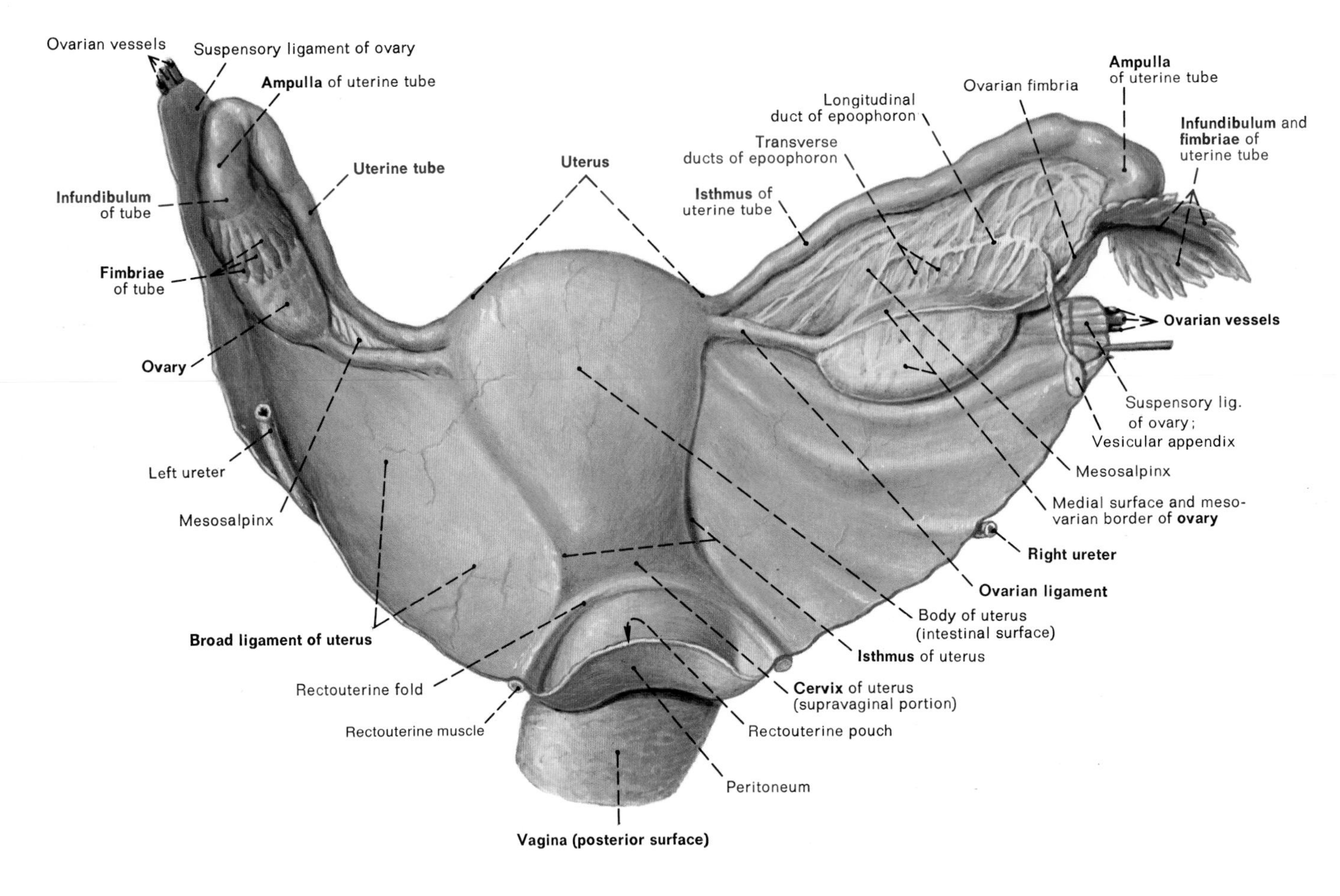

Fig. 26–14: Posterior view of the reproductive viscera of the female pelvis.

the uterine tube into the peritoneal cavity. These two small openings are believed to be the reason for peritonitis being more common in women than in men.

The **ovaries** are the primary sex organs of the female reproductive system. The ovary is an oval body whose lateral pole is embraced by the fimbriae of the uterine tube (Figs. 26–14 and 26–16). A fibrous cord, called the **round ligament of the ovary** (**ovarian ligament**), extends from the medial pole of each ovary to the side of the uterus (Figs. 26–14, 26–15, and 26–16).

The Peritoneal Reflections about the Uterus, the Uterine Tubes, and the Ovaries

In the midline region of the female pelvic cavity, the peritoneum which covers the anterior surface of the middle third of the rectum is continuous inferiorly with the peritoneum that covers the uppermost posterior surface of the vagina (Fig. 26–13). This peritoneal reflection from the rectum onto the vagina lines the lowest part of a pouch in the floor of the female pelvic cavity called the **rectouterine pouch** (the **pouch of Douglas**). The layer of peritoneum that covers the uppermost posterior surface of the vagina extends upward on the posterior surface of the uterus, passes over the fundus of the uterus, extends downward on the anterior surface of the uterus, and finally reflects onto the superior surface of the urinary bladder; peritoneum thus covers all the surfaces of the body of the uterus and the supravaginal part of the cervix, except for thin slits on the lateral surfaces of these parts of the uterus. The peritoneal reflection from the uterus onto the bladder lines the lowest part of a pouch in the floor of the female pelvic cavity called the **vesicouterine pouch.**

The reflections of peritoneum from the rectum onto the vagina and uterus and from the uterus onto the bladder extend across the entire breadth of the female pelvic cavity (in other words, the peritoneal reflections extend to the lateral pelvic walls) (Fig. 26–13). On each

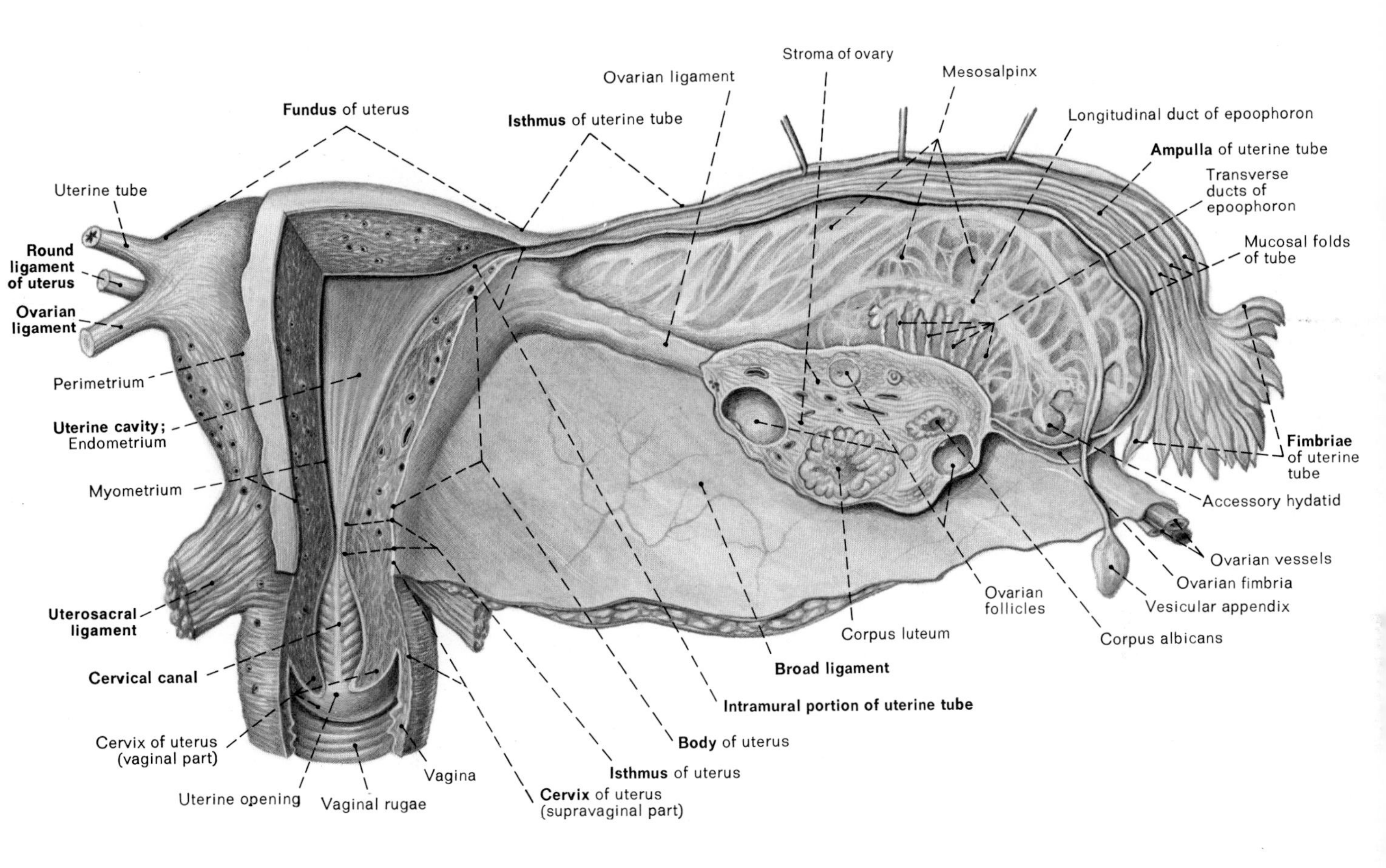

Fig. 26–15: A coronal sectional view of the uterus, uterine tube, and ovary.

side of the uterus, a broad layer of peritoneum (extending from the side of the uterus to the lateral pelvic wall) ascends from the floor of the pelvis, drapes over the uterine tube, and then descends onto the floor of the pelvis; the ascending and decending layers of peritoneum together are called a **broad ligament of the uterus** (Figs. 26–14 and 26–17). The round ligament of the ovary and the round ligament of the uterus course through the extraperitoneal space sandwiched between the ascending and descending peritoneum layers of the broad ligament (Fig. 26–17).

Each ovary is attached by a peritoneal ligament called the **mesovarium** to the posterior peritoneum layer of the broad ligament of the uterus (Fig. 26–17). The ovary, however, is not covered with peritoneum. The ovary bears a layer of cuboidal epithelium that is continuous with the mesothelium of the peritoneum at the edge of the mesovarium. The part of each broad ligament which lies below the uterine tube but above the mesovarium and the round ligament of the ovary is called the **mesosalpinx** (Figs. 26–14, 26–15 and 26–17). The part of each broad ligament that extends from the ovary to the lateral pelvic wall is called the **suspensory ligament of the ovary** (Figs. 26–14 and 26–16).

The Supports of the Uterus

The principal supports of the uterus are the paired levator ani muscles and two paired condensations of endopelvic fascia attached to the cervix and the vault of the vagina. The **cardinal, or transverse cervical, ligaments** extend from the lateral pelvic walls to the cervix and vagina along the lowest margin of the extraperitoneal space within the broad ligaments of the uterus (Figs. 26–17 and 26–18). The cardinal ligaments help stabilize the midline position of the cervix and the vault of the vagina.

The **uterosacral ligaments** arise from the lower end of the sacrum and extend anteriorly around the sides of the rectum to attach to the cervix and vagina

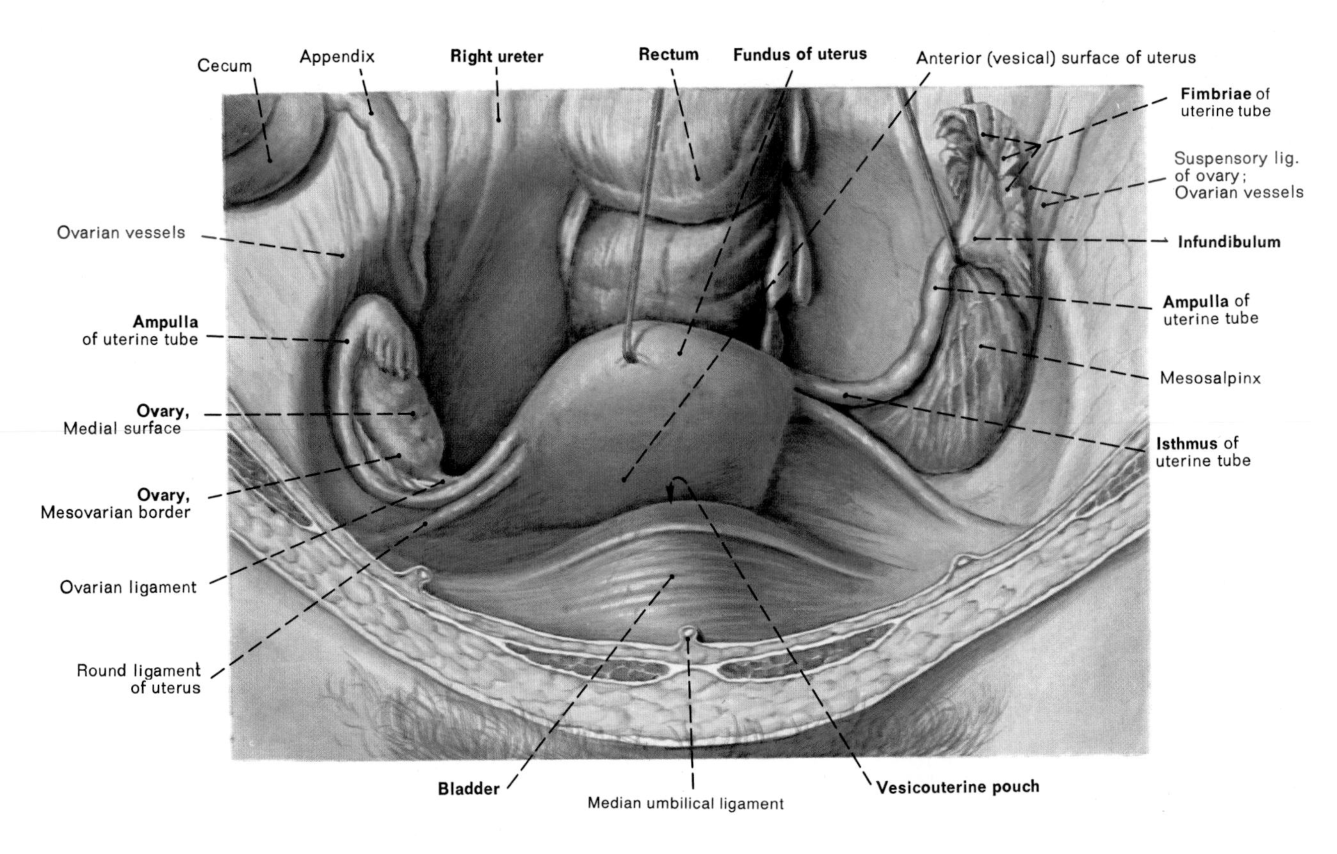

Fig. 26–16: Anterosuperior view of the female pelvic organs.

(Fig. 26–15 shows the attachment of the left uterosacral ligament to the cervix and vagina). The uterosacral ligaments have smooth muscle tissue called the rectouterine muscle (Fig. 26–14). On each side, the uterosacral ligament forms a fascial shelf that projects from the lateral wall of the pelvis toward the midline; the fascial shelf is covered by a fold of peritoneum called a **rectouterine fold** (Figs. 26–13 and 26–14). The uterosacral ligaments securely tether the cervix to the sacrum, and help stabilize the 90° angle between the longitudinal axes of the vagina and uterus.

The Arterial Supply of the Vagina, Uterus, Uterine Tubes, and Ovaries

Four paired arteries collectively supply the vagina, uterus, uterine tubes, and ovaries (Fig. 26–19):

(1) The **paired ovarian arteries** arise from the abdominal aorta just below the origins of the renal arteries. As the ovarian artery descends into the pelvis by crossing over the external iliac vessels at the pelvic brim, it is closely associated with the ureter. The ovarian artery extends through the extraperitoneal space of the suspensory ligament of the ovary and then gives rise to (a) branches that extend through the mesovarium to supply the ovary and (b) branches that extend medially through the broad ligament to supply the uterine tube and fundus of the uterus.

(2) The **paired uterine arteries** arise on each side from the anterior division of the internal iliac artery, and extend medially atop the floor of the pelvis within the cardinal ligament to reach the side of the uterus (near the lateral fornix of the vagina). As the uterine artery extends toward the uterus, it passes over the ureter. The uterine artery gives rise to branches that ascend the uterus to supply the body of the uterus and to anastomose with the vaginal branches of the ovarian artery near the entrance of the uterine tube into the uterus. The uterine artery also gives rise to branches which descend along the vagina to supply the upper part of the vagina.

(3) The **paired vaginal arteries** are each a branch of the anterior division of the internal iliac artery. The vaginal artery gives rise to branches which descend along the vagina to supply the lower part of the vagina. The vaginal artery also gives rise to branches which ascend along the vagina to supply the upper part of the vagina and to anastomose with branches of the uterine artery.

(4) The **paired internal pudendal arteries** are each a branch of the anterior division of the internal iliac artery. The internal pudendal artery gives rise to (a) branches which supply the lowest part of the vagina and (b) branches which ascend the vagina to anastomose with branches of the vaginal artery.

The Lymphatic Drainage of the Vagina, Uterus, Uterine Tubes, and Ovaries

Lymph collected from the lower third of the vagina drains into the horizontal group of superficial inguinal nodes. Lymph collected from the middle third of the vagina drains into internal iliac nodes. Lymph collected from the upper third of the vagina drains into external and internal iliac nodes.

Most of the lymph collected from the cervix and body of the uterus drains into internal iliac nodes. Lymph collected from the fundus of the uterus, the uterine tubes, and the ovaries drains into aortic nodes at the level of the body of the first lumbar vertebra.

The Sensory Innervation of the Uterus and Vagina

The sensory fibers innervating the body of the uterus enter the spinal cord at those spinal cord segments that provide preganglionic sympathetic innervation for the body of the uterus (specifically, the sensory fibers enter at the T11-L1 levels). Disease or injury of the body of the uterus produces visceral pain that is most commonly poorly localized to the hypogastric region. Disease or injury of the body of the uterus may refer pain to parts of the T11-L1 dermatomes.

The sensory fibers innervating the cervix of the uterus and the upper part of the vagina enter the spinal cord at those spinal cord segments that provide preganglionic parasympathetic innervation (through the pelvic splanchnic nerves) for the uterus and the upper part of the vagina (specifically, the sensory fibers enter at the S2, S3, and S4 levels). Disease or injury of the cervix of the uterus or the upper part of the vagina produces visceral pain that is usually poorly localized to the lumbosacral region. Disease or injury of the cervix of the uterus or the upper part of the vagina may refer pain to parts of the S2, S3, and S4 dermatomes.

The left and right pudendal nerves provide sensory innervation for the lower part of the vagina.

THE REPRODUCTIVE VISCERA OF THE MALE PELVIS

The **vas deferens** (**ductus deferens**) conducts sperm from the tail of the epididymis to the origin of the ejaculatory duct. The vas deferens extends from the tail of the epididymis to the origin of the ejaculatory duct by first ascending through the scrotum, then traversing the inguinal canal, and finally coursing posteriorly along the lateral wall of the pelvis to the posterior surface of the bladder. In Figure 26–9, the segment of the vas deferens in the scrotum is labelled the scrotal part of the ductus deferens, the segment in the inguinal canal is labelled the inguinal part of the vas deferens, and the segment in the pelvic cavity is labelled the pelvic part of the duct. Upon emerging from the inguinal canal at the deep inguinal ring, the vas deferens curves around the lateral side of the inferior epigastric artery near the artery's origin from the external iliac artery (Fig. 26–6). As the vas deferens extends medially to reach the superolateral corner of the posterior surface of the bladder, it crosses over the ureter (Fig. 26–6). The vas deferens ends at its union with the duct of the seminal vesicle to form the ejaculatory duct.

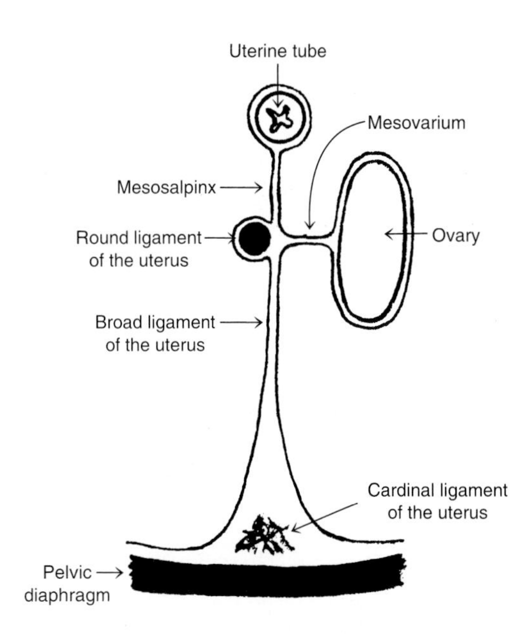

Fig. 26–17: Diagram of a sagittal sectional view of the tissues covered by the broad ligament of the uterus in a female pelvis.

The **seminal vesicles** lie against the base of the bladder (Fig. 26–6). These highly coiled, tubular glands add secretions to semen.

The **ejaculatory ducts** extend from their origin near the neck of the bladder anteroinferiorly through the prostate gland (Fig. 26–9). They open into the prostatic urethra (Fig. 26–11).

The **prostate gland** has a superior **base** and an inferior **apex.** The base faces the undersurface of the neck of the bladder; the apex rests atop the superior

CLINICAL PANEL V-3

Acute Prostatitis

Description: Acute prostatitis is acute inflammation of the prostate from infection.

Common Symptoms: Common symptoms of acute prostatitis include dysuria, increased urgency and frequency of urination, perineal and/or low back pain or discomfort, and fever. The pain upon urination may be a burning sensation resulting from the stimulation of thermal receptors in the inflamed prostatic urethra during the passage of urine. Voiding may be difficult because a decrease in the size of the lumen of the prostatic urethra. Stimulation of sensory fibers in the inflamed prostatic urethra accounts for the increased urgency and frequency of urination. Pain or discomfort in the perineum and/or lower back represents referred pain.

Common Signs: Digital examination of the rectum commonly reveals an enlarged, tender, and warm prostate that feels boggy (sponge-like).

layer of fascia of the UG diaphragm. The prostate transmits the initial segment of the urethra in the male, the prostatic urethra (Fig. 26–11). The gland is incompletely divided into lobes by the prostatic urethra and ejaculatory ducts. During sexual climax, the prostate adds secretions to the semen through 20 to 30 ductules which open into the prostatic urethra.

The prostate is supplied by branches of the **inferior vesical** and **middle rectal arteries.** Venous outflow drains into a venous plexus which lies superficial to the fibrous capsule of the gland. The **prostatic venous plexus** not only drains into the internal iliac veins, but also the **vesical plexus** (the venous plexus around the bladder) and the **vertebral venous plexuses.** The communications between the prostatic and vertebral venous plexuses are believed to be responsible for the metastasis of prostatic neoplastic cells to vertebral bodies. Most of the lymph collected from the prostate drains into internal iliac nodes.

The first and second lumbar splanchnic nerves provide preganglionic sympathetic innervation for the prostate, and the pelvic splanchnic nerves provide the preganglionic parasympathetic innervation. The sensory fibers that innervate the prostate enter the spinal cord at those spinal cord segments that provide preganglionic sympathetic and preganglionic parasympathetic innervation for the gland (in other words, the sensory fibers enter at the L1, L2, S2, S3, and S4 levels). Disease of the prostate may refer pain to parts of the L1, L2, S2, S3, and S4 dermatomes.

Digital examination of the rectum, which offers palpation of the posterior surface of the prostate, can provide relatively early diagnosis of a prostatic neoplasm.

THE ANAL REGION OF THE PERINEUM

The anal region of the perineum is traversed in the midline by the anal canal. The left and right halves of the anal region are called the **left** and **right ischiorectal fossae** (Fig. 26–20). The ischiorectal fossae are filled with fat which is readily displaced when feces pass through the anal canal. Each ischiorectal fossa is a wedge-shaped space whose superomedial border is the origin of levator ani from the tendinous arch of obturator internus. Each fossa is bordered laterally by obturator internus and the ischial tuberosity. Medially, each fossa is bounded by the downward sloping levator ani and the external anal sphincter. Posteriorly, each fossa extends as far as the sacrotuberous ligament and the lesser sciatic foramen.

On each side, the **pudendal nerve** and the **internal pudendal artery** enter the ischiorectal fossa by the lesser sciatic foramen. The nerve and artery extend anteriorly through the fossa in a fascial sheath which rests against the medial surface of obturator internus. The passageway formed by this fascial sheath is called the **pudendal canal** (Fig. 26–20).

THE RECTUM AND ANAL CANAL

The rectum and anal canal are the terminal segments of the digestive tract. Whereas the rectum is a pelvic organ, the anal canal traverses the perineum. Passage of feces into the rectum stimulates the desire to defecate; the musculature associated with the anal canal controls the timing of rectal evacuation.

As the rectum descends from the rectosigmoid junction to the anorectal junction, it presents an anterior, concave curvature called the sacral flexure (Fig. 26–9). The forward pull of the puborectalis parts of the levator ani muscles at the anorectal junction gives the junction a posterior, concave curvature called the perineal flexure of the rectum (Fig. 26–9).

The rectum also exhibits three lateral curvatures along its length; the uppermost and lowermost bulge to the left, while the middle bulges to the right. The upper margin of each lateral curvature is demarcated internally by a transverse, crescentic fold which extends into the lumen of the rectum; these permanent folds are called the **superior, middle,** and **inferior rectal valves** (Figure 26–21 displays the superior and middle rectal valves). It has been speculated that the rectal valves provide for the passage of flatus through a feces-laden rectum.

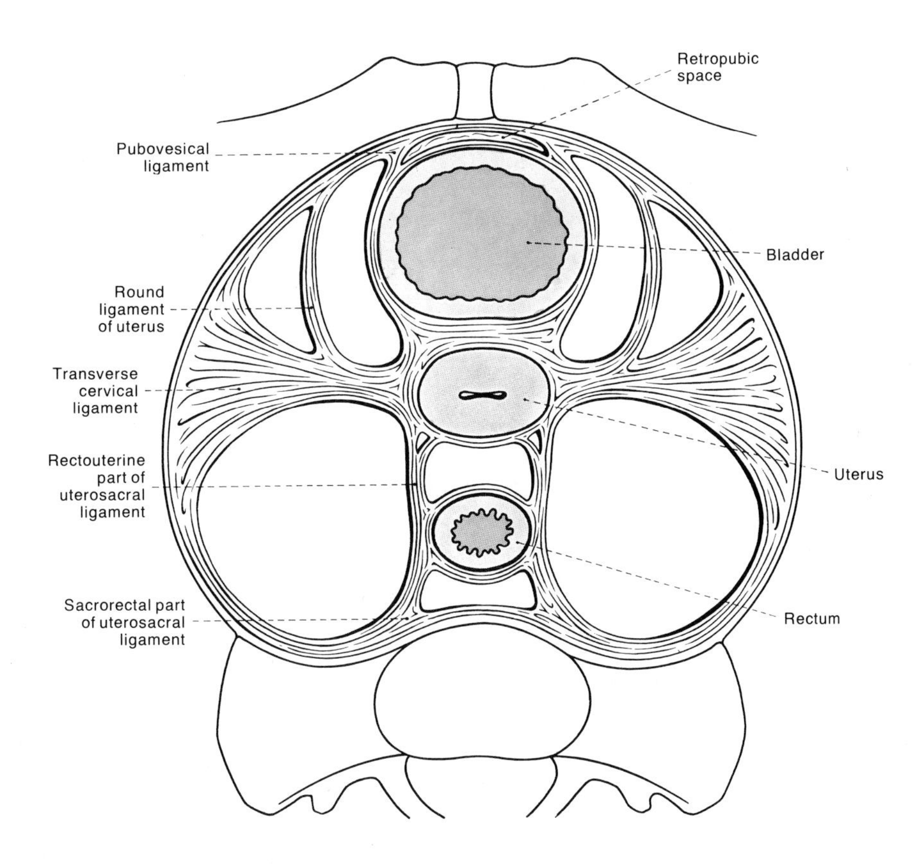

Fig. 26–18: Diagram of a superior view of the transverse cervical and uterosacral ligaments of the female pelvis.

The lower part of the rectum is more dilated and more distensible than the upper part. The lower part of the rectum is called the **rectal ampulla** (Fig. 26–21). The middle rectal valve marks the upper margin of the rectal ampulla.

The anal canal extends posteriorly and inferiorly from its origin at the anorectal junction to its end at the anus. Three sets of muscle fibers contribute to its control of the timing of fecal and flatus evacuation:

The **internal anal sphincter** consists of the thickened circular layer of muscularis externa extending from the lowermost part of the rectum through the upper two-thirds of the anal canal (Fig. 26–21). The internal anal sphincter is thus composed of smooth muscle fibers. The internal anal sphincter has a sharply defined lower border called the **intermuscular (intersphincteric) groove.** The intermuscular groove is palpable upon digital examination of the anal canal.

The **external anal sphincter** is a thick layer of skeletal muscle fibers which surrounds the entire anal canal (Figs. 26–2A, 26–3, 26–7, 26–9, 26–21, and 26–22). The midregion of the external anal sphincter consists mostly of muscle fibers that extend past the lateral sides of the anal canal; these muscle fibers insert anteriorly onto the central tendon of the perineum and posteriorly onto the coccyx (Figs. 26–2A, 26–3, and 26–7). The anteroposterior orientation of these muscle fibers compresses the lumen of the anal canal into a slit-like, longitudinal fissure.

The muscle fibers of the **puborectalis parts of the levator ani muscles** interdigitiate extensively with the muscle fibers of the deepest part of the external anal sphincter (Fig. 26–22) to form a prominent muscular ring, called the **anorectal ring,** at the upper margin of the anal canal. The anorectal ring is palpable upon digital examination of the anal canal.

Fecal continence requires tonic contraction of the internal anal sphincter, external anal sphincter, and the puborectalis parts of the levator ani muscles. The passage of feces into the rectum stimulates the urge to defecate. Distention of the rectal wall by fecal matter elicits a reflexive relaxation of the internal anal sphincter and, in a toilet-trained individual, voluntary contraction of the external anal sphincter and puborectalis parts of the levator ani muscles. At a time suitable for defecation, voluntary relaxation of the external anal sphincter results in (1) a lateral spreading of the anal canal from the level of the anus up to the level of the intersphincteric groove and (2) decreased tension along the wall of the remainder of the anal canal. Voluntary relaxation of the puborectalis parts of levator ani decreases the curvature of the perineal flexure; the course of the anal canal thus becomes more in line with that of the rectal ampulla. Increased parasympathetic stimulation of the rectum increases peristaltic activity, and contraction of the diaphragm and anterolateral abdominal wall muscles is used to increase abdominopelvic cavity pressure upon the fecal contents in the rectum.

The Arterial Supply of the Rectum and Anal Canal

Two unpaired and two paired arteries supply the rectum and anal canal (Fig. 26–23):

The **superior rectal artery** supplies the rectum and upper half of the anal canal; it is almost the sole arterial supply of the mucosal lining of these digestive tract segments.

The **paired middle rectal arteries** supply mainly the muscularis externa layers of the rectum and anal canal. The middle rectal artery is a branch of the anterior division of the internal iliac artery.

The **paired inferior rectal arteries** supply the lower half of the anal canal as well as the external and internal anal sphincters. The inferior rectal artery arises from the internal pudendal artery as it courses through the perineum.

The **middle sacral artery** supplies the posterior parts of the anorectal junction and the anal canal. The middle sacral artery arises from the abdominal aorta above its terminal bifurcation.

The Venous Drainage of the Rectum and Anal Canal

Whereas the mucosa of the rectum and upper half of the anal canal is drained by tributaries of the **superior rectal vein,** the mucosa of the lower half of the anal canal is drained by tributaries of the **inferior rectal veins.** The border between the upper and lower halves of the anal canal is thus a region of **portal-systemic**

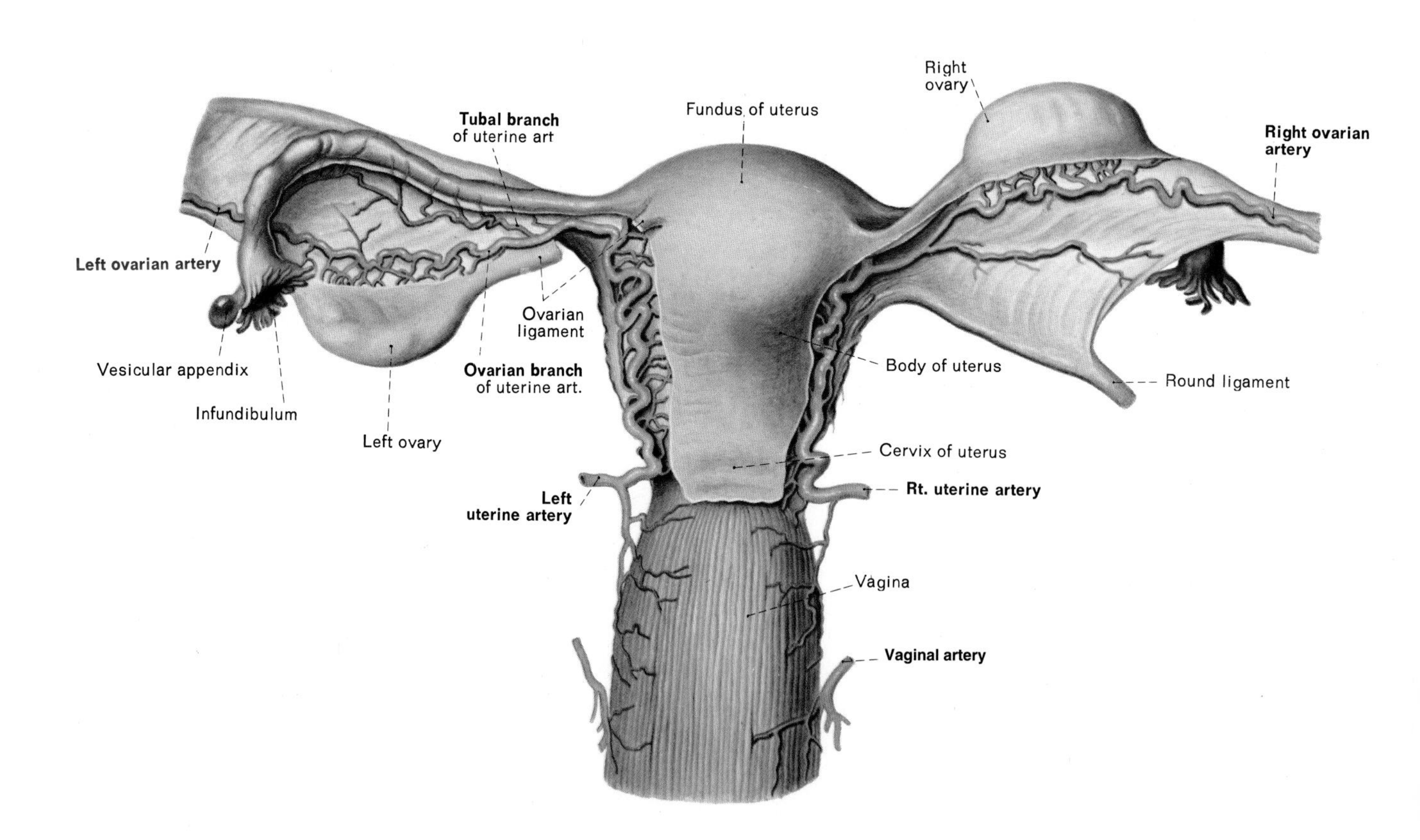

Fig. 26–19: Diagram of the arterial supply of the female reproductive viscera.

anastomoses. Varicosities of the superior rectal tributaries in the submucosa of the upper half of the anal canal are called internal hemorrhoids; varicosities of the inferior rectal tributaries in the submucosa of the lower half of the anal canal are called external hemorrhoids. The **middle rectal veins** drain mostly the muscular walls of the rectum and anal canal.

The superior rectal tributaries in the submucosa of the upper half of the anal canal raise the mucosal lining into longitudinal folds called **anal columns** (Fig. 26–21). Short, transverse folds of mucosal lining called **anal valves** extend between the inferior ends of adjacent anal columns. Each anal valve forms a recess, called an **anal sinus,** at the base of the space between two adjacent anal columns (Fig. 26–21). The anal valves collectively form a transverse fold which encircles the anal canal at its upper margin; this transverse fold is called the **anorectal (pectinate) line.**

The Lymphatic Drainage of the Rectum and Anal Canal

Lymph collected from the upper half of the rectum drains into the inferior mesenteric nodes. Lymph collected from the lower half of the rectum and upper half of the anal canal drains into internal iliac nodes. Lymph drained from the lower half of the anal canal drains into the horizontal group of superficial inguinal nodes.

The Innervation of the Rectum and Anal Canal

The first and second lumbar splanchnic nerves provide preganglionic sympathetic innervation for the rectum and the upper half of the anal canal, and the pelvic plexuses transmit the postganglionic sympathetic fibers to these viscera. The pelvic splanchnic

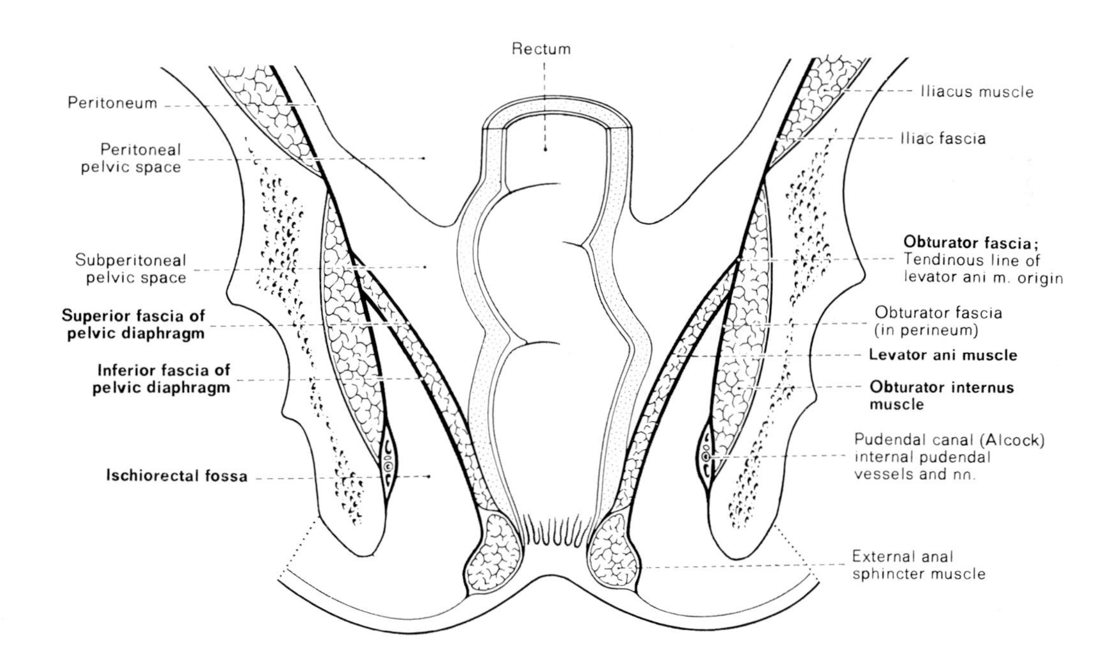

Fig. 26–20: Diagram of a coronal section through the male pelvis and perineum.

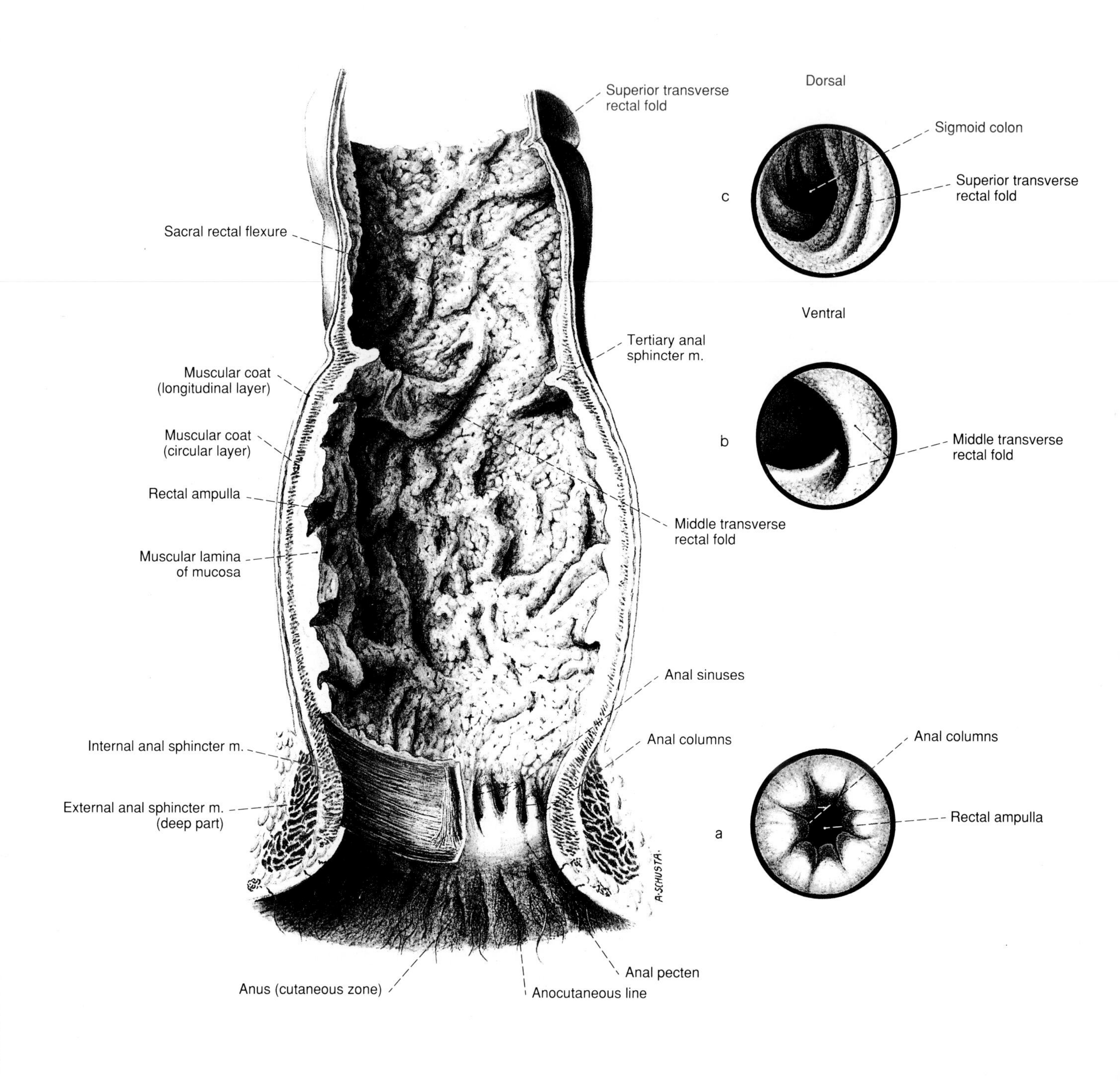

a = Anal canal

b = Rectal ampulla near the middle tranverse rectal fold (Kohlrausch)

c = Sacral flexure near the superior transverse rectal fold

Fig. 26–21: Posterior view of the rectal mucosa lining the anterior wall of the rectum. The three circular illustrations show proctoscope views of the rectal mucosa at different levels.

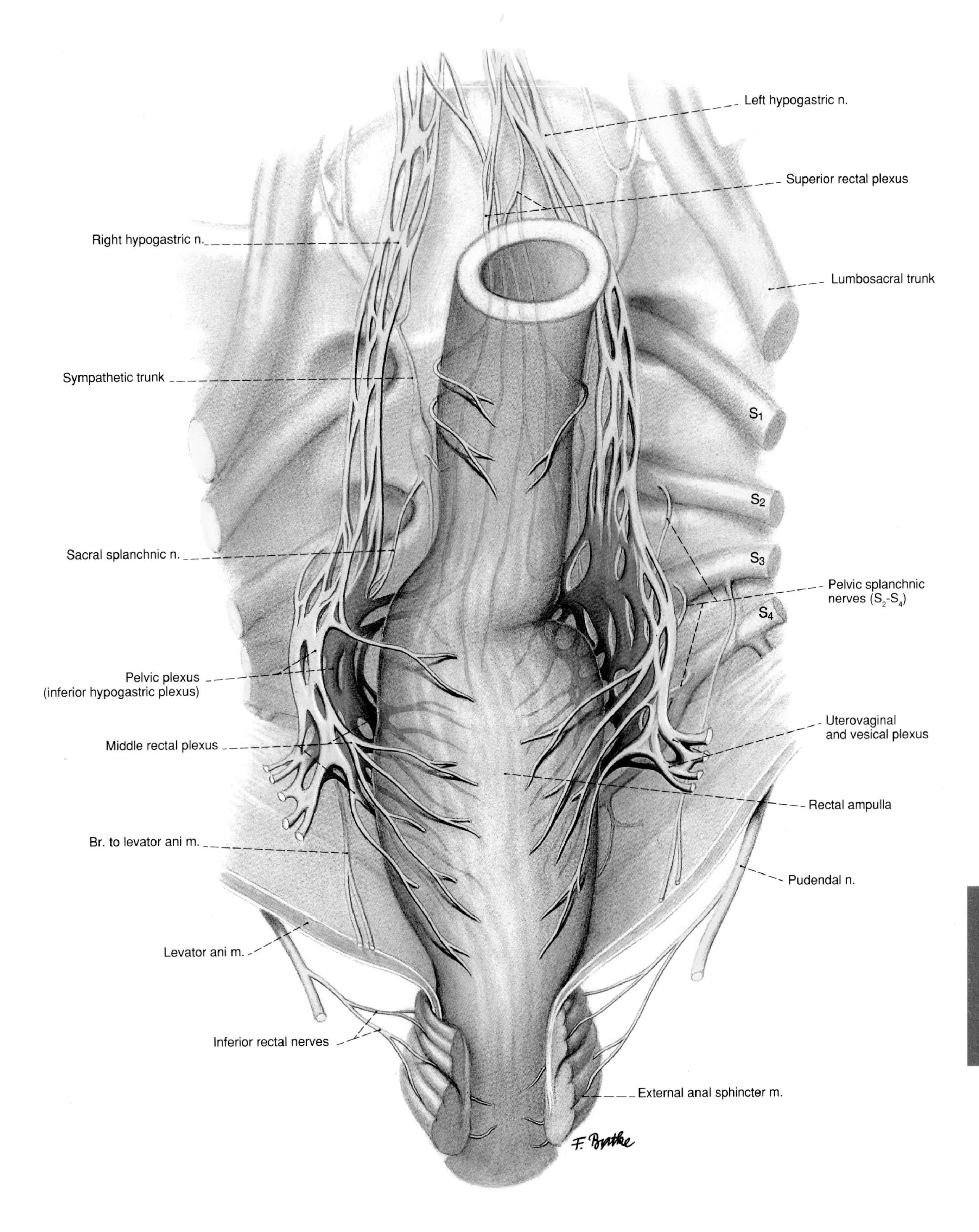

Fig. 26–22: Schematic illustration of an anterior view of the sensory innervation of the rectum and anal canal.

nerves provide the preganglionic parasympathetic innervation for the rectum and upper half of the anal canal.

The sensory fibers that innervate the rectum and upper half of the anal canal enter the spinal cord at those spinal cord segments that provide preganglionic sympathetic and preganglionic parasympathetic innervation for these viscera (in other words, the sensory fibers enter at the L1, L2, S2, S3, and S4 levels). Disease or injury of the rectum and upper half of the anal canal produces visceral pain that is, most commonly, poorly localized to the lumbosacral region. Disease or injury of these viscera may refer pain to parts of the L1, L2, S2, S3, and S4 dermatomes.

The left and right pudendal nerves provide sensory innervation for the lower half of the anal canal.

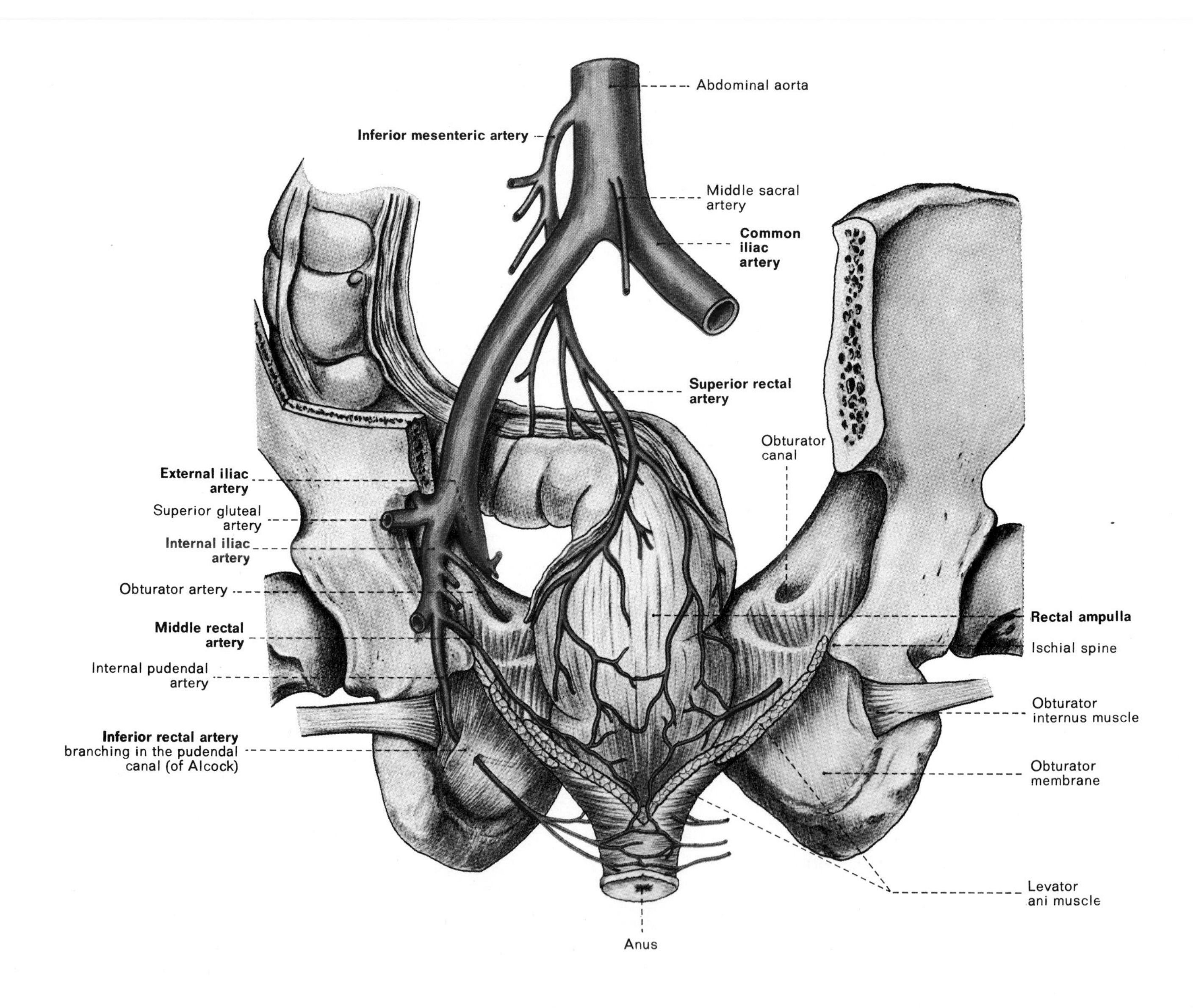

Fig. 26–23: Posterior view of the arterial supply of the rectum and anal canal.

CASE **V.7**

The Case of Mahmoud Ravazi

INITIAL PRESENTATION AND APPEARANCE OF THE PATIENT

A 67 year-old, white man named Mahmoud Ravazi walks into the ambulatory clinic at 2:00 PM. He appears uncomfortable and sits with the trunk of his body bent forward.

QUESTIONS ASKED OF THE PATIENT

What is bothering you? I have a pain in my belly.

Where does it hurt? Over here [the patient places his right hand over the umbilical region of his abdomen].

When did the pain start? Yesterday evening, after dinner.

Do you recall what you were doing when the pain started? Yes, I had just sat down to read the newspaper.

How would you describe the pain? It's a deep ache that never goes away. *[The patient's answer suggests pain of visceral nature.]*

Has the nature of the pain or its location changed since yesterday? No, except that it seems to be getting worse.

Have you ever had a pain like this before? No, never.

Is there anything that makes the pain worse? No.

Is there anything that relieves the pain? No.

Is there anything else that is also bothering you, or that you think may be associated with your abdominal pain? I don't know. I don't think so.

Have you recently felt nauseous or vomited? No.

Have there been any recent changes in your bowel habits or bowel movements? No.

Have you had any recent problems with urination? No.

Have you felt feverish? No.

Do you know any individuals around you, such as members of your family or friends, who have also recently suffered from abdominal pain? No.

Have you had any recent illnesses or injuries? No.

Are you taking any medications for previous illnesses? No. *[The examiner finds the patient alert and fully cooperative during the interview.]*

PHYSICAL EXAMINATION OF THE PATIENT

Vital signs: Blood pressure
Lying supine: 125/75 left arm and 120/75 right arm
Standing: 120/75 left arm and 120/75 right arm
Pulse: 77
Rhythm: regular
Temperature: 98.6°F.
Respiratory rate: 22
Height: 5′8″
Weight: 135 lbs.
HEENT Examination: Normal
Lungs: Normal
Cardiovascular Examination: Auscultation reveals bilateral femoral bruits. Dorsalis pedis pulses are bilaterally absent and posterior tibial pulses are bilaterally weak.
Abdomen: Inspection, auscultation, percussion, and palpation of the abdomen are normal except for the following findings: Palpation reveals a pulsatile, epigastric mass. The patient also exhibits guarding but no rebound tenderness.
Genitourinary Examination: Normal except for an enlarged prostate.
Musculoskeletal Examination: Normal
Neurologic Examination: Normal

INITIAL ASSESSMENT OF THE PATIENT'S CONDITION

The patient is suffering from acute pain in the umbilical region associated with a pulsatile epigastric mass and cardiovascular abnormalities in the lower limbs.

ANATOMIC BASIS OF THE PATIENT'S HISTORY AND PHYSICAL EXAMINATION

(1) **The patient's abdominal pain appears to be umbilical visceral pain.** This presentation suggests disease or disorder of one or more of the viscera supplied by the superior mesenteric artery.

(2) **The finding of bilateral femoral bruits indicates a constriction or aneurysm of the aorta or both common iliac-external iliac-femoral arterial trunks.** Normal blood flow through arteries is streamlined and therefore silent. Turbulent blood flow through arteries produces sounds called bruits. A bruit indicates the presence of an upstream constriction or aneurysm of the artery. However, it should be noted that not all constrictions or aneurysms produce bruits.

(3) **The bilaterally weak distal pulses in the lower limbs in association with systolic and diastolic blood pressures in the normal range and distal pulses of normal strength in the upper limbs suggest a constriction or aneurysm of (a) the aorta distal to the origin of the left subclavian artery from the aortic arch or (b) both common iliac-external iliac-femoral arterial trunks.**

(4) **The pulsatile epigastric mass indicates a large abdominal aortic aneurysm.**

CLINICAL REASONING PROCESS

The acute onset of intense abdominal pain in an elderly individual absent of any symptoms of GI disease or disorder suggests an expanding abdominal aortic aneurysm, intestinal ischemia, pancreatitis, or acute renal disorder. The finding of a pulsatile midline mass is virtually diagnostic of an expanding abdominal aortic aneurysm. The pulsatile mass is encountered in the epigastric or umbilical regions, as the abdominal aorta generally bifurcates into the common iliac arteries at a level within the umbilical region.

Abdominal aortic aneurysms can be asymptomatic for several years as they progressively enlarge. An acute increase in the rate of enlargement elicits continuous abdominal pain that is generally midline in location and visceral in nature; the pain frequently radiates to the back.

RADIOLOGIC EVALUATION AND FINAL RESOLUTION OF THE PATIENT'S CONDITION

Ultrasonography revealed an abdominal aortic aneurysm measuring approximately 7 cm in diameter. The patient was admitted for immediate surgical repair of the aneurysm.

At surgery, a 7.5 cm fusiform aortic aneurysm involving the inferior mesenteric artery was found. The inferior mesenteric artery was ligated and the aneurysm repaired with a prosthetic graft. Blood supply to the distal parts of the large bowel was not compromised following ligation of the inferior mesenteric artery because of collateral supply provided to the marginal artery of Drummond by branches of the superior mesenteric artery.

CASE **V.8**

The Case of Karen Hanson

INITIAL APPEARANCE AND PRESENTATION OF THE PATIENT

A 27 year-old white woman named Karen Hanson has made an appointment to seek treatment for a painful abdomen. The patient appears slightly uncomfortable. When standing, she bends her body slightly forward and to the right.

QUESTIONS ASKED OF THE PATIENT

How can I help you? I have a pain in my belly which has been getting worse.

Do you have pain now? Yes.

Can you show me or describe to me where you feel pain? [The patient points to the RUQ of her abdomen.]

When did you first notice the pain? It started 4 days ago. *[The patient's answer indicates that the patient is suffering from (a) an acute disease or condition or (b) a disease or condition that has become symptomatic only recently.]*

Do you recall what you were doing when the pain started? Yes. The pain started in the afternoon while I was seated at work. I'm a computer programmer.

How would you describe the pain? Most of the time it feels like a deep ache. *[The patient's description suggests visceral pain (a dull ache).]*

Has the nature of the pain or its location changed at any time during the past 4 days? The location of the pain hasn't changed. But it has been getting worse each day, and since yesterday, the pain sometimes feels sharp, like a knife cutting me. *[The patient's answer indicates recent onset of somatic pain (a sharp pain).]*

Have you found anything that makes the pain worse? Yes. The pain is worse when I take a deep breath.

Have you found anything that relieves the pain? No, not really.

Have you ever had pains like this before in the upper right part of your abdomen? No.

Is there anything else that is also bothering you? Well, I have felt feverish during the past 3 days. *[The report of fever suggests the presence of an inflammatory process, and thus prompts the examiner to inquire about symptoms common to disease of the body's major organ systems.]*

Have you had any respiratory problems, such as a cough or difficulty in catching your breath? No.

Have you had any recent digestive problems, such as heartburn, nausea, or vomiting? No.

Have you noticed any recent change in your bowel habits, such as diarrhea or constipation? No.

Have you had any recent problems with urination? No.

Have you had any recent menstrual problems? Well, my last menstrual period, which was 5 days ago, was a little on the heavy side. And I've noticed some yellowish vaginal discharge in the last few days, but I haven't had any discomfort or pain. *[The patient indicates that she had menorrhagia (an increased volume of menstrual discharge) during her last menstrual period. The report of a yellowish vaginal discharge suggests an infectious process of the lower genital tract. (In a female, the lower genital tract consists of the vagina and cervix of the uterus. The upper genital tract consists of the body of the uterus, the uterine tubes, and the ovaries.)]*

Have your menstrual periods been normal and regular prior to this last abnormal period? Yes.

Have you ever had a sexually transmitted disease? No.

Do you practice birth control? Yes, I have an IUD [an intrauterine device]. It was inserted about a year ago.

Are you currently having sexual relations with anyone? Yes. I met a guy about 3 months ago, and we started having sexual relations about a month ago.

Are you taking any medications for previous injuries or illnesses? No. *[The examiner finds the patient alert and fully cooperative during the interview.]*

PHYSICAL EXAMINATION OF THE PATIENT

Vital Signs: Blood pressure
Lying supine: 115/70 left arm and 115/70 right arm
Standing: 115/70 left arm and 115/70 right arm
Pulse: 75
Rhythm: regular
Temperature: 100.5°F
Respiratory rate: 19
Height: 5′9″
Weight: 135 lbs.

[The examiner recognizes that the patient is febrile (has an elevated temperature, as recorded by an oral thermometer). The normal range of oral temperatures is from 96.4°F (35.8°C) to 99.1°F (37.3°C).]

HEENT Examination: Normal
Lungs: Normal
Cardiovascular Examination: Normal
Abdomen: Inspection, auscultation, percussion, and palpation of the abdomen are normal except for the following findings: Deep palpation reveals a firm, regular, smooth but tender liver edge palpable below the right costal margin on inspiration. Fist percussion applied to the lower ribs above the right costal margin elicits an increase in RUQ pain. Deep palpation elicits localized tenderness in RLQ and LLQ. Light palpation elicits localized tenderness in the RUQ only.
Pertinent normal findings include the following: The gallbladder is not palpable, and the test for Murphy's sign is negative.
Genitourinary Examination: Inspection of the external genitalia and bi-manual examination of the pelvis reveal the following findings: The introitus (entrance) to the vagina shows the presence of vaginal discharge. The vaginal walls and cervix are erythematous. The cervix is mucopurulent and appears nulliparous. The cervix is firm and mobile but tender.
Bi-manual examination finds the body of the uterus small and globular but tender. The adnexa on both sides are mobile and palpable but tender. [When used in reference to the female reproductive tract, the term adnexa refers to the uterine tubes and ovaries and their associated vascular and connective tissues.]
Musculoskeletal Examination: Normal
Neurologic Examination: Normal

INITIAL ASSESSMENT OF THE PATIENT'S CONDITION

The patient appears to be suffering from progressive worsening of an acute disease or condition associated with inflammation of one or more viscera of the RUQ and the genital tract.

ANATOMIC BASIS OF THE PATIENT'S HISTORY AND PHYSICAL EXAMINATION

(1) **The patient's liver is tender (as demonstrated by deep palpation).** The finding of a tender liver explains the exacerbation of RUQ pain upon deep inspiration or fist percussion to the right costal margin. This is because any force (such as that delivered from fist percussion to the right costal margin or from the downward thrust of the diaphragm during deep inspiration) that puts increased pressure upon a tender liver will produce upper abdominal pain.

(2) **The RUQ tenderness elicited upon light palpation represents rebound tenderness. This presentation indicates inflammation of the parietal peritoneum lining the anterolateral abdominal wall of the RUQ. The patient's RUQ pain is partially or entirely somatic.** The inflammation of the RUQ parietal peritoneum could have been acquired from direct contact with inflamed RUQ abdominal viscera or from migration of inflammatory agents from the thorax, the other abdominal quadrants, or the pelvis. The absence of RUQ abdominal wall rigidity suggests that the parietal peritoneum inflammation is not marked at this time.

(3) **The tenderness of the cervix, the erythema of the intravaginal cervix and vaginal walls, and the presence of a mucopurulent discharge collectively indicate infectious inflammation of the lower genital tract.**

(4) **The tenderness of the body of the uterus and its adnexa indicate inflammation of the upper genital tract.**

(5) **The RLQ and LLQ tenderness to deep palpation suggests inflammation and/or distention of one or more lower quadrant abdominopelvic viscera and/or inflammation of the parietal peritoneum lining the posterior abdominal wall in the lower quadrants.**

(6) **The patient's moderately elevated temperature indicates the presence of an inflammatory process.**

INTERMEDIATE EVALUATION OF THE PATIENT'S CONDITION

The patient appears to be suffering from infectious inflammation of the lower and upper genital tract, inflammation of RUQ parietal peritoneum, and a tender liver.

CLINICAL REASONING PROCESS AND FINAL RESOLUTION OF THE PATIENT'S CONDITION

The patient's chief symptom of RUQ pain initially focused the examiner's attention during the taking of the history to the possibility of hepatobiliary disease (disease of the liver, gallbladder, and/or extrahepatic biliary ducts) and even right lower lobar viral pneumonia. The early questioning in the history, however, did not reveal evidence of symptoms typical of hepatobiliary disease. The patient's report of fever suggested the possibility of hepatitis, but she did not report other constitutional symptoms typically manifested by hepatitis [symptoms such as nausea, vomiting, anorexia (loss of appetite), fatigue, and arthralgia and/or myalgia (joint and/or muscle pain)]. The patient also did not give a history of biliary colic, which would have indicated the presence of gallstones (refer to Case V-4 for a discussion of biliary colic). The possibility of viral pneumonia was diminished by the absence of a non-productive cough and malaise or fatigue.

The patient's account of a mucopurulent vaginal discharge subsequently focused the examiner's attention to the possibility of genital tract disease or disorder. The patient's report of her use of an intrauterine device and her recent menstrual period make it very unlikely that she is suffering from complications of an ectopic pregnancy. However, upon completion of the history but prior to the physical exam, the examiner was still considering conditions such as a ruptured ovarian cyst or twisted adnexa as possible etiologies of the patient's condition.

A primary care physician or obstetrician/gynecologist would recognize that the patient's signs and symptoms are characterisitc of pelvic inflammatory disease complicated by perihepatitis. The expression **pelvic inflammatory disease (PID)** refers to infectious inflammation of the upper female reproductive tract (the pelvic parts of the reproductive tract) from dissemination of an infection of the intravaginal part (perineal part) of the cervix of the uterus. PID complicated by perihepatitis is referred to by an eponym: the Fitz-Hugh-Curtis syndrome. The Fitz-Hugh-Curtis syndrome is a condition in which infectious agents responsible for PID disseminate from the upper genital tract (generally one or both uterine tubes) to the liver, and there provoke inflammation of the fibrous capsule of the liver but not of its parenchyma (in other words, the inflammation is confined to the fibrous capsule only). The dissemination can occur through lymphatics, hematogenous routes, or a transperitoneal route (the transperitoneal route probably involves the migration of the infectious agents from the pelvis to the liver by the right lateral paracolic gutter).

A gynecologist would advise hospitalization of the patient because of her fever and RUQ rebound tenderness. The physician would recommend a pregnancy test, various blood tests (including a white blood cell count), and microbiologic work-up of a cervical swab. Laparoscopy would be advisable if the collective evidence was not as supportive of a diagnosis of PID (as it is in this case). Laparoscopy is used to obtain visual confirmation of PID and to collect peritoneal fluid for microbiologic work-up. Treatment involves antibiotic therapy, removal of the IUD (as it is believed to augment dissemination of infectious agents from the lower to the upper genital tract), and consultation and treatment of the patient's male sex partner.

EXPLANATION OF THE PATIENT'S CONDITION

The patient became infected with *Chlamydia trachomatis* (a bacterial pathogen) as a result of sexual intercourse with her male sex partner (who is not aware that he has a sexually transmissible disease). The IUD provided a structural tract for the ascent of the bacteria from the lower to the upper genital tract. The erythema of the vaginal walls and intravaginal cervix is a consequence of infectious inflammation. The mucopurulent vaginal discharge is a product of immunologic response. Infectious inflammation of the body of the cervix and the uterine tubes is responsible for their tenderness upon bi-manual pelvic examination. The infectious inflammation of the body of the uterus and the uterine tubes is also the anatomic basis of the RLQ and LLQ tenderness to deep palpation.

The bacteria extended from the pelvis to the perihepatic area through ascent along the right lateral paracolic gutter. Infectious inflammation of the liver capsule has elicited visceral pain (that in this case is not centered about the midline but in the RUQ). Acute distention of the inflamed liver capsule by deep inspiration and fist percussion to the right costal margin intensifies the visceral pain. Infectious inflammation has extended to the parietal peritoneum of the anterolateral abdominal wall region anterior to the liver, and this is the anatomic basis of the RUQ rebound tenderness.

CASE **V.9**

The Case of Jack Mullen

INITIAL APPEARANCE AND PRESENTATION OF THE PATIENT

A 66 year-old white man named Jack Mullen has made an appointment to seek treatment for a painful back. The patient is overweight and appears uncomfortable.

QUESTIONS ASKED OF THE PATIENT

How can I help you? I've been having pain in the lower part of my back for about two months.

Do you have back pain now? Yes.

Can you show me or describe to me where you feel pain? Back here. [The patient places his right hand over the lumbar region of the vertebral column.]

Do you recall what you were doing when the pain started? No. All I know is that pain started gradually about two months ago.

Do you recall injuring yourself before the back pain started? No.

How would you describe the pain? It just feels like an ache in my back.

Has the nature of the pain or its location changed at any time during the past two months? No, not really.

Have you found anything that makes the pain worse? No, not really. The pain just seems a bit worse at the end of a long day.

Have you found that the back pain becomes worse if you move or bend the back in a particular way? No.

Have you found that the back pain becomes momentarily worse when you cough, sneeze, or bear down during a bowel movement? No.

Have you found anything that relieves the pain? Well, if I lie down, the pain seems a bit better, but it's still there.

Is there anything else that is also bothering you? No. *[At this juncture in the history, the examiner assesses the patient's condition as chronic low back pain of 2 months duration and insidious onset. There is no evidence to indicate that the low back pain is (a) a consequence of an injury, (b) aggravated by movement or weight-bearing, or (c) associated with other symptoms. The examiner appreciates that the patient may not recognize the presence of associated symptoms, and therefore decides to inquire of the presence of specific symptoms, symptoms typical of disease or dysfunction of various organ systems.]*

Have you had any recent injuries or illnesses? No.

Do you have any respiratory problems, such as a cough or shortness of breath? Well, I've had a morning cough for about 4 years. I generally cough every morning to clear my lungs.

What generally is the color of the sputum? It's yellowish or brownish.

Do you ever cough up blood? Oh, once or twice I've seen a speck of blood in the sputum I've coughed up in the morning.

Do you smoke? Yes. I smoke 2 or 3 cigars a day. I've been smoking cigars for about the past 10 years. Before that, I used to smoke about a pack of cigarettes a day. I smoked cigarettes for about 20 years.

Do you have any digestive problems, such as heartburn or indigestion? Yes. I have had some problems with heartburn and indigestion for a number of years. I have found that antacids help the heartburn and the belching of gas. I haven't been too concerned about the heartburn because it comes and goes.

Have you had any recent changes in your bowel habits, such as diarrhea or constipation? No.

Have you had any recent problems with urination? Well, I do have trouble starting and stopping. The urine will sometimes dribble out before I can get to the bathroom.

Is it ever painful to urinate? No. *[The presence of painful urination would indicate inflammation of the lower urinary tract.]*

Have you found that you have to get up frequently at night to urinate? Yes. I go about 3 or 4 times a night. I guess my bladder must be weak because I don't generally pass much urine.

Has there been any recent change in your weight? Yes. I've lost about 10 pounds in the last 2 months.

Are you taking any medications for previous injuries or illnesses? No. *[The examiner finds the patient alert and fully cooperative during the interview.]*

PHYSICAL EXAMINATION OF THE PATIENT

Vital Signs: Blood pressure
Lying supine: 135/80 left arm and 135/80 right arm
Standing: 135/80 left arm and 135/80 right arm
Pulse: 75
Rhythm: regular
Temperature: 98.4°F.
Respiratory rate: 19
Height: 5′9″
Weight: 225 lbs.
HEENT Examination: Normal
Lungs: Inspection, palpation, percussion, and auscultation of the lungs are normal except for a few expiratory rhonchi throughout both lungs. The rhonchi are diminished upon coughing.
Cardiovascular Examination: Normal
Abdomen: Inspection, auscultation, percussion, and palpation of the abdomen are normal except for the following findings: The rectal exam reveals a hard, irregular painless nodule in the right lateral lobe of the prostate and obliteration of the midline furrow.
Genitourinary Examination: Normal
Musculoskeletal Examination: Normal
Neurologic Examination: Normal

INITIAL ASSESSMENT OF THE PATIENT'S CONDITION

The patient appears to be suffering from chronic low back pain, urgency (the sudden desire to urinate), dysuria (difficulty with urination), and nocturia (excessive urination at night). The patient also has a hard, painless prostatic nodule.

ANATOMIC BASIS OF THE PATIENT'S HISTORY AND PHYSICAL EXAMINATION

(1) **The patient's dysuria and nocturia suggest an altered configuration of the urethrovesical junction (the junction of the neck of the urinary bladder with the prostatic urethra).** Alterations in the configuration of the urethrovesical junction reduce the efficacy of the circumferentially-oriented connective tissue fibers in the neck of the bladder to restrict the passage of urine into the urethra. Passage of urine into the urethral segment immediately distal to the neck of the bladder stimulates sensory fibers that provoke an intense urge to urinate. It is through this mechanism that an altered configuration of the urethrovesical junction can cause urgency, nocturia, and difficulty in terminating urine flow at the end of micturition.

(2) **The hard, irregular painless nodule in the right lateral lobe of the prostate indicates cancer of the prostate.** The posterior aspect of the prostate is palpable upon rectal exam. A midline furrow on the posterior surface of the prostate divides the palpable posterior aspect of the gland into left and right halves which are called lateral lobes. Palpation of the lateral lobes can detect the hard, painless nodules characteristic of cancer of the prostate before metastasis occurs.

Palpation of the prostate can also detect **benign prostatic hypertrophy,** a condition common among older men in which the prostate's glandular tissue undergoes benign hypertrophy. The generalized glandular enlargement can lead to obstruction of the prostatic urethra and/or alteration of the configuration of the urethrovesical junction. Common symptoms of benign prostatic hypertrophy include the symptoms of lower urinary tract dysfunction encountered in this case: urgency, dysuria, and nocturia.

(3) **The expiratory rhonchi indicate the presence of narrowed conducting airways.** Rhonchi are adventitious (abnormal) respiratory sounds produced by turbulent air flow through narrowed and/or irregular conducting airways. Causes of such narrowing and/or irregularity include excessive bronchial secretions, inflammatory swelling of airway walls, bronchiolar spasm, luminal encroachment by a tumor, or external compression by enlarged lymph nodes.

INTERMEDIATE EVALUATION OF THE PATIENT'S CONDITION

The patient appears to be suffering from cancer of the prostate.

CLINICAL REASONING PROCESS

The etiology of low back pain (pain in the lumbosacral region of the vertebral column) can be divided into mechanical versus nonmechanical causes. Mechanical causes are disorders intrinsic to the vertebral column produced by excessive strain on or degenerative changes in its musculoskeletal framework. Mechanical causes include musculotendinous or ligamentous strains, herniated intervertebral discs, degenerative changes of the vertebral column's synovial and cartilaginous joints, and spondylosis (stiffening or fixation of vertebral column joints as a result of the formation of fibrous or bony structures between articular surfaces). Mechanical causes typically produce low back pain that is affected by vertebral column movement and/or has a radicular component. Radicular pain is pain produced by compression and/or inflammation of the roots of one or more spinal nerves. Radicular pain is frequently perceived as radiating (extending) along the courses of those nerves transmitting nerve fibers to and from the involved spinal nerves.

Chronic low back pain that is (a) not associated with radicular pain and (b) essentially unaffected by weight-bearing or vertebral column movement is **OMINOUS.** Here the letters of the acronym represent seven major causes of nonmechanical, low back pain: **O**steomyelitis (infectious inflammation of bone), **M**etabolic bone disease (such as Paget's disease), **I**nflammatory disease [such as spondylitis (inflammation of one or more vertebrae)], **N**eoplasm, **O**ther condition (such as abscess), **U**nstable vertebral column [such as that produced by spondylolisthesis (anterior movement of a lumbar vertebra relative to the underlying lumbar vertebra or sacrum)], and **S**pinal canal disease (such as a spinal cord tumor).

In this case, the examiner began to more actively pursue information regarding non-mechanical causes of low back pain after learning that the patient's low back pain did not appear to have a radicular component to it nor to be aggravated by vertebral column movements. Given the patient's age and sex, the finding of problems with urinary retention prompted the examiner to more seriously consider disease or disorder of the urinary bladder and/or prostate.

RADIOLOGIC EVALUATION AND FINAL RESOLUTION OF THE PATIENT'S CONDITION

PA and lateral chest films did not show any abnormalities in the mediastinum or lung fields. However, osteoblastic lesions (lesions of increased radiodensity) were observed in the bodies and pedicles of the tenth, eleventh, and twelfth thoracic vertebrae. An AP erect abdominal film did not show any abnormalities except for osteoblastic lesions in the bodies and pedicles of the lumbar vertebrae, the sacrum, and the ileal wings of the coxal bones. A nuclear medicine study using a radiolabelled phosphate complex revealed increased activity (increased uptake of the radiolabel) in the left eleventh rib, right twelfth rib, lower thoracic vertebrae, all the lumbar vertebrae, the sacrum, and the ileal wings of the coxal bones.

Based upon the findings of the history, physical exam, and radiologic procedures, a primary care physician or oncologist would diagnose the patient's condition as metastatic cancer of the prostate [metastatic cancer is cancer that has spread (metastasized) beyond the site of origin]. Further examination would involve biopsy of the prostate and analysis for blood-borne markers suggestive or indicative of prostate cancer. Depending upon the results of the prostate biopsy (and assuming a positive diagnosis of cancer of the prostate), an oncologist would recommend some combination of surgical, radiotherapeutic and chemotherapeutic procedures.

CHRONOLOGY OF THE PATIENT'S CONDITION

The patient is suffering from metastatic cancer of the prostate. The enlargement of the neoplastic nodule in the right lateral lobe during the past year distorted the configuration of the urethrovesical junction to the extent that it produced the onset of urgency, nocturia, and dysuria about eight months previously.

The valveless communications among the prostatic venous plexus and the vertebral venous plexuses provided vascular routes of metastasis to the vertebral volumn, the coxal bones, and the ribs. Metastases became established in the parts of these bones bearing hematopoetic marrow (the bodies and pedicles of the vertebra, the shafts of the ribs, and the ileal wings of the coxal bones). Increased deposition of bone tissue at the metastatic sites accounts for the radiographic evidence of osteoblastic areas (sites of increased radiodensity) and the nuclear medicine evidence of areas of increased deposition of radiolabelled phosphate complex. The patient's constant low back pain is pain elicited by the expanding metastases in the bodies and pedicles of the patient's lumbar vertebrae and the microfractures that have resulted from the disorganization of bony trabeculae in the bodies of the lumbar vertebrae.

Lymphatics have also transported neoplastic cells to internal iliac, common iliac, and para-aortic nodes, and resulted in enlargement of these lymph nodes.

The patient's chronic productive cough (it produces sputum) and expiratory rhonchi are manifestations of chronic bronchitis (chronic inflammation of the conducting airways of the lungs). The chronic bronchitis is a result of the patient's extensive history of smoking tobacco products.

RECOMMENDED REFERENCES FOR ADDITIONAL INFORMATION ON THE VISCERA OF THE PELVIS AND PERINEUM

Jones III, H. W., Wentz, A. C., and L. S. Burnett, *Novak's Textbook of Gynecology,* Williams & Wilkins, Baltimore, 1988: *Pages 5 and 6 in Chapter 1 provide a brief discussion of the topics relevant to a gynecologic history. Pages 10–12 in Chapter 1 offer a description of a pelvic examination.*

DeGowin, R. L., Jochimsen, P. R., and E. O. Theilen, *DeGowin & DeGowin's Bedside Diagnostic Examination,* 5th ed., Macmillan Publishing Co., New York, 1987: *Pages 590 through 597 offer a discussion of common signs and symptoms of urinary problems.*

SECTION **VI**

Head and Neck

CHAPTER **27**

The Neck

The neck is bordered inferiorly by the thoracic inlet and superiorly by the base of the skull and the floor of the mouth.

THE BONY SKELETON OF THE NECK

The Cervical Vertebrae

The seven cervical vertebrae form most of the skeletal framework of the neck (Fig. 27–1). The cervical vertebrae exhibit a curved alignment which is concave posteriorly in the median plane; this secondary cervical curvature in the vertebral column aids the cervical vertebrae in their support of the head.

The transverse processes of the cervical vertebrae each bear a vertically-oriented passageway called a **transverse foramen** (Fig. 27–2). The transverse foramina of the upper six cervical vertebrae transmit on each side the **vertebral artery** and its venae comitantes (Fig. 27–3) and postganglionic sympathetic fibers. The transverse foramina of the seventh cervical vertebra transmit only the venae comitantes.

The second through sixth cervical vertebrae are distinguished from all other vertebrae in that their spinous processes are **bifid** (the posterior part of the spinous process is partially cleft into two ends) (Fig. 27–2).

A prominent, sturdy ligament called the **ligamentum nuchae** joins the spinous processes of the cervical vertebrae. The posterior edge of the ligamentum nuchae extends from the external protuberance of the skull's occipital bone to the tip of the seventh cervical vertebra's spinous process (Fig. 27–4). The seventh cervical vertebra is called the **vertebra prominence** because it has the most prominent, palpable spinous process of all the cervical vertebrae (Fig. 27–1). The spinous process of the seventh cervical vertebra is commonly the highest prominent spinous process in the midline of the back.

The first cervical vertebra is called the **atlas** (after the mythologic Greek titan who supported the earth and heavens) because it helps support the skull (Fig. 27–1). The atlas has neither a body nor a spinous process (Fig. 27–5). It is basically a bony ring with an anterior arch and a posterior arch joined by two lateral masses.

The second cervical vertebra is called the **axis** because it provides a pivot around which the atlas can rotate (Fig. 27–6). This pivot is a tooth-like process called the **dens** (**odontoid process**) which projects superiorly from the body of the axis. In the vertebral column, the dens extends up to the level of the atlas (Fig. 27–7), where it is encircled anteriorly by the atlas's anterior arch and posteriorly by a strong, curved ligament called the **transverse ligament of the atlas** (which extends transversely between the ends of the atlas's anterior arch) (Fig. 27–4). The transverse ligament is part of a cross-shaped ligament appropriately called the **cruciform ligament;** the superior band of this cruciform ligament extends superiorly to the occipital bone, and the inferior band extends inferiorly to the third cervical vertebra (Fig. 27–8). The dens is attached to the undersurface of the occipital bone by a bilateral pair of sturdy ligaments called alar ligaments and a midline ligament called the apical ligament (the apical ligament is not illustrated in Fig. 27–8).

It is **unequivocally imperative** that the head be immoblized in a cervical collar if it is suspected that trauma to the neck has torn one or more of the ligaments which retain the dens in close apposition to the anterior arch of the atlas. The tearing of one or more of these ligaments may render the dens relatively free to move posteriorly within the vertebral canal, and thereby crush the spinal cord. If the dens crushes the spinal cord at or above the level of the C2 roots, death ensues within a few minutes as a result of denervation of the muscles of respiration [the diaphragm receives its motor innervation from the anterior rami of C3, C4, and C5 (through the phrenic nerves); the intercostal muscles receive their innervation from the anterior rami of T1-T11].

The atlas articulates with both the **occipital bone of the skull** above and the axis below. The atlas forms a pair of synovial joints called the **left** and **right atlanto-occipital joints** with the occipital bone; in these joints, the superior articular facets of the atlas articulate with the condyles on the undersurface of the occipital bone. The atlas forms three synovial joints called the **left lateral, median,** and **right lateral atlanto-axial joints** with the axis; the inferior articular facets of the atlas articulate with the superior articular facets of the axis in the lateral joints (Fig. 27–8), and the anteri-

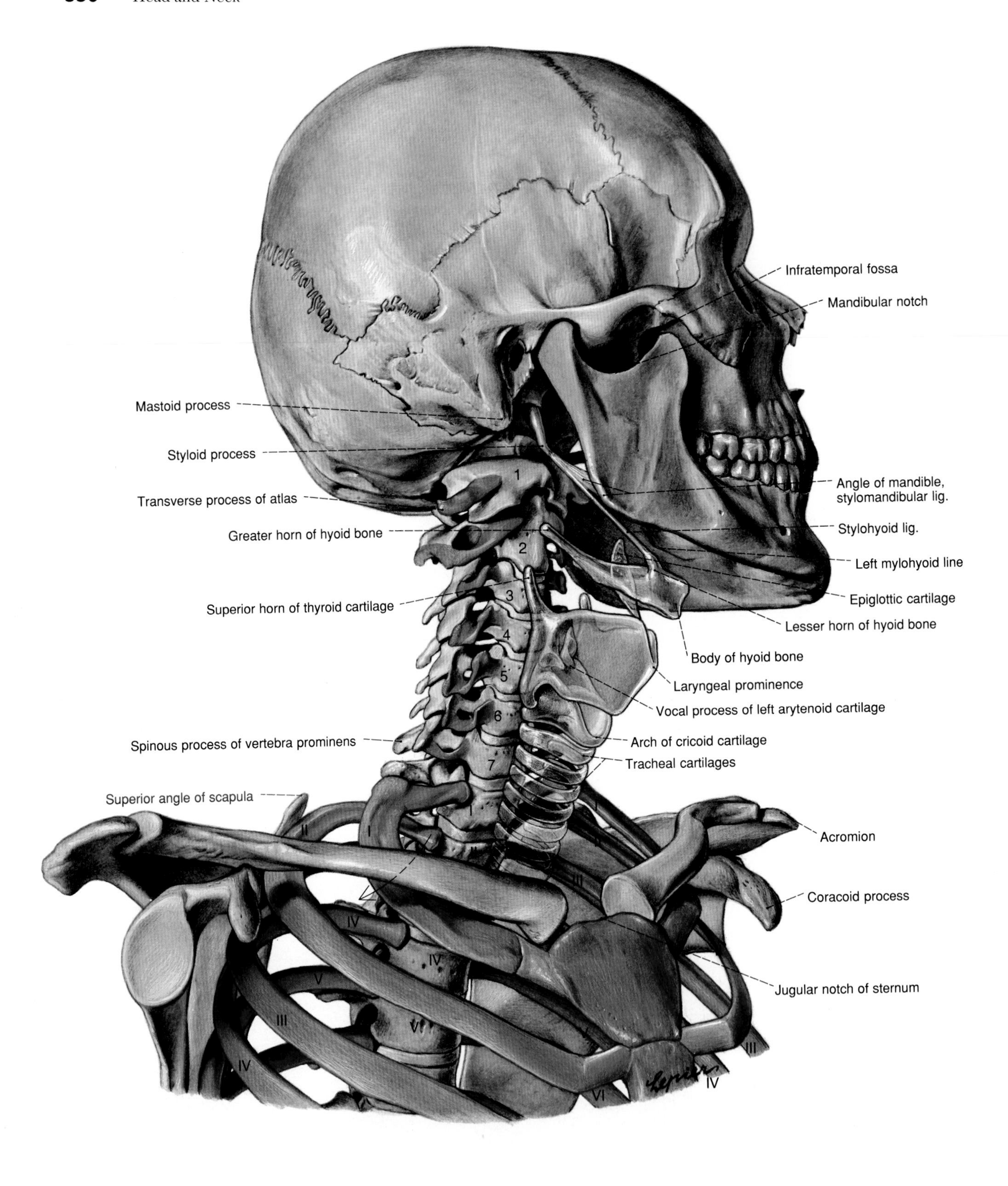

1–7 = Cervical vertebra 1–7
I–VI = Ribs and thoracic vertebrae 1–6

Fig. 27–1: Anterolateral view of the skull, cervical vertebrae, laryngeal and tracheal cartilages, and upper rib cage.

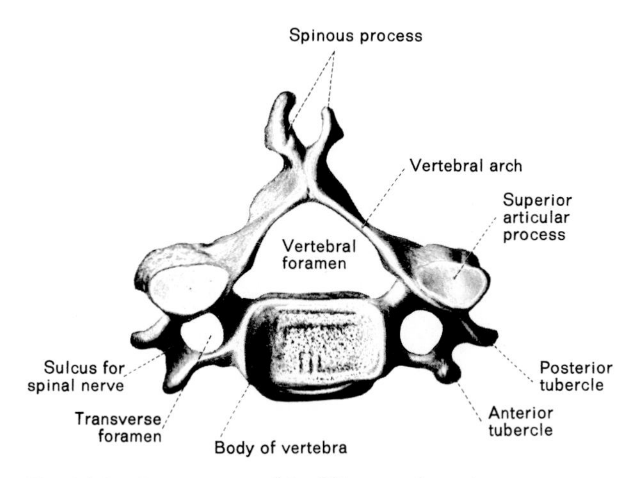

Fig. 27–2: Superior view of the fifth cervical vertebra.

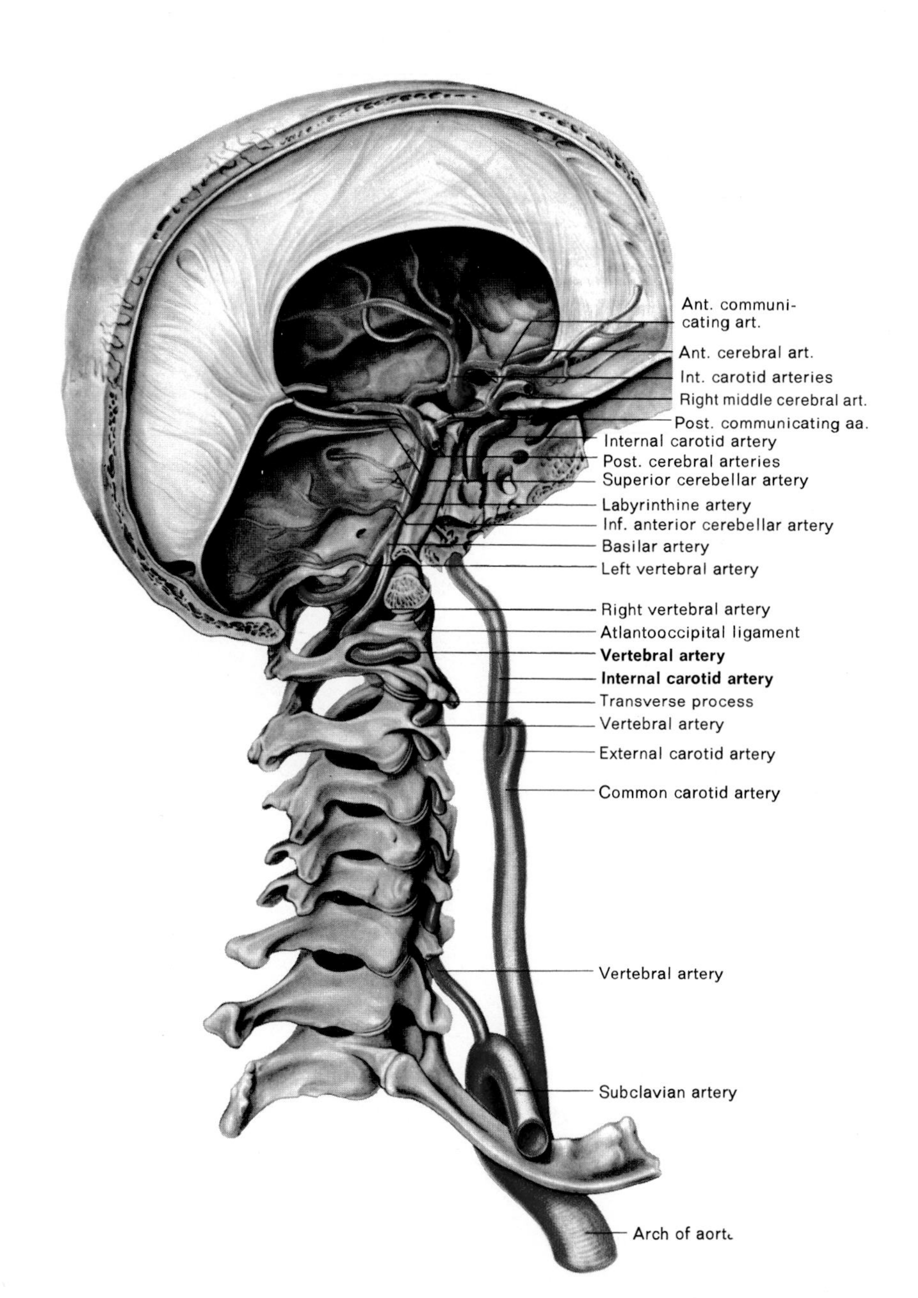

Fig. 27–3: Posterolateral view of the vertebral and internal carotid arteries.

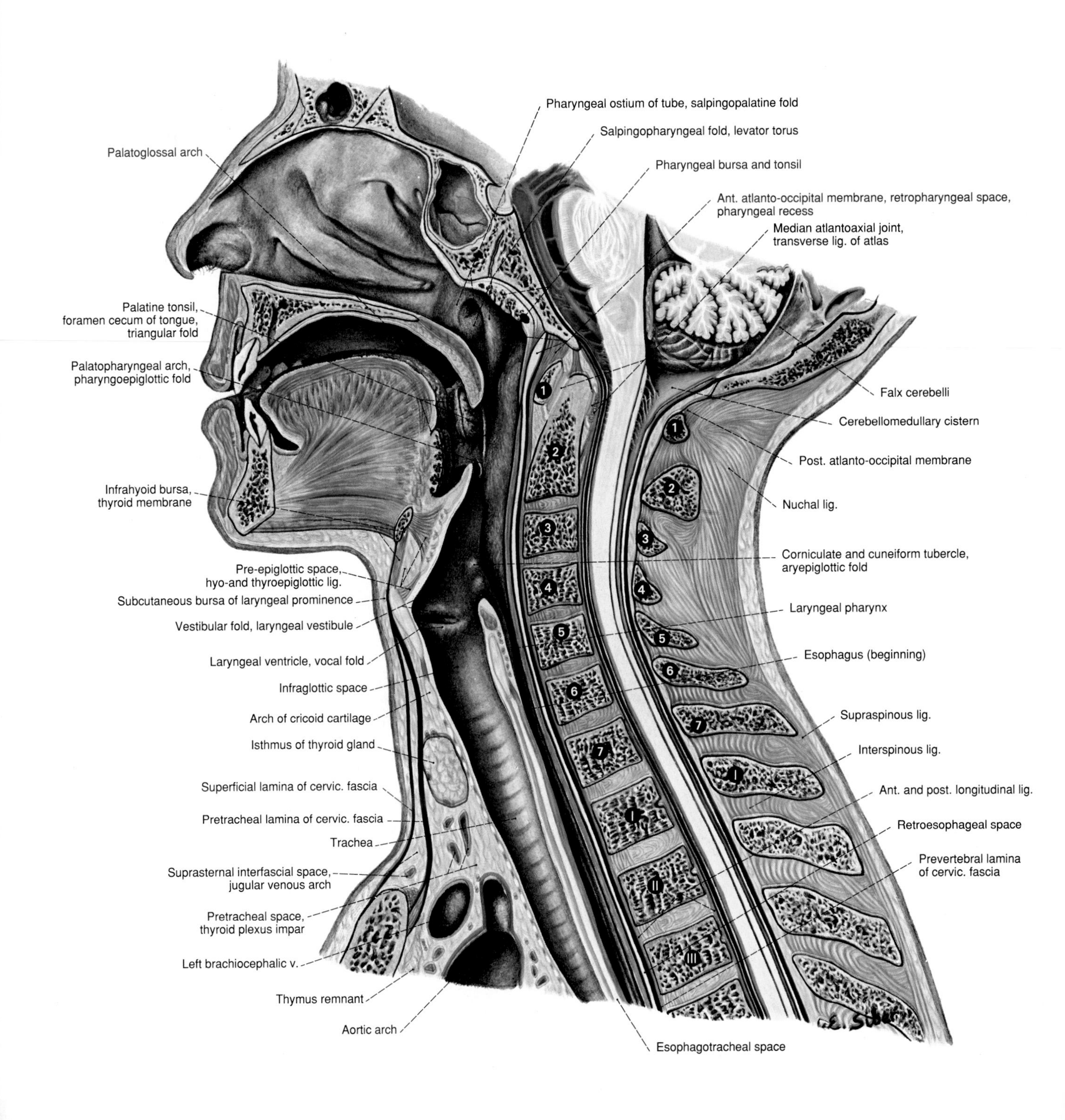

1–7 = Cervical vertebra 1–7
I–III = Thoracic vertebra 1–3

Fig. 27–4: Semischematic illustration of the median sectional view of the neck and adjoining parts of the head and chest.

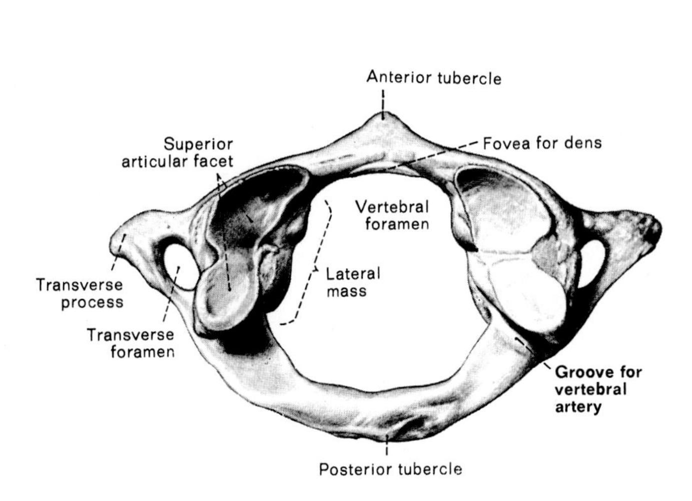

Fig. 27–5: Superior view of the atlas.

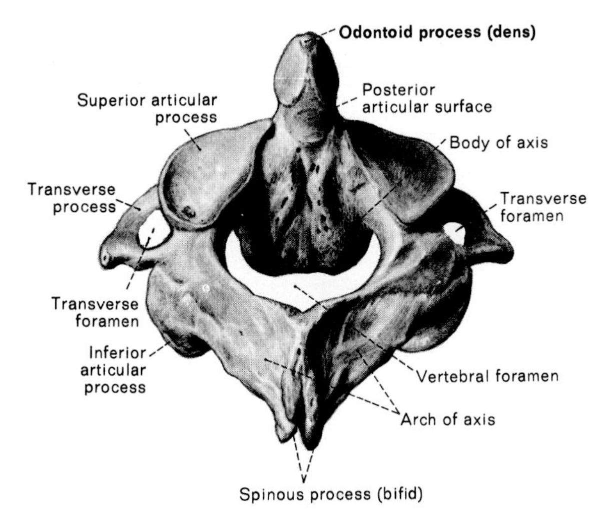

Fig. 27–6: Posterosuperior view of the axis.

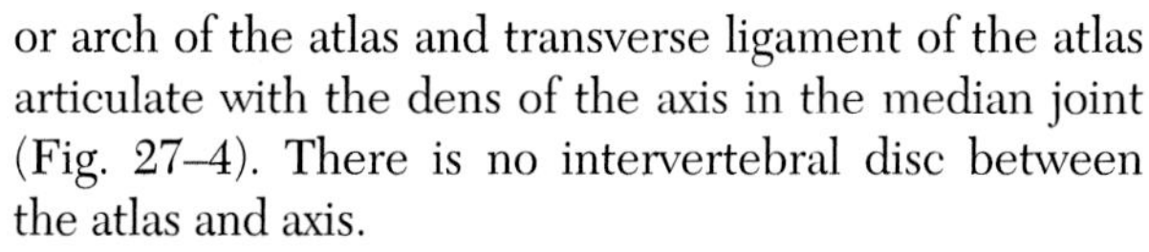

or arch of the atlas and transverse ligament of the atlas articulate with the dens of the axis in the median joint (Fig. 27–4). There is no intervertebral disc between the atlas and axis.

When the head is nodded forward and backward, much of the movement occurs within the atlanto-occipital joints (the occipital condyles rock forward and backward atop the superior articular facets of the atlas). By contrast, when the head is rotated to either side, much of the movement occurs within the atlanto-axial joints (the skull and atlas rotate in union about the dens).

The Hyoid Bone

The hyoid bone is a **U**-shaped bone which can be felt in the deep angle between the chin and the anterior part of the neck (Fig. 27–1). It lies approximately at the level of the body of the fourth cervical vertebra (Figure 27–1 shows it lying at the level of the body of the third cervical vertebra). The anterior, curved part of the bone is called the **body.** On each side, the body ends posteriorly at the point where it gives rise to a long process (a **greater horn**) which projects backward and a short process (a **lesser horn**) which projects upward (Fig. 27–1).

The hyoid bone serves as both a base for the tongue and a support from which the larynx is suspended. The larynx, which rests atop the trachea, is composed largely of cartilages joined by fibroelastic membranes, ligaments, and muscles. The largest cartilage of the larynx is the shield-shaped **thyroid cartilage** (Fig. 27–1). A fibroelastic membrane called the **thyrohyoid membrane** suspends the thyroid cartilage from the hyoid bone above.

THE MUSCLES OF THE NECK

The muscles of the anterior part of the neck may be classified functionally into four groups: four muscles that flex the neck, four muscles that raise the hyoid, four muscles that lower the hyoid, and a muscle of facial expression. Tables 27–1 to 27–3 list for quick reference the origin, insertion, nerve supply, and actions of these neck muscles except for the muscle of facial expression.

Sternocleidomastoid is the most prominent muscle in the anterior part of the neck (Fig. 27–9). It is named for three of its four attachments. The muscle has two heads of origin: one head arises from the sternum and the other from the clavicle. The muscle extends posterosuperiorly from its origins to insert onto the mastoid process of the temporal bone and an external ridge of the occipital bone.

Its action tilts the head toward the ipsilateral side concurrently with rotation of the head toward the contralateral side; the net effect is that the face is turned to the contralateral side and directed upward. When both sternocleidomastoids act in concert, they anteriorly flex the neck.

Injury to a newborn's sternocleidomastoids is one of the major potential complications of vaginal delivery. If one of a neonate's sternocleidomastoid muscles becomes fibrotic and shortened as a result of injury during birth, the common bearing of the individual's head will be that the face is turned to the contralateral side and directed upward. The physical condition produced is called **congenital torticollis** (which literally means "a twisted neck dating from birth").

Sternocleidomastoid's prominence and placement in the neck is the basis for its use as an anatomic bisector

Table 27–1
Muscles of the Neck which Flex the Neck

Muscle	Origin	Insertion	Nerve Supply	Actions
Sternocleidomastoid	Manubrium of the sternum and the medial third of the clavicle	Mastoid process of the temporal bone and the lateral half of the superior nuchal line of the occipital bone	Spinal part of the accessory and C2 and C3 fibers	Turns face to the opposite side and directs it upward
Scalenus anterior	Transverse processes of the 3rd, 4th, 5th and 6th cervical vertebrae	Upper surface of the first rib	C4, C5, and C6 fibers	Laterally flexes the neck or raises the first rib
Scalenus medius	Transverse processes of the upper 6 cervical vertebrae	Upper surface of the first rib	C3, C4, C5, C6, C7 and C8 fibers	Laterally flexes the neck or raises the first rib
Scalenus posterior	Transverse processes of the 4th, 5th and 6th cervical vertebrae	Upper surface of the second rib	C6, C7 and C8 fibers	Laterally flexes the neck or raises the second rib

Table 27–2
The Suprahyoid Muscles of the Neck

Muscle	Origin	Insertion	Nerve Supply	Actions
Mylohyoid	Mylohyoid line of the medial surface of the mandible	Contralateral mylohyoid and hyoid	Mandibular division of the trigeminal (via mylohyoid nerve, a branch of the inferior alveolar nerve)	Raises the hyoid, floor of the mouth, and tongue
Digastric	Posterior belly from mastoid process of the temporal bone; anterior belly from the digastric fossa of the mandible	Intermediate tendon and body and greater horn of the hyoid	Posterior belly by facial; anterior belly by mandibular division of the trigeminal	Raises the hyoid and lowers the mandible
Stylohyoid	Styloid process of the temporal bone	Body of the hyoid	Facial	Raises and retracts the hyoid
Geniohyoid	Inferior mental spine of the mandible	Body of the hyoid	C1 fibers	Raises the hyoid and pulls it forward

of each side of the neck into anterior and posterior triangles. The **anterior triangle of the neck** is bordered anteriorly by the midline of the neck, posteriorly by the anterior border of sternocleidomastoid, and superiorly by the lower margin of the body of the mandible (in Figure 27–9, the region of the neck anterior to the sternocleidomastoid muscle represents the anterior triangle of the neck). The **posterior triangle of the neck** is bordered anteriorly by the posterior border of sternocleidomastoid, posteriorly by the anterior border of trapezius, and inferiorly by the middle third of the clavicle (in Figure 27–9, the triangular region of the neck posterior to the sternocleidomastoid muscle, anterior to the trapezius muscle, and superior to the clavicle represents the posterior triangle of the neck).

Scaleni anterior, medius, and posterior extend from cervical vertebrae to the first two ribs (Fig. 27–10). They function as lateral flexors of the neck and accessory muscles of respiration.

Table 27–3
The Infrahyoid Muscles of the Neck

Muscle	Origin	Insertion	Nerve Supply	Actions
Sternothyroid	Manubrium of the sternum	Oblique line of the thyroid cartilage	C1, C2 and C3 fibers	Lowers the larynx
Thyrohyoid	Oblique line of the thyroid cartilage	Body and greater horn of the hyoid	C1 fibers	Lowers the hyoid and raises the larynx
Sternohyoid	Manubrium of the sternum and medial end of the clavicle	Body of the hyoid	C1, C2 and C3 fibers	Lowers the hyoid
Omohyoid	Superior border of the scapula	Body of the hyoid	C1, C2 and C3 fibers	Lowers and retracts the hyoid

Scalenus anterior is an important feature in the neck because of its relationships to the subclavian vessels, the brachial plexus, and the phrenic nerve. As the subclavian artery and the more proximal parts of the brachial plexus extend over the first rib en route to the axillary region, they pass behind scalenus anterior. The subclavian vein extends over the first rib en route to the cervical region by passing in front of scalenus anterior. The phrenic nerve descends in the neck from its C3, C4, and C5 origins by passing anterior to scalenus anterior.

The muscles attached to the hyoid are used to move the tongue and larynx during the acts of swallowing and phonation (production of sound). The suprahyoid muscles collectively suspend the hyoid from the cranium and mandible. The infrahyoid muscles collectively tether the hyoid inferiorly to the manubrium, clavicle, and scapula.

Mylohyoid originates from a gently curved ridge on the medial surface of the mandible that extends from the **symphysis menti** to the position of the last molar tooth (the symphysis menti is the border of fusion between the left and right halves of the neonatal mandible) (Fig. 27–11). The most posterior fibers in each muscle extend medially to attach onto the body of the hyoid. All other fibers in each muscle meet their

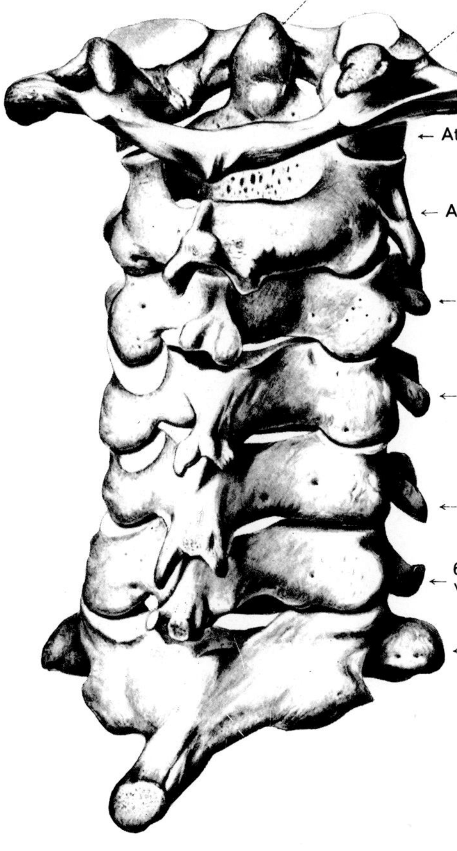

Fig. 27–7: Posterior aspect of the cervical vertebrae.

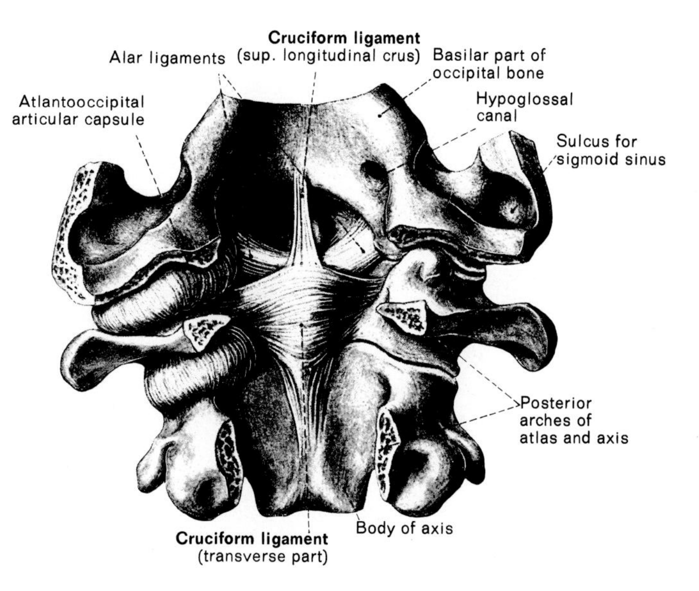

Fig. 27–8: Posterior view of the atlanto-occipital and atlanto-axial joints. The posterior arches of the atlas and axis have been removed to exhibit the cruciform ligament.

4–8 = Ventral rami of C-4 to C-8 nerves
I = Ventral ramus of T-1 nerve

L. c. = Cricothyroid ligament
M. h. th. = Thyrohyoid membrane
I, II = 1st and 2nd ribs

C. cr. = Cricoid cartilage
C. th. = Thyroid cartilage
H = Hyoid bone

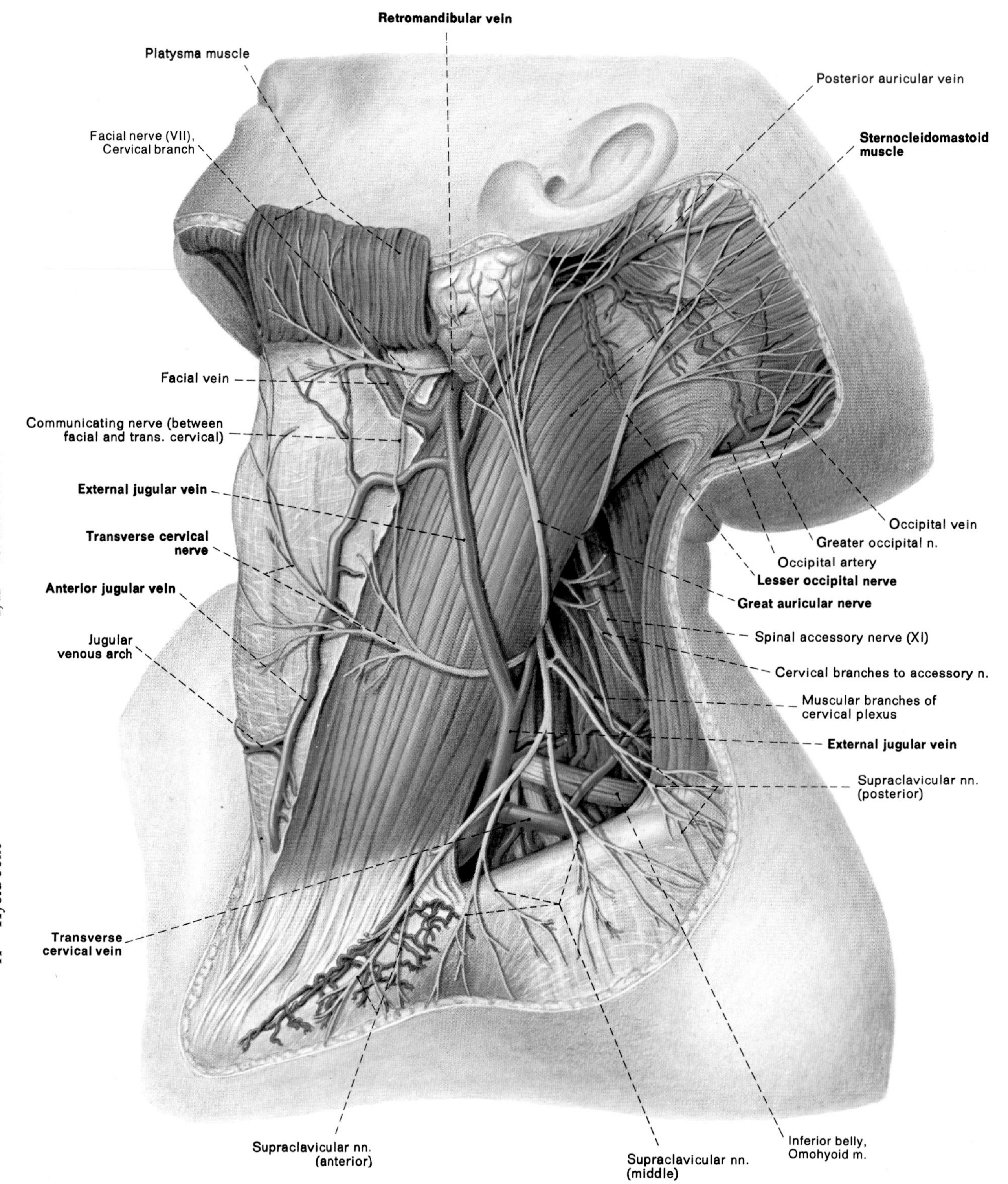

Fig. 27–9: Lateral view of a relatively superficial dissection of the neck. The skin and superficial fascia of the neck have been removed. All of the platysma muscle except for a reflected upper part has also been removed.

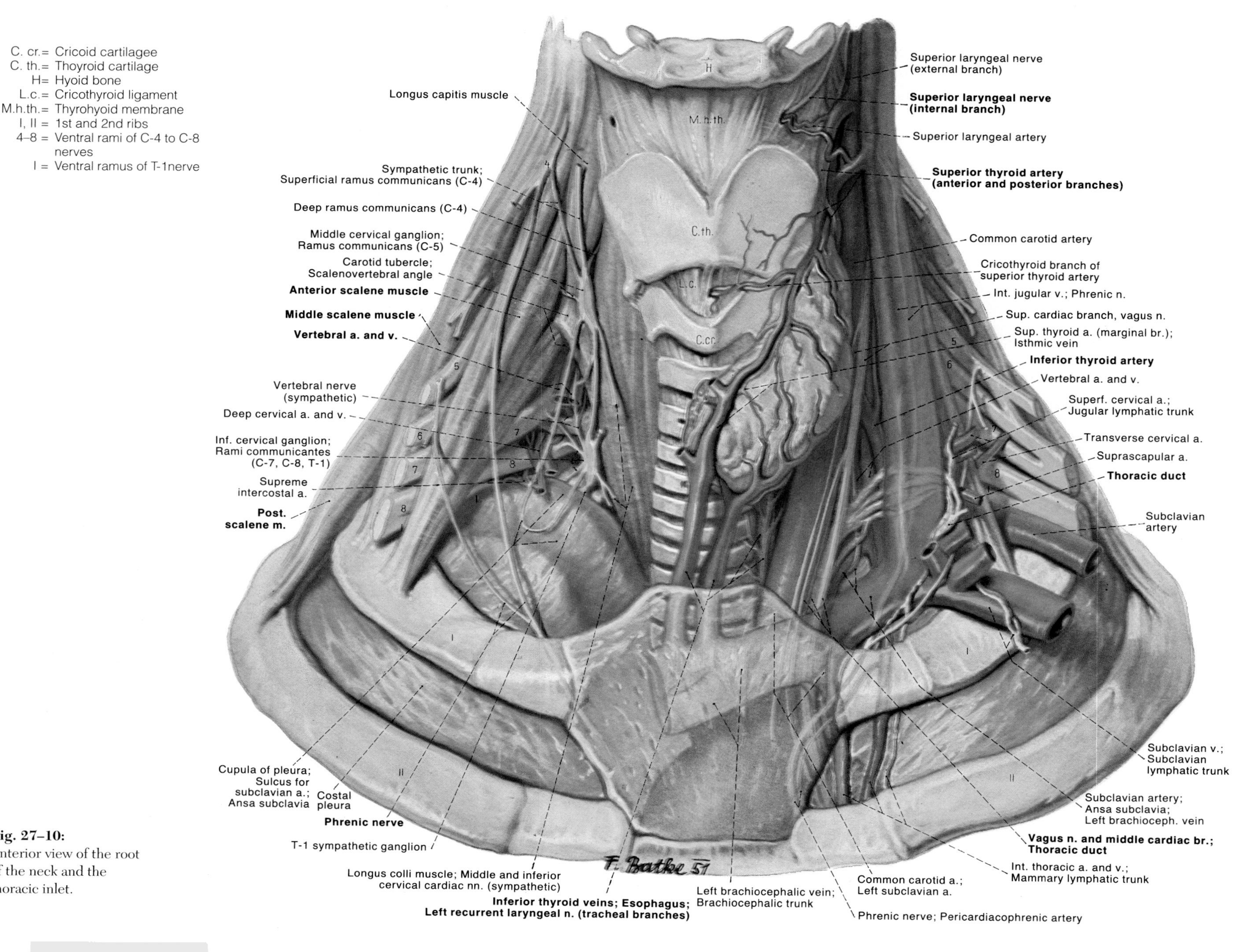

C. cr.= Cricoid cartilagee
C. th.= Thoyroid cartilage
H= Hyoid bone
L.c.= Cricothyroid ligament
M.h.th.= Thyrohyoid membrane
I, II = 1st and 2nd ribs
4–8 = Ventral rami of C-4 to C-8 nerves
I = Ventral ramus of T-1nerve

Fig. 27–10: Anterior view of the root of the neck and the thoracic inlet.

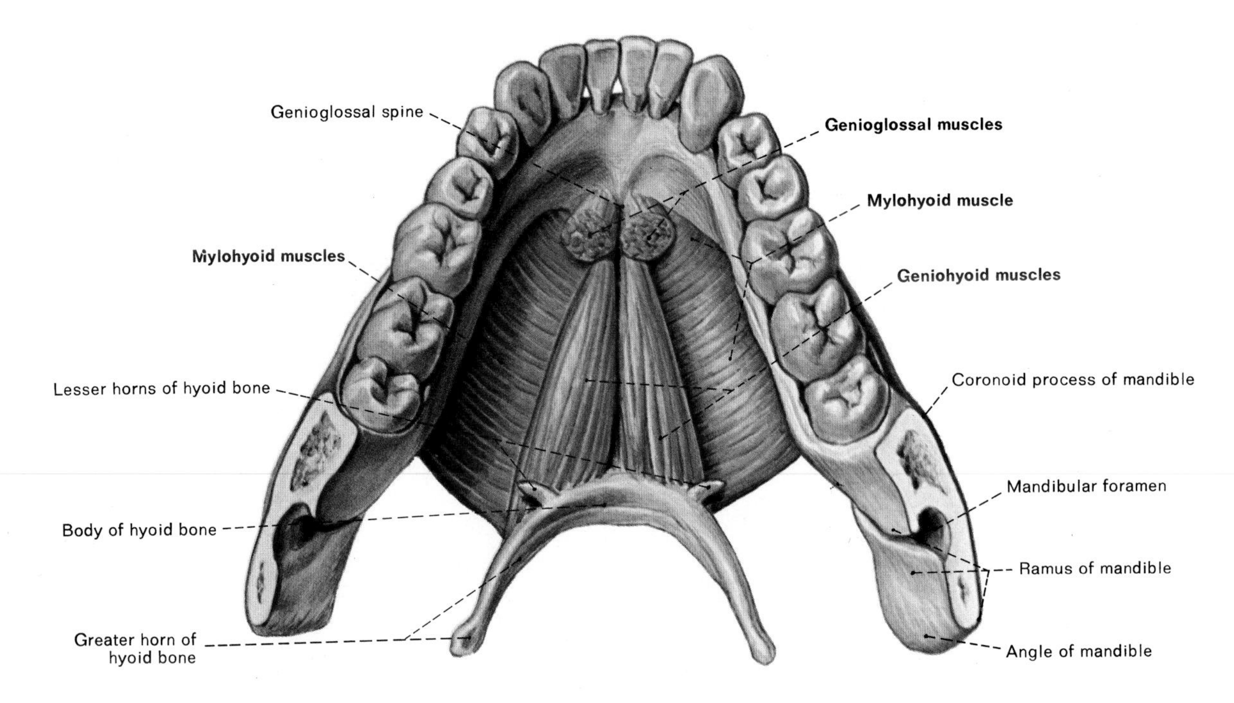

Fig. 27–11: Superior view of the mylohyoid and geniohyoid muscles.

counterparts along a median fibrous raphe (seam) which extends from the symphysis menti to the hyoid (Fig. 27–12). It is in this fashion that the paired mylohyoids form a muscular floor of the mouth.

Digastric is the suprahyoid muscle with two bellies joined by an intermediate tendon (Fig. 27–12). The posterior belly originates from the mastoid process of the temporal bone, and the anterior belly originates from the lower border of the mandible near the symphysis menti. The intermediate tendon is bound to the body of the hyoid by a loop of deep fascia. The two-bellied digastric and the overlying, lower margin of the mandible define a small area in the anterior triangle of the neck called the **submandibular triangle.** The anterior belly of digastric also borders another small area in the anterior triangle of the neck called the **submental triangle;** the submental triangle is bordered medially by the midline of the neck, laterally by the anterior belly of digastric, and inferiorly by the hyoid bone.

Stylohyoid extends from the styloid process of the temporal bone to the body of the hyoid (Fig. 27–12). Stylohyoid lies directly superior to the posterior belly of digastric.

Geniohyoid extends backward from the inner surface of the mandible near the midline to the body of the hyoid (Fig. 27–11). Geniohyoid overlies mylohyoid in the floor of the mouth.

Sternothyroid is a strap-shaped muscle extending from the manubrium of the sternum to the thyroid cartilage of the larynx (Fig. 27–12). Sternothyroid superficially covers the lateral lobe of the thyroid gland (Fig. 27–12).

Thyrohyoid extends from the thyroid cartilage of the larynx to the body of the hyoid (Fig. 27–12).

Sternohyoid is a strap-shaped muscle which extends from the clavicle and the manubrium of the sternum to the body of the hyoid (Fig. 27–12). Sternohyoid superficially covers sternothyroid and thyrohyoid.

Omohyoid is the infrahyoid muscle with two bellies joined by an intermediate tendon (Fig. 27–12). The inferior belly originates from the upper border of the scapula, and the superior belly inserts onto the body of the hyoid. The intermediate tendon is tethered down to the clavicle and first rib by a loop of deep fascia. The superior belly of the omohyoid borders a triangular area in the anterior triangle of the neck called the **carotid triangle;** this area is bordered anteriorly by the superior belly of omohyoid, posteriorly by the anterior border of sternocleidomastoid, and superiorly by the posterior belly of digastric (Figure 27–12 provides an anterior view of the borders of the right carotid triangle).

Platysma is a very thin muscle that originates from the deep fascia overlying the clavicular origin of pectoralis major (Fig. 27–13). It extends through the superficial fascia of the neck to insert onto the lower margin of the mandible and the superficial fascia and skin about the mouth. Platysma increases tension in the skin of the neck, and helps lower the corner of the mouth. Platysma is termed a muscle of facial expression because it can affect the expression of one's face. Platysma is innervated by the facial nerve. **The facial nerve is the 7th cranial nerve.**

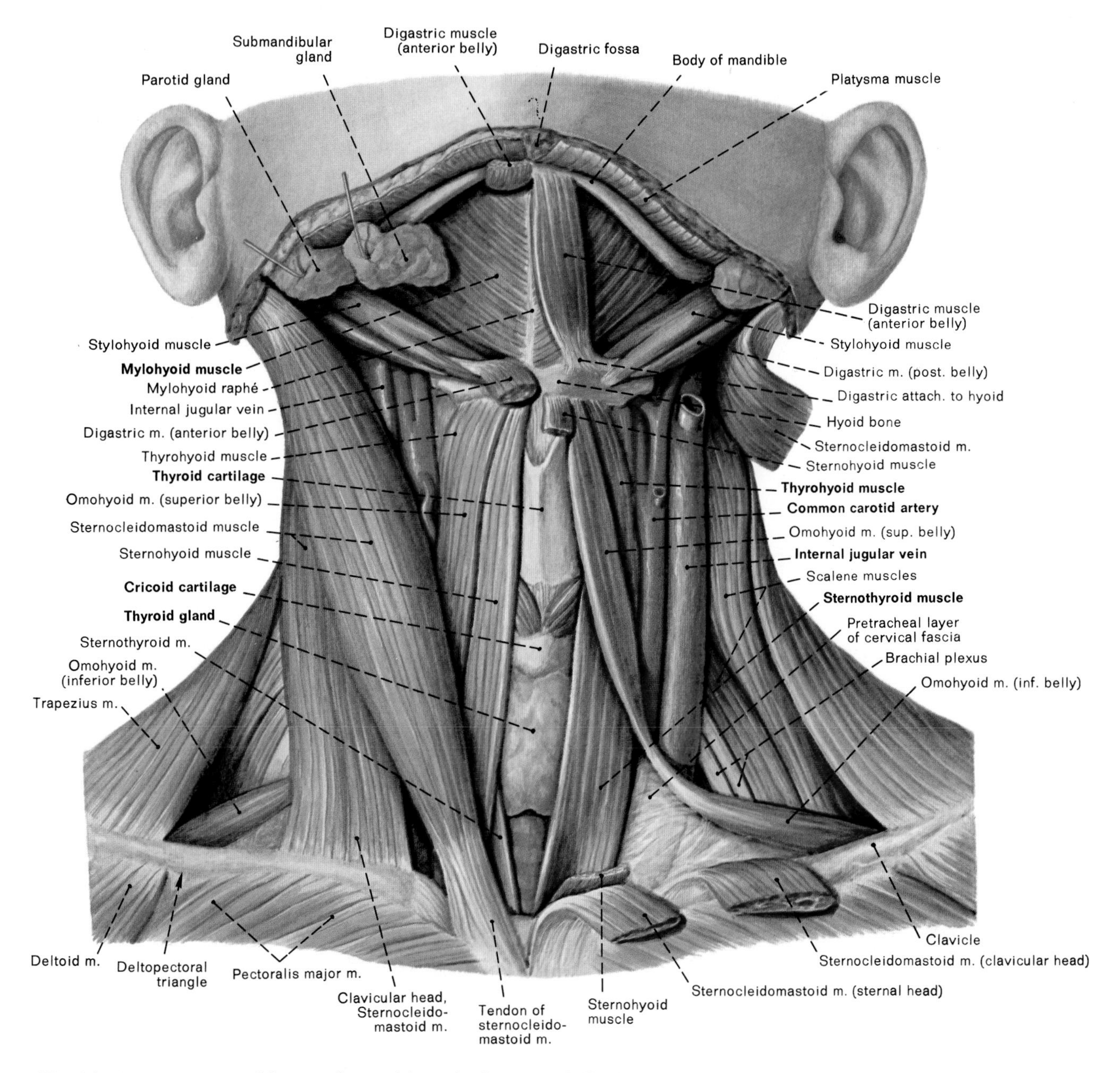

Fig. 27–12: Anterior view of the musculature of the neck. The anterior belly of the right digastric has been removed. The left sternocleidomastoid and sternohyoid muscles have been transected and the left submandibular gland removed.

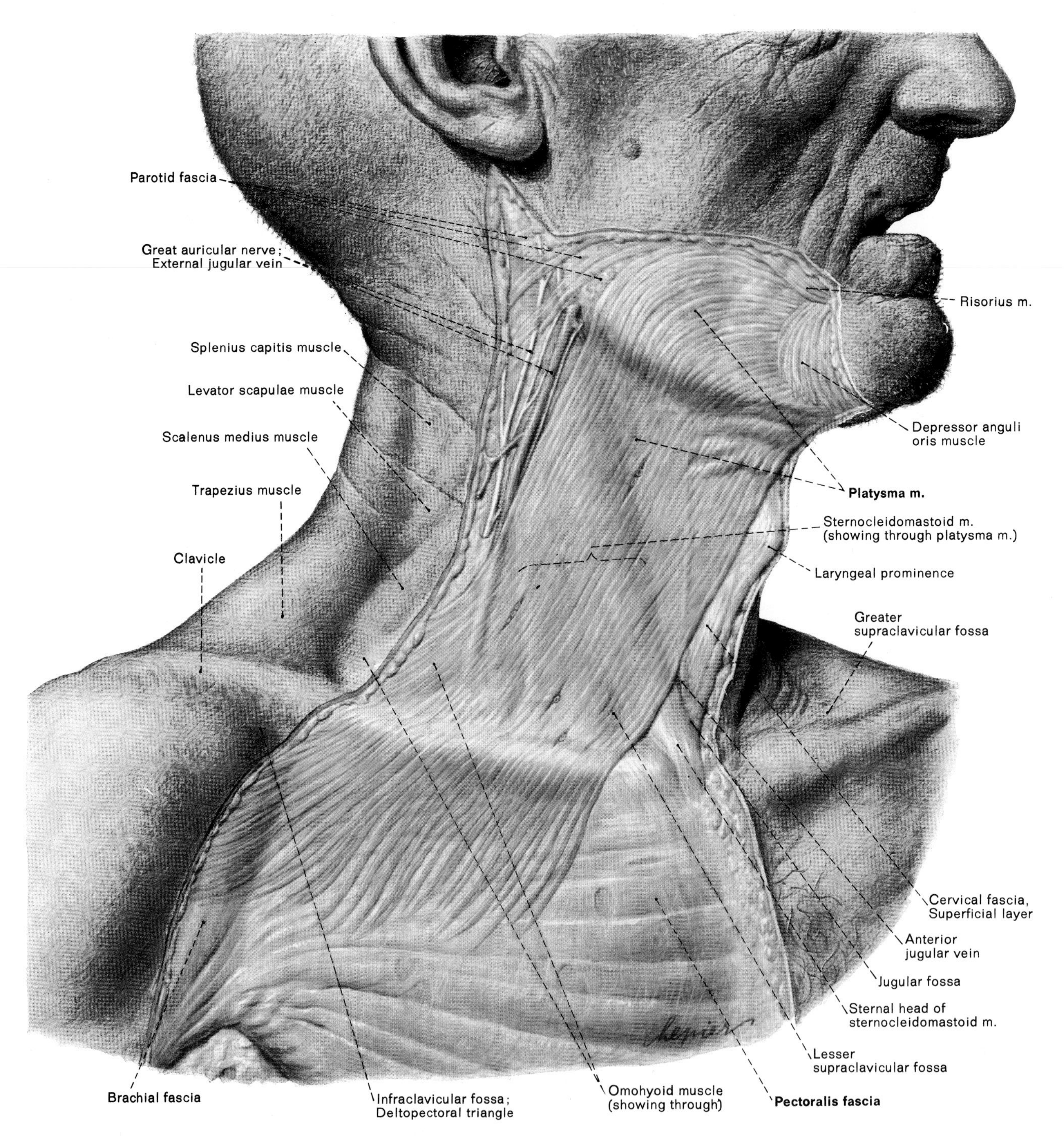

Fig. 27–13: Dissection of the right platysma muscle.

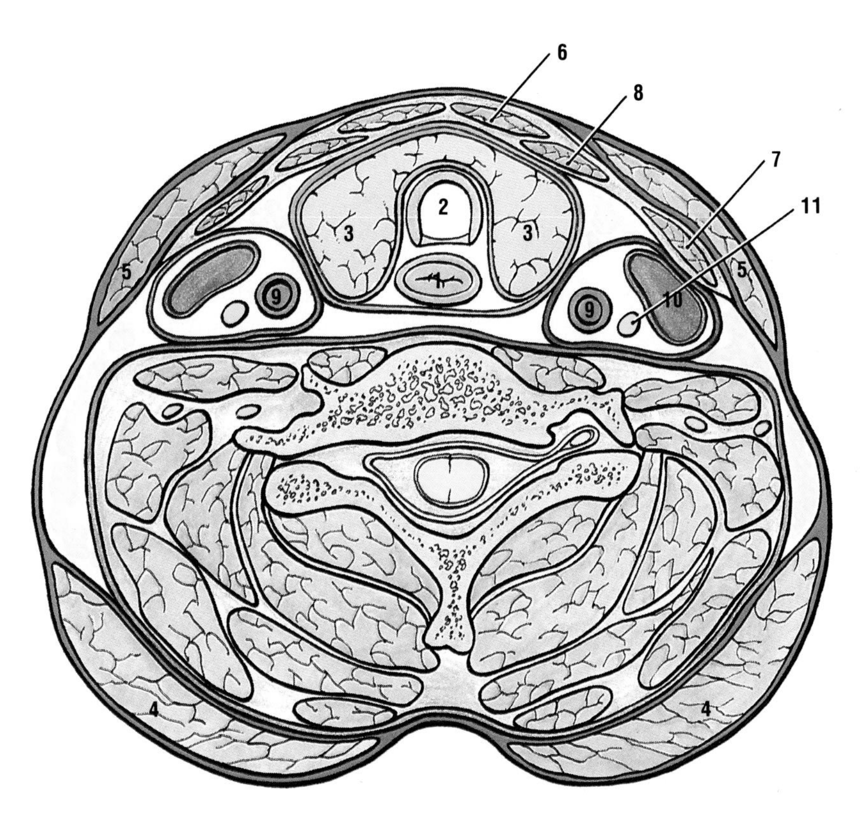

1 - **Esophagus,** *2* - **Trachea,** *3* - **Thyroid gland,** *4* - **Trapezius,** *5* - **Sternocleidomastoid,** *6* - **Sternohyoid,** *7* - **Omohyoid,** *8* - **Sternothyroid,** *9* - **Common carotid artery,** *10* - **Internal jugular vein,** *11* - **Vagus nerve.**

Fig. 27–14: Diagram of a horizontal section through the neck at the level of the body of the seventh cervical vertebra showing the investing (red), prevertebral (blue), and pretracheal (orange) layers of the deep cervical fascia and also the carotid sheath (purple).

THE DEEP CERVICAL FASCIA

There are three layers of deep cervical fascia. They generally limit the routes by which untreated infections disperse through the neck. Microbes spread across the breadth of these layers more often than penetrating them.

(1) The **investing layer** is the most superficial deep fascial layer (Fig. 27–14). It is a fascial collar which surrounds the cervical vertebrae and all the neck viscera and musculature (except the platysma). Its upper and lower margins form continuous seals of attachment to the upper and lower skeletal boundaries of the neck (from the occipital, temporal, and zygomatic bones and mandible above to the manubrium, clavicle, and scapula below). It tends to restrict abscesses immediately deep to it from spreading to the surface.

Anterolaterally, the investing layer forms the roof of the anterior and posterior triangles of the neck (Fig. 27–14). Posteriorly, it adheres to the length of the ligamentum nuchae. It divides to invest the superficial and deep surfaces of the two largest neck muscles (sternocleidomastoid and trapezius) and two salivary glands (the parotid and submandibular glands).

The spinal part of the accessory nerve descends through the neck by first extending beneath sternocleidomastoid (where it innervates the muscle from its deep surface), then coursing obliquely through the posterior triangle of the neck, and finally extending beneath trapezius to innervate it from its deep surface (Fig. 27–9). The spinal part of the accessory nerve is susceptible to injury by relatively superficial neck lacerations along its course through the posterior triangle of the neck, where it is covered only by skin, subcutaneous fascia, and the investing layer of the deep cervical fascia.

(2) The **prevertebral layer** is the deepest of the deep fascial layers (Fig. 27–14). It extends from the base of the skull down into the superior mediastinum (down to the level of the third thoracic vertebra). It

envelops the cervical vertebrae, the scaleni muscles, and the prevertebral and postvertebral muscles of the neck. In the lower part of the neck, the prevertebral layer gives rise to the **axillary sheath,** the tubular fascial sheath that surrounds the subclavian artery, subclavian vein, and the more proximal parts of the brachial plexus as they extend through the neck towards the axilla.

The prevertebral layer forms the floor of the posterior triangle of the neck. When an abscess deep to the prevertebral layer bursts through the layer, it most commonly bursts through the layer in the region on each side where it forms the floor of the posterior triangle.

A very loose connective tissue called the **alar fascia** separates the anterior, midline region of the prevertebral layer from the **buccopharyngeal fascia** which envelops the pharynx and esophagus. The alar fascia permits the up-and-down movements of the pharynx and esophagus which occur during deglutition (swallowing). The space immediately posterior to the alar fascia is called the **retropharyngeal space** (the alar fascia and the retropharyngeal space are not shown in Figure 27–14; both are located in the thin space where the posterior aspect of the pretracheal fascia faces the anterior aspect of the prevertebral fascia). Abscesses within the retropharyngeal space may either bulge anteriorly into the pharynx (or esophagus) or track inferiorly into the mediastinum.

(3) The **pretracheal layer** envelops the thyroid gland and covers the anterolateral surfaces of the larynx and trachea and the lateral surfaces of the esophagus (Fig. 27–14). The pretracheal layer extends from the hyoid bone down to the fibrous pericardium. Infections which exit the larynx or the cervical parts of the trachea or esophagus are bordered anteriorly by the pretracheal layer and posteriorly by the prevertebral layer, and can thus track inferiorly into the mediastinum.

The **carotid sheath** is a tubular sheath of deep cervical fascia that envelops the **common and internal carotid arteries,** the **internal jugular vein,** the **vagus nerve** (**the vagus nerve is the tenth cranial nerve**) and some of the **deep cervical lymph nodes** (Fig. 27–14). The cervical parts of the sympathetic trunks lie embedded in loose connective tissue directly posterior to the carotid sheaths.

THE CERVICAL PLEXUS

The cervical plexus arises from the anterior rami of C1–C4. The major branches of the cervical plexus, their segmental derivation, and distribution are as follows:

The C1 Fibers Which Innervate Geniohyoid and Thyrohyoid

C1 supplies motor fibers to one suprahyoid and all four infrahyoid muscles. All these fibers extend along the hypoglossal nerve **(the hypoglossal nerve is the twelfth cranial nerve)** for part of their course through the neck. The motor fibers which innnervate geniohyoid and thyrohyoid are the last C1 fibers to leave the hypoglossal nerve (the nerve labelled the "nerve to thyrohyoid m." in Figure 27–15 represents the bundle of C1 motor fibers that innervate thyrohyoid; note that at the gross level of resolution, the nerve to the thyrohyoid muscle appears to be a branch of the hypoglossal nerve).

The C1, C2, and C3 Fibers Which Innervate Sternohyoid, Omohyoid, and Sternothyroid

The first C1 motor fibers to leave the hypoglossal nerve are those which, in combination with C2 and C3 fibers, innervate the three infrahyoid muscles **S**ternohyoid, **O**mohyoid, and **S**ternothyroid (which may be mnemonically remembered as the **SOS** muscles of the neck). The C1 fibers for the SOS muscles leave the hypoglossal nerve and initially descend in the neck as a distinct, slender bundle of fibers called the **superior root of the ansa cervicalis** (Fig. 27–15). The C2 and C3 fibers for the SOS muscles descend for a part of their course in the neck as a separate slender bundle of fibers called the **inferior root of the ansa cervicalis** (Fig. 27–15). In the lower part of the neck, the two bundles meet to form a **U**-shaped nerve loop called the **ansa cervicalis** (the term ansa, which is derived from a Latin word meaning handle, is synonymous in anatomic names with loop) (Fig. 27–15). Because, at the gross level, the bundle of C1 fibers appears to be a branch of the hypoglossal nerve (Fig. 27–15), the superior root of the ansa cervicalis previously was referred to as the **descendens hypoglossi.** The ansa cervicalis lies either embedded within or immediately superficial to the carotid sheath.

The Phrenic Nerve to the Diaphragm

The phrenic nerve is derived from C3, C4, and C5 fibers; C4 provides most of the fibers. The phrenic nerve lies directly anterior to scalenus anterior as the nerve descends through the lower part of the neck (Fig. 27–15). As the phrenic nerve descends through the lowest part of the neck and the superior mediastinum, it lies posterior or posterolateral to the internal jugular vein (Figs. 27–10 and 27–15) before passing posterior to the subclavian vein at or near its union with the internal jugular vein to form the brachiocephalic vein (Fig. 27–10). The close proximity of the

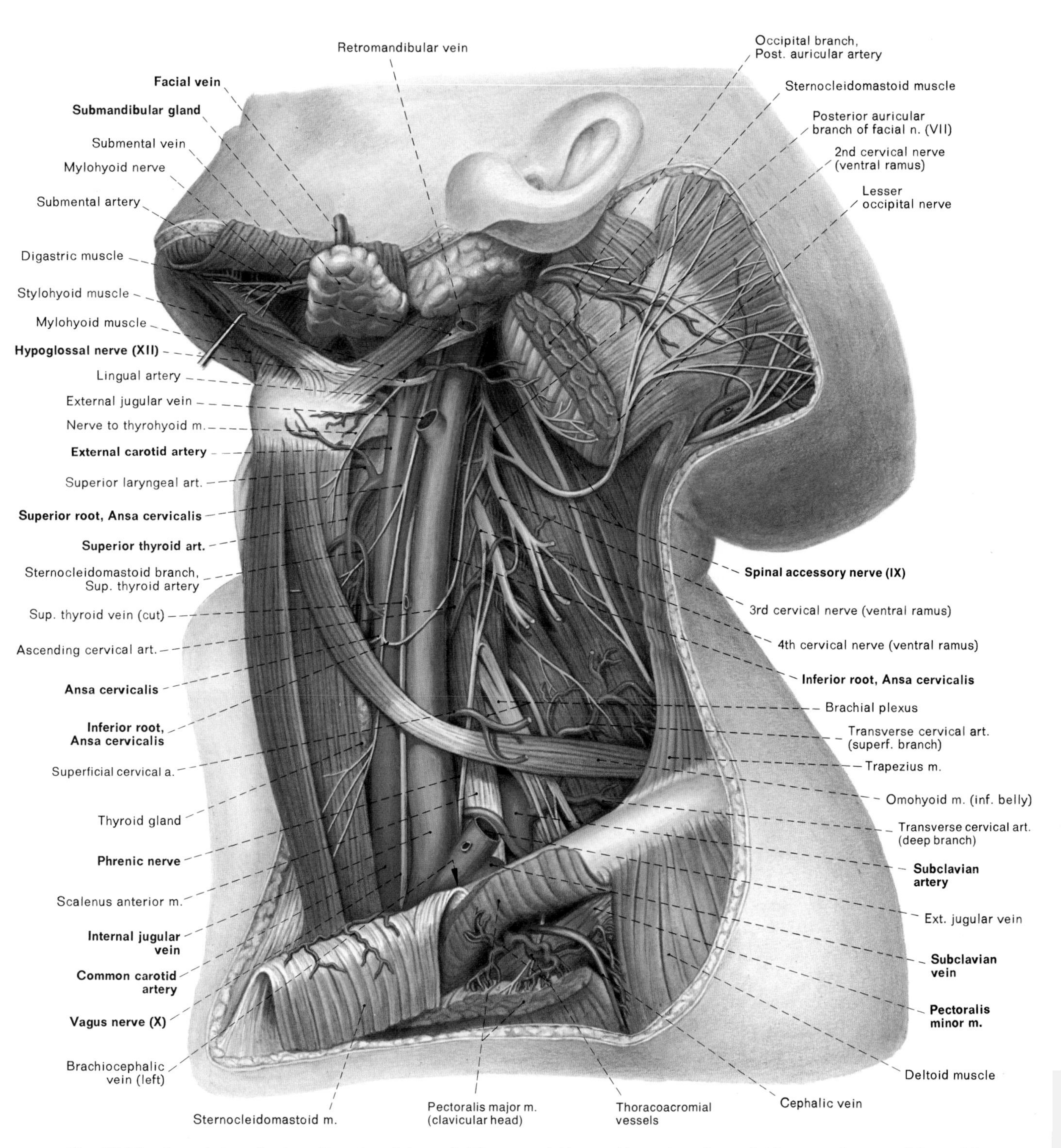

Fig. 27–15: Lateral view of a deep dissection of the neck. The sternocleidomastoid muscle and superficial veins and nerves of the neck have been been removed, the facial vein transected, and the superficial part of the submandibular gland reflected upward.

phrenic nerve to these veins renders the nerve susceptible to injury during catheterization of these veins near the thoracic inlet.

The Sensory Cutaneous Nerves of the Cervical Plexus

The cervical plexus gives rise to four sensory cutaneous nerves. All four nerves emerge in the posterior triangle of the neck at the midway point of the posterior border of sternocleidomastoid, and then extend from this point to their respective cutaneous regions of distribution (Fig. 27–9). The lesser occipital nerve (C2) provides sensory innervation in particular to the strip of the scalp immediately posterior to the outer ear. The great auricular nerve (C2 & C3) provides sensory innervation in particular to the skin overlying the angle of the mandible, the parotid gland, and the mastoid process of the temporal bone. The transverse cervical nerve (C2 & C3) provides sensory innervation in particular to almost all the skin overlying the anterior triangle of the neck. The supraclavicular nerves (C3 & C4) provide sensory innervation in particular to the skin overlying the top of the shoulder.

THE CERVICAL PART OF THE SYMPATHETIC TRUNK

The parts of the sympathetic trunks that extend through the neck lie anterolaterally to the cervical vertebrae. There are generally three cervical sympathetic ganglia, called the **superior, middle, and inferior cervical ganglia.** The superior cervical ganglion, which lies at the level of the atlas and axis, is generally the largest cervical ganglion. The inferior cervical ganglion (Fig. 27–10), which lies at the level of the seventh cervical vertebra, is frequently fused with the first thoracic ganglion to form a roughly star-shaped ganglion called a **stellate ganglion.**

On each side of the neck, there are at least two trunks of fibers connecting the lower two cervical ganglia. The more anterior trunk descends from the middle cervical ganglion and passes in front of the first part of the subclavian artery before looping under the artery and then ascending back up to the inferior cervical or stellate ganglion. This nerve loop is called the **ansa subclavia** (removal of the right subclavian artery in Figure 27–10 provides an unhindered view of the right ansa subclavia).

Because the cervical spinal cord segments do not contribute preganglionic fibers to the sympathetic trunks, no white rami communicantes extend from the cervical spinal nerves to the cervical ganglia. The T1–T5 spinal cord segments provide the preganglionic fibers ascending within the cervical parts of the sympathetic trunks. These preganglionic fibers synapse within one of the cervical ganglia.

Each cervical spinal nerve receives a gray ramus communicans (conveying postganglionic fibers) from one of the cervical ganglia. C1–C4 each receive a gray ramus communicans from the superior ganglion, C5 and C6 each receive a gray ramus communicans from the middle ganglion, and C7 and C8 each receive a gray ramus communicans from the inferior or stellate ganglion.

Many postganglionic fibers exit the superior ganglion to extend along the internal and external carotid arteries and their branches. These postganglionic fibers from the superior cervical ganglion supply mainly the smooth muscle tissue in the walls of the branches of the internal and external carotid arteries. Finally, there are also some postganglionic fibers which exit medially from all three cervical ganglia on each side to descend into the superior mediastinum and contribute to the cardiac plexus; stimulation of these sympathetic fibers increases the rate of the heartbeat.

Transection in the neck of one of the sympathetic trunks or surgical removal of one or more of the cervical ganglia on one side produces a syndrome of conditions called Horner's syndrome. **Horner's syndrome is characterized by ptosis (drooping of the upper eyelid), meiosis (a constricted pupil), and anhydrosis (absence of sweating on the affected side of the face).** Ptosis results because both smooth and skeletal muscle fibers are responsible for elevating the upper eyelid, and the smooth muscle fibers are sympathetically innervated. Meiosis results because whereas the muscle which dilates the pupil (dilator pupillae) is sympathetically innervated, the muscle which constricts the pupil (sphincter pupillae) is parasympathetically innervated. Anhydrosis results because the sweat glands in the skin of the face are sympathetically innervated.

THE ARTERIES OF THE NECK

The Subclavian Artery and its Branches

The right subclavian artery begins posterior to the right sternoclavicular joint as one of the two terminal branches of the brachiocephalic artery. The left subclavian artery begins in the superior mediastinum as the third branch of the aortic arch (Fig. 18–5); the artery enters the root of the neck by passing posterior to the left sternoclavicular joint.

Each subclavian artery courses through the root of the neck by arching over the apex of the lung, then passing behind scalenus anterior, and finally extending over the first rib (Fig. 27–10). The subclavian artery becomes the axillary artery at the lateral border of the first rib. The relationship of the subclavian artery to scalenus anterior divides the artery into three parts: the first part is medial to the muscle, the second part is

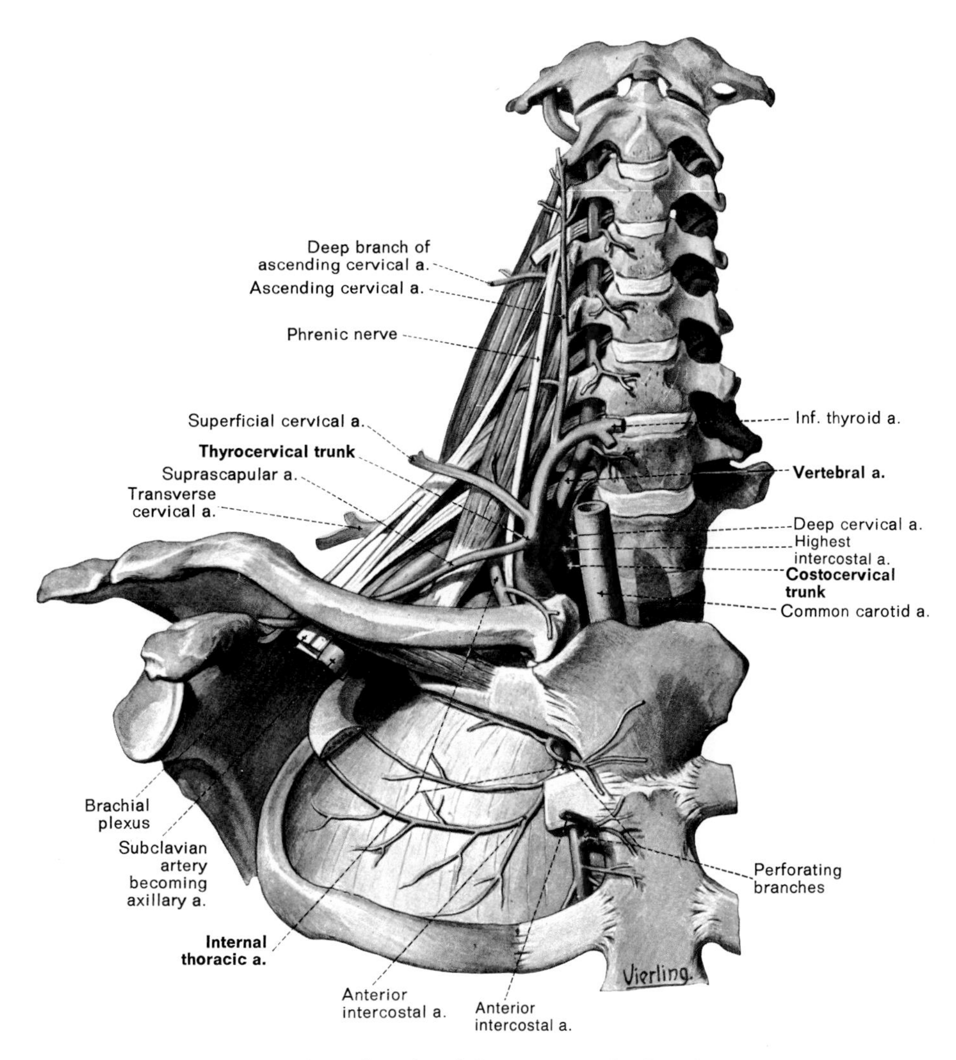

Fig. 27–16: The right subclavian artery and its branches.

posterior to the muscle, and the third part is lateral to the muscle.

The **internal thoracic (mammary) artery** originates from the first part of the subclavian artery (Fig. 27–16). The artery descends from its origin by extending alongside the deep surface of the chest wall (just lateral to the sternum). On each side, the artery gives rise to the anterior intercostal arteries for the upper five intercostal spaces before terminating at its birfurcation into the musculophrenic and superior epigastric arteries.

The **vertebral artery** is the largest branch of the first part of the subclavian artery. It ascends from its origin by passing through the transverse foramina of the upper six cervical vertebrae (Fig. 27–3). Upon passing through the transverse foramen of the atlas, the artery turns first posteriorly and then medially to pierce the dura mater and arachnoid mater of the vertebral canal. Upon entering the subarachnoid space of the vertebral canal, the artery turns upward and enters the cranial cavity by passing through the foramen magnum.

In the neck, the vertebral artery supplies deep muscles and the spinal cord and its meninges. In the cranial cavity, the artery gives rise to branches which supply the brainstem, cerebellum, and the temporal and occipital lobes of the cerebrum.

The **thyrocervical trunk** is a relatively short branch of the first part of the subclavian artery (Fig. 27–16). It gives rise to three arteries: **the inferior thyroid, suprascapular,** and **transverse cervical arteries.** The paired inferior thyroid arteries supply the thyroid gland and its closely associated parathyroid glands. The suprascapular and transverse cervical arteries supply the shoulder region, and their branches unite with branches of the subscapular artery to form an anastomotic network which encircles the scapula and provides collateral circulation across the shoulder joint (Fig. 8–1).

The **costocervical trunk** is the only branch of the second part of the subclavian artery (Fig. 27–16). It is a relatively short arterial trunk which bifurcates into the **superior intercostal and deep cervical arteries.** The superior intercostal artery gives rise to the two uppermost posterior intercostal arteries, and the deep cervical artery supplies deep structures in the neck.

The Common Carotid Artery and its Branches

The right common carotid artery begins posterior to the right sternoclavicular joint as one of the two terminal branches of the brachiocephalic artery. The left common carotid artery begins in the superior mediastinum as the second branch of the aortic arch (Fig. 18–5); the artery enters the root of the neck by passing posterior to the left sternoclavicular joint.

Both arteries extend superiorly from the root of the neck to the upper border of the thyroid cartilage of the larynx. Here each artery bifurcates into its terminal branches: **the internal** and **external carotid arteries.** These terminal branches are the only branches of the common carotid artery (Fig. 27–3). The pulse of the common carotid artery can be readily palpated in the carotid triangle of the neck against the side of the thyroid cartilage, immediately deep to the anterior border of sternocleidomastoid.

The origin of the internal carotid artery commonly exhibits a dilatation called the **carotid sinus** (the carotid sinus may also be located at the termination of the common carotid artery). Each carotid sinus bears a high concentration of **baroreceptors** (pressure receptors). Each carotid sinus is innervated primarily by the glossopharyngeal nerve and to a lesser extent by the vagus nerve and the sympathetic division of the autonomic nervous system. **The glossopharyngeal nerve is the ninth cranial nerve.** When the baroreceptors of the carotid sinus are subjected to a sudden change in blood pressure, their response initiates an autonomic reflex which restores the blood pressure to normal levels; the restoration occurs by regulation of arteriolar constriction and the rate of the heartbeat.

Physical massage of the carotid sinuses is occasionally used in the differential diagnosis and treatment of tachycardia (an elevated heartbeat rate). Massage of one of the carotid sinuses for 5 to 10 sec. will produce increases in blood pressure sufficient to stimulate the sinus's baroreceptors. This stimulation initiates a reflex vagal stimulation of the cardiac plexus, and some tachycardias respond to such vagal stimulation.

However, it must be noted that carotid massage in very elderly patients may cause cardiac arrest. Moreover, carotid massage should not be attempted in patients with occlusive carotid arterial disease, as the massage may loosen an atheromatous plaque and produce a cerebral embolism.

On each side of the neck, a small mass of tissue called the **carotid body** lies near the origins of the internal and external carotid arteries. Each carotid body is supplied by small nutrient branches from the external carotid artery; these nutrient arteries provide an extremely high blood flow through the carotid body (2 liters/100 g of tissue per minute; by comparison, blood flow through the heart and kidney are respectively only 54 and 420 ml per 100 g tissue per minute). Each carotid body bears a high concentration of **chemoreceptors** (similar chemoreceptors are also found in analogous bodies, called **aortic bodies,** located near the aortic arch). The chemoreceptors of the carotid body are innervated primarily by the glossopharyngeal nerve and to a lesser extent by the vagus nerve and the sympathetic division of the autonomic nervous system (the chemoreceptors of the aortic bodies are innervated primarily by the vagus nerves). When the chemoreceptors in the carotid and aortic bodies are exposed to an oxygen pressure decrease and/or a carbon dioxide pressure increase in the blood, their response initiates an autonomic reflex that restores these blood gases to normal levels; the restoration occurs through increases in the rates of respiration and the heartbeat.

Each internal carotid artery ascends from its origin to the base of the skull, where it enters the cranial cavity by the carotid canal (Fig. 27–3). The internal carotid artery does not give rise to any branches in the neck, and is enveloped along its entire length by the carotid sheath. The paired internal carotid arteries supply most of the cerebrum and the tissues of the forehead, orbital cavities, and nose.

Each external carotid artery ascends from its origin by passing deep to the posterior belly of digastric (Fig. 27–15). As it then ascends posterior to the ramus of the mandible (Fig. 27–17), it passes either through or medial to the parotid gland. The artery ends at the level of the neck of the mandible by bifurcating into the superficial temporal and maxillary arteries (Fig. 27–17).

The external carotid arteries are the main supply for the tissues of the face and neck. Each artery gives rise to six major branches prior to its terminal bifurcation: three major branches emerge from the artery's anterior surface and the other three major branches from the artery's posterior surface.

The **superior thyroid artery** is the first branch that emerges from the anterior surface of the external carotid artery (Figs. 27–15 and 27–17). The artery supplies the thyroid gland (Fig. 27–10) and its closely associated parathyroid glands. The left and right superior thyroid arteries are joined by anastomoses that can provide effective collateral circulation across the midline.

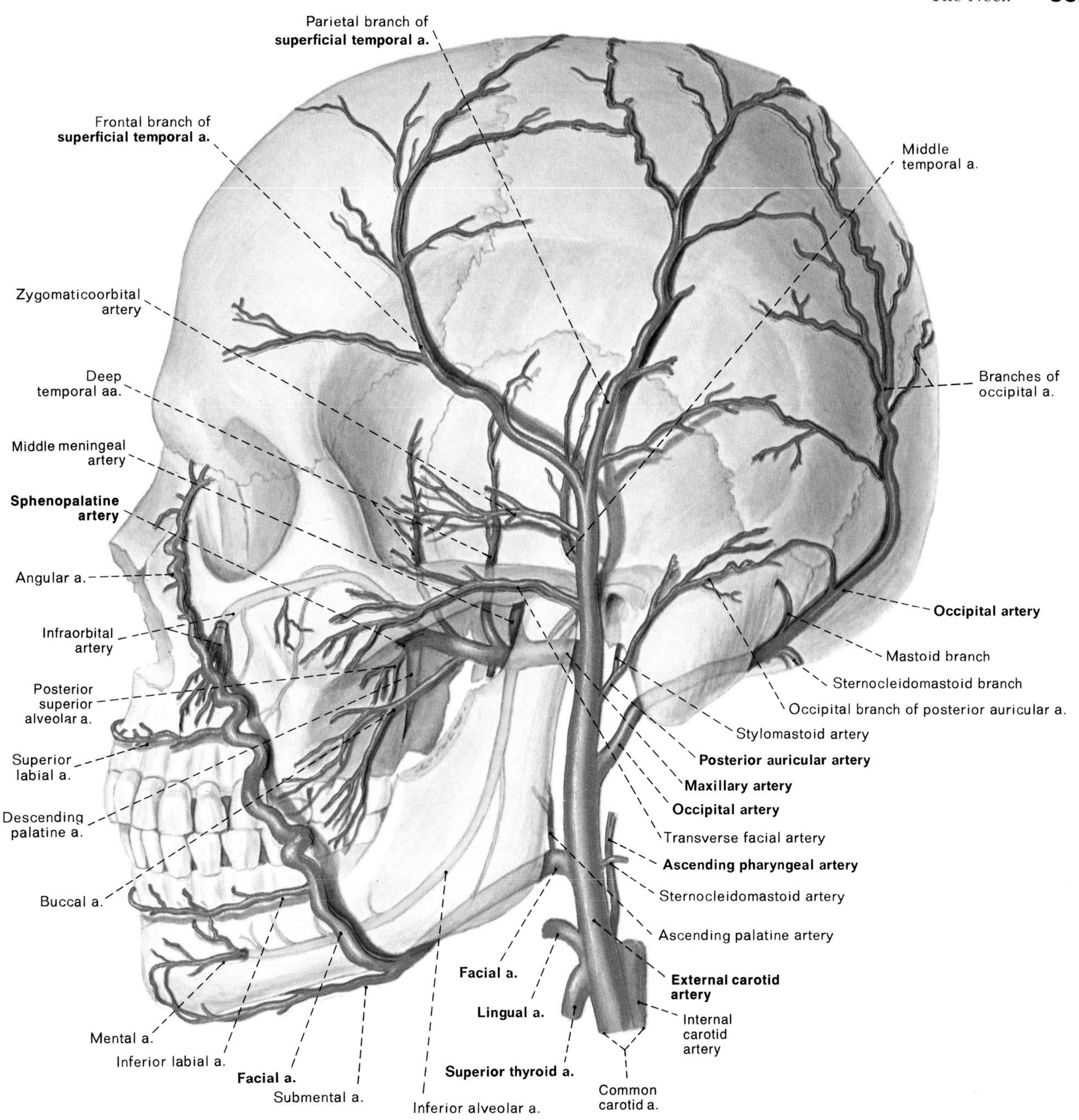

Fig. 27–17: The external carotid artery and its branches in the skull.

The **lingual artery** is the second highest branch which emerges from the anterior surface of the external carotid artery (Figs. 27–15 and 27–17). It supplies the tongue and the floor of the mouth.

The **facial artery** is commonly the most superior of the branches that extend from the anterior surface of the external carotid artery (Fig. 27–17). The pulse of the facial artery is palpable at the point where the artery crosses the lower margin of the body of the mandible (Fig. 27–17). It supplies the muscles of the face, submandibular gland, palatine tonsil, and soft palate.

The ascending pharyngeal, occipital, and posterior auricular arteries are the three major branches which emerge from the posterior surface of the external carotid artery (Fig. 27–17). The pharyngeal artery supplies the muscles of the pharynx, the occipital artery supplies the back of the scalp, and the posterior auricular artery supplies the auricle and the back of the scalp. The left and right occipital arteries are joined by anastomoses that can provide effective collateral circulation across the midline.

The **superficial temporal artery** is a terminal branch of the external carotid artery. The pulse of the superficial temporal artery is palpable in front of the ear at the point where the artery ascends superficially over the zygomatic process of the temporal bone (Fig. 27–17). The artery supplies the parotid gland, temporomandibular joint, auricle, and temporal region of the scalp.

The **maxillary artery** is a terminal branch of the external carotid artery. It extends anteriorly through the infratemporal fossa, giving rise to branches which supply deep tissues in the upper part of the face. Its major branches and their distribution are as follows: the middle meningeal artery supplies the cranial dura mater, the inferior alveolar artery supplies the lower jaw and its teeth, the deep temporal branches supply temporalis, the masseteric artery supplies the masseter, the posterior superior alveolar artery supplies the upper jaw and its teeth, the greater palatine artery supplies the soft palate and the palatine tonsil, and the sphenopalatine artery supplies the frontal, maxillary, ethmoid, and sphenoid sinuses (Figure 27–17 displays all these branches except the masseteric and greater palatine arteries).

THE VEINS OF THE NECK

The internal and external jugular veins are the major veins of the neck.

The Internal Jugular Vein

The internal jugular veins conduct toward the heart much of the blood drained from the brain. The veins also receive tributaries in the neck which collect blood from the face and neck. On each side, the largest tributary is the major vein of the face, the **facial vein** (Fig. 27–18).

The internal jugular veins begin in the jugular foramina of the base of the skull as the direct continuations of the **sigmoid sinuses** (Fig. 27–18). The sigmoid sinuses are the largest of the intracranial venous channels that collect blood drained from the brain.

The carotid sheath envelops the internal jugular vein along its entire descent through the neck (Fig. 27–14). From the base of the skull down to the upper border of the thyroid cartilage, the internal jugular vein lies immediately lateral to the internal carotid artery. From the upper border of the thyroid cartilage down to the root of the neck, the vein lies immediately lateral to the common carotid artery (Fig. 27–14). The vein ends immediately above the thoracic inlet by uniting with the subclavian vein to form the brachiocephalic vein (Fig. 27–18).

The External Jugular Vein and its Tributaries

The external jugular veins conduct back toward the heart blood drained from superficial tissues of the head and neck and deep tissues of the upper part of the face. On each side, the external jugular vein begins behind the angle of the mandible at the point where a posterior division of the retromandibular vein joins the posterior auricular vein (Fig. 27–9). The retromandibular vein is, as its name indicates, a vein which lies behind the ramus of the mandible (Fig. 27–18); it begins as the union of the superficial temporal and maxillary veins. The posterior auricular vein is a superficial vein which descends on the side of the head behind the ear.

The external jugular vein passes obliquely over sternocleidomastoid as the vein descends through the superficial fascia of the neck (Fig. 27–9). Upon entering the lower part of the posterior triangle of the neck, the external jugular vein commonly receives the following tributaries: the posterior external jugular vein (a superficial vein descending from the back of the head), the venae comitantes of the suprascapular and transverse cervical arteries, and the anterior jugular vein (all these tributaries except the posterior external jugular vein are displayed in Figure 27–18). The anterior jugular veins are paired superficial veins which begin in the submental triangles of the neck and descend to the suprasternal region; in the suprasternal region, the paired veins are joined together by a vein called the jugular venous arch (Fig. 27–9).

The external jugular vein ends by first passing through the investing layer of deep cervical fascia in the lower part of the posterior triangle of the neck and then extending deeper to join the subclavian vein (Figure 27–18 displays the end of the external jugular vein but not its passage through the investing layer of deep cervical fascia). It is very important to note that the investing layer of deep cervical fascia is attached to the margin of the vein at the site where the vein pierces the fascia. This fascial attachment tends to always keep the vein open. Consequently, if the external jugular vein is cut anywhere above its attachment to the investing layer of deep cervical fascia, there is the risk that a pulmonary air embolism will occur as a result of air's being sucked into the vein during inspiration.

The internal and external jugular veins, subclavian veins, brachiocephalic veins, and superior vena cava are the major venous trunks that extend inferiorly through the neck and upper thorax to conduct blood into the right atrium of the heart. When a healthy adult is standing or seated upright, blood normally fills these venous trunks up to a level about 2 to 3 cm above the sternal angle. Under these conditions, blood fills the entirety of the superior vena cava, brachiocephalic veins, and subclavian veins but only the lower parts of the jugular veins. The height to which the jugular veins are filled with blood is proportional to right atrial pressure.

Accordingly, the jugular veins can be used as manometers (pressure meters) of right atrial pressure.

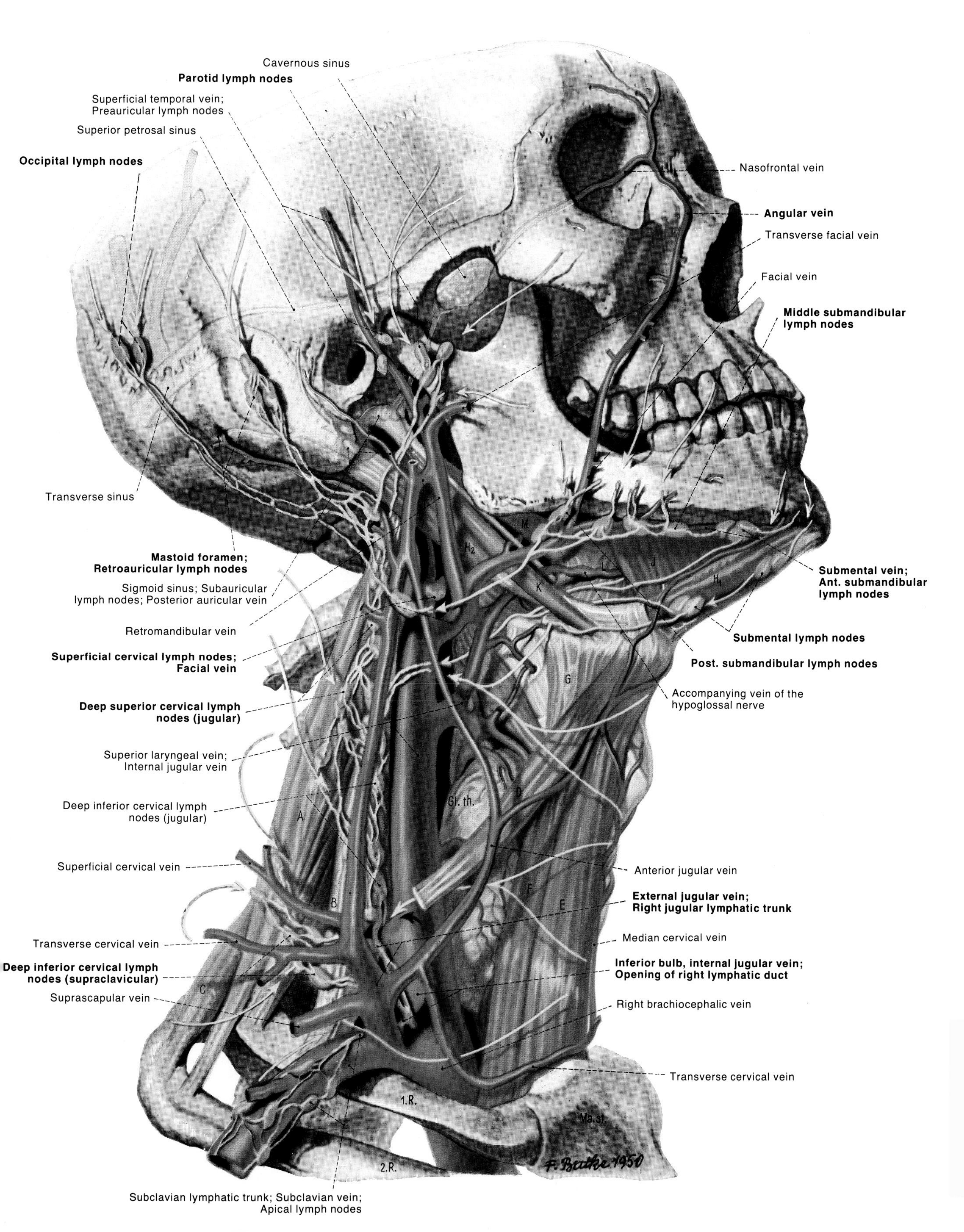

Fig. 27–18: Lateral view of the major veins, superficial lymph nodes, and lymphatic vessels of the lower head, face and neck.

Right atrial pressure is frequently called **central venous pressure,** because the blood pressure of the right atrium approximates that of the large systemic veins converging upon the right atrium. There are, however, three diagnostic points to remember when using jugular veins as manometers of central venous pressure.

(1) The right internal jugular vein is the most appropriate jugular vein to select for the measurement of central venous pressure, because it is the jugular vein which is most closely co-aligned with the superior vena cava. The angular union of the left internal jugular vein with the left brachiocephalic vein and the angular unions of the external jugular veins with the subclavian veins make the blood heights in these jugular veins less reliable monitors of central venous pressure.

(2) During each heartbeat, there are three transient increases in right atrial pressure. The first pulsatile increase (which is called the **a** wave) occurs when contraction of the atrial musculature increases atrial pressure. The second increase (which is called the **c** wave) occurs during the early moments of systole, when the increasing pressure in the contracting right ventricle bulges the tricuspid valve into the right atrium. The third pulsatile increase (which is called the **v** wave) occurs as blood flowing into the right atrium begins to bulge the heart chamber during the late moments of ventricular contraction.

The pulse-like increases in right atrial pressure which occur during each heartbeat are transmitted in a retrograde (backward) fashion through the blood in the internal and external jugular veins. When the jugular venous pulses reach the meniscus (the curved, upper surface) of the blood in each jugular vein, they produce fluctuations in the level of the meniscus, and these meniscal fluctuations, in turn, produce up-and-down movements of the overlying skin. These skin movements over each jugular vein are generally the best indicator of the height to which blood fills the vein. The skin movements are optimally observed by viewing the silhouette of the skin overlying the vein. The silhouette of the skin overlying the right internal jugular vein is best viewed from the patient's left side.

(3) In patients with thin necks, the pulsations of the carotid arteries may also produce skin movements. Carotid arterial pulses differ from jugular venous pulses in their compressiblity and palpability. Jugular venous pulses are compressible (pressure applied to the supraclavicular region of the neck will generally abolish jugular pulses) and are generally not palpable. Carotid arterial pulses, on the other hand, are palpable but not compressible. An alternative method for discriminating between jugular and carotid pulses is to note the inward-versus-outward direction of the largest fastest movements of the pulses. Whereas the largest fastest movement of jugular venous pulses is inward, the largest fastest movement of carotid arterial pulses is outward.

Inspection of jugular venous pulses during a physical exam should begin with the patient in a supine position. Some distention of the external jugular veins should be observed in all patients when lying down. This is simply because the jugular veins and the right atrium are all at about the same level when a person is lying down, and engorgement of external jugular veins generally is easily observable. Failure to observe distended external jugular veins suggests a markedly decreased central venous pressure, due to circulatory shock, dehydration, or hemorrhagic blood loss.

Assuming that the external jugular veins are either partly or completely distended with the patient in a supine position, the patient's head, neck, and trunk are next elevated above the horizontal plane sufficiently to lower the height of the jugular pulses to a level which is below the angle of the mandible but above the clavicle (a 30 to 45° elevation is generally sufficient). In a healthy adult, pulsatile activity will be visible near the lower ends of the jugular veins, commonly up to a vertical distance of 2 to 3 cm above the level of the sternal angle. This is because the average right atrial pressure in a healthy person is about 7 to 8 cm water (the center of the right atrium is about 5 cm below the level of the sternal angle in an average adult).

The most common cause of markedly higher jugular pulses is right ventricular failure. The expression "right ventricular failure" refers to any condition in which the right ventricle fails to eject during systole all of the blood it receives during diastole; the resulting damming of blood in the right atrium raises central venous pressure and the heights to which blood fills the jugular veins.

THE DEEP CERVICAL LYMPH NODES

Almost all the lymph drained from the superficial and deep tissues of the head and neck ultimately passes through the neck's deep nodes. On each side, these nodes lie strung alongside the length of the internal jugular vein (Fig. 27–18). The nodes lie either embedded within the carotid sheath or external to it. The efferent lymphatics from the deep cervical nodes anastomose on each side to form a jugular trunk. The right jugular trunk generally empties into the right brachiocephalic vein, near the vein's origin. The right jugular trunk may also join the right subclavian and bronchomediastinal trunks to form a right lymphatic duct, which, in turn, empties directly into the right brachiocephalic vein. The left jugular trunk empties into either the thoracic duct or the left brachiocephalic vein.

Physicians commonly refer to the deep cervical nodes clustered at the levels of the intermediate tendons of the digastric and omohyoid muscles as respectively the jugulo-digastric and jugulo-omohyoid groups. The **jugulo-digastric group** is the first cluster of deep cervical nodes to receive lymph drained from the posterior third of the tongue. The **jugulo-omohyoid group** is the first cluster of deep cervical nodes to receive lymph drained from the anterior two-thirds of the tongue.

THE THYROID GLAND AND THE PARATHYROID GLANDS

The thyroid gland consists of a pair of conical lobes joined across the midline by a band of glandular tissue called the **isthmus** (Fig. 27–19). Each lobe lies superficial and anterolateral to the larynx and the uppermost part of the trachea; the upper pole of each lobe extends up to the oblique line of the thyroid cartilage, and the lower pole extends down to the fourth or fifth tracheal ring. The isthmus lies in front of the trachea, generally at the levels of the second, third, and fourth tracheal rings.

In about 40% of the population, an accessory lobe of glandular tissue, called the **pyramidal lobe,** extends superiorly from the isthmus towards the hyoid. The pyramidal lobe is attached to the hyoid with a band of fibrous or muscular tissue. If the band is muscular, it is called the levator glandulae thyroidae.

The superior and inferior thyroid arteries supply the thyroid gland (Fig. 27–20). In some individuals, a thyroid artery arises from the aortic arch or brachiocephalic trunk; this thyroid artery is called the **thyroid ima artery** (ima means lowest). Branches of the thyroid arteries anastomose with each other extensively both on the surface and within the substance of the gland.

The thyroid gland is drained on each side by three veins called the **superior, middle,** and **inferior thyroid veins.** Whereas the superior and middle thyroid veins are tributaries of the internal jugular vein, the inferior thyroid vein is a tributary of the left brachiocephalic vein (Figure 27–10 shows the left superior and inferior thyroid veins but not the left middle thyroid vein).

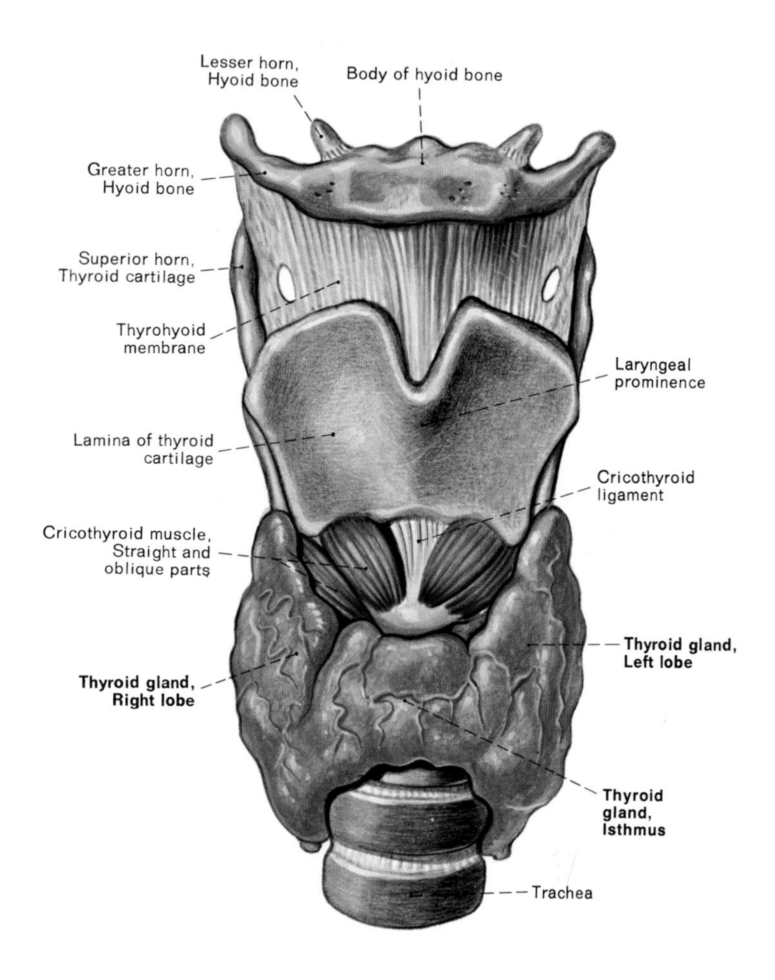

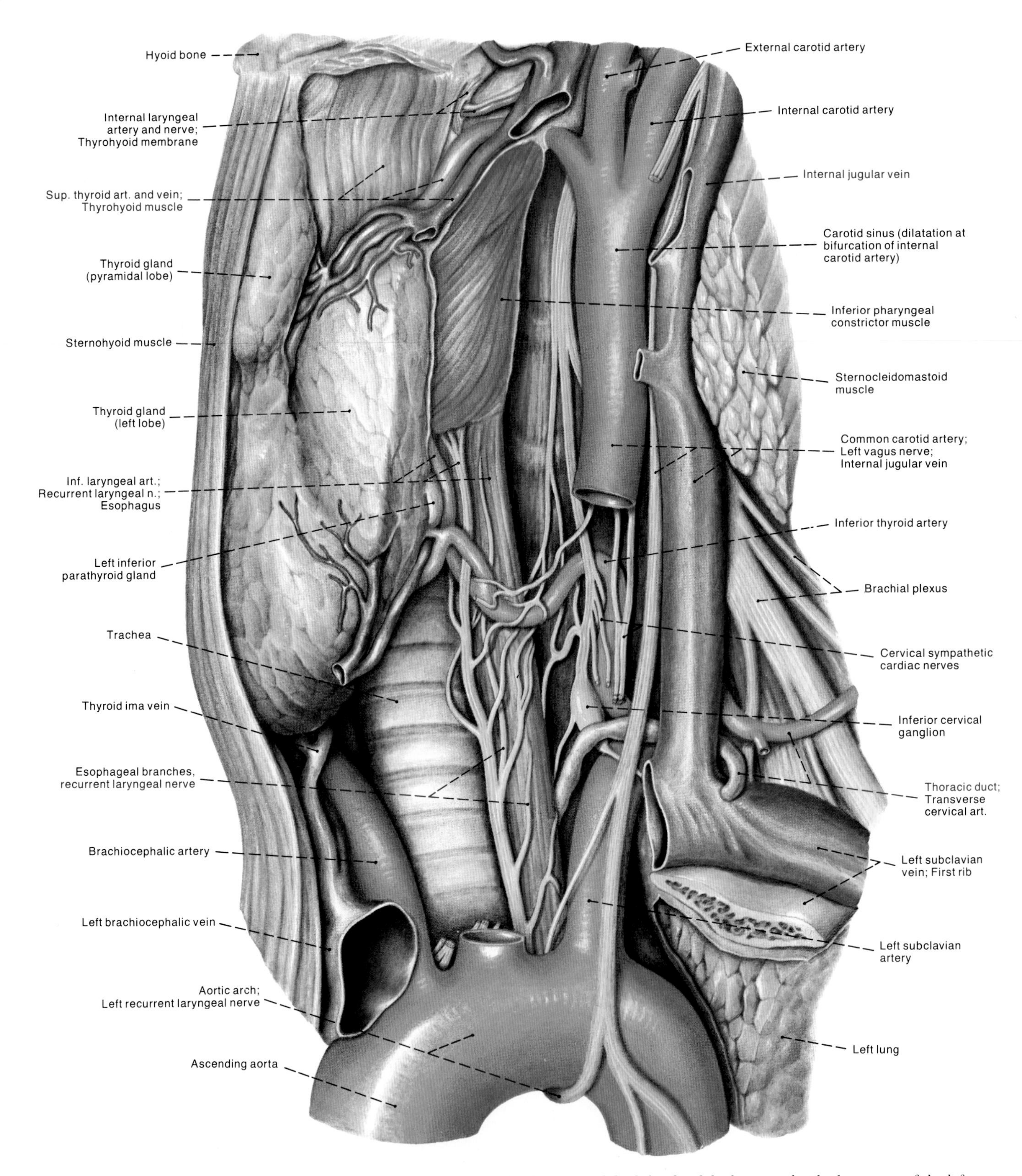

Fig. 27-20: Anterolateral view of the nerves and blood vessels in the deep part of the left side of the lower neck. The lower part of the left common carotid artery has been removed. Additionally, the left brachiocephalic vein has been removed from its point of origin at the junction of the left subclavian and internal jugular veins to the site at which the thyroid ima vein joins it.

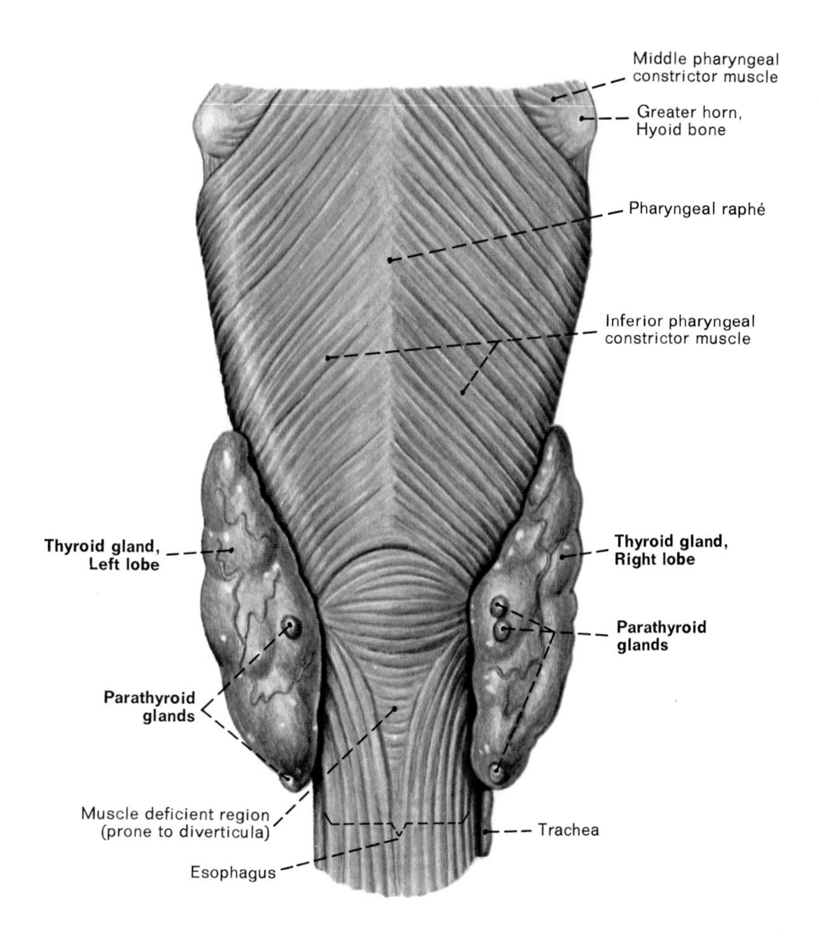

Fig. 27–21: Posterior view of the thyroid gland and its relationships to the pharynx and parathyroid glands.

Almost all the lymph drained from the thyroid gland is collected by lymphatics afferent to the deep cervical nodes (some lymphatics may drain directly into the thoracic duct). Carcinoma of the thyroid gland is the most common cause of enlarged, deep cervical nodes at the level of the isthmus of the thyroid gland.

The number of **parathyroid glands** and their locations are variable among the population. Most individuals have four small parathyroid glands: 80% of the population has four glands, 17% has fewer than four, and the remaining 3% has more. There is a superior and an inferior gland on each side (Fig. 27–21). The superior parathyroid gland generally lies behind the midregion of the lateral lobe of the thyroid gland, and the inferior parathyroid gland usually lies behind the inferior pole of the lateral lobe. The parathyroid glands are supplied by the superior and inferior thyroid arteries.

The parathyroid glands are almost always preserved during partial or total thyroidectomies. This is because a total lack of parathyroid function produces a severe hypocalcemia (a very low level of calcium in the blood) which must be treated throughout the remainder of the patient's life. The yellowish-brown color of the parathyroid glands aids in their identification from the adjacent, reddish-brown thyroid tissue.

CASE **VI.1**

The Case of Marjorie Patterson

INITIAL PRESENTATION AND APPEARANCE OF THE PATIENT

A 52 year-old white woman named Marjorie Patterson comes into the ambulatory clinic at 11:00 AM. She appears comfortable and relaxed.

QUESTIONS ASKED OF THE PATIENT

What is it that has brought you to the clinic? I'm having pains in my arm.

Can you point to where you have pains? No, not exactly. My pains are not always in the same place.

Can you indicate then the areas in your arm where you have pains? Yes [the patient slides the dorsum of her right hand over the medial aspects of her left arm, forearm, and hand].

Do you have pains in any one of these areas more often than the other areas? No.

Do you recall when these pains started? Oh, I would say about six months ago or longer. At first, I thought they would go away, because the pains come and go. But in the last two months, I have had the pains more frequently and they're more intense. *[The patient's account of the temporal profile of the upper limb pains suggests an insidious onset and progressive worsening of the condition.]*

How would you describe your pains? They're hard to describe; sometimes they feel like an ache. At other times, it feels like there's a bunch of small pins sticking me. *[The patient's answer indicates that the disease or disorder responsible for her condition can elicit pain or paresthesia (abnormal sensation).]*

Is there any time during the day or night when you get these pains more frequently than at other times? Yes. Many times I get the pains when I'm asleep in the early morning, because they will wake me up. I have also noticed that I tend to get them when I paint; I started painting with oils about a year ago, and, now that you mention it, I realize that many times the pains start after I have been painting for about a half hour or so.

Is there anything that relieves the pains? Yes. I have found that moving my arm like this [the patient circumducts her left arm] gets rid of the pains temporarily.

Is there anything that makes the pains worse? No.

Is there anything else that is also bothering you, or that you think may be associated with your arm pains? Well, yes. I sometimes feel numbness in the areas of my arm where I also get the pains. And, within the last two months, I've noticed that my left hand will get cold when I have the pains. *[The patient's answer indicates that the disease or disorder responsible for her left upper limb pain can also elicit hypesthesia or anesthesia (partial or complete numbness) and may be associated with blood circulatory disturbances.]*

Are you left or right-handed? Left-handed.

Have you noticed any clumsiness with activities done by your left hand? No.

Have you had any recent injuries or illnesses? No.

Are you taking any medications for previous injuries or illnesses? No. *[The examiner finds the patient alert and fully cooperative during the interview.]*

PHYSICAL EXAMINATION OF THE PATIENT

Vital signs: Blood pressure
Lying supine: 130/75 left arm and 125/75 right arm
Standing: 125/70 left arm and 130/75 right arm
Pulse: 74
Rhythm: regular
Temperature: 98.5°F.
Respiratory rate: 17
Height: 5′9″
Weight: 150 lbs.
HEENT Examination: Normal
Lungs: Normal
Cardiovascular Examination: The Adson maneuver with the left upper limb diminishes the left radial pulse. The Adson maneuver with the right upper limb does not affect the right radial pulse.
Abdomen: Normal

Genitourinary Examination: Normal
Musculoskeletal Examination: Normal
Neurologic Examination: The Adson maneuver with the left upper limb elicits a pins-and-needles sensation along the medial aspects of the left forearm and arm. The Adson maneuver with the right upper limb does not elicit any sensory effects.

INITIAL EVALUATION OF THE PATIENT'S CONDITION

The patient is suffering from intermittent episodes of left upper limb pain, hypesthesia or anesthesia, and paresthesia (pins-and-needles sensation) in association with blood circulatory disturbances.

ANATOMIC BASIS OF THE PATIENT'S HISTORY AND PHYSICAL EXAMINATION

(1) **The apparent contribution of certain daily activities or body positions (sleeping and painting) to the patient's left upper limb pain and the relief provided by a specific motion of the upper limb (circumduction of the arm) suggests a musculoskeletal disorder as the basis of the patient's upper limb pain.**

(2) **The co-location of pain, numbness, and paresthesia along the medial aspect of the patient's left upper limb indicates involvement of nerves derived from the medial cord of the brachial plexus.** Branches of the intercostobrachial nerve, the medial cutaneous nerve of the arm, the medial cutaneous nerve of the forearm, and the ulnar nerve provide cutaneous innervation of the medial aspect of the upper limb (Fig. 4–12, E and F). All of these nerves except the intercostobrachial nerve are derived from the medial cord of the brachial plexus. The intercostobrachial nerve is a sensory branch of the second intercostal nerve. The neuronal cell bodies for these cutaneous sensory fibers reside in the dorsal root ganglia of C8, T1, and T2 (Fig. 4–12, E and F).

(3) **The recent concurrence in the onset of coldness in the hand with the onset of left upper limb pain and paresthesia suggests a common etiology for the neurologic and cardiovascular symptoms.**

(4) **The positive Adson test indicates scalenus anticus syndrome.** The Adson maneuver is conducted as follows:

(i) As the patient sits with the head and neck in the anatomic position and the forearms resting pronated on the thighs, the examiner first establishes a baseline radial pulse during deep inspiration (that is, when the patient takes a deep breath of air inward and holds it in).

(ii) The examiner then monitors the radial pulse during deep inspiration with the patient's head extended backward and turned to the side being tested. The head extension and rotation maneuver narrows the interval between the scalenus anterior and medius muscles and thus manifests or increases any compression of the subclavian artery and/or brachial plexus within the interval. Maintenance of deep inspiration imposes increased traction on the subclavian artery and the brachial plexus through elevation of the first rib.

(iii) The test is positive if the head extension and rotation maneuver diminishes or obliterates the radial pulse **and** also reproduces or aggravates the patient's neurologic symptoms. A positive test suggests that the patient's neurologic symptoms are attributable to entrapment and compression of the subclavian artery and brachial plexus between the scalenus anterior and medius muscles. This condition is known as the scalenus anticus syndrome (scalenus anticus is the old-fashioned name for scalenus anterior).

CLINICAL REASONING PROCESS

Upper limb pain (in the absence of injury) can be the result of a disease or disorder of the musculoskeletal, vascular, and/or nervous tissues of the upper limb, one or more parts of the central nervous system, or one or more viscera of the neck, thorax, and/or abdomen. In this case, the two chief findings which markedly limit the diagnostic possibilities are (1) that certain daily activities or body positions appear to contribute to the onset of upper limb pain and (2) that a specific upper limb movement alleviates the pain. These two findings suggest that a disease or disorder of the musculoskeletal system is the basis of the patient's condition.

The physical examination is remarkable for the absence of any pain upon active or passive movements at any the left upper limb's joints or upon isometric testing of the muscles which act across these joints. There is also no pain upon active or passive movements of the patient's head and neck. Collectively, these negative findings suggest that if a musculoskeletal disease or disorder is responsible for the patient's condition, then the pain is not of musculoskeletal origin (in other words, the pain does not represent stimulation of pain fibers in skeletal muscles, ligaments, joints, or bones).

In the absence of any evidence indicating disease or injury of musculoskeletal tissues, it becomes reasonable to consider impingement or entrapment syndromes as a basis for upper or lower limb pain that is provoked and/or palliated by body movement or position. In this case, since the upper limb pain (a) occurs at sites ranging from the upper arm to the hand and (b) is associated with cutaneous nerves derived from the C8 and T1 roots of the brachial plexus, the most likely sites of possible impingement or entrapment appear to be sites in the neck or axilla.

The most common impingement or entrapment syndromes in the neck are cervical spondylosis (osteoarthritis of the cervical spine) and herniation of a cervical intervertebral disc. In cervical spondylosis, degenerative changes in the joints between the cervical vertebrae result in narrowing of the intervertebral foramina and the formation of periarticular osteophytes (bony outgrowths about the joints). These degenerative changes are asymptomatic unless they produce impingement of the roots of the cervical spinal nerves within the intervertebral foramina. If neurologic symptoms appear, head and neck movements, especially extension and rotation to the painful side, commonly reproduce or exacerbate the symptoms.

A herniated intervertebral disc is one which is partly protruded outward beyond its normal location between the bodies of two adjacent vertebrae. A posterolateral herniation may impinge upon the roots of a spinal nerve. If neurologic symptoms result from herniation of a cervical intervertebral disc, the symptoms are frequently confined to the motor and sensory distribution of just one cervical spinal nerve. The symptoms are commonly exacerbated if the examiner hyperextends the patient's head and neck and applies downward pressure on the head.

The most common impingement or entrapment syndromes in or about the axilla are the thoracic outlet syndromes: the hyperabduction, costoclavicular, cervical rib, and scalenus anticus syndromes. The general features of thoracic outlet syndromes and discussions of the hyperabduction and costoclavicular syndromes are presented in Chapter 8.

The cervical rib syndrome is a consequence of a congenital bony or fibrous extension of the lateral aspect of the seventh cervical vertebra. The extension may compress the subclavian artery and/or the divisions of the brachial plexus because the neurovascular bundle arches over the extension as the bundle exits the neck region. Cervical ribs occur in approximately 1% of the population, but less than 10% of cervical ribs become symptomatic. A cervical rib frequently subjects the divisions of the lower trunk of the brachial plexus to the greatest degree of compression, and thus frequently elicits neurologic symptoms associated with the nerves derived from the lower trunk. An oblique radiograph of the cervical spine provides an appropriate perspective for detecting the presence of a bony cervical rib.

RADIOGRAPHIC EVALUATION AND FINAL RESOLUTION OF THE PATIENT'S CONDITION

An oblique radiograph of the cervical spine showed evidence of cervical spondylosis (periarticular osteophytes) but did not show any evidence of a bony cervical rib. The positive Adson test, in combination with the absence of any evidence for other impingement or entrapment syndromes in the neck or axilla, indicates that the patient is suffering from scalenus anticus syndrome. Conservative treatment of the patient would require physical therapy.

EXPLANATION OF THE PATIENT'S CONDITION

The patient is suffering from the scalenus anticus syndrome as a consequence of connective tissue changes in the interval between the left scalenus anterior and medius muscles. Raynaud's phenomenon accounts for the coldness in the patient's left hand (refer to Chapter 11 for a discussion of Raynaud's phenomenon).

CASE **VI.2**

The Case of Catherine Donnelly

INITIAL PRESENTATION AND APPEARANCE OF THE PATIENT

A 42 year-old white woman named Catherine Donnelly has made an appointment to seek evaluation of a swelling in her neck. She appears comfortable and relaxed.

QUESTIONS ASKED OF THE PATIENT

How can I help you? I would like for you to look at this swelling in my neck.

When did you first notice the swelling? Well, I didn't know that I had the swelling until my husband saw it there 3 days ago. *[The duration of the presence of the cervical (neck) swelling is an important factor in the determination of the most likely etiologies. Unfortunately, the patient is unaware of the acute versus chronic presence of the swelling.]*

Where is the swelling located? [The patient uses the tips of her left index and middle fingers to mark a site in the lower region of the neck on the right side.] *[The examiner asks this question not only to have the patient mark the site of the swelling, but also to quickly establish whether the swelling is located in a side of the neck versus the midline. A midline versus side location is another important factor in the determination of the most likely etiologies.]*

How would you describe the swelling? It feels round and sort of firm. *[The examiner wants the patient to describe in her own words the physical characteristics of the swelling, in part to insure that the examiner will be able to accurately identify and palpate the swelling in question.]*

Do you recall bruising or injuring your neck before your husband noticed the swelling? No. *[This answer suggests that the swelling is not a bruise from a recent injury.]*

Does the swelling hurt? No.

Does the swelling hurt if you touch it? No. *[At this juncture during the history-taking, the examiner's knowledge of five key characteristics of the cervical swelling (it is a painless, non-tender swelling in the side of the neck that has been present for at least three days and is probably not a hematoma from trauma) permits the examiner to tentatively conclude that the swelling is most likely an enlarged lymph node, a neoplastic lesion, or a thyroid nodule. Enlargement of cervical lymph nodes can be due to infection (either local or systemic), metastatic cancer, or lymphoma. Because infection is the most common cause of cervical lymphadenopathy, the examiner decides to inquire if the patient has had any recent symptoms of either local or systemic infection. However, the examiner recognizes that lymph nodes enlarged by infectious inflammation are frequently tender.]*

Have you recently had a cold or sore throat? No.

Have you had any recent problems with your teeth or gums or any recent dental work? No.

Have you had a recent ear infection? No.

Have you recently had any flu-like symptoms, such as a fever or a feeling of being run-down and tired? No. *[Given the absence of symptoms of a local or systemic infection, the examiner next decides to inquire if the patient has any symptoms that might herald malignancy of a head or neck tissue or lymphoma.]*

Have you found any other swellings in your neck? No.

Have you had any pains in your head or neck? No, other than the occasional headache.

Have you had any sores in your mouth that don't seem to heal? No.

Have you noticed any change in the pitch or clarity of your voice? No. *[A change in the pitch or clarity of an individual's voice indicates either a lesion intrinsic to the larynx or a lesion that involves one or more of the nerves that innervate the muscles of the larynx.]*

Do you have any problems swallowing solid or liquid food? No. *[Dysphagia (difficulty in swallowing) indicates dysfunction of the pharynx and/or esophagus.]*

Do you have any trouble breathing? No. *[In the context of this case, a positive reply would have prompted the examiner to consider the possiblity that* *the dyspnea (difficulty in breathing) is due to obstruction of the upper respiratory tract by a cervical or mediastinal lesion.]*

[Given the negative replies to all of the preceding questions, the examiner now decides to inquire if the patient has any symptoms typical of either thyrotoxicosis or hypothyroidism. Thyrotoxicosis is the condition which results from the exposure of the body tissues to significantly elevated levels of thyroid hormones; there are a number of different diseases which can produce thyrotoxicosis. Hypothyroidism is the condition which results from the exposure of the body tissues to inadequate levels of thyroid hormones. There are a number of clinical entities which can result in the thyroid gland producing inadequate levels of its hormones. The examiner's questions concern the symptoms commonly produced by elevated or inadequate blood plasma levels of the thyroid hormones triiodothyronine (T_3) and thyroxine (T_4).]

Are you intolerant of the heat or cold; that is, do you find that you are uncomfortably warmer or colder than others in a room? No. *[Increased levels of T_3 and T_4 elevate the basal metabolic rate of body tissues; the concurrent increase in body heat production promotes increased perspiration and renders the individual less tolerant of heat. Decreased levels of T_3 and T_4 result in a reduction of the basal metabolic rate of body tissues; the concurrent decrease in body heat production renders the individual less tolerant of cold.]*

Have you recently lost or gained any weight? No. In fact, my weight has been fairly constant for the past few years now.

Has there been any recent change in your appetite? No. *[The increased metabolic rate that occurs with thyrotoxicosis can manifest weight loss and/or increased appetite. The decreased metabolic rate that occurs with hypothyroidism can result in weight gain.]*

Has there been any change in the nature or frequency of your bowel movements? No. *[Thyrotoxicosis increases gastrointestinal motility, which may produce more frequent bowel movements. Hypothyroidism decreases gastrointestinal motility, the product of which may be constipation.]*

Have you noticed any recent tendency to become tired more easily? No.

Have you noticed any recent weakness in your muscles? For example, do you find it more difficult to get up from a chair or to climb stairs? No. *[Thyrotoxicosis may produce muscle weakness as a result of increased protein catabolism.]*

Have you recently found yourself more nervous or more easily irritated with daily problems? No. *[Thyrotoxicosis may result in a dissociation of thought processes, which manifests itself as emotional lability.]*

Has there been any change in the regularity or flow of your menstrual periods? No. *[In the female, increased levels of T_3 and T_4 may cause oligomenorrhea (reduced menstrual bleeding). Decreased levels of T_3 and T_4 may produce polymenorrhea (frequent menstrual bleeding) and/or menorrhagia (excessive menstrual bleeding).]*

Have any regions of your head or neck ever been subjected to radiation therapy? No. *[Exposure of the head and neck to ionizing radiation predisposes an individual to thyroid cancer or hypothyroidism.]*

Have you ever had any problems with your thyroid or parathyroid glands? No, not that I know of.

Is there any history in your family of thyroid problems or any other glandular, hormonal, or endocrine problems? No. *[Family members may share a heritable disorder in which thyroid tumors occur in association with other endocrine tumors. Tumors or hyperplasia of the adrenal and parathyroid glands are frequently associated with thyroid tumors.]*

Are you taking any medications for previous injuries or illnesses? No. *[The examiner finds the patient alert and fully cooperative during the interview.}*

PHYSICAL EXAMINATION OF THE PATIENT

Vital signs: Blood pressure
Lying supine: 125/65 left arm and 125/65 right arm
Standing: 125/65 left arm and 125/65 right arm
Pulse: 71
Rhythm: regular
Temperature: 98.7°F.
Respiratory rate: 16
Height: 5′9″
Weight: 160 lbs.
[The normal heart rate is a pertinent normal finding, since thyrotoxicosis is almost always associated with tachycardia (an increased heart rate). Bradycardia (a decreased heart rate) may occur with hypothyroidism.]

HEENT Examination: Examination of the head, eyes, ears, nose, and throat are normal except for the following findings: Palpation reveals a 2 cm × 2 cm discretely bordered, firm, nontender, round mass in the right side of the neck whose center lies 3 cm inferior to the level of the laryngeal prominence (the Adam's apple). The mass moves first up and then down during deglutition (the act of swallowing).

Pertinent normal findings include the absence of any enlarged cervical or supraclavicular lymph nodes and a thyroid gland within the normal size range (with size assessed by inspection and palpation).

Lungs: Normal
Cardiovascular Examination: Normal
Abdomen: Normal
Genitourinary Examination: Normal

Musculoskeletal Examination: Normal
Neurologic Examination: Normal

[The absence of hyporeflexia in tendon reflex tests (such as the biceps brachii and quadriceps femoris tendon reflex tests) is a pertinent normal finding because severe hypothyroidism is almost always associated with hyporeflexia. Hyperreflexia may occur with thyrotoxicosis. The absence of muscle tremor is another important normal finding, since a fine muscle tremor is a characteristic sign of thyrotoxicosis. The fine muscle tremor has a characteristic frequency of 10 to 15 cycles per second; it can be demonstrated by assessing the vibrations in a sheet of paper laid on the volar surfaces of the patient's extended fingers.]

Integumentary Examination: The patient's skin has a normal consistency and temperature and is not moist.

[These normal findings relative to the patient's integument are pertinent because thyrotoxicosis and hypothyroidism are frequently associated with changes in skin texture and temperature. The skin of individuals with thyrotoxicosis is frequently soft, warm, moist and velvety in texture. The skin of individuals with hypothyroidism is frequently coarse or rough, dry, and thickened or edematous; alopecia (loss of body hair) may also be an associated finding.]

INITIAL EVALUATION OF THE PATIENT'S CONDITION

The history and physical examination shows that the patient has a painless, non-tender mass in the right side of the lower neck.

ANATOMIC BASIS OF THE PATIENT'S HISTORY AND PHYSICAL EXAMINATION

(1) The location of the cervical mass and its upward-and-downward movement during deglutition indicates that the mass is either a thyroid nodule or a mass tethered to the larynx and trachea.

The pretracheal layer of deep cervical fascia (Fig. 27–14) and the sternothyroid muscles tether the thyroid gland posteriorly to the larynx and trachea. The thyroid gland thus moves with the larynx and trachea during deglutition (the act of swallowing) and phonation (the production of sound by the larynx). This association aids in the physical diagnosis of cervical swellings. A cervical swelling is not a thyroid nodule (a focal enlargement of the thyroid gland) if it does not move first up and then down when the patient swallows.

Diffuse enlargements (diffuse goiters) of the lateral lobes of the thyroid gland cannot extend superiorly beyond the insertion sites of the overlying sternothyroids to the oblique lines of the thyroid cartilage. As diffuse goiters enlarge, they may stretch the overlying sternothyroids, compress the underlying trachea and esophagus, or extend inferiorly into the superior mediastinum, where they may compress the brachiocephalic veins.

(2) The patient does not have any signs or symptoms of either thyrotoxicosis or hypothyroidism.

INTERMEDIATE EVALUATION OF THE PATIENT'S CONDITION

The patient appears to be a euthyroid individual with a thyroid nodule. The term euthyroid refers to the condition in which an individual's thyroid gland is producing its hormones at normal levels.

It is important to appreciate, however, that euthyroidism does not necessarily imply the absence of thyroid disease. This is because there are benign thyroid follicular tumors which produce thyroid hormones autonomously [independently of the stimulation provided by the anterior pituitary hormone TSH (thyroid stimulating hormone)]. The normal thyroid glandular tissue responds to the presence of these tumors by reducing thyroid hormone production so as to initially maintain a euthyroid state. Progressive enlargement of these tumors ultimately produces thyrotoxicosis.

CLINICAL REASONING PROCESS

This case illustrates the general kinds of information a physician seeks from adult patients who present with a solitary cervical mass. The most common types of solitary cervical masses in adult patients are enlarged lymph nodes, neoplastic lesions, and thyroid nodules. In this case, the upward-and-downward movement of the cervical mass during deglutition points to a focal thyroid mass.

In consideration of the patient's condition, a primary care physician or endocrinologist would consider four principal types of thyroid lesions: a thyroid adenoma, multinodular goiter, a thyroid cyst, or thyroid carcinoma. Thyroid adenomas are well-delineated, encapsulated, benign neoplasms of thyroid tissue. A multinodular goiter is a thyroid gland bearing multiple foci of hyperplastic thyroid tissue. Thyroid cysts are round to oval, fluid-filled lesions that frequently result from necrosis of neoplastic thyroid tissue, either benign or malignant.

Further evaluation of the patient's solitary cervical nodule requires scintillation scanning of radiolabelled iodide uptake by the thyroid gland. The thyroid gland actively takes up iodide from the blood plasma and covalently binds it to tyrosyl residues of the glycoprotein thyroglobulin; the biosynthesis of thyroglobulin and its iodination occur within the follicular cells of the thyroid gland. Thyroglobulin is stored in the follicles of the thyroid gland until it is enzymatically hydrolyzed to produce the thyroid hormones T_3 and T_4. The amount of iodide taken up by a region of the thyroid is thus an approximate measure of its overall rate of T_3 and T_4 production. Measurement of radiolabelled iodide uptake by scintillation scanning affords assessment at the gross level of resolution of the uniformity of hormone production within the thyroid gland. Thyroid scintillation scanning is performed 24 hours following intravenous administration of radiolabelled iodide.

Assessment of the uniformity of hormone production within the thyroid gland can also be conducted with the anion pertechnetate. The thyroid actively takes up pertechnetate in a fashion similar to that of its active uptake of iodide. However, because pertechnetate rapidly diffuses out of the thyroid gland and does not become organically bound to thyroglobulin, radiolabelled pertechnetate uptake can be measured by scintillation scanning within 20 minutes of its intravenous administration.

Scintillation scanning of either radiolabelled iodide or pertechnetate uptake shows nodules to be hyperfunctioning (to have taken up more radiolabel than the surrounding glandular tissue), isofunctioning (to have taken up radiolabel to an extent indistinguishable from that taken up by surrounding glandular tissue), or hypofunctioning (to have taken up less radiolabel than the surrounding glandular tissue). Although most thyroid carcinomas form hypofunctioning foci, some appear as isofunctioning. Scintillation scanning of radiolabelled iodide or pertechnetate uptake thus cannot effectively exclude the possibility of thyroid carcinoma in a patient with a solitary thyroid nodule (however, the likelihood is extremely low if the nodule is hyperfunctioning). The principal benefit of scintillation scanning is detection of functionally abnormal foci in addition to the palpable lesion.

Complete evaluation of the patient's solitary cervical nodule also requires measurement of blood plasma levels of thyroid hormones, ultrasonography, and fine needle aspiration biopsy.

RADIOGRAPHIC EVALUATION AND FINAL RESOLUTION OF THE PATIENT'S CONDITION

Serum levels of T_3 and T_4 were within the normal range. Scintillation scanning of radiolabelled pertechnetate uptake revealed uniform distribution of the radiolabel throughout the thyroid gland except for the region corresponding to the palpable cervical mass, which exhibited decreased radiolabel. Ultrasonography demonstrated that the mass was solid rather than cystic (fluid-filled). Fine needle aspiration biopsy of the hypofunctioning nodule revealed a solid, benign mass of follicular cells. The thyroid adenoma was surgically removed.

RECOMMENDED REFERENCES FOR ADDITIONAL INFORMATION ON THE NECK

Slaby, F., and E. R. Jacobs, *Radiographic Anatomy*, Harwal Publishing Co., Malvern, PA, 1990: *Pages 185 through 187 in Chapter 5 outline the major anatomic features of the lateral neck film. The outline describes the lateral neck film in terms of the images cast by two major sets of anatomically distinct structures: (1) the bones of the neck and (2) the soft tissues of the neck.*

DeGowin, R. L., Jochimsen, P. R., and E. O. Theilen, *DeGowin & DeGowin's Bedside Diagnostic Examination*, 5th ed., Macmillan Publishing Co., New York, 1987: *Pages 209 through 211 present a discussion of the physical examination for thyroid disorders.*

Jarvis, C., *Physical Examination and Health Assessment*, W. B. Saunders Co., Philadelphia, 1992: *Pages 291 through 294 present a discussion of the physical examination of the thyroid gland.*

CHAPTER **28**

The Cranial Cavity

The **skull** is the skeleton of the head and face. The term **cranium** refers to all the bones of the skull except the mandible (which is the bone of the lower jaw). The term **calvaria** refers to the domelike, upper portion of the cranium. The remaining, lower portion of the cranium is called the **base of the skull.** The space enclosed by the calvaria and the base of the skull is called the **cranial cavity.** The upper part of the cranial cavity is termed the **cranial vault.**

The cranial cavity houses the brain. The reader is referred to neuroanatomy texts for descriptions of the gross structure of the brain and organization of neuronal pathways in the central nervous system. This chapter focuses on the skeletal framework of the cranial cavity, the cranial meninges, and the dural venous sinuses.

THE CALVARIA

The calvaria of an adult consists of the upper parts of four bones: the unpaired **frontal bone** anteriorly, the paired **parietal bones** on the sides, and the unpaired **occipital bone** posteriorly (Fig. 28–1). The articulations among these bones are united by fibrous joints called **sutures.** The suture in the midline between the paired parietal bones is called the **sagittal suture.** The suture between the frontal bone and the paired parietal bones is called the **coronal suture;** the point of union between the coronal and sagittal sutures is called the **bregma.** The suture between the occipital bone and the paired parietal bones is called the **lambdoid suture;** the point of union between the lambdoid and sagittal sutures is called the **lambda.**

The bones of the calvaria all consist of outer and inner plates of compact bone (called **tables**) separated by a middle cancellous layer called the **diploë.** Because the diploë bears hematopoietic marrow throughout adulthood, it always remains a potential site of secondary metastases. Small valveless veins called **diploic veins** drain the bones of the calvaria. Veins called **emissary veins** traverse the bones of the calvaria completely and connect the veins of the scalp with the intracranial, dural venous sinuses.

The bones of the calvaria ossify through intramembranous ossification. The occipital and paired parietal bones each ossify radially outward from a single ossification center; the ossification of the frontal bone begins with a bilateral pair of ossification centers. At birth, the bones are only partially ossified, and they are separated by fibrous tissue membranes. These intervening membranes permit the bones of the calvaria to slide past and even overlap each other during vaginal delivery; these movements mold the cranial vault into a shape which eases the passage of the neonate's head through the birth canal. Some realignment of the bones usually occurs during the first week postnatally.

In the anterior part of a neonate's calvaria, the left and right halves of the frontal bone and the paired parietal bones are separated by a diamond-shaped membrane called the **anterior fontanelle** (Fig. 28–2). In the posterior part of the neonatal calvaria, the occipital bone and the paired parietal bones are separated by a triangular membrane called the **posterior fontanelle.** The tension of these so-called "soft spots" can be used as a manometer of intracranial pressure. The posterior and anterior fontanelles become impalpable after 12 and 18 months, respectively.

THE SCALP

The scalp covers the calvaria. The scalp consists of five soft tissue layers; the combination of the first letters of the names of these layers spells the word **scalp:** **s**kin, **c**onnective tissue, **a**poneurosis of epicranius, **l**oose connective tissue, and **p**ericranium (Figure 28–3 illustrates the three most superficial layers of the scalp). **Epicranius** (**occipitofrontalis**) is a muscle with anterior and posterior muscular bellies joined by an intermediate aponeurotic tendon called the **galea aponeurotica** (Fig. 28–4). The posterior (occipital) belly originates from the occipital and temporal bones, and the anterior (frontal) belly is attached to the superficial fascia of the forehead and upper eyelid. Epicranius can raise the eyebrow and the skin of the forehead. Epicranius is a muscle of facial expression, and it is innervated by the facial nerve. The pericranium is the periosteum of the outer table of the bones of the calvaria.

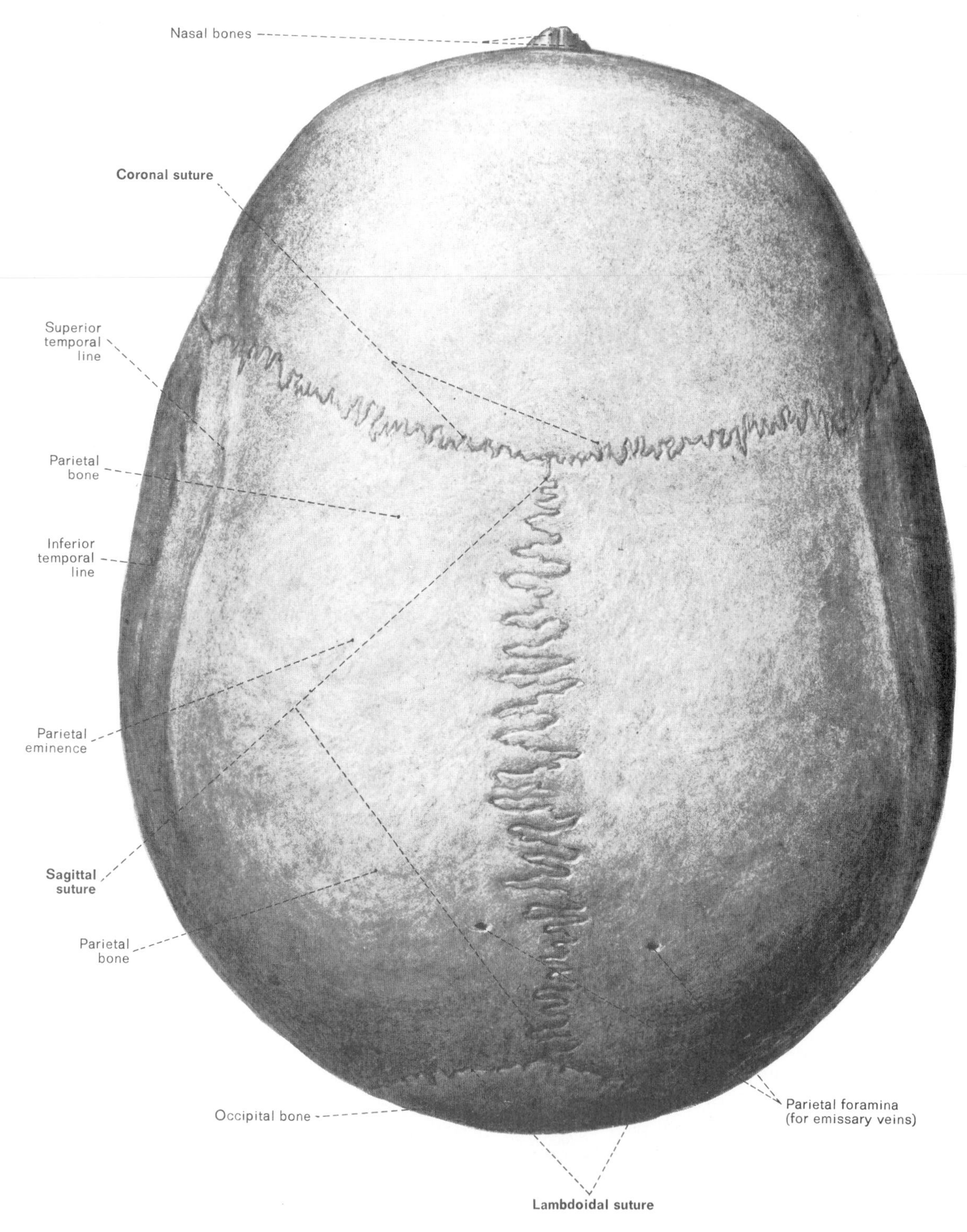

Fig. 28–1: Superior view of the adult calvaria.

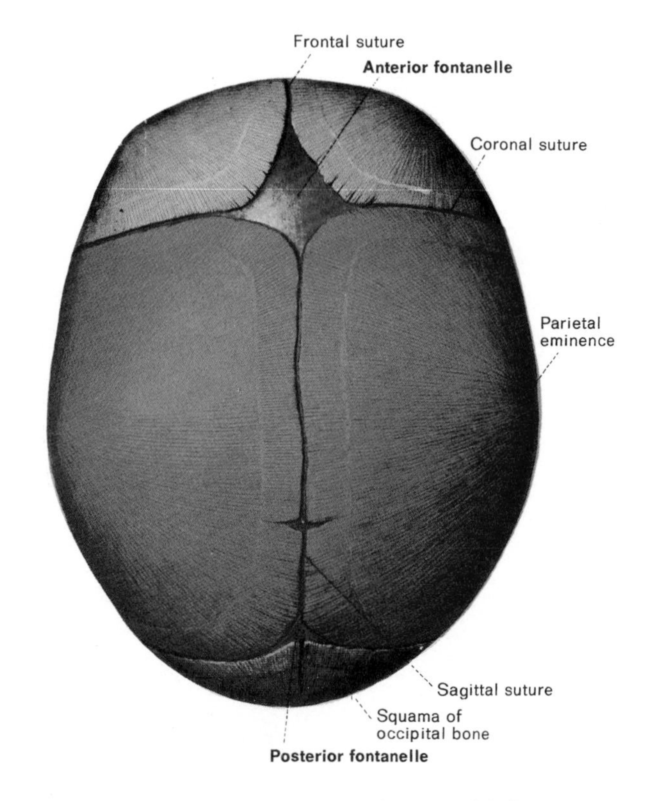

Fig. 28–2: Superior view of a neonatal skull.

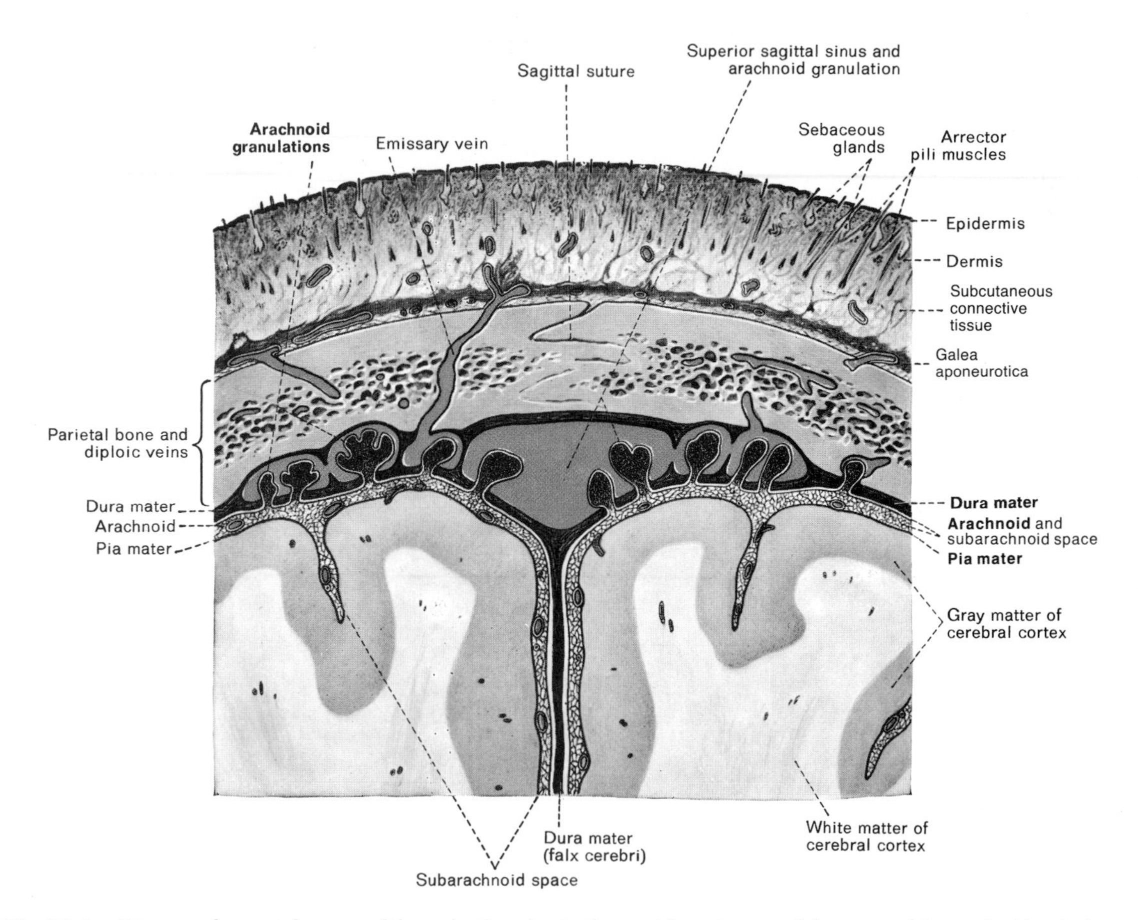

Fig. 28–3: Diagram of a coronal section of the scalp, the calvaria, the cranial meninges, and the cortex of the cerebral hemispheres.

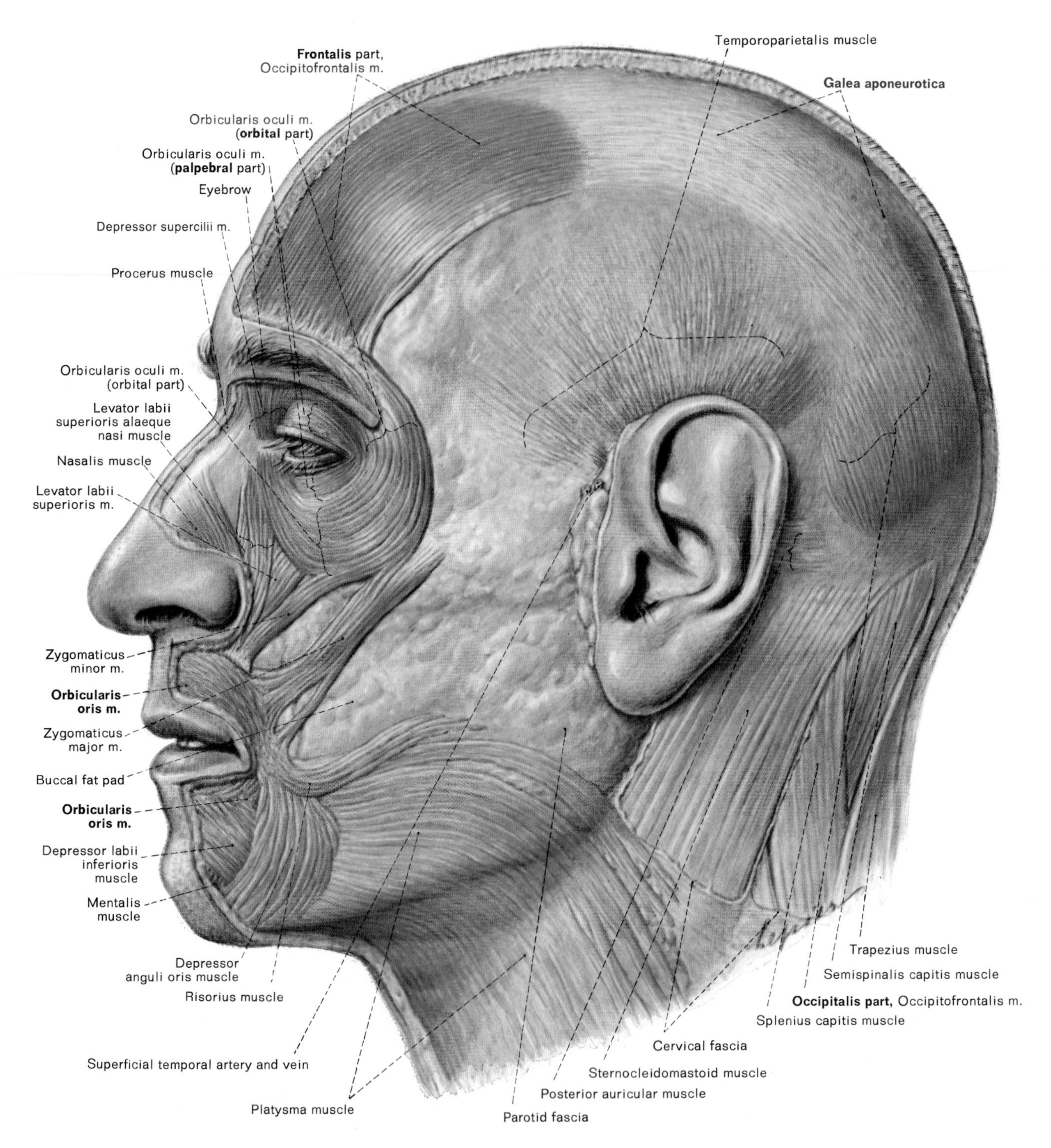

Fig. 28–4: Lateral view of the muscles of facial expression and the superficial posterior cervical muscles.

THE INTERIOR OF THE BASE OF THE SKULL

The bones that border the interior of the base of the skull include the lower parts of the bones of the calvaria, the unpaired sphenoid and ethmoid bones, and the paired temporal bones.

The Sphenoid

The sphenoid is the central bone of the base of the skull (Figs. 28–5 and 28–6). The bone is named for the wedge shape it presents when viewed either anteriorly or posteriorly (Fig. 28–7). The central part of the sphenoid is called the **body,** and it houses the **left** and **right sphenoid sinuses,** two of the sinuses of the nasal cavities (Fig. 28–8). The paired wing-like projections that extend laterally from the upper part of the body are called the **lesser wings,** and the paired wing-like projections that extend laterally from the lower part of the body are called the **greater wings** (Figs. 28–6 and 28–7). The **pterygoid processes** are the paired processes that extend inferiorly from the regions at which the greater wings are united to the body; as each pterygoid process extends inferiorly, it splits into two plates called the medial and lateral pterygoid plates (Fig. 28–7).

The superior surface of the body of the sphenoid bears a hollow called the **hypophyseal fossa,** because the **hypophysis cerebri (pituitary gland)** resides within it (Figs. 28–5 and 28–6). The hollow is also called the **sella turcica** (Turkish saddle), because of its shape. The transverse ridge which anteriorly borders the sella turcica is called the **tuberculum sellae** (which literally means "tubercle of the sella turcica"). The transverse ridge which posteriorly borders the sella turcica is called the **dorsum sellae,** and its lateral ends are called the **posterior clinoid processes** (clinoid means bed-like). The opposing **anterior clinoid processes** are the posteromedial angles of the lesser wings of the sphenoid.

The superior surface of the body of the sphenoid also bears a transverse groove called the **chiasmatic (optic) groove** lying in front of the ridgelike tuberculum sellae (Fig. 28–6). The lateral ends of the chiasmatic groove transmit the optic nerves as they extend from the optic chiasma toward the orbital cavities (the cavities which house the eyeballs).

The Ethmoid

The ethmoid has a vertical, bony plate called the **perpendicular plate** which lies in the median plane of the head (Fig. 29–1). The upper end of the perpendicular plate resembles a cock's comb, and hence is called the **crista galli** (which literally means the "cock's crest") (Fig. 29–1). The lower end of the perpendicular plate contributes to the nasal septum (the nasal septum is the part bony-part cartilaginous septum which separates the paired nasal cavities).

The ethmoid has a horizontal, bony plate which, since it is speckled with numerous, small perforations, is called the **cribriform plate** (the term cribriform is derived from the Latin word for sieve) (Fig. 28–5). The left and right halves of the cribriform plate form the roofs of the paired nasal cavities.

The cribriform plate bears on each side a labyrinth of bone-encased air cells. These air cells collectively represent the ethmoid paranasal sinus of each nasal cavity; each ethmoid sinus is divided into three subgroups of air cells called the **anterior, middle,** and **posterior ethmoid sinuses** (Figure 28–8 shows the anterior and posterior ethmoid sinuses but not the middle ethmoid sinuses). On each side, two downward-curving, bony shelves called the **superior** and **middle conchae** project medially from the air cell labyrinth (the term concha is derived from the Latin word for shell) (Fig. 29–2). The paired air cell labyrinths and paired superior and middle conchae of the ethmoid contribute to the lateral walls of the nasal cavities.

The Temporal Bone

The temporal bone is commonly divided into four parts:

(1) The **squamous part** is so named because it is the flat, scalelike part of the bone (Fig. 28–9). The squamous parts of the paired temporal bones underlie much of the sides of the head (the "temples") (Fig. 28–10). A spearlike process called the **zygomatic process** projects forward from the outer surface of the squamous part (Figs. 28–9 and 28–10); the zygomatic process helps shape the side of the cheek.

(2) The **petrous part** is so named because it is the rocklike part of the bone (Fig. 28–9). The petrous part houses the internal ear and the mastoid air cells; the mastoid air cells open directly into the tympanic cavity of the middle ear (Fig. 30–2B). The carotid canal passes through the petrous part.

(3) The **tympanic part** is so named because it forms the bony frame around the **tympanic membrane** (the "eardrum"). The tympanic part houses the external acoustic meatus.

(4) The **styloid process** is the pointed part of the temporal bone which extends anteroinferiorly from the tympanic part (Figs. 28–9 and 28–10). The styloid process serves as the origin of three muscles (stylohyoid, styloglossus, and stylopharyngeus).

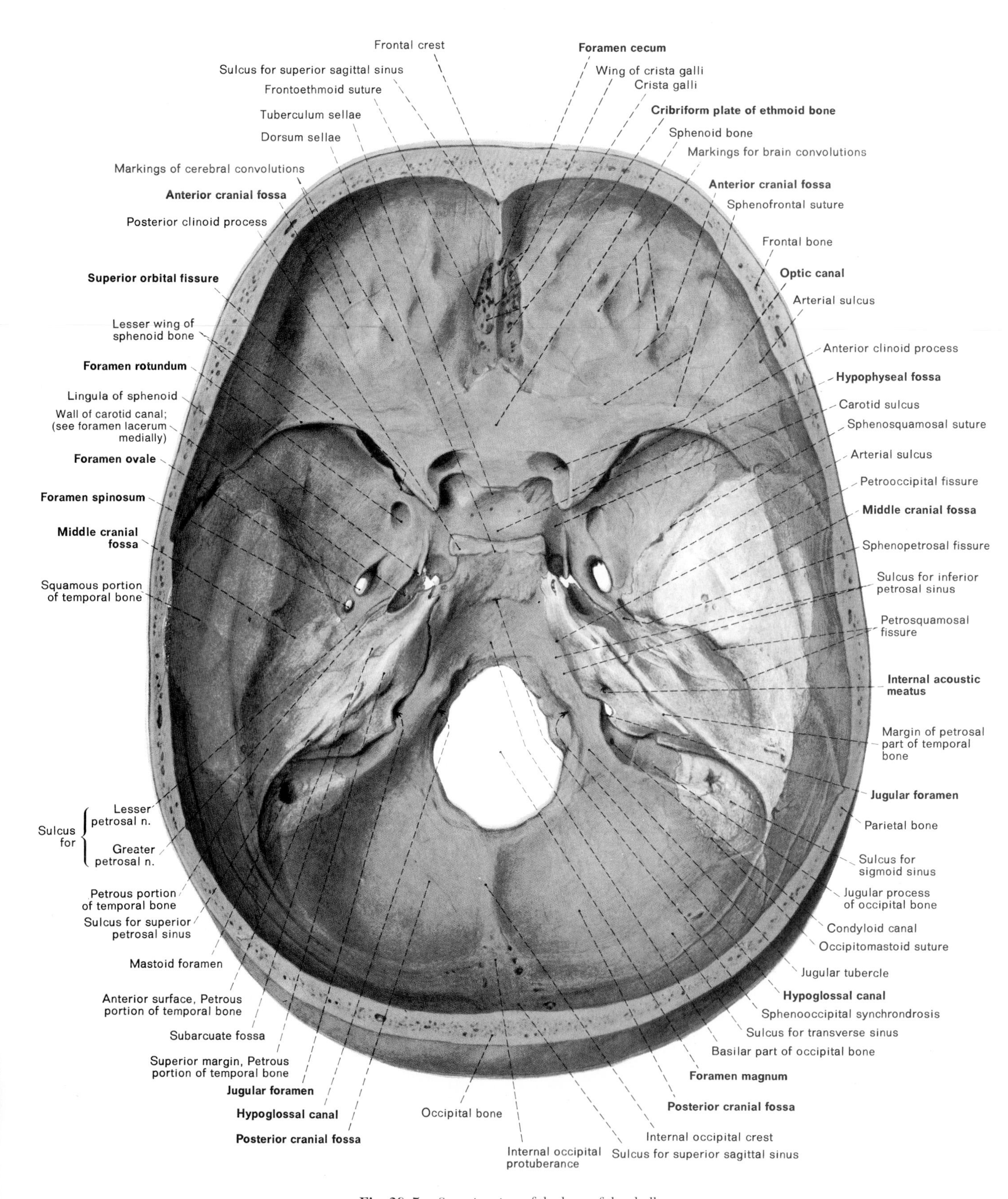

Fig. 28–5: Superior view of the base of the skull.

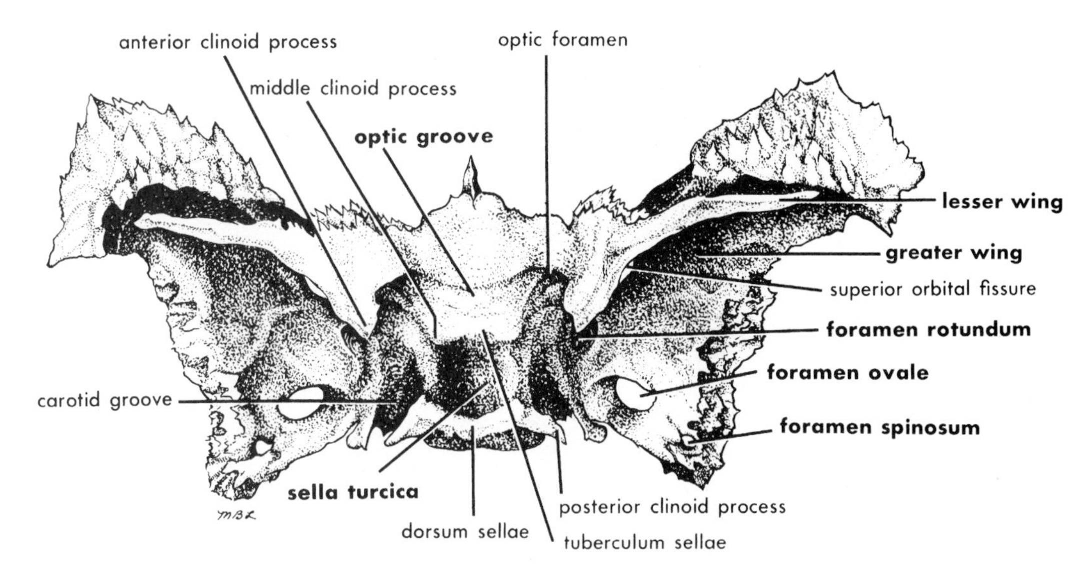

Fig. 28–6: Superior view of the sphenoid bone.

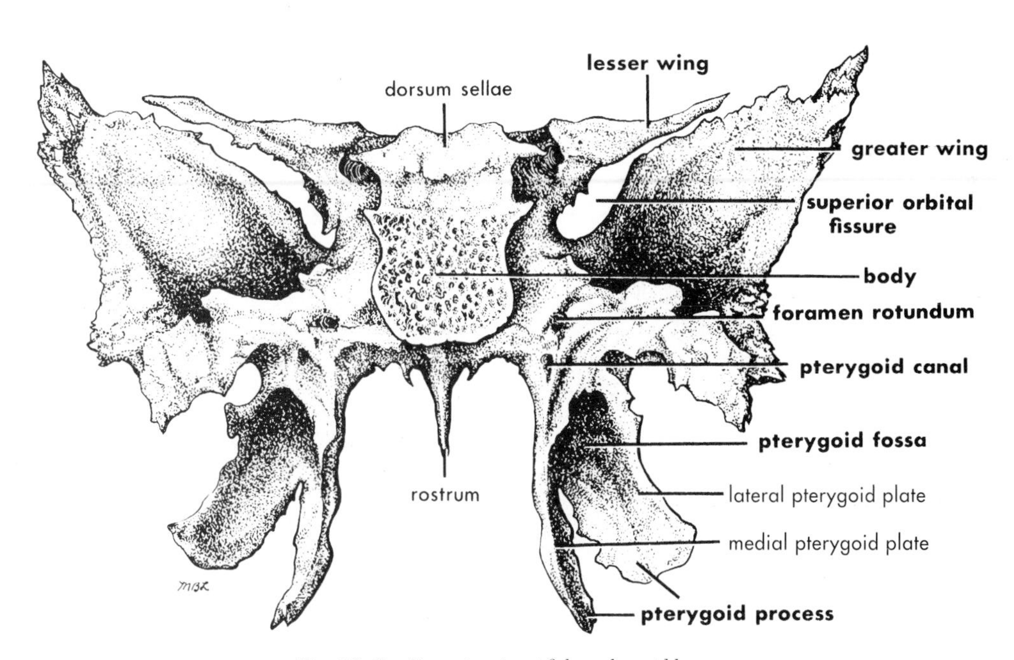

Fig. 28–7: Posterior view of the sphenoid bone.

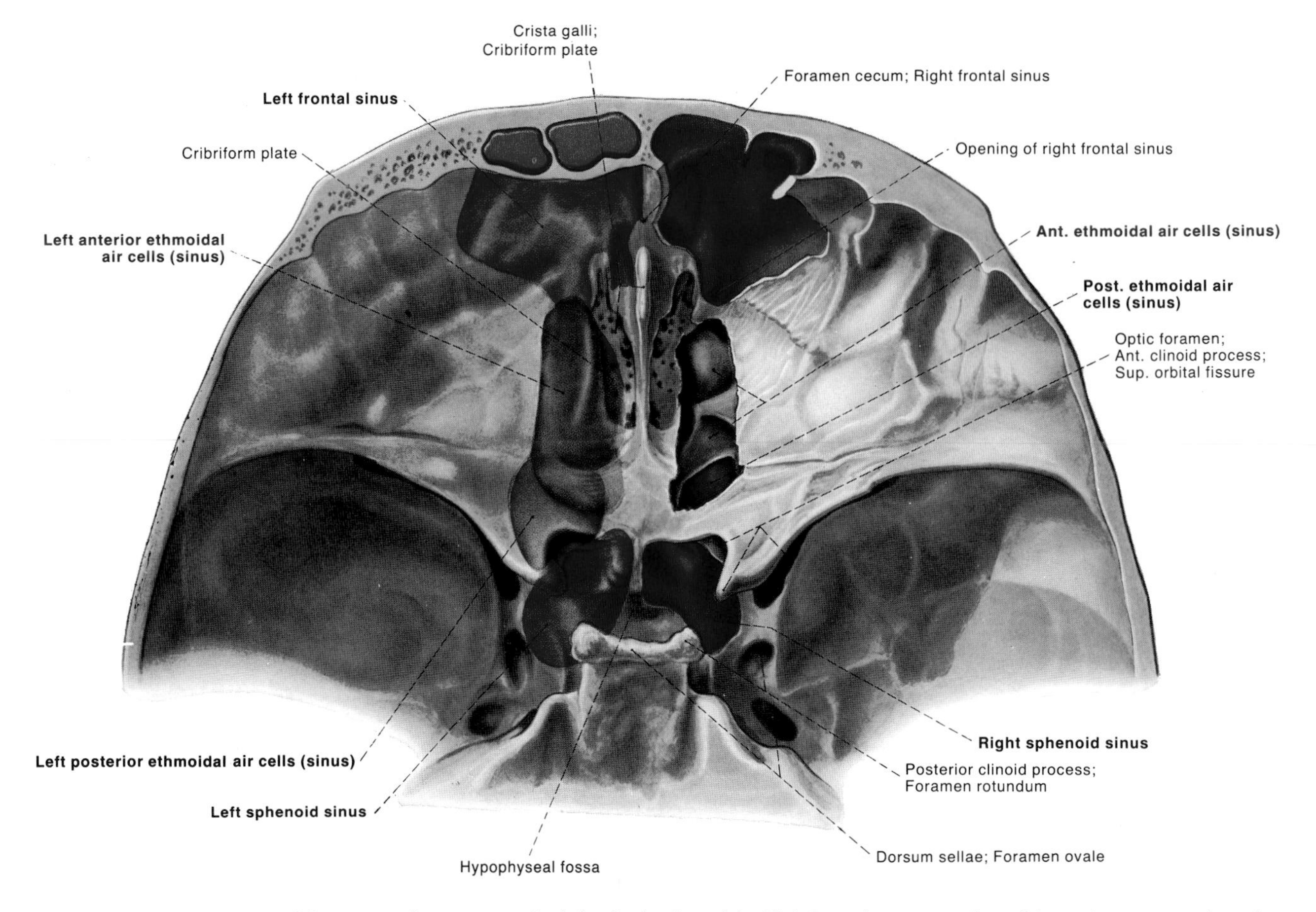

Fig. 28–8: Superior view of the paranasal sinuses. On the left side the frontal (reddish brown), anterior ethmoid (green), posterior ethmoid (lavender), and sphenoid (brown) sinuses are projected onto the bones of the base of the skull. In the right side portions of the frontal, ethmoid, and sphenoid bones have been removed to bring the sinuses into view from above.

The Frontal Bone

The frontal bone underlies the forehead (Fig. 28–10). The upper part of the bone curves downward and forward from the calvaria to the upper margins (supraorbital margins) of the orbital cavities. The frontal bone presents in the midline a small elevation called the glabella; the glabella underlies the area of the forehead between the medial ends of the eyebrows.

The frontal bone has a pair of horizontally oriented plates which extend backward from the supraorbital margins; these plates are called the **orbital plates** because they form most of the bony roofs of the orbital cavities (in Figure 28–5, the orbital plates are the parts of the frontal bone bearing markings for brain convolutions). The frontal bone also houses the frontal sinuses, one of the sinus pairs of the nasal cavities (Fig. 28–8). The frontal sinuses lie opposite each other immediately above and behind the medial ends of the supraorbital margins.

The Occipital Bone

The occipital bone forms the lower, back part of the base of the skull (Fig. 28–10). The condyles on its undersurface articulate with the superior articular facets of the atlas.

The Cranial Fossae

The internal surface of the base of the skull forms the triple-tiered floor of the cranial cavity (Fig. 28–5). Since each tier bears a major pair of depressions, the tiers are referred to as fossae, and named the anterior, middle, and posterior cranial fossae.

The floor of the cranial fossae contains openings by which the 12 cranial nerves extend between the cranial cavity and the neck or extracranial regions of the head (these openings are discussed in the following sections on the cranial fossae). The cranial nerves all emerge from or extend toward the undersurface of the brain

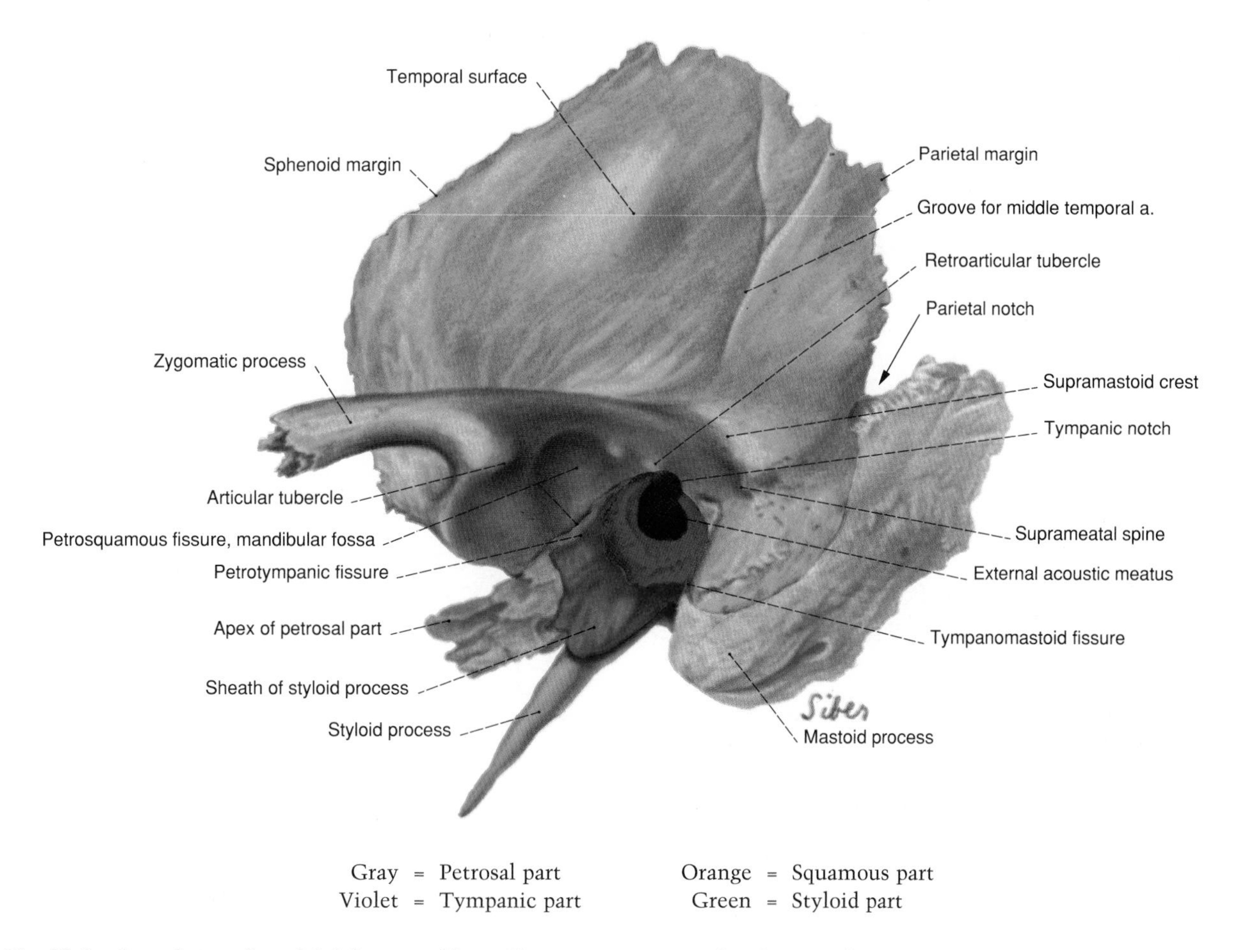

Fig. 28–9: Lateral view of an adult left temporal bone. The squamous part is colored orange, the petrous part gray, the tympanic part violet, and the styloid process green.

(Fig. 28–11). The olfactory tracts are connected to the cerebral hemispheres. The optic nerves are connected to the diencephalon (a small region of the brain located between the cerebral hemispheres) (the diencephalon is not shown in Figure 28–11). All the other cranial nerves are connected to the brainstem, the part of the brain that connects the cerebral hemispheres, diencephalon, and cerebellar hemispheres with the spinal cord. Figure 28–11 displays the pons and medulla oblongata, which are the largest parts of the brainstem. This figure also displays the **semilunar ganglion of the trigeminal nerve;** this ganglion contains the cell bodies of all the sensory fibers of the trigeminal nerve (the trigeminal nerve is the fifth cranial nerve).

The **anterior cranial fossa** is the highest of the cranial fossae. The fossa is formed by the orbital plates of the frontal bone, the cribriform plate and crista galli of the ethmoid, and the lesser wings and anterior part of the body of the sphenoid (Fig. 28–5). The posterior edges of the lesser wings of the sphenoid form most of the bony ridge which separates the anterior cranial fossa from the middle cranial fossa.

The floor of the anterior cranial fossa transmits neuronal elements of the olfactory nerve **(the olfactory nerve is the first cranial nerve)** (the neuronal elements labelled "I" in Figure 28–12 represent nerve fibers of the olfactory nerve). The frontal lobes of the cerebral hemispheres rest upon the floor of the anterior cranial fossa. The anterior cranial fossa overlies the paired orbital and nasal cavities and frontal and ethmoid sinuses (Figure 28–8 shows the frontal and ethmoid sinuses beneath the floor of the anterior cranial fossa, and Figure 28–12 shows the left orbital cavity beneath the left side of the floor of the anterior cranial fossa).

The **middle cranial fossa** is formed by the greater wings and body of the sphenoid and the petrous and squamous parts of the temporal bones (Fig. 28–5). On each side, the superior border of the petrous part of the temporal bone marks the posterior boundary of the middle cranial fossa.

The lateral floor areas of the middle cranial fossa lodge the temporal lobes of the cerebral hemispheres and overlie (on each side) the inner ear, the middle ear

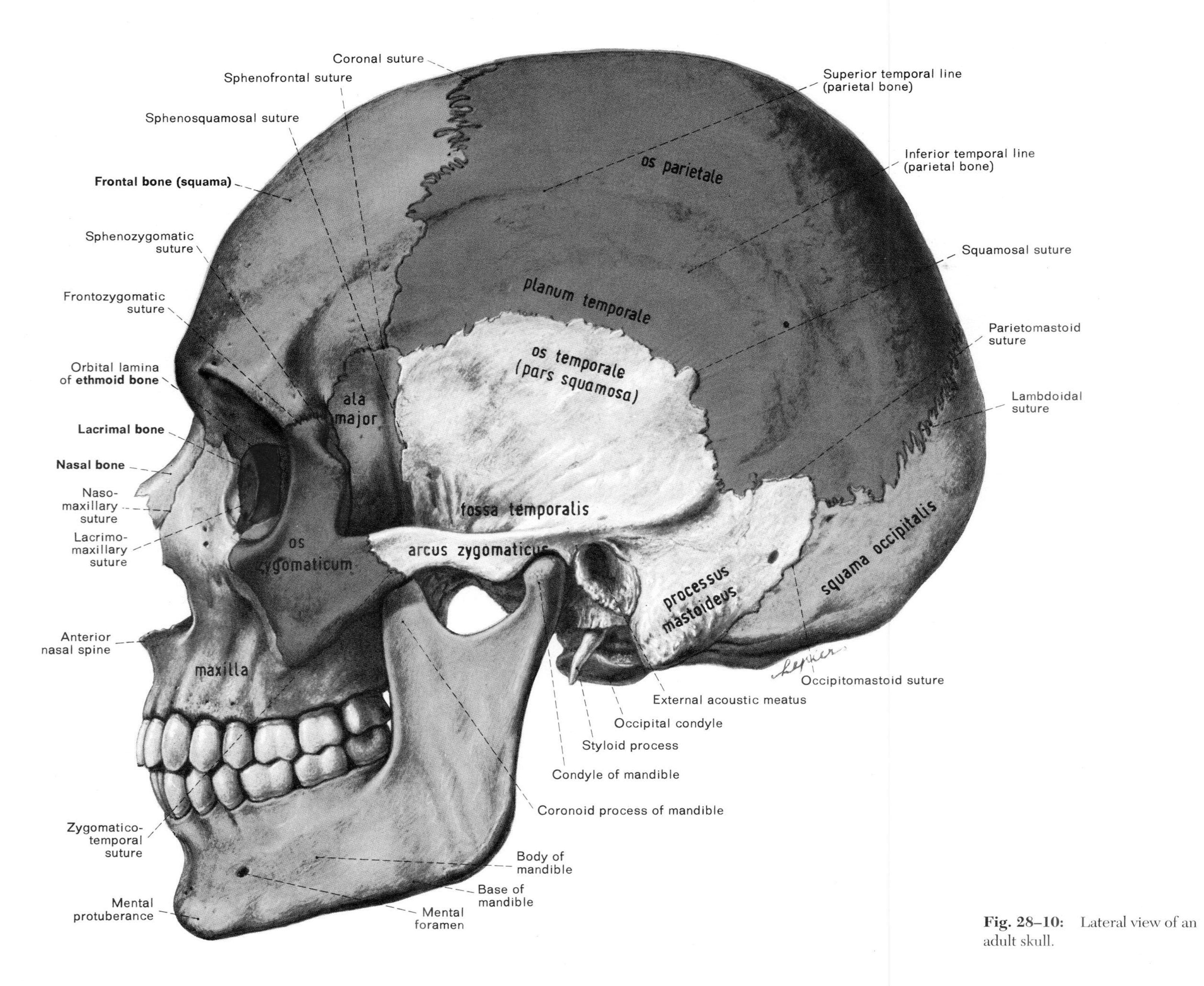

Fig. 28–10: Lateral view of an adult skull.

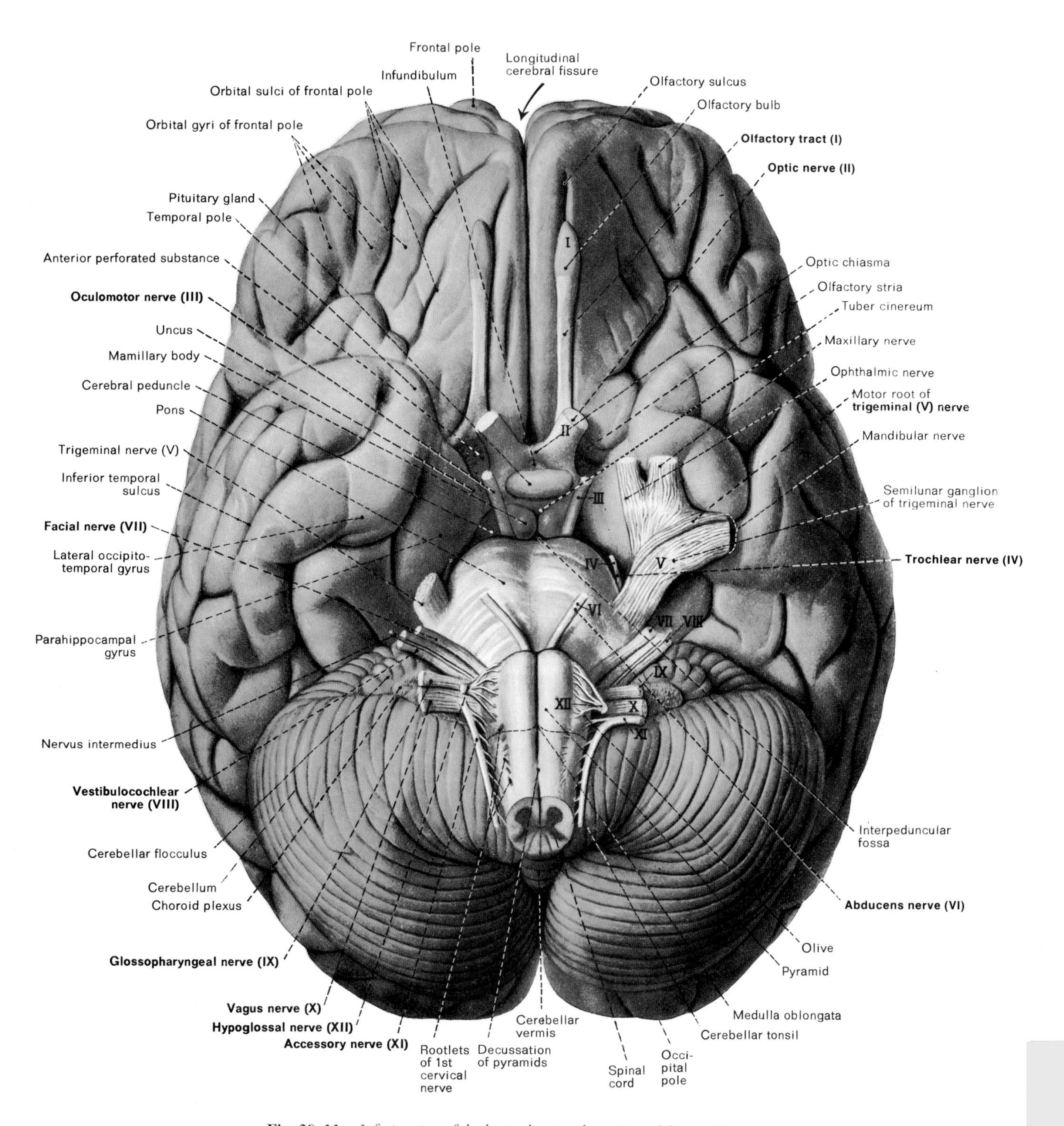

Fig. 28–11: Inferior view of the brain showing the origins of the cranial nerves.

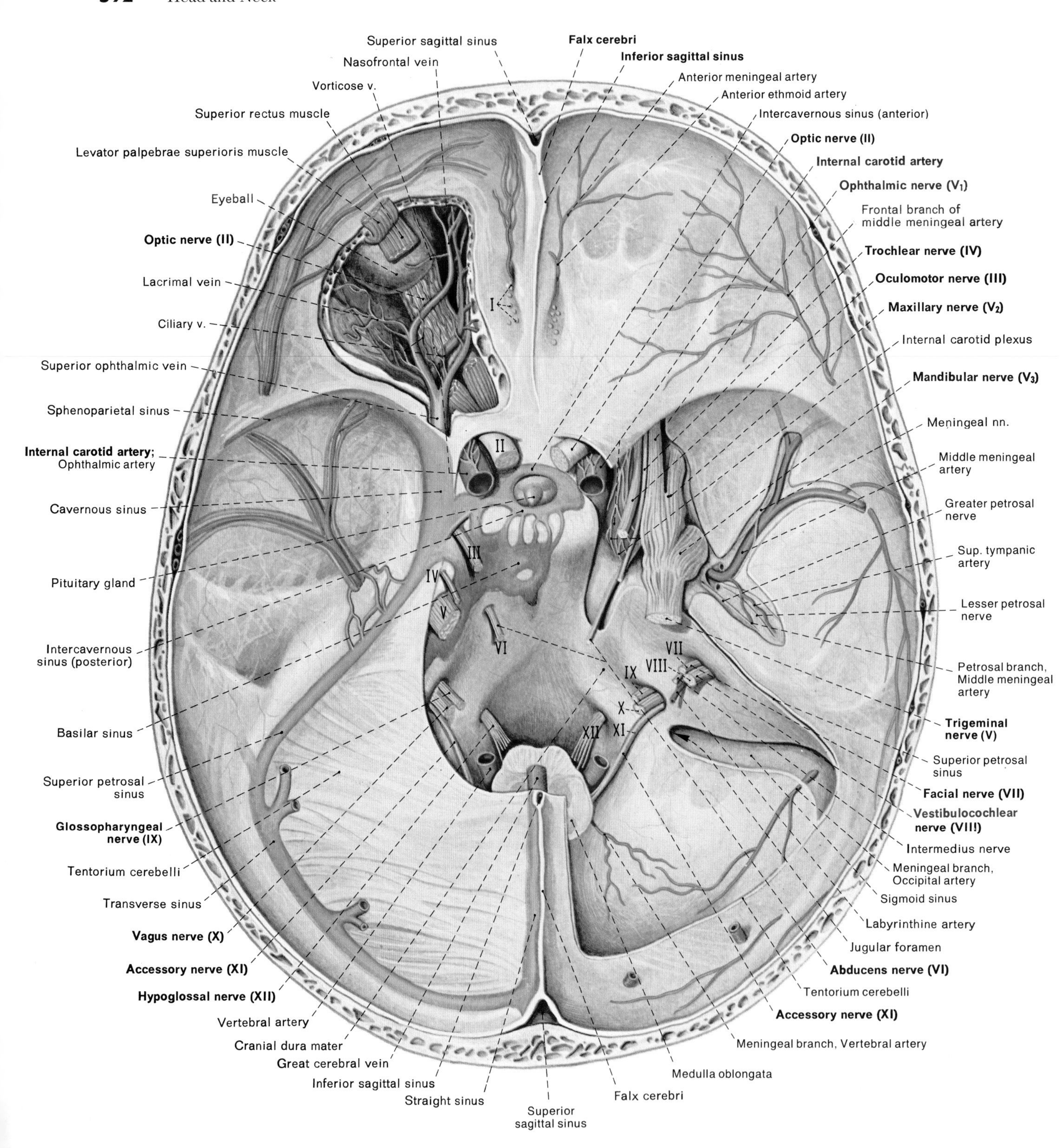

Fig. 28–12: Superior view of blood vessels, nerves and dura mater of the base of the skull. The dura mater and the orbital plate of the left frontal bone have been chipped away to expose structures in the left orbital cavity.

and associated mastoid air cells, and the infratemporal region. In the central floor area of the middle cranial fossa, the pituitary gland lies suspended within the hypophyseal fossa (Fig. 28–12). The central floor area overlies the sphenoid sinuses (Fig. 28–8).

The floor of the middle cranial fossa bears a number of major openings that transmit structures to or from the cranial cavity:

(a) The **optic canal** is a passageway in the lesser wing of the sphenoid. It transmits the optic nerve and the ophthalmic artery between the orbital and cranial cavities **(the optic nerve is the second cranial nerve).** (Compare Figures 28–5 and 28–12).

(b) The **superior orbital fissure** is a passageway between the greater and lesser wings of the sphenoid (Fig. 28–7). It transmits the oculomotor nerve, the trochlear nerve, the ophthalmic division of the trigeminal nerve, the abducent nerve, and the superior and inferior ophthalmic veins between the orbital and cranial cavities **(the oculomotor nerve is the third cranial nerve, the trochlear nerve is the fourth cranial nerve, the trigeminal nerve is the fifth cranial nerve, and the abducent nerve is the sixth cranial nerve; the trigeminal nerve has three divisions called the ophthalmic, maxillary, and mandibular divisions).** (Compare Figures 28–5 and 28–12; the inferior ophthalmic vein is not displayed in Fig. 28–12).

(c) The **foramen rotundum** is an opening in the greater wing of the sphenoid (Fig. 28–6). It transmits the maxillary division of the trigeminal nerve from the pterygopalatine fossa to the cranial cavity. (Compare Figs. 28–5 and 28–12). The **pterygopalatine fossa** is a deep fossa of the face named for its posterior wall (the pterygoid process of the sphenoid) and medial wall (the palatine bone).

(d) The **foramen ovale** is an opening in the greater wing of the sphenoid (Fig. 28–6). It transmits the mandibular division of the trigeminal nerve between the cranial cavity and infratemporal fossa (compare Figs. 28–5 and 28–12).

(e) The **foramen spinosum** is an opening in the greater wing of the sphenoid (Fig. 28–6). It transmits the **middle meningeal artery** from the infratemporal fossa to the cranial cavity (compare Figs. 28–5 and 28–12).

The **posterior cranial fossa** is formed by the occipital bone and the petrous parts of the temporal bones (Fig. 28–5). The inner, concave surface of the occipital bone forms most of the floor of the fossa; on each side, the posterior surface of the petrous part of the temporal bone forms the anterolateral wall of the fossa. The posterior cranial fossa lodges the cerebellar hemispheres, the pons, and the medulla oblongata.

The floor of the posterior cranial fossa bears a number of major openings that transmit structures to or from the cranial cavity:

(a) The **internal acoustic meatus** is a passageway through the petrous part of the temporal bone. It transmits the facial nerve between the cranial cavity and the facial canal and the vestibulocochlear nerve from the inner ear to the cranial cavity **(the facial nerve is the seventh cranial nerve, and the vestibulocochlear nerve is the eighth cranial nerve).** (Compare Figures 28–5 and 28–12).

(b) The **jugular foramen** is an opening between the occipital bone and the petrous part of the temporal bone. It transmits the glossopharyngeal nerve, the vagus nerve, and the cranial and spinal parts of the accessory nerve between the cranial cavity and the neck **(the glossopharyngeal nerve is the ninth cranial nerve, the vagus nerve is the tenth cranial nerve, and the accessory nerve is the eleventh cranial nerve).** (Compare Figs. 28–5 and 28–12). The jugular foramen is also the site at which the internal jugular vein begins as a continuation of the sigmoid sinus.

(c) The **hypoglossal canal** is a passageway through the occipital bone. It transmits the hypoglossal nerve between the cranial cavity and the neck **(the hypoglossal nerve is the twelfth cranial nerve).** (Compare Figures 28–5 and 28–12).

(d) The **foramen magnum** (which is the largest opening in the base of the skull) is an opening in the occipital bone. It transmits the medulla oblongata, the spinal part of the accessory nerve on each side, and the vertebral artery on each side between the cranial cavity and the cervical region. (Compare Figures 28–5 and 28–12).

THE CRANIAL MENINGES AND THE DURAL VENOUS SINUSES

The brain and the spinal cord are enveloped by three connective tissue membranes called meninges. The spinal meninges (the meninges that envelop the spinal cord) are direct continuations of the cranial meninges (the meninges that surround the brain). For both the brain and the spinal cord, the outermost meningeal layer is called the **dura mater,** the middle layer the **arachnoid mater,** and the innermost layer the **pia mater.**

The cranial dura mater (the dura mater about the brain) is very firmly attached in most places to the inner periosteum of the bones of the cranial cavity (Fig. 28–3). However, there are four margins where the cranial dura mater folds in to form taut connective tissue septa which separate and support parts of the brain. Two of the dural septa lie in the median plane (midline) of the body: the **falx cerebri** extends down-

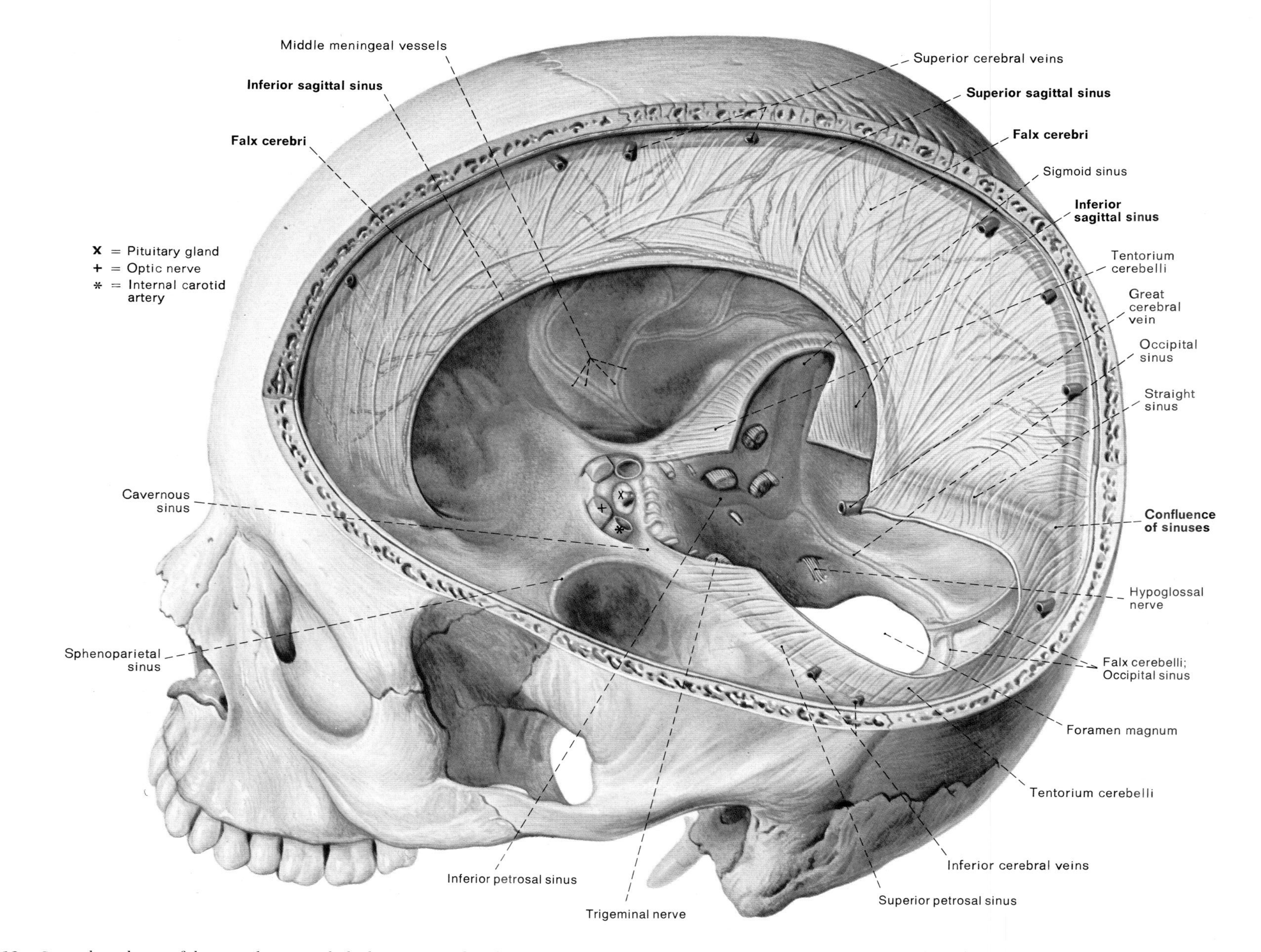

Fig. 28–13: Superolateral view of the cranial cavity with the brain removed to show the cranial dura mater infoldings and the dural venous sinuses. Most of the tentorium cerebelli on the left and part of it on the right have been cut away to show the posterior cranial fossa.

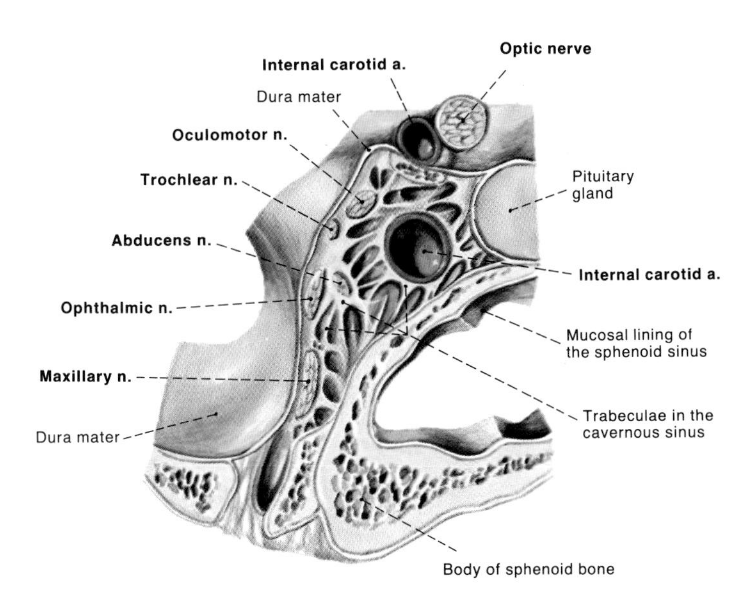

Fig. 28–14: Coronal section through the left cavernous sinus.

ward from the calvaria to form a partial septum between the cerebral hemispheres, and the **falx cerebelli** extends upward and forward from the floor of the posterior cranial fossa to form a partial septum between the cerebellar hemispheres (Fig. 28–13). The other two dural septa extend transversely through (i.e., lie horizontally within) the cranial cavity: the **tentorium cerebelli** extends anteriorly from the occipital bone to form a partial septum between the cerebellar hemispheres and the occipital lobes of the cerebral hemispheres (Figs. 28–12 and 28–13), and the **diaphragma sellae** extends inward from the bony margins encircling the hypophyseal fossa to form a dural roof over the pituitary gland (the pale blue area around the pituitary gland in Figure 28–12 represents the dural venous sinuses of the diaphragma sellae). In the posterior part of the cranial cavity, the falx cerebri and the falx cerebelli join the tentorium cerebelli in the midline.

The dural venous sinuses are the endothelium-lined, valveless, venous channels of the cranial cavity that extend along margins of separation between the cranial dural mater and the inner periosteum lining the cranial cavity. Many of the dural venous sinuses extend along the margins of the dural septa. The dural venous sinuses receive blood from the brain, and ultimately drain most of it into the internal jugular veins.

The **superior and inferior sagittal sinuses** extend along the superior and inferior margins, respectively, of the falx cerebri (Fig. 28–13). The posterior ends of the superior and inferior sagittal sinuses are connected by the **straight sinus** (which extends along the margin of attachment between the falx cerebri and tentorium cerebelli) (Figs. 28–12 and 28–13). The dilated union of the superior sagittal sinus with the straight sinus is called the **confluence of sinuses** (Fig. 28–13). The **transverse sinuses** emerge laterally from either the superior sagittal sinus (near its posterior end) or the confluence of sinuses and extend along the posterior margin of the tentorium cerebelli (Fig. 28–12). At the border between the floors of the middle and posterior cranial fossae, the transverse sinus on each side becomes continuous with the **sigmoid sinus;** the sigmoid sinus follows an **S**-shaped course on the floor of the posterior cranial fossa down to the jugular foramen. At the jugular foramen, the sigmoid sinuses become continuous with the **internal jugular veins.**

The **cavernous sinuses** are a pair of small dural venous sinuses lying on either side of the hypophyseal fossa (Fig. 28–12). Segments of important structures extend through either the interior or lateral wall of each cavernous sinus. Segments of the oculomotor nerve, ophthalmic and maxillary divisions of the trigeminal nerve, and trochlear nerve extend through the lateral wall of the cavernous sinus (Fig. 28–14). The interior of the cavernous sinus is traversed by segments of the internal carotid artery and abducent nerve (Fig. 28–14). The left and right cavernous sinuses are connected by **intercavernous sinuses** that lie anterior and posterior to the pituitary gland (Fig. 28–12).

The **superior petrosal sinus** on each side connects the cavernous sinus with the transverse sinus, near its union with the sigmoid sinus (Fig. 28–13). The **inferior petrosal sinus** on each side connects the cavernous sinus with the internal jugular vein, near its origin from the sigmoid sinus (Fig. 28–13).

Most of the cerebrospinal fluid (CSF) about the brain drains into venous blood through protrusions of the arachnoid mater (**arachnoid villi or arachnoid granulations**) in the walls of dural venous sinuses (Fig. 28–3). The superior sagittal sinus generally bears numerous arachnoid villi.

CHAPTER 29

The Nasal and Orbital Cavities

The nasal cavities serve to clean, humidify, and warm the air inhaled through the nostrils. The nasal cavities open posteriorly into the uppermost part of the pharynx (the nasopharynx) (Fig. 27–4). The nasopharynx receives the air that passes through the nasal cavities and channels it down into the lower parts of the pharynx towards the larynx.

The orbital cavities house the eyeballs, the muscles that move them, and the lacrimal glands. The orbital cavities underlie the anterior cranial fossa of the cranial cavity and overlie the maxillary sinuses of the maxillary bones.

THE NASAL CAVITIES

The Skeletal and Cartilaginous Foundation of the Nasal Cavities

Each nasal cavity has a roof, medial wall, lateral wall, and floor. The skeletal and cartilaginous foundation of these walls is as follows:

The **cribriform plate of the ethmoid** forms the roof of each nasal cavity (Fig. 28–5).

The **nasal septum** is the shared medial wall of the nasal cavities (Fig. 29–1). The nasal septum consists of the **septal cartilage** anteriorly, the **perpendicular plate of the ethmoid** posterosuperiorly, and the **vomer** posteroinferiorly. The vomer is an unpaired bone of the skull, and it contributes principally to the skeletal foundation of the nasal septum.

The lateral wall of each nasal cavity bears three bony shelves called the conchae that project medially into the nasal cavity (Figs. 29–2 and 31–3). The **superior** and **middle conchae** are parts of the ethmoid. The **inferior concha** on each side is an individual bone of the skull, and it contributes principally to the skeletal foundation of the part of the nasal cavity's lateral wall for which it is named.

The **palatine process of the maxillary bone** and the **horizontal plate of the palatine** form the floor of each nasal cavity (Fig. 29–2). The paired maxillary bones of the skull form the upper jaw (Fig. 28–10). The paired palatine bones of the skull lie in the interior of the skull, bordered by the paired maxillary bones anteriorly and the pterygoid processes of the sphenoid posteriorly.

The nasal conchae divide each nasal cavity into four curved, anteroposterior passageways. The **superior, middle,** and **inferior meatuses** extend, respectively, beneath the superior, middle, and inferior conchae (Fig. 29–2). The **sphenoethmoidal recess** passes above the superior concha. The nasal conchae serve as turbinates which baffle the air that is breathed through the meatuses and the sphenoethmoidal recesses of the nasal cavities. They also increase the surface area within the nasal cavities with which inspired air comes into contact. The baffled air is cleansed, humidified, and warmed as it passes over the sticky, mucus-lined, nasal mucosa.

Four paired sets of **paranasal sinuses** communicate with the nasal cavities. The **maxillary sinus** (in the maxillary bone) is the largest paranasal sinus; it drains its mucus into the middle meatus (Fig. 29–3). The **frontal sinus** (in the frontal bone) drains its mucus into the middle meatus. The ethmoid houses the **anterior, middle, and posterior ethmoid sinuses.** The anterior and middle ethmoid sinuses drain into the middle meatus, and the posterior ethmoid sinus drains into the superior meatus (Fig. 29–3). The **sphenoid sinus** (in the body of the sphenoid) lies anteroinferior to the hypophyseal fossa, and drains its mucus into the sphenoethmoidal recess (Fig. 29–3).

The Innervation of the Nasal Cavity

The innervation of the nasal cavity involves cranial nerves I, V, and VII, the pterygopalatine ganglion, and postganglionic sympathetic fibers from the superior cervical ganglion.

Cranial Nerve I: The olfactory nerve arises from first-order neurons (sensory neurons sensitive to smell) which lie within the mucosal lining of the uppermost region of the nasal cavity (Fig. 29–4). The afferent fibers of the first-order neurons extend from the nasal cavity into the cranial cavity by passing through perforations in the cribriform plate of the ethmoid. Upon entering the cranial cavity, the afferent fibers of the first-order neurons extend up into the **olfactory bulb** to synapse with second-order neurons (neurons that are second in order in transmitting olfactory sensation

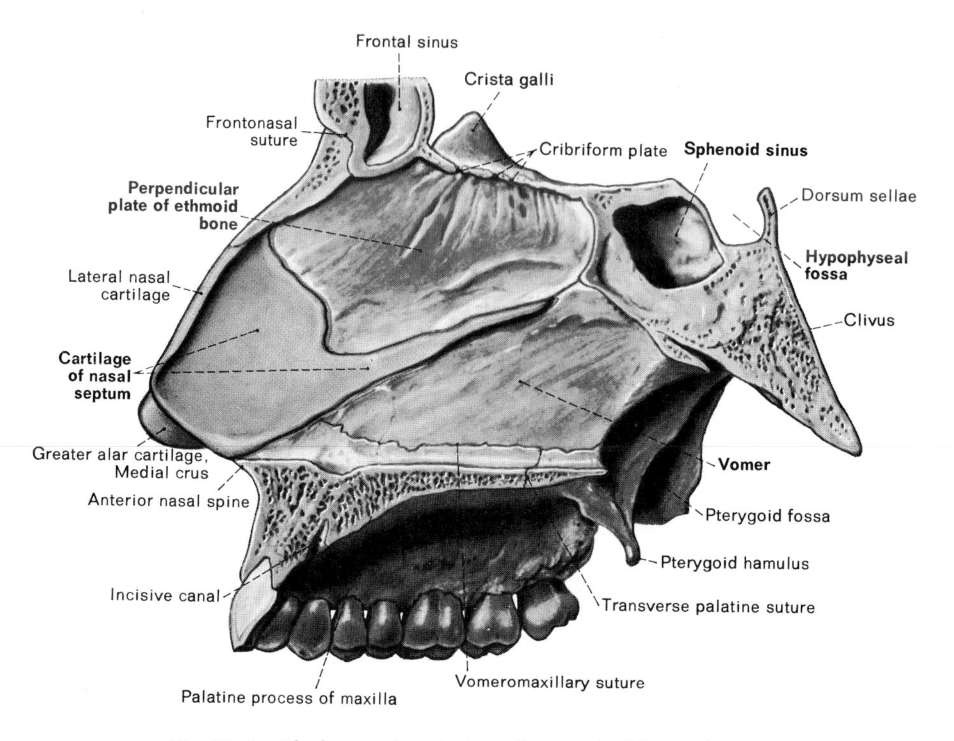

Fig. 29–1: The bony and cartilaginous framework of the nasal septum.

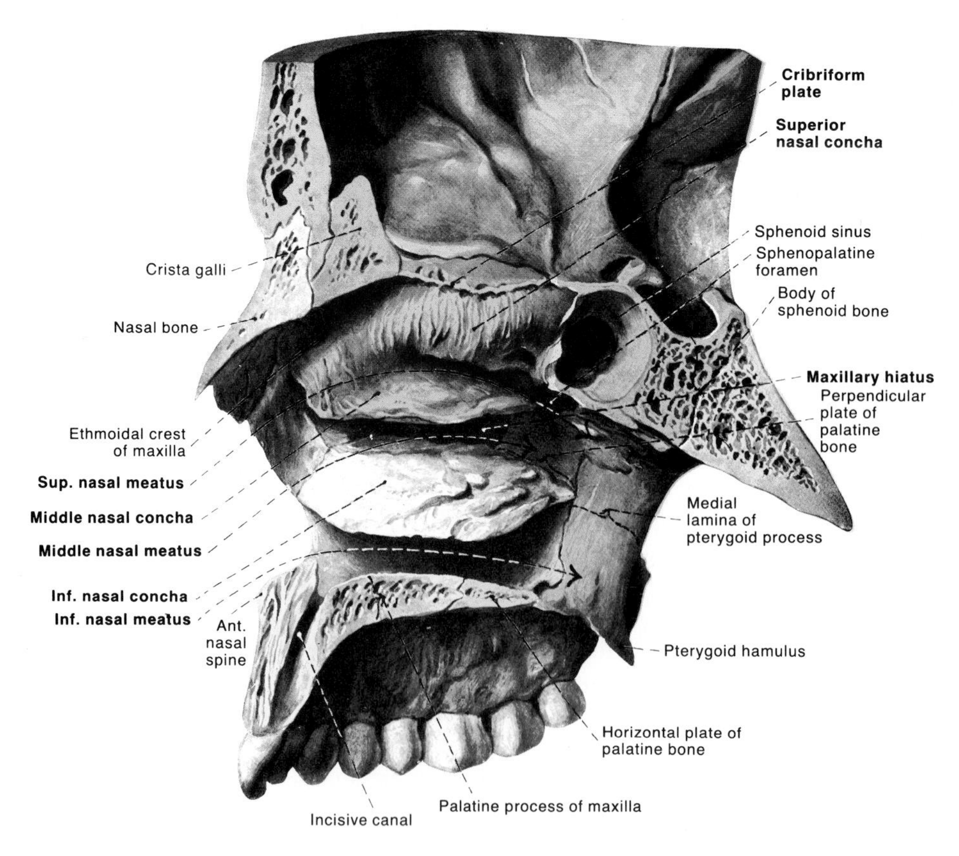

Fig. 29–2: The bony lateral wall of the right nasal cavity.

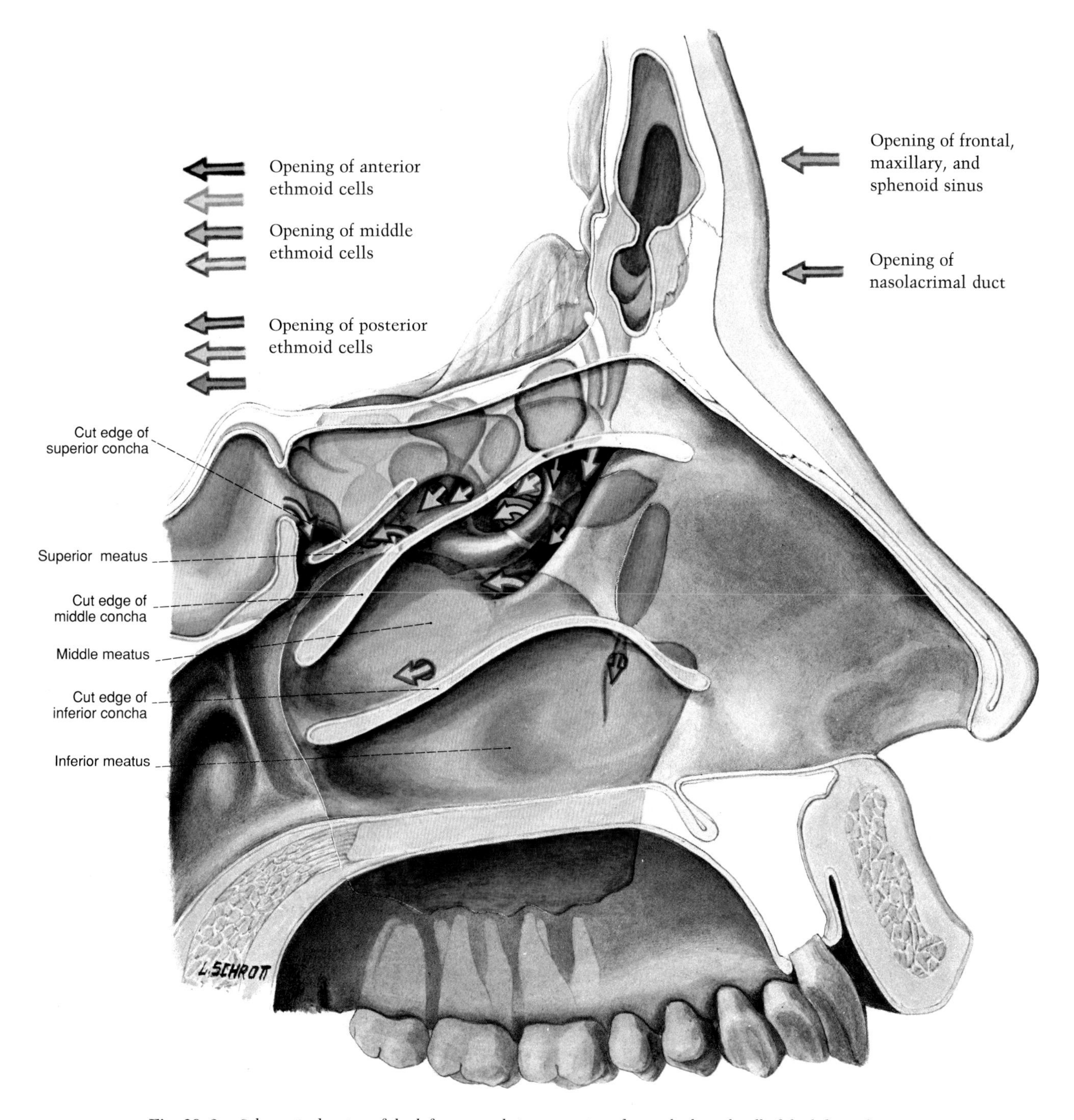

Fig. 29–3: Schematic drawing of the left paranasal sinuses projected onto the lateral wall of the left nasal cavity.

to the brain). The afferent fibers of the second-order neurons extend posteriorly through the **olfactory tract** to enter the brain (Fig. 28–11).

The Pterygopalatine Ganglion: The pterygopalatine ganglion is a ganglion of the parasympathetic division of the autonomic nervous system. It is attached to the maxillary division of the trigeminal nerve in the pterygopalatine fossa (Fig. 29–5) [the pterygopalatine fossa is a deep fossa of the face named for its posterior wall (the pterygoid process of the sphenoid) and medial wall (the palatine bone); the maxillary division of the trigeminal nerve extends from the cranial cavity into the pterygopalatine fossa by passing through the foramen rotundum (Figs. 28–12 and 29–5)]. The ganglion contains cell bodies of postganglionic parasympathetic neurons (Fig. 29–5). These cell bodies provide postganglionic parasympathetic secretomotor fibers to the mucosal glands of the nasal cavity. These postganglionic parasympathetic secretomotor fibers are transmitted by branches of the pterygopalatine ganglion, such as

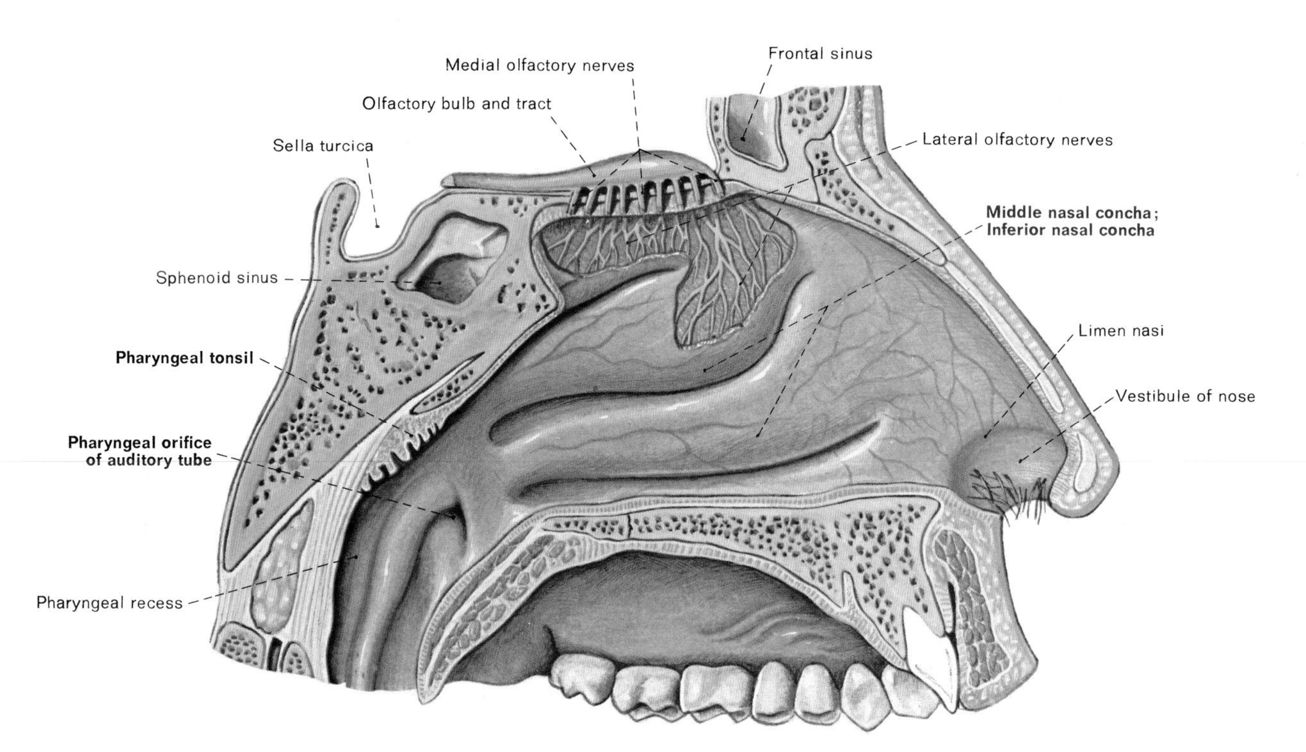

Fig. 29–4: The lateral wall of the left nasal cavity. The mucous membrane overlying fibers of the olfactory nerve has been removed.

the nasopalatine and greater and palatine nerves (Fig. 29–5).

Postganglionic Sympathetic Fibers: The pterygopalatine ganglion also transmits postganglionic sympathetic fibers that innervate tissues of the nasal epithelium. These fibers originate from postganglionic sympathetic neurons of the **superior cervical ganglion** (the most superior ganglion of the sympathetic trunk) (Fig. 29–5). The fibers reach the pterygopalatine fossa by extending first within the **internal carotid plexus** (which is the plexus of autonomic fibers that extends along the length of the internal carotid artery) (Fig. 29–6), then within the **deep petrosal nerve** (which is a bundle of autonomic fibers derived from the internal carotid plexus), and finally within the **nerve of the pterygoid canal** (Fig. 29–5). [The nerve of the pterygoid canal is formed by the union of the deep petrosal and greater petrosal nerves (Fig. 29–5). The pterygoid canal is a passageway in the pterygoid process of the sphenoid.] The postganglionic sympathetic fibers pass through the pterygopalatine ganglion and reach the nasal epithelium through the branches of the ganglion.

Cranial Nerve V: Branches of the ophthalmic and maxillary divisions of the trigeminal nerve provide sensory innervation to the epithelial lining of, respectively, the anterior one-third and the posterior two-thirds of the nasal cavity.

Cranial Nerve VII: The **facial nerve,** through its **greater petrosal nerve,** provides the preganglionic parasympathetic fibers which synapse within the pterygopalatine ganglion (Fig. 29–5).

Blood Supply and Venous and Lymphatic Drainage of the Nasal Mucosa

The sphenopalatine and greater palatine branches of the maxillary artery provide most of the blood supply to the nasal mucosa. Branches of the opththalmic artery (an intracranial branch of the internal carotid artery) and the facial artery contribute blood supply to the anterior part of the nasal mucosa. Venous drainage of the nasal mucosa includes drainage by the sphenopalatine vein (an indirect tributary of the maxillary vein), the facial vein, and the superior and inferior ophthalmic veins (the ophthalmic veins are tributaries of the intracranial cavernous sinus).

The lymphatics from the anterior part of the nasal mucosa are afferent to submandibular lymph nodes. The lymphatics from the remainder of the nasal mucosa are afferent to upper deep cervical nodes.

CLINICAL PANEL VI.1

Acute Paranasal Sinusitis

Description: Acute paranasal sinusitis is acute inflammation of a paranasal sinus with infection.

Common Symptoms: The patient commonly has nasal congestion, nasal discharge, and a cough. Acute maxillary sinusitis commonly presents with pain in the cheek and one or more upper molar teeth (the upper molar teeth may ache because their roots come into contact with the mucous membrane lining the floor of the maxillary sinus). Acute frontal sinusitis typically presents with pain in the forehead region immediately superomedial to the eyeball; pain may be referred to the occiput (the back of the head) or the anterior aspect of the temporal region. Acute ethmoid sinusitis commonly presents with pain between the eyes or posterior to the eye on the affected side; pain may be referred to the vertex (the most superior point of the skull) or the upper posterior aspect of the temporal region. Acute sphenoid sinusitis generally presents with retroorbital pain (pain behind the eye); pain may be referred to the vertex or occiput. Any maneuver that increases engorgement of the nasal mucosal vasculature (such as bending the head forward and downward) typically worsens the pain of acute paranasal sinusitis.

Common Signs: Acutely inflamed maxillary and frontal sinuses are generally tender to percussion. Transillumination of the interior of these sinuses frequently reveals opacity.

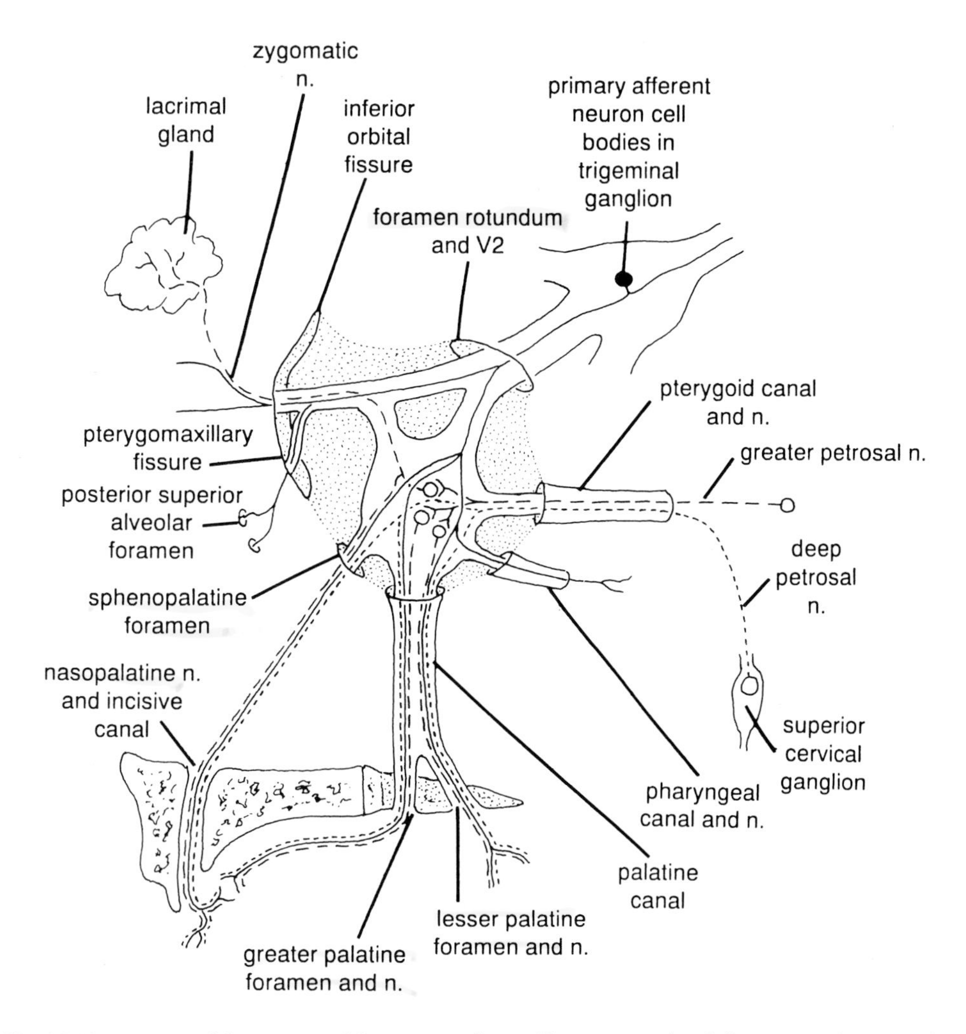

Fig. 29–5: Diagram of the origins and destinations of nerve fibers associated with the pterygopalatine ganglion.

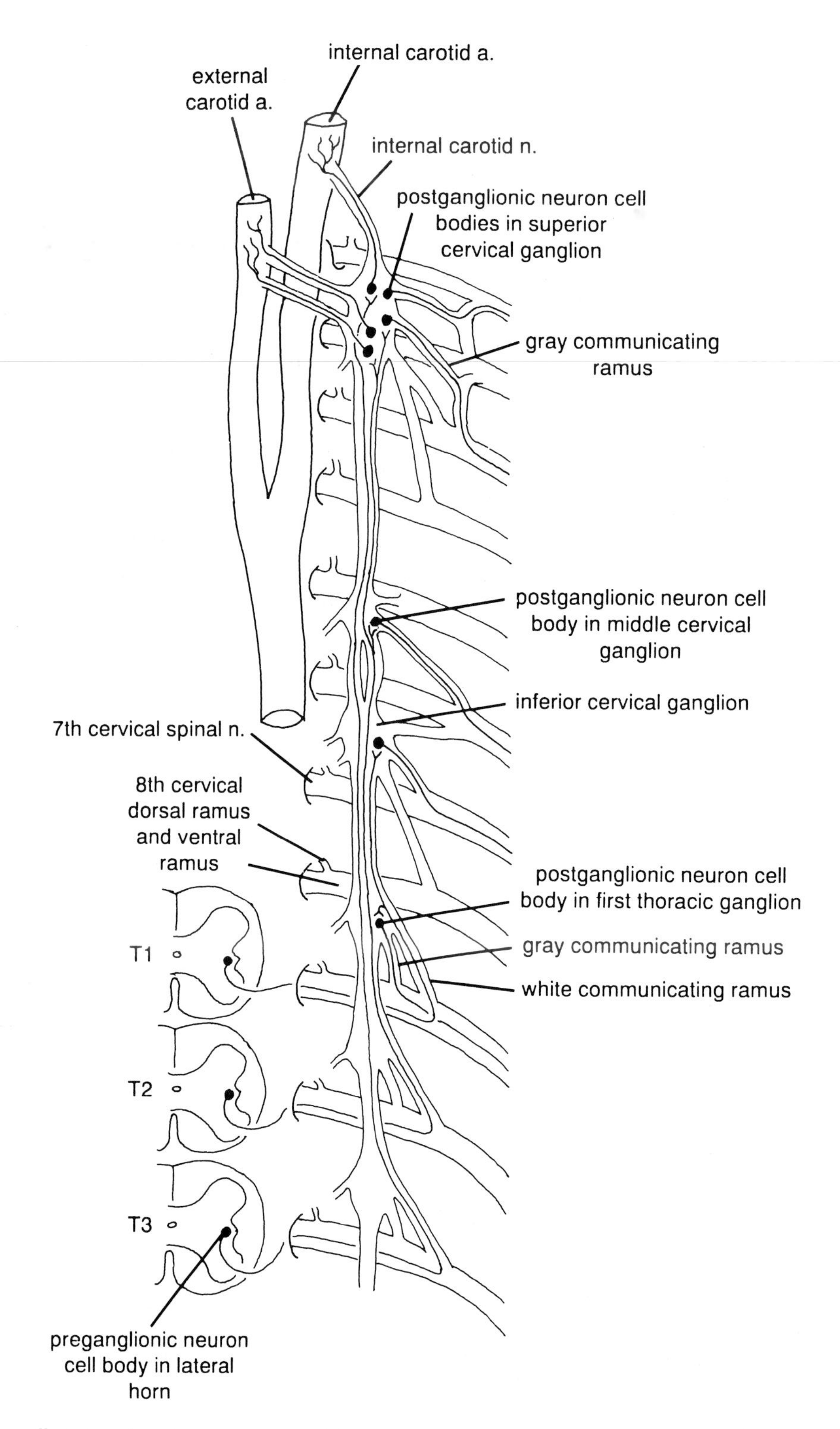

Fig. 29–6: Diagram illustrating the origins and courses of the preganglionic and postganglionic sympathetic fibers associated with the sympathetic innervation of the head and neck.

THE ORBITAL CAVITIES

The Walls of the Orbit

The walls of the orbital cavity resemble a four-sided pyramid lying on one of its sides; the apex of this pyramidal skeletal foundation projects posteromedially and the base faces anteriorly (Fig. 29–7). The **orbital plate of the frontal bone** forms most of the roof of each orbital cavity (the lesser wing of the sphenoid contributes to the posterior aspect of the roof). The **orbital plate of the maxillary bone** forms most of the floor of each orbital cavity (the orbital process of the palatine bone and part of the zygomatic bone also contribute to the floor). Proceeding from the anterior rim to the posterior limit, the medial wall is constructed from the **frontal process of the maxillary bone,** the **lacrimal bone,** the **orbital plate of the ethmoid,** and the **body of the sphenoid** (surrounding the optic canal). The lateral wall is formed anteriorly by the **frontal process of the zygomatic (malar) bone** and posteriorly by the greater wing of the sphenoid. The zygomatic (malar) bone on each side of the skull underlies the prominence of the cheek (Fig. 28–10).

The Eyeball

The eyeball has three tissue layers (Fig. 29–8). The transparent **cornea** forms the most anterior part of the outermost tissue layer of the eyeball, and the white, fibrous **sclera** forms the remainder of the outermost tissue layer (the visible, white part of the eye represents the anterior margin of the eye's sclera). The cornea and sclera are continuous with each other at the circular **sclerocorneal junction (limbus)**.

The middle tissue layer is a vascular layer that transmits the larger blood vessels of the eye's outer tissue layers; it consists of the iris, ciliary body, and choroid. The **iris** (the annular, colored part of the eye) is the most anterior part of the middle layer; the iris lies directly anterior to the lens of the eye (Fig. 29–8). The iris surrounds the **pupil,** the dark circular aperture of the eye. The iris contains two muscles composed of smooth muscle fibers: whereas the smooth muscle fibers of **sphincter pupillae** are circularly-oriented within the iris, the smooth muscle fibers of **dilator pupillae** are radially-oriented (Figs. 29–8 and 29–9). Sphincter and dilator pupillae are antagonists of each other: whereas sphincter pupillae acts to decrease the

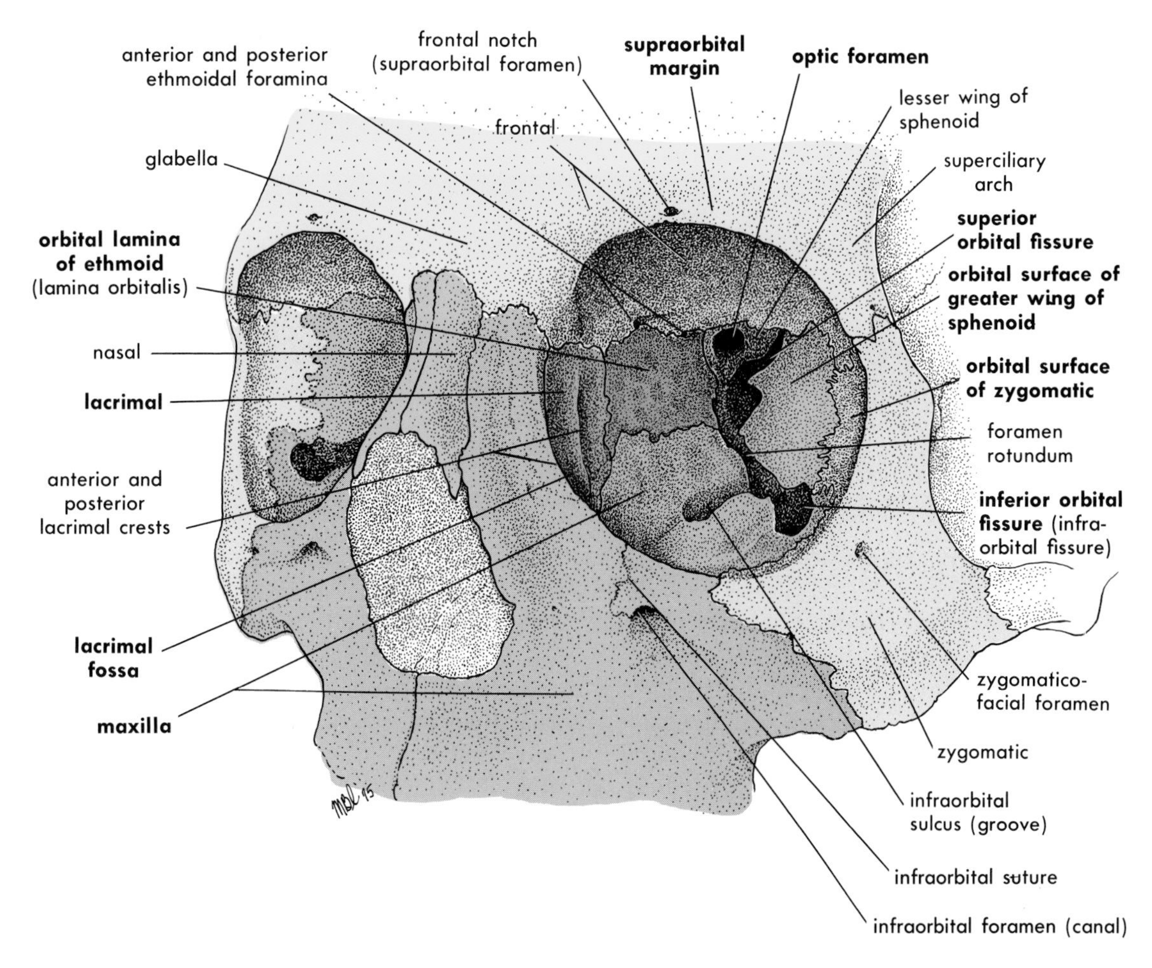

Fig. 29–7: Anterolateral view of the bony foundation of the left orbital cavity.

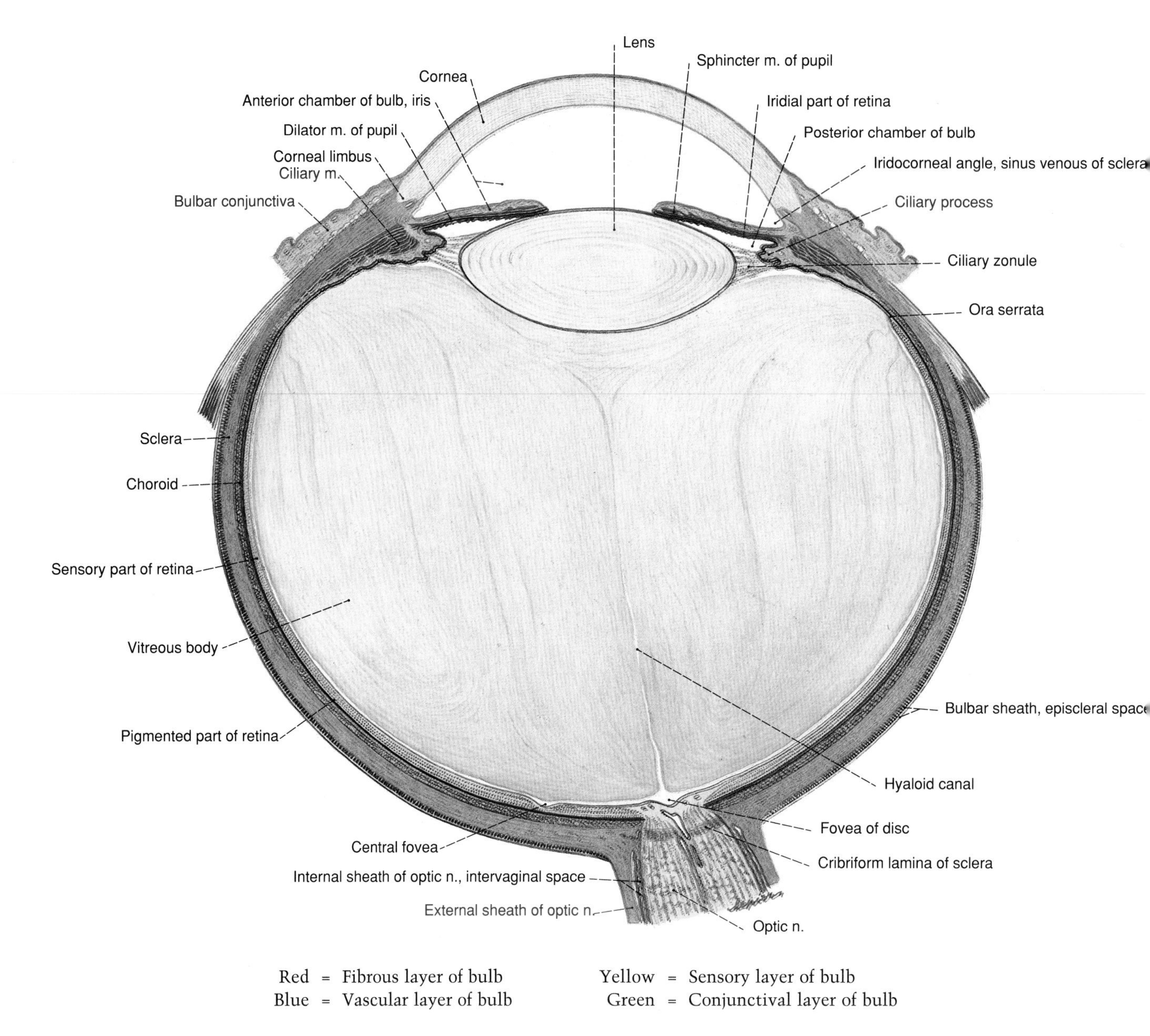

Fig. 29–8: Horizontal section of the left eyeball through the optic nerve and optic disc. The outer, fibrous layer of the eyeball is colored red, the middle, vascular layer blue, and the inner sensory layer yellow. The bulbar conjunctiva lining the visible part of the eyeball's sclera is colored green.

size of the pupil, dilator pupillae serves to increase the size of the pupil.

The **ciliary body** is an annular body of tissue in the anterior portion of the eye that encircles and suspends the lens of the eye; it is the part of the eyeball's middle tissue layer that is continuous anteriorly with the outer margin of the iris (Fig. 29–9). Finger-like extensions from the inner margin of the ciliary body (the **ciliary processes**) suspend the lens with fine ligamentous strands called the **suspensory ligaments of the lens** (**zonular fibers**) (Fig. 29–9). The ciliary processes and their suspensory ligaments stretch the lens in such a fashion so as to diminish the convexity of its anterior and posterior surfaces.

The ciliary body contains circularly-oriented and meridionally-oriented (radially-oriented) smooth muscle fibers collectively called **ciliaris,** or the **ciliary muscle** (the meridionally-oriented muscle fibers originate from the sclera near the sclerocorneal junction and insert into the posterior part of the ciliary body) (Fig. 29–9). When the circularly-oriented and meridionally-oriented smooth muscle fibers of ciliaris concentrically contract, they collectively draw the ciliary processes inward toward the optic (anteroposterior) axis of the eye; this action lessens the tension exerted on the lens through its suspensory ligaments, thus per-

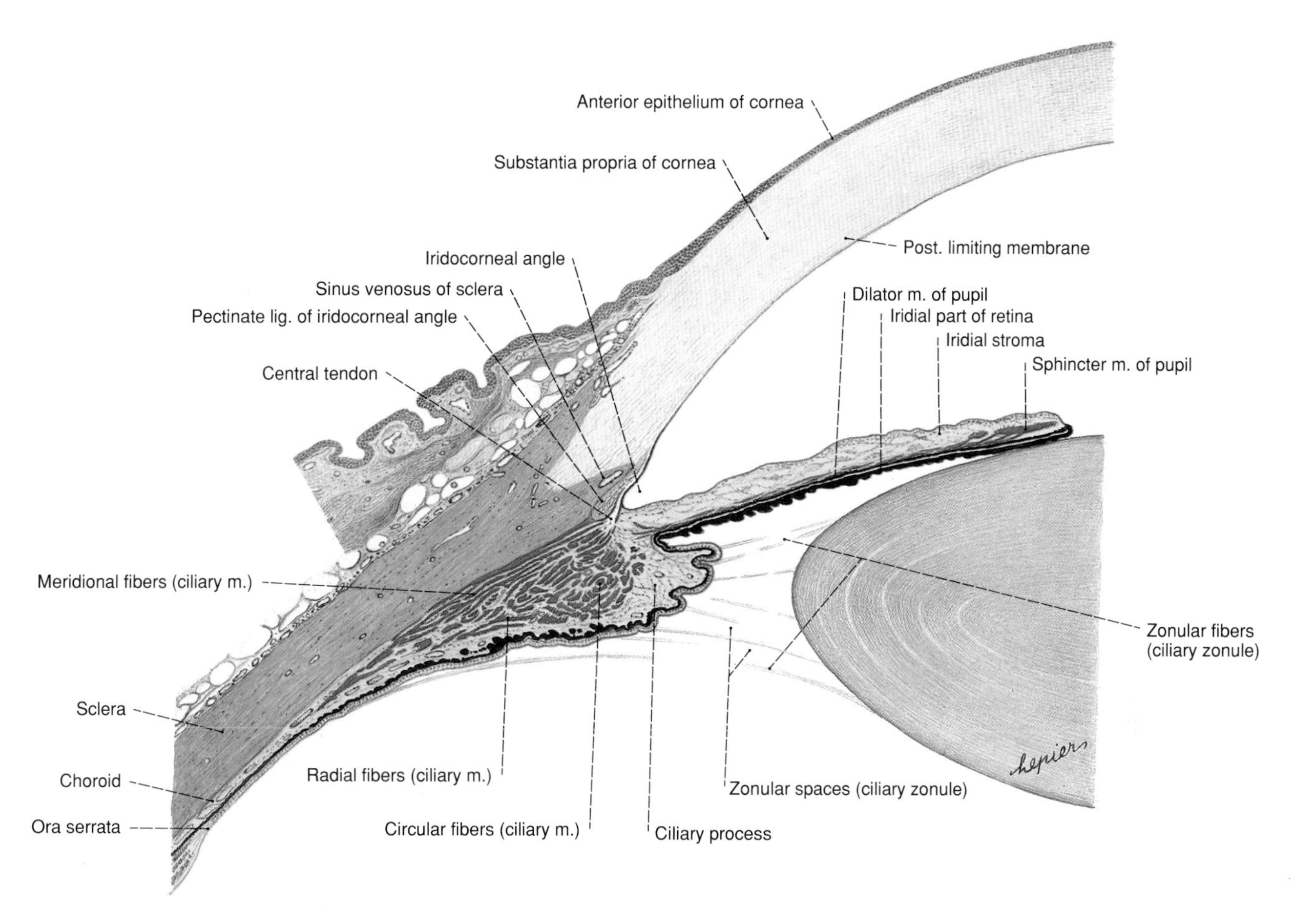

Fig. 29–9: Horizontal section through the anterior part of the eyeball. The outer, fibrous layer of the eyeball is colored red, the middle, vascular layer blue, and the inner sensory layer yellow. The bulbar conjunctiva lining the visible part of the eyeball's sclera is colored green.

mitting the lens to assume a more globular shape (the increased convexity of its anterior and posterior surfaces allows the lens to focus light rays from closer objects onto the retina).

The ciliary processes of the ciliary body also secrete the fluid **(aqueous humor)** that fills the cavity in the anterior portion of the eye between the cornea and the lens; the parts of this cavity which lie anterior and posterior to the iris are called, respectively, the **anterior and posterior chambers of the eye** (aqueous humor fills both chambers) (Fig. 29–8).

The most posterior part of the middle vascular layer of the eye is called the **choroid** (Figs. 29–8 and 29–9). The choroid is continuous at its anterior margin with the posterior margin of the ciliary body.

The **retina** forms the innermost tissue layer of the eye (Fig. 29–8). It contains the cell bodies of the photo-sensitive receptor cells (rods and cones) of the eye and the other neurons called bipolar and ganglion neurons whose afferent fibers form most of the optic nerve. The retina envelops the large cavity in the posterior portion of the eye (the vitreous chamber) that accounts for about four-fifths of the eyeball's interior; the vitreous chamber is filled with a translucent, gelatinous substance called the **vitreous body** (Fig. 29–8).

The ophthalmoscope is used to examine the **optic fundus** (the posterior portion of the interior of the eye). Ophthalmoscopic examination shows the region of continuity between the retina and the optic nerve as a lightly pigmented, circular elevation called the **optic disc** (the optic disc lies medial and slightly superior to the center of the optic fundus) (Fig. 29–10). Branches of the **central artery of the retina** radiate outward from the optic disc and tributaries of the **central vein of the retina** converge upon the optic disc. A round dark region called the **macula lutea** lies near the center of the optic fundus; the photo-sensitive neurons of the macula lutea provide the greatest visual acuity of the retina.

The Muscles that Move the Eyeball

Six muscles direct the movements of the eyeball. The actions of these muscles are described by how they move the eyeball from its primary position. In the eyeball's primary position, the cornea faces directly anteriorly (the eye looks straight ahead).

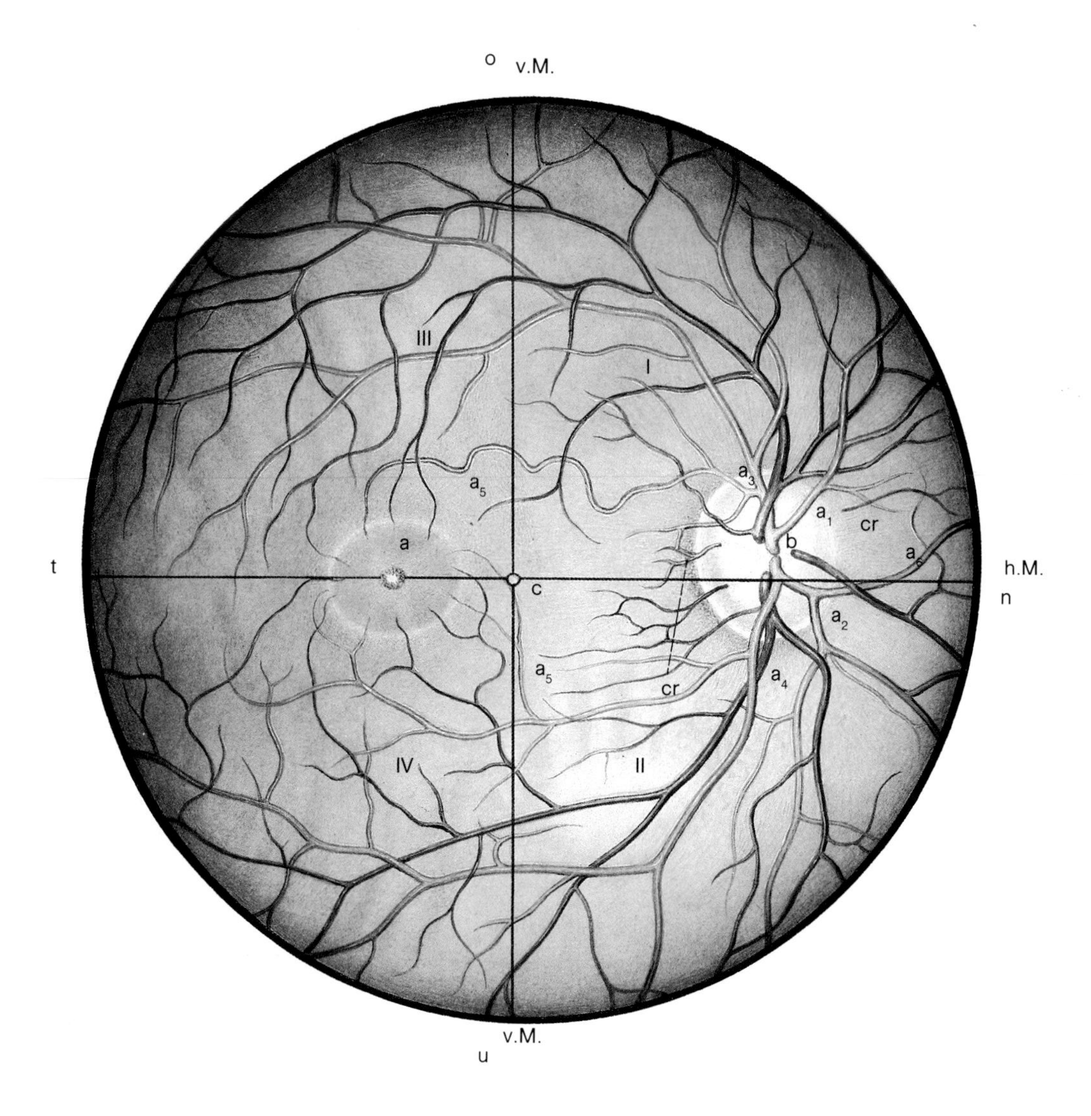

a = Macula, fovea centralis	a_5 = Macular arterioles	t = Temporal side of fundus
a_1 = Superior nasal retinal arteriole	a_6 = Medial arteriole of retina	u = Lower side of fundus
a_2 = Inferior nasal retinal arteriole	b = Optic disc	v.M. = Meridian
a_3 = Superior temporal retinal arteriole	c = Midpoint of the fundus	I = Upper nasal quadrant
a_4 = Inferior temporal retinal arteriole	cr = Cilioretinal anastomoses	II = Lower nasal quadrant
	h.M. = Equator	III = Upper temporal quadrant
	n = Nasal side of fundus	IV = Lower temporal quadrant
	o = Upper side of fundus	

Fig. 29–10: The optic fundus.

The four rectus muscles **(superior rectus, inferior rectus, medial rectus,** and **lateral rectus)** originate from a ring of fibrous connective tissue (common tendinous ring) that encircles the optic canal and middle portion of the superior orbital fissure (Fig. 29–11). The four rectus muscles all insert onto the sclera of the eye just posterior to the sclerocorneal junction (Fig. 29–12). The medial rectus muscle adducts the eyeball: it moves the eyeball so that the cornea faces directly medially (in this position, the eye looks inward through the medial corner of the eye, and thus looks toward the nose). The lateral rectus muscle abducts the eyeball: it moves the eyeball so that the cornea faces directly laterally (in this position, the eye looks outward through the lateral corner of the eye). The superior rectus moves the eyeball so that the cornea faces superomedially (in this position, the eye looks both upward and inward). The inferior rectus moves the eyeball so that the cornea faces inferomedially (in this position, the eye looks both downward and inward).

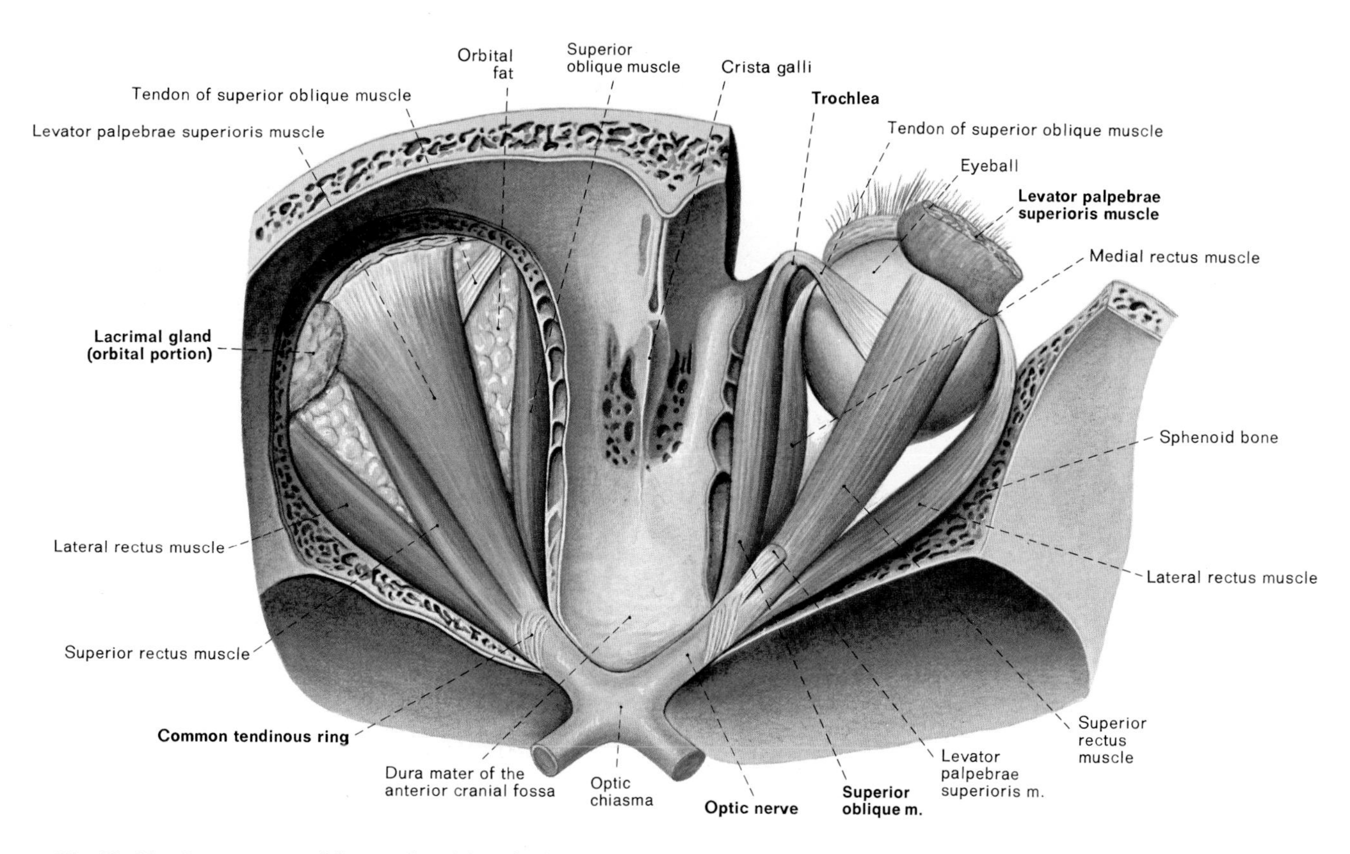

Fig. 29–11: Superior view of the muscles of the orbital cavity. The orbital plates of the frontal bone have been removed to provide a superior view of the orbital cavities. On the right side, levator palpebrae superioris has been resected and the orbital fat removed to expose the ocular muscles.

Superior oblique originates from the body of the sphenoid superomedial to the optic canal (Fig. 29–11). Its muscular belly extends anteriorly through the superomedial aspect of the orbital cavity (Fig. 29–12) before giving rise to a tendon which loops through a fibrocartilaginous trochlea (pulley) attached to the frontal bone (Figs. 29–11 and 29–12). Upon passing through the trochlea, superior oblique's tendon extends posterolaterally over the eyeball and inserts onto the sclera at the posterolateral aspect of the superior surface of the eyeball (Fig. 29–11). Superior oblique moves the eyeball so that the cornea faces inferolaterally (in this position, the eye looks both downward and outward).

Inferior oblique originates from the anteromedial aspect in the floor of the orbit (Fig. 29–12). Its muscle belly extends posterolaterally beneath the eyeball before giving rise to a tendon which inserts onto the sclera at the posterolateral aspect of the inferior surface of the eyeball. Inferior oblique moves the eyeball so that the cornea faces superolaterally (in this position, the eye looks both upward and outward).

The activity of medial rectus is tested by asking the patient to look toward the nose. With the cornea facing directly medially, the activities of the superior and inferior obliques are tested by asking the patient to look, respectively, downward and upward. The activity of lateral rectus is tested by asking the patient to look outward through the lateral corner of the eye. With the cornea facing directly laterally, the activities of the superior and inferior recti are tested by asking the patient to look, respectively, upward and downward.

Note that in testing for the activities of the superior and inferior recti and superior and inferior obliques, an examiner does **not** request the patient to move the eyeball in a manner that corresponds to the action of each of these four muscles. When the examiner requests the patient to look toward the nose, the patient positions the eyeball so that the axis extending from the cornea to the eyeball's center becomes approximately parallel to (a) the axis extending from superior oblique's trochlea to the muscle's insertion onto the eyeball and (b) the axis extending from inferior oblique's origin to its insertion onto the eyeball. In this orientation (with the cornea facing directly medially), superior oblique becomes the only muscle whose concentric contraction can direct the cornea to face downward, and inferior oblique becomes the only muscle whose concentric contraction can direct the cornea to face upward. This is the anatomic basis for requesting the patient to look at the nose to test the activities of the superior and inferior obliques.

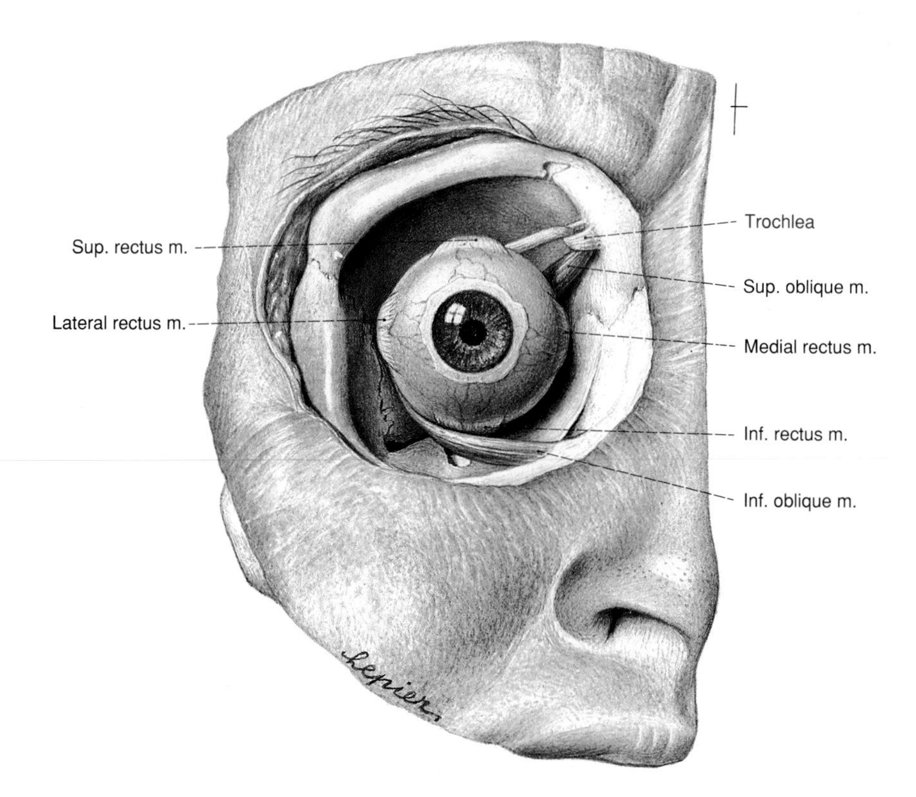

Fig. 29–12: Anterior view of the extraocular muscles of the right orbital cavity.

When the examiner requests the patient to look outward through the lateral corner of the eye, the patient positions the eyeball so that the axis extending from the cornea to the eyeball's center becomes approximately parallel to (a) the axis extending from superior rectus's origin to its insertion onto the eyeball and (b) the axis extending from inferior rectus's origin to its insertion onto the eyeball. In this orientation (with the cornea facing directly laterally), superior rectus becomes the only muscle whose concentric contraction can direct the cornea to face upward, and inferior rectus becomes the only muscle whose concentric contraction can direct the cornea to face downward. This is the anatomic basis for requesting the patient to look out through the lateral corner of the eye in order to test the activities of the superior and inferior recti.

The Eyelids

The upper and lower eyelids are tissue folds extending, respectively, from the upper and lower parts of the anterior bony rim of the orbital cavity. The area between the free margins of the eyelids is called the **palpebral fissure.** The upper and lower eyelids meet each other at the **medial and lateral canthi** (at the medial and lateral corners) of the eye.

Each eyelid has a tough connective tissue foundation consisting of the orbital septum and a tarsal plate (or tarsus). The **orbital septum** is a fascial extension of the periosteum lining the anterior bony rim of the orbital cavity; the orbital septum extends into and through the upper part of the upper eyelid and the lower part of the lower eyelid (Figs. 29–13 and 29–14). The **tarsal plate (tarsus)** in each eyelid extends from the inner margin of the orbital septum to the free margin of the eyelid. The tarsal plates of the eyelids are attached to the anterior bony rim of the orbital cavity at the medial and lateral canthi by tendinous bands called, respectively, the **medial and lateral palpebral ligaments** (Fig. 29–13).

The most central muscle fibers of **orbicularis oculi** lie within the eyelids anterior to the orbital septum and the tarsal plates (the more peripheral muscle fibers of orbicularis oculi lie within the superficial fascia deep to the skin surrounding the eyelids) (Figs. 28–4 and 29–14). The circularly-oriented muscle fibers of oribcularis oculi act to close the eyelids (to bring their free margins into contact with each other). Orbicularis oculi is a muscle of facial expression.

Levator palpebrae superioris is the muscle responsible for raising the upper eyelid; it consists of both skeletal and smooth muscle fibers. It originates from the lesser wing of the sphenoid, extends forward within the orbital cavity (superior to superior rectus) (Fig. 29–11), and inserts onto the tarsal plate of the upper eyelid (Fig. 29–14).

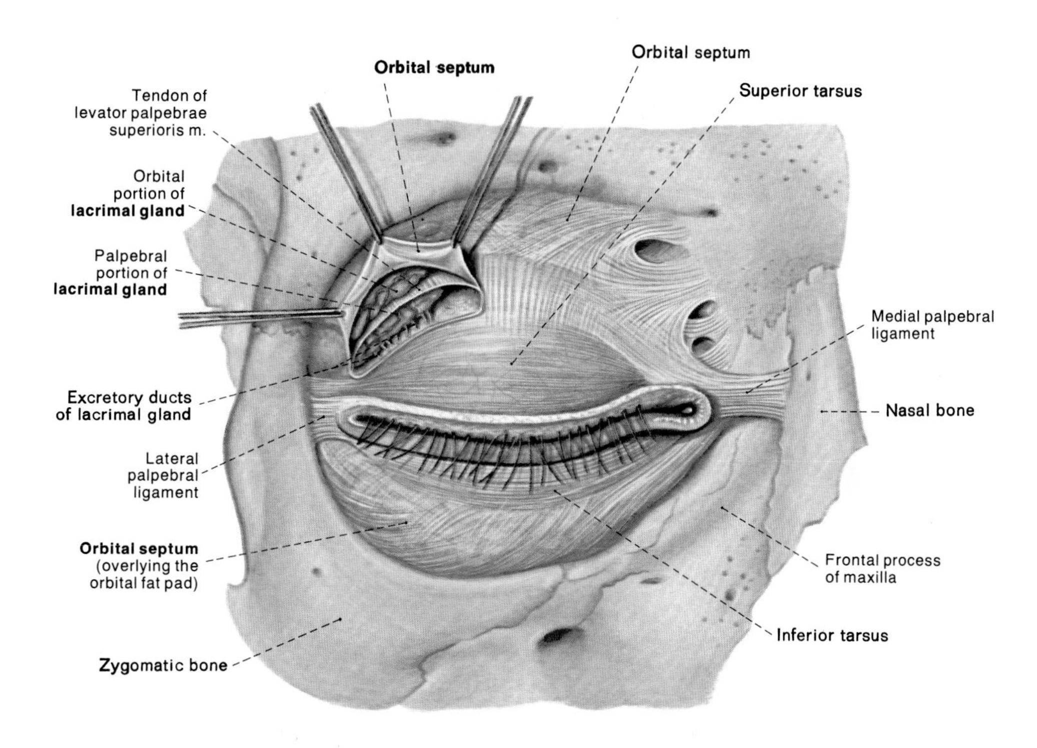

Fig. 29–13: Anterior view of the orbital septum, tarsal plates, palpebral ligaments, and lacrimal gland of the right eye.

The Conjunctival Sac

The conjunctival sac is the space between the visible, anterior portion of the eyeball and the apposing, posterior portions of the eyelids (Figure 29–14 shows a sagittal sectional view of the upper part of the conjunctival sac). The conjunctival sac is lined by a mucous membrane. The posterior portion of this mucous membrane is called the **bulbar conjunctiva;** it lines the visible part of the eyeball's sclera and its epithelium is continuous at the sclerocorneal junction with the epithelium lining the anterior surface of the cornea. The anterior portion of the mucous membrane lining the conjunctival sac is called the **palpebral conjunctiva;** it lines the posterior surfaces of the eyelids. The bulbar and palpebral conjunctiva are continuous with each other at the outer margin of the visible part of the eyeball's sclera. The spaces within the conjunctival sac at the upper and lower margins of reflection between the bulbar and palpebral conjunctiva are called, respectively, the **superior** and **inferior fornices of the conjunctival sac** (Figure 29–14 shows the superior fornix of the conjunctival sac).

The space within the conjunctival sac is lubricated by the watery secretion of the lacrimal gland. The **lacrimal gland** lies in the superolateral aspect of the orbital cavity (Fig. 29–13); its excretory ducts open into the superior fornix of the conjunctival sac. Eyelid movements spread lacrimal fluid inferomedially across the front of the eyeball and ensure that it covers all surfaces of the conjunctival sac. Lacrimal fluid that collects at the medial canthus drains into capillary-sized canals (the **superior** and **inferior lacrimal canaliculi**) in the medial ends of the eyelids (Fig. 29–15). Orbicularis oculi muscle fibers are attached to the lacrimal canaliculi. Concentric contraction of these muscle fibers dilates the lacrimal canaliculi, emptying, in the process, lacrimal fluid collected by the lacrimal canaliculi into the **lacrimal sac.** The lacrimal sac is drained by the **nasolacrimal duct,** which, in turn, drains into the inferior meatus of the nasal cavity (Fig. 29–3).

The Innervation of the Orbital Cavity

The innervation of the orbital cavity involves cranial nerves II to VII, the ciliary ganglion, the pterygopalatine ganglion, and postganglionic sympathetic fibers from the superior cervical ganglion.

Cranial Nerve II: The afferent fibers of the ganglion neurons in the retina of the eyeball converge at the back of the eyeball (specifically, the optic disc) to form the optic nerve (Fig. 29–8). The optic canal transmits the optic nerve from the orbital cavity to the cranial cavity (Fig. 28–12). Upon entering the cranial cavity, the paired optic nerves join to form the optic chiasma, from which the optic tracts extend posterolaterally to the brain (Fig. 29–16). In the optic chiasma, the afferent fibers from the nasal half of each eyeball's retina cross to the opposite side to accompany the afferent fibers from the temporal half of the other eyeball's retina in the contralateral optic tract. In other

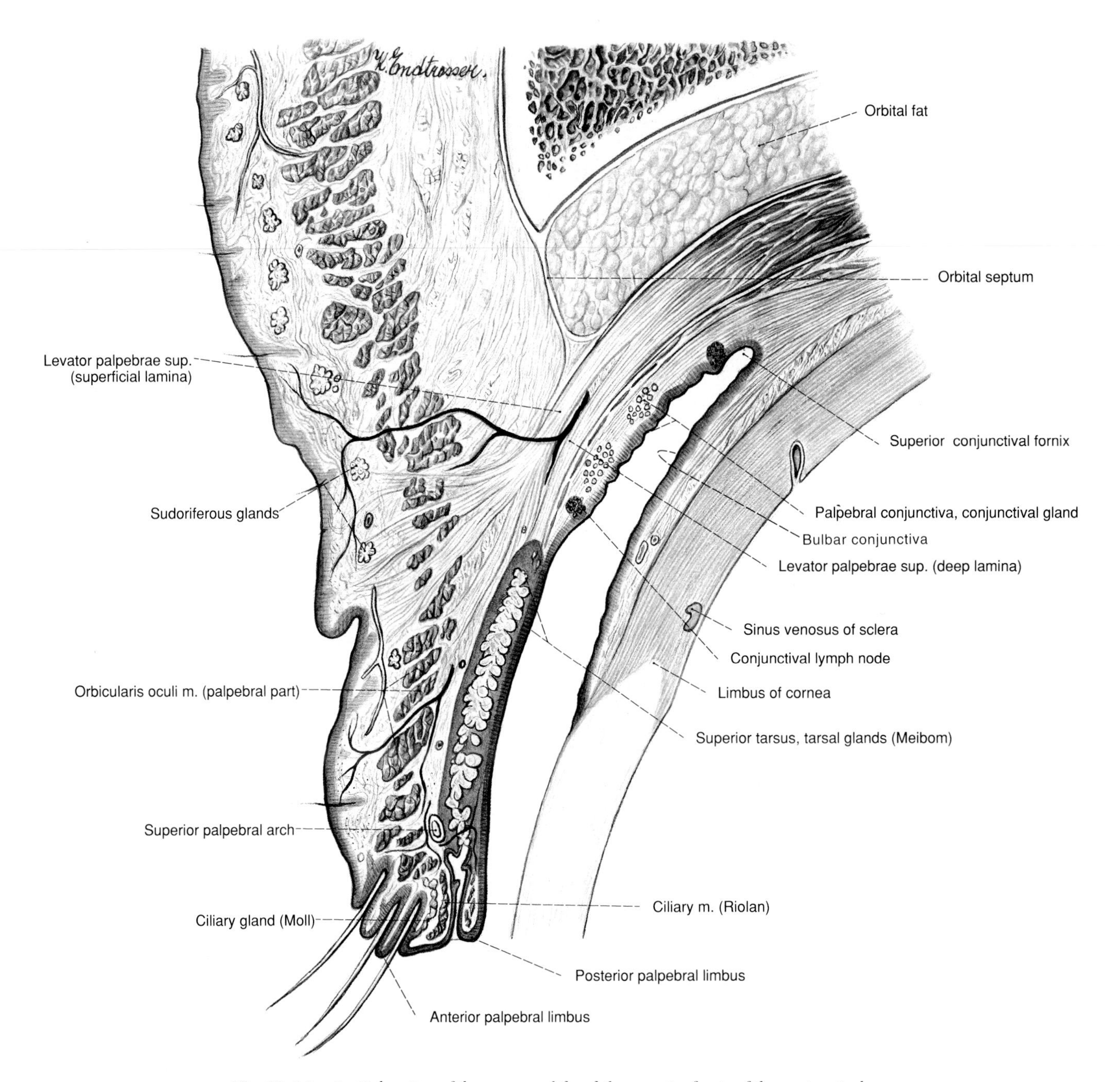

Fig. 29–14: Sagittal section of the upper eyelid and the superior fornix of the conjunctival sac.

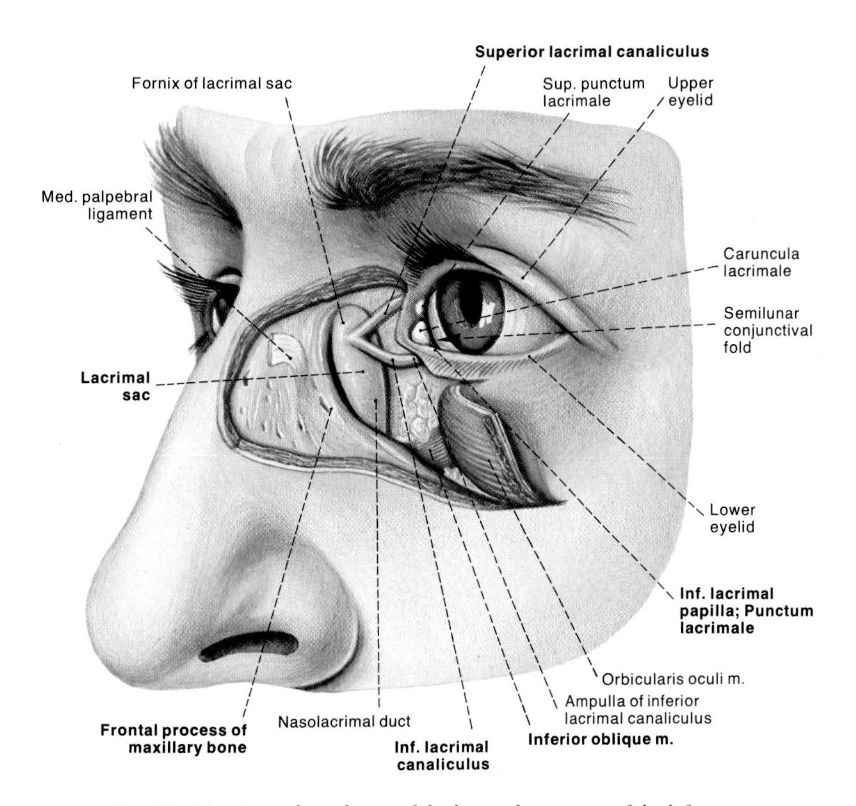

Fig. 29–15: Anterolateral view of the lacrimal apparatus of the left eye.

words, the left optic tract transmits afferent fibers from the left half of the retina of each eyeball, and the right optic tract transmits afferent fibers from the right half of the retina of each eyeball.

Cranial Nerve III: The oculomotor nerve (Fig. 29–17) innervates superior rectus, medial rectus, inferior rectus, inferior oblique, and the skeletal muscle fibers of levator palpebrae superioris. The oculomotor nerve also provides the preganglionic parasympathetic fibers which synapse within the ciliary ganglion (Fig. 29–18).

The Ciliary Ganglion: The ciliary ganglion is a ganglion of the parasympathetic division of the autonomic nervous system. It is located in the orbital cavity behind the eyeball, attached to the oculomotor nerve (Fig. 29–17). The ciliary ganglion provides postganglionic parasympathetic fibers which innervate ciliaris and sphincter pupillae (Fig. 29–18); the **short ciliary nerves** transmit these fibers from the ganglion to the eyeball (Fig. 29–17).

Cranial Nerve IV: The trochlear nerve (Fig. 29–17) innervates superior oblique.

Cranial Nerve V: The orbital cavity transmits the three branches of the ophthalmic division of the trigeminal nerve: the **nasociliary, frontal,** and **lacrimal nerves** (Fig. 29–17). Collectively, these branches provide sensory innervation for the epithelial lining of the anterior one-third of the nasal cavity, the skin of the uppermost part of the face (specifically, that covering the forehead, the upper eyelids, and the bridge and tip of the nose) (Fig. 29–19), the bulbar conjunctiva, the corneal epithelium, and the palpebral conjunctiva of the upper eyelid.

The maxillary division of the trigeminal nerve provides sensory innervation for the skin and palpebral conjunctiva of the lower eyelid (Fig. 29–19).

Cranial Nerve VI: The abducent nerve (Fig. 29–17) innervates lateral rectus.

The Pterygopalatine Ganglion: The pterygopalatine ganglion provides postganglionic parasympathetic secretomotor fibers for the lacrimal gland; the zygomatic nerve of the maxillary division of the trigeminal nerve and its branches and the lacrimal nerve of the ophthalmic division of the trigeminal nerve transmit these fibers from the ganglion to the gland (Fig. 29–20).

Cranial Nerve VII: The facial nerve innervates orbicularis oculi. The facial nerve, through its greater petrosal nerve, also provides the preganglionic parasympathetic fibers which synapse within the pterygopalatine ganglion (Figs. 29–5 and 29–20).

Postganglionic Sympathetic Fibers: Postganglionic sympathetic fibers innervate dilator pupillae. These fibers emerge from the superior cervical ganglion (Fig. 29–6), extend within the internal carotid plexus, and finally reach the eyeball by the **long ciliary nerve** (a branch of the nasociliary nerve) and/or a branch to the ciliary ganglion and subsequent continu-

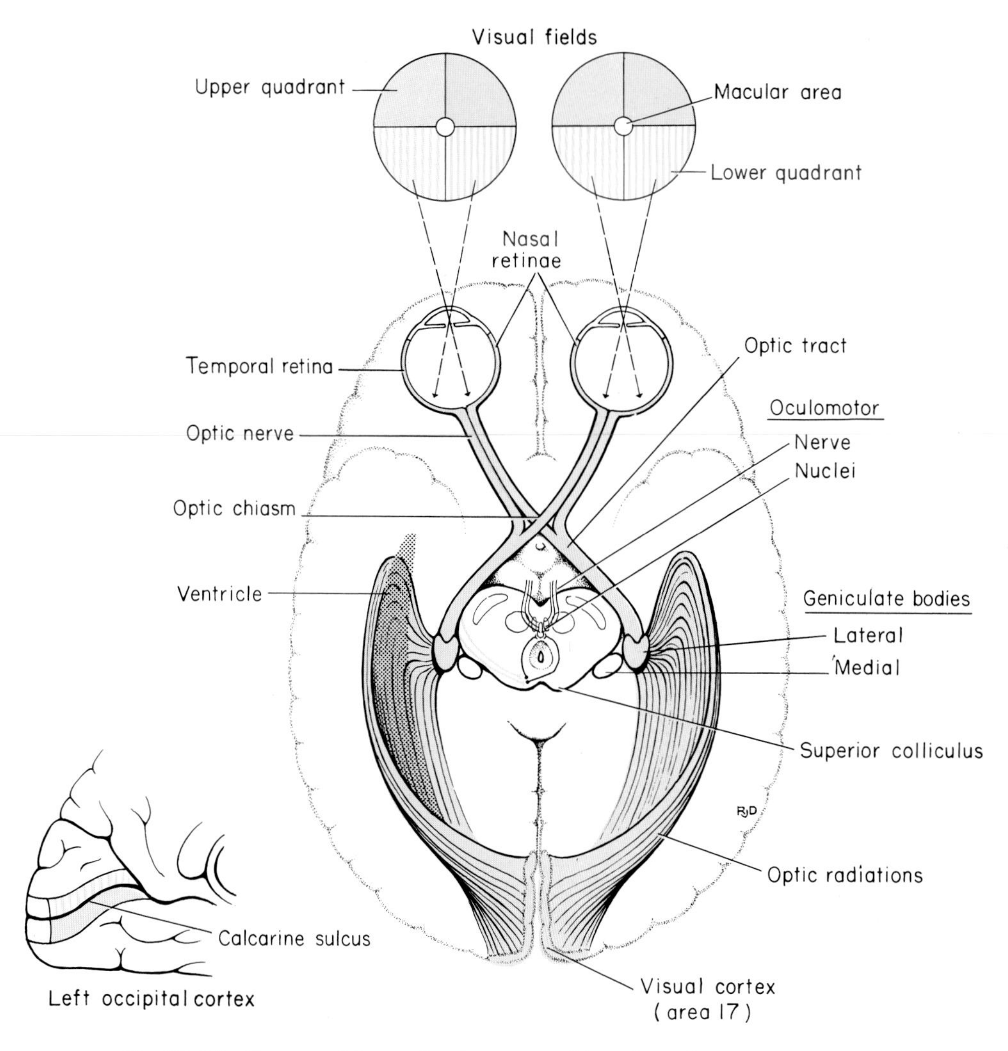

Fig. 29–16: Diagram of the visual pathways viewed from the undersurface of the brain.

ation within the short ciliary nerves. Postganglionic sympathetic fibers also innervate the smooth muscle fibers of levator palpebrae superioris.

The Corneal Reflex Test

Stimulation of the sensory fibers of the corneal epithelium elicits a reflexive concentric contraction of the orbicularis oculi that quickly shuts the eyelids. This reflex can be tested during a physical examination by the examiner's holding a wisp of cotton out of view at the side of the patient's head and then bringing the wisp of cotton inward from the lateral corner of the patient's eye into contact with the corneal epithelium. The corneal reflex test involves sensory fibers of the ophthalmic division of the trigeminal nerve and motor fibers of the facial nerve.

Blood Supply and Venous and Lymphatic Drainage of Orbital Cavity Tissues

The ophthalmic artery is the chief source of blood supply to the eyeball and other tissues of the orbital cavity. The ophthalmic artery arises from the internal carotid artery as it extends beyond the cavernous sinus. The optic canal transmits the ophthalmic artery from the cranial cavity to the orbital cavity. The central artery of the retina, which is the sole source of blood supply to the retina, is a branch of the ophthalmic artery. The superior and inferior ophthalmic veins receive most of the venous drainage from orbital cavity tissues. The superior and inferior ophthalmic veins are tributaries of the cavernous sinus.

The lymphatics from the medial aspect of the eyelids are afferent to submandibular lymph nodes. The lymphatics from the lateral aspect of the eyelids are afferent to parotid nodes.

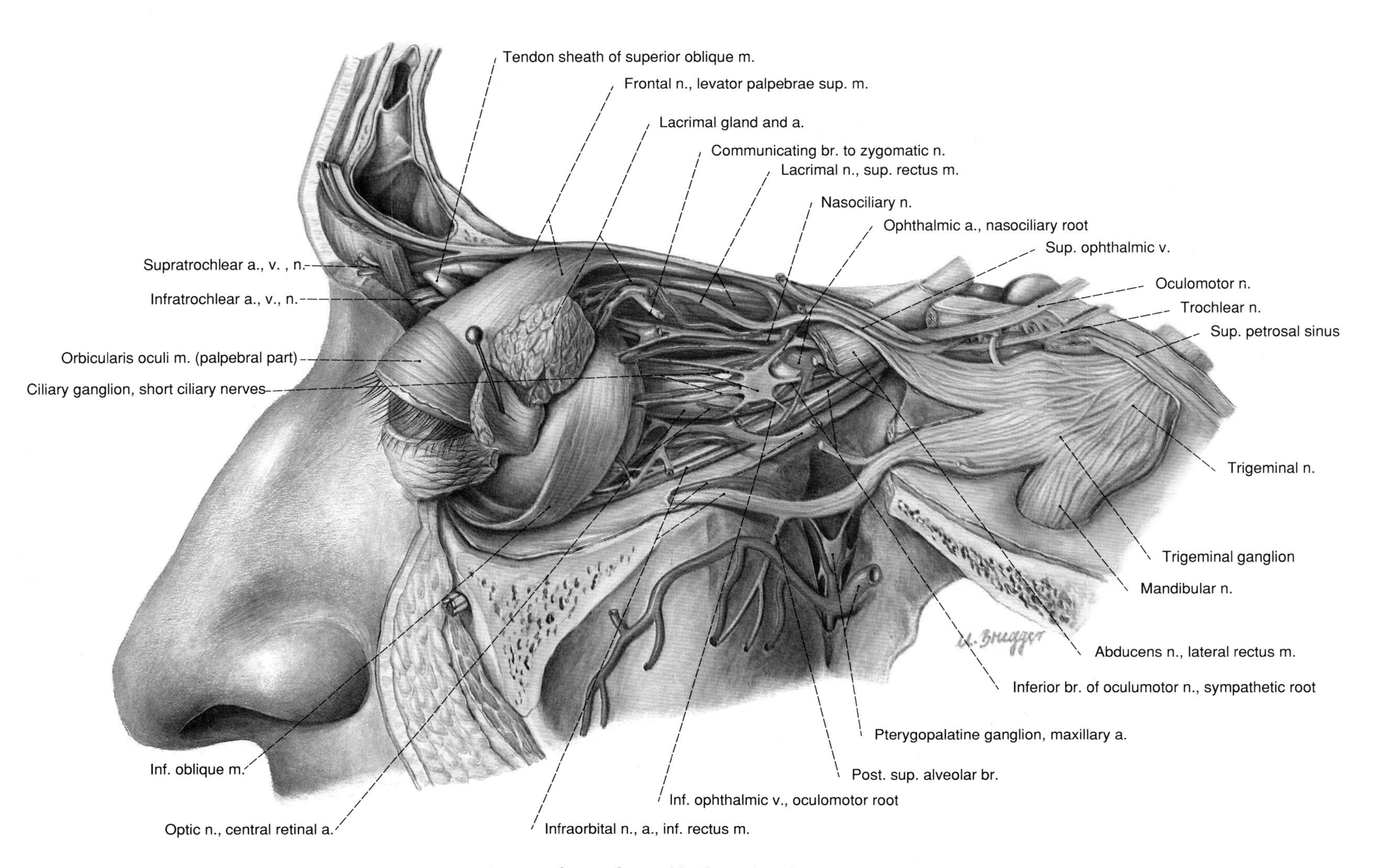

Fig. 29–17: Lateral view of major blood vessels and nerves of the left orbital cavity.

CLINICAL PANEL VI.2

Cavernous Sinus Thrombophlebitis

Description: Cavernous sinus thrombophlebitis is an infectious inflammation of the cavernous sinus with thrombus formation. This condition is frequently a complication of nasal cavity and/or paranasal sinus infection. The condition is associated with significant morbidity and mortality because of the rapid extension of the infection to the cranial meninges (which produces meningitis) and the systemic circulation (which results in systemic septicemia).

Early Signs and Symptoms: The patient becomes acutely ill with high fever, eye pain, and headache. Examination frequently reveals periorbital edema, chemosis (edema of the bulbar conjunctiva), papilledema (edema of the optic disc), proptosis (protrusion of the eyeball), and ocular palsy involving one, two, or all three of the cranial nerves (oculomotor, trochlear, and abducent) which innervate the extraocular muscles. The tributaries of the cavernous sinus on each side include the valveless superior and inferior ophthalmic veins of the orbital cavity (Fig. 29–17) and the central vein of the retina. The periorbital edema, chemosis, papilledema, and proptosis are all a consequence of obstruction of the flow of blood through these veins by the cavernous sinus thrombus or thrombi. The oculomotor, trochlear, and abducent nerves are susceptible to inflammation because of their extension through either the lateral wall or the interior of the inflamed cavernous sinus (Fig. 28–14). The intercavernous sinuses provide routes of easy extension of the infection to both cavernous sinuses, and thus bilateral involvement of both eyes is a frequent finding.

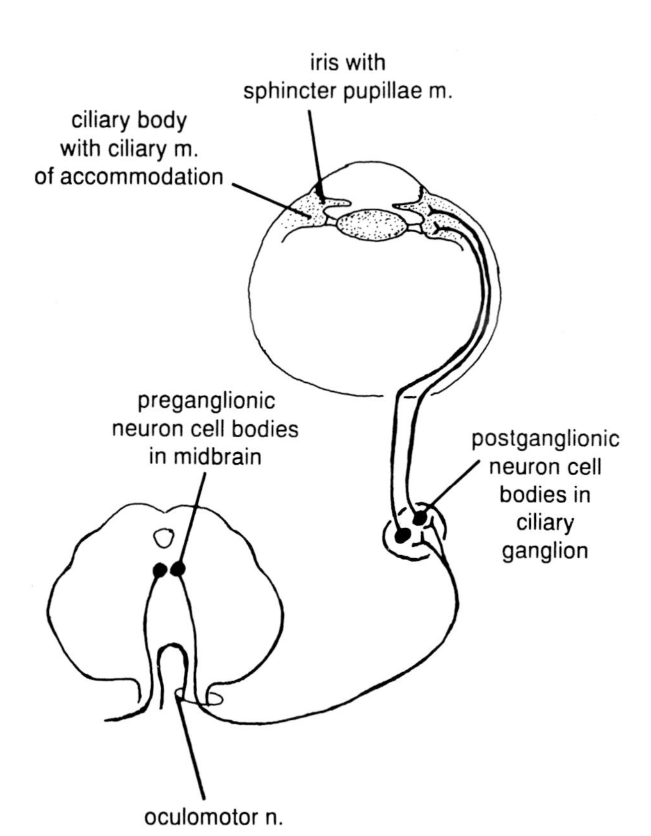

Fig. 29–18: Diagram of the origins and destinations of nerve fibers associated with the ciliary ganglion.

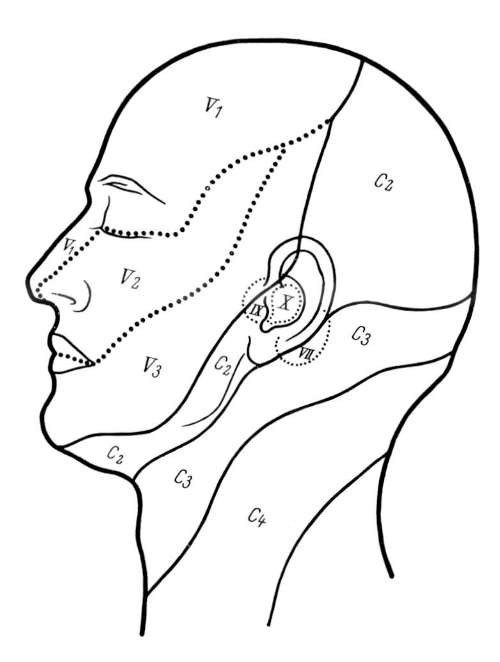

Fig. 29–19: Diagram showing the cutaneous innervation of the head and neck. Branches of the ophthalmic (V1), maxillary (V2), and mandibular (V3) divisions of the trigeminal nerve provide cutaneous innervation for anterior and lateral aspects of the head and face. Branches of the second, third, and fourth cervical spinal nerves (C2, C3, and C4) provide cutaneous innervation for posterior and lateral aspects of the head and neck. Branches of the facial (VII), glossopharyngeal (IX), and vagus (X) nerves provide cutaneous innervation for small areas of skin around the ear.

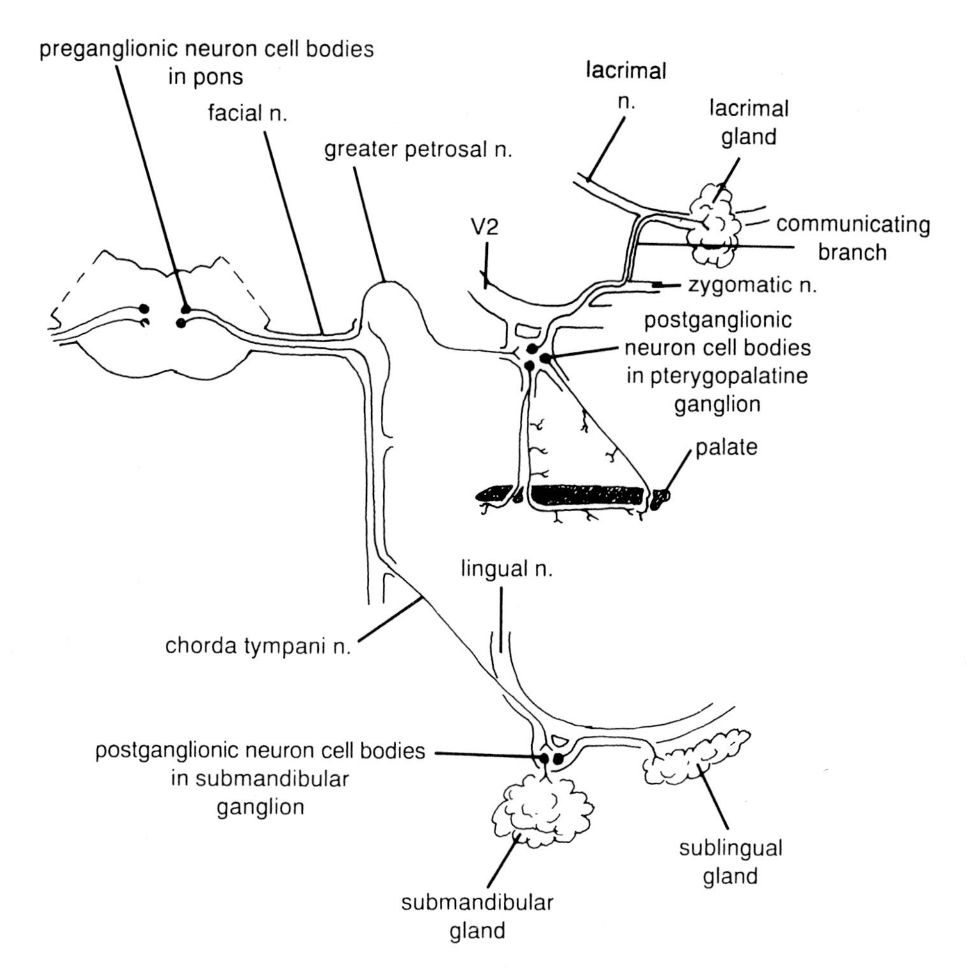

Fig. 29–20: Diagram of the distribution of the parasympathetic innervation provided by the facial nerve.

CASE **VI.3**

The Case of Duane Carlson

INITIAL PRESENTATION AND APPEARANCE OF THE PATIENT

Mr. and Mrs. Carlson have made an appointment for their son Duane, a 3 year-old white male with swollen left upper and lower eyelids. Duane is a small, thin, boy. He sits quietly on his mother's lap, clutching a towel and a stuffed teddy bear. He appears apprehensive and is only partially soothed by his mother's quiet words of comfort.

QUESTIONS ASKED OF THE PATIENT'S PARENTS

What can you tell me about Duane's swollen eye? Well, we first noticed some swelling around his left eye two days ago when we picked him up from preschool. When we talked with his teacher about his eye, she said she noticed it was a little red and puffy around noontime that day, but that she didn't know of any injury to his eye. His eye has gotten worse each day to the point where, as you can see, it's swollen shut. We don't think that his swollen eye hurts him, because he has not been crying about it. However, he won't let us touch the eye. *[The parents' remarks indicate first that the patient's swollen eye does not appear to be the result of a physical injury. Second, the swelling has progressively worsened during the past two days to the extent that it has swollen the eyelids shut. In the absence of any apparent cause for the swelling of the eyelids, the examiner decides to ask if the patient has had any other recent medical problems, because one or more of these problems may be the basis for the swollen eye.]*

Is there anything else that you know may be bothering Duane? Not anything that we can think of other than he's had a fever since last night. However, he has not been himself since yesterday; he's quieter and wants to be left alone. He does seem to act more like himself after Tylenol (acetaminophen), which we've been giving him for his fever. *[The parents' report of fever suggests that the patient is suffering from an infection, the most common cause of fever in young children. The parents' report of a change in the patient's personality (he's quieter and wants to be left alone) warrants serious consideration. Young children frequently express a feeling of poor health by a decline in their general level of physical and mental activity and/or a reduction in their food intake. In this case, however, in which it is possible that the patient has an infection within or about the left orbital cavity, the examiner is particularly mindful of signs or symptoms (such as a change in personality, irritability, lethargy, or malaise) which commonly herald intracranial extension of an infectious process.]*

Has he complained of a headache? No.

Have you seen or has he complained of any muscle spasms or awkward movements involving his hands or feet? No.

Has he had any nausea or vomiting? No. *[Episodes of muscle spasms or awkward limb movements might signify seizures. In this case, evidence of headache, seizures, nausea, or vomiting would warrant consideration of infection of the central nervous system.]*

Has he had any recent injuries or illnesses? He had a runny nose this past week for about 3 or 4 days, but it pretty much stopped yesterday. *[The parents' remarks indicate that an infection of the nasal cavities and/or paranasal sinuses preceded and may be concurrent with the patient's swollen eye problem.]*

Has he ever had an eye problem like this before? No.

Has there been any tearing from his eye? Yes. *[The examiner asked this question to acquire the parents' impression of the level of lacrimation in the left eye. Absence of tearing would suggest involvement of the lacrimal gland in the pathologic process affecting the left eye.]*

Has he had any other illnesses or injuries since birth? He was a full-term baby; he weighed 7 lbs. 8 oz. at birth. He was never really sick until he started school 3 months ago. Since then, he's had an ear infection once and now this swollen eye.

Has he received all his immunizations? Yes, we've been careful to see that he's received all his immunizations. *[Childhood immunizations tend to limit the variety of infectious pathogens to which a child is susceptible.]*

Have you been giving him any medications for his runny nose and fever other than Tylenol? No. *[The examiner finds the patient's parents alert and fully cooperative during the interview.]*

PHYSICAL EXAMINATION OF THE PATIENT

Vital signs: Blood pressure: 95/65
Pulse: 94
Rhythm: regular
Temperature: 101.7°F.
Respiratory rate: 27
Height: 94 cm.
Weight: 13.5 kg.

[A pediatrician would recognize that except for the temperature, the patient's vital signs are within normal limits. The patient's weight and height are respectively in the 25th and 50th percentiles for boys 3 years of age.]

HEENT Examination: Examination of the head, eyes, ears, nose, and throat are normal except for the following findings: The left eyelids are edematous and mildly erythematous on both their cutaneous and conjunctival surfaces (the edema limits the extent to which the eyelids can be everted for observation). There is no induration of the cutaneous surfaces of the left eyelids nor excessive warmth. Palpation of the erythematous area about the left eye is minimally painful. The patient has difficulty breathing through the nose because of congestion. Halitosis is pronounced. A few submandibular and deep cervical lymph nodes are enlarged but soft and nontender.

Pertinent normal findings include the following observations: The patient reports vision with the left eye (the patient is able to identify and describe small toy figures based upon observations made with the left eye). There is no evidence of eyeball displacement within the left orbital cavity. There appears to be full range of painless, ocular movements in both eyes. The pupils are round, equal, and reactive to light. The sclera are not hyperemic. The levels of lacrimation in both eyes appear equivalent. There is no evidence of dental caries or pharyngeal inflammation. The outer ear canals and tympanic membranes are clear, and the malleus is visible through the tympanic membranes bilaterally.
[The pertinent normal observations of the left eye indicate that the pathologic process responsible for marked swelling of the eyelids has not significantly affected vision, the actions of the extraocular muscles, reflexive pupillary constriction in response to light, or tear secretion by the lacrimal gland. There is also no evidence of conjunctivitis (inflammation of the sclera). Many of the tissues of the orbital cavity thus appear to be spared from the pathologic process.]

Lungs: Normal
Cardiovascular Examination: Normal
Abdomen: Normal
Genitourinary Examination: Normal
Musculoskeletal Examination: Normal
Neurologic Examination: Normal. Pertinent normal findings are (a) absence of pain or rigidity upon active and passive flexion and extension of the head and (b) absence of positive Brudzinski's and Kernig's signs.

[Brudzinski's and Kernig's tests are tests that stretch the spinal meninges enveloping the spinal cord and its spinal nerve roots. Both tests are conducted with the patient initially lying supine on an examination table with the patient's hands folded behind the head. In Brudzinski's test, the examiner requests the patient to raise the head, an act which stretches the spinal meninges from the cervical part of the vertebral column. If such stretching produces pain in the head, neck or back, the active anterior flexion of the head will elicit a reflexive contraction of the lower limbs at the hip and knee joints. The reflexive movements at the hip and knee joints constitute a positive sign for Brudzinski's test, because they represent a reflexive attempt by the patient to minimize tension in the spinal meninges through flexion of the thighs and legs. In Kernig's test, the patient slowly raises one of the lower limbs, taking care to keep the leg fully extended at the knee. This action stretches the spinal meninges from the lumbosacral part of the vertebral column. If such stretching produces pain in the head, neck, or back, and the pain can be relieved by flexion of the leg at the knee, then the test is judged positive. In this case, positive Brudzinski's and Kernig's signs would suggest meningitis (infectious inflammation of the meninges).]

INITIAL ASSESSMENT OF THE PATIENT'S CONDITION

The patient appears to have an infection the manifestations of which are fever, swollen left eyelids, nasal congestion, and halitosis.

ANATOMIC BASIS OF THE PATIENT'S HISTORY AND PHYSICAL EXAMINATION

(1) The patient's elevated temperature suggests the presence of an entrenched infection. Infection is the most common cause of fever in a young child.

(2) The patient's halitosis suggests sinusitis (infectious inflammation of one or more of the paranasal sinuses). Sinusitis is the most likely basis of halitosis in a young child without dental caries, pharyngitis, or a foreign body trapped in the nasal passageways. Suspicion of sinusitis in this case is strengthened by the facts that the patient (a) had rhinorrhea antecedent to the swelling of the left eyelids and (b) currently has nasal congestion.

(3) **The physical characteristics of the left eyelids indicates that they are either the site of an incipient infection or indirectly involved with an entrenched infection of tissues which share common channels of venous and/or lymphatic drainage.** Although the left eyelids are markedly edematous and mildly erythematous, there is no evidence of extreme tenderness, induration, or excessive warmth.

A progressive, untreated infection of a cutaneous area and the superficial fascia underlying it commonly elicits a **cellulitis** (inflammatory response) of marked edema, erythema, and warmth. The infected site becomes exquisitely tender to palpation. The physical signs of infection are extreme because the site of the infection is subcutaneous, and thus subject to direct visual and physical examination. The cellular death in the region surrounding the site of the infection soon renders the cutaneous area indurated (hardened).

When examining the eyelids for evidence of a cellulitis, it is important to visually examine both the cutaneous and conjunctival surfaces (edema may limit the extent to which the eyelids can be everted for observation of the conjunctival surfaces). This is because the orbital septa and tarsal plates of the eyelids effectively serve as mechanical barriers to the spread of infections within the eyelids. An infection of the postseptal tissues of an eyelid will tend to spread coronally outward through the postseptal tissues before piercing the orbital septum and/or tarsal plate to infect the preseptal tissues. The preseptal tissues of an eyelid are the tissues anterior to the orbital septum and tarsal plate; the postseptal tissues lie posterior to the orbital septum and tarsal plate. An edematous and mildly erythematous cutaneous surface of an eyelid thus sometimes can be the only physical evidence of a raging cellulitis in the postseptal tissues.

In this patient, the physical characteristics of the left eyelids indicate that the eyelids are not a site of an entrenched infection, because both the cutaneous and conjunctival surfaces do not show any evidence of a cellulitis. The left eyelids thus do not appear to be the source of the infection that presumably has made the patient febrile. It therefore follows that the left eyelids are either the site of an incipient infection or indirectly involved with an entrenched infection of tissues which share common channels of venous and/or lymphatic drainage. The anatomic basis for the latter possibility is as follows: A progressive, untreated infection of a tissue ultimately leads to congestion or obstruction of the tissue's lymphatic and venous drainage. The resulting pressure increases in the lymphatics and veins draining the infected tissue are transmitted in a retrograde fashion back toward all the tributaries of these vascular channels. Consequently, any uninfected tissue which shares common channels for lymphatic and venous drainage with the infected tissue soon becomes edematous and erythematous.

INTERMEDIATE EVALUATION OF THE PATIENT'S CONDITION

The limited evidence from the history and physical examination suggests that the patient's fever and swollen left eyelids are manifestations of an occult abscess or empyema of one or more of the paranasal sinuses.

CLINICAL REASONING PROCESS

In evaluating this case, a pediatrician would know that sinusitis in young children most commonly involves the ethmoid and maxillary sinuses. The pediatrician would also know that the tissues of the nasal and orbital cavities share common channels for lymphatic and venous drainage. Because there is reasonably strong evidence (halitosis) for an occult sinusitis and no evidence for a raging cellulitis of the left eyelids, a pediatrician would regard the marked edema and slight erythema of the left eyelids to be most likely the product of congested or obstructed venous and/or lymphatic drainage from an abscess or empyema of one or more of the ethmoid and maxillary sinuses.

If untreated, this condition can rapidly progress to infections in or about the orbital and cranial cavities. **Periorbital cellulitis** (infection of the preseptal tissues of the eyelids) is the mildest of the potential complications. Periorbital cellulitis is indicated if the cutaneous surfaces of the eyelids acquire a violaceous hue and become indurated and extremely warm and tender. **Orbital cellulitis** (infection of the tissues of the orbital cavity) can result from direct extension of the infection through the orbital plate of the ethmoid (if an ethmoid sinusitis is present) or the orbital plate of the maxilla (if a maxillary sinusitis is present). The valveless superior and inferior ophthalmic veins can serve as conduits for intracranial infection, because the ophthalmic veins receive some of the venous drainage of the nasal cavities and paranasal sinuses and are directly continuous with the cavernous sinuses. Intracranial extension of the infection can produce any of a number of dire complications, such as **meningitis** (inflammation of the meninges), **epidural or subdural abscesses** (abscesses immediately superficial or immediately deep to the dura mater), and **cavernous sinus thrombosis** (formation of thrombi in the cavernous sinuses).

Other possible but less likely disorders that a pediatrician might consider at this stage in the evaluation are dacryocystitis and dacryoadenitis. **Dacryocystitis** (infection of the lacrimal duct) is indicated by marked edema, erythema, and tenderness beneath the medial canthus. **Dacryoadenitis** (infection of the lacrimal gland) is indicated by marked edema and mild erythema involving primarily the lateral part of the upper eyelid. A pediatrician would not consider **conjunctivitis** because the sclera of the patient's left eye is not hyperemic.

RADIOLOGIC EVALUATION AND FINAL RESOLUTION OF THE PATIENT'S CONDITION

CT scans of the patient's head and neck showed a small subperiosteal region with a radiodensity of water in the left middle ethmoid sinus. The areas of pneumatization (the interior cavities of the air cells) of the left ethmoid sinuses showed increased radiodensity, and the air cells were outlined by poorly defined bony margins. There was no evidence of a subperiosteal abscess or of a suppurative process in the cranial cavity or left orbital cavity. The maxillary sinuses appeared normal. [The paranasal sinuses develop throughout childhood and adolescence through the progressive evagination of pneumatized (air-filled) cells from the nasal cavity. The maxillary and ethmoid sinuses are the only paranasal sinuses partially pneumatized at birth.]

These radiologic findings indicate a diagnosis of ethmoid sinusitis complicated by an ethmoid abscess and ethmoid osteitis (infection of the ethmoid air cells complicated by infection of the compact bone lining the walls of the ethmoid air cells). The subperiosteal region of water radiodensity represents an abscess. The increased radiodensity of the areas of pneumatization is caused by the effusion from and swelling of the mucous membrane lining the ethmoid air cells. The lack of definition of the bony outlines of the air cells is a consequence of demineralization of the compact bone forming the walls of the air cells.

Management of this case would involve immediate parenteral antibiotic therapy with oral and per nasal administration of antihistamine. The antihistamine is administered to decrease the inflammatory swelling of the ethmoid sinuses, which, in turn, should result in an increase in blood supply (with its antibiotics) to the ethmoid sinuses. The patient must be closely monitored during the first 48 hours of this treatment to assess the efficacy of the combined antibiotic-antihistamine therapy to resolve the ethmoid sinusitis, abscess, and osteitis. If the patient's condition does not begin to improve, surgical extirpation of the ethmoid abscess is required.

THE CHRONOLOGY OF THE PATIENT'S CONDITION

The patient's problems began seven days previously with inflammatory obstruction of the nasal passageways complicated by ethmoidal sinusitis (specifically, bacterial infection of the ethmoid sinuses). During the first four days of this upper respiratory infection, the patient experienced rhinorrhea and a low grade fever.

The swelling of the mucous membrane lining the ethmoid air cells fully obstructed mucus drainage from the ethmoid sinuses beginning two days previously. The cessation of most of the patient's rhinorrhea was a result of this obstruction. The entrapment of bacteria-laden mucus secretions within the left ethmoid sinuses increased pressures within the sinuses sufficient to severely compromise blood supply. The resultant necrosis and the accompanying expanding infectious process prompted establishment of the abscess and extension of the ethmoidal infection into (a) the bony walls of the ethmoid air cells and (b) the veins and lymphatics draining the ethmoid sinuses. Inflammation of these veins and lymphatics led to their obstruction, which in turn led to impairment of venous and lymphatic drainage from the left eyelids. The left eyelids became edematous and mildly erythematous due to their diminished venous and lymphatic drainage.

RECOMMENDED REFERENCES FOR ADDITIONAL INFORMATION ON THE NASAL AND ORBITAL CAVITIES

Slaby, F., and E. R. Jacobs, *Radiographic Anatomy*, Harwal Publishing Co., Malvern, PA, 1990: *Pages 177 through 179 in Chapter 5 outline the major anatomic features of the Waters' and lateral radiographic views of the face. The Waters' view provides relatively unobstructed views of the frontal maxillary sinuses. The lateral view provides partially superimposed views of all four major sets of paired sinuses.*

Coles, W. H., *Ophthalmology*, Williams & Wilkins, Baltimore, 1989: *Pages 65 through 87 in Chapter 4 present a discussion of the ophthalmologic findings associated with frequently encountered diseases and disorders of the retina.*

CHAPTER 30

The Ear and the Temporal, Infratemporal, and Parotid Regions

The ear houses the organs for equilibrium and hearing. The vestibulocochlear nerve serves both sensory modalities: the vestibular portion of the nerve is associated with the sense of head position and movement, and the cochlear portion of the nerve is associated with hearing.

The temporal, infratemporal, and parotid regions are on the side of the head. The temporal region is the superficial region on the side of the head, which lies superior to the zygomatic arch. The infratemporal region is the deep region in the side of the head which lies deep and inferior to the zygomatic arch. The parotid region is the superficial region on the side of the head which lies directly posterior to the ramus of the mandible.

THE EAR

The ear is divided into outer, middle, and inner parts. The outer and middle parts and the cochlea of the inner part provide for the collection and perception of sound waves for hearing. The semicircular canals of the inner part provide for the perception of head movement and position.

The Outer Ear

The outer ear consists of the **auricle** (the external, fleshy part of the ear) (Fig. 30–1) and the **external acoustic meatus** (Fig. 30–2). The auricle collects the sound vibrations in the air that impinge upon an individual, and the external acoustic meatus conducts the vibrations to the **tympanic membrane** (eardrum). The external acoustic meatus has an **S**-shaped course (Fig. 30–2) which can be partly straightened in an adult for otoscopic examination by pulling the auricle upward and back. The course of the external acoustic meatus can be partly straightened because the lateral third of the passageway has a flexible, cartilaginous foundation (Fig. 30–2). The medial two-thirds of the passageway has a bony foundation) (Fig. 30–3).

The Middle Ear

The middle ear is a cavity within the petrous part of the temporal bone (Fig. 30–3). This cavity is called the **tympanic cavity,** and, by convention, it is described as bordered by six walls: lateral, anterior, medial, and posterior walls, a roof, and a floor. The roof of the tympanic cavity underlies the temporal lobe of the cerebral hemisphere, and the floor overlies the internal jugular vein.

The tympanic membrane (eardrum) forms most of the lateral wall of the tympanic cavity (Fig. 30–4). Otoscopic examination of the lateral surface of the tympanic membrane reveals the impression of the manubrium of the malleus on the tympanic membrane (Fig. 30–5).

The medial wall of the tympanic cavity separates the cavity from the inner ear. The medial wall has two openings (the oval window and round window) to the bony labyrinth of the inner ear (Figs. 30–6 and 30–10). The **oval window (vestibular window)** opens into the vestibule of the inner ear, and it is closed by the base of the stapes. The **round window (cochlear window)** opens into the cochlea, and it is closed by the secondary tympanic membrane. The central region of the medial wall bears a bulge called the **promontory** formed by the first turn of the cochlea.

An articulated arch of three bony ossicles (the **malleus, incus,** and **stapes**) extends across the breadth of the tympanic cavity (from its lateral to its medial wall) (Fig. 30–7). Fine ligaments suspend the ossicles in place within the tympanic cavity (Fig. 30–4). The **manubrium** (handle) **of the** hammer-shaped **malleus** is attached to the tympanic membrane (Fig. 30–7). The **head of the malleus** articulates (through a synovial joint) with the **body of the** anvil-shaped **incus.** The lenticular process at the end of the **long crus (limb) of the incus** articulates (through a synovial joint) with the **head of the** stirrup-shaped

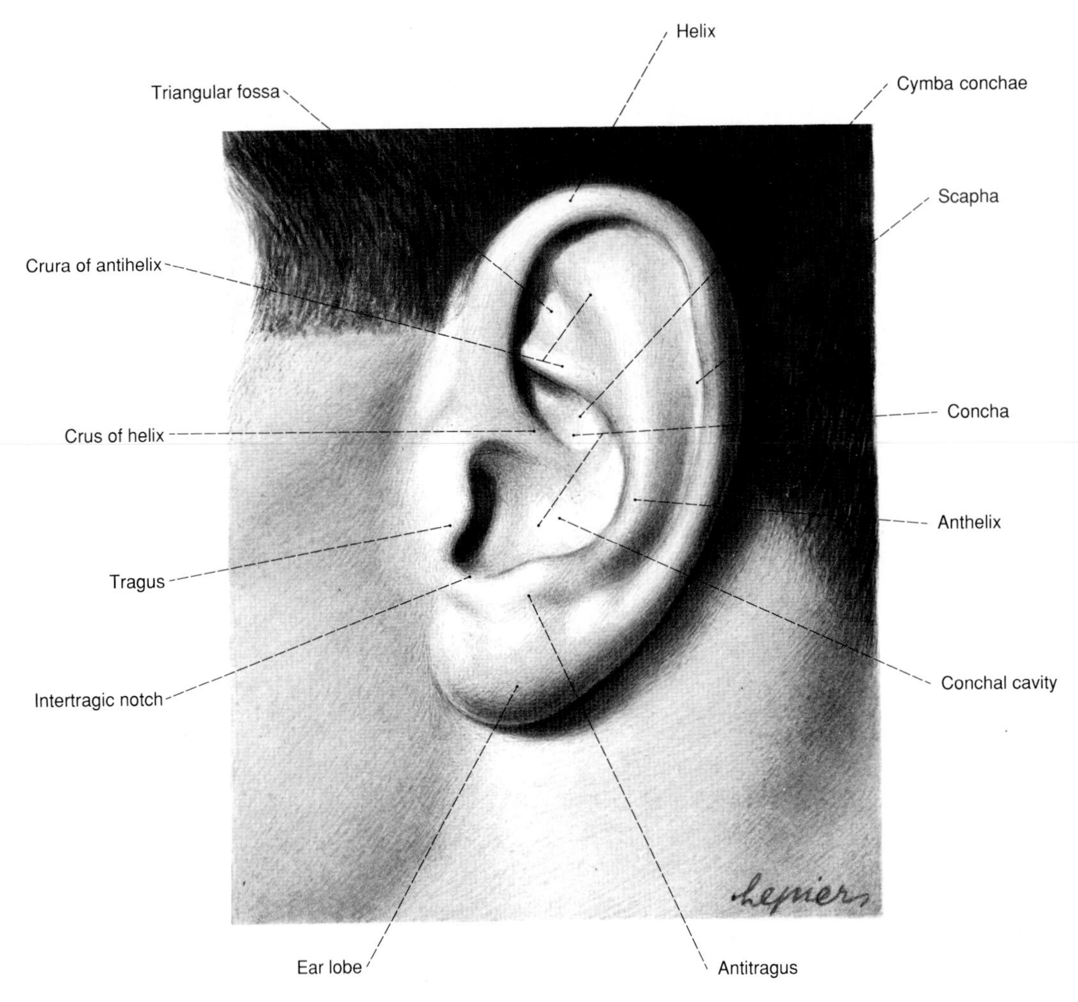

Fig. 30–1: The left auricle.

stapes. The **base of the stapes** fills the oval window (fenestra vestibuli) in the bony labyrinth of the inner ear.

The malleus, incus, and stapes serve as a mechanical couple which converts the vibrations of the tympanic membrane into vibrations of the perilymph of the inner ear. Medial movement of the tympanic membrane and the attached manubrium of the malleus sets in motion the following chain of movements (Fig. 30–8): The malleus rotates in such a fashion that the medial movement of its manubrium is accompanied by lateral movement of its head; the body of the incus moves laterally with the head of the manubrium. The incus rotates in such a fashion that the lateral movement of its body is accompanied by medial movement of its long crus; the medial movement of the long crus of the incus swings the base of the stapes medially within the oval window. All these movements are reversed when the tympanic membrane moves laterally. Movements of the base of the stapes within the oval window generate vibrations in the perilymph of the inner ear which travel to the membrane (the secondary tympanic membrane) which encloses the round window.

The tympanic cavity has two muscles (tensor tympani and stapedius) which limit the movements of the malleus, incus, and stapes. Extremely loud sounds elicit a simultaneous, reflexive contraction of the muscles called the **tympanic reflex. Tensor tympani** originates from the anterior wall of the tympanic cavity and inserts onto the manubrium of the malleus (Fig. 30–4). Tensor tympani acts to tense the tympanic membrane, and thus dampens the amplitude of its vibrations; its actions also push the base of the stapes medially within the oval window. **Stapedius** originates from the posterior wall of the tympanic cavity, and inserts onto the neck of the stapes (Fig. 30–9). Stapedius pulls the stapes posteriorly; this action pulls the base of the stapes laterally within the oval window, and thus opposes the action of tensor tympani upon the stapes during the tympanic reflex.

The **auditory tube** is a passageway that connects the tympanic cavity with the **nasopharynx** (the uppermost part of the pharynx). The auditory tube opens

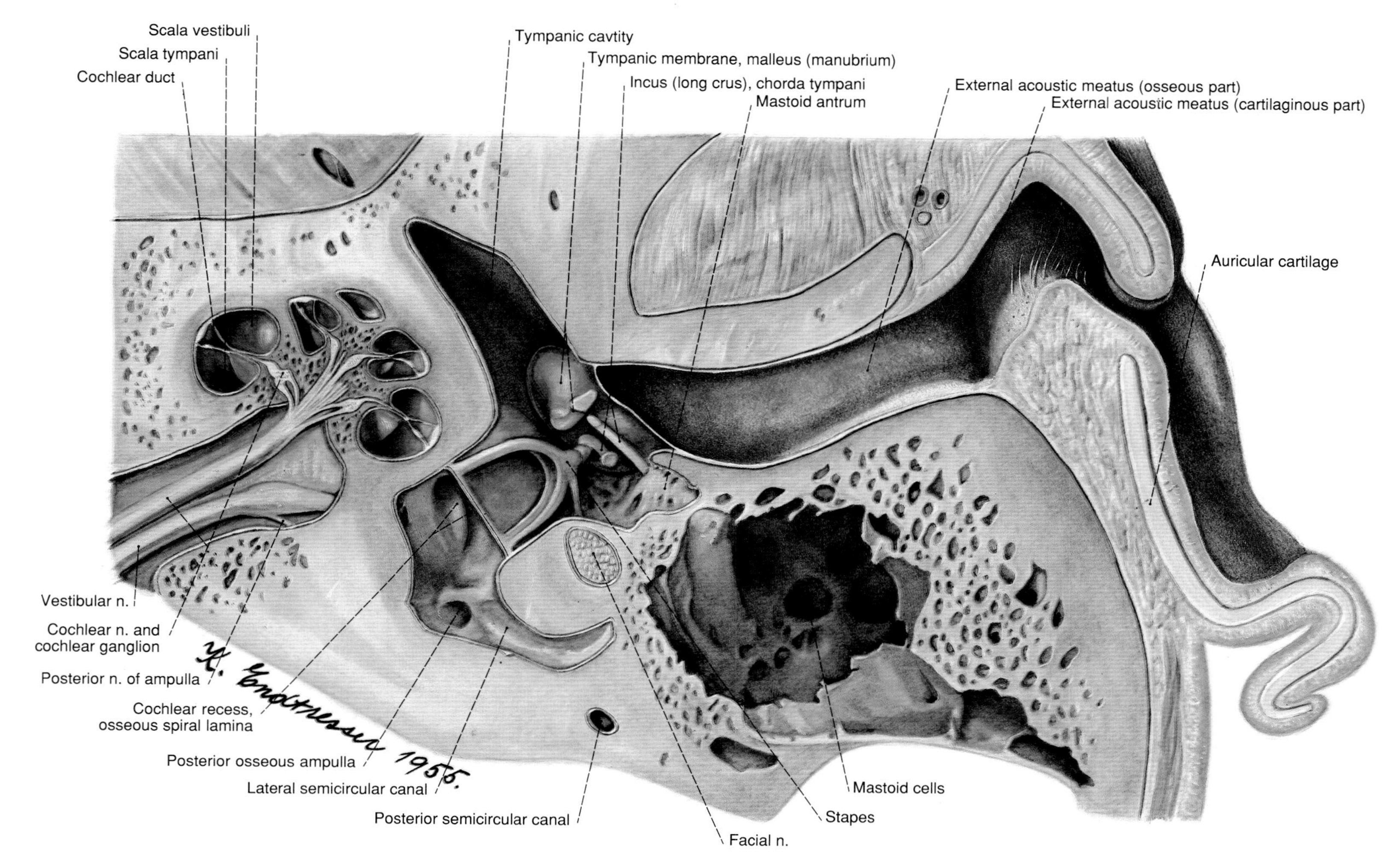

Fig. 30–2: Horizontal section of the right ear. Parts of the malleus and incus are removed.

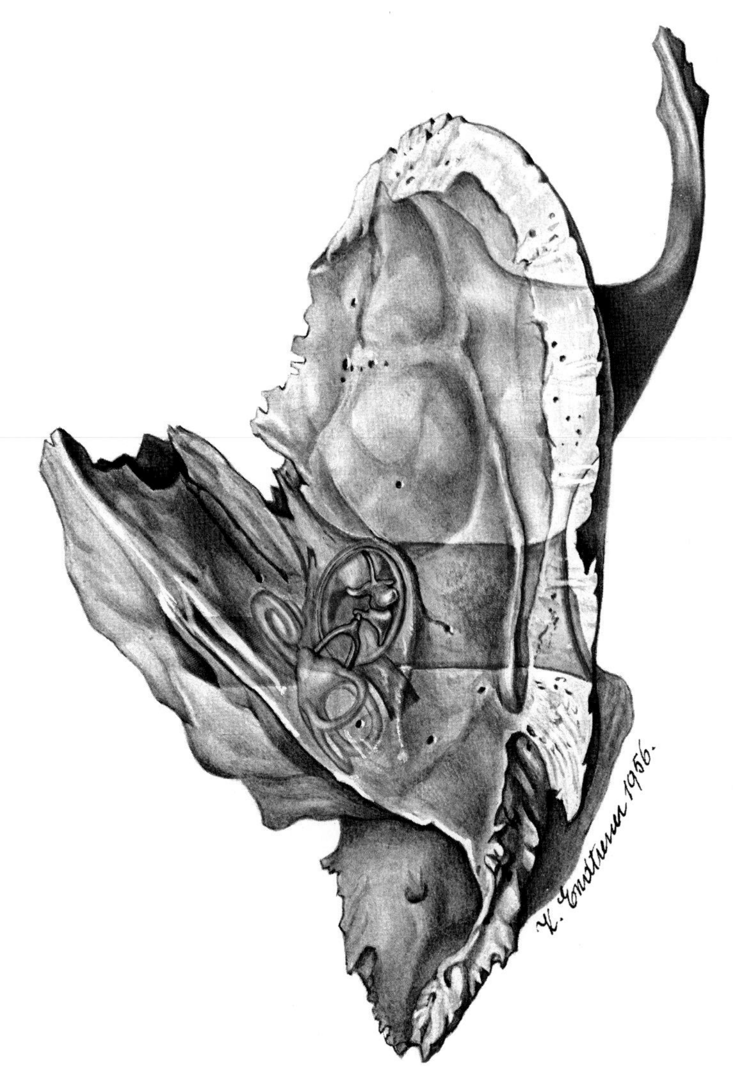

Fig. 30–3: Superior view of the parts of the right ear seen through a semitransparent temporal bone. The outer ear is colored red, the middle ear green, and the inner ear blue.

into the tympanic cavity via its anterior wall (Figs. 30–4 and 30–9). The auditory tube is normally closed. During the act of swallowing, however, the auditory tube is momentarily opened when a muscle of the roof of the mouth (tensor veli palatini) and a muscle of the pharynx (salpingopharyngeus) pull down upon the cartilaginous foundation of the medial portion of the auditory tube; the momentary opening provides for equilibrium of air pressure across the tympanic membrane.

The **mastoid process of the temporal bone** contains a number of mucous-membrane-lined air cells (Fig. 30–9), the largest and most superior of which is called the **mastoid antrum** (Fig. 30–6). The mastoid antrum and the other **mastoid air cells** all communicate with each other. The mastoid antrum communicates with the tympanic cavity through an opening in its posterior wall.

The Inner Ear

The inner ear consists of a membranous labyrinth of communicating sacs and ducts suspended within a bony labyrinth of chambers and canals in the petrous part of the temporal bone (Fig. 30–10). The bony labyrinth is divided into three major parts (Fig. 30–3): The **cochlea** is the snail-shaped anterior part of the bony labyrinth. The round window opens into the cochlea. The **vestibule** is the central part of the bony labyrinth; the oval window opens into the vestibule. The three **semicircular canals** form the posterior part of the bony labyrinth.

The bony labyrinth is lined by a serous membrane which secretes a fluid called **perilymph.** The perilymph bathes the external surface of the membranous labyrinth that lies suspended in the bony labyrinth

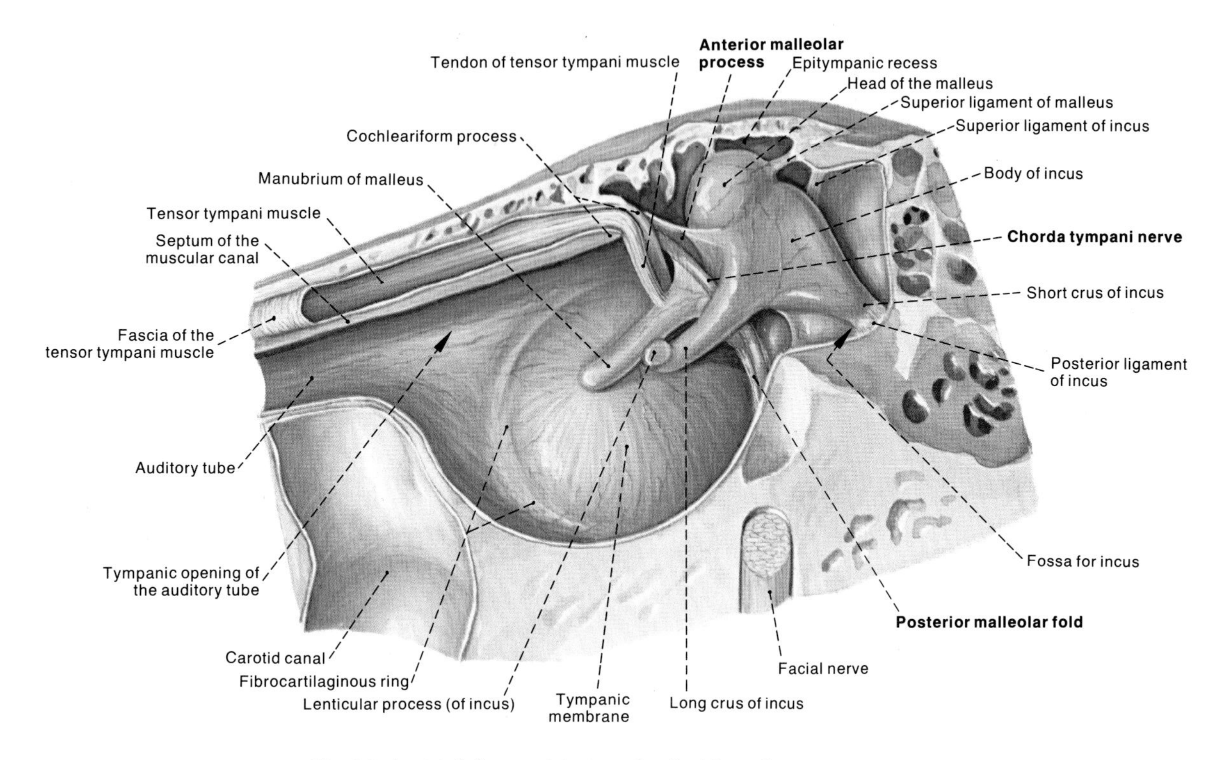

Fig. 30–4: Medial view of the lateral wall of the right tympanic cavity.

(Fig. 30–10). The fluid which fills the interior of the membranous labryrinth is called **endolymph.**

The membranous labyrinth is divided into three major parts (Fig. 30–10): The spiralled **cochlear duct** lies within the cochlea. The cochlear duct contains the **organ of Corti,** the end receptor organ of hearing. The two membranous sacs called the **utricle** and **saccule** represent a second major part of the membranous labyrinth; the utricle and saccule lie within the vestible. The three semicircular ducts (the **anterior, posterior,** and **lateral semicircular ducts**) that lie within the semicircular canals form the third major part of the membranous labyrinth. The utricle, saccule, and ampullae at the ends of the semicircular ducts contain sensors of movement.

The Innervation of the Ear

The innervation of the ear involves cranial nerves V, VII, VIII, IX, and X.

Cranial Nerve V: The auriculotemporal branch of the mandibular division of the trigeminal nerve provides sensory innervation for skin covering part of the outer surface of the tympanic membrane and parts of the external auditory meatus and auricle (Figure 29–19 shows the regions of the auricular skin which receive sensory innervation through the auriculotemporal branch of the mandibular division of the trigeminal nerve). A branch of the mandibular division also innervates tensor tympani.

Cranial Nerve VII: The facial nerve innervates stapedius. It also provides sensory innervation for skin covering parts of the external acoustic meatus, auricle, and scalp immediately posterior to the auricle. Figure 29–19 shows the parts of the auricle and scalp whose skin receives sensory innervation from the facial nerve.

Cranial Nerve VIII: The vestibular portion of the vestibulocochlear nerve is associated with the sense of head position and movement, and the cochlear portion of the nerve is associated with hearing.

Cranial Nerve IX: The glossopharyngeal nerve provides sensory innervation for the mucous membrane covering the inner surface of the mastoid air cells, tympanic cavity, and auditory tube. It also provides sensory innervation for skin covering parts of the auricle and the side of the face immediately anterior to the auricle (Fig. 29–19).

Cranial Nerve X: The vagus nerve exits the cranial cavity and enters the neck through the jugular foramen (Fig. 28–12). The nerve bears a superior ganglion near the level of the jugular foramen and an inferior ganglion immediately below the jugular foramen. Both ganglia are sensory ganglia and contain neuronal cell bodies for the nerve's sensory fibers. The superior ganglion gives rise to auricular branches that provide sensory innervation for skin covering the outer surface of the tympanic membrane and parts of the external auditory meatus and auricle. Figure 29–19 shows the part of the auricle whose skin receives sensory innervation from the auricular branches of the vagus nerve. If

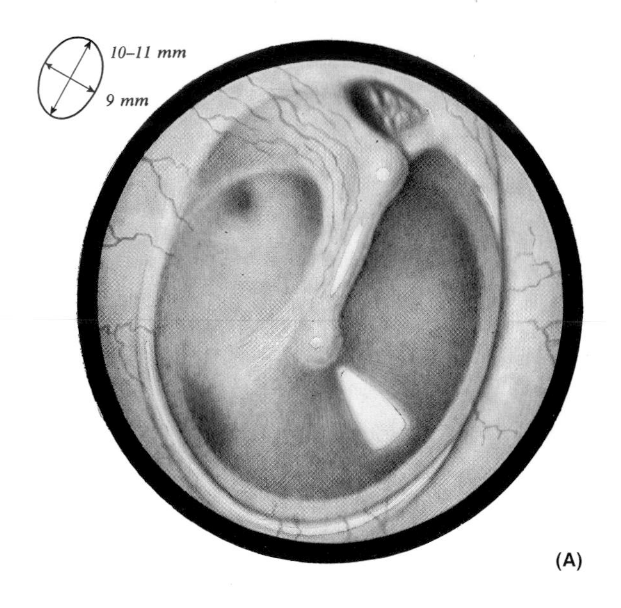

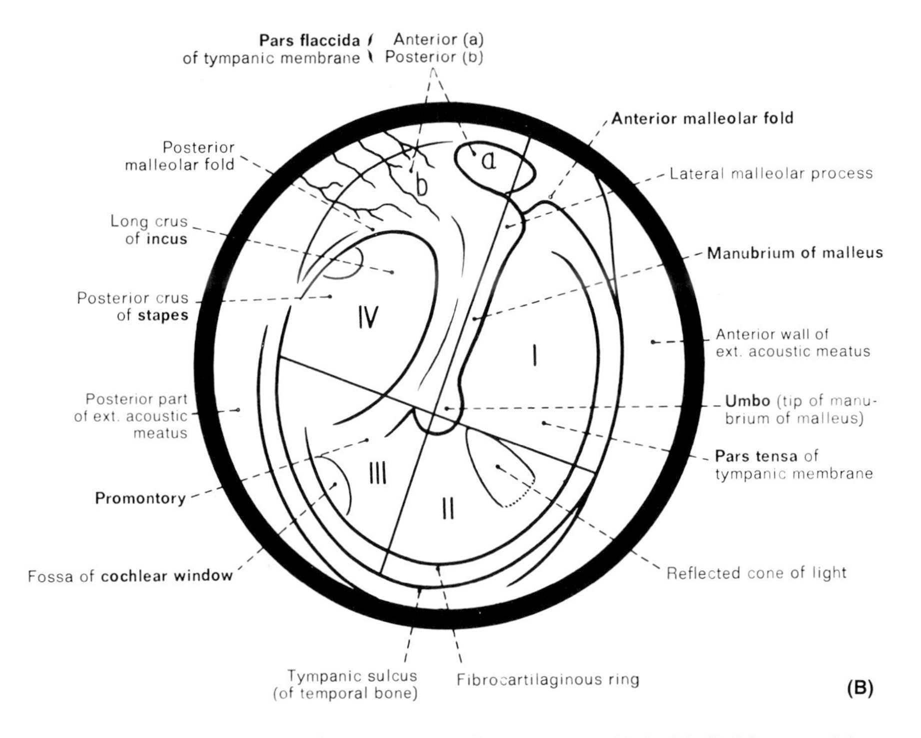

Fig. 30–5: (A) Right tympanic membrane as seen with an otoscope in a living person and (B) a labelled diagram of the image displayed in (A).

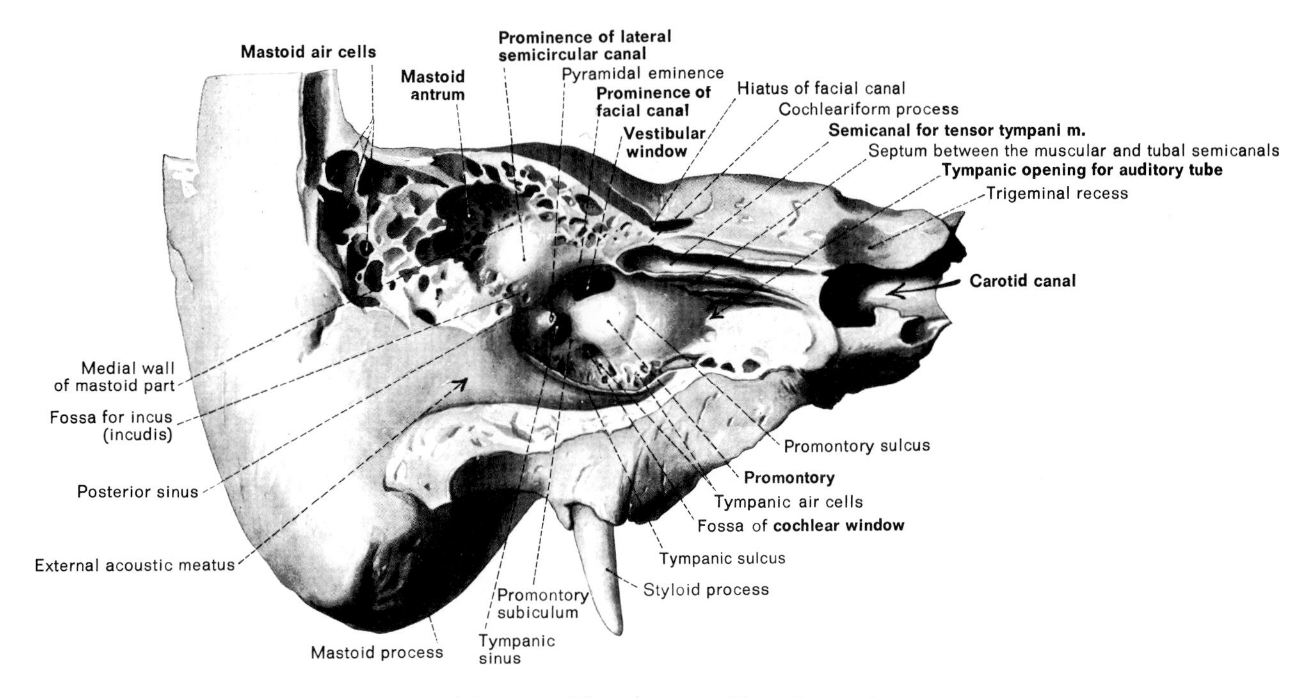

Fig. 30–6: Lateral dissection of the right temporal bone showing the tympanic cavity.

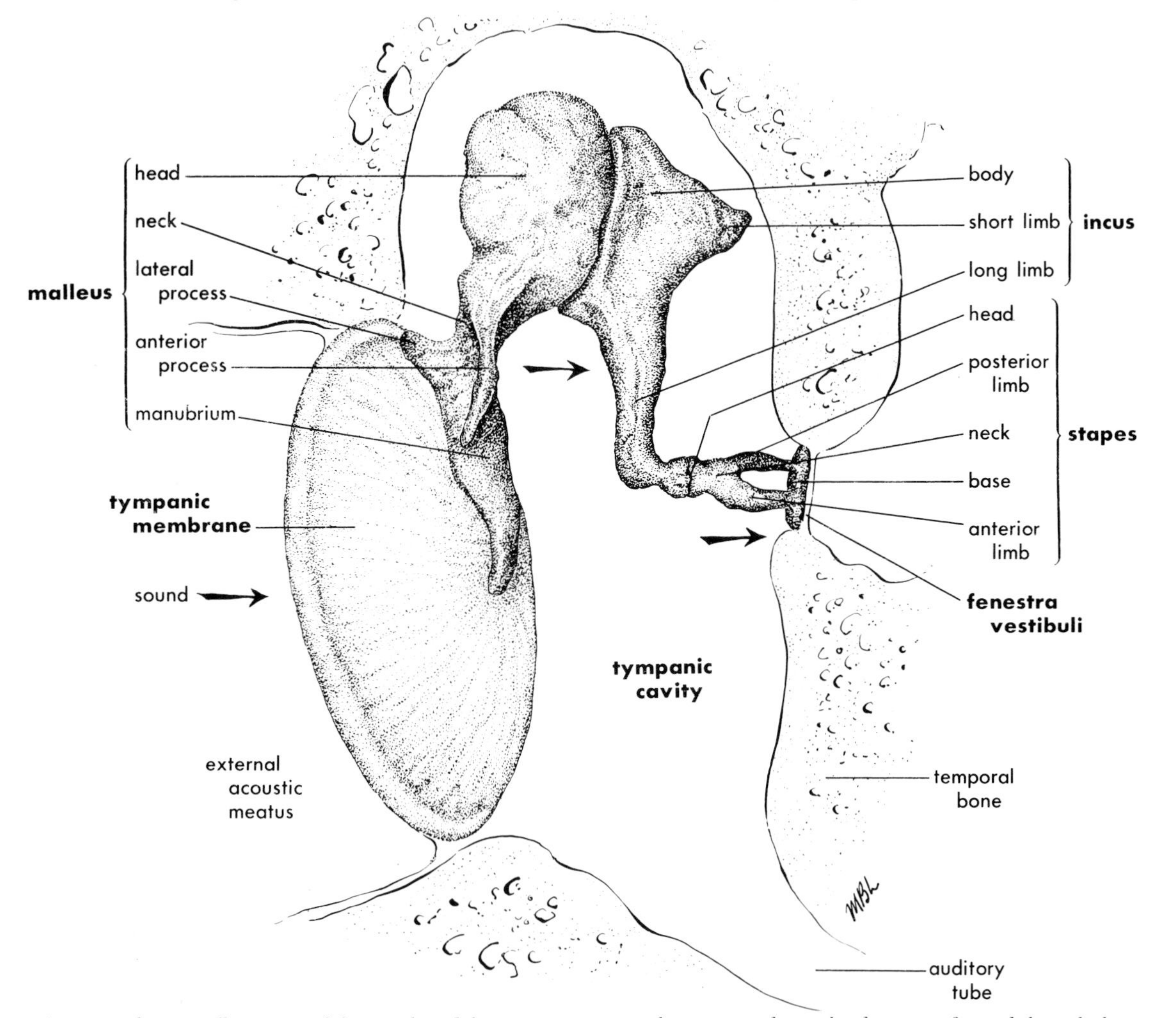

Fig. 30–7: Semischematic illustration of the ossicles of the tympanic cavity. The arrows indicate the direction of sound through the external acoustic meatus and its impact upon the movement of the long limb of the incus and the stapes during inward (medially directed) movement of the tympanic membrane.

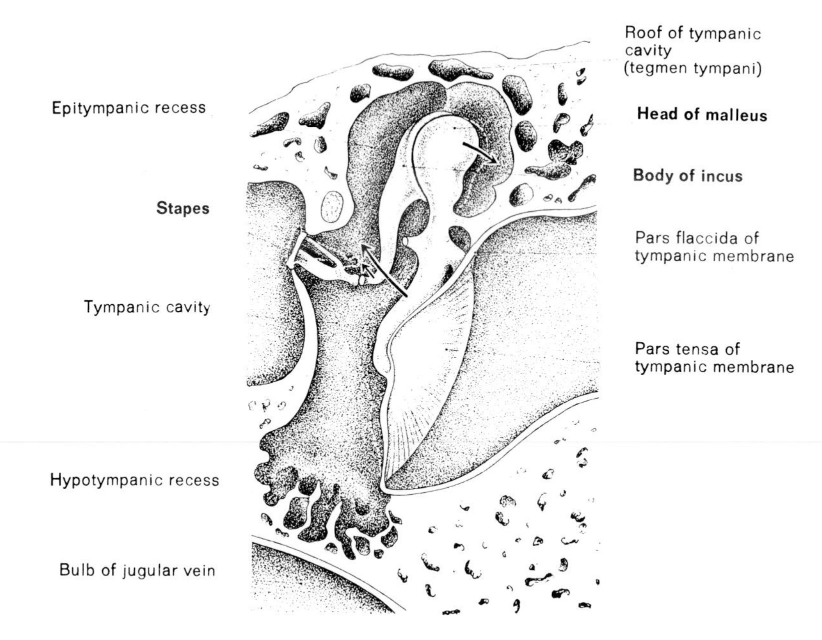

Fig. 30–8: Frontal section of the left ear through the ossicles of the tympanic cavity. The arrows indicate the direction of movement of the head of the malleus, body of the incus, long crus of the incus, and stapes during inward (medially directed) movement of the tympanic membrane.

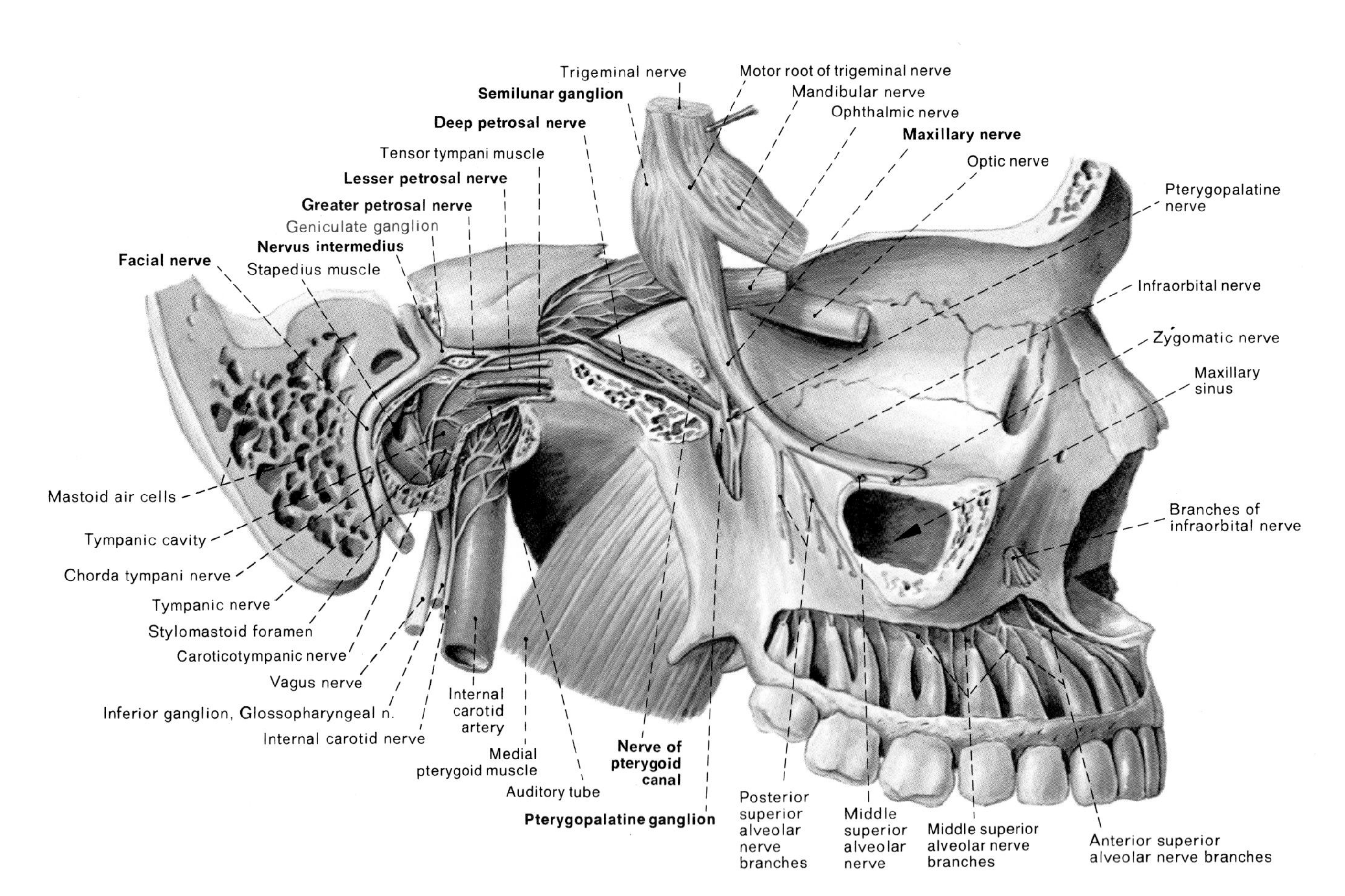

Fig. 30–9: The course of the facial nerve through the facial canal and the origins of the greater and lesser petrosal nerves.

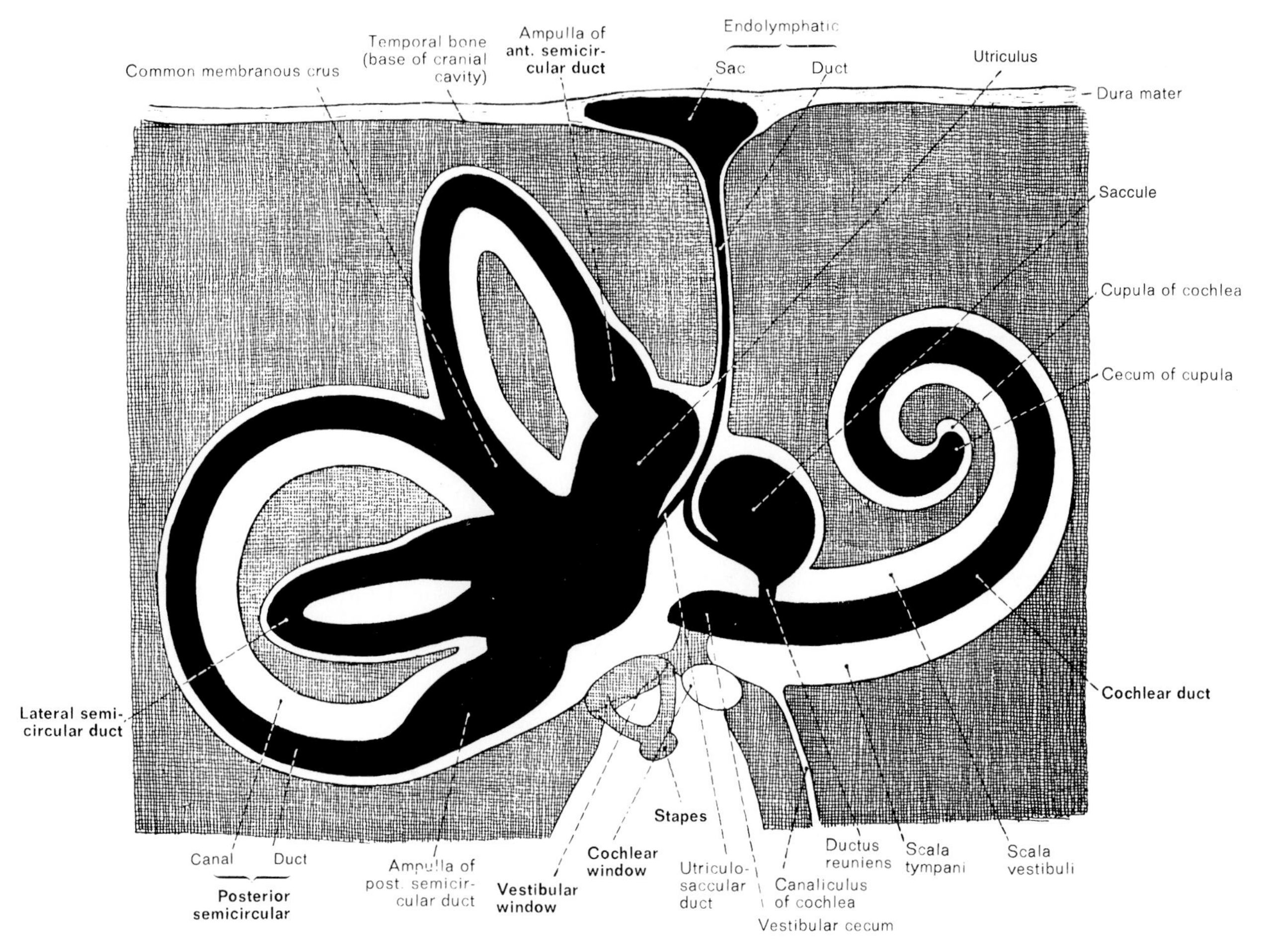

Fig. 30–10: Diagram of the membranous labyrinth suspended within the bony labyrinth of the inner ear. The petrous part of the temporal bone surrounding the bony labyrinth is shown as cross hatched. The perilymphatic spaces are diagrammed as white. The interior of the membranous labyrinth is diagrammed as black.

these vagal sensory fibers are irritated during rough cleaning of the external auditory meatus, they initiate the gag reflex. In elderly individuals, such irritation has on occasion caused heart arrest, as a consequence of reflex stimulation of the vagal preganglionic parasympathetic fibers to the cardiac plexuses.

The Course of the Facial Nerve through the Temporal Bone

The facial nerve passes through the temporal bone in extending from the cranial cavity to the face. This intratemporal course, which is closely related to the inner ear and middle ear, is as follows:

The facial nerve emerges from the brainstem in the form of two nerve fiber bundles, a motor root and a sensory root. The sensory root is called the **nervus intermedius.** The motor and sensory roots exit the cranial cavity through the internal acoustic meatus (which is a passageway through the petrous part of the temporal bone) (Fig. 28–12). At the bottom of the internal acoustic meatus, the roots of the facial nerve enter a passageway called the **facial canal.** The facial canal directs the roots of the facial nerve laterally over the vestibule of the inner ear to the medial wall of the tympanic cavity, at which point the roots converge upon the geniculate ganglion (Fig. 30–9). The **geniculate ganglion** is the sensory ganglion of the facial nerve and contains the neuronal cell bodies for its sensory fibers (Fig. 30–11).

From the geniculate ganglion, the facial canal directs the facial nerve first posteriorly over the promontory and then sharply inferiorly posterior to the tympanic cavity (the facial nerve's downward course here lies within the medial wall of the opening that joins the mastoid antrum with the tympanic cavity) (Figs. 30–2 and 30–9). The facial canal finally directs the facial nerve inferiorly through the temporal bone to exit through a foramen called the **stylomastoid foramen** (so named because it is located between the styloid and mastoid processes of the temporal bone) (Fig. 30–9).

The facial nerve gives rise to three important branches during its intratemporal course. (1) The first branch is the **greater petrosal nerve,** which arises

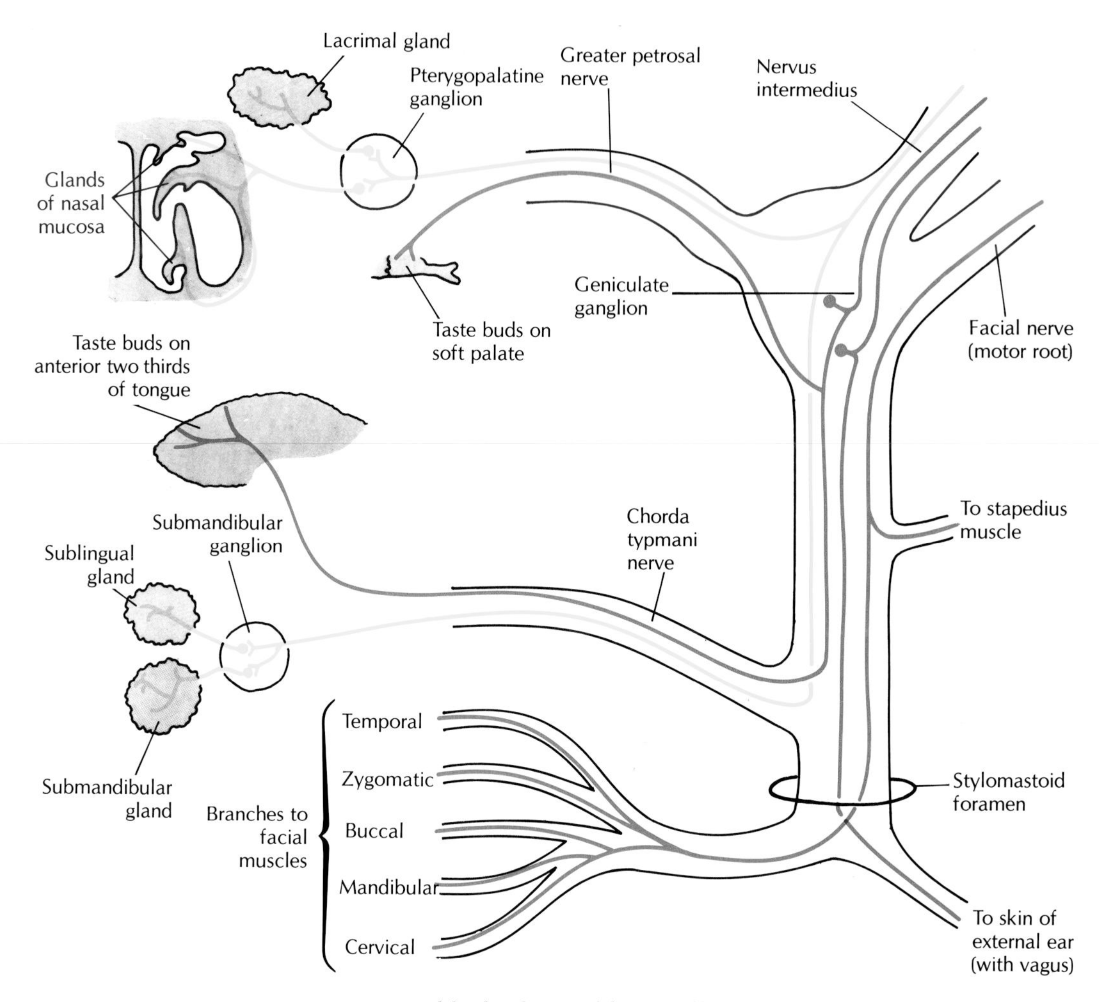

Fig. 30–11: Diagram of the distribution of the nerve fibers of the facial nerve.

from the geniculate ganglion (Fig. 30–11). The greater petrosal nerve has preganglionic parasympathetic fibers which synapse within the pterygopalatine ganglion (Figs. 29–5, 29–20, and 30–11). (2) The second branch is the nerve to stapedius (Fig. 30–11). (3) The third branch is the **chorda tympani** (Fig. 30–11), which extends anterosuperiorly from the facial nerve to enter the tympanic cavity by an opening in its posterior wall (Fig. 30–4). In the tympanic cavity, the chorda tympani crosses the medial surfaces of the tympanic membrane and the manubrium of the malleus as it extends anteriorly to enter the temporal bone again (Fig. 30–4). It exits the temporal bone and enters the infratemporal fossa through an opening called the petrotympanic fissure. The chorda tympani joins the lingual nerve to provide preganglionic parasympathetic fibers for the submandibular ganglion and taste fibers for the anterior two-thirds of the tongue (Fig. 30–11). The submandibular ganglion is discussed in Chapter 31 in the section on the innervation of the oral cavity.

The Course and Distribution of Glossopharyngeal Fibers through the Temporal Bone

The glossopharyngeal nerve exits the cranial cavity and enters the neck by the jugular foramen (Fig. 28–12). The nerve bears a superior ganglion near the level of the jugular foramen and an inferior ganglion immediately below the jugular foramen. Both ganglia are sensory ganglia and contain neuronal cell bodies for the nerve's sensory fibers.

The inferior ganglion gives rise to the **tympanic nerve,** which ascends in the neck to enter the tympanic cavity through an opening in the cavity's floor (Fig. 30–9). In the tympanic cavity, the tympanic nerve gives rise to a nerve plexus called the **tympanic plexus** (some facial nerve fibers also contribute to the tympanic plexus). The tympanic plexus provides sensory innervation for the mucous membrane covering the inner surface of the mastoid air cells, tympanic cavity, and auditory tube.

Many fibers of the tympanic plexus combine to form the **lesser petrosal nerve** (Fig. 30–9), a nerve which extends from the tympanic cavity of the temporal bone into the middle cranial fossa and then through the foramen ovale or an unnamed opening into the infratemporal fossa. The lesser petrosal nerve provides preganglionic parasympathetic fibers for the otic ganglion. The otic ganglion is discussed later in this chapter in the section on the parotid gland.

THE TEMPORAL, INFRATEMPORAL AND PAROTID REGIONS

The temporal region is the superficial region on the side of the head that extends from the **superior temporal line** above to the **zygomatic arch** below. The zygomatic arch consists of the zygomatic process of the temporal bone and the temporal process of the zygomatic bone (Fig. 30–12). The infratemporal region is the region on the side of the head deep to the ramus of the mandible, superficial to the lateral pterygoid plate of the sphenoid, and inferior to the floor of the middle cranial fossa (Fig. 30–12). The parotid region is chiefly that region lying between the ramus of the mandible anteriorly and the mastoid process of the temporal bone posteriorly.

The side of the skull underlying the temporal region bears an **H**-shaped set of sutures (Fig. 30–12). The sutures among the frontal bone, parietal bone, squamous part of the temporal bone, and greater wing of the sphenoid form the **H**-shaped set of sutures. The area around the crossbar of the **H**-shaped set of sutures is called the **pterion.**

The pterion overlies part of the course of the anterior branch of the middle meningeal artery within the cranial cavity. This anatomic relationship is important for the following reason: The close relationship of the anterior branch of the middle meningeal artery to the relatively thin plates of bone that form the **H**-shaped set of sutures about the pterion renders the arterial branch susceptible to injury from trauma to the side of the skull. Trauma to the side of the skull can result in rupture of the arterial branch and/or its accompanying vein. An extradural hemorrhage results if the rupture site lies superficial to the dura mater. Such an extradural hemorrhage almost always requires prompt identification and mangement to avoid a fatal outcome.

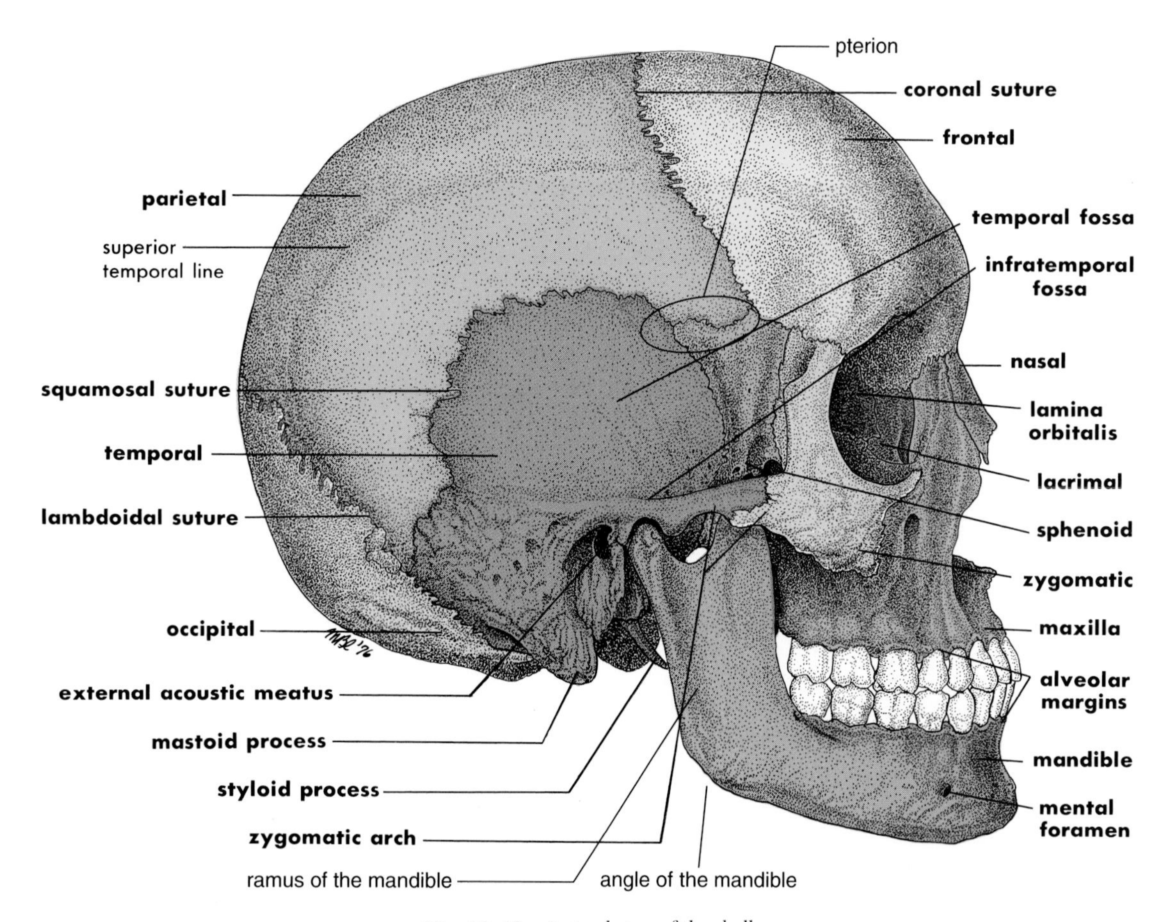

Fig. 30–12: Lateral view of the skull.

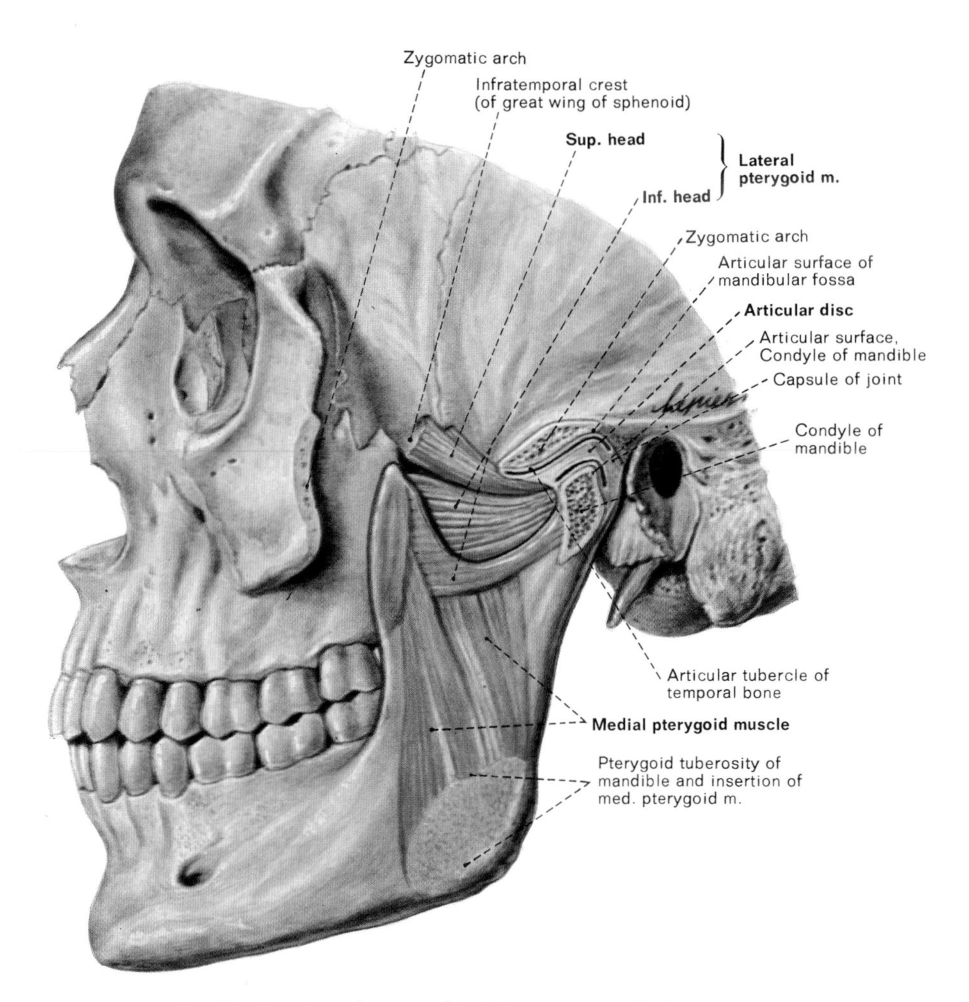

Fig. 30–13: Sagittal section of the left temporomandibular joint (TMJ).

The Mandible

The mandible forms the lower jaw. The ramus of the mandible borders the infratemporal fossa laterally (Fig. 30–12).

The mandible supports the lower dental arch. In an adult, the lower and upper dental arches each consist of 4 incisors, 2 canines, 4 premolars, and 6 molars.

The Temporomandibular Joint (TMJ)

The TMJ is a synovial joint that joins the **head of the condyle of the mandible** and the **mandibular fossa of the zygomatic process of the temporal bone** (Fig. 30–13). A disc of fibrous cartilage called the **articular disc** divides the synovial cavity of the TMJ into upper and lower cavities.

The mandible can be **rotated upward** and **downward (elevated** or **depressed)** at its paired TMJs; such rotation mainly involves a hinge motion in the lower cavity of the paired TMJs. The mandible can also be **moved forward** and **backward (protracted** and **retracted)** at its paired TMJs; such movement mainly involves a gliding motion in the upper cavity of the paired TMJs (in other words, protraction and retraction of the mandible mainly involves gliding of the mandibular heads and the articular discs forward and backward beneath the mandibular fossa of the zygomatic process of the temporal bone).

A slight opening of the jaws from the closed position happens from the downward rotation motion in the lower cavity of the TMJs. Further opening of the jaws involves both the downward rotation motion in the lower cavity of the TMJs and the forward gliding motion in the upper cavity of the TMJs (compare Figures 30–14, A and B).

The Muscles of Mastication

The major muscles of the temporal and infratemporal regions are the muscles of mastication (the muscles of chewing). The four muscles of mastication move the mandible at its temporomandibular joints:

Temporalis is a fan-shaped muscle which extends from the temporal fossa of the skull to the coronoid process of the mandible (Fig. 30–15). It can raise and retract the mandible.

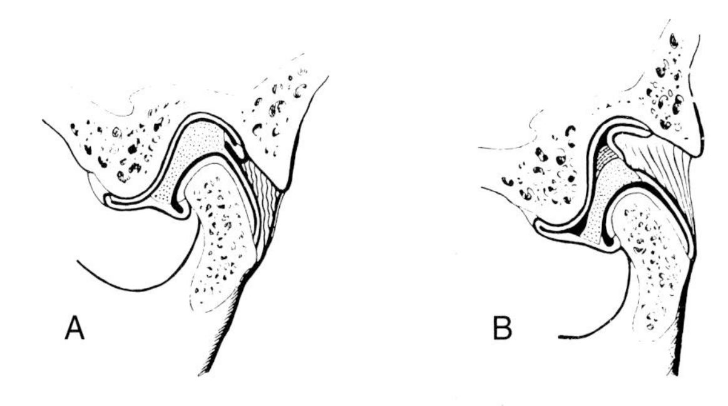

Fig. 30–14: Diagrams showing the configuration of the TMJ in (A) the closed and (B) the wide open positions of the jaws.

Masseter overlies much of the ramus of the mandible. Its fibers extend inferoposteriorly from the zygomatic arch to the lateral surface of the coronoid process, ramus, and angle of the mandible (Fig. 30–16). It can raise the mandible.

Medial pterygoid lies deep to the ramus of the mandible (Fig. 30–13). Its fibers extend inferoposteriorly mainly from the lateral ptyergoid plate of the sphenoid to the medial surface of the angle of the mandible. It can protract and raise the mandible.

Lateral pterygoid lies deep to the ramus of the mandible (Fig. 30–13). Its fibers extend posteriorly from the lateral pterygoid plate and greater wing of the sphenoid to the fibrous capsule and articular disc of the TMJ. It can protract and lower the mandible.

The medial and lateral pterygoids on each side produce chewing movements by alternate concentric contraction with the pterygoids on the contralateral side. Such alternate concentric contraction protrudes the mandible on the side of the concentrically contracting pterygoids at the same time that the mandible is retracting on the side of the eccentrically contracting pterygoids; this action grinds the upper and lower dental arches against each other.

The stretch receptors and muscle fibers of the muscles of mastication are innervated by branches of the mandibular division of the trigeminal nerve. The integrity of the sensory and motor supply to the muscles of mastication can be assessed during a physical exam by the jaw jerk test. An examiner conducts the **jaw jerk test** by first requesting the patient to slightly open the jaws and then gently tapping the patient's lower jaw with a reflex hammer. The abrupt stimulation of the stretch receptors of the muscles of mastication elicits a deep tendon reflex in which the temporalis, masseter, and medial pterygoid muscles slightly raise the mandible.

THE PAROTID GLAND

The parotid gland is the largest salivary gland. It has superficial and deep parts: the former lies superficial to the masseter muscle (Fig. 30–16); its deep part lies between the ramus of the mandible anteriorly and the external acoustic meatus posteriorly.

The **parotid duct (Stenson's duct)** emerges from the anterior border of the gland (Fig. 30–16). It extends anteriorly across the superficial surface of masseter and then turns medially to pierce buccinator and open into the mouth opposite the upper second molar.

A number of structures pass through or lie within the parenchyma of the parotid gland:

The terminal part of the facial nerve: Upon emerging in the face by passing through the stylomastoid foramen, the facial nerve extends anteriorly through the parotid gland and ramifies into its temporal, zygomatic, buccal, mandibular, and cervical branches (Fig. 30–17). These branches collectively innervate all the muscles of facial expression.

The retromandibular vein: The retromandibular vein forms within the gland from the union of the superficial temporal and maxillary veins.

The bifurcation of the external carotid artery: The external carotid artery terminally bifurcates within the gland into the superficial temporal and maxillary arteries.

Parotid lymph nodes: A number of lymph nodes called the parotid lymph nodes lie embedded within the parotid gland.

Stimulation of salivary secretion by the parotid gland involves the otic ganglion, part of the parasympathetic division of the autonomic nervous system. The **otic ganglion** is located in the infratemporal fossa directly beneath the foramen ovale. The ganglion contains cell bodies of postganglionic parasympathetic neurons (Fig. 30–18). The cell bodies provide postganglionic parasympathetic secretomotor fibers for the parotid gland; the **auriculotemporal nerve** (a branch of the mandibular division of the trigeminal nerve) transmits these fibers from the ganglion to the gland. The glossopharyngeal nerve, through its **lesser petrosal nerve,** provides preganglionic parasympathetic fibers which synapse within the otic ganglion.

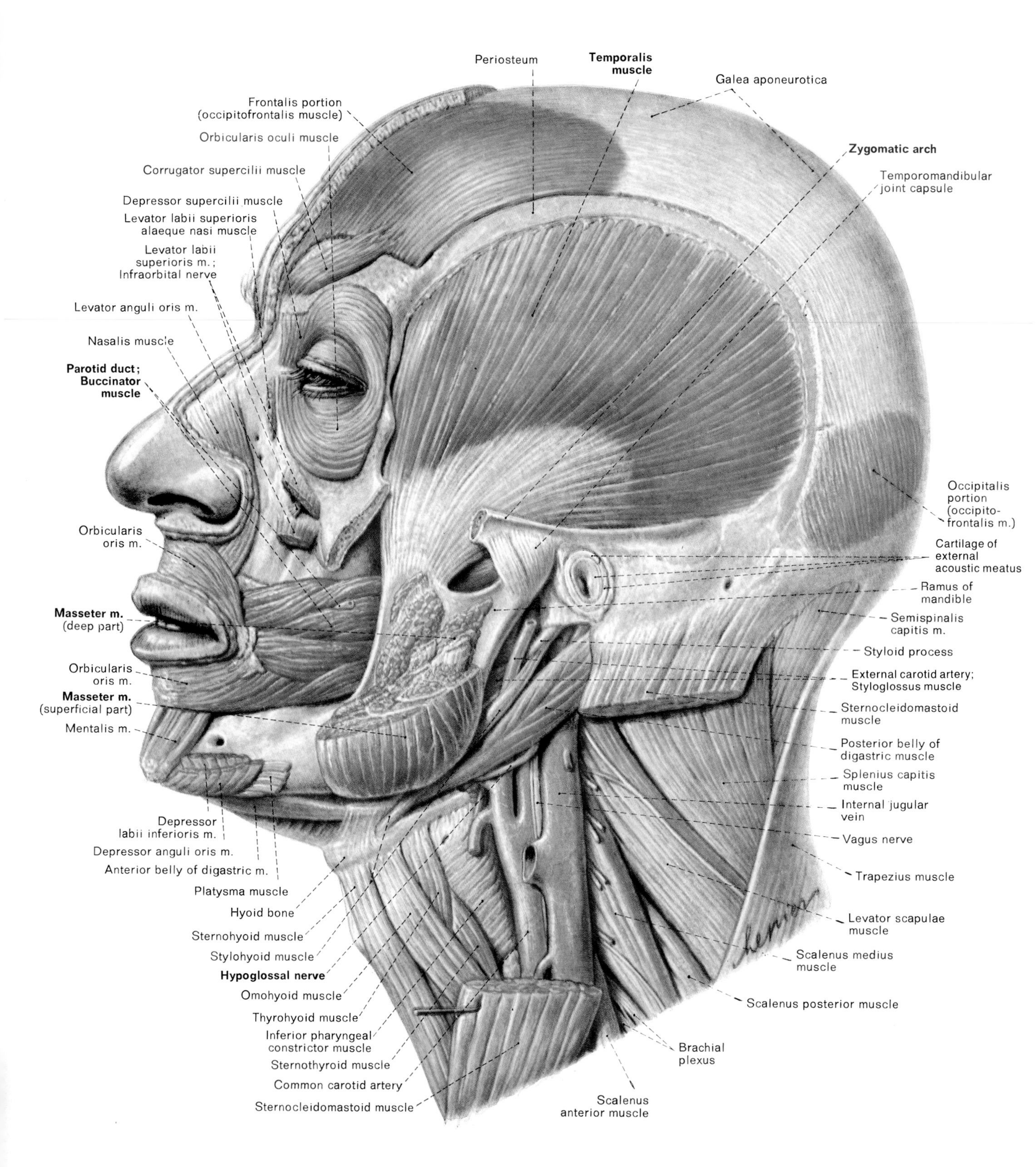

Fig. 30–15: Lateral view of the left temporalis muscle. The auricle, zygomatic arch, and bulk of the masseter muscle have been removed.

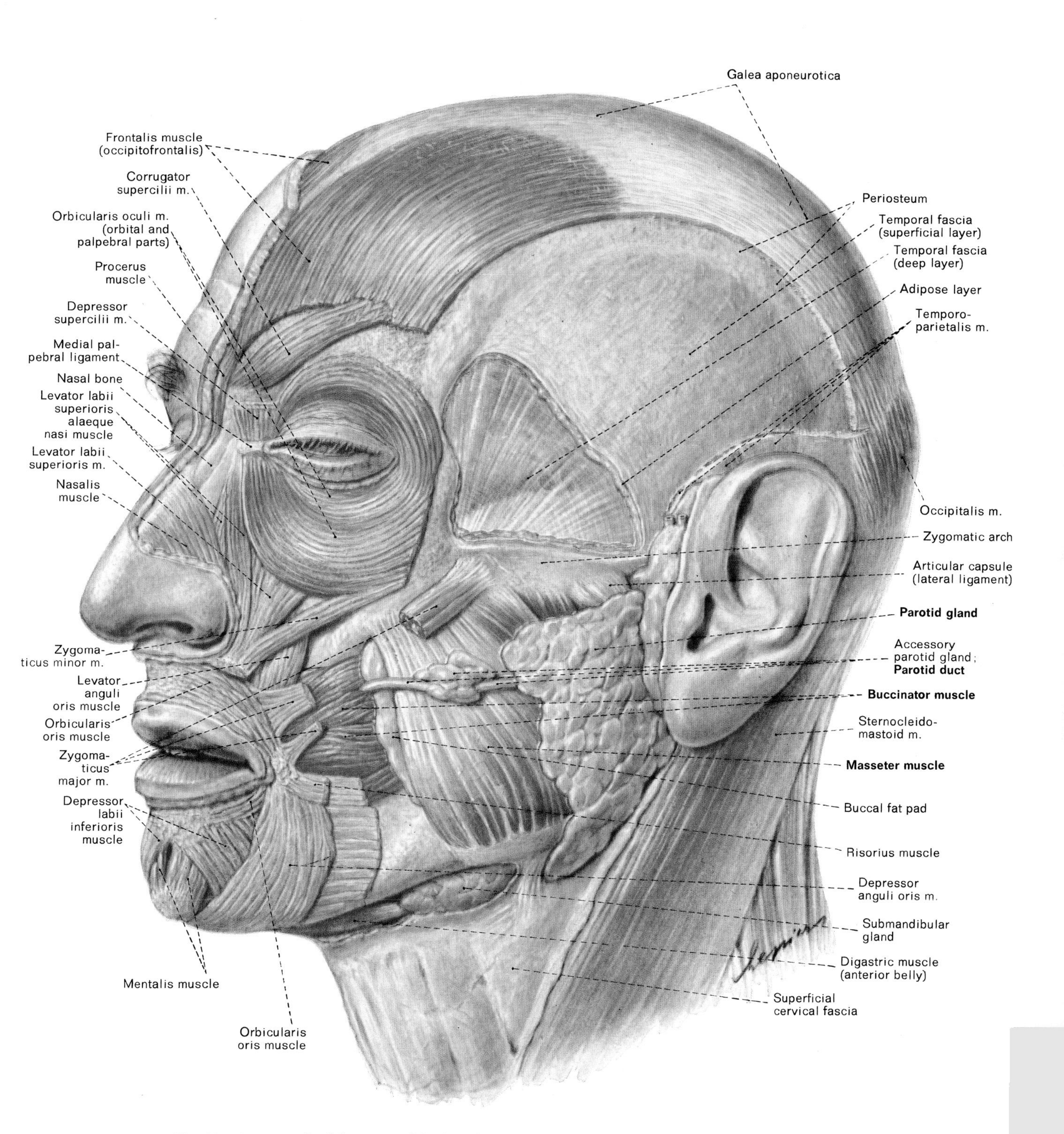

Fig. 30–16: Superficial dissection of the face showing the parotid gland and duct and the masseter muscle.

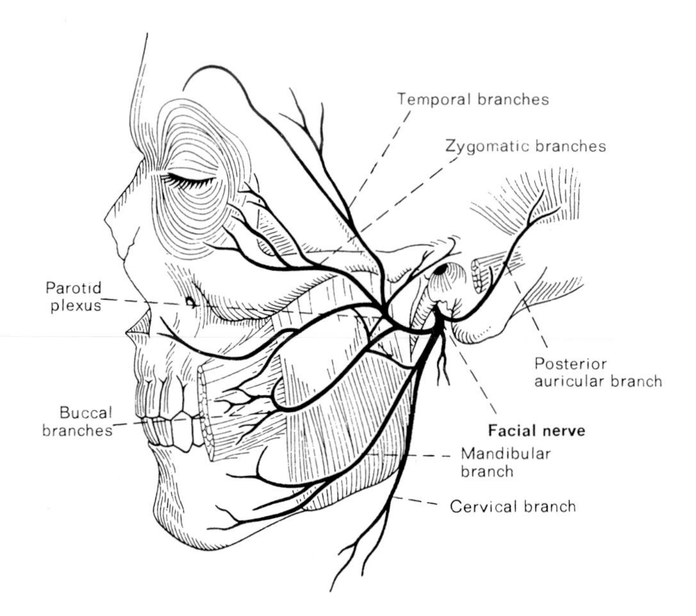

Fig. 30–17: Diagram of the branches of the facial nerve which collectively innervate all the muscles of facial expression.

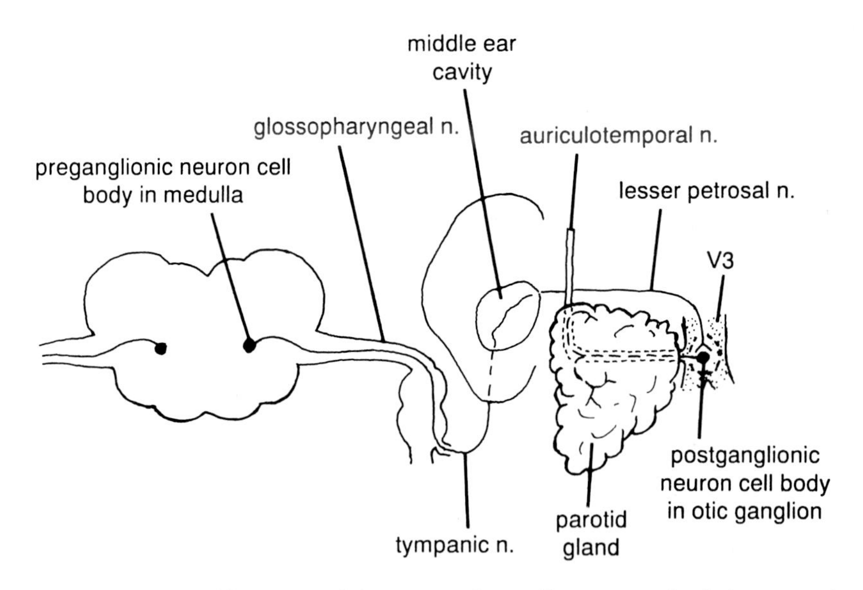

Fig. 30–18: Diagram of the origins and destinations of nerve fibers associated with the otic ganglion.

CASE **VI.4**

The Case of Noryu Takanami

INITIAL PRESENTATION AND APPEARANCE OF THE PATIENT

Mr. and Mrs. Takanami have made an appointment for their 4 year-old son Noryu. Noryu is a small, thin boy. He sits quietly on his mother's lap, holding a model space ship. The examiner observes that Noryu's right ear projects more anteroinferiorly than the left and that Noryu's facial expression is asymmetric when he smiles weakly in response to the examiner's greeting.

QUESTIONS ASKED OF THE PATIENT'S PARENTS

What seems to be the problem with Noryu? We're worried because he's had a fever for the past five days. His problems seemed to start about ten days ago when he caught a cold. Since he had a little bit of a fever, we kept him home from daycare and hoped that his cold would clear up in a few days. At first, his cold didn't seem to bother him much. Even though his nose was running and he complained that he couldn't breathe through his nose, he seemed otherwise healthy and full of energy. And, then, just when he seemed to be getting over the cold, he developed a fever that he's now had for the past five days.

What was his temperature when you last measured it? We took his temperature this morning just before coming to see you, and it was 103°. *[Infection is the most common cause of fever in young children. The parents' account of the patient's health during the past 10 days indicates that the patient has had an infection of the upper respiratory tract. At this point, the examiner embarks upon a series of questions designed to identify the most likely source of the patient's fever.]*

Has there been any change in Noryu's behavior or activities in the last 5 days? He's not been himself for the past 2 days. He's quiet and wants to be left alone. He also seems to have lost much of his appetite. We can only get him to drink fruit juice or milk. *[The examiner poses this question partly because young children frequently express a feeling of poor health with a decline in their general level of physical and mental activity and/or a reduction in their food intake. However, in this case, in which it is likely that an infection associated with the upper respiratory tract is responsible for the patient's persistent fever, the examiner is particularly mindful of signs or symptoms (such as a change in personality, irritability, lethargy, or malaise) which commonly herald intracranial extension of an infectious process.]*

Has he complained of a headache? No.

Have you seen or has he complained of any muscle spasms or awkward movements involving his hands or feet? No.

Has he had any nausea or vomiting? No. *[Episodes of muscle spasms or awkward limb movements might signify seizures. In this case, evidence of headache, seizures, nausea, or vomiting would warrant consideration of infection of the central nervous system.]*

Does Noryu still have any symptoms of a cold? No, other than his nose seems to be a bit congested. *[This answer suggests that the nasal cavities and/or one or more of the paranasal sinuses are inflamed and congested.]*

Has he complained of a sore throat? No. *[A positive response to this question would suggest painful inflammation of the pharynx.]*

Has he complained of an earache? Not in the past two days. But he did complain of an earache in his right ear about three or four days ago, and we noticed then that he would frequently rub his ear. But since then, he hasn't said anything about an earache.

I noticed when I first looked at Noryu that his right ear seems to stick out more than his left ear. Is this normal or something that has recently happened? No, it's not normal. We noticed the same thing yesterday morning, and we also noticed that the skin behind his ear is red and swollen. *[The presence of otalgia three or four days previously, the anteroinferior displacement of the right pinna, and the parents' description of erythematous and edematous skin behind the right ear all point to the right ear as a likely site of infection.]*

Has Noryu complained of a loss of hearing? No.

Has he complained of dizziness? No.

Have you seen or has he complained of clumsiness? No. *[The examiner poses these questions to determine if there is any evidence of disturbance of either cochlear or vestibular function. In this case, dizziness and/or clumsiness would suggest extension of the right ear infection to the semicircular canals or the cranial cavity.]*

Does he have a cough? No.

Has he complained of difficulty in breathing? No. *[The examiner asks these questions because coughing and dyspnea are common symptoms of respiratory tract disease.]*

Has he had any diarrhea? No. *[The examiner asks this question because diarrhea is a common symptom of infection of the gastrointestinal tract.]*

Has there been any change in his urination habits? No. *[The examiner poses this question to determine if the patient exhibits any symptoms (such as painful urination, an increase in the frequency of urination, or a change in the appearance of the urine) common to infection of the genitourinary tract.]*

Has there been any changes in his skin? No.

Has he complained of pain in his joints? No. *[The examiner asks these questions to determine if the patient exhibits any symptoms of localized or generalized infection of the integumentary and musculoskeletal systems.]*

Has he had any other recent illnesses or injuries? No.

Have you given him any medications for cold or fever? We didn't give him anything for his cold, but we have given him Tylenol for the fever. He does seem to act more like himself after Tylenol. *[The examiner finds the patient's parents to be alert and fully cooperative during the interview.]*

PHYSICAL EXAMINATION OF THE PATIENT

Vital signs: Blood pressure: 110/75
Pulse: 100
Rhythm: regular
Temperature: 103.5°F.
Respiratory rate: 29
Height: 109 cm.
Weight: 19.5 kgs.

[A pediatrician would recognize that except for the temperature, the patient's vital signs are within normal limits. The patient's weight and height are in the 90th percentile for boys 4 years of age.]

HEENT Examination: Examination of the head, eyes, ears, nose, and throat are normal except for the following findings: The right pinna is displaced anteroinferiorly; its posterior surface is erythematous and edematous. The skin overlying the right side of the skull immediately posterosuperior to the pinna is swollen, erythematous, and extremely tender to palpation. The right external acoustic meatus is erythematous, contains a purulent discharge, and exhibits a sagging posterosuperior wall. The right tympanic membrane is hyperemic, perforated, and opaque.

Wrinkling of the forehead does not occur as rapidly on the right side as it does on the left side when the patient looks upward. The right palpebral fissure is slightly wider than the left palpebral fissure. Closure of the right eyelids occurs slightly more slowly and less forcefully than that of the left eyelids. The right buccal angle does not retract as much as the left buccal angle when the patient smiles.

The parotid and upper deep cervical lymph nodes on the right side are enlarged, regularly shaped, firm, and tender.

The patient can hear a quietly ticking watch 20 cm. from the left ear but not the right ear. The patient can hear the snap of a finger 20 cm. from either ear.

A pertinent normal finding is that the level of lacrimation in the right eye appears to equal that in the left eye. The patient does not cooperate with an attempt to ascertain salivary secretion from the right submandibular duct.

Lungs: Normal
Cardiovascular Examination: Normal
Abdomen: Normal
Genitourinary Examination: Normal
Musculoskeletal Examination: Normal
Neurologic Examination: Normal. Pertinent normal findings are the absence of (a) positive Brudzinski's and Kernig's signs and (b) nystagmus.

[A positive Brudzinski's or Kernig's sign would suggest meningitis (see the case of Duane Carlson, Case VI.3). In this case, evidence of nystagmus would suggest infectious inflammation of the semicircular canals in the patient's right ear.]

INITIAL ASSESSMENT OF THE PATIENT'S CONDITION

The patient's chief medical problem appears to be an infection involving the tissues within and about the right ear.

ANATOMIC BASIS OF THE PATIENT'S HISTORY AND PHYSICAL EXAMINATION

(1) **The patient's elevated temperature suggests the presence of an entrenched infection.** Infection is the most common cause of fever in a young child.

(2) **Examination of the right ear reveals marked inflammation of the pinna, external acoustic meatus, and tympanic membrane.** The presence of a purulent discharge in the right external acoustic meatus suggests that the inflammation has occurred in response to infection by bacterial and/or viral agents. The inflammation and perforation of the tympanic membrane indicates that the infection involves tissues of the middle ear as well as those of the outer ear.

(3) **The inflammatory swelling of the tissues behind the right pinna, the anteroinferior displacement of the right pinna, and the sagging of the posterosuperior wall of the right external meatus indicate a swelling of the right mastoid process and the tissues about it.** In a young child, the mastoid process of the temporal bone lies deep to the soft tissues directly behind the pinna and immediately posterosuperior to the external acoustic meatus and tympanic cavity. Marked swelling of the mastoid process and the soft tissues about it in a young child will therefore exert (a) an anteroinferiorly-directed pressure against the external acoustic meatus (which could anteroinferiorly displace the outer cartilaginous part of the meatus and the attached pinna and/or depress the posterosuperior wall of the meatus) and (b) a laterally-directed pressure against the back of the ear (which could turn the pinna to face more directly anteriorly). The displacement of the pinna and the distortion of the external acoustic meatus thus suggest that the infection within and about the right ear also involves the mastoid process.

(4) **The weakness and slowness of movements involving the skin of the forehead, the eyelids, and the lips indicate dysfunction of the muscles of facial expression on the right side of the face.** Epicranius is the muscle of facial expression chiefly responsible for wrinkling the forehead. Orbicularis oculi is the muscle chiefly responsible for forcible closure of the eyelids; its tonus is a chief determinant of the width of the palpebral fissure. Zygomaticus major, buccinator, and risorius all retract the buccal angle during smiling. The facial nerve innervates all the muscles of facial expression.

As previously discussed, the facial nerve extends first posteriorly through the upper medial wall of the tympanic cavity and then inferiorly through the posterior wall of the tympanic cavity as it courses within the facial canal of the temporal bone. The facial nerve's descent through the posterior wall of the tympanic cavity places it directly between the tympanic cavity anteriorly and the mastoid air cells posteriorly, and thus susceptible to impairment by an infection involving both the middle ear and the air cells of the mastoid process. The dysfunction of the muscles of facial expression on the right side of the face thus suggests that the infection within and about the right ear additionally involves the facial nerve at one or more points along its course within the facial canal. The point or points of involvement appear to be distal to the origin of the greater petrosal nerve, as there appears to no asymmetry with respect to the levels of lacrimation in the left and right eyes. The patient's refusal to cooperate with observation of salivary secretion from the right submandibular duct negates the attempt to determine chorda tympani function.

(5) **The physical characteristics of the parotid and upper deep cervical nodes on the right side of the face indicate an immunologic response to infectious agents.** Lymph nodes which are mounting an immunological response to infectious agents are typically enlarged but regularly shaped, tender, relatively mobile, and firm. The immunologic response by the parotid and upper deep cervical nodes suggests lymphatic dissemination of the infectious agents responsible for the infection in and about the right ear. The marked swelling of postauricular tissues on the right side masks the physical characteristics of the postauricular nodes.

(6) **The ability to hear the snapping of a finger but not the more quiet ticks of a watch on the right side indicate a partial loss of hearing in the right ear.**

INTERMEDIATE EVALUATION OF THE PATIENT'S CONDITION

The patient is suffering from an infection involving the middle and outer parts of the right ear, the adjoining mastoid process, and the right facial nerve. This infection appears to be the source of the patient's persistent fever of five days duration.

CLINICAL REASONING PROCESS

In considering this case, a pediatrician would conclude that the patient has **otitis media** (inflammation of the middle ear) complicated by **mastoiditis** (inflammation of the mastoid process) and partial paralysis of the muscles of facial expression. Otitis media is indicated by the purulent otorrhea in association with the inflamed and perforated tympanic membrane. The findings suggest that a purulent effusion of the tympanic cavity has generated sufficient pressure to rupture the tympanic membrane. It should be noted, however, that these findings are not characteristic for all cases of otitis media. For example, otitis media can occur without purulent effusion in the tympanic cavity. Mastoiditis is indicated by the marked postauricular swelling, anteroinferior displacement of the pinna, and sagging of the posterosuperior wall of the external acoustic meatus.

The principal issue to be resolved is the extent of the infection within the mastoid process. In patients with acute otitis media, there is almost always concurrent inflammation of the mucous membrane lining the mastoid air cells. The interior of the mastoid air cells in such patients shows increased radiodensity in CT scans; this clouding of the interior of the air cells represents effusion from and swelling of the mucous membrane lining. Progression of the mastoid infection from this early stage can result in **mastoid periosteitis** (infection of the periosteum lining the outer surface of the mastoid process) and/or **mastoid osteitis** (infection of the bony trabeculae of the mastoid air cells). Mastoid periosteitis can occur in the absence of mastoid osteitis when the emissary veins of the mastoid process serve as conduits for the pathogens from the interior of the mastoid air cells to the mastoid periosteum and overlying tissue layers of the scalp. Extension of the infection beyond the mastoid process can produce **labyrinthitis** (inflammation of the semicircular canals) or intracranial complications such as **meningitis** (inflammation of the meninges), an **epidural abscess,** a **subdural empyema** (a pus-filled cavity between the dura mater and arachnoid mater), or **dural venous sinus thrombosis** (formation of thrombi in a dural venous sinus, generally the lateral or sigmoid sinus).

RADIOLOGIC EVALUATION AND FINAL RESOLUTION OF THE PATIENT'S CONDITION

CT scans of the patient's head and neck showed distortion and poor definition of the outline of the mastoid air cells in the right ear. The areas of pneumatization in the right mastoid process (the interior cavities of the air cells) exhibited increased radiodensity and were outlined by poorly defined bony margins. There was no evidence of a subperiosteal abscess or of a suppurative process in the cranial cavity.

These radiologic findings indicate a diagnosis of acute mastoiditis osteitis. The increased radiodensity of the areas of pneumatization is caused by the effusion from and swelling of the mucous membrane lining the air cells. The lack of definition of the bony outlines of the air cells is a consequence of demineralization of the compact bone forming the walls of the air cells.

Management of this case would involve immediate parenteral antibiotic therapy promptly followed by mastoidectomy (surgical extirpation of the mastoid air cells). Drains are inserted into the mastoid cavity and tympanic cavity to provide for external drainage and ventilation.

THE CHRONOLOGY OF THE PATIENT'S CONDITION

The patient's problems began ten days before with inflammatory obstruction of the nasal passageways complicated by maxillary and ethmoidal sinusitis (specifically, bacterial infection of the maxillary and ethmoid sinuses). During the first three days of this upper respiratory infection, there were several occasions when bacteria-laden, nasal secretions were insufflated into the right tympanic cavity during the act of swallowing.

Two factors combined to produce the insufflation of nasal secretions. (1) Because of a slight asymmetric formation of cartilage in the auditory tubes, the patient's right auditory tube did not have sufficient cartilage to adequately support tensor veli palatini's capacity to open the auditory tube. This impairment in the active opening mechanism for the right auditory tube resulted in a higher than normal negative pressure within the right tympanic cavity (because of sustained absorption of gases from the tympanic cavity by its membranous lining). (2) Nasal congestion prevented equilibration of nasopharyngeal air pressure with the exterior through the nasal passageways. Consequently, whenever the nasopharynx was closed by the raising of the soft palate during the act of swallowing, there occurred a transient initial increase in nasopharyngeal air pressure (above that of atmospheric pressure) quickly followed by a transient decrease in nasopharyngeal air pressure (below that of atmospheric pressure). The imposition of a transient increase in air pressure at the pharyngeal end of the auditory tube in combination with the high negative pressure at the tympanic end generated a momentary pressure difference along the closed segment of the auditory tube sufficient to inject bacteria-laden nasal secretions into the tympanic cavity. Drainage of these secretions back into the nasopharynx was restricted by the impairment in the active opening mechanism for the auditory tube.

The entrapment of the bacteria-laden nasal secretions within the right tympanic cavity led to acute otitis media (and attendant mastoiditis) six days previously. The accumulation of a high-pressure purulent effusion within the tympanic cavity was coincident with the appearance of otalgia (earache) and fever. The high-pressure purulent effusion markedly restricted blood flow to the walls of the mastoid air cells, resulting in demineralization of the bony septa. Ear pain was markedly diminished three days before, after perforation of the tympanic membrane. The purulent otorrhea represents discharge from the tympanic cavity. Partial paralysis of the muscles of facial expression on the right side occurred because of extension of the infection into the facial canal along its course downward through the posterior wall of the tympanic cavity.

RECOMMENDED REFERENCES FOR ADDITIONAL INFORMATION OF THE EAR

DeGowin, R. L., Jochimsen, P. R., and E. O. Theilen, *DeGowin & DeGowin's Bedside Diagnostic Examination,* 5th ed., Macmillan Publishing Co., New York, 1987: *Pages 192 and 193 present a discussion of Weber's and Rinne's testing for assessing sensorineural versus conductive hearing loss. Sensorineural loss is a consequence of a disorder involving the cochlear portion of the vestibulocochlear nerve, and conductive loss is a consequence of interference with the conduction of physical vibrations through the outer and middle parts of the ear.*

Ballenger, J. J., *Diseases of the Nose, Throat, Ear, Head, and Neck,* 14th ed., Lea & Febiger, Philadelphia, 1991: *Pages 1104 through 1108 in Chapter 34 provide a discussion of acute inflammatory diseases of the middle ear.*

CHAPTER 31

The Oral Cavity and the Pharynx

The oral cavity is divisible into two parts: the oral cavity proper and the vestibule. The **oral cavity proper** is the region within the mouth that is anterolaterally bordered by the upper and lower dental arches. The **vestibule** of the mouth is the region superficial to the upper and lower dental arches; the vestibule is bordered on its sides by the cheeks and anteriorly by the upper and lower lips.

The pharynx is a funnel-shaped, muscular tube which extends from the undersurface of the base of the skull down through the neck to the esophagus (Fig. 31–8). The pharynx is divisible into three parts which, proceeding from the uppermost to the lowest, are called the **nasopharynx, oropharynx,** and **laryngopharynx** (or **hypopharynx**). The posterior openings of the nasal cavities open into the anterior aspect of the nasopharynx, and the medial ends of the auditory tubes open into the sides of the nasopharynx (Fig. 31–3). The palatoglossal arches mark the boundary where the oral cavity opens into the anterior aspects of the oropharynx. The **laryngeal inlet** faces the anterior aspect of the hypopharynx.

THE ORAL CAVITY

The Muscles around the Mouth

The muscles around the mouth are muscles of facial expression, and thus are innervated by the facial nerve. The muscle fibers of orbicularis oris encircle the lips (Figs. 28–4 and 30–15); their primary action is to close the mouth at the lips. Buccinator is the chief muscle of the cheek (Fig. 30–15); it acts to compress the cheek against the dental arches.

Levator anguli oris and depressor anguli oris repectively act to raise and lower the corner of the mouth (Fig. 30–16). Levator labii superioris and depressor labii inferioris respectively act to raise the upper lip and lower the lower lip. Zygomaticus major pulls the corner of the mouth superolaterally.

The Tongue

The tongue rests on the floor of the mouth. Its dorsal surface is divided into anterior and posterior parts by a V-shaped groove called the **sulcus terminalis** (Fig. 32–9). The anterior two-thirds of the tongue lies in the oral cavity proper, and the posterior one-third faces the oral part of the pharynx.

The tongue consists of intrinsic and extrinsic muscles. The intrinsic muscles are confined to the tongue itself, and function to change the shape and configuration of the tongue. The extrinsic muscles of the tongue originate from sites above and below the tongue; they function to move the tongue within and outside of the mouth. There are four pairs of intrinsic muscles (vertical, transverse, superior longitudinal, and inferior longitudinal) and four pairs of extrinsic muscles (genioglossus, hyoglossus, styloglossus, and palatoglossus).

Genioglossus originates from the posterior surface of the midline portion of the body of the mandible; its muscle fibers radiate posterosuperiorly through the tongue (Fig. 31–1). Genioglossus is the only extrinsic muscle that can extend the tongue out from the mouth. When genioglossus acts to protrude the tip of the tongue from the mouth, it also deviates the tip of the tongue to the contralateral side. Genioglossus can also lower the tongue within the mouth.

Hyoglossus originates from the hyoid; its muscle fibers extend anterosuperiorly through the tongue (Fig. 31–1). Hyoglossus can retract and lower the tongue within the mouth.

Styloglossus originates from the styloid process of the temporal bone; its muscle fibers extend anteroinferiorly through the tongue (Fig. 31–1). Styloglossus can retract and raise the tongue within the mouth.

Palatoglossus originates from the **soft palate** (the pliable, posterior part of the roof of the mouth); its muscle fibers extend anteroinferiorly through the tongue (Fig. 31–1). Palatoglossus can retract and raise the tongue within the mouth. Palatoglossus forms the substance of a soft tissue arch (the **palatoglossal arch**) in the mouth that extends downward behind the last molars from the soft palate into the tongue (Figs. 31–1 and 31–2). The palatoglossal arches represent the boundary between the oral cavity proper and the oropharynx.

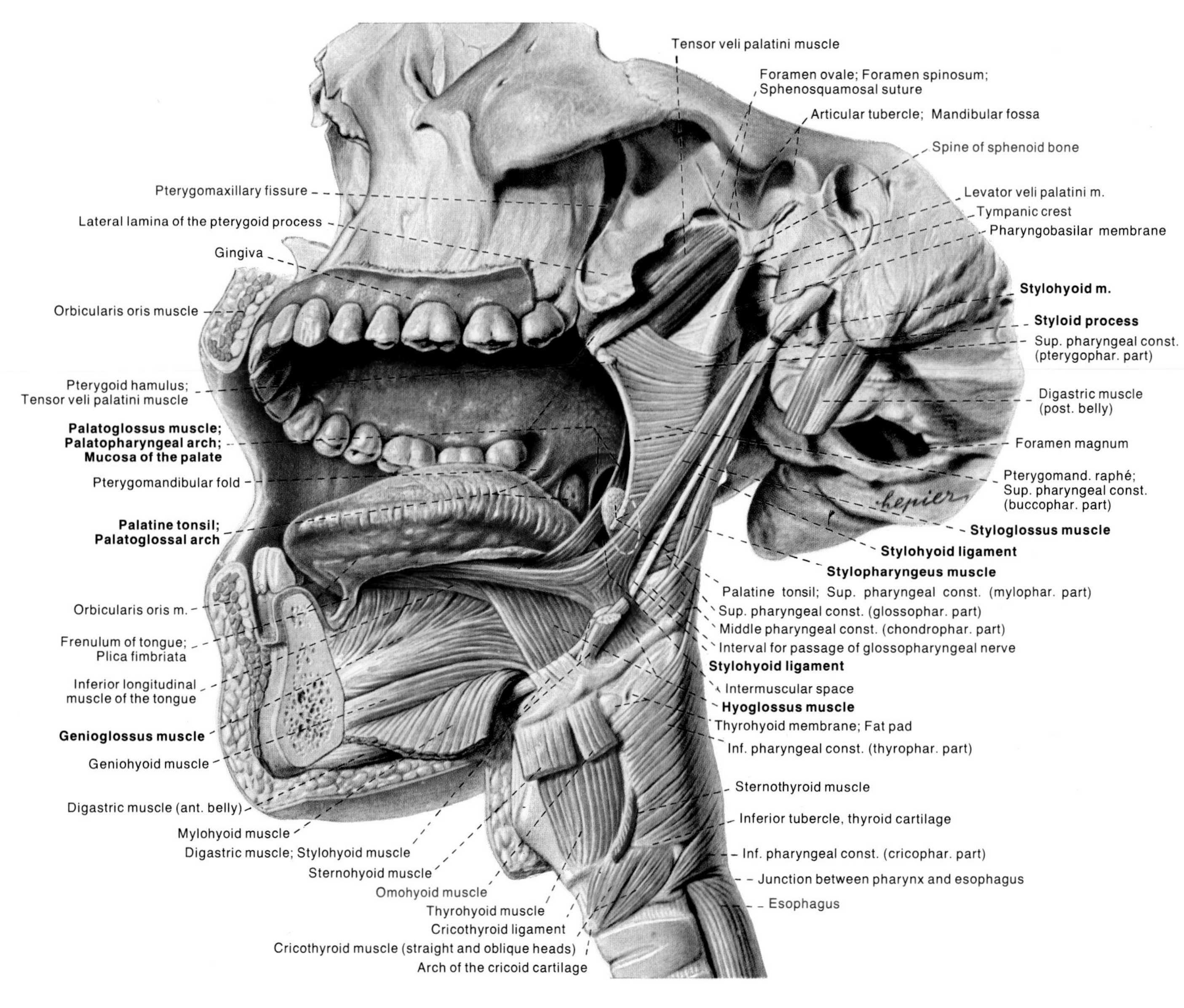

Fig. 31–1: Lateral view of the lingual musculature and the pharyngeal constrictors. The superficial and deep facial structures on the left side and the left half of the mandible have been removed.

The Palate

The anterior part of the roof of the mouth has a bony foundation, and thus is called the **hard palate** (Fig. 31–3). The **palatine processes of the maxillae** and the **horizontal plates of the palatine bones** form the bony foundation of the hard palate (Fig. 29–2).

The posterior part of the roof of the mouth has a fibromuscular foundation, and thus is called the **soft palate** (Fig. 31–3). The conical projection of soft tissues that hangs downward from the posterior margin of the soft palate in the midline is called the **uvula** (Figs. 31–2 and 31–3).

Five muscles are associated with the soft palate. Two of the muscles (tensor veli palatini and palatoglossus) act on the anterior portion of the soft palate; the other three muscles (levator veli palatini, musculus uvula, and palatopharyngeus) act on the posterior portion of the soft palate.

Tensor veli palatini descends from its origins (the pterygoid process of the sphenoid and the lateral aspect of the medial end of the auditory tube) to give rise to a tendon which hooks medially around a bony projection (the pterygoid hamulus) on the undersurface of the sphenoid (Figs. 31–1 and 31–4). Upon hooking around the pterygoid hamulus, the tendon extends medially and horizontally into the anterior portion of the soft palate. Tensor veli palatini acts to tense the anterior portion of the soft palate and to open the auditory tube.

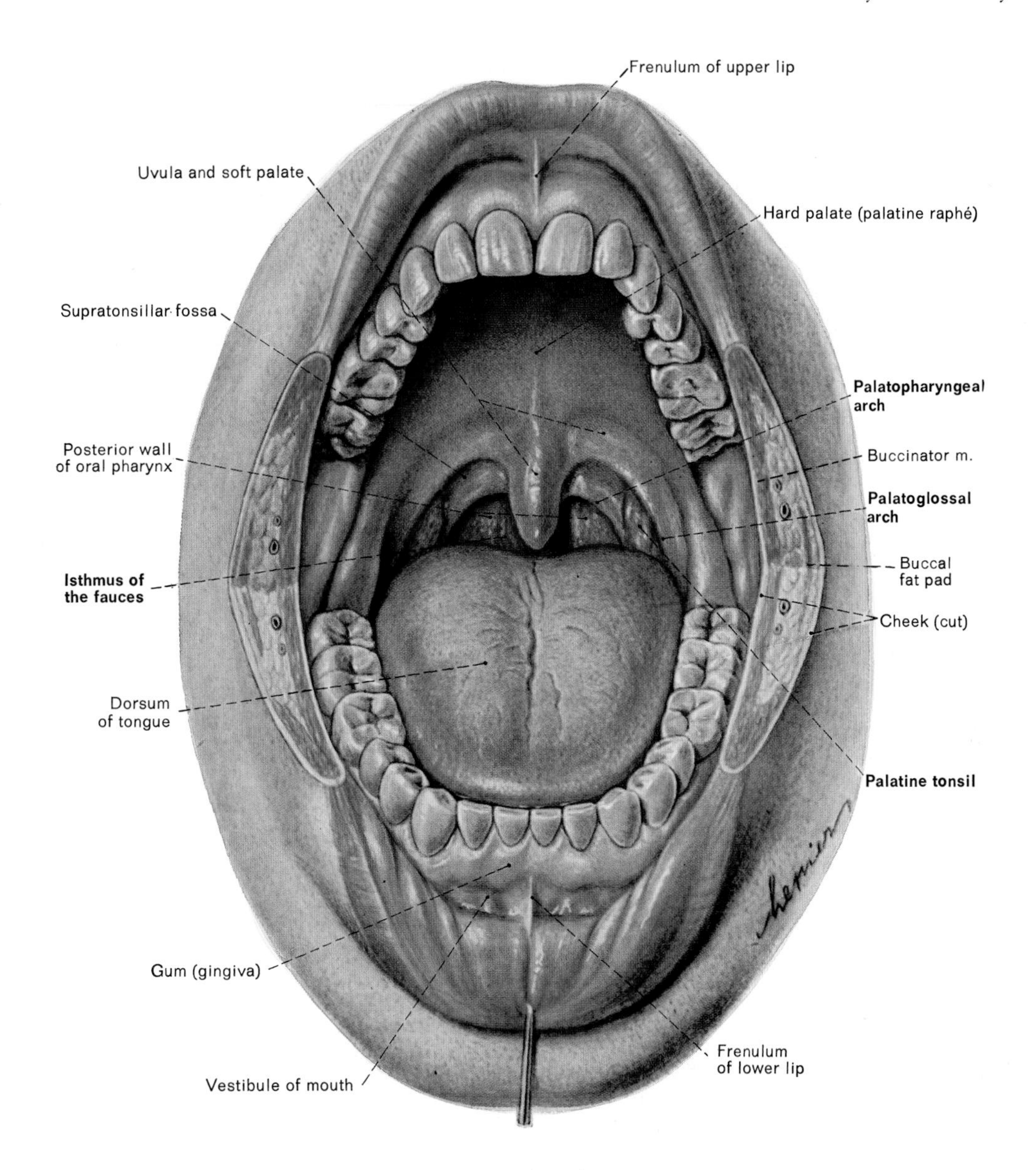

Fig. 31–2: The oral cavity.

Levator veli palatini descends from the undersurface of the petrous part of the temporal bone into the posterior portion of the soft palate (Figs. 31–1 and 31–4). Levator veli palatini acts to elevate the posterior portion of the soft palate.

Musculus uvula originates from the posterior margin of the bony foundation of the hard palate and extends posteriorly into the uvula (Fig. 31–4). The muscle on each side acts to raise the uvula and to pull it to the ipsilateral side.

Palatopharyngeus descends from the soft palate into the lateral wall of the pharynx (Fig. 31–4). Palatopharyngeus acts to raise the pharynx. Palatopharyngeus underlies a soft tissue arch (the **palatopharyngeal arch**) that extends downward from the soft palate behind the palatoglossal arch (Figs. 31–2 and 31–3).

The Submandibular and Sublingual Glands

The submandibular and sublingual glands lie near the floor of the mouth. The submandibular gland is the second largest salivary gland. Its superficial and deep parts lie, respectively, beneath and above the floor of the mouth (Fig. 31–5). The **submandibular duct** opens into the oral cavity beneath the tongue. The submandibular duct and deep part of the gland curve around the posterior margin of the mylohyoid muscle.

The sublingual gland is the smallest major salivary gland; it lies beneath the tongue (Fig. 31–5). Its secretions are released into the oral cavity through a variable number of small ducts.

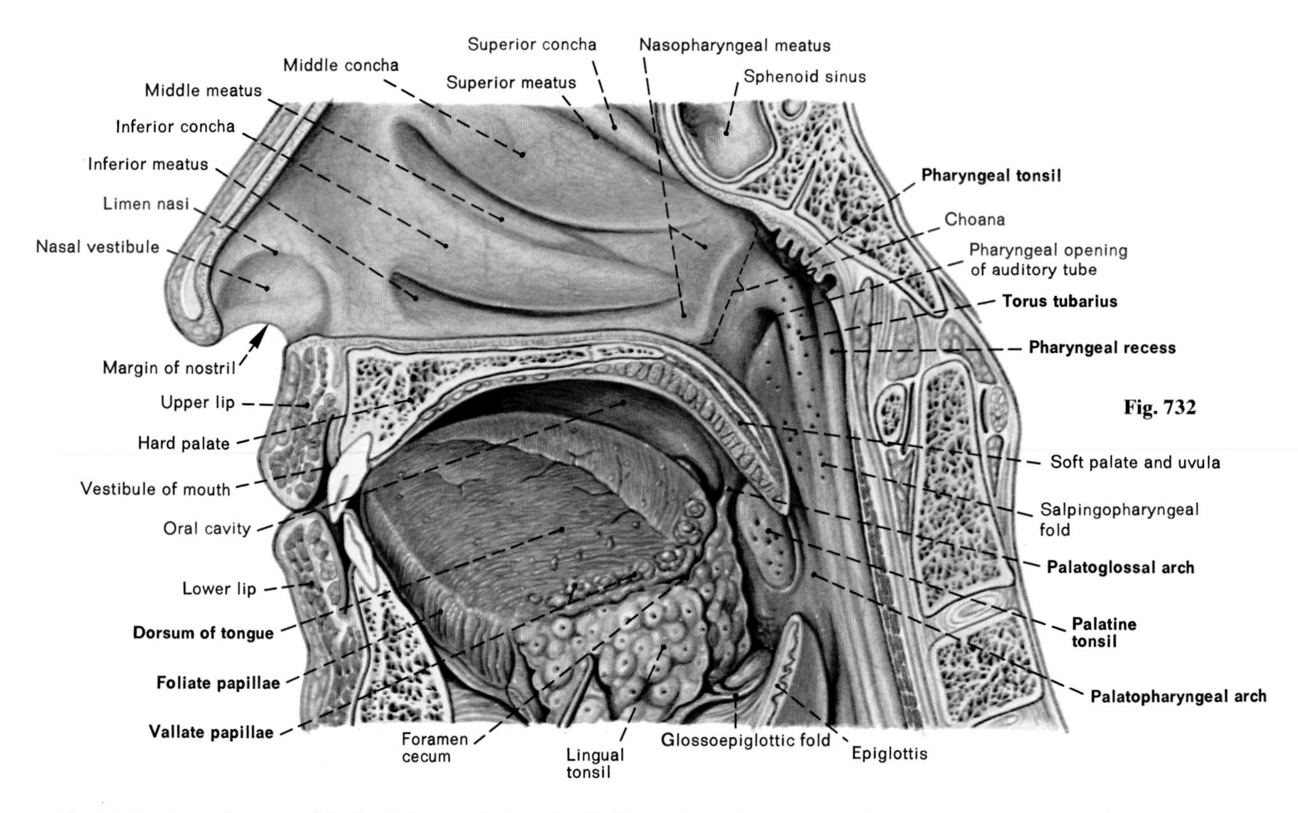

Fig. 31–3: Sagittal section of the head showing the lateral wall of the right nasal cavity, the oral cavity, the nasopharynx and the oropharynx.

The Innervation of the Oral Cavity

The innervation of the oral cavity involves cranial nerves V, VII, and IX to XII and the submandibular ganglion.

Cranial Nerve V: The maxillary division of the trigeminal nerve exits the cranial cavity and enters the pterygopalatine fossa by passing through the foramen rotundum (Fig. 28–12). The maxillary division extends from the pterygopalatine fossa to a groove in the floor of the orbital cavity called the infraorbital groove; here the maxillary division is called the **infraorbital nerve** (Figs. 30–9 and 31–6). The pterygopalatine fossa branches of the maxillary division and the infraorbital nerve branches collectively provide sensory innervation for the upper dental arch, the mucous membrane of the upper gum and upper lip, and all of the mucous membrane of the palate except that of the uvula. Cutaneous branches of the infraorbital nerve provide sensory innervation for the skin of the midregion of the face (specifically, that covering the lower eyelid, overlying the maxillary bone, upper dental arch, and upper gum, and covering the upper lip) (Fig. 29–19).

The mandibular division of the trigeminal nerve exits the cranial cavity and enters the infratemporal fossa by passing through the foramen ovale (Fig. 28–12). In the infratemporal fossa, the mandibular division gives rise to four major sensory branches: the **auriculotemporal nerve, inferior alveolar nerve, lingual nerve,** and **buccal nerve** (Fig. 31–7). The auriculotemporal nerve passes anterior to the external acoustic meatus as it ascends toward the scalp on the side of the head (Fig. 31–7). The inferior alveolar nerve enters the mandible through the mandibular foramen in the deep surface of the ramus of the mandible in route to its forward extension within the body of the mandible (Fig. 31–6). A terminal cutaneous branch of the inferior alveolar nerve, called the **mental nerve** (Fig. 31–6), exits the body of the mandible by the mental foramen. The lingual nerve extends anteroinferiorly deep to the mucous membrane lining the floor of the mouth (Fig. 31–7). The buccal nerve extends anteriorly through the cheek (Fig. 31–7). These sensory branches of the mandibular division collectively provide sensory innervation for the lower dental arch, the mucous membrane of the lower gum and lower lip, and all of the mucous membrane of the anterior two-thirds of the tongue and the floor of the mouth underlying the anterior two-thirds of the tongue. They also provide sensory innervation for the skin of the lowest region of the face (specifically, that overlying the lower margin of the body of the mandible, lower dental arch, and lower gum and covering the lower lip) (Fig. 29–19).

In the oral cavity, the mandibular division of the trigeminal nerve also innervates tensor veli palatini.

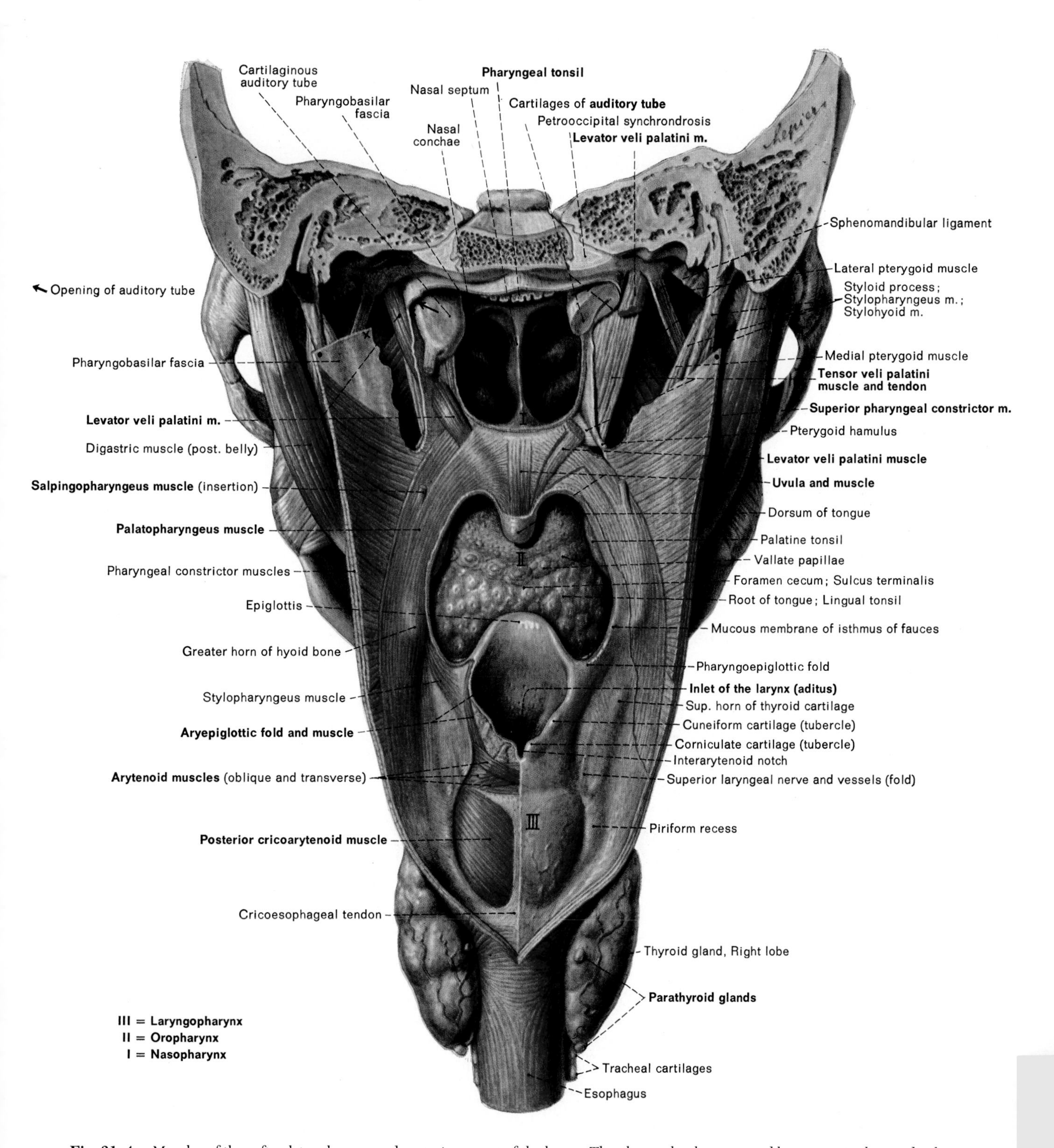

Fig. 31–4: Muscles of the soft palate, pharynx, and posterior aspect of the larynx. The pharynx has been opened by a posterior longitudinal incision and the mucous membrane has been removed from the soft palate, pharynx, and left side of the posterior aspect of the larynx.

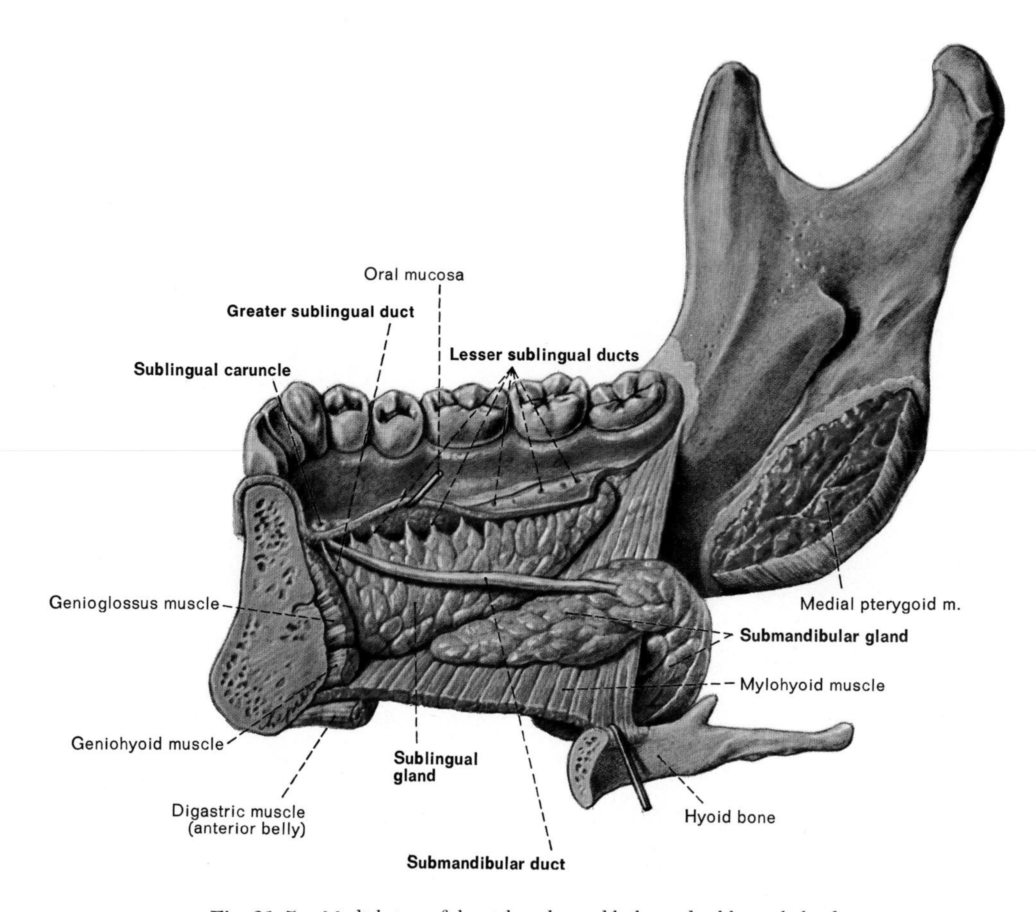

Fig. 31–5: Medial view of the right submandibular and sublingual glands.

The Submandibular Ganglion: The submandibular ganglion is a ganglion of the parasympathetic division of the autonomic nervous system. It is located beneath the mucous membrane of the floor of the mouth, where it is attached to the lingual nerve (Fig. 31–7). The ganglion contains cell bodies of postganglionic parasympathetic neurons. The cell bodies provide postganglionic parasympathetic secretomotor fibers for the submandibular and sublingual glands (Fig. 29–20); the lingual nerve transmits these fibers from the ganglion to the glands.

Cranial Nerve VII: The facial nerve, through its chorda tympani branch, provides preganglionic parasympathetic fibers which synapse within the submandibular ganglion. The chorda tympani joins the lingual nerve in the infratemporal fossa (Fig. 31–7), and the lingual nerve then transmits the chorda tympani's preganglionic parasympathetic fibers to the ganglion (Fig. 29–20).

The chorda tympani also provides taste sensation for the anterior two-thirds of the tongue (Fig. 30–11). The lingual nerve transmits these taste sensation fibers from the anterior two-thirds of the tongue to the chorda tympani. The neuronal cell bodies for the taste sensation fibers are located in the geniculate ganglion of the facial nerve (Fig. 30–11).

Cranial Nerve IX: The glossopharyngeal nerve provides sensory innervation for the mucous membrane of the uvula and the posterior third of the tongue. It also provides taste sensation for the posterior third of the tongue. The neuronal cell bodies for these general and taste sensation fibers of the oral cavity are located in the inferior ganglion of the glossopharyngeal nerve.

Cranial Nerves X and XI: Levator veli palatini, musculus uvula, palatopharyngeus, and palatoglossus are innervated by motor nerve fibers of the cranial part of the accessory nerve. The motor nerve fibers at first descend in the neck within the vagus nerve and are then transmitted by branches of the vagus nerve to the **pharyngeal plexus** (the pharyngeal plexus is the motor and sensory nerve plexus of the pharynx).

Cranial Nerve XII: The hypoglossal nerve innervates all the intrinsic and extrinsic muscles of the tongue except palatoglossus.

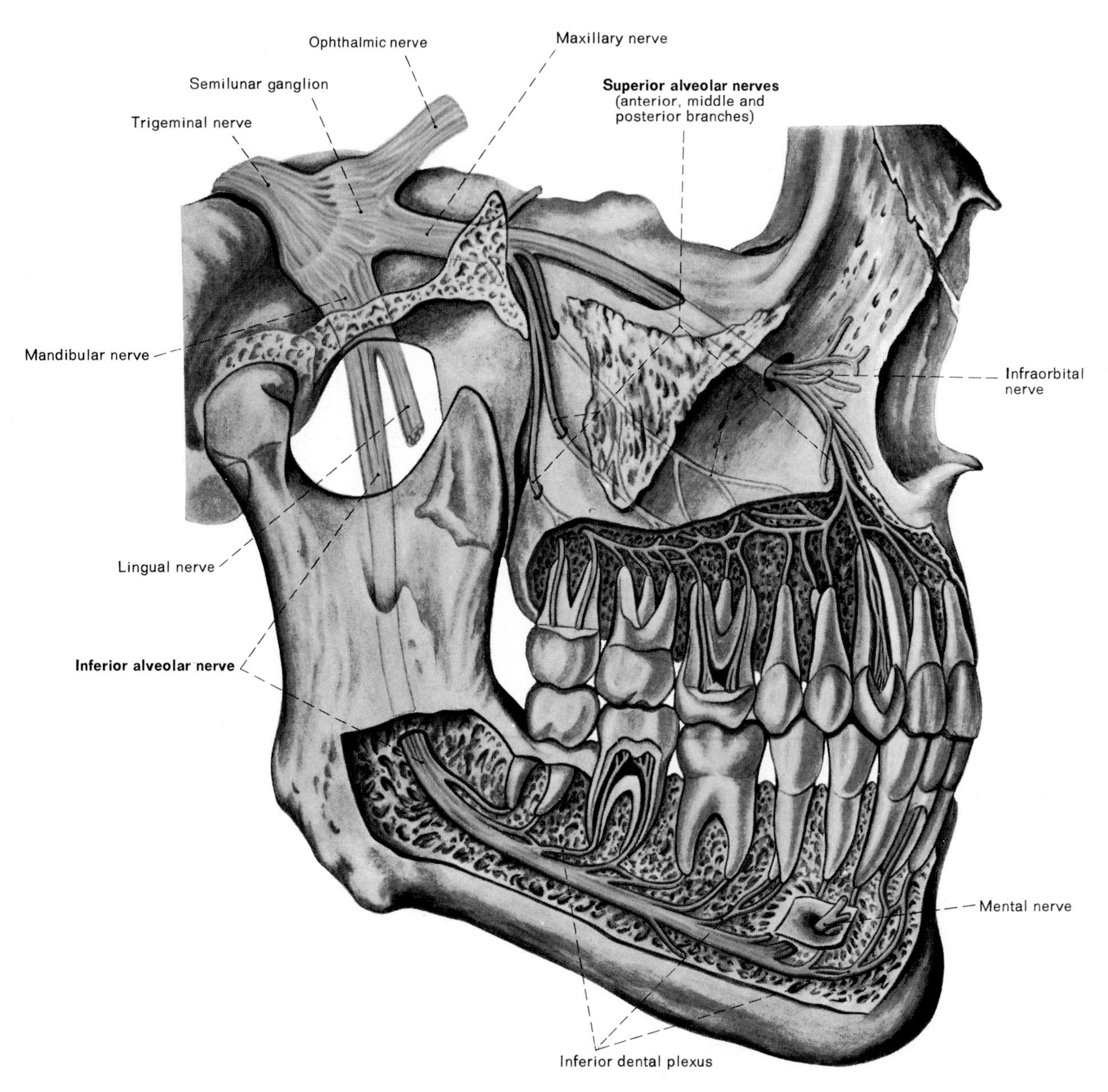

Fig. 31–6: Nerves of the upper and lower teeth.

Blood Supply and Venous and Lymphatic Drainage of the Tongue

The lingual branch of the external carotid artery provides most of the blood supply to the tongue. The tonsillar and ascending palatine branches of the facial artery supply the posterior part of the tongue.

The dorsum and sides of the tongue are drained by the dorsal lingual veins. The **dorsal lingual veins** are tributaries of the **lingual veins,** which are the veins that accompany the lingual artery. The lingual veins are tributaries of the internal jugular vein.

The inferior surface of the tongue is drained on each side by the deep lingual vein. The **deep lingual vein** begins near the tip of the tongue and extends posteriorly through the tongue immediately deep to the mucous membrane lining the inferior surface of the tongue. The tributaries of the deep lingual vein provide rapid absorption of drugs administered orally for dissolution in the oral cavity between the tongue and the floor of the mouth. The deep lingual vein joins a vein from the sublingual salivary gland to form a vein that accompanies the hypoglossal nerve; this latter vein ends by joining the internal jugular, facial, or lingual vein.

The lymphatics from the tip of the tongue are afferent to the submental lymph nodes. Refer to Chapter 32 for a discussion of the lymph nodes of the head and neck. The lymphatics from the lateral aspects of the remainder of the anterior two-thirds of the tongue are af-

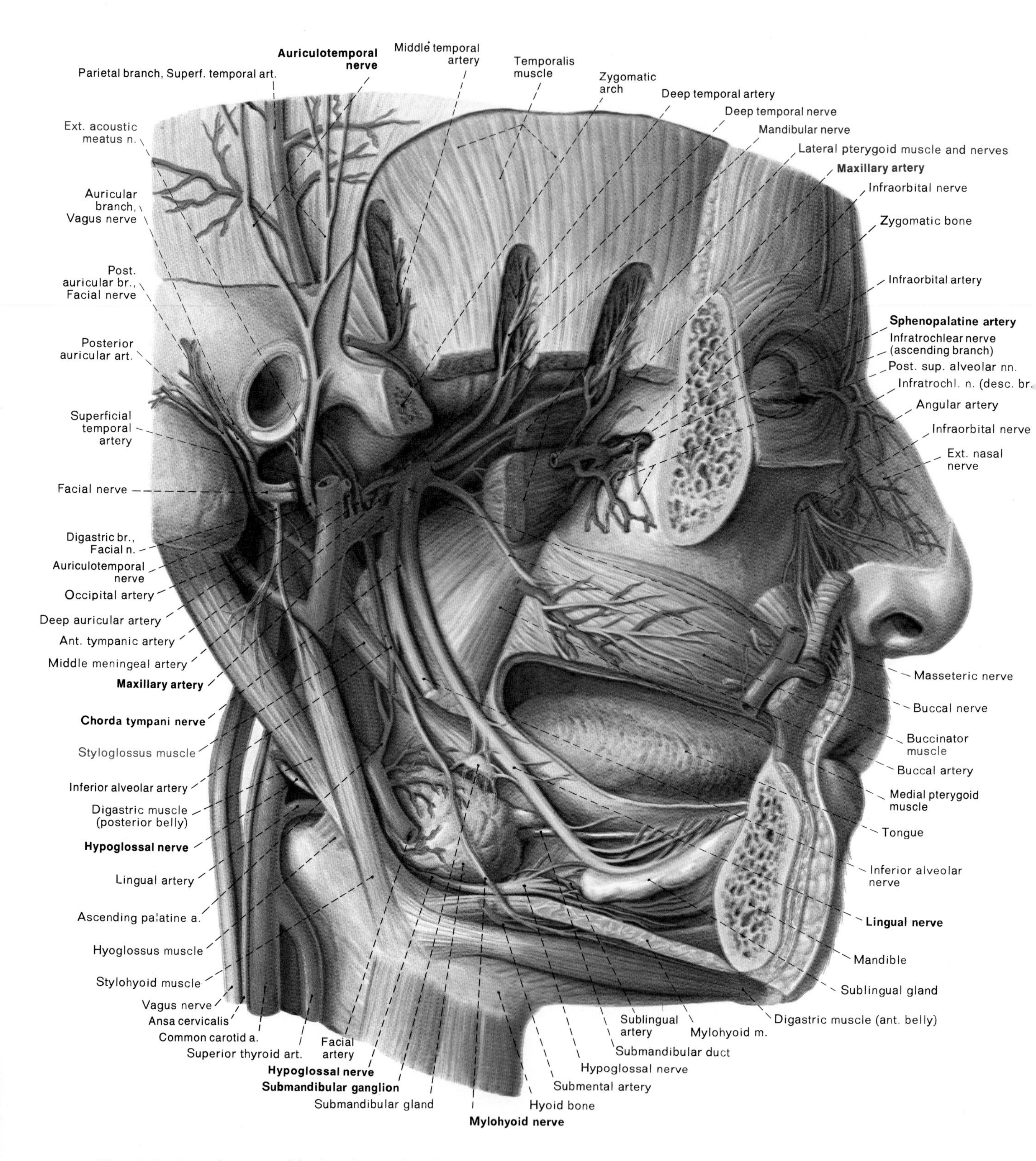

Fig. 31–7: Deep dissection of the face showing the infratemporal region and the branches of the mandibular division of the trigeminal nerve.

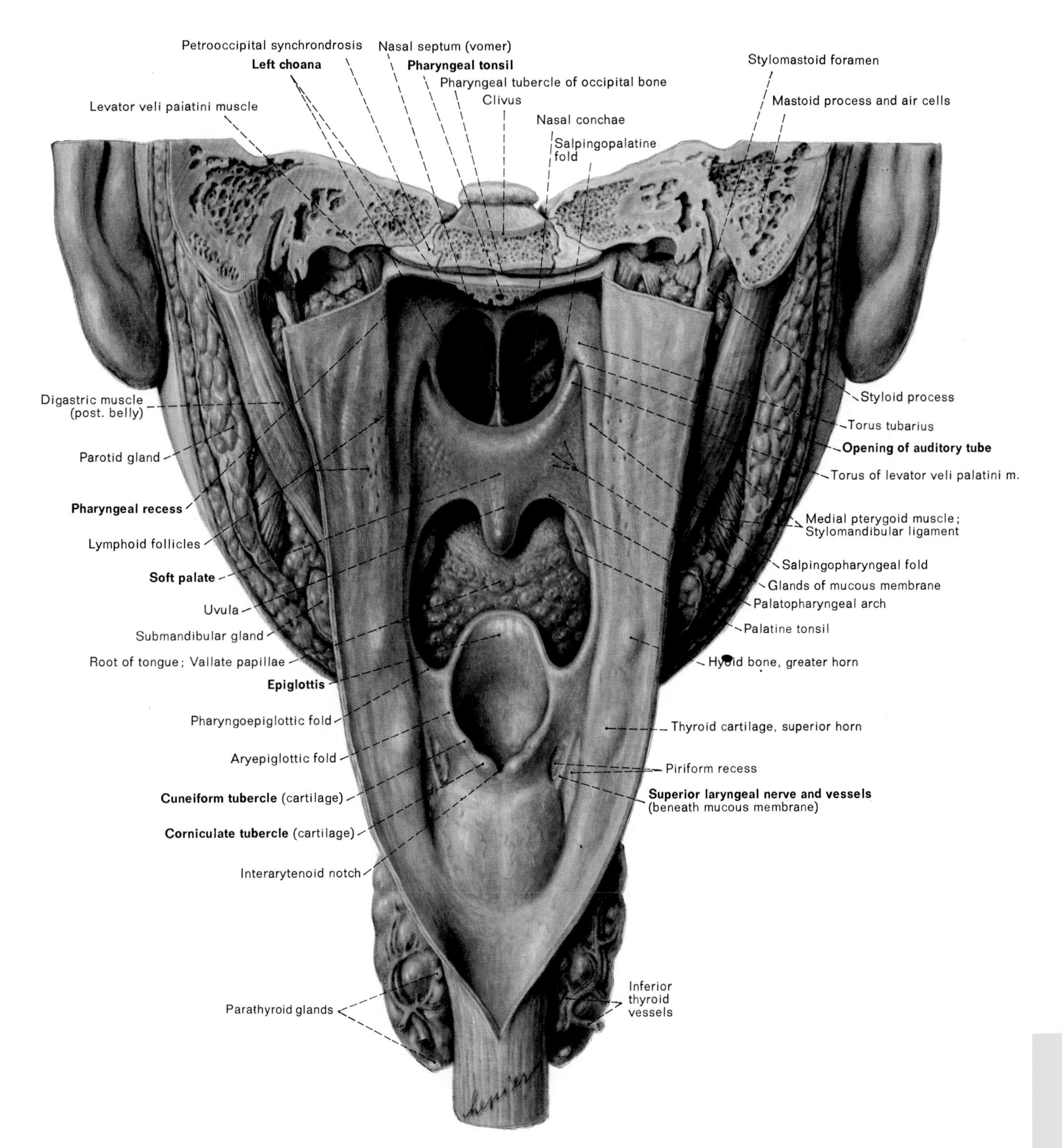

Fig. 31–8: Posterior view of the interior of the pharynx. The pharynx has been opened by a posterior longitudinal incision.

ferent to the submandibular nodes. The lymphatics from the medial parts of the remainder of the anterior two-thirds of the tongue are afferent to deep cervical nodes in the lower part of the neck. The lymphatics from the posterior third of the tongue are afferent to deep cervical nodes in the upper part of the neck. Whereas the lymphatics of the anterior two-thirds of the tongue generally drain toward nodes in the same side of the neck, the lymphatics from the posterior third of the tongue drain toward upper deep cervical nodes on both sides of the neck.

THE PHARYNX

The Muscles of the Pharynx

The pharynx has three muscles (the superior, middle, and inferior constrictors) the muscle fibers of which are circumferentially oriented within the walls of the pharynx (Fig. 31–9) and three muscles (stylopharyngeus, salpingopharyngeus, and palatopharyngeus) the muscle fibers of which are longitudinally oriented within the walls of the pharynx (Fig. 31–4). The three constrictors can narrow the lumen of the pharynx, and the three longitudinally oriented muscles can raise the pharynx [the pharynx moves upward and downward during deglutition (the act of swallowing) and phonation (the act of producing sounds)]. Salpingopharyngeus can also open the auditory tube.

The muscle fibers of each constrictor fuse with the muscle fibers of its contralateral counterpart along the midline of the posterior wall of the pharynx. The muscle fibers of the **superior constrictor** extend backward from the pterygomandibular raphe (a connective tissue seam that extends along the posterior margin of the buccinator) (Figs. 31–1 and 31–9). The muscle fibers of the **middle constrictor** extend backward from the hyoid. The muscle fibers of the **inferior constrictor** extend backward from the side of the thyroid and cricoid cartilages of the larynx.

The muscle fibers of **stylopharyngeus, salpingopharyngeus,** and **palatopharyngeus** extend downward from their origins into the lateral wall of the pharynx (Fig. 31–4). Stylopharyngeus and salpingopharyngeus originate, respectively, from the styloid process of the temporal bone and the medial end of the auditory tube.

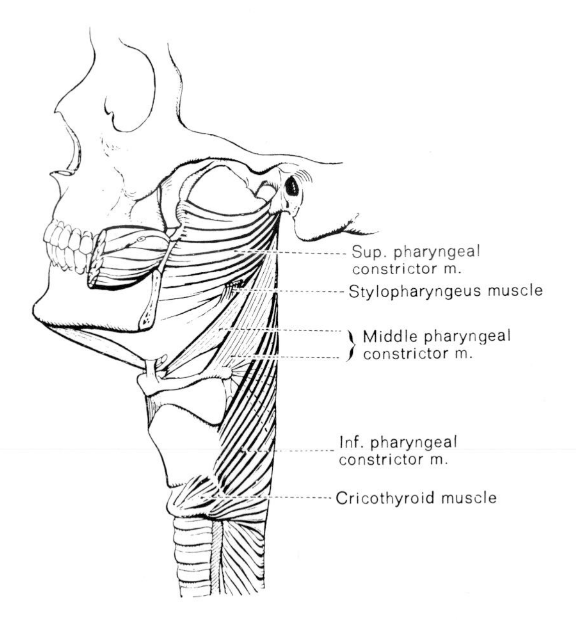

Fig. 31–9: Diagram of a lateral view of the pharyngeal constrictor muscles.

The Innervation of the Pharynx

The innervation of the pharynx involves cranial nerves V and IX–XI.

Cranial Nerve V: The maxillary division of the trigeminal nerve provides sensory innervation for the mucous membrane lining the upper part of the nasopharynx.

Cranial Nerve IX: The glossopharyngeal nerve provides sensory innervation for the mucous membrane lining all the parts of the pharynx except the upper part of the nasopharynx. The glossopharyngeal nerve also innervates stylopharyngeus.

Cranial Nerves X and XI: The vagus nerve provides sensory innervation for the mucous membrane lining all the parts of the pharynx except the upper part of the nasopharynx.

All the muscles of the pharynx except for stylopharyngeus are innervated by motor nerve fibers of the cranial part of the accessory nerve. The motor nerve fibers at first descend in the neck within the vagus nerve and are then transmitted by branches of the vagus nerve to the pharyngeal plexus.

The Gag Reflex

The placement of any object (a broad wooden applicator is generally used during a physical examination) onto the surface of the posterior third of the tongue or the posterior surfaces of the oropharynx of a normal individual elicits a reflexive, concerted contraction of pharyngeal and laryngeal muscles that is designed to expel the object from the oral cavity. This is called the gag reflex. It occurs when objects come into contact with the surface of the posterior third of the tongue, the oropharynx, and/or the hypopharynx in the absence of the act of swallowing. Such contact, which occurs normally during the act of swallowing, does not elicit the gag reflex if it occurs as part of the act of swallowing.

The gag reflex involves primarily sensory fibers of the glossopharyngeal nerve and motor fibers of the cranial part of the accessory nerve (these latter fibers are distributed as branches of the vagus nerve).

Aggregates of Lymphoid Tissue in the Pharynx

Numerous aggregates of lymphoid tissue reside in the walls of the pharynx. The major aggregates are called the tonsils. The **palatine tonsil** is located on each side in the fossa (the **palatine fossa**) lying between the palatoglossal and palatopharyngeal arches (Figs. 31–3) and 31–8). The **pharyngeal tonsil** is located in the posterosuperior corner of the nasopharynx (Fig. 31–3). The **tubal tonsil** on each side resides within the auditory tube (near its medial end).The tubal tonsil is not displayed in Figure 31–3.

RECOMMENDED REFERENCE FOR ADDITIONAL INFORMATION ON THE OROPHARYNX

Ballenger, J. J., *Diseases of the Nose, Throat, Ear, Head, and Neck*, 14th ed., Lea & Febiger, Philadelphia, 1991: *Pages 243 through 258 of Chapter 16 present discussins of the signs and symptoms of common diseases of the oropharynx.*

CHAPTER **32**

The Larynx and the Lymph Nodes of the Head and Neck

The larynx functions as both a respiratory sphincter and the most important organ involved in the production of the human voice. Its most important physical relationships are that it lies beneath and behind the root of the tongue, surmounts the trachea, and communicates posteriorly with the hypopharynx (Fig. 32–1). The side walls of the hypopharynx are attached anteriorly to the sides of the larynx.

The lymph nodes of the head and neck include the deep cervical nodes along the internal jugular vein and a ring of nodes that encircles the junction of the head with the neck.

THE LARYNX

The Cartilaginous Skeleton of the Larynx

The larynx has a skeleton composed of three unpaired cartilages (the cricoid, thyroid, and epiglottic cartilages) and three paired cartilages (the arytenoid, cuneiform, and corniculate cartilages).

The **cricoid cartilage** is the most inferior of the laryngeal cartilages; it lies atop the first tracheal ring (Figs. 32–2 and 32–3). The cricoid cartilage is shaped like a signet ring. The signet part, which forms the posterior face of the cartilage, is called the **lamina;** the narrow, anterior part of the cartilage is called the **arch.**

The **thyroid cartilage** is the largest of the laryngeal cartilages (Figs. 32–2 and 32–3). It anteriorly shields much of the interior of the larynx (indeed, the name thyroid is derived from a Greek expression meaning shieldlike) (Fig. 32–4). The cartilage consists of a pair of large, curved laminae fused anteriorly in the midline. The upper end of this border of fusion between the two laminae is called the **laryngeal prominence,** because it is the part of the larynx which forms the most prominent, subcutaneous landmark of the larynx in the neck (the laryngeal prominence is labelled the superior thyroid notch in Figures 32–3 and 32–4). After puberty, the laryngeal prominence is more acute and relatively larger in males than in females; this sexual difference is the basis for the prominence's colloquial name, the "Adam's apple."

A fibrous membrane called the **thyrohyoid membrane** suspends the thyroid cartilage from the hyoid above (Fig. 32–2). A pair of synovial joints called the cricothyroid joints hinge the inferior horns of the thyroid cartilage onto the cricoid cartilage below (Figs. 32–2 and 32–3).

The **epiglottic cartilage** is a leaf-shaped cartilage (Fig. 32–3) embedded in the uppermost anterior wall of the larynx (Figs. 32–1 and 32–5). The uppermost anterior wall of the larynx is called the **epiglottis** because it takes the shape of the broad, upper part of the epiglottic cartilage (Fig. 31–4). The epiglottis is appropriately named, for it overhangs the glottis below it (the term **glottis** refers to the vocal folds and the space between their free margins).

The **arytenoid cartilages** are a pair of pyramidal cartilages which lie atop the lateral slopes of the lamina of the cricoid cartilage (Figs. 32–3 and 32–5). Two stout processes called the vocal and muscular processes project, respectively, anteriorly and laterally from the base of each arytenoid cartilage (Fig. 32–4). The vocal processes are so named because they contribute to the posterior ends of the vocal folds. The **vocal folds** are the pair of tissue folds which generate the sounds produced by the larynx. The muscular processes are named for their attachment to many of the intrinsic muscles of the larynx.

The base of each arytenoid cartilage articulates by a synovial joint with the shoulder of the lamina of the underlying cricoid cartilage (Fig. 32–3). This cricoarytenoid joint provides for two kinds of movements by the arytenoid cartilage: it can rotate atop the lamina and glide (both anteroposteriorly and transversely) over the lamina's shoulder.

The **corniculate cartilages** (which are named for their diminutive horn shape) lie atop the apices of the arytenoid cartilages (Figs. 32–3 and 32–4).

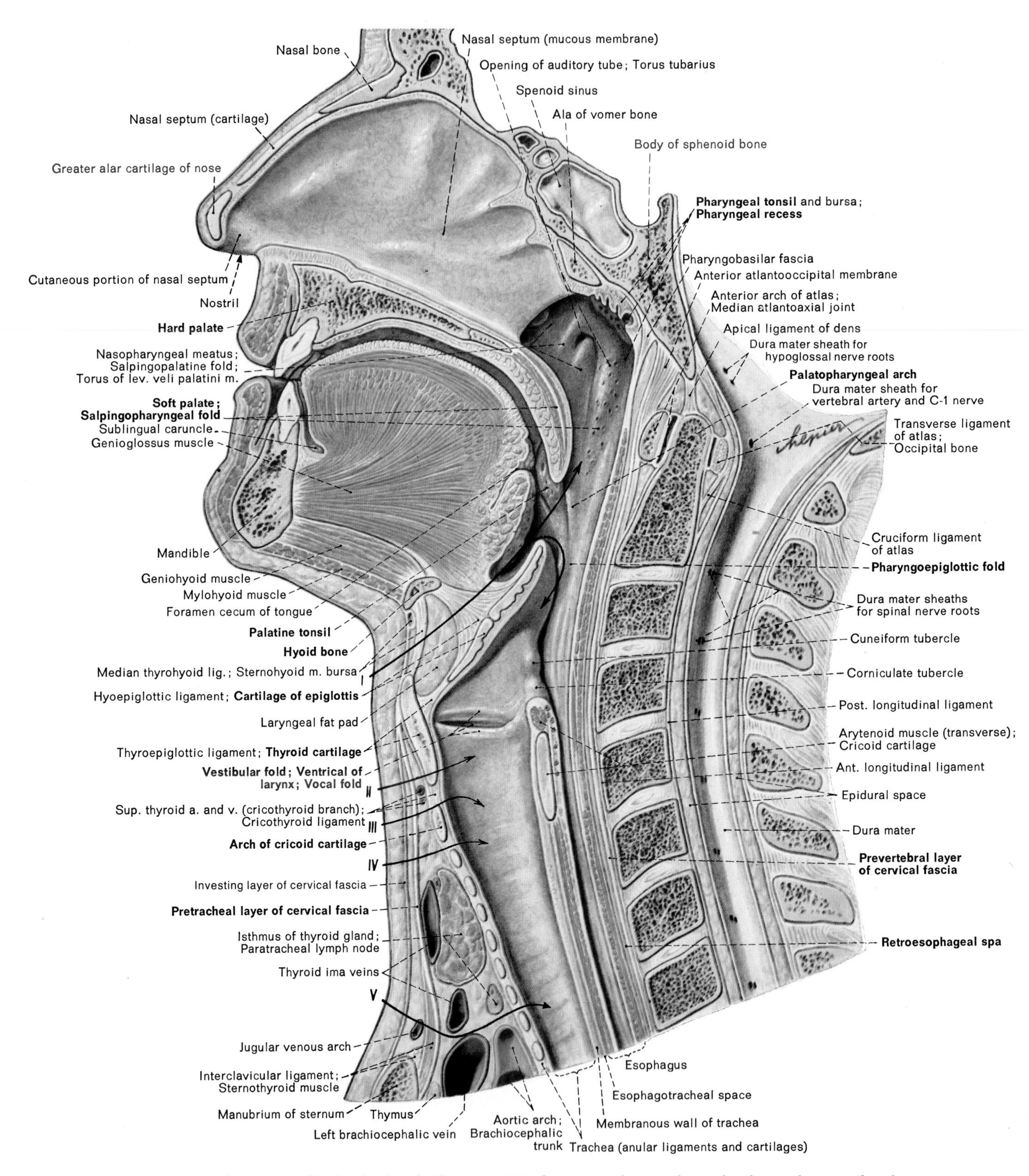

Fig. 32–1: Median section of the head and neck. The arrows I-V indicate surgical approaches to the pharynx, larynx, and trachea.

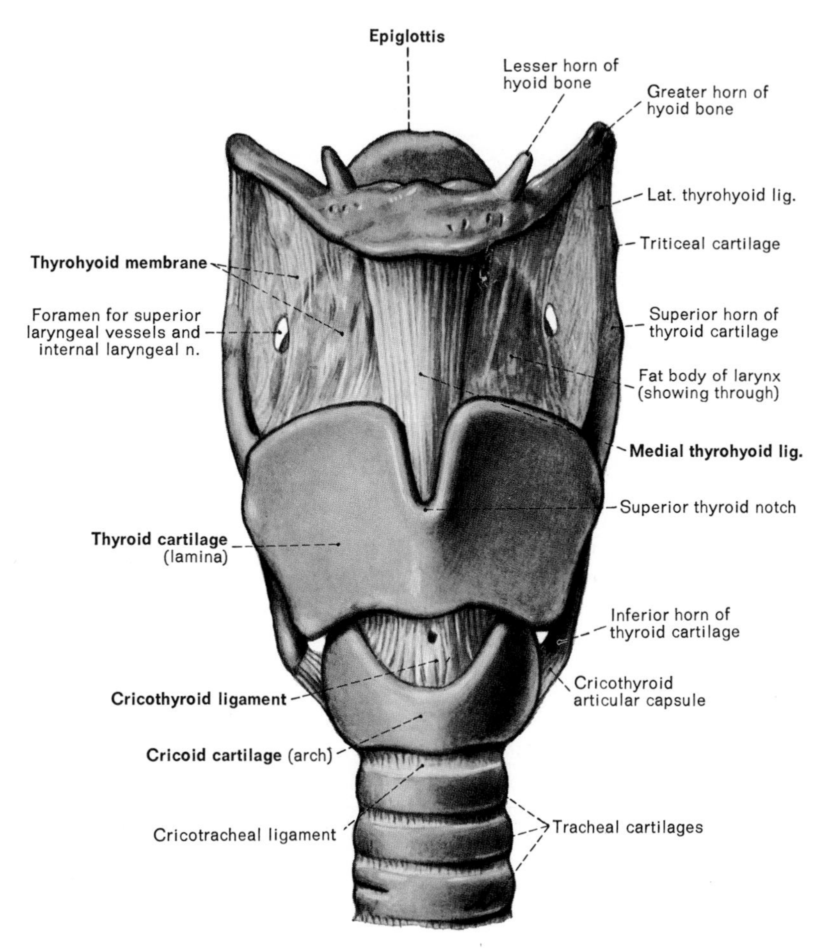

Fig. 32–2: Anterior view of the cartilages and ligaments of the larynx.

The **cuneiform cartilages** are named for their wedge shape. They are embedded along the upper margins of the pair of **aryepiglottic folds** (which are the folds of tissue that extend from the sides of the epiglottic cartilage to the arytenoid cartilages) (Fig. 32–6).

The Interior of the Larynx

The term laryngoscopy refers to any procedure by which the interior of the larynx is examined. Laryngoscopy may be performed with a laryngeal mirror (Fig. 32–7). The anterior part of the tongue may be grasped and gently pulled forward out of the oral cavity so as to minimize the extent to which the posterior part of the tongue overlies the epiglottis and the laryngeal inlet. Laryngoscopy affords a view of the interior of the larynx and the lingual and pharyngeal surfaces which surround the laryngeal inlet. The following material describes the laryngoscopic view of the larynx and the surrounding lingual and pharyngeal surfaces, beginning with the most superior and then sequentially proceeding to the most inferior features.

The broad, curled, upper end of the epiglottis presents as the most superior structure of the larynx (Fig. 32–8). The gently curved edge of the upper end of the epiglottis marks the anterior boundary of the laryngeal inlet.

The epiglottis's upward and forward projection from behind and below the root of the tongue (Fig. 32–1) creates a space between the anterior surface of the epiglottis and the posterior surface of the tongue (Fig. 32–9). The space is bordered by a pair of tissue folds extending from the sides of the epiglottis to the posterior lateral margins of the tongue; these folds are called the **lateral glossoepiglottic folds.** A midline tissue fold called the **median glossoepiglottic fold** divides the space into a pair of smaller spaces called the **valleculae.**

The next lowest laryngeal structures are a pair of tissue folds that extend downward and backward from the sides of the epiglottis onto the apices of the arytenoid cartilages. These folds are called the **aryepiglottic folds,** and they form the lateral boundaries of the laryngeal inlet (Fig. 32–8). The corniculate

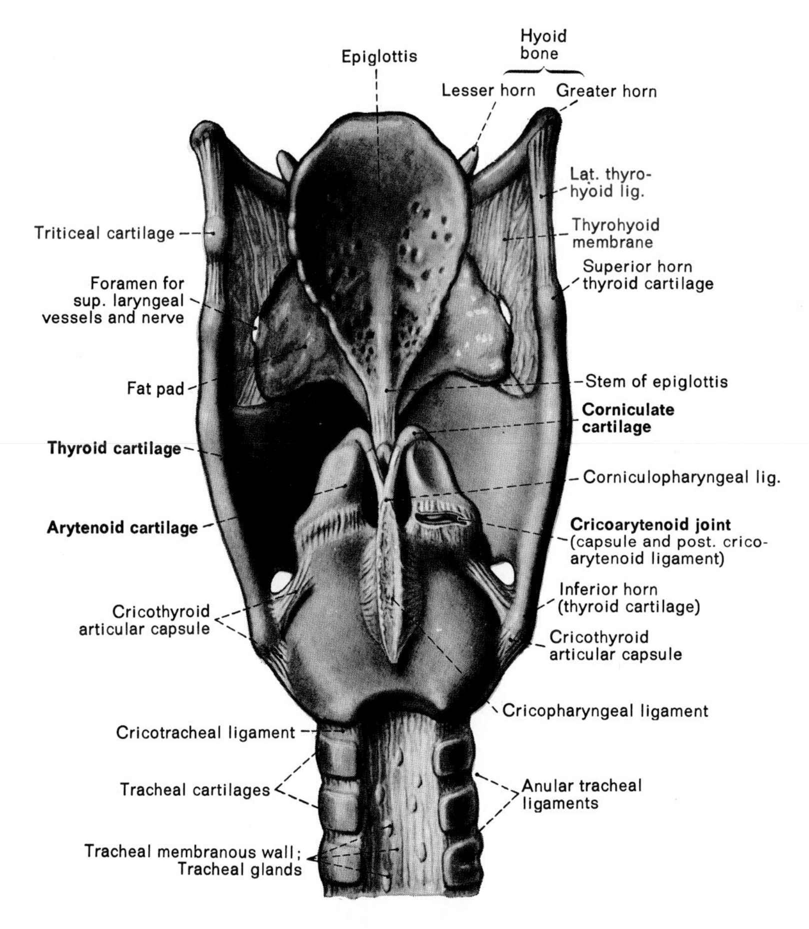

Fig. 32–3: Posterior view of the cartilages and ligaments of the larynx.

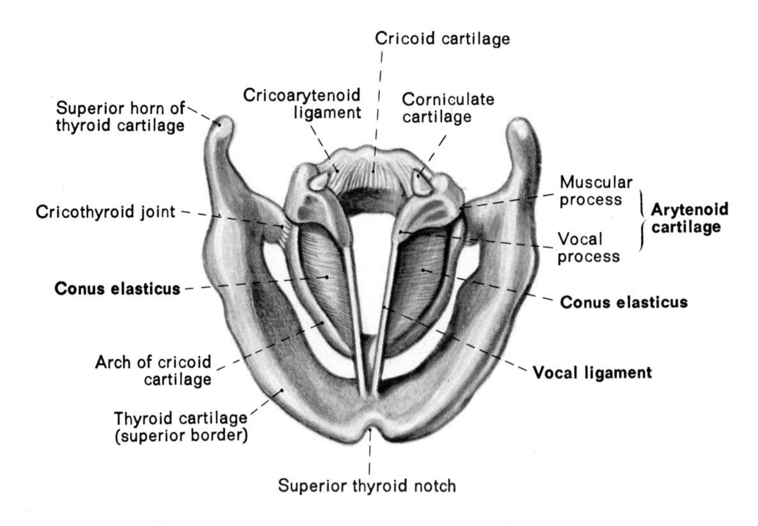

Fig. 32–4: Superior view of the thyroid, cricoid, and arytenoid cartilages and the vocal ligaments.

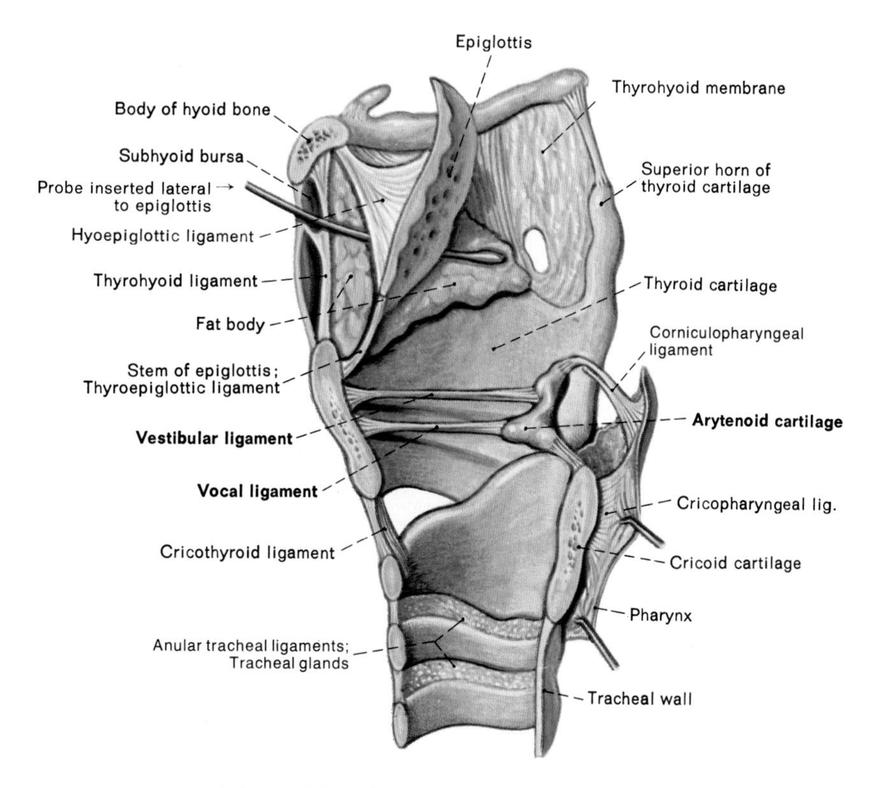

Fig. 32–5: Medial view of the right half of the cartilages and ligaments of the larynx.

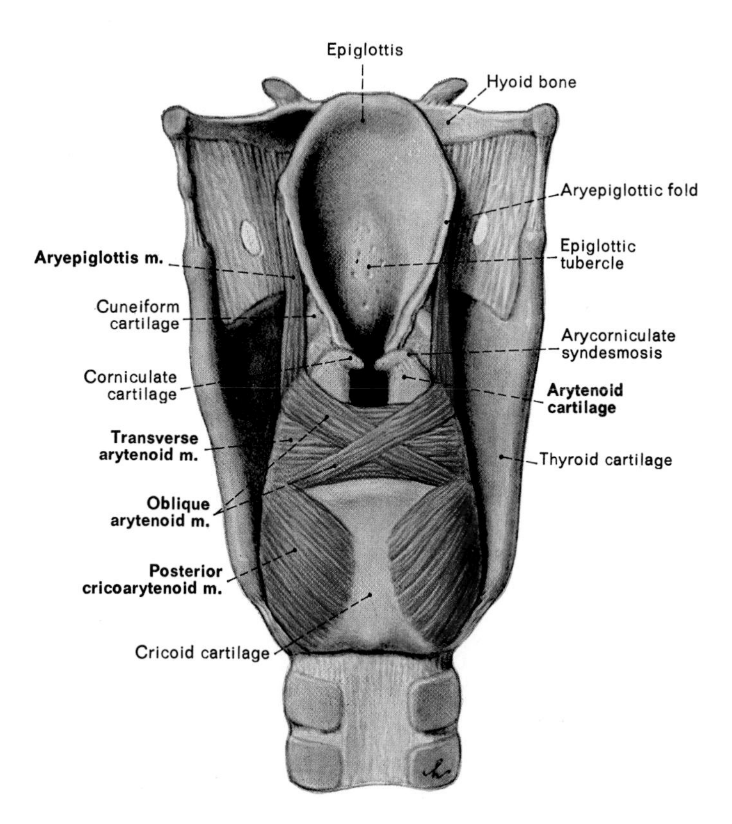

Fig. 32–6: Posterior view of the intrinsic muscles of the larynx.

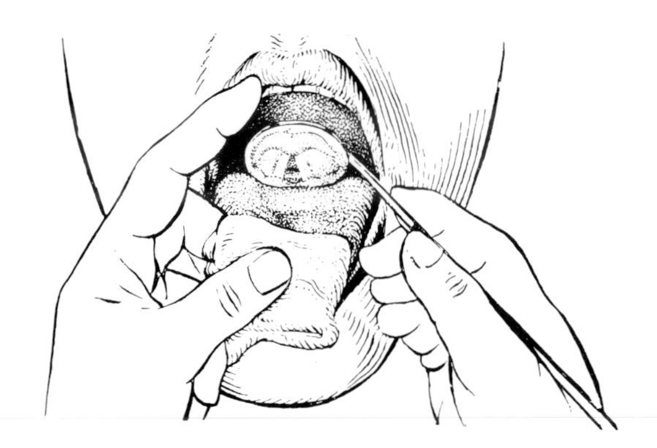

Fig. 32–7: Diagram illustrating indirect visualization of the larynx with a laryngeal mirror.

and cuneiform cartilages lie within the aryepiglottic folds, and form visible elevations in the upper margins of the folds (Figs. 31–4 and 31–8).

The anterior attachments of the side walls of the hypopharynx to the sides of the larynx form, on each side, a relatively deep recess called a **piriform recess** between the aryepiglottic fold and the side wall of the hypopharynx (Figs. 31–4 and 31–8). The piriform recesses are thus paired, lateral recesses in the lumen of the hypopharynx.

It is very common for a foreign object (usually the proverbial chicken or fish bone) to become entrapped within one of the piriform recesses. This is because when food passes through the hypopharynx during swallowing, it is forced in part to stream through the piriform recesses. Removal of foreign objects from a piriform recess risks damage to either the superior laryngeal nerve or its internal laryngeal branch, because both nerves course immediately deep to the mucous membrane lining the floor of the recess.

The first major internal feature of the larynx encountered below the level of the laryngeal inlet is a pair of relatively thick, pink tissue folds called the **ventricular,** or **vestibular, folds** (Fig. 32–8). The ventricular folds curve downward into the lumen of the larynx from the side walls of the larynx to the arytenoid cartilages (Fig. 32–10). The midline space between the free margins of the ventricular folds is called the **rima vestibuli.** The interior region of the larynx that extends from the laryngeal inlet above to the ventricular folds below is called the **vestibule.**

The ventricular folds serve mainly as an expiratory sphincter. When you "hold your breath" during forceful contraction of the expiratory muscles, you tightly adduct (bring tightly together) the downward-curving, free margins of the ventricular folds. This action forms a sphincter atop the trachea which entraps inspired air.

Immediately below the ventricular folds lie another pair of folds which project into the lumen of the larynx from its side walls (Figs. 32–8 and 32–10). These folds are called the **vocal folds,** and, like the ventricular folds, they too are attached anteriorly to the thyroid cartilage and posteriorly to the arytenoid cartilages. The free margin of each vocal fold bears a ligamentous strand called the **vocal ligament** (Fig. 32–10). The midline space between the free margins of the vocal folds is called the **rima glottidis,** or simply the **glottis.** Since the rima vestibuli is larger than the glottis during quiet respiration, the ventricular and vocal folds are both observed during laryngoscopic examination (Fig. 32–8). The small indented space between the ventricular fold and vocal fold on each side of the larynx is called a **ventricle** (Fig. 32–10).

The vocal folds serve two functions:

(1) They are the source of the sounds which emanate from the larynx. The vocal folds produce audible vibrations when their free margins are closely apposed and air is forcibly expired past them in an intermittent fashion.
(2) They serve as the major inspiratory sphincter of the larynx. Tight adduction of the upward-curving, free margins of the vocal folds forms an effective sphincter against inspiration.

The lowest interior region of the larynx is called the **infraglottic region** (Fig. 32–10). It is continuous inferiorly with the superior end of the trachea.

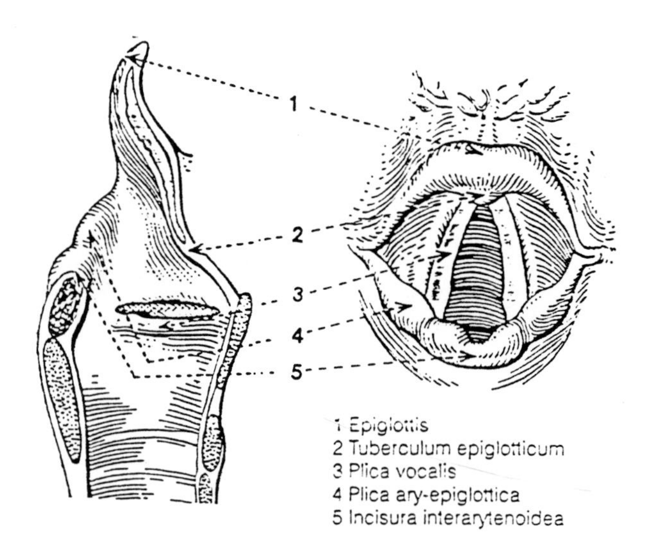

Fig. 32–8: Diagram of the laryngoscopic view of the larynx.

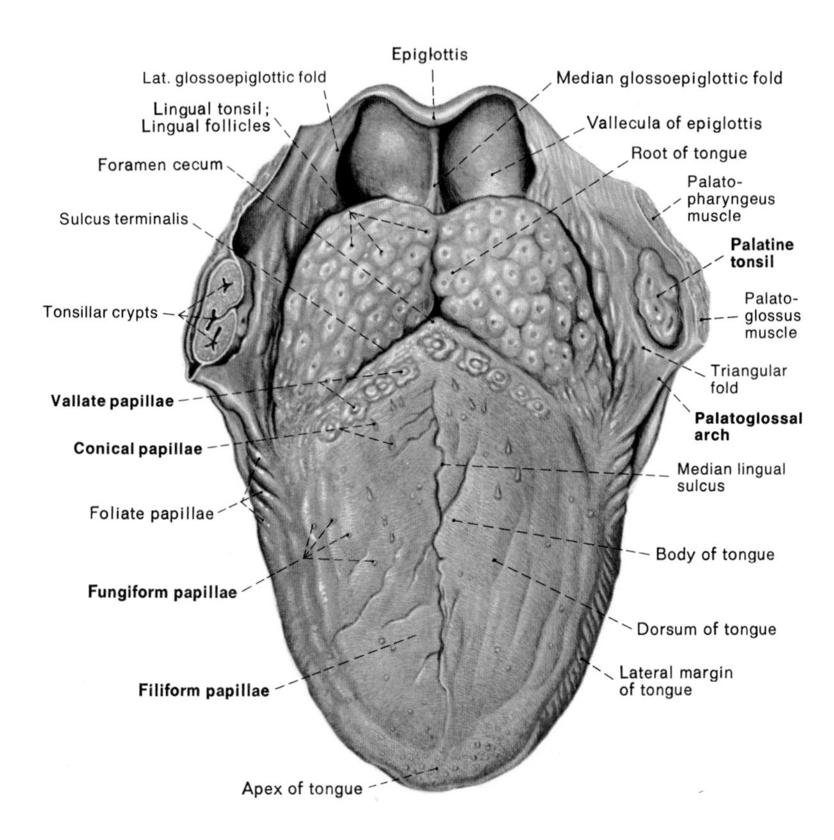

Fig. 32–9: Superior view of the tongue and the valleculae.

The Muscles of the Larynx

The larynx is said to have two sets of muscles: an extrinsic set and an intrinsic set. The extrinsic muscles consist of those muscles which, by virtue of their attachment to the hyoid or the pharynx, raise and lower the larynx during phonation and deglutition. The extrinsic muscles thus consist of all the hyoid muscles and the three elevator muscles of the pharynx (stylopharyngeus, salpingopharyngeus, and palatopharyngeus).

The intrinsic muscles of the larynx consist of those muscles which, by virtue of their attachment to the cartilages and ligaments of the larynx, modify the sizes of the laryngeal inlet, the rima vestibuli and the glottis and vary the tension within the vocal folds. In the following discussion, only the major actions of each intrinsic muscle will be noted; this is because the specific actions of some intrinsic muscles are still in dispute.

The **posterior cricoarytenoids** originate from the posterior surface of the lamina of the cricoid cartilage, and extend upward and laterally to insert onto the posterolateral surface of the muscular processes of the arytenoid cartilages (Figs. 32–6 and 32–11). The posterior cricoarytenoids are the only muscles which can abduct the vocal folds (move the vocal folds apart laterally). The posterior cricoarytenoids abduct the vocal folds in two ways: (1) They laterally rotate the arytenoid cartilages atop the lamina of the cricoid cartilage (they rotate the vocal processes away from the midline). (2) They posterolaterally slide the arytenoid cartilages down the shoulders of the lamina of the cricoid cartilage (thereby increasing the distance between the paired arytenoid cartilages).

The **lateral cricoarytenoids** are the prime adductors of the vocal folds; their actions are most directly antagonistic to those of the posterior cricoarytenoids (Fig. 32–11).

The **transverse arytenoid** is the only unpaired, intrinsic laryngeal muscle; its fibers extend transversely between the posterior surfaces of the arytenoid cartilages (Figures 32–6 and 32–11). The **oblique arytenoids** are superficial to the transverse arytenoid; each oblique arytenoid originates from the posterior surface of an arytenoid cartilage, and extends obliquely upward to insert onto the apex of the contralateral arytenoid cartilage. Some fibers continue even further, extending through the aryepiglottic fold of the contralateral arytenoid cartilage to reach the side of the epiglottic cartilage; this subset of oblique arytenoid fibers is called the **aryepiglottic muscle** (Figs. 32–10 and 32–11.

The transverse and oblique arytenoids act in concert during deglutition to close the laryngeal inlet, the rima vestibuli, and the glottis. The posteromedial tugs of the

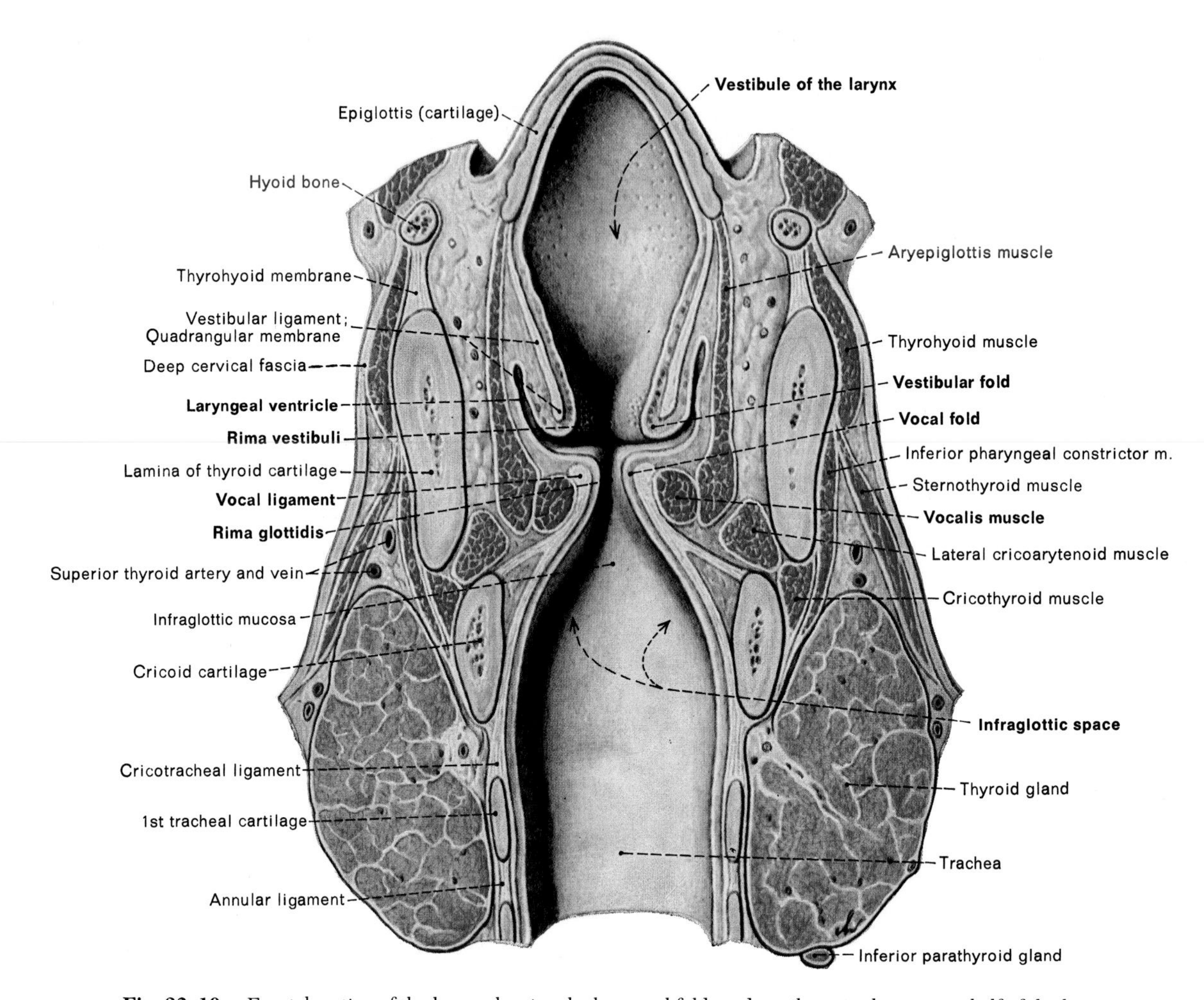

Fig. 32–10: Frontal section of the larynx showing the laryngeal folds and cartilages in the anterior half of the larynx.

oblique arytenoids upon the aryepiglottic folds draw the folds medially together and help pull the epiglottis backward over the glottis; these movements help close the laryngeal inlet. The oblique arytenoids also assist the transverse arytenoid in bringing the arytenoid cartilages together at the midline; this movement helps close the rima vestibuli and glottis.

The **cricothyroids** are the intrinsic laryngeal muscles primarily responsible for lengthening, thinning and tensing the vocal folds (Fig. 32–12). They exert these actions mainly by rotating the cricoid cartilage upward and backward around the axis that passes through the cricothyroid joints (this upward and backward rocking movement increases the distance between the front of the thyroid cartilage and the tips of the vocal processes of the arytenoid cartilages).

The **thyroarytenoids** extend anteroposteriorly through the ventricular and vocal folds. The thyroarytenoids shorten and relax the vocal folds by pulling the arytenoid cartilages anteriorly and the thyroid cartilage posteriorly. The **vocalis muscle** is the name given to the deep part of each thyroarytenoid that extends through the vocal fold immediately lateral to the vocal ligament (Fig. 32–10). The paired vocales play a major role in regulating the tension within the vocal folds.

The Innervation of the Larynx

The larynx receives all its sensory and motor supply from the **superior** and **recurrent laryngeal nerves** (Fig. 27–10). The superior and recurrent laryngeal nerves are branches of the vagus nerve. On each side, the superior laryngeal nerve divides into the **internal** and **external laryngeal nerves** as it approaches the larynx.

The internal laryngeal nerves provide the sensory supply to the mucous membrane of the larynx above the level of the vocal folds; the recurrent laryngeal nerves provide the sensory supply to the mucous membrane below the level of the vocal folds.

With the exception of the cricothyroids, all the intrinsic muscles of the larynx are innervated by the recurrent laryngeal nerves. The cricothyroids are innervated by the external laryngeal nerves. The motor fibers to the intrinsic muscles of the larynx (in both the recurrent and external laryngeal nerves) are from the

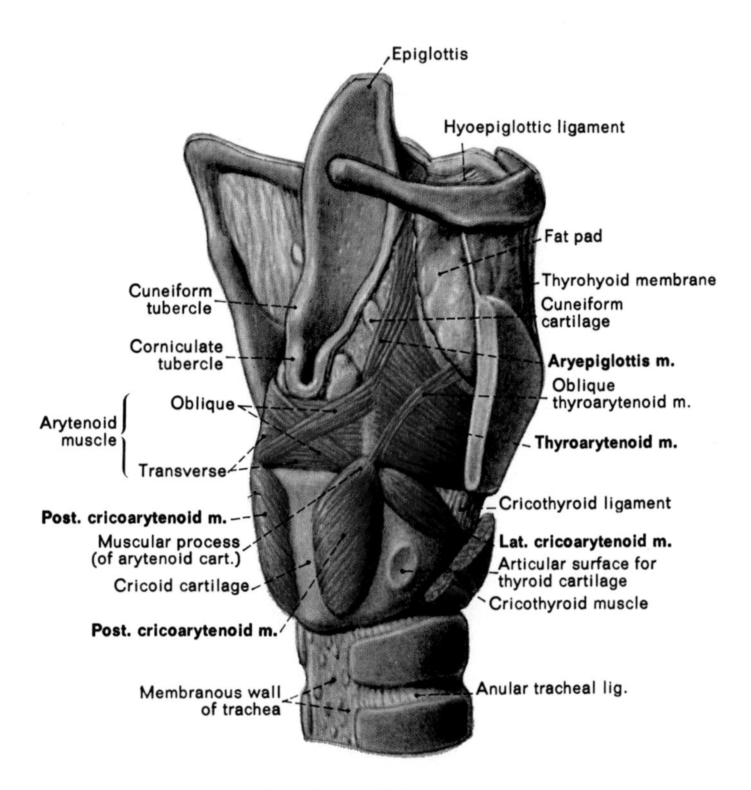

Fig. 32–11: Posterolateral view of the intrinsic muscles of the larynx.

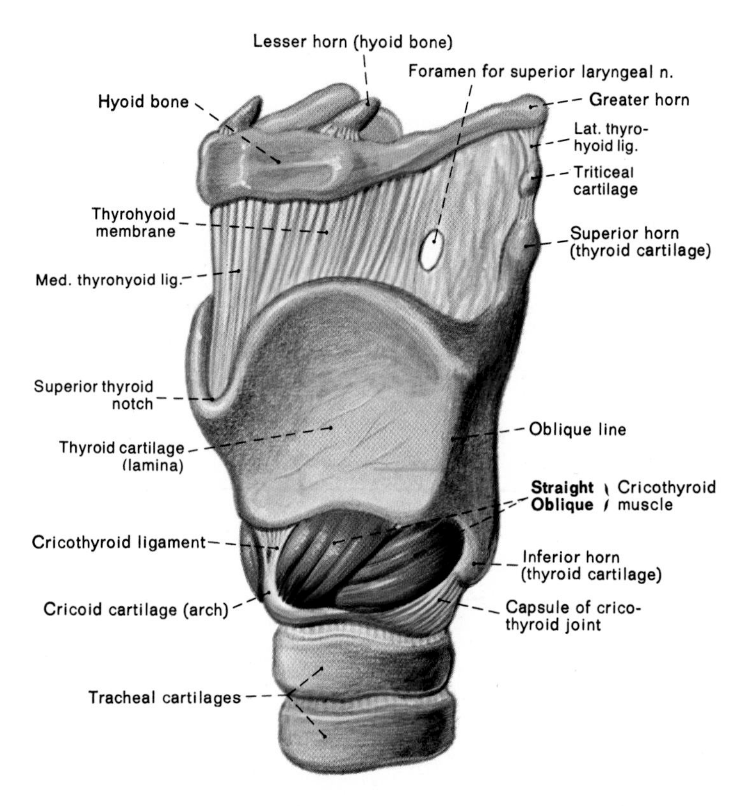

Fig. 32–12: Anterolateral view of the cricothyroid muscle.

cranial part of the accessory nerve (which joins the vagus nerve in the uppermost part of the neck).

The ligation of the superior and inferior thyroid arteries during thyroidectomy (surgical removal of the thyroid gland) places the external laryngeal and recurrent laryngeal nerves at risk of inadvertent injury. This is because part of the route of the superior thyroid artery on each side is closely associated with the external laryngeal nerve (Fig. 27–10), and the terminal part of the route of the inferior thyroid artery on each side is closely associated with the recurrent laryngeal nerve (Fig. 27–20).

The Roles of the Vestibular and Vocal Folds in the Valsalva Maneuver

The sphincteric actions by the vestibular and vocal folds are integral to the Valsalva maneuver. This maneuver is designed to temporarily increase the pressures within both the thoracic and abdominopelvic cavities, and it consists of three steps conducted in quick succession:

(1) The lungs are inflated by deep inspiration, during which the vestibular and vocal folds are abducted widely (are pulled apart widely).
(2) At the end of the deep inspiration, both pairs of folds are tightly adducted.
(3) The anterolateral abdominal wall muscles are then strongly concentrically contracted to increase pressures within both the thoracic and abdominopelvic cavities. The diaphragm transmits the increased abdominopelvic cavity pressure to the thoracic cavity.

The Valsalva maneuver can serve several purposes:

(1) It is almost always used to assist defecation or micturition. In these activities, the maneuver is performed in conjunction with relaxation of the external anal sphincter or the sphincter urethrae. During either of these activities, the vestibular and vocal folds temporarily remain tightly adducted so as to maximize the abdominopelvic cavity pressure which can be brought to bear down upon the rectum or bladder.
(2) Because the Valsalva maneuver is so commonly associated with defecation and micturition (which are activities promoted by parasympathetic reactions), employment of the maneuver elicits in most individuals a parasympathetic reaction. Employment of the maneuver may, therefore, resolve an episode of sinus tachycardia (an accelerated heartbeat caused by an increased rate of depolarization of the SA node) through stimulation of the vagal parasympathetic fibers innervating the cardiac plexuses.

CLINICAL PANEL VI.3

Aspiration of A Large Foreign Body

Description: The common setting for the aspiration of a large foreign body by an older child, adolescent, or adult is during eating. A large piece of a solid food (such as meat) may be accidentally aspirated through the laryngeal inlet into the vestibule of the larynx, where it becomes entrapped above the vestibular folds. Such entrapment may completely seal off the upper respiratory tract. Realizing that breathing is impossible, the afflicted individual typically bolts upward from the chair and grasps the throat. The mouth is open and the individual is speechless (because the larynx is blocked). The individual will die within five minutes unless the Heimlich maneuver is successfully performed.

The Heimlich Maneuver: Position your body directly behind the suffocating individual and tightly wrap your arms around the individual's upper abdomen. Make a fist with one of your hands, and turn that hand so that the flexed interphalangeal joint of its thumb is directed slightly upward into the individual's epigastric region. Grasp your fist with the other hand, and then both suddenly and very forcefully drive your fist upward and backward into the individual's epigastrium. This sudden and forceful maneuver (the Heimlich maneuver) imparts a sudden and forceful upward thrust on the individual's diaphragm. This is because the maneuver (a) imparts a sudden upward and backward thrust upon the viscera in the individual's epigastrium and (b) suddenly compresses the abdominopelvic cavity. The upward thrust of the diaphragm generally increases the air pressure within the respiratory trees of the individual's lungs to such an extent that the force of the increased air pressure propels the foodstuff from the laryngeal vestibule out through the individual's open mouth. The maneuver sometimes must be performed repeatedly to dislodge the foreign body.

(3) The maneuver is used to sneeze or cough. When sneezing or coughing, a quick Valsalva maneuver is used to raise thoracic cavity pressure, and both pairs of folds are then abducted rapidly and widely to release explosively high-pressure air from the lungs into either the nasal cavities (and so produce a sneeze) or the oral cavity (and thereby produce a cough).

(4) The maneuver is used during weight lifting. In this activity, the maneuver is performed in conjunction with concentric contraction of both the external anal sphincter and the sphincter urethrae. The object of the maneuver during weight lifting is to transform mechanically the thoracic and abdominopelvic cavities into a rigid beam. This "beam" bears some of the force required to lift a weight, and thus reduces the magnitude of the force which must be borne by the intervertebral discs and vertebral bodies of the vertebral column.

THE LYMPH NODES OF THE HEAD AND NECK

All the lymp collected from head and neck tissues drains toward the deep cervical nodes, either directly from the tissues or indirectly after traversing an outlying group of nodes.

The junction of the head with the neck is encircled by a pericervical collar of lymph nodes. The nodes which form this collar are clustered into the following groups (Fig. 27–18):

The **occipital nodes** drain the back of the scalp.

The **retroauricular (mastoid) nodes** drain the scalp posterosuperior to the auricle, the auricle, and external acoustic meatus.

The **parotid nodes** drain the scalp anterosuperior to the auricle, the forehead, eyelids, cheek, auricle, external acoustic meatus, middle ear, and parotid gland.

The **submandibular nodes** drain the anterior part of the scalp, the eyelids, cheek, nose, lips, the frontal, maxillary, and ethmoid sinuses, the upper and lower teeth, anterior two-thirds of the tongue, floor of the mouth, and gums.

The **submental nodes** drain the tip of the tongue, floor of the mouth beneath the tip of the tongue, lower incisors, and midline part of the lower lip.

The deep cervical nodes clustered at the levels of the intermediate tendons of the digastric and omohyoid muscles are respectively called the **jugulo-digastric** and **jugulo-omohyoid nodes.** The jugulo-digastric nodes are the first deep cervical nodes to receive lymph drained from the posterior third of the tongue; the jugulo-omohyoid nodes are among the first deep cervical nodes to receive lymph drained from the anterior two-thirds of the tongue.

CASE **VI.5**

The Case of Matthew Garson

INITIAL PRESENTATION AND APPEARANCE OF THE PATIENT

A 19 year-old black man, Matthew Garson, has made an appointment to seek treatment for a sore throat and difficulty in breathing. Upon entering the examination room, the examiner observes that the patient appears extremely anxious, is seated bolt upright, with the neck extended and the head held forward, and frequently expectorates oral secretions into disposable tissues.

QUESTIONS ASKED OF THE PATIENT

What can I do for you? [The patient replies in a soft, raspy voice.] I would like for you to look at my throat; it's so sore that I'm having trouble breathing. If I don't keep my head up like this and breathe slowly, I find it difficult to breathe.

How long has your throat been sore? Since yesterday morning. [As it is 3:00 PM, the patient's answer indicates that his throat has been painful for at least 27 hours.]

How long have you had difficulty in breathing? Since this morning.

Your voice sounds hoarse. Is this your normal voice? No. I started getting hoarse last night, and it's just been getting worse ever since. *[A complaint of a sore throat suggests pharyngitis (inflammation of the pharynx). Hoarseness (a harshening of the voice) suggests laryngitis (inflammation of the larynx] and/or a noninflammatory disorder of the larynx (any disorder which would cause partial or complete paralysis of one or more of the intrinsic muscles of the larynx). Inflammation of the upper respiratory tract may be caused by infectious and/or chemical agents. In this case, the examiner considers it likely that the patient's sore throat and hoarseness share a common cause, and, therefore, that the patient is suffering from inflammation of both pharyngeal and laryngeal tissues. The patient's account of his problems suggest the upper respiratory tract inflammation originated in the pharynx and spread to the larynx within 6 to 12 hours.*

In this case, dyspnea (difficulty in breathing) is an ominous symptom. Given the likelihood that the patient is suffering from an inflammation of the upper respiratory tract, the finding of dyspnea raises the possibility that the inflammation is producing a narrowing or obstruction of the upper respiratory tract. Such a narrowing or obstruction is obviously a life-threatening condition.]

INITIAL ASSESSMENT OF THE PATIENT'S CONDITION

Even though the taking of the patient's history has only begun, the examiner already tentatively concludes that the patient's chief medical problem is a critical, life-threatening condition: progressive acute obstruction of the upper respiratory tract, presumably from inflammation. The patient's comment that he needs to keep his head and neck extended in order to facilitate breathing makes the examiner acutely aware that certain parts of the physical examination may have to be deferred because of their potential to completely obstruct the patient's upper airway. The examiner decides to ask just a few more questions and conduct a limited physical examination before deciding on a prompt course of action.

CONTINUATION OF THE PATIENT'S HISTORY

Is there anything else that is also bothering you? Yes. It hurts and it's hard for me to swallow. I've also found that I have difficulty in getting my breath at all if I swallow.

Do you mean you can't breathe when you're swallowing or after you swallow? After I swallow. In the last few hours, I've found two or three times that after I swallow, I'll try to get air into my lungs but nothing happens. It's like I'm suffocating for a few seconds. *[The examiner recognizes that odynophagia (pain upon swallowing) and dyphagia (difficulty upon swallowing) are common symptoms of pharyngitis. The description of obstructive apnea (absence of breathing* *due to obstruction) in association with the act of swallowing heightens even further the examiner's awareness that the patient appears extremely susceptible to sudden total obstruction of the upper airway.]*

Does your tongue feel swollen? No. *[A swollen tongue could be contributing to the patient's dyspnea and dysphagia. The examiner asks this question to avoid asking the patient to open his mouth and stick out his tongue for direct observation. The examiner is concerned that such an activity could provoke a sudden total obstruction of the upper airway.]*

Do you have a fever or chills? I haven't taken my temperature, but I do feel feverish.

Do you have a cough? No. *[Fevers and coughs are important but non-specific symptoms.]*

Have you had a cold, flu, or flu-like symptoms recently? No.

Have you had any recent illnesses or injuries? No.

Are you taking any medications? No.

Now, this next question I'm about to ask, I'm asking it because I want to have as much correct information as possible about your sore throat and difficulty with breathing. Have you taken any illegal drugs? No. *[The examiner inquires about illegal drug use because the patient is a young adult, one of the age groups most commonly involved with illegal drug use, and certain illegal drugs can produce upper respiratory disorders.]*

Have you recently come into contact with any irritating fumes or gases? No. *[This question addresses the possiblity that the patient's upper airway came into contact with noxious gases that provoked a marked inflammation of pharyngeal and laryngeal tissues.]*

Have you been bitten by an animal or had an animal lick an open wound or sore on your body? No. *[The examiner recognizes that the patient exhibits some of the symptoms of rabies (such as a sore throat, dysphagia, and excessive salivation) during the acute encephalitis stage of the disease (the stage of acute inflammation of the brain). The question addresses the most common means by which the rabies virus is transmitted from a rabid animal to a human. The examiner finds the patient to be alert and fully cooperative during the interview.]*

PHYSICAL EXAMINATION OF THE PATIENT

Vital signs: Blood pressure
Standing: 115/65 left arm and 115/65 right arm
Pulse: 80
Rhythm: regular
Temperature: 102.2°F.
Respiratory rate: 27
Height: 6′0″
Weight: 160 lbs.
HEENT Examination: Deferred, except for the observation of nasal flaring and suprasternal retraction during inspiration.
Lungs: Inspection of the chest and auscultation and percussion of the lung fields are normal except for the presence of stridor chiefly upon inspiration.
Cardiovascular Examination: Deferred
Abdomen: Deferred
Genitourinary Examination: Deferred
Musculoskeletal Examination: Deferred
Neurologic Examination: Deferred

ANATOMIC BASIS OF THE PATIENT'S HISTORY AND PHYSICAL EXAMINATION

(1) **The patient's sore throat, odynophagia, dysphagia, and drooling collectively indicate pharyngitis.** Given the absence of any evidence for chemical irritation of the pharynx, it appears likely that bacterial and/or viral agents are responsible for the pharyngitis. The pharyngeal pain is caused in large part by the acute edematous distention of the pharyngeal mucosa. The swelling of the pharyngeal wall tissues has narrowed the pharyngeal lumen, and this narrowing accounts for the dysphagia. The drooling is the product of excessive salivation (if the infectious process has also involved one or more of the salivary glands) and/or diminished swallowing of saliva because of the odynophagia.

(2) **Hoarseness is an indicator of inflammation or a disorder of the larynx.** In this case, it is likely that the patient is suffering from infectious inflammation of the larynx. Inflammatory swelling of the laryngeal mucosa (especially that overlying the vocal folds) interferes with the normal movement of the vocal folds, producing distortion, harshness and diminished intensity of an individual's voice.

(3) **Stridor is an indicator of partial obstruction of the upper airway.** Stridor is loud, harsh, high-pitched breathing sounds, akin to the blowing of the wind. It is generated by the turbulent passage of air through a narrowed segment of the upper airway. It is most pronounced during the inspiratory phase of respiration.

The patient's comment that quiet respiration minimizes the dyspnea is comprehensible in terms of Bernoulli's principle. This states that when gas flows through a passageway (and friction effects are negligible), the velocity of the gas flow is inversely proportional to the pressure exerted by the gas against the wall of the passageway. Application of Bernoulli's principle to this case suggests that when air flows through the patient's partially obstructed upper airway, the velocity of air flow is greatest and the pressure lowest along the most stenosed (narrowest) segment. Forced inspiration generates a higher velocity of air flow through the most stenosed segment than quiet respiration, and thus a lower intraluminal pressure. The lower the intraluminal pressure across the interval of the most stenosed segment, the greater the tendency of the swollen tissues lining the segment to approximate each other. Accordingly, the patient finds it easier to breathe with quiet inspiration than with forced inspiration.

(4) **The nasal flaring and retraction of the suprasternal fossa at the base of the neck indicate labored breathing.** The nasal flaring is due to the action of nasalis (Fig. 30–16), a muscle of facial expression whose action is to widen the nasal aperture. The concentric contraction of the accessory muscles of respiration in the neck (sternocleidomastoid and the scaleni muscles) account for the retraction of the suprasternal fossa.

INTERMEDIATE EVALUATION OF THE PATIENT'S CONDITION

The patient appears to be suffering from acute infectious obstruction in or about the larynx.

CLINICAL REASONING PROCESS

In considering this case, a primary or emergency care physician would recognize that acute infectious obstruction of the upper respiratory tract can occur at one or more of the following levels: oropharynx, laryngopharynx, vestibule of the larynx, glottis, infraglottic region of the larynx, trachea, and primary bronchi. In the older adolescent and adult population, however, only two disease entities are likely to be responsible for the upper airway obstruction: **epiglottitis** (inflammation of the epiglottis and aryepiglottic folds) or **croup (laryngotracheobronchitis)** (inflammation of the infraglottic region of the larynx, the trachea, and the primary bronchi). In this case, a primary or emergency care physician would tentatively consider croup to be an unlikely possibility, because the bark-like cough characteristic of croup is conspicuously absent.

Cases of epiglottitis and croup are most commonly encountered among infants and school-age children. In the pediatric population, whose body parts are comparatively small, each of these disease entities produces such marked inflammatory swelling that obstruction of the upper airway is a common and potentially life-threatening complication. Other disease entities which are a frequent cause of upper airway obstruction in the pediatric population include uvelitis (inflammation of the uvula), peritonsillar abscesses, and retropharyngeal abscesses.

In either a pediatric or adult patient where acute infection of the upper airway is hypothesized to be the case of moderate to severe dyspnea, **the patient should not be submitted to any procedure which makes breathing more difficult.** In particular, it is inappropriate to attempt visualization of the oral cavity and pharynx with a tongue depressor or visualization of the larynx with a laryngeal mirror, because either instrument may elicit the gag reflex, which in turn may cause edematous tissues at the site or sites of partial obstruction to adhere to each other and produce a sudden total obstruction of the upper airway. **Without unduly alarming the patient, it becomes imperative to rapidly assemble medical personnel with the necessary skills and equipment to establish an emergency airway.** Clinical judgment is required to decide whether to attempt direct and/or radiographic visualization of the patient's oral cavity, pharynx, and larynx prior to attempting endotracheal intubation (i.e., placing a semirigid tube through the pharynx and larynx into the trachea in order to secure an open airway).

RADIOLOGIC EVALUATION AND FINAL RESOLUTION OF THE PATIENT'S CONDITION

The patient in this case was examined by a primary care physician in an out-patient clinic. Upon making the intermediate evaluation of the patient's condition, the physician assembled an appropriate medical team and then ordered AP and lateral radiographs of neck soft tissues **stat** (to be done immediately). The medical team accompanied the patient to the radiology suite and comforted him as a radiologist examined the neck films. The lateral neck film showed an enlarged, rounded, and blunted epiglottic shadow, whose outline was poorly defined. The radiologist confirmed that these radiographic findings are diagnostic of epiglottitis. The marked swelling of the epiglottis accounts for the enlargement, roundness, bluntness, and poor definition of its radiographic shadow in a lateral neck film.

The patient was admitted to the intensive care unit for endotracheal intubation and appropriate antibiotic treatment. A markedly edematous and erythematous epiglottis was observed at the time of endotracheal intubation. The marked inflammation of the epiglottis and surrounding laryngeal and pharyngeal tissues should begin to resolve within 24 hours of antibiotic treatment. The patient is extubated when endoscopy confirms resolution of the inflammatory process.

THE CHRONOLOGY OF THE PATIENT'S CONDITION

The patient's problems began 36 hours previously with infectious inflammation of the pharynx. Accordingly, when he awoke yesterday morning, he had a sore throat. The infection spread inferiorly to the larynx, and elicited an inflammatory response sufficient to produce hoarseness by last evening.

Continued swelling of the epiglottis and the aryepiglottic folds during the night produced the dyspnea experienced by the patient this morning. The patient soon learned by trial-and-error that sitting bolt upright with the neck extended and the head held forward minimized the dyspnea. This position minimizes the dyspnea because it lessens the extent to which the tongue presses backward and downward against the swollen epiglottis; any backward and downward push on the swollen epiglottis markedly constricts the laryngeal inlet. The patient began drooling because of diminished swallowing of salivary secretions; subconscious swallowing of salivary secretions declined because of the odynophagia and dysphagia.

The patient experienced three episodes of obstructive apnea during the last four hours when he swallowed. On each of these occasions, the raising of the larynx and the downward and backward movement of the epiglottis over the laryngeal inlet during the act of swallowing led to momentary adherence of the markedly edematous epiglottis with the similarly edematous aryepiglottic folds. The epiglottis remained momentarily stuck over the laryngeal inlet after the larynx descended to its resting position in the neck, so producing the episode of obstructive apnea.

Case List

Section V: The Abdomen, Pelvis and Perineum

Section VI: The Head and Neck

Credits

The following figures were reproduced with permission:

From Abeloff, D.: *Medical Art: Graphics for Use,* Williams & Wilkins, Baltimore, 1982, Figures I–A, I–B.

From Adams, J.C., and Hamblen, D.L.: *Outline of Orthopaedics,* 11th ed., Churchill-Livingstone, Edinburgh, 1990, Figure 13–12.

From Agur, A.: *Grant's Atlas of Anatomy,* 9th ed., Williams & Wilkins, Baltimore, 1991, Figures 7–6, 10–18, 10–19, 10–20, 10–21, 10–22, 10–23, 10–24, 10–25, 18–15, 27–14.

From Backhouse, K.M., and Hutchings, R.T.: *Color Atlas of Surface Anatomy,* Williams & Wilkins, Baltimore, 1986, Figures 10–1, 10–5, 10–33, 10–34, 11–1, 14–5, 15–7.

From Barr, M.L., and Kiernan, J.A.: *The Human Nervous System. An Anatomical Viewpoint,* 5th ed., J.B. Lippincott, Philadelphia, 1988, Figure 30–11.

From Barrett, C.P., Poliakoff, S.J., and Holder, L.E.: *Primer of Sectional Anatomy with MRI and CT correlation,* Williams and Wilkins, Baltimore, 1990, Figures 6–4, 6–5.

From Bartles, H., and Bartles, R.: *Physiologie, Lehrbuch and Atlas,* Abb. v. Marks, G.3., Oberarb, Aufl. Urban & Schwarzenberg, Munich, 1987, Figure 1–2.

From Bates, B.: *A Guide to Physical Examination and History Taking,* 4th ed., J.B. Lippincott, Philadelphia, 1991, Figure 7–26.

From Carpenter, M.B., and Sutin, Jerome: *Human Neuroanatomy,* 3rd ed., Williams & Wilkins, Baltimore, 1983, Figures 4–8, 4–9, 4–10, 4–14, 4–16, 4–19, 4–20, 29–16.

From Clemente, C.D.: *Anatomy, A Regional Atlas of the Human Body,* 3rd ed., Urban & Schwarzenberg, Munich, 1987, Figures 3–5, 4–6, 4–7, 4–15, 7–13, 7–14, 7–15, 7–16, 7–17, 7–18, 7–19, 7–20, 8–2, 8–8, 8–10, 9–8, 9–11, 9–12, 9–13, 9–14, 9–15, 9–16, 10–10, 10–11, 10–14, 10–16, 10–26, 10–27, 10–28, 10–29, 10–30, 10–31, 10–32, 11–2, 11–3, 11–4, 11–5, 11–10, 12–3, 12–4, 12–7, 12–8, 12–19, 12–20, 12–24, 12–26, 13–1, 13–2, 13–4, 13–6, 13–7, 13–8, 13–13, 13–15, 13–17, 13–18, 13–20, 13–22, 13–23, 13–24, 14–1, 14–14, 14–15, 14–16, 14–18, 14–19, 14–21, 14–25, 14–27, 14–29, 15–3, 15–4, 15–8, 15–9, 15–10, 15–11, 15–12, 15–13, 16–2, 16–3, 16–8, 16–9, 16–13, 17–2, 17–3, 17–5, 17–6, 18–1, 18–2, 18–3, 18–5, 18–6, 18–13, 18–14, 19–2, 19–4, 19–5, 20–4, 20–5, 21–3, 21–4, 21–5, 21–6, 21–8, 21–9, 22–1, 22–2, 22–3, 22–4, 22–5, 22–6, 22–7, 22–10, 22–11, 22–13, 23–2, 24–2, 25–1, 25–2, 26–3, 26–4, 26–5, 26–6, 26–7, 26–8, 26–10, 26–11, 26–12, 26–13, 26–14, 26–15, 26–16, 26–18, 26–19, 26–20, 26–23, 27–2, 27–3, 27–5, 27–6, 27–7, 27–8, 27–9, 27–10, 27–11, 27–12, 27–13, 27–15, 27–16, 27–17, 27–18, 27–19, 27–20, 27–21, 28–1, 28–2, 28–3, 28–4, 28–5, 28–8, 28–10, 28–11, 28–12, 28–13, 28–14, 29–1, 29–2, 29–4, 29–11, 29–13, 29–15, 29–19, 30–4, 30–5, 30–6, 30–8, 30–9, 30–10, 30–13, 30–14, 30–15, 30–16, 30–17, 31–1, 31–2, 31–3, 31–4, 31–5, 31–6, 31–7, 31–8, 31–9, 32–1, 32–2, 32–3, 32–4, 32–5, 32–6, 32–9, 32–10, 32–11, 32–12.

From Crouch, J.E.: *Functional Human Anatomy,* 4th ed., Lea & Febiger, Philadelphia, 1985, Figures 17–7, 19–3, 23–1, 24–3, 26–1, 28–6, 28–7, 29–7, 30–7, 30–12.

From Dandy, D.J.: *Essential Orthopaedics and Trauma,* Churchill Livingstone, Edinburgh, 1989, Figure 8–5.

From Freedman, M.: *Clinical Imaging. An Introduction to the Role of Imaging in Clinical Practice,* Churchill Livingstone, New York, 1988, Figure 6–3.

From Gardner, M.: *Basic Anatomy of the Head and Neck,* Lea & Febiger, Philadelphia, 1992, Figures 29–5, 29–6, 29–18, 29–20, 30–18.

From George, R.B., Light, R.W., Matthay, M.A., and Matthay, R.A.: *Chest Medicine: Essentials of Pulmonary and Critical Care Medicine,* 2nd ed., Williams & Wilkins, Baltimore, 1990, Figure 17–8.

From Greenspan, A.: *Orthopedic Radiology. A Practical Approach,* J.B. Lippincott, Philadelphia, 1988, Figure 14–9.

From Hall-Craggs, E.C.B.: *Anatomy as a Basis for Clinical Medicine,* 2nd Ed., Urban & Schwarzenberg, Munich, 1990, Figures 2–1, 3–2, 3–3, 3–4, 3–6, 4–2, 4–3, 4–4, 4–5, 4–11, 4–12, 4–17, 4–18, 5–1, 5–2, 5–3, 6–6, 7–3, 7–4, 7–5, 8–1, 8–3, 8–4, 9–1, 9–2, 9–3, 9–6, 9–7, 9–9, 9–10, 10–6, 10–15, 11–6, 12–25, 13–5, 13–9, 13–10, 13–21, 14–6, 14–7, 14–10, 14–11, 14–12, 14–13, 14–28, 15–1, 15–2, 15–16.

From Langman, J., and Woerdeman, M.W.: *Atlas of Medical Anatomy,* W.B. Saunders, Philadelphia, 1978, Figures 12–10, 12–18, 20–3, 21–7.

From Lawrence, P., Bell, R.M., and Dayton, M.T.: *Essentials of General Surgery,* 2nd ed., Williams & Wilkins, Baltimore, 1992, Figure 23–5.

From Lippert, H.: *Lehrbuch Anatomie,* Urban & Schwarzenberg, Munich, 1990, Figures 1–1, 1–3, 3–1, 4–1, 7–1, 7–2, 7–7, 10–2, 11–7, 12–2, 19–1, 20–1, 20–2, 32–7, 32–8.

From Magee, D.J.: *Orthopedic Physical Assessment,* 2nd ed., W.B. Saunders, Philadelphia, 1992, Figures 8–6, 8–7, 11–8, 11–9.

From Melloni, J.L., Dox, I., Melloni, H.P., and Melloni, B.J.: *Melloni's Illustrated Review of Human Anatomy,* J.B. Lippincott, Philadelphia, 1988, Figures 22–12, 23–3, 23–4, 25–3.

From Morrissy, R.T.: *Lovell and Winter's Pediatric Orthopedics,* 3rd ed., J.B. Lippincott, Philadelphia, 1990, Figure 13–3.

From Norkin, C.C., and Lavangie, P.K.: *Joint Structure and Function. A Comprehensive Analysis,* 2nd ed., F.A. Davis, Philadelphia, 1992, Figures 12–1, 12–22, 12–23, 13–11, 14–3, 14–4, 14–23, 14–24, 15–5, 15–6.

From Oski, F.A., DeAngelis, C.D., Feigin, R.D., and Warshaw, J.D.: *Principles and Practices of Pediatrics,* J.B. Lippincott, Philadelphia, 1990, Figure 12–16.

From Pernkopf, E.: *Atlas of Topographical and Applied Human Anatomy,* 3rd ed. Edited by W. Platzer.

Translated by H. Monsen. Urban & Schwarzenberg, Munich, 1990, Figures 4–13, 8–9, 10–12, 10–13, 10–17, 12–5, 12–6, 12–9, 12–11, 13–25, 13–26, 16–4, 17–4, 18–8, 18–9, 18–11, 18–12, 18–16, 22–8, 22–9, 24–4, 24–5, 24–6, 26–2, 26–9, 26–21, 26–22, 27–1, 27–4, 28–9, 29–3, 29–8, 29–9, 29–10, 29–12, 29–14, 29–17, 30–1, 30–2, 30–3.

From Rothman, R.H., and Simeone, F.A.: *The Spine,* 3rd ed., W.B. Saunders, Philadelphia, 1992, Figure 15–15.

From Salter, R.B.: *Textbook of Disorders and Injuries of the Musculoskeletal System.* Williams & Wilkins, 1970, Baltimore, Figure 9–21.

From Shields, T.W.: *General Thoracic Surgery,* 3rd ed., Lea & Febiger, Philadelphia, 1989, Figures 6–1, 6–2, 16–5, 16–6, 16–7, 17–9, 17–10.

From Slaby, F., and Jacobs, E.R.: *Radiographic Anatomy,* Harwal, Philadelphia, 1990, Figures 6–1, 6–2, 7–21, 7–22, 9–4, 9–5, 9–19, 9–20, 10–3, 10–4, 13–27, 13–28, 13–29, 13–30, 14–2, 14–8, 17–12, 17–13, 17–14.

From Sobotta, J.: *Atlas of Human Anatomy,* 10th English ed. Edited by H. Ferner and J. Staubesand. Translated and edited by W. J. Hild. Urban & Schwarzenberg, Munich, 1983, Figures 12–12, 12–13, 24–1.

From Swash, M: *Hutchinson's Clinical Methods,* 19th ed., Balliere Tindall, London, 1989, Figures 7–11, 12–15.

From Thorek, P: *Anatomy in Surgery,* 3rd ed., Springer-Verlag, New York, 1985, Figures 16–10, 16–14, 17–1, 18–4.

From Tweitmyer, A., and McCracken, T.: *Coloring Guide to Regional Human Anatomy,* 2nd ed., Lea & Febiger, Philadelphia, 1992, Figure 18–10.

From Warwick, P.L., and Williams, R.: *Gray's Anatomy,* 36th ed., Churchill-Livingstone, Edinburgh, 1980, Figure 12–14.

From Zitelli, B.J., and Davis, H.W.: *Atlas of Pediatric Physical Diagnosis.* Mosby, St. Louis, 1987, Figures 9–17, 12–7.

Index

Page numbers printed in *italics* indicate figures; page numbers followed by a 't' indicate tables; page numbers followed by 'CP' indicate discussion in a clinical panel; and page numbers followed by 'CC' indicate discussion of a clinical case.

B

D

G

I

J

K

N